Response Surfaces, Mixtures, and Ridge Analyses

THE WILEY BICENTENNIAL—KNOWLEDGE FOR GENERATIONS

*E*ach generation has its unique needs and aspirations. When Charles Wiley first opened his small printing shop in lower Manhattan in 1807, it was a generation of boundless potential searching for an identity. And we were there, helping to define a new American literary tradition. Over half a century later, in the midst of the Second Industrial Revolution, it was a generation focused on building the future. Once again, we were there, supplying the critical scientific, technical, and engineering knowledge that helped frame the world. Throughout the 20th Century, and into the new millennium, nations began to reach out beyond their own borders and a new international community was born. Wiley was there, expanding its operations around the world to enable a global exchange of ideas, opinions, and know-how.

For 200 years, Wiley has been an integral part of each generation's journey, enabling the flow of information and understanding necessary to meet their needs and fulfill their aspirations. Today, bold new technologies are changing the way we live and learn. Wiley will be there, providing you the must-have knowledge you need to imagine new worlds, new possibilities, and new opportunities.

Generations come and go, but you can always count on Wiley to provide you the knowledge you need, when and where you need it!

WILLIAM J. PESCE
PRESIDENT AND CHIEF EXECUTIVE OFFICER

PETER BOOTH WILEY
CHAIRMAN OF THE BOARD

Response Surfaces, Mixtures, and Ridge Analyses

Second Edition

GEORGE E. P. BOX
Departments of Industrial Engineering and Statistics
University of Wisconsin
Madison, Wisconsin

NORMAN R. DRAPER
Department of Statistics
University of Wisconsin
Madison, Wisconsin

WILEY-INTERSCIENCE
A JOHN WILEY & SONS, INC., PUBLICATION

Library of Congress Cataloging in Publication Data:

Box, George E. P.
    Response surfaces, mixtures, and ridge analyses / George E. P. Box, Norman R. Draper.
--2nd ed.
        p. cm.
    Includes bibliographical references and index.
    ISBN 978-0-470-05357-7 (cloth)
    1. Experimental design. 2. Response surfaces (Statistics) 3. Mixture distributions
(Probability theory) 4. Ridge regression (Statistics) I. Draper, Norman Richard. II. Title.
III. Title: Empirical model-building and response surfaces.

QA279.B67   2007
519.5′—dc22   2006043975

Printed in the United States of America

10 9 8 7 6 5 4 3 2 1

Contents

Preface to the Second Edition

This book, entitled *Response Surfaces, Mixtures, and Ridge Analyses* to reflect new and expanded coverage of topics, is a successor volume to the 1987 book *Empirical Model Building and Response Surfaces* by the same authors.

The 1987 book contained very little about ridge analysis of second order response surfaces; at that time, the method was quite limited. However, recent advances have turned ridge analysis into a very useful practical tool that can easily be applied in a variety of circumstances. Any number of predictor variables can be analyzed with ease, the selected starting point of the ridge paths can now be anywhere in the experimental space, and the method can now be applied to cases where one, two, or more linear restrictions apply to the predictor variables. This has particularly important relevance for mixture experiments, where at least one linear restriction (most often $x_1 + x_2 + \cdots + x_q = 1$) *always* applies, and where other linear restrictions may be added stage by stage. A particular type of such addition comes about when the initial ridge path hits a boundary, defined by a linear function of the predictor variables, and the path must then be followed further on this boundary. The additional restriction is simply added to any previous ones to form a new set of linear restrictions, whereupon the modified ridge path can be easily recalculated and followed to an optimal location or the next restriction, whichever occurs first.

To present and integrate this new material, we have added six new chapters. One (Chapter 12) describes standard ridge analysis, three (Chapters 16–18) discuss techniques relevant to mixtures, one (Chapter 19) is about ridge regression when there are linear restrictions (including, or not including, mixture models, as the case may be) and one (Chapter 20) explains how standard canonical reduction of second order response surfaces can be achieved in a general case, if desired, when linear restrictions are present. Other additions and changes have been made where appropriate—the most extensive of these are in Chapter 5—and the opportunity to correct known errors has been taken. One chapter of the 1987 book, the former Chapter 12, has been deleted entirely. A number of new exercises have been added. Many of the exercises contain authentic published data, with the source provided. Nearly all the exercises have solutions; a few additional data sets are given with no solutions, in response to requests for this in the past. As a whole, the exercises form an extremely important component of the text, because they provide many supplementary examples of response surface use. An extensive bibliography is also provided. It contains articles and books mentioned in the text, plus a much wider selection on response surfaces generally.

The methods described in Chapters 19 and 20 arise from joint research work performed by N. R. Draper and F. Pukelsheim. In connection with this research, N. R. Draper is grateful to the Alexander von Humboldt-Stiftung for support through a Max-Planck-Award for cooperative research with Professor Friedrich Pukelsheim at the University of Augsburg, Germany. The methods described in Chapter 18 arise from joint research work performed by Professor Philip Prescott and N. R. Draper. N. R. Draper is extremely grateful to both of his coworkers, not only for their provision of complete research facilities during many visits at the Universities of Augsburg and Southampton, U. K., but also for their greatly valued friendship, extended over many years.

For more than 40 years, starting with Bea Shube, working with John Wiley has been a pleasure. For this book, Editor Steve Quigley and Associate Editor Susanne Steitz Filler provided continual encouragement and help until our manuscript was ready. Production Editor Lisa Van Horn steered us smoothly and patiently through the complications of production. Brookhaven Typesetting Services brought murmurs of admiration. We are exceedingly grateful to them all.

Contacting the Authors

If you wish to comment on any aspect of this book (e.g., to point out an error, to enquire about an ambiguity, to provide suggestions for exercises, and so on) please email to draper@stat.wisc.edu or write to Professor Norman R. Draper, Statistics Department, University of Wisconsin, 1300 University Avenue, Madison, WI, 53706-1532. Experience has shown that, in spite of the enormous care that goes into producing a book like this, errors can occur. When they do, we like to minimize any adverse effects from their presence. If you would like to receive a list of "errors found to date" via email, please ask.

GEORGE E. P. BOX
NORMAN R. DRAPER

Madison, Wisconsin
January, 2007

CHAPTER 1

Introduction to Response Surface Methodology

In this chapter we discuss, in a preliminary way, some general philosophy necessary to the understanding of the theory and practice of response surface methodology (RSM). Many of the points made here receive more detailed treatment in later chapters.

1.1. RESPONSE SURFACE METHODOLOGY (RSM)

The mechanism of some scientific scientific phenomena are understood sufficiently well that useful mathematical models that flow from the physical mechanism can be written down. Although a number of important statistical problems arise in the building and study of such models, they are not considered in this book. Instead, the methods we discuss will be appropriate to the study of phenomena that are presently not sufficiently well understood to permit the mechanistic approach.

Response surface methodology comprises a group of statistical techniques for empirical model building and model exploitation. By careful design and analysis of experiments, it seeks to relate a *response*, or *output* variable, to the levels of a number of *predictors*, or *input* variables, that affect it.

The variables studied will depend on the specific field of application. For example, response in a chemical investigation might be *yield* of sulfuric acid, and the input variables affecting this yield might be the *pressure* and *temperature* of the reaction. In a psychological experiment, an investigator might want to find out how a test *score* (output) achieved by certain subjects depended upon (a) the *duration* (input 1) of the period during which they studied the relevant material and (b) the *delay* (input 2) between study and test. In mathematical language, in this latter case, we can say that the investigator is interested in a presumed *functional relationship*

$$\eta = f(\xi_1, \xi_2), \qquad (1.1.1)$$

Response Surfaces, Mixtures, and Ridge Analyses, Second Edition. By G. E. P. Box and N. R. Draper
Copyright © 2007 John Wiley & Sons, Inc.

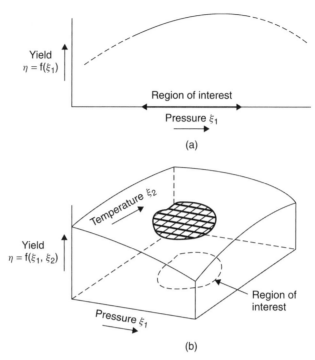

Figure 1.1. (*a*) A response curve. (*b*) A response surface.

which expresses the response or output score, η (Greek letter eta) as a function of two input variables *duration*, ξ_1, and *delay* ξ_2 (Greek letters xi). Both our examples involve two inputs, but, in general, we shall have not two, but k, input variables $\xi_1, \xi_2, \ldots, \xi_k$, and the functional relationship between the mean response and the levels of the k inputs can then be written

$$\eta = f(\xi_1, \xi_2, \ldots, \xi_k). \qquad (1.1.2)$$

More compactly, if $\boldsymbol{\xi}$ denotes a column vector* with elements $\xi_1, \xi_2, \ldots, \xi_k$, the mean response function may be written as

$$\eta = f(\boldsymbol{\xi}). \qquad (1.1.3)$$

If there is only one input variable ξ_1, we can relate an output η to a single input ξ_1 by a *response curve* such as that in Figure 1.1*a*. If we have two inputs ξ_1 and ξ_2 and we draw, in three-dimensional space, a graph of η against ξ_1 and ξ_2, we obtain a *response surface* such as that shown in Figure 1.1*b*. When k ξ's are involved and k is greater than 2, we shall still talk of a response surface in the $(k + 1)$-dimensional space of the variables even though only sectional representation of the surface is possible in the three-dimensional space actually available to us.

The operation of the system with the k inputs adjusted to some definite set of levels $\boldsymbol{\xi}$ is referred to as an *experimental run*. If repeated runs are made at the

---

*The word vector is used here only as a shorthand word referring to all k elements.

same conditions ξ, the measured response will vary because of measurement errors, observational errors, and basic variability in the experimental material. Therefore, we regard η as the mean response at particular conditions $(\xi_1, \ldots, \xi_k)$. An actual observed response result, which we call y, would fall in some statistical distribution about its mean value η. We say that the expected value of y equals η; that is, $E(y) = \eta$. In any particular run, we shall refer to the discrepancy $y - \eta$ between the observed value y and the hypothetical mean value η as the *error* and denote it by ε (Greek epsilon). In general then, our object is to investigate certain aspects of a functional relationship affected by error and expressed as

$$y = f(\xi) + \varepsilon. \tag{1.1.4}$$

Extending the Scope of Graphical Procedures

The underlying relationship between η and a single-input variable ξ can often be represented by a smooth curve like that in Figure 1.1a. Even though the exact nature of the function f was unknown, a smooth curve such as a polynomial acting as a kind of mathematical French curve might be used to represent the function *locally*. Similarly, the relationship between η and two inputs ξ_1 and ξ_2 can often be represented by a smooth surface like that in Figure 1.1b; and a suitable polynomial, acting as a two-dimensional French curve, could be used to approximate it locally. Mathematical French curves, like other French curves, are, of course, not expected to provide a global representation of the function but only a local approximation over some limited region of current interest.

The empirical use of graphical techniques is an important tool in the solution of many scientific problems. The basic objective of the use of response surface methods is to extend this type of empirical procedure to cases where there are more than just one or two inputs. In these higher-dimensional cases, simple graphical methods are inadequate, but we can still fit graduating polynomials and, by studying the features of these fitted empirical functions, greatly extend the range of application of empirical procedures.

So that the reader can better appreciate where response surface methodology fits in the overall picture of scientific experimentation, we now briefly survey some aspects of scientific method. A more extensive review of the role of statistics in the scientific method is given by Box (1976).

1.2. INDETERMINANCY OF EXPERIMENTATION

At first sight, the conduct of an experimental investigation seems to be a highly arbitrary and uncertain process. For example, suppose that we collected 10 teams* of experimenters competent in a particular field of science or technology, locked each team in a separate room, presented them all with the same general scientific problem, and asked each team to submit a plan that could lead to the problem's solution. It is virtually certain that no two teams would present the same plan.

*Or 10 individual experimenters.

Consider, in particular, some of the questions which would arise in planning an initial set of experiments, but on which the teams would be unlikely to agree.

1. *Which input variables* $\xi_1, \xi_2, \ldots$, *should be studied?* If it were a chemical reaction that was being studied, for example, it might be that most investigators would regard, say, temperature and pressure as being important but that there might be a diversity of opinion about which should be included among other input variables such as the initial rate of addition of the reactant, the ratio of certain catalysts, the agitation rate, and so on. Similar and perhaps even stronger disagreement might occur in a psychological experiment.

2. *Should the input variables* ξ *be examined in their original form, or should transformed input variables be employed?* An input variable ξ, such as energy, may produce a linear increase in a response η, such as perceived loudness of noise, when ξ is varied on a *geometrical* scale—or, equivalently, when $\ln \xi$ is varied on a linear scale. It is simpler to express such a relationship therefore by first transforming the input ξ to its logarithm. Another input might be related to the response η by an inverse square law, suggesting an inverse square transformation ξ^{-2}. Other examples can be found leading to transformations such as the square root $\xi^{1/2}$, the inverse square root $\xi^{-1/2}$, and the reciprocal ξ^{-1}. A choice of a transformation for a single variable is often called a *choice of metric* for that variable. More generally, a transformation on the input variables can involve two or more of the original inputs. Suppose, for example, that the amounts ζ_1 and ζ_2 (Greek zeta) of two nitrogenous fertilizers were being investigated. Rather than employing ζ_1 and ζ_2 themselves as the input variables, their sum $\xi_1 = \zeta_1 + \zeta_2$, the total amount of nitrogenous fertilizer, and their ratio $\xi_2 = \zeta_1/\zeta_2$ might be used if it were likely that the response relationship in terms of ξ_1 and ξ_2 could be more simply expressed. In some instances, the theory of dimensionless groups can be used to indicate appropriate transformations of this kind, but usually the best choice of metrics and transformations is not clear-cut and, initially at least, will be the subject of conflicting opinion.

3. *How should the response be measured?* It is often far from clear how the response should be defined. For example, in a study to improve a water treatment plant, the biochemical oxygen demand (BOD) of the effluent would often be regarded by experts as a natural measurement of purity. However, it is possible to have an effluent with zero BOD which is lethal to both bacteria and humans. Thus the appropriate response depends on the end use of the response and will often be the subject of debate.

In question 2, we discussed transformation of the input variables. Transformation of the response variable is also possible, and the appropriate metric for the response is not always clear. In subsequent chapters we show how transformation in the response can lead both to simpler representation and to simpler assumptions.

4. *At which levels of a given input variable* ξ *should experiments be run?* Suppose temperature is an important input. One experimenter might feel that, for the particular system under study, experiments covering temperatures over the range 100–140°C should be made. Another experimenter, believing the system to be very sensitive to temperature, might choose a range from 115°C to 125°C. A third

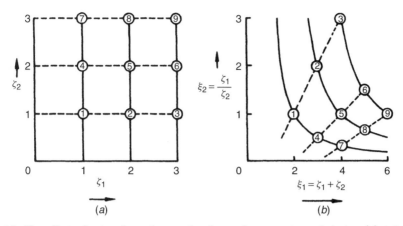

Figure 1.2. The effect of a transformation on the shape of an experimental design: (*a*) A balanced design in the (ζ_1, ζ_2) space. (*b*) The same design in the $\{(\zeta_1 + \zeta_2), \zeta_1/\zeta_2\}$ space.

experimenter believing that considerably higher temperatures would be needed, might pick a range from 140°C to 180°C. In practice, of course, not one but several inputs must be considered simultaneously. If the temperature is to be changed over a 20° range, for example, what is a suitable* "commensurate" change for concentration? In brief, the investigator must choose not only a location for his experiments but also an appropriate scaling for each of the variables.

5. *How complex a model is necessary in a particular situation?* This question is, of course, related to questions 2, 3, and 4. By definition, the more appropriately transformations and metrics are chosen, the simpler is the model that may be employed. Also, the more extensive the region of interest is in the ξ space, the more complex will be the model needed. Thus again we see that this question is a hazy one and is intimately bound up with questions considered under 2, 3, and 4.

6. *How shall we choose qualitative variables?* The foregoing discussion is posed entirely in terms of *quantitative* inputs such as temperature and concentration. Similar indeterminacies occur in experimentation with *qualitative* inputs such as type of raw material, type of catalyst, identity of operator, and variety of seed. For example, if we are to compare three kinds of seed, which varieties should be employed? Should the tested varieties include the cheapest, that believed to give the greatest yield, and that which is most popular? The answer depends on the objectives of the investigation, and that, partially at least, depends on opinion.

7. *What experimental arrangement (experimental design) should be used?* This question is intimately bound up with all of the questions already studied. For instance, consider the hypothetical experiment of question 2. If the amounts ζ_1 and ζ_2 of the two nitrogen fertilizers are each tested at three levels in all nine possible combinations, the experimental arrangement is that of Figure 1.2*a*.

---

*A "rule of thumb" sometimes used by chemists is that a 10°C change in temperature roughly doubles the rate of a chemical reaction. Thus, a commensurate change for concentration in this example would consist of quadrupling the concentration, However, here again, the effect of varying concentration differs for different chemical systems and, in particular, for the reaction order.

However, if the same experimental plan is set out in terms of total nitrogen $\xi_1 = \zeta_1 + \zeta_2$ and nitrogen ratio $\xi_2 = \zeta_1/\zeta_2$, we obtain the very unbalanced arrangement of Figure 1.2*b*. In general, the question of "design optimality" is intimately bound up with arbitrary choices of metric transformation, and the size and shape of the region to be studied.

In brief then, the investigator deals with a number of entities whose natures are necessarily matters of opinion. Among these are (a) the identity of the space of the inputs and outputs in which the experiments should be conducted, (b) the scales, metrics, and transformations in which the variables should be measured, and (c) the location of the region of interest, the specification of the model over it, and the experimental arrangement that should be used to explore the region of interest.

1.3. ITERATIVE NATURE OF THE EXPERIMENTAL LEARNING PROCESS

Faced with so many indeterminancies and uncertainties, the reader might easily despair of a successful outcome of any kind. His spirits may be sustained, however, by the thought that these difficulties have nothing to do with statistical methods per se and that they are hurdles that confront and have always confronted all experimenters. In spite of them, practical experimentation is frequently successful. How does this come about? The situation appears more hopeful when we remember that any group of experimental runs is usually only *part of an iterative sequence* and that an investigational strategy should be aimed at the overall furthering of knowledge rather than just the success of any individual group of trials. Our problem is to so organize matters that we are likely in due course to be led to the right conclusions even though our initial choices of the region of interest, the metrics, the transformations, and levels of the input variables may not all be good. Our strategy must be such as will allow any poor initial choices to be rectified as we proceed. Obviously, the path to success is not unique (although it may seem so to the first investigator in a field). Therefore, it is not *uniqueness* of the path that we should try to achieve, but rather the probable and rapid convergence of an iterative sequence to the right conclusions.

This iterative process of learning by experience can be roughly formalized. It consists essentially of the successive and repeated use of the sequence

CONJECTURE—DESIGN—EXPERIMENT—ANALYSIS

as illustrated in Figure 1.3. Here, the words "design" and "analysis" do not refer only to *statistical* design and analysis. By "design," we mean the synthesis of a suitable experiment to test, estimate, and develop a current conjectured model. By "analysis" we mean the treatment of the experimental results, leading to either (a) verification of a postulated model and the working out of its consequences or (b) the forming of a new or modified conjuncture. These complementary processes are used many times during an investigation, and, by their intelligent alteration, the experimenter's knowledge of the system becomes steadily greater.

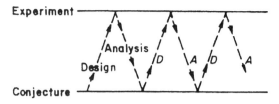

Figure 1.3. The iterative nature of experimentation.

The indeterminacy of the learning process, but the ultimate possibility of its convergence, is familiarly exemplified by the playing of the game of 20 questions. Ten players engaged independently in playing 20 questions all might succeed eventually in discovering that the object in question was the left ear of the Statue of Liberty, but they would almost certainly reach that conclusion by different routes. Let us consider how to do it.

A player may be supplied initially with the information that the object is "mineral." This limits consideration to a certain class of objects which he has to reduce to one as rapidly as possible. When the game is played so that each question concerns a choice between two specific alternatives, the best design of experiment (each "experiment" here is a question asked) is one which will classify the possible objects into two approximately equiprobable groups. A good (but not uniquely good) experiment would thus consist of asking." Is it metal or stone?", a poor experiment would consist of asking "Is it the stone in my ring?" The experiment having been performed and the data (in the form of an answer) having become available, the player would now analyze the reply in the light of any relevant prior knowledge he possessed and would then form some new or modified conjecture as to the nature of the object in question. To the question "Is it metal or stone?" the answer "stone" might conjure up in his mind, on the one hand, buildings, monuments, and so on, and, on the other hand, small stones, both precious (such as diamonds, rubies) and nonprecious (such as pebbles). The player might feel that a good way to discriminate between these two types of objects in the light of information then available would be the question "Is it larger or smaller than this room?" The answer to this question would now result in new analysis which would raise new conjectures and give rise to a new question and so the game would continue.

As another example of iterative investigation, consider a detective employed in solving a mystery. Some data in the form of a body, and the facts that the door was locked on the inside and that Mr. X would benefit from the will, are already known or have been discovered. These data lead to conjectures on the part of the investigator which in turn lead to certain data-producing actions, or "experiments." As readers of detective novels will realize, the sequence of events which finally leads to the detection of the culprit is by no means unique. Alternatively good (but by no means unique) experiments might be a visit to Baker Street subway station to question the ticket collector about a one-armed man with a scar under his left eye, or a visit to the Manor House to find if the flower bed beneath Lady Cynthia's window shows a tell-tale footprint. The skill of the detective is partially measured

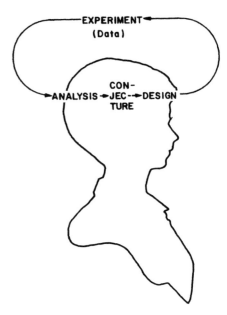

Figure 1.4. The iterative experimental process in relation to the experimenter.

by his ability to conceive at each stage, appropriate conjectures and to select those experiments which, in the light of these conjectures, will best illuminate current aspects of the case. He pursues a strategy which (he hopes) will cause him to follow *one* of the paths leading to the unmasking of the assassin (preferably before the latter has had time to flee the country).

It is through the iterative process of "learning as we go" that the problem of determinancy which we mention may be resolved. We shall find out that, in the application of the investigative processes described in this book, a multidimensional iteration occurs in which modification in the *location* of the experimental runs, in *scaling*, in *transformation*, and in the *complexity* of the contemplated model all can occur. While we cannot, and should not, attempt to ensure uniqueness for the investigation, we can ensure that the crucial components of good design and analysis on which convergence depends are organized so as to illuminate and stimulate the ideas of the experimenter as effectively as possible, and so lead him quickly along one (nonunique) path to the truth. Figure 1.4 illustrates this general idea.

The truly iterative nature of some investigations may sometimes be obscured by the length of time taken by each iterative cycle. In this case it may be possible to see the wider iteration only by "stepping back" and examining what occurs over months or years. In this wider context, iteration may skip from one investigator to another, even from one country to another, and its phases may be very long. Even in this situation, however, it is important to bear in mind that the important consideration is the *overall* acquisition of knowledge.

What is being said is not, of course, peculiar to response surface methodology. "Traditional" designs such as randomized blocks, latin squares, and factorial designs have, since their inception, been used by statisticians as building blocks in iterative learning sequences. The possibility of rapid convergence of such a sequence depends, to an important extent on (a) the efficiency of these designs and their associated analyses and (b) their ability to illuminate and stimulate the ideas of the investigator, This notion was inherent in R. A. Fisher's* general attitude toward the use of these designs in scientific investigation.

1.4. SOME CLASSES OF PROBLEMS (WHICH, HOW, WHY)

Within the foregoing iterative context, we now consider some specific classes of scientific problems and the characteristics of models used in solving these problems.

Mechanistic and Empirical Models

It is helpful in this discussion to distinguish between empirical models and mechanistic models. We consider first what we might mean by a purely mechanistic model. Suppose that, in the study of some physical phenomenon, we know enough of its physical mechanism to *deduce* the form of the functional relationship linking the mean value η of the output to the levels ξ of the inputs via an expression

$$E(y) = \eta = f(\xi, \theta), \tag{1.4.1}$$

where $\xi = (\xi_1, \xi_2, \ldots, \xi_k)'$ is a set of input variables, measuring, for example, initial concentrations of reactants, temperatures, and pressures, and where $\theta = (\theta_1, \theta_2, \ldots, \theta_p)'$ represents a set of physical parameters measuring such things as activation energies, diffusion coefficients, and thermal conductivities. Then we would say that Eq. (1.4.1) represented a mechanistic model.

In practice, some or all of the θ's would need to be estimated from the data. Also, mechanistic knowledge might most naturally be expressed by a set of differential equations or integral equations for which (1.4.1) was the solution.

Often, however, the necessary physical knowledge of the system is absent or incomplete, and consequently no mechanistic model is available. In these circumstances, it could often be realistically assumed that the relationship between η and ξ would be smooth and, consequently, that $f(\xi, \theta)$ could be locally *approximated* (over limited ranges of the experimental variables ξ) by an interpolation function $g(\xi, \beta)$, such as a polynomial. In this latter expression, the β's, which are the elements of β, would be coefficients in the interpolation function. They would be related to, but be distinct from, the parameters (the θ's) of the physical system. The interpolation function $g(\xi, \beta)$ could provide a local empirical model for the system and, as we have said, would act simply as a mathematical French curve.

Now the theoretical mechanistic model $\eta = f(\xi, \theta)$ and the purely empirical model $\eta \simeq g(\xi, \beta)$, as defined above, represent extremes. The former would be appropriate in the extreme case where a great deal was accurately known about the system, and the latter would be appropriate in the other extreme case, where nothing could be assumed except that the response surface was locally smooth.

---

*Sir Ronald Fisher (1890–1962), the famous British statistician, was responsible for many of the basic ideas of experimental design. His books *Statistical Methods for Research Workers* and *The Design of Experiments* are classic.

Table 1.1. Some Scientific Problems

Supposed Unknown	Objective	Descriptive Name	Stage
f ξ θ	Determine the subset ξ of important variables from a given larger set Ξ of potentially important variables	Screening variables	Which
f θ	Determine empirically the effects of the known input variables ξ	Empirical model building	How
f θ	Determine a local interpolation approximation $g(\xi, \beta)$ to $f(\xi, \theta)$	Response surface methodology	
f θ	Determine f	Mechanistic model building	Why
θ	Determine θ	Mechanistic model fitting	

The situation existing in most real investigations is somewhere in between and, as experimentation proceeds and we gain information, the situation can change. Because real problems occur at almost all points between the extremes mentioned above, a variety of statistical tools is needed to cope with them.

The state of ignorance in which we begin our experimental work, along with the state of comparative knowledge to which we wish to be brought, will determine our approach. It must be realized, of course, that no real problem ever *quite* fits any prearranged category. With this proviso in mind, it is nevertheless helpful to distinguish the basic *types* of problems shown in Table 1.1. These are categorized in terms of what is unknown about the "true mechanistic model." For reference purposes we have given each type of problem a descriptive name, as well as a briefer but more pointed "stage" name which makes it clear what stage of investigation is involved.

At the WHICH stage, our object is to determine *which* ones of all the suggested input variables have a significant and important effect upon the response variable. At the HOW stage we would like to learn more about the pattern of the response behavior as changes are made in these important variables. At the WHY stage we attempt to find the mechanistic reasons for the behavior we have observed in the previous stage. (In practice, the stages would usually overlap.)

We now briefly sketch some of the procedures that may be employed to deal with these problems. Our book will then discuss more fully the use of response surface methods for empirical model building.

Screening (WHICH Stage)

It often happens at the beginning of an investigation that there is rather a long list of variables $\xi_1, \xi_2, \ldots,$ which could be of importance in affecting η. One way to reduce the list to manageable size is to sit down with the investigator (the biologist, chemist, psychologist, etc.) and ask him to pick out the variables he believes to be most important. To press this too far, however, is dangerous because, not infrequently, a variable initially believed unimportant turns out to have a major effect. A good compromise is to employ a preliminary screening design such as a two-level fractional factorial (see Chapter 5) to pick out variables worthy of further study.

In one investigation, for instance, the original list of variables $\xi_1, \xi_2, \ldots,$ that might have affected the response η contained 11 candidates. Three of these were, after careful thought, eliminated as being sufficiently unimportant to be safely ignored. A 16-run two level fractional factorial design was run on the remaining eight variables, and four of the eight were designated as probably influential over the ranges studied. Three of these four had already been selected by the investigator as likely to be critical, confirming his judgment. The fourth was unexpected and turned out to be of great importance. Screening designs are often carried out sequentially in small blocks and are very effective when performed in this way. For additional details see the papers by Box and Hunter (1961a, b, especially pp. 334–337) or see Box et al. (2005).

Empirical Model Building (HOW Stage)

When input variables are quantitative and the experimental error is not too large compared with the range covered by the observed responses (see Section 8.2), it may be profitable to attempt to *estimate* the response function within some area of immediate interest. In many problems, the form of the true response function $f(\xi, \theta)$ is unknown and cannot economically be obtained but may be capable of being locally approximated by a polynomial or some other type of graduating function, $g(\xi, \beta)$, say. Suitable experimental designs for this purpose have been developed. The essentially iterative nature of *response surface methodology* (RSM) would ensure that, as the investigation proceeded, it would be possible to learn about (a) the amount of replication needed to achieve sufficient precision, (b) the location of the experimental region of most interest, (c) appropriate scalings and transformations for the input and output variables, and (d) the degree of complexity of an approximating function, and hence of the designs, needed at various stages. These matters are discussed by Box et al. (2005) and, more fully, in this book.

Mechanistic Model Building (WHY Stage)

If it were possible, we might like to use the *true* functional form $f(\xi, \theta)$ to represent the response, rather than to approximate it by a graduating function. In some problems, we can hope to achieve useful working mechanistic models which, at least, take account of the *principal* features of the mechanism. These models often are most naturally expressed in terms of differential equations or other nonexplicit forms, but modern developments in computing facilities and in the theory of nonlinear design and estimation have made it possible to cope with the resulting problems. A mechanistic model has the following advantages:

   1. It contributes to our scientific understanding of the phenomenon under study.

2. It usually provides a better basis for extrapolation (at least to conditions worthy of further experimental investigation if not through the entire ranges of all input variables).

3. It tends to be parsimonious (i.e., frugal) in the use of parameters and to provide better estimates of the response.

Results from fitting mechanistic models have sometimes been disappointing because not enough attention has been given to discovering what *is* an appropriate model form. It is easy to collect data that never "place the postulated model in jeopardy" and so it is common (e.g., in chemical engineering) to find different research groups each advocating a different model for the same phenomenon and each proffering data that "prove" their claim. In such cases, methods that discriminate between the various candidate models must be applied.

Comment

As we have said, the lines between the WHICH, the HOW, and the WHY stages of investigation are often ill-defined in practice. Thus, toward the end of a (WHICH) screening investigation, for example, we may employ a fractional factorial design that is then used as the first of a series of building blocks in a response surface study (HOW). In other circumstances (see, e.g., Box and Youle, 1955), careful study of an empirical fitted response surface at the HOW stage can generate ideas about the possible underlying mechanism (WHY). The present text is concerned mostly with empirical model building—that is, the HOW stage of experimentation—and specifically with what is now called *response surface methodology* or RSM. The various aspects of RSM will be dealt with in the chapters that follow.

1.5. NEED FOR EXPERIMENTAL DESIGN

It sometimes happens (e.g., in investigations of industrial plant processes) that a large amount of past operating data is available. It may then be urged that no experimentation is actually needed because it ought to be possible to extract information relating changes in a response of interest to changes that have occurred *naturally* in the input variables. Such investigations are often valuable as preliminary studies, but the existence of such data rarely eliminates the need for further planned experimentation. There are several reasons for this. For example:

1. Important input variables affecting the response are often the very ones that are not varied.

2. Relations between the response variable and various input variables may be induced by unrecorded "lurking" variables that affect both the response and the input variables. These can give rise to "nonsense correlations."

3. Historical operating data often contain gaps and omit important ancillary information.

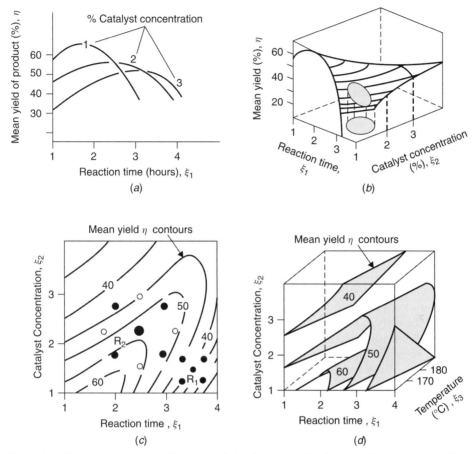

Figure 1.5. Geometrical representations of a relationship connecting the mean yield of product (%), η, with reaction time, ξ_1, catalyst concentration (%), ξ_2, and temperature (°C), ξ_3.

1.6. GEOMETRIC REPRESENTATION OF RESPONSE RELATIONSHIPS

Our ability to think about systems with multiple inputs and outputs is greatly enhanced by geometrical representations and, in particular, by the use of contour diagrams. Some examples are given in Figure 1.5. The curves of Figure 1.5a for example, represent a relationship between yield of product η and reaction ξ_1 at three different levels of catalyst concentration ξ_2.

In Figure 1.5b, a three-dimensional representation is given, and the three curves in Figure 1.5a are now sections of the response surface in Figure 1.5b.

If we were to slice off portions of the response surface at various horizontal levels of the mean yield, η, and then project the outlines of the slices onto the (ξ_1, ξ_2) base of Figure 1.5b, we would obtain the *contours* (or *contour lines*) in Figure 1.5c. We call Figure 1.5c a *contour diagram*. A contour diagram has the advantage that a relationship between a response η and two predictors ξ_1 and ξ_2 (i.e., a relationship that involves three variables) can be represented in only two-dimensional space. Figure 1.5d is also a contour diagram but one in which a third dimension, representing a third predictor variable, temperature, has been added, and three-dimensional *contour surfaces* drawn. If we took a section, at a selected, fixed temperature, of the three-dimensional representation in Figure 1.5d, we would obtain a two-dimensional contour diagram of the form of Figure 1.5c. We see that, with only *three* dimensions at our disposal, we can actually think about the relationship between *four* variables $\eta, \xi_1, \xi_2, \xi_3$. Because the contour diagrams are drawn in the space of the input variables, the ξ's, each potential experimental run will appear simply as a point in this input space. A special selected pattern of points chosen to investigate a response function relationship is called an *experimental design*. Two such experimental design patterns are shown in the regions R_1 and R_2 in Figure 1.5c.

1.7. THREE KINDS OF APPLICATIONS

Response surface methods have proved valuable in the solution of a large variety of problems. We now outline three common classes of such problems using an industrial example for illustration.

A particular type of extruded plastic film should possess a number of properties, one of which is high transparency. This property, regarded as a response η_1, will be affected by such input variables as the screw speed of extrusion, ξ_1, and the temperature of the extruder barrel, ξ_2.

1. Approximate Mapping of a Surface Within a Limited Region

Suppose we were exploring the capability of an extruding machine used in actual manufacture. Such a machine would usually be designed to run at a particular set of conditions, but moderate adjustments in the speed ξ_1 and temperature ξ_2 might be possible. If the unknown response function, $f(\xi_1, \xi_2)$ could be graduated by, say, a polynomial $g(\xi_1, \xi_2)$ in the limited region $R(\xi_1, \xi_2)$ over which the machine could be run, we could approximately predict the transparency of the product for any specified adjustment of the machine.

2. Choice of Operating Conditions to Achieve Desired Specifications

It frequently happens that we are interested in more than one property of the product. For example, in the case of the extruded film, we might be interested in tear strength η_2 in addition to transparency η_1,

It is convenient to visualize the situation by thinking of the contours of η_1 and η_2 superimposed in the ξ_1 and ξ_2 space (see Figure 1.6). For clarity, we can think of the contours of η_1 as being drawn in red ink and those of η_2 drawn in green ink. Actual diagrams and models in which two or more sets of contours are superimposed are in fact often used in elucidating possibilities for industrial systems

Suppose we desired a high transparency $\eta_1 > \eta_{10}$ and a high tear strength $\eta_2 > \eta_{20}$, where η_{10} and η_{20} are specified values. Then, if the responses and tear

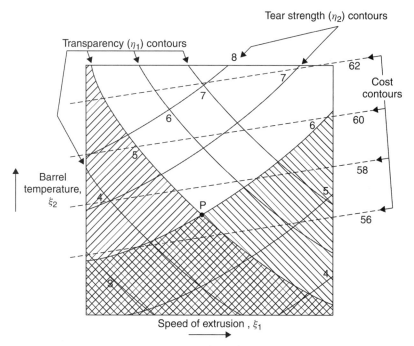

Figure 1.6. The unshaded region contains values (ξ_1, ξ_2) in which the desired product with transparency $\eta_1 > 5$ and tear strength $\eta_2 > 6$ is obtained.

strength were known to be graduated, respectively, by

$$\eta_1 = g_1(\xi_1, \xi_2), \qquad \eta_2 = g_2(\xi_1, \xi_2),$$

we could draw the appropriate critical contours $\eta_1 = \eta_{10}$, $\eta_2 = \eta_{20}$ on the same figure and hence determine the region in the (ξ_1, ξ_2) space in which *both* inequalities were satisfied and so *both* responses attained satisfactory values. Figure 1.6 illustrates how this might be done. The shading indicates the region in the space of the input variables in which an undesirable product would be obtained. Transparencies greater than 5 and tear strengths greater than 6 can be obtained within the unshaded region.

In practice, of course, there may be more than two predictors and/or more than two response variables. One response that would often need to be examined is the overall average cost associated with any particular set of operating conditions. Thus we might seek to obtain a product which satisfied the specification $\xi_1 > \xi_{10}$, $\xi_2 > \xi_{20}$ *and* had the smallest possible cost η_3. We see from Figure 1.6 that manufacture at conditions corresponding to the point P would just satisfy the specifications and, at the same time, lead to a minimum cost.

The formal description for the problem above has many aspects closely related to linear and nonlinear programming problems. The situation we consider here, however, is more complicated, because the relationships between the η's and ξ's are not given but must be estimated.

3. Search for Optimal Conditions

In the examples above, we imply that reasonably satisfactory manufacturing conditions have already been determined. However, at an earlier stage of process development, work would have been carried out in the laboratory and pilot plant to determine suitable conditions. Frequently, there would be a large region in the space of the variables ξ_1 and ξ_2 where the process was operable. An important problem is that of finding, within this operability region, the *best* operating conditions. Let us suppose that the merit of any particular set (ξ_1, ξ_2) of input conditions can be assessed in terms of a response η representing, say, an overall measure of profitability. The problem would be that of locating, and determining the characteristics of, the optimum conditions.

In the chapters that follow, we shall discuss further the details of response methodology and how the various applications of it can be made.

CHAPTER 2

The Use Of Graduating Functions

2.1. APPROXIMATING RESPONSE FUNCTIONS

Basic Underlying Relationship

We have said that investigations of physical, chemical, and biological systems are often concerned with the elucidation of some functional relationship

$$E(y) = \eta = f(\xi_1, \xi_2, \ldots, \xi_k)$$
$$= f(\boldsymbol{\xi}) \tag{2.1.1}$$

connecting the expected value (i.e., the mean value) of a response y such as the yield of a product with k quantitative variables $\xi_1, \xi_2, \ldots, \xi_k$ such as temperature, time, pressure, concentration, and so on. In some problems, a number of different responses may be of interest, as for example, percentage yield y_1, purity y_2, viscosity y_3, and so on. For each of these, there will be a separate functional relationship. In this book we assume that the natures of these various functional relationships are not known and that local polynomial approximations are used instead.

In what follows it is convenient not to have to deal with the actual numerical measures of the variables ξ_i, but instead to work with coded or "standardized" variables x_i. For example, if at some stage of an investigation we defined the current region of interest for ξ_i to be $\xi_{io} \pm S_i$, where ξ_{io} is the center of the region, then it would be convenient to define an equivalent working variable x_i where

$$x_i = \frac{\xi_i - \xi_{io}}{S_i}.$$

Thus, for example, if ξ_i is temperature and the current region of interest is $115 \pm 10°$, then for any setting of temperature we have the equivalent coded value $x_i = (\xi_i - 115)/10$. We note that the coded quantities x_i are simply convenient

Response Surfaces, Mixtures, and Ridge Analyses, Second Edition. By G. E. P. Box and N. R. Draper
Copyright © 2007 John Wiley & Sons, Inc.

17

linear transformations of the original ξ_i, and so expressions containing the x_i can always be readily rewritten in terms of the ξ_i.

Polynomial Approximations

In general, a polynomial in the coded inputs $x_1, x_2, \ldots, x_k$ is a function which is a linear aggregate (or combination) of powers and products of the x's. A term in the polynomial is said to be of order j (or degree j) if it contains the product of j of the x's (some of which may be repeated). Thus terms involving x_1^3, $x_1 x_2^2$, and $x_1 x_2 x_3$ would all be said to be of order 3 (or degree 3). A polynomial is said to be of order d, or degree d, if the term(s) of highest order in it is (are) of order or degree d. Thus, if $k = 2$ and if x_1 and x_2 denote two coded inputs, the general polynomial can be written

$$g(\mathbf{x}, \boldsymbol{\beta}) = \beta_0 + (\beta_1 x_1 + \beta_2 x_2) + (\beta_{11} x_1^2 + \beta_{22} x_2^2 + \beta_{12} x_1 x_2)$$
$$+ (\beta_{111} x_1^3 + \beta_{222} x_2^3 + \beta_{112} x_1^2 x_2 + \beta_{122} x_1 x_2^2) + (\beta_{1111} x_1^4 \ldots)$$
$$+ \text{etc.,} \tag{2.1.2}$$

where terms of the same order are bracketed for convenience. Note that the subscript notation is chosen so that each β coefficient can be easily identified with its corresponding x term. For example, β_{122} is the coefficient of $x_1 x_2 x_2$, that is, $x_1 x_2^2$. In the expression (2.1.2), the β's are coefficients of (empirical) *parameters* which, in practice, have to be estimated from the data.

As is seen from Table 2.1, the number of such parameters increases rapidly as the number, k, of the input variables and the degree, d, of the polynomial are both increased.

A polynomial expression of degree d can be thought of as a Taylor's series expansion of the true underlying theoretical function $f(\boldsymbol{\xi})$ truncated after terms of dth order. The following will usually be true:

1. The higher the degree of the approximating function, the more closely the Taylor series can approximate the true function.
2. The smaller the region R over which the approximation needs to be made, the better is the approximation possible with a polynomial function of given degree.

Table 2.1. Number of Coefficients in Polynomials of Degree d Involving k Inputs

Number of Inputs, k	Degree of Polynomial, d			
	1	2	3	4
	Planar	Quadratic	Cubic	Quartic
2	3	6	10	15
3	4	10	20	35
4	5	15	35	70
5	6	21	56	126

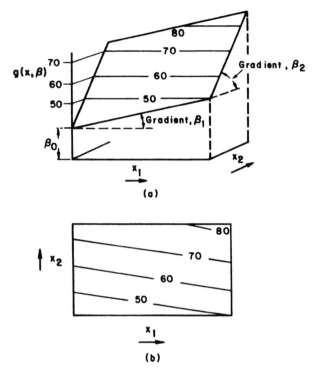

Figure 2.1. (*a*) Geometrical representation of a first degree polynomial (planar approximation) $g(\mathbf{x}, \boldsymbol{\beta})$ $= \beta_0 + \beta_1 x_1 + \beta_2 x_2$. (*b*) The equal height contours of $g(\mathbf{x}, \boldsymbol{\beta})$ projected onto the (x_1, x_2) plane.

In practice, we can often proceed by supposing that, over limited regions of the factor space, a polynomial of only first or second degree might adequately represent the true function.

First-Degree (or First-Order) Approximation

In Eq. (2.1.2) the first set of parentheses contains first-order terms. If we truncated the expression at this point, we should have the first-degree polynomial approximation for $k = 2$ predictor variables, x_1 and x_2,

$$g(\mathbf{x}, \boldsymbol{\beta}) = \beta_0 + \beta_1 x_1 + \beta_2 x_2, \qquad (2.1.3)$$

capable of representing a tilted plane. The height and tilt of the plane are determined by the coefficients β_0, β_1, and β_2. Specifically (as in Figure 2.1a) β_0 is the intercept of the plane with the g axis at the origin of x_1 and x_2 and β_1 and β_2 are the gradients (slopes) in the directions x_1 and x_2. The height contours of such a plane would be equally spaced parallel straight lines. Some of these contours are shown on the plane in Figure 2.1a, and projected onto the (x_1, x_2) plane in Figure 2.1b. (Some possible height readings are attached to the contours to help the visualization.) We shall discuss such approximations in considerably more detail in Chapter 6.

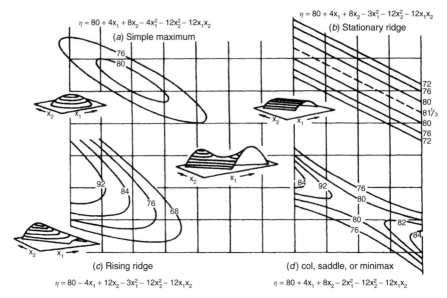

Figure 2.2. Some examples of types of surfaces defined by second-order polynomials in two predictor variables x_1 and x_2.

Second-Degree (or Second-Order) Approximation

If we truncated the expression (2.1.2) at the second set of parentheses, we would have the polynomial approximation of second degree for $k = 2$ predictor variables, x_1, and x_2,

$$g(\mathbf{x}, \boldsymbol{\beta}) = \beta_0 + \beta_1 x_1 + \beta_2 x_2 + \beta_{11} x_1^2 + \beta_{22} x_2^2 + \beta_{12} x_1 x_2. \qquad (2.1.4)$$

This defines what is called a general second-order (or quadratic) surface, here in two variables x_1 and x_2 only.

Figure 2.2 illustrates how, by suitable choices of the coefficients, the second-order surface in x_1 and x_2 can take on a variety of useful shapes. Both contour plots and associated surfaces are shown.

A simple maximum is shown in Figure 2.2a, and a stationary or flat ridge is shown in Figure 2.2b. Figure 2.2c shows a rising ridge, and Figure 2.2d shows what is variously called a col, saddle, or minimax. Although even locally the true underlying model cannot be expected to correspond *exactly* with such forms, nevertheless the main features of a true surface could often be approximated by one of these forms. We shall discuss approximations in considerably more detail in Chapters 9–11.

Relationship Between the Approximating Polynomial and the "True" Underlying Response Surface

To appreciate the potential usefulness of such empirical approximations, consider Figure 2.3. This shows a theoretical response surface in which the yield of a

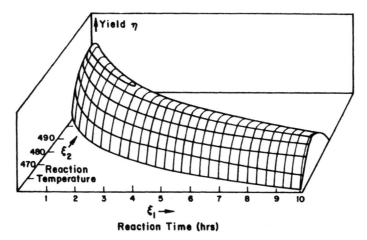

Figure 2.3. A theoretical response surface showing a relationship between response η (yield of product) as a function of predictors ξ_1 (reaction time in hours), and ξ_2 (reaction temperature in degrees Kelvin).

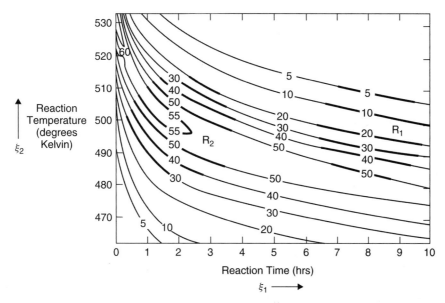

Figure 2.4. Yield contours of a theoretical response surface, showing the relationship between response η (yield of product) as a function of predictors ξ_1 (reaction time in hours), and ξ_2 (reaction temperature in degrees Kelvin).

product η is represented as a function of time of reaction in hours, ξ_1, and the absolute temperature of reaction, ξ_2, in degrees Kelvin. A contour diagram corresponding to this surface is shown in Figure 2.4. [The theoretical function shown in Figures 2.3 and 2.4 arises from a particular physical theory about the manner in which the reaction of interest occurred. This theory leads to a set of differential equations whose solution is the function shown. The function itself,

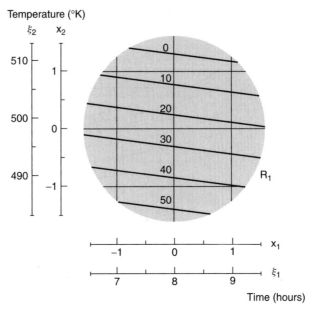

Figure 2.5. Local representation in region R_1 of the response surface in Figure 2.4. The contours plotted are those of the first-degree polynomial $24.5 - 2x_1 - 18.5x_2$, where $x_1 = \xi_1 - 8$, and $x_2 = (\xi_2 - 498)/10$.

$\eta = f(\xi_1, \xi_2; \theta_1, \theta_2, \theta_3, \theta_4)$, which depends on two predictor variables ξ_1 and ξ_2 and on four physical constants, $\theta_1, \theta_2, \theta_3, \theta_4$, is given in Appendix 2A. For our present purposes, it is sufficient to appreciate that, underlying any system, there is a mechanism which, if fully understood, could be represented by a theoretical relationship $f(\mathbf{\xi}, \mathbf{\theta})$.]

If we *knew* the *form* of the theoretical relationship (that is, the nature of the function f) but did not know the values of the parameters θ, then usually it would be best to fit this theoretical function directly to the data. If, as we suppose here, we *do not* know the form of the function, then the relationship can often be usefully approximated, over a limited region, by a simple polynomial function. The purpose of this approximation is not to represent the true underlying relationship everywhere, but merely to graduate it locally. This is illustrated in Figure 2.5, which should be compared with Figure 2.4. We see that is is possible to represent the main features of the true relationship moderately well over the region labeled R_1 by the first degree polynomial

$$\eta \simeq 24.5 - 2x_1 - 18.5x_2,$$

in which the input variables are represented through the convenient coding

$$x_1 = (\xi_1 - 8)/1, \qquad x_2 = (\xi_2 - 498)/10.$$

The approximation is far from perfect. In particular, it is of too simple a kind to represent the uneven spacing of the theoretical contours. However, it does convey

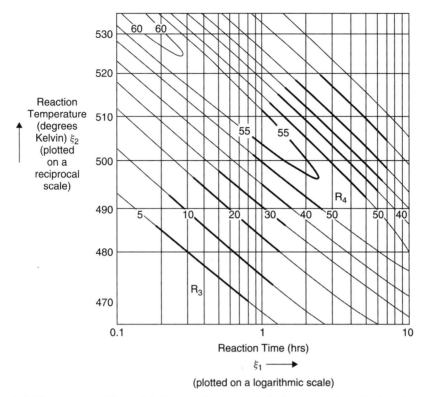

Figure 2.6. The contours of Figure 2.4 after transformation; ξ_1 is plotted on a logarithmic scale, and ξ_2 is plotted on a reciprocal scale.

the correct impression that the main feature of the surface in the region is a large negative gradient in temperature. Furthermore, the inevitable experimental error met in practice might make dubious the usefulness of greater elaboration at this stage.

Over the region R_2 in Figure 2.4, a second-degree polynomial representing a surface like that of Figure 2.2c could, again, very approximately represent the function locally.

Improvement in Approximation Obtainable by Transformation

As we mentioned in Chapter 1, a considerable improvement in representational capability can often be obtained by allowing the possibility of transformations in the inputs and outputs. For illustration, we reconsider the kinetic example introduced earlier. Figure 2.6 is a contour diagram of the same function as plotted in Figure 2.4. However, in Figure 2.6, reaction time ξ_1 is plotted on a log scale, and temperature in degrees Kelvin ξ_2 is plotted on a reciprocal scale.* It is easy to see

*The reciprocal axis is reversed to allow more ready comparison between Figures 2.4 and 2.6. *Over the range studied*, reciprocal temperature is almost a linear function of temperature so that, for this example, the improved approximation is almost totally due to plotting reaction time on a log scale.

Table 2.2. Data on Worsted Yarn: Cycles to Failure Under Various Loading Conditions

Length of Test Specimen (mm), ξ_1	Amplitude of Load Cycle (mm), ξ_2	Load (g), ξ_3	Cycles to Failure, Y	Antilog of $\hat{y}$
250	8	40	674	692
350	8	40	3636	3890
250	10	40	170	178
350	10	40	1140	1000
250	8	50	292	309
350	8	50	2000	1738
250	10	50	90	94
350	10	50	360	447

that considerably closer approximations by first- and second-degree polynomials would be possible after these transformations had been made. In particular, a fitted plane in the region R_3 and a fitted quadratic surface in R_4 would now represent the transformed function quite closely.

2.2. AN EXAMPLE

We now give an example which has been chosen (a) to illustrate, in an elementary way, some of the considerations already mentioned and (b) to motivate the work of the chapters which follow.

An Experiment in Worsted Yarn

Table 2.2 shows a partial* listing of data from an unpublished report to the Technical Committee, International Wool Textile Organization, by Dr. A. Barella and Dr. A. Sust. These numbers were obtained in a textile investigation of the behavior of worsted yarn under cycles of repeated loading. The final column should be ignored for the moment. Because of the wide range of variation of Y (from 90 to 3636 in this data set), it is more natural to consider an analysis in terms of $y = \log Y$ (the base 10 is used) rather than in terms of Y itself. Also the input or predictor variables are more conveniently used in the coded forms

$$x_1 = (\xi_1 - 300)/50, \qquad x_2 = (\xi_2 - 9)/1, \qquad x_3 = (\xi_3 - 45)/5. \quad (2.2.1)$$

The transformed data are shown in Table 2.3 (Again, ignore the $\hat{y}$ column for the moment.) We shall show later that the transformed data can be fitted very well by a simple first-degree polynomial in x_1, x_2, and x_3 of form

$$g(\mathbf{x}, \boldsymbol{\beta}) = \beta_0 + \beta_1 x_1 + \beta_2 x_2 + \beta_3 x_3, \quad (2.2.2)$$

*The complete set of data, along with a more comprehensive analysis, is given in Chapter 7. These data were quoted and analyzed by Box and Cox (1964).

Table 2.3. Coded Data on Worsted Yarn

Specimen Length (Coded), x_1	Amplitude (Coded), x_2	Load (Coded), x_3	Log Cycles to Failure, y	Fitted Value Log Cycles, $\hat{y}$
−1	−1	−1	2.83	2.84
1	−1	−1	3.56	3.59
−1	1	−1	2.23	2.25
1	1	−1	3.06	3.00
−1	−1	1	2.47	2.49
1	−1	1	3.30	3.24
−1	1	1	1.95	1.90
1	1	1	2.56	2.65

where we estimate the coefficients $\beta_0, \beta_1, \beta_2, \beta_3$ from the data. If we denote the respective estimates by b_0, b_1, b_2, b_3 and the *fitted value*—that is, the response obtained from the fitted equation at a general point (x_1, x_2, x_3)—by $\hat{y}$, then

$$\hat{y} = b_0 + b_1 x_1 + b_2 x_2 + b_3 x_3. \tag{2.2.3}$$

For this particular set of eight data values, we show later that estimates of the β coefficients, evaluated via the method of least squares, are

$$
\begin{aligned}
b_0 &= 2.745 \pm 0.025, \\
b_1 &= 0.375 \pm 0.025, \\
b_2 &= -0.295 \pm 0.025, \\
b_3 &= -0.175 \pm 0.025.
\end{aligned}
\tag{2.2.4}
$$

(The numbers following the $\pm$ signs are the standard errors of the estimates.) Thus the fitted equation derived from the data is

$$\hat{y} = 2.745 + 0.375x_1 - 0.295x_2 - 0.175x_3. \tag{2.2.5}$$

For comparison, the $\hat{y}$'s evaluated from this fitted equation by substituting the appropriate x values are shown beside the actual observed y's in Table 2.3. The antilogs of these $\hat{y}$'s are shown in Table 2.2, opposite the corresponding observed values of the Y's. It will be seen that there is a close agreement between each Y, the actual number of cycles to failure observed, and the corresponding antilog $\hat{y}$ predicted by the fitted equation.

Figure 2.7a shows contours of the fitted equation for $\hat{y}$ in the x_1, x_2, x_3 space computed from the linear expression Eq. (2.2.5). In Figure 2.7b, these same scales and contours have been relabeled in terms of the original variables $\xi_1, \xi_2,$ and ξ_3, and the original observations are shown at the corners of the cube.

A first inspection of Figure 2.7 certainly suggests that a first-degree approximation in terms of the log response provides an excellent representation of the relationship between the response variable, cycles to failure, and the input variables, length, amplitude, and load.

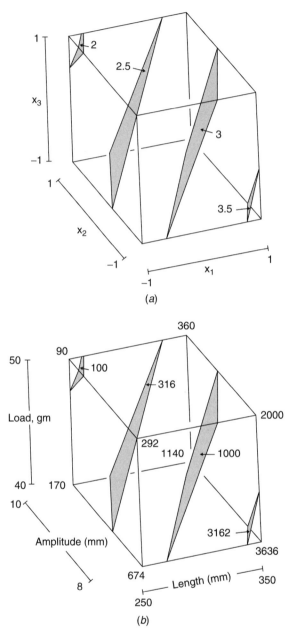

Figure 2.7. (*a*) Contours of fitted surface $\hat{y} = 2.745 + 0.375x_1 - 0.295x_2 - 0.175x_3$, with the transformation $y = \log Y$ applied to the original Y observations. (*b*) Contours of fitted surface as in (*a*) but labeled in original units of Y; the actual observations are shown at the cube corners.

If this is so, then the relationship could be of great practical value.

1. It could indicate the direction and magnitude of change in response when changes are made in each of the three input variables.

2. It could allow calculations to be made of the response at intermediate conditions not actually tested.

3. It could show the direction in which to move if we wish to change the input levels so as to increase the response "cycles to failure" as much as possible.

However, this example raises a number of questions. Some of these we now set out, and we also indicate where in this book these questions will be answered.

SOME QUESTIONS THAT ARISE

1. *Least Squares*. The estimated expression $\hat{y} = 2.745 + 0.375x_1 - 0.295x_2 - 0.175x_3$ was obtained by fitting a first-degree polynomial of general form (2.2.2) to the data of Table 2.3 by the method of least squares. What *is* the method of least squares? What are its assumptions, and how may it be generally applied to the fitting of such relationships? (See Chapter 3.)
2. *Standard Errors of Coefficients*. If relationships of this kind are to be intelligently employed, we need measures of precision of the estimated coefficients. In Eqs. (2.2.4), standard errors, which are estimates of the standard deviations of the coefficients, were obtained from the data. How can such estimates be calculated? (See Chapter 3.)
3. *Using First-Order Response Models*. How may we use such expressions to ascertain a direction of increasing response, perhaps pointing the way toward a maximum response? (See Chapter 6.)
4. *Higher-Order Expressions*. How may we fit and use higher-order polynomial models—in particular, general second-order expressions? How may fitted expressions such as in 1 be used to indicate sets of conditions for which *several* responses achieve desired specifications. (See Chapters 7, 9, 10, and 11.)
5. *Adequacy of Fit*. In approximating an unknown theoretical function empirically, we need to be able to check whether or not a given degree of approximation is adequate. How can analysis of variance (ANOVA) and the examination of residuals (the discrepancies between observed and fitted values) help to check adequacy of fit? (See Chapters 3 and 7.)
6. *Designs*. A simple arrangement of experiments was employed in the above example in which all eight combinations of two levels of the input variables were run. This is called a 2^3 *factorial design*. What designs are suitable for fitting polynomials of first and second degrees? (See Chapters 4, 5, 15, and 13.)
7. *Transformations*. By making a log transformation of the output variable Y above, a very simple first-degree relationship could be employed. How can such transformations be found in general? (See Chapters 7 and 8.)

The motivation for topics discussed in chapters not specifically mentioned above will become clearer as we proceed.

APPENDIX 2A. A THEORETICAL RESPONSE FUNCTION

The function plotted in Figures 2.3 and 2.4 is given by

$$\eta = k_1\{\exp(-k_1\xi_1) - \exp(-k_2\xi_1)\}/(k_1 - k_2), \qquad (2A.1)$$

where

$$k_1 = \theta_1 \exp\{-\theta_2/\xi_2\}, \qquad k_2 = \theta_3\exp\{-\theta_4/\xi_2\},$$

$$\xi_1 = \text{reaction time (h)}, \tag{2A.2}$$

$$\xi_2 = \text{reaction temperature } (^\circ K) = T + 273,$$

where T is temperature in degrees Celsius. Expression (2A.1) arises if η is the fractional yield of the intermediate product B in a consecutive chemical reaction $A \to B \to C$ subject to first-order kinetics, with the dependence of the rates k, k_2 on temperature following the Arrhenius law. The actual parameter values which were used to give Figures 2.3 and 2.4 are

$$\theta_1 = 5.0259 \times 10^{16}, \qquad \theta_2 = 1.9279 \times 10^4,$$
$$\theta_3 = 1.3862 \times 10^{14}, \qquad \theta_4 = 1.6819 \times 10^4. \tag{A.3}$$

CHAPTER 3

Least Squares for Response Surface Work

Readers with an understanding of least squares (regression analysis) may wish to skip this chapter entirely, or simply dip into it as necessary. The chapter does not, of course, represent a complete treatment of least squares. For that, the reader should consult one of the standard texts.

3.1. THE METHOD OF LEAST SQUARES

We saw in Chapter 2 that some method was needed for fitting empirical functions to data. The fitting procedure we employ is the *method of least squares*. Given certain assumptions which are later discussed and which often approximately describe real situations, the method has a number of desirable properties.

We shall present the basic results needed, omitting proofs and illustrating with examples as we proceed. Because matrices are necessary to develop these methods efficiently, we provide a brief review of the required matrix algebra in Appendix 3C.

Least-Squares Estimates

We have said in Chapter 2 [see Eq. (2.1.1)] that the investigator often wishes to elucidate some model

$$y = f(\xi, \theta) + \varepsilon, \tag{3.1.1}$$

where

$$E(y) = \eta = f(\xi, \theta) \tag{3.1.2}$$

is the mean level of the response y which is affected by k variables $(\xi_1, \xi_2, \ldots, \xi_k)$ $= \xi'$. The model involves, in addition, p parameters $(\theta_1, \theta_2, \ldots, \theta_p) = \theta'$, and ε is an experimental error. To examine this model, the experimenter may make a series of experimental runs at n different sets of conditions $\boldsymbol{\xi}_1, \boldsymbol{\xi}_2, \ldots, \boldsymbol{\xi}_n$, observing the corresponding values of the response, $y_1, y_2, \ldots, y_n$. Two important questions that arise are:

1. Does the postulated model adequately represent the data?
2. Assuming that the model does adequately represent the data, what are the best estimates of the parameters in $\boldsymbol{\theta}$?

Somewhat paradoxically we need to study the second of these two questions first, leaving the first question until later.

Suppose some specific values were chosen for the parameters on some basis or another. This would enable us to compute $f(\boldsymbol{\xi}, \boldsymbol{\theta})$ for each of the experimental runs $\boldsymbol{\xi}_1, \boldsymbol{\xi}_2, \ldots, \boldsymbol{\xi}_n$ and hence to obtain the n discrepancies $\{y_1 - f(\boldsymbol{\xi}_1, \boldsymbol{\theta})\}, \{y_2 - f(\boldsymbol{\xi}_2, \boldsymbol{\theta})\}, \ldots, \{y_n - f(\boldsymbol{\xi}_n, \boldsymbol{\theta})\}$. The method of least squares selects, as the best estimate of $\boldsymbol{\theta}$, the value that makes the sum of squares of these discrepancies, namely

$$S(\boldsymbol{\theta}) = \sum_{u=1}^{n} \{y_u - f(\boldsymbol{\xi}_u, \boldsymbol{\theta})\}^2, \qquad (3.1.3)$$

as small as possible. $S(\boldsymbol{\theta})$ is called the *sum of squares function*. For any given choice of the p parameters in $\boldsymbol{\theta}$, there will be a specific value of $S(\boldsymbol{\theta})$. The minimizing choice of $\boldsymbol{\theta}$, its *least-squares estimate*, is denoted by $\hat{\boldsymbol{\theta}}$; the corresponding (minimized) value of $S(\boldsymbol{\theta})$ is thus $S(\hat{\boldsymbol{\theta}})$.

Are the least-squares estimates of the θ's "good" estimates? In general, their goodness depends on the nature of the distribution of the errors. Specifically, we shall see later that least-squares estimates would be appropriate if it could be assumed that the experimental errors $\varepsilon_u = y_u - \eta_u$, $u = 1, 2, \ldots, n$, were *statistically independent*, with *constant variance*, and were *normally distributed*. These "standard assumptions" and their relevance are discussed later; for the moment, we concentrate on the problem of how to calculate least-squares estimates.

3.2. LINEAR MODELS

A great simplification in the computation of the least-squares estimates occurs when the response function is linear in the parameters—that is, of the form

$$\eta = f(\boldsymbol{\xi}, \boldsymbol{\theta}) = \theta_1 z_1 + \theta_2 z_2 + \cdots + \theta_p z_p. \qquad (3.2.1)$$

In this expression, the z's are known constants and, in practice, are known functions of the experimental conditions $\xi_1, \xi_2, \ldots, \xi_k$ or, equivalently, of their coded forms $x_i = (\xi_i - \xi_{i0})/S_i$, $i = 1, 2, \ldots, k$, where ξ_{i0} and S_i are suitable location and scale factors, respectively. Adding the experimental error $\varepsilon = y - \eta$, we thus have the model

$$y = \theta_1 z_1 + \theta_2 z_2 + \cdots + \theta_p z_p + \varepsilon. \qquad (3.2.2)$$

Any model of this form is said to be a *linear model*—that is, *linear in the parameters*. Linear models are more widely applicable than might at first appear, as will be clear from the following examples.

Example 3.1. An experimenter believes that the electrical conductivity y of cotton fiber depends on the humidity ξ and that, over the range of humidity of interest, an approximately linear relationship, obscured by experimental error, will exist. It is convenient to transform linearly a variable such as humidity ξ to coded form $x = (\xi - \xi_0)/S$, where ξ_0 is some convenient origin and S is a scale factor. The model may then be written

$$\eta = \beta_0 + \beta_1 x, \tag{3.2.3}$$

where β_0 represents the conductivity at $x = 0$ (i.e., at $\xi = \xi_0$) and β_1 represents the increase in conductivity per unit increase in x (i.e., per S units of humidity). In this example, there is $k = 1$ predictor variable, namely the humidity x, and there are $p = 2$ parameters, β_0 and β_1. The response function can be written in the general form (3.2.1) as

$$\eta = \theta_1 z_1 + \theta_2 z_2 \tag{3.2.4}$$

by setting $z_1 = 1$, $z_2 = x = (\xi - \xi_0)/S$, $\theta_1 = \beta_0$, and $\theta_2 = \beta_1$.

Example 3.2. Suppose, in the foregoing example, that the relationship between the mean electrical conductivity and the humidity was expected to be curved. This might be provided for by adding a term in x^2 so that the postulated response function was

$$\eta = \beta_0 + \beta_1 x + \beta_{11} x^2. \tag{3.2.5}$$

Identification with the general form

$$\eta = \theta_1 z_1 + \theta_2 z_2 + \theta_3 z_3 \tag{3.2.6}$$

is provided by setting $z_1 = 1$, $z_2 = x$, $z_3 = x^2$, $\theta_1 = \beta_0$, $\theta_2 = \beta_1$, and $\theta_3 = \beta_{11}$. Note that this is an example of a function which is linear in the parameters but nonlinear (in fact quadratic) in x. It is clear that we can cover any of the polynomial models discussed in Chapter 2 using the general linear model formulation.

Example 3.3. It is postulated that the presence of a certain additive in gasoline increases gas mileage y. A number of trials are run on a standard Volkswagen engine *with* and *without* the additive. The experimenter might employ the response function

$$\eta = \beta_0 + \beta_1 x \tag{3.2.7}$$

in which x is used simply to denote *presence* or *absence* of the additive. Thus

$$
\begin{aligned}
x = 0 \quad &\text{if additive absent} \\
x = 1 \quad &\text{if additive present}
\end{aligned}
\tag{3.2.8}
$$

and then β_0 represents the mean gas mileage with no additive while β_1 is the incremental effect of the additive. Here $k = 1$; the single predictor variable x does not have a continuous scale but can take only the values 0 or 1. Such a variable is sometimes called an *indicator*, or *dummy*, *variable*. There are $p = 2$ parameters, β_0 and β_1, and the model function in the general form (3.2.1) is as in Eq. (3.2.4).

Dummy variables are very useful to represent the effects of qualitative variables such as operator, day on which experiment is conducted, type of raw material, and so on.

An Algorithm for All Linear Least-Squares Problems

Suppose that we are in a situation where the response function

$$\eta = \theta_1 z_1 + \theta_2 z_2 + \cdots + \theta_p z_p \tag{3.2.9}$$

has been postulated and experimental conditions $(z_{1u}, z_{2u}, \ldots, z_{pu})$, $u = 1, 2, \ldots, n$, have been run, yielding observations $y_1, y_2, \ldots, y_n$. Then the model relates the observations to the known z_{iu}'s and the unknown θ_i's by n equations

$$y_1 = \theta_1 z_{11} + \theta_2 z_{21} + \cdots + \theta_p z_{p1} + \varepsilon_1,$$

$$y_2 = \theta_1 z_{12} + \theta_2 z_{22} + \cdots + \theta_p z_{p2} + \varepsilon_2,$$

$$\cdots$$

$$\cdots$$

$$y_n = \theta_1 z_{1n} + \theta_2 z_{2n} + \cdots + \theta_p z_{pn} + \varepsilon_n. \tag{3.2.10}$$

These can be written in matrix form as

$$\mathbf{y} = \mathbf{Z}\boldsymbol{\theta} + \boldsymbol{\varepsilon}, \tag{3.2.11}$$

where

$$\mathbf{y} = \begin{bmatrix} y_1 \\ y_2 \\ \vdots \\ y_n \end{bmatrix}, \quad \mathbf{Z} = \begin{bmatrix} z_{11} & z_{21} & \cdots & z_{p1} \\ z_{12} & z_{22} & \cdots & z_{p2} \\ \vdots & \vdots & & \vdots \\ z_{1n} & z_{2n} & \cdots & z_{pn} \end{bmatrix}, \quad \boldsymbol{\theta} = \begin{bmatrix} \theta_1 \\ \theta_2 \\ \vdots \\ \theta_p \end{bmatrix}, \quad \boldsymbol{\varepsilon} = \begin{bmatrix} \varepsilon_1 \\ \varepsilon_2 \\ \vdots \\ \varepsilon_n \end{bmatrix}$$

$$n \times 1 \qquad\qquad n \times p \qquad\qquad\qquad p \times 1 \qquad n \times 1$$

$$\tag{3.2.12}$$

and where the dimensions of the vectors and matrices are given beneath them. The sum of squares function Eq. (3.1.3) is

$$S(\boldsymbol{\theta}) = \sum_{u=1}^{n} \left(y_u - \theta_q z_{1u} - \theta_2 z_{2u} - \cdots - \theta_p z_{pu} \right)^2, \tag{3.2.13}$$

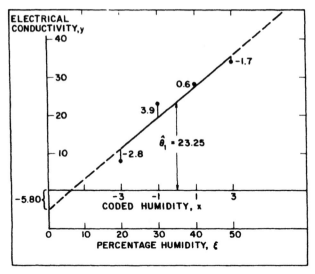

Figure 3.1. Conductivity—humidity example. Plot of the data and the fitted line.

or in matrix format,

$$S(\theta) = (y - Z\theta)'(y - Z\theta). \qquad (3.2.14)$$

We show later that the least-squares estimates $(\hat{\theta}_1, \hat{\theta}_2, \ldots, \hat{\theta}_p) = \hat{\theta}'$ which minimize $S(\theta)$ are given by the solutions of the *p normal equations*

$$\hat{\theta}_1 \Sigma z_1^2 + \hat{\theta}_2 \Sigma z_1 z_2 + \cdots + \hat{\theta}_p \Sigma z_1 z_p = \Sigma z_1 y,$$

$$\hat{\theta}_1 \Sigma z_2 z_1 + \hat{\theta}_2 \Sigma z_2^2 + \cdots + \hat{\theta}_p \Sigma z_2 z_p = \Sigma z_2 y,$$

$$\cdots \qquad (3.2.15)$$

$$\cdots$$

$$\hat{\theta}_1 \Sigma z_p z_1 + \hat{\theta}_2 \Sigma z_p z_2 + \cdots + \hat{\theta}_p \Sigma z_p^2 = \Sigma z_p y,$$

where the abbreviated notation indicates that, for example,

$$\Sigma z_1 z_2 = \sum_{u=1}^{n} z_{1u} z_{2u} \quad \text{and} \quad \Sigma z_1 y = \sum_{u=1}^{n} z_{1u} y_u. \qquad (3.2.16)$$

Example 3.4. Consider the following data which relate to Example 3.1.

Humidity percent, ξ:	20	30	40	50
Coded humidity, $x = (\xi - 35)/5$:	−3	−1	1	3
Observed electrical conductivity, y:	8	23	28	34

We postulate the relationship $y = \beta_0 + \beta_1 x + \varepsilon$ or $y = \theta_1 z_1 + \theta_2 z_2 + \varepsilon$, where $z_1 = 1$, $z_2 = x$, $\theta_1 = \beta_0$, and $\theta_2 = \beta_1$. Then $\Sigma z_1^2 = 4$, $\Sigma z_1 z_2 = 0$, $\Sigma z_2^2 = 20$, $\Sigma z_1 y = 93$, $\Sigma z_2 y = 83$, and the normal equations are

$$
\begin{aligned}
4\hat{\theta}_1 \qquad &= 93, \\
20\hat{\theta}_2 &= 83,
\end{aligned}
\tag{3.2.17}
$$

whence $\hat{\theta}_1 = 23.25$ and $\hat{\theta}_2 = 4.15$. The fitted least-squares straight line is thus given by

$$
\hat{y} = 23.25 + 4.15x,
\tag{3.2.18}
$$

where $\hat{y}$ denotes a *fitted* or *estimated value*. The data points and the fitted line are plotted in Figure 3.1. Note that the line looks sensible in relation to the points, providing a modest, but important, visual check that the calculations are correct.

3.3. MATRIX FORMULAS FOR LEAST SQUARES

In the general case, the p normal equations (3.2.15) are conveniently written in matrix form as

$$
\mathbf{Z}'\mathbf{Z}\hat{\boldsymbol{\theta}} = \mathbf{Z}'\mathbf{y},
\tag{3.3.1}
$$

where $\mathbf{Z}'$ (of dimension $p \times n$) is the transpose of $\mathbf{Z}$. Thus, for the conductivity experiment,

$$
\mathbf{y} = \begin{bmatrix} 8 \\ 23 \\ 28 \\ 34 \end{bmatrix}, \quad
\mathbf{Z} = \begin{bmatrix} 1 & -3 \\ 1 & -1 \\ 1 & 1 \\ 1 & 3 \end{bmatrix},
$$

$$
\mathbf{Z}'\mathbf{y} = \begin{bmatrix} 1 & 1 & 1 & 1 \\ -3 & -1 & 1 & 3 \end{bmatrix} \begin{bmatrix} 8 \\ 23 \\ 28 \\ 34 \end{bmatrix} = \begin{bmatrix} 93 \\ 83 \end{bmatrix},
$$

$$
\mathbf{Z}'\mathbf{Z} = \begin{bmatrix} 1 & 1 & 1 & 1 \\ -3 & -1 & 1 & 3 \end{bmatrix} \begin{bmatrix} 1 & -3 \\ 1 & -1 \\ 1 & 1 \\ 1 & 3 \end{bmatrix} = \begin{bmatrix} 4 & 0 \\ 0 & 20 \end{bmatrix}.
\tag{3.3.2}
$$

Thus the normal equations in matrix form (3.3.1) become

$$
\begin{bmatrix} 4 & 0 \\ 0 & 20 \end{bmatrix} \begin{bmatrix} \hat{\theta}_1 \\ \hat{\theta}_2 \end{bmatrix} = \begin{bmatrix} 93 \\ 83 \end{bmatrix},
\tag{3.3.3}
$$

which, of course, are the same equations as (3.2.17).

Conditions on the Experimental Design Necessary to Ensure that the θ's Can Be Separately Estimated

A schedule of experimental conditions is referred to as an *experimental design*. Thus, in the conductivity example, the set of conditions for humidity, $\xi = 20, 30, 40, 50$, is the experimental design. The design determines the $\mathbf{Z}$ matrix. We here consider some elementary points about the choice of an experimental design. Consider again the linear model

$$y = \theta_1 z_1 + \theta_2 z_2 + \varepsilon, \tag{3.3.4}$$

where y might, for example, refer to the rate of reaction and z_1 and z_2 might be percentages of two catalysts A and B affecting this rate. Suppose an experimental design were chosen in which the levels of z_1 and z_2 happened to be proportional to one another, so that, for every one of the runs made, $z_2 = \delta z_1$. Thus, for $\delta = 2$, for example, the percentage of catalyst B would be twice that of catalyst A, in every run. We might then have

$$\mathbf{Z} = \begin{bmatrix} 1 & 2 \\ 2 & 4 \\ 3 & 6 \\ 4 & 8 \end{bmatrix}. \tag{3.3.5}$$

In this case the model could equally well be written as

$$\begin{aligned} y &= \theta_1 z_1 + \theta_2 \delta z_1 + \varepsilon \\ &= (\theta_1 + \delta\theta_2) z_1 + \varepsilon \\ &= \delta^{-1}(\theta_1 + \delta\theta_2) z_2 + \varepsilon. \end{aligned} \tag{3.3.6}$$

The normal equations for θ_1 and θ_2 can be written down but they would not provide a unique solution for estimates of θ_1 and θ_2, which could not be estimated separately. We could estimate only the linear combination $\theta_1 + \delta\theta_2$. The reason is that, when $z_2 = \delta z_1$, changes associated with z_1 (catalyst A) are completely indistinguishable from changes associated with z_2 (catalyst B). Now $z_2 = \delta z_1$ implies that $z_2 - \delta z_1 = 0$. In general, the pathological case occurs whenever there is an exact linear relationship of the form $\alpha_1 z_1 + \alpha_2 z_2 = 0$ (for our example, $\alpha_1 = -\delta$, $\alpha_2 = 1$) linking the columns of the matrix $\mathbf{Z}$. In general, an $n \times p$ matrix $\mathbf{Z}$ (where $n \geq p$) is said to have *full* (column) *rank p* if there are no linear relationships of the form

$$\alpha_1 z_1 + \alpha_2 z_2 + \cdots + \alpha_p z_p = 0 \tag{3.3.7}$$

linking the elements of its columns, and the matrix $\mathbf{Z}'\mathbf{Z}$ is then said to be *nonsingular*. If, instead, there are $q > 0$ independent linear relationships, among the columns of $\mathbf{Z}$, then $\mathbf{Z}$ is said to have rank $p - q$ and $\mathbf{Z}'\mathbf{Z}$ is said to be *singular*. Whenever $\mathbf{Z}$ is of rank $p - q < p$, it will not be possible to estimate the p θ's separately but only $p - q$ linear functions of them.

To enable the p parameters of the model to be estimated separately, we must therefore employ an experimental design that provides an $n \times p$ matrix $\mathbf{Z}$ of full column rank. The matrix $\mathbf{Z'Z}$ will then be nonsingular and will possess an inverse. In such a case, the solution to the normal equations (3.3.1) may be written

$$\hat{\boldsymbol{\theta}} = (\mathbf{Z'Z})^{-1}\mathbf{Z'y}. \tag{3.3.8}$$

Using our example once again to illustrate this, we find

$$(\mathbf{Z'Z})^{-1} = \begin{bmatrix} \frac{1}{4} & 0 \\ 0 & \frac{1}{20} \end{bmatrix} \tag{3.3.9}$$

$$\begin{bmatrix} \hat{\theta}_1 \\ \hat{\theta}_2 \end{bmatrix} = \begin{bmatrix} \frac{1}{4} & 0 \\ 0 & \frac{1}{20} \end{bmatrix}\begin{bmatrix} 93 \\ 83 \end{bmatrix} = \begin{bmatrix} 23.25 \\ 4.15 \end{bmatrix}. \tag{3.3.10}$$

Meaning of the Fitted Constants

Consider now the meaning of the estimated parameters in the fitted line

$$\hat{y} = 23.25 + 4.15x. \tag{3.3.11}$$

Because $x = (\xi - 35)/5$, we can rewrite this equally well in terms of ξ, the percentage humidity, as

$$\hat{y} = -5.80 + 0.83\xi. \tag{3.3.12}$$

Notice that, in Figure 3.1, the estimated constant 23.25 in Eq. (3.3.11) is the value of $\hat{y}$ (the intercept) when $x = 0$. Similarly, in the form of Eq. (3.3.12), the intercept when $\xi = 0$ is -5.80. Outside the range of humidities actually tested (20 to 50%) the fitted line is shown as a broken line in Figure 3.1, because we do not have data to check its approximate validity outside that range. In particular, the intercept -5.80 should be regarded merely as a construction point in drawing the line over the experimental range. (Obviously the idea that electrical conductivity has a negative value at zero humidity is nonsense.)

The second estimated parameter in the fitted equation measures the gradient of the line in the units of measurement employed. Thus $\hat{y}$ increases by 4.15 units for each unit change in x or, equivalently, by $0.83 = 4.15/5$ units for each unit change in ξ.

Calculation of Residuals

Using Eq. (3.3.11), we can calculate, for each x value, corresponding values of $\hat{y}$ and of $e = y - \hat{y}$ as shown in Table 3.1. The reader should note that $\Sigma ez_1 = 0$ and $\Sigma ez_2 = 0$, namely, $-2.8 + 3.9 + 0.6 - 1.7 = 0$ and $-3(-2.8) - 1(3.9) + 1(0.6) + 3(-1.7) = 0$. These results provide a check on the calculations and are fundamental properties which are, in theory, exactly true. (In practice, they are valid only to the level of the rounding error in the calculations.) As we shall see in the next section, all least-squares estimates are such that the residuals have zero sum of

Table 3.1. Calculation of Residuals for Our Examples

z_1	$z_2 = x$	y	$\hat{y}$	$e = y - \hat{y}$
1	-3	8	10.8	-2.8
1	-1	23	19.1	3.9
1	1	28	27.4	0.6
1	3	34	35.7	-1.7
Sums		93	93.0	0

Table 3.2. A Portion of the Barella and Sust Data on Failure of Worsted Yarn; the Response Is log(Cycles to Failure)

x_1	x_2	x_3	y	$\hat{y}$	$e = y - \hat{y}$
-1	-1	-1	2.83	2.84	-0.01
1	-1	-1	3.56	3.59	-0.03
-1	1	-1	2.23	2.25	-0.02
1	1	-1	3.06	3.00	0.06
-1	-1	1	2.47	2.49	-0.02
1	-1	1	3.30	3.24	0.06
-1	1	1	1.95	1.90	0.05
1	1	1	2.56	2.65	-0.09

products with each one of the z's. It is this property, formally stated, which produces *normal* equations.

The residuals are marked in Figure 3.1; they are the vertical distances between the observed data points y and the corresponding values $\hat{y}$ on the fitted straight line. As we discuss later, they provide all the information available from the data on the adequacy of fit of the model.

Example 3.5. For a further illustration of least-squares calculations, consider Table 3.2 which shows again the eight observations already discussed in Section 2.2. To these were fitted the model

$$y = \beta_0 + \beta_1 x_1 + \beta_2 x_2 + \beta_3 x_3 + \varepsilon. \qquad (3.3.13)$$

In the notation of Eq. (3.1.4), $p = 4$, $z_1 = 1$, $z_2 = x_1$, $z_3 = x_2$, $z_4 = x_3$, $\theta_1 = \beta_0$, $\theta_2 = \beta_1$, $\theta_3 = \beta_2$, and $\theta_4 = \beta_3$. The model can then be written as

$$\mathbf{y} = \mathbf{Z}\boldsymbol{\theta} + \boldsymbol{\varepsilon} \qquad (3.3.14)$$

with

$$
\mathbf{y} = \begin{bmatrix} 2.83 \\ 3.56 \\ 2.23 \\ 3.06 \\ 2.47 \\ 3.30 \\ 1.95 \\ 2.56 \end{bmatrix}, \quad \mathbf{Z} = \begin{bmatrix} 1 & -1 & -1 & -1 \\ 1 & 1 & -1 & -1 \\ 1 & -1 & 1 & -1 \\ 1 & 1 & 1 & -1 \\ 1 & -1 & -1 & 1 \\ 1 & 1 & -1 & 1 \\ 1 & -1 & 1 & 1 \\ 1 & 1 & 1 & 1 \end{bmatrix},
$$
$$
8 \times 1 \qquad\qquad\qquad 8 \times 4
$$

$$
\boldsymbol{\theta} = \begin{bmatrix} \beta_0 \\ \beta_1 \\ \beta_2 \\ \beta_3 \end{bmatrix}, \quad \boldsymbol{\varepsilon} = \begin{bmatrix} \varepsilon_1 \\ \varepsilon_2 \\ \varepsilon_3 \\ \varepsilon_4 \\ \varepsilon_5 \\ \varepsilon_6 \\ \varepsilon_7 \\ \varepsilon_8 \end{bmatrix}, \quad \mathbf{Z'Z} = \begin{bmatrix} 8 & 0 & 0 & 0 \\ 0 & 8 & 0 & 0 \\ 0 & 0 & 8 & 0 \\ 0 & 0 & 0 & 8 \end{bmatrix},
$$
$$
4 \times 1 \qquad\qquad 8 \times 1
$$

$$
(\mathbf{Z'Z})^{-1} = \begin{bmatrix} \frac{1}{8} & 0 & 0 & 0 \\ 0 & \frac{1}{8} & 0 & 0 \\ 0 & 0 & \frac{1}{8} & 0 \\ 0 & 0 & 0 & \frac{1}{8} \end{bmatrix}, \quad \mathbf{Z'y} = \begin{bmatrix} 21.96 \\ 3.00 \\ -2.36 \\ -1.40 \end{bmatrix}.
$$

Thus

$$
\hat{\boldsymbol{\theta}} = \begin{bmatrix} \frac{1}{8} & 0 & 0 & 0 \\ 0 & \frac{1}{8} & 0 & 0 \\ 0 & 0 & \frac{1}{8} & 0 \\ 0 & 0 & 0 & \frac{1}{8} \end{bmatrix} \begin{bmatrix} 21.96 \\ 3.00 \\ -2.36 \\ -1.40 \end{bmatrix} = \begin{bmatrix} 2.745 \\ 0.375 \\ -0.295 \\ -0.175 \end{bmatrix}. \tag{3.3.15}
$$

The fitted equation is therefore

$$
\hat{y} = 2.745 + 0.375x_1 - 0.295x_2 - 0.175x_3. \tag{3.3.16}
$$

The contours of the fitted plane in terms of the coded variables x_1, x_2, x_3 were shown in Figure 2.7a; in terms of the original variables ξ_1 (length), ξ_2 (amplitude), ξ_3 (load), the contours appear in Figure 2.7b. The fitted values and residuals are shown in the last two columns of Table 3.2. We shall discuss this example further as we proceed.

Table 3.3. Dyestuff Yields at Various Reaction Temperatures, Fitted Values, and Residuals

Reaction Temperature	x	Yield (%), y	$\hat{y}$	e
56	-4	45.9	46.0	-0.1
60	0	79.8	78.6	1.2
61	1	78.9	80.5	-1.6
63	3	77.1	76.6	0.5
65	5	62.5	62.6	-0.1

Exercise. Confirm that the set of residuals is orthogonal to the set of elements in every column of the $\mathbf{Z}$ matrix, namely, that $\Sigma ez_1 = 0$, $\Sigma ez_2 = 0$, $\Sigma ez_3 = 0$, and $\Sigma ez_4 = 0$. □

Example 3.6. Table 3.3 contains data obtained in a laboratory experiment to examine the change in percentage yield y of a certain dyestuff as the reaction temperature was changed. The experimenter believed that the yield would pass through a maximum in this temperature range, and that the true relationship could perhaps be graduated by a quadratic (or second-order) expression in the variable temperature,

$$y = \beta_0 + \beta_1 x + \beta_{11} x^2 + \varepsilon,$$

where $x = (\text{temperature} - 60)$. Thus, for this example, $p = 3$, $z_1 = 1$, $z_2 = x$, $z_3 = x^2$, $\theta_1 = \beta_0$, $\theta_2 = \beta_1$, and $\theta_3 = \beta_{11}$. The model is of the form (3.2.11) with

$$
\mathbf{y} = \begin{bmatrix} 45.9 \\ 79.8 \\ 78.9 \\ 77.1 \\ 62.5 \end{bmatrix}, \quad
\mathbf{Z} = \begin{bmatrix} 1 & -4 & 16 \\ 1 & 0 & 0 \\ 1 & 1 & 1 \\ 1 & 3 & 9 \\ 1 & 5 & 25 \end{bmatrix}, \quad
\boldsymbol{\theta} = \begin{bmatrix} \beta_0 \\ \beta_1 \\ \beta_{11} \end{bmatrix}, \quad
\boldsymbol{\varepsilon} = \begin{bmatrix} \varepsilon_1 \\ \varepsilon_2 \\ \varepsilon_3 \\ \varepsilon_4 \\ \varepsilon_5 \end{bmatrix}, \quad (3.3.17)
$$

$$
\mathbf{Z'Z} = \begin{bmatrix} 5 & 5 & 51 \\ 5 & 51 & 89 \\ 51 & 89 & 963 \end{bmatrix},
$$

$$
(\mathbf{Z'Z})^{-1} = \begin{bmatrix} 0.435321 & -0.002917 & -0.022785 \\ -0.002917 & 0.023398 & -0.002008 \\ -0.022785 & -0.002008 & 0.002431 \end{bmatrix},
$$

$$
\mathbf{Z'y} = \begin{bmatrix} 344.2 \\ 439.1 \\ 3069.7 \end{bmatrix}, \quad
\hat{\boldsymbol{\theta}} = (\mathbf{Z'Z})^{-1}\mathbf{Z'y} = \begin{bmatrix} 78.614709 \\ 3.106313 \\ -1.263149 \end{bmatrix}. \quad (3.3.18)
$$

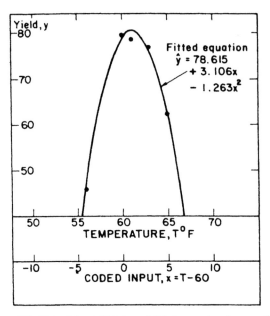

Figure 3.2. Plot of dyestuff data and fitted second-order equation.

The fitted equation is thus

$$\hat{y} = 78.615 + 3.106x - 1.263x^2. \tag{3.3.19}$$

A plot of the original data and the fitted second-order curve is shown in Figure 3.2. The fitted values and residuals appear in the last two columns of Table 3.3.

3.4. GEOMETRY OF LEAST SQUARES

A deep understanding of least squares and its associated analysis is possible using elementary ideas of coordinate geometry.

Least Squares with One Regressor

Consider the simple model

$$y = \beta + \varepsilon, \tag{3.4.1}$$

which expresses the fact that y is varying about an unknown mean β. Suppose we have just three observations of y, $\mathbf{y} = (4, 1, 1)'$. Then, as we have seen, the model can be written

$$\mathbf{y} = \mathbf{z}_1 \boldsymbol{\theta} + \boldsymbol{\varepsilon} \tag{3.4.2}$$

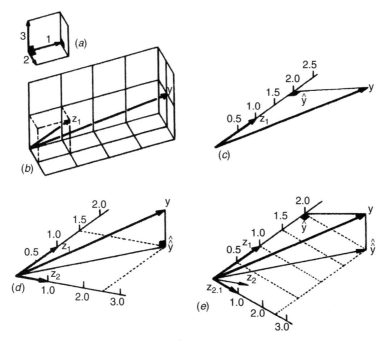

Figure 3.3. The geometry of least-squares estimation for a simple example involving three observations.

with $\mathbf{z}_1 = (1, 1, 1)'$ and $\boldsymbol{\theta} = \theta = \beta$. That is,

$$\begin{bmatrix} 4 \\ 1 \\ 1 \end{bmatrix} = \begin{bmatrix} 1 \\ 1 \\ 1 \end{bmatrix} \theta + \begin{bmatrix} \varepsilon_1 \\ \varepsilon_2 \\ \varepsilon_3 \end{bmatrix}. \tag{3.4.3}$$

Now the vectors $\mathbf{y}$ and $\mathbf{z}_1$ may be represented by points in three-dimensional space. When there are n observations, in general, we shall need an n-dimensional space, but it is possible to argue directly by analogy. Suppose we agree that the first, second, and third coordinates of every vector will be measured along the three axes marked 1, 2, and 3, respectively, in Figure 3.3a. Then we can construct a three-dimensional grid representing "three-dimensional graph paper" like that shown in Figure 3.3b with unit cubes being the three-dimensional analogy of unit squares on ordinary paper. On such a grid, we can represent a three-element vector $\mathbf{y} = (y_1, y_2, y_3)'$, say, by a line* joining the origin $(0, 0, 0)$ to the point with coordinates (y_1, y_2, y_3). In Figure 3.3b we shown the vectors $\mathbf{z}_1 = (1, 1, 1)'$ and $\mathbf{y} = (4, 1, 1)'$ which appear in Eq. (3.4.3). Recall three basic facts of coordinate

*A vector has magnitude and direction but holds no particular *position* in space. Any other line having the same length and pointing in the same direction represents the same vector.

geometry:

1. The squared distance between any two points $U = (u_1, u_2, u_3)$ and $V = (v_1, v_2, v_3)$ is $(v_1 - u_1)^2 + (v_2 - u_2)^2 + (v_3 - u_3)^2$, the sum of squares of the differences of their coordinates. Thus, in particular, the squared length of a vector is equal to the sum of squares of its elements. Thus, the length of the vector $\mathbf{y} = (4, 1, 1)'$ is $|\mathbf{y}| = (4^2 + 1^2 + 1^2)^{1/2} = (18)^{1/2} = 3(2)^{1/2}$ and the length of $\mathbf{z} = (1, 1, 1)'$ is $|\mathbf{z}| = (1^2 + 1^2 + 1^2)^{1/2} = 3^{1/2}$.

2. If ϕ is the angle between two vectors $\mathbf{u} = (u_1, u_2, u_3)'$ and $\mathbf{v} = (v_1, v_2, v_3)'$,

$$\cos \phi = \frac{u_1 v_1 + u_2 v_2 + u_3 v_3}{\left(u_1^2 + u_2^2 + u_3^2\right)^{1/2}\left(v_1^2 + v_2^2 + v_3^2\right)^{1/2}}$$

$$= \frac{\Sigma uv}{\left\{(\Sigma u^2)(\Sigma v^2)\right\}^{1/2}}$$

$$= \frac{\mathbf{u}'\mathbf{v}}{\left\{(\mathbf{u}'\mathbf{u})(\mathbf{v}'\mathbf{v})\right\}^{1/2}}. \tag{3.4.4}$$

It follows that, if $\mathbf{u}'\mathbf{v} = 0$, the vectors $\mathbf{u}$ and $\mathbf{v}$ are at right angles to each other, that is, are orthogonal.

3. Multiplication of a vector by a constant multiplies all the element of that vector by the constant. Thus, for example, $2.5\mathbf{z}_1$ is the vector $(2.5, 2.5, 2.5)'$ whose endpoint is marked 2.5 in Figure 3.3c and which is in the same direction as $\mathbf{z}_1$ but is 2.5 times as long.

Least-Squares Estimates

Let us now look at Figure 3.3c. The vector $\hat{\mathbf{y}} = \mathbf{z}_1 \hat{\theta}$ is the point in the direction of z_1 which makes

$$\Sigma e^2 = e_1^2 + e_2^2 + e_3^2 = \left(y_1 - \hat{y}_1\right)^2 + \left(y_2 - \hat{y}_2\right)^2 + \left(y_3 - \hat{y}_3\right)^2 \tag{3.4.5}$$

as small as possible, that is, the point that makes the distance $|\mathbf{e}| = |\mathbf{y} - \hat{\mathbf{y}}|$ as small as possible. Thus $\hat{\theta}$ must be chosen so that the vector $\mathbf{y} - \hat{\mathbf{y}}$ is at right angles to the vector $\mathbf{z}_1$, as indicated in Figure 3.3c. It follows that

$$\mathbf{z}_1'(\mathbf{y} - \hat{\mathbf{y}}) = 0,$$

or

$$\mathbf{z}_1'\left(\mathbf{y} - \mathbf{z}_1 \hat{\theta}\right) = 0, \tag{3.4.6}$$

that is,

$$\Sigma z_1 y - \hat{\theta} \Sigma z_1^2 = 0.$$

This is the single normal equation for this example and corresponds to Eqs. (3.2.15) and (3.3.1) for $p = 1$. The solution is clearly

$$\hat{\theta} = \left(\Sigma z_1^2\right)^{-1} \Sigma z_1 y. \tag{3.4.7}$$

For our numerical example, $\Sigma z_1^2 = 3$, $\Sigma z_1 y = 6$, and hence $\hat{\theta} = \bar{y} = 2$. Thus the fitted model obtained by the method of least squares is

$$\hat{\mathbf{y}} = 2\mathbf{z}_1 \tag{3.4.8}$$

as is evident by inspection of Figure 3.3c. The relationships between the original observations, the fitted values given by Eq. (3.4.8), and the residuals $y - \hat{y}$ are expressed by

$$\mathbf{y} = \hat{\mathbf{y}} + (\mathbf{y} - \hat{\mathbf{y}}),$$

$$\begin{bmatrix} 4 \\ 1 \\ 1 \end{bmatrix} = \begin{bmatrix} 2 \\ 2 \\ 2 \end{bmatrix} + \begin{bmatrix} 2 \\ -1 \\ -1 \end{bmatrix}. \tag{3.4.9}$$

3.5. ANALYSIS OF VARIANCE FOR ONE REGRESSOR

Suppose that special interest were associated with some particular value of θ, say $\theta_0 = 0.5$, and we wanted to check whether the null hypothesis $H_0: \theta = \theta_0 = 0.5$ was plausible. If this null hypothesis were true, the mean observation vector would be given by

$$\boldsymbol{\eta}_0 = \mathbf{z}_1 \theta_0, \tag{3.5.1}$$

or

$$\begin{bmatrix} 0.5 \\ 0.5 \\ 0.5 \end{bmatrix} = \begin{bmatrix} 1 \\ 1 \\ 1 \end{bmatrix} 0.5. \tag{3.5.2}$$

The appropriate observation breakdown

$$\mathbf{y} - \boldsymbol{\eta}_0 = (\hat{\mathbf{y}} - \boldsymbol{\eta}_0) + (\mathbf{y} - \hat{\mathbf{y}}) \tag{3.5.3}$$

would then be

$$\begin{bmatrix} 3.5 \\ 0.5 \\ 0.5 \end{bmatrix} = \begin{bmatrix} 1.5 \\ 1.5 \\ 1.5 \end{bmatrix} + \begin{bmatrix} 2 \\ -1 \\ -1 \end{bmatrix}. \tag{3.5.4}$$

Associated with this breakdown is the analysis of variance in Table 3.4.

We shall discuss the use of this table later. For the moment, notice that it is essentially a bookkeeping analysis. The sums of squares represent the squared lengths of the vectors, and the corresponding degrees of freedom indicate the number of dimensions in which the various vectors are free to move. To see this, consider the situation *before the data became available*. The data $\mathbf{y}$ could lie anywhere in the three-dimensional space so that $\mathbf{y} - \boldsymbol{\eta}_0$ has three degrees of freedom (3 df). However, the model (3.4.2) says that, whatever the data, the systematic part $\hat{\mathbf{y}} - \boldsymbol{\eta}_0 = (\hat{\theta} - \theta_0)\mathbf{z}_1$ of $\mathbf{y} - \boldsymbol{\eta}_0$ *must* lie in the direction of $\mathbf{z}_1$ so

Table 3.4. Analysis of Variance Associated with the Null Hypothesis $H_0: \theta = 0.5 \ (= \theta_0)$

Source	Degrees of Freedom (df)	Sum of Squares ($=$ length^2) (SS)	Mean Square (MS)	F	Expected Value of Mean Square, E(MS)
Model	1	$\|\hat{\mathbf{y}} - \boldsymbol{\eta}_0\|^2 = (\hat{\theta} - \theta_0)^2 \Sigma z_1^2$ = 6.75	6.75	$F_0 = 2.25$	$\sigma^2 + (\theta - \theta_0)^2 \Sigma z_1^2$
Residual	2	$\|\mathbf{y} - \hat{\mathbf{y}}\|^2 = \Sigma(y - \hat{\theta} z_1)^2$ = 6.00	3.00		σ^2
Total	3	$\|\mathbf{y} - \boldsymbol{\eta}_0\|^2 = \Sigma(y - \eta_0)^2$ = 12.75			

that $\hat{\mathbf{y}} - \boldsymbol{\eta}_0$ has only one degree of freedom (1 df). Finally, whatever the data, the residual vector must be perpendicular to $\mathbf{z}_1$ and is therefore free to move in two dimensions and thus has two degrees of freedom (2 df). The sum of the degrees of freedom for model ($\hat{\mathbf{y}} - \boldsymbol{\eta}_0$) and for residual ($\mathbf{y} - \hat{\mathbf{y}}$) is equal to n (i.e., to that for total $\mathbf{y} - \boldsymbol{\eta}_0$).

The sums of squares in Table 3.4 are the squared lengths of the vectors $\hat{\mathbf{y}} - \boldsymbol{\eta}_0$, $\mathbf{y} - \hat{\mathbf{y}}$, and $\mathbf{y} - \boldsymbol{\eta}_0$. It follows from Pythagoras's theorem that the sums of squares for the model and residual add up to give the total sum of squares.

Now what can be said about the null hypothesis that $\theta = \theta_0 = 0.5$? Look at the analysis of variance table (Table 3.4) and at Figure 3.4. The component $\|\hat{\mathbf{y}} - \boldsymbol{\eta}_0\|^2 = (\hat{\theta} - \theta_0)^2 \Sigma z^2$ is a measure of the discrepancy between the postulated model $\boldsymbol{\eta}_0 = \mathbf{z}_1 \theta_0$ and the estimated model $\hat{\mathbf{y}} = \mathbf{z}_1 \hat{\theta}$. Under the standard assumptions mentioned earlier, it can be shown that the expected value of the sum of squares for the model component assuming the model true is $(\theta - \theta_0)^2 \Sigma z_1^2 + \sigma^2$, while that for the residual component is $2\sigma^2$ (or, in general, $\nu_2 \sigma^2$, where ν_2 is the number of degrees of freedom of the residuals). A natural measure of discrepancy from the null hypothesis $\theta = \theta_0$ is therefore

$$F_0 = \frac{\|\hat{\mathbf{y}} - \boldsymbol{\eta}_0\|^2/1}{\|\mathbf{y} - \hat{\mathbf{y}}\|^2/2} = \frac{\text{mean square for model component}}{\text{mean square for residual}}. \qquad (3.5.5)$$

If the null hypothesis $\theta = \theta_0$ were true, then both the numerator and the denominator of F_0 would estimate the same σ^2. A value of F_0 greater than 1 suggests the existence of $\theta - \theta_0$ and hence of a possible departure from the null hypothesis that $\theta = \theta_0$. Now consider the angle ϕ_0, shown on Figure 3.4a, between the vector $\mathbf{y} - \boldsymbol{\eta}_0$ and the $\mathbf{z}_1$ vector. We see that

$$F_0 = 2 \cot^2 \phi_0. \qquad (3.5.6)$$

Thus the smaller the ϕ_0 (and the larger is F_0), the more we shall be led to doubt the null hypothesis. Now on the standard assumptions, the errors $\boldsymbol{\varepsilon} = \mathbf{y} - \boldsymbol{\eta}$ are

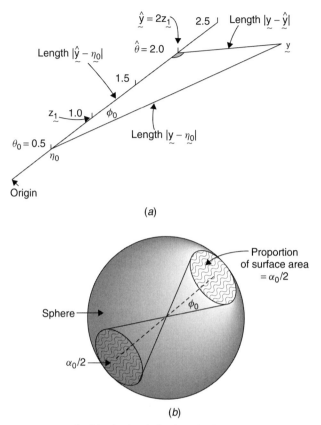

Figure 3.4. Geometry connected with check of the hypothesis $\theta = 0.5$ for the model $y = \theta z + \varepsilon$. (*a*) The sum of squares breakup. (*b*) The significance level as a surface area.

identically, independently, and normally distributed, so that

$$p(\mathbf{y} - \boldsymbol{\eta}) = \text{constant } \sigma^{-3} \exp \left\{ \frac{-\frac{1}{2} \sum_{u} (y_u - \eta_u)^2}{\sigma^2} \right\}$$

$$= f\left\{ \sum_{u} (y_u - \eta_u)^2 \right\}. \tag{3.5.7}$$

Thus, the three-dimensional distribution for the y's has spherical contours and, if the null hypothesis is true and $\boldsymbol{\eta} = \boldsymbol{\eta}_0$, the contours of probability density will be spheres centered at $\boldsymbol{\eta}_0$. To determine how often an angle as small as ϕ_0 will occur by chance is evidently equivalent to asking "How large are the two caps of a sphere subtended by a cone of rotation of angle ϕ_0 compared to the area of the sphere's surface?" (see Figure 3.4*b*).

Calculations equivalent to the above have been made to obtain the tables of the percentage points of the F distribution given in this book. For a series of upper-tail reference probabilities, denoted by α and called *significance levels*, the fixed numbers $F_\alpha(\nu_1, \nu_2)$ are given such that $\text{prob}\{F(\nu_1, \nu_2) \geq F_\alpha(\nu_1, \nu_2)\} = \alpha$, where $F(\nu_1, \nu_2)$ denotes the random variable of an F distribution with ν_1 and ν_2 df. The F_0 value obtained in the analysis of variance table is compared to this set of reference $F_\alpha(\nu_1, \nu_2)$'s. For our example, $F_0 = 2.25$. The tables show that, for $\nu_1 = 1$, $\nu_2 = 2$ df, this value would be exceeded by chance with a probability exceeding 25% (i.e., with an $\alpha > 0.25$) so that the value $\theta = \theta_0 = 0.5$ is not discredited by the data at even the $\alpha = 0.25$ level.

An important special case arises when the null hypothesis is $H_0: \theta = \theta_0 = 0$, implying that there is no relationship between y and the regressor z_1 or, equivalently in this case, that the mean of the data is zero.

Exercise. Compute the analysis of variance table for the data when $\theta_0 = 0$ (i.e., when $\boldsymbol{\eta}_0 = \mathbf{0}$). (*Answer*: $12 + 6 = 18$, same df.)   □

3.6. LEAST SQUARES FOR TWO REGRESSORS

Our previous model, Eq. (3.4.1), said that y could be represented by a mean value $\beta = \theta$ plus an error. We now suppose instead that, for the same observations $\mathbf{y}' = (4, 1, 1)$, it was believed that there might be systematic deviations from the mean associated with the humidity in the laboratory. Suppose the coded levels of humidity were 0.8, 0.2, and -0.4 in the three experimental runs. Thus, the model would now be the equation of a straight line

$$y = \beta_0 + \beta_1 x + \varepsilon, \tag{3.6.1}$$

or

$$\mathbf{y} = \mathbf{z}_1 \theta_1 + \mathbf{z}_2 \theta_2 + \boldsymbol{\varepsilon} \tag{3.6.2}$$

with $\mathbf{y} = (4, 1, 1)'$, $\theta_1 = \beta_0$, $\theta_2 = \beta_1$, $\mathbf{z}_1 = (1, 1, 1)'$, and $\mathbf{z}_2 = (0.8, 0.2, -0.4)'$.

Least-Squares Estimates

Now examine Figure 3.3d. Whereas the previous model function $\boldsymbol{\eta} = \mathbf{z}_1 \theta_1$ with $\mathbf{z}_1 = (1, 1, 1)'$ said that the vector $\boldsymbol{\eta}$ lay on the equiangular line, the revised model

$$\boldsymbol{\eta} = \mathbf{z}_1 \theta_1 + \mathbf{z}_2 \theta_2, \tag{3.6.3}$$

says that $\boldsymbol{\eta} = E(\mathbf{y})$ is in the plane defined by linear combinations of the vectors $\mathbf{z}_1, \mathbf{z}_2$. Note that (because $\mathbf{z}_1' \mathbf{z}_2 = \Sigma z_1 z_2 = 0.6 \neq 0$) the vectors $\mathbf{z}_1$ and $\mathbf{z}_2$ are *not* at right angles. However, a point on the plane corresponding to any choice of θ_1 and θ_2 could be found by imagining the parameter plane to be covered by an oblique "graph paper" grid on which the basic element was not a square as on ordinary graph paper but a parallelogram having $\mathbf{z}_1$ and $\mathbf{z}_2$ for two of its sides.

The least-squares values $\hat{\theta}_1, \hat{\theta}_2$ which produce a vector

$$\hat{\mathbf{y}} = \mathbf{z}_1 \hat{\theta}_1 + \mathbf{z}_2 \hat{\theta}_2 \tag{3.6.4}$$

in the parameter plane are those which make the squared length $\Sigma(y - \hat{y})^2 = |\mathbf{y} - \hat{\mathbf{y}}|^2$ of the residual vector as small as possible. Thus $\hat{\mathbf{y}}$ is the foot of the perpendicular from the end of the vector $\mathbf{y}$ onto the plane defined by $\mathbf{z}_1$ and $\mathbf{z}_2$. The normal equations now express the fact that $\mathbf{y} - \hat{\mathbf{y}}$ must be perpendicular (i.e., normal) to both $\mathbf{z}_1$ and $\mathbf{z}_2$. Therefore,

$$\mathbf{z}_1'(\mathbf{y} - \hat{\mathbf{y}}) = 0$$
$$\mathbf{z}_2'(\mathbf{y} - \hat{\mathbf{y}}) = 0 \tag{3.6.5}$$

or, equivalently,

$$\Sigma z_1\left(y - \hat{\theta}_1 z_1 - \hat{\theta}_2 z_2\right) = 0$$
$$\Sigma z_2\left(y - \hat{\theta}_1 z_1 - \hat{\theta}_2 z_2\right) = 0, \tag{3.6.6}$$

yielding the normal equations (3.2.15) with $p = 2$. In matrix format,

$$\mathbf{y} = \begin{bmatrix} 4 \\ 1 \\ 1 \end{bmatrix}, \quad \mathbf{Z}' = \begin{bmatrix} 1 & 1 & 1 \\ 0.8 & 0.2 & -0.4 \end{bmatrix}, \quad \hat{\boldsymbol{\theta}} = \begin{bmatrix} \hat{\theta}_1 \\ \hat{\theta}_2 \end{bmatrix}. \tag{3.6.7}$$

The normal equations are then

$$\mathbf{Z}'\left(\mathbf{y} - \mathbf{Z}\hat{\boldsymbol{\theta}}\right) = \mathbf{0}, \tag{3.6.8}$$

which yields the equations (3.3.1) with solution (3.3.8). For our example

$$\mathbf{Z}'\mathbf{Z} = \begin{bmatrix} 3.00 & 0.60 \\ 0.60 & 0.84 \end{bmatrix}, \quad (\mathbf{Z}'\mathbf{Z})^{-1} = \tfrac{1}{18}\begin{bmatrix} 7 & -5 \\ -5 & 25 \end{bmatrix},$$

$$\mathbf{Z}'\mathbf{y} = \begin{bmatrix} 6 \\ 3 \end{bmatrix}, \quad \hat{\boldsymbol{\theta}} = \begin{bmatrix} \hat{\theta}_1 \\ \hat{\theta}_2 \end{bmatrix} = \begin{bmatrix} 1.5 \\ 2.5 \end{bmatrix}. \tag{3.6.9}$$

The least-squares fit resulting from a model of form Eq. (3.6.1) is thus

$$\hat{y} = 1.5 + 2.5x. \tag{3.6.10}$$

Inserting successively the three values of x in the data provides fitted values 3.5, 2.0, 0.5 and leads to the observation breakdown, exemplified in Figure 3.3d:

$$\mathbf{y} = \hat{\mathbf{y}} + (\mathbf{y} - \hat{\mathbf{y}}) \tag{3.6.11}$$

$$\begin{bmatrix} 4 \\ 1 \\ 1 \end{bmatrix} = \begin{bmatrix} 3.5 \\ 2.0 \\ 0.5 \end{bmatrix} + \begin{bmatrix} 0.5 \\ -1.0 \\ 0.5 \end{bmatrix}. \tag{3.6.12}$$

Let us now consider the two fitted models together.

$$p = 1, \quad \hat{y} = 2.0z_1, \qquad \text{i.e.,} \quad \hat{y} = 2,$$

$$p = 2, \quad \hat{\hat{y}} = 1.5z_1 + 2.5z_2, \quad \text{i.e.,} \quad \hat{\hat{y}} = 1.5 + 2.5x. \qquad (3.6.13)$$

The introduction of the second explanatory variable $x = z_2$ has resulted in the coefficient of z_1 changing from 2.0 to 1.5. To see why, look at Figures 3.3c and 3.3d. The projection of y on the line of z_1 is $\hat{y} = 2z_1$, but the projection of y on the plane of z_1 and z_2 is $\hat{\hat{y}} = 1.5z_1 + 2.5z_2$. However, had z_1 and z_2 been orthogonal —that is, $z_1'z_2 = 0$—the coefficient of z_1 would *not* have changed. This is easily seen from the geometry. Alternatively, we can see that it is true algebraically, for in this case the off-diagonal elements of $Z'Z$ would have been zero and the normal equations for $\hat{\theta}_1$ and $\hat{\theta}_2$ could have been solved independently. We should not be surprised by this change because the coefficient of z_1 has a different meaning in the two cases.

3.7. GEOMETRY OF THE ANALYSIS OF VARIANCE FOR TWO REGRESSORS

Now suppose special interest were associated with a null hypothesis involving particular values of the parameters, say $\theta_1 = \theta_{10} = 0.5$, $\theta_2 = \theta_{20} = 1.0$. If this null hypothesis were true, the mean observation vector $\boldsymbol{\eta}_0$ would be representable as

$$\boldsymbol{\eta}_0 = \theta_{10}z_1 + \theta_{20}z_2, \qquad (3.7.1)$$

or

$$\begin{bmatrix} 1.3 \\ 0.7 \\ 0.1 \end{bmatrix} = 0.5\begin{bmatrix} 1 \\ 1 \\ 1 \end{bmatrix} + 1.0\begin{bmatrix} 0.8 \\ 0.2 \\ -0.4 \end{bmatrix}. \qquad (3.7.2)$$

Thus we could write

$$\mathbf{y} - \boldsymbol{\eta}_0 = (\hat{\hat{\mathbf{y}}} - \boldsymbol{\eta}_0) + (\mathbf{y} - \hat{\hat{\mathbf{y}}}),$$

$$\begin{bmatrix} 2.7 \\ 0.3 \\ 0.9 \end{bmatrix} = \begin{bmatrix} 2.2 \\ 1.3 \\ 0.4 \end{bmatrix} + \begin{bmatrix} 0.5 \\ -1.0 \\ 0.5 \end{bmatrix}. \qquad (3.7.3)$$

The geometry of this breakdown will be clear from Figure 3.5a. The associated analysis of variance is given in Table 3.5. Again, the plausibility of the null hypothesis $\theta_1 = 0.5$, $\theta_2 = 1.0$ can be assessed by comparing mean squares via the F statistic

$$F_0 = \frac{|\hat{\hat{\mathbf{y}}} - \boldsymbol{\eta}_0|^2/2}{|\mathbf{y} - \hat{\hat{\mathbf{y}}}|^2/1} = 2.23. \qquad (3.7.4)$$

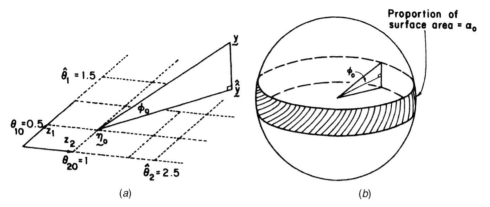

Figure 3.5. Geometry connected with check of the hypothesis $\theta_1 = 0.5$, $\theta_2 = 1.0$ for the model $y = z_1\theta_1 + z_2\theta_2 + \varepsilon$. (a) The sum of squares breakup. (b) The hypothesis probability as surface area.

Table 3.5. Analysis of Variance Split-Up Appropriate for Testing the Null Hypothesis
$\theta_1 = \theta_{10} = 0.5, \theta_2 = \theta_{20} = 1.0$

Source	df	SS		MS	F
Model z_1 and z_2	2	$|\hat{\mathbf{y}} - \mathbf{\eta}_0|^2 = \Sigma\{(\theta_1 - \theta_{10})z_1 + (\theta_2 - \theta_{20})z_2\}^2$	$= 6.69$	3.345	2.23
Residual	1	$|\mathbf{y} - \hat{\mathbf{y}}|^2 = \Sigma(y - \hat{\theta}_1 z_1 - \hat{\theta}_2 z_2)^2$	$= 1.50$	1.50	
Total	3	$|\mathbf{y} - \mathbf{\eta}_0|^2 = \Sigma(y - \eta_0)^2$	$= 8.19$		

As before, $F_0 = (\nu_2/\nu_1)\cot^2 \phi_0$, where $\nu_1 = 2$, $\nu_2 = 1$ are the degrees of freedom for numerator and denominator, respectively. If the null hypothesis is true, then the data are generated by a spherical normal distribution centered at $\mathbf{\eta}_0$ and the probability that $F > F_0$ is the ratio of the area of the shaded band in Figure 3.5b compared with the area of the whole sphere.

3.8. ORTHOGONALIZING THE SECOND REGRESSOR, EXTRA SUM OF SQUARES PRINCIPLE

In this example, the vector $\mathbf{z}_2$ is not orthogonal to $\mathbf{z}_1$. However, we can find the component $\mathbf{z}_{2\cdot1}$ of $\mathbf{z}_2$ that *is* orthogonal to $\mathbf{z}_1$ and rewrite the response function in terms of $\mathbf{z}_1$ and $\mathbf{z}_{2\cdot1}$. To obtain $\mathbf{z}_{2\cdot1}$, we use the least-squares property that a residual vector is orthogonal to the space in which the predictor variables lie. Temporarily regarding $\mathbf{z}_2$ as the "response" vector and $\mathbf{z}_1$ as the predictor variable, we obtain $\hat{\mathbf{z}}_2 = 0.2\mathbf{z}_1$ by least squares and hence the residual vector is

$$\mathbf{z}_{2\cdot1} = \mathbf{z}_2 - \hat{\mathbf{z}}_2 = \mathbf{z}_2 - 0.2\mathbf{z}_1. \tag{3.8.1}$$

Thus,

$$\mathbf{z}_{2\cdot1} = (0.6, 0, -0.6)'. \tag{3.8.2}$$

Now rewrite the model function in the form

$$\boldsymbol{\eta} = (\theta_1 + 0.2\theta_2)\mathbf{z}_1 + \theta_2(\mathbf{z}_2 - 0.2\mathbf{z}_1),$$

that is,

$$\boldsymbol{\eta} = \theta\mathbf{z}_1 + \theta_2\mathbf{z}_{2\cdot1}. \tag{3.8.3}$$

For this form of the model

$$(\mathbf{Z'Z})^{-1} = \begin{bmatrix} 3.00 & 0 \\ 0 & 0.72 \end{bmatrix}^{-1} = \tfrac{1}{18}\begin{bmatrix} 6 & 0 \\ 0 & 25 \end{bmatrix},$$

$$\mathbf{Z'y} = \begin{bmatrix} 6 \\ 1.8 \end{bmatrix}, \quad \text{and} \quad \hat{\boldsymbol{\theta}} = \begin{bmatrix} \hat{\theta} \\ \hat{\theta}_2 \end{bmatrix} = \begin{bmatrix} 2 \\ 2.5 \end{bmatrix}. \tag{3.8.4}$$

We can now compare our various fitted equations:

1. $\hat{y} = 2z_1$, for the model with one parameter
2a. $\hat{y} = 1.5z_1 + 2.5z_2$ ⎫ for the two forms of the model with
2b. $\hat{y} = 2.0z_1 + 2.5z_{2\cdot1}$ ⎭ two parameters

Note the following points:

1. Because $\mathbf{z}_1$ and $\mathbf{z}_{2\cdot1}$ are orthogonal, the coefficient of z_1 in model 2b is the same as the coefficient 2 of z_1 in model 1, which uses only a single regressor.
2. The coefficient of $z_{2\cdot1}$ in model 2b is the same as that for z_2 in 2a.

These facts repay study because they generalize directly in a manner to be discussed shortly.

The geometry of what we have just done is illustrated in Figure 3.3e. The vector $\mathbf{z}_1$ and the component $\mathbf{z}_{2\cdot1}$ of $\mathbf{z}_2$, which is orthogonal to $\mathbf{z}_1$, can be used just as well as $\mathbf{z}_1$ and $\mathbf{z}_2$ to define the plane that constitutes the response function locus. In the nonorthogonal coordinates defined by the basis vectors $\mathbf{z}_1$ and $\mathbf{z}_2$, the point $\hat{y}$ has coordinates $\hat{\theta}_1 = 1.5$, $\hat{\theta}_2 = 2.5$; in the orthogonal coordinates defined by basis vectors $\mathbf{z}_1$ and $\mathbf{z}_{2\cdot1}$, the *same point* $\hat{y}$ has coordinates $\hat{\theta} = 2$, $\hat{\theta}_2 = 2.5$.

Analysis of Variance Associated with Augmentation of the Model

Problems often arise where the efficacy of including an additional variable z_2 (or, as we shall see later, of including several additional variables) is under consideration. The appropriate null hypothesis is then that $\theta_2 = 0$. It is immaterial what value we choose for θ. Let it be such that $\boldsymbol{\eta}_0 = \mathbf{z}_1\theta$. Then the data may be broken down as follows:

$$\mathbf{y} - \boldsymbol{\eta}_0 = \hat{\mathbf{y}} - \boldsymbol{\eta}_0 + (\hat{\hat{\mathbf{y}}} - \hat{\mathbf{y}}) + (\mathbf{y} - \hat{\mathbf{y}}). \tag{3.8.5}$$

Table 3.6. Analysis of Variance Table for Our Example with $\theta = 0$ ($\boldsymbol{\eta_0} = 0$), Showing The Orthogonal Contribution Produced by Augmentation of the Model

Source	df	SS
Response function with z_1 only	1	$\lvert \hat{\mathbf{y}} - \boldsymbol{\eta_0} \rvert^2 = (\hat{\theta} - \theta_0)^2 \Sigma z_1^2 = 12.0$
Extra due to z_2 (given z_1)	1	$\lvert \hat{\hat{\mathbf{y}}} - \hat{\mathbf{y}} \rvert^2 = \hat{\theta}_2^2 \Sigma z_{2\cdot1}^2 = 4.5$
Residual	1	$\lvert \mathbf{y} - \hat{\hat{\mathbf{y}}} \rvert^2 = \Sigma(y - \hat{\hat{y}})^2 = 1.5$
Total	3	$\lvert \mathbf{y} - \boldsymbol{\eta_0} \rvert^2 = \Sigma(y - \eta_0)^2 = 18.0$

More specifically, we may as well set $\theta = 0$ so that $\boldsymbol{\eta_0} = \mathbf{0}$ and, for our example,

$$\begin{bmatrix} 4 \\ 1 \\ 1 \end{bmatrix} = \begin{bmatrix} 2 \\ 2 \\ 2 \end{bmatrix} + \begin{bmatrix} 1.5 \\ 0 \\ -1.5 \end{bmatrix} + \begin{bmatrix} 0.5 \\ -1.0 \\ 0.5 \end{bmatrix}. \tag{3.8.6}$$

The three vectors on the right are all orthogonal to one another leading to the analysis of variance of Table 3.6. The relationship in the sum of squares column may be arrived at by noting that, because

$$\hat{\mathbf{y}} = \mathbf{z}_1 \hat{\theta} \quad \text{and} \quad \hat{\hat{\mathbf{y}}} - \hat{\mathbf{y}} = \mathbf{z}_1 \hat{\theta} + \mathbf{z}_{2\cdot1} \hat{\theta}_2 - \mathbf{z}_1 \hat{\theta} = \mathbf{z}_{2\cdot1} \hat{\theta}_2, \tag{3.8.7}$$

the equations $\mathbf{y} = \hat{\mathbf{y}} + (\hat{\hat{\mathbf{y}}} - \hat{\mathbf{y}}) + (\mathbf{y} - \hat{\hat{\mathbf{y}}})$ can be written

$$\mathbf{y} = \mathbf{z}_1 \hat{\theta} + \mathbf{z}_{2\cdot1} \hat{\theta}_2 + \left(\mathbf{y} - \hat{\hat{\mathbf{y}}} \right). \tag{3.8.8}$$

The three vectors on the right are orthogonal, so that

$$\lvert \mathbf{y} \rvert^2 = \hat{\theta}^2 \lvert \mathbf{z}_1 \rvert^2 + \hat{\theta}_2^2 \lvert \mathbf{z}_{2\cdot1} \rvert^2 + \lvert \mathbf{y} - \hat{\hat{\mathbf{y}}} \rvert^2 \tag{3.8.9}$$

and the sum of squares split-up of Table 3.6 is immediately obtained.

The Extra Sum of Squares Principle

Consider the extra sum of squares due to z_2 in the analysis of variance, Table 3.6. We have

$$\lvert \hat{\hat{\mathbf{y}}} - \hat{\mathbf{y}} \rvert^2 = \lvert \hat{\hat{\mathbf{y}}} \rvert^2 - \lvert \hat{\mathbf{y}} \rvert^2$$
$$= (\text{SS for } z_1 \text{ and } z_2) - (\text{SS for } z_1 \text{ alone}). \tag{3.8.10}$$

Alternatively and equivalently,

$$\lvert \hat{\hat{\mathbf{y}}} - \hat{\mathbf{y}} \rvert^2 = \lvert \mathbf{y} - \hat{\mathbf{y}} \rvert^2 - \lvert \mathbf{y} - \hat{\hat{\mathbf{y}}} \rvert^2$$
$$= (\text{residual SS with } z_1 \text{ only}) - (\text{residual SS with } z_1 \text{ and } z_2). \tag{3.8.11}$$

We see that, to obtain the "extra SS in the analysis of variance associated with the orthogonal sum of squares," we do not need to calculate $\mathbf{z}_{2 \cdot 1}$ explicitly. All that is needed is to fit each model in turn and obtain the difference in the residual sums of squares. More generally, if we have a linear model including any set of regressors defined by the columns of matrix $\mathbf{Z}_1$, say, and we are considering a more elaborate model with additional regressors represented by the columns of $\mathbf{Z}_2$, then a sum of squares associated with the additional regressors, given the others, can always be found by first fitting the model with both sets of regressors $(\mathbf{Z}_1, \mathbf{Z}_2)$ and then fitting the simpler model with $\mathbf{Z}_1$ only. The required extra sum of squares is then obtained either as

$$(\text{regression SS for model with } \mathbf{Z}_1, \mathbf{Z}_2)$$

$$-(\text{regression SS for model with } \mathbf{Z}_1 \text{ only}) \qquad (3.8.12)$$

or as

$$(\text{residual SS for model with } \mathbf{Z}_1 \text{ only})$$

$$-(\text{residual SS for model with } \mathbf{Z}_1 \text{ and } \mathbf{Z}_2). \qquad (3.8.13)$$

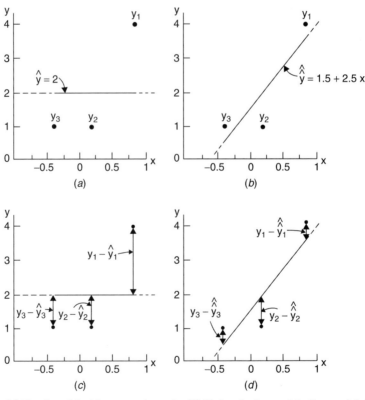

Figure 3.6. (*a*) Fitted model with mean value only. (*b*) Fitting sloping straight-line model. (*c*) Split-up of the observations into fitted values and residuals for the mean only model. (*d*) Split-up for the sloping straight-line model.

Another Way to Consider the Analysis

To see what has been done from a slightly different point of view, we can examine Figure 3.6 which shows a plot of the data in relation to the two different models. The first fitted model is the horizontal straight line $\hat{y} = 2$ in Figure 3.6a. The second fitted model is the sloping line $\hat{y} = 1.5 + 2.5x$ in Figure 3.6b; $\hat{\theta}_1 = 1.5$ is the value of $\hat{y}$ when $x = 0$ and $\hat{\theta}_2 = 2.5$ is the gradient (or slope) of the line. Figures 3.6c and 3.6d show the partitioning of the observations into fitted values and residuals.

Consider next the meaning associated with the breakdown

$$|\mathbf{y}|^2 = |\hat{\mathbf{y}}|^2 + |\hat{\hat{\mathbf{y}}} - \hat{\mathbf{y}}|^2 + |\mathbf{y} - \hat{\hat{\mathbf{y}}}|^2, \tag{3.8.14}$$

that is,

$$\Sigma y^2 = \Sigma \hat{y}^2 + \Sigma\left(\hat{\hat{y}} - \hat{y}\right)^2 + \Sigma\left(y - \hat{\hat{y}}\right)^2. \tag{3.8.15}$$

The analysis of variance of Table 3.4 but with $\theta_0 = 0$ says that the total sum of squares $y_1^2 + y_2^2 + y_3^2 = \Sigma y^2$ can be split into a model component $\hat{y}_1^2 + \hat{y}_2^2 + \hat{y}_3^2$ and a residual component $(y_1 - \hat{y}_1)^2 + (y_2 - \hat{y}_2)^2 + (y_3 - \hat{y}_3)^2$. Because $\hat{y} = z_1\hat{\theta} = \hat{\theta} = \bar{y}$, the model component measures the deviation of the sample mean from zero. For the analysis of the sloping-line model, the model component $\hat{\hat{y}}_1^2 + \hat{\hat{y}}_2^2 + \hat{\hat{y}}_3^2$ measures the deviation of the fitted straight-line values from zero and leaves a residual component $(y_1 - \hat{\hat{y}}_1)^2 + (y_2 - \hat{\hat{y}}_2)^2 + (y_3 - \hat{\hat{y}}_3)^2$. In the analysis of variance of Table 3.6, the component $\hat{\hat{y}}$ is split up into the parts $\hat{y} + (\hat{\hat{y}} - \hat{y}) = \bar{y} + \hat{\theta}_2 z_{2\cdot1} = \bar{y} + \hat{\theta}_2(x - \bar{x})$. The first represents the effect of the average and the second the deviations of the fitted values given by the straight line from the fitted average.

3.9. GENERALIZATION TO p REGRESSORS

All these ideas can be readily generalized to any number of observations n and any number of parameters p. The development is most easily carried out using matrix algebra but the reader should keep in mind, as a guide, the geometrical interpretations set out for $n = 3$ above.

Least Squares and the Normal Equations

The n relations implicit in the response function may be written

$$\boldsymbol{\eta} = \mathbf{Z}\boldsymbol{\theta}, \tag{3.9.1}$$

where it is assumed that $\mathbf{Z}$ is of full column rank p and hence that $\mathbf{Z}'\mathbf{Z}$ is positive definite and invertible. Let $\hat{\boldsymbol{\theta}}$ be the vector of estimates given by the normal equations

$$(\mathbf{y} - \hat{\mathbf{y}})'\mathbf{Z} = \mathbf{0}, \tag{3.9.2}$$

that is,

$$\left(\mathbf{y} - \mathbf{Z}\hat{\boldsymbol{\theta}}\right)'\mathbf{Z} = \mathbf{0}. \tag{3.9.3}$$

It is clear that $\hat{\boldsymbol{\theta}}$ provides least-squares estimates from the following argument. The sum of squares function is

$$
\begin{aligned}
S(\boldsymbol{\theta}) &= (\mathbf{y} - \boldsymbol{\eta})'(\mathbf{y} - \boldsymbol{\eta}) \\
&= (\mathbf{y} - \hat{\mathbf{y}})'(\mathbf{y} - \hat{\mathbf{y}}) + (\hat{\mathbf{y}} - \boldsymbol{\eta})'(\hat{\mathbf{y}} - \boldsymbol{\eta})
\end{aligned} \tag{3.9.4}
$$

because the cross-product is zero from the normal equations. Thus

$$S(\boldsymbol{\theta}) = S(\hat{\boldsymbol{\theta}}) + (\hat{\boldsymbol{\theta}} - \boldsymbol{\theta})'\mathbf{Z}'\mathbf{Z}(\hat{\boldsymbol{\theta}} - \boldsymbol{\theta}). \tag{3.9.5}$$

However, because $\mathbf{Z}'\mathbf{Z}$ is positive definite, $S(\boldsymbol{\theta})$ is minimized when $\boldsymbol{\theta} = \hat{\boldsymbol{\theta}}$. Thus the normal equations solution

$$\hat{\boldsymbol{\theta}} = (\mathbf{Z}'\mathbf{Z})^{-1}\mathbf{Z}'\mathbf{y} \tag{3.9.6}$$

will always produce least-squares estimates.

Analysis of Variance

The breakdown (3.9.4) can be conveniently set out in the form of Table 3.7.

Orthogonal Breakdown

Suppose that the p regressors represented by the p columns of $\mathbf{Z}$ split naturally into two sets $\mathbf{Z}_1, \mathbf{Z}_2$ of p_1, p_2 columns, respectively, so that the response function $\boldsymbol{\eta} = \mathbf{Z}\boldsymbol{\theta}$ can also be written

$$\boldsymbol{\eta} = \mathbf{Z}_1\boldsymbol{\theta}_1 + \mathbf{Z}_2\boldsymbol{\theta}_2. \tag{3.9.7}$$

The simpler model function $\boldsymbol{\eta} = \mathbf{Z}_1\boldsymbol{\theta}_1$ might be one which it was hoped might be adequate and $\mathbf{Z}_2\boldsymbol{\theta}_2$ might represent further terms which perhaps would have to be added if the terms of $\mathbf{Z}_1\boldsymbol{\theta}_1$ were inadequate to represent the response. Suppose further (at first) that $\mathbf{Z}_1$ and $\mathbf{Z}_2$ were orthogonal so that $\mathbf{Z}_1'\mathbf{Z}_2 = \mathbf{0}$. Then the equation $\hat{\boldsymbol{\theta}} = (\mathbf{Z}'\mathbf{Z})^{-1}\mathbf{Z}'\mathbf{y}$ which provides the least-squares estimator $\hat{\boldsymbol{\theta}}$ of $\boldsymbol{\theta}$ splits into the two parts

$$\hat{\boldsymbol{\theta}}_1 = (\mathbf{Z}_1'\mathbf{Z}_1)^{-1}\mathbf{Z}_1'\mathbf{y} \quad \text{and} \quad \hat{\boldsymbol{\theta}}_2 = (\mathbf{Z}_2'\mathbf{Z}_2)^{-1}\mathbf{Z}_2'\mathbf{y}. \tag{3.9.8}$$

Table 3.7. Analysis of Variance Table for the Case of General Regression

Source	df	SS
Response function	p	$\lvert\hat{\mathbf{y}} - \boldsymbol{\eta}\rvert^2 = (\hat{\boldsymbol{\theta}} - \boldsymbol{\theta})'\mathbf{Z}'\mathbf{Z}(\hat{\boldsymbol{\theta}} - \boldsymbol{\theta})$
Residual	$n - p$	$\lvert\mathbf{y} - \hat{\mathbf{y}}\rvert^2 = \Sigma(y - \hat{y})^2$
Total	n	$\lvert\mathbf{y} - \boldsymbol{\eta}\rvert^2 = \Sigma(y - \eta)^2$

Table 3.8. Analysis of Variance Table when Model Splits Up into Orthogonal Parts (i.e., $Z_1'Z_2 = 0$)

Source	df	SS
Response function (Z_1 only)	p_1	$(\hat{\boldsymbol{\theta}}_1 - \boldsymbol{\theta}_1)'Z_1'Z_1(\hat{\boldsymbol{\theta}}_1 - \boldsymbol{\theta}_1)$
Extra for Z_2	p_2	$(\hat{\boldsymbol{\theta}}_2 - \boldsymbol{\theta}_2)'Z_2'Z_2(\hat{\boldsymbol{\theta}}_2 - \boldsymbol{\theta}_2)$
Residual	$n - p_1 - p_2$	$(y - \hat{y})'(y - \hat{y}) = \Sigma(y - \hat{y})^2$
Total	n	$(y - \boldsymbol{\eta})'(y - \boldsymbol{\eta}) = \Sigma(y - \eta)^2$

Moreover, the sum of squares for the response function of the more elaborate model may be written as

$$(\boldsymbol{\theta} - \hat{\boldsymbol{\theta}})'Z'Z(\boldsymbol{\theta} - \hat{\boldsymbol{\theta}}) = (\boldsymbol{\theta}_1 - \hat{\boldsymbol{\theta}}_1)Z_1'Z_1(\boldsymbol{\theta}_1 - \hat{\boldsymbol{\theta}}_1) + (\boldsymbol{\theta}_2 - \hat{\boldsymbol{\theta}}_2)Z_2'Z_2(\boldsymbol{\theta}_2 - \hat{\boldsymbol{\theta}}_2)$$

$$(3.9.9)$$

and the associated analysis of variance is thus as in Table 3.8.

Orthogonalization when $Z_1'Z_2 \neq 0$

Consider the model function $\boldsymbol{\eta} = Z_1\boldsymbol{\theta}_1$. An important fact about least-squares estimates, which follows from the normal equations, is that the vector of residuals

$$y - \hat{y} = y - Z_1\hat{\boldsymbol{\theta}} \qquad (3.9.10)$$

is orthogonal to every one of the columns of Z_1. This thus provides a general method for determining that component of a vector y which is orthogonal to a space of p_1 dimensions whose basis vectors are the columns of a matrix Z_1. The required orthogonal component, $y_{\cdot 1}$ say, is given by

$$\begin{aligned} y_{\cdot 1} &= y - \hat{y} = y - Z_1\hat{\boldsymbol{\theta}}_1 \\ &= y - Z_1(Z_1'Z_1)^{-1}Z_1'y \\ &= (I - R_1)y \qquad (3.9.11) \end{aligned}$$

say, where

$$R_1 = Z_1(Z_1'Z_1)^{-1}Z_1'. \qquad (3.9.12)$$

Now, instead of the single-column vector y, consider the p_2 columns of Z_2 in the same role. A matrix

$$Z_{2\cdot 1} = Z_2 - \hat{Z}_2 \qquad (3.9.13)$$

with every one of its p_2 columns orthogonal to every one of the p_1 columns of Z_1 can be obtained via

$$\begin{aligned} Z_{2\cdot 1} &= (I - R_1)Z_2 \\ &= Z_2 - Z_1A, \qquad (3.9.14) \end{aligned}$$

Table 3.9. Analysis of Variance for Orthogonalized General Model

Source	df	SS
Response function Z_1 only	p_1	$(\theta - \hat{\theta})'Z_1'Z_1(\theta - \hat{\theta})$
Extra for Z_2	p_2	$(\theta_2 - \hat{\theta}_2)Z_{2\cdot1}'Z_{2\cdot1}(\theta_2 - \hat{\theta}_2)$
Residual	$n - p_1 - p_2$	$(y - \hat{y})'(y - \hat{y}) = \Sigma(y - \hat{y})^2$
Total	n	$(y - \eta)'(y - \eta) = \Sigma(y - \eta)^2$

where

$$A = (Z_1'Z_1)^{-1}Z_1'Z_2 \qquad (3.9.15)$$

is a $p_1 \times p_2$ matrix of coefficients obtained by regressing each of the p_2 columns of Z_2 onto all the p_1 columns of Z_1. This matrix A is sometimes referred to as the *alias matrix*, or the *bias matrix*.

The Orthogonalized Model and the Extra Sum of Squares Procedure in This Context

In the situation where Z_1 and Z_2 are not orthogonal, we can write the response function in the form

$$y = Z_1(\theta_1 + A\theta_2) + (Z_2 - Z_1A)\theta_2 + \varepsilon$$
$$= Z_1\theta + Z_{2\cdot1}\theta_2 + \varepsilon, \qquad (3.9.16)$$

where $\theta = \theta_1 + A\theta_2$. The corresponding analysis of variance table is shown in Table 3.9.

The following points should be noticed:

1. The vector of estimates $\hat{\theta}$ obtained by fitting $y = Z_1\theta + Z_{2\cdot1}\theta_2 + \varepsilon$ will be identical to the one that would be obtained by fitting the model $y = Z_1\theta + \varepsilon$ which contained only Z_1.

2. If the additional regressors in Z_2 are desired in the model, the vector of estimates $\hat{\theta}_2$ will be the same if they are obtained by least squares from the full model (3.9.7), or from the model $y = Z_{2\cdot1}\theta_2 + \varepsilon$.

   (Points 1 and 2 are ensured by the orthogonality of the matrices Z_1 and $Z_{2\cdot1}$, i.e., $Z_1'Z_{2\cdot1} = 0$.)

3. If the additional regressors in Z_2 are required, that is, if $\theta_2 \neq 0$, then $\hat{\theta}$ will not provide an unbiased estimate of θ_1 but, rather, of the combination $\theta_1 + A\theta_2$.

4. The need for the regressors in Z_2 to be in the response function will be indicated by the size of the sum of squares "extra for Z_2" with p_2 df, when θ_2 is set equal to 0 in that sum of squares.

5. The "extra for Z_2" sum of squares may be obtained by first fitting the simpler model $y = Z_1\theta_1 + \varepsilon$ and then the more elaborate model $y = Z_1\theta_1 + Z_2\theta_2 + \varepsilon$ and calculating the difference in the model sums of squares or, alternatively, in the residual sums of squares.

3.10. BIAS IN LEAST-SQUARES ESTIMATORS ARISING FROM AN INADEQUATE MODEL

Suppose that it had been decided to fit the model* $y = Z_1\theta_1 + \varepsilon$ but the true model which should have been fitted was $y = Z_1\theta_1 + Z_2\theta_2 + \varepsilon$. We would have estimated θ_1 by

$$\hat{\theta}_1 = (Z_1'Z_1)^{-1}Z_1'y, \tag{3.10.1}$$

but, under the true model,

$$E(\hat{\theta}_1) = (Z_1'Z_1)^{-1}Z_1'E(y)$$

$$= (Z_1'Z_1)^{-1}Z_1'(Z_1\theta_1 + Z_2\theta_2)$$

$$= \theta_1 + A\theta_2, \tag{3.10.2}$$

where

$$A = (Z_1'Z_1)^{-1}Z_1'Z_2 \tag{3.10.3}$$

is the so-called *bias* or *alias* matrix. Equation (3.10.2) tells us that, unless $A = 0$, $\hat{\theta}_1$ will estimate not θ_1, but a combination of θ_1 and θ_2. Now $A = 0$ only if $Z_1'Z_2 = 0$, when all the regressors in Z_1 are orthogonal to all the regressors in Z_2.

Example 3.7. For illustration, consider again the data of Table 3.3, but suppose that, instead of fitting the "true" quadratic model $y = \beta_0 + \beta_1 x + \beta_{11} x^2 + \varepsilon$, we had fitted the straight-line model $y = \beta_0 + \beta_1 x + \varepsilon$. Figure 3.7 shows the least-squares fitted straight line $\hat{y} = 66.777 + 2.063x$, which fits the data very poorly. It is obvious that the intercept $b_0 = 66.777$, which is the predicted value of y at $x = 0$, and the slope $b_1 = 2.063$ will have expectations contaminated by the true value of the omitted curvature parameter, β_{11}. The exact nature of this contamination can be found by calculating the matrix A. We find

$$Z_1 = \begin{bmatrix} 1 & -4 \\ 1 & 0 \\ 1 & 1 \\ 1 & 3 \\ 1 & 5 \end{bmatrix}, \quad \theta_1 = \begin{bmatrix} \beta_0 \\ \beta_1 \end{bmatrix}, \quad Z_2 = \begin{bmatrix} 16 \\ 0 \\ 1 \\ 9 \\ 25 \end{bmatrix}, \quad \theta_2 = \beta_{11}. \tag{3.10.4}$$

*As before, we use the notation θ for the parameters of a general model and β's for specific cases.

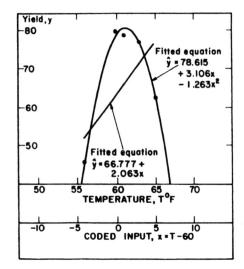

Figure 3.7. Plot of dyestuff data and fitted first- and second-order equation.

Thus the bias matrix is

$$\mathbf{A} = (\mathbf{Z}_1'\mathbf{Z}_1)^{-1}\mathbf{Z}_1'\mathbf{Z}_2 = \begin{bmatrix} 5 & 5 \\ 5 & 51 \end{bmatrix}^{-1} \begin{bmatrix} 51 \\ 89 \end{bmatrix}$$

$$= \frac{1}{230} \begin{bmatrix} 51 & -5 \\ -5 & 5 \end{bmatrix} \begin{bmatrix} 51 \\ 89 \end{bmatrix}$$

$$= \frac{1}{230} \begin{bmatrix} 2156 \\ 190 \end{bmatrix}$$

$$= \begin{bmatrix} 9.374 \\ 0.826 \end{bmatrix}. \tag{3.10.5}$$

Hence

$$E(b_0) = \beta_0 + 9.374\beta_{11},$$
$$E(b_1) = \beta_1 + 0.826\beta_{11}. \tag{3.10.6}$$

We see that $b_0 = 66.777$ is not an unbiased estimate of β_0 in the true quadratic model but is instead an estimate of $\beta_0 + 9.374\beta_{11}$. Similarly, $b_1 = 2.063$ is an estimate of $\beta_1 + 0.826\beta_{11}$, rather than of β_1.

When, as in this example, it is possible to estimate the coefficients of the true model separately, the linear relationships indicated by the alias structure will also hold for the least-squares estimates.* For example, recall that the estimated quadratic equation was

$$\hat{y} = 78.615 + 3.106x - 1.263x^2 \tag{3.10.7}$$

---

*This fact is a direct consequence of the Gauss-Markov theorem (see Appendix 3B).

while the estimated straight-line equation was

$$\hat{y} = 66.777 + 2.063x.$$

It is readily confirmed that (to within rounding error)

$$66.777 = 78.615 + 9.374(-1.263),$$
$$2.063 = 3.106 + 0.826(-1.263). \qquad (3.10.8)$$

Example 3.8.   Aliases for a 2^{5-2} design. Consider the 2^{5-2} fractional factorial design* with generating relation $I = 124 = 135$ whose levels are given by the second through sixth columns of the $\mathbf{Z}_1$ matrix below. Suppose the fitted first-degree polynomial was

$$\hat{y} = b_0 + b_1 x_1 + b_2 x_2 + b_3 x_3 + b_4 x_4 + b_5 x_5 \qquad (3.10.9)$$

when, in fact, to obtain an adequate representation over the region covered by the x's, we would need a second-degree polynomial model

$$
\begin{aligned}
y = \; & \beta_0 + \beta_1 x_1 + \beta_2 x_2 + \beta_3 x_3 + \beta_4 x_4 + \beta_5 x_5 \\
& + \beta_{11} x_1^2 + \beta_{22} x_2^2 + \beta_{33} x_3^2 + \beta_{44} x_4^2 + \beta_{55} x_5^2 \\
& + \beta_{12} x_1 x_2 + \beta_{13} x_1 x_3 + \beta_{14} x_1 x_4 + \beta_{15} x_1 x_5 + \beta_{23} x_2 x_3 \\
& + \beta_{24} x_2 x_4 + \beta_{25} x_2 x_5 + \beta_{34} x_3 x_4 + \beta_{35} x_3 x_5 + \beta_{45} x_4 x_5 + \varepsilon.
\end{aligned}
\qquad (3.10.10)
$$

What will be the expected values of $b_0, b_1, \ldots, b_5$? We have

$$
\mathbf{Z}_1 =
\begin{array}{c@{\quad}ccccc}
& 0 & 1 & 2 & 3 & 4 & 5 \\
& \begin{bmatrix}1 & -1 & -1 & -1 & 1 & 1 \\
1 & 1 & -1 & -1 & -1 & -1 \\
1 & -1 & 1 & -1 & -1 & 1 \\
1 & 1 & 1 & -1 & 1 & -1 \\
1 & -1 & -1 & 1 & 1 & -1 \\
1 & 1 & -1 & 1 & -1 & 1 \\
1 & -1 & 1 & 1 & -1 & -1 \\
1 & 1 & 1 & 1 & 1 & 1 \end{bmatrix}
\end{array},
$$

$$
\mathbf{Z}_2 =
\begin{array}{c}
\begin{smallmatrix}11 & 22 & 33 & 44 & 55 & 12 & 13 & 14 & 15 & 23 & 24 & 25 & 34 & 35 & 45\end{smallmatrix} \\
\begin{bmatrix}
1 & 1 & 1 & 1 & 1 & 1 & 1 & -1 & -1 & 1 & -1 & -1 & -1 & -1 & 1 \\
1 & 1 & 1 & 1 & 1 & -1 & -1 & -1 & -1 & 1 & 1 & 1 & 1 & 1 & 1 \\
1 & 1 & 1 & 1 & 1 & -1 & 1 & 1 & -1 & -1 & -1 & 1 & 1 & -1 & -1 \\
1 & 1 & 1 & 1 & 1 & 1 & -1 & 1 & -1 & -1 & 1 & -1 & -1 & 1 & -1 \\
1 & 1 & 1 & 1 & 1 & 1 & -1 & -1 & 1 & -1 & -1 & 1 & 1 & -1 & -1 \\
1 & 1 & 1 & 1 & 1 & -1 & 1 & -1 & 1 & -1 & 1 & -1 & -1 & 1 & -1 \\
1 & 1 & 1 & 1 & 1 & -1 & -1 & 1 & 1 & 1 & -1 & -1 & -1 & -1 & 1 \\
1 & 1 & 1 & 1 & 1 & 1 & 1 & 1 & 1 & 1 & 1 & 1 & 1 & 1 & 1
\end{bmatrix}
\end{array}
$$

$$\boldsymbol{\theta}_1 = (\beta_0, \beta_1, \beta_2, \beta_3, \beta_4, \beta_5)',$$

$$\boldsymbol{\theta}_2 = (\beta_{11}, \beta_{22}, \beta_{33}, \beta_{44}, \beta_{55}, \beta_{12}, \beta_{13}, \beta_{14}, \beta_{15}, \beta_{23}, \beta_{24}, \beta_{25}, \beta_{34}, \beta_{35}, \beta_{45})',$$

*Designs of this type are discussed in Chapter 5.

$$\mathbf{A} = (\mathbf{Z}_1'\mathbf{Z}_1)^{-1}\mathbf{Z}_1'\mathbf{Z}_2$$

$$= \begin{bmatrix} 1 & 1 & 1 & 1 & 1 & & & & & \\ & & & & & & & 1 & & 1 \\ & & & & & & 1 & & & \\ & & & & & & & 1 & & \\ & & & & & 1 & & & & \\ & & & & & & 1 & & & \end{bmatrix}, \qquad (3.10.11)$$

where empty spaces represent zeros. Then

$$E(b_0) = \beta_0 + \beta_{11} + \beta_{22} + \beta_{33} + \beta_{44} + \beta_{55},$$
$$E(b_1) = \beta_1 + \beta_{24} + \beta_{35},$$
$$E(b_2) = \beta_2 + \beta_{14},$$
$$E(b_3) = \beta_3 + \beta_{15}, \qquad (3.10.12)$$
$$E(b_4) = \beta_4 + \beta_{12},$$
$$E(b_5) = \beta_5 + \beta_{13}.$$

Thus all estimators are aliased unless the β_{ij} shown are all zero.

Notice that the aliasing which occurs in this design cannot be further elucidated unless additional experimental runs are performed. For example, the columns in $\mathbf{Z}_1$ and $\mathbf{Z}_2$ associated with x_1, x_2x_4, and x_3x_5 are *identical*. Here we can only say, for instance, that unless the unknown coefficients β_{24} and β_{35} are zero, then b_1 will be a biased estimate of β_1. Because of the inadequacy of the design, no possibility exists here to estimate the second degree model and so to unbias (or de-alias) the estimates. (In the previous example, this could be done.)

3.11. PURE ERROR AND LACK OF FIT

Genuine Replicates

It is frequently useful to employ experimental arrangements in which two or more runs are made at an identical set of values of the input variables $x_1, x_2, \ldots, x_k$. If these runs can be made in such a way that they are subject to all the sources of error that beset runs made at different conditions, we call them *genuine replicates.**

Genuine replicates are very valuable because differences in the response y between them can provide an estimate of the error variance no matter what the true model may be. For r genuine replicates in which $\mathbf{x}$ is fixed, the model can be written

$$y_u = f(\mathbf{x}, \boldsymbol{\theta}) + \varepsilon_u, \qquad u = 1, 2, \ldots, r, \qquad (3.11.1)$$

*Some care is needed to achieve genuine replicated runs. In particular, a group of such runs should normally not be run consecutively but should be randomly ordered. Replicate runs must be subject to all the usual setup errors, sampling errors, and analytical errors which affect runs made at different conditions. Failure to achieve this will typically cause underestimation of the error and will invalidate the analysis.

implying

$$\bar{y} = f(\mathbf{x}, \boldsymbol{\theta}) + \bar{\varepsilon}, \tag{3.11.2}$$

and

$$\sum_{u=1}^{r} (y_u - \bar{y})^2 = \sum_{u=1}^{r} (\varepsilon_u - \bar{\varepsilon})^2, \tag{3.11.3}$$

whatever the model function $f(\mathbf{x}, \boldsymbol{\theta})$ may be. Thus, if the ε_u are independent random variables with variance σ^2,

$$E\left\{ \sum_{u=1}^{r} (y_u - \bar{y})^2 \right\} = E\left\{ \sum_{u=1}^{r} (\varepsilon_u - \bar{\varepsilon})^2 \right\} = (r-1)\sigma^2 \tag{3.11.4}$$

and, furthermore, if the $\varepsilon_u \sim N(0, \sigma^2)$ and are independent, both sides of (3.11.4) are distributed as $\sigma^2 \chi^2_{r-1}$ variables.

If we have m such sets of replicated runs with r_i runs in the ith set made at $\mathbf{x}_i$, the individual internal sums of squares may be pooled together to form a *pure error sum of squares* having, as its degrees of freedom, the sum of the separate degrees of freedom. Thus the (total) *pure error sum of squares is*

$$\sum_{i=1}^{m} \sum_{u=1}^{r_i} (y_{iu} - \bar{y}_i)^2 \tag{3.11.5}$$

with degrees of freedom

$$\sum_{i=1}^{m} (r_i - 1) = \sum_{i=1}^{m} r_i - m. \tag{3.11.6}$$

When $r_i = 2$, the simpler formula

$$\sum_{u=1}^{2} (y_{iu} - \bar{y}_i)^2 = \frac{1}{2}(y_{i1} - y_{i2})^2 \tag{3.11.7}$$

can be used; such a component has 1 df.

When any model is fitted to data, the pure error sum of squares is always *part* of the residual sum of squares. The residual from the uth observation at $\mathbf{x}_i$ is

$$y_{iu} - \hat{y}_i = (y_{iu} - \bar{y}_i) - (\hat{y}_i - \bar{y}_i); \tag{3.11.8}$$

on squaring both sides and summing,

$$\sum_{i=1}^{m} \sum_{u=1}^{r_i} (y_{iu} - \hat{y}_i)^2 = \sum_{i=1}^{m} \sum_{u=1}^{r_i} (y_{iu} - \bar{y}_i)^2 + \sum_{i=1}^{m} r_i(\hat{y}_i - \bar{y}_i)^2, \tag{3.11.9}$$

that is,

$$\text{residual SS} = \text{pure error SS} + \text{lack of fit SS}, \tag{3.11.10}$$

Table 3.10. Analysis of Variance Table Showing the Entry of Pure Error

Source	SS	df	MS
b_0	$n\bar{y}^2$	1	
Additional model terms	$\hat{\boldsymbol{\theta}}'\mathbf{Z}'\mathbf{y} - n\bar{y}^2$	$p - 1$	MS_M
Lack of fit	$\displaystyle\sum_{i=1}^{m} r_i(\hat{y}_i - \bar{y}_i)^2$	$m - p$	MS_l
Pure error	$\displaystyle\sum_{i=1}^{m}\sum_{u=1}^{r_i} (y_{iu} - \bar{y}_i)^2$	$n - m$	MS_e
Total	$\displaystyle\sum_{i=1}^{m}\sum_{u=1}^{r_i} y_{iu}^2$	n	

the cross-product vanishing in the summation over u for each i. The last term on the right is called the lack of fit sum of squares because it is a measure of the discrepancy between the model prediction $\hat{y}_i$ and the average $\bar{y}_i$ of the replicated runs made at the ith set of experimental conditions.

The corresponding equation for degrees of freedom is

$$\sum_{i=1}^{m} r_i - p = \sum_{i=1}^{m} (r_i - 1) + m - p, \qquad (3.11.11)$$

where p is the number of parameters in the model. Obviously, we must have $m \geq p$; that is, there must be at least as many or more *distinct* $\mathbf{x}$ values at which y's are observed than there are parameters to be estimated. If $m = p$, there will be no sum of squares nor degrees of freedom for lack of fit, since the model will always "fit perfectly."

An analysis of variance table may now be constructed as in Table 3.10. In this and similar tables, the mean squares, denoted by MS, are obtained by dividing each sum of squares by the corresponding number of degrees of freedom (df).

Test for Lack of Fit

It is always true that

$$E(\mathrm{MS}_e) = \sigma^2. \qquad (3.11.12)$$

To evaluate the expected value of MS_L, assume that the model to be fitted is

$$\mathbf{y} = \mathbf{Z}\boldsymbol{\theta} + \boldsymbol{\varepsilon} \qquad (3.11.13)$$

but the true model is

$$\mathbf{y} = \mathbf{Z}\boldsymbol{\theta} + \mathbf{Z}^*\boldsymbol{\theta}^* + \boldsymbol{\varepsilon}. \qquad (3.11.14)$$

Then, with

$$\mathbf{A} = (\mathbf{Z}'\mathbf{Z})^{-1}\mathbf{Z}'\mathbf{Z}^* \qquad (3.11.15)$$

being the alias matrix,

$$E(\mathrm{MS}_L) = \sigma^2 + \boldsymbol{\theta}^{*\prime}(\mathbf{Z}^* - \mathbf{ZA})^\prime(\mathbf{Z}^* - \mathbf{ZA})\boldsymbol{\theta}^*/(m - p). \quad (3.11.16)$$

Thus, provided $\mathbf{Z}^* - \mathbf{ZA}$ is non-null, $E(\mathrm{MS}_L)$ will be inflated when $\boldsymbol{\theta}^*$ is nonzero. If $\varepsilon \sim N(0, \sigma^2)$ and $\boldsymbol{\theta}^* = \mathbf{0}$, it can be shown that the observed F ratio

$$F = \mathrm{MS}_L/\mathrm{MS}_e \quad (3.11.17)$$

follows an $F(m - p, n - m)$ distribution. Thus we can test the null hypothesis $H_0: \boldsymbol{\theta}^* = \mathbf{0}$ versus the alternative $H_1: \boldsymbol{\theta}^* \neq \mathbf{0}$ by comparing $\mathrm{MS}_L/\mathrm{MS}_e$ to a suitable upper percentage point of the $F(m - p, n - m)$ distribution.

A large and significant value of F discredits the fitted model. In most cases, this would initiate a search for a more adequate model. The nature of the inadequacy would first be sought by analyzing the residuals, and remedial measures, perhaps involving transformation of y or of one or more of the x's or possibly the use of a radically different model, would then be taken. Occasionally, and particularly when there is a very large amount of data, investigation might show that the deficient model is, nevertheless, sufficient for the purpose at hand and therefore may be used with proper caution. (Remember that all models are wrong; the practical question is how wrong do they have to be to not be useful.)

The R^2 Statistic

The statistic

$$R^2 = \left(\hat{\boldsymbol{\theta}}^\prime\mathbf{Z}^\prime\mathbf{y} - n\bar{y}^2\right)/(\mathbf{y}^\prime\mathbf{y} - n\bar{y}^2) \quad (3.11.18)$$

represents the fraction of the variation about the mean that is explained by the fitted model. It is often used as an overall measure of the fit attained. Note that no model can explain pure error, so that the maximum possible value of R^2 is

$$\max R^2 = \{(\mathbf{y}^\prime\mathbf{y} - n\bar{y}^2) - \text{pure error SS}\}/(\mathbf{y}^\prime\mathbf{y} - n\bar{y}^2). \quad (3.11.19)$$

3.12. CONFIDENCE INTERVALS AND CONFIDENCE REGIONS

Confidence Contours

We have seen that we can obtain least-squares estimates for the parameters in the model $\mathbf{y} = \mathbf{Z}\boldsymbol{\theta} + \boldsymbol{\varepsilon}$ by evaluating $\hat{\boldsymbol{\theta}} = (\mathbf{Z}^\prime\mathbf{Z})^{-1}\mathbf{Z}^\prime\mathbf{y} = \mathbf{Ty}$, say, where $\mathbf{T} = (\mathbf{Z}^\prime\mathbf{Z})^{-1}\mathbf{Z}^\prime$ is the linear transformation matrix which converts the vector $\mathbf{y}$ into the vector of linear functions $\hat{\boldsymbol{\theta}} = \mathbf{Ty}$. Now the variance–covariance matrix of $\hat{\boldsymbol{\theta}}$ is given by

$$\begin{aligned}
\mathbf{V}(\hat{\boldsymbol{\theta}}) = \mathbf{V}(\mathbf{Ty}) = \mathbf{TV}(\mathbf{y})\mathbf{T}^\prime &= \mathbf{TI}\sigma^2\mathbf{T}^\prime \\
&= \mathbf{TT}^\prime\sigma^2 \\
&= (\mathbf{Z}^\prime\mathbf{Z})^{-1}\sigma^2, \quad (3.12.1)
\end{aligned}$$

and in fact it is true that, if the model is correct,

$$\hat{\boldsymbol{\theta}} \sim N\left(\boldsymbol{\theta}, (\mathbf{Z}'\mathbf{Z})^{-1}\sigma^2\right). \tag{3.12.2}$$

As a consequence of this result, it can be shown that the quadratic form

$$(\boldsymbol{\theta} - \hat{\boldsymbol{\theta}})'\mathbf{Z}'\mathbf{Z}(\boldsymbol{\theta} - \hat{\boldsymbol{\theta}}) \sim \sigma^2 \chi^2(p), \tag{3.12.3}$$

where $\chi^2(p)$ denotes a chi-squared variable with p df and where p, is, as before, the number of parameters in $\boldsymbol{\theta}$. Furthermore, if s^2 is an independent estimate of σ^2 with ν df, such that

$$\nu s^2 \sim \sigma^2 \chi^2(\nu), \tag{3.12.4}$$

$$(\boldsymbol{\theta} - \hat{\boldsymbol{\theta}})'\mathbf{Z}'\mathbf{Z}(\boldsymbol{\theta} - \hat{\boldsymbol{\theta}})/(ps^2) \sim F(p, \nu), \tag{3.12.5}$$

where $F(p, \nu)$ denotes an F variable with p and ν df, because the ratio of two independent χ^2 variables each divided by their respective degrees of freedom is an F variable.

In those examples where the residual mean square with $(n - p)$ df is used to supply an estimate of error, $\nu = n - p$.

Suppose that $F_\alpha(p, \nu)$ denotes the α-significance level of the F distribution with p and $\nu = n - p$ df. Then the equation

$$(\boldsymbol{\theta} - \hat{\boldsymbol{\theta}})'\mathbf{Z}'\mathbf{Z}(\boldsymbol{\theta} - \hat{\boldsymbol{\theta}})/(ps^2) = F_\alpha(p, \nu) \tag{3.12.6}$$

defines an ellipsoidal region in the parameter space which we call a "$1 - \alpha$ confidence region for $\boldsymbol{\theta}$." This region includes the true value of $\boldsymbol{\theta}$ with probability $1 - \alpha$ in the sense that, if we imagine the model generating an infinite sequence of sets of y's for the same $\mathbf{Z}$ values, with similar calculations performed for each set, then a proportion $1 - \alpha$ of the regions so generated would actually contain the true point $\boldsymbol{\theta}$.

Individual (also called marginal) intervals for the separate θ's can be obtained as follows. The variances of $\hat{\theta}_1, \hat{\theta}_2, \ldots, \hat{\theta}_p$ are given by

$$V(\hat{\theta}_i) = c^{ii}\sigma^2, \tag{3.12.7}$$

where c^{ii} denotes the ith diagonal term of $(\mathbf{Z}'\mathbf{Z})^{-1}$. Then

$$(\hat{\theta}_i - \theta_i)/(s^2 c^{ii})^{1/2} \tag{3.12.8}$$

has a t distribution with ν df, and a $1 - \alpha$ confidence interval for θ_i is given by

$$\hat{\theta}_i \pm t_{\alpha/2}(\nu)(s^2 c^{ii})^{1/2}, \tag{3.12.9}$$

where $t_{\alpha/2}(\nu)$ denotes the upper $\frac{1}{2}\alpha$ percentage point of the $t(\nu)$ distribution. Confidence intervals of this type are easily obtainable and useful, but they do not draw attention to the correlations between the various $\hat{\theta}_i$'s. These correla-

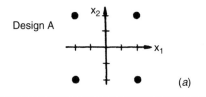

Design A

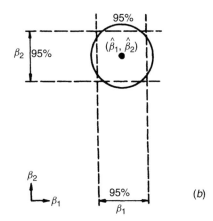

Figure 3.8. (a) Orthogonal design A. (b) The resulting 0.95 joint confidence region and 0.95 individual confidence intervals.

tions became large if the corresponding columns of the $\mathbf{Z}$ matrix are highly nonorthogonal.

The situation can be understood by considering the joint estimation of the two parameters β_1 and β_2 in the first-order model

$$y = \beta_0 + \beta_1 x_1 + \beta_2 x_2 + \varepsilon. \qquad (3.12.10)$$

We shall suppose, without affecting the point at issue, that $V(\varepsilon) = \sigma^2$ is known and is equal to 1.

Design A   (An Orthogonal Design)

We shall see later that efficient estimates of the parameters in the above model can be obtained by using the 2^2 factorial design of Figure 3.8a for which the levels of x_1 and x_2 are

x_1	x_2
-1	-1
1	-1
-1	1
1	1

so that $\mathbf{Z'Z} = \begin{bmatrix} 4 & 0 & 0 \\ 0 & 4 & 0 \\ 0 & 0 & 4 \end{bmatrix} = 4I_3. \qquad (3.12.11)$

With $\sigma^2 = 1$, the covariance matrix for the estimates $\hat{\beta}_0, \hat{\beta}_1, \hat{\beta}_2$ of $\beta_0, \beta_1,$ and β_2 is then

$$(\mathbf{Z'Z})^{-1}\sigma^2 = \begin{bmatrix} \frac{1}{4} & 0 & 0 \\ 0 & \frac{1}{4} & 0 \\ 0 & 0 & \frac{1}{4} \end{bmatrix} \qquad (3.12.12)$$

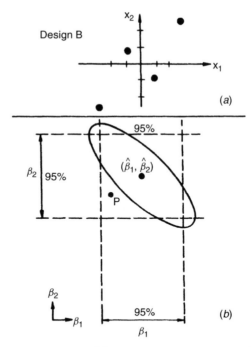

Figure 3.9. (*a*) Nonorthogonal design B. (*b*) The resulting 0.95 joint confidence region and 0.95 individual confidence intervals.

and the (circular) joint 0.95 confidence region for β_1 and β_2 is centered at the point $(\hat{\beta}_1, \hat{\beta}_2)$ and consists of all points (β_1, β_2) such that

$$4\left(\beta_1 - \hat{\beta}_1\right)^2 + 4\left(\beta_2 - \hat{\beta}_2\right)^2 \le 5.99 = \chi_{0.05}^2(2). \qquad (3.12.13)$$

Individual (marginal) intervals for β_1 and β_2 are given by

$$4\left(\beta_1 - \hat{\beta}_1\right)^2 = 3.84 = \chi_{0.05}^2(1) \quad \text{and} \quad 4\left(\beta_2 - \hat{\beta}_2\right)^2 = 3.84. \quad (3.12.14)$$

The joint and individual intervals are shown in Figure 3.8*b*.

Design B   (A Nonorthogonal Design)

Now suppose we use a different design in x_1 and x_2 as shown in Figure 3.9*a* for which the levels of x_1 and x_2 are

x_1	x_2
-1.34	-1.34
0.45	-0.45
-0.45	0.45
1.34	1.34

so that $\mathbf{Z'Z} = \begin{bmatrix} 4 & 0 & 0 \\ 0 & 4 & 3.2 \\ 0 & 3.2 & 4 \end{bmatrix}$.   (3.12.15)

The levels of x_1 and x_2 have been chosen so that $\bar{x}_i = 0$ and $\sigma_{x_i} = \{\Sigma x_i^2/4\}^{1/2}$ are the same for both designs A and B. For design B, given that $\sigma^2 = 1$, the covariance matrix of $\hat{\beta}_0, \hat{\beta}_1, \hat{\beta}_2$ is

$$(\mathbf{Z'Z})^{-1}\sigma^2 = \begin{bmatrix} \frac{1}{4} & 0 & 0 \\ 0 & \frac{25}{36} & -\frac{5}{9} \\ 0 & -\frac{5}{9} & \frac{25}{36} \end{bmatrix} \qquad (3.12.16)$$

and the (elliptical) joint 0.95 confidence region for β_1 and β_2 is centered at $(\hat{\beta}_1, \hat{\beta}_2)$ and consists of all points (β_1, β_2) such that

$$4\left(\beta_1 - \hat{\beta}_1\right)^2 + 4\left(\beta_2 - \hat{\beta}_2\right)^2 + 6.4\left(\beta_1 - \hat{\beta}_1\right)\left(\beta_2 - \hat{\beta}_2\right) \le 5.99. \quad (3.12.17)$$

Individual (marginal) intervals for β_1 and β_2 are given by

$$\tfrac{25}{36}\left(\beta_1 - \hat{\beta}_1\right)^2 = 3.84 \quad \text{and} \quad \tfrac{25}{36}\left(\beta_1 - \hat{\beta}_1\right)^2 = 3.84. \qquad (3.12.18)$$

The joint and individual intervals are shown in Figure 3.9b.

Notice that the orthogonal design is much more desirable than the nonorthogonal design in the senses that:

1. The area of the joint confidence region is much smaller.

2. The lengths of the individual confidence intervals are much smaller.

Correlation of the Parameter Estimates

Now consider Figures 3.8 and 3.9 together. Compare, first of all, the marginal intervals and the joint interval in Figure 3.9b for the nonorthogonal design. Consider a pair of parameter values (β_{10}, β_{20}) corresponding to the point P. We see that, although β_{10} falls within the marginal interval for β_1 and β_{20} falls within the marginal interval for β_2, the point P (β_{10}, β_{20}) itself falls outside the joint interval. The meaning to be associated with this is that, although the value β_{10} is acceptable for some values of β_2, it is not acceptable for the particular value β_{20}. In general, to understand the joint acceptability of values for a group of parameters, it is necessary to consider the joint region and these are not at all easy to visualize when we have more than two or three parameters. Figure 3.8b shows how this difficulty is considerably lessened, but not eliminated, by using an orthogonal design. Such designs lead to circular contours (for two parameters), spherical contours (for three parameters), or hyperspherical contours (for more parameters).

Greater Precision of Estimates from Orthogonal Designs

A further point which is illustrated by comparing Figures 3.8b and 3.9b is the greater precision of estimates obtained from orthogonal designs. Notice that, if such comparisons are to be made fairly, it is necessary to scale, to the same units, the designs to be compared. This has been done in this example by choosing $\Sigma z_1^2 = \Sigma x_2^2 = 4$ for both designs.

3.13. ROBUST ESTIMATION, MAXIMUM LIKELIHOOD, AND LEAST SQUARES

The method of maximum likelihood is known to produce estimates having desirable properties (see, for example, Johnson and Leone, 1977, Vol. 1, pp. 212–214). Thus least squares is often justified by the argument that the least-squares estimators of the elements of $\boldsymbol{\theta}$ are the maximum likelihood estimators when the errors $\varepsilon_1, \varepsilon_2, \ldots, \varepsilon_n$ are distributed $\boldsymbol{\varepsilon} \sim N(\mathbf{0}, \mathbf{I}\sigma^2)$. This can be seen as follows. Since the joint probability distribution function of the ε_u is, for $-\infty \le \varepsilon_u \le \infty$.

$$p(\boldsymbol{\varepsilon}) \propto \sigma^{-n} \exp\left\{ \frac{-\boldsymbol{\varepsilon}'\boldsymbol{\varepsilon}}{2\sigma^2} \right\}, \tag{3.13.1}$$

and since

$$\mathbf{y} = \mathbf{Z}\boldsymbol{\theta} + \boldsymbol{\varepsilon}, \tag{3.13.2}$$

the likelihood function is

$$L(\boldsymbol{\theta}|\mathbf{y}) \propto \sigma^{-n} \exp\left\{ \frac{-(\mathbf{y} - \mathbf{Z}\boldsymbol{\theta})'(\mathbf{y} - \mathbf{Z}\boldsymbol{\theta})}{2\sigma^2} \right\}$$

$$= \sigma^{-n} \exp\left\{ \frac{-S(\boldsymbol{\theta})}{2\sigma^2} \right\}. \tag{3.13.3}$$

This is maximized with respect to $\boldsymbol{\theta}$ when the sum of squares function $S(\boldsymbol{\theta}) = (\mathbf{y} - \mathbf{Z}\boldsymbol{\theta})'(\mathbf{y} - \mathbf{Z}\boldsymbol{\theta})$ is minimized with respect to $\boldsymbol{\theta}$. Thus, the least-squares estimators are maximum likelihood estimators when $\boldsymbol{\varepsilon} \sim N(\mathbf{0}, \mathbf{I}\sigma^2)$, as stated. This and a similar Bayesian justification make it clear that the method of least squares is fully appropriate when the errors can be assumed (1) to be statistically independent, (2) to have constant variance σ^2, and (3) to be normally distributed. We can embrace all these assumptions by saying that, when they are true, the errors are *spherically normally distributed*. Conversely, if we know the errors to have some other distributional form, least-squares estimates of parameters would be inappropriate but we could use maximum likelihood to indicate what would then be suitable. It is instructive to consider some of the possibilities so offered.

Correlated Errors with Nonconstant Variance

Suppose we assume that, while the n errors $\varepsilon_1, \varepsilon_2, \ldots, \varepsilon_n$ are normally distributed, they have a covariance matrix $\mathbf{W}^{-1}\sigma^2$, where $\mathbf{W}^{-1}$ is known but σ^2 is not known. (Under such a setup, the previous model, in which the errors are supposed independent and with constant variance, corresponds to the special case $\mathbf{W}^{-1} = \mathbf{I}$.) Then

$$L(\boldsymbol{\theta}|\mathbf{y}) \propto \sigma^{-n} |\mathbf{W}|^{1/2} \exp\left\{ \frac{-Q(\boldsymbol{\theta})}{2\sigma^2} \right\}, \tag{3.13.4}$$

where

$$Q(\boldsymbol{\theta}) = (\mathbf{y} - \mathbf{Z}\boldsymbol{\theta})'\mathbf{W}(\mathbf{y} - \mathbf{Z}\boldsymbol{\theta}). \tag{3.13.5}$$

The likelihood is maximized with respect to θ when $Q(\theta)$ is minimized. Also, writing $\mathbf{W} = \{w_{tu}\}$, we see that the maximum likelihood estimates of θ are obtained by minimizing a *weighted* sum of *products* of discrepancies,

$$Q(\theta) = \sum_{t=1}^{n} \sum_{u=1}^{n} w_{tu}(y_t - \mathbf{z}_t'\theta)(y_u - \mathbf{z}_u'\theta), \qquad (3.13.6)$$

where $\mathbf{z}_t'$ is the tth row of $\mathbf{Z}$, rather than minimizing the sum of squares

$$S(\theta) = \sum_{u=1}^{n} (y_u - \mathbf{z}_u'\theta)^2. \qquad (3.13.7)$$

It is easily shown that these estimates are such that

$$\hat{\theta} = (\mathbf{Z}'\mathbf{W}\mathbf{Z})^{-1}\mathbf{Z}'\mathbf{W}\mathbf{Y} \qquad (3.13.8)$$

with

$$V(\hat{\theta}) = (\mathbf{Z}'\mathbf{W}\mathbf{Z})^{-1}\sigma^2. \qquad (3.13.9)$$

Also an unbiased estimates s^2 of σ^2 is provided by $Q(\hat{\theta})/(n-p)$, where p is the number of parameters.

An interesting way to view these estimates is as follows. Any symmetric positive-definite matrix $\mathbf{W}$ can be written in the form

$$\mathbf{W} = (\mathbf{W}^{1/2})'\mathbf{W}^{1/2} \qquad (3.13.10)$$

say, where the choice of the matrix designated as $\mathbf{W}^{1/2}$ is not unique. Now define pseudovariables $\dot{\mathbf{Z}} = \mathbf{W}^{1/2}\mathbf{Z}$ and $\dot{\mathbf{y}} = \mathbf{W}^{1/2}\mathbf{y}$, and we find that

$$\hat{\theta} = (\dot{\mathbf{Z}}'\dot{\mathbf{Z}})^{-1}\dot{\mathbf{Z}}'\dot{\mathbf{y}} \qquad (3.13.11)$$

with

$$V(\hat{\theta}) = (\dot{\mathbf{Z}}'\dot{\mathbf{Z}})^{-1}\sigma^2, \qquad (3.13.12)$$

which are the standard least-squares formulas for the pseudovariates $\dot{\mathbf{Z}}$ and $\dot{\mathbf{y}}$.

Autocorrelated Errors—Time Series Analysis

Frequently, business and economic data occur as time series—for example, as monthly values of the gross national product and of unemployment. In investigations aimed at discovering the relationship between such series, even though the assumption of constant error variance may be plausible, the assumption of independent errors will most often not be. It is natural for data obtained serially to be correlated serially—that is, for each error to be correlated with its near neighbors.

The use of ordinary least squares in these circumstances can lead to nonsensical results [see, e.g., Coen et al., (1969) and Box and Newbold (1971)].

Legitimate analysis using time series analysis (see, for example, Box and Jenkins, 1976) in effect uses the data to estimate $\mathbf{W}$ appropriately.

Independent Errors with Nonconstant Variance—Weighted Least Squares

Another important special case occurs when it can be assumed that errors are independent but do not have constant variance so that

$$V(y_u) = \frac{\sigma^2}{w_u}. \qquad (3.13.13)$$

As before, the w_u are supposed known, but σ^2 is not or, equivalently, the relative variances of the y_u's are supposed known. In this case

$$\mathbf{W} = \text{diag}(w_1, w_2, \ldots, w_n), \qquad (3.13.14)$$

and the maximum likelihood estimates minimize the *weighted* sum of squares

$$Q_d(\mathbf{\theta}) = \sum_{u=1}^{n} w_u (y_u - \mathbf{z}_u'\mathbf{\theta})^2. \qquad (3.13.16)$$

Pseudovariates are obtained (in this special case) by writing $\dot{z}_{iu} = w_u^{1/2} z_{iu}$ and $\dot{y}_u = w_u^{1/2} y_u$.

For simplicity, we illustrate when $p = 2$, so that the model is then

$$y_u = \theta_1 z_{1u} + \theta_2 z_{2u} + \varepsilon_u,$$

with $V(y_u) = \sigma^2/w_u$. Then writing, for example, $\Sigma w z_1 z_2$ for $\Sigma_{u=1}^{n} w_u z_{1u} z_{2u}$, we obtain the weighted least-squares estimates of θ_1 and θ_2 as the solutions of the normal equations

$$\hat{\theta}_1 \Sigma w z_1^2 + \hat{\theta}_2 \Sigma w z_1 z_2 = \Sigma w z_1 y,$$
$$\hat{\theta}_1 \Sigma w z_1 z_2 + \hat{\theta}_2 \Sigma w z_2^2 = \Sigma w z_2 y. \qquad (3.13.17)$$

The residuals are given by

$$r_u = y_u - \hat{\theta}_1 z_{1u} - \hat{\theta}_2 z_{2u}. \qquad (3.13.18)$$

The estimate of the experimental error variance σ^2 is

$$\sigma^2 = s^2 = \frac{\displaystyle\sum_{u=1}^{n} w_u r_u^2}{(n - p)} \qquad (3.13.19)$$

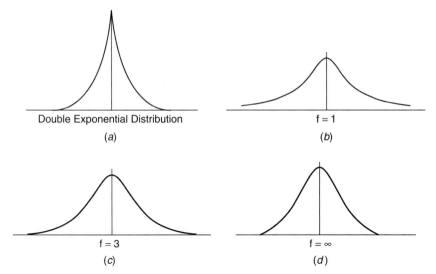

Figure 3.10. (a) The double exponential distribution. (b, c, d) t distributions, $f = 1, 3, \infty$ (normal).

and the covariance matrix for the estimates is given by

$$
\mathbf{V}(\hat{\boldsymbol{\theta}}) = \begin{bmatrix} \Sigma w z_1^2 & \Sigma w z_1 z_2 \\ \Sigma w z_1 z_2 & \Sigma w z_2^2 \end{bmatrix}^{-1} \sigma^2, \tag{3.13.20}
$$

where $\hat{\boldsymbol{\theta}} = (\hat{\theta}_1, \hat{\theta}_2)'$.

Non-Normality

There are of course an infinite number of ways in which a distribution can be non-normal. It has been suggested in particular that error distributions in which extreme deviations occur more frequently than with the normal distribution often arise. Such distributions are said to be "heavy-tailed" or "leptokurtic." For such an error distribution, the maximum likelihood estimates differ from the least-squares estimates. For example, one highly leptokurtic error distribution is the "double exponential" shown in Figure 3.10a. This has the form

$$
p(\varepsilon) \propto \sigma^{-1} \exp\left\{\left|\frac{\varepsilon}{2\sigma}\right|\right\},
$$

where σ is an appropriate scale parameter. For this distribution, the maximum likelihood estimates are those obtained by minimizing the sum of absolute deviations

$$
\sum_{u=1}^{n} |y_u - \mathbf{z}_u' \boldsymbol{\theta}|. \tag{3.13.21}
$$

Suppose the model is simply $y_u = \theta + \varepsilon_u$. Then the least-squares estimator of θ, appropriate if the error distribution were normal, would be $\hat{\theta} = \bar{y}$, the sample average. However, for the heavy-tailed double exponential distribution the maximum likelihood estimator that minimizes (3.13.21) is the *sample median*. Notice that this estimator is much less sensitive to the extreme deviations that occur with the double exponential distribution.

Heavy tailed error distributions may be represented somewhat more realistically by t distributions for which

$$p(\varepsilon) \propto \frac{1}{\sigma}\left(1 + \frac{\varepsilon^2}{f\sigma^2}\right)^{-(1/2)(f+1)},$$

where again σ is a scale parameter. Examples of t distributions with $f = 1, 3$, and ∞ df are shown in Figures 3.10*b*, 3.10*c* and 3.10*d*.

The t distribution with $f = 1$ is extremely heavy tailed and is sometimes called the Cauchy distribution. The tendency to leptokurtosis decreases as f increases and, for $f = \infty$, the distribution is exactly the normal distribution.

Effect of Bad Values

It sometimes happens that data are afflicted with one or more bad values (also called "rogue" observations). An atypical value of this kind can arise, for example, from an unrecognized mistake in conducting an experimental run. Now least-squares estimates, which are appropriate under standard normal theory which presupposes that bad values *never* occur, can be excessively influenced by bad values. By employing alternative assumptions, which more closely model the true situation, more appropriate estimates are obtained which place less emphasis on outlying observations.

For example, the double exponential distribution of Figure 3.10*a* has much heavier tails than the normal, and, as already observed, the maximum likelihood estimate of the population mean is the sample median rather than the sample average. Obviously, the median is much less likely to be affected by outlying observations than is the sample average, and so it is more robust against the possibility that the distribution has heavy tails.

Robustification Using Iteratively Reweighted Least Squares
To guard against misleading estimates caused by non-normal heavy-tailed distributions and occasional bad values, the results for a number of maximum likelihood analyses may be compared using different distributional assumptions. Unfortunately, however, the direct evaluation of maximum likelihood estimates from non-normal distributions can become complicated.

An ingenious method for obtaining the required maximum likelihood estimates for a wide class of non-normal distributions employs weighted least squares iteratively. The general argument is set out in Appendix 3A. We illustrate for the important special case when it is supposed that, rather than follow the normal distribution, the errors follow a t distribution having f df. It then turns out that appropriate estimates $\hat{\theta}$ and s^2 for θ and σ^2 may be obtained by iteratively

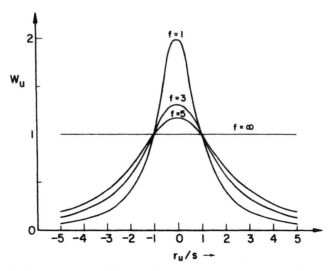

Figure 3.11. Weight functions appropriate when the error distribution is a t distribution with degrees of freedom $f = 1, 3, 5, \infty$.

reweighted least squares with weights given by

$$w_u = \frac{f + 1}{f + (r_u/s)^2}.$$

(3.13.22)

In Figure 3.11, this weight function is plotted for $f = 1, 3, 5, \infty$. Recall that, when $f = \infty$, the t distribution becomes the normal distribution; the weights are then uniform and equal to unity, corresponding to the employment of ordinary least squares. As the value of f decreases (corresponding to the assumption that the error distribution is more heavy-tailed than the normal), the down-weighting associated with large residuals occurs.

Iterative Reweighting

Because estimates of unknown parameters appear in the weights (3.13.22), the weighted least-squares calculations must be performed iteratively. A suitable scheme is as follows:

The iteration is started by obtaining preliminary estimates $(\hat{\boldsymbol{\theta}}^{(0)}, s^{(0)})$ in some way (for example, by ordinary least squares). From these, residuals $r_u^{(0)}$ and weights $w_u^{(0)}$ may be calculated from which new weighted estimates $(\hat{\boldsymbol{\theta}}^{(1)}, s^{(1)})$ are computed. These new estimates give rise to new weights and so on, until the process converges. The procedure is called iteratively reweighted least squares, or IRLS for short. (See Appendix 3A.)

Choice of f

For a given set of data we will almost certainly not know what value of f is appropriate. A useful practical procedure is to carry through the analysis for a few different values of f (for example, 1, 3, and ∞) and see how the estimates are affected.

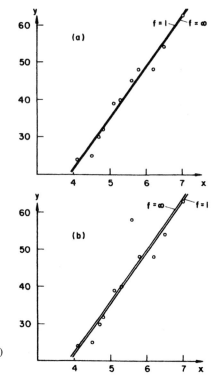

Figure 3.12. IRLS estimation on (*a*) original data and (*b*) data with outlier.

Table 3.11. (a) A Set of Observations. (b) The Same Set but with 13 Units Added to y_7

Observation Number	(a)		(b)	
	x	y	x	y
1	4.1	24	4.1	24
2	4.5	25	4.5	25
3	4.7	30	4.7	30
4	4.8	32	4.8	32
5	5.1	39	5.1	39
6	5.3	40	5.3	40
7	5.6	45	5.6	58
8	5.8	48	5.8	48
9	6.2	48	6.2	48
10	6.5	54	6.5	54
11	7.0	63	7.0	63

An Example

In Figure 3.12*a* and Table 3.11 we show a set of data which appears to follow a straight line relationship

$$y_u = \theta_0 + \theta_1(x_u - \bar{x}) + \varepsilon.$$

Figure 3.12*b* shows the same data but with 13 units added to the value of *y* for the seventh observation making it depart markedly from the linear relationship.

Table 3.12. IRLS Estimates (Standard Errors), and Observation Weights for the Example

f	Data (a)				Data (b)			
	∞	5	3	1	∞	5	3	1
$\hat{\theta}_0$	40.73(0.65)	40.85(0.61)	40.89(0.59)	41.00(0.54)	41.91(1.56)	41.30(1.19)	41.21(1.09)	41.13(0.90)
$\hat{\theta}_1$	13.54(0.76)	13.54(0.70)	13.55(0.68)	13.60(0.61)	13.83(1.83)	13.66(1.35)	13.63(1.23)	13.61(0.99)
Observation	Weights				Weights			
1	1	1.11	1.17	1.36	1	1.19	1.30	1.75
2	1	0.79	0.73	0.62	1	1.00	0.98	0.93
3	1	1.11	1.15	1.25	1	1.16	1.25	1.56
4	1	1.18	1.29	1.67	1	1.19	1.31	1.83
5	1	0.93	0.90	0.86	1	1.13	1.18	1.28
6	1	1.16	1.26	1.68	1	1.20	1.33	1.92
7	1	1.04	1.06	1.13	1	0.52	0.44	0.31
8	1	0.99	0.99	1.01	1	1.16	1.24	1.46
9	1	0.79	0.73	0.60	1	0.98	0.97	0.92
10	1	1.05	1.05	1.01	1	1.12	1.18	1.35
11	1	1.14	1.23	1.63	1	1.20	1.33	1.92

The IRLS estimates, with their standard errors in parentheses and the observation weights, are shown in Table 3.12.

Two of the fitted straight lines for $f = 1$ and $f = \infty$ are shown in Figure 3.12. By considering the weights for observation 7 we can see how the influence of this observation is successively discounted as f becomes smaller. In this example it is easy to appreciate the situation by inspection of the graphs. With the more complicated models that are usually employed in this book, this is not possible. However, useful information can still be obtained by noticing the manner in which the estimates of θ and the weights change as f changes.

It is important to keep an open mind concerning the conclusions to be drawn from application of these methods. Robustification merely points out that certain observations are unusual given particular model assumptions. This does not necessarily mean that these observations are unimportant. The history of science abounds with examples where it was the *discrepant* observation that led to a new discovery. Also there are many situations where there are different plausible explanations for the same data. Robust estimation might indicate that model A was applicable if an observation y_m were discounted, but that a more complicated model B would accommodate this particular observation y_m. Distinguishing between such alternative possibilities is a matter for careful consideration and analysis and, in particular, for further experimentation in an attempt to separate these possibilities.

APPENDIX 3A. ITERATIVELY REWEIGHTED LEAST SQUARES

Consider any error distribution function $p(\varepsilon_u)$ which can be written in the form

$$p(\varepsilon_u) \propto \sigma^{-1} g\left\{ \left(\frac{\varepsilon_u}{\Sigma} \right)^2 \right\},$$

where σ is a scale parameter, $g\{\cdot\}$ denotes a functional form, and $\varepsilon_u = y_u - \Sigma_i \theta_i z_{iu}$. Then the log likelihood for θ and σ^2, given a sample $\mathbf{y}$ of n observations, may be written

$$l(\theta, \sigma^2|\mathbf{y}) = \text{constant} - \frac{1}{2} n \ln \sigma^2 + \sum_u \ln g\left\{ \left(\frac{\varepsilon_u}{\sigma} \right)^2 \right\}.$$

Now write

$$-2\left[\frac{\partial \ln g\{(\varepsilon_u/\sigma)^2\}}{\partial(\varepsilon_u/\sigma)^2} \right] = w_u(\theta, \sigma^2) = w_u,$$

where the notation $w_u(\theta, \sigma^2)$ is a reminder that w_u is a function of both θ and of σ^2. On differentiating the log likelihood, we obtain

$$\frac{\partial l}{\partial \theta_j} = \sigma^{-2} \sum_u w_u \left(y_u - \sum_i \theta_i z_{iu} \right) z_{ju}, \qquad j = 1, 2, \ldots, p,$$

$$\frac{\partial l}{\partial \sigma^2} = -\frac{1}{2} n \sigma^{-2} + \frac{1}{2} \sigma^{-4} \sum_u w_u \left(y_u - \sum_i \theta_i z_{iu} \right)^2.$$

On equating the derivatives to zero, we obtain the maximum likelihood estimates $\hat{\boldsymbol{\theta}}$ and $\hat{\sigma}^2$ as the solution of the equations

$$\sum_u w_u \left(y_u - \sum_i \hat{\theta}_i z_{iu} \right) z_{ju} = 0, \qquad j = 1, 2, \ldots, p, \qquad (3A.1)$$

$$s^2 = \hat{\Sigma}^2 = \frac{\sum w_u \left(y_u - \Sigma \hat{\theta}_i z_{iu} \right)^2}{n}, \qquad (3A.2)$$

where $w_u = w_u(\hat{\boldsymbol{\theta}}, s^2)$.

Equations (3A.1) have the form of the normal equations for weighted least squares while Eq. (3A.2) with n replaced by $(n - p)$ has the form of the unbiased estimate of σ^2 supplied by weighted least squares [see Eqs. (3.13.17), (3.13.18), (3.13.19), and (3.13.20)]. The equations may be solved iteratively as described earlier. In particular, if the errors are distributed in a t distribution scaled by a parameter σ, then

$$g\left\{ \left(\frac{\varepsilon_u}{\sigma} \right)^2 \right\} = \left\{ 1 + \frac{\varepsilon_u^2}{(f\sigma^2)} \right\}^{-(1/2)(f+1)}$$

and then $w_u = (f + 1)/\{f + (r_u/s)^2\}$, where r_u is the residual $y_u - \sum_i \hat{\theta}_i z_{iu}$.

APPENDIX 3B. JUSTIFICATION OF LEAST SQUARES BY THE GAUSS–MARKOV THEOREM; ROBUSTNESS

An alternative widely used justification for least-squares fitting of linear models employs the Gauss–Markov theorem. This states that, for the model $\mathbf{y} = \mathbf{Z}\boldsymbol{\theta} + \boldsymbol{\varepsilon}$ with the elements $\varepsilon_1, \varepsilon_2, \ldots, \varepsilon_n$ of $\boldsymbol{\varepsilon}$ pairwise uncorrelated and having equal variances σ^2, the least-squares estimators of the p parameters in $\boldsymbol{\theta}$ have, individually, the smallest variances of all linear unbiased estimators of these parameters.

A linear estimator T is one which is of the form $T = l_1 y_1 + l_2 y_2 + \cdots + l_n y_n$, where the l's are selected constants. An estimator T_j of θ_j, $j = 1, 2, \ldots, p$, is unbiased if $E(T_j) = \theta_j$.

The minimum variance property is appealing and does not depend on normality. It might at first be thought that it would automatically endow ordinary least-squares estimators with robust properties. Unfortunately, this is not necessarily the case. To see why, we must consider the relationship between variance, bias, and mean square error. Suppose the true value of a parameter is θ and an estimator of it (not necessarily linear in the data) is $\tilde{\theta} = f(y_1, y_2, \ldots, y_n) = f(\mathbf{y})$. Now what we require in an estimator is that it be in some sense close to the true value. In particular, we might choose estimators with small *mean square error* (MSE), that is, ones such that

$$\text{MSE}(\hat{\theta}) = E(\tilde{\theta} - \theta)^2$$

is small. Now write

$$\tilde{\theta} - \theta = \tilde{\theta} - E(\tilde{\theta}) + E(\tilde{\theta}) - \theta,$$

where $E(\tilde{\theta})$ is the mean value of $\tilde{\theta}$. Then

$$
\begin{aligned}
\text{MSE}(\tilde{\theta}) &= E(\tilde{\theta} - \theta)^2 \\
&= E\{\tilde{\theta} - E(\tilde{\theta})\}^2 + \{E(\tilde{\theta}) - \theta\}^2 \\
&\quad + 2E\{\tilde{\theta} - E(\tilde{\theta})\}\{E(\tilde{\theta}) - \theta\}.
\end{aligned}
$$

The last term is zero, $E\{\tilde{\theta} - E(\tilde{\theta})\}^2$ is the variance of $\tilde{\theta}$, and $E(\tilde{\theta}) - \theta$ is the bias of $\tilde{\theta}$, so that

$$
\text{MSE}(\tilde{\theta}) = \text{variance } \{\tilde{\theta}\} + \{\text{bias } \tilde{\theta}\}^2.
$$

The Gauss–Markov theorem says that if $\tilde{\theta} = f(\mathbf{y})$ is a *linear* function of the y's, and *if* we set the bias term equal to zero, then the variance of $\tilde{\theta}$ (which becomes, with these restrictions, the MSE) will be minimized. Now it turns out (perhaps rather unexpectedly) that, if these restrictions are relaxed and we allow nonlinear functions of the data and biased estimators to be used, then estimators with smaller MSE are possible. That is to say, by making the bias term nonzero, a more than compensating *decrease* in variance can be obtained. Robust estimators are examples of such estimators. Other examples are Stein's shrinkage estimators and Hoerl and Kennard's ridge regression estimators.

APPENDIX 3C. MATRIX THEORY

Matrix, Vector, Scalar

A $p \times q$ matrix $\mathbf{M}$ is a rectangular array of numbers containing p rows and q columns written

$$
\mathbf{M} = \begin{bmatrix}
m_{11} & m_{12} & \cdots & m_{1q} \\
m_{21} & m_{22} & \cdots & m_{2q} \\
\cdots & & & \\
m_{p1} & m_{p2} & \cdots & m_{pq}
\end{bmatrix}
$$

For example,

$$
\mathbf{A} = \begin{bmatrix}
4 & 1 & 3 & 7 \\
-1 & 0 & 2 & 2 \\
6 & 5 & -2 & 1
\end{bmatrix}
$$

is a 3×4 matrix. The plural of *matrix* is *matrices*. A "matrix" with only one row is called a *row vector*; a "matrix" with only one column is called a *column vector*. For example, if

$$
\mathbf{a} = [1, 6, 3, 2, 1], \qquad \mathbf{b} = \begin{bmatrix} -1 \\ 0 \\ 1 \end{bmatrix},
$$

then **a** is a row vector of length five and **b** is a column vector of length three. A 1×1 "vector" is an ordinary number or *scalar*.

Equality

Two matrices are equal if and only if their dimensions are identical and they have exactly the same entries in the same positions. Thus a matrix equality implies as many individual equalities as there are terms in the matrices set equal.

Sum and Difference

The sum (or difference) of two matrices is the matrix each of whose elements is the sum (or difference) of the corresponding elements of the matrices added (or subtracted). For example,

$$
\begin{bmatrix} 7 & 6 & 9 \\ 4 & 2 & 1 \\ 6 & 5 & 3 \\ 2 & 1 & 4 \end{bmatrix} - \begin{bmatrix} 1 & 2 & 4 \\ -1 & 3 & -2 \\ 6 & 2 & 1 \\ 7 & 0 & 2 \end{bmatrix} = \begin{bmatrix} 6 & 4 & 5 \\ 5 & -1 & 3 \\ 0 & 3 & 2 \\ -5 & 1 & 2 \end{bmatrix}.
$$

The matrices must be of exactly the same dimensions for addition or subtraction to be carried out. Otherwise the operations are not defined.

Transpose

The transpose of a matrix **M** is a matrix **M'** whose rows are the columns of **M** and whose columns are the rows of **M** in the same original order. Thus, for **M** and **A** as defined above,

$$
\mathbf{M'} = \begin{bmatrix} m_{11} & m_{21} & \cdots & m_{p1} \\ m_{12} & m_{22} & \cdots & m_{p2} \\ \cdots & & & \\ m_{1q} & m_{2q} & \cdots & m_{pq} \end{bmatrix}
$$

$$
\mathbf{A'} = \begin{bmatrix} 4 & -1 & 6 \\ 1 & 0 & 5 \\ 3 & 2 & -2 \\ 7 & 2 & 1 \end{bmatrix}.
$$

Note that the transpose notation enables us to write, for example,

$$
\mathbf{b'} = (-1, 0, 1) \quad \text{or, alternatively,} \quad \mathbf{b} = (-1, 0, 1)'.
$$

Note: The parentheses around a matrix or vector can be square-ended or curved. Often, capital letters are used to denote matrices, whereas lowercase letters are used to denote vectors.

Symmetry

A matrix $\mathbf{M}$ is said to be *symmetric* if $\mathbf{M}' = \mathbf{M}$.

Trace of a Square Matrix

The trace of a square matrix is the sum of the elements on its main diagonal. So, if $p = q$ in $\mathbf{M}$ above, then

$$\text{trace } \mathbf{M} = m_{11} + m_{22} + \cdots + m_{pp}.$$

Multiplication

Suppose we have two matrices, $\mathbf{A}$, which is $p \times q$, and $\mathbf{B}$, which is $r \times s$. They are *conformable* for the product $\mathbf{C} = \mathbf{AB}$ only if $q = r$. The resulting product is then a $p \times s$ matrix, the multiplication procedure being defined as follows: If

$$\mathbf{A} = \begin{bmatrix} a_{11} & a_{12} & \cdots & a_{1q} \\ a_{21} & a_{22} & \cdots & a_{2q} \\ \cdots & & & \\ a_{p1} & a_{p2} & \cdots & a_{pq} \end{bmatrix}, \qquad \mathbf{B} = \begin{bmatrix} b_{11} & b_{12} & \cdots & b_{1s} \\ b_{21} & b_{22} & \cdots & b_{2s} \\ \cdots & & & \\ b_{q1} & b_{q2} & \cdots & b_{qs} \end{bmatrix},$$

$$\quad\quad\quad p \times q \quad\quad\quad\quad\quad\quad\quad\quad q \times s$$

then the product

$$\mathbf{AB} = \mathbf{C} = \begin{bmatrix} c_{11} & c_{12} & \cdots & c_{1s} \\ c_{21} & c_{22} & \cdots & c_{2s} \\ \cdots & & & \\ c_{p1} & c_{p2} & \cdots & c_{ps} \end{bmatrix}$$

$$p \times s$$

is such that

$$c_{ij} = \sum_{l=1}^{q} a_{il} b_{lj},$$

that is, the entry in the ith row and jth column of $\mathbf{C}$ is the *inner product* (the element by element cross-product) of the ith row of $\mathbf{A}$ with the jth column of $\mathbf{B}$. For example,

$$\begin{bmatrix} 1 & 2 & 1 \\ -1 & 3 & 0 \end{bmatrix} \begin{bmatrix} 1 & 2 & 3 \\ 4 & 0 & -1 \\ -2 & 1 & 3 \end{bmatrix}$$

$$\quad\quad 2 \times 3 \quad\quad\quad\quad 3 \times 3$$

$$= \begin{bmatrix} 1(1) + 2(4) + 1(-2) & 1(2) + 2(0) + 1(1) & 1(3) + 2(-1) + 1(3) \\ -1(1) + 3(4) + 0(-2) & -1(2) + 3(0) + 0(1) & -1(3) + 3(-1) + 0(3) \end{bmatrix}$$

$$= \begin{bmatrix} 7 & 3 & 4 \\ 11 & -2 & -6 \end{bmatrix}.$$

$$\quad 2 \times 3$$

We say that, in the product **AB**, we have *premultiplied* **B** by **A** or we have *postmultiplied* **A** by **B**. Note that, in general, **AB** and **BA**, even if both products are permissible (conformable), do not lead to the same result. In a matrix multiplication, the order in which the matrices are arranged is crucially important, whereas the order of the numbers in a scalar product is irrelevant.

When several matrices and/or vectors are multiplied together, the product should be carried out in the way that leads to the least work. For example, the product

$$\underset{p \times p}{\mathbf{W}} \ \underset{p \times n}{\mathbf{Z}'} \ \underset{n \times 1}{\mathbf{y}}$$

could be carried out as $(\mathbf{WZ'})\mathbf{y}$, or as $\mathbf{W}(\mathbf{Z'y})$, where the parenthesized product is evaluated first. In the first case we would have to carry out pn p-length cross-products and p n-length cross-products; in the second case, p p-length and p n-length, clearly a saving in effort.

Special Matrices and Vectors

We define

$$\mathbf{I}_n = \begin{bmatrix} 1 & 0 & 0 & \cdots & 0 \\ 0 & 1 & 0 & \cdots & 0 \\ 0 & 0 & 1 & \cdots & 0 \\ \cdots & & & & \\ 0 & 0 & 0 & \cdots & 1 \end{bmatrix}$$

a square $n \times n$ matrix with 1's on the diagonal, 0's elsewhere as the *unit matrix* or *identity matrix*. This fulfills the same role as the number 1 in ordinary arithmetic. If the size of $\mathbf{I}_n$ is clear from the context, the subscript n is often omitted. We further use **0** to denote a vector

$$\mathbf{0} = (0, 0, \ldots, 0)'$$

or a matrix

$$\mathbf{0} = \begin{bmatrix} 0 & 0 & \cdots & 0 \\ 0 & 0 & \cdots & 0 \\ \cdots & & & \\ 0 & 0 & \cdots & 0 \end{bmatrix},$$

all of whose values are zeros; the actual size of **0** is usually clear from the context. We also define

$$\mathbf{j} = (1, 1, \ldots, 1)',$$

a vector of all 1's; the size of **j** is either specified or is clear in context.

Orthogonality

A vector $\mathbf{a} = (a_1, a_2, \ldots, a_n)'$ is said to be *orthogonal* to a vector $\mathbf{b} = (b_1, b_2, \ldots, b_n)'$ if the sum of products of their elements is zero, that is, if

$$\sum_{i=1}^{n} a_i b_i = \mathbf{a'b} = \mathbf{b'a} = 0.$$

Inverse Matrix

The inverse $\mathbf{M}^{-1}$ of a square matrix $\mathbf{M}$ is the unique matrix such that

$$\mathbf{M}^{-1}\mathbf{M} = \mathbf{I} = \mathbf{M}\mathbf{M}^{-1}.$$

The columns $\mathbf{m}_1, \mathbf{m}_2, \ldots, \mathbf{m}_n$ of an $n \times n$ matrix are *linearly dependent* if there exist constants $\lambda_1, \lambda_1, \ldots, \lambda_n$, not all zero, such that

$$\lambda_1 \mathbf{m}_1 + \lambda_2 \mathbf{m}_2 + \cdots + \lambda_n \mathbf{m}_n = \mathbf{0}$$

and similarly for rows. A square matrix some of whose rows (or some of whose columns) are linearly dependent is said to be *singular* and does not possess an inverse. A square matrix that is not singular is said to be *nonsingular* and can be inverted.

   If $\mathbf{M}$ is symmetric, so is $\mathbf{M}^{-1}$.

Obtaining an Inverse

The process of matrix inversion is a relatively complicated one and is best appreciated by considering an example. Suppose we wish to obtain the inverse $\mathbf{M}^{-1}$ of the matrix

$$\mathbf{M} = \begin{bmatrix} 3 & 4 & 5 \\ 1 & 2 & 6 \\ 7 & 1 & 9 \end{bmatrix}.$$

Let

$$\mathbf{M}^{-1} = \begin{bmatrix} a & b & c \\ d & e & f \\ g & h & k \end{bmatrix}.$$

Then we must find $(a, b, c, \ldots, h, k)$ so that

$$\begin{bmatrix} a & b & c \\ d & e & f \\ g & h & k \end{bmatrix} \begin{bmatrix} 3 & 4 & 5 \\ 1 & 2 & 6 \\ 7 & 1 & 9 \end{bmatrix} = \begin{bmatrix} 1 & 0 & 0 \\ 0 & 1 & 0 \\ 0 & 0 & 1 \end{bmatrix},$$

that is, so that

$$
\begin{array}{lll}
3a + b + 7c = 1, & 3d + e + 7f = 0, & 3g + h + 7k = 0, \\
4a + 2b + c = 0, & 4d + 2e + f = 1, & 4g + 2h + k = 0, \\
5a + 6b + 9c = 0, & 5d + 6e + 9f = 0, & 5g + 6h + 9k = 1.
\end{array}
$$

Solving these three sets of three linear simultaneous equations yields

$$\mathbf{M}^{-1} = \begin{bmatrix} \frac{12}{103} & -\frac{31}{103} & \frac{14}{103} \\ \frac{33}{103} & -\frac{8}{103} & -\frac{13}{103} \\ -\frac{13}{103} & \frac{25}{103} & \frac{2}{103} \end{bmatrix} = \frac{1}{103} \begin{bmatrix} 12 & -31 & 14 \\ 33 & -8 & -13 \\ -13 & 25 & 2 \end{bmatrix}.$$

(Note the removal of a common factor, explained below.) In general, for an $n \times n$ matrix there will be n sets of n simultaneous linear equations. Accelerated methods for inverting matrices adapted specifically for use with electronic computers permit inverses to be obtained with great speed, even for large matrices.

Determinants

An important quantity associated with a square matrix is its *determinant*. Determinants occur naturally in the solution of linear simultaneous equations and in the inversion of matrices. For a 2×2 matrix

$$\mathbf{M} = \begin{bmatrix} a & b \\ c & d \end{bmatrix}$$

the determinant is defined as

$$\det \mathbf{M} = \begin{vmatrix} a & b \\ c & d \end{vmatrix} = ad - bc.$$

For a 3×3 matrix

$$\begin{bmatrix} a & b & c \\ d & e & f \\ g & h & k \end{bmatrix}$$

it is

$$a \begin{vmatrix} e & f \\ h & k \end{vmatrix} - b \begin{vmatrix} d & f \\ g & k \end{vmatrix} + c \begin{vmatrix} d & e \\ g & h \end{vmatrix} = aek - afh - bdk + bfg + cdh - ceg.$$

Notice that we expand by the first row, multiplying a by the determinant of the matrix left when we cross out the row and column containing a, multiplying b by the determinant of the matrix left when we cross out the row and column containing b, and multiplying c by the determinant of the matrix left when we cross out the row and column containing c. We also attach alternate signs $+, -, +$ to these three terms, counting from the top left-hand corner element: $+$ to a, $-$ to b, $+$ to c, and so on, alternately, if there were more elements in the first row.

In fact, the determinant can be written down as an expansion of *any* row or column by the same technique. The signs to be attached are counted $+ - + -$, and so on, from the top left corner element alternating either along row or column (but *not* diagonally). In other words the signs

$$\begin{bmatrix} + & - & + \\ - & + & - \\ + & - & + \end{bmatrix}$$

are attached and any row or column is used to write down the determinant. For example, using the second row we have

$$-d \begin{vmatrix} b & c \\ h & k \end{vmatrix} + e \begin{vmatrix} a & c \\ g & k \end{vmatrix} - f \begin{vmatrix} a & b \\ g & h \end{vmatrix}$$

to obtain the same result as before.

The same principle is used to get the determinant of any matrix. Any row or column is used for the expansion and we multiply each element of the row or column by:

1. Its appropriate sign, counted as above.
2. The determinant of the submatrix obtained by deletion of the row and column in which the element of the original matrix stands.

Determinants arise in the inversion of matrices as follows. The inverse $\mathbf{M}^{-1}$ may be obtained by first replacing each element m_{ij} of the original matrix $\mathbf{M}$ by an element calculated as follows:

1. Find the determinant of the submatrix obtained by crossing out the row and column of $\mathbf{M}$ in which m_{ij} stands.
2. Attach a sign from the $+ - + -$ count, as above.
3. Divide by the determinant of $\mathbf{M}$.

When all elements of $\mathbf{M}$ have been replaced, *transpose the resulting matrix*. The transpose will be $\mathbf{M}^{-1}$.

The reader might like to check these rules by showing that

$$\mathbf{M}^{-1} = \begin{bmatrix} a & b \\ c & d \end{bmatrix}^{-1} = \begin{bmatrix} d/D & -b/D \\ -c/D & a/D \end{bmatrix},$$

where $D = ad - bc$ is the determinant of $\mathbf{M}$; and that

$$\mathbf{Q}^{-1} = \begin{bmatrix} a & b & c \\ d & e & f \\ g & h & k \end{bmatrix}^{-1} = \begin{bmatrix} A & B & C \\ D & E & F \\ G & H & K \end{bmatrix},$$

where

$$A = (ek - fh)/Z, \qquad B = -(bk - ch)/Z, \qquad C = (bf - ce)/Z,$$
$$D = -(dk - fg)/Z, \qquad E = (ak - cg)/Z, \qquad F = -(af - cd)/Z,$$
$$G = (dh - eg)/Z, \qquad H = -(ah - bg)/Z, \qquad K = (ae - bd)/Z$$

and where

$$Z = aek + bfg + cdh - afh - bdk - ceg$$

is the determinant of $\mathbf{Q}$. Note that if $\mathbf{M}$ is symmetric (so that $b = c$), $\mathbf{M}^{-1}$ is also symmetric. Also, if $\mathbf{Q}$ is symmetric (so that $b = d$, $c = g$, $f = h$), then $\mathbf{Q}^{-1}$ is also symmetric because then $B = D$, $C = G$, and $F = H$.

Common Factors

If *every* element of a matrix has a common factor, it can be taken outside the matrix. Conversely, if a matrix is multiplied by a constant c, every element of the matrix is multiplied by c. For example,

$$\begin{bmatrix} 4 & 6 & -2 \\ 8 & 6 & 2 \end{bmatrix} = 2\begin{bmatrix} 2 & 3 & -1 \\ 4 & 3 & 1 \end{bmatrix}.$$

Note that if a matrix is square and of size $p \times p$, and if c is a common factor, then the determinant of the matrix has a factor c^p, not just c. For example,

$$\begin{vmatrix} 4 & 6 \\ 8 & 6 \end{vmatrix} = 2^2\begin{vmatrix} 2 & 3 \\ 4 & 3 \end{vmatrix} = 2^2(6 - 12) = -24.$$

Eigenvalues and Eigenvectors

If $\mathbf{M}$ is a $p \times p$ square matrix with elements m_{ij}, the *eigenvalues* of $\mathbf{M}$ are the p roots of the determinantal equation

$$\det(\mathbf{M} - \lambda\mathbf{I}) = 0$$

which is a polynomial of the pth degree in λ. Suppose $\lambda_1 \geq \lambda_2 \geq \cdots \geq \lambda_p$ are the eigenvalues. Then the $p \times 1$ vector $\mathbf{v}_i$ such that

$$\mathbf{M}\mathbf{v}_i = \lambda_i\mathbf{v}_i$$

(separately for each and every one of the λ_i) is called the eigenvector associated with λ_i. These quantities are needed to perform the *canonical reduction* of a second order fitted response surface in Chapters 10 and 11, and to carry out ridge analysis of response surfaces. Computer programs are widely available for obtaining eigenvalues and eigenvectors; their evaluation otherwise is extremely tedious.

Kronecker Product

Suppose we have two matrices $\mathbf{A}$ (size p rows and q columns with elements a_{ij}) and $\mathbf{B}(r \times s$ with elements b_{ij}). Then the *Kronecker product* (sometimes called the *direct product*) of $\mathbf{A}$ and $\mathbf{B}$, *In that order*, *is*

$$\mathbf{A} \otimes \mathbf{B} = \begin{bmatrix} a_{11}\mathbf{B} & a_{12}\mathbf{B} & \cdots & a_{1q}\mathbf{B} \\ a_{21}\mathbf{B} & a_{22}\mathbf{B} & \cdots & a_{2q}\mathbf{B} \\ \cdots & \cdots & & \cdots \\ a_{p1}\mathbf{B} & a_{p2}\mathbf{B} & \cdots & a_{pq}\mathbf{B} \end{bmatrix},$$

and is of size $pr \times qs$. Some properties are:

- $(\mathbf{A} \otimes \mathbf{B})' = \mathbf{A}' \otimes \mathbf{B}'$
- $(\mathbf{A} \otimes \mathbf{B})(\mathbf{C} \otimes \mathbf{D}) = \mathbf{AC} \otimes \mathbf{BD}$
- trace$(\mathbf{A} \otimes \mathbf{B})$ = trace$(\mathbf{A}) \times$ trace$(\mathbf{B})$, if both $\mathbf{A}$ and $\mathbf{B}$ are square
- $(\mathbf{A} \otimes \mathbf{B})^{-1} = \mathbf{A}^{-1} \otimes \mathbf{B}^{-1}$, if both $\mathbf{A}$ and $\mathbf{B}$ are square and nonsingular
- $(\mathbf{A} + \mathbf{C}) \otimes \mathbf{B} = \mathbf{A} \otimes \mathbf{B} + \mathbf{C} \otimes \mathbf{B}$, if $\mathbf{A}$ and $\mathbf{C}$ are the same size.

Special Case ($p = 1, r = 1$)
Suppose $\mathbf{A} = (a_1, a_2, \ldots, a_q)$ and $\mathbf{B} = (b_1, b_2, \ldots, b_s)$. Then

$$\mathbf{A} \otimes \mathbf{B} = (a_1\mathbf{B}, a_2\mathbf{B}, \ldots, a_q\mathbf{B})$$

$$= (a_1b_1, a_1b_2, \ldots, a_1b_s; a_2b_1, a_2b_2, \ldots, a_2b_s; \ldots; a_qb_1, a_qb_2, \ldots, a_qb_s),$$

which is a row vector of length qs.

APPENDIX 3D. NONLINEAR ESTIMATION

Nonlinear Least Squares

Our discussion in Chapter 3 on fitting linear models leaves open the question of how to proceed when the model we wish to fit is intrinsically nonlinear. We discuss the matter briefly here. Suppose we wish to fit the general model

$$y_u = f(\boldsymbol{\xi}_u, \boldsymbol{\theta}) + \varepsilon_u, \tag{3D.1}$$

where f is some known function, $\boldsymbol{\xi}$ is a vector of predictor variables, $\xi_1, \xi_2, \ldots, \xi_k$, $\boldsymbol{\theta} = (\theta_1, \theta_2, \ldots, \theta_p)'$ is a vector of parameters to be estimated, ε is a random error from a distribution with mean zero and unknown variance σ^2, and the subscript $u = 1, 2, \ldots, n$ ranges over the n observations. We also suppose that the errors are normally distributed, so that $\varepsilon_u \sim N(0, \sigma^2)$.

To obtain least-squares estimates, we must find the value of $\boldsymbol{\theta}$ which minimizes the sum of squares

$$S(\boldsymbol{\theta}) = \sum_{u=1}^{n} \{y_u - f(\boldsymbol{\xi}_u, \boldsymbol{\theta})\}^2. \tag{3D.2}$$

Equating first derivatives with respect to the θ's to zero yields

$$\sum_{u=1}^{n} \{y_u - f(\boldsymbol{\xi}_u, \boldsymbol{\theta})\}\left\{\frac{\partial f}{\partial \theta_i}\right\} = 0, \qquad i = 1, 2, \ldots, p. \tag{3D.3}$$

Because these normal equations are nonlinear in $\boldsymbol{\theta}$, they are not amenable to easy solution. We now describe an alternative minimizing procedure based on local linearization originally suggested by Gauss.

Linearization (or Gauss) Procedure

If the function $f(\boldsymbol{\xi}_u, \boldsymbol{\theta})$ is expanded as a Taylor series in the elements $\boldsymbol{\theta}$ to first-order terms about some guessed value $\boldsymbol{\theta}_0$, we can use linear least-squares theory to give a solution $\boldsymbol{\theta}_1$ for this *linearized* model. We can then repeat the procedure with $\boldsymbol{\theta}_1$ in place of $\boldsymbol{\theta}_0$, to give rise to $\boldsymbol{\theta}_2$, and so on. Under favorable conditions, successive iterations will converge to the least-squares estimate $\hat{\boldsymbol{\theta}}$. In practice they do not always so converge. (For further technical details of this procedure, see Draper and Smith, 1998, Chapter 24.)

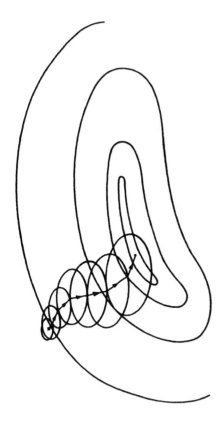

Figure 3D.1. Typical iterations in a Marquardt compromise procedure.

The Marquardt Compromise Procedure

The crude Gauss procedure outlined above has a tendency to "overshoot," that is, to go beyond points θ where smaller $S(\theta)$ values exist to points where larger $S(\theta)$ values occur. While recovery from "overshoot" sometimes occurs on the next iteration, sometimes it does not, and the iterations may continue to diverge. To avoid this problem, Levenberg (1944) suggested restraining the next iteration to points on an ellipsoid about the current best estimate. As the current iteration gets closer and closer to the minimizing value $\hat{\theta}$, overshoot becomes less likely and the dimensions of the ellipsoid can be enlarged. This is the procedure Marquardt (1963) implemented. In the appropriately scaled standard error coordinates, two constants are chosen, one to set the spherical radius and one to use as an enlarging factor, as the iterations proceed; in the various programmed versions, the selection is done automatically. Figure 3D.1 shows how the iterations might proceed in a particular case.

One justification for the procedure is as follows: Initially, when steepest ascent is likely to work well, the procedure follows the path of steepest ascent orthogonal to the contours; later when linearization is likely to work well, nearer to $\hat{\theta}$, it becomes more and more like linearization. (Thus the Marquardt method *compromises* between these two approaches; hence its name.) Whether Marquardt's is actually the best procedure from this point of view has been questioned. However, an advantage of the Marquardt compromise is that it overcomes problems that arise when there is ill-conditioning, that is, when the sum of squares surface $S(\boldsymbol{\theta})$ is approximated locally by an attenuated ridge type of surface.

Initial Values

In order to actually carry out any of the iterative procedures needed to minimize $S(\boldsymbol{\theta})$, one has to begin with a set of *initial guesses* $\boldsymbol{\theta}_0 = (\theta_{10}, \theta_{20}, \ldots, \theta_{p0})'$ of the parameters. The "better" these initial guesses are, the faster the iterative procedure will converge. Thus, efforts to obtain good initial values are often worthwhile. Approximate graphical methods are frequently appropriate, and other methods based on picking p selected runs and solving for the θ's when the model is made to pass through these points exactly can sometimes be employed. However, it must also be said that even with extremely poor initial guesses, convergence can sometimes occur surprisingly fast, so that it is not *always* necessary to spend a great deal of effort selecting starting values.

Confidence Contours

By linearizing $f(\boldsymbol{\xi}, \boldsymbol{\theta})$ about $\hat{\boldsymbol{\theta}}$, and appealing to results that are true for the linear model case, we obtain an approximate $100(1 - \alpha)\%$ elliptical confidence region in the form

$$(\boldsymbol{\theta} - \hat{\boldsymbol{\theta}})'\hat{\mathbf{Z}}'\hat{\mathbf{Z}}(\boldsymbol{\theta} - \hat{\boldsymbol{\theta}}) \leq ps^2 F(p, n - p, 1 - \alpha), \qquad (3\text{D}.4)$$

where $\hat{\mathbf{Z}}$ is an $n \times p$ matrix whose element in the uth row and ith column is

$$\left.\frac{\partial f(\boldsymbol{\xi}_u, \boldsymbol{\theta})}{\partial \theta_i}\right|_{\substack{\text{evaluated} \\ \text{at } \boldsymbol{\theta} = \hat{\boldsymbol{\theta}}}} \qquad (3\text{D}.5)$$

and where

$$s^2 = \frac{S(\hat{\boldsymbol{\theta}})}{(n - p)}. \qquad (3\text{D}.6)$$

Ways of checking the adequacy of this linear approximation have been considered, as have methods for choosing appropriate transformations of the parameters which can often greatly improve the adequacy of the linear approximation. For additional reading see, for example, Bates and Watts (1980), Beale (1960), and Box (1960b).

For wider reading on nonlinear estimation, see Draper and Smith (1998, Chapter 24) and Bates and Watts (1988).

APPENDIX 3E. RESULTS INVOLVING V(ŷ)

We have

$$\hat{\mathbf{y}} = \mathbf{Z}\hat{\boldsymbol{\theta}} = \mathbf{Z}(\mathbf{Z}'\mathbf{Z})^{-1}\mathbf{Z}'\mathbf{y}$$

$$= \mathbf{Ry}, \qquad (3\text{E}.1)$$

say, where $\mathbf{R} = \mathbf{Z}(\mathbf{Z}'\mathbf{Z})^{-1}\mathbf{Z}'$. Note that $\mathbf{R}$ is symmetric ($\mathbf{R}' = \mathbf{R}$) and idempotent ($\mathbf{R}^2 = \mathbf{RR} = \mathbf{R}$). It follows that

$$V(\hat{\mathbf{y}}) = \mathbf{R}'V(\mathbf{y})\mathbf{R}$$

$$= \mathbf{R}(\mathbf{I}\sigma^2)\mathbf{R}$$

$$= \mathbf{R}\sigma^2. \tag{3E.2}$$

Thus $V(\hat{y}_i)$, $i = 1, 2, \ldots, n$, is the ith diagonal term of $\mathbf{R}\sigma^2$. Consider the sum

$$\sum_{i=1}^{n} V(\hat{y}_i) = \text{trace}(\mathbf{R}\sigma^2) = \sigma^2 \text{ trace } \mathbf{R}, \tag{3E.3}$$

where "trace" means "take the sum of all the diagonal elements of the square matrix indicated." Now it is true that $\text{trace}(\mathbf{AB}) = \text{trace}(\mathbf{BA})$. So, taking $\mathbf{A} = \mathbf{Z}$ and $\mathbf{B} = (\mathbf{Z}'\mathbf{Z})^{-1}\mathbf{Z}'$, we see that

$$\text{trace } \mathbf{R} = \text{trace}(\mathbf{Z}(\mathbf{Z}'\mathbf{Z})^{-1}\mathbf{Z}') = \text{trace}\{(\mathbf{Z}'\mathbf{Z})^{-1}\mathbf{Z}'\mathbf{Z}\}$$

$$= \text{trace}(\mathbf{I}_p)$$

$$= p, \tag{3E.4}$$

the dimension of $\mathbf{Z}'\mathbf{Z}$. It follows that

$$n^{-1}\sum_{i=1}^{n} V(\hat{y}_i) = \frac{p\sigma^2}{n}. \tag{3E.5}$$

This means that the average variance of $\hat{y}$ over a set of points used for a regression calculation is $p\sigma^2/n$, and is thus fixed when p, n, and σ^2 are fixed.

EXERCISES

3.1. The data below arose in a test of certain semiconductor memory devices.

Supply voltage during "write" operation, ξ: 25.00 25.05 25.10 25.15 25.20
Retention time (h $\times$ 10^{-4}), Y: 1.55   2.36   3.93   7.11 13.52

(a) Plot the data, fit the model $Y = \beta_0 + \beta_1\xi + \varepsilon$, and find the fitted values and residuals. Confirm that $\Sigma Y_i^2 = \Sigma\hat{Y}_i^2 + \Sigma(Y_i - \hat{Y}_i)^2$, within rounding error.

(b) Let ξ be coded to $x = (\xi - \xi_0)/S$. What are suitable values for ξ_0 and S?

(c) Fit the model $Y = \beta_0 + \beta_1 x + \varepsilon$ and find the fitted values and residuals.

(d) Which is preferable, model (a) or model (c)? Why?

(e) What do the residuals indicate?

Table E3.4

z_1	z_2	y
3	6	176
4	4	192
5	7	262
6	3	230
7	6	308
8	4	312

(f) It is now suggested that the analysis should have been done *not* with Y but with $y = \log Y$. Fit $y = \beta_0 + \beta_1 x + \varepsilon$ and draw the fitted line on a plot of y versus x.

(g) Find the fitted values y_i and the residuals $y_i - \hat{y}_i$ in (f) and show that the vectors **y**, **ŷ**, and **y** − **ŷ** form a right-angled triangle with internal angles of $2°$ (approximately), $90°$, and $88°$ (approximately).

(h) For the model in (f), test the hypothesis $(\beta_0, \beta_1) = (0.6, 0.25)$. What do you conclude?

3.2. A colleague who reads the above exercise suggests that "it might have been better to fit a quadratic $Y = \beta_0 + \beta_1 x + \beta_2 x^2 + \varepsilon$ to the original Y data." Carry out the analysis. Do you agree with her? Explain.

3.3. Consider the model $y = z\theta + \varepsilon$ where the data values of z are in $\mathbf{z}' = (1, 1, 1, 1, 1)$ and the corresponding y observations are in $\mathbf{y}' = (11, 8, 9, 10, 7)$. Estimate θ and test $H_0: \theta = 8$ versus $H_1: \theta \neq 8$.

3.4. The data in Table E3.4 were obtained from an experiment to determine the effect of two gasoline additives (whose percentages are x_1 and x_2) on the "cold start ignition time" in seconds, Y, of a test vehicle. (We work with $z_1 = 10x_1$, $z_2 = 10x_2$, and $y = 100Y$ to remove decimals in what follows.)

(a) Fit the model $y = \theta_1 z_1 + \varepsilon$ by least squares. (What important assumption is being made here?)

(b) Fit $y = \theta_1 z_1 + \theta_2 z_2 + \varepsilon$.

(c) Determine $z_{2.1}$ and fit $y = \theta z_1 + \theta_2 z_{2.1} + \varepsilon$. How are $\hat{\theta}_1$, $\hat{\theta}_2$, and $\hat{\theta}$ related?

(d) Construct an analysis of variance table.

(e) If model (a) is fitted but model (b) is "true," what is $E(\hat{\theta}_1)$? What relationship does this have to your answer in (c)?

(f) Construct a 95% joint confidence region for θ_1 and θ_2.

3.5. (Source: Inactivation of adrenaline and nonadrenaline by human and other mammalian liver in vitro, by W. A. Bain and J. E. Batty, *British Journal of Pharmacology and Chemotherapy*, **11**, 1956, 52–57.) The data in Table E3.5 are $n = 14$ epinephrine (adrenaline) concentrations Y (erg/mL) for five "times in the lower tissues," X (min), coded to x.

Table E3.5

X	x	Y			ΣY
6	−2	30.0	28.6	28.5	87.1
18	−1	8.9	8.0	10.8	27.7
30	0	4.1	—	4.7	8.8
42	1	1.8	2.6	2.2	6.6
54	2	0.8	0.6	1.0	2.4
					132.6

(a) What is the coding?

(b) Assume for the moment that the data consist of 14 independent observations. Fit $Y = \beta_0 + \beta_1 x + \varepsilon$, but show that lack of fit exists. What alternative model would you recommend?

(c) Fit a suitable alternative model.

(d) You now look up the original paper and find that each Y column in the table is a separate experiment, in each of which samples were taken successively in time from the same tube. Might this affect your analysis? If so, how and why?

3.6. The postulated straight-line model $y = \beta_0 + \beta_1 x + \varepsilon$ is to be fitted by least squares to observations $x_1, x_2, \ldots, x_n$. The first three moments of the x's are defined by

$$\bar{x} = \sum_i x_i / n, \qquad c = \sum_i x_i^2 / n, \qquad d = \sum_i x_i^3 / n,$$

where all summations are over $1 = 1, 2, \ldots, n$. Show that, if the model $y = \beta_0 + \beta_1 x + \beta_{11} x^2 + \varepsilon$ is feared, but the straight line is fitted, then

$$E(b_0) = \beta_0 + \left\{ (c^2 - \bar{x}d)/(c - \bar{x}^2) \right\} \beta_{11}$$

$$E(b_1) = \beta_1 + \left\{ (d - \bar{x}c)/(c - \bar{x}^2) \right\} \beta_{11}.$$

(a) Is it possible to choose a viable set of x's so that both estimates are unbiased? Explain.

(b) Is it possible to choose a viable set of x's so that b_1 is unbiased? If yes, provide a (relatively) simple way to achieve this.

3.7. Remember the following useful facts about eigenvalues:

(a) The trace of a square matrix is equal to the sum of its eigenvalues.

(b) The determinant of a square matrix is equal to the product of its eigenvalues.

Check out these facts on a small example, using symbols or numbers.

CHAPTER 4

Factorial Designs at Two Levels

4.1. THE VALUE OF FACTORIAL DESIGNS

In the model-building process, the step of model specification or identification, as a result of which a model worthy of being entertained is put forward, is a somewhat tenuous one. It is a creative step in which the human mind, taking account of what is known about the system under study, must be allowed to interact freely with the data, making comparisons, seeking similarities, differences, trends, and so on. A class of experimental designs called "factorials" greatly facilitates this process.

We obtain a complete *factorial design* in k factors by choosing n_1 levels of factor 1, n_2 levels of factor 2, ..., and n_k levels of factor k and then selecting the $n = n_1 \times n_2 \times \cdots \times n_k$ runs obtained by taking all possible combinations of the levels selected. Figure 4.1a shows a $3 \times 2 \times 2$ factorial design in the factors:

Temperature	(three levels: 180°C, 190°C, 200°C)
Pressure	(two levels: 70 psi, 90 psi)
Catalyst	(two levels: type A, type B)

Suppose the numbers shown in Figure 4.1b are the measured amounts of an impurity which are produced at the various sets of conditions. By simply looking at the figure, an observer can make many different and informative comparisons. For example, suppose we knew that the experimental error was small (that the standard deviation σ was less than 1, say). Then the "horizontal" comparisons $(7 \to 8 \to 10)$, $(6 \to 8 \to 11)$, $(3 \to 5 \to 5)$, $(3 \to 4 \to 5)$ would suggest that, over the ranges studied, increasing temperature produced increasing impurity.

Also, the "vertical" comparisons $(7 \to 6)$, $(8 \to 8)$, $(10 \to 11)$, $(3 \to 3)$, $(5 \to 4)$, $(5 \to 5)$ would suggest that there is little to choose between the two pressures examined. The six comparisons between the two types of catalysts would suggest that, over the ranges tested, B produces less impurity than A. Furthermore, the increase of impurity with increased temperature is less with B than with A (so that

Response Surfaces, Mixtures, and Ridge Analyses, Second Edition. By G. E. P. Box and N. R. Draper
Copyright © 2007 John Wiley & Sons, Inc.

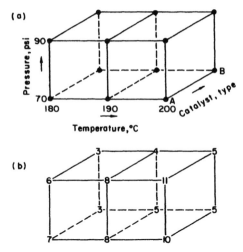

Figure 4.1. A $3 \times 2 \times 2$ factorial design. (*a*) The points of the design. (*b*) Impurity values observed at the 12 design point locations.

there is a temperature–catalyst interaction). The possibility is also seen that perhaps a fixed change in temperature produces a fixed *proportionate* response so that it is possible that the logarithm of the response is linearly and additively affected by temperature.

All of the above ideas can, of course, be subjected to more formal examination. It is the ability of the factorial design to suggest ideas to the investigator that we wish to emphasize.

In the above example, the factorial design is used to study two quantitative factors or variables (temperature and pressure) and one qualitative variable (type of catalyst). Most of this book will be concerned only with the study of quantitative variables. Factorial designs possess the following desirable properties:

1. They allow multitudes of comparisons to be made and so facilitate model creation and criticism.
2. They provide highly efficient estimates of constants (parameters)—that is, estimates whose variances are as small, or nearly as small, as those that could be produced by any design occupying the same space.
3. They give rise to simple calculations.

On the mistaken supposition that simplicity of calculation is the *only* reason for their use, it is sometimes suggested that factorial designs are no longer needed because computers have made even complex calculations easy. Also it may be argued that "optimal" designs may now be computed (see Chapter 14 for details) which do not necessarily give nice patterns and yet, *if the model is assumed known*, will also give highly efficient estimates. However, this argument discounts the importance of allowing the mind to see patterns in the data at the model specification stage.

4.2. TWO-LEVEL FACTORIALS

Of particular importance for our purpose are the "two-level" factorial designs in which each variable occurs at just two levels. Such designs are especially useful at the exploratory stage of an investigation, when not very much is known about a system and the model is still to be identified. As we shall see in later examples, the initial two-level pattern can be a first building block in developing structures of many different sorts. A model-identification technique of great value that may be used in association with this type of design is the plotting of coefficients on normal probability paper (see Daniel, 1959, 1976). For additional details see Appendix 4B.

For a first illustration, we return to the 2^3 (i.e., $2 \times 2 \times 2$) factorial design introduced in Chapter 2 (see Table 2.2). This design was used to study $Y = $ cycles to failure for combinations of three variables, each tested at two levels as follows.

		Coded Levels	
	x_i	-1	1
Length of test specimen (mm)	ξ_1	250	350
Amplitude of load cycle (mm)	ξ_2	8	10
Load (g)	ξ_3	40	50

As discussed in Chapter 2, it is convenient to code the lower and upper levels as x_1, x_2, x_3, taking the values -1 for the lower level and 1 for the upper level. Thus, for this example,

$$x_1 = \frac{(\xi_1 - 300)}{50}, \qquad x_2 = \frac{(\xi_2 - 9)}{1}, \qquad \text{and} \qquad x_3 = \frac{(\xi_3 - 45)}{5}. \quad (4.2.1)$$

The 2^3 design is shown in Table 4.1 in both uncoded and coded form, together with the response values $y = $ cycles to failure and the values of $y = \log Y$. (The logarithms are taken to base 10.)

Table 4.1. Textile Data. A 2^3 Factorial Design in Uncoded and Coded Units with Response Y, and $y = \log Y$ Values

Uncoded			Coded			Cycles	
Specimen Length (mm), ξ_1	Amplitude (mm), ξ_2	Load (g), ξ_3	Specimen Length, x_1	Amplitude x_2	Load, x_3	to Failure, Y	$y = \log Y$
250	8	40	-1	-1	-1	674	2.83
350	8	40	1	-1	-1	3636	3.56
250	10	40	-1	1	-1	170	2.23
350	10	40	1	1	-1	1140	3.06
250	8	50	-1	-1	1	292	2.47
350	8	50	1	-1	1	2000	3.30
250	10	50	-1	1	1	90	1.95
350	10	50	1	1	1	360	2.56

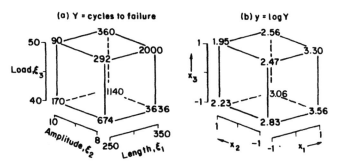

Figure 4.2. Textile data. A 2^3 design shown geometrically in (*a*) uncoded and (*b*) coded units, with responses (*a*) Y = cycles to failure and (*b*) y = log Y.

Figure 4.2 shows the eight runs set out in the space of the three variables where the design points appear as the vertices of a cube. On the vertices of the left-hand cube are shown the values of the response Y, which is the number of cycles to failure, and on the right-hand cube are shown the corresponding transformed values y = log Y. The geometrical display again illustrates the virtues of the factorial design. It is very easy to appreciate the many comparisons that can be made. In particular, one can see the consistent increases that occur in Y with increase in length (ξ_1) and the reductions that occur with increase in amplitude (ξ_2) and with increase in load (ξ_3). Moreover, the differences in magnitudes of these changes largely disappear after log transformation, indicating the regularity that occurs in the *proportional* increases in this example.

In general, a 2^k factorial design consists of all the 2^k runs (points) with levels (coordinates)

$$(x_1, x_2, \ldots, x_k) = (\pm 1, \pm 1, \ldots, \pm 1), \tag{4.2.2}$$

where every possible combination of $\pm$ signs is selected in turn. Geometrically the design consists of the vertices of a hypercube in k dimensions. For purposes of analysis, it is convenient to list the runs in *standard order*, not in the order (usually randomized) in which they were made. This standard order is obtained by writing alternate $-$ and $+$ signs in the column headed x_1, alternate pairs $- -$, $+ +$, in the x_2 column, alternate fours $- - - -$, $+ + + +$, in the x_3 column, and so on. For illustration, the runs are set out in standard order in Table 4.1.

Analysis of the Factorial Design

The main effect of a given variable as defined by Yates (1937) is the average difference in the level of response as one moves from the low to the high level of that variable. For illustration, consider the calculation of the main effects for

$y = \log Y$. A glance at Figure 4.2b shows that the main effect indicated by 1, of variable x_1, is*

$$1 \leftarrow \tfrac{1}{4}(3.56 + 3.06 + 3.30 + 2.56) - \tfrac{1}{4}(2.83 + 2.23 + 2.47 + 1.95)$$

$$= 0.75. \tag{4.2.3}$$

Similar calculations provide the results

$$2 \leftarrow -0.59, \tag{4.2.4}$$

$$3 \leftarrow -0.35. \tag{4.2.5}$$

A valuable property of the factorial design is that it makes possible not only the calculation of main effects (i.e., average effects) but also the calculation of interaction effects between variables.

Two variables, say 1 and 3, are said to interact (in their effect on the response) if the effect of 1 is different at the two different levels of 3. In the textile example, if we confine attention for the moment to the first four runs in which x_3 is at its lower level, the main (average) effect of 1 is

$$(1|x_3 = -1) \leftarrow \tfrac{1}{2}(3.56 + 3.06) - \tfrac{1}{2}(2.83 + 2.23) = 0.78. \tag{4.2.6}$$

However, for the last four runs, with x_3 at its upper level, the main effect of 1 is

$$(1|x_3 = 1) \leftarrow \tfrac{1}{2}(3.30 + 2.56) - \tfrac{1}{2}(2.47 + 1.95) = 0.72. \tag{4.2.7}$$

The interaction between variables 1 and 3 is defined as half the difference between the main effect of 1 at the upper level of x_3 and the main effect of 1 at the lower level of x_3. This interaction is denoted by the symbol 13, so that

$$13 = \tfrac{1}{2}\{(1|x_3 = 1) - (1|x_3 = -1)\} \leftarrow \tfrac{1}{2}\{0.72 - 0.78\} = -0.03. \tag{4.2.8}$$

(There is no ambiguity in this definition in the sense that interchanging the roles of variables 1 and 3 does not change the value of the interaction.)

Exercise.   Evaluate $\tfrac{1}{2}\{(3|x_1 = 1) - (3|x_1 = -1)\}$ and show that it produces a result identical to the 13 value already obtained.   □

---

*Here and in Chapter 5, we use the arrow pointing to the left ($\leftarrow$) to mean "is estimated by," so that Eq. (4.2.3) is read as "the main effect of variable 1 is estimated by the number 0.75." Similarly, the reverse notation "0.75 $\rightarrow$ 1" is read as "0.75 is an estimate of the main effect of variable 1."

Via similar calculations we obtain the other two two-factor interactions as

$$12 \leftarrow -0.03, \tag{4.2.9}$$

$$23 \leftarrow -0.04. \tag{4.2.10}$$

Finally, the interaction 12 might be different at different levels of the variable 3. To check this we could evaluate and compare

$$(12|x_3 = -1) \leftarrow \tfrac{1}{2}\{(3.06 - 2.23) - (3.56 - 2.83)\} = 0.05,$$

$$(12|x_3 = 1) \leftarrow \tfrac{1}{2}\{(2.56 - 1.95) - (3.30 - 2.47)\} = -0.11. \tag{4.2.11}$$

Half the difference between these quantities is called the 123 interaction, that is,

$$123 = \tfrac{1}{2}\{(12|x_3 = 1) - (12|x_3 = -1)\} \leftarrow -0.08. \tag{4.2.12}$$

Exercise. Show that the same result occurs if we define the 123 interaction as half the difference between the interaction 23 at levels of $x_1 = 1$ and $x_1 = -1$, or as half the difference between the interaction 13 at levels of $x_2 = 1$ and $x_2 = -1$. Thus the three-way interaction 123 is unambiguous.   □

There is an easier systematic way of making these calculations using the columns of signs of Table 4.2. Consider the calculation of the main effect 1 as in Eq. (4.2.3). This could be written

$$1 \leftarrow \tfrac{1}{4}(-y_1 + y_2 - y_3 + y_4 - y_5 + y_6 - y_7 + y_8). \tag{4.2.13}$$

In other words, the main effect 1 would be obtained by multiplying the column of data **y** by the column of $-$ and $+$ signs in the column labeled **1** in Table 4.2 and dividing by the divisor 4 indicated there. Note that the divisors are the numbers of $+$ signs in the corresponding columns.

Table 4.2. Columns of Signs and the Divisors for Systematically Obtaining the Factorial Effects in a 2^3 Factorial Design

I	1	2	3	12	13	23	123	y
+	−	−	−	+	+	+	−	2.83
+	+	−	−	−	−	+	+	3.56
+	−	+	−	−	+	−	+	2.23
+	+	+	−	+	−	−	−	3.06
+	−	−	+	+	−	−	+	2.47
+	+	−	+	−	+	−	−	3.30
+	−	+	+	−	−	+	−	1.95
+	+	+	+	+	+	+	+	2.56
Divisor 8	4	4	4	4	4	4	4	

Similarly, it can readily be confirmed that

$$2 \leftarrow \tfrac{1}{4}(-y_1 - y_2 + y_3 + y_4 - y_5 - y_6 + y_7 + y_8), \tag{4.2.14}$$

where the signs and divisor are taken from the column of Table 4.2 labeled **2**. The main effect of 3 is similarly obtained.

If the operations leading to the calculation of the two-factor interaction 13 in Eq. (4.2.8) are similarly analyzed, it will be found that

$$13 \leftarrow \tfrac{1}{4}(y_1 - y_2 + y_3 - y_4 - y_5 + y_6 - y_7 + y_8). \tag{4.2.15}$$

This effect can be mechanically computed from the **13** column in the table; the signs in that column are simply the results of taking the products of the signs in the **1** and **3** columns, line by line. The interaction effects 12 and 23 are calculated in exactly similar fashion using their corresponding columns.

Finally, if the calculations leading to the 123 interaction are analyzed, we find that

$$123 \leftarrow \tfrac{1}{4}(-y_1 + y_2 + y_3 - y_4 + y_5 - y_6 - y_7 + y_8). \tag{4.2.16}$$

The signs are those shown in the **123** column in Table 4.2, and this column is formed by taking the triple product of signs from the **1**, **2**, and **3** columns.

The table of signs is completed by the addition of a column of plus signs, shown at the left with the heading **I**. This is needed for obtaining the average of the response values, and its divisor (the number of plus signs in the column) is 8 rather than 4 as in the other columns.

A table such as Table 4.2 is very easily constructed for any 2^k factorial design as follows. Begin with a column of 1's of length 2^k. The next k columns are ± 1 signs for the design written down in standard order. Columns $12, 13, \ldots, 123, \ldots,$ **123...k**, $(2^k - k - 1)$ in number, are then obtained by multiplying signs, row by row, in the way indicated by the headings. At the bottom are written the divisors, 2^k for the first column, and 2^{k-1} for all the others.

The table of signs is very useful for understanding the nature of the various factorial effects. However, a quicker method of obtaining the effects mechanically is available, due to Yates; this is described in Appendix 4A.

Variance and Standard Errors of Effects for 2^k Designs

For a complete 2^k design, if $V(y) = \sigma^2$,

$$V(\text{grand mean}) = \frac{\sigma^2}{2^k},$$
$$V(\text{effect}) = \frac{4\sigma^2}{2^k}. \tag{4.2.17}$$

If the responses are not "y's" but are instead "$\bar{y}$'s," the averages of (say) r y-observations, then

$$V(\text{grand mean}) = \frac{\sigma^2}{n}, \qquad V(\text{effect}) = \frac{4\sigma^2}{n}, \qquad (4.2.17a)$$

where n is the total number of y-observations. Here $n = r2^k$. Formulas (4.2.17a) are also generally applicable to 2^{k-p} fractional factorial designs (see Chapter 5).

In practice, we shall need to obtain an estimate s^2 of the experimental error variance σ^2. We shall consider methods for doing this as we proceed. For the textile data, suppose that an estimate $s^2 = 0.0050$ were available. Then

$$\hat{V}(\bar{y}) = 0.000625, \qquad \hat{V}(\text{effect}) = 0.0025, \qquad (4.2.18)$$

and the corresponding standard errors are the square roots

$$s(\bar{y}) = 0.025, \qquad s(\text{effect}) = 0.05. \qquad (4.2.19)$$

The complete table of effects and their standard errors is as follows:

$$
\begin{aligned}
I &\leftarrow \bar{y} = 2.745 \pm 0.025, \\
1 &\leftarrow \quad 0.75 \pm 0.05, \\
2 &\leftarrow -0.59 \pm 0.05, \\
3 &\leftarrow -0.35 \pm 0.05, \\
12 &\leftarrow -0.03 \pm 0.05, \\
13 &\leftarrow -0.03 \pm 0.05, \\
23 &\leftarrow -0.04 \pm 0.05, \\
123 &\leftarrow -0.08 \pm 0.05.
\end{aligned}
\qquad (4.2.20)
$$

Effects and Regression Coefficients

If we fit a first-degree polynomial to the textile data, as was done in Eq. (2.2.4), we obtain

$$\hat{y} = \underset{(0.025)}{2.745} + \underset{(0.025)}{0.375x_1} - \underset{(0.025)}{0.295x_2} - \underset{(0.025)}{0.175x_3}. \qquad (4.2.21)$$

where the numbers in parentheses are the standard errors of the coefficients. Notice that the estimated regression coefficients $b_1 = 0.375$, $b_2 = -0.295$, $b_3 = -0.175$ and their standard errors are exactly half the main effects and standard errors just calculated. The factor of one-half arises because an effect was defined as the difference in response on moving from the -1 level to the $+1$ level of a given variable x_i, which corresponds to the change in y when x_i is changed by *two* units. The regression coefficient b_i is, of course, the change in y when x_i is changed by *one* unit.

Table 4.3. Coding of Variables in the 2^6 Example

Variables	ξ_i	Coded Levels, x_i		x_i in Terms of ξ_i
		-1	$+1$	
Polysulfide index	ξ_1	6	7	$x_1 = (\xi_1 - 6.5)/0.5$
Reflux rate	ξ_2	150	170	$x_2 = (\xi_2 - 160)/10$
Moles polysulfide	ξ_3	1.8	2.4	$x_3 = (\xi_3 - 2.1)/0.3$
Time (min)	ξ_4	24	36	$x_4 = (\xi_4 - 30)/6$
Solvent (cm^3)	ξ_5	30	42	$x_5 = (\xi_5 - 36)/6$
Temperature (°C)	ξ_6	120	130	$x_6 = (\xi_6 - 125)/5$

Table 4.4. A 2^6 Factorial Design and Resulting Observations of Strength, Hue, and Brightness

			Design Levels					Responses	
Actual Run Order	Polysulfide Index, x_1	Reflux Rate, x_2	Moles Polysulfide, x_3	Time, (min), x_4	Solvent (cm^3), x_5	Temperature (°C), x_6	Strength, y_1	Hue, y_2	Brightness, y_3
26	$-$	$-$	$-$	$-$	$-$	$-$	3.4	15	36
3	$+$	$-$	$-$	$-$	$-$	$-$	9.7	5	35
11	$-$	$+$	$-$	$-$	$-$	$-$	7.4	23	37
5	$+$	$+$	$-$	$-$	$-$	$-$	10.6	8	34
42	$-$	$-$	$+$	$-$	$-$	$-$	6.5	20	30
18	$+$	$-$	$+$	$-$	$-$	$-$	7.9	9	32
41	$-$	$+$	$+$	$-$	$-$	$-$	10.3	13	28
14	$+$	$+$	$+$	$-$	$-$	$-$	9.5	5	38
17	$-$	$-$	$-$	$+$	$-$	$-$	14.3	23	40
27	$+$	$-$	$-$	$+$	$-$	$-$	10.5	1	32
19	$-$	$+$	$-$	$+$	$-$	$-$	7.8	11	32
56	$+$	$+$	$-$	$+$	$-$	$-$	17.2	5	28
23	$-$	$-$	$+$	$+$	$-$	$-$	9.4	15	34
8	$+$	$-$	$+$	$+$	$-$	$-$	12.1	8	26
32	$-$	$+$	$+$	$+$	$-$	$-$	9.5	15	30
7	$+$	$+$	$+$	$+$	$-$	$-$	15.8	1	28
46	$-$	$-$	$-$	$-$	$+$	$-$	8.3	22	40
13	$+$	$-$	$-$	$-$	$+$	$-$	8.0	8	30
58	$-$	$+$	$-$	$-$	$+$	$-$	7.9	16	35
38	$+$	$+$	$-$	$-$	$+$	$-$	10.7	7	35
43	$-$	$-$	$+$	$-$	$+$	$-$	7.2	25	32
55	$+$	$-$	$+$	$-$	$+$	$-$	7.2	5	35
6	$-$	$+$	$+$	$-$	$+$	$-$	7.9	17	36
64	$+$	$+$	$+$	$-$	$+$	$-$	10.2	8	32
22	$-$	$-$	$-$	$+$	$+$	$-$	10.3	10	20
4	$+$	$-$	$-$	$+$	$+$	$-$	9.9	3	35
16	$-$	$+$	$-$	$+$	$+$	$-$	7.4	22	35
47	$+$	$+$	$-$	$+$	$+$	$-$	10.5	6	28
63	$-$	$-$	$+$	$+$	$+$	$-$	9.6	24	27
51	$+$	$-$	$+$	$+$	$+$	$-$	15.1	4	36
20	$-$	$+$	$+$	$+$	$+$	$-$	8.7	10	36
29	$+$	$+$	$+$	$+$	$+$	$-$	12.1	5	35

Table 4.4 (*Continued*)

Actual Run Order	Design Levels						Responses		
	Polysulfide Index, x_1	Reflux Rate, x_2	Moles Polysulfide, x_3	Time (min), x_4	Solvent (cm^3), x_5	Temperature (°C), x_6	Strength, y_1	Hue, y_2	Brightness, y_3
62	−	−	−	−	−	+	12.6	32	32
1	+	−	−	−	−	+	10.5	10	34
37	−	+	−	−	−	+	11.3	28	30
61	+	+	−	−	−	+	10.6	18	24
44	−	−	+	−	−	+	8.1	22	30
24	+	−	+	−	−	+	12.5	31	20
59	−	+	+	−	−	+	11.1	17	32
60	+	+	+	−	−	+	12.9	16	25
35	−	−	−	+	−	+	14.6	38	20
50	+	−	−	+	−	+	12.7	12	20
48	−	+	−	+	−	+	10.8	34	22
36	+	+	−	+	−	+	17.1	19	35
21	−	−	+	+	−	+	13.6	12	26
9	+	−	+	+	−	+	14.6	14	15
33	−	+	+	+	−	+	13.3	25	19
57	+	+	+	+	−	+	14.4	16	24
10	−	−	−	−	+	+	11.0	31	22
39	+	−	−	−	+	+	12.5	14	23
25	−	+	−	−	+	+	8.9	23	22
40	+	+	−	−	+	+	13.1	23	18
30	−	−	+	−	+	+	7.6	28	20
31	+	−	+	−	+	+	8.6	20	20
28	−	+	+	−	+	+	11.8	18	20
49	+	+	+	−	+	+	12.4	11	36
52	−	−	−	+	+	+	13.4	39	20
15	+	−	−	+	+	+	14.6	30	11
34	−	+	−	+	+	+	14.9	31	20
53	+	+	−	+	+	+	11.8	6	35
2	−	−	+	+	+	+	15.6	33	16
12	+	−	+	+	+	+	12.8	23	32
45	−	+	+	+	+	+	13.5	31	20
54	+	+	+	+	+	+	15.8	11	20
			64 observations Σy_2				711.9	1085	1810
			64 observations Σy				8443.41	24,421	54,340

4.3. A 2^6 DESIGN USED IN A STUDY OF DYESTUFFS MANUFACTURE

In the manufacture of a certain dyestuff, it is important to:

1. Obtain a product of a desired hue and brightness, and of maximum strength.
2. Know what changes in the process variables should be made if the requirements of the customer (as expressed by hue and brightness of the product) should change.
3. Know what sorts of changes in the process variables might compensate for given departures from specification.

Table 4.5. Effects Ranked in Order for Strength, Hue, and Brightness

Strength, y_1		Hue, y_2		Brightness, y_3	
Effect Name	Effect Value	Effect Name	Effect Value	Effect Name	Effect Value
Mean	11.12	Mean	16.95	Mean	28.28
4	2.98	1	−11.28	6	−8.87
6	2.69	6	10.84	4	−3.00
1	1.75	2456	−2.84	2346	−2.87
1245	−1.16	135	−2.78	12346	−2.81
12	0.89	2	−2.72	123456	−2.81
24	−0.86	13	2.66	234	−2.56
16	−0.82	2346	2.53	35	2.44
124	0.82	1235	2.47	1245	−2.37
1346	−0.77	46	2.34	1456	−2.37
12346	0.74	136	2.34	1356	2.19
345	0.72	256	−2.22	15	2.12
13456	0.71	12345	2.22	1246	1.94
2	0.70	1236	−2.16	56	−1.88
12345	0.70	234	2.09	126	1.75
123	−0.65	1245	−1.97	135	1.62
1235	0.65	3	−1.91	2456	−1.62
1356	−0.65	23	−1.91	125	−1.56
456	0.63	25	−1.91	1235	−1.56
23	0.60	36	−1.84	2	1.50
256	0.57	14	−1.78	1234	−1.50
146	−0.56	146	−1.72	5	−1.44
12356	0.50	23456	1.72	456	1.31
1456	−0.47	2345	−1.59	25	1.25
12456	0.47	1246	−1.47	146	1.25
5	−0.42	1356	−1.47	1256	1.25
15	−0.42	124	−1.41	134	−1.19
2346	−0.41	456	1.41	145	1.19
34	0.40	5	1.34	45	1.12
134	0.40	3456	1.34	14	1.06
234	−0.40	123	−1.22	26	1.06
2356	0.39	26	−1.16	246	1.00
236	0.35	345	1.16	3	−0.94
1345	0.34	45	1.09	12	0.94
123456	0.34	126	−1.09	16	0.94
25	−0.33	12346	1.09	136	−0.94
145	−0.33	236	1.03	156	0.94
245	−0.33	245	−1.03	346	−0.94
45	−0.32	15	−0.97	3456	−0.94
1246	−0.32	346	−0.97	235	−0.88
246	0.30	123456	−0.84	24	0.81
1236	0.29	16	0.78	2345	−0.81
3456	−0.28	34	−0.78	13	0.75
56	0.27	35	0.78	236	−0.69
125	−0.27	1256	−0.78	12345	−0.50
126	−0.25	12	0.72	12356	−0.50

Table 4.5. (*Continued*)

Strength, y_1		Hue, y_2		Brightness, y_3	
Effect Name	Effect Value	Effect Name	Effect Value	Effect Name	Effect Value
2345	−0.24	12456	−0.66	245	0.44
356	−0.23	156	−0.53	256	−0.44
36	−0.22	1456	−0.53	1	0.38
1234	−0.22	2356	0.47	34	0.38
26	−0.18	56	0.41	345	0.38
346	0.17	4	−0.34	1236	−0.38
14	0.15	1345	0.34	23456	−0.38
13	0.14	12356	−0.34	46	−0.31
23456	0.14	1234	0.28	23	0.25
235	0.12	1346	−0.28	1346	0.25
136	0.11	246	0.22	13456	−0.25
3	0.10	356	0.22	36	0.13
156	0.10	24	0.16	123	0.06
35	0.08	125	0.16	1345	0.06
135	0.07	235	0.16	2356	−0.06
2456	0.03	13456	0.16	12456	0.06
46	0.02	134	0.03	124	0
1256	0.02	145	0.03	356	0

(These problems are essentially those we have listed in Section 1.7 as "Choice of operating conditions to achieve desired specifications" and "Approximate mapping of a surface within a limited region.") The particular process under study was one whose chemical mechanism was not well understood, so that an empirical approach was mandatory.

Six variables were suggested as possibly of importance in affecting strength, hue, and brightness. They were ξ_1 = polysulfide index, ξ_2 = reflux rate, ξ_3 = moles of polysulfide, ξ_4 = reaction time, ξ_5 = amount of solvent, and ξ_6 = reaction temperature. The ranges over which they were changed in the design, along with the details of their coding to x's are shown in Table 4.3 The design employed was a 2^6 full factorial design.

Model Specification

Data from past manufacturing experience had indicated that effects of potential importance were not likely to be large compared with the (rather high) experimental error. Such evidence as existed suggested that, over the range of the region of interest, effects might be roughly linear, and, in particular, no response was likely to be close to a maximum or other turning point. This was later checked by making additional runs both at the center of the 2^6 design and also at more extreme conditions along the six axial directions. No evidence of curved relationships was found, and we shall here use only the data from the principal part of the design, namely the complete 2^6 factorial. The results from these 64 runs (which were actually performed in random order) are shown in Table 4.4.

The preliminary statistical analysis for model specification consisted of calculating the factorial effects (see Table 4.5 and Appendix 4A) shown plotted on the

NORMAL PLOTS OF EFFECTS

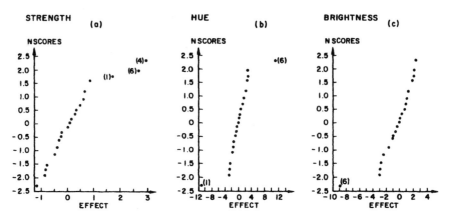

Figure 4.3. Normal plots for (a) strength, (b) hue, (c) brightness. (The effects plotted are those of orders 1, 2, 3, 4, 8, 12, ..., 60, 61, 62, and 63.)

Table 4.6. The 64 Observations on Each of Three Responses Grouped in Sets of Eight to Reflect Their Dependence on the Variables x_1 = Polysulfide Index, x_4 = Time, and x_6 = Temperature

x_1 x_4 x_6	x_2, x_3, x_5								Row Averages	Row Totals
	− − −	+ − −	− + −	+ + −	− − +	+ − +	− + +	+ + +		
				Strength						
− − −	3.4	7.4	6.5	10.3	8.3	7.9	7.2	7.9	7.3625	58.9
+ − −	9.7	10.6	7.9	9.5	8.0	10.7	7.2	10.2	9.2250	73.8
− + −	14.3	7.8	9.4	9.5	10.3	7.4	9.6	8.7	9.6250	77.0
+ + −	10.5	17.2	12.1	15.8	9.9	10.5	15.1	12.1	12.9000	103.2
− − +	12.6	11.3	8.1	11.1	11.0	8.9	7.6	11.8	10.3000	82.4
+ − +	10.5	10.6	12.5	12.9	12.5	13.1	8.6	12.4	11.6375	93.1
− + +	14.6	10.8	13.6	13.3	13.4	14.9	15.6	13.5	13.7125	109.7
+ + +	12.7	17.1	14.6	14.4	14.6	11.8	12.8	15.8	14.2250	113.8
				Hue						
− − −	15	23	20	13	22	16	25	17	18.875	151
+ − −	5	8	9	5	8	7	5	8	6.875	55
− + −	23	11	15	15	10	22	24	10	16.250	130
+ + −	1	5	8	1	3	6	4	5	4.125	33
− − +	32	28	22	17	31	23	28	18	24.875	199
+ − +	10	18	31	16	14	23	20	11	17.875	143
− + +	38	34	12	25	39	31	33	31	30.375	243
+ + +	12	19	14	16	30	6	23	11	16.375	131
				Brightness						
− − −	36	37	30	28	40	35	32	36	34.250	274
+ − −	35	34	32	38	30	35	35	32	33.875	271
− + −	40	32	34	30	20	35	27	36	31.750	254
+ + −	32	28	26	28	35	28	36	35	31.000	248
− − +	32	30	30	32	22	22	20	20	26.000	208
+ − +	34	24	20	25	23	18	20	36	25.000	200
− + +	20	22	26	19	20	20	16	20	20.375	163
+ + +	20	35	15	24	11	35	32	20	24.000	192

probability scales of Figure 4.3. For clarity, in the center of each plot only every fourth point is inserted. A brief discussion of the normal plotting of factorial effects is given in Appendix 4B. From this preliminary analysis, it appeared that the data were adequately explained in terms of linear effects in only three of the variables x_1, x_4, and x_6—that is, in polysulfide index, reaction time, and reaction temperature, respectively. The preliminary identification thus suggested that linear models might be tentatively entertained in the variables x_1, x_4, and x_6 with the residual variation ascribed principally to random error or "noise."

On this basis the data could be tentatively analyzed as if they came from a 2^3 factorial in x_1, x_4, and x_6 performed eight times over. The observations, appropriately rearranged to reflect this, are set out again in Table 4.6. The totals on the right of this table are used to facilitate the subsequent tentative analysis.

Model Fitting

The linear equations in x_1, x_4, and x_6 fitted by least squares to all 64 observations and rounded to two decimal places are*:

$$\text{Strength: } \hat{y}_1 = 11.12 + 0.87x_1 + 1.49x_4 + 1.35x_6 \qquad (4.3.1)$$
$$(0.24) \quad (0.24) \quad (0.24) \quad (0.24)$$

$$\text{Hue: } \hat{y}_2 = 16.95 - 5.64x_1 - 0.17x_4 + 5.42x_6 \qquad (4.3.2)$$
$$(0.74) \quad (0.74) \quad (0.74) \quad (0.74)$$

$$\text{Brightness: } \hat{y}_3 = 28.28 + 0.19x_1 - 1.50x_4 - 4.44x_6 \qquad (4.3.3)$$
$$(0.67) \quad (0.67) \quad (0.67) \quad (0.67)$$

The appropriate analyses of variance are set out in Table 4.7 The estimates of the error standard deviations obtained from the residual sums of squares after fitting these equations are, respectively,

$$s_1 = 1.90, \qquad s_2 = 5.93, \qquad s_3 = 5.39. \qquad (4.3.4)$$

These estimates of the standard deviations might be biased somewhat upward because a number of small main effects and interactions have been ignored, and perhaps biased slightly downward because of the effect of selection; that is, only the large estimates were taken to be of real effects. However, these are the figures we have used in estimating the standard errors of the coefficients shown in parentheses beneath the coefficients.

---

*Not all the estimated regression coefficients b_1, b_4, and b_6 in Eqs. (4.3.1) to (4.3.3) are significantly different from zero, but the values shown are the best estimates available in the subspace of greatest interest. We do not substitute values of zero for the nonsignificant coefficients, since these would not be the best estimates. It might be thought that this judgment is somewhat contradictory, since variables x_2, x_3, and x_5 have been dropped completely. In fact, however, we can simply regard the fitted equations as the best estimates in the three-dimensional subspace of the full six-dimensional space in which x_2, x_3, and x_5 are at their average values. Our treatment implies only that, had we prepared contour diagrams for the responses as functions of x_1, x_4, and x_6 for *other* levels of x_2, x_3, and x_5 over their relevant ranges, the appearances of the various diagrams would not be appreciably different.

Table 4.7. Analysis of Variance Tables, for Strength, Hue, and Brightness

	Strength			
Source of Variation	SS	df	MS	F Ratio
Total SS = Σy^2	8,443.4100	64		
Correction factor,				
SS due to $b_0 = (\Sigma y)^2/64$	7,918.7752	1		
Corrected Total SS	524.6348	63		
Due to $b_1 = b_1\Sigma x_1 y = (\Sigma x_1 y)^2/\Sigma x_1^2$	48.8252	1	48.8252	13.47[a]
Due to $b_4 = b_4\Sigma x_4 y = (\Sigma x_4 y)^2/\Sigma x_4^2$	142.5039	1	142.5039	39.32[a]
Due to $b_6 = b_6\Sigma x_6 y = (\Sigma x_6 y)^2/\Sigma x_6^2$	115.8314	1	115.8314	31.96[a]
Residual	217.4743	60	3.6246	
			$s_1 = 1.9038$	

$\Sigma y = 711.9$; $\Sigma x_1 y = 55.9$; $\Sigma x_4 y = 95.5$; $\Sigma x_6 y = 86.1$

	Hue			
Source of Variation	SS	df	MS	F Ratio
Total SS = Σy^2	24,421.0000	64		
SS due to $b_0 = (\Sigma y)^2/64$	18,394.1406	1		
Corrected Total	SS	6,026.8594	63	6,026.8594
Due to $b_1 = b_1\Sigma x_1 y = (\Sigma x_1 y)^2/\Sigma x_1^2$	2,036.2656	1	2,036.2656	57.98[a]
Due to $b_4 = b_4\Sigma x_4 y = (\Sigma x_4 y)^2/\Sigma x_4^2$	1.8906	1	1.8906	0.05
Due to $b_6 = b_6\Sigma x_6 y = (\Sigma x_6 y)^2/\Sigma x_6^2$	1,881.3906	1	1,881.3906	53.57[a]
Residual	2,107.3126	60	35.1219	
			$s_2 = 5.9264$	

$\Sigma y = 1085$; $\Sigma x_1 y = -361$; $\Sigma x_4 y = -11$; $\Sigma x_6 y = 347$

	Brightness			
Source of Variation	SS	df	MS	F Ratio
Total SS = Σy^2	54,340.0000	64		
SS due to b_0	51,189.0625	1		
Corrected Total SS	3,150.9375	63		
Due to b_1	2.2500	1	2.2500	0.08
Due to b_4	144.0000	1	144.0000	4.95[a]
Due to b_6	1260.2500	1	1260.2500	43.35[a]
Residual	1744.4375	60	29.0740	
			$s_3 = 5.3920$	

$\Sigma y = 1810$; $\Sigma x_1 y = 12$; $\Sigma x_4 y = -96$; $\Sigma x_6 y = -284$

[a]Significant at $\alpha = 0.05$ or smaller α-level; $P[F(1.60) \geq 4.00] = 0.05$.

Table 4.8. Residuals from Individual and from Average Observations, Fitted Values $\hat{y}$ Obtained by Using Eqs. (4.3.1) to (4.3.3), and Time Order of Observations[a]

x_1 x_4 x_6	x_2, x_3, x_5								From Average	$\hat{y}$
	− − −	+ − −	− + −	+ + −	− − +	+ − +	− + +	+ + +		
				Strength						
− − −	−4.0	0	−0.9	2.9	0.9	0.5	−0.2	0.5	0	7.4
+ − −	0.5	1.4	−1.3	0.3	−1.2	1.5	−2.0	1.0	0	9.2
− + −	3.9	−2.6	−1.0	−0.9	−0.1	−3.0	−0.8	−1.7	−0.8	10.4
+ + −	−1.6	5.1	0	3.7	−2.2	−1.6	3.0	0	0.8	12.1
− − +	2.5	1.2	−2.0	1.0	0.9	−1.2	−2.5	1.7	0.2	10.1
+ − +	−1.4	−1.3	0.6	1.0	0.6	1.2	−3.3	0.5	−0.3	11.9
− + +	1.5	−2.3	0.5	0.2	0.3	1.8	2.5	0.4	0.6	13.1
+ + +	−2.1	2.3	−0.2	−0.4	−0.2	−3.0	−2.0	1.0	−0.6	14.8
				Hue						
− − −	−2	6	3	−4	5	−1	8	0	2	17
+ − −	−1	2	3	−1	2	1	−1	2	1	6
− + −	6	−6	−2	−2	−7	5	7	−7	−1	17
+ + −	−5	−1	2	−5	−3	0	−2	−1	−2	6
− − +	4	0	−6	−11	3	−5	0	−10	−3	28
+ − +	−7	1	14	−1	−3	6	3	−6	1	17
− + +	10	6	−16	−3	11	3	5	3	2	28
+ + +	−5	2	−3	−1	13	−11	6	−6	−1	17
				Brightness						
− − −	2	−3	−4	−6	6	1	−2	2	0	34
+ − −	1	0	−2	4	−4	1	1	−2	0	34
− + −	9	1	3	−1	−11	4	−4	5	1	31
+ + −	1	−3	−5	−3	4	−3	5	4	0	31
− − +	7	5	5	7	−3	−3	−5	−5	1	25
+ − +	8	−2	−6	−1	−3	−8	−6	10	−1	26
− + +	−2	0	4	−3	−2	−2	−6	−2	−2	22
+ + +	−3	12	−8	1	−12	12	9	−3	1	23
				Time Order						
− − −	26	11	42	41	46	58	43	6		
+ − −	3	5	18	14	13	38	55	64		
− + −	17	19	23	32	22	16	63	20		
+ + −	27	56	8	7	4	47	51	29		
− − +	62	37	44	59	10	25	30	28		
+ − +	1	61	24	60	39	40	31	49		
− + +	35	48	21	33	52	34	2	45		
+ + +	50	36	9	57	15	53	12	54		

[a]The layout is similar to that of Table 4.6, and all figures are rounded to the length of the original observations.

4.4. DIAGNOSTIC CHECKING OF THE FITTED MODELS,
2^6 DYESTUFFS EXAMPLE

Table 4.8 shows the residuals from the models (4.3.1) to (4.3.3). The time order of the observations is also shown. Figures 4.4a–c show residuals for the fitted models plotted in various ways. There is some evidence that the variance of the response hue (y_2) is not homogeneous (see Figure 4.4b, the residuals versus $\hat{y}_2$ plot). However, in general, it appears that the tentative assumption that the changes in the responses are associated mainly with the linear effects of the three variables polysulfide index (x_1), reaction time (x_4), and reaction temperature (x_6) is borne out.

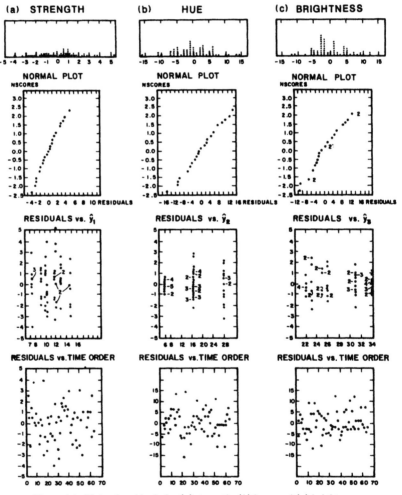

Figure 4.4. Plots of residuals for (a) strength, (b) hue, and (c) brightness.

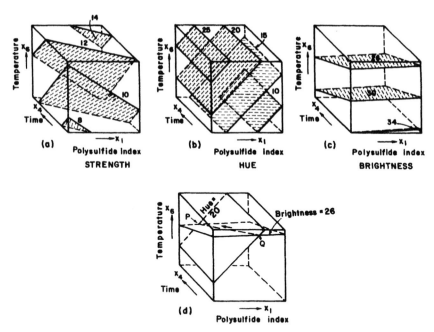

Figure 4.5. Contours of (*a*) strength, (*b*) hue, (*c*) brightness with (*d*) intersection of the hue = 20 and brightness = 26 planes.

4.5. RESPONSE SURFACE ANALYSIS OF THE 2^6 DESIGN DATA

We now consider the model's implications for the manufacturing problem, assuming that, to a sufficient approximation, the fitted models are adequate to represent the data. The immediate objective of these experiments was to determine operating conditions, if any such existed, at which a standard hue of 20 units and a standard brightness of 26 units could be obtained with maximum strength. A second objective was to understand the way in which the three responses were affected by the variables. Such an understanding would allow future modifications to be made in the process as necessary. Such modifications might be needed, for example, to (a) prepare special consignments of dye whose hue and brightness had to differ from current specifications and (b) allow intelligent compensatory adjustments to be made when unknown disturbances had caused the response to depart temporarily from the standard. What was needed in this latter situation in essence was some kind of "navigation chart."

The contours of the fitted equations are shown in Figures 4.5*a*–*c* in cubes whose corners are the design points.

Figure 4.5*d* shows, together, the predicted contour planes for brightness and hue equal to their desired levels of 26 units and 20 units, respectively; the line *PQ* along which these two planes intersect has predicted strengths ranging from about 11.08 at *Q* to about 12.46 at *P*, as Table 4.9 indicates. Approximately, then, we expect that, anywhere along the line of intersection, satisfactory brightness and hue would be obtained and that higher strengths would result by moving in the

Table 4.9. Predicted Strengths at Points Along the Line PQ in Figure 4.5d

	Coordinates on line PQ			
	x_1	x_4	x_6	Predicted Strength, $\hat{y}_1$
Q	0.32	-1.00	0.87	11.08
	0.09	-0.39	0.65	11.50
	-0.05	0.00	0.51	11.77
	-0.18	0.34	0.39	12.00
P	-0.42	1.00	0.16	12.46

direction indicated by the arrow in the figure. The estimated difference in strengths at the points P and Q in Figure 4.5d is given by

$$\hat{y}_P - \hat{y}_Q = b_1(x_{1P} - x_{1Q}) + b_4(x_{4P} - x_{4Q}) + b_6(x_{6P} - x_{6Q})$$

$$= 12.46 - 11.08$$

$$= 1.38 \tag{4.5.1}$$

with a variance of

$$V(\hat{y}_P - \hat{y}_Q) = \left\{(x_{1P} - x_{1Q})^2 + (x_{4P} - x_{4Q})^2 + (x_{6P} - x_{6Q})^2\right\} V(b_i)$$

$$= \overline{PQ}^2 V(b_i), \tag{4.5.2}$$

where $V(b_i) = \sigma^2/n$, and $\overline{PQ}^2$ is the squared distance between the points P and Q in the scale of the x's. Therefore the standard deviation of $(\hat{y}_P - \hat{y}_Q) = 1.38$ is $\overline{PQ}\sigma/n^{1/2}$, which we can evaluate as

$$\left\{(-0.42 - 0.32)^2 + (1 + 1)^2 + (0.16 - 0.87)^2\right\}^{1/2} \sigma/64^{1/2} = 0.2809\sigma. \tag{4.5.3}$$

From Table 4.7 we substitute $s_1 = 1.9038$ for σ, giving a standard error of about 0.53. The difference $\hat{y}_P - \hat{y}_Q = 1.38$ is thus about 2.6 times its standard error and we conclude that the strength is significantly higher at the position P than at the position Q.

[We note, in passing, that this design happens to be what is called a rotatable first-order design (see Chapter 14) and therefore has the interesting property that the variance of differences in estimated yields between two points is simply a function of the distance between the two points concerned (see Box and Draper, 1980).]

It is obvious that our analysis can be only an approximate one, since the data on which it is based are only approximate. However, on the assumption that the form of the prediction equation is not seriously at fault, we can readily calculate confidence limits at the points P and Q, or at any other set of values, for $E(y_1)$, $E(y_2)$, and $E(y_3)$. For any of the three responses, at a general point (x_1, x_4, x_6), we can write

$$V(\hat{y}_i) = V(b_0) + x_1^2 V(b_1) + x_4^2 V(b_4) + x_6^2 V(b_6)$$
$$= \sigma_i^2 \left[1 + x_1^2 + x_4^2 + x_6^2 \right] / 64, \tag{4.5.4}$$

where the σ_i^2 is the appropriate one for the response desired, and is estimated by s_i^2 from the appropriate part of Table 4.7. Thus, for example, the estimated variance of $\hat{y}_1$ at the point P with coordinates $(-0.42, 1.00, 0.16)$ is

$$V(\hat{y}_1) = \frac{3.6246(2.202)}{64} = 0.1247, \tag{4.5.5}$$

where we have substituted for σ_1^2, $s_1^2 = 3.6246$ based on 60 df from Table 4.7. Taking the square root, we then have to two decimal places that

$$\text{standard error } (\hat{y}_1) = 0.35. \tag{4.5.6}$$

Proceeding in exactly similar fashion and using $t_{0.025}(60) = 2.00$, we obtain at P and Q 95% confidence limits for $E(y_1)$, $E(y_2)$ and $E(y_3)$ as follows:

	At Point P	At Point Q
Yield	12.46 ± 0.70	11.08 ± 0.80
Hue	20 ± 2.20	20 ± 2.51
Brightness	26 ± 2.00	26 ± 2.28

(In Appendix 4C, methods are given for constructing confidence regions both for contour planes and for their lines of intersection.)

Confirmatory runs made in the regions indicated by this analysis verified the general conclusions given above. A three-dimensional model with contours of strength, hue, and brightness shown by different colors was constructed and proved to be of great value in improving the product and in day-to-day operation of the process.

In this investigation we were somewhat fortunate in finding that first-order expressions were sufficient to represent the three response functions. The area of application of such methods is greatly widened by allowing the possibility of nonlinear transformation in either the response or predictor variables or, sometimes, in both. Quite frequently, although the relationships may not be adequately represented by first-order expressions in the original variables, approximate first-order representation can be obtained by employing, for example, the logarithmic or inverse transformation. Methods whereby appropriate transformations may be determined are discussed later (see Chapters 8 and 13).

APPENDIX 4A.   YATES' METHOD FOR OBTAINING THE FACTORIAL EFFECTS FOR A TWO-LEVEL DESIGN

Yates' method is best understood by considering an example. Table 4A.1 shows the calculations for the textile data previously discussed in Chapter 4. Column C_1 is

Table 4A.1. Yates' Method for Obtaining Factorial Effects

Row	Setting of Variable x_1	x_2	x_3	Observations, y_u	C_1	C_2	C_3	Divisor	Effect Value	Name
1	−1	−1	−1	2.83	6.39	11.68	21.96	8	2.745	I
2	1	−1	−1	3.56	5.29	10.28	3.00	4	0.75	1
3	−1	1	−1	2.23	5.77	1.56	−2.36	4	−0.59	2
4	1	1	−1	3.06	4.51	1.44	−0.12	4	−0.03	12
5	−1	−1	1	2.47	0.73	−1.10	−1.40	4	−0.35	3
6	1	−1	1	3.30	0.83	−1.26	−0.12	4	−0.03	13
7	−1	1	1	1.95	0.83	0.10	−0.16	4	−0.04	23
8	1	1	1	2.56	0.61	−0.22	−0.32	4	−0.08	123

developed from the y_u column by these operations:

$$\text{In row 1 of } C_1 \text{ place the entry } y_2 + y_1,$$
$$2 \qquad\qquad y_4 + y_3,$$
$$3 \qquad\qquad y_6 + y_5,$$
$$4 \qquad\qquad y_8 + y_7,$$
$$5 \qquad\qquad y_2 - y_1,$$
$$6 \qquad\qquad y_4 - y_3,$$
$$7 \qquad\qquad y_6 - y_5,$$
$$\text{In row 8 of } C_1 \text{ place the entry } y_8 - y_7.$$

The identical operations performed on the C_1 column now give rise to the C_2 column, and the latter is used to calculate the C_3 column in the same manner. Each entry in the C_3 column is then divided by a divisor. The first divisor is the number of runs in the design, here 8. All the other divisors are half of this. The results of this mechanical procedure are the effects of the variables in the standard order corresponding to the plus signs in the variable settings; here, for example, the order is $I, 1, 2, 12, 3, 13, 23, 123$. Rearranging this order, we obtain the estimates already given.

Grand mean	$I \leftarrow \quad 2.745 \quad (= \bar{y})$
Main effects	$1 \leftarrow \quad 0.75$
	$2 \leftarrow -0.59$
	$3 \leftarrow -0.35$
Two-factor interactions	$12 \leftarrow -0.03$
	$13 \leftarrow -0.03$
	$23 \leftarrow -0.04$
Three-factor interaction	$123 \leftarrow -0.08$

Yates' method may be applied to any two-level 2^k factorial design by carrying out the sum and difference operations on k columns. (For its implementation on 2^{k-p} fractional factorial designs, see Section 5.5.)

APPENDIX 4B. NORMAL PLOTS ON PROBABILITY PAPER

When two-level factorial and fractional factorial designs are run, it is frequently the case that no independent estimate of error is available. A useful technique that is often effective in distinguishing real effects from noise employs *Normal probability paper*, for example, Keuffel and Esser 46-8003, or Codex 3227, or Team 3211 or Dietzgen 340-PS-90. Computerized versions are widely available.

What Is Normal Probability Paper?

Observations y having a normal distribution with mean η and variance σ^2 have the probability density function

$$f(y) = \frac{1}{\sigma\sqrt{2\pi}} e^{-(y-\eta)^2/(2\sigma^2)}. \tag{4B.1}$$

The corresponding cumulative density function is

$$F(y) = \int_{-\infty}^{y} \frac{1}{\sigma\sqrt{2\pi}} e^{-(t-\eta)^2/(2\sigma^2)}\, dt. \tag{4B.2}$$

Both the normal density function and its cumulative are plotted in Figure 4B.1. We see that the cumulative density function $F(y)$ is the area under the probability density function $f(y)$ from $-\infty$ to y, and that $0 \le F(y) \le 1$. The normal cumulative distribution function is an "ogive," that is, an "S" shaped curve that begins at $(-\infty, 0)$ and passes through points such as $(\eta - \sigma, 0.1587)$, $(\eta, 0.5)$, $(\eta + \sigma, 0.8413)$, and $(\infty, 1.0)$, and in general through the points $\{\eta + u\sigma, F(\eta + u\sigma)\}$ where u is

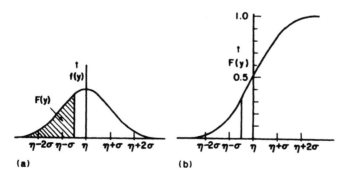

Figure 4B.1. Plots of (*a*) the normal density function $f(y)$ and (*b*) the cumulative normal density function $F(y)$. Note, for example, that prob($y \le \eta - 0.5\sigma$) = 0.3085 is the area shaded in (*a*) and is the vertical height marked in (*b*).

called the *normal score*. Normal probability paper shown in Figure 4B.2 is arranged so that the ogive curve $F(y)$ is made into a straight line by stretching outward from the center symmetrically the vertical dimension of the paper. Shown on the vertical axis are both the probability $F(y) = F(\eta + u\sigma)$ and the normal score u.

We illustrate use of the probability paper by drawing a random sample of 20 observations from a normal distribution with mean $\eta = 7$ and variance $\sigma^2 = 0.5$. The sample is rearranged in ascending order in Table 4B.1 and plotted as a dot

Table 4B.1. The 20 Ordered Observations $y_{(i)}$ and the Corresponding Values of $F(i) = (i - \frac{1}{2}) / 20$ for $i = 1, 2, \ldots, 20$

$y_{(i)}$	$F(i)$	$y_{(i)}$	$F(i)$
5.66	0.025	6.97	0.525
6.03	0.075	7.04	0.575
6.04	0.125	7.08	0.625
6.38	0.175	7.10	0.675
6.38	0.225	7.24	0.725
6.46	0.275	7.33	0.775
6.59	0.325	7.39	0.825
6.64	0.375	7.51	0.875
6.88	0.425	7.94	0.925
6.91	0.475	8.08	0.975

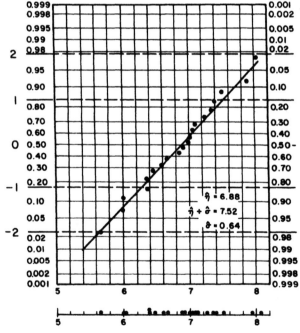

Figure 4B.2. Plots of 20 randomly drawn observations from a normal distribution with mean $\eta = 7.0$ and standard deviation $\sigma = 0.707$.

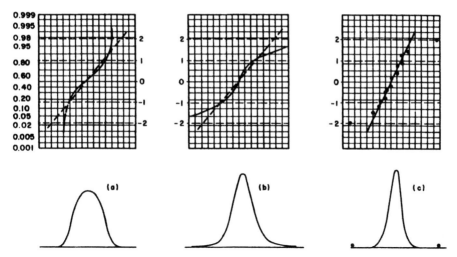

Figure 4B.3. (*a*) Light-tailed distribution. (*b*) Heavy-tailed distribution. (*c*) Normal distribution with two outliers.

diagram at the bottom of Figure 4B.2. With the *i*th in order, $y_{(i)}$, is associated

$$F(i) = F(y_{(i)}) = \frac{\left(i - \frac{1}{2}\right)}{n}, \tag{4B.3}$$

evaluated in Table 4B.1 for $n = 20$. The points $\{y_{(i)}, F(i)\}$ are then plotted on the probability paper in Figure 4B.2.

On the assumption that the data are a sample from a normal distribution, an "eye-fitted" straight line can be used to provide rough estimates of the mean and standard deviation. The estimate of the mean η is simply the abscissa value associated with $F(y) = 0.50$. To obtain an estimate of σ, recall that prob($y \le \eta + \sigma$) = 0.8413. The abscissa values for $F(y) = 0.50$ and $F(y) = 0.8413$ are, respectively, 6.88 and 7.52; thus for this set of data, $\hat{\eta} = 6.88$ and $\hat{\sigma} = 0.64 = 7.52 - 6.88$.

If the parent distribution were not normal, the cumulative distribution would not plot as a straight line on normal probability paper. Figures 4B.3*a* and 4B.3*b* show the curved lines that result from a light-tailed distribution and a heavy-tailed distribution, respectively. Figure 4B.3*c* shows a plot for a sample in which all observations are random drawings from a normal distribution except for two outliers.

Application to Estimates from 2^k Designs

If all k factors in a two-level factorial or fractional factorial design are without influence, all the $2^k - 1$ estimates of effects and interactions would be expected to be approximately normal with mean zero and variance $\{4\sigma^2/n\}$. Thus, if a normal plot of the estimated effects approximates a straight line, there would be no reason to believe that any of the true effects were nonzero. By contrast, points that fall well off the line to the right at the top and to the left at the bottom would suggest

Table 4B.2. Ordered Estimates from a 2^4 Design

Rank	Effect Name	Estimate	Rank	Effect Name	Estimate
1	2	−4.22	9	124	0.72
2	13	−2.49	10	12	0.91
3	234	−1.58	11	4	1.01
4	24	−1.18	12	123	1.20
5	1	−0.80	13	34	1.49
6	23	−0.80	14	1234	1.52
7	14	−0.58	15	3	3.71
8	134	0.40			

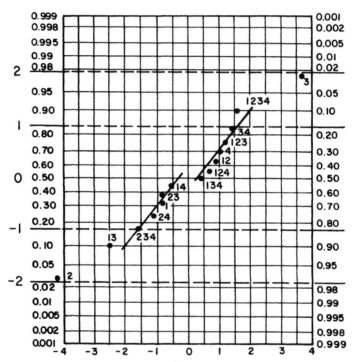

Figure 4B.4. Effect of an aberrant observation on a normal plot of factorial effect estimates from a 2^4 design.

the existence of real effects. Examples of how this works in practice are seen in Figure 4.3.

Detection of Aberrant Observations

Consider the estimates from a 2^4 factorial design in Table 4B.2. A normal plot of these is shown in Figure 4B.4. The data suggest, though not strongly, that the main effect 2 (and possibly 3 also) is distinguishable from noise, but this plot also exhibits another striking characteristic. The inlying values, which should represent

Table 4B3. Table of Signs for the Analysis of a 2^4 Design. Matching the Signs of Biased Effects with the Run that Probably Causes the Bias

Sign	−	−	+	+	+	−	−	−	−	+	+	+	(+)	−	+
Estimate	−0.80	−4.22	3.71	1.01	0.91	−2.49	−0.58	−0.80	−1.18	1.49	1.20	0.72	0.40	−1.58	1.52
	1	2	3	4	12	13	14	23	24	34	123	124	134	234	1234
1	−	−	−	−	+	+	+	+	+	+	−	−	−	−	+
2	+	−	−	−	−	−	−	+	+	+	+	+	+	−	−
3	−	+	−	−	−	+	+	−	−	+	+	+	−	+	−
4	+	+	−	−	+	−	−	−	−	+	−	−	+	+	+
5	−	−	+	−	+	−	+	−	+	−	+	−	+	+	−
6	+	−	+	−	−	+	−	−	+	−	−	+	−	+	+
7	−	+	+	−	−	−	+	+	−	−	−	+	+	−	+
8	+	+	+	−	+	+	−	+	−	−	+	−	−	−	−
9	−	−	−	+	+	+	−	+	−	−	−	+	+	+	−
10	+	−	−	+	−	−	+	+	−	−	+	−	−	+	+
11	−	+	−	+	−	+	−	−	+	−	+	−	+	−	+
12	+	+	−	+	+	−	+	−	+	−	−	+	−	−	−
13	−	−	+	+	+	−	−	−	−	+	+	+	−	−	+
14	+	−	+	+	−	+	+	−	−	+	−	−	+	−	−
15	−	+	+	+	−	−	−	+	+	+	−	−	−	+	−
16	+	+	+	+	+	+	+	+	+	+	+	+	+	+	+

117

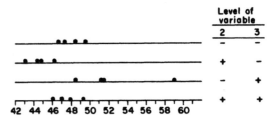

Figure 4B.5. Plot of replicate values when the 2^4 design is considered as a replicated 2^2 design in factors 2 and 3.

just noise, appear to be fitted not by one line, but by two parallel straight lines which "break apart" close to the zero value on the abscissa, and thus divide positive from negative effect estimates.

Now consider Table 4B.3, the body of which shows the $\pm$ signs which determine the effects, set out in the usual format. Suppose an observation were aberrant, for example, suppose observation 1 were too big by an amount δ. Then, from the table of signs, we see that effects 1, 2, 3, 4, 123, 124, 134, 234 would all be reduced by an amount $\delta/8$ and the remaining effects 12, 13, 14, 23, 24, 34, 1234 would all be increased by $\delta/8$. If we regard all the effects with the exception of the main effects 2 and 3 as possibly arising from noise, and consider the signs of the effects that remain, we see that, except for 134, these signs correspond to observation 13, suggesting that this observation is in error, and possibly too large. (Sometimes the complements of the signs make the match up, indicating a "too small" observation.)

Further, if we assume that factors x_2 and x_3 *are* the only ones of possible importance, the 2^4 design then becomes a replicated 2^2 in these factors. The replicate values are plotted in Figure 4B.5, from which it again seems likely that the thirteenth observation of 59.15 is in error.

APPENDIX 4C. CONFIDENCE REGIONS FOR CONTOUR PLANES
(see Section 4.5)

We can construct confidence regions for the contour planes as follows. Consider, for example, the equation for the contour plane of the desired level of hue, that is,

$$20 = 16.95 - 5.64x_1 - 0.17x_4 + 5.42x_6.$$

Let

$$Z_H \equiv 20 - 16.95 + 5.64x_1 + 0.17x_4 - 5.42x_6.$$

For any particular point (x_1, x_4, x_6) which lies on the "true" contour plane for a response of 20 in hue, we have

$$E(Z_H) = 0.$$

Also, we know that the estimate of variance of Z_H is

$$\hat{\sigma}_{Z_H}^2 = \left(1 + x_1^2 + x_4^2 + x_6^2\right)s_2^2/64.$$

Thus a confidence region for all points (x_1, x_4, x_6) which satisfy hue $= 20$ is given by

$$-t_{0.025}(\nu)\hat{\sigma}_{Z_H} \leq Z_H \leq t_{0.025}(\nu)\hat{\sigma}_{Z_H}$$

where $t_{0.025}(\nu)$ is the upper $97\frac{1}{2}\%$ point of a t distribution with ν df and where $\nu = 60$ is the number of degrees of freedom on which $s_2^2 = 35.1219$ (from Table 4.7) is based.

It follows that the (x_1, x_4, x_6) boundary for the confidence region for the hue $= 20$ contour plane is defined by

$$(20 - 16.95 + 5.64x_1 + 0.17x_4 - 5.42x_6)^2$$

$$= \left(1 + x_1^2 + x_4^2 + x_6^2\right)2^2(35.1219)/64$$

or

$$(3.05 + 5.64x_1 + 0.17x_4 - 5.42x_6)^2 = 2.195\left(1 + x_1^2 + x_4^2 + x_6^2\right).$$

This equation represents two hyperbolic surfaces which "sandwich" the estimated hue $= 20$ plane and go further away from it as we travel further away from the point $(\bar{x}_1, \bar{x}_4, \bar{x}_6)$. Points (x_1, x_4, x_6) that make left-hand side (LHS) $<$ right-hand side (RHS) lie within the confidence region. Points for which LHS $>$ RHS lie outside it.

EXERCISES

4.1. Explain briefly, in a few sentences, Yates' method of evaluating effects in a 2^k design.

4.2. A 2^4 design gave rise to the following contrasts in standard order (1, 2, 12, 3, 13, 23, 123, 4, 14, etc.): 18, 21, 3, 6, 0, 4, -1, 24, -8, 11, -9, 3, 1, 1, -2. Use a normal plot technique to analyze these data.

Note: A *contrast* is a linear combination of the y's whose coefficients add to zero. Thus the factorial design estimates of main effects and of interactions are all contrasts, but $\bar{y}$ is not.

4.3. A 2^4 factorial design yielded the following contrasts:

$$
\begin{array}{lll}
A \leftarrow 0.4 & AB \leftarrow 16.7 & ABC \leftarrow -0.1 \\
B \leftarrow -7.6 & AC \leftarrow 3.1 & ABD \leftarrow -4.7 \\
C \leftarrow 14.1 & AD \leftarrow 5.2 & ACD \leftarrow 7.7 \\
D \leftarrow 66.7 & BC \leftarrow 8.3 & BCD \leftarrow -2.3 \\
 & BD \leftarrow -3.6 & ABCD \leftarrow 3.9 \\
 & CD \leftarrow 14.3 &
\end{array}
$$

Interpret the results, employing a normal plot technique.

4.4. **(a)** If a full 2^k experiment is performed and each factor has levels -1 and 1, what are the relationships between the estimates of (i) the main effects of the k factors, and (ii) the coefficients β_i in a first-order model.

(b) How, in the same situation, are two-factor interactions represented in a regression model?

4.5. Imagine that individual trials comprising a 2^3 factorial design are run sequentially in standard order (in complete disregard for randomization !) while, simultaneously, a fourth variable disregarded by the experimenter takes the values shown in Table E4.5.

Table E4.5

x_1	x_2	x_3	x_4
$-$	$-$	$-$	-7
$+$	$-$	$-$	5
$-$	$+$	$-$	7
$+$	$+$	$-$	3
$-$	$-$	$+$	-3
$+$	$-$	$+$	-7
$-$	$+$	$+$	-5
$+$	$+$	$+$	7

If the first-order model $y = \beta_0 + \beta_1 x_1 + \beta_2 x_2 + \beta_3 x_3 + \varepsilon$ is fitted by least squares, determine the bias in the estimates of b_0, b_1, b_2, and b_3 of β_0, β_1, β_2, and β_3 induced by the failure to add a term $\beta_4 x_4$ to the model.

4.6. Use Yates' algorithm to evaluate the effects from the observations in Table E4.6, obtained from a 2^3 factorial design.

4.7. Imagine a 2^k factorial design with a center point replicated r times. What effect do these additional center observations have on the estimation of the factorial effects and on the overall mean?

Table E4.6

x_1	x_2	x_3	y
−	−	−	13
+	−	−	11
−	+	−	9
+	+	−	5
−	−	+	11
+	−	+	8
−	+	+	9
+	+	+	7

4.8. The primary purpose of quenching a steel is to produce martensite. The transition from austenite to martensite depends upon the steel's composition, which affects the temperature of the martensite formation. It has long been believed that carbon and other alloying elements lower the temperature range in which martensite forms. A preliminary 2^3 factorial study to determine the effects of carbon, manganese, and nickel on "martensite start temperature" produced the data in Table E4.8.

Table E4.8

Test Conditions			Start Temperature
%C	%Mn	%Ni	(°F)
0.1	0.2	0.1	940
0.9	0.2	0.1	330
0.1	1.4	0.1	820
0.9	1.4	0.1	250
0.1	0.2	0.2	840
0.9	0.2	0.2	280
0.1	1.4	0.2	760
0.9	1.4	0.2	200

(a) Code the predictor variables (i.e., the test conditions).

(b) Draw a diagram of the design and mark the response observations (i.e., the start temperatures) on it.

(c) Estimate all the main effects, two-factor interactions, and the three-factor interaction.

(d) Assume that the standard deviation of the response observations is $\sigma = 65$. Which estimates in (c) are large compared with their standard errors?

(e) What are the practical conclusions of your analysis?

Table E4.9

X_1	X_2	X_3	y
110	2500	2	37.7
110	1000	2	25.6
60	2500	2	22.2
110	1000	7.3	36.6
110	1000	7.3	34.4
110	2500	7.3	49.1
60	1000	2	12.5
60	2500	2	21.0
110	2500	7.3	46.0
60	2500	7.3	33.5
110	1000	2	28.2
110	2500	2	39.2
60	1000	2	12.2
60	1000	7.3	19.5
60	2500	7.3	35.0
60	1000	7.3	20.8

4.9. (Adapted from data provided by Mario Karfakis.) The data in Table E4.9 resulted from a study of the effect of three factors

$$X_1 = \text{revolutions per minute (rpm)},$$

$$X_2 = \text{downward thrust (weight), (lb)},$$

$$X_3 = \text{drilling fluid flow rate},$$

on the rate of penetration, y, of a drill. The observations are listed in the random order in which they were taken. Code the X's, and identify the type of design used. Then carry out a suitable analysis and list the practical conclusions you reach.

4.10. Four variables were examined for their effects on the yield y of a chemical process:

$$x_1 = \text{catalyst concentration},$$

$$x_2 = \text{NaOH concentration},$$

$$x_3 = \text{level of agitation},$$

$$x_4 = \text{temperature}.$$

A 2^4 factorial design was employed. The trials were run in random order. Interpret the data, given in Table E4.10.

Table E4.10

1	2	3	4	y	1	2	3	4	y
−	+	−	+	61	+	−	+	+	84
+	−	−	+	70	+	+	+	−	41
−	−	−	−	46	−	−	+	−	67
−	−	+	+	87	+	−	−	−	49
+	+	+	+	62	+	−	−	+	59
+	−	+	−	64	−	+	−	−	36
+	+	−	−	38	−	−	−	+	68
−	+	+	−	38	−	+	+	+	62

4.11. A manufacturer of paper bags wishes to check tear resistance, y, for which she has a long-used numerical scale. She examines three factors at each of two coded levels, namely,

$$x_1 = \text{type of paper,}$$

$$x_2 = \text{humidity,}$$

$$x_3 = \text{direction of tear,}$$

and obtains three replicate observations as shown in Table E4.11. (The runs have been rearranged for convenience from the random order in which they were performed.) The total sum of squares for the 24 observations is 496.87. Perform a factorial analysis on the data, estimate σ^2 from the repeat runs, and formulate conclusions about the factors.

Table E4.11

x_1	x_2	x_3	Tear Resistance			Row Sums
−	−	−	3.8	3.1	2.2	9.1
+	−	−	6.6	8.0	6.8	21.4
−	+	−	3.4	1.7	3.8	8.9
+	+	−	6.8	8.2	6.0	21.0
−	−	+	2.3	3.1	0.7	6.1
+	−	+	4.7	3.5	4.4	12.6
−	+	+	2.1	1.1	3.6	6.8
+	+	+	4.2	4.7	2.9	11.8
						97.7

4.12. Look again at the data of Exercise 4.9. Fit the first-order model $y = \beta_0 + \beta_1 x_1 + \beta_2 x_2 + \beta_3 x_3 + \varepsilon$ by least squares, and perform the usual analyses including an overall test for lack of fit. Your conclusions should be consistent with those from the factorial analysis.

4.13. To the data of Exercise 4.10 fit the model

$$y = \beta_0 + \beta_2 x_2 + \beta_3 x_3 + \beta_4 x_4 + \beta_{23} x_2 x_3 + \beta_{24} x_2 x_4 + \beta_{34} x_3 x_4 + \varepsilon.$$

Provide the associated analysis of variance table, estimate σ^2, and check the fitted coefficients for significance.

4.14. Look at the data of Exercise 4.11. We could, formally, fit a regression equation to these data but they are much more suitably analyzed by factorial methods. Why?

4.15. **(a)** Fit the model $y = \beta_0 + \beta_1 x_1 + \beta_2 x_2 + \beta_3 x_3 + \varepsilon$ to the data given in Table E4.15. Test for lack of fit. What final model would you adopt, if any?

(b) This design has 12 observations; exhibit the vectors c_i for 12 meaningful orthogonal contrasts which exist and which provide a complete transformation of the original data set $y_1, y_2, \ldots, y_{12}$.

Table E4.15

Run Order	x_1	x_2	x_3	y
3	−1	−1	−1	49
12	1	−1	−1	48
11	−1	1	−1	53
8	1	1	−1	54
4	−1	−1	1	57
5	1	−1	1	58
9	−1	1	1	61
7	1	1	1	55
6	0	0	0	55
2	0	0	0	54
1	0	0	0	56
10	0	0	0	54

4.16. (*Source:* Maximum data through a statistical design, by C. D. Chang, O. K. Kononenko, and R. E. Franklin, Jr., *Industrial and Engineering Chemistry*, **52**, November 1960, 939–942. Material reproduced and adapted by permission of the copyright holder, the American Chemical Society. The data were obtained under a grant from the Sugar Research Foundation, Inc., now the World Sugar Research Organization, Ltd.) Table E4.16 shows the results obtained from a 2^4 factorial design plus six center points. Four response variables were observed. The trials were performed in random order but are listed in standard order. Estimate all factorial main effects and all interac-

Table E4.16. A 2^4 Factorial Design with Six Center Points; Four Responses are Shown

Trial Number	Factor Levels				Responses			
	x_1, NH$_3$	x_2, T	x_3, H$_2$O	x_4, P	y_1, PDA	y_2, DMP	y_3, PD	y_4, R
1	−1	−1	−1	−1	1.8	58.2	24.7	84.7
2	1	−1	−1	−1	4.3	23.4	45.5	73.2
3	−1	1	−1	−1	0.4	21.9	8.6	30.9
4	1	1	−1	−1	0.7	21.8	9.1	31.6
5	−1	−1	1	−1	0.3	14.3	75.5	90.1
6	1	−1	1	−1	4.5	6.3	86.5	96.3
7	−1	1	1	−1	0.0	4.5	10.0	14.5
8	1	1	1	−1	1.6	21.8	50.1	73.5
9	−1	−1	−1	1	1.3	46.7	43.3	91.3
10	1	−1	−1	1	4.2	53.2	39.7	97.1
11	−1	1	−1	1	1.9	23.7	5.4	31.0
12	1	1	−1	1	0.7	40.3	9.7	50.7
13	−1	−1	1	1	0.0	7.5	78.8	86.3
14	1	−1	1	1	2.3	13.3	77.8	93.4
15	−1	1	1	1	0.8	49.3	21.1	71.2
16	1	1	1	1	7.3	20.1	37.8	65.2
17[a]	0	0	0	0	5.0	32.8	45.1	82.8
Pure error estimates of σ_i^2, $s_i^2 =$					1.08	9.24	15.37	8.18

[a]Average yield of six trials.

tions for every response, and evaluate the appropriate standard errors. Which variables appear to be effective? So far, the six center points have been used only to estimate the four σ_i^2, the variances of the four responses. Calculate the values of the contrasts

$$CC_i = (\text{average response at factorial points})$$

$$- (\text{average response at center points})$$

and determine their standard errors. Which responses show evidence of pure quadratic curvature? (*Note:* In general if there are k factors, such a contrast has expectation

$$E(CC) = \beta_{11} + \beta_{22} + \cdots + \beta_{kk}.$$

Thus if the CC contrast is compared to its standard error, a test for H_0: $\Sigma \beta_{ii} = 0$ versus H_1: $\Sigma \beta_{ii} \neq 0$ may be made.) Suppose you fitted a first-order model to each response. Would the center point average response be counted as one data point or as six data points. (The answer is six; why?) A reviewer of the original article suggested that the responses might have been transformed via

$$\omega_i = \sin^{-1}\left(\frac{y_i}{100}\right)^{1/2},$$

which is a transformation often appropriate when the data are percentages, as here, or proportions (when the division by 100 would not be needed). Is such a transformation useful in the sense that it renders a simpler interpretation of the factors' effects? Transform all the response values and repeat the factorial analysis to find out. (The y_2 data will be used again in Exercise 11.16 and will be combined there with additional data, for a second-order surface fitting exercise.)

4.17. The following estimates of effects resulted from a 2^5 experiment:

$$
\begin{array}{lll}
E \leftarrow -224 & BCD \leftarrow -18 & BE \leftarrow 29 \\
C \leftarrow -153 & ABDE \leftarrow -14 & DE \leftarrow 30 \\
ABCDE \leftarrow -77 & D \leftarrow -9 & ABCE \leftarrow 31 \\
ACE \leftarrow -58 & B \leftarrow -6 & BCE \leftarrow 39 \\
AD \leftarrow -54 & CD \leftarrow -4 & ACDE \leftarrow 47 \\
BC \leftarrow -53 & ABC \leftarrow 0 & AC \leftarrow 53 \\
ABD \leftarrow -34 & AE \leftarrow 2 & ABCD \leftarrow 58 \\
ACD \leftarrow -33 & BD \leftarrow 7 & AB \leftarrow 64 \\
BDE \leftarrow -28 & CDE \leftarrow 12 & CE \leftarrow 83 \\
ABE \leftarrow -22 & BCDE \leftarrow 16 & A \leftarrow 190 \\
 & ADE \leftarrow 21 &
\end{array}
$$

(a) Plot these contrasts on normal probability paper.

(b) Which effects do you think are distinguishable from noise?

(c) Replot the remaining contrasts.

(d) Estimate the standard deviation of the contrasts from plot (c). See p. 131.

[Partial answers: (b) E, C, A. (d) Roughly 40.]

4.18. *Source:* A case study of the use of an experimental design in preventing shorts in nickel-cadmium cells, by S. Ophir, U. El-Gad, and M. Schneider, *Journal of Quality Technology*, **20**(1), 1988, 44—150.) In order to overcome problems caused by the high incidence of faulty batteries during the manufacturing process, a special study was initiated in the plant. After a mechanical assembly fault had been corrected and an Ishikawa fishbone diagram (essentially a pictorial listing of possible sources of the problem) had been constructed, seven factors were selected for further attention. They were divided into two groups of three and four, and a 2^3 factorial design was performed on the three. These three were: (1) method of sintering, that is, of pressing the electrodes (new method or current method); (2) electrode separator (thick or thin); and (3) rolling pin (thick or thin). With these factors in standard order and with the two levels coded to $(-1, 1)$ or, more briefly, to $(-, +)$, the percentages of faulty battery cells manufactured in the experiments were as follows:

1	2	3	y
−	−	−	0.0
+	−	−	2.0
−	+	−	2.0
+	+	−	2.0
−	−	+	0.0
+	−	+	2.0
−	+	+	2.0
+	+	+	4.9

Perform a factorial analysis of these data. What do you think of the experimenters' conclusion that variable 3 was not a crucial one, but that the new sintering method and a thick separator should be adopted for future runs?

Next, a 2^{4-1} design, of a type described in Chapter 5, was run on the other set of four factors, but in doing this, it became clear that changing two of these variables would not be feasible under production conditions. This led to a final 2^2 design being run on the remaining variables: (4) order of rolling electrodes (negative first, positive first) and (5) direction of rolling (lower edge during impregnation rolled first, or the reverse). The results were:

4	5	y
−	−	0.8
+	−	7.2
−	+	1.4
+	+	3.9

What overall conclusion did the investigators reach, do you think?

4.19. *Source:* (Using DOE to determine AA battery life, by E. Wasiloff and C. Hargitt, *Quality Progress*, March 1999, 67–71.) The two authors, both model car race enthusiasts attending Farmington High School, MI, tested two types of batteries in a 2^3 factorial design experiment. The three factors and their codings were:

1. Battery type: Low cost ($x_1 = -1$), high cost ($x_1 = 1$).
2. Connector contacts: Tamiya standard ($x_2 = -1$), Dean gold-plated ($x_2 = 1$).
3. Initial battery temperature: Cold ($x_3 = -1$), ambient ($x_3 = 1$).

The response of interest, y, was the time to discharge, in minutes, of the batteries. The resultant data are shown in Table E4.19. Analyze the data and

Table E4.19. Battery Life Data

Run	x_1	x_2	x_3	y
1	1	1	1	493
2	1	1	−1	490
3	1	−1	1	489
4	1	−1	−1	612
5	−1	1	1	94
6	−1	1	−1	75
7	−1	−1	1	93
8	−1	−1	−1	72

Table E4.20. Data from a 2^4 Full Factorial Design with Four Added Center Points

No.	x_1	x_2	x_3	x_4	y
1	−1	−1	−1	−1	39.8
2	1	−1	−1	−1	61.4
3	−1	1	−1	−1	44.4
4	1	1	−1	−1	67.3
5	−1	−1	1	−1	44.6
6	1	−1	1	−1	62.5
7	−1	1	1	−1	52.1
8	1	1	1	−1	69.1
9	−1	−1	−1	1	39.5
10	1	−1	−1	1	60.6
11	−1	1	−1	1	46.7
12	1	1	−1	1	66.5
13	−1	−1	1	1	49.3
14	1	−1	1	1	61.3
15	−1	1	1	1	50.2
16	1	1	1	1	70.9
17	0	0	0	0	57.3
18	0	0	0	0	50.6
19	0	0	0	0	53.5
20	0	0	0	0	52.7

form conclusions about how battery life depends on the three factors and their interactions.

4.20. The 20 observations in Table E4.20 arose from a 2^4 full factorial design with four added center points, performed to examine the effects of four predictor variables on a measure of tire performance. Analyze these data and report on the results you find.

CHAPTER 5

Blocking and Fractionating 2^k Factorial Designs

5.1. BLOCKING THE 2^6 DESIGN

In the 2^6 design used above for illustration, the runs were made in a random order; the design was *fully randomized*. However, it might have been much better to run the arrangement in a series of *randomized blocks*. Suppose, for example, we expected trouble because of appreciable inhomogeneity of the raw material. Suppose further that a blender was available, large enough to produce homogeneous blends of raw material sufficient for making eight batches—that is, for conducting 8 of the 64 runs. Then the 64-run experiment might have been arranged using 8 blends (blocks) of homogeneous raw material within each of which 8 runs were made. The difficulty would be that differences between blends could bias the estimates of effects. R. A. Fisher found a way of circumventing this biasing problem by making the (often reasonable) assumption that the effects of blocks would be approximately additive—that is, that the differences between blends would cause the response results simply to be raised or lowered by a fixed (but unknown) amount.

Two Blocks

We introduce the method in a simpler context. Suppose first that we would like to make the 64 runs in two blocks (I and II) each with 32 runs. (Such an arrangement would be appropriate if we had a blender large enough to provide homogeneous raw material for 32 runs.) Consider the partial table of signs, Table 5.1, which shows the 2^6 design laid out in standard order. Suppose now that we allocate runs for which the product column **123456** has a minus sign to block I, along with runs for which **123456** has a plus sign to block II. Then, following Fisher, we say that the block effect (denoted by B) is completely *confounded* with the 123456 interaction, that is,

$$\mathbf{B} = \mathbf{123456}. \qquad (5.1.1)$$

Response Surfaces, Mixtures, and Ridge Analyses, Second Edition. By G. E. P. Box and N. R. Draper
Copyright © 2007 John Wiley & Sons, Inc.

Table 5.1. Blocking the 2^6 Design into Two Blocks Using B = 123456

Run Number	Variable Number						123456	Block
	1	**2**	**3**	**4**	**5**	**6**		
1	−	−	−	−	−	−	+	II
2	+	−	−	−	−	−	−	I
3	−	+	−	−	−	−	−	I
4	+	+	−	−	−	−	+	II
5	−	−	+	−	−	−	−	I
6	+	−	+	−	−	−	+	II
7	−	+	+	−	−	−	+	II
8	+	+	+	−	−	−	−	I
9	−	−	−	+	−	−	−	I
10	+	−	−	+	−	−	+	II
11	−	+	−	+	−	−	+	II
12	+	+	−	+	−	−	−	I
13	−	−	+	+	−	−	+	II
14	+	−	+	+	−	−	−	I
⋮	⋮	⋮	⋮	⋮	⋮	⋮	⋮	⋮
63	−	+	+	+	+	+	−	I
64	+	+	+	+	+	+	+	II

Thus, we lose the ability to estimate (at least with the same accuracy) the 123456 interaction. However, in most practical contexts, this very high-order interaction will probably be negligible in size and of little interest.

Now the 2^k factorial designs have an important orthogonal* property whereby a sequence of signs corresponding to a particular effect is orthogonal to every other such sequence. The important implication of this in the present context is that altering the apparent 123456 effect by superimposing on it the difference between blocks (blends) does not change the estimate of any of the other effects.

Exercise.   Add 3 to each of the runs for which elements of the **123456** column are negative and subtract 7 from each of the runs for which they are positive, and confirm by recalculation that the estimates of all effects except for 123456 are unchanged.   □

Four Blocks; Eight Blocks

Let us now get a little more ambitious and try to arrange the design in four blocks each containing 16 runs. This can be done by confounding two high-order interactions with block contrasts. Suppose, as a first shot, we associate the contrasts 123456 and 23456 with blocks. (This might seem reasonable since we shall wish to confound interactions of the highest possible order with blocks.) Thus the *blocking*

---

*Two columns of ± signs, or two columns of numbers, are said to be *orthogonal* if the sum of the cross-products of corresponding signs, or of corresponding numbers, is zero. The geometrical implication is that the two columns represent two vectors in a space which are orthogonal—that is, at right angles—to each other.

generators for our design will be

$$\mathbf{B}_1 = \mathbf{123456}, \qquad \mathbf{B}_2 = \mathbf{23456}, \qquad (5.1.2)$$

say. The runs would then be allocated to the four blocks according to the following scheme:

$$\mathbf{B}_2 = 23456$$

$$(5.1.3)$$

That is, runs would be allocated to the four blocks I, II, III, and IV as the signs associated with the columns **123456** and **23456** took the values $(--)(-+)$ $(+-)$ and $(++)$. Unfortunately, a serious difficulty occurs with this arrangement. There are, of course, 3 df (independent contrasts) among the four blocks of 16 runs each. If we associate $\mathbf{B}_1$ and $\mathbf{B}_2$ with two of these contrasts, the third must be the interaction $\mathbf{B}_1 \times \mathbf{B}_2 = \mathbf{B}_1 \mathbf{B}_2$. However, then the interaction $\mathbf{B}_1 \mathbf{B}_2$ will be confounded with the "interaction" between **123456** and **23456**. TO see what this interaction is, consider the following table of signs in standard order. (We write them as rows to conserve space.)

123456	$+--+$	$-++-$	$-++-$	$\cdots$	$-+$
23456	$--++$	$++--$	$++--$	$\cdots$	$++$
product	$-+-+$	$-+-+$	$-+-+$	$\cdots$	$-+$
of				
above				

We see that the interaction between **123456** and **23456** is the main effect **1**. This arrangement would thus be a very poor one, because the main effect of the first variable would be confounded with block differences. To make a better choice, we need to understand how to calculate interactions between complex effects. To see how to do this, we note that if we take any set of signs for any effect (say 1) and multiply those signs by the signs of the *same* effect, we obtain a row of $+$'s which we denote by the *identity* **I**.

1	$-+-+$	$-+-+$	$-+-+$	$\cdots$	$-+$
1	$-+-+$	$-+-+$	$-+-+$	$\cdots$	$-+$
$1 \times 1 = 1^2 = \mathbf{I}$	$++++$	$++++$	$++++$	$\cdots$	$++$

Thus, using the multiplication sign to imply the multiplication of the signs in corresponding positions in two rows (or two columns if the design is arranged vertically, as would be usual), we can write

$$1 \times 1 = 1^2 = \mathbf{I}, \qquad 2 \times 2 = 2^2 = \mathbf{I}, \qquad 12 \times 12 = 1^2 2^2 = \mathbf{I}^2 = \mathbf{I}, \quad (5.1.4)$$

and so on. Also, multiplication of any contrast by the identity $\mathbf{I}$ leaves the contrast unchanged. Thus

$$1 \times \mathbf{I} = 1, \qquad 23 \times \mathbf{I} = 23, \qquad 12345 \times \mathbf{I} = 12345, \qquad (5.1.5)$$

and so on. Applying this rule in the case of the design above, we have that

$$\mathbf{B}_1\mathbf{B}_2 = 123456 \times 23456 = 1^2 2 3^2 4^2 5^2 6^2 = 1 \times \mathbf{I}^5 = 1, \qquad (5.1.6)$$

indicating that $B_1 B_2$ and 1 are confounded.

An arrangement in four blocks of 16 runs so that all the interactions confounded with blocks are of the highest possible order may be obtained by using as generators two four-factor interactions in which only two symbols overlap. For example, if we choose

$$\mathbf{B}_1 = 1234$$
$$\mathbf{B}_2 = 3456,$$

then

$$\mathbf{B}_1\mathbf{B}_2 = 1256. \qquad (5.1.7)$$

We can now write out, with the design arranged in standard order, as usual, rows of signs for $\mathbf{B}_1$ and $\mathbf{B}_2$ and allocate the runs to the four blocks corresponding to the sign combinations $(\mathbf{B}_1, \mathbf{B}_2) = (-,-), (-,+), (+,-), (+,+)$, as follows:

$\mathbf{B}_1 = 1234$	$+\ -\ -\ +$	$-\ +\ +\ -$	$-\ +\ +\ -$	$\cdots$	$-\ +$
$\mathbf{B}_2 = 3456$	$+\ +\ +\ +$	$-\ -\ -\ -$	$-\ -\ -\ -$	$\cdots$	$+\ +$
block number	IV II II IV	I III III I	I III III I	$\cdots$	II IV

On the assumption that the blocks contribute only additive effects, the main effects, two-factor interactions, and three-factor interactions will all remain unconfounded with any block effect.

If, as would often be acceptable, we are prepared to confound some three-factor interactions, we can split the 64 runs into eight blocks each containing eight runs. For example, choosing the generators

$$\mathbf{B}_1 = 1234, \qquad \mathbf{B}_2 = 3456, \qquad \mathbf{B}_3 = 136, \qquad (5.1.8)$$

we obtain the following confounding pattern for the seven contrasts among the eight blocks:

$$\mathbf{B}_1 = 1234$$
$$\mathbf{B}_2 = 3456$$
$$\mathbf{B}_1\mathbf{B}_2 = 1256$$
$$\mathbf{B}_3 = 136 \qquad (5.1.9)$$
$$\mathbf{B}_1\mathbf{B}_3 = 246$$
$$\mathbf{B}_2\mathbf{B}_3 = 1\ 45$$
$$\mathbf{B}_1\mathbf{B}_2\mathbf{B}_3 = \ 23\ 5.$$

Table 5.2. Useful Blocking Arrangements for 2^k Factorial Designs

k = Number of Variables	Block Size	Block Generator	Interactions Confounded with Blocks
3	4	$B_1 = 123$	123
	2	$B_1 = 12, B_2 = 13$	12, 13, 23
4	8	$B_1 = 1234$	1234
	4	$B_1 = 124, B_2 = 134$	124, 134, 23
	2	$B_1 = 12, B_2 = 23, B_3 = 34$	12, 23, 34, 13, 1234, 24, 14
5	16	$B_1 = 12345$	12345
	8	$B_1 = 123, B_2 = 345$	123, 345, 1245
	4	$B_1 = 125, B_2 = 235, B_3 = 345$	125, 235, 345, 13, 1234, 24, 145
	2	$B_1 = 12, B_2 = 13, B_3 = 34, B_4 = 45$	12, 13, 34, 45, 23, 1234, 1245, 14, 1345 35, 24, 2345, 1235, 15, 25 that is, all 2fi and 4fi[a]
6	32	$B_1 = 123456$	123456
	16	$B_1 = 1236, B_2 = 3456$	1236, 3456, 1245
	8	$B_1 = 135, B_2 = 1256, B_3 = 1234$	135, 1256, 1234, 236, 245, 3456, 146
	4	$B_1 = 126, B_2 = 136, B_3 = 346, B_4 = 456$	126, 136, 346, 456, 23, 1234, 1245 14, 1345, 35, 246, 23456, 12356, 156, 25
	2	$B_1 = 12, B_2 = 23, B_3 = 34, B_4 = 45, B_5 = 56$	All 2fi, 4fi, and 6fi
7	64	$B_1 = 1234567$	1234567
	32	$B_1 = 12367, B_2 = 34567$	12367, 34567, 1245
	16	$B_1 = 123, B_2 = 456, B_3 = 167$	123, 456, 167, 123456, 2367, 1457, 23457
	8	$B_1 = 1234, B_2 = 567, B_3 = 345, B_4 = 147$	1234, 567, 345, 147, 1234567, 125, 237, 3467, 1456, 1357, 1267, 2356, 2457, 136, 246
	4	$B_1 = 127, B_2 = 237, B_3 = 347, B_4 = 457, B_5 = 567$	127, 237, 347, 457, 567, 13, 1234, 1245, 1256, 24, 2345, 2356, 3456, 35, 1234567, 46, 147, 13457, 13567 12467, 12357, 257, 24567, 23467, 367, 15, 1456 1346, 1236, 26, 167
	2	$B_1 = 12, B_2 = 23, B_3 = 34, B_4 = 45, B_5 = 56, B_6 = 67$	All 2fi, 4fi, and 6fi

[a]"fi" is an abbreviation for "factor interaction"; thus, for example, 2fi means two-factor interaction.

Source: Table 5.2 is reproduced and adapted with permission from the table on pp. 346–347 of *Statistics for Experimenters*: *An Introduction to Design, Data Analysis and Model Building*, by G. E. P. Box, W. G. Hunter, and J. S. Hunter, published by John Wiley & Sons, Inc., New York, 1978.

On the assumption that blocks have only additive effects, no main effect or two-factor interaction is confounded with any block effect.

Exercise. Allocate the 64 runs of a 2^6 design to eight blocks using the generators $B_1 = 1346$, $B_2 = 1235$, and $B_3 = 156$ and write down the confounding pattern for the seven contrasts among the eight blocks. Is any main effect or two-factor interaction confounded with blocks? (No.)   □

A short table of useful confounding arrangements adapted from Box et al. (1978) is given in Table 5.2. Other tables will be found in Davies (1954), Clatworthy et al. (1957), and McLean and Anderson (1984).

5.2. FRACTIONATING THE 2^6 DESIGN

To study the dyestuffs problem in Chapter 4, we used a 2^6 factorial design that required 64 runs. As it turned out, only main effects of three of the variables could be distinguished from the noise. Such a discovery naturally leads to the speculation as to whether so many runs were really needed. In this particular example, the standard deviation σ of the experimental error was rather large compared with the size of the effects of interest, so the full 64 runs *were* probably needed. However, if σ had been smaller, a design containing a smaller number of runs than the full $2^6 = 64$ might have proved adequate. A class of two-level designs requiring fewer runs than the full 2^k factorials is that of the 2^{k-p} *fractional factorials*.

Half-Fractions of the 2^6 Factorial

To illustrate the nature and the utility of such designs, consider again the problem of blocking the 2^6 factorial into two blocks, each of 32 runs, with the six-factor interaction **123456** ($= \mathbf{B}$) used as the defining (or blocking) contrast. (See Section 5.1.) Let us temporarily augment our previous notation a little by using the symbol $\mathbf{I}_{32}$ to refer to a sequence of 32 plus signs. For each of the 32 runs in the first block, the signs of the six factors will multiply to give -1 and for each of the 32 runs in the second block, the signs will multiply to give $+1$. Thus,

$$\mathbf{123456} = -\mathbf{I}_{32} \quad \text{for block I,}$$

$$\mathbf{123456} = \mathbf{I}_{32} \quad \text{for block II.} \tag{5.2.1}$$

Suppose now that, instead of performing all the 64 runs, we carried out only the 32 runs in, say, block II. These are the runs for which elements of **123456** are $+$ in Table 5.1. Let us set out a table of signs for the calculation of main effects, two-factor interactions, three-, four-, five-, and six-factor interactions, just as for a full factorial design but using only block II runs, as shown in Table 5.3.

We could certainly use the table of signs to compute all the 63 main effects and interactions. Nevertheless, something would seem to be amiss because we can scarcely expect to compute 63 independent effects from only 32 observations. The explanation lies in the duplication that occurs in the table. We see, as expected, that the signs in the **123456** column are all identical to those of $\mathbf{I}_{32}$ (because we have deliberately arranged this). Notice also, however, that the signs for the interaction column **23456** are identical to those for the column **1**, or symbolically **1** = **23456**. Additionally, **2** = **13456**, and so on. Let us denote l_1, l_2, and so on, the various linear contrasts obtained by adding together all the response values with plus signs in the **1, 2**, and so on, column and subtracting all those with minus signs, and dividing by the appropriate column divisor. Also* let us use an arrow pointing to the right ($\rightarrow$) to mean "is an estimate of." Then, for example, l_1 provides an estimate not just of the main effect of 1, nor of the interaction 23456, but of the sum of these, that is,

$$l_1 \rightarrow 1 + 23456. \tag{5.2.2}$$

To establish this, we need only write out the formulas for the estimates of 1 and of 23456 from the full 2^6 and add the two together. It will be found that addition and cancellation leaves only the observations in block II with the signs of column **1** in

---

*As in Chapter 4, the arrow pointing to the left ($\leftarrow$) will mean "is estimated by."

Table 5.3. Columns of Signs for Calculation of Effects in a Half-Fraction of a 2^6 Design

Run Number from Table 5.1	I_{32}	Main Effects						Two-Factor Interactions					Five-Factor Interactions			Six-Factor Interaction
		1	2	3	4	5	6	12	13	14	...	...	12456	13456	23456	123456
1	+	−	−	−	−	−	−	+	+	+	...	...	−	−	−	+
4	+	+	+	−	−	−	−	+	−	−			−	+	+	+
6	+	+	−	+	−	−	−	−	+	−			+	−	+	+
7	+	−	+	+	−	−	−	−	−	+			+	+	−	+
10	+	+	−	−	+	−	−	−	−	+			−	+	−	+
11	+	−	+	−	+	−	−	+	+	−			−	−	+	+
13	+	−	−	+	+	−	−	+	−	−			+	−	−	+
...	...	...									...	...				
...	...	...									...	...				
64	+	+	+	+	+	+	+	+	+	+	...	...	+	+	+	+
Divisor	32	16	16	16	16	16	16	16	16	16	...	...	16	16	16	16

Table 5.3 attached, divided by the divisor 16. In similar fashion,

$$l_2 \rightarrow 2 + 13456 \qquad (5.2.3)$$

and so on. The device that enables us quickly to identify all these *aliases* is supplied by the generator for this half-fraction of the 2^6 design which we can write as

$$\mathbf{I} = \mathbf{123456}, \qquad (5.2.4)$$

(where now, for simplicity, we write $\mathbf{I}$ for $\mathbf{I}_{32}$). Equation (5.2.4) simply tells us to pick out runs from the full 64 whose product of signs **123456** is plus, same as the $\mathbf{I}$ column. Multiplying both sides of Eq. (5.2.4) by **1**, we have $\mathbf{1} = \mathbf{23456}$ which implies that $l_1 \rightarrow 1 + 23456$.

All other existing alias relationships are obtained similarly; for example, on multiplying Eq. (5.2.4) by **123** we obtain

$$\mathbf{123} = \mathbf{456}, \qquad \text{implying that } l_{123} \rightarrow 123 + 456. \qquad (5.2.5)$$

(Note that each l can be denoted by two equivalent names; for example, $l_{123} = l_{456}$.)

For quantitative variables, the various interactions $12, 345, \ldots$, reflect the effects of mixed derivatives of the response function η, such as

$$\frac{\partial^2 \eta}{\partial x_1\, \partial x_2}, \frac{\partial^3 \eta}{\partial x_3\, \partial x_4\, \partial x_5}, \ldots.$$

It would frequently be reasonable to assume that derivatives higher than the second could be ignored, that is, in a Taylor's series expansion of the response function, terms of third and of higher orders could be assumed to be negligible within the local region of experimentation. In these circumstances, we could, correspondingly, ignore interactions of three or more factors in the factorial analysis of the data. *Under such an assumption*, no confounding would in fact occur in the half-fraction $\mathbf{I} = \mathbf{123456}$ of the 2^6 design above. When σ is not too large, so that fewer runs than a full factorial will provide adequate accuracy, fractional factorials are frequently very useful, particularly in experimental situations with more than three or four variables.

For our illustration above, we chose the half-fraction defined by $\mathbf{I} = \mathbf{123456}$, but we could equally well have used the other half-fraction defined by $-\mathbf{I} = \mathbf{123456}$ or, equivalently, $\mathbf{I} = -\mathbf{123456}$. In this case, the columns **1** and $-\mathbf{23456}$ will be identical and, denoting the linear contrasts defined by columns $\mathbf{1}, \mathbf{2}, \ldots$, and so on by $l_1', l_2', \ldots$, and so on, we shall find that

$$l_1' \rightarrow 1 - 23456, \qquad l_2' \rightarrow 2 - 13456, \ldots, \qquad (5.2.6)$$

and so on. Notice that, if we perform *both* halves of the 2^6 design, we can recover all the individual main effects and interactions by averaging or halving and differencing. For example,

$$\tfrac{1}{2}(l_1 + l_1') \rightarrow \tfrac{1}{2}(1 + 23456 + 1 - 23456) = 1 \qquad (5.2.7)$$

and this would be identical to the estimate supplied by the full factorial. Also, similarly,

$$\tfrac{1}{2}(l_1 - l_1') \rightarrow 23456, \qquad\qquad (5.2.8)$$

again identical to that supplied by the full factorial.

A one-half replicate of a full factorial in k factors, that is, $\frac{1}{2}$ of a 2^k, or a 2^{-1} fraction of a 2^k, is often referred to as a 2^{k-1} fractional factorial. Thus the fractions we have described above are both 2^{6-1} fractional factorial designs.

Quarter Fractions of the 2^6 Factorial

Suppose now we ran only 16 of the 64 runs of the full 2^6 design corresponding to one block of the design when it is divided into four blocks (as in Section 5.1). This *quarter replicate* would be referred to as a 2^{6-2} fractional factorial design. There are four blocks from which to choose. Suppose we use the one for which $(\mathbf{B}_1, \mathbf{B}_2) = (+, +)$. Then for all the runs in this block,

$$\mathbf{B}_1 = \mathbf{I} = \mathbf{1234}, \qquad \mathbf{B}_2 = \mathbf{I} = \mathbf{3456}, \qquad \mathbf{B}_1\mathbf{B}_2 = \mathbf{I} = \mathbf{1256}. \qquad (5.2.9)$$

We can multiply through this defining relation by $\mathbf{1, 2, \ldots, 12, 13, \ldots}$ to give the alias pattern. For example,

$$\mathbf{1} = \mathbf{234} = \mathbf{13456} = \mathbf{256},$$
$$\mathbf{2} = \mathbf{134} = \mathbf{23456} = \mathbf{156},$$
$$\cdots \qquad\qquad (5.2.10)$$
$$\mathbf{12} = \mathbf{34} = \mathbf{123456} = \mathbf{56},$$
$$\cdots$$

and so on. Ignoring interactions involving four or more factors we see that the design allows estimates of 16 combinations of main effects and interactions such as, for example,

$$l_1 \rightarrow 1 + (234 + 256),$$
$$l_2 \rightarrow 2 + (134 + 156),$$
$$l_3 \rightarrow 3 + (124 + 456),$$
$$l_4 \rightarrow 4 + (123 + 356),$$
$$l_5 \rightarrow 5 + (346 + 126),$$
$$l_6 \rightarrow 6 + (345 + 125), \qquad (5.2.11)$$
$$l_{12} \rightarrow 12 + 34 + 56,$$
$$l_{13} \rightarrow 13 + 24,$$
$$l_{14} \rightarrow 14 + 23,$$
$$\cdots$$

(Note that the l subscripts can be those of any of the aliased effects. For example, $l_{12} = l_{34} = l_{56}$, and so on. Typically, we choose subscripts which occur earlier in the ordering $1, 2, \ldots, 12, 13, \ldots, 123, \ldots$, eliminating aliased duplicates, but this is not essential.)

The estimation equations show that all main effects are confounded with third-order (or ignored higher-order) interactions, not with two-factor interactions or other main effects. Two-factor interactions are confounded with other two-factor interactions. The defining relation must indicate all the independent generators and their products in twos, threes, $\ldots$, all of which produce the identify **I**. For example, suppose we had used the 2^{6-2} design defined by $(\mathbf{B}_1, \mathbf{B}_2) = (-, +)$, we would have had

$$\mathbf{B}_1 = -\mathbf{I} = \mathbf{1234}, \qquad \mathbf{B}_2 = \mathbf{I} = \mathbf{3456}, \qquad \mathbf{B}_1\mathbf{B}_2 = -\mathbf{I} = \mathbf{1256}, \quad (5.2.12)$$

which leads to the defining relation

$$\mathbf{I} = -\mathbf{1234} = \mathbf{3456} = -\mathbf{1256}, \tag{5.2.13}$$

and means that the estimates obtained would be as follows:

$$\begin{aligned} l_1' &\to 1 - 234 - 256, \\ l_2' &\to 2 - 134 - 156, \\ &\cdots \end{aligned} \tag{5.2.14}$$

and so on, ignoring interactions of four or more factors.

The variance of any of the estimates in (5.2.11) or (5.2.14) is

$$V(l) = 4\sigma^2/n \tag{5.2.15}$$

where n is the total number of individual observations (with variance σ^2) contributing to l. This formula applies to estimates throughout this chapter.

5.3. RESOLUTION OF A 2^{k-p} FACTORIAL DESIGN

The resolution R of a 2^{k-p} fractional factorial design is the length of the shortest word in the defining relation. For example, the 2^{6-1} design with defining relation **I** = **123456** is of resolution VI. (Note that the resolution is expressed in Roman numerals.) The quarter fraction 2^{6-2} design with defining relation **I** = **1234** = **3456** = **1256** is of resolution IV, and so on.

In this book we shall frequently use polynomial approximations of first or second degree. First-order designs to obtain data to estimate the coefficients in a first degree equation should therefore be of resolution at least III. This will ensure that no main effect is aliased with any other main effect, although main effects may (if R = III) be aliased with two-factor interactions. On the strict assumption that the response can be represented by a first degree equation, these two-factor interactions will all be zero. A resolution IV design provides added security because then no main effect would be aliased with either main effects or two-factor interactions.

Table 5.4. A Saturated, Resolution III, 2^{7-4} Design

I	1	2	3	4 = 12	5 = 13	6 = 23	7 = 123
+	−	−	−	+	+	+	−
+	+	−	−	−	−	+	+
+	−	+	−	−	+	−	+
+	+	+	−	+	−	−	−
+	−	−	+	+	−	−	+
+	+	−	+	−	+	−	−
+	−	+	+	−	−	+	−
+	+	+	+	+	+	+	+

We shall see later that second-order designs used to fit second degree polynomials can be formed by a suitable augmentation of two-level factorial or fractional factorial designs.

5.4. CONSTRUCTION OF 2^{k-p} DESIGNS OF RESOLUTION III AND IV

A design of resolution III can be constructed by *saturating* all the interactions of a full factorial design with additional factors. We illustrate this with an initial 2^3 design which can be saturated to produce a one-sixteenth replicate of a 2^7 design, namely a 2^{7-4} design. Consider again the columns of signs of Table 4.2 for the 2^3 factorial, now reproduced as Table 5.4.

If we use the **12** column to accommodate factor **4**, the **13** column similarly for factor **5**, the **23** column for **6**, and the **123** column for **7**, we have made the equivalences

$$4 = 12, \qquad 5 = 13, \qquad 6 = 23, \qquad \text{and} \qquad 7 = 123. \qquad (5.4.1)$$

Thus the generators of the design are

$$I = 124 = 135 = 236 = 1237. \qquad (5.4.2)$$

By multiplying these generators together in all possible ways, we obtain the defining relation for this 2^{7-4} design, and this leads to the alias relationships shown below. In this tabulation, we have ignored all interactions involving more than two factors for the sake of simplicity.

$$l_1 \rightarrow 1 + 24 + 35 + 67,$$
$$l_2 \rightarrow 2 + 14 + 36 + 57,$$
$$l_3 \rightarrow 3 + 15 + 26 + 47,$$
$$l_4 \rightarrow 4 + 12 + 56 + 37, \qquad (5.4.3)$$
$$l_5 \rightarrow 5 + 13 + 46 + 27,$$
$$l_6 \rightarrow 6 + 23 + 45 + 17,$$
$$l_7 \rightarrow 7 + 34 + 25 + 16.$$

Eight-Run Designs for Fewer than Seven Variables

When the number of variables examined is smaller than seven, useful designs can also be obtained by omitting one or more of the columns of the 2^{7-4} design or, more accurately, not using those columns for a variable. (The column is still used for estimating combinations of interactions but no longer specifies the levels of a variable.) The choice of which columns to drop can make use of the judgment of the investigator that only certain specific interactions are likely to occur. For example, suppose we have five factors and a particular one of these, which we designate as 1, is thought likely to interact with two others, 2 and 3, say; apart from this no other interactions are likely. Thus we would seek a design which can estimate the main effects of five factors, including 1, 2, and 3, and also the interactions 12 and 13, assuming all other interactions are zero. Looking at Table 5.4, we see that the required design is obtained by associating our five factors with the columns **1**, **2**, **3**, **6**, and **7**. If our assumptions are correct, then the linear combinations will provide the following estimates:

$$l_1 \to 1, \quad l_4 \to 12, \quad l_6 \to 6,$$
$$l_2 \to 2, \quad l_5 \to 13, \quad l_7 \to 7. \quad\quad (5.4.4)$$
$$l_3 \to 3,$$

For four variables, we can use variables 1, 2, 3, and 7 to obtain a design of resolution IV.

Adding Further Fractions

Sometimes, it will happen that the results from a highly fractionated design are ambiguous. In such a case we may wish to run a further fraction or fractions which will resolve the uncertainties created by the first. There are many possibilities; the following illustrations will indicate the general idea. We assume that the first fraction is the 2^{7-4} given by $\mathbf{I} = \mathbf{124} = \mathbf{135} = \mathbf{236} = \mathbf{1237}$ with estimates as in Eq. (5.4.3).

Possibility I
Suppose that, from a view of the results of the first fraction, it seems likely that a particular factor is dominating, and we decide we would like to add a further fraction and estimate the main effect of the dominating factor and all its interactions with other factors, clear of two-factor aliases. We can achieve this by choosing the second fraction whose generators are like those of the first fraction but with the sign of the variable of interest reversed everywhere. For example, if variable **1** is the designated variable we use

$$\mathbf{I} = -\mathbf{124} = -\mathbf{135} = \mathbf{236} = -\mathbf{1237} \quad\quad (5.4.5)$$

and obtain the individual runs of the design by writing an initial 2^3 factorial in variables **1**, **2**, and **3** and then generating columns $4 = -\mathbf{12}, 5 = -\mathbf{13}, 6 = \mathbf{23}$, and

$7 = -123$. From the second fraction alone we can then estimate

$$l_1' \rightarrow 1 - 24 - 35 - 67,$$

$$l_2' \rightarrow 2 - 14 + 36 + 57,$$

$$l_3' \rightarrow 3 - 15 + 26 + 47,$$

$$l_4' \rightarrow 4 - 12 + 56 + 37, \qquad (5.4.6)$$

$$l_5' \rightarrow 5 - 13 + 46 + 27,$$

$$l_6' \rightarrow 6 + 23 + 45 - 17,$$

$$l_7' \rightarrow 7 + 34 + 25 - 16,$$

which has the same basic alias structure as Eq. (5.4.3) but with certain signs reversed, as shown. It will now be seen that by combining the results from both fractions we can compute the estimates:

$$\tfrac{1}{2}(l_1 - l_1') \rightarrow 24 + 35 + 67, \qquad \tfrac{1}{2}(l_1 + l_1') \rightarrow 1,$$

$$\tfrac{1}{2}(l_2 - l_2') \rightarrow 14, \qquad \tfrac{1}{2}(l_2 + l_2') \rightarrow 2 + 36 + 57,$$

$$\tfrac{1}{2}(l_3 - l_3') \rightarrow 15, \qquad \tfrac{1}{2}(l_3 + l_3') \rightarrow 3 + 26 + 47,$$

$$\tfrac{1}{2}(l_4 - l_4') \rightarrow 12, \qquad \tfrac{1}{2}(l_4 + l_4') \rightarrow 4 + 56 + 37, \quad (5.4.7)$$

$$\tfrac{1}{2}(l_5 - l_5') \rightarrow 13, \qquad \tfrac{1}{2}(l_5 + l_5') \rightarrow 5 + 46 + 27,$$

$$\tfrac{1}{2}(l_6 - l_6') \rightarrow 17, \qquad \tfrac{1}{2}(l_6 + l_6') \rightarrow 6 + 23 + 45,$$

$$\tfrac{1}{2}(l_7 - l_7') \rightarrow 16, \qquad \tfrac{1}{2}(l_7 + l_7') \rightarrow 7 + 34 + 25.$$

Thus we see that, provided our assumption that interactions between more than two factors are negligible is true, we can isolate estimates of the main effect of factor 1 and all its interactions with other factors free from aliases.

Possibility 2

Suppose instead that we wished to run a second fraction that would provide all main effects clear of *all* two-factor interactions. This can be done by choosing a second fraction whose generators have the signs of all the variables reversed compared with the first one. For example, from Eq. (5.4.2) this procedure gives the fraction

$$\mathbf{I} = -\mathbf{124} = -\mathbf{135} = -\mathbf{236} = \mathbf{1237} \qquad (5.4.8)$$

whose columns are obtained by writing down a 2^3 design in the initial columns **1**, **2**, and **3** and then setting $4 = -12$, $5 = -13$, $6 = -23$, and $7 = 123$. Such a design produces the estimates

$$l'_1 \rightarrow 1 - 24 - 35 - 67,$$

$$l'_2 \rightarrow 2 - 14 - 36 - 57,$$

$$l'_3 \rightarrow 3 - 15 - 26 - 47,$$

$$l'_4 \rightarrow 4 - 12 - 56 - 37, \qquad (5.4.9)$$

$$l'_5 \rightarrow 5 - 13 - 46 - 27,$$

$$l'_6 \rightarrow 6 - 23 - 45 - 17,$$

$$l'_7 \rightarrow 7 - 34 - 25 - 16.$$

Combining these estimates with those of Eq. (5.4.3) allows us to separate the main effects from the groups of two-factor interactions as follows:

$$\tfrac{1}{2}(l_1 + l'_1) \rightarrow 1, \qquad \tfrac{1}{2}(l_1 - l'_1) \rightarrow 24 + 35 + 67,$$

$$\tfrac{1}{2}(l_2 + l'_2) \rightarrow 2, \qquad \tfrac{1}{2}(l_2 - l'_2) \rightarrow 14 + 36 + 57, \qquad (5.4.10)$$

$$\ldots \qquad\qquad \ldots$$

and so on.

Designs of Resolution IV via "Foldover"

The foregoing combination design is of resolution IV, and so the method of switching all signs provides a method of building up a resolution IV design from one of resolution III. The switching can be carried out directly on the sign pattern as follows. Suppose **A** represents the block of signs shown in Table 5.4 (ignoring the **I** column for the moment) for the 2^{7-4} design **I = 124 = 135 = 236 = 1237**. Then the signs

$$\begin{matrix} \mathbf{A} \\ -\mathbf{A} \end{matrix} \qquad (5.4.11)$$

provide 16 lines of signs which give a 2^{7-3} design of resolution IV. The $-\mathbf{A}$ block will be exactly the second set of runs as in "Possibility 2" above but will occur in a different order. Such a design is usually called a *foldover* design because we have "folded over" the signs—that is, switched them all. We can actually do better than this and obtain a 2^{8-4} design of resolution IV by associating an eighth factor with the initial $\mathbf{I}_8$ column of ones. The folded design can then be written

$$\begin{matrix} \text{runs } 1-8 \\ \text{runs } 9-16 \end{matrix} \begin{bmatrix} \mathbf{A} & \mathbf{I}_8 \\ -\mathbf{A} & -\mathbf{I}_8 \end{bmatrix}. \qquad (5.4.12)$$

The design matrix of the 2^{8-4} resolution IV design obtained in this manner is shown in Table 5.5. Equation (5.4.13) shows the estimates that can be obtained

Table 5.5. A 2^{8-4} Design of Resolution IV, Obtained Here by the Foldover Technique

Run	Mold Temperature 1	Moisture Content 2	Holding Pressure 3	Cavity Thickness 4	Booster Pressure 5	Cycle Time 6	Gate Size 7	Screw Speed 8	Shrinkage y
1	−	−	−	+	+	+	−	+	14.0
2	+	−	−	−	−	+	+	+	16.8
3	−	+	−	−	+	−	+	+	15.0
4	+	+	−	+	−	−	−	+	15.4
5	−	−	+	+	−	−	+	+	27.6
6	+	−	+	−	+	−	−	+	24.0
7	−	+	+	−	−	+	−	+	27.4
8	+	+	+	+	+	+	+	+	22.6
9	+	+	+	−	−	−	+	−	22.3
10	−	+	+	+	+	−	−	−	17.1
11	+	−	+	+	−	+	−	−	21.5
12	−	−	+	−	+	+	+	−	17.5
13	+	+	−	−	+	+	−	−	15.9
14	−	+	−	+	−	+	+	−	21.9
15	+	−	−	+	+	−	+	−	16.7
16	−	−	−	−	−	−	−	−	20.3

Source: Table 5.5 is reproduced and adapted with permission from the table on p. 402 of *Statistics for Experimenters: An Introduction to Design, Data Analysis and Model Building*, by G. E. P. Box, W. G. Hunter, and J. S. Hunter, published by John Wiley & Sons, Inc., New York, 1978.

from the columns whose subscripts are shown in the l's.

$$l_1 \rightarrow 1,$$
$$l_2 \rightarrow 2,$$
$$l_3 \rightarrow 3,$$
$$l_4 \rightarrow 4,$$
$$l_5 \rightarrow 5,$$
$$l_6 \rightarrow 6,$$
$$l_7 \rightarrow 7,$$
$$l_8 \rightarrow 8,$$
$$l_{12} \rightarrow 12 + 37 + 48 + 56 \quad (B_1),$$
$$l_{13} \rightarrow 13 + 27 + 46 + 58 \quad (B_2),$$
$$l_{14} \rightarrow 14 + 28 + 36 + 57 \quad (B_3),$$
$$l_{15} \rightarrow 15 + 26 + 38 + 47 \quad (B_1 B_2 B_3),$$
$$l_{16} \rightarrow 16 + 25 + 34 + 78 \quad (B_2 B_3),$$
$$l_{17} \rightarrow 17 + 23 + 68 + 45 \quad (B_1 B_2),$$
$$l_{18} \rightarrow 18 + 24 + 35 + 67 \quad (B_1 B_3).$$

(5.4.13)

Exercise.   Use the data shown in Table 5.5 and obtain the numerical estimates for the combinations of effects in Eq. (5.4.13). They are, in order, -0.7, -0.1, 5.5, -0.3, -3.8, -0.1, 0.6, 1.2, -0.6, 0.9, -0.4, 4.6, -0.3, -0.2, and -0.6. What do you conclude about the factors as a result of examining these figures? For additional details see Box et al. (1978, p. 399).   □

The foldover method of Eq. (5.4.12) is completely general and will provide a resolution IV design whenever **A** represents a block of signs designating a resolution III design.

Exercise.   Write down the 2^{3-1} resolution III design given by $\mathbf{I} = \mathbf{123}$. Associate a fourth variable with $\mathbf{I}_4$. Fold over the entire design. Confirm that the combination design is the resolution IV design $\mathbf{I} = \mathbf{1234}$.   □

In the parentheses in Eq. (5.4.13) we show how to generate an interesting blocking arrangement which allows the 2^{8-4} design to be run in eight blocks of two runs each, without confounding main effects with blocks. The way in which this is done will be clear from Table 5.6 which shows three blocking variables associated with selected interactions. The eight possible sign combinations of the blocking variables provide the eight blocks. It is of interest to note that the two runs within any block are mirror images of each other. All the signs of any one run are reversed in its companion run.

5.5. COMBINATION OF DESIGNS FROM THE SAME FAMILY

By changing the signs of one or more of the generators that define any given fractional factorial design, we obtain a family of fractional factorial designs (Box and Hunter, 1961a). For example,

$$\mathbf{I} = \pm\mathbf{124} = \pm\mathbf{135} = \pm\mathbf{236} = \pm\mathbf{1237}$$

provides a family of sixteen 2_{III}^{7-4} fractional factorials, obtained from the $2^4 = 16$ possible sign choices. Clearly, knowledge of the defining relation of any one fraction in a family enables us to obtain the defining relation of any other family member fraction quite quickly. Suppose we combine two members of the same family. What type fraction, and of what resolution, is the combination design?

The basic rule for obtaining the defining relation of a combination of two designs from the same family is simple to state, but can be tedious to do. One simply writes down the defining relation for each fraction (which will be the same words, apart from signs) and selects those words that have the same sign in each. For example, if we combine the design (D_1 say)

$$\mathbf{I} = \mathbf{124} = \mathbf{135} = \mathbf{236} = \mathbf{1237}$$

with (D_2 say)

$$\mathbf{I} = -\mathbf{124} = -\mathbf{135} = \mathbf{236} = -\mathbf{1237},$$

Table 5.6. A 2^{8-4}_{IV} Design in Eight Blocks of Size Two

| | 2^{8-4}_{IV} Design | | | | | | | | Block Variable | | |
Run	1	2	3	4	5	6	7	8	B₁ 12	B₂ 13	B₃ 14
1	−	−	+	+	+	+	−	+	+	+	−
2	+	−	−	−	−	+	+	+	−	−	−
3	−	+	−	−	+	−	+	+	−	+	+
4	+	+	+	+	−	−	−	+	+	−	+
5	−	−	+	+	−	−	+	+	+	−	−
6	+	−	−	−	+	−	−	+	−	+	−
7	−	+	+	−	−	+	−	+	−	−	+
8	+	+	+	+	+	+	+	+	+	+	+
9	+	+	−	−	−	−	+	−	+	+	−
10	−	+	+	+	+	−	−	−	−	−	−
11	+	−	+	+	−	+	−	−	−	+	+
12	−	−	−	−	+	+	+	−	+	−	+
13	+	+	−	−	+	+	−	−	+	−	−
14	−	+	+	+	−	+	+	−	−	+	−
15	+	−	−	+	+	−	+	−	−	−	+
16	−	−	−	−	−	−	−	−	+	+	+

Design Rearranged in Eight Blocks

Block	1	2	3	4	5	6	7	8	B₁	B₂	B₃	Run
1	+	−	−	−	−	+	+	+	−	−	−	2
	−	+	+	+	+	−	−	−	−	−	−	10
2	−	−	+	+	−	−	+	+	+	−	−	5
	+	+	−	−	+	+	−	−	+	−	−	13
3	+	−	−	−	+	−	−	+	−	+	−	6
	−	+	+	+	−	+	+	−	−	+	−	14
4	−	−	+	+	+	+	−	+	+	+	−	1
	+	+	−	−	−	−	+	−	+	+	−	9
5	−	+	+	−	−	+	−	+	−	−	+	7
	+	−	−	+	+	−	+	−	−	−	+	15
6	+	+	+	+	−	−	−	+	+	−	+	4
	−	−	−	−	+	+	+	−	+	−	+	12
7	−	+	−	−	+	−	+	+	−	+	+	3
	+	−	+	+	−	+	−	−	−	+	+	11
8	+	+	+	+	+	+	+	+	+	+	+	8
	−	−	−	−	−	−	−	−	+	+	+	16

Source: Table 5.6 is reproduced and adapted with permission from the table on p. 406 of *Statistics for Experimenters: An Introduction to Design, Data Analysis and Model Building,* by G. E. P. Box, W. G. Hunter, and J. S. Hunter, published by John Wiley & Sons, Inc., New York, 1978.

we compare the two defining relations

$$\begin{aligned}
\mathbf{I} = \;&\pm\mathbf{124} = \pm\mathbf{135} = \mathbf{236} = \pm\mathbf{1237} \\
&= \mathbf{2345} = \pm\mathbf{1346} = \mathbf{347} = \pm\mathbf{1256} = \mathbf{257} = \pm\mathbf{167} \\
&= \mathbf{456} \;\; = \pm\mathbf{1457} = \mathbf{2467} = \mathbf{3567} \\
&= \pm\mathbf{1234567},
\end{aligned}$$

where the top sign refers to D_1 and the bottom sign refers to D_2. The common part is obviously

$$\mathbf{I} = \mathbf{236} = \mathbf{2345} = \mathbf{347} = \mathbf{257} = \mathbf{456} = \mathbf{2467} = \mathbf{3567},$$

which is a resolution III design that can be generated, for example, by $\mathbf{I} = \mathbf{236} = \mathbf{347} = \mathbf{257}$. Thus the design is a 2_{III}^{7-3}. Thinking somewhat symbolically, we can say that we "averaged" the two individual defining relations. (For other examples, see Exercises 5.8 to 5.10.) Generally, if we combine any two of the 2^p members of a 2^{k-p} family, we obtain a 2^{k-p+1} design.

An alternative and quicker way of getting generators for the combined design is to use the following rule, due to Box and Hunter (1961a, pp. 328–329).

The U–L Rule for Combining Fractions from the Same Family

Suppose, for two designs of the same family, we have two generating relations that use exactly the same words, although the signs of those words will vary. Suppose the designs have L generators common and of the same (or LIKE) sign and $U = p - L$ generators which are the same apart from a sign change—that is, of UNLIKE sign. Then the L generators of like sign, together with $U - 1$ generators arising from independent even products of the U generators of unlike signs, provide $L + U - 1 = p - 1$ generators for the combined design.

Example. Using the example above, we see that D_1 and D_2 have $L = 1$ word of like sign (**236**) and $U = 3$ words of unlike sign. Two independent even products of these latter are $(\pm\mathbf{135})(\pm\mathbf{1237}) = \mathbf{257}$ and $(\pm\mathbf{124})(\pm\mathbf{1237}) = \mathbf{347}$. The third possible product $(\pm\mathbf{124})(\pm\mathbf{135}) = \mathbf{2345}$ is not independent of the first two since it is the product of them, namely $(\mathbf{257})(\mathbf{347})$.

Exercise. Show that when the two 2_{III}^{5-2} designs $\mathbf{I} = \mathbf{124} = \mathbf{135}$ and $\mathbf{I} = -\mathbf{124} = -\mathbf{135}$ are combined, the joint design is $\mathbf{I} = \mathbf{2345}$, namely a 2_{IV}^{5-1}.   □

Note carefully that the two sets of generators must be identical words except for signs. Also all the even products must be taken from one design or the other, not from a mixture of both. In the example above, we can take the product $(\mathbf{135})(\mathbf{1237})$ or $(-\mathbf{135})(-\mathbf{1237})$ but must not cross over by $(\mathbf{135})(-\mathbf{1237})$, an error to be avoided.

More generally, whenever 2^r of the designs from a family of 2^{k-p} designs are combined, the result will be a 2^{k-p+r} design, and its generators can be obtained either by applying the U–L rule r times or by finding the common parts of all the 2^r defining relations and then finding generators which will give rise to those common parts.

The device of combining fractions from the same family is an extremely useful one and provides a very flexible system for concentrating the direction of an investigation. The choice of a second fraction to combine with a first can be made *after* the results of the first fraction have been examined. The choice can then be made in such a way that as many as possible of the questions raised by the first fraction can be resolved.

Exercise. Show that the combination of the two 2_{III}^{7-4} designs $I = 124 = 135 = 236 = 1237$ and $I = -124 = -135 = -236 = 1237$ leads to the 2_{IV}^{7-3} design $I = 1237 = 2345 = 1346$.   □

Exercise. Show that the combination of the two 2_{III}^{7-4} designs $I = 124 = 135 = 236 = 1237 = 8$ and $I = -124 = -135 = -236 = 1237 = -8$ is the 2_{IV}^{8-4} design $I = 1237 = 2345 = 1346 = 1248$. (This result applies in the foldover example in Section 5.4.)   □

Combinations, More Generally

For any given fractional factorial design, the question arises as to what is the best way to "fold it over" in a more general manner, that is, add a second design from the same family but with one or more sign changes in the generators compared to those of the initial design. (Our previous foldover involved changing all signs.) When two such pieces are combined, there are better and worse designs, depending on the objectives of the experimenter. In our earlier example, we began with a 2_{III}^{7-4} design generated by $I = 124 = 135 = 236 = 1237$ and added another 2_{III}^{7-4}, $I = -124 = -135 = 236 = -1237$ to obtain a 2_{III}^{7-3} design, $I = 236 = 347 = 257$. We can characterize the second fraction with respect to the first by saying that we have "changed the sign of **1**" in the original generators. (Less economically, we could also say that we changed the signs of **4**, **5**, and **7**, which would give the same design, but with the runs in a different order; and so on.) What is the best way to choose a second fraction? It depends, of course, on which effects the experimenter regards as important to separate out. The general question of which sign changes would produce higher resolution was considered by Li and Lin (2003a, b). We give an example from their Table 2 (2003a).

Design D_1: $I = 125 = 1346(= 23456)$, 2_{III}^{6-2}.
Design D_2: $I = -125 = 1346(= -23456)$, 2_{III}^{6-2}, changing the sign of **5**.
$D_1 + D_2$: $I = 1346$. 2_{IV}^{6-1}.

Design D_2: $I = -125 = -1346(= 23456)$, 2_{III}^{6-2}, changing the signs of **5, 6**.
$D_1 + D_2$: $I = 23456$. 2_V^{6-1}.

We see that the second combination is somewhat better than the first in the sense that it is of higher resolution, giving clear estimates of *all* 2fi's, whereas the first confounds three pairs (13 + 46), (14 + 36) and (16 + 34). For more on this topic, see Li and Lin (2003a, b) and Fang et al. (2003).

Semifolding

It is also possible to "semifold" designs—that is, fold over half of the initial designs, in order to elucidate specific features from the initial fraction. While this technique has been known for many years [e.g., see Addelman (1961) and Daniel (1962)], a more recent useful investigation was given by Mee and Peralta (2000).

5.6. SCREENING, USING 2^{k-p} DESIGNS (INVOLVING PROJECTIONS TO LOWER DIMENSIONS)

In some experimental situations, one might wish to look at many factors, even though it is anticipated that only a *few* of those factors will have an effect on the response. However, it is usually not known beforehand which those few may be! (Such a situation is often called making an assumption of *effect sparsity*.) A typical basic analysis when a screening design is used is to evaluate the estimates of effects associated with all of the columns of the design and then seek to retain only those columns associated with large effects, projecting the design down to those dimensions for further investigation. This may suggest the performance of additional runs to resolve ambiguities of interpretation seen in the analysis so far. (These extra runs could be made using only the factors that appear to be important, or perhaps all factors, as seems sensible and/or feasible.) Therefore, before we choose and run a screening design, it is useful to know what sorts of designs can arise from it when it is projected down to various fewer numbers of factors. (In another context, we might also want to simplify a setup design if we decided, before running it, that the introduction of a certain variable was pointless, for example, because it requires equipment that is unavailable. We might then wish to see what sort of design remains if this variable is eliminated.)

What does a 2^{k-p} design become if variables are dropped? The basic method is simply stated: Write out the defining relation, cross out any word that contains the variable(s) to be omitted, and choose generators that give rise to the new defining relation. There is also a rule that has similarities to the $U-L$ rule described above:

Rule: Suppose d variables are dropped from a 2^{k-p} fractional factorial design generated by p given generators. Suppose that q of these generators contain the dropped variable(s), so that $p - q$ do not. The reduced design will have at least $p - d$ generators and may have more. They are obtained by selecting the $p - q$ original generators which do not contain the dropped variables together with as many as possible (but not fewer than $q - d$) additional independent generators which *do not* contain dropped variables, formed from products of the q generators which *do* contain the dropped variables.

Example 1.   Consider the 2^{7-4}_{III} design $\mathbf{I} = \mathbf{124} = \mathbf{135} = \mathbf{236} = \mathbf{1237}$. What design results if variable **1** is dropped? Thus, $d = 1$. Three generators contain **1** and there is $p - q = 4 - 3 = 1$ which does not. So **236** will be a generator of the new design and we need at least $q - d = 3 - 1 = 2$ independent products of the other generators that do not have **1** in them. There are three possible such products, namely $(\mathbf{124})(\mathbf{135}) = \mathbf{2345}$, $(\mathbf{124})(\mathbf{1237}) = \mathbf{347}$, and $(\mathbf{135})(\mathbf{1237}) = \mathbf{257}$, each of which is the product of the other two, so only two are needed. For example, the design can be generated from $\mathbf{I} = \mathbf{236} = \mathbf{347} = \mathbf{257}$. This is a 2^{6-3}_{III} design in six variables numbered 2–7.

Exercise. Obtain the same result by writing out the defining relation of the original design, crossing out any word that contains the variable **1**, and then choosing generators that give rise to the new defining relation. ☐

Example 2. Consider from Table 5.15 the 2_{III}^{6-3} design which has defining relationship **I** = **124** = **135** = **236** = **2345** = **1346** = **1256** = **456**. What reduced designs result from deleting three variables? Crossing out factors 4, 5, and 6 in the defining relation, for example, leaves no unaffected words remaining. This means that the original design, when projected into the dimensions of factors 1, 2, and 3 must be a full factorial. (This is obvious anyway, because this design is often generated by starting with the 8 runs of a 2^3 design in factors 1, 2, and 3 and assigned the three variables 4, 5, and 6 to interaction columns 12, 13, and 23, respectively.) Of the $\binom{6}{3} = 20$ ways of deleting three of the six dimensions, 16 retentions leave full factorials and only 4 retentions, in the dimensions 124, 135, 236, and 456, lead to doubled-up (replicated) 2^{3-1} designs. Note that these four specific triplets are all in the defining relation, so that this result is not surprising.

Example 3. Consider another example taken from Table 5.15, the 16-runs 2_{IV}^{6-2} design with defining relation **I** = **1235** = **2346** = **1456**. If we project this design into various choices of three dimensions by crossing out subsets of (the remaining) three numbers, we find that *every* subset leaves no defining relation words standing. So the resulting projections must all be 2^3 factorials. However, because the original design has 16 runs, each 2^3 factorial is necessarily replicated, or doubled up, because we have to account for all of the available 16 runs. What happens if we cross out only two numbers from the original design? This can be done in $\binom{6}{2} = 15$ ways; 12 ways result in a full 2^4 projection, and three ways leave a replicated 2^{4-1} design. These three ways are as follows:

Deletion of 1 and 5 leaves a replicated 2^{4-1} design such that **I** = **2346**.
Deletion of 2 and 3 leaves a replicated 2^{4-1} design such that **I** = **1456**.
Deletion of 4 and 6 leaves a replicated 2^{4-1} design such that **I** = **1235**.

More generally, Cheng (1995b) has pointed out that, for there to be a full factorial in four factors remaining from all projections of 2^{k-p} designs, defining words of length three and four must be absent from the defining relation. This ensures that all main effects and two-factor interactions in the reduced design are unconfounded with one another in all four-dimensional projections.

Exercise. If the 2_{IV}^{8-3} design of Table 5.15, generated by **I** = **1236** = **1247** = **23458**, is projected into all $\binom{8}{4} = 70$ possible choices of four dimensions, which three choices of those dimensions will leave four copies of a 2_{IV}^{4-1} design? What will the other 67 choices leave?

Answers: The full defining relation, developed by taking all products of the three generators, is **I** = **1236** = **1247** = **23458** = **3467** = **14568** = **13578** = **25678**. So the three choices of surviving dimensions 1236, 1247, and 3467 will each leave 2_{IV}^{4-1} designs, which must occur four times over because the original design has 32 runs and the projected 2_{IV}^{4-1} designs have only eight each; the $\binom{8}{4} - 3 = 67$ other choices necessarily give a 2^4 (16-run) design which must occur twice, again to account for the original 32 runs available.   □

Exercise. If the 2_V^{8-2} design shown in Table 5.15 is projected into d dimensions, for what values of d will the result *always* result in a full factorial?

   Answer: The defining relation is $\mathbf{I} = \mathbf{12347} = \mathbf{12568} = \mathbf{345678}$ and the smallest word of this is of length five, so some projections into five and higher dimensions will remain as fractional factorials. Thus the answer is $d \le 4$.   □

Screening Designs in Robustness Studies

Another context in which screening designs may be useful is where the objective is to subject a product to variation in many factors in the expectation that the results will show *no* significant variation in response! If that expectation seemed to be confirmed by the data obtained, this would enable the experimenter to assert that the product under test is robust and unaffected by changes in the variables examined. Sensible variables to check in this context might be, for example, humidity, temperature, speed, stress, and so on, namely conditions likely to be experienced by the product in normal use.

5.7. COMPLETE FACTORIALS WITHIN FRACTIONAL FACTORIAL DESIGNS

The idea of deleting variables leads naturally to the following question. Suppose we wish to examine a large number of variables k in a screening situation in which we expect only a smaller subset of $k - d$ variables to be important. Again, we do not know which variables they will be. Whichever they are, we should like to have information on their main effects and 2fi's. Ideally, we need a fractional factorial in all k variables such that if we delete *any* d variables, a full factorial in the remaining $k - d$ variables remains. For what values of k and d is this possible? A basic rule (to be expanded below) is the following.

   Rule. A 2^{k-p} design of resolution R provides a complete factorial design in any subset of $R - 1$ variables if the other $k - R + 1$ variables are deleted. The design remaining may be single or replicated, depending on the values of k and R.

The rule is easily confirmed as follows. A design of resolution R has words of length R or higher in its defining relation. If, after deletion of $k - R + 1$ variables, only $R - 1$ variables remain, all words in the defining relation must have been struck out. The design necessarily provides a full factorial in all the remaining variables. Whether the full factorial design is replicated or not depends on the values of k and R.

   Example 1. If any one variable is dropped from a four-run 2_{III}^{3-1} design, generator $\mathbf{I} = \mathbf{123}$, the design reduces to a single 2^2 factorial design in the remaining two variables.

Example 2. If any five variables are dropped from an eight-run 2_{III}^{7-4} design, the design reduces to a replicated 2^2 factorial in the remaining two variables. The replication occurs because eight runs are available and the reduced 2^2 design occupies only four runs, so it is necessarily replicated.

After appreciating these examples, we can re-formulate the rule more generally as:

Rule. If a 2^{k-p} design of resolution R contains $2^{R-1} \times r$ runs, then it provides a complete factorial design, replicated r times, in any subset of $R - 1$ variables if the other $k - R + 1$ variables are deleted.

Example 3. The 2_{IV}^{8-4} design $\mathbf{I = 1235 = 1246 = 1347 = 2348}$ has 16 runs and $R = 4$, and $16 = 2^{R-1} \times 2$. Thus if any five variables are dropped, the reduced design is a replicated 2^3 design in the remaining three variables. Because there are $\binom{8}{3} = 56$ ways of choosing three factors out of eight, this design provides rather remarkable coverage if, in fact, only three variables out of eight have real effects. Not only will there be estimates of all three main effects, three two-factor interactions, and the three-factor interaction, but also σ^2 can be tentatively estimated from the pseudo-repeat runs so created. If a prior estimate of σ^2 exists, we then have a useful check on the nullity of the other five factors.

Example 4. If any four variables are dropped from a 16-run 2_{III}^{6-2} design, the reduced design is a 2^2 design in the remaining two variables, four times over.

Example 5. If any six variables are dropped from a 32-run 2_{IV}^{9-4} design, the reduced design is a 2^3 design in the remaining three variables, four times over.

Deletion of $(k - R)$ Variables from a 2_R^{k-p}

When we delete only $(k - R)$ variables, instead of $(k - R + 1)$ from a 2_R^{k-p} design, we cannot be assured that a full factorial design will be left in the remaining variables. The key to the situation is this rule:

Rule. If $k - R$ variables are deleted from a design of resolution R, the reduced design will provide in the remaining R variables either a full factorial design or a replicated half-fractional factorial design which itself may be repeated, depending on the values of k and R. If the R variables that are not deleted form a word of the original defining relation, half fractions will be obtained. Otherwise full factorials will occur.

Example 1. Consider the 2_{III}^{7-4} with generators $\mathbf{I = 124 = 135 = 236 = 1237}$. Its defining relation is $\mathbf{I = 124 = 135 = 236 = 1237 = 2345 = 1346 = 347 = 1256 = 257 = 167 = 456 = 1457 = 2467 = 3567 = 1234567}$. In this, there are seven words of length three, namely $\mathbf{124, 135, 236, 347, 257, 167}$. Now there are $\binom{7}{3} = 35$ ways of selecting $k - R = 7 - 3 = 4$ variables to delete. If the remaining (nondeleted) variables form one of the seven words, the reduced design is a half fraction, replicated. The other 28 sets of three remaining variables form full 2^3 factorials. This is an excellent guarantee from an eight-run design!

Example 2. Consider the 2_{IV}^{8-4} design with generators $\mathbf{I} = \mathbf{1235} = \mathbf{1246}$ $= \mathbf{1347} = \mathbf{2348}$. Its defining relation is $\mathbf{I} = \mathbf{1235} = \mathbf{1246} = \mathbf{1347} = \mathbf{2348} = \mathbf{3456}$ $= \mathbf{2457} = \mathbf{1458} = \mathbf{2367} = \mathbf{1368} = \mathbf{1278} = \mathbf{1567} = \mathbf{2568} = \mathbf{3578} = \mathbf{4678} = \mathbf{12345678}$. There are $\binom{8}{4} = 70$ ways of choosing four variables out of eight. Suppose four variables are deleted from the design. Then in the *remaining* four variables, the reduced design will be

(a) a full factorial if the remaining variables *do not* form a word in the defining relation (56 possibilities).

(b) a replicated half-fraction of resolution IV if the remaining variables *do* form a word in the defining relation (14 possibilities).

All in all, this is an excellent guarantee for 16 runs.

5.8. PLACKETT AND BURMAN DESIGNS FOR $n = 12$ TO 60 (BUT NOT 52)

As we have seen, saturated resolution III designs for up to $k = n - 1$ factors in $n = 2^q$ runs can be generated from any "core" 2^q factorial design by associating new variables with some or all of the interaction columns of the core design. Thus, designs of this type for numbers of runs, n, equal to powers of 2 are easily generated. For intermediate values that are multiples of four but *not* powers of two, such as $n = 12, 20, 24, 28, 36, \ldots$, saturated designs given by Plackett and Burman (1946) can be used. (In fact, that paper also covers $n = 2^q$ cases.) In all, Plackett and Burman (1946, pp. 323–324) provide designs for n equal to a multiple of four for $n \leq 100$, except 92. For $n = 92$, see Baumert et al. (1962); also see Draper (1985a). All such designs are resolution III with complicated alias structures. Full foldovers of them, with reversals of all signs, produce resolution IV designs. The mirror image pairs produced by foldover may then be deployed in blocks of size two, if desired.

Table 5.7. Plackett and Burman Design Matrix for $n = 12$

Run No.	\multicolumn{11}{c}{Variable and Column Number}

Run No.	1	2	3	4	5	6	7	8	9	10	11
1	+	+	−	+	+	+	−	−	−	+	−
2	−	+	+	−	+	+	+	−	−	−	+
3	+	−	+	+	−	+	+	+	−	−	−
4	−	+	−	+	+	−	+	+	+	−	−
5	−	−	+	−	+	+	−	+	+	+	−
6	−	−	−	+	−	+	+	−	+	+	+
7	+	−	−	−	+	−	+	+	−	+	+
8	+	+	−	−	−	+	−	+	+	−	+
9	+	+	+	−	−	−	+	−	+	+	−
10	−	+	+	+	−	−	−	+	−	+	+
11	+	−	+	+	+	−	−	−	+	−	+
12	−	−	−	−	−	−	−	−	−	−	−

We provide here the bases for Plackett and Burman designs for $n = 12$, 16, 20, 24, 28, 36, 40, 44, 48, 56, and 60. (We omit 52 simply because of the space needed to construct it.) When $n = 12$, 16, 20, 24, 32, 36, 44, 48, or 60, write down, as the first row of the design matrix, the line of $n − 1$ signs given below. Permute these cyclically in the second row, taking the extreme right-hand sign into the first position and moving all other signs to the right. Continue this until $(n − 1)$ rows have been obtained. Add an nth row consisting of all minus signs. We now have n runs (rows) in $n − 1$ variables (columns). All columns are orthogonal to one another. To examine fewer than $n − 1$ variables, t say, use *any* t of the $n − 1$ columns. Table 5.7 shows this construction in full for $n = 12$.

```
n = 12:   + + − + + + − − − + −
n = 16:   + + + + − + − + + − − + − − −
n = 20:   + + − − + + + + − + − + − − − − + + −
n = 24:   + + + + + − + − + + − − + + − − + − + − − − −
n = 32:   − − − − +   − + − + +   + − + + −   − − + + +   + + − − +
          + − + − −   +
n = 36:   − + − + +   + − − − +   + + + + −   + + + − −   + − − − −
          + − + − +   + − − + −
n = 44:   + + − − +   − + − − +   + + − + +   + + + − −   − + − + +
          + − − − −   − + − − −   + + − + −   + + −
n = 48:   + + + + +   − + + + +   − − + − +   − + + + −   − + − − +
          + − + + −   − − + − +   − + + − −   − − + − −   − −
n = 60:   + + − + +   + − + − +   − − + − −   + + + − +   + + + − −
          + + + + +   − − − − −   + + − − −   − + − − −   + + − + +
          − + − + −   − − + −
```

For all such designs, estimates of the effects are obtained in the usual way, by taking the cross product of a column with the column of responses and dividing by a divisor $n/2$ (e.g., 6 for the $n = 12$ example above). If a regression program is used, the regression coefficients are half of the effects, as usual, because the divisor is doubled to n in the regression calculation (12 for the example above).

When $n = 28$, a different construction is used. Plackett and Burman provide three 9×9 blocks of signs, A, B, and C, given in Table 5.8. These, set side by side, form nine rows of width $k = 27$. Cyclic permutation to BCA and then CAB gives a total of 27 rows, and a final row of minuses completes the $n = 28$ experimental runs. When $n = 40$, 56, the designs are obtained by "doubling" the designs for $n = 20$, 28, respectively. "Doubling" is interpreted as follows. Suppose we denote the design matrix for $n = 20$ or 28 by **A**. Then the design matrix for $n = 40$ or 56 is

$$\begin{bmatrix} \mathbf{A} & \mathbf{A} & \mathbf{I} \\ \mathbf{A} & -\mathbf{A} & -\mathbf{I} \end{bmatrix}$$

where **I** denotes a column of + signs with the same number of rows as **A**. As always, $t < n − 1$ variables may be examined by using any t of the columns.

The method for writing down the design for $n = 52$ is similar to that for $n = 28$ except that a single first line and column must be written down, and five 10×10

Table 5.8. Three Construction Blocks for a Plackett and Burman Design that Examines 27 Variables in 28 Runs

Construction block A

+	−	+	+	+	+	−	−	−
+	+	−	+	+	+	−	−	−
−	+	+	+	+	+	−	−	−
−	−	−	+	−	+	+	+	+
−	−	−	+	+	−	+	+	+
−	−	−	−	+	+	+	+	+
+	+	+	−	−	−	+	−	+
+	+	+	−	−	−	+	+	−
+	+	+	−	−	−	−	+	+

Construction block B

−	+	−	−	−	+	−	−	+
−	−	+	+	−	−	+	−	−
+	−	−	−	+	−	−	+	−
−	−	+	−	+	−	−	−	+
+	−	−	−	−	+	+	−	−
−	+	−	+	−	−	−	+	−
−	−	+	−	−	+	−	+	−
+	−	−	+	−	−	−	−	+
−	+	−	−	+	−	+	−	−

Construction block C

+	+	−	+	−	+	+	−	+
−	+	+	+	+	−	+	+	−
+	−	+	−	+	+	−	+	+
+	−	+	+	+	−	+	−	+
+	+	−	−	+	+	+	+	−
−	+	+	+	−	+	−	+	+
+	−	+	+	−	+	+	+	−
+	+	−	+	+	−	−	+	+
−	+	+	−	+	+	+	−	+

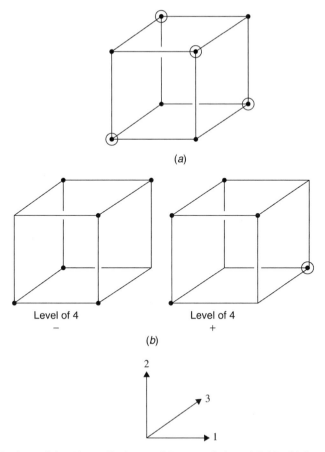

Figure 5.1. Projections of the 12-run Plackett and Burman design of Table 5.9 into (*a*) dimensions $(1, 2, 3)$ and (*b*) dimensions $(1, 2, 3, 4)$. Note that the two portions of (*b*), when merged left to right, give (*a*).

blocks of signs are permuted in cyclic order. The details are given by Plackett and Burman (1946, p. 323). In all, Plackett and Burman (1946, pp. 323–324) provide designs for n equal to a multiple of 4 for $n \le 100$, except 92. For $n = 92$, see Baumert et al. (1962). Also see Draper (1985a).

5.9. SCREENING, USING PLACKETT AND BURMAN DESIGNS (INVOLVING PROJECTIONS TO LOWER DIMENSIONS)

The Plackett and Burman designs are excellent for screening experiments. We first look at projections of the useful and versatile 12-run design shown in Table 5.7. Projection down into the dimensions of any two factors always leads to a 2^2 factorial, three times over. Projection into three factors always gives a full 2^3 design (8 runs) plus a 2^{3-1} design (4 runs). See Figure 5.1a for the specific projection into the dimensions of columns 1, 2 and 3; more generally, see Lin and Draper (1992) and Box and Bisgaard (1993).

Projection into any four dimensions also leads to only one type of result. We get 11 runs of a 2^4 design plus one run repeated, as indicated for factors 1, 2, 3, and 4 in Figure 5.1b. Notice that such reduced designs have certain symmetries, in spite of the fact that there are only 12 points. There are three points on each of the opposite faces of the two cubes in all choices of three dimensions, for example. By adding five more runs, it is possible to complete a 2^4 design in the selected factors, a useful possibility. If this is to be done, the question arises as to what fixed levels the "ignored" factors should be set. Most experimenters would set them at their central values on the basis that these factors were tentatively judged to be not important ones when they were shifted from their central values; this is often the "most neutral" decision in the circumstances. Other experimenters would choose differently, depending on circumstances and experimental convenience. Returning to Figure 5.1 for a moment, we can appreciate that, if the two portions of Figure 5.1b were merged by pushing them together accordion-wise, Figure 5.1a would be created. The types of projections shown in Figures 5.1a and 5.1b always occur, but the axes need to be relabeled by variable and perhaps in their ($-$ to $+$) directions, according to the particular factor space into which the original design is projected.

There are two distinct types of projection of the 12-run Plackett and Burman design into five dimensions. Which type occurs depends on which set of five factors are retained. Standardized versions are shown in Table 5.9(a) and (b). By standardized versions, we mean versions that characterize the two types; specific projections may have all the signs in one or more columns changed and/or the runs and/or the columns may occur in a different order. However, one type always has a repeat run as in (a), and the other has a pair of mirror image runs as in (b). In all, there are 462 possible projections into five dimensions; 66 are of type (a) and 396 are of type (b). By adding two more runs to type (a), we can complete an eight-run 2_{III}^{5-2} design; alternatively, six runs can be added to complete a 2_V^{5-1} design. Adding two more runs to type (b) will complete a 2_{III}^{5-2} design, and adding eight more will complete a 2_{IV}^{5-1} design; adding 10 more will give a 2_V^{5-1} design. The runs in either (a) or (b), whether projected down from a larger design or not, can also be used as the basis for so-called *small composite designs* for fitting a second-order response surface in five factors by adding suitable *star* runs. Details are given in Chapter 15.

Projections of Larger Plackett and Burman Two-Level Designs

The projections of larger two-level Plackett and Burman designs have been examined by a number of authors. Some relevant references are Box and Tyssedal (1996), Deng and Tang (1999), Evangelaras, Georgiou, and Koukovinos (2003, 2004), Evangelaras and Koukovinos (2004a, b, 2006), Ito et al. (1981), Kimura (1989), Lin and Draper (1992, 1993, 1995), and Wang and Wu (1995). Note that projections of two-level designs can be used to investigate main effects and interactions of the factors retained, but cannot support a full quadratic model, for which at least three levels are mandatory. However, there are also economical three-level designs that can be used for screening when a second-order fit is desirable. These are discussed in Chapter 15, Section 15.5.

Table 5.9. The Two Characteristic Projections of a 12-Run Plackett and Burman Design into Five Dimensions

(a) With a pair of repeat runs

	1	2	3	4	5
1	−	−	−	−	−
2	−	−	−	−	−
3	−	−	+	+	+
4	−	+	−	+	+
5	−	+	+	−	+
6	−	+	+	+	−
7	+	−	−	+	+
8	+	−	+	−	+
9	+	−	+	+	−
10	+	+	−	−	+
11	+	+	−	+	−
12	+	+	+	−	−

(b) With a mirror-image pair of runs

	1	2	3	4	5
1	−	−	−	−	−
2	+	+	+	+	+
3	−	−	+	+	+
4	−	+	+	−	+
5	+	−	−	+	+
6	+	+	−	+	−
7	+	+	+	−	−
8	+	−	+	−	−
9	+	−	−	−	+
10	−	+	−	+	−
11	−	+	−	−	+
12	−	−	+	+	−

Supersaturated Designs

Designs that examine more factors than there are runs are called *supersaturated designs*. Such designs are not commonly used for screening, unless one can assume that only a very few of many variables will affect the response. For some comments on this, see the letter exchange in Lin (1995b). Such designs may also be useful in robustness studies, of course. Some selected references are Abraham et al. (1999), Booth and Cox (1962), Chen and Lin (1998), Cheng (1997), Iida (1994), Li and Lin (2003a), Li and Wu (1997), Lin (1993a, 1995a), Nguyen (1996a), Tang and Wu (1997), Westfall et al. (1998), Yamada and Lin (1997, 1999, 2002), Yamada and Matsui (2002), and Yamada et al. (1999, 2006).

5.10. EFFICIENT ESTIMATION OF MAIN EFFECTS AND TWO-FACTOR INTERACTIONS USING RELATIVELY SMALL TWO-LEVEL DESIGNS

Consider the investigation of k factors for which it is desired to estimate all k main effects and all $k(k-1)/2$ two-factor interactions efficiently, under the (usually tentative) assumption that " ≥ 3 fi = 0," that is, under the assumption that all interactions between three or more factors can be ignored. How few runs are

Table 5.10. Rechtschaffner's 22-Run Design for $k = 6$

	x_1	x_2	x_3	x_4	x_5	x_6
1	-1	-1	-1	-1	-1	-1
2	-1	1	1	1	1	1
3	1	-1	1	1	1	1
4	1	1	-1	1	1	1
5	1	1	1	-1	1	1
6	1	1	1	1	-1	1
7	1	1	1	1	1	-1
8	1	1	-1	-1	-1	-1
9	1	-1	1	-1	-1	-1
10	1	-1	-1	1	-1	-1
11	1	-1	-1	-1	1	-1
12	1	-1	-1	-1	-1	1
13	-1	1	1	-1	-1	-1
14	-1	1	-1	1	-1	-1
15	-1	1	-1	-1	1	-1
16	-1	1	-1	-1	-1	1
17	-1	-1	1	1	-1	-1
18	-1	-1	1	-1	1	-1
19	-1	-1	1	-1	-1	1
20	-1	-1	-1	1	1	-1
21	-1	-1	-1	1	-1	1
22	-1	-1	-1	-1	1	1

Run 1: all levels -1.
Runs 2–7: one level -1.
Runs 8–22: four levels -1.

needed using two-level designs? This is a complicated question, studied by Mee (2004), for $5 \leq k \leq 15$, and we summarize his conclusions in what follows. Counting an overall mean level, we need to estimate $m = 1 + k + k(k - 1)/2$ quantities in all, so at least that many runs are essential. Mee's recommended designs require n runs for k factors as shown in the following display:

Factors, k:	2	3	4	5	6	7	8	9	10	11	12	13	14	15
Estimates, m:	4	7	11	16	22	29	37	46	56	67	79	92	106	121
Runs, n:	4	8	16	16	22	29	48	64	64	96	128	128	128	128

The factorial main effects and two factor interaction estimates can be calculated using standard regression techniques by fitting the model:

$$y = \beta_0 + \beta_1 x_1 + \beta_2 x_2 + \cdots + \beta_k x_k + \beta_{12} x_1 x_2 + \beta_{13} x_1 x_3 + \cdots + \beta_{k-1,k} x_{k-1} x_k + \varepsilon$$

$$(5.10.1)$$

We now list the specific recommended designs. Note that, in general, there are many of these designs with different sign combinations, which can be produced by deciding on which factors, if any, one associates the $(-1, 1)$ levels with (high, low) rather than the more conventional (low, high). In particular, all the signs can be reversed. This is sometimes useful when a specific run shown in the tabled design cannot be achieved. For example, in some experiments, it is not practical to hold all factors at their minimum levels, or all at their maximum levels, simultaneously.

List of Recommended Designs

$2 \leq k \leq 4$: Use the full 2^k design.

$k = 5$: Use a 2_V^{5-1} resolution V design defined by $I = \mathbf{12345}$ or $I = -\mathbf{12345}$.

$k = 6$: Use Rechtschaffner's (1967, p. 570) saturated 22-run design, shown in Table 5.10. See also Mee (2004, p. 406).

$k = 7$: Use Tobias's (1966) saturated 29-run design, shown in Table 5.11.

$k = 8$: Use John's (1961, 1962) three-quarters of a 2_V^{8-2} design with 48 runs, shown in Table 5.12 [see also Addelman (1961) and Mee (2004, p. 405)]. Each of the three-quarters is a 2_{III}^{8-4} fractional factorial; see John (1962, p. 176).

$k = 9$: Look at the 64-run design recommended for $k = 10$ in Table 5.13, and omit the column for factor 2.

$k = 10$: Use the 64-run design in Table 5.13.

$k = 11$: Use Peter John's three-quarter replicate of a 2_V^{11-4} design with 96 runs. This design can be run in three blocks of 32 runs each.

$k = 12-15$: Use a 128-run nonregular design, omitting 3 columns for $k = 12$, omitting 2 columns for $k = 13$, and omitting 1 column for $k = 14$.

Designs for $k = 11-15$ are not given here. For more detail on those and for other related matters, see Mee (2004).

Table 5.11. Tobias's (1966) Saturated 29-Run Design for $k = 7$

	x_1	x_2	x_3	x_4	x_5	x_6	x_7
1	-1	-1	-1	-1	-1	-1	1
2	-1	-1	-1	-1	-1	1	-1
3	-1	-1	-1	-1	1	-1	-1
4	-1	-1	-1	1	-1	-1	-1
5	-1	-1	1	-1	-1	-1	-1
6	-1	1	-1	-1	-1	-1	-1
7	1	-1	-1	-1	-1	-1	-1
8	1	1	1	1	1	1	1
9	-1	-1	1	1	1	1	1
10	-1	1	-1	1	1	1	1
11	-1	1	1	-1	1	1	1
12	-1	1	1	1	-1	1	1
13	-1	1	1	1	1	-1	1
14	-1	1	1	1	1	1	-1
15	1	-1	-1	1	1	1	1
16	1	-1	1	-1	1	1	1
17	1	-1	1	1	-1	1	1
18	1	-1	1	1	1	-1	1
19	1	-1	1	1	1	1	-1
20	1	1	1	1	-1	-1	-1
21	1	1	1	-1	1	-1	-1
22	1	1	1	-1	-1	1	-1
23	1	1	1	-1	-1	-1	1
24	1	1	-1	1	1	-1	-1
25	1	1	-1	1	-1	1	-1
26	1	1	-1	1	-1	-1	1
27	1	1	-1	-1	-1	1	1
28	1	1	-1	-1	1	-1	1
29	1	1	-1	-1	1	1	-1

5.11. DESIGNS OF RESOLUTION V AND OF HIGHER RESOLUTION

When five or more factors are being considered, very useful designs of resolution V or higher are provided by half-fractions, the 2^{k-1} designs. For example, the 2^{5-1} design with generator $\mathbf{I} = \mathbf{12345}$ (i.e., we set $\mathbf{5} = \mathbf{1234}$ after writing down an initial 2^4 design in $\mathbf{1}$, $\mathbf{2}$, $\mathbf{3}$, and $\mathbf{4}$) is a 16-run design in which main effects are confounded only with four-factor interactions and two-factor interactions are confounded only with three-factor interactions. Thus, if the assumption that interactions between more than two factors are negligible is correct, the design allows unbiased estimation of all main effects and two-factor interactions. Designs of resolution five or more are of particular value as main building blocks of *composite designs* (presented later in this book) which allow estimation of all the terms in a second-degree polynomial approximation. For eight factors or more, one-quarter replicates can produce designs of resolution V. Some useful resolution V designs and associated blocking arrangements are shown in Table 5.14. Note that the design resolution has been attached as a subscript, for example, 2_V^{8-2}.

Table 5.12. Peter John's 48-Run Design for $k = 8$

	x_1	x_2	x_3	x_4	x_5	x_6	x_7	x_8
1	-1	-1	-1	-1	-1	1	1	-1
2	1	-1	-1	-1	1	1	-1	-1
3	-1	1	-1	-1	1	1	-1	1
4	1	1	-1	-1	-1	1	1	1
5	-1	-1	1	-1	-1	-1	-1	-1
6	1	-1	1	-1	1	-1	1	-1
7	-1	1	1	-1	1	-1	1	1
8	1	1	1	-1	-1	-1	-1	1
9	-1	-1	-1	1	-1	-1	-1	1
10	1	-1	-1	1	1	-1	1	1
11	-1	1	-1	1	1	-1	1	-1
12	1	1	-1	1	-1	-1	-1	-1
13	-1	-1	1	1	-1	1	1	1
14	1	-1	1	1	1	1	-1	1
15	-1	1	1	1	1	1	-1	-1
16	1	1	1	1	-1	1	1	-1
17	-1	-1	-1	-1	1	-1	1	-1
18	1	-1	-1	-1	-1	-1	-1	-1
19	-1	1	-1	-1	-1	-1	-1	1
20	1	1	-1	-1	1	-1	1	1
21	-1	-1	1	-1	1	1	-1	-1
22	1	-1	1	-1	-1	1	1	-1
23	-1	1	1	-1	-1	1	1	1
24	1	1	1	-1	1	1	-1	1
25	-1	-1	-1	1	1	1	-1	1
26	1	-1	-1	1	-1	1	1	1
27	-1	1	-1	1	-1	1	1	-1
28	1	1	-1	1	1	1	-1	-1
29	-1	-1	1	1	1	-1	1	1
30	1	-1	1	1	-1	-1	-1	1
31	-1	1	1	1	-1	-1	-1	-1
32	1	1	1	1	1	-1	1	-1
33	-1	-1	-1	-1	-1	-1	1	1
34	1	-1	-1	-1	1	-1	-1	1
35	-1	1	-1	-1	1	-1	-1	-1
36	1	1	-1	-1	-1	-1	1	-1
37	-1	-1	1	-1	-1	1	-1	1
38	1	-1	1	-1	1	1	1	1
39	-1	1	1	-1	1	1	1	-1
40	1	1	1	-1	-1	1	-1	-1
41	-1	-1	-1	1	-1	1	-1	-1
42	1	-1	-1	1	1	1	1	-1
43	-1	1	-1	1	1	1	1	1
44	1	1	-1	1	-1	1	-1	1
45	-1	-1	1	1	-1	-1	1	-1
46	1	-1	1	1	1	-1	-1	-1
47	-1	1	1	1	1	-1	-1	1
48	1	1	1	1	-1	-1	1	1

Table 5.13. Sixty-Four-Run Design for $k = 10$ and $k = 9$ (Omit x_2)

	x_1	x_2	x_3	x_4	x_5	x_6	x_7	x_8	x_9	x_{10}
1	-1	-1	-1	-1	1	-1	-1	-1	-1	1
2	-1	-1	-1	-1	1	-1	-1	-1	1	-1
3	-1	-1	-1	-1	1	1	1	1	-1	-1
4	-1	-1	-1	-1	1	1	1	1	1	1
5	-1	-1	-1	1	-1	-1	-1	-1	-1	1
6	-1	-1	-1	1	-1	-1	-1	1	-1	-1
7	-1	-1	-1	1	-1	1	1	-1	1	-1
8	-1	-1	-1	1	-1	1	1	1	1	1
9	-1	-1	1	-1	-1	-1	1	-1	1	1
10	-1	-1	1	-1	-1	-1	1	1	1	-1
11	-1	-1	1	-1	-1	1	-1	-1	-1	-1
12	-1	-1	1	-1	-1	1	-1	1	-1	1
13	-1	-1	1	1	-1	-1	1	-1	-1	-1
14	-1	-1	1	1	-1	-1	1	-1	-1	1
15	-1	-1	1	1	1	1	-1	1	-1	1
16	-1	-1	1	1	1	1	-1	1	1	-1
17	-1	1	-1	-1	-1	-1	1	-1	-1	-1
18	-1	1	-1	-1	-1	-1	1	1	-1	1
19	-1	1	-1	-1	-1	1	-1	-1	1	1
20	-1	1	-1	-1	-1	1	-1	1	1	-1
21	-1	1	-1	1	1	-1	1	1	-1	1
22	-1	1	-1	1	1	-1	1	1	1	-1
23	-1	1	-1	1	1	1	-1	-1	-1	-1
24	-1	1	-1	1	1	1	-1	-1	1	1
25	-1	1	1	-1	1	-1	-1	1	-1	-1
26	-1	1	1	-1	1	-1	-1	1	1	1
27	-1	1	1	-1	1	1	1	-1	-1	1
28	-1	1	1	-1	1	1	1	-1	1	-1
29	-1	1	1	1	-1	-1	-1	-1	1	-1
30	-1	1	1	1	-1	-1	-1	1	1	1
31	-1	1	1	1	-1	1	1	-1	-1	1
32	-1	1	1	1	-1	1	1	1	-1	-1
33	1	-1	-1	-1	-1	-1	-1	1	1	1
34	1	-1	-1	-1	-1	-1	1	1	1	-1
35	1	-1	-1	-1	-1	1	-1	-1	-1	-1
36	1	-1	-1	-1	-1	1	1	-1	-1	1
37	1	-1	-1	1	1	-1	-1	1	1	1
38	1	-1	-1	1	1	-1	1	-1	-1	-1
39	1	-1	-1	1	1	1	-1	1	1	-1
40	1	-1	-1	1	1	1	1	-1	-1	1
41	1	-1	1	-1	1	-1	-1	-1	1	-1
42	1	-1	1	-1	1	-1	1	1	-1	1
43	1	-1	1	-1	1	1	-1	-1	1	1
44	1	-1	1	-1	1	1	1	1	-1	-1
45	1	-1	1	1	-1	-1	-1	1	-1	-1
46	1	-1	1	1	-1	-1	1	1	-1	1
47	1	-1	1	1	-1	1	-1	-1	1	1
48	1	-1	1	1	-1	1	1	-1	1	-1
49	1	1	-1	-1	1	-1	-1	1	-1	-1

Table 5.13. (*Continued*)

	x_1	x_2	x_3	x_4	x_5	x_6	x_7	x_8	x_9	x_{10}
50	1	1	−1	−1	1	−1	1	−1	1	1
51	1	1	−1	−1	1	1	−1	1	−1	1
52	1	1	−1	−1	1	1	1	−1	1	−1
53	1	1	−1	1	−1	−1	−1	−1	1	−1
54	1	1	−1	1	−1	−1	1	−1	1	1
55	1	1	−1	1	−1	1	−1	1	−1	1
56	1	1	−1	1	−1	1	1	1	−1	−1
57	1	1	1	−1	−1	−1	−1	−1	−1	1
58	1	1	1	−1	−1	−1	1	−1	−1	−1
59	1	1	1	−1	−1	1	−1	1	1	−1
60	1	1	1	−1	−1	1	1	1	1	1
61	1	1	1	1	1	−1	−1	−1	−1	1
62	1	1	1	1	1	−1	1	1	1	−1
63	1	1	1	1	1	1	−1	−1	−1	−1
64	1	1	1	1	1	1	1	1	1	1

Table 5.14. Construction and Blocking of Some Designs of Resolution V and Higher so that No Main Effect or Two-Factor Interaction Is Confounded with Any Other Main Effect or Two-Factor Interaction

(1) Number of Variables	(2) Number of Runs	(3) Degree of Fractionation	(4) Type of Design	(5) Method of Introducing "New" Factors	(6) Blocking (with No Main Effect or 2fi Confounded)	(7) Method of Introducing Blocks
5	16	$\frac{1}{2}$	2_V^{5-1}	$\pm 5 = 1234$	Not available	
6	32	$\frac{1}{2}$	2_{VI}^{6-1}	$\pm 6 = 12345$	Two blocks of 16 runs	$B_1 = 123$
7	64	$\frac{1}{2}$	2_{VII}^{7-1}	$\pm 7 = 123456$	Eight blocks of 8 runs	$B_1 = 1357$ $B_2 = 1256$ $B_3 = 1234$
8	64	$\frac{1}{4}$	2_V^{8-2}	$\pm 7 = 1234$ $\pm 8 = 1256$	Four blocks of 16 runs	$B_1 = 135$ $B_2 = 348$
9	128	$\frac{1}{4}$	2_{VI}^{9-2}	$\pm 8 = 13467$ $\pm 9 = 23567$	Eight blocks of 16 runs	$B_1 = 138$ $B_2 = 129$ $B_3 = 789$
10	128	$\frac{1}{8}$	2_V^{10-3}	$\pm 8 = 1237$ $\pm 9 = 2345$ $\pm \overline{10} = 1346$	Eight blocks of 16 runs	$B_1 = 149$ $B_2 = 12\overline{10}$ $B_3 = 89\overline{10}$
11	128	$\frac{1}{16}$	2_V^{11-4}	$\pm 8 = 1237$ $\pm 9 = 2345$ $\pm \overline{10} = 1346$ $\pm \overline{11} = 1234567$	Eight blocks of 16 runs	$B_1 = 149$ $B_2 = 12\overline{10}$ $B_3 = 89\overline{10}$

Source: Table 5.14 is reproduced and adapted with permission from the table on p. 408 of *Statistics for Experimenters*: *An Introduction to Design, Data Analysis and Model Building*, by G. E. P. Box, W. G. Hunter, and J. S. Hunter, published by John Wiley & Sons, Inc., New York, 1978.

Table 5.15. Two-Level Fractional Designs for k Variables and N Runs (Numbers

N	Number of Variables k				
	3	4	5	6	7
4	2_{III}^{3-1} $\pm 3 = 12$				
8	2^2	2_{IV}^{4-1} $\pm 4 = 123$	2_{III}^{5-2} $\pm 4 = 12$ $\pm 5 = 13$	2_{III}^{6-3} $\pm 4 = 12$ $\pm 5 = 13$ $\pm 6 = 23$	2_{III}^{7-4} $\pm 4 = 12$ $\pm 5 = 13$ $\pm 6 = 23$ $\pm 7 = 123$
16	2^3	2^4 2 times	2_V^{5-1} $\pm 5 = 1234$	2_{IV}^{6-2} $\pm 5 = 123$ $\pm 6 = 234$	2_{IV}^{7-3} $\pm 5 = 123$ $\pm 6 = 234$ $\pm 7 = 134$
32	2^3 4 times	2^4 2 times	2^5	2_{VI}^{6-1} $\pm 6 = 12345$	2_{IV}^{7-2} $\pm 6 = 1234$ $\pm 7 = 1245$
64	2^3 8 times	2^4 4 times	2^5 2 times	2^6	2_{VII}^{7-1} $\pm 7 = 123456$
128	2^3 16 times	2^4 8 times	2^5 4 times	2^6 2 times	2^7
	↖ (16)	↖ (8)	↖ (4)	↖ (2)	↖ (1)

Source: Table 5.15 is reproduced and adapted with permission from the table on p. 272 of *Statistics for Experimenters,* by G. E. P. Box, J. S. Hunter, and W. G. Hunter, published by John Wiley & Sons, Inc., New

in Parentheses Represent Replication)

	Number of Variables k			
	8	9	10	11
4				
8				
16	2_{IV}^{8-4} $\pm 5 = 234$ $\pm 6 = 134$ $\pm 7 = 123$ $\pm 8 = 124$	2_{III}^{9-5} $\pm 5 = 123$ $\pm 6 = 234$ $\pm 7 = 134$ $\pm 8 = 124$ $\pm 9 = 1234$	2_{III}^{10-6} $\pm 5 = 123$ $\pm 6 = 234$ $\pm 7 = 134$ $\pm 8 = 124$ $\pm 9 = 1234$ $\pm \overline{10} = 12$	2_{III}^{11-7} $\pm 5 = 123$ $\pm 6 = 234$ $\pm 7 = 134$ $\pm 8 = 124$ $\pm 9 = 1234$ $\pm \overline{10} = 12$ $\pm \overline{11} = 13$
32	2_{IV}^{8-3} $\pm 6 = 123$ $\pm 7 = 124$ $\pm 8 = 2345$	2_{IV}^{9-4} $\pm 6 = 2345$ $\pm 7 = 1345$ $\pm 8 = 1245$ $\pm 9 = 1235$	2_{IV}^{10-5} $\pm 6 = 1234$ $\pm 7 = 1235$ $\pm 8 = 1245$ $\pm 9 = 1345$ $\pm \overline{10} = 2345$	2_{IV}^{11-6} $\pm 6 = 123$ $\pm 7 = 234$ $\pm 8 = 345$ $\pm 9 = 134$ $\pm \overline{10} = 145$ $\pm \overline{11} = 245$
64	2_{V}^{8-2} $\pm 7 = 1234$ $\pm 8 = 1256$	2_{IV}^{9-3} $\pm 7 = 1234$ $\pm 8 = 1356$ $\pm 9 = 3456$	2_{IV}^{10-4} $\pm 7 = 2346$ $\pm 8 = 1346$ $\pm 9 = 1245$ $\pm \overline{10} = 1235$	2_{IV}^{11-5} $\pm 7 = 345$ $\pm 8 = 1234$ $\pm 9 = 126$ $\pm \overline{10} = 2456$ $\pm \overline{11} = 1456$
128	2_{VIII}^{8-1} $\pm 8 = 1234567$	2_{VI}^{9-2} $\pm 8 = 13467$ $\pm 9 = 23567$	2_{V}^{10-3} $\pm 8 = 1237$ $\pm 9 = 2345$ $\pm \overline{10} = 1346$	2_{V}^{11-4} $\pm 8 = 1237$ $\pm 9 = 2345$ $\pm \overline{10} = 1346$ $\pm \overline{11} = 1234567$

↖ $(\frac{1}{128})$

↖ $(\frac{1}{64})$

↖ $(\frac{1}{32})$

↖ (1)　　↖ $(\frac{1}{2})$　　↖ $(\frac{1}{4})$　　↖ $(\frac{1}{8})$　　↖ $(\frac{1}{16})$

York, 2005. For additional details see that book and "Minimum aberration 2^{k-p} designs," by A. Fries and W. G. Hunter, *Technometrics*, **22**, 1980, 601–608. The numbers in parentheses show the replication factors in a "north–west" direction.

A Table of 2^{k-p} Designs

Extensive tables of 2^{k-p} designs of various resolutions are available; see, for example, Clatworthy et al. (1957). A short but useful listing is given in Table 5.15.

Calculation of Effects Using Yates' Method

In the foregoing discussion, we have shown how two-level fractional factorial designs requiring $2^{k-p} = 2^q$ runs may be generated by first writing down a complete 2^q factorial as a "core" design and then using interaction columns of the 2^q design to accommodate additional factors. To analyze the resulting fractional factorial, Yates' method may be used to compute the $2^q - 1$ contrasts *from the complete core factorial* in the usual way. The string of effects and interactions estimated by each of the $2^q - 1$ contrasts is then obtained from the alias pattern for the design (see also Berger, 1972).

5.12. APPLICATION OF FRACTIONAL FACTORIAL DESIGNS TO RESPONSE SURFACE METHODOLOGY

Suppose the expected response $E(y) = \eta$ is a function of k predictor variables $x_1, x_2, \ldots, x_k$, coded so that the center of the region of interest is at the origin $(0, 0, \ldots, 0)$. Consider a Taylor's series expansion

$$E(y) = \eta = \eta_0 + \sum_{i=1}^{k} \left[\frac{\partial \eta}{\partial x_i} \right]_0 x_i + \frac{1}{2} \sum_{i=1}^{k} \sum_{j=1}^{k} \left[\frac{\partial^2 \eta}{\partial x_i \, \partial x_j} \right]_0 x_i x_j + \cdots, \quad (5.12.1)$$

where the subscript zero indicates evaluation at the origin $(0, 0, \ldots, 0)$. If we ignore terms of order higher than the first, the expansion yields the first degree (or first-order, or planar) approximation

$$\eta = \beta_0 + \sum_{i=1}^{k} \beta_i x_i. \quad (5.12.2)$$

If, in addition, we retain terms of second degree, we obtain the second-degree (or second-order, or quadratic) approximation

$$\eta = \beta_0 + \sum_{i=1}^{k} \beta_i x_i + \sum_{i=1}^{k} \sum_{j \geq i}^{k} \beta_{ij} x_i x_j. \quad (5.12.3)$$

We shall use these approximations a great deal in subsequent chapters and will refer to designs suitable for collecting data to estimate the β parameters in Eqs. (5.12.2) and (5.12.3) as *first-order designs* and *second-order designs*, respectively.

Clearly, if two-level fractional factorial designs are to be used as first-order designs, they must have resolution of a least III so that, if model (5.12.2) is valid, the main effect contrasts of the design will provide estimates of all the β_i unconfounded with one another.

When a two-level design is used as a building block in second-order composite designs, it is usually* of resolution at least V. Then, if model (5.12.3) is valid, the main effect and two-factor interaction contrasts of the two-level design provide estimates of all main effect coefficients β_i and of all interaction coefficients β_{ij} ($i \neq j$), unconfounded with one another. (See also Sections 9.2, 13.8, 14.3, and 15.3.)

Because we are never sure about the adequacy of an approximation of selected degree (or order), we shall need to check lack of fit as well as to estimate the model parameters. Resolution III designs that are not saturated can allow estimation of suspect interactions. Also (as we shall see in Chapter 9) the addition of center points makes possible a general curvature check. Resolution IV designs provide safer first-order designs because first-order coefficients will not then be confused with any two-factor interactions. Although the latter are ignored in a first-order fitting, they may yet exist; a resolution IV design will enable combinations of them to be estimated as a check of lack of fit, if desired.

5.13. PLOTTING EFFECTS FROM FRACTIONAL FACTORIALS ON PROBABILITY PAPER

To distinguish real effects from noise, estimates of effects from fractional factorial designs may be plotted on normal probability paper (see Appendix 4B) in just the same way as for full factorial designs.

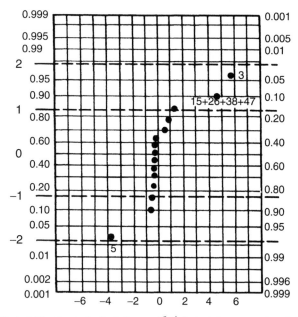

Figure 5.2. Plot of 15 estimated effects from a 2^{8-4} factorial on normal probability paper.

*However, also see Section 15.5.

Numerical Example, 2_{IV}^{8-4} Design

Sixteen observations were obtained from a 2_{IV}^{8-4} design given by Box et al. (1978, Table 12.11). The mean of these was $\bar{y} = 19.75$ and there were 15 estimates in addition, namely of:

1. Eight main effects: -0.7, -0.1, 5.5, -0.3, -3.8, -0.1, 0.6, 1.2.
2. Seven sets of four two-factor interactions: -0.6, 0.9, -0.2, -0.4, -0.6, -0.3, 4.6.

These effects are plotted on a normal probability scale in Figure 5.2. The plot suggests that two main effects (of variables 3 and 5) and one combination of two-factor interactions ($15 + 26 + 38 + 47$) are distinguishable from the noise.

EXERCISES

5.1. Are the following statements true or false, for a 2_R^{k-p} design?
   (a) $k \geq R$
   (b) $k \geq R + p$
   Give, in a few words, reasons for your answers.

5.2. A 2_{III}^{5-2} design with generators $\mathbf{I} = \mathbf{1234} = \mathbf{135}$ is run.
   (a) If it is *known* that factors 1, 2, and 3 all act independently of one another, and that factors 3, 4, and 5 all act independently of one another, what (ignoring interactions of three or more factors) can we estimate?
   (b) Can you suggest a better 2_{III}^{5-2} design for these factors, or is the one given the best available?

5.3. For the construction of a one-quarter replicate of a 2^5 factorial, that is, a 2^{5-2} factorial, two generating relations are proposed:
   (a) $\mathbf{I} = \mathbf{ABCDE} = \mathbf{BCDE}$
   (b) $\mathbf{I} = \mathbf{ABC} = \mathbf{BCDE}$
   Which design would you prefer, and why?

5.4. (a) Using the principle of foldover, show how the 2_{III}^{7-4} design generated by

$$\mathbf{I} = \mathbf{1234} = \mathbf{125} = \mathbf{146} = \mathbf{247}$$

   can provide a 2_{IV}^{8-4} design.
   (b) What are the generators of the new design?
   (c) If this second design is again folded, but the $\mathbf{I}$ column is *not* folded with it this time, what type is the final design obtained?

5.5. Consider the 2_{IV}^{8-4} design generated by $\mathbf{I} = \mathbf{1235} = \mathbf{1246} = \mathbf{1347} = \mathbf{2348}$.
   (a) By using a column that normally provides an estimate of the sum of four two-factor interactions, split the design into two blocks.

(b) Retaining two-factor interactions of all kinds, show what estimates are available.

(c) If the block variable does not interact, show what estimates are available.

5.6. (a) Obtain a design for examining six variables in four blocks each containing four runs.

(b) What are the generators for the design (including blocking)?

(c) Find the defining relation (including blocking).

(d) Show what estimates can be found, assuming interactions between three or more factors are zero and that block variables do not interact with ordinary variables.

(*Hint:* Consider the 2_{IV}^{6-2} design generated by $I = 1235 = 1246$. Associate $B_1 = 12$, $B_2 = 13$, so that $B_1 B_2 = 23$. The design is then generated by $I = 1235 = 1246 = 12B_1 = 13B_2$, including blocking. In developing the estimates, remember that although $B_1 B_2 5$, for example, looks like a *three*-factor interaction, it must be regarded as a *two*-factor interaction between the blocking variable $B_1 B_2$ and the ordinary variable **5**.)

(e) Is the choice suggested in the hint a good one? Explain the reasons for your answer. If you decide that better choices exist, provide one.

5.7. (a) Obtain a design for examining six variables in eight blocks each containing two runs so that the estimates of main effects are not influenced by block effects.

(b) What are the generators for the design (including blocking)?

(c) Find the defining relation.

(d) Show what estimates can be found assuming interactions between three or more factors are zero and that block variables do not interact with ordinary variables.

(*Hint:* Consider the 2_{IV}^{6-2} design generated by $I = 1235 = 1246$. Associate $B_1 = 12$, $B_2 = 13$ (so that $B_1 B_2 = 23$), and $B_3 = 14$. Then $B_1 B_3 = 24$, $B_2 B_3 = 34$, and $B_1 B_2 B_3 = 1234$. The blocks are now defined. Remember that, for example, $B_1 B_2 B_3 5$ has the status of a two-factor interaction.)

5.8. An experimenter begins an investigation with a 2^{5-2} design using generators $4 = 123$ and $5 = 23$. Later he performs the eight foldover runs obtained by reversing all the signs in the five-variable design table.

(a) What is the resolution of the combined 16-run design? What generators would produce this design?

(b) What better design might the experimenter have used had he known in advance that 16 runs would be made?

5.9. Construct a 16-run two-level fractional factorial design in six factors such that all main effects and all two-factor interactions associated with factor **1** are estimable.

5.10. If the design $I = 124 = 135 = 236$ is followed by the design in which variables **1** and **2** have their signs switched, what is the alias structure of the 16-run design so formed?

5.11. An experimenter needs a 16-run two-level fractional factorial design that will estimate the main effects $1, 2, \ldots, 6$ of six variables as well as the interactions 12, 123, 13, 14, 15, 16, 23, 24 between them, assuming all interactions of three or more factors can be ignored except for 123. Does such a design exist? If so, what is its defining relation?

5.12. Verify that the 2^{8-4} resolution IV design can be used in all the following alternative ways. (Assume that all interactions between three or more factors can be neglected.)

(a) As a "main effects" design to obtain the main effects of eight factors clear of all two factor interactions.

(b) As in (a) but blocked into two, or four, or eight blocks.

(c) If there are three major variables, all of whose main effects, two-factor interactions, and three-factor interaction are needed, and eight minor variables all of whose interactions can be neglected, the design can be used to estimate all these. (*Hint:* Look at Eq. (5.4.13) and associate each $\mathbf{B}_i$ with a major variable $\mathbf{M}_i$, $i = 1, 2, 3$. Remember that all the two-factor interactions involving numbers, e.g., 12, 13, etc., are zero. Take the numbered variables as the minor variables.)

(d) As in (c) but with eight minor variables, two major variables, and in two blocks of eight runs each. (*Hint:* Replace $\mathbf{B}_1$, $\mathbf{B}_2$ and $\mathbf{B}_3$ in Eq. (5.4.13) by $\mathbf{M}_1$, $\mathbf{M}_2$, and $\mathbf{B}$.)

(e) As a design for eight minor variables and one major variable in four blocks of four runs each.

(f) As a screening design that will provide a complete factorial, twice over, in *any* set of three variables, if the other five variables have no effect. (*Hint:* If five variables are not effective we can "cross them out" in the defining relationship that leads to the alias structure. Since no words of length three or smaller appear in this—because the design is of resolution IV—all words are "deleted," no matter which five variables are chosen, implying that a full factorial design remains in the other three variables. Because the original design had 16 runs, the 2^3 is necessarily replicated.)

(g) As a screening design that will provide a complete factorial in some sets of four variables, or a 2^{4-1}, twice over, in all other sets of four variables, provided the remaining four variables *not* in the considered set have no effect. (*Hint:* If the chosen four variables form a word in the defining relationship, a replicated 2^{4-1} occurs. Otherwise we get a 2^4.)

5.13. Select a set of suitable generators for a 2^{10-3}_{V} design and write out the defining relation which provides the alias structure. Select a suitable blocking generator for division into two blocks. (*Hint:* See Table 5.7.)

5.14. A 2^{5-1}_{V} design was used to examine five factors A, B, C, D, and E. It was assumed that all interactions involving three or more variables could be neglected. From the resulting data, the following estimates (rounded to the nearest unit) were obtained:

$A \leftarrow$	0	$AB \leftarrow$	17	$BD \leftarrow$	-4
$B \leftarrow$	-8	$AC \leftarrow$	3	$BE \leftarrow$	8
$C \leftarrow$	14	$AD \leftarrow$	5	$CD \leftarrow$	14
$D \leftarrow$	67	$AE \leftarrow$	-2	$CE \leftarrow$	-5
$E \leftarrow$	4	$BC \leftarrow$	8	$DE \leftarrow$	0

Plot these results on normal probability paper and identify effects that appear to be important. Remove these, replot the rest, and estimate σ, where $V(y_i) = \sigma^2$. See Appendix 4B for additional details.

5.15. Identify the design given in Table E5.15 and perform an appropriate analysis. Assume $V(y_i) = \sigma^2 = 2$.

Table E5.15

x_1	x_2	x_3	x_4	y
1	-1	-1	-1	105
-1	1	-1	-1	107
-1	-1	1	-1	102
-1	-1	-1	1	104
1	1	1	-1	114
1	1	-1	1	111
1	-1	1	1	105
-1	1	1	1	107

What factors appear to influence the response? What features need to be further elucidated? If you were allowed to perform just four more runs, what would they be? Why?

5.16. What design is given in Table E5.16? Perform an appropriate factorial analysis of the data.

Table E5.16

x_1	x_2	x_3	x_4	x_5	y
-1	-1	1	-1	-1	18, 15
1	-1	-1	-1	1	16, 18
-1	1	-1	-1	1	18, 17
1	1	1	-1	-1	15, 16
-1	-1	1	1	1	17, 19
1	-1	-1	1	-1	31, 30
-1	1	-1	1	-1	18, 18
1	1	1	1	1	30, 27

Would you like to have additional data to clarify what you have seen in this experiment? If four more runs were authorized what would you suggest? Why? If eight more runs were authorized what would you suggest? Why?

5.17. Consider a 2_V^{5-1} fractional factorial design defined by $\mathbf{I} = -\mathbf{12345}$. Suppose we wish to choose two blocking variables $\mathbf{B}_1$ and $\mathbf{B}_2$ so that the design can be divided into four blocks. Can this be done in such a way that all main effects and all two-factor interactions are *not* confounded with blocks? If your answer is yes, write out the design and show what estimates can be made.

5.18. (a) You have convinced an experimenter that she should use a 2^{7-1} design. Each run takes about 20 min, and 16 runs is the most she wishes to attempt during one day's work. Thus she wishes the design divided into four blocks of 16 runs each. Assume that the design will be generated using $\mathbf{I} = \mathbf{1234567}$ and blocked using blocking generators $\mathbf{B}_1 = \mathbf{1357}$, $\mathbf{B}_2 = \mathbf{1256}$. Write out the 16 runs in the $(\mathbf{B}_1, \mathbf{B}_2) = (+, +)$ block.

(b) After these 16 runs have been completed, the experimental unit suffers a breakdown. It will not be repaired until after the monthly meeting at which the experimenter has to make a report. Assuming that all interactions involving three or more of the seven factors are negligible, and that blocks do not interact with the factors, write out the 16 quantities that can be estimated using only the 16 runs currently available.

5.19. (*Source:* Toby J. Mitchell.) Ozzie Cadenza, holder of a Ph.D. in Statistics from the University of Wisconsin, and now owner and manager of Ozzie's Bar and Grill, recently decided to study the factors that influence the amount of business done at his bar.

At first, he did not know which factors were important and which were not, but he drew up the following list of six, which he decided to investigate by means of a fractional factorial experiment:

1. The amount of lighting in the bar.
2. The presence of free potato chips and chip dip at the bar.
3. The volume of the juke box.
4. The presence of Ozzie's favorite customer, a young lady by the name of Rapunzel Freeny. Miss Freeny was a real life-of-the-party type, continually chatting with the customers, passing around the potato chips, and so on, all of which made Ozzie feel that she had a real effect on the amount of his bar business.
5. The presence of a band of roving gypsies, who had formed a musical group called the Roving Gypsy Band and who had been hired by Ozzie to play a limited engagement there.
6. The effect of the particular bartender who happened to be on duty. There were originally three bartenders, Tom, Dick, and Harry, but Harry was fired so that each factor in the experiment would have only two levels.

Plus and minus levels were assigned to these six factors as follows:

	−	+
1.	Lights are dim	Lights are bright
2.	No chips are at the bar	Chips are at the bar
3.	Juke box is playing softly	Juke box is blaring loudly
4.	Miss Freeny stays at home	Miss Freeny is there
5.	Gypsies are not there	Gypsies are there
6.	Tom is the bartender	Dick is the bartender

Ozzie decided to perform one "run" every Friday night during the cocktail hour (4:30 to 6:30 P.M.). He thought he should try a fractional factorial with as few runs as possible, since he was never quite sure just when the Roving Gypsy Band would pack up and leave. He finally decided to use a member of the family of 2_{III}^{6-3} designs with principal generating relation $\mathbf{I} = \mathbf{124} = \mathbf{135} = \mathbf{236}$. He had wanted to find a resolution III design which would be such that the juke box would never be blaring away while the gypsies were playing but found this requirement to be impossible.

(a) Why?

He did insist, however, that in *no* run of the experiment could variables **1**, **3**, and **5** attain their plus levels simultaneously. This restriction was made necessary by the annoying tendency of all the lights to fuse whenever the gypsies plugged in their electric zither at the same time the lights and the juke box were on full blast. Note that this restriction made it impossible for the principal member of the chosen family to be used.

(b) What members of the given family *does* this restriction allow?

Ozzie settled on the generating relation. $\mathbf{I} = \mathbf{124} = -\mathbf{135} = \mathbf{236}$. The design matrix and the "response" (income in dollars) corresponding to each run are given in Table E5.19.

(c) Assuming third- and higher-order interactions are negligible, write down the estimates obtained from this experiment and tell what they estimate.

(d) In a few sentences, tell what fraction you would perform next and why.

Table E5.19

1	2	3	4 = 12	5 = −13	6 = 23	y
−	−	−	+	−	+	265
+	−	−	−	+	+	155
−	+	−	−	−	−	135
+	+	−	+	+	−	205
−	−	+	+	+	−	195
+	−	+	−	−	−	205
−	+	+	−	+	+	125
+	+	+	+	−	+	315

Table E5.20

Run Number	\multicolumn{24}{c}{Factor Number}																								Response y
	1	2	3	4	5	6	7	8	9	10	11	12	13	14	15	16	17	18	19	20	21	22	23	24	
1	+	+	+	−	−	−	+	+	+	+	+	−	+	−	−	+	+	+	+	+	−	−	−	+	133
2	+	+	−	−	−	−	+	+	+	+	+	+	+	−	+	−	+	−	−	+	+	+	+	−	49
3	+	−	−	+	−	−	−	+	−	+	−	−	+	−	+	+	−	+	+	−	+	+	−	−	62
4	−	−	+	+	+	+	−	−	−	+	−	+	−	+	+	+	+	+	+	+	+	+	+	+	45
5	+	+	−	−	+	+	−	−	+	+	+	+	+	+	−	−	+	−	−	+	−	−	+	+	88
6	+	+	−	+	+	+	−	−	−	+	−	+	+	−	+	+	+	+	−	+	−	−	−	−	52
7	−	−	+	+	+	+	−	−	−	−	−	−	+	−	+	−	−	−	−	−	+	+	+	−	300
8	−	−	+	+	+	+	+	−	+	−	+	−	+	+	−	−	+	+	+	+	+	+	+	+	56
9	−	+	+	+	+	+	+	+	+	+	−	+	+	+	+	+	−	+	+	−	−	+	+	−	47
10	+	−	−	−	+	−	+	−	−	−	−	−	−	+	−	+	+	+	+	−	+	−	−	−	88
11	+	−	+	+	+	−	+	+	+	+	+	−	−	+	−	+	−	+	−	+	+	+	−	+	116
12	−	−	+	+	−	−	−	+	−	+	−	−	−	−	−	−	+	−	+	−	−	−	−	−	83
13	−	−	−	−	+	+	−	−	+	+	+	+	+	−	+	−	+	+	+	+	−	−	+	+	193
14	+	+	+	−	+	+	−	+	+	+	+	+	+	+	+	+	−	+	+	−	+	+	+	+	230
15	−	+	+	+	+	+	+	+	+	+	+	−	−	+	+	+	+	+	+	+	+	+	+	+	51
16	+	−	+	−	−	−	−	+	−	−	+	+	+	−	+	−	−	−	−	−	−	−	−	+	82
17	−	+	+	−	−	−	−	−	−	−	+	+	+	−	+	+	−	+	+	+	+	+	−	+	32
18	+	+	−	+	+	+	+	+	+	−	+	+	−	−	−	+	−	+	+	+	+	+	+	+	58
19	+	+	−	+	+	+	+	+	+	+	+	+	+	+	+	+	−	+	+	+	+	+	+	+	201
20	+	−	+	+	−	−	−	−	−	−	+	−	+	+	+	+	−	−	−	+	+	−	−	−	56
21	+	+	+	+	−	−	−	−	−	−	+	+	+	+	+	+	−	−	−	+	+	+	+	+	97
22	−	+	−	+	+	+	+	+	+	−	−	+	+	+	−	−	−	−	−	+	+	+	+	+	53
23	+	+	−	−	+	+	+	+	+	−	−	+	+	+	−	+	−	−	−	−	+	+	+	+	276
24	+	+	−	+	+	+	+	+	+	−	−	+	+	+	+	+	+	−	−	+	+	+	+	+	145
25	+	−	+	+	−	+	+	+	+	−	−	+	+	+	−	−	−	−	−	+	+	−	−	−	130
26	−	+	−	−	+	+	−	+	−	−	−	+	−	+	+	+	−	−	−	+	−	+	+	−	55
27	+	+	+	+	−	+	+	+	+	+	+	+	−	+	−	+	−	−	−	+	−	+	+	−	160
28	−	−	−	+	−	+	−	+	−	+	+	−	+	+	−	−	−	−	−	−	−	−	−	−	127
CP	−365	−281	−95	511	175	−165	−189	415	213	199	−51	−221	−321	−433	1209	−321	−599	−191	−49	−683	117	−451	−167	−173	3065
EF	−26	−20	−7	37	13	−12	−14	30	15	14	−4	−16	−23	−31	−86	−23	−43	−14	−4	−49	−8	−32	−12	−12	109

[a] The level of run 8, factor 20 is here shown as +. In the source reference, the lower level appears; that seems to be a typographical error.

Being partial to Miss Freeny and encouraged by the results of the first fraction, Ozzie chose a second fraction which would give unaliased estimates of her and each of her interactions. The results of this second fraction, given with variables **1**, **2**, and **3** in standard order, were 135, 165, 285, 175, 205, 195, 295, 145.

(e) Write down the estimates obtained by combining the results of both fractions.

(f) In the light of these results, was the choice of the second fraction a wise one?

(g) Offer a brief conjecture which might explain the presence and direction of the interactions involving Miss Freeny.

5.20. (*Source:* Designed experiments, by K. R. Williams, *Rubber Age*, **100**, August 1968, 65–71.) Table E5.20 shows a 28-run Plackett and Burman type design for 24 variables together with the responses obtained from each run. Except for the last column, the CP row (cross-product) is the result of taking CPs of each column of signs with the y values and the EF row ("effect") is the CP row divided by 14 and rounded. In the last column, $\Sigma y = 3065$ and $\bar{y} = 109$. Confirm one or more of the effects calculations. If, a priori, it were believed that at least one-third of the factors were ineffective, but not known which factors these were, which factors would be regarded as effective ones?

5.21. A 2^3 factorial design has been partially replicated as shown in Table E5.21. Analyze the data using regression methods, with a "full factorial analysis" type model.

Table E5.21

x_1	x_2	x_3	y
−	−	−	46
+	−	−	61, 57
−	+	−	37, 45
+	+	−	56
−	−	+	37, 41
+	−	+	68
−	+	+	33
+	+	+	66, 68

5.22. (*Source:* Sucrose-modified phenolic resins as plywood adhesives, by C. D. Chang and O. K. Kononenko, *Adhesives Age*, **5**(7), July 1962, 36–40. Material reproduced and adapted with the permission of *Adhesives Age*. The data were obtained under a grant from the Sugar Research Foundation, Inc., now the World Sugar Research Organization, Ltd.) The object of a research study was to improve both the dry and wet strip-shear strength in pounds per square inch (psi) of Douglas fir plywood glued with a resin modified with sugar. Here, we shall examine only the dry strength data, however.

Table E5.22a. Levels of the Six Predictor Variables Examined

Variable (Units), Designation	Level		Coding
	-1	1	
Sucrose (g), S	43	71	$x_1 = (S - 57)/14$
Paraform (g), P	30	42	$x_2 = (P - 36)/6$
NaOH (g), N	6	10	$x_3 = (N - 8)/2$
Water (g), W	16	20	$x_4 = (W - 18)/2$
Temperature maximum (°C), T	80	90	$x_5 = (T - 85)/5$
Time at T°C, t	25	35	$x_6 = (t - 30)/5$

Table E5.22b. The Design and the Response Values

x_1	x_2	x_3	x_4	x_5	x_6	y
-1	-1	-1	-1	-1	-1	162^a
1	-1	-1	-1	1	-1	146
-1	1	-1	-1	1	1	182
1	1	-1	-1	-1	1	133
-1	-1	1	-1	1	1	228
1	-1	1	-1	-1	1	143
-1	1	1	-1	-1	-1	223
1	1	1	-1	1	-1	172
-1	-1	-1	1	-1	1	168
1	-1	-1	1	1	1	128
-1	1	-1	1	1	-1	175
1	1	-1	1	-1	-1	186
-1	-1	1	1	1	-1	197
1	-1	1	1	-1	-1	175
-1	1	1	1	-1	1	196
1	1	1	1	1	1	173

^{a}Shown as 172 in the source reference; however, according to Dr. O. K. Kononenko, the 172 may be a typographical error.

The first part of the investigation involved six reaction variables $x_1, x_2, \ldots, x_6$, each examined at two levels coded to -1 and 1. The variables are shown in Table E5.22a, together with the coding used for each variable.

A 16-run 2_{IV}^{6-2} design generated by $\mathbf{I} = \mathbf{1235} = \mathbf{2346}$ (so that $\mathbf{I} = \mathbf{1456}$, also) was performed, and it provided the responses in Table E5.22b. (The y values are average values of six to eight samples per run; the fact that this provides observations whose variances may vary slightly is ignored in what follows.) Five additional center-point runs at $(0, 0, \ldots, 0)$ provided a pure error estimate $s_e^2 = 95.30$ for $V(y) = \sigma^2$.

Perform a factorial analysis on these data, assuming all interactions involving three or more factors can be neglected, and so confirm the results in Table E5.22c. Note, in the left column of that table, that "other" denotes groups of interactions of order three or more. These should represent "error" if interactions between three or more factors are negligible, and so

Table E5.22c. The Estimated Factorial Effects from the
2^{6-2} Experiment

Effect Name	Estimate
1	-34.375^a
2	11.625
3	28.375^a
4	1.125
5	1.875
6	-10.625
12 + 35	6.375
13 + 25	-10.875
14 + 56	15.875^a
15 + 23 + 46	-6.375
16 + 45	-14.875^a
24 + 36	3.875
26 + 34	-7.375
other (124 +)	6.125
other (134 +)	6.875
s.e.	4.88
2.78 (s.e.)	13.57

^{a}Significant at $\alpha = 0.05$.

should not be statistically significant under this tentative assumption. The designation (124 +) stands for (124 + 345 + 136 + 2456) while (134 +) represents (134 + 245 + 126 + 3456). The superscript a's indicate estimated effects that exceed 2.78(s.e.) in modulus. (The standard error is based on five average readings, and thus on 4 df. Ninety-five percent of the t_4 distribution lies within ± 2.78 so that estimates exceeding 2.78 (s.e.) in modulus are those whose true values are tentatively judged to be nonzero.)

At this stage, it was decided to explore outward on a path of steepest ascent, ignoring the two significant interactions and this was successful, leading to the following tentative conclusions:

1. Variables x_5 and x_6 have shown little effect and may be dropped in future runs.

2. A good balance of dry and wet strength values was achieved around $(S, P, N, W) = (47, 36.6, 8.8, 23.8)$, and this point will be the center for a subsequent experiment.

3. A 27-run three-level rotatable design with blocking, equivalent to a rotated "cube plus star" design (see Section 15.4 and/or Box and Behnken, 1960a) with added center points, will be performed next, and a second-order surface will be fitted to the y values thus obtained. (For the continuation of this work, see Exercise 7.23. For steepest ascent, see Chapter 6.)

5.23. (Source: A first-order five variable cutting-tool temperature equation and chip equivalent, by S. M. Wu and R. N. Meyer, Journal of the Engineering

Table E5.23. Coded Cutting Conditions and Temperature Results

Testing Order	x_1	x_2	x_3	x_4	x_5	Temperature Results T_1	T_2
4	-1	-1	-1	-1	1	462	498
16	1	-1	-1	-1	-1	743	736
9	-1	1	-1	-1	-1	714	681
8	1	1	-1	-1	1	1070	1059
11	-1	-1	1	-1	-1	474	486
5	1	-1	1	-1	1	832	810
1	-1	1	1	-1	1	764	756
15	1	1	1	-1	-1	1087	1063
14	-1	-1	-1	1	-1	522	520
6	1	-1	-1	1	1	854	828
2	-1	1	-1	1	1	773	756
10	1	1	-1	1	-1	1068	1063
7	-1	-1	1	1	1	572	574
12	1	-1	1	1	-1	831	856
13	-1	1	1	1	-1	819	813
3	1	1	1	1	1	1104	1092

Industry, ASME Transactions Series B, **87**, 1965, 395–400. Adapted with the permission of the copyright holder, The American Society of Mechanical Engineers, Copyright © 1965.) Fit a first-order model $y = \beta_0 + \beta_1 x_1 + \cdots + \beta_5 x_5 + \varepsilon$ by least squares to the data in Table E5.23. The design is a 2_V^{5-1} design generated by $x_1 x_2 x_3 x_4 x_5 = 1$ where the x's are coded predictor variables. The actual response is T (temperature in °F) but the model should be fitted in terms of $y = \ln T$. Treat the two values at each set of conditions as repeat runs for pure error purposes. Test the model for lack of fit and, if none is revealed, test if all x's are needed. Decide what model is suitable for these data, check the residuals, especially against the testing order, and interpret what you have done. (Some complications stemming from blocking have been ignored in this exercise modification; see the source reference for additional details.)

5.24. Is it possible to obtain a 2_R^{k-p} design which provides clear estimates of the main effects of nine variables in 16 runs, assuming all interactions involving three or more variables are zero? If yes, provide the design. If no, provide the best 16-run design in the circumstances and specify how many two-factor interactions would need to be zero to achieve the requirement of clear estimates of main effects.

5.25. Variables **4** and **7** are dropped from the 2_{III}^{7-4} design generated by **I** = **124** = **235** = **136** = **1237**. What is the resulting design?

5.26. The data in Table E5.26 were obtained from the records of a medium-sized dairy company seeking to improve sales. Explain what design is being used, analyze the data, and suggest what sensible next step(s) should be considered by the company.

Table E5.26

1	2	3	4	5	Response
Daily Deliveries?	Roundsman Wears Uniform?	Older Man?	City Round?	Delivery Before Noon?	Sales Rating
No	No	Yes	Yes	No	4
No	No	Yes	No	Yes	3
Yes	Yes	Yes	Yes	Yes	0
No	Yes	No	Yes	No	3
Yes	No	No	Yes	Yes	0
Yes	Yes	Yes	No	No	4
Yes	No	No	No	No	4
No	Yes	No	No	Yes	3

Table E5.27

Variable	$(-1, 1)$
1: Foreman	(absent, present)
2: Sex of packer	(man, woman)
3: Time of day	(morning, afternoon)
4: Temperature	(normal, high)
5: Music	(none, piped in)
6: Age of packer	(under 25, 25 or over)
7: Factory location	(Los Angeles, New York)

5.27 **(a)** A 2_{III}^{7-4} design with generators $I = -126 = 135 = -237 = -1234$ is used in a factory investigation of the time taken to pack 100 standard items. The variables investigated are given in Table E5.27. With variables **1**, **2**, and **3** in standard order, the observed packing times were, in minutes, 46.1, 55.4, 44.1, 58.7, 56.3, 18.9, 46.4, 16.4. Perform a standard analysis and state tentative conclusions.

(b) It was decided to perform a second fractional factorial, obtained from the first by reversing the signs of all variables. For these eight experiments, the packing times were (with **1**, **2** and **3** in order $+ + +$, $- + +$, $+ - +$, $- - +$, $+ + -$, $- + -$, $+ - -$, $- - -$), 41.8, 40.1, 61.5, 37.0, 22.9, 34.1, 17.7, 42.7. Perform a standard analysis.

(c) Combine the two fractions given above to obtain estimates of main effects and two factor interactions. Comment on the results. If you were the factory owner, what would you do?

5.28. An industrial spy is disturbed while photographing experimental records. On later inspection, he discovers he has only seven runs of a 2_R^{k-p} design, given in Table E5.28. How can he estimate the effects from the incomplete data? Do this. The information from these tests is vital to the spy's employer who is prepared to run another 2^{k-p} block if the results from the photographed data indicate this to be a worthwhile endeavor. Should additional tests be run? If so, what would you recommend?

Table E5.28

Variable Number						
1	**2**	**3**	**4**	**5**	**6**	Response, y
+	+	−	−	−	−	29
−	+	+	−	+	−	27
+	−	−	+	+	−	42
−	+	−	+	−	+	0
−	−	−	−	+	+	30
+	+	+	+	+	+	0
−	−	+	+	−	−	39

5.29. Construct the Plackett and Burman design for $n = 20$.

5.30. Construct the Plackett and Burman design for $n = 8$ and identify it as a 2_{III}^{k-p} design. (*Hint:* Generate the design from the seven signs $+ + + - + - -$ and choose column numbers $1, 2, \ldots, 7$ to correspond to the order of the given signs. Rearrange the rows by setting columns 1, 2, and 3 in standard order. Then see how columns 4, 5, 6, and 7 may be obtained from products of columns 1, 2, and 3. Hence provide a set of generators for the design.

5.31. (*Source:* John Charlton.) Consider the standard form 12-run Plackett and Burman design displayed in Table 5.7. This design was used to examine four factors, allocated to columns 1–4, in a study of water quality. The observations, in the run order of the table, were as follows: 700, 3600, 949, 1700, 1895, 2293, 1361, 2140, 550, 3600, 1627, 1100. Previous experience indicated that the factors should exert no significant effects. However, it was reported that factors 1, 2, and 3 were all significant! Enquiries elicited that a model $y = \beta_0 + \beta_1 x_1 + \beta_2 x_2 + \beta_3 x_3 + \beta_4 x_4 + \varepsilon$ had been fitted to the data and that the regression program had shown significant values for the estimated coefficients of these variables.

(a) Verify this calculation.

(b) Now reanalyze the data by fitting a first-order model (with intercept added) to all 11 columns and plot all the contrasts in a normal probability plot, being sure not to plot the intercept, which is not a contrast. What do you conclude now?

(c) In view of what happens in (b), use the bias formula $E(\mathbf{b}) = \boldsymbol{\beta} + (\mathbf{X}'\mathbf{X})^{-1}\mathbf{X}'\mathbf{X}_2\boldsymbol{\beta}_2 = \boldsymbol{\beta} + \mathbf{A}\boldsymbol{\beta}_2$ to find out what combinations of main effects and two-factor interactions are being estimated by the 11 columns apart from the mean estimate. In this notation, $\boldsymbol{\beta}$ contains the main effect parameters, the β_i, $i = 1, 2, 3, 4$, while $\boldsymbol{\beta}_2$ contains the six β_{ij} with subscripts in order 12, 13, 14, 23, 24, 34, as usual.

(d) Now fit a model with four main effects and six two factor interactions by least squares. Make a normal plot of the regression coefficients as an approximate analysis, even though the estimates are correlated. What is your conclusion now?

Table E5.32. Data from a Five-Factor, 16-Run Factorial Experiment

Row	x_1	x_2	x_3	x_4	x_5	y
1	-1	-1	-1	-1	-1	6.92
2	1	-1	-1	-1	1	6.30
3	1	-1	-1	1	-1	7.70
4	-1	-1	-1	1	1	17.97
5	1	-1	1	-1	-1	6.33
6	-1	-1	1	-1	1	7.83
7	-1	-1	1	1	-1	8.26
8	1	-1	1	1	1	18.14
9	1	1	-1	-1	-1	6.24
10	-1	1	-1	-1	1	7.72
11	-1	1	-1	1	-1	7.13
12	1	1	-1	1	1	18.86
13	-1	1	1	-1	-1	6.29
14	1	1	1	-1	1	7.62
15	1	1	1	1	-1	7.36
16	-1	1	1	1	1	11.26

5.32. The data in Table E5.32 come from a five-factor 16-run factorial experiment. Analyze these data and provide some sensible and useful conclusions about the data, the main effects of the five factors, and their interactions. Note that there is an additional problem, beyond the standard analysis, that you will need to solve.

5.33. (*Source:* Warren Porter, Professor of Zoology, University of Wisconsin, Madison, WI.) The 94 observations in Table E5.33a arose from a screening experiment to investigate the influences of eight factors on the growth of mice. The factors, with their lower (coded -1), middle (0), and upper ($+1$) levels are shown in Table 5.33b. The initial planning envisioned observations obtained in the following manner. A 16 run 2_{IV}^{8-4} fractional factorial design using the ± 1 levels was set up in eight blocks of size 2. This design was replicated, making 32 runs. To each of the 16 blocks was added a center point run, bringing the number of runs to $32 + 16 = 48$. Four mice were allocated to each of these. Thus the initial planning was to work with 192 mice. In the event, a number of mice died during the experiment. Table E5.33a contains, in the y column, the weights in grams of the 94 that survived. The initial neat balance of the experiment was thus disturbed, although much information remained. In the table, variables $x_1, x_2, \ldots, x_8$ are the factors being investigated and variables x_9, x_{10}, and x_{11} are blocking variables, whose eight possible combinations of signs separate the runs of each 2_{IV}^{8-4} fractional factorial into eight blocks. If the eight blocks are ordered so that the ± 1 levels of x_9, x_{10}, and x_{11} are listed in the standard order of a 2^3 factorial design, the numbers of runs within the blocks are 13, 6, 11, 15, 14, 11, 12, and 12, totaling 94 runs in all. In the reduced design of Table E5.33a, blocks are no longer orthogonal to the factors, nor to two-factor interactions. Variable x_{12} is a dummy variable used to indicate replicate

Table E5.33a. Ninety-Four Observations of the Mouse Experiment

	x_1	x_2	x_3	x_4	x_5	x_6	x_7	x_8	x_9	x_{10}	x_{11}	x_{12}	y
1	-1	-1	-1	-1	-1	-1	-1	-1	1	1	1	-1	3.6
2	1	-1	-1	-1	1	1	1	-1	-1	1	1	-1	7.6
3	1	-1	-1	-1	1	1	1	-1	-1	1	1	1	8.1
4	1	-1	-1	-1	1	1	1	-1	-1	1	1	1	7.1
5	-1	1	-1	-1	1	1	-1	1	1	-1	1	-1	11.8
6	-1	1	-1	-1	1	1	-1	1	1	-1	1	-1	11.6
7	-1	1	-1	-1	1	1	-1	1	1	-1	1	1	9.2
8	-1	1	-1	-1	1	1	-1	1	1	-1	1	1	9.0
9	1	1	-1	-1	-1	-1	1	1	-1	-1	1	-1	6.6
10	1	1	-1	-1	-1	-1	1	1	-1	-1	1	-1	7.3
11	1	1	-1	-1	-1	-1	1	1	-1	-1	1	1	6.5
12	-1	-1	1	-1	1	-1	1	1	1	1	-1	-1	10.6
13	-1	-1	1	-1	1	-1	1	1	1	1	-1	-1	10.5
14	1	-1	1	-1	-1	1	-1	1	-1	1	-1	-1	9.2
15	1	-1	1	-1	-1	1	-1	1	-1	1	-1	-1	10.2
16	1	-1	1	-1	-1	1	-1	1	-1	1	-1	1	3.5
17	-1	1	1	-1	-1	1	1	-1	1	-1	-1	-1	6.7
18	-1	1	1	-1	-1	1	1	-1	1	-1	-1	-1	6.2
19	-1	1	1	-1	-1	1	1	-1	1	-1	-1	1	5.8
20	-1	1	1	-1	-1	1	1	-1	1	-1	-1	1	6.2
21	-1	-1	-1	1	-1	1	1	1	-1	-1	-1	-1	5.5
22	-1	-1	-1	1	-1	1	1	1	-1	-1	-1	-1	4.8
23	-1	-1	-1	1	-1	1	1	1	-1	-1	-1	-1	5.5
24	-1	-1	-1	1	-1	1	1	1	-1	-1	-1	1	6.3
25	-1	-1	-1	1	-1	1	1	1	-1	-1	-1	1	5.7
26	-1	-1	-1	1	-1	1	1	1	-1	-1	-1	1	6.2
27	-1	-1	-1	1	-1	1	1	1	-1	-1	-1	1	5.0
28	-1	1	-1	1	1	-1	1	-1	-1	1	-1	-1	6.7
29	-1	1	-1	1	1	-1	1	-1	-1	1	-1	-1	6.1
30	-1	1	-1	1	1	-1	1	-1	-1	1	-1	-1	6.6
31	-1	1	-1	1	1	-1	1	-1	-1	1	-1	-1	6.3
32	-1	1	-1	1	1	-1	1	-1	-1	1	-1	1	6.6
33	-1	1	-1	1	1	-1	1	-1	-1	1	-1	1	6.7
34	-1	1	-1	1	1	-1	1	-1	-1	1	-1	1	6.9
35	1	1	-1	1	-1	1	-1	-1	1	1	-1	-1	4.8
36	1	1	-1	1	-1	1	-1	-1	1	1	-1	-1	4.5
37	1	1	-1	1	-1	1	-1	-1	1	1	-1	-1	4.7
38	1	1	-1	1	-1	1	-1	-1	1	1	-1	-1	4.3
39	1	1	-1	1	-1	1	-1	-1	1	1	-1	1	8.3
40	1	1	-1	1	-1	1	-1	-1	1	1	-1	1	8.3
41	1	1	-1	1	-1	1	-1	-1	1	1	-1	1	8.7
42	-1	-1	1	1	1	1	-1	-1	-1	-1	1	-1	6.8
43	-1	-1	1	1	1	1	-1	-1	-1	-1	1	-1	7.0
44	-1	-1	1	1	1	1	-1	-1	-1	-1	1	1	5.0
45	-1	-1	1	1	1	1	-1	-1	-1	-1	1	1	5.2
46	-1	-1	1	1	1	1	-1	-1	-1	-1	1	1	5.1
47	-1	-1	1	1	1	1	-1	-1	-1	-1	1	1	4.8
48	1	-1	1	1	-1	-1	1	-1	1	-1	1	-1	3.8
49	-1	1	1	1	-1	-1	-1	1	-1	1	1	-1	5.5

Table E5.33a (*Continued*)

	x_1	x_2	x_3	x_4	x_5	x_6	x_7	x_8	x_9	x_{10}	x_{11}	x_{12}	y
50	−1	1	1	1	−1	−1	−1	1	−1	1	1	−1	6.5
51	1	1	1	1	1	1	1	1	1	1	1	−1	9.1
52	1	1	1	1	1	1	1	1	1	1	1	−1	8.0
53	1	1	1	1	1	1	1	1	1	1	1	−1	9.4
54	1	1	1	1	1	1	1	1	1	1	1	−1	8.5
55	1	1	1	1	1	1	1	1	1	1	1	1	8.0
56	1	1	1	1	1	1	1	1	1	1	1	1	9.0
57	1	1	1	1	1	1	1	1	1	1	1	1	8.1
58	1	1	1	1	1	1	1	1	1	1	1	1	7.9
59	0	0	0	0	0	0	0	0	−1	−1	−1	−1	7.5
60	0	0	0	0	0	0	0	0	−1	−1	−1	−1	7.0
61	0	0	0	0	0	0	0	0	−1	−1	−1	−1	7.6
62	0	0	0	0	0	0	0	0	−1	−1	−1	1	7.0
63	0	0	0	0	0	0	0	0	−1	−1	−1	1	7.0
64	0	0	0	0	0	0	0	0	−1	−1	−1	1	6.6
65	0	0	0	0	0	0	0	0	1	−1	−1	1	8.8
66	0	0	0	0	0	0	0	0	1	−1	−1	1	8.5
67	0	0	0	0	0	0	0	0	−1	1	−1	1	9.7
68	0	0	0	0	0	0	0	0	1	1	−1	−1	8.8
69	0	0	0	0	0	0	0	0	1	1	−1	−1	7.9
70	0	0	0	0	0	0	0	0	1	1	−1	−1	7.3
71	0	0	0	0	0	0	0	0	1	1	−1	1	8.2
72	0	0	0	0	0	0	0	0	1	1	−1	1	9.0
73	0	0	0	0	0	0	0	0	1	1	−1	1	9.0
74	0	0	0	0	0	0	0	0	−1	−1	1	−1	7.3
75	0	0	0	0	0	0	0	0	−1	−1	1	−1	7.3
76	0	0	0	0	0	0	0	0	−1	−1	1	−1	8.0
77	0	0	0	0	0	0	0	0	−1	−1	1	1	5.8
78	0	0	0	0	0	0	0	0	−1	−1	1	1	6.2
79	0	0	0	0	0	0	0	0	−1	−1	1	1	6.8
80	0	0	0	0	0	0	0	0	1	−1	1	−1	8.6
81	0	0	0	0	0	0	0	0	1	−1	1	−1	9.1
82	0	0	0	0	0	0	0	0	1	−1	1	−1	8.5
83	0	0	0	0	0	0	0	0	1	−1	1	1	7.0
84	0	0	0	0	0	0	0	0	1	−1	1	1	7.5
85	0	0	0	0	0	0	0	0	1	−1	1	1	7.8
86	0	0	0	0	0	0	0	0	−1	1	1	−1	9.4
87	0	0	0	0	0	0	0	0	−1	1	1	−1	9.5
88	0	0	0	0	0	0	0	0	−1	1	1	−1	8.8
89	0	0	0	0	0	0	0	0	−1	1	1	1	6.2
90	0	0	0	0	0	0	0	0	−1	1	1	1	6.2
91	0	0	0	0	0	0	0	0	−1	1	1	1	6.5
92	0	0	0	0	0	0	0	0	1	1	1	1	7.8
93	0	0	0	0	0	0	0	0	1	1	1	1	7.9
94	0	0	0	0	0	0	0	0	1	1	1	1	8.1

Table E5.33b. Levels of Eight Factors

Factor	-1 Level	0 Level	1 Level
1. Time dependence of temperature	Steady	$\frac{1}{2}$ oscillating	Oscillating
2. Temperature, °C	11	17.5	23
3. Relative humidity	90%	50%	20%
4. Number of young	2	3	4
5. Nestbox	No	No roof	Yes
6. Food, quantity	80% ad lib	90% ad lib[a]	ad lib[a]
7. Food, quality	No sprouts	$\frac{1}{2}$ normal	Sprouts
8. Water available	80% ad lib	90% ad lib[a]	ad lib[a]

[a]ad lib indicates amounts normally consumed when no limits were imposed.

runs when the levels of the other factors remain fixed. The signs designate first $(-)$ and second $(+)$ replicates within blocks, and the variation within such sets of observations allows a pure error calculation to be made via standard regression methods.

Examine these data and answer the question: In this large screening experiment, which variables appear to affect the response significantly? Was the blocking effective? What sort of experiment would you consider doing next, if any?

5.34. (*Source:* Improve molded part quality, by M. B. de V. Azerado, S. S. da Silva, and K. Rekab, *Quality Progress*, July 2003, 72–76.) An injection molding process for a plastic closure for medical infusion bottles has "complex geometry plus many functional properties." The authors focused on one important functional response, the force, y, needed to open the closure mechanism. Seven variables A, B, . . . , G that might influence the response were investigated at two selected levels each, as shown in Table E5.34, after coding to -1 and 1. The design is a 2_{III}^{7-4} fractional factorial, generated by **I = ABD = ACE = BCF = ABCG**. Three response readings of y were obtained at each set of conditions. Analyze the data and build a reduced predictive regression model that contains the variables that appear to be important. Do the residuals from this model look satisfactory?

Table E5.34. Data from an Injection Molding Process

Run No.	Inject Speed A	Mold Temp. B	Melt Temp. C	Hold Pressure D	Hold Time E	Cool Time F	Eject Speed G	y_1	y_2	y_3
1	1	1	1	1	1	1	1	64.33	73.43	70.95
2	-1	1	1	-1	-1	1	-1	42.77	41.15	39.49
3	1	-1	1	-1	1	-1	-1	71.62	78.44	73.96
4	-1	-1	1	1	-1	-1	1	65.51	62.48	59.05
5	1	1	-1	1	-1	-1	-1	63.02	64.12	62.67
6	-1	1	-1	-1	1	-1	1	44.12	46.46	32.33
7	1	-1	-1	-1	-1	1	1	68.59	70.89	71.53
8	-1	-1	-1	1	1	1	-1	41.04	44.02	41.89

5.35. (*Source:* Augmented ruggedness testing to prevent failures, by M. J. Anderson, *Quality Progress*, May 2003, 38–45.) The investigator, who had previously used premixes to bake bread, conducted an experiment to investigate whether bread could be made as well, or better, using basic primary ingredients. His selected four investigational factors, and their codings, were:

1. Liquid: Water (-1), milk (1).
2. Oil: Butter (-1), margarine (1).
3. Flour: Regular flour (-1), bread flour (1).
4. Yeast: Regular yeast (-1), bread yeast (1).

The experiment chosen was a 2_{IV}^{4-1} design defined by $I = 1234$, which associates all main effects with 3fi's and also associates pairs of 2fi's. The response measured, modified from the original choice, was simply 1 (good, the bread rises) and 0 (bad, it doesn't rise). Three runs were repeated, as shown in Table E5.35, but the repeat responses were identical, so that the pure error estimate was zero. Analyze these data and form conclusions about the contributions of the variables under investigation.

Table E5.35. A 2^3 Factorial Design for Bread Making with Three Runs Replicated

Setting	x_1	x_2	x_3	x_4	y
1	-1	-1	-1	-1	0, 0
2	1	-1	-1	1	1, 1
3	-1	1	-1	1	0, 0
4	1	1	-1	-1	1
5	-1	-1	1	1	1
6	1	-1	1	-1	1
7	-1	1	1	-1	1
8	1	1	1	1	1

Four more runs were then performed using a "semifold" technique (Mee and Peralta, 2000). They resulted as follows:

Setting	x_1	x_2	x_3	x_4	y
9	-1	-1	-1	1	0
10	1	-1	-1	-1	1
11	-1	1	-1	-1	0
12	1	1	-1	1	1

The four added runs are, in fact, very like runs 5–8, but with the signs of x_3 reversed. Whereas, in runs 1–8, we had $x_1 x_3 = x_2 x_4$, in runs 9–12, $x_1 x_3 = -x_2 x_4$. This means that we can now estimate both of these two-factor interactions individually, using all 15 observations together. What are your conclusions now, incorporating the extra information?

Table E5.36. Data from a 2_{III}^{7-3} Fractional Factorial Experiment

Run	x_1	x_2	x_3	x_4	x_5	x_6	x_7	y
1	-1	-1	-1	-1	-1	-1	-1	58.7
2	1	-1	-1	-1	1	1	1	79.7
3	-1	1	-1	-1	1	1	-1	90.7
4	1	1	-1	-1	-1	-1	1	69.0
5	-1	-1	1	-1	1	-1	1	59.1
6	1	-1	1	-1	-1	1	-1	60.5
7	-1	1	1	-1	-1	1	1	71.7
8	1	1	1	-1	1	-1	-1	69.0
9	-1	-1	-1	1	-1	1	1	60.2
10	1	-1	-1	1	1	-1	-1	58.9
11	-1	1	-1	1	1	-1	1	76.8
12	1	1	-1	1	-1	1	-1	74.3
13	-1	-1	1	1	1	1	-1	79.1
14	1	-1	1	1	-1	-1	1	57.1
15	-1	1	1	1	-1	-1	-1	73.1
16	1	1	1	1	1	1	1	90.4

5.36. A 2_{III}^{7-3} design generated by $\mathbf{I = 1235 = 1246 = 1347}$ is used to screen seven variables, of which it is believed that four, at most, will be of importance. Under these tentative assumptions, analyze the resultant data in Table E5.36.

CHAPTER 6

The Use of Steepest Ascent
to Achieve Process Improvement

If we attempted to fit an empirical function such as a polynomial over the whole *operability region* (that is, over the whole region within which the studied system could be operated), a very complex function would usually be needed. The fitting of such a function could involve an excessive number of experiments. The exploration of the whole operability region is, however, almost never a feasible or sensible objective. First, the extent of this region is almost never known, and, second, the experimenter can often safely dismiss as unprofitable whole areas of the region where experiments could, theoretically, be conducted. What he usually wants to do, with good reason, is to explore a smaller subregion *of interest* which often changes as the investigation progresses. He can expect that, within this smaller subregion, a fairly simple graduating equation will be representationally adequate.

Now the mathematical procedure of fitting selects, from all possible surfaces of the degree fitted, that which approximates the responses *at the experimental points* most closely (in the least-squares sense). The features of the fitted surface at points remote from the region of the experimental design need not, and usually would not, bear any resemblance to the features of the actual surface. We can therefore expect our approximation to be useful only in the immediate neighborhood of the current experimental region. If, for example, we were already close to a maximum, we might represent its main features approximately by fitting, to a suitable locally placed set of experimental points, an equation of only second degree, possibly in transformed variables. However, particularly if the system were being investigated for the first time, starting conditions would often not be very close to such a maximum. Thus, the experimenter often first needs some *preliminary procedure*, to bring him to a suitable point where the second-degree equation can most usefully be employed. *One* such preliminary procedure is the one-factor-at-a-time method. An alternative which, in the authors' opinion, is usually more effective and economical in experiments (at least in the fields of application in

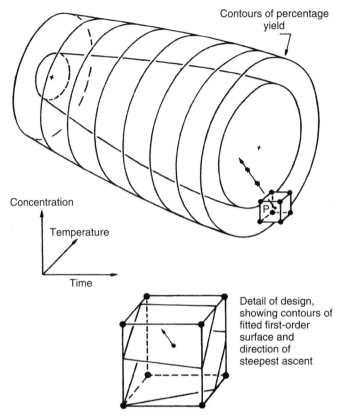

Contours of percentage yield

Concentration

Temperature

Time

Detail of design, showing contours of fitted first-order surface and direction of steepest ascent

Figure 6.1. A stage of the steepest ascent procedure.

which it has been used) is the "steepest ascent" method. This preliminary procedure, which may later be followed by the fitting of an equation of second degree (possibly in transformed variables), is the basis of a sequential experimentation method which we discuss in this chapter and which has proved extremely effective in a wide variety of applications.

6.1. WHAT STEEPEST ASCENT DOES

For illustration, consider the investigation of a chemical system whose percentage yield depends on the three predictor variables time, temperature, and concentration; a possible situation is shown in Figure 6.1. The overall objective is to find settings of the variables which will result in improved yields. Suppose we are at an early stage of investigation when considerable improvement is possible. Then the planar contours of a first-degree equation can be expected to provide a fair approximation in the immediate region of a point such as P, which is far from the optimum. A direction at right angles to these contour planes is the direction of *steepest ascent*, if it points toward higher yield values. (The opposite direction is that of *steepest descent*, needed for *reducing*, and thus improving, a response such as "impurity" for example.) In practice, the first-degree approximating equation must be estimated via experiments. This may be done by running a design such as that indicated by the dots around the point P.

Exploratory runs performed along the path of steepest ascent and indicated by stars in Figure 6.1 will normally show increasing yield along the path. The best point found, or an interpolated estimated maximum point on the path, could then be made the base for a new first-order design from which a further advance might be possible.

The direction of ascent we shall calculate is the steepest when the response surface is scaled in the units of the design as it is, for example, in Figure 6.1. The effects of scale changes are discussed in Appendix 6A.

It will usually be found that, after one or perhaps two applications of steepest ascent, first-order effects will no longer dominate and the first-order approximation will be inadequate. At this stage, further steepest ascent progress will not be possible and more sophisticated second-order methods, discussed in Chapters 7 and 9, should be applied. The steepest ascent technique is particularly effective for the initial investigation of a new process. Such an investigation typically takes place on the laboratory scale.

For an already operating process arrived at after much development work, designs performed in the vicinity of the currently known best process may show hardly any first-order effects at all, because these have already been exploited to obtain the current conditions. In this situation then, no initial large gains may be possible using steepest ascent, but, again, the more sophisticated second-order methods of Chapter 9 may still lead to substantial improvement.

6.2. AN EXAMPLE. IMPROVEMENT OF A PROCESS USED IN DRUG MANUFACTURE

In laboratory experiments on a reaction taking place between four reagents A, B, C, and D, the amount of A was kept constant, but five factors—time of reaction (ξ_1), temperature of reaction (ξ_2), amount of B (ξ_3), amount of C (ξ_4), and the amount of D (ξ_5)—were varied as in Table 6.1. A 2_V^{5-1} design with **I = 12345** was used and the design and resulting percentage yields, y, were as shown on the left of Table 6.2.

Table 6.1. Factors ξ_i and Coded Levels x_i

	Selected Levels of x_i		
ξ_i	Lower(-1)	Upper($+1$)	
ξ_1	6 h	10 h	so that $x_1 = (\xi_1 - 8)/2$
ξ_2	85°C	90°C	so that $x_2 = (\xi_2 - 87.5)/2.5$
ξ_3	30 cm^3	60 cm^3	so that $x_3 = (\xi_3 - 45)/15$
ξ_4	90 cm^3	115 cm^3	so that $x_4 = (\xi_4 - 102.5)/12.5$
ξ_5	40 g	50 g	so that $x_5 = (\xi_5 - 45)/5$

Table 6.2. 2^{5-1} Design, Responses, and the Yates' Analysis

Run Order and Number $x_1 x_2 x_3 x_4 x_5$	y	Yates' Analysis				Half-Effect (5)^a	Effect Name
		(1)	(2)	(3)	(4)		
16 − − − − +	51.8	108.1	213.2	420.3	914.8	$\bar{y} = 57.1750$	0
2 + − − − −	56.3	105.1	207.1	494.5	−53.6	−3.3500	1
10 − + − − −	56.8	112.1	242.0	−19.5	−34.6	−2.1625	2
1 + + − − +	48.3	95.0	252.5	−34.1	2.2	0.1375	12
14 − − + − −	62.3	122.1	−4.0	−20.1	4.4	0.2750	3
8 + − + − +	49.8	119.9	−15.5	−14.5	−12.0	−0.7500	13
9 − + + − +	49.0	132.4	−16.8	−3.5	−24.2	−1.5125	23
7 + + + − −	46.0	120.1	−17.3	5.7	−30.6	−1.9125	45
4 − − − + −	72.6	4.5	−3.0	−6.1	74.2	4.6375	4
15 + − − + +	49.5	−8.5	−17.1	−10.5	−14.6	−0.9125	14
13 − + − + +	56.8	−12.5	−2.2	−11.5	5.6	0.3500	24
3 + + − + −	63.1	−3.0	−12.3	−0.5	9.2	0.5750	35
12 − − + + +	64.6	−23.1	−13.0	−14.1	16.6	1.0375	34
6 + − + + −	67.8	6.3	9.5	−10.1	11.0	0.6875	25
5 − + + + −	70.3	3.2	29.4	22.5	4.0	0.2500	15
11 + + + + +	49.8	−20.5	−23.7	−53.1	−75.6	−4.7250	5

^{a}Except for $\bar{y}$.

On the assumption that the second-degree polynomial

$$y = \beta_0 + \sum_{i=1}^{5} \beta_i x_i + \sum_{i=1}^{5} \beta_{ii} x_i^2 + \sum_{i<j}^{5\,5} \beta_{ij} x_i x_j + \varepsilon$$

is adequate to represent the response locally, least-squares estimates of the five first-order coefficients β_i and the 10 two-factor interaction coefficients β_{ij} are

$$b_1 = -3.35, \quad b_{12} = 0.14, \quad b_{24} = 0.35,$$
$$b_2 = -2.16, \quad b_{13} = -0.75, \quad b_{25} = 0.69,$$
$$b_3 = 0.28, \quad b_{14} = -0.92, \quad b_{34} = 1.04,$$
$$b_4 = 4.64, \quad b_{15} = 0.25, \quad b_{35} = 0.58,$$
$$b_5 = -4.73, \quad b_{23} = -1.51, \quad b_{45} = -1.91.$$

It will be recalled that, because each of the x variables has been allocated the coded levels -1 and $+1$, the least-squares values of the b's will be one-half of the values obtained via Yates' method in Table 6.2. In this experiment, some of the estimated second-order effects are not particularly small compared with those of first order. On the whole, however, the first-order effects are considerably larger in magnitude than those of second order, as evidenced, for example, by the analysis of variance in Table 6.3. As an approximation, it was decided to assume tentatively that second degree effects could, for the present, be ignored and to combine the second-order effects to provide an "estimate of error" based on 10 df. From this, the standard error of each estimate is $(15.08/16)^{1/2} = 0.971$. Note that

Table 6.3. First- and Second-Order Contributions to Corrected Total Sum of Squares

Source	df	SS	MS
First order	5	956.905	191.38
Second order	10	150.785	15.08
Total, corrected	15	1107.690	

the approximation is conservative in the sense that, if second-order effects are not negligible, this calculation will tend to *inflate* the estimate of error variance. Under these assumptions we have

$$b_1 = -3.35 \pm 0.97,$$

$$b_2 = -2.16 \pm 0.97,$$

$$b_3 = 0.28 \pm 0.97,$$

$$b_4 = 4.64 \pm 0.97,$$

$$b_5 = -4.73 \pm 0.97.$$

The estimated direction of steepest ascent, when the variables are scaled in units of the design, then follows the vector of coefficient values $[-3.35, -2.16, 0.28, 4.64, -4.73]$. The length of this vector is $[(-3.35)^2 + (-2.16)^2 + \cdots + (-4.73)^2]^{1/2} = 7.74$. A vector of unit length in the direction of steepest ascent therefore has coordinates $-3.35/7.74 = -0.43$, $-2.16/7.74 = -0.28$, and so on, that is, $[-0.43, -0.28, 0.04, 0.60, -0.61]$. If we follow the steepest ascent path, then, for a change of -0.43 units of the first variable we should change the second variable by -0.28 units, change the third variable by $+0.04$ units, and so on. Exploratory runs were chosen at the conditions shown in Table 6.4, obtained by multiplying the unit length vector by 2, 4, 6, and 8, respectively. These particular multiples were chosen because the first of these is close to the periphery of the experimental region. It corresponds quite closely with run number 4, which, in fact, gave the highest recorded yield of 72.6, while the spacing separating the runs E_2, E_4, E_6, and E_8 is about the same as that separating the upper and lower levels of the factorial design. Note that the decoded conditions may be obtained by inverting the

Table 6.4. Points on the Path of Steepest Ascent in Coded and Decoded Units

	Coded Conditions					Decoded Conditions					Observed
	x_1	x_2	x_3	x_4	x_5	Time	Temperature	Amount B	Amount C	Amount D	Yields
E_2:	−0.86	−0.56	0.08	1.20	−1.22	6.3	86.1	46.2	117.5	38.9	72.6 (run 4)
E_4:	−1.72	−1.12	0.16	2.40	−2.44	4.6	84.7	47.4	132.5	32.8	85.1
E_6:	−2.58	−1.68	0.24	3.60	−3.66	2.8	83.3	48.6	147.5	26.7	82.4
E_8:	−3.44	−2.24	0.32	4.80	−4.88	1.1	81.9	49.8	162.5	20.6	80.8

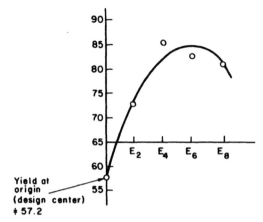

Figure 6.2. Yields found along the path of steepest ascent. The yield at the origin is approximated by the average 57.2 of all the 16 runs. The steepest ascent procedure has thus been very effective, moving us to a new set of conditions yielding over 80%.

formulas in Table 6.1 so that

$$\xi_1 = 2x_1 + 8,$$
$$\xi_2 = 2.5x_2 + 87.5,$$
$$\xi_3 = 15x_3 + 45,$$
$$\xi_4 = 12.5x_4 + 102.5,$$
$$\xi_5 = 5x_5 + 45.$$

Because of its closeness to the conditions of run number 4, the experiment E_2 was not actually run. The observed yields in Table 6.4 are plotted in Figure 6.2, from which it was concluded that a conditional maximum along the path of steepest ascent was located close to E_6.

6.3. GENERAL COMMENTS ON STEEPEST ASCENT

Design

It will be remarked that in the example, the fractional factorial employed was a 2_V^{5-1}, a design of resolution V, requiring 16 runs. It might have been possible to make progress with an eight-run design, but in this case the experimenters decided on a somewhat conservative course, partly because of their feeling that the first-order (i.e., main) effects might not be particularly large compared with the experimental error.

Steepest Ascent or Descent?

In this example, the *objective function* was yield and the objective was to increase it. Other objective functions that might be important in chemical investigations would be *unit cost* and *level of impurity*, both of which we would wish to *decrease* rather than increase. This would require a path of steepest *descent*, given by changing all the signs from those of the path of steepest ascent at any point. Alternatively, the same effect can be achieved by changing the sign of the objective function and following the path of steepest ascent. Thus, essentially, all problems can be set up as *ascent* problems.

When Will Ascent Methods Work?

If we are looking for, say, a maximum yield, it would be wasteful to investigate *in any detail* regions of *low* yield. Typically, the main features in such regions are the first-order (i.e., main) effects which can point a direction of ascent up a surface. Once first-order effects become small compared to those of second order, or compared to the error, or to both, it may be necessary to switch to a second-order approximation. The appropriate designs can then be chosen to cover somewhat larger regions.

Checks on the adequacy of the first-order approximation or, equivalently, on the need for second order terms are supplied by

1. Examining individual interaction contrasts, which are not used for estimation of first-order effects.
2. An overall curvature check.

A useful overall curvature check is supplied by adding a number of center points to a two-level fractional factorial or factorial design. The order in which the complete design, including center points, is carried out should be random. Thus, a set of randomly replicated experimental runs from which a pure error estimate of the basic variation in the data can be calculated is obtained. The overall curvature check is supplied by the contrast

$$c = \{\text{average response in two-level factorial runs}\}$$
$$- \{\text{average response for runs at the center}\}.$$

It may be shown that, if the true response function is of second degree, c supplies an unbiased estimate of the sum of the pure quadratic effects, namely

$$c \rightarrow \sum_{i=1}^{k} \beta_{ii}.$$

For a surface that contained a minimax, the β_{ii} would be of different signs and theoretically $E(c)$ could be zero even when large curvatures β_{ii} occurred. However, minimax surfaces are rare in practice, and typically the β_{ii} will be of the same sign.

There will nevertheless be intermediate situations where it is unclear whether or not the first-order approximation is good enough to allow progress via steepest ascent. In such cases, it is usually worthwhile to calculate the path of steepest ascent anyway and to make a few trials along it to find out.

A discussion of the effect of experimental error on the estimation of the direction of steepest ascent is given in Section 6.4.

The First-Order Design as a Building Block

Suppose that, possibly after previous steepest ascent application(s), a new first-order design has just been completed. The appropriate checks, possibly augmented by tentative local exploration, indicate that no further advance by steepest ascent is likely and that we need to proceed with a second-order exploration. It will often be possible to incorporate the first-order design just completed as a building block of the second-order design now required. Consider, for instance, the chemical process example discussed above. Suppose the 2^{5-1} design had shown large second-order effects (i.e., two-factor interactions) and that these appeared to be reasonably well estimated. Then the addition of further points to the first-order design in a manner we describe in more detail in Sections 9.2 and 13.8 could convert it into a second-order design suitable for fitting a full second-degree equation. The results from this second-order fit might then permit further response improvement.

Steepest Ascent as a Precursor to Further Investigation

Steepest ascent is rarely an end in itself. It is mainly of value as a preliminary procedure, to move the region of interest to a neighborhood of improved response, worthy of more thorough investigation.

Steepest Ascent Classroom Simulation

The steepest ascent procedure can be realistically simulated for teaching purposes (see, for example, Mead and Freeman, 1973).

6.4. A CONFIDENCE REGION FOR THE DIRECTION OF STEEPEST ASCENT

In the example of Section 6.2 the first-order coefficients were estimated as

$$b_1 = -3.35 \pm 0.97,$$
$$b_2 = -2.16 \pm 0.97,$$
$$b_3 = 0.28 \pm 0.97, \qquad (6.4.1)$$
$$b_4 = 4.64 \pm 0.97,$$
$$b_5 = -4.73 \pm 0.97,$$

where the standard error of each estimate (0.97) was based on an estimated variance having 10 df. Retaining now the convention that we consider the variables as scaled in the units of the design, the vector

$$(-3.35, -2.16, 0.28, 4.64, -4.73) \qquad (6.4.2)$$

provides the estimated best direction of advance. Assuming that a first-degree response equation provides an adequate model, we may now ask how much in error due to sampling variation this direction might be. This question can be answered in a conventional way by obtaining a confidence region for the direction of steepest ascent (Box, 1954, p. 211). This confidence region will turn out to be a cone (or hypercone depending on the number of dimensions involved) whose axis is the estimated direction vector.

Suppose there are k variables and we seek a confidence region for the *true* direction of steepest ascent as defined by its direction cosines $\delta_1, \delta_2, \ldots, \delta_i, \ldots, \delta_k$. If $b_1, b_2, \ldots, b_k$ are the estimated first-degree effects, the expected values of these quantities are proportional to $\delta_1, \delta_2, \ldots, \delta_k$ so that

$$E(b_i) = \gamma \delta_i, \qquad i = 1, 2, \ldots, k, \qquad (6.4.3)$$

where γ is some constant. Now if we think of this relationship as a regression model in which the b_i are responses and the δ_i are the levels of a single-predictor variable, then γ is the "regression coefficient" of $b_1, \ldots, b_k$ on $\delta_1, \delta_2, \ldots, \delta_k$. The required region is supplied by those elements $\delta_1, \delta_2, \ldots, \delta_k$ which just fail to make the residual mean square significant compared with $V(b_i) = \sigma_b^2$ at some desired level α. That is, for those δ's which satisfy

$$\frac{\left\{ \sum b_i^2 - \left(\sum_{i=1}^{k} b_i \delta_i \right)^2 \Big/ \sum \delta_i^2 \right\} \Big/ (k-1)}{s_b^2} \leq F_\alpha(k-1, \nu_b), \qquad (6.4.4)$$

where s_b^2 is an estimate of σ_b^2 and ν_b is the number of degrees of freedom on which this estimate s_b^2 is based. For the particular case where $k = 2$, we have an application of Fieller's theorem (Fieller, 1955), of which the present development is one extension. Because all the quantities in the foregoing inequality are known except for the values of the δ's, this expression defines a set of acceptable δ's, thus defining a set of acceptable vectors and, hence, a confidence region for the direction of steepest ascent.

As an illustration we can again use the process example of Section 6.2. For simplicity and for the time being, we ignore the last two variables and treat the problem as if there had been only three variables ($k = 3$). The confidence region then takes the form of a cone in three dimensions about the estimated direction of steepest ascent as in Figure 6.3. For this reduced example, we have

$$\sum b_i^2 = (-3.35)^2 + (-2.16)^2 + (0.28)^2 = 15.9665,$$

$$\sum b_i \delta_i = -3.35 \delta_1 - 2.16 \delta_2 + 0.28 \delta_3,$$

$$\sum \delta_i^2 = 1, \qquad (6.4.5)$$

$$(k-1) = 2, \qquad s_b = 0.97, \qquad F_{0.05}(2, 10) = 4.10,$$

and the confidence region is thus defined by the cone with apex at the origin and such that all points $(\delta_1, \delta_2, \delta_3)$ a unit distance away from the origin satisfy

$$15.9665 - (-3.35 \delta_1 - 2.16 \delta_2 + 0.28 \delta_3)^2 \leq 2 \times 0.97^2 \times 4.10, \qquad (6.4.6)$$

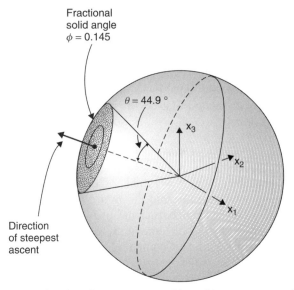

Figure 6.3. Direction of steepest ascent with confidence cone around it.

that is,

$$(-3.35\delta_1 - 2.16\delta_2 + 0.28\delta_3)^2 \geq 8.2511. \qquad (6.4.7)$$

When there are more than three variables, the preparation of a diagram like that of Figure 6.3 is not possible. However, in practical applications, all we usually need to know is whether the direction of steepest ascent has been determined accurately enough for us to proceed. A good indication is supplied by the magnitude of the solid angle of the confidence cone about the estimated vector. Consider again Figure 6.3. Suppose the cap on the unit sphere centered at the origin has a surface which is a fraction ϕ of the total area of the sphere; then we shall say that the confidence cone subtends a fractional solid angle ϕ. Associated with the confidence cone, we might also consider another angle, the semiplane angle θ at the vertex between a line on the surface of the cone at the origin and the axis of the cone. For a given number of dimensions, there will be a one-to-one correspondence between the fractional solid angle ϕ and the semiplane angle θ at the vertex.

This correspondence is given by tables of the t distribution. These tables are equivalent to a listing of the fractional solid angle 2ϕ (which corresponds to the double-tailed probability in the tables) in terms of the function $t = (k - 1)^{1/2} \cot \theta$. The fractional solid angle ϕ may therefore be obtained by inverse interpolation in the tables of t. Specifically, in terms of the quantities we have defined,

$$\sin \theta = \left\{ \frac{(k - 1)s_b^2 F_\alpha(k - 1, \nu_b)}{\sum_{i=1}^{k} b_i^2} \right\}^{1/2} \qquad (6.4.8)$$

and

$$t_{2\phi}(k - 1) = (k - 1)^{1/2} \cot \theta$$

$$= \left\{ \frac{\sum\limits_{i=1}^{k} b_i^2}{s_b^2 F_\alpha(k - 1, \nu_b)} - (k - 1) \right\}^{1/2}. \tag{6.4.9}$$

In the simplified example above in which we have pretended that there were only three variables we have

$$\sin^2\theta = \frac{\{2F_{0.05}(2, 10)\}0.97^2}{15.9665} = \frac{2(4.10)(0.97^2)}{(15.9665)} = 0.4832, \tag{6.4.10}$$

whence $\theta = 44.0°$. Thus, the confidence region for this example consists of a cone about the direction of steepest ascent with a semiplane angle at the vertex of $\theta = 44.0°$. Applying Eq. (6.4.9) with $\theta = 44.0°$, $k = 3$, we have, since $\cot \theta = 1.034$,

$$t_{2\phi}(2) = 1.462. \tag{6.4.11}$$

Interpolation in the t tables gives $\phi = 0.141$ approximately. Thus the 95% confidence cone excludes about 85.9% of the possible directions of advance.

This kind of statement is available to us; however, many variables are involved. Thus, without necessarily being able to visualize the situation geometrically, we can appreciate the size of the confidence region simply by calculating ϕ.

For the original five-variable example, we would set $k = 5$, $\nu_2 = 10$ to obtain

$$t_{2\phi}(4) = \sqrt{\frac{59.8690}{0.97 \times 3.48} - 4} = 3.78. \tag{6.4.12}$$

Interpolation in the t tables yields $\phi = 0.01$ approximately. Thus, in this instance the 95% confidence cone for the direction of steepest ascent excludes about 99% of possible directions of advance. The appropriate direction is, therefore, known with some considerable accuracy.

6.5. STEEPEST ASCENT SUBJECT TO A CONSTRAINT

In some problems, we cannot proceed very far in the direction of steepest ascent before some constraint is encountered. It is then of interest to explore the modified direction of steepest ascent subject to the condition that the constraint is not violated. For example, in Figure 6.4 we show a 2^3 factorial experiment in the space of x_1, x_2, and x_3. As usual, x_1, x_2, and x_3 are standardized and relate to real variables such as temperature, ξ_1, time, ξ_2, and so on, by such equations as $x_i = (\xi_i - \xi_{i0})/S_i$. It may happen that the path of steepest ascent leads to a region of inoperability. Suppose that, at least as a local approximation, the inoperability

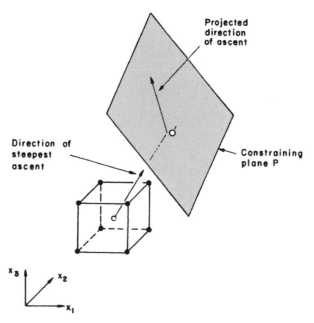

Figure 6.4. Steepest ascent subject to a constraint.

region is bounded by the plane

$$a_0 + a_1 x_1 + a_2 x_2 + \cdots + a_k x_k = 0. \tag{6.5.1}$$

A vector perpendicular to this plane is then $\mathbf{a} = (a_1, a_2, \ldots, a_k)'$. Let the unmodified vector of steepest ascent be given by $\mathbf{b} = (b_1, b_2, \ldots, b_k)'$. Then the ith element of the modified vector direction of steepest ascent is

$$e_i = b_i - c a_i \qquad i = 1, 2, \ldots, k, \tag{6.5.2}$$

where

$$c = \frac{\sum b_i a_i}{\sum a_i^2}. \tag{6.5.3}$$

It will be recognized that the e_i are the residuals obtained by regressing the elements of the steepest ascent vector on to the coefficients which represent the plane, the quantity c being the appropriate estimated regression coefficient. The above calculation gives the *direction* of the modified vector only. We also need to calculate the point O' where the initial vector of steepest ascent hits the constraining plane P and the new direction takes over. Clearly, any point on the initial direction of steepest ascent will have coordinates

$$x_1 = \lambda b_1, \qquad x_2 = \lambda b_2, \qquad \ldots, x_k = \lambda b_k, \tag{6.5.4}$$

for some λ. Consequently, the required point is that which, while on the path of steepest ascent, also lies on the plane $a_0 + a_1 x_1 + a_2 x_2 + \cdots + a_k x_k = 0$. Thus, it corresponds to the value λ_0 which satisfies $a_0 + (a_1 b_1 + a_2 b_2 + \cdots + a_k b_k)\lambda_0 = 0$, that is,

$$\lambda_0 = \frac{-a_0}{\sum a_i b_i}. \tag{6.5.5}$$

An Example

The activity of a certain chemical mixture depends upon the proportion of three major ingredients A, B, and C. The mixture usually also contains an inactive diluent which is added to bring the total volume to a fixed value. A 2^3 factorial design was run to determine the effects of the ingredients on the activity y of the mixture. The lower and upper levels of the ingredients A, B, and C were as follows:

Ingredients	Amount (cm^3)	
	Lower	**Upper**
A	15	25
B	10	30
C	20	40

$$\tag{6.5.6}$$

In each case, sufficient diluent was added to bring up the total volume to 100 cm^3. Thus our standardized variables are

$$x_1 = \frac{(A - 20)}{5}, \qquad x_2 = \frac{(B - 20)}{10}, \qquad x_3 = \frac{(C - 30)}{10}. \tag{6.5.7}$$

In this example, the interactions were found to be fairly small compared with the main effects and their standard errors, and the activity y of the resulting mixture was adequately represented by the linear relationship

$$\hat{y} = 12.2 + 3.8x_1 + 2.6x_2 + 1.9x_3. \tag{6.5.8}$$

Thus, the estimated direction of steepest ascent runs from the origin $x_1 = x_2 = x_3 = 0$ along the vector $(2.8, 2.6, 1.9)$. In the original variables, the origin corresponds to the mixture of 20 cm^3 of A, 20 cm^3 of B, and 30 cm^3 of C and, if we follow the direction of steepest ascent from this point, we shall clearly need to keep increasing the amounts of all the active ingredients. However, we are working with a system having the constraint

$$A + B + C \leq 100. \tag{6.5.9}$$

Because $A = 20 + 5x_1$, $B = 20 + 10x_2$, and $C = 30 + 10x_3$, this implies that

$$20 + 5x_1 + 20 + 10x_2 + 30 + 10x_3 \leq 100. \tag{6.5.10}$$

Thus, the constraining plane has the equation

$$-30 + 5x_1 + 10x_2 + 10x_3 = 0. \tag{6.5.11}$$

It follows that

$$\lambda_0 = -(-30)/\{5(3.8) + 10(2.6) + 10(1.9)\} = 0.469. \qquad (6.5.12)$$

Thus the path of steepest ascent just meets the constraining plane at the point where

$$x_1 = 0.469(3.8) = 1.78, \qquad x_2 = 0.469(2.6) = 1.22, \qquad x_3 = 0.469(1.9) = 0.89,$$
$$(6.5.13)$$

that is, at the point O' with coded coordinates $(1.78, 1.22, 0.89)$. To check, we note that

$$5(1.78) + 10(1.22) + 10(0.89) = 30.00 \qquad (6.5.14)$$

so that this point is therefore just on the constraining plane. We now wish to calculate the modified direction of steepest ascent, which is the projection of this direction in the plane $-30 + 5x_1 + 10x_2 + 10x_3 = 0$. Regressing the initial steepest ascent vector on to the coefficients of the constraining plane, we find that

$$c = \frac{3.8(5) + 2.6(10) + 1.9(10)}{5^2 + 10^2 + 10^2} = 0.284. \qquad (6.5.15)$$

Thus the modified vector direction of steepest ascent from O' has elements

$$e_1 = 3.8 - 0.284(5) = 2.38,$$
$$e_2 = 2.6 - 0.284(10) = -0.24, \qquad (6.5.16)$$
$$e_3 = 1.9 - 0.284(10) = -0.94.$$

Further progress in this direction of constrained steepest ascent will therefore lie along the line having coded coordinates

$$x_1 = 1.78 + \mu(2.38), \qquad x_2 = 1.22 - \mu(0.24), \qquad x_3 = 0.89 - \mu(0.94),$$
$$(6.5.17)$$

where μ is some suitably chosen multiplier.

Some specimen calculations along the unmodified direction of steepest ascent and then along the modified direction of steepest ascent are shown in Table 6.5.

The first set of conditions listed in the table is for the center point. Conditions two and three lie on the steepest ascent path $x_1 = 3.8\lambda$, $x_2 = 2.6\lambda$, and $x_3 = 1.9\lambda$, where λ is 0.2 and 0.4. At the fourth set of conditions, the steepest ascent vector reaches the constraining plane ($\lambda = \lambda_0 = 0.469$). The remaining conditions listed are on the modified direction of steepest ascent defined in the foregoing display with $\mu = 0.2$ and 0.4. It will be noticed that the predicted activity continues to rise, but at a considerably reduced rate, on the modified path of ascent.

The elements comprising the vector of the path of steepest ascent are, of course, often subject to fairly large errors and, in any case, the linear approximation is probably inaccurate at points not close to the region of the design. The type

Table 6.5. Points on the Original and Modified Directions of Steepest Ascent

Point Number	Total (cm^3)	x_1	x_2	x_3	Predicted Activity	Comments
1	70	0	0	0	12.2	Center O
2	82.8	0.76	0.52	0.38	17.2 ⎫	Unconstrained
3	95.6	1.52	1.04	0.76	22.1 ⎭	path
4	100.0	1.78	1.22	0.89	23.8	Point O'
5	100.0	2.26	1.17	0.70	25.2 ⎫	Constrained
6	99.95	2.73	1.12	0.51	26.5 ⎭	path

of calculation given above should not therefore be regarded as an exact prediction but rather as supplying an experimental path worthy of further exploration. In some cases the constraint is not known exactly *a priori*, as in this example, but must also be estimated. It often happens that, in a given set of experimental runs, a number of responses may be measured, one of which, the principal response function, represents an objective function such as yield, which we desire to maximize, and the others are measures of physical characteristics of the product—purity, color, viscosity, odor, and so on. In those cases where the objective of the experiment is to maximize the principal response, the constraining relations may themselves be estimated from the auxiliary responses. We have seen an example of this situation earlier where we were required to maximize the strength of a dyestuff, subject to the attainment of suitable levels of hue and brightness.

To understand constraining relations between three variables, it is very worthwhile to examine appropriate sets of three-dimensional models. Computer graphics have now greatly simplified such exploration. One can say, of course, that the geometrical models do no more than express the related algebra, and so consequently algebraic manipulation without the accompanying models is enough. In practice, however, it seems that there is much less likelihood of missing important aspects of the problem and of making outright mistakes, if graphical and geometrical illustrations are used whenever possible.

For an excellent pictorial representation of steepest ascent using five-factor simplex designs, see Gardiner and Cowser (1961).

APPENDIX 6A. EFFECTS OF CHANGES IN SCALES ON STEEPEST ASCENT

In this book, we have tried to emphasize that the outcome of an investigation depends critically on a number of crucial questions which are not directly dependent on the data and which the experimenter must decide from knowledge and experience. Moreover, these decisions are typically modified as the investigation unfolds. For example, the experimenter decides *which* factors to include, the *region* of the factor space in which the experiments are to be carried out, and the *relative scales* of measurement of the factors. Such questions are quite outside the competence of the statistician, although he ought always to ask appropriately probing questions to ensure that the experimenter has properly considered all the relevant issues, and the statistician may need to point out, after the data are available, that certain suppositions of the experimenter appear to be of doubtful validity. Granted all this, it is the role of the statistician to design the best experiment within the framework currently believed to be most appropriate by the experimenter and, in particular, in the scaling of the variables the experimenter regards as most applicable at any given time. It has to be remembered that a subject as concrete and mathematically satisfying as experimental design is actually embedded in a morass of uncertainty, uncertainty due to the possibilities that the experimenter might choose wrong variables, might explore the wrong region, or might use scaling that was inappropriate. However, the gloomy view that successful experimentation is a matter of purest luck is lightened by two circumstances: (1) Experimenters often *do* know a very great deal about the system they are studying, and (2) because most investigations are conducted sequentially, the experimenter does not need to guess exactly right but need only guess sufficiently right to place himself on one of the many possible paths that will lead adaptively to a satisfactory answer. In particular, the correction of any grossly unsuitable scaling of the predictor variables will normally occur as the experimental iteration proceeds.

The problem of scaling the variables in steepest ascent is one area where the general indeterminacy of experimentation does not lie conveniently hidden but makes itself manifest. In order to appreciate the problem, consider the following: Suppose the scale factors that one experimenter adopts are $s_1, s_2, \ldots, s_k$ and that the scale factors adopted by a second experimenter are $s_1' = a_1 s_1, s_2' = a_2 s_2, \ldots, s_k' = a_k s_k$. (For example, if in a two-level factorial or fractional factorial design, one experimenter changed temperature by $s_1 = 20°$ and another by $s_1' = 10°$, then $a_1 = s_1'/s_1 = 0.5$.) Suppose further that the direction of steepest ascent as measured in the first experimenter's scales has direction cosines proportional to the elements $1, 1, \ldots, 1$. (We can make this assumption without loss of generality because we could always rotate the axes of our x space so that the direction, whatever it may be, lies in this particular direction in new coordinates.) In this case, the direction of steepest ascent appropriate to the second experimenter would have direction cosines proportional to $a_1^2, a_2^2, \ldots, a_k^2$. Figure 6A.1 shows the situation for $k = 3$.

The fact that the direction of steepest ascent was not invariant to scale change was pointed out in the original response surface paper by Box and Wilson (1951). In the discussion of the paper, this point was taken up by N. L. Johnson, who emphasized the dependence of the elements of the direction on the square of the amount by which the scale factor was altered. In this same discussion it was also argued that it might be better simply to rely on the signs of the effects, which are of course scale invariant.

We can, to some extent, answer the question of how much *additional* useful information is contained in the direction of steepest ascent compared with using only the signs of the effects by examining a brief, unpublished investigation undertaken by G. E. P. Box and G. A. Coutie some years ago.

Let us choose some measure $G(|a|)$ of how widely different the second experimenter's scales are from those adopted by the first experimenter. Because multiplying all the a's by some constant factor will not change the actual situation, we need to choose, for G, some suitable homogeneous function of degree zero in the a's. One such suitable function is $G = |a|_{max} / |a|_{min}$. If we adopt this crite-

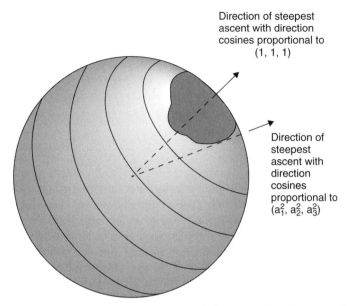

Direction of steepest
ascent with direction
cosines proportional to
(1, 1, 1)

Direction of
steepest
ascent with
direction
cosines
proportional to
(a_1^2, a_2^2, a_3^2)

Figure 6A.1. Two directions of steepest ascent, one of which traces a closed curve on the unit sphere about the other (see text).

rion, we shall regard all values of the a's that produce the same value, G_0 say, of G as representing a degree of scaling equally discordant with the original one. Such values of the a's will trace out a closed curve on the unit sphere as shown in Figure 6A.1. Thus, a measure of how much uncertainty is introduced in the direction of steepest ascent for a scaling as discordant as, or less discordant than, that represented by $G = G_0$ will be represented by the content of the closed curve bounded by $G = G_0$ on the unit sphere, taken as a proportion of the whole surface of the unit sphere. In the three-dimensional case of Figure 6A.1, this is simply proportional to the surface of the cap enclosed by the curve $G = G_0$ on the unit sphere.

Now let us write $a_i^2 = y_i$, then $G^4 = a_{max}^4/a_{min}^4 = y_{max}^2/y_{min}^2$ and the inequality $G < G_0$ is equivalent to the inequality $G^4 < G_0^4$. It can be shown that the required value is

$$\phi = \frac{1}{(2\pi)^{k/2}} \int_R \exp\left\{ -\tfrac{1}{2} \sum_{i=1}^{k} y_i^2 \right\} dy, \qquad (6A.1)$$

where the integral is taken over the region R such that $y_{max}^2/y_{min}^2 < G_0^4$. This is the same integral as would arise in a comparison of variances each having 1 df using Hartley's (1950) F_{max} criterion. H. A. David (1952) gives the upper and lower 5% and 1% points of F_{max} for a set of k variance estimates each based on ν df for $\nu \geq 2$. The test required here is for $\nu = 1$, which is not given in the tables presumably because it is usually of very little importance. The appropriate values for this integral have, however, been calculated for $\nu = 1$ to give Table 6A.1.

As an example of the use of Table 6A.1, consider the following, in which we shall leave aside the question of sampling error. Suppose we were concerned with $k = 5$ variables. Let us further suppose that experimenters may be expected to

Table 6A.1. Table of the Reduction of the "Percentage of Uncertainty" Due to Changes of Scale in a Particular *n*-tant

Value of G	Values of k				
	2	3	4	5	6
$\sqrt{2}$	41.0	15.5	5.6	2.0	0.7
2	68.8	45.4	28.0	16.7	11.6
3	85.7	69.8	55.8	43.9	33.8
4	93.8	83.8	75.1	67.7	61.4

differ in their ideas about scaling by a factor G representing the ratio of the largest to the smallest modifying factor. Finally, let us suppose a reference experimenter has so scaled the variables that the elements of the steepest ascent direction are proportional to $1, 1, 1, 1, 1$ in the scaling adopted. A knowledge of all the signs will reduce the uncertainty concerning the direction of advance by a factor of $1/2^5 = 1/32 = 0.03125$. If $G = 2$, the uncertainty will be reduced still further by a factor of 0.167 (obtained from Table 6A.1 with $k = 5$, $G = 2$), giving an overall uncertainty of $0.167(0.03125) = 0.0052$ for $G = 2$; by a factor of 0.439 if $G = 3$, giving an overall uncertainty of $0.439(0.03125) = 0.014$; and by a factor of 0.677 for $G = 4$, giving an overall uncertainty of $0.677(0.03125) = 0.021$. Our certainty of knowledge of the direction is thus 1 minus these figures or 0.9948, 0.986, and 0.979 for $G = 2$, 3, and 4, respectively.

In summary, the present authors feel that the direction of steepest ascent correctly distills the first-degree information concerning possible directions of improvement, by combining the experimenter's prior knowledge concerning the nature of the response relationship appropriately with the available data. This direction does and should change for different experimenters because different prior opinions must enter the problem.

APPENDIX 6B. PLACEMENT OF FIRST-ORDER DESIGNS WITHIN SPHERICAL AND CUBOIDAL REGIONS OF INTEREST

First-order designs are arrangements of experimental points that give rise to data to which a first-order (planar) model can be fitted. Each factor must of course have (at least) two levels, and so the two-level factorial or fractional factorial designs of Chapter 5 are obvious candidates. Designs of resolution $R = 4$ or higher are recommended to avoid aliasing first-order terms with second-order terms; often, center points are added because this permits an overall test for second-order curvature; see Sections 6.3 and 13.8. In deciding to run such a design, one is faced with the question, "How widely should the points be spaced in the experimental region being considered?" We offer here some useful guidelines based on work by Box and Draper (1959, 1963); see also the theoretical detail in Chapter 13.

We first assume that we have a specified region of interest in the coded predictor variables. Experimenters often express this region as a sphere or a cuboid, and these will be the applications we discuss here. We shall assume the sphere has radius 1 and the cuboid has width 2 in each coded x-dimension and that both are centered at the origin. These regions have moments obtained by calculus methods, integrating quantities of the form $x_1^a x_2^b \ldots x_k^k/n$ over the respective region and dividing the result by a volume scaling factor, obtained by integrating 1 over the same region. Similarly, given a selection of points that form a design, we can evaluate corresponding design moments by averaging quantities of the form $x_1^a x_2^b \ldots x_k^k/n$ over the design points.

We now state two very general theoretical results, and then explain why they help us in practice.

1. Suppose there are no experimental errors in the observations, and the only errors in fitting the data to a model of degree d_1 arise from the fact that we are fitting a model of too low a degree. Assume that the true model is of degree $d_2 > d_1$. We thus say that there is no *variance error*, but only *bias error*. Then the best design will be obtained by making the moments of the design equal to the moments of the region up to and including order $d_1 + d_2$. Such a design is called an *all-bias design*. [We note that these conditions are "sufficient" ones. The "necessary and sufficient" conditions are given in Box and Draper (1963, p. 339.)]

2. Suppose, alternatively, there is no bias error; that is, the model fitted is of an appropriate order (so that $d_2 = d_1$). Then the best design will be obtained by making the moments of the design as large as possible, thus spreading out the design points as far as the experimental limitations allow. Such a design is called an *all-variance design*.

In practice, neither extreme is usually appropriate. However, studies (e.g., Draper and Lawrence, 1965a) have shown that the following compromise is often a sensible and practical one. If $d_2 = d_1 + 1$ (that is, we fit a polynomial of order d_1 but are somewhat worried that we might need a polynomial of one order higher at worst), first select an appropriate all-bias design as in paragraph 1 above; then expand the linear dimensions by about 10–20%. This result is based on the idea that, when both variance and bias errors exist, *such a design is appropriate for situations where the two kinds of error are roughly of the same size.*

Spherical Region

For a spherical region of interest centered at the origin, for example, all odd-order region moments, including the first and third, are zero, and the pure second-order region moment has value $1/(k + 2)$. So if we wish to use a two-level factorial or fractional factorial of resolution $R \geq 4$ to fit a plane, we need to match up only the second-order moments, because the first and third ($d_1 + d_2 = 3$ here) are already zero. Suppose the design has n_a factorial type points $(\pm a, \pm a, \ldots, \pm a)$ and suppose further that we add n_0 center points; then, for the all-bias design we need to set $n_a a^2/(n_a + n_0) = 1/(k + 2)$—that is, to solve

$$a^2 = (1 + n_0/n_a)/(k + 2).$$

Some example values of a are given in the display below for $n_0 = 0$, 1, 2, 3, and 4.

k	n_a	$n_0 = 0$	$n_0 = 1$	$n_0 = 2$	$n_0 = 3$	$n_0 = 4$
2	4	0.500	0.559	0.612	0.661	0.707
3	8	0.447	0.474	0.500	0.524	0.548
4	8	0.408	0.433	0.456	0.479	0.500
4	16	0.408	0.421	0.433	0.445	0.456
5	16	0.378	0.390	0.401	0.412	0.423
5	32	0.378	0.384	0.390	0.395	0.401

Next, expanding these designs by, say, 15% leads to the following suggested values of a.

k	n_a	$n_0 = 0$	$n_0 = 1$	$n_0 = 2$	$n_0 = 3$	$n_0 = 4$
2	4	0.575	0.643	0.704	0.761	0.813
3	8	0.514	0.545	0.575	0.603	0.630
4	8	0.469	0.498	0.524	0.551	0.575
4	16	0.469	0.484	0.498	0.512	0.525
5	16	0.435	0.448	0.461	0.474	0.486
5	32	0.435	0.441	0.449	0.455	0.461

Thus, for example, for a 2^3 design with three center points, the factorial points are at $(\pm 0.603, \pm 0.603)$ in relation to a sphere of radius 1.

Cuboidal Region

A cuboidal region centered at the origin with sides parallel to the axes also has zero value odd-order region moments, and the pure second-order region moment is $\frac{1}{3}$ for all k. We first solve $a^2 = (1 + n_0/n_a)/3$. When $n_0 = 0$, $a = 0.577$ for all k. For $n_0 = 0$, 1, 2, 3, and 4, we have the values displayed below.

k	n_a	$n_0 = 0$	$n_0 = 1$	$n_0 = 2$	$n_0 = 3$	$n_0 = 4$
2	4	0.577	0.645	0.707	0.764	0.816
3	8	0.577	0.612	0.645	0.677	0.707
4	8	0.577	0.612	0.645	0.677	0.707
4	16	0.577	0.595	0.612	0.629	0.645
5	16	0.577	0.595	0.612	0.629	0.645
5	32	0.577	0.586	0.595	0.604	0.612

Multiplying these a values by 1.15 then provides the following suggested values of a.

k	n_a	$n_0 = 0$	$n_0 = 1$	$n_0 = 2$	$n_0 = 3$	$n_0 = 4$
2	4	0.664	0.742	0.813	0.878	0.939
3	8	0.664	0.704	0.742	0.779	0.813
4	8	0.664	0.704	0.742	0.779	0.813
4	16	0.664	0.684	0.704	0.724	0.742
5	16	0.664	0.684	0.704	0.724	0.742
5	32	0.664	0.674	0.684	0.694	0.704

Note that for both region shapes, as center points are added, the factorial design portion expands because more information is being provided at the center of the design.

The values shown for both regions should not be rigidly interpreted, but instead used to give a general feel for how spread out a factorial design should be in relation to the regions of interest assumed above. In use, the designs would usually be re-coded so that the design levels are -1 and 1. Here, the a values are those appropriate in relation to a sphere of radius 1 and a cuboid of side 2, both centered at the origin.

Applications of these methods to second-order designs will be found in Appendix 7H.

EXERCISES

6.1. An experimenter begins a steepest ascent procedure on two variables (X_1, X_2) at the current central point $(90, 20)$ and performs five runs with the response results provided:

X_1	80	100	80	100	90
X_2	10	10	30	30	20
y	11	0	29	6	12

Code (X_1, X_2) sensibly to variables (x_1, x_2) and fit a first-order (planar) model $\hat{y} = b_0 + b_1 x_1 + b_2 x_2$ to the data. Determine the direction of steepest ascent. The experimenter now performs six more runs:

X_1	64.5	47.5	39	30.5	43.25	34.75
X_2	38	50	56	62	53	59
y	43	58	72	62	65	68

Which of these runs lie on the path of steepest ascent you determined earlier? The experimenter decides now to combine the two runs $(X_1, X_2, y) = (43.25, 53, 65)$ and $(34.75, 59, 68)$ with six more, namely these:

X_1	34.75	43.25	39	39	39	39
X_2	53	59	56	56	56	56
y	71	68	71	72	72	73

Fit a first-order model to those eight runs and use the repeat observations to test for lack of fit. Plot, as points on a diagram, all the runs performed so far, with their y values attached. What should the experimenter do next? Go off in a new direction of steepest ascent? Fit a second-order surface? Or what?

6.2. Two coded variables $x_1 = (X_1 - 99)/10$, $x_2 = (X_2 - 17)/20$ are examined, and the data of Table E6.2 are obtained. Is the point $(X_1, X_2) = (59, 67)$ on the path of steepest ascent?

Table E6.2

x_1	x_2	y
-1	-1	11
1	-1	2
-1	1	29
1	1	6
0	0	12

6.3. In a steepest ascent investigation to examine the effects of three variables, the initial design of Table E6.3 was employed.

The model $\hat{y} = 4.7 + 0.25x_1 + x_2 + 0.5x_3$ is fitted. The following restriction exists:

$$4X_1 + 5X_2 + 6X_3 \le 371.$$

Is the point $(X_1, X_2, X_3) = (36.40, 27.10, 15.00)$ on the restricted path of steepest ascent? (Allow for the fact that round-off errors may occur unless adequate figures are retained.)

Table E6.3

	Uncoded			Coded		
X_1	X_2	X_3	x_1	x_2	x_3	
30	10	6	-1	-1	-1	
40	10	6	1	-1	-1	
30	20	6	-1	1	-1	
40	20	6	1	1	-1	
30	10	16	-1	-1	1	
40	10	16	1	-1	1	
30	20	16	-1	1	1	
40	20	16	1	1	1	
35	15	11	0	0	0	
35	15	11	0	0	0	
35	15	11	0	0	0	
35	15	11	0	0	0	

6.4. A mixture of three ingredients X_1, X_2, X_3 (plus inerts to make up to 100%) is being examined. Steepest ascent techniques are being applied to improve the response variable, percentage yield. The following fitted equation is obtained from the first set of runs made:

$$\hat{y} = 9.925 - 4.10x_1 - 9.25x_2 + 4.90x_3,$$

where

$$x_1 = X_1 - 8, \qquad x_2 = X_2 - 14, \qquad x_3 = 2(X_3 - 7).$$

If the restriction $X_1 + X_2 + X_3 \geq 26.4$ applies, is the point $X_1 = 7.11$, $X_2 = 11.29$, $X_3 = 8.00$ on the path of steepest ascent? (There will undoubtedly be a little rounding error in your calculations, remember.)

6.5. A four-dimensional steepest ascent path is obtained from the fitted coefficients $b_1 = 1$, $b_2 = 2$, $b_3 = 3$, $b_4 = 2$. If we have to keep inside or on the restricting boundary

$$-4 + x_1 + x_2 + x_3 + x_4 = 0,$$

answer this question: Is the point $(X_1, X_2, X_3, X_4) = (-5/2, 10, 45/2, 10)$ on the adjusted path of steepest ascent? Assume $x_i = (X_i - 5)/5$, $i = 1, 2, 3, 4$.

6.6. A manufacturer of circular saw blades has two empirical first-order equations for blade life in hours (η_1, observed as y_1) and for unit blade cost in cents (η_2; y_2) in terms of the manufacturing time (coded to x_1) and the hardness of the steel used (coded to x_2). These are

$$\hat{y}_1 = 20 + x_1 + x_2,$$
$$\hat{y}_2 = 50 + 4x_1 + 2x_2,$$

and they are valid only for $-2 \leq x_i \leq 2$. If the unit blade cost must be kept below 54 cents and the blade life must exceed 21.5 h, can the manufacturer produce blades? If yes, explain how he must operate. If no, explain why he cannot.

CHAPTER 7

Fitting Second-Order Models

In Chapter 6 we saw how, in appropriate situations, a fitted first-degree approximation to the response function could be exploited to produce process improvement. In this chapter, we begin to consider the use of quadratic (second-order) approximating functions.

7.1. FITTING AND CHECKING SECOND-DEGREE POLYNOMIAL GRADUATING FUNCTIONS

To fix ideas, we discuss in more detail an example first introduced in Chapter 2 concerning the behavior of worsted yarn under cycles of repeated loading. For simplicity in the earlier analysis, only 8 of the full set of 27 runs were utilized. These 8 runs are distinguished by superscript b's in Table 7.1, where the complete data are set out.

Reconsideration of the Textile Example

It will be recalled that, in this example, the three inputs were

$$\xi_1 = \text{the length of test specimen (mm)},$$

$$\xi_2 = \text{the amplitude of load cycle (mm)},$$

$$\xi_3 = \text{the load (g)},$$

which were conveniently coded in terms of

$$x_1 = \frac{(\xi_1 - 300)}{50}, \quad x_2 = (\xi_2 - 9), \quad x_3 = \frac{(\xi_3 - 45)}{5}. \quad (7.1.1)$$

The output (or response) of interest was $Y =$ number of cycles to failure. However the analysis was conducted in terms of $y = \log_{10} Y$ and the first-degree approximat-

---

Response Surfaces, Mixtures, and Ridge Analyses, Second Edition. By G. E. P. Box and N. R. Draper

211

Table 7.1. Textile Data for 3 × 3 × 3 Factorial Design, from an Unpublished Report to the Technical Committee, International Wool Textile Organization by Dr. A. Barella and Dr. A. Sust

	ξ_1 = length of specimen (mm) $x_1 = (\xi_1 - 300)/50$		ξ_2 = amplitude of load cycle (mm) $x_2 = (\xi_2 - 9)/1$	ξ_3 = load (g) $x_3 = (\xi_3 - 45)/5$	
Run Number[a]	Length, x_1	Amplitude, x_2	Load, x_3	Y = cycles to failure	$y = (\log_{10} Y)$
1[b]	−1	−1	−1	674	2.83
2	0	−1	−1	1414	3.15
3[b]	1	−1	−1	3636	3.56
4	−1	0	−1	338	2.53
5	0	0	−1	1022	3.01
6	1	0	−1	1568	3.19
7[b]	−1	1	−1	170	2.23
8	0	1	−1	442	2.65
9[b]	1	1	−1	1140	3.06
10	−1	−1	0	370	2.57
11	0	−1	0	1198	3.08
12	1	−1	0	3184	3.50
13	−1	0	0	266	2.42
14	0	0	0	620	2.79
15	1	0	0	1070	3.03
16	−1	1	0	118	2.07
17	0	1	0	332	2.52
18	1	1	0	884	2.95
19[b]	−1	−1	1	292	2.47
20	0	−1	1	634	2.80
21[b]	1	−1	1	2000	3.30
22	−1	0	1	210	2.32
23	0	0	1	438	2.64
24	1	0	1	566	2.75
25[b]	−1	1	1	90	1.95
26	0	1	1	220	2.34
27[b]	1	1	1	360	2.56
Total sum of squares				ΣY^2 = 40,260,625,	Σy^2 = 208.68 (27 df)

[a] Run number is used for identification only and does not indicate the order in which the runs were actually made.
[b] Runs employed in earlier analysis.

ing equation fitted in Chapter 2 was

$$\hat{y} = 2.745 + 0.375x_1 - 0.295x_2 - 0.175x_3, \tag{7.1.2}$$

which (see Table 2.3) gave fitted values agreeing very closely with the eight observations considered.

Structure of Empirical Models

Let us recapitulate certain ideas, using this example for illustration. There presumably exists some true physical relationship between the expectation $E(y) = \eta$ of the output y and the three inputs ξ_1, ξ_2, and ξ_3 via physical constants $\boldsymbol{\theta}$. The nature of this true *expectation function*

$$\eta = E(y) = f(\xi_1, \xi_2, \xi_3, \boldsymbol{\theta}) = f(\boldsymbol{\xi}, \boldsymbol{\theta}) \tag{7.1.3}$$

for this system is, however, unknown to us. In particular, we do not know the true functional form f, nor do we know the nature of the constants $\boldsymbol{\theta}$.

We therefore replace $f(\boldsymbol{\xi}, \boldsymbol{\theta})$ by a graduating function $g(\mathbf{x}, \boldsymbol{\beta})$ which we hope can approximate it locally. In the analysis of Chapter 2, it was supposed that the expected value of $y = \log Y$ could be represented by a first-degree polynomial graduating function

$$g_1(\mathbf{x}, \boldsymbol{\beta}) = \beta_0 + \beta_1 x_1 + \beta_2 x_2 + \beta_3 x_3, \qquad (7.1.4)$$

where each x_i was a linear coding $x_i = (\xi_i - \xi_{i0})/S_i$ of an input ξ_i. It should be remembered that, in general, the x_i's could be any *functions* $x_i = f_i(\xi_1, \xi_2, \xi_3)$, $i = 1, 2, \ldots$, of some or all of the inputs which the experimenter feels are appropriate. In some engineering applications, for instance, they might be suitably chosen dimensionless groups.

The graduating function approximates the expectation function so that

$$E(y) = \eta \simeq g(\mathbf{x}, \boldsymbol{\beta}), \qquad (7.1.5)$$

which establishes an approximate link between $E(y)$ and $\boldsymbol{\xi}$ the expected output and the inputs. In practice, we do not know the mean value $E(y)$ of the output response for any particular choice of the inputs. We have only an observation y (or sometimes a number of replicated observations) subject to error $\varepsilon = y - E(y)$. Although, in any given case, the actual error ε is unknown, we suppose it to arise from a fixed distribution which we refer to as the error distribution. We shall assume that (possibly after suitable transformation of the response), the errors are to a sufficient approximation distributed normally and independently of one another with the same variance σ^2. The observed output response y is linked to the inputs in $\boldsymbol{\xi}$ in two stages. The link between y and $E(y)$ occurs via the error distribution, and the link between $E(y)$ and $\boldsymbol{\xi}$ via the graduating function:

$$y \xrightarrow{\text{error distribution}} E(y) \xrightarrow{\text{graduating function}} \xi.$$

Thus, there are two sources of error:

1. Random error $\varepsilon = y - E(y)$,
2. Systematic error, or bias, $E(y) - g(\mathbf{x}, \boldsymbol{\beta})$, arising from the inability of the graduating function $g(\mathbf{x}, \boldsymbol{\beta})$ to exactly match the expectation function $E(y) = f(\boldsymbol{\xi}, \boldsymbol{\theta})$.

As we shall see later, it is important to keep in mind that both kinds of errors are involved. This consideration is helpful in deciding how precise a graduating function needs to be. In particular, there is little point in straining hard to reduce the level of systematic errors in the estimated response function much below that induced by random errors.

Analysis of Variance for Textile Example (Logged Data)

In our present consideration of the textile example, we now employ the whole set of data from the 27 runs set out in Table 7.1. For the time being, we continue to

Table 7.2. Textile Data. Polynomials of Zero-, First-, and Second-Degree Fitted to Logged Data, with Standard Errors of Coefficients Indicated by $\pm$ Values

Zero-Degree Polynomial

Estimated response function is $\qquad$ $\hat{y} = \bar{y} = 2.751$
$$\pm 0.016$$

Regression sum of squares is $\qquad$ $S_0 = (\Sigma y)^2 / n = 204.2975 \ (1 \ df)$

First-Degree Polynomial

Estimated response function is $\qquad$ $\hat{y} = \quad 2.751 + 0.362 x_1 - 0.274 x_2 - 0.171 x_3$
$$\pm 0.016 \pm 0.020 \quad \pm 0.020 \quad \pm 0.020$$

Regression sum of squares is $\qquad$ $S_1 = 208.5292 \ (4 \ df)$

Second-Degree Polynomial

Estimated response function is $\qquad$ $\hat{y} = \quad 2.786 + 0.362 x_1 - 0.274 x_2 - 0.171 x_3$
$$\pm 0.043 \pm 0.020 \quad \pm 0.020 \quad \pm 0.020$$

$$-0.037 x_1^2 + 0.013 x_2^2 - 0.029 x_3^2$$
$$\pm 0.034 \quad \pm 0.034 \quad \pm 0.034$$

$$-0.014 x_1 x_2 - 0.029 x_1 x_3 - 0.010 x_2 x_3$$
$$\pm 0.024 \quad \quad \pm 0.024 \quad \quad \pm 0.024$$

Regression sum of squares is $\qquad$ $S_2 = 208.5573 \ (10 \ df)$

work in terms of $y = \log Y$, the log of the number of cycles to failure. Later in this chapter we discuss in some detail questions concerning such data transformations.

The design used in this investigation was a $3 \times 3 \times 3$ factorial. That is, each of three levels of the three inputs were run in all combinations. The provision of three levels allows in particular a general quadratic polynomial to be fitted, and the results of doing this as well as of fitting first- and zero-degree polynomials to the complete set of data are set out in Table 7.2. The $\pm$ limits below the estimated coefficients are $\pm$ their standard errors, calculated from the square roots of the diagonal terms of the matrix $(\mathbf{Z'Z})^{-1} s^2$ (see Appendix 7A). We use the estimate $s^2 = 0.0071$, obtained from the residual for the second degree model, in all calculations to obtain consistency between the three sets of standard errors. Also shown in Table 7.2 are the regression sums of squares and degrees of freedom associated with the fits of successively higher order. From these, the extra sums of squares $S_1 - S_0$, $S_2 - S_1$, and their associated degrees of freedom have been calculated, to yield the analysis of variance of Table 7.3.

A question of immediate interest is whether the fitted first-degree equation provides an adequate representation of the response function. In this design, there was no replication and therefore no estimate of pure error with which the mean square associated with second-degree terms might be compared. However, on the *assumption* that an adequate model is supplied by a polynomial of *at most* second degree the residual mean square provides an estimate of error. The F ratio

$$\frac{\text{mean square for added second-degree terms}}{\text{mean square for residual}} = \frac{0.0047}{0.0071} = 0.66 \quad (7.1.6)$$

Table 7.3. Analysis of Variance for Textile Example (Logged Data)

Source	SS	df	MS
Mean (zero-degree polynomial)	204.2975	1	
Added first-order terms	4.2317	3	1.4106
Added second-order terms	$\left.\begin{array}{l} 0.0281 \\ 0.1214 \end{array}\right\} 0.1495$	$\left.\begin{array}{l} 6 \\ 17 \end{array}\right\} 23$	$\left.\begin{array}{l} 0.0047 \\ 0.0071 \end{array}\right\} 0.0065$
Residual			
Total	208.6787	27	

provides no reason to doubt the adequacy of the first-degree polynomial representation. At this stage of our analysis, we are thus *tentatively* entertaining the fitted first-degree model

$$\hat{y} = 2.751 + 0.362x_1 - 0.274x_2 - 0.171x_3. \qquad (7.1.7)$$

On this basis, the residual mean square obtained from fitting the first-degree model provides an estimate of σ^2 of

$$\frac{(0.0281 + 0.1214)}{(6 + 17)} = 0.0065 \qquad (7.1.8)$$

based on 23 df. This leads to standard error values slightly smaller than those previously quoted. Note that our present fitted model based on all 27 observations agrees very well with that fitted in Chapter 2 [see Eq. (7.1.2)], which, however, used only 8 of the 27 observations.

Choice of Metric for the Output Response

We shall return to this example once again when we come to consider, in Chapter 8, the general question of *estimating* appropriate transformations for the output and input variables to ensure maximum simplicity in the model. For the moment, we take note of the fact that the analysis has so far been conducted in terms of the *logarithm*, $y = \log Y$, of the number of cycles to failure rather than in terms of Y itself. We have already noted that, in choosing a transformation or "metric" in terms of which the output (response) is best expressed, the implications for the *error distribution* must always be kept in mind. The estimation procedure of (unweighted) least squares is efficient, provided that the standard normal theory assumptions are roughly true for the error distribution implied by the choice of model. Of major importance among these standard assumptions is the supposition that the observations have (approximately) constant variance. In Box and Cox (1964), it was pointed out that, for this particular example, the logarithmic transformation is a sensible choice on prior grounds. The number of cycles to failure has a range zero to infinity and the actual data values extend from $Y = 90$ to $Y = 3636$. It is much more likely with such data that the *percentage* errors made at different sets of experimental conditions, rather than the absolute errors, will be roughly comparable. This is equivalent to saying that, as a first guess at least,

$y = \log Y$ (which has a range from minus infinity to plus infinity) might be expected to have constant variance, in which case the standard least-squares procedure would be efficient for a model of the form

$$\log Y = y \simeq g(\mathbf{x}, \boldsymbol{\beta}) + \varepsilon. \tag{7.1.9}$$

Some further checks on the adequacy of the first degree equation in $\log Y$ are provided by examination of the residuals calculated in Table 7.4. Figure 7.1a shows

Table 7.4. Observations, Fitted Values, and Residuals for the Unlogged Data Y and Logged Data $y = \log Y$. In each case, the first-order model form $\beta_0 + \beta_1 x_1 + \beta_2 x_2 + \beta_3 x_3 + \varepsilon$ and the second-order model form $\beta_0 + \beta_1 x_1 + \beta_2 x_2 + \beta_3 x_3 + \beta_{11} x_1^2 + \beta_{22} x_2^2 + \beta_{33} x_3^2 + \beta_{12} x_1 x_2 + \beta_{13} x_1 x_3 + \beta_{23} x_2 x_3 + \varepsilon$ have been fitted to the response data. (Note: Residuals are calculated from the fitted model before round-off of coefficients.)

| | | \multicolumn{4}{c}{Cycles to Failure} | \multicolumn{4}{c}{$\log_{10}$ (Cycles to Failure)} |
| | | First-degree Model | | Second-Degree Model | | First-Degree Model | | | Second-Degree Model | |
Run Number	Y	$\hat{Y}$	$Y - \hat{Y}$	$\hat{Y}$	$Y - \hat{Y}$	y	$\hat{y}$	$y - \hat{y}$	$\hat{y}$	$y - \hat{y}$
1	674	1048	−374	654	20	2.83	2.83	0.00	2.76	0.07
2	1414	1708	−294	1768	−354	3.15	3.20	−0.05	3.21	−0.06
3	3636	2368	1268	3358	278	3.56	3.56	0.00	3.57	−0.01
4	338	512	−174	156	182	2.53	2.56	−0.03	2.50	0.03
5	1022	1172	−150	813	209	3.01	2.92	0.09	2.93	0.08
6	1568	1832	−264	1947	−379	3.19	3.28	−0.09	3.28	−0.09
7	170	−24	194	210	−39	2.23	2.29	−0.06	2.26	−0.03
8	442	636	−194	409	32	2.65	2.65	0.00	2.68	−0.03
9	1140	1296	−156	1088	52	3.06	3.01	0.05	3.02	0.04
10	370	737	−367	484	−114	2.57	2.66	−0.09	2.66	−0.09
11	1198	1397	−199	1362	−164	3.08	3.02	0.06	3.07	0.01
12	3184	2057	1127	2717	467	3.50	3.39	0.11	3.41	0.09
13	266	201	65	129	137	2.42	2.39	0.03	2.39	0.03
14	620	861	−241	551	69	2.79	2.75	0.04	2.79	0.00
15	1070	1521	−451	1449	−379	3.03	3.11	−0.08	3.11	−0.08
16	118	−335	453	326	−208	2.07	2.12	−0.05	2.14	−0.07
17	332	325	7	290	42	2.52	2.48	0.04	2.53	−0.01
18	884	985	−101	733	151	2.95	2.84	0.11	2.84	0.11
19	292	426	−134	218	74	2.47	2.49	−0.02	2.50	−0.03
20	634	1086	−452	860	−226	2.80	2.85	−0.05	2.88	−0.08
21	2000	1746	254	1980	20	3.30	3.22	0.08	3.19	0.11
22	210	−109	319	6	204	2.32	2.22	0.10	2.22	0.10
23	438	551	−113	192	246	2.64	2.58	0.06	2.59	0.05
24	566	1211	−645	855	−289	2.75	2.94	−0.19	2.88	−0.13
25	90	−645	735	345	−255	1.95	1.94	0.01	1.96	−0.01
26	220	15	205	74	146	2.34	2.31	0.03	2.32	0.02
27	360	675	−315	281	79	2.56	2.67	−0.11	2.60	−0.04
Check totals (should be zero, within rounding)			3		1			−0.01		0.02

LOGGED DATA

(a)   Residuals Overall

(b)   Residuals vs. $\hat{y}$.

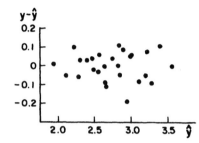

Figure 7.1. Plots of residuals $y_u - \hat{y}_u$ from first-degree model fitted to logged data. (a) Residuals overall, (b) Residuals versus $\hat{y}_u$.

the residuals $y_u - \hat{y}_u$ from the fitted first-degree model plotted in a dot diagram. Figure 7.1b shows the $y_u - \hat{y}_u$ plotted against $\hat{y}_u$. If the standard deviation σ_y increased as the mean value $E(y)$ increased, we should expect the range of the residuals to increase as $\hat{y}_u$ increased, giving rise to a triangular-shaped scatter. The plot in Figure 7.1b shows no tendency of this sort, however. If the y's were not suitably transformed, we should expect to see a quadratic tendency in Figure 7.1b; no such tendency is evident.

To see how this quadratic tendency *can* occur, suppose in general that some model form

$$\beta_0 + g(\mathbf{x}, \boldsymbol{\beta}) \tag{7.1.10}$$

is being considered, where $g(\mathbf{x}, \boldsymbol{\beta})$ is some graduating function; for instance in the textile example, $g(\mathbf{x}, \boldsymbol{\beta}) = \beta_1 x_1 + \beta_2 x_2 + \beta_3 x_3$. Suppose that the model form would be more nearly appropriate after suitable data transformation. That is to say, a better approximation would be provided by the model form

$$\eta^{(\lambda)} = \beta_0 + g(\mathbf{x}, \boldsymbol{\beta}) \tag{7.1.11}$$

where $\eta^{(\lambda)}$ is a parametric transformation of η, such as a power transformation, for example. In these circumstances, we can employ, over moderate ranges of η, the approximation

$$\eta^{(\lambda)} = \alpha_0 + \alpha_1 \eta - \alpha_2 \eta^2. \tag{7.1.12}$$

Table 7.5. Textile Data: Polynomials of Zero-, First-, and Second-Degree Fitted to Original (Unlogged) Data

---

Zero-Degree Polynomial

Estimated response function is	$\hat{Y} = \bar{Y} = 861.3$ ± 52.3
Regression sum of squares is	$S_0 = 20{,}031.2 \times 10^3$ (1 df)

First-Degree Polynomial

Estimated response function is	$\hat{Y} = 861.3 + 660.0x_1 - 535.9x_2 - 310.8x_3$ $\pm 52.3 \quad \pm 64.1 \qquad \pm 64.1 \qquad \pm 64.1$
Regression sum of squares is	$S_1 = 34{,}779.7 \times 10^3$ (4 df)

Second-Degree Polynomial

Estimated response function is	$\hat{Y} = \quad 550.7 + 660.0x_1 - 535.9x_2 - 310.8x_3$ $\pm 138.4 \quad \pm 64.1 \qquad \pm 64.1 \qquad \pm 64.1$ $+ 238.7x_1^2 + 275.7x_2^2 - 48.3x_3^2$ $\pm 111.0 \qquad \pm 111.0 \qquad \pm 111.0$ $- 456.5x_1x_2 - 235.7x_1x_3 + 143.0x_2x_3$ $\pm 78.5 \qquad \pm 78.5 \qquad \pm 78.5$
Regression sum of squares is	$S_2 = 39{,}004.0 \times 10^3$ (10 df)

Then, after equating (7.1.11) and (7.1.12), dividing through by α_1, and renaming the parameters, the model form becomes, approximately,

$$\eta = \alpha\eta^2 + \beta_0' + g(\mathbf{x}, \boldsymbol{\beta}'). \tag{7.1.13}$$

This relationship applies for each set of observations so that

$$\eta_u - \beta_0' - g(\mathbf{x}_u, \boldsymbol{\beta}') = \alpha\eta_u^2, \quad u = 1, 2, \ldots, n. \tag{7.1.14}$$

We can obtain rough estimates of the left-hand side of (7.1.14) by fitting the model $y_u = \beta_0' + g(\mathbf{x}_u, \boldsymbol{\beta}') + \varepsilon$ by least squares and finding the residuals $y_u - \hat{y}_u$. Also the same $\hat{y}_u$'s will provide estimates of the η_u's. Then, if $y = \log Y$ were an unsuitable transformation, we should expect a quadratic relationship to show up between the $y_u - \hat{y}_u$ and the $\hat{y}_u$ because they would reflect the relationship indicated in (7.1.14). No such tendency appears, however, as Figure 7.1b shows.

An Analysis of the Unlogged Data

To illustrate the importance of appropriate transformation and the value of the various residual checks, we consider what would have happened if the logarithmic transformation had *not* been used in this example. Table 7.5 summarizes the results of fitting polynomials of degrees 0, 1, and 2 to the original (unlogged) data Y. The corresponding analysis of variance for Y is shown in Table 7.6. If, as before, the assumption were made that an adequate model was supplied by a

Table 7.6. Analysis of Variance for Unlogged Textile Data

Source	$SS \times 10^{-3}$	df	$MS \times 10^{-3}$
Mean (zero-degree polynomial)	20,031.2	1	
Added first-order terms	14,748.5	3	4,916.2
Added second-order terms	4,224.3	6	704.1
Residual	1,256.6	17	73.9
Total	40,260.6	27	

polynomial of *at most* second degree, then the residual mean square will provide an estimate of error. The ratio

$$\frac{\text{mean square for added second-order terms}}{\text{mean square for residual}} = 9.52 \qquad (7.1.15)$$

is now much larger than for the logged data, and reference to the F table with 6 and 17 df shows it to be highly significant. Thus, for the unlogged data, second-degree terms are needed to represent curvature previously taken account of by the log transformation, and the fitted equation now takes the more complicated second-degree form

$$\hat{Y} = 550.7 + 660.0x_1 - 535.9x_2 - 310.8x_3$$
$$+ 238.7x_1^2 + 275.7x_2^2 - 48.3x_3^2$$
$$- 456.5x_1x_2 - 235.7x_1x_3 + 143.0x_2x_3. \qquad (7.1.16)$$

Notice that this more complicated representation in terms of Y is less satisfactory in accounting for the variation in the data than the simple first-order representation in $y = \log Y$. Specifically, the F value corresponding to the ratio (regression mean square)/(residual mean square) is 7.6 times larger for the logged data employing a first-order graduating function than for the unlogged data employing a second-order graduating function. In this example it may also be shown (Box and Fung, 1983) that the reduction factor in the variance of estimation of the response caused by employing the transformed (logged) response in combination with the simpler (first-degree) model, is greater than 1 at every point in the design. At one point it is as high as 308. The values of Y, $\hat{Y}$, and $Y - \hat{Y}$ are shown in Table 7.4, both for the first-degree fitted model and for the second-degree fitted model. One feature immediately apparent from the fitted first-degree model in the original metric is that it does not make physical sense, because four of the calculated values of $\hat{Y}$, representing the "number of cycles to failure," are negative. The need for a transformation of the original observations Y is confirmed by the quadratic tendency evident in the plots of $Y - \hat{Y}$ versus $\hat{Y}$ shown in Figures 7.2b and 7.3b for the first degree and the second-degree models. Evidently the need for transformation is not totally eliminated by the introduction of second-order terms.

UNLOGGED DATA

First-Degree Model   Second-Degree Model

(a)

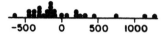

(b)

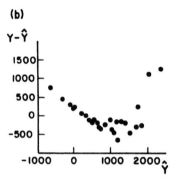

Figure 7.2. Plots of residuals $Y_u - \hat{Y}_u$ from first-degree model fitted to unlogged data; (*a*) Residuals overall, (*b*) Residuals versus $\hat{Y}_u$.

Figure 7.3. Plots of residuals $Y_u - \hat{Y}_u$ from second-degree model fitted to unlogged data; (*a*) Residuals overall, (*b*) Residuals versus $\hat{Y}_u$.

7.2. A COMPREHENSIVE CHECK OF THE NEED FOR TRANSFORMATION IN THE RESPONSE VARIABLE

A comprehensive check of the need for transformation is provided [see Atkinson (1973b)] by the appropriate predictive score function

$$g = \sum_{u=1}^{n} \frac{z_u r_u}{s}, \tag{7.2.1}$$

where

$$z_u = Y_u \left\{ 1 - \ln\left(\frac{Y_u}{\dot{Y}}\right) \right\}, \qquad r_u = \frac{(Y_u - \hat{Y}_u)}{s}, \qquad s^2 = \sum_{u=1}^{n} \frac{(Y_u - \hat{Y}_u)^2}{(n-p)}, \tag{7.2.2}$$

and where ln denotes logarithms to the base e, n is the number of observations, p is the number of parameters in the model, and

$$\ln \dot{Y} = \frac{\left(\sum_{u=1}^{n} \ln Y_u\right)}{n}, \tag{7.2.3}$$

so that $\dot{Y}$ is the geometric mean of the data.

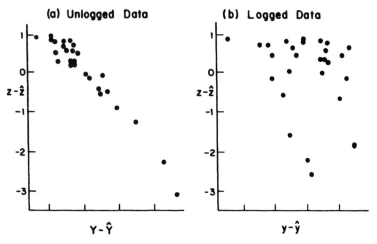

Figure 7.4. Plot of residuals from score function statistic versus corresponding residuals from the data, (*a*) unlogged, (*b*) logged, after fitting first-degree polynomial model.

Equivalently, a graphical check is provided by plotting (a) the residuals $z_u - \hat{z}_u$ obtained after fitting the constructed variable z_u to a model of the desired order or type against (b) the residuals obtained from fitting the same form of model to the response of current interest (Y or $y = \log_{10} Y$ as the case may be). The plots for the first-degree models for unlogged and logged data are given in Figures 7.4*a* and 7.4*b*. The strong dependence shown in the "unlogged" plot clearly indicates the need for transformation. In the logged data plot, this dependence has disappeared, indicating the success of the log transformation actually made.

It is instructive to consider the relationship between the score function plot Figure 7.4*a* and the earlier plot of residuals overall and versus $\hat{Y}$ for the first-degree model, in Figures 7.2*a* and 7.2*b*. These earlier plots are made in terms of nonstandardized quantities, but we could recast them in terms of the standardized residuals $r_u = (Y_u - \hat{Y}_u)/s$ and standardized deviations of calculated values from the sample average $c_u = (\hat{Y}_u - \bar{Y})/s$ by making origin and scale changes. The shapes would not alter. In these plots the need for transformation may be evidenced by:

1. Skewness in the overall plot of the residuals r_u.
2. A tendency for the spread of the residuals r_u to become larger or smaller as c_u increases or decreases.
3. A quadratic tendency in a plot of residuals r_u versus c_u.

As was noted by Anscombe (1961) and by Anscombe and Tukey (1963), appropriate checking functions for these three characteristics are, respectively,

$$\text{(a) } \Sigma r_u^3 = T_{30},$$

$$\text{(b) } \Sigma r_u^2 c_u = T_{21}, \qquad\qquad (7.2.4)$$

$$\text{(c) } \Sigma r_u c_u^2 = T_{12}.$$

An additional checking function that would measure skewness is

$$d = \frac{(\bar{Y} - \dot{Y})}{s}, \tag{7.2.5}$$

the standardized difference between the arithmetic and geometric means for the original data.

Now it can be shown that z_u is closely approximated by the quadratic expression

$$z_u \doteq \dot{Y} + B(Y_u - \dot{Y})^2, \tag{7.2.6}$$

where B is a constant that can be evaluated. Thus, after writing

$$\frac{(Y_u - \dot{Y})}{s} = r_u + c_u + d, \tag{7.2.7}$$

we see that

$$g \doteq Bs \sum_{u=1}^{n} (r_u + c_u + d)^2 r_u \tag{7.2.8}$$

$$\alpha \left\{ 2(n - p)d + T_{30} \right\} + 2T_{21} + T_{12}, \tag{7.2.9}$$

after reduction, using the facts that $\Sigma r_u = 0$, for any linear model containing a β_0 term, $\Sigma r_u \hat{Y}_u = 0$ for any linear model, and $(n - p)s^2 = \Sigma(Y_u - \hat{Y}_u)^2$. Thus the score function plot appropriately brings together all the evidence for transformation.

An exact test for one type of possible discrepancy mentioned above, namely (3) a quadratic tendency in a plot of residuals r_u versus c_u, can be carried out, as shown by Tukey (1949); see also Anscombe (1961). The test is usually referred to as "Tukey's one degree of freedom test for additivity" and is described in Appendix 7B.

7.3. INTERIM SUMMARY

1. Although we have used this example to illustrate the mechanism of fitting a second-degree model, we have also taken the opportunity to illustrate that the mechanical application of the technique can be dangerous (as can the mechanical application of any technique).

2. In the original analysis of these data, Barella and Sust had fitted a second-degree polynomial expression to the unlogged data Y. The above analysis (see also Box and Cox, 1964) shows that a simpler and more precise representation is possible by first transforming the data via $y = \log Y$, after which a first degree polynomial adequately expresses the relationship. With the response measured in the y metric, all six of the second-order parameters $\beta_{11}, \beta_{22}, \beta_{33}, \beta_{12}, \beta_{13}, \beta_{23}$ can be dispensed with, and the representation by the first-degree equation in $\log Y$ is

much better than that of the second-degree equation in Y itself. Indeed, a very large gain in accuracy of *estimation of the response* is achieved by using the simpler model.

3. Although, in this example, the need to fit a second degree expression is avoided by an appropriate use of a transformation, we shall later consider examples where a function of at least second degree in the inputs is essential to represent the dependence of output on inputs. This is especially so when we wish to represent a multidimensional maximum or minimum in the response.

4. The example illustrates how appropriate plotting of residuals can help to diagnose model inadequacy (and, in particular, the need for a transformation).

5. We later discuss how an appropriate transformation may be *estimated* from the data itself. When such an analysis is conducted for the textile data, it selects essentially the logarithmic transformation. However, we emphasize that the choice of $\log Y$ rather than Y is a natural one* and is also suggested by physical considerations.

7.4. FURTHER EXAMINATION OF ADEQUACY OF FIT; HIGHER-ORDER TERMS

The textile example has allowed us to examine a number of general questions concerning the fitting of empirical functions and the use of transformations. We shall now employ this same example to illustrate some further general points. The analysis has so far been conducted on the assumption that, at most, an equation of second degree can represent functional dependence between the response and the inputs. It is natural to ask how the analyses would be affected if terms of third order were introduced, that is, if the "true" model were of the form

$$
\begin{aligned}
y = \beta_0 &+ \beta_1 x_1 + \beta_2 x_2 + \beta_3 x_3 + \beta_{11} x_1^2 + \beta_{22} x_2^2 + \beta_{33} x_3^2 + \beta_{12} x_1 x_2 \\
&+ \beta_{13} x_1 x_3 + \beta_{23} x_2 x_3 + \big\{ \beta_{111} x_1^3 + \beta_{222} x_2^3 + \beta_{333} x_3^3 + \beta_{122} x_1 x_2^2 \\
&\quad + \beta_{112} x_1^2 x_2 + \beta_{133} x_1 x_3^2 + \beta_{113} x_1^2 x_3 \\
&\quad + \beta_{233} x_2 x_3^2 + \beta_{223} x_2^2 x_3 + \beta_{123} x_1 x_2 x_3 \big\} + \varepsilon. \quad (7.4.1)
\end{aligned}
$$

For the particular, but important, 3^3 design used in the textile experiment, we shall consider three distinct but closely related questions:

1. How would coefficient estimates derived on the assumed adequacy of a second-order model be biased by the existence of third-order terms?

---

*We have sometimes been dismayed to find that the engineer newly introduced to statistical methods may put statistics and engineering in different compartments of his mind and feel that when he is doing statistics, he no longer needs to be an engineer. This is of course not so. All his engineering knowledge, hunches, and cunning concerning such matters as choice of variables and transformation of variables must still be employed in conjunction with statistical design and analysis. Statistical methods used with good engineering know-how make a powerful combination, but poor engineering combined with mechanical and unimaginative use of inadequately understood statistical methods can be disastrous.

2. How may the need for third-order terms be evidenced?

3. If rcal third-order terms exist, to what extent can we estimate them?

Biases in Second-Degree Model Estimates Produced by Existence of Third-Order Terms

We show in Appendix 7C that, for the 3^3 design of the textile experiment, all the estimates in the second-degree model except those of the first-order terms are unbiased by third-order terms. For the first-order terms, we find

$$E(b_1) = \beta_1 + \beta_{111} + \tfrac{2}{3}\{\beta_{122} + \beta_{133}\},$$

$$E(b_2) = \beta_2 + \beta_{222} + \tfrac{2}{3}\{\beta_{112} + \beta_{233}\}, \qquad (7.4.2)$$

$$E(b_3) = \beta_3 + \beta_{333} + \tfrac{2}{3}\{\beta_{113} + \beta_{223}\}.$$

Existence and Estimation of Third-Order Terms

We can check for overall existence of third-order terms by separating an appropriate component sum of squares in the analysis of variance table. However, a difficulty arises. For any particular regressor x_i, the 3^3 design employs only three levels which we can denote by the coded values $(-1, 0, 1)$. The added regressor x_i^3 also has the same three levels $(-1, 0, 1)$, and so the first-order and pure cubic effects (β_i and β_{iii}) cannot be separately estimated, for $i = 1, 2, 3$. The remaining seven third-order effects $\beta_{112}, \beta_{122}, \beta_{113}, \beta_{133}, \beta_{223}, \beta_{233}, \beta_{123}$ are, however, all estimable. We therefore proceed in this case by fitting the third-order expression (7.4.1) but *omitting* the regressors x_1^3, x_2^3, and x_3^3. Details will be found in Appendix 7D. An extra sum of squares based on 7 df corresponding to "added third-order terms" may be constructed, and its magnitude can supply some idea of the necessity for representation by a third-order polynomial.

If this type of analysis indicates that the third-order terms we can estimate appear to be important, we might consider the possibility of appropriately transforming the response and/or predictor variables (Box and Draper, 1982, 1983; Draper, 1983) or, failing this, of running a more elaborate design which allowed all such third-order terms to be estimated for the original response and predictors.

The analyses of variance for both logged and unlogged data are given for the textile example in Table 7.7. Since the need for transformation of the unlogged response has been demonstrated, it is hardly surprising to find an F ratio of $138.1/29.0 = 4.76$, indicating the need for third-order terms. It will be seen, however, that a similar F ratio $0.0125/0.0034 = 3.68$ is also found for *the logged data*, and this must be considered more carefully.

For the logged data the sum of squares associated with the 7 df for third-order terms is 0.0877. This sum of squares may be split into the component parts shown in Table 7.8.

The main contribution to the total arises from b_{122}, which implies that the pure quadratic contribution for predictor variable x_2 changes as the level of x_1 changes. It must be remembered that anomalous effects of this kind are sometimes produced by an individual outlying value. Examination of the residuals does not indicate that this is the case here, however. The effect is exhibited in Figure 7.5 by

Table 7.7. Analysis of Variance of Textile Data Showing Effect of Third-Order Terms

Source	Logged SS	Unlogged SS $\times 10^{-3}$	df	Logged Data MS	Unlogged Data MS $\times 10^{-3}$
Mean	204.2975	20,031.2	1	204.2975	20,031.2
Added first-order terms	4.2317	14,748.5	3	1.4106	4,916.2
Added second-order terms	0.0281	4,224.3	6	0.0047	704.1
Added third-order terms	0.0877	966.9	7	0.0125	138.1
Residual	0.0336	289.8	10	0.0034	29.0

Logged Data: $F = 3.68$ (for third-order terms vs residual)

Unlogged Data: $F = 4.76$

Table 7.8. Logged Data. Further Analysis of the Extra Sum of Squares for Third-Order Terms

Source	SS	df
b_{122}	0.0552	1
b_{112}	0.0038	1
b_{113}	0.0000	1
b_{133}	0.0156	1
b_{233}	0.0003	1
b_{223}	0.0000	1
b_{123}	0.0128	1
Total third order	0.0877	7

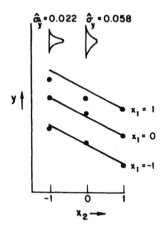

$\hat{\sigma}_y = 0.022$ $\hat{\sigma}_y = 0.058$

y

$x_1 = 1$

$x_1 = 0$

$x_1 = -1$

$x_2 \longrightarrow$

Figure 27.5. The b_{122} effect. Also shown are normal distributions: (i) with standard deviation $\hat{\sigma}_y = 0.058$, (ii) with standard deviation $\hat{\sigma}_y = 0.022$.

plotting, for each level of x_1, the response at each level of x_2 averaged over all levels of x_3. The straight lines in the figure are corresponding averages of values $\hat{y}$ calculated from the fitted first-degree equation. Also shown in the same figure are normal distributions with standard deviations $\hat{\sigma}_y = 0.058$ and $\hat{\sigma}_{\hat{y}} = 0.022$. The numerical value $\hat{\sigma}_y^2 = 0.0034$ has been obtained from the last line of the analysis of variance of Table 7.7 and is an estimate of the experimental error variance. Also $\hat{\sigma}_{\hat{y}}^2 = 0.0034 \times \frac{4}{27}$ is an estimate of the *average* variance of the $\hat{y}$'s (see Appendix 3E). It is clear that, although the b_{122} effect is clearly visible, it is quite small in relation to the more important characteristics exhibited by the data, namely the first-order effects, which stand out very clearly. Thus, in this example, even though the existence of a significant third-order effect can be demonstrated, for most purposes the convenient first-degree model in $y = \log Y$ would adequately describe the system under study over the ranges considered.

Table 7.9. Box–Behnken Second-Order Design in Four Variables with Two Sets of Data from (a) Bacon (1970) and (b) Box and Behnken (1960a)

Block Number	Coded Levels of Variables x_1	x_2	x_3	x_4	(a) Bacon Data	(b) Box–Behnken Data
1	−1	−1	0	0	84.5	84.7
	1	−1	0	0	62.9	93.3
	−1	1	0	0	90.7	84.2
	1	1	0	0	63.2	86.1
	0	0	−1	−1	70.9	85.7
	0	0	1	−1	69.2	96.4
	0	0	−1	1	80.1	88.1
	0	0	1	1	79.8	81.8
	0	0	0	0	75.1	93.8
2	−1	0	0	−1	81.8	89.4
	1	0	0	−1	61.8	88.7
	−1	0	0	1	92.2	77.8
	1	0	0	1	70.7	80.9
	0	−1	−1	0	72.4	80.9
	0	1	−1	0	76.4	79.8
	0	−1	1	0	71.9	86.8
	0	1	1	0	74.9	79.0
	0	0	0	0	74.5	87.3
3	0	−1	0	−1	69.4	86.1
	0	1	0	−1	71.2	87.9
	0	−1	0	1	77.3	85.1
	0	1	0	1	81.1	76.4
	−1	0	−1	0	88.0	79.7
	1	0	−1	0	63.0	92.5
	−1	0	1	0	86.7	89.4
	1	0	1	0	65.0	86.9
	0	0	0	0	73.0	90.7

We are reminded that:

1. Just as lack of a significant effect does not demonstrate that no effect exists but only that, for the particular experiment under study, none can be demonstrated, so the presence of a significant effect does not necessarily imply an important effect that needs to be taken account of in the model.

Table 7.10a. Estimates of Coefficients for Various Values of f, for Bacon (1970) Data; Standard Errors Are Given in Parentheses

Coefficient Estimated	Value of f			
	∞	5	3	1
β_0	74.20 (0.62)	74.25 (0.58)	74.27 (0.56)	74.33 (0.50)
β_1	-11.44 (0.31)	-11.48 (0.31)	-11.49 (0.30)	-11.50 (0.30)
β_2	1.59 (0.31)	1.59 (0.29)	1.59 (0.27)	1.60 (0.24)
β_3	-0.28 (0.31)	-0.29 (0.28)	-0.29 (0.27)	-0.30 (0.23)
β_4	4.74 (0.31)	4.72 (0.29)	4.72 (0.28)	4.71 (0.25)
β_{11}	1.50 (0.47)	1.50 (0.45)	1.50 (0.44)	1.48 (0.41)
β_{22}	-0.33 (0.47)	-0.33 (0.43)	-0.33 (0.42)	-0.34 (0.38)
β_{33}	-0.03 (0.47)	-0.07 (0.43)	-0.09 (0.41)	-0.15 (0.37)
β_{44}	0.88 (0.47)	0.85 (0.44)	0.85 (0.43)	0.82 (0.40)
β_{12}	-1.48 (0.54)	-1.48 (0.53)	-1.48 (0.53)	-1.48 (0.52)
β_{13}	0.83 (0.54)	0.85 (0.50)	0.86 (0.48)	0.88 (0.45)
β_{14}	-0.38 (0.54)	-0.37 (0.57)	-0.37 (0.58)	-0.37 (0.60)
β_{23}	-0.25 (0.54)	-0.25 (0.48)	-0.25 (0.45)	-0.26 (0.38)
β_{24}	0.50 (0.54)	0.48 (0.48)	0.47 (0.46)	0.44 (0.40)
β_{34}	0.35 (0.54)	0.34 (0.48)	0.33 (0.46)	0.30 (0.40)

Table 7.10b. Estimates of Coefficients for Various Values of f, for Box and Behnken (1960a) Data; Standard Errors Are Given in Parentheses

Coefficient Estimated	Value of f			
	∞	5	3	1
β_0	96.60 (0.84)	90.59 (0.64)	90.58 (0.59)	90.53 (0.48)
β_1	1.93 (0.42)	2.32 (0.33)	2.30 (0.30)	2.42 (0.23)
β_2	-1.96 (0.42)	-1.96 (0.31)	-1.96 (0.28)	-1.95 (0.22)
β_3	1.13 (0.42)	1.14 (0.31)	1.13 (0.28)	1.12 (0.22)
β_4	-3.68 (0.42)	-3.37 (0.33)	-3.31 (0.30)	-3.19 (0.23)
β_{11}	-1.42 (0.63)	-1.47 (0.50)	-1.49 (0.46)	-1.52 (0.39)
β_{22}	-4.33 (0.63)	-4.29 (0.48)	-4.27 (0.44)	4.22 (0.36)
β_{33}	-2.24 (0.63)	-2.21 (0.48)	-2.19 (0.44)	-2.14 (0.36)
β_{44}	-2.58 (0.63)	-2.63 (0.49)	-2.65 (0.46)	-2.69 (0.39)
β_{12}	-1.68 (0.73)	-1.67 (0.55)	-1.67 (0.49)	-1.67 (0.38)
β_{13}	-3.83 (0.73)	-3.82 (0.54)	-3.82 (0.49)	-3.82 (0.37)
β_{14}	0.95 (0.73)	-0.76 (0.70)	-0.68 (0.69)	-0.49 (0.68)
β_{23}	-1.68 (0.73)	-1.67 (0.54)	-1.66 (0.49)	-1.64 (0.40)
β_{24}	-2.63 (0.73)	-2.63 (0.54)	-2.63 (0.49)	-2.63 (0.38)
β_{34}	-4.25 (0.73)	-4.25 (0.54)	-4.25 (0.48)	-4.25 (0.36)

Table 7.11a. Values of Weights w_u for Various Values of f, Bacon (1970) Data

Observation	Value of f			
Number	∞	5	3	1
1	1	0.99	1.00	1.03
2	1	0.85	0.80	0.70
3	1	0.96	0.96	0.99
4	1	0.87	0.83	0.74
5	1	1.19	1.32	1.90
6	1	1.02	1.03	1.03
7	1	1.12	1.18	1.38
8	1	1.20	1.33	2.00
9	1	1.00	1.01	1.08
10	1	0.75	0.68	0.57
11	1	0.85	0.81	0.72
12	1	0.82	0.77	0.67
13	1	0.78	0.72	0.62
14	1	1.20	1.33	1.98
15	1	1.11	1.17	1.36
16	1	1.09	1.14	1.27
17	1	1.20	1.32	1.95
18	1	1.19	1.31	1.92
19	1	1.00	1.00	0.98
20	1	1.17	1.28	1.81
21	1	1.20	1.33	2.00
22	1	1.12	1.19	1.41
23	1	1.18	1.28	1.65
24	1	1.05	1.08	1.17
25	1	0.91	0.87	0.78
26	1	1.10	1.13	1.20
27	1	0.90	0.86	0.78

2. If wider ranges of the variables were being studied in a subsequent trial, these high-order effects might well become important and this possibility would have to be borne in mind.

7.5. ROBUST METHODS IN FITTING GRADUATING FUNCTIONS

As we explained in Chapter 3, occasional "faulty" observations can be expected to occur in practical experimental circumstances. They could arise, for example, from copying errors, mistakes in measurement, or mistakes in setting variable levels. Robust procedures can both minimize the effect of faulty values as well as help to identify them. A number of different robust estimation procedures have been proposed. These differ in their details, but usually not very much in their conclusions. Here, we employ the iteratively reweighted least-squares method introduced in Section 3.13 with the weight function (3.13.22), namely,

$$w_u = \frac{(f+1)}{\left(f + r_u^2/s^2\right)}, \tag{7.5.1}$$

where, at each iteration, the r_u are the residuals and s^2 is the residual mean square taken from the previous stage, and f is a selected constant. (Below, we choose $f = \infty$, 5, 3, and 1.)

For illustration, we consider two sets of data, each obtained when studying four predictor variables using a particular kind of three-level arrangement called a Box–Behnken design (see Chapter 15). In general, this class of second-order designs requires many fewer runs than the complete three-level factorials. As indicated in Table 7.9, the four-factor design may be run in three orthogonal blocks. Two sets of data were obtained in quite separate investigations at different times by different experimenters. One set is taken from Bacon (1970) and the other from Box and Behnken (1960a).

Our purpose here is only to illustrate the estimation of the model coefficients using a robust analysis, not to discuss the resulting fitted equations. Tables 7.10a and 7.10b show the estimates of the coefficients of a second degree polynomial and (in parentheses) their standard errors using the weight function (7.5.1) with $f = \infty$, 5, 3, and 1, and with block effects eliminated, for (a) the Bacon data and (b) the Box–Behnken data. The corresponding weights used are shown in Tables 7.11a and 7.11b.

Table 7.11b. Values of Weights w_u for Various Values of f, Box and Behnken (1960a) Data

Observation Number	Value of f			
	∞	5	3	1
1	1	1.06	1.11	1.39
2	1	1.16	1.28	1.85
3	1	1.09	1.16	1.50
4	1	1.14	1.25	1.78
5	1	1.16	1.21	1.67
6	1	1.09	1.16	1.57
7	1	1.16	1.28	1.90
8	1	1.18	1.31	1.95
9	1	1.01	1.00	0.93
10	1	0.43	0.34	0.22
11	1	1.20	1.32	1.88
12	1	1.66	1.26	1.62
13	1	0.51	0.43	0.32
14	1	1.17	1.28	1.72
15	1	1.15	1.22	1.39
16	1	1.08	1.09	1.07
17	1	1.20	1.33	1.94
18	1	1.05	1.07	1.15
19	1	1.08	1.15	1.48
20	1	1.20	1.33	1.72
21	1	1.20	1.33	1.78
22	1	1.07	1.13	1.41
23	1	1.17	1.30	1.96
24	1	1.09	1.15	1.49
25	1	1.10	1.17	1.56
26	1	1.16	1.29	1.92
27	1	1.20	1.33	1.95

Table 7A.1. Matrix Z_s of Regressors for Second-Order Model and the Data in Unlogged and Logged Form

1	x_1	x_2	x_3	x_1^2	x_2^2	x_3^2	x_1x_2	x_1x_3	x_2x_3	Y	$y = \log Y$
1	−1	−1	−1	1	1	1	1	1	1	674	2.83
1	·	−1	−1	·	1	1	·	·	1	1414	3.15
1	1	−1	−1	1	1	1	−1	−1	1	3636	3.56
1	−1	·	−1	1	·	1	·	1	·	338	2.53
1	·	·	−1	·	·	1	·	·	·	1022	3.01
1	1	·	−1	1	·	1	·	−1	·	1568	3.19
1	−1	1	−1	1	1	1	−1	1	−1	170	2.23
1	·	1	−1	·	1	1	·	·	−1	442	2.65
1	1	1	−1	1	1	1	1	−1	−1	1140	3.06
1	−1	−1	·	1	1	·	1	·	·	370	2.57
1	·	−1	·	·	1	·	·	·	·	1198	3.08
1	1	−1	·	1	1	·	−1	·	·	3184	3.50
1	−1	·	·	1	·	·	·	·	·	266	2.42
1	·	·	·	·	·	·	·	·	·	620	2.79
1	1	·	·	1	·	·	·	·	·	1070	3.03
1	−1	1	·	1	1	·	−1	·	·	118	2.07
1	·	1	·	·	1	·	·	·	·	332	2.52
1	1	1	·	1	1	·	1	·	·	884	2.95
1	−1	−1	1	1	1	1	1	−1	−1	292	2.47
1	·	−1	1	·	1	1	·	·	−1	634	2.80
1	1	−1	1	1	1	1	−1	1	−1	2000	3.30
1	-1	·	1	1	·	1	·	−1	·	210	2.32
1	·	·	1	·	·	1	·	·	·	438	2.64
1	1	·	1	1	·	1	·	1	·	566	2.75
1	−1	1	1	1	1	1	−1	−1	1	90	1.95
1	·	1	1	·	1	1	·	·	1	220	2.34
1	1	1	1	1	1	1	1	1	1	360	2.56

Note that, for the Bacon data, the changes that occur in the estimates and their standard errors are slight. For the Box–Behnken data, they are much more pronounced, however; in particular, note the large change in the estimate of β_{14}. In Table 7.11b we see, for the Box–Behnken data, the very large down-weightings of runs 10 and 13 that occur as f is made small. This strongly suggests that observations 10 and 13 may be faulty.

Note. The following appendixes make use of various results in least-squares theory. A discussion of least squares is given in Chapter 3.

APPENDIX 7A. FITTING OF FIRST- AND SECOND-DEGREE MODELS

The matrix Z_s of regressors for the second-order (or quadratic) model and the response data, both unlogged and logged, are shown in Table 7A.1. The matrix Z_f of regressors for the first-order model consists of the first four columns of Z_s. Matrices needed for the fitting of the first- and second-order models are shown in Tables 7A.2 and 7A.3.

Table 7A.2. Fitting of the First-Degree Model

$$\mathbf{Z}_f'\mathbf{Z}_f \qquad\qquad\qquad (\mathbf{Z}_f'\mathbf{Z}_f)^{-1}$$

$$
\begin{bmatrix}
27 & \cdot & \cdot & \cdot \\
\cdot & 18 & \cdot & \cdot \\
\cdot & \cdot & 18 & \cdot \\
\cdot & \cdot & \cdot & 18
\end{bmatrix}
\qquad\qquad
\begin{bmatrix}
\frac{1}{27} & \cdot & \cdot & \cdot \\
\cdot & \frac{1}{18} & \cdot & \cdot \\
\cdot & \cdot & \frac{1}{18} & \cdot \\
\cdot & \cdot & \cdot & \frac{1}{18}
\end{bmatrix}
$$

$\mathbf{Z}_f'\mathbf{Y}$	$\mathbf{Z}_f'\mathbf{y}$	$\mathbf{b}_Y$	$\mathbf{b}_y$
$\begin{bmatrix} 23{,}256 \\ 11{,}880 \\ -9{,}646 \\ -5{,}594 \end{bmatrix}$	$\begin{bmatrix} 74.27 \\ 6.51 \\ -4.93 \\ -3.08 \end{bmatrix}$	$\begin{bmatrix} 861.3 \\ 660.0 \\ -535.9 \\ -310.8 \end{bmatrix}$	$\begin{bmatrix} 2.751 \\ 0.362 \\ -0.274 \\ -0.171 \end{bmatrix}$

The matrix $\mathbf{Z}_s'\mathbf{Z}_s$ has a pattern of a special kind frequently encountered in the fitting of second-degree polynomial models. Suppose in general that there are k variables $x_1, x_2, \ldots, x_k$ and that the columns of the $n \times \frac{1}{2}(k+1)(k+2)$ matrix $\mathbf{Z}_s$ are rearranged so that the first $k+1$ columns are those corresponding to

$$1, x_1^2, x_2^2, \ldots, x_k^2.$$

Denote the matrix with rearranged columns by $\dot{\mathbf{Z}}$. Now suppose that $\dot{\mathbf{Z}}'\dot{\mathbf{Z}}$ is partitioned after the first $k+1$ rows and columns. Then it is easily seen that the $\dot{\mathbf{Z}}'\dot{\mathbf{Z}}$ derived from $\mathbf{Z}_s$ in Table 7A.3 is of the form

$$\dot{\mathbf{Z}}'\dot{\mathbf{Z}} = \left[\begin{array}{c|c} \mathbf{M} & \mathbf{0} \\ \hline \mathbf{0}' & \mathbf{P} \end{array}\right], \tag{7A.1}$$

where $\mathbf{M}$ is a $k+1$ by $k+1$ matrix, $\mathbf{0}$ is a null matrix of size $(k+1) \times \frac{1}{2}k(k+1)$, and $\mathbf{P}$ is $\frac{1}{2}k(k+1) \times \frac{1}{2}k(k+1)$ and diagonal. The inverse is thus of the form

$$(\dot{\mathbf{Z}}'\dot{\mathbf{Z}})^{-1} = \left[\begin{array}{c|c} \mathbf{M}^{-1} & \mathbf{0} \\ \hline \mathbf{0}' & \mathbf{P}^{-1} \end{array}\right] \tag{7A.2}$$

and the elements of the diagonal matrix $\mathbf{P}^{-1}$ are the reciprocals of the corresponding elements in the matrix $\mathbf{P}$. Furthermore, $\mathbf{M}$ is a patterned matrix of the general form

$$
\mathbf{M} = \begin{array}{c}
\begin{array}{cccccc} 0 & 11 & 22 & 33 & & kk \end{array} \\
\begin{bmatrix}
d & e & e & e & \cdots & e \\
e & f & g & g & \cdots & g \\
e & g & f & g & \cdots & g \\
\vdots & & & & & \vdots \\
e & g & g & g & \cdots & f
\end{bmatrix}
\begin{array}{c} 0 \\ 11 \\ 22 \\ \vdots \\ kk. \end{array}
\end{array}
\tag{7A.3}
$$

Table 7A.3. Fitting of the Second-Degree Model

$$\mathbf{Z}_s'\mathbf{Z}_s$$

	0	1	2	3	11	22	33	12	13	23
0	27	·	·	·	18	18	18	·	·	·
1	·	18	·	·	·	·	·	·	·	·
2	·	·	18	·	·	·	·	·	·	·
3	·	·	·	18	·	·	·	·	·	·
11	18	·	·	·	18	12	12	·	·	·
22	18	·	·	·	12	18	12	·	·	·
33	18	·	·	·	12	12	18	·	·	·
12	·	·	·	·	·	·	·	12	·	·
13	·	·	·	·	·	·	·	·	12	·
23	·	·	·	·	·	·	·	·	·	12

$$(\mathbf{Z}_s'\mathbf{Z}_s)^{-1}$$

$$\frac{1}{108}\begin{bmatrix}
28 & · & · & · & -12 & -12 & -12 & · & · & · \\
· & 6 & · & · & · & · & · & · & · & · \\
· & · & 6 & · & · & · & · & · & · & · \\
· & · & · & 6 & · & · & · & · & · & · \\
-12 & · & · & · & 18 & · & · & · & · & · \\
-12 & · & · & · & · & 18 & · & · & · & · \\
-12 & · & · & · & · & · & 18 & · & · & · \\
· & · & · & · & · & · & · & 9 & · & · \\
· & · & · & · & · & · & · & · & 9 & · \\
· & · & · & · & · & · & · & · & · & 9
\end{bmatrix}$$

$\mathbf{Z}_s'\mathbf{Y}$	$\mathbf{Z}_s'\mathbf{y}$	$\mathbf{b}_Y$	$\mathbf{b}_y$
23,256	74.27	550.6	2.786
11,880	6.51	660.0	0.362
−9,646	−4.93	−535.9	−0.274
−5,594	−3.08	−310.8	−0.171
16,936	49.29	238.7	−0.037
17,158	49.59	275.7	0.013
15,214	49.34	−48.3	−0.029
−5,478	−0.17	−456.5	−0.014
−2,828	−0.35	−235.7	−0.029
1,716	−0.12	143.0	−0.010

It is readily confirmed that the inverse follows a similar pattern with elements denoted by the corresponding capital letters

$$D = H^{-1}\{f + (k - 1)g\}(f - g),$$

$$E = -H^{-1}e(f - g),$$

$$F = H^{-1}\{df + (k - 2)\,dg - (k - 1)e^2\}, \tag{7A.4}$$

$$G = H^{-1}(e^2 - dg),$$

$$H = (f - g)\{df + (k - 1)\,dg - ke^2\}$$

replacing their lowercase counterparts.

The matrices resulting from the three-level factorial design employed in the present chapter are of even more specialized form than those so far discussed. For these designs, the elements of $\mathbf{M}$ are

$$d = n, \qquad e = \frac{2n}{3}, \qquad f = \frac{2n}{3}, \qquad g = \frac{4n}{9},$$

where $n = 3^k$, whence the elements of $\mathbf{M}^{-1}$ are

$$D = \frac{(2k+1)}{n}, \qquad E = \frac{-3}{n}, \qquad F = \frac{9}{2n}, \qquad G = 0.$$

This particular result may also be obtained more directly.

For the case $k = 3$ the resulting inverse is that given in Table 7A.3, after the appropriate rearrangement of rows and columns back to their original positions.

APPENDIX 7B. CHECKING FOR NONADDITIVITY WITH TUKEY'S TEST

We have seen how a graphical test for nonadditivity looks for a quadratic tendency in a plot of $y - \hat{y}$ versus $\hat{y}$. A formal exact test for such a tendency was proposed by Tukey (1949); see also Anscombe (1961).

To perform the test, first fit the model in the ordinary way to obtain residuals $y - \hat{y}$ and calculated values $\hat{y}$. Now refit using the augmented model

$$y_u = \beta_0' + g(\mathbf{x}_u, \boldsymbol{\beta}') + \alpha q_u + \varepsilon_u \tag{7B.1}$$

where $\beta_0' + g(\mathbf{x}_u, \boldsymbol{\beta}')$ is the original model form but with parameters renamed by the addition of primes, and where

$$q_u = \hat{y}_u^2. \tag{7B.2}$$

[For additional detail, see the text around Eqs. (7.1.10) through (7.1.14).] The extra sum of squares for the estimate $\hat{\alpha}$ given the other parameters is associated with Tukey's "one degree of freedom for nonadditivity." It may be compared with the new residual mean square, which now has 1 df fewer, to provide an exact F test.

For the textile example, the special symmetry of the original design permits the use of a more direct method of performing the calculations, as follows:

After the initial fit, $q_u = \hat{y}_u^2$ is computed and this is treated as a "pseudo-response" and is fitted to the (original) model form to give residuals $q_u - \hat{q}_u$. The required extra sum of squares corresponding to Tukey's 1 df is then

$$S_\alpha = \frac{\left\{ \Sigma(y_u - \hat{y}_u)(q_u - \hat{q}_u) \right\}^2}{\Sigma(q_u - \hat{q}_u)^2}.$$

The new residual sum of squares will be $S_R - S_\alpha$ and it will have 1 df fewer than S_R. The appropriate analysis of variance breakup of the residual sum of squares is shown in Table 7B.1. For the textile data we obtain the results in Tables 7B.2 through 7B.5. The analysis confirms that, while highly significant nonadditivity is found for the original data Y, use of $y = \log Y$ causes the nonadditivity to disappear.

Table 7B.1. Adjustment of Analysis of Variance Table to Check Nonadditivity

Source	SS	df	MS	F
Nonadditivity	$S_\alpha = \hat{\alpha}^2 \Sigma(q_u - \hat{q}_u)^2$	1	MS_α	$MS_\alpha / MS_{R'}$
Adjusted residual	$S_{R'} = S_R - S_\alpha$	$n - p - 1$	$MS_{R'}$	
Residual	$S_R = \Sigma(y_u - \hat{y}_u)^2$	$n - p$		

Table 7B.2. Checking for Nonadditivity. First-Degree Model Fitted to Unlogged Data Y

Source	SS $\times 10^{-3}$	df	MS $\times 10^{-3}$	F
Nonadditivity	4037.0	1	4037.0	61.5
Adjusted residual	1444.0	22	65.6	
Residual	5481.0	23		

Table 7B.3. Checking for Nonadditivity. Second-Degree Model Fitted to Unlogged Data Y

Source	SS $\times 10^{-3}$	df	MS $\times 10^{-3}$	F
Nonadditivity	704.3	1	704.3	20.4
Adjusted residual	552.4	16	34.5	
Residual	1256.7	17		

Table 7B.4. Checking for Nonadditivity. First-Degree Model Fitted to Logged Data y

Source	SS	df	MS	F
Nonadditivity	0.00213	1	0.00213	0.32
Adjusted residual	0.14735	22	0.00670	
Residual	0.14948	23		

Table 7B.5. Checking for Nonadditivity. Second-Degree Model Fitted to Logged Data y

Source	SS	df	MS	F
Nonadditivity	0.00233	1	0.00233	0.32
Adjusted residual	0.11814	16	0.00738	
Residual	0.12047	17		

APPENDIX 7C. THIRD-ORDER BIASES IN ESTIMATES FOR SECOND-ORDER MODEL, USING THE 3^3 DESIGN

The least-squares estimator

$$\mathbf{b}_1 = (\mathbf{Z}_1'\mathbf{Z}_1)^{-1}\mathbf{Z}_1'\mathbf{y} \qquad (7C.1)$$

Table 7C.1. Bias Calculations for the Textile Example

$\mathbf{Z}_2$

111	222	333	122	112	133	113	233	223	123
−1	−1	−1	−1	−1	−1	−1	−1	−1	−1
·	−1	−1	·	·	·	·	−1	−1	·
1	−1	−1	1	−1	1	−1	−1	−1	1
−1	·	−1	·	·	−1	−1	·	·	·
·	·	−1	·	·	·	·	·	·	·
1	·	−1	·	·	1	−1	·	·	·
−1	1	−1	−1	1	−1	−1	1	−1	1
·	1	−1	·	·	·	·	1	−1	·
1	1	−1	1	1	1	−1	1	−1	−1
−1	−1	·	−1	−1	·	·	·	·	·
·	−1	·	·	·	·	·	·	·	·
1	−1	·	1	−1	·	·	·	·	·
−1	·	·	·	·	·	·	·	·	·
·	·	·	·	·	·	·	·	·	·
1	·	·	·	·	·	·	·	·	·
−1	1	·	−1	1	·	·	·	·	·
·	1	·	·	·	·	·	·	·	·
1	1	·	1	1	·	·	·	·	·
−1	−1	1	−1	−1	−1	1	−1	1	1
·	−1	1	·	·	·	·	−1	1	·
1	−1	1	1	−1	1	1	−1	1	−1
−1	·	1	·	·	−1	1	·	·	·
·	·	1	·	·	·	·	·	·	·
1	·	1	·	·	1	1	·	·	·
−1	1	1	−1	1	−1	1	1	1	−1
·	1	1	·	·	·	·	1	1	·
1	1	1	1	1	1	1	1	1	1

$\mathbf{Z}_1'\mathbf{Z}_2$

	111	222	333	122	112	133	113	233	223	123
0	·	·	·	·	·	·	·	·	·	·
1	18	·	·	12	·	12	·	·	·	·
2	·	18	·	·	12	·	·	12	·	·
3	·	·	18	·	·	·	12	·	12	·
11	·	·	·	·	·	·	·	·	·	·
22	·	·	·	·	·	·	·	·	·	·
33	·	·	·	·	·	·	·	·	·	·
12	·	·	·	·	·	·	·	·	·	·
13	·	·	·	·	·	·	·	·	·	·
23	·	·	·	·	·	·	·	·	·	·

$\mathbf{A} = (\mathbf{Z}_1'\mathbf{Z}_1)^{-1}\mathbf{Z}_1'\mathbf{Z}_2$

	111	222	333	122	112	133	113	233	223	123
0	·	·	·	·	·	·	·	·	·	·
1	1	·	·	2/3	·	2/3	·	·	·	·
2	·	1	·	·	2/3	·	·	2/3	·	·
3	·	·	1	·	·	·	2/3	·	2/3	·
11	·	·	·	·	·	·	·	·	·	·
22	·	·	·	·	·	·	·	·	·	·
33	·	·	·	·	·	·	·	·	·	·
12	·	·	·	·	·	·	·	·	·	·
13	·	·	·	·	·	·	·	·	·	·
23	·	·	·	·	·	·	·	·	·	·

associated with the postulated model

$$\mathbf{y} = \mathbf{Z}_1 \boldsymbol{\beta}_1 + \boldsymbol{\varepsilon} \qquad (7C.2)$$

is [with the minimal assumption $E(\boldsymbol{\varepsilon}) = 0$] unbiased, so that

$$E(\mathbf{b}_1) = \boldsymbol{\beta}_1. \qquad (7C.3)$$

If, however, the estimator (7C.1) appropriate to the model (7C.2) is employed, when in fact the correct model is

$$\mathbf{y} = \mathbf{Z}_1 \boldsymbol{\beta}_1 + \mathbf{Z}_2 \boldsymbol{\beta}_2 + \boldsymbol{\varepsilon}, \qquad (7C.4)$$

then, in general,

$$E(\mathbf{b}_1) = \boldsymbol{\beta}_1 + \mathbf{A}\boldsymbol{\beta}_2, \qquad (7C.5)$$

that is, the estimator $\mathbf{b}_1$ is now biased, and the matrix $\mathbf{A}$ of bias coefficients is (see Section 3.10)

$$\mathbf{A} = (\mathbf{Z}_1'\mathbf{Z}_1)^{-1}\mathbf{Z}_1'\mathbf{Z}_2. \qquad (7C.6)$$

We now illustrate this general analysis for the 3^3 design used in the textile example. Here, $\mathbf{Z}_1$ is the matrix for the second degree polynomial model, called $\mathbf{Z}_s$ and enumerated in Table 7A.1, whose rows are appropriate values of

$$\{1, x_1, x_2, x_3, x_1^2, x_2^2, x_3^2, x_1 x_2, x_1 x_3, x_2 x_3\}, \qquad (7C.7)$$

while $\mathbf{Z}_2$, given in Table 7C.1 is the matrix whose rows are appropriate values of the additional third-order terms

$$\{x_1^3, x_2^3, x_3^3, x_1 x_2^2, x_1^2 x_2, x_1 x_3^2, x_1^2 x_3, x_2 x_3^2, x_2^2 x_3, x_1 x_2 x_3\}. \qquad (7C.8)$$

Then it is also seen that the matrix $\mathbf{Z}_1'\mathbf{Z}_2$ of cross-products between columns of $\mathbf{Z}_1$ and $\mathbf{Z}_2$ is that given in Table 7C.1 from which, after premultiplying by the inverse $(\mathbf{Z}_1'\mathbf{Z}_1)^{-1}$ [denoted by $(\mathbf{Z}_s'\mathbf{Z}_s)^{-1}$ in Table 7A.3], we obtain the bias matrix $\mathbf{A}$, yielding the bias relationships of Eq. (7.4.2).

APPENDIX 7D. ANALYSIS OF VARIANCE TO DETECT POSSIBLE THIRD-ORDER TERMS

To perform the analysis, we would first need to fit the full third-degree polynomial model

$$\mathbf{y} = \mathbf{Z}_1 \boldsymbol{\beta}_1 + \mathbf{Z}_2 \boldsymbol{\beta}_2 + \boldsymbol{\varepsilon} \qquad (7D.1)$$

in which $\mathbf{Z}_1 \boldsymbol{\beta}_1$ includes terms up to second order and $\mathbf{Z}_2 \boldsymbol{\beta}_2$ terms of third order, and then find the extra sum of squares due to fitting the third-order terms. An apparent difficulty for the 3^3 design is that at least four levels of a given variable are needed to estimate pure cubic coefficients $\beta_{111}, \beta_{222}, \beta_{333}$, and we have only three levels. However, any third-order coefficient of the form β_{ijj} is estimable because it measures the change in quadratic curvature associated with the jth variable as the ith variable is changed and quadratic curvature in any one input can certainly be estimated at each of the three levels of any other input.

(*Note:* Here and elsewhere, the generic "β_{ijj}" includes terms such as β_{113}.)

Orthogonalization of the Model

We can understand the situation more clearly if we proceed as follows. It has been noted in Chapter 3 that, by subtracting $\mathbf{Z}_1\mathbf{A}\boldsymbol{\beta}_2$ from the second term and adding it to the first in Eq. (7D.1), the model may be equivalently written so that terms of

Table 7D.1. Matrix $\mathbf{Z}_{2\cdot 1}$ for Estimation of Third-Order Terms

$$\mathbf{Z}_{2\cdot 1} = \mathbf{Z}_2 - \mathbf{Z}_1\mathbf{A}$$

111	222	333	122	112	133	113	233	223	123
·	·	·	$-\frac{1}{3}$	$-\frac{1}{3}$	$-\frac{1}{3}$	$-\frac{1}{3}$	$-\frac{1}{3}$	$-\frac{1}{3}$	-1
·	·	·	·	$\frac{2}{3}$	·	$\frac{2}{3}$	$-\frac{1}{3}$	$-\frac{1}{3}$	·
·	·	·	$\frac{1}{3}$	$-\frac{1}{3}$	$\frac{1}{3}$	$-\frac{1}{3}$	$-\frac{1}{3}$	$-\frac{1}{3}$	1
·	·	·	$\frac{2}{3}$	·	$-\frac{1}{3}$	$-\frac{1}{3}$	·	$\frac{2}{3}$	·
·	·	·	·	·	·	$\frac{2}{3}$	·	$\frac{2}{3}$	·
·	·	·	$-\frac{2}{3}$	·	$\frac{1}{3}$	$-\frac{1}{3}$	·	$\frac{2}{3}$	·
·	·	·	$-\frac{1}{3}$	$\frac{1}{3}$	$-\frac{1}{3}$	$-\frac{1}{3}$	$\frac{1}{3}$	$-\frac{1}{3}$	1
·	·	·	·	$-\frac{2}{3}$	·	$\frac{2}{3}$	$\frac{1}{3}$	$-\frac{1}{3}$	·
·	·	·	$\frac{1}{3}$	$\frac{1}{3}$	$\frac{1}{3}$	$-\frac{1}{3}$	$\frac{1}{3}$	$-\frac{1}{3}$	-1
·	·	·	$-\frac{1}{3}$	$-\frac{1}{3}$	$\frac{2}{3}$	·	$\frac{2}{3}$	·	·
·	·	·	·	$\frac{2}{3}$	·	·	$\frac{2}{3}$	·	·
·	·	·	$\frac{1}{3}$	$-\frac{1}{3}$	$-\frac{2}{3}$	·	$\frac{2}{3}$	·	·
·	·	·	$\frac{2}{3}$	·	$\frac{2}{3}$	·	·	·	·
·	·	·	·	·	·	·	·	·	·
·	·	·	$-\frac{2}{3}$	·	$\frac{2}{3}$	·	·	·	·
·	·	·	$-\frac{1}{3}$	$\frac{1}{3}$	$\frac{2}{3}$	·	$-\frac{2}{3}$	·	·
·	·	·	·	$-\frac{2}{3}$	·	·	$-\frac{2}{3}$	·	·
·	·	·	$\frac{1}{3}$	$\frac{1}{3}$	$-\frac{2}{3}$	·	$-\frac{2}{3}$	·	·
·	·	·	$-\frac{1}{3}$	$-\frac{1}{3}$	$-\frac{1}{3}$	$\frac{1}{3}$	$-\frac{1}{3}$	$\frac{1}{3}$	1
·	·	·	·	$\frac{2}{3}$	·	$-\frac{2}{3}$	$-\frac{1}{3}$	$\frac{1}{3}$	·
·	·	·	$\frac{1}{3}$	$-\frac{1}{3}$	$\frac{1}{3}$	$\frac{1}{3}$	$-\frac{1}{3}$	$\frac{1}{3}$	-1
·	·	·	$\frac{2}{3}$	·	$-\frac{1}{3}$	$\frac{1}{3}$	·	$-\frac{2}{3}$	·
·	·	·	·	·	·	$-\frac{2}{3}$	·	$-\frac{2}{3}$	·
·	·	·	$-\frac{2}{3}$	·	$\frac{1}{3}$	$\frac{1}{3}$	·	$-\frac{2}{3}$	·
·	·	·	$-\frac{1}{3}$	$\frac{1}{3}$	$-\frac{1}{3}$	$\frac{1}{3}$	$\frac{1}{3}$	$\frac{1}{3}$	-1
·	·	·	·	$-\frac{2}{3}$	·	$-\frac{2}{3}$	$\frac{1}{3}$	$\frac{1}{3}$	·
·	·	·	$\frac{1}{3}$	$\frac{1}{3}$	$\frac{1}{3}$	$\frac{1}{3}$	$\frac{1}{3}$	$\frac{1}{3}$	1

third order are orthogonal to the lower-order terms. We thus rewrite the model (7D.1) as

$$\mathbf{y} = \mathbf{Z}_1(\boldsymbol{\beta}_1 + \mathbf{A}\boldsymbol{\beta}_2) + (\mathbf{Z}_2 - \mathbf{Z}_1\mathbf{A})\boldsymbol{\beta}_2 + \boldsymbol{\varepsilon}. \tag{7D.2}$$

When the third-degree model is written in this form, it is seen that $\mathbf{y}$ is to be regressed onto $\mathbf{Z}_1$ and also onto the part, $\mathbf{Z}_2 - \mathbf{Z}_1\mathbf{A} = \mathbf{Z}_{2\cdot1}$, of $\mathbf{Z}_2$ which is orthogonal to $\mathbf{Z}_1$. Regression of $\mathbf{y}$ onto $\mathbf{Z}_1$ (with the third-order model assumed to be true) provides the vector of estimates $\mathbf{b}_1 = (\mathbf{Z}_1'\mathbf{Z}_1)^{-1}\mathbf{Z}_1'\mathbf{y}$ of Eq. (7C.1) which, in Eq. (7D.2), are correctly shown, not as estimates of $\boldsymbol{\beta}_1$ alone, but as estimates of the biased combination of vectors $\boldsymbol{\beta}_1 + \mathbf{A}\boldsymbol{\beta}_2$; see Eq. (7C.5). Regression of $\mathbf{y}$ onto $\mathbf{Z}_{2\cdot1} = \mathbf{Z}_2 - \mathbf{Z}_1\mathbf{A}$ produces the normal equations $\mathbf{Z}_{2\cdot1}'\mathbf{Z}_{2\cdot1}\mathbf{b}_2 = \mathbf{Z}_{2\cdot1}'\mathbf{y}$. In circumstances where the $\mathbf{Z}_{2\cdot1}'\mathbf{Z}_{2\cdot1}$ matrix is nonsingular, this would give the unbiased estimator $\mathbf{b}_2 = (\mathbf{Z}_{2\cdot1}'\mathbf{Z}_{2\cdot1})^{-1}\mathbf{Z}_{2\cdot1}'\mathbf{y}$ of $\boldsymbol{\beta}_2$. It would follow from the Gauss–Markov theorem that, provided all of the β coefficients are estimable, the unbiased least-squares estimates of the coefficients $\boldsymbol{\beta}_1$ obtained by direct fitting of the third-order model (7D.1) could be recovered from the calculation

$$\hat{\boldsymbol{\beta}}_1 = \mathbf{b}_1 - \mathbf{A}\mathbf{b}_2. \tag{7D.3}$$

It is obvious not only that the coefficients $\beta_{111}, \beta_{222}, \beta_{333}$ cannot be estimated but also that β_{jjj} is wholly confounded with the linear coefficient β_j. This is so for any three-level factorial confined to levels $(-1, 0, 1)$, for then $x_j^3 \equiv x_j$. The fact that $\beta_{111}, \beta_{222}, \beta_{333}$ are not separately estimable shows up, as expected, in the matrix $\mathbf{Z}_{2\cdot1}$ given in Table 7D.1, where it will be noted that the first three columns consist entirely of zeros. However, it is also clear from $\mathbf{Z}_{2\cdot1}$ that seven estimates of β_{122}, $\beta_{112}, \beta_{133}, \beta_{113}, \beta_{233}, \beta_{223}$, and β_{123} can be obtained, and the extra sum of squares associated with these can certainly be calculated.

APPENDIX 7E. ORTHOGONALIZATION OF EFFECTS FOR THREE-LEVEL DESIGNS

Suppose a single input ξ_1 has three equally spaced levels, coded as $x_1 = -1, 0, 1$, and a quadratic model

$$y = \beta_0 x_0 + \beta_1 x_1 + \beta_{11} x_1^2 + \varepsilon \tag{7E.1}$$

is fitted, where $x_0 = 1$. If we write the model in matrix form as $\mathbf{y} = \mathbf{Z}\boldsymbol{\beta} + \boldsymbol{\varepsilon}$ the $\mathbf{Z}$ matrix takes the form

$$\mathbf{Z} = \begin{matrix} x_0 & x_1 & x_1^2 \\ \begin{bmatrix} 1 & -1 & 1 \\ 1 & \cdot & \cdot \\ 1 & 1 & 1 \end{bmatrix} \end{matrix}. \tag{7E.2}$$

The third column is not orthogonal to the first, but we can recast the model to achieve a $\mathbf{Z}$ matrix with orthogonal columns by the following procedure. We regress the x_1^2 column (as "response") on to the x_0 column (as "predictor") and take residuals to give a new column. The new third column consists of elements of the form $x_{1u}^2 - \bar{x_1^2}$, that is, of the form $x_{1u}^2 - \frac{2}{3}$, and to avoid fractions we shall, in fact, use $3x_{1u}^2 - 2$. We can adjust the model to take account of this change to obtain

$$y = \left(\beta_0 + \tfrac{2}{3}\beta_{11} \right)x_0 + \beta_1 x_1 + \tfrac{1}{3}\beta_{11}\left(3x_1^2 - 2\right) + \varepsilon \tag{7E.3}$$

$$= \alpha_0 x_0 + \alpha_1 x_1 + \alpha_2 x^{(2)} + \varepsilon \tag{7E.4}$$

say, with an obvious notation, for which the $\mathbf{Z}$ matrix, which we call $\mathbf{Z}_0$, is shown on the left, below.

$$
\begin{array}{ccc}
x_0 & x_1 & x^{(2)} = 3x_1^2 - 2
\end{array}
$$

$$
\mathbf{Z}_0 = \begin{bmatrix} 1 & -1 & 1 \\ 1 & \cdot & -2 \\ 1 & 1 & 1 \end{bmatrix}, \; \mathbf{y} = \begin{bmatrix} y_1 \\ y_2 \\ y_3 \end{bmatrix} = \begin{bmatrix} 1 \\ 6 \\ 5 \end{bmatrix}. \tag{7E.5}
$$

Now suppose we have a set of data like $\mathbf{y}$ on the right, above. Then, by regressing the columns of $\mathbf{Z}_0$ onto $\mathbf{y}$, we obtain the mean, linear, and quadratic orthogonal components of $\mathbf{y}$, that is, the least-squares estimators a_0, a_1 and a_2 of α_0, α_1, and α_2, and we can also obtain the corresponding sums of squares. If $\mathbf{z}_i$ is a particular column of $\mathbf{Z}_0$ then we simply calculate

$$a_i = \frac{\mathbf{z}_i'\mathbf{y}}{\mathbf{z}_i'\mathbf{z}_i} \quad \text{and} \quad \mathrm{SS}(a_i) = \frac{(\mathbf{z}_i'\mathbf{y})^2}{\mathbf{z}_i'\mathbf{z}_i}. \tag{7E.6}$$

For the data set on the right of (7E.5), we can thus evaluate

$$a_0 = \text{mean} = \frac{1 + 6 + 5}{3} = 4,$$

$$a_1 = \text{linear component} = \frac{(1 \times -1) + (1 \times 5)}{2} = 2, \tag{7E.7}$$

$$a_2 = \text{quadratic component} = \frac{(1 \times 1) + (-2 \times 6) + (1 \times 5)}{6} = -1.$$

Thus the fitted equation is

$$\hat{y} = 4 + 2x_1 - 1\left(3x_1^2 - 2\right), \tag{7E.8}$$

which can of course be readily rearranged into the form of the original quadratic model to give the fitted equation

$$\hat{y} = 6 + 2x_1 - 3x_1^2. \tag{7E.9}$$

Table 7E.1. A 9×9 Orthogonalized Matrix Z_0 for Two Components x_1 and x_2

Order of terms	0	1		2			3		4
	1	x_1	x_2	$x_1^{(2)}$	$x_1 x_2$	$x_2^{(2)}$	$x_1^{(2)} x_2$	$x_1 x_2^{(2)}$	$x_1^{(2)} x_2^{(2)}$

$$Z_0 = \begin{bmatrix} 1 & -1 & -1 & 1 & 1 & 1 & -1 & -1 & 1 \\ 1 & \cdot & -1 & -2 & \cdot & 1 & 2 & \cdot & -2 \\ 1 & 1 & -1 & 1 & -1 & 1 & -1 & 1 & 1 \\ 1 & -1 & \cdot & 1 & \cdot & -2 & \cdot & 2 & -2 \\ 1 & \cdot & \cdot & -2 & \cdot & -2 & \cdot & \cdot & 4 \\ 1 & 1 & \cdot & 1 & \cdot & -2 & \cdot & -2 & -2 \\ 1 & -1 & 1 & 1 & -1 & 1 & 1 & -1 & 1 \\ 1 & \cdot & 1 & -2 & \cdot & 1 & -2 & \cdot & -2 \\ 1 & 1 & 1 & 1 & 1 & 1 & 1 & 1 & 1 \end{bmatrix}$$

For these data, the total sum of squares is $1^2 + 6^2 + 5^2 = 62$ and the component sums of squares are as follows.

Source	SS	df
Mean	$12^2/3 = 48$	1
Added linear	$4^2/2 = 8$	1
Added quadratic	$6^2/6 = 6$	1
Total	62	3

It is clearly simpler to work with the set of orthogonal variables 1, x_1, and $x_1^{(2)} = 3x_1^2 - 2$ than with the set 1, x_1 and x_1^2, and this simplicity is maintained however many inputs $x_1, x_2, \ldots, x_k$, are included within the factorial, because the balanced design ensures that the orthogonality is maintained across variables. Thus, for two inputs coded as x_1 and x_2, there are nine orthogonal components. These are indicated in Table 7E.1. The curly brackets indicate terms of different orders as indicated by the figures $0, 1, \ldots, 4$ at the heads of the columns. The appropriate adjustments must, of course, be made in the model. Because all components are orthogonal, all their individual contributions to the sums of squares are, again, given by the simple equation (7E.6).

When effects of different kinds, such as linear, pure quadratic, interaction, and so on, occur, they will, in general have different variances. To enable them to be jointly assessed using a normal probability plot, they need to be rescaled to have the same variance. A convenient choice is the rescaling that produces standardized effects A_i whose variances each equal the experimental error variance σ^2. Because the unscaled effect $a_i = z_i' y / z_i' z_i$ has variance $\sigma^2 / z_i' z_i$, we set

$$\text{standardized effect } A_i = (z_i' z_i)^{1/2} a_i = z_i' y / (z_i' z_i)^{1/2}.$$

Equivalently, for any effect which has 1 df, A_i is the square root of the mean square for a_i in the analysis of variance table. This is $\pm\{SS(a_i)/1\}^{1/2}$ with the sign attached taken to be that of a_i.

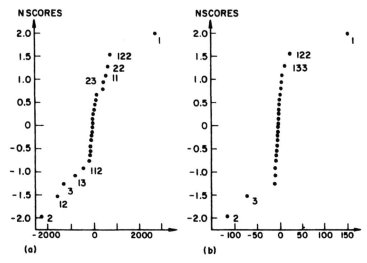

Figure 7F.1. Normal plots of root mean square components for (*a*) unlogged data and (*b*) logged data $\times 100$.

APPENDIX 7F. ORTHOGONAL EFFECTS ASSOCIATED WITH INDIVIDUAL DEGREES OF FREEDOM AND RELATED NORMAL PLOTS

Via details in Appendix 7E, the three-level factorial design used in the textile investigation provided an analysis of variance table in which contributions for individual degrees of freedom associated with the various effects could readily be isolated. It was thus possible to calculate standardized effects and to plot them on normal probability paper (see Appendix 4B) as shown in Figure 7F.1 for both unlogged and logged data. For the logged data (*b*), first-order effects are clearly of paramount importance. For the unlogged data (*a*), however, the 12 and 13 interactions, for example, and various other interactions as well, are large and do not appear to be attributable to random error alone. These interactions can readily be explained therefore as *transformable* ones, that is, they arise only because of an inappropriate choice of metric.

For didactic purposes, we have introduced the analysis of individual components only after fitting some postulated models. In many investigations, particularly if identification of the model were not as clear-cut as it is with this textile example, the analysis into individual degrees of freedom, and the plotting of components on normal probability paper, is a sensible place to start, rather than to finish.

APPENDIX 7G. YATES' ALGORITHM FOR 3^k FACTORIAL DESIGNS

Yates' algorithm can be used to calculate effects and their corresponding sums of squares for 3^k designs. We illustrate this in Tables 7G.1 and 7G.2 with the 3^3 factorial textile data for unlogged and logged values.

The 27 observations are shown in columns (1) of these tables in the standard order exemplified in Table 7.1 (p. 212). There are nine sets of three observations.

Table 7G.1. Yates' Algorithm for Textile Data (Unlogged Data)

(1)	(2)	(3)	(4)	Effect Name (5)	Divisor (6)	Orthogonal Effect (7)	MS (8)	$\sqrt{(8)}$
674	5724	10,404	23,256	000	27	861.33	20,031,168	
1414	2928	8,042	11,880	100	18	660.0	7,840,800	2800.1
3636	1752	4,810	4,296	$1^2 00$	54	79.56	341,771	584.6
338	4752	5,162	−9,646	020	18	−535.89	5,169,184	2273.6
1022	1956	4,384	−5,478	120	12	−456.5	2,500,707	1581.4
1568	1334	2,334	−2,890	$1^2 20$	36	−80.28	232,003	481.7
170	2926	1,770	4,962	$02^2 0$	54	91.89	455,953	675.2
442	1214	1,592	4,710	$12^2 0$	36	130.83	616,225	785.0
1140	670	934	4,722	$1^2 2^2 0$	108	43.72	206,456	454.4
370	2962	−3,972	−5,594	003	18	−310.78	1,738,491	1318.5
1198	1230	−3,418	−2,828	103	12	−235.67	666,465	816.4
3184	970	−2,256	−836	$1^2 03$	36	−23.22	19,414	139.3
266	2814	−1,992	1,716	023	12	143.0	245,388	495.4
620	804	−2,048	554	123	8	69.25	38,365	195.9
1070	766	−1,438	42	$1^2 23$	24	1.75	74	8.6
118	1708	−1,056	−452	$02^2 3$	36	−12.56	5,675	75.3
332	356	−820	−206	$12^2 3$	24	−8.58	1,768	42.0
884	270	−1,014	−950	$1^2 2^2 3$	72	−13.19	12,535	112.0
292	1482	1,620	−870	003^2	54	−16.11	14,017	118.4
634	−138	2,174	−1,272	103^2	36	−35.33	44,944	212.0
2000	426	1,168	−480	$1^2 03^2$	108	−4.44	2,133	46.2
210	1158	1,472	608	023^2	36	16.89	10,268	101.3
438	96	1,972	666	123^2	24	27.75	18,482	136.0
566	338	1,266	−430	$1^2 23^2$	72	−5.97	2,568	50.7
90	1024	2,184	−1.560	$02^2 3^2$	108	−14.44	22,533	150.1
220	−100	1,304	−1,206	$12^2 3^2$	72	−16.75	20,201	142.1
360	10	1,234	810	$1^2 2^2 3^2$	216	3.75	3,038	55.1

Let w_1, w_2 and w_3 denote the first, second, and third observations, respectively, in one such set. Then the analysis is conducted in terms of the following three operations: (0) $w_1 + w_2 + w_3$, (i) $w_3 - w_1$, (i^2) $w_1 - 2w_2 + w_3$.

The first operation, characterized by (0), corresponds to computing the mean of the three values, the second (i) to computing the linear effect, and the third (i^2) to computing the quadratic effect. (Appropriate divisors are introduced in a later stage of the calculation.)

The operation (0) is first performed on each of the nine sets and the results are entered in column (2) *as its first nine entries*. Then, the operation (i) is performed on each of the nine sets of three observations in column (1), and the nine results are set down in order in column (2), forming its second set of nine entries. Finally, the operation (i^2) is carried out on each of the nine sets in column (1), producing the third set of nine entries of column (2).

Column (3) is obtained from column (2) in exactly the same manner; then column (4) is obtained similarly from column (3). (In general, this operation is repeated as many times as there are factors.) The first entry in column (4) is now

Table 7G.2. Yates' Algorithm for Textile Data (Logged Data)

(1)	(2)	(3)	(4)	Effect Name (5)	Divisor (6)	Orthogonal Effect (7)	MS $\times 10^4$ (8)	$\sqrt{(8)}$
2.83	9.54	26.21	74.27	0 0 0	27	2.75	2,042,975.15	
3.15	8.73	24.93	6.51	1 0 0	18	0.36	23,544.50	153.44
3.56	7.94	23.13	−0.67	1^2 0 0	54	−0.01	83.13	9.11
2.53	9.15	2.22	−4.93	0 2 0	18	−0.27	13,502.72	116.20
3.01	8.24	2.42	−0.17	1 2 0	12	−0.01	24.08	4.90
3.19	7.54	1.87	−0.37	1^2 2 0	36	−0.01	38.03	6.16
2.23	8.57	−0.22	0.23	0 2^2 0	54	0.00	9.80	3.13
2.65	7.71	−0.24	1.41	1 2^2 0	36	0.04	552.25	23.50
3.06	6.85	−0.21	1.25	1^2 2^2 0	108	0.01	144.68	12.02
2.57	0.73	−1.60	−3.08	0 0 3	18	−0.17	5,270.22	72.59
3.08	0.66	−1.61	−0.35	1 0 3	12	−0.03	102.08	10.10
3.50	0.83	−1.72	0.01	1^2 0 3	36	0.00	0.03	0.17
2.42	0.93	0.10	−0.12	0 2 3	12	−0.01	12.00	3.46
2.79	0.61	−0.05	−0.32	1 2 3	8	−0.04	128.00	11.31
3.03	0.88	−0.22	−0.24	1^2 2 3	24	−0.01	24.00	4.89
2.07	0.83	−0.10	−0.02	0 2^2 3	36	−0.00	0.11	0.33
2.52	0.43	0.07	0.34	1 2^2 3	24	0.01	48.17	6.94
2.95	0.61	−0.34	−0.26	1^2 2^2 3	72	−0.00	9.39	3.06
2.47	0.09	0.02	−0.52	0 0 3^2	54	−0.01	50.07	7.07
2.80	−0.30	0.21	−0.75	1 0 3^2	36	−0.02	156.25	12.50
3.30	−0.01	0.00	0.05	1^2 0 3^2	108	0.00	0.23	0.47
2.32	−0.09	0.24	−0.10	0 2 3^2	36	−0.00	2.78	1.66
2.64	−0.13	0.59	−0.02	1 2 3^2	24	−0.00	0.17	0.41
2.75	−0.02	0.58	−0.58	1^2 2 3^2	72	−0.01	46.72	6.83
1.95	0.17	0.68	−0.40	0 2^2 3^2	108	−0.00	14.81	3.84
2.34	−0.21	0.15	−0.36	1 2^2 3^2	72	−0.01	18.00	4.24
2.56	−0.17	0.42	0.80	1^2 2^2 3^2	216	0.00	29.63	5.44

the sum of all 27 observations, and the remaining entries are sums of products $z_i'y$, corresponding to the various main effects and interactions whose identity is indicated in column (5). Thus, for example, $1 2^2 3$ is the interaction between the linear effects of x_1 and x_3 and the quadratic effect of x_2. The sequence of effects follows the same standard order as used for the design set out in Table 7.1 (p. 212) with $-1, 0, 1$ in the ith column replaced by $0, i, i^2$.

The (orthogonalized) effects are obtained by dividing column (4) by column (6), while the mean squares in column (8) are obtained by squaring the entries in column (4) and dividing by the corresponding divisor in column (6). The divisors of column (6) are obtained from appropriate products of the sum of squared coefficients of the w's in the operations $(0), (i), (i^2)$. The last mentioned are:

$$\text{for } (0): \quad 1^2 + 1^2 + 1^2 = 3,$$

$$\text{for } (i): \quad 1^2 + 0^2 + 1^2 = 2,$$

$$\text{for } (i^2): \quad 1^2 + 2^2 + 1^2 = 6.$$

Then, for example, the divisor for the effect $[02^23]$—that is, the interaction between the quadratic effect of x_2 and the linear effect of x_3—is $3 \times 6 \times 2 = 36$.

To illustrate how the orthogonal effects may be employed, suppose it is assumed that a polynomial of degree 2 will provide an adequate representation. This is equivalent to supposing that all terms of the third and higher order are zero. Recalling that the appropriate orthogonal second degree contrast is $(3x_i^2 - 2)$, we may thus write down the second-degree function fitted to (for example) the unlogged data as follows.

$$\hat{Y} = 861.3 + 660.0x_1 - 535.9x_2 - 310.8x_3 + 79.6\left[3x_1^2 - 2\right]$$

$$+ 91.9\left[3x_2^2 - 2\right] - 16.1\left[3x_3^2 - 2\right] - 456.5x_1x_2 - 235.7x_1x_3 + 143.0x_2x_3.$$

$$(7G.1)$$

On collecting terms, this agrees, apart from small rounding discrepancies, with the fitted second-degree equation of Table 7A.3.

Normal Plots

The standardized components $\mathbf{z}_i'\mathbf{y}/(\mathbf{z}_i'\mathbf{z}_i)^{1/2}$ described in Appendix 7E are the square roots of the mean squares given in column (8) of Table 7G.2 for the logged data $y = \log Y$. Similar quantities $\mathbf{z}'\mathbf{Y}/(\mathbf{z}_i'\mathbf{z}_i)^{1/2}$ for the unlogged data Y are given in Table 7G.1. As already mentioned in Appendix 7F, these components may be plotted on normal probability paper in exactly the same manner as was done for similar quantities in the case of the 2^k factorial. This was done in Figure 7F.1. Notice that a slightly different and alternative notation is in use there, for example (122) for (12^20).

APPENDIX 7H. DESIRABLE FEATURES AND PLACEMENT OF SECOND ORDER DESIGNS FOR SPHERICAL AND CUBOIDAL REGIONS OF INTEREST.

Second-order designs are arrangements of experimental points that give rise to data to which a second-order (quadratic) model can be fitted. Each factor must of course have (at least) three levels, and so three-level factorial or fractional factorial designs are obvious candidates. However, there are other useful designs. In this section, we consider how the guidelines applied in Appendix 6B (to first-order designs based on two-level factorials and fractional factorials of resolution ≥ 4) would be interpreted in the context of second-order designs. We again consider the work by Box and Draper (1959, 1963); see also the theoretical detail in Chapter 13.

Again, we assume that we have a specified region of interest in the coded predictor variables. Experimenters often express this region as a sphere or a cuboid, and these will be the applications we discuss here. We shall assume that the sphere has radius 1 and the cuboid has width 2 in each coded x-dimension and that both are centered at the origin. These regions have moments obtained by calculus methods, integrating quantities of the form $x_1^a x_2^b \ldots x_k^k/n$ over the respective region and dividing the result by a volume scaling factor, obtained by integrating 1 over the same region. Similarly, given a selection of points that form a design, we can evaluate corresponding design moments by averaging quantities of the form $x_1^a x_2^b \ldots x_k^k/n$ over the design points.

As was previously done in Appendix 6B, we state two general theoretical results and then explain why they help us in second-order situations.

1. Suppose there are no experimental errors in the observations, and the only errors in fitting the data to a model of degree d_1 arise from the fact that we are fitting a model of too low a degree. Assume that the true model is of degree $d_2 > d_1$. We thus say that there is no *variance error*, but only *bias error*. Then the best design will be obtained by making the moments of the design equal to the moments of the region up to and including order $d_1 + d_2$. Such a design is called an *all-bias design*. (We again note that these conditions are "sufficient" ones. The "necessary and sufficient" conditions are given in Box and Draper, 1963, p. 339.)

2. Suppose, alternatively, there is no bias error; that is, the model fitted is of an appropriate order (so that $d_2 = d_1$). Then the best design will be obtained by making the moments of the design as large as possible, thus spreading out the design points as far as the experimental limitations allow. Such a design is called an *all-variance design*.

In practice, neither extreme is usually appropriate. However, studies (e.g., Draper and Lawrence, 1965a) have shown that the following compromise is often a sensible and practical one. If $d_2 = d_1 + 1$ (that is, we fit a polynomial of order d_1 but are somewhat worried that we might need a polynomial of one order higher at worst), first select an appropriate all-bias design as in paragraph in 1 above; then expand the linear dimensions by about 10–20%. This result is based on the idea that, when both variance and bias errors exist, *such a design is appropriate for situations where the two kinds of error are roughly of the same size.*

In the present context, we take $d_1 = 2$ and $d_2 = 3$. It follows that, for an all-bias design, we need to match the moments of the design to the moments of the region up to and including order 5.

Spherical Region

For a spherical region of interest, for example, all odd-order region moments, including the first, third, and fifth, are zero. The pure second-order region moment has value $1/(k + 2)$. The pure fourth-order moment is $3/\{(k + 2)(k + 4)\}$ and the mixed fourth-order moment is $1/\{(k + 2)(k + 4)\}$. So, if we start off with a design that has the symmetries necessary to make the odd order first, third, and fifth design moments zero, we need to match up only the second- and fourth-order moments, because the first, third, and fifth moments ($d_1 + d_2 = 5$ here) are already zero. Suppose we consider designs made up of two types of point sets; a two-level factorial alone will not provide enough levels for fitting a second-order model. The first point set consists of n_a factorial type points $(\pm a, \pm a, \ldots, \pm a)$

with resolution $R \geq 4$, often referred to as a "cube" in this role, no matter what the value of k. The second point set consists of a "star," defined as k pairs of axial points of the form $(\pm b, 0, 0, \ldots, 0), (0, \pm b, 0, \ldots, 0), \ldots, (0, 0, \ldots, 0, \pm b)$. Such a design is called a central composite design. Suppose further that we add n_0 center points. Then, for the all-bias design, we need to set $\{n_a a^2 + 2b^2\}/(n_a + 2k + n_0) = 1/(k + 2)$ for the pure second-order moment condition, and $\{n_a a^4 + 2b^4\}/(n_a + 2k + n_0) = 3/\{(k + 2)(k + 4)\}$ for the pure fourth-order moment condition. For the mixed fourth-order condition, we set $n_a a^4/(n_a + 2k + n_0) = 1/\{(k + 2)(k + 4)\}$. Manipulation of these equations lead to the solution, successively, as

$$n_0 = \left(n_a^{1/2} + 2\right)^2 (k + 2)/(k + 4) - n_a - 2k,$$

$$a = \left\{(n_a + 2k + n_0)/[n_a(k + 2)(k + 4)]\right\}^{1/4},$$

$$b = a \times n_a^{1/4}.$$

All-bias solutions for selected values of k and n are shown below. Note that the solution provides mostly noninteger values for n_0. In practice of course, such values need to be rounded to integers; we suggest rounding up in all cases. Moreover, note that the ratio $b/a = n_a^{1/4}$. This causes the design to be *rotatable*, which means that the contours of the variance function of the predicted values $\hat{y}$ are spherical, providing equal information on spheres centered at the origin. This is usually regarded as a very desirable (though not absolutely essential) design characteristic; for more on this, see Sections 14.3 and 15.3.

k	n_a	n_0	a	b	b/a
2	4	2.67	0.577	0.816	1.414
3	8	2.65	0.494	0.831	1.682
4	8	1.49	0.462	0.777	1.682
4	16	3.00	0.433	0.866	2
5	16	2.00	0.408	0.816	2
6	16	0.80	0.387	0.775	2
6	32	2.90	0.368	0.875	2.378

Next, expanding these designs by (say) 15% and rounding up n_0 to the next integer leads to the following suggested values.

k	n_a	n_0	a	b	b/a
2	4	3	0.664	0.939	1.414
3	8	3	0.568	0.955	1.682
4	8	2	0.531	0.893	1.682
4	16	3	0.498	0.996	2
5	16	2	0.469	0.939	2
6	16	1	0.445	0.891	2
6	32	3	0.423	1.006	2.378

Cuboidal Region

For a cuboidal region of interest, of side 2 units and centered at the origin, all odd-order region moments, including the first, third, and fifth are zero. The pure second-order region moment has value $\frac{1}{3}$ for all k. The pure fourth-order moment is $\frac{1}{5}$, and the mixed fourth-order moment is $\frac{1}{9}$. So, if we start off with a design that has the symmetries necessary to make the odd-order first, third, and fifth design moments zero, we need to match up only the second- and fourth-order moments, because the first, third, and fifth moments $(d_1 + d_2 = 5$ here) are already zero. Suppose we again consider the cube plus star plus n_0 center points central composite design, used above. Then for the all-bias design, we need to set $\{n_a a^2 + 2b^2\}/(n_a + 2k + n_0) = \frac{1}{3}$ for the pure second-order moment condition, and $\{n_a a^4 + 2b^4\}/(n_a + 2k + n_0) = \frac{1}{5}$ for the pure fourth-order moment condition. For the mixed fourth-order condition, we set $n_a a^4/(n_a + 2k + n_0) = \frac{1}{9}$. Manipulation of these equations leads to the solution, successively, as

$$n_0 = 4(2n_a/5)^{1/2} + 1.6 - 2k,$$

$$N = n_a + 2k + n_0,$$

$$a = (N/9n_a)^{1/4},$$

$$b = (2N/45)^{1/4}.$$

All-bias solutions for selected values of k and n are shown below. Note that the solution provides noninteger value for n_0. In practice, of course, such values should be rounded; we suggest rounding up in all cases.

k	n_a	n_0	a	b
2	4	2.66	0.738	0.830
3	8	2.76	0.695	0.929
4	8	0.76	0.695	0.929
4	16	3.72	0.662	1.054
5	16	1.72	0.662	1.054
6	16	−0.28	0.662	1.054
6	32	3.91	0.639	1.208

Expanding these designs by 15% and rounding up the n_0 values to the next integer leads to the following suggested values.

k	n_a	n_0	a	b
2	4	3	0.848	0.954
3	8	3	0.799	1.068
4	8	1	0.799	1.068
4	16	4	0.762	1.212
5	16	2	0.762	1.212
6	16	0	0.762	1.212
6	32	4	0.734	1.389

The values shown for both regions should not be rigidly interpreted, but instead used to give a general feel for how spread out a composite design should be in relation to the regions of interest assumed above. In use, the designs would usually be re-coded so that the factorial design levels are -1 and 1. Here, the two sets of a and b values are those appropriate respectively in relation to a sphere of radius 1 and a cuboid of side 2, both centered at the origin.

Applications of these methods to first-order designs will be found in Appendix 6B.

EXERCISES

7.1. Consider the 3^2 factorial design and data shown below.

x_1	x_2	y	
-1	-1	2.57	$n = 9$
0	-1	3.08	$\Sigma y = 24.93$
1	-1	3.50	$\bar{y} = 2.770$
-1	0	2.42	$\Sigma y^2 = 70.50$
0	0	2.79	
1	0	3.03	
-1	1	2.07	
0	1	2.52	
1	1	2.95	

(a) Write down the first-order model, normal equations, fitted model, and associated analysis of variance table, and evaluate the residuals.

(b) Examine your results and make a judgment as to the adequacy of the fitted model.

7.2. Consider again the 3^2 design and data in the foregoing exercise.

(a) Write down the $\mathbf{Z}$ matrix associated with a *second*-order polynomial model and also the corresponding $\mathbf{Z'Z}$ matrix.

(b) Rewrite the second-order model so that $\mathbf{Z'Z}$ becomes a diagonal matrix. (*Hint:* Replace x_1^2 by $x_1^2 - \frac{2}{3}$, where $\Sigma x_{1u}^2/n = \frac{2}{3}$. Add a term $2\beta_{11}/3$ to β_0 to compensate. Do the same for x_2^2.)

7.3. Fit the second-order model to the data given in Exercise 7.1 and construct the associated analysis of variance table. [Use either Exercise 7.2(a) or 7.2(b); your answer should be the same in the end.]

7.4. Using the data and design given in Exercise 7.1, and a first-order model, construct the 1 df test for nonadditivity. (See Appendix 7B.)

7.5. Consider again the 3^2 factorial design and data given in Exercise 7.1.

(a) Write down the standard third-order polynomial model.

(b) Which third-order parameters can be separately estimated? Which cannot?

(c) Develop the appropriate alias matrix given that a second-order model has been fitted, and that a third-order model might be appropriate.

(d) Are your results in (c) consistent with what you found in (b)? (*Hint:* Do they have to be?)

7.6. (a) Using the fitted first-order model from Exercise 7.1, construct separate 95% confidence limits for each parameter in the model.

(b) Predict the response, and determine 95% confidence limits for the true mean value of the response, at the point $x_1 = 0.7$, $x_2 = 0.5$.

(c) Construct the joint 90% confidence region for β_1 and β_2.

(d) Note that $(0.95)(0.95) = 0.9025 \simeq 0.90$. Compare the region defined by the two pairs of limits in (a) with the region in (c). Comment on what you see.

(e) Show that the variance $V[y(\mathbf{x})]$ of the predicted response $\hat{y}(\mathbf{x})$ at $\mathbf{x} = (x_1, x_2)'$ is constant on circles (or radius ρ, say) centered at the point $x_1 = 0$, $x_2 = 0$.

Table E7.7

x_1	x_2	x_3	y	
−1	−1	−1	159	$n = 27$
0	−1	−1	395	$\Sigma y = 7145$
1	−1	−1	149	
−1	0	−1	25	$\bar{y} = 264.63$
0	0	−1	255	
1	0	−1	251	$\Sigma y^2 = 2{,}328{,}793$
−1	1	−1	184	
0	1	−1	363	
1	1	−1	378	
−1	−1	0	260	
0	−1	0	454	
1	−1	0	112	
−1	0	0	98	
0	0	0	422	
1	0	0	270	
−1	1	0	237	
0	1	0	362	
1	1	0	363	
−1	−1	1	146	
0	−1	1	417	
1	−1	1	150	
−1	0	1	103	
0	0	1	455	
1	0	1	172	
−1	1	1	195	
0	1	1	492	
1	1	1	278	

Note: The "observations" y shown in Table E7.7 are in fact the sums of pairs of repeats, but we are ignoring this for the purposes of our question. For additional detail involving the original data, please refer to the Davies reference.

7.7. The accompanying 3^3 factorial design and data (Table E7.7) are taken from *Design and Analysis of Industrial Experiments*, edited by O. L. Davies, Oliver and Boyd, Edinburgh, 1956, p. 333. A limp-cover revised edition was published in 1978 by the Longman Group, New York.

  (a) Fit a full second-order model to these data.

  (b) Determine the residuals and analyze them.

  (c) Make the 1 df test for nonadditivity. (See Appendix 7B.)

  (d) Using the algorithm for the three-level factorial, separately estimate all possible individual degree of freedom estimates, and plot the results on normal probability paper. (See Appendix 4B.)

  (e) Interpret your analysis.

7.8. The 3^3 factorial design of Table E7.8 arose in connection with an investigation by A. J. Feuell and R. E. Wagg, Statistical methods in detergency investigations, *Research*, **2**, 1949, 334–337; see p. 335. (*Note: Research* later became *Research Applied in Industry*. The data were also reported again in *Statistical Analysis in Chemistry and the Chemical Industry* by C. A. Bennett

Table E7.8

x_1	x_2	x_3	y	
−1	−1	−1	106	
0	−1	−1	198	
1	−1	−1	270	
−1	0	−1	197	
0	0	−1	329	
1	0	−1	361	
−1	1	−1	223	
0	1	−1	320	
1	1	−1	321	
−1	−1	0	149	
0	−1	0	243	
1	−1	0	315	$n = 27$
−1	0	0	255	$\Sigma y = 8068$
0	0	0	364	
1	0	0	390	
−1	1	0	294	$\bar{y} = 298.81$
0	1	0	410	
1	1	0	415	
−1	−1	1	182	$\Sigma y^2 = 2{,}609{,}558$
0	−1	1	232	
1	−1	1	340	
−1	0	1	259	
0	0	1	389	
1	0	1	406	
−1	1	1	297	
0	1	1	416	
1	1	1	387	

and N. L. Franklin, John Wiley & Sons, New York, 1954, p. 509.) The response was a measure of washing performance (percentage reflectance in terms of white standards) and the response data were multiplied by 10 to produce the y values shown in the table. The three predictor variables were detergent level coded to x_1, sodium carbonate (x_2), and sodium carboxymethyl cellulose level (x_3). Using the 3^k algorithm, construct an appropriate model and an analysis of variance table for these data, and plot the individual 1-df effects on normal probability paper.

7.9. The 3^3 factorial design, replicated three times, in Table E7.9 is from a study made to investigate the roles of three printing machine variables, speed (coded as x_1), pressure (x_2), and distance (x_3), upon the application of coloring inks onto package labels. Perform as complete an analysis as possible on these data.

Table E7.9

			Replicates			
x_1	x_2	x_3	I	II	III	Totals
-1	-1	-1	34	10	28	72
0	-1	-1	115	116	130	361
1	-1	-1	192	186	263	641
-1	0	-1	82	88	88	258
0	0	-1	44	178	188	410
1	0	-1	322	350	350	1022
-1	1	-1	141	110	86	337
0	1	-1	259	251	259	769
1	1	-1	290	280	245	815
-1	-1	0	81	81	81	243
0	-1	0	90	122	93	305
1	-1	0	319	376	376	1071
-1	0	0	180	180	154	514
0	0	0	372	372	372	1116
1	0	0	541	568	396	1505
-1	1	0	288	192	312	792
0	1	0	432	336	513	1281
1	1	0	713	725	754	2192
-1	-1	1	364	99	199	662
0	-1	1	232	221	266	719
1	-1	1	408	415	443	1266
-1	0	1	182	233	182	597
0	0	1	507	515	434	1456
1	0	1	846	535	640	2021
-1	1	1	236	126	168	530
0	1	1	660	440	403	1503
1	1	1	878	991	1161	3030
			8808	8096	8584	25488

Table E7.10

%A	%S	x_1	x_2	y
2	14	−1	−1	338, 344
4	14	0	−1	258, 272
6	14	1	−1	320, 334
2	16	−1	0	264, 290
4	16	0	0	242, 207
6	16	1	0	308, 310
2	18	−1	1	332, 325
4	18	0	1	258, 233
6	18	1	1	336, 350

Table E7.11

Run	x_1	x_2	y
1	−1	−1	33
2	1	−1	27
3	−1	1	28
4	1	1	52
5	0	0	45
6	0	0	41
7	$\sqrt{2}$	0	33
8	$-\sqrt{2}$	0	26
9	0	$\sqrt{2}$	37
10	0	$-\sqrt{2}$	25
11	0	0	42
12	0	0	46

7.10. An investigation to examine the effects of adding sulphur (S) and asphalt (A) in the formation of sand aggregate pavements resulted in the 3^2 factorial design, replicated, shown in Table E7.10. The response y is the failure stress in pounds per square inch. Perform a complete analysis of these data.

7.11. Consider the data in Table E7.11.

(a) Plot the data.

(b) Fit a first-order model by least squares and do the appropriate analysis on it. Comment.

(c) Now fit a second-order model $y = \beta_0 + \beta_1 x_1 + \beta_2 x_2 + \beta_{11} x_1^2 + \beta_{22} x_2^2 + \beta_{12} x_1 x_2 + \varepsilon$ to the data. Do the appropriate analysis and comment.

(d) Add 37 to observations 7–12. Repeat the work of part (c) and make a list of all the differences between your original (c) results and the new ones. What is the implication of the differences and the similarities you observe?

7.12. Analyze the data below.

x_1	x_2	y
	Block 1	
-1	-1	73
1	-1	85
-1	1	83
1	1	79
0	0	80
0	0	81
0	0	79
		$\overline{560}$
	Block 2	
$-\sqrt{2}$	0	67
$\sqrt{2}$	0	73
0	$-\sqrt{2}$	69
0	$\sqrt{2}$	71
0	0	71
0	0	69
0	0	70
		$\overline{490}$

Facts:

$$\begin{bmatrix} 7 & 4 & 4 \\ 4 & 6 & 2 \\ 4 & 2 & 6 \end{bmatrix}^{-1} = \frac{1}{48}\begin{bmatrix} 16 & -8 & -8 \\ -8 & 13 & 1 \\ -8 & 1 & 13 \end{bmatrix}$$

$$\Sigma y = 1050 \qquad N\bar{y}^2 = 78{,}750$$

$$\Sigma y^2 = 79{,}208 \qquad \Sigma(y - \bar{y})^2 = 458$$

7.13. (*Source:* Statistical design of experiments for process development of MBT, by S. A. Frankel, *Rubber Age*, **89**, 1961, June, 453–458. Material reproduced and adapted by permission of the author, of his employer American Cyanamid Company, and of *Rubber Age*.) The diagram shows 25 hypothesized observa-

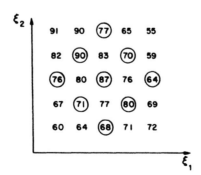

Table E7.14

x_1	x_2	y
-1	-1	130
1	-1	121
-1	1	110
1	1	125
-1.5	0	107
1.5	0	117
0	-1.5	112
0	1.5	111
0	0	149, 152, 148, 153

tions of an experimental program for investigation of two factors ξ_1 and ξ_2. Assume each ξ_i has its levels coded to $x_i = -2, -1, 0, 1, 2$. The nine encircled points are those of a composite design with four cube points at $(\pm 1, \pm 1)$, four star points at $(\pm 2, 0)$, $(0, \pm 2)$ and one center point $(0, 0)$.

Fit a second-order model to the original 25 data points and then just to the nine composite design data points. Perform a full analysis, and predict all 25 responses for each fit. Compare the two sets of results; what do you conclude?

7.14. Fit a second-order model to the data shown in Table E7.14 and provide an analysis of variance table.

Draw the contours of the function $V(\hat{y}(\mathbf{x}))$ within a square of side length 4 whose sides are parallel to the (x_1, x_2) axes and whose center is at $(x_1, x_2) = (0, 0)$.

7.15. (*Source:* Copyright © 1960. TAPPI. Reprinted and adapted from L. H. Phifer and J. B. Maginnis, Dry ashing of pulp and factors which influence it, *Tappi*, **43**(1), 1960, 38–44, with permission.) Ash values of paper pulp were widely used as a measure of the level of inorganic impurities present in the pulp. A study was carried out to examine the effects of two variables, temperature T in degrees Celsius and time t in hours, on the ash values of four specific pulps. The results for one of these, "pulp A," are shown in Table E7.15. The coded predictor variables shown are

$$x_1 = \frac{(T - 775)}{115}, \qquad x_2 = \frac{(t - 3)}{1.5},$$

and the response y is (dry ash value in %) $\times 10^3$, the factor being attached simply to remove decimals. The data were taken in the run order $1, 2, \ldots, 10$ indicated at the left of the table.

Perform a second-order surface fitting on these data and investigate (as appropriate) lack of fit, whether or not both x_1 and x_2 should be retained, and whether or not a quadratic model is needed. Plot contours for the model

Table E7.15

Run Order	x_1	x_2	y
9	-1	-1	211, 213
5	1	-1	92, 88
6	-1	1	216, 212
7	1	1	99, 83
10	-1.5	0	222, 217
3	1.5	0	48, 62
4	0	-1.5	168, 175
2	0	1.5	179, 151
1	0	0	122, 157
8	0	0	175, 146

you decide to retain and examine the residuals in the usual ways, including against run order. What do you conclude?

If it were difficult to control the temperature variable, how would you feel about the use of the method to compare various pulps, based on what you can see in this particular set of data?

7.16. (*Source:* Statistical design of experiments for process development of MBT, by S. A. Frankel, *Rubber Age*, **89**, 1961, June, 453–458. Material reproduced and adapted by permission of the author, of his employer American Cyanamid Company, and of *Rubber Age*.) In a study of the manufacture of mercapto-benzothiazole by the autoclave reaction of aniline, carbon bisulfide, and sulfur at 250°C without medium, two input variables were varied. These were ξ_1 = reaction time in hours, and ξ_2 = reaction temperature in degrees Celsius, coded to $x_1 = (\xi_1 - 12)/8$ and $x_2 = (\xi_2 - 250)/30$. The response was the percentage yield of the reaction.

First, runs 1–12 shown in Table E7.16 were performed, and a second-order surface was fitted, namely,

$$\hat{y} = 82.17 - 1.01x_1 - 8.61x_2 + 1.40x_1^2 - 8.76x_2^2 - 7.20x_1x_2.$$

The contours of this surface are shown in Figure E7.16a. Repeat this fitting for yourself, carrying out the usual tests for lack of fit, first- and second-order terms, and so on, and confirm the correctness of the fitted equation and of Figure E7.16a.

High yields were desirable. As a result of examining the surface above, five confirmatory runs, numbers 13–17, were performed. Are the five new response values consistent with the results that the second-order surface predicts? If so, refit the second-order surface to all 17 observations to obtain

$$\hat{y} = 81.11 - 2.09x_1 - 6.76x_2 + 1.15x_1^2 - 5.41x_2^2 - 9.49x_1x_2,$$

Table E7.16

Run Number	x_1	x_2	y
1	-1	0	83.8
2	1	0	81.7
3	0	0	82.4
4	0	0	82.9
5	0	-1	84.7
6	0	1	57.9
7	0	0	81.2
8	$-\theta$	$-\theta$	81.3
9	$-\theta$	θ	83.1
10	θ	$-\theta$	85.3
11	θ	θ	72.7
12	-1	0	82.0
13	-1	1	82.1
14	1	-1	88.5
15	-1	1.3	84.4
16	-1.25	0.5	85.2
17	-1.25	1	83.2

Note: $\theta = 2^{-1/2}$ in runs 8–11.

and carry out the usual analyses. Confirm that the surface shown in Figure E7.16*b* is an appropriate representation of this equation. What practical information has this investigation provided?

[*Note:* A $100(1 - \alpha)\%$ confidence prediction interval for individual observations at (x_{10}, x_{20}) is given by

$$\hat{y}_0 \pm t_{\nu, \alpha/2}\{1 + \mathbf{z}_0'(\mathbf{X'X})^{-1}\mathbf{z}_0\}^{1/2},$$

where $\hat{y}_0$ is the predicted value at (x_{10}, x_{20}), where $t_{\nu, \alpha/2}$ is the percentage point of the t distribution with ν df which leaves $\alpha/2$ in the upper tail, where $\mathbf{z}_0' = (1, x_{10}, x_{20}, x_{10}^2, x_{20}^2, x_{10}x_{20})$, and where $\mathbf{X}$ is the usual $\mathbf{X}$ matrix which occurs in the least-squares fitting of the second-order model.]

7.17. (*Source:* An exploratory study of Taylor's tool life equation by power transformations, by S. M. Wu, D. S. Ermer, and W. J. Hill, University of Wisconsin. Changes have been made to the original data as follows. The original x_1 is divided by 100, the original x_2 has been multiplied by 100. Responses are rounded and repeat observations appear on the same line.) Fit a second-order model to the data in Table E7.17. Test for lack of fit and, if no lack of fit is evident, test whether second-order terms are needed and whether any of the x's can be dropped. What predictive model would you use?

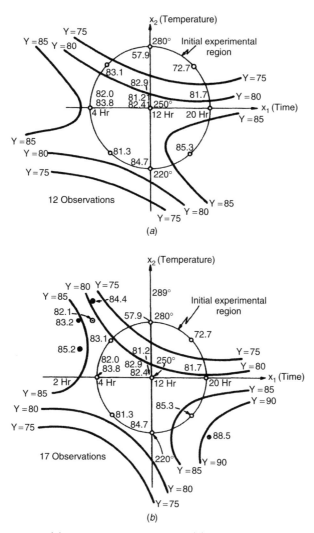

Figure E7.16. (a) Surface fitted to runs 1–12. (b) Surface fitted to runs 1–17.

Table E7.17

x_1	x_2			y	
7	1.57	2.54	2.78	3.10	3.50
5	1.57	8.45	9.13	9.33	10.67
8	1.725	0	−0.51	−1.49	−2.08
7	1.725	0	0.86	1.87	2.15
6	1.725	3.77	4.25	4.73	5.42
5	1.725	7.43	8.08	8.78	10.29
4	1.725	10.99	11.33	11.47	12.06
6	2.20	0.86	1.87	2.15	2.15
4.5	2.20	5.80	6.38	6.80	7.22

7.18. [*Sources:* (1) Maximization of potato yield under constraint, by H. P. Hermanson, *Agronomy Journal*, **57**, 1965, 210–213. Adapted by permission of the American Society of Agronomy. (2) The fertility of some Minnesota peat soils, by H. P. Hermanson, Ph.D. Thesis, University of Minnesota, 1961, 154 pp. (order No. 61-5846, University Microfilms, Ann Arbor, MI).] In 1959 a peat soil fertility trial was carried out in Aitkin County, Minnesota, using Red Pontiac potatoes. Three predictor variables N = nitrogen, P = phosphorus, and K = potassium, all measured in pounds per acre were examined in a dodecahedron design for their effects upon the response Y = yield in units of 100 pounds per acre. (This soil–crop situation required appreciable potassium fertilizer but the response to nitrogen and phosphorus is strongly seasonably dependent. A frost terminated growth early that year.)

A second-order response function was fitted to the data and was slightly modified by removal of the N^2 term to provide the response equation

$$\hat{Y} = 123 - 0.476N + 4.77P + 0.917K$$

$$- 0.0936P^2 - 0.00278K^2$$

$$- 0.000670NP + 0.00100NK - 0.00192PK. \qquad (1)$$

Important in potatoes is the specific gravity G, which must lie above a certain minimum value specified by the customer to satisfy the quality specifications of chippers. The function

$$10^3(G - 1) = 62.7 - 0.0680K \qquad (2)$$

relates the specific gravity to the fertilizer rate.

The design that gave rise to the data which led to (1) was a dodecahedron type. All noncentral design points were on the surface

$$(0.0222N - c)^2 + (0.0758P - c)^2 + (0.0133K - c)^2 = c^2, \qquad (3)$$

where $c = (1 + 5^{1/2})/2 = 1.618$.

Given that we must have $G > 1.055$, determine the combination of (N, P, K) that results in the maximum response $\hat{Y}$ determined by (1).

Is this point within or on the region defined by (3)? If such a restriction were necessary, how would one determine the best fertilizer combination?

7.19. (*Source:* Variable shear rate viscosity of SBR-filler-plasticizer systems, by G. C. Derringer, *Rubber Chemistry and Technology*, **47**, 1974, 825–836.) Experimental studies were carried out to relate the Mooney viscosity η of various rubber compounds to three predictor variables ξ_1 = silica filler, phr, ξ_2 = oil filler, phr, and ξ_3 = shear rate, s^{-1}. (phr = parts per hundred parts of rubber.) In the accompanying tables (Tables E7.19a and E7.19b), x_1 and

Table E7.19a

Coded Levels of		y Value When ξ_3 Has Value			
x_1	x_2	3	75	1500	3000
-1	-1	13.71	11.01	8.32	7.71
1	-1	14.15	11.35	8.85	—
-1	1	12.87	10.40	—	7.30
1	1	13.53	10.80	8.20	7.74
-1.4	0	12.99	10.68	8.05	7.36
1.4	0	13.89	11.04	8.51	8.10
0	-1.4	14.16	11.33	8.75	8.22
0	1.4	12.90	10.57	8.00	7.45
0	0	13.75	10.96	8.36	7.86
0	0	13.66	10.94	8.38	7.80
0	0	13.86	10.95	8.40	7.80
0	0	13.63	10.88	8.35	7.80
0	0	13.74	—	8.36	—

Additional run to be added to those above: $(x_1, x_2, \xi_3, y) = (-1, 1, 1550, 7.88)$.

Table E7.19b

Coded Levels of		y Value When ξ_3 Has Value				
x_1	x_2	3	7.5	75	750	3000
-0.86	-0.86		12.58		8.80	
0.86	-0.86		13.12		9.09	
-0.86	0.86		11.97		8.46	
0.86	0.86		12.24		8.63	
-1.4	0			10.51		
1.4	0			10.74		
0	-1.4			11.22		
0	1.4			10.24		
0	0	13.06		10.73		7.60
0	0			10.69		
0	0			10.65		
0	0			10.67		
0	0			10.69		
0	0			10.64		

x_2 are the coded variables

$$x_1 = \frac{(\xi_1 - 60)}{15}, \qquad x_2 = \frac{(\xi_2 - 21)}{15},$$

and the responses y are the observed values of $\ln \eta$. Fit, to each of these sets of data, the incomplete cubic model

$$y = \beta_0 + \beta_1 x_1 + \beta_2 x_2 + \beta_3 x_3 + \beta_{33} x_3^2 + \beta_{13} x_1 x_3 + \beta_{23} x_2 x_3$$
$$+ \beta_{133} x_1 x_3^2 + \beta_{233} x_2 x_3^2 + \varepsilon,$$

where $x_3 = \ln(\xi_3 + 1)$, and carry out appropriate tests for the validity and usefulness of the fitted surface. Note that the model can be rewritten in the form

$$y = \left(\beta_0 + \beta_3 x_3 + \beta_{33} x_3^2 \right) + \left(\beta_1 + \beta_{13} x_3 + \beta_{133} x_3^2 \right) x_1$$
$$+ \left(\beta_2 + \beta_{23} x_3 + \beta_{233} x_3^2 \right) x_2 + \varepsilon.$$

This shows how the model was conceived, namely as a "plane in x_1 and x_2" but with coefficients which are all quadratic functions of x_3.

7.20. An experimenter finds the following response surfaces for (a) yield, percent:

$$\hat{y}_1 = 80 + 4x_1 + 8x_2 - 4x_1^2 - 12x_2^2 - 12x_1 x_2,$$

and (b) unit cost per pound:

$$\hat{y}_2 = 80 + 4x_1 + 8x_2 - 2x_1^2 - 12x_2^2 - 12x_1 x_2.$$

If he must operate at yields not less than 80% and costs not more than 76 cents per pound to make a profit, what values of the predictors x_1 and x_2 are profitable ones?

7.21. Fit a second-order response surface to the coded data in Table E7.21 and provide a full analysis.

7.22. (*Source:* Cutting-tool temperature-predicting equation by response surface methodology, by S. M. Wu and R. N. Meyer, *Journal of the Engineering Industry, ASME Transactions, Series B,* **86**, 1964, 150–156. Adapted with the

Table E7.21

x_1	x_2	x_3	y
-1	-1	-1	10.6
1	-1	-1	11.1
-1	1	-1	11.9
1	1	-1	13.0
-1	-1	1	19.7
1	-1	1	18.1
-1	1	1	18.7
1	1	1	22.1
-2	0	0	14.2, 15.3
2	0	0	17.5, 16.6
0	-2	0	12.7, 13.4
0	2	0	16.5, 17.6
0	0	-2	8.4, 9.0
0	0	2	23.8, 23.3
0	0	0	16.0, 15.5, 15.6, 16.2

Table E7.22a. Coded Input Values and Temperatures Observed

Trial Number	Test Sequence Number		Coded Cutting Conditions			Response Values, T (°F)	
	First 24 Tests	Second 24 Tests	x_1	x_2	x_3	First 24 Tests	Second 24 Tests
1	7	12	-1	-1	-1	409	398
2	1	1	1	-1	-1	711	709
3	4	3	-1	1	-1	599	604
4	9	11	1	1	-1	1025	1016
5	3	6	-1	-1	1	474	498
6	8	9	1	-1	1	856	851
7	11	8	-1	1	1	700	714
8	6	5	1	1	1	1086	1087
9	2	4	0	0	0	728	742
10	5	2	0	0	0	728	730
11	10	10	0	0	0	716	722
12	12	7	0	0	0	738	722
13	17	13	$-2^{1/2}$	0	0	498	509
14	18	14	$2^{1/2}$	0	0	1063	1059
15	14	16	0	$-2^{1/2}$	0	565	558
16	13	17	0	$2^{1/2}$	0	917	915
17	16	15	0	0	$-2^{1/2}$	633	626
18	15	18	0	0	$2^{1/2}$	785	779
19	22	22	$-2^{1/2}$	0	0	486	509
20	24	19	$2^{1/2}$	0	0	1070	1072
21	19	21	0	$-2^{1/2}$	0	559	555
22	21	23	0	$2^{1/2}$	0	917	917
23	23	20	0	0	$-2^{1/2}$	624	612
24	20	24	0	0	$2^{1/2}$	792	789

permission of the copyright holder, The American Society of Mechanical Engineers, copyright © 1964.) Table E7.22a shows the results obtained from a cube plus doubled star plus four center points design run on three variables, speed V, feed f, and depth of cut d, which were varied in a cutting tool investigation. The coded input variables are $x_1 = (\ln V - \ln 70)/(\ln 70 - \ln 35)$, $x_2 = (\ln f - \ln 0.010)/(\ln 0.010 - \ln 0.005)$, and $x_3 = (\ln d - \ln 0.040)/(\ln 0.040 - \ln 0.020)$. The desired response variable is $y = \ln T$, where $T°$ is the temperature in degrees Fahrenheit. Two sets of observations were obtained, so that 48 sets of (x_1, x_2, x_3, y) are available in all. The tests within each set were run in random order as denoted by the "test sequence numbers" given in the source reference and reproduced in Table E7.22a.

Fit a second-order equation to the 48 pieces of data and provide an analysis of variance table. Test for lack of fit. (Assume that the four repeat runs at the center and the six pairs of star points provide 9 df for pure error *within each set*, and so 18 in all. Assume that there may be a block difference between the "first 24 tests" and the "second 24 tests" and take out a sum of

Table E7.22b. Confirmatory Runs

x_1	x_2	x_3	Observed Temperature(s) (°F)
1.28010	−0.46642	−1.23447	843, 832
−0.80735	0.76553	0.80737	711, 709
0.77761	−1.24129	−0.79837	658, 667
0.19265	0.26303	1.32194	866, 873
−1.37440	1.13750	0.0	609

Note: The x_1 and x_2 coordinates of the fifth confirmatory run have been corrected; in the source reference they do not match the V and f levels recorded.

squares for blocks.) Check the residuals, particularly against the sequence numbers of the runs.

Use your fitted equation to predict the responses for the nine confirmatory runs in Table E7.22b, and compare your predicted temperatures with those observed.

7.23. (*Source:* Sucrose-modified phenolic resins as plywood adhesives, by C. D. Chang and O. K. Kononenko, *Adhesives Age*, **5**(7), July 1962, 36–40. Material reproduced and adapted with the permission of *Adhesives Age*. The data were obtained under a grant from the Sugar Research Foundation, Inc., now the World Sugar Research Organization, Ltd.) This work is a continuation of that in Exercise 5.22. Four variables: (1) sucrose in grams, S, coded to $x_1 = (S − 47)/5$; (2) paraform in grams, P, coded to $x_2 = (P − 36.6)/3$; (3) NaOH in grams, N, coded to $x_3 = (N − 8.8)/2$; and (4) water in grams, W, coded to $x_4 = (W − 23.8)/7$ were examined for their effect on the dry strip-shear strength in pounds per square inch (psi) of Douglas fir plywood glued with a resin modified with sugar. Three coded levels, $−1, 0, 1$ were used for each factor, and 27 runs were made. These constituted a four-factor cube plus star rotatable design with three center points, but one not in the conventional orientation. (For additional explanation see Section 15.4 and/or Box and Behnken, 1960a.) The design and the response observations are shown in Table E7.23a. Note that each block contains one center point and that blocks are orthogonal; that is, the blocking will not affect the estimation of a second-order model. For the conditions for this, see Section 15.3. A separate set of repeat runs gave rise to a pure error estimate $s_3^2 = 50.97$.

Confirm that the tabled data may be fitted to a full second-order equation to provide

$$\hat{y} = 1.99 − 0.25x_1 − 7.667x_2 + 8.833x_3 + 5.417x_4$$

$$− 3.167x_1^2 − 6.792x_2^2 − 1.292x_3^2 + 13.833x_4^2$$

$$+ 6.5x_1x_2 − 3x_1x_3 − 4.75x_1x_4 − 10.5x_2x_3 − 6x_2x_4 + 7x_3x_4.$$

Furthermore, obtain the analysis of variance table given as Table E7.23b, and show that the lack of fit F statistic, $F(10, \nu) = 3.33$, is not significant if

Table E7.23a. A Three-Level, Second-Order, Response Surface Design in Four Factors Divided into Three Orthogonal Blocks

Block	x_1	x_2	x_3	x_4	y	Σy
1	-1	-1	0	0	203	
	1	-1	0	0	189	
	-1	1	0	0	191	
	1	1	0	0	203	
	0	0	-1	-1	211	
	0	0	1	-1	206	
	0	0	-1	1	218	
	0	0	1	1	241	
	0	0	0	0	211	1873
2	-1	0	0	-1	178	
	1	0	0	-1	196	
	-1	0	0	1	210	
	1	0	0	1	209	
	0	-1	-1	0	178	
	0	1	-1	0	169	
	0	-1	1	0	211	
	0	1	1	0	160	
	0	0	0	0	184	1695
3	0	-1	0	-1	218	
	0	1	0	-1	213	
	0	-1	0	1	219	
	0	1	0	1	190	
	-1	0	-1	0	184	
	1	0	-1	0	181	
	-1	0	1	0	222	
	1	0	1	0	207	
	0	0	0	0	202	1836
						5404

Table E7.23b. Analysis of Variance for the Fitted Second-Degree Equation

Source	SS	df	MS	F
Blocks	1960.5	2	980.25	5.77
First order	1944.5	4	486.13	2.86
Second order given b_0	3110.8	10	311.08	1.83
Residual	1697.6	10	169.76	
Total, corrected	8713.4	26		

$\nu \leq 8$. (This number ν is not specified in the paper, but the authors remark on p. 30 that the quadratic fit is adequate. It is most likely that five runs were made, as happened in an earlier part of the study.)

We see that only blocks are significant, indicating that the variation between blocks of runs is higher than any variation attributable to the factors x_1, x_2, x_3, and x_4. After three more exploratory trials to confirm this

Table E7.24a

Block	x_1	x_2	x_3	Yield (lb/plot), y Replicate 1	Replicate 2
1	d	c	0	10.6	7.8
	0	$-d$	c	11.5	6.9
	$-c$	0	$-d$	9.7	7.9
	1	-1	-1	9.1	5.1
	0	0	0	11.0	7.4
	0	0	0	10.3	6.6
2	d	$-c$	0	9.1	5.2
	0	d	$-c$	9.8	4.7
	$-c$	0	d	8.5	7.8
	1	1	1	10.3	7.6
	0	0	0	12.1	7.0
	0	0	0	9.3	7.5
3	$-d$	c	0	10.7	7.5
	0	$-d$	$-c$	11.4	4.4
	c	0	d	8.2	5.5
	-1	-1	1	6.2	5.1
	0	0	0	10.8	8.4
	0	0	0	11.3	7.2
4	$-d$	$-c$	0	8.9	4.8
	0	d	c	10.7	7.4
	c	0	$-d$	10.3	7.8
	-1	1	-1	10.4	7.5
	0	0	0	11.3	6.4
	0	0	0	11.1	9.1
5	-1	1	1	10.4	8.7
	-1	-1	-1	11.3	6.0
	1	1	-1	11.3	7.9
	1	-1	1	9.4	6.7
	0	0	0	10.8	7.6
	0	0	0	10.7	7.3
Totals				306.5	206.8

Note: $c = \frac{1}{2}(5^{1/2} + 1) = 1.618$; $d = \frac{1}{2}(5^{1/2} - 1) = 0.618 = c^{-1}$.

situation, the authors concluded that "it is apparent that a broad plateau has been reached" and selected the point $(S, P, N, W) = (44.6, 34, 47.1, 16.3)$ as the "optimal conditions." They reconfirmed this point with a final run.

7.24. (*Source:* Adapted from: An agronomically useful three-factor response surface design based on dodecahedron symmetry, by H. P. Hermanson, C. E. Gates, J. W. Chapman, and R. S. Farnham, *Agronomy Journal*, **56**, 1964, 14–17, by permission of the American Society of Agronomy.) Table E7.24a shows a 30-point design divided into five blocks. Two replicate values are obtained for each treatment combination, making 60 observations in all.

Table E7.24b

Source	SS	df	MS	F
Replicates	165.67	1	165.67	143.69[a]
First order	19.99	3	6.663	5.78[a]
Second order b_0	12.72	6	2.12	1.84
Lack of fit	47.34	39	1.214	1.32
Pure error	9.17	10	0.917	
Total, corrected	254.89	59		

[a]Significant at $\alpha = 0.05$ or lower α.

(a) Confirm that the design is a rotatable dodecahedron plus 10 center points, 2 per block, divided into orthogonal blocks; that is, the blocks are orthogonal to a second-order model matrix. Confirm also that "replicates" are orthogonal to the second-order model. (See Sections 14.3 and 15.3.)

(b) Obtain the fitted (by least squares) model $\hat{y} = \frac{1}{2}\{18.32 - 0.0552x_1 + 1.4063x_2 + 0.1155x_3 - 0.6225x_1^2 - 0.7075x_2^2 - 0.4850x_3^2 - 0.1667x_1x_2 - 0.1917x_1x_3 + 0.4167x_2x_3\} + 1.66z$. *Note:* (1) $z = 1$ for "replicate 1," $z = -1$ for "replicate 2." (2) The portion $\{\ldots\}$ of the model is what is obtained if a second-order surface is fitted to totals $\{$"y for replicate 1"+ "y for replicate 2"$\}$—for example, to $\{10.6 + 7.8\} = 18.4$, and so on.

(c) Confirm the analysis of variance, Table E7.24b. Note the following points, (1) Pure error is obtained only from pairs of center points within the same block *and* within the same replicate. There are 10 such pairs, for example, $(11.0, 10.3)$, $(7.4, 6.6)$, $(12.1, 9.3), \ldots, (7.6, 7.3)$. (2) The "replicates" sum of squares is given by $(306.5)^2/30 + (206.8)^2/30 - (513.3)^2/60 = 4556.95 - 4391.28 = 165.67$. (3) Certain subtleties involving "block strips SS" (blocks) and "replicate by block interaction SS" have been ignored; see the source reference for additional details. (These details would have needed investigation if lack of fit had shown up.) (4) The pooled estimate of σ^2 given by $s^2 = (47.34 + 9.17)/(39 + 10) = 1.153$ is used to obtain all F values except that for lack of fit; the latter F value is not significant; hence the validity of the pooling.

(d) Confirm that a first-order model explains $R^2 = 0.7284$ of the variation about the mean and that the addition of the (nonsignificant) second-order terms raises this to $R^2 = 0.7812$.

(e) Check the residuals from both first- and second-order models. Would you feel happy about using the fitted first-order model $\hat{y} = 8.555 + 1.66z - 0.052x_1 + 1.4063x_2 + 0.1155x_3$? Why or why not?

7.25. (*Source:* Chemical process improvement by response surface methods, by P. W. Tidwell, *Industrial and Engineering Chemistry*, **52**, 1960, June, 510–512. Adapted by permission of the copyright holder, the American Chemical Society.) Fit a second-order model to each set of y data in Table E7.25. Test

Table E7.25

x_1	x_2	x_3	y_1	y_2
− 1	− 1	− 1	62.1	15.79
1	− 1	− 1	61.3	12.03
− 1	1	− 1	17.8	11.91
1	1	− 1	68.8	12.37
− 1	− 1	1	61.3	15.77
1	− 1	1	61.0	11.92
− 1	1	1	16.6	11.58
1	1	1	66.5	12.27
− 2	0	0	8.9	11.26
2	0	0	67.0	10.96
0	− 1.65	0	62.4	13.68
0	1.65	0	45.1	12.68
0	0	− 2.4	66.8	14.02
0	0	2.4	66.5	13.80
0	0	0	64.6	13.70
0	0	0	65.5	13.71
0	0	0	64.6	13.90
0	0	0	65.1	14.08

for lack of fit and, if no lack of fit is evident, test whether second-order terms are needed and whether any of the x's can be dropped. What predictive model would you use?

7.26. (*Source:* A statistical experimental design and analysis of the extraction of silica from quartz by digestion in sodium hydroxide solutions, by R. L. Stone, S. M. Wu, and T. D. Tiemann, *Transactions of the Society of Mining Engineers*, **232**, June 1965, 115–124. Adapted with the permission of the Society of Mining Engineers of AIME, copyright © 1965.) Table E7.26a shows the values of four coded variables x_1, x_2, x_3, x_4 and the response y in an experiment to determine the effects of four predictor variables, quartz size, time, temperature, and sodium hydroxide concentration on the extraction of silica from quartz. The actual predictor variables have been logged to the base e and coded, and the response is ln(extraction).

Consider runs 1–20 as one block and runs 21–28 as a second block. Is this design orthogonally blocked? Fit a second-order surface to these data, adding a blocking variable whose levels are $z = -\frac{2}{7}$ for runs $1 - 20$ and $z = \frac{5}{7}$ for runs 21–28. Evaluate an analysis of variance table of the form indicated in Table E7.26b and also examine the usefulness of the individual x's as well as the need for quadratic terms. What model would you adopt? Examine the residuals, including a plot of residuals against test order. State your conclusions. (For orthogonal blocking, see Section 15.3.)

7.27. (*Source:* Use of statistics and computers in the selection of optimum food processing techniques, by A. M. Swanson, J. J. Geissler, P. J. Magnino, Jr., and J. A. Swanson, *Food Technology*, **21**(11), 1967, 99–102. Adapted with the

Table E7.26a

Test Order	x_1	x_2	x_3	x_4	y
1	−1	−1	−1	−1	4.190
20	1	−1	−1	−1	5.333
11	−1	1	−1	−1	4.812
5	1	1	−1	−1	5.886
17	−1	−1	1	−1	5.333
3	1	−1	1	−1	6.225
9	−1	1	1	−1	5.905
12	1	1	1	−1	6.574
15	−1	−1	−1	1	4.533
6	1	−1	−1	1	5.642
8	−1	1	−1	1	5.199
13	1	1	−1	1	6.094
4	−1	−1	1	1	5.628
19	1	−1	1	1	6.373
14	−1	1	1	1	6.127
2	1	1	1	1	6.702
10	0	0	0	0	5.775
7	0	0	0	0	5.802
18	0	0	0	0	5.781
16	0	0	0	0	5.796
25	−2	0	0	0	4.615
21	2	0	0	0	6.690
28	0	−2	0	0	5.220
23	0	2	0	0	6.385
27	0	0	−2	0	4.796
24	0	0	2	0	6.693
26	0	0	0	−2	5.416
22	0	0	0	2	6.182

Table E7.26b

Source	df
b_0	1
Blocks	1
First order given blocks	4
Second order given b_0, blocks	10
Lack of fit	9
Pure error	3
Total	28

Table E7.27. Data from a Sterilized, Concentrated, Baby Formula Experiment

Run Reference Number	Levels of Coded Predictors					Response Values (subscript = storage period in months)			
	x_1	x_2	x_3	x_4	x_5	y_0	y_3	y_6	y_9
1	−1	−1	−1	−1	−1	9.8	7.5	12.5	41.5
26	1	−1	−1	−1	1	30.2	35.0	22.5	45.0
5	−1	1	−1	−1	1	17.5	17.5	12.5	20.0
22	1	1	−1	−1	−1	12.5	10.0	7.5	12.5
3	−1	−1	1	−1	1	512.5	1950.0	2070.0	3030.0
24	1	−1	1	−1	−1	655.0	670.0	450.0	1700.0
7	−1	1	1	−1	−1	342.5	262.5	410.0	322.5
20	1	1	1	−1	1	1020.0	1050.0	970.0	1230.0
2	−1	−1	−1	1	1	82.5	145.0	162.5	145.0
25	1	−1	−1	1	−1	19.0	22.0	17.5	25.0
6	−1	1	−1	1	−1	9.3	5.8	5.0	12.5
21	1	1	−1	1	1	27.5	22.5	15.0	20.0
4	−1	−1	1	1	−1	270.0	237.5	337.5	717.5
23	1	−1	1	1	1	282.5	710.0	650.0	547.5
8	−1	1	1	1	1	172.5	237.5	210.0	190.0
19	1	1	1	1	−1	172.5	155.0	257.5	435.0
9	−1	0	0	0	0	45.8	52.5	62.5	57.5
27	1	0	0	0	0	77.5	62.5	70.0	113.8
11	0	−1	0	0	0	195.8	262.5	252.5	276.3
12	0	1	0	0	0	33.0	22.5	15.0	27.5
13	0	0	−1	0	0	20.0	15.0	17.5	17.5
14	0	0	1	0	0	337.5	117.5	105.0	177.5
15	0	0	0	−1	0	70.0	147.5	60.0	147.5
16	0	0	0	1	0	83.8	62.5	132.5	105.0
17	0	0	0	0	−2	40.0	40.0	22.5	60.0
18	0	0	0	0	2	287.5	450.0	482.5	495.0
10	0	0	0	0	0	67.5	77.5	45.0	107.5

permission of the copyright holder, the Institute of Food Technologists.) The data in Table E7.27 are obtained from a larger experiment on a sterilized, concentrated, baby formula. The purpose of the experiments was to select conditions which would provide a formula with acceptable storage stability for over a year at room temperature, and which would withstand terminal sterilization once the concentrate had been diluted with water. The five factors, or predictor variables, and their ranges of interest were as follows:

1. Preheating of the milk solids for 25 min (175–205°F).
2. Addition of sodium polyphosphate (0–0.14%).
3. Addition of sodium alginate (0–0.3%).
4. Addition of lecithin (0–1.5%).
5. Addition of carrageenan (0.444–0.032%).

Three levels were chosen for predictor variables 1–4, five levels for variable 5, and the coding shown in the table was adopted for these levels. The basic design is a 2_V^{5-1} fractional factorial design ($\mathbf{I} = -\mathbf{12345}$) plus five pairs of axial points at the extreme levels, plus one center point—that is, a cube plus star (unbalanced in x_5) plus center point. The response y_t, $t = 0, 3, 6, 9$ is the Brookfield viscosity in centipoise after t months, immediately after opening and without any agitation. To be acceptable, the product viscosity had to be less than 1000 cP. (Other responses measured, but not given in the paper, related to storage stability and the results of terminal stabilization. In fact, y_t was a subsidiary response and optimization was performed on the other responses first with a check on y_t to confirm that it was acceptable. However, for the purposes of this exercise, we regard it as the primary response.)

Fit, to each column of response values shown, a full second-order surface in $x_1, x_2, \ldots, x_5$. Provide an analysis of variance table with as complete a breakup as possible and test for lack of fit, and for second- and first-order significant regression, as is possible.

Evaluate the residuals and examine them. In particular, arrange the residuals in run order (shown on the left of the table) and see if any effects from that ordering are apparent.

Consider the following question: Would it be a good idea to transform the response? Apply the Box–Cox transformation technique to find the "best" transformation in the family (see Section 8.4)

$$z = \frac{(y^\lambda - 1)}{\lambda}, \qquad \lambda \neq 0$$

$$= \ln y, \qquad \lambda = 0,$$

and if a value of λ different from unity is indicated, make the transformation, and carry out all the analyses mentioned in the foregoing. What do you conclude, now?

If the viscosity has to be kept within the range $800 \leq y_t \leq 1000$, what levels of the x_i are permissible? How about for $y_t \leq 200$?

7.28. (*Source:* Sequential method for estimating response surfaces, by G. C. Derringer, *Industrial and Engineering Chemistry*, **61**, 1969, 6–13. Adapted by permission of the copyright holder, the American Chemical Society.) Table E7.28 shows data from a two-factor, third-order, rotatable, orthogonally blocked, response surface design consisting of a circle of eight points, a circle of seven points, and six center points in the coded (x_1, x_2) space. Also, shown are the values of six response variables $y_1, \ldots, y_6$. (Full descriptive details of the experiment will be found in the source paper and are not repeated here.) The orthogonal blocking conditions are satisfied for first- and third-order terms by the nature of the design, and the second-order orthogonal blocking condition determines the radius of the seven-point circle given the radius of the eight-point circle and the number of center points used. Specifically, here, $r_2^2 = 2(7)(8)/\{14(7)\} = 8/7 = (1.069)^2$.

Table E7.28

Run	Block	x_1	x_2	y_1	y_2	y_3	y_4	y_5	y_6
1	1	1	1	1140	3040	3670	570	9.4	14.5
2	1	1	-1	590	1750	3430	720	13.4	20.0
3	1	-1	1	910	2390	3790	660	7.5	13.0
4	1	-1	-1	530	1500	2940	710	11.9	30.5
5	1	0	0	900	2410	3800	660	12.0	16.0
6	1	0	0	880	2410	3860	660	10.6	15.0
7	1	0	0	870	2330	3830	670	11.3	15.0
8	2	1.414	0	1040	2760	3670	600	12.7	17.5
9	2	-1.414	0	650	1830	3570	710	8.2	18.0
10	2	0	1.414	990	2560	3770	630	8.9	13.0
11	2	0	-1.414	440	1300	3040	730	20.7	41.5
12	2	0	0	840	2240	3870	690	11.2	16.0
13	2	0	0	860	2270	3690	660	11.9	16.0
14	2	0	0	860	2260	3820	680	11.9	16.5
15	3	0.667	0.836	870	2400	3440	620	10.1	14.5
16	3	-0.238	1.042	730	2000	3560	660	7.9	13.5
17	3	-0.963	0.464	570	1590	3200	700	9.4	15.0
18	3	-0.963	-0.464	490	1350	2860	710	10.5	21.0
19	3	-0.238	-1.042	460	1290	3010	740	14.5	28.0
20	3	0.667	-0.836	620	1880	3510	690	14.7	22.0
21	3	1.069	0	870	2440	3660	630	12.3	17.5

Fit a third-order polynomial individually to each response, perform tests to determine what order of surface is needed in each case, and plot each fitted response within the rectangle $-1.75 \le x_i \le 1.75$, $i = 1, 2$. Your model will require two blocking variables and your analysis of variance table should contain a 2-df sum of squares for blocks. Suitable values for the blocking variables are $Z_1 = -1, 0, 1$ and $Z_2 = 1, -2, 1$ for blocks 1, 2, 3, respectively, and the model should contain extra terms $\alpha_1 Z_1 + \alpha_2 Z_2$. [Note: The Z_1 and Z_2 values are those of the orthogonal polynomials for three levels. See Appendix 19A for their normalized forms, that is forms in which their sums of squares add to 1.]

CHAPTER 8

Adequacy of Estimation and the Use of Transformation

Usually, the metrics in which data are recorded are chosen merely as a matter of convenience in measurement. As the textile example in the foregoing chapter illustrated, they need not be those in which the system is most simply modeled. To achieve simplicity, transformations may be applied to the response, or to the predictor variables, or both. We begin by discussing the effects of such transformations.

8.1. WHAT CAN TRANSFORMATIONS DO?

Nonlinear transformations such as the square root, log, and reciprocal of some necessarily positive response Y have the effect of expanding the scale at one part of the range and contracting it at another. This is illustrated in Table 8.1 and Figure 8.1 where, for Y covering the range $1-4$, the effects of square root, log, inverse square root, and reciprocal transformations are illustrated. These are all examples* of simple power transformations characterized by $y = Y^\alpha$. Transformations such as those illustrated in which α is less than unity have the effect of contracting the range at high values and may be called contractive transformations. Power transformations with α greater than unity have the reverse effect and may be called expansive. Contractive transformations of the kind illustrated in Figure 8.1 are most often needed in practical problems. Two facts about nonlinear transformations which should be remembered are:

1. The effect of the transformation is greater the greater the ratio Y_{max}/Y_{min} considered.
2. The effect of a power transformation $y = U^\alpha$ is greater the more α differs from unity.

*The log transformation can be regarded as a special case of the power transformation $y = Y^\alpha$ with $\alpha \to 0$, as explained in Section 8.4.

Response Surfaces, Mixtures, and Ridge Analyses, Second Edition. By G. E. P. Box and N. R. Draper
Copyright © 2007 John Wiley & Sons, Inc.

Table 8.1. Numerical Values of Contractive Transformations

Y	1	1.5	2	2.5	3	3.5	4
$Y^{1/2}$	1	1.225	1.414	1.581	1.732	1.871	2
$\log_{10} Y$	0	0.176	0.301	0.398	0.477	0.544	0.602
$Y^{-1/2}$	1	0.816	0.707	0.632	0.577	0.535	0.500
Y^{-1}	1	0.667	0.500	0.400	0.333	0.286	0.250

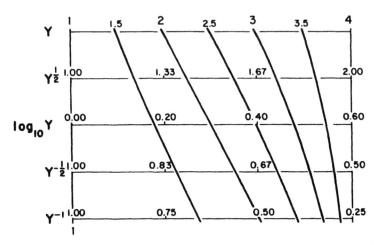

Figure 8.1. Some contractive transformations plotted over the range from $Y = 1$ to $Y = 4$. The scale for each transformed variable is shown on the appropriate horizontal line.

Note: When using the logarithmic transformation, either logarithms to the base 10 or to the base e can be used. They differ only by a constant factor, so that any subsequent analysis is unaffected except for this difference in scale. When we specifically wish to denote logarithms to the base e, we use the notation ln.

It is supposed in the above that Y is referred to a natural origin and is necessarily positive over the range considered. Thus, if we are interested in the concentration or amount of a particular ingredient in a chemical reaction, the natural origin would usually be zero. For temperature (T) measured in degrees Celsius, the functionally important origin is, most often, absolute zero ($-273°C$) and temperatures are thus usually expressed in degrees Kelvin ($T + 273$). For example, the Arrhenius equation postulates that the logarithm of the rate of a chemical reaction is a linear function of the reciprocal of the absolute temperature, namely that

$$\log(\text{rate}) = \alpha + \beta(T + 273)^{-1}.$$

If temperatures are measured in degrees Celsius, the above implies that, before the transformation (to reciprocals), a constant 273 is added to the temperature. Sometimes, an *empirically chosen* constant μ is subtracted from a response Y. This produces a transformation of greater nonlinearity if μ is positive, as is usually the case. For example, for any fixed α, the effect of the transformation $(Y - \mu)^{\alpha}$ will

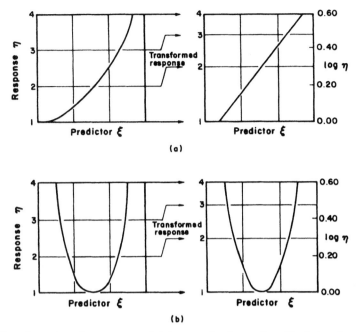

Figure 8.2. Transformation of a response η. (a) An example where a logarithmic transformation of the response variable η produces a straight line. (b) An example where a logarithmic transformation of the response variable η produces a quadratic curve.

depend on $(Y - \mu)_{\max}/(Y - \mu)_{\min}$. Thus, with the restriction $0 < \mu < Y_{\min}$, the larger μ is taken, the greater the effect induced by the transformation.

Effects of Simple Transformations on the Functional Relationships $\eta = f(\xi)$

For the moment, let us set aside the important question of the effect of transformation on distributional assumptions and consider only their effect on the functional relationship $E(Y) = \eta = f(\xi)$.

Figure 8.2a illustrates a situation where a relationship between η and ξ is linearized by making a contractive transformation in the response η. The logarithmic transformation is actually used, so that $\log \eta = \alpha + \beta\xi$ and the original relationship is of the form $\eta = AB^{\xi}$. The same transformation applied in Figure 8.2b produces a quadratic relationship. Thus $\log \eta = \alpha + \beta\xi + \gamma\xi^2$ and the original relationship is of the form $Ae^{B(\xi - \xi_0)^2}$.

Figure 8.3a shows how a contractive transformation (log, again) of the *predictor variable* ξ linearizes a curved relationship. We see that a contractive transformation on the response η can linearize a "convex downward" curve (Figure 8.2a) and a contractive transformation on the predictor ξ can linearize a "convex upward" curve. A further implication which can easily be confirmed is that, where linearity can be produced by a contractive transformation on η, a similar result might be obtained, at least approximately, by an expansive transformation in ξ. The opposite is also true.

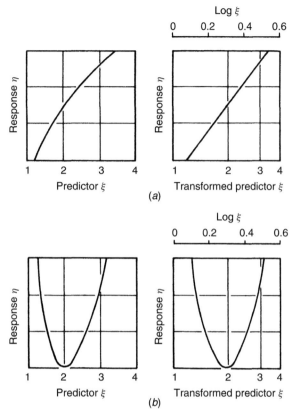

Figure 8.3. Transformation of a predictor variable ξ. (*a*) An example where a logarithmic transformation of the predictor variable ξ produces a straight line. (*b*) An example where a logarithmic transformation of the predictor variable ξ produces a quadratic curve.

Figure 8.3*b* shows the same contractive transformation of the *predictor variable* ξ, producing a quadratic curve. The quadratic curves in the second parts of Figures 8.2*b* and 8.3*b* are chosen to be identical within their respective frameworks, although the axial scales are different in the two cases. By examination and comparison of Figures 8.2*b* and 8.3*b*, we can appreciate that a quadratic function may be used to model an unsymmetric maximum *only* if ξ is transformed. Transformation of η can produce a sharper or flatter peaked curve, but *not* an unsymmetric one.

For η depending on more than one predictor variable, Figures 8.4*a* and 8.4*b* illustrate the fact that transformation of η relabels contours without changing their shape. In the illustration of Figure 8.4*a*, log η is linear in ξ_1 and ξ_2. Notice that this implies that the contours of *any* other function of η (in particular η itself) will have parallel straight line contours. However, in this particular illustration, it is only for log η that equal intervals in the transformed responses will be represented by equal distances in space. Thus only log η is *linear* in ξ_1 and ξ_2. Similarly in Figure 8.4*b*, log η is quadratic in ξ_1 and ξ_2 and the elliptic contours are also contours of η or any other function of η. But only log η is quadratic in ξ_1 and ξ_2.

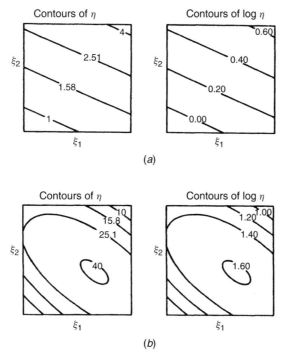

Figure 8.4. Transformation of a response η which depends on two predictor variables ξ_1 and ξ_2. (a) An example where a logarithmic transformation of the response variable η produces a planar surface. (b) An example where a logarithmic transformation of the response variable η produces a quadratic surface.

When η is monotonic in ξ_1 and ξ_2 but its contours are not straight lines, it may be possible by transformations of ξ_1 and ξ_2 to produce a planar surface as in Figure 8.5a but transformation of η alone clearly cannot. Similarly, when a set of closed contours about a maximum are not ellipses, it may, nevertheless, be possible by suitable transformations of ξ_1 and ξ_2 to produce a quadratic surface, as in Figure 8.5b. It is clearly not possible to do this by transformation of η alone.

Effect of Transformation of the Response on Distributional Assumptions

In practice, we do not know the expected response values (the η's); transformation must be carried out on the observed response values (the y's) which contain error. Thus, transformation of the response affects the error structure.

To illustrate how this occurs, consider an experiment designed to study how the size of the red blood cells of a rabbit were affected by two dietary substances added in amounts ξ_1 and ξ_2 to the feed. Suppose that, over the immediate region of interest $R(\boldsymbol{\xi})$, the amounts ξ_1 and ξ_2 of these two additives affected the radius y of the cells approximately linearly and that, for this measurement, the usual normal theory assumptions were approximately valid. Then y_u, the cell radius measured in the uth experiment, can be written

$$y_u = \eta_u + \varepsilon_u \tag{8.1.1}$$

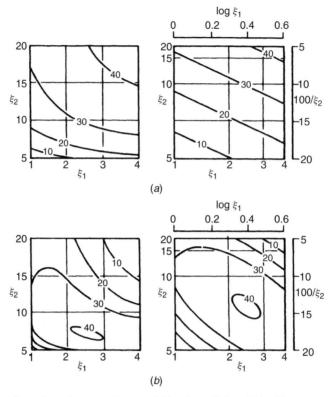

Figure 8.5. Transformation of two predictor variables ξ_1 and ξ_2 which affect a response η. (*a*) An example where transformation of two predictor variables ξ_1 and ξ_2 produces a planar surface. (*b*) An example where transformation of two predictor variables ξ_1 and ξ_2 produces a quadratic surface.

with

$$\eta_u = \beta_0 + \beta_1 \xi_{1u} + \beta_2 \xi_{2u}.$$

Now in practice it might be most convenient to measure the area A_u of a cell or, equivalently, $A_u / \pi = Y_u$, the radius squared. But if

$$Y_u = y_u^2,$$

then, using (8.1.1),

$$Y_u = \eta_u^2 + 2\eta_u \varepsilon_u + \varepsilon_u^2. \tag{8.1.2}$$

To convert (8.1.2) into a form comparable with (8.1.1), we must write

$$Y_u = \eta_u' + \varepsilon_u',$$

where

$$\eta_u' = \eta_u^2 + \sigma_\varepsilon^2 \quad \text{and} \quad \varepsilon_u' = 2\eta_u \varepsilon_u + \varepsilon_u^2 - \sigma_\varepsilon^2,$$

so that, using the normality assumption where necessary,

$$E(\varepsilon'_u) = 0 \quad \text{and} \quad V(\varepsilon'_u) = 2\sigma_\varepsilon^2(2\eta_u^2 + \sigma_\varepsilon^2) = 2\sigma_\varepsilon^2(2\eta'_u - \sigma_\varepsilon^2).$$

Thus, if modeling were carried out directly on the untransformed cell area, we would need a second-degree equation to express the relationship between the response and the predictor variables. In addition, the errors ε'_u would have (a) a variance which was linearly dependent on the mean level η'_u of the response and (b) a distribution which was non-normal. However, a (square root) transformation of Y would yield a linear model with normal errors having constant variance.

From the above, it is clear that if we regard the response Y as principally subject to error while the predictor variables are relatively error-free, the transformation on Y will not only affect the form of the functional relationship $\eta = f(\xi)$ but also the error structure. On the other hand, transformation of the ξ's will, on these assumptions, not affect the error structure at all. There will be examples such as the (artificial) blood cell example above and also the (real) textile example already considered, where transformation of the response can simultaneously simplify the functional relationship and improve compliance with the distributional assumptions. Other examples will occur where this is not possible. In some examples, where distribution assumptions are already adequate, appropriate simplification may be possible by transformation of the predictor variables ξ alone. In other examples, it may be necessary to transform both the response and the predictor variables to obtain simplicity of functional form and a satisfactory distribution of error.

With the possibility of simplification via transformation ever present, the task of obtaining empirically an adequate functional representation is not a trivial one; we consider various aspects of the matter in the next section. We begin by considering a measure of the adequacy of estimation of a fitted function which is related to a certain F ratio in the analysis of variance.

8.2. A MEASURE OF ADEQUACY OF FUNCTION ESTIMATION

Attempts to interpret relationships that have been inadequately estimated are likely to prove misleading and sometimes disastrous. For example, the detailed behavior of plotted contours from a fitted second-degree surface ought not to be taken seriously on the basis of mere statistical significance of the "overall regression" versus "residual" F test. This section is intended to throw light on the following questions: "When is a fitted function sufficiently well estimated to permit useful interpretations?" and "Should adequacy be judged on the size of F, R^2, or what else?" We follow an argument due to Box and Wetz (1973; also see Draper and Smith, 1998, pp. 243–250).

We can illustrate adequately the questions at issue in terms of a single-input ξ. Suppose that observations of a response y are made at n different settings of the input ξ. Figure 8.6a shows a plot of $n = 8$ observations $y_1, y_2, \ldots, y_8$ against the known levels $\xi_1, \xi_2, \ldots, \xi_8$, of the single-input ξ. Suppose further that $\eta = g(\xi, \boldsymbol{\beta})$ is an appropriate graduating function (such as a polynomial in ξ) linear in p adjustable parameters in addition to the mean and that the curve in Figure 8.6b

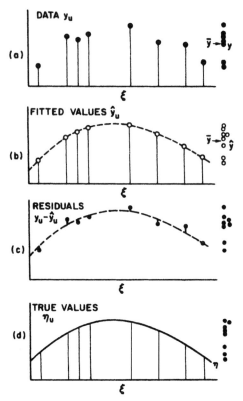

Figure 8.6. Data y_u, fitted values $\hat{y}_u$, ,residuals $y_u - \hat{y}_u$, and true values η_u for a relationship involving one input ξ. Also shown are the empirical distributions of y's, $\hat{y}$'s, $y - \hat{y}$'s, and η's, the sum of whose squares enter the analysis of variance table.

represents this function fitted to the data by least squares. The values that the fitted function takes at $\xi = \xi_1, \xi = \xi_2, \ldots, \xi = \xi_8$ are denoted by $\hat{y}_1, \hat{y}_2, \ldots, \hat{y}_8$ and are indicated in Figure 8.6b by open circles. Figure 8.6c shows the residuals $y_1 - \hat{y}_1, y_2 - \hat{y}_2, \ldots, y_8 - \hat{y}_8$ which measure the discrepancies between observation and fitted function at the chosen values of ξ. Figure 8.6d shows the true (expected) values, η. We know from Chapter 3 that certain identities exist between the sums of squares of these quantities. In particular, taking squares about the grand average $\bar{y}$ provides

$$\text{total sum of squares} = (\text{regression sum of squares})$$
$$+ (\text{residual sum of squares}),$$

or

$$\Sigma(y_u - \bar{y})^2 = \Sigma(\hat{y}_u - \bar{y})^2 + \Sigma(y_u - \hat{y}_u)^2. \qquad (8.2.1)$$

The corresponding identity in the degrees of freedom is

$$n - 1 = p + (n - p - 1). \qquad (8.2.2)$$

Using these identities, an appropriate analysis of variance table may be written down. For the purpose of this discussion, we shall assume for the time being that the relationship $E(y) = \eta = g(\xi, \beta)$ is representationally adequate. Then the expected values of the regression sum of squares about the mean, which is denoted by SS_R, and of the residual sum of squares, which is denoted by SS_D, are given by

$$E(\text{SS}_R) = E\{\text{regression SS}|b_0\}$$

$$= E\left\{\Sigma(\hat{y}_u - \bar{y})^2\right\} = \Sigma(\eta_u - \bar{\eta})^2 + p\sigma^2, \qquad (8.2.3)$$

$$E(\text{SS}_D) = E\{\text{residual SS}\}$$

$$= E\left\{\Sigma(y_u - \hat{y}_u)^2\right\} = (n - p - 1)\sigma^2. \qquad (8.2.4)$$

Now consider Figure 8.6d. Because of estimation errors, the estimated function $\hat{y} = g(\xi, \hat{\beta})$ differs somewhat from the true function $\eta = g(\xi, \beta)$. The values $\eta_1, \eta_2, \ldots, \eta_8$, in Figure 8.6$d$, provide a "sampling" of the function over the range actually observed and the mean square deviation $\Sigma(\eta_u - \bar{\eta})^2/n$ is a measure of the real change in η accounted for by the functional dependence of η on the input variable ξ. (It is the mean squared deviation of the distribution of the given η_u.) We want to compare this with a measure of the uncertainty with which $\eta - \bar{\eta}$ is estimated. Such a measure is the average of the expected values of the squares of discrepancies between the estimated deviations $\hat{y}_u - \bar{y}$ and "true" deviations $\eta_u - \bar{\eta}$ measured at the n (= 8, here) values of ξ.

We denote this averaged mean square error of estimate by $\sigma^2_{\hat{y}-\bar{y}}$ so that

$$\sigma^2_{\hat{y}-\bar{y}} = \left[E\{(\hat{y}_1 - \bar{y}) - (\eta_1 - \bar{\eta})\}^2 + \cdots \right.$$

$$\left. + E\{(\hat{y}_8 - \bar{y}) - (\eta_8 - \bar{\eta})\}^2\right]/n. \qquad (8.2.5)$$

A natural measure of estimation adequacy is, therefore, the quantity

$$Q^2 = \frac{\Sigma(\eta_u - \bar{\eta})^2/n}{\sigma^2_{\hat{y}-\bar{y}}}. \qquad (8.2.6)$$

In Appendix 8A, we show that, irrespective of the experimental design, $\sigma^2_{\hat{y}-\bar{y}} = p\sigma^2/n$. Thus, an alternative form to (8.2.6) is

$$Q^2 = \frac{\Sigma(\eta_u - \bar{\eta})^2/n}{p\sigma^2/n} = \frac{\Sigma(\eta_u - \bar{\eta})^2/p}{\sigma^2}. \qquad (8.2.7)$$

The quantity Q^2 is familiar in classical statistics as a *measure of noncentrality*, and Q may also be regarded as a measure of *signal-to-noise ratio*. Now, on the assumptions made, the residual mean square $s^2 = \text{SS}_D/(n - p - 1)$ provides an unbiased estimator of σ^2 so that, from (8.2.3), $(\text{SS}_R - ps^2)/n$ provides an unbiased estimator of $\Sigma(\eta_u - \bar{\eta})^2/n$ and ps^2/n provides an unbiased estimator of

Table 8.2. Values of Estimation Adequacy Measure $\hat{Q}^2 = F - 1$ for Textile Data

Response used	$y = \log Y$	Y
Linear equation	217.0^a	65.5^b
Quadratic equation	65.7^b	27.5^b

[a] Via residual from appropriate planar equation.
[b] Via residual from appropriate quadratic equation.

$p\sigma^2/n = \sigma^2_{\hat{y}-\bar{y}}$. Thus, an estimator having the property that the expected values of numerator and denominator are those of Q^2 is

$$\hat{Q}^2 = \frac{(SS_R - ps^2)/n}{ps^2/n} = \frac{SS_R/p}{s^2} - 1 = F - 1, \tag{8.2.8}$$

where F is the usual ratio of regression to residual mean squares.

As an illustration of the use of the measure of estimation adequacy $\hat{Q}^2 = F - 1$, we employ it to assess the effects of transformation on the textile data already analyzed in Chapter 7. The various values of $\hat{Q}^2$ are shown in Table 8.2. In the original analysis of these data, the investigators, Barella and Sust, fitted a quadratic function to the untransformed Y. This yielded a value of $\hat{Q}^2$ of 27.5. We see that, for this example, a sevenfold increase in $\hat{Q}^2$ ($217.0/27.5 = 7.9$) is possible by use of the simpler straight line equation fitted to $y = \log Y$. Evidently the improvement may be credited to the following facts:

1. The method of (unweighted) least squares is an efficient estimation procedure only if the data have constant variance. A change from Y to $y = \log Y$ apparently makes the constant variance assumption approximately valid.
2. There are, in addition to the mean, nine parameters in the quadratic model but only three in the linear model. Thus, in the log metric, there are six redundant parameters that should improve estimation efficiency by an additional factor of $\frac{10}{4} = 2.5$.

When Is a Response Function Adequately Estimated?

After an analysis of variance has been performed, the ratio F = (regression mean square)/(error mean square) is customarily computed and compared to an F-distribution percentage point, appropriate on the assumption that the errors ε are normally distributed with constant variance. Significance of this F ratio contradicts the assertion that the noncentrality measure $Q^2 = \{\Sigma(\eta_u - \bar{\eta})^2/p\}/\sigma^2$ is *zero*, and suggests that changes of some sort are detectable in the response η. However, to permit the assertion that we have a reasonable *estimate* of the response function, much more than mere statistical significance is required.

Now consider Figures 8.6c and 8.6d. The quantity $\{\Sigma(\eta_u - \bar{\eta})^2/n\}^{1/2}$ is a root mean square measure of *change* in the function $\eta = g(\xi, \beta)$, while $\sigma_{\hat{y}-\bar{y}}$ is a root mean square measure of the average *error of estimate* of that same function. Q is the ratio of these two quantities. Before we could assert that a worthwhile *estimate* of the response function had been obtained, we would need to ensure that the change in the true function over the region of ξ considered was reasonably large compared with the error of estimate in that region. Thus we would need to be assured that Q had some minimum value Q_0.

To make assertions of this kind, we must consider the distribution of $F = \hat{Q}^2 + 1$ when the noncentrality measure is nonzero. This distribution (called the noncentral F distribution) is known and has been tabulated, but, for our present purposes, it will be sufficient to use a crude approximation. Very roughly, over the range of values of interest, to assert that $Q > Q_0$ requires that the relevant significance point obtained from the F table be multiplied* by $Q_0^2 + 1$. Thus, to ensure that Q is greater than 3, we should require that the ratio (mean square regression)/(mean square residual) exceed its normal F significance level by a factor of 10.

The general conclusion to be drawn is that adequacy of estimation of a function ought to be judged by the size of $F = $ (regression mean square)/(residual mean square). The value of F should exceed the chosen tabulated significance level by a factor of $Q_0^2 + 1$, which we suggest should be about 10.

An alternative and, in spirit, equivalent procedure is to compare the range of the n fitted $\hat{y}$'s with their average standard error $(ps^2/n)^{1/2}$.

Application to the Textile Data

For the textile data, for the first-degree polynomial fit, $p = 3$, $n - p - 1 = 23$ and the 5% level of $F = 3.03$. Applying the above rule, we should require, therefore, that the F value actually achieved exceeded 30. In fact, for the logged data, it is 217.

Conversely, the value $F = 217$ implies that, for the example, $Q \approx (217 - 1)^{1/2} \approx 14.7$, which is a very satisfactory value.

Inadequacy of Models for Unlogged Textile Data

The discussion above has been carried out as if, in all the cases considered, we could believe in the adequacy of the fitted functions and of the normal theory models implied by the least-squares analysis. In fact, as we saw in Chapter 7, if the analysis of the textile example had been conducted in terms of the unlogged data, its inadequacy would have been seen as soon as the residuals were plotted.

8.3. CHOOSING THE RESPONSE METRIC TO VALIDATE DISTRIBUTION ASSUMPTIONS

In considering the question of an appropriate choice of response metric to validate distribution assumptions, the first thing to notice is that a *linear* transformation of Y, say to $y = (Y - k_0)/k_1$ where the k's are constants, will be without effect on

---

*The argument involves approximation of the appropriate noncentral F by $(1 + Q_0^2)$ times a central F. This would in fact be exact if the model were a random-effects model. When, as here, the model is a fixed-effects one, the result is approximate. However, the approximation appears to be conservative; see Patnaik (1949).

these assumptions since all that has happened is a rescaling and relocation of the graduating function.

[Purely for *computational* convenience it is nevertheless useful sometimes to make a linear transformation (or "coding") of the output Y just as similar linear codings are often usefully applied to the inputs. For instance, in a study of the manner in which the specific gravity Y of an aqueous solution was affected by certain input variables, the range of variation in Y over the whole set of experimental conditions was from 1.0027 to 1.0086. The data invite the coding $y = 10,000(Y - 1)$. The range of variation in the coded output y's is then from 27 to 86. Such a coding reduces worries about misplaced decimals, rounding errors, and computer overflow, and, when the calculations have been completed, the inverse transformation $Y = 1 + 0.0001y$ restores the calculated response to the original units.]

Although linear transformation of the response has no effect on the shape of the error distribution, *nonlinear* transformations such as $y = \log Y$, $y = \log(Y - k)$, and $y = Y^\lambda$ can, in suitable circumstances, produce marked changes in distributional shape and in variance homogeneity.

When Do Data Transformations Have an Important Effect on Distribution Assumptions?

For data that are necessarily positive, the choice of metric for Y becomes important when the ratio Y_{max}/Y_{min} is large. For if this ratio is small, moderate nonlinear transformations of Y such as $\log Y, Y^{1/2}, Y^{-1}$ are nearly linear over the relevant range and make little difference. On the other hand, when the ratio is large, transformation can bring profound changes. To see what is involved, consider Figure 8.7. Suppose that $y = \log Y$ has a normal distribution and constant variance independent of its expectation. Then the identical distributions A, B, and C on the $y = \log Y$ axis will correspond with the very different distributions A', B', and C' on the Y axis. In this illustration, the distributions A' and C' have been centered on the values $Y = 90$ and $Y = 3600$, the extreme values covered by the textile data. It is seen that normality in $y = \log Y$ will induce somewhat skewed distributions in Y because of *local* curvature in the plot of Y versus y; however, such induced non-normality will be of only moderate importance. What is more important is the dramatic difference in the *standard deviation* of the distributions A' and C' because of "global" differences in gradient. In fact, in our illustration, C' has a standard deviation $3600/90 = 40$ times as large as that of A'. The variance of the distribution of C' is thus 1600 times as large as that of the distribution A'! Equivalently, the weight of an observation from A' is 1600 times as large as that of an observation from C'. This means that one observation from A' is as accurate as the mean of 1600 from C' and is so treated by the logarithmic analysis. However, if ordinary least squares is applied to Y instead of to $\log Y$, these observations of vastly different accuracy are treated as having equal weights! Notice also that the distributions B' and C', centered at 2500 and 3600, have standard deviations in proportion to $3600/2500 = 1.44$, and therefore variances in the ratio of about 2 to 1, so that, by comparison, this discrepancy is a minor one.

To sum up: Over the range that encompasses B' and C', the curve relating Y and $\log Y$ is roughly linear and the choice between Y and $\log Y$ is not very

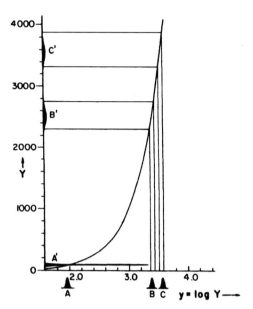

Figure 8.7. The logarithmic transformation and its effect on distributional assumptions.

important. By contrast, the range encompassing A' and C' is highly nonlinear and, for data such as we actually have in the textile example, for which $Y_{max}/Y_{min} = 3600/90 = 40$, careful choice of transformation is essential.

For further explanation, we suppose that some nonlinear transformation $y = h(Y)$ can be approximated (see Figure 8.8) by the first two terms of its Taylor's series* expansion about some value $Y = Y_0$. Thus, approximately

$$y = h(Y) = h_0 + h_1(Y - Y_0) + \tfrac{1}{2}h_2(Y - Y_0)^2, \qquad (8.3.1)$$

where

$$h_0 = h(Y_0), \qquad h_1 = \frac{dh(Y)}{dY}\bigg|_{Y=Y_0}, \qquad h_2 = \frac{d^2h(Y)}{dY^2}\bigg|_{Y=Y_0}. \qquad (8.3.2)$$

In this formula, h_1 measures the slope at Y_0, while h_2 measures the quadratic curvature at Y_0. It is important to notice that, while the distortion of the distribution shape produced by transformation is largely due to the curvature h_2, the change in spread produced by the transformation is associated with the change in gradient h_1, so that approximately,

$$\sigma_y \simeq h_1 \sigma_Y. \qquad (8.3.3)$$

*The formula simply says that, for a smooth function $h(Y)$, the value at some Y close to Y_0 is given by its value $h_0 = h(Y_0)$ at Y_0, plus a term to allow for the rate h_1 at which the function is changing at Y_0, plus a further term to allow for the rate h_2 at which the rate is itself changing and so on.

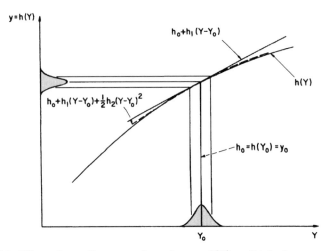

Figure 8.8. Effects of a nonlinear transformation $y = h(Y)$ on distribution assumptions.

We can thus write, for the ratio of the standard deviations of y at the extremes of the range of the data,

$$\frac{\sigma_{y_{\max}}}{\sigma_{y_{\min}}} \simeq \left(\frac{h_1(Y_{\max})}{h_1(Y_{\min})} \right) \left(\frac{\sigma_{Y_{\max}}}{\sigma_{Y_{\min}}} \right).$$

For the textile data, for example, if the standard deviation is roughly constant for $y = h(y) = \log Y$, then $h_1 = Y^{-1}$ and

$$\frac{\sigma_{Y_{\max}}}{\sigma_{Y_{\min}}} \simeq \frac{3600}{90} = 40.$$

Variance Stabilizing Transformations

We have seen that an important requirement for the efficiency and validity of unweighted least-squares estimation is that, to a rough approximation, the variance of the response Y is independent of the magnitude of $E(Y) = \eta$.

Suppose this is not so and that the standard deviation σ_Y of Y is some function $F(\eta)$ of the mean value η, of Y. Then, referring to the argument of the foregoing section, a transformed variable or metric $h(Y)$ for which the variance will be approximately stable is such that

$$\frac{dh(Y)}{dY} \propto \frac{1}{F(Y)}. \tag{8.3.4}$$

This result leads, in particular, to the important transformations shown in Table 8.3. In the textile example, an assumption that the *percentage* error was roughly constant, so that $\sigma_Y \propto \eta$, would yield the metric $y = \log Y$.

Table 8.3. Appropriate Variance Stabilizing Transformations when $\sigma_Y = F(\eta)$

Nature of Dependence $\sigma_Y = F(\eta)$			Variance Stabilizing Transformation
$\sigma_Y \propto \eta^k$		$(Y \geq 0)$	Y^{1-k}
and in particular			
$\sigma_Y \propto \eta^{1/2}$	(Poisson)	$(Y \geq 0)$	$Y^{1/2}$
$\sigma_Y \propto \eta$		$(Y \geq 0)$	$\log Y$
$\sigma_Y \propto \eta^2$		$(Y \geq 0)$	Y^{-1}
$\sigma_Y \propto \eta^{1/2}(1 - \eta)^{1/2}$	(binomial)	$(0 \leq Y \leq 1)$	$\sin^{-1}(Y^{1/2})$
$\sigma_Y \propto (1 - \eta)^{1/2}\eta$	(negative binomial)	$(0 \leq Y \leq 1)$	$(1 - Y)^{1/2} - (1 - Y)^{3/2}/3$
$\sigma_Y \propto (1 - \eta^2)^{-2}$		$(-1 \leq Y \leq 1)$	$\log\{(1 + Y)/(1 - Y)\}$

In practice, the nature of the dependence between σ_Y and η may be suggested by theoretical considerations or by preliminary empirical analysis, or by both.

Examples of Transformations Derived from Theoretical Considerations

A relationship between σ_Y and Y can sometimes be deduced by considering the theoretical distribution of Y.

Poisson Distribution

Suppose the observed response Y was the *frequency* with which an event occurred, such as a blood cell count. Then it might be expected that, under constant experimental conditions ξ, Y would vary from one count to another in a Poisson distribution

$$P(Y) = \frac{\eta^Y}{Y!}e^{-\eta} \tag{8.3.5}$$

for which

$$\sigma_Y \propto \eta^{1/2}. \tag{8.3.6}$$

Referring to Table 8.3, we see that the appropriate transformation is the square root transformation. The effect of the transformation would be such that if we now change the experimental conditions ξ, resulting in changes in $\eta = f(\xi)$, the variance of $y = Y^{1/2}$ would, nevertheless, remain approximately constant. Incidentally, the transformation would also render the distributions at various conditions ξ more nearly normal.

Binomial Distribution

As a further example, suppose that Y was the proportion of "successes" out of a fixed number of trials such as the proportion of 50 tested resistors with resistances less than 1 ohm. Then Y might be expected to follow a binomial distribution. If the probability of a success is η and is constant for all resistors at a given set of experimental conditions, then

$$p(Y) \propto (\eta)^{nY}(1 - \eta)^{n(1-Y)} \qquad (8.3.7)$$

for which

$$\sigma_Y \propto \eta^{1/2}(1 - \eta)^{1/2}. \qquad (8.3.8)$$

From Table 8.3, we see that the appropriate transformation for which the variance would be nearly stable and independent of the true probability of success η is $\sin^{-1}(Y^{1/2})$.

Transformations Derived from Empirical Considerations

When it happens that replicate runs are made at a number of experimental conditions, a preliminary fit can be made in what is judged to be the best metric Y and then the sample variance s_ξ^2 and the mean $\bar{Y}_\xi$ may be calculated for each set of conditions ξ. The possible dependence of s_ξ^2 on the mean $\bar{Y}_\xi$ may then be judged empirically by graphical methods.

In particular, if it is assumed that σ_Y is proportional to some power of η, so that

$$\sigma_Y \propto \eta^k, \qquad (8.3.9)$$

and so

$$\log \sigma_Y = c + k \log \eta, \qquad (8.3.10)$$

then the slope of a plot of $\log s_\xi$ against $\log \bar{Y}_\xi$ can provide an estimate for k.

Simplicity of Response Functions and Validity of Distribution Assumptions

In the past, transformation has sometimes been carried out purely to simplify the form of the response function and insufficient consideration has been given to the vital question of how such transformations affect error assumptions, which alone determine efficient choice of the estimation procedure. For example, in physical chemistry, rate equations involving reciprocal representations, such as

$$\eta = 1/(\theta_0 + \theta_1 \xi_1 + \theta_2 \xi_2), \qquad (8.3.11)$$

frequently arise. The constants (the θ's) in such expressions are often estimated by making a reciprocal transformation of the response function and fitting the resulting linear expression to the reciprocals of the response data by least squares, implying a model

$$y = Y^{-1} = \theta_0 + \theta_1 \xi_1 + \theta_2 \xi_2 + \varepsilon. \qquad (8.3.12)$$

If, for this reciprocal scale, the normal theory constant variance assumptions were valid for ε, ordinary least squares would produce efficient estimates with *correct* standard errors, and further analysis resting on the validity of derived F and t

distributions would be appropriate. However, suppose, on the other hand, that the normal theory error assumptions, and particularly the assumptions about constancy of variance, had been true for the model

$$Y = 1/(\theta_0 + \theta_1\xi_1 + \theta_2\xi_2) + \varepsilon'. \tag{8.3.13}$$

Then, for Y's covering a wide range, least-squares fitting of the reciprocal function (8.3.12) could result in very inefficient estimators and incorrect standard errors. By contrast, direct fitting of the nonlinear model (8.3.13) using *nonlinear* least squares would yield efficient estimates. (This option is discussed in Appendix 3D.) Alternatively, an appropriate weighted linear least-squares analysis of the linearized form (Box and Hill, 1974) could be employed to produce a close approximation to the nonlinear fit.

Further Analysis of the Textile Data

While transformations employed merely to achieve a linear model can often seriously distort distributional assumptions, they may also occasionally make the usual assumptions more nearly applicable. For illustration, consider once more the textile data of Table 7.1, following the discussion of Box and Cox (1964). In many fields of technology, power relationships are common. For the textile example, such a relationship would postulate that cycles to failure was, apart from error, proportional to

$$\gamma_0 \xi_1^{\gamma_1}\xi_2^{\gamma_2}\xi_3^{\gamma_3}, \tag{8.3.14}$$

where ξ_1 = length of test specimen (mm), ξ_2 = amplitude of load cycle (mm), and ξ_3 = load (g). Now we know already that the normal assumptions appear to be approximately true for $y = \log Y$. After taking logarithms, the power relationship yields the model

$$y = \log\gamma_0 + \gamma_1\log\xi_1 + \gamma_2\log\xi_2 + \gamma_3\log\xi_3 + \varepsilon. \tag{8.3.15}$$

We have already seen that the distributional assumptions appear valid for $y = \log Y$ rather than for Y itself. However, the model (8.3.15) is linear in the *logs of the input* and not in the inputs themselves which is the form we fitted in coded form (see Table 7.2) in the previous analysis. Since we know that the equation in the unlogged inputs fits very well, this might at first lead us to question whether logged inputs would be worth further consideration. However, in this particular example, the inputs (*unlike the output*), cover quite narrow ranges. The largest value of ξ_{max}/ξ_{min} is found for the first input ξ_1 (length of test specimen) for which $\xi_{1\,max}/\xi_{1\,min} = 350/250 = 1.4$. Over this small a range, the log transformation is practically linear, as we see from Figure 8.7. Thus, for these data, the fact that we get a good fit with the unlogged ξ's *guarantees* a fit which is nearly as good (or possibly slightly better) for the logged ξ's. In fact, if model (8.3.15) is estimated by least squares, an excellent fit is obtained with

$$\hat{\gamma}_1 = 4.96 \pm 0.20, \qquad \hat{\gamma}_2 = -5.27 \pm 0.30, \qquad \hat{\gamma}_3 = -3.15 \pm 0.30. \tag{8.3.16}$$

There is an advantage in moving as close as possible toward a model which might conceivably have more theoretical justification, and it sometimes happens that interesting possibilities are suggested by the results of what began as an empirical analysis, as we now illustrate. Since in Eq. (8.3.16), $\hat{\gamma}_2 \simeq -\hat{\gamma}_1$, the quantity $\log \xi_2 - \log \xi_1 = \log(\xi_2/\xi_1)$ is suggested as being of importance. Furthermore, $\xi_2/\xi_1 = \xi_4$ (say) is the fractional amplitude and is suggested by dimensional considerations. Finally, then the simple relationship

$$\text{cycles to failure} = k\xi_4^{-5}\xi_3^{-3} \tag{8.3.17}$$

is obtained, which represents the data very well. Thus our iterative analysis leads finally to a very simple representation.

8.4. ESTIMATING RESPONSE TRANSFORMATION FROM THE DATA

The textile example illustrates how it is sometimes possible to simplify greatly the problem of empirical representation by making a suitable transformation on the response

Consider again the general question of the choice of transformation $y = h(Y)$ for an output variable Y. If possible, such a transformation should be chosen so that the model

$$y = g(\xi, \beta) + \varepsilon \tag{8.4.1}$$

provides an accurate and easily understood representation over the region $R(\xi)$ of immediate interest and can be readily fitted. Ideally, then, the transformation might be chosen to achieve these features:

1. $g(\xi, \beta)$ should have a simple and parsimonious form (requiring a minimum number of parameters).
2. The error variance should be approximately constant.
3. The error distributions should be approximately normal.

As has been illustrated above, scientific knowledge of the system, common sense, and the use of simple plots can greatly help in an appropriate choice. A further tool (to be used in conjunction with the above) is the likelihood function, which can help determine a suitable transformation from the data themselves.

Although the method is applicable to any class of parametric transformations, we illustrate it here for the important power transformations where (supposing the data values Y are all positive) $y = Y^\lambda$. These include, for example,

$\lambda = 1$	$\lambda = \frac{1}{2}$	$\lambda = 0$	$\lambda = -\frac{1}{2}$	$\lambda = -1$
$y = Y$	$y = Y^{1/2}$	$y = \log Y$	$y = 1/Y^{1/2}$	$1/Y.$
no	square	logarithm	reciprocal	reciprocal
transformation	root		square root	

Since $Y^0 = 1$ for all Y, an explanation is in order as to why the log transformation is associated with the value $\lambda = 0$. Because we can write

$$Y^\lambda = e^{\lambda \ln Y} \simeq 1 + \lambda \ln Y + \tfrac{1}{2}\lambda^2 (\ln Y)^2 + \cdots, \qquad (8.4.2)$$

we have that

$$\lim_{\lambda \to 0} \left(\frac{Y^\lambda - 1}{\lambda} \right) = \ln Y. \qquad (8.4.3)$$

In essence, what this says is that if we take a small positive or negative power of Y (say $Y^{0.01}$ or $Y^{-0.01}$), it will plot against $\log Y$ very nearly as a straight line, and linearity will be more and more nearly achieved the smaller the power we take.*

Use of the Likelihood Function in Selecting a Power Transformation

We begin by describing the mechanics of the method using the textile data for illustration. Suppose it is postulated that some power transformation $y = Y^\lambda$ exists, in terms of which the graduating function $g(\boldsymbol{\xi}, \boldsymbol{\beta})$ in (8.4.1) has some specific simple form, and in terms of which the normal theory assumptions apply. The idea is to employ the method of maximum likelihood to estimate the transformation parameter λ at the same time as the parameters $\boldsymbol{\beta}$ of the model are estimated. The necessary steps (Box and Cox, 1964) are as follows:

1. Compute the geometric mean $\dot{Y}$ of all the data from $\ln \dot{Y} = n^{-1}\Sigma \ln Y_u$.
2. For a series of suitable values of λ, compute the transformed values $Y_u^{(\lambda)}$ using the form

$$Y_u^{(\lambda)} = \begin{cases} (Y_u^\lambda - 1)/(\lambda \dot{Y}^{\lambda-1}), & \text{if } \lambda \neq 0 \\ \dot{Y} \ln Y_u, & \text{if } \lambda = 0. \end{cases} \qquad (8.4.4)$$

3. Fit the parsimonious model $g(\boldsymbol{\xi}, \boldsymbol{\beta})$ to $Y^{(\lambda)}$ by least squares, and record the residual sum of squares $S(\lambda)$, for each chosen value of λ.
4. Plot $\ln S(\lambda)$ against λ. The value of λ which makes $\ln S(\lambda)$, and so $S(\lambda)$, smallest is the maximum likelihood value $\hat{\lambda}$.
5. An approximate $100(1 - \alpha)\%$ confidence interval for λ is given by determining graphically the two values of λ for which

$$\ln S(\lambda) - \ln S(\hat{\lambda}) = \frac{\chi_\alpha^2(1)}{\nu_r}, \qquad (8.4.5)$$

---

*Use of logarithms to the base e (ln) is needed for Eq. (8.4.2), but logarithms to any base (e.g., 10) can be used in the transformation. The difference amounts only to a constant factor.

where ν_r is the number of residual degrees of freedom, and $\chi_\alpha^2(1)$ is the upper α significance point of χ^2 with 1^* df.

We illustrate the method by applying it to the untransformed textile data Y given in Table 7.1. Our object is to estimate a λ for which the errors, $y_u - g(\boldsymbol{\xi}_u, \boldsymbol{\beta})$, of the transformed observations y_u from the simple linear graduating function

$$g(\boldsymbol{\xi}, \boldsymbol{\beta}) = \beta_0 + \beta_1 \xi_1 + \beta_2 \xi_2 + \beta_3 \xi_3, \tag{8.4.6}$$

most nearly follow normal theory assumptions. We find that:

1. The geometric mean for the 27 observations of Table 7.1 is

$$\dot{Y} = 562.34.$$

2. The transformed data values for any given value of λ are given by

$$Y_u^{(\lambda)} = \begin{cases} \lambda^{-1}(562.34)^{1-\lambda}(Y_u^\lambda - 1), & \text{if } \lambda \neq 0, \\ (562.34)\ln Y_u, & \text{if } \lambda = 0. \end{cases}$$

3. If the linear model above is fitted to each set of transformed values $Y_u^{(\lambda)}$, $u = 1, 2, \ldots, n$, for each of the following values of λ, the residual sums of squares and their natural logarithms are as given:

λ	-1.0	-0.8	-0.6	-0.4	-0.2	0.0
$S(\lambda)$	3.9955	2.1396	1.1035	0.5478	0.2920	0.2519
$\ln S(\lambda)$	1.3852	0.7606	0.0985	-0.6018	-1.2310	-1.3787
	0.2	0.4	0.6	0.8	1.0	
	0.4115	0.8178	1.5968	2.9978	5.4810	
	-0.8879	-0.2011	0.4680	1.0979	1.7013	

4. The plot of $\ln S(\lambda)$ against λ yields a minimum value at about $\hat{\lambda} = -0.06$. (Figure 8.9 shows the situation around the minimum.)

5. An approximate 95% confidence interval for λ is obtained by computing

$$\frac{\chi_{0.05}^2(1)}{\nu_r} = \frac{3.84}{23} = 0.167$$

and observing the two values of λ at which the horizontal line $\ln S(\hat{\lambda}) + 0.167$ cuts the $\ln S(\lambda)$ curve. In the present case, the 95% confidence interval extends from about $\lambda = -0.20$ to 0.08.

---

*In general the transformation could involve q unknown parameters (instead of a single λ as here). In this case, the approximate expression (8.4.5) would still hold but the chi-squared table would be entered with q df.

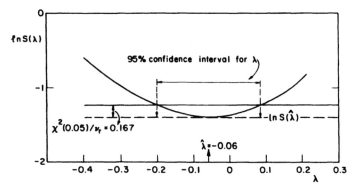

Figure 8.9. A plot of $\ln S(\lambda)$ versus λ in the neighborhood of the minimum $\ln S(\lambda)$, for the textile data.

Our analysis confirms the conclusions previously reached on a common sense basis, and reinforces them. We see, for example, that the values $\lambda = 1$ (no transformation), $\lambda = \frac{1}{2}$ (square root transformation), and $\lambda = -\frac{1}{2}$ (reciprocal square root transformation) are all rejected by the analysis. Plausible values for λ are closely confined to a region near to $\lambda = 0$ (log transformation).

We remind ourselves once more that data transformation is of potential importance in this example because the ratio $Y_{\max}/Y_{\min} = 3636/90 \simeq 40$ is very large. As we have said before, when $Y_{\max}/Y_{\min}$ is not large, transformations such as log, reciprocal, and square root behave locally almost like linear transformations and so do not greatly affect the analysis. If, when $Y_{\max}/Y_{\min}$ is not large, the likelihood analysis is nevertheless carried through, a very wide confidence interval for λ will be obtained, indicating that it will make little difference which transformation is used.

Other Transformations

The method may be applied not only to simple power transformations but much more generally.

Suppose we consider some class of transformations y of Y which involves a vector of parameters, $\boldsymbol{\lambda}$. Suppose these transformations are monotonic functions of Y for each value of $\boldsymbol{\lambda}$ over the admissible range. Then the analysis is made with the transformed values

$$Y_u^{(\lambda)} = \frac{y_u}{J^{1/n}}, \tag{8.4.7}$$

where

$$J = \prod_{u=1}^{n} \left| \frac{\partial y_u}{\partial Y_u} \right|. \tag{8.4.8}$$

For example, for the power transformations with shifted origin, $(Y + \lambda_2)^{\lambda_1}$, it is convenient to work with the form $y = \{(Y + \lambda_2)^{\lambda_1} - 1\}/\lambda_1$ which takes the value $\ln(Y + \lambda_2)$ when λ_1 is zero. The analysis is thus made in terms of

$$Y^{(\lambda)} = \frac{(Y + \lambda_2)^{\lambda_1} - 1}{\lambda_1 (Y \dot{+} \lambda_2)^{\lambda_1 - 1}}, \tag{8.4.9}$$

where $Y \dot{+} \lambda_2$ is the geometric mean of the n values $Y_u + \lambda_2$, $u = 1, 2, \ldots, n$.

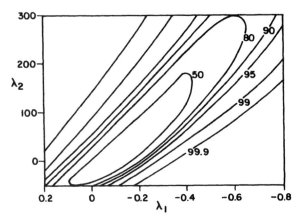

Figure 8.10. Checking the value of the transformation $(Y + \lambda_2)^{\lambda_1}$ for the textile data. Contours shown are those of the natural logarithm of the residual sum of squares $\ln S(\lambda_1, \lambda_2)$, labeled with their approximate confidence probability. [Reproduced from Box and Cox (1964), by permission of the Royal Statistical Society, London.]

Box and Cox (1964) illustrated the use of this type of transformation by applying it to the textile data of Table 7.1. The function (8.4.6) was fitted by least squares to the transformed data, which this time depended on two parameters, λ_1 and λ_2. For a selected grid of points (λ_1, λ_2), the values of $\ln S(\lambda_1, \lambda_2)$, which is the logarithm of the residual sum of squares for the fitted model, were calculated. This grid of values was used to construct Figure 8.10. The contours in that figure have been labeled to indicate the approximate probability associated with the confidence regions they encompass. There is no indication that a nonzero value of λ_2 is needed for these data, which confirms the validity of our previous analysis.

8.5. TRANSFORMATION OF THE PREDICTOR VARIABLES

So far, we have written mostly about transformation of the response variable. In some cases, as we have illustrated earlier in this chapter, needed simplification may be obtained by transforming some or all of the individual predictor variables. Occasionally, transformation of both response *and* predictor variables is necessary. We now discuss certain types of transformation of the predictor variables. Let $\xi_1^{(\alpha_1)}, \xi_2^{(\alpha_2)}, \ldots,$ denote individual parametric transformations of the predictor variables $\xi_1, \xi_2, \ldots,$ each dependent on a single parameter $\alpha_1, \alpha_2, \ldots,$ respectively, and consider a model of the form

$$y = g\left(\xi_1^{(\alpha_1)}, \xi_2^{(\alpha_2)}, \ldots, \xi_k^{(\alpha_k)}; \boldsymbol{\beta} \right) + \varepsilon. \tag{8.5.1}$$

For illustration, Figure 8.11*a* shows contours of the function

$$\eta = 25 + 5 \ln \xi_1 + 100\xi_2^{-1} + 6\xi_3^{1/2}, \tag{8.5.2}$$

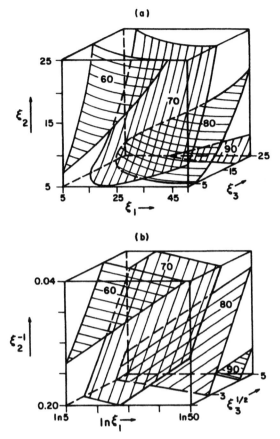

Figure 8.11. A three variable response function (*a*) before, and (*b*) after, transformation. [Reproduced from Box and Tidwell (1962), by permission of the American Statistical Association.]

over a cubical region in the ξ space, while Figure 8.11*b* shows contours of the same function plotted in terms of the transformed variables $\ln \xi_1$, ξ_2^{-1}, and $\xi_3^{1/2}$. If a polynomial were directly fitted to the ξ's, an equation of at least second degree would be needed to obtain a suitable fit, but in terms of the transformed variables the equation is first order. What we should like to do, in general, is to recognize when suitable transformation of the predictors will help us attain this sort of simplification. Transformation of the individual predictor variables should be considered whenever:

1. There is theoretical or empirical evidence to suggest that model simplification may occur as a result.
2. Transformation of some sort is clearly called for, but the error structure is satisfactory in the original response metric.

An appropriate transformation will frequently be suggested by theoretical considerations. The Arrhenius law, which connects the logarithm of the rate η of a chemical reaction to the absolute temperature ξ, is of the form

$$\ln \eta = \beta_0 + \beta_1 \xi^{-1}, \tag{8.5.3}$$

where β_0 and β_1 are constants., Thus if, as might well be the case, the usual normal theory distribution assumptions were expected to be true for the logarithm $y = \ln Y$ of the observed rate Y and if also a wide temperature range were being used so that choice of transformation was important, then a linear expression in the reciprocal of the temperature ξ should be entertained, and fitted by least squares, to the y values. Again, suppose the object of interest was the time period η of a pendulum as a function of the length ξ of the string from which the plumb bob was suspended. For a "perfect" pendulum, physical theory says that

$$\eta = \beta\sqrt{\xi}, \tag{8.5.4}$$

where β is a constant. In this instance then, if normal theory assumptions could be expected to apply to observations of the time period, then it would be natural to entertain a linear model in the square root of ξ.

Detecting the Need for, and Estimating, Power Transformations in the Predictor Variables, Using Only Linear Least-Squares Techniques

Suppose that the response η is a function of the k predictor variables in $\boldsymbol{\xi} = (\xi_1, \xi_2, \ldots, \xi_k)'$ and that it is desired to fit some graduating function g by least squares. Suppose further that, while the predictor variables have already been transformed to what is currently believed to be their most appropriate forms, it is possible, nevertheless, that they may require further transformation to powers $\xi_1^{\alpha_1}, \xi_2^{\alpha_2}, \ldots, \xi_k^{\alpha_k}$. Then the true response for the uth observation is a function

$$\eta_u = g(\xi_{1u}^{\alpha_1}, \xi_{2u}^{\alpha_2}, \ldots, \xi_{ku}^{\alpha_k}; \boldsymbol{\beta}^*), \tag{8.5.5}$$

where $\boldsymbol{\beta}^* = (\beta_0^*, \beta_1^*, \beta_2^*, \ldots, \beta_p^*)'$, say, is a vector of parameters. Following Box and Tidwell (1962), we can expand η_u in a Taylor's series to first order about the values $\alpha_1 = 1, \alpha_2 = 1, \ldots, \alpha_k = 1$ to give, instead of the original model $Y_u = \eta_u + \varepsilon_u$, the approximation

$$Y_u = g(\boldsymbol{\xi}_u, \boldsymbol{\beta}^*) + \sum_{j=1}^{k} (\alpha_j - 1)\left[\frac{\partial \eta_u}{\partial \xi_j}\right]_{\text{all } \alpha_i = 1} (\xi_{ju} \ln \xi_{ju}) + \varepsilon_u, \tag{8.5.6}$$

which can also be written as

$$Y_u - g(\boldsymbol{\xi}_u, \boldsymbol{\beta}^*) = \sum_{j=1}^{k} (\alpha_j - 1)Z_{ju} + \varepsilon_u, \tag{8.5.7}$$

where

$$Z_{ju} = \left[\frac{\partial \eta_u}{\partial \xi_j}\right]_{\text{all } \alpha_i = 1} (\xi_{ju} \ln \xi_{ju}). \tag{8.5.8}$$

Detection of Need for Transformation
Now if we fit the model

$$Y_u = g(\boldsymbol{\xi}_u, \boldsymbol{\beta}^*) + \varepsilon_u \tag{8.5.9}$$

by least squares and take the residuals $e_u = Y_u - \hat{Y}_u$, we shall have estimates of the left-hand sides of (8.5.7). Moreover, the same $\hat{Y}_u$ values can be used to give us the

$$\hat{Z}_{ju} = \frac{\partial \hat{Y}_u}{\partial \xi_j}(\xi_{ju}\ln \xi_{ju}), \tag{8.5.10}$$

which will be estimates of the Z_{ju} in (8.5.8). Thus, we can expect that, approximately, the need for transformation of the ξ_j will be indicated by a significant planar relationship apparent in plots of the e_u versus the $\hat{Z}_{ju}$, $j = 1, 2, \ldots, k$.

Estimating the Transformation
To get estimates of the α_j we can proceed as follows:

1. Fit (8.5.9) and use it to obtain the estimates $\hat{Z}_{ju}$ via (8.5.10).
2. Fit, by least squares, the model

$$Y_u = g(\boldsymbol{\xi}_u, \boldsymbol{\beta}^*) + \sum_{j=1}^{k}(\alpha_j - 1)\hat{Z}_{ju} + \varepsilon_u, \tag{8.5.11}$$

which is derived from (8.5.7) by replacing Z_{ju} by $\hat{Z}_{ju}$.
3. See which $(\hat{\alpha}_j - 1)$ are large compared to their respective standard errors and use those $\hat{\alpha}_j$ to transform ξ_j to $\xi_j^{\hat{\alpha}_j}$.
4. If desired, repeat 1–3 with the transformed variables in the roles of the corresponding untransformed variables.

This iterative process should produce improved estimates of the α's and should converge fairly quickly in most cases.

In practice, we often work, not with the ξ_{ju}, but with scaled variables $x_{ju} = (\xi_{ju} - \xi_{j0})/S_j$ so that $\xi_{ju} = \xi_{j0} + S_j x_{ju}$. Use of the scaled variables often prevents an escalation in size of the numbers involved in the calculation as the iterations proceed. If, upon rescaling, $g(\boldsymbol{\xi}_u, \boldsymbol{\beta}^*)$ becomes $g(\mathbf{x}_u, \boldsymbol{\beta})$, then we obtain, for first- and second-order polynomial models, the following.

First-Order Polynomial

$$g(\boldsymbol{\xi}_u, \boldsymbol{\beta}^*) = \beta_0^* + \sum_j \beta_j^* \xi_{ju}, \tag{8.5.12}$$

$$g(\mathbf{x}_u, \boldsymbol{\beta}) = \beta_0 + \sum_j \beta_j x_{ju}, \tag{8.5.13}$$

so that $\beta_j = S_j \beta_j^*$, and $\beta_0 = \beta_0^* + \sum_j \beta_j^* \xi_{j0}$. It follows that

$$\hat{Z}_{ju} = \hat{\beta}_j^* \xi_{ju} \ln \xi_{ju} \tag{8.5.14}$$

$$= S_j^{-1} \hat{\beta}_j (\xi_{j0} + S_j x_{ju}) \ln(\xi_{j0} + S_j x_{ju}). \tag{8.5.15}$$

Second-Degree Polynomial

$$g(\boldsymbol{\xi}_u, \boldsymbol{\beta}^*) = \beta_0^* + \sum_j \beta_j^* \xi_{ju} + \sum\sum_{i \geq j} \beta_{ij}^* \xi_{iu} \xi_{ju}, \tag{8.5.16}$$

$$g(\mathbf{x}_u, \boldsymbol{\beta}) = \beta_0 + \sum_j \beta_j x_{ju} + \sum\sum_{i \geq j} \beta_{ij} x_{iu} x_{ju}, \tag{8.5.17}$$

so that

$$\beta_{ij} = S_i S_j \beta_{ij}^*, \qquad \beta_j = S_j \left(\beta_j^* + \sum_i \beta_{ij}^* \xi_{i0} \right),$$

$$\beta_0 = \beta_0^* + \sum_j \beta_j^* \xi_{j0} + \sum\sum_{i \geq j} \beta_{ij}^* \xi_{i0} \xi_{j0}. \tag{8.5.18}$$

It follows that

$$\hat{Z}_{ju} = \left\{ \hat{\beta}_j^* + 2 \hat{\beta}_{jj}^* \xi_{ju} + \sum_{i \neq j} \hat{\beta}_{ij}^* \xi_{iu} \right\} \xi_{ju} \ln \xi_{ju} \tag{8.5.19}$$

$$= S_j^{-1} \left\{ \left(\hat{\beta}_j - \sum_i S_i^{-1} \hat{\beta}_{ij} \xi_{i0} + 2 S_j^{-1} \hat{\beta}_{jj} \xi_{j0} + \sum_{i \neq j} S_i^{-1} \hat{\beta}_{ij} \xi_{i0} \right) \right.$$

$$\left. + 2 \hat{\beta}_{jj} x_{ju} + \sum_{i \neq j} \hat{\beta}_{ij} x_{iu} \right\} (\xi_{j0} + S_j x_{ju}) \ln(\xi_{j0} + S_j x_{ju}). \tag{8.5.20}$$

To illustrate these ideas, we consider an example with a single predictor variable ξ, taken from Box and Tidwell (1962).

Example: Yield of Polymer as a Function of Catalyst Charge

An experimenter observes the quantity of a polymer formed in 1 hour, under steady-state conditions, for various (equally spaced) charges of a catalyst. The results are shown in Table 8.4 and are plotted in Figure 8.12a.

The need for a linearizing transformation is, of course, obvious in this case, and certainly one could guess a transformation for ξ which would produce near linearity. Figure 8.12b shows that $\xi^{1/2}$ *is* such a transformation. However, for illustration, we shall proceed according to the method outlined above. We want eventually to fit a straight line model

$$Y_u = \beta_0 + \beta_1 \xi_u^\alpha + \varepsilon_u \tag{8.5.21}$$

Table 8.4. Polymer Formed for Various Catalyst Charges

Grams of Polymer Formed, Y_u	Grams of Catalyst Charged to Reactor, ξ_u	Values of $(13.729)\xi_u \ln \xi_u = \hat{Z}_u$
28.45	0.5	-4.76
47.78	1.5	8.34
64.35	2.5	31.45
74.67	3.5	60.20
83.65	4.5	92.92

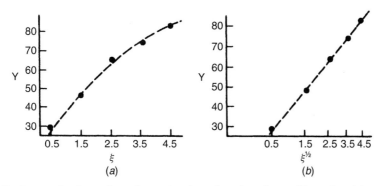

Figure 8.12. Grams of polymer formed as a function of catalyst charge. [Reproduced from Box and Tidwell (1962), by permission of the American Statistical Association.]

with the best α. We first fit (8.5.9) which, for our example, is (8.5.21) with $\alpha = 1$. This provides the fitted equation

$$\hat{Y}_u = 25.47 + 13.729\xi_u \qquad (8.5.22)$$

with derivative $\partial \hat{Y}_u / \partial \xi_u = 13.729$. We now construct the variable

$$\hat{Z}_u = 13.729\xi_u \ln \xi_u \qquad (8.5.23)$$

from (8.5.8); the values of $\hat{Z}_u$ are shown in Table 8.4. We then fit, by least squares, the model (8.5.11) which is

$$Y_u = \beta_0 + \beta_1 \xi_u + (\alpha - 1)\hat{Z}_u + \varepsilon_u. \qquad (8.5.24)$$

The fitted equation is

$$\hat{Y}_u = 11.222 + 28.286\xi_u - 0.589\hat{Z}_u. \qquad (8.5.25)$$

It follows that $\hat{\alpha} = 0.41$. The standard error of $\hat{\alpha}$ is 0.08, so that transformation is clearly indicated. We could now do additional iterations to improve the estimate of α, but it is clear that a value of $\alpha = 0.5$ is both convenient and within the range of reasonable values for the parameter. Furthermore, if we use $\alpha = 0.5$, a satisfactory linearization is achieved, as Figure 8.12*b* has already shown. Thus the transformation analysis could sensibly be terminated at this point and the model

$$Y_u = \beta_0 + \beta_1 \xi_u^{1/2} + \varepsilon_u \qquad (8.5.26)$$

could now be fitted and validated in the usual manner.

Transformations in the Predictor Variables Via Nonlinear Estimation

The procedure that we have given above is simple and useful, particularly when only one iteration is adequate for the purpose at hand. As we have intimated, the procedure could be the first step in a series of iterations which would lead to improved estimates of the α_j. When a standard nonlinear least-squares program is available, however, the above calculations can be bypassed and the nonlinear model

$$Y_u = g(\xi_{1u}^{\alpha_1}, \xi_{2u}^{\alpha_2}, \ldots, \xi_{ku}^{\alpha_k}; \boldsymbol{\beta}) + \varepsilon_u \qquad (8.5.27)$$

can be fitted directly.

APPENDIX 8A. PROOF OF A RESULT USED IN SECTION 8.2*

We can write $\sigma_{\hat{y}-\bar{y}}^2$ as $1/n$ times S, where

$$S = E\{(\hat{\mathbf{y}} - \bar{y}\mathbf{j}) - (\boldsymbol{\eta} - \bar{\eta}\mathbf{j})\}'\{(\hat{\mathbf{y}} - \bar{y}\mathbf{j}) - (\boldsymbol{\eta} - \bar{\eta}\mathbf{j})\}.$$

The vector $\mathbf{j} = (1, 1, \ldots, 1)'$ is of length n. Now

$$\hat{\mathbf{y}} - \bar{y}\mathbf{j} = \mathbf{R}\mathbf{y} - \frac{1}{n}\mathbf{j}\mathbf{j}'\mathbf{y} = \left(\mathbf{R} - \frac{1}{n}\mathbf{j}\mathbf{j}'\right)\mathbf{y},$$

where

$$\mathbf{R} = \mathbf{X}(\mathbf{X}'\mathbf{X})^{-1}\mathbf{X}'$$

and

$$\boldsymbol{\eta} - \bar{\eta}\mathbf{j} = \left(\mathbf{I} - \frac{1}{n}\mathbf{j}\mathbf{j}'\right)\boldsymbol{\eta},$$

where $\boldsymbol{\eta} = \mathbf{X}\boldsymbol{\beta}$. Also $\mathbf{R}$, $[\mathbf{R} - (1/n)\mathbf{j}\mathbf{j}']$ and $[\mathbf{I} - (1/n)\mathbf{j}\mathbf{j}']$ are all idempotent; that is, $\mathbf{R}\mathbf{R} = \mathbf{R}$, and so on. Thus we can write

$$S = E\left\{\mathbf{y}'\left(\mathbf{R} - \frac{1}{n}\mathbf{j}\mathbf{j}'\right)\mathbf{y} - 2\mathbf{y}'\left(\mathbf{R} - \frac{1}{n}\mathbf{j}\mathbf{j}'\right)\left(\mathbf{I} - \frac{1}{n}\mathbf{j}\mathbf{j}'\right)\boldsymbol{\eta} + \boldsymbol{\eta}'\left(\mathbf{I} - \frac{1}{n}\mathbf{j}\mathbf{j}'\right)\boldsymbol{\eta}\right\}.$$

*See Section 8.2 below Eq. (8.2.6).

Provided that the model has an intercept term β_0, $\mathbf{X'j}$ is the first column of $\mathbf{X'X}$ whereupon it follows that

$$\mathbf{Rj} = \mathbf{X(X'X)}^{-1}\mathbf{X'j} = \mathbf{X}(1,0,0,\ldots,0)' = \mathbf{j},$$

whereupon

$$S = E\left\{ \mathbf{y'}\left(\mathbf{R} - \frac{1}{n}\mathbf{jj'}\right)\mathbf{y} - 2\mathbf{y'}\left(\mathbf{R} - \frac{1}{n}\mathbf{jj'}\right)\boldsymbol{\eta} + \boldsymbol{\eta'}\left(\mathbf{I} - \frac{1}{n}\mathbf{jj'}\right)\boldsymbol{\eta}\right\}$$

$$= \boldsymbol{\eta'}\left(\mathbf{R} - \frac{1}{n}\mathbf{jj'}\right)\boldsymbol{\eta} + \sigma^2\, \text{trace}\left(\mathbf{R} - \frac{1}{n}\mathbf{jj'}\right) - 2\boldsymbol{\eta'}\left(\mathbf{R} - \frac{1}{n}\mathbf{jj'}\right)\boldsymbol{\eta}$$

$$+ \boldsymbol{\eta'}\left(\mathbf{I} - \frac{1}{n}\mathbf{jj'}\right)\boldsymbol{\eta}$$

[using the fact that $E(\mathbf{y'Ay}) = \boldsymbol{\eta'A\eta} + \text{trace}(\boldsymbol{\Sigma A})$ where $E(\mathbf{y}) = \boldsymbol{\eta}$, $V(\mathbf{y}) = \boldsymbol{\Sigma}$]

$$= \boldsymbol{\eta'}(\mathbf{I} - \mathbf{R})\boldsymbol{\eta} + \sigma^2\, \text{trace}\left(\mathbf{R} - \frac{1}{n}\mathbf{jj'}\right)$$

$$= \sigma^2 p$$

because the first term vanishes due to the fact that $(\mathbf{I} - \mathbf{R})\mathbf{X} = \mathbf{0}$ and the second term is σ^2 times $\text{trace}\{\mathbf{X(X'X)}^{-1}\mathbf{X'} - (1/n)\mathbf{jj'}\} = \text{trace}\{(\mathbf{X'X})^{-1}\mathbf{X'X}\} - (1/n)\text{trace}\,\mathbf{j'j}$, using the fact that $\text{trace}(\mathbf{AB}) = \text{trace}(\mathbf{BA})$, which reduces to $(p + 1) - 1 = p$, assuming there are p parameters in the model in addition to the constant term. Dividing by $1/n$ gives us $\sigma^2_{\hat{y}-\bar{y}} = p\sigma^2/n$ as required.

EXERCISES

8.1. A measure of the uncertainty associated with some event E is provided by its probability, $\text{prob}(E) = \theta$ where θ is restricted to the range $0 \le \theta \le 1$. A popular transformation of this measure of uncertainty is provided by the *odds* on an event where $\text{odds}(E) \equiv \phi = \theta/(1 - \theta)$. A further transformation is provided by the $\log(\text{odds}) = \log\{\theta/(1 - \theta)\}$. Plot, on a piece of rectangular graph paper, curves which will transform θ to ϕ to $\log \phi$.

8.2. You are given that a contour of the true response η is the following function of two variables r and θ:

$$r^2 + 4r \cos \theta + 6r \sin \theta = 12.$$

Transform this equation by $x = r \cos \theta$ and $y = r \sin \theta$, and so determine the shape of the contour in the (x, y) coordinate system.

8.3. Given the textile data of Table 7.1, fit the model

$$y \equiv \log Y = \gamma + \gamma_1 \log \xi_3 + \gamma_2 \log \xi_4 + \varepsilon$$

Table E8.4. Mooney Viscosities $y_1(\mathrm{ML}_4)$ and $y_2(\mathrm{MS}_4)$ at 100°C for Various Combinations of Coded Variables[a] x_1 and x_2

x_1	x_2	y_1	y_2
-1	-1	58	28
-1	1	41	18
1	-1	—	93
1	1	106	54
0	$-2^{1/2}$	104	57
0	$2^{1/2}$	56	30
$-2^{1/2}$	0	38	20
$2^{1/2}$	0	—	100
0	0	97	41
0	0	74	40
0	0	78	42
0	0	83	45
0	0	90	48

[a] $x_1 = (X_1 - 40)/20$, $x_2 = (X_2 - 20)/10$, where X_1 = level of Silica A, H_i-Sil 233, PPG Industries, and X_2 = level of Aromatic oil, Sundex 790, Sun Oil Co., in NBR Hycar 1052, B. F. Goodrich Chemical Co.

where $\xi_4 = \xi_2/\xi_1$, by least squares, and test the null hypothesis H_0: $\gamma_1 = -3$, $\gamma_2 = -5$ versus H_1: some violation of H_0. Is the remark around Eq. (8.3.17) justified, do you think? [*Hint:* On a standard regression program, it may be easier to fit $(y + 3 \log \xi_3 + 5 \log \xi_4) = \gamma + \delta_1 \log \xi_3 + \delta_2 \log \xi_4 + \varepsilon$, where $\delta_1 = \gamma_1 + 3$, $\delta_2 = \gamma_2 + 5$, and then test H_0: $\delta_1 = \delta_2 = 0$.]

8.4. (*Source:* An empirical model for viscosity of filled and plasticized elastomer compounds, by G. C. Derringer, *Journal of Applied Polymer Science*, **18**, 1974, 1083–1101. Copyright © 1974, John Wiley & Sons, adapted with permission.) Table E8.4 shows data for two responses y_1 and y_2 at various levels of two inputs. For each response individually, choose the best value of λ in the transformation $w = (y^\lambda - 1)/\lambda$ for $\lambda \neq 0$, $w = \ln y$ for $\lambda = 0$ to enable the model $w = \beta_0 + \beta_1 x_1 + \beta_2 x_2 + \varepsilon$ to be well fitted to the data by least squares. After choosing the best transformation parameter estimate $\hat{\lambda}$ in each case, carry out all the usual regression analyses including examination of the residuals. [*Note:* Once λ is chosen, the response y^λ may be used for the subsequent analysis instead of $(y^\lambda - 1)/\lambda$, if $\lambda \neq 0$. Or, if $\lambda = 0$, logarithms to any base, not necessarily e, can be used if desired. Such alternative choices do not affect the basic analysis, just the size of the numbers involved in the calculation. These remarks apply to all such transformation problems.]

8.5. (*Source:* An empirical model for viscosity of filled and plasticized elastomer compounds, by G. C. Derringer, *Journal of Applied Polymer Science*, **18**, 1974, 1083–1101. Copyright © 1974, John Wiley & Sons, adapted with permission.) Choose the best value of λ in the transformation $w = (y^\lambda - 1)/\lambda$ for $\lambda \neq 0$, $w = \ln y$ for $\lambda = 0$, to enable the model $w = \beta_0 + \beta_1 x_1 + \beta_2 x_2 + \beta_3 x_3 + \varepsilon$ to be well fitted to the data of Table E8.5. After choosing the best transforma-

Table E8.5. Mooney Viscosity $y(ML_4)$ at 100°C for Various Combinations of Coded Variables[a] x_1, x_2, and x_3

x_1	x_2	x_3	y
-1	-1	1	73
-1	1	-1	78
1	-1	-1	95
1	1	1	121
1.5	0	0	114
-1.5	0	0	66
0	1.5	0	96
0	-1.5	0	80
0	0	1.5	84
0	0	-1.5	79
0	0	0	89
0	0	0	86

[a]$x_i = (X_i - 10)/5$, $i = 1, 2, 3$, where X_1 = level of Silica A, Hi-Sil 233, PPG Industries, X_2 = level of Silica B, Hi-Sil EP, PPG Industries, and X_3 = level of N774, Cabot Corp.

Table E8.6. Feuell and Wagg (1949) Detergent Data

x_1	x_2	x_3	y
-1	-1	-1	106
0	-1	-1	198
1	-1	-1	270
-1	0	-1	197
0	0	-1	329
1	0	-1	361
-1	1	-1	223
0	1	-1	320
1	1	-1	321
-1	-1	0	149
0	-1	0	243
1	-1	0	315
-1	0	0	255
0	0	0	364
1	0	0	390
-1	1	0	294
0	1	0	410
1	1	0	415
-1	-1	1	182
0	-1	1	232
1	-1	1	340
-1	0	1	259
0	0	1	389
1	0	1	406
-1	1	1	297
0	1	1	416
1	1	1	387

Table E8.7. Data from Brownlee (1953)

W	B	H	S	y
−1	−1	−1	−1	4.2
−1	−1	1	−1	2.7
−1	1	−1	−1	1.4
−1	1	1	−1	1.3
1	−1	−1	−1	3.3
1	−1	1	−1	3.5
1	1	−1	−1	1.3
1	1	1	−1	1.9
−1	−1	−1	1	2.2
−1	−1	1	1	3.1
−1	1	−1	1	4.0
−1	1	1	1	4.1
1	−1	−1	1	5.0
1	−1	1	1	3.0
1	1	−1	1	2.2
1	1	1	1	2.6

tion parameter estimate $\hat{\lambda}$, carry out all the usual regression analyses including examination of the residuals. (See note attached to foregoing problem.)

8.6. (*Source:* Statistical methods in detergency investigations, by A. J. Feuell and R. E. Wagg, *Research*, **2**, 1949, 334–337; see p. 335. [*Note: Research* later became *Research Applied in Industry*. The data were also reported again in *Statistical Analysis in Chemistry and the Chemical Industry* by C. A. Bennett and N. L. Franklin, John Wiley & Sons, Inc., New York, 1954, p. 509.] The response in a detergent study was a measure of washing performance (percentage reflectance in terms of white standards) and the response data were multiplied by 10 to produce the 27 values of y shown in Table E8.6. The three predictor variables were detergent level (coded to x_1), sodium carbonate (x_2), and sodium carboxymethyl cellulose level (x_3). Consider the fit of a full quadratic equation to these data, and investigate whether a transformation on the response y would improve the fit to such a model and, perhaps, make the use of quadratic terms unnecessary.

8.7. (*Sources:* Quality quandaries: Not all models are polynomials, by S. Bisgaard, C. A. Vivacqua, and L. S. de Pinho, *Quality Engineering*, 2005, **17**, 181–186; data from *Industrial Experimentation*, 4th edition, by K. A. Brownlee, 1949, reprinted 1953, Chemical Publishing Co, p. 95.) Table E8.7 shows the original Brownlee data, obtained from a 2^4 experiment. Analyze these data and make tentative conclusions about the influence of the factors W, B, H, and S. Noting that the ratio (maximum y)/(minimum y) = 3.846, seek a suitable transformation of the data and reanalyze it. Redo your analysis and reassess your conclusions.

CHAPTER 9

Exploration of Maxima and Ridge Systems with Second-Order Response Surfaces

9.1. THE NEED FOR SECOND-ORDER SURFACES

In earlier chapters we have considered the use of polynomial graduating functions to represent locally the true expectation function $f(\xi, \theta)$. Thus it was supposed that an estimate $\hat{\beta}$ of a vector of empirical constants β could be found such that, over the region in the ξ space of immediate interest,

$$g(\xi, \hat{\beta}) \simeq f(\xi, \theta), \tag{9.1.1}$$

where $g(\xi, \beta)$ was a polynomial of first or second degree.

In Chapter 7, an example was discussed where the observed response Y of interest was the number of cycles to failure of the tested textile specimen and where the variables ξ_1, ξ_2, and ξ_3 were length, amplitude, and load, respectively. It turned out in this example that, if we wished to represent $E(Y)$ directly by a polynomial approximation, then an expression of at least second degree in the ξ's was needed. However, if we first applied a simple transformation to the response (specifically, by use of $y = \log Y$ instead of Y), an expression of only first degree in the ξ's produced an adequate approximation. In general, if, over the region of interest considered, the observed response is curved but smoothly monotonic in each of the elements of ξ, it will sometimes be possible (and where it is so, preferable) to employ a first-degree expression in transformed variables, rather than a second-degree expression in untransformed variables. Such a simplified representation may be possible either by transforming the output Y, as above, or by transforming some or all of the input ξ's, or by transforming both ξ's and Y simultaneously.

Response Surfaces, Mixtures, and Ridge Analyses, Second Edition. By G. E. P. Box and N. R. Draper
Copyright © 2007 John Wiley & Sons, Inc.

However, when the object is to approximate the response function in a region in which it has a turning point, we shall need a polynomial of at least second degree.* Transformed metrics may, nevertheless, make possible the use of a polynomial of lower degree than would otherwise be needed in untransformed metrics. Thus, for example, we can often avoid the need for a third-degree polynomial by employing a second-degree approximation in transformed variables.

As we have seen, the number of terms required by an approximating polynomial increases rapidly as the degree of the polynomial increases. Also, when N runs are used and the polynomial model has p parameters, the average variance of estimation is

$$\bar{V}(\hat{y}) = N^{-1} \sum_{u=1}^{N} V(\hat{y}_u) = p\sigma^2/N. \tag{9.1.2}$$

Thus, solely on the basis of consideration of the variance of the predicted values, the lower-order approximating model should always be preferred, or, stated more forcefully, movement to higher-order approximating polynomial models should be resisted until all efforts at keeping to the lower-order approximations have been expended.

9.2. EXAMPLE: REPRESENTATION OF POLYMER ELASTICITY AS A QUADRATIC FUNCTION OF THREE OPERATING VARIABLES

For illustration, we describe the elucidation of the nature of a maximal region for a polymer elasticity experiment. In this example, we encounter a number of issues which will be taken up in more detail in later chapters.

A Central Composite Design

The design employed at this stage of the investigation was a second-order central *composite design*. Such a design consists of a two-level factorial or fractional factorial (chosen so as to allow the estimation of all first-order and two-factor interaction terms) augmented with further points which allow pure quadratic effects to be estimated also. For this design, shown in Figure 9.1 for $k = 3$ inputs, the additional experimental points consist of center points plus "star" points arranged along the axes of the variables and symmetrically positioned with respect to the factorial cube. We discuss such designs in more detail in Chapter 15.

First Set of Runs

The objective of this investigation was to explore reaction conditions yielding close to maximal elasticity for a certain polymer. At this stage of the enquiry, the variables of importance had been narrowed down to three: the percentage concen-

---

*Functions other than polynomials can, of course, be used to produce curves containing maxima. One such function that is sometimes useful for a single input x, is

$$\ln Y = \alpha + \beta \ln x + \gamma x.$$

With $\beta > 0$ and $\gamma < 0$, this produces a curve containing an asymmetric maximum.

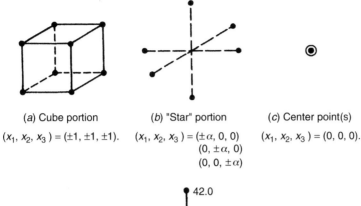

(a) Cube portion
$(x_1, x_2, x_3) = (\pm 1, \pm 1, \pm 1)$.

(b) "Star" portion
$(x_1, x_2, x_3) = (\pm \alpha, 0, 0)$
$(0, \pm \alpha, 0)$
$(0, 0, \pm \alpha)$

(c) Center point(s)
$(x_1, x_2, x_3) = (0, 0, 0)$.

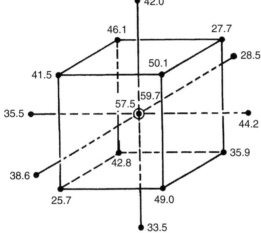

(d) Cube plus star plus center point(s) for
$k = 3$, with polymer elasticity data
attached.

Figure 9.1. A composite design in three dimensions ($k = 3$). (a) Cube portion, $(x_1, x_2, x_3) = (\pm 1, \pm 1, \pm 1)$. (b) "Star" portion, $(x_1, x_2, x_3) = (\pm \alpha, 0, 0)$, $(0, \pm \alpha, 0)$, $(0, 0, \pm \alpha)$. (c) Center point(s), $(x_1, x_2, x_3) = (0, 0, 0)$. (d) Cube plus star plus center point(s) for $k = 3$, with polymer elasticity data attached.

trations C_1 and C_2 of two constituents in the reaction mixture, and the reaction temperature T. Earlier groups of experiments using factorials and steepest ascent had produced conditions yielding a measured elasticity of about 55 units, and a rough estimate of the experimental error standard deviation of 2 units. A further 2^3 factorial straddling these "so far best conditions" was run using the pairs of levels shown in Table 9.1.

Thus, via the standard factorial coding

$$x_1 = \frac{(C_1 - 18)}{3}, \qquad x_2 = \frac{(C_2 - 2.7)}{0.4}, \qquad x_3 = \frac{(T - 145)}{10}. \qquad (9.2.1)$$

The experimental results are given in Table 9.2. If we were to fit a polynomial of first degree to these data we would obtain

$$\hat{y} = \underset{\pm 0.71}{39.85} + \underset{\pm 0.71}{0.83 x_1} - \underset{\pm 0.71}{1.73 x_2} + \underset{\pm 0.71}{1.49 x_3}. \qquad (9.2.2)$$

Table 9.1. Coded and Uncoded Levels of Three Predictor Variables

Predictor Variables	Levels			
	-1	$+1$	Midlevel	Semirange
Percentage concentration of constituent 1, C_1	15	21	18	3
Percentage concentration of constituent 2, C_2	2.3	3.1	2.7	0.4
Temperature (°C), T	135	155	145	10

Table 9.2. Results from a 2^3 Factorial Design

x_1	x_2	x_3	y
-1	-1	-1	25.74
1	-1	-1	48.98
-1	1	-1	42.78
1	1	-1	35.94
-1	-1	1	41.50
1	-1	1	50.10
-1	1	1	46.06
1	1	1	27.70

where the standard errors of the coefficients are evaluated on the assumption that $\sigma = 2$. If we attempt to fit a second-degree polynomial to the present data, the pure quadratic effects are not of course separately estimable, because the design contains only two levels of each of the predictor variables. In fact, the quadratic effects appear as aliases of the constant term, as indicated in Table 9.3. It is now seen that those estimates of second-order coefficients which are available, namely, b_{12}, b_{13}, b_{23}, are large not only compared with their own standard errors, but also compared with the estimates b_1, b_2, b_3 of terms of first order. From this analysis, it is evident therefore that we have moved into a region where the first-degree polynomial no longer provides a good approximation. Further experimentation to allow estimation of all second-order effects (pure quadratic terms as well as interaction terms) is needed to clarify the situation.

Table 9.3. The Estimates from the 2^3 Design and Their Expected Values Under the Assumption that the True Model Is the Complete Quadratic

Estimated Coefficients $\pm$ Standard Errors	Expected Value on the Assumption that the True Model Is a Second-Degree Polynomial
$b_0 = 39.85 \pm 0.71$	$\beta_0 + \beta_{11} + \beta_{22} + \beta_{33}$
$b_1 = 0.83 \pm 0.71$	β_1
$b_2 = -1.73 \pm 0.71$	β_2
$b_3 = 1.49 \pm 0.71$	β_3
$b_{12} = -7.13 \pm 0.71$	β_{12}
$b_{13} = -3.27 \pm 0.71$	β_{13}
$b_{23} = -2.73 \pm 0.71$	β_{23}

Table 9.4. Results from the Added Star and Center Point Runs

		Uncoded Levels			Coded Levels		Response
	$C_1\%$	$C_2\%$	$T°C$	x_1	x_2	x_3	y
Center points	18	2.7	145	0	0	0	57.52
	18	2.7	145	0	0	0	59.68
	12	2.7	145	−2	0	0	35.50
	24	2.7	145	2	0	0	44.18
Star points	18	1.9	145	0	−2	0	38.58
	18	3.5	145	0	2	0	28.46
	18	2.7	125	0	0	−2	33.50
	18	2.7	165	0	0	2	42.02

Second Set of Runs

The three-dimensional star of experimental points and two center points were therefore symmetrically added to the cube at this stage. The two sets of experiments together form the composite design of Figure 9.1d. The experimental results from this second group of experiments are shown in Table 9.4. The least-squares calculations for the combined set of 16 results are shown in Table 9.5. In this table, the rows of the 16 by 10 matrix $\mathbf{Z}_1$ contain the elements 1, x_1, x_2, x_3, x_1^2, x_2^2, x_3^2, $x_1 x_2$, $x_1 x_3$, $x_2 x_3$ corresponding (row by row) to the 16 sets of experimental conditions. The 10 by 1 vector of estimates $\mathbf{b}_1$ with elements b_0, b_1, b_2, b_3, b_{11}, b_{22}, b_{33}, b_{12}, b_{13}, b_{23} is given by

$$\mathbf{b}_1 = (\mathbf{Z}_1'\mathbf{Z}_1)^{-1}\mathbf{Z}_1'\mathbf{y}, \tag{9.2.3}$$

and the standard errors of the various estimates are the square roots of the corresponding diagonal elements of $(\mathbf{Z}_1'\mathbf{Z}_1)^{-1}\sigma^2$. Carrying through these calculations, we obtain the fitted second-degree polynomial as

$$\hat{y} = 57.3 + \underset{\pm 0.5}{1.5x_1} - \underset{\pm 0.5}{2.1x_2} + \underset{\pm 0.5}{1.8x_3} - \underset{\pm 0.5}{4.7x_1^2} - \underset{\pm 0.5}{6.3x_2^2}$$

$$- \underset{\pm 0.5}{5.2x_3^2} - \underset{\pm 0.7}{7.1x_1 x_2} - \underset{\pm 0.7}{3.3x_1 x_3} - \underset{\pm 0.7}{2.7x_2 x_3}, \tag{9.2.4}$$

where, once again, the standard errors of the coefficients (shown below the corresponding coefficients) are computed on the assumption that $\sigma = 2$. An analysis of variance table is shown in Table 9.6.

Removal of the Block Difference

In this particular investigation, more than a week elapsed between the performance of the first set of eight runs (the 2^3 factorial "cube") and of the second set of eight runs (the "star" plus two center points). There were, therefore, opportunities for systematic differences to occur between the first and second sets of eight runs, possibly associated with slight changes in the raw materials employed, in the standard solutions used in the analyses, or in the setting of the instruments.

Table 9.5. Least-Squares Calculations for Polymer Elasticity Example

$\mathbf{Z}_1$

1	x_1	x_2	x_3	x_1^2	x_2^2	x_3^2	x_1x_2	x_1x_3	x_2x_3
1	−1	−1	−1	1	1	1	1	1	1
1	1	−1	−1	1	1	1	−1	−1	1
1	−1	1	−1	1	1	1	−1	1	−1
1	1	1	−1	1	1	1	1	−1	−1
1	−1	−1	1	1	1	1	1	−1	−1
1	1	−1	1	1	1	1	−1	1	−1
1	−1	1	1	1	1	1	−1	−1	1
1	1	1	1	1	1	1	1	1	1
1	.	.	.	.	.	.	.	.	.
1	.	.	.	.	.	.	.	.	.
1	−2	.	.	4	.	.	.	.	.
1	2	.	.	4	.	.	.	.	.
1	.	−2	.	.	4	.	.	.	.
1	.	2	.	.	4	.	.	.	.
1	.	.	−2	.	.	4	.	.	.
1	.	.	2	.	.	4	.	.	.

$\mathbf{Z}_2^*$				$\mathbf{Z}_2^0$		y
$x_1^{(3)}$	$x_2^{(3)}$	$x_3^{(3)}$	$x_1x_2x_3$	x_B	x_E	
1	1	1	−1	−1	.	25.74
−1	1	1	1	−1	.	48.98
1	−1	1	1	−1	.	42.78
−1	−1	1	−1	−1	.	35.94
1	1	−1	1	−1	.	41.50
−1	1	−1	−1	−1	.	50.10
1	−1	−1	−1	−1	.	46.06
−1	−1	−1	1	−1	.	27.70
.	.	.	.	1	−1	57.52
.	.	.	.	1	1	59.68
−2	.	.	.	1	.	35.50
2	.	.	.	1	.	44.18
.	−2	.	.	1	.	38.58
.	2	.	.	1	.	28.46
.	.	−2	.	1	.	33.50
.	.	2	.	1	.	42.02

Following the terminology introduced in Chapter 5, we shall say that the experiment was run in two *blocks* of eight runs.

A systematic difference between two blocks may be estimated and eliminated as follows. If all 16 runs had been performed under the conditions of the first block, suppose the constant term in the second-degree polynomial would have been $\beta_0 - \delta$, while if all runs had been performed under the conditions of the second block, suppose the constant term would have been $\beta_0 + \delta$. Then the true mean

Table 9.5. (*Continued*)

$\mathbf{Z}_1'\mathbf{Z}_1$ and $\mathbf{Z}_1'\mathbf{y}$

	0	1	2	3	11	22	33	12	13	23	$\mathbf{Z}_1'\mathbf{y}$
0	16	·	·	·	16	16	16	·	·	·	658.24
1	·	16	·	·	·	·	·	·	·	·	24.00
2	·	·	16	·	·	·	·	·	·	·	−34.08
3	·	·	·	16	·	·	·	·	·	·	28.96
11	16	·	·	·	40	8	8	·	·	·	637.52
22	16	·	·	·	8	40	8	·	·	·	586.96
33	16	·	·	·	8	8	40	·	·	·	620.88
12	·	·	·	·	·	·	·	8	·	·	−57.04
13	·	·	·	·	·	·	·	·	8	·	−26.16
23	·	·	·	·	·	·	·	·	·	8	−21.84

$32(\mathbf{Z}_1'\mathbf{Z}_1)^{-1}$ and $\mathbf{b}_1$

	0	1	2	3	11	22	33	12	13	23	$\mathbf{b}_1$
0	14	·	·	·	−4	−4	−4	·	·	·	57.31
1	·	2	·	·	·	·	·	·	·	·	1.50
2	·	·	2	·	·	·	·	·	·	·	−2.13
3	·	·	·	2	·	·	·	·	·	·	1.81
11	−4	·	·	·	2	1	1	·	·	·	−4.69
22	−4	·	·	·	1	2	1	·	·	·	−6.27
33	−4	·	·	·	1	1	2	·	·	·	−5.21
12	·	·	·	·	·	·	·	4	·	·	−7.13
13	·	·	·	·	·	·	·	·	4	·	−3.27
23	·	·	·	·	·	·	·	·	·	4	−2.73

Table 9.6. Analysis of Variance for Polymer Elasticity Example

Source	SS	df	MS
Mean (zero-degree polynomial)	27,078.0	1	
Added first-order terms (given b_0)			
Added second-order terms (given zero- and first-order terms)	161.0 ⎫ 1,290.6 ⎭ 1.451.6	3 ⎫ 6 ⎭ 9	53.7 ⎫ 215.1 ⎭ 161.3
Residual	41.6	6	6.9
Total	28,571.2	16	

difference between blocks is 2δ and the model becomes

$$y = g(\mathbf{x}, \boldsymbol{\beta}) + x_B\delta + \varepsilon, \qquad (9.2.5)$$

where the indicator variable x_B (the *blocking variable* shown in Table 9.5 in the first column of $\mathbf{Z}_2^0$) is -1 in the first block and $+1$ in the second block. A rather remarkable circumstance will be noted for this particular design. The indicator variable x_B is orthogonal to each of the columns of $\mathbf{Z}_1$. It follows (see Section 3.9)

Table 9.7. Further Analysis of the Residual Sum of Squares in the Polymer Elasticity Example

Source	SS	df	MS	
Blocks	26.6	1	26.6	$F = 8.9$
Residual, after removal of blocks	15.0	5	3.0	
Residual, before removal of blocks	41.6	6		

that

$$\hat{\delta} = \frac{\left(\sum\limits_{j=1}^{16} x_{Bj} y_j \right)}{\sum\limits_{j=1}^{16} x_{Bj}^2} = 1.29. \tag{9.2.6}$$

Moreover, we can separate out from the residual sum of squares a "blocks" contribution with 1 df given by (see also pp. 492–495)

$$16\hat{\delta}^2 = 26.63. \tag{9.2.7}$$

Thus, the 6 df labeled "residual" in the analysis of variance of Table 9.6 can be further analyzed as indicated in Table 9.7. The contribution of blocks in this example is clearly quite substantial. The estimated standard deviation computed from the residual mean square is now $s = 1.73$, close to the value $\sigma = 2$ assumed previously.

Orthogonal Blocking and Sequential Design

Because the blocking is orthogonal in this example, the removal of the contribution associated with blocks does not change the estimated coefficients in the fitted function (9.2.4). However, that portion of the original residual sum of squares accounted for by the systematic block difference is removed, and the accuracy of the experiment is correspondingly increased.

The Importance of Blocking in the Sequential Assembly of Designs

Blocking, and where possible, orthogonal blocking is of great importance in sequential experimentation. When the 2^3 factorial corresponding to the first half of this design was planned, several possibilities were open for the succeeding step. If the first-order effects had been fairly large (a) compared with their standard errors *and* (b) compared with the estimated interaction terms, then an application of steepest ascent would n appropriate and this might have moved the center of interest to a new location possibly yielding higher elasticity where a further design would have been run. In the event, as we have seen, an inspection of the set of eight factorial runs made it seem likely that a second-degree exploration in the immediate neighborhood was necessary. The second part of the design corresponding to the star with center points was added with the knowledge that the second-degree polynomial equation could now be conveniently estimated, even though a change in level could have occurred between the two blocks of runs. This possibility of *sequential assembly* of different kinds of design building blocks as the need for them is demonstrated is a very valuable one and we discuss it in more detail in Section 15.3.

9.3. EXAMINATION OF THE FITTED SURFACE

The residual variance of 3.0 obtained after removing blocks is based on 5 df and agrees quite well with previous expectation. We shall show later how it is possible to analyze this residual variance further and to isolate measures of adequacy of fit. Here, these checks do not show any abnormalities, and we proceed on the basis that the fit is adequate. If the residual variance of 3.0 is used as an estimate of error, the overall F value associated with the complete second-degree regression equation is $161.3/3.0 = 53.8$. The tabled 5% value of F for 9 and 5 df is 4.8. The observed value of F exceeds by more than 10 times its 5% significance level, and application of the rough guideline introduced in Section 8.2 indicates that, in this example, we would probably have something worthwhile to interpret from the fitted surface. For this example, where there are just three predictor variables, we can easily visualize what the second degree fitted equation (9.2.4) is telling us, by mapping elasticity contours in the space of two of the predictor variables for a series of chosen values of the third. In Figures 9.2a to 9.2e, the fitted contours of elasticity are drawn in the space of C_1 and C_2 (x_1 and x_2) for five values of the temperature T (x_3). The fitted surface shows a maximum, with some factor dependence between the variables. This dependence is such that a higher concentration C_1 is, to some extent, compensated by a lower concentration C_2, and higher temperatures T are, to some extent, compensated by lower levels of C_1.

Location of the Maximum of the Fitted Surface

The maximum of the fitted surface may be located by differentiating the fitted second degree polynomial with respect to each predictor variable and equating derivatives to zero. For instance, for $k = 3$, differentiation with respect to x_1, x_2, and x_3 of

$$\hat{y} = b_0 + b_1 x_1 + b_2 x_2 + b_3 x_3 + b_{11} x_1^2 + b_{22} x_2^2 + b_{33} x_3^2$$
$$+ b_{12} x_1 x_2 + b_{13} x_1 x_3 + b_{23} x_2 x_3, \tag{9.3.1}$$

yields the equations

$$2b_{11} x_1 + b_{12} x_2 + b_{13} x_3 = -b_1,$$
$$b_{12} x_1 + 2b_{22} x_2 + b_{23} x_3 = -b_2, \tag{9.3.2}$$
$$b_{13} x_1 + b_{23} x_2 + 2b_{33} x_3 = -b_3,$$

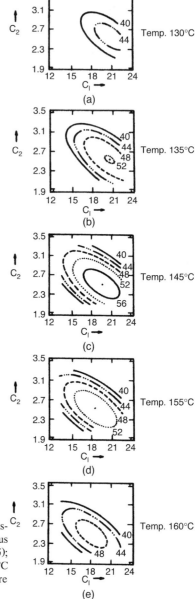

Figure 9.2. The fitted surface: Contours show polymer elasticity as a function of concentrations C_1 and C_2 for various values of temperature T. (a) Temperature 130°C ($x_3 = -1.5$); (b) temperature 135°C ($x_3 = -1.0$); (c) temperature 145°C ($x_3 = 0$); (d) temperature 155°C ($x_3 = 1.0$); (e) temperature 160°C ($x_3 = 1.5$).

from which the form for general k is obvious. The main diagonal coefficients on the left are twice the estimated pure quadratic effects, while the off-diagonal coefficients are the interaction effects. The right-hand sides are the estimated linear effects with signs reversed. For the present example, after changing signs, we obtain

$$9.38x_1 + 7.13x_2 + 3.27x_3 = 1.50,$$
$$7.13x_1 + 12.54x_2 + 2.73x_3 = -2.13, \qquad (9.3.3)$$
$$3.27x_1 + 2.73x_2 + 10.42x_3 = 1.81.$$

Solving these equations yields

$$x_1^* = 0.460, \qquad x_2^* = -0.465, \qquad x_3^* = 0.151 \qquad (9.3.4)$$

for the coordinates of the maximum of the fitted surface, or in terms of the original predictor variables,

$$C_1^* = 19.4\%, \qquad C_2^* = 2.5\%, \qquad T^* = 146.5°C. \qquad (9.3.5)$$

Also, it is easily shown that $\hat{y}^*$, the value given by the fitted equation at this maximum point, is

$$\hat{y}^* = b_0 + \tfrac{1}{2}(b_1 x_1^* + b_2 x_2^* + b_3 x_3^*). \qquad (9.3.6)$$

Thus, for the present example,

$$\hat{y}^* = 58.29.$$

On the assumption of the adequacy of the second-degree equation, a confidence region for the maximum point may be computed (for details, see Box and Hunter, 1954).

9.4. INVESTIGATION OF ADEQUACY OF FIT: ISOLATION OF RESIDUAL DEGREES OF FREEDOM FOR THE COMPOSITE DESIGN

Before accepting that the fitted second-degree equation provides a satisfactory approximation to the true surface, one must of course consider possible evidence for lack of fit. In this example, there are only five residual degrees of freedom, but nevertheless they can provide some information about lack of fit.

In general, suppose N observations are fitted to a linear model containing p parameters. Then the fitting process itself will introduce p linear relationships among the N residuals $y - \hat{y}$. If N is large with respect to p, then the effect of this induced dependence among the residuals will be slight, and the plotting techniques employed to examine residuals will be useful in revealing any inadequacies in the model. As p becomes larger and, in particular, as it approaches N, patterns caused by the induced dependencies become dominant and can mask those due to model inadequacies. However, as we saw for the factorial design in Section 7.4, it may be possible to obtain information on the adequacy of fit by isolating and identifying individual residual degrees of freedom associated with feared model inadequacies. In particular, when considering the fitting of a polynomial of a given degree, it is natural to be concerned with the possibility that, in fact, a polynomial of higher degree may be needed. This immediately focuses attention on the characteristics of the estimates when the feared alternative model applies, but the simpler assumed model has been fitted. In these circumstances, contemplation of the fitted model embedded in a more complex one makes it possible to answer two kinds of questions.

1. To what extent are the original estimates of the coefficients biased if the more complex model is true?

2. What are appropriate checking functions to warn of the possible need for a more complex model?

Both questions are critically affected by the particular choice of design.

For illustration, we shall consider the possibility that a third-degree polynomial might be needed in the elasticity example.

Bias Characteristics of the Design

We may write the extended (third-order) polynomial model in the form

$$y = Z_1\beta_1 + Z_2\beta_2 + \varepsilon,$$

or, arguing as in Section 3.9, we can write it in the orthogonalized form

$$y = Z_1(\beta_1 + A\beta_2) + (Z_2 - Z_1A)\beta_2 + \varepsilon, \tag{9.4.1}$$

where $Z_1\beta_1$ includes all terms up to and including second order, and $Z_2\beta_2$ includes all terms of third order.

The alias (or bias) matrix A (Table 9.8) shows that only the estimates of first-order terms are biased by third-order terms. In fact, if a third-order model is appropriate and if b_1, b_2, and b_3 are our previous least-squares estimates (assuming a second-order model has been fitted), then

$$E(b_1) = \beta_1 + 2.5\beta_{111} + 0.5\beta_{122} + 0.5\beta_{133},$$

$$E(b_2) = \beta_2 + 2.5\beta_{222} + 0.5\beta_{112} + 0.5\beta_{233}, \tag{9.4.2}$$

$$E(b_3) = \beta_3 + 2.5\beta_{333} + 0.5\beta_{113} + 0.5\beta_{223}.$$

Checking Functions for the Design

Examination of the matrix $Z_{2\cdot1} = Z_2 - Z_1A$ in Table 9.8 reveals a rather remarkable circumstance. Although the matrix has 10 columns, only 4 of these are independent and these 4 are all simple multiples of the 4 columns of Z_2^* in Table 9.5. Consider, for example, the first column (labeled 111) of the matrix $Z_{2\cdot1}$. We can write this as

$$1.5x_1^{(3)}, \tag{9.4.3}$$

where

$$x_1^{(3)} = (1, -1, 1, -1, 1, -1, 1, -1, 0, 0, -2, 2, 0, 0, 0, 0)' \tag{9.4.4}$$

is the first column of Z_2^*. This vector is, at first sight, rather like the x_1 vector which is the second column in the matrix Z_1. Notice, however, that the signs of the first eight elements of $x_1^{(3)}$ are the reverse of those of x_1, while the signs associated with the elements $(-2, 2)$ are the same as those of x_1. In spite of their similar form, x_1 and $x_1^{(3)}$ are, of course, orthogonal. Further inspection shows that the elements of the columns labeled 122 and 133 in $Z_{2\cdot1}$ are those of $-0.5x_1^{(3)}$ in each case.

Table 9.8. Calculations Leading to the Alias Matrix for the Polymer Elasticity Example

$$\mathbf{Z}_2$$

111	222	333	122	112	133	113	233	223	123
−1	−1	−1	−1	−1	−1	−1	−1	−1	−1
1	−1	−1	1	−1	1	−1	−1	−1	1
−1	1	−1	−1	1	−1	−1	1	−1	1
1	1	−1	1	1	1	−1	1	−1	−1
−1	−1	1	−1	−1	−1	1	−1	1	1
1	−1	1	1	−1	1	1	−1	1	−1
−1	1	1	−1	1	−1	1	1	1	−1
1	1	1	1	1	1	1	1	1	1
.	.	.	.	.	.	.	.	.	.
.	.	.	.	.	.	.	.	.	.
−8	.	.	.	.	.	.	.	.	.
8	.	.	.	.	.	.	.	.	.
.	−8	.	.	.	.	.	.	.	.
.	8	.	.	.	.	.	.	.	.
.	.	−8	.	.	.	.	.	.	.
.	.	8	.	.	.	.	.	.	.

$$\mathbf{Z}_1'\mathbf{Z}_2$$

	111	222	333	122	112	133	113	233	223	123
0	.	.	.	.	.	.	.	.	.	.
1	40	.	.	8	.	8	.	.	.	.
2	.	40	.	.	8	.	.	8	.	.
3	.	.	40	.	.	.	8	.	8	.
11	.	.	.	.	.	.	.	.	.	.
22	.	.	.	.	.	.	.	.	.	.
33	.	.	.	.	.	.	.	.	.	.
12	.	.	.	.	.	.	.	.	.	.
13	.	.	.	.	.	.	.	.	.	.
23	.	.	.	.	.	.	.	.	.	.

We now define a normalized linear combination l_{111} of the observations which uses the elements of $\mathbf{x}_1^{(3)}$ as coefficients. This is

$$l_{111} = \frac{\mathbf{x}_1^{(3)'}\mathbf{y}}{\mathbf{x}_1^{(3)'}\mathbf{x}_1^{(3)}} = \frac{\mathbf{x}_1^{(3)'}\mathbf{y}}{16}. \qquad (9.4.5)$$

Then

$$E(l_{111}) = 1.5\beta_{111} - 0.5\beta_{122} - 0.5\beta_{133}. \qquad (9.4.6)$$

Similarly, we can define estimates l_{222}, l_{333}, and obtain

$$E(l_{222}) = 1.5\beta_{222} - 0.5\beta_{112} - 0.5\beta_{233},$$
$$E(l_{333}) = 1.5\beta_{333} - 0.5\beta_{113} - 0.5\beta_{223}. \qquad (9.4.7)$$

Table 9.8. (*Continued*)

$$\mathbf{Z}_{2\cdot1} = \mathbf{Z}_2 - \mathbf{Z}_1\mathbf{A}$$

111	222	333	122	112	133	113	233	223	123
1.5	1.5	1.5	−0.5	−0.5	−0.5	−0.5	−0.5	−0.5	−1
−1.5	1.5	1.5	0.5	−0.5	0.5	−0.5	−0.5	−0.5	1
1.5	−1.5	1.5	−0.5	0.5	−0.5	−0.5	0.5	−0.5	1
−1.5	−1.5	1.5	0.5	0.5	0.5	−0.5	0.5	−0.5	−1
1.5	1.5	−1.5	−0.5	−0.5	−0.5	0.5	−0.5	0.5	1
−1.5	1.5	−1.5	0.5	−0.5	0.5	0.5	−0.5	0.5	−1
1.5	−1.5	−1.5	−0.5	0.5	−0.5	0.5	0.5	0.5	−1
−1.5	−1.5	−1.5	0.5	0.5	0.5	0.5	0.5	0.5	1
.	.	.	.	.	.	.	.	.	.
−3	.	.	1	.	1	.	.	.	.
3	.	.	−1	.	−1	.	.	.	.
.	−3	.	.	1	.	.	1	.	.
.	3	.	.	−1	.	.	−1	.	.
.	.	−3	.	.	.	1	.	1	.
.	.	3	.	.	.	−1	.	−1	.

$$\mathbf{A} = (\mathbf{Z}_1'\mathbf{Z}_1)^{-1}\mathbf{Z}_1'\mathbf{Z}_2$$

	111	222	333	122	112	133	113	233	223	123
0	.	.	.	.	.	.	.	.	.	.
1	2.5	.	.	0.5	.	0.5	.	.	.	.
2	.	2.5	.	.	0.5	.	.	0.5	.	.
3	.	.	2.5	.	.	.	0.5	.	0.5	.
11	.	.	.	.	.	.	.	.	.	.
22	.	.	.	.	.	.	.	.	.	.
33	.	.	.	.	.	.	.	.	.	.
12	.	.	.	.	.	.	.	.	.	.
13	.	.	.	.	.	.	.	.	.	.
23	.	.	.	.	.	.	.	.	.	.

A parallel definition of

$$l_{123} = \Sigma x_1 x_2 x_3 y / 8 \tag{9.4.8}$$

will give us

$$E(l_{123}) = \beta_{123}. \tag{9.4.9}$$

Thus, although we cannot obtain estimate of each of the third-order effects individually, with this design we can isolate certain linear combinations of them (that is, certain alias groups). The size of these combinations can indicate particular directions in which there may be lack of fit. In particular, suppose that l_{jjj} were excessively large. This could indicate, for example, that a transformation of x_j might be needed to obtain an adequate representation using a second-degree equation. This transformation aspect is discussed in more detail in Section 13.8.

Table 9.9. Analysis of the Residual Sum of Squares into Individual Degrees of Freedom for the Elasticity Example

Source	SS	df	MS
Blocks	26.6	1	26.6
111	7.2	1	7.2
222	2.6	1	2.6
333	1.6	1	1.6
123	1.2	1	1.2
Pure error	2.3	1	2.3
Residual sum of squares	41.6[a]	6	

[a] The slight discrepancy in addition is due to round-off error.

Complete Breakup of the Residual Sum of Squares in the Elasticity Example

There are 16 runs in the composite design of the elasticity example. Ten degrees of freedom are used in the estimation of the second-degree polynomial. Of the remaining 6 df, one is used for blocking and one is a pure error comparison in which the two center points are compared. These are associated with the two vectors in $\mathbf{Z}_2^0$ of Table 9.5. Four degrees of freedom remain, and these can be associated with possible lack of fit from neglected third-order terms (or alternatively with the need for transformation of the variables). We notice that the columns of $\mathbf{Z}_2^*$ and $\mathbf{Z}_2^0$ are all orthogonal to one another and to the columns of $\mathbf{Z}_1$ so that, for the elasticity example, we can work out each contribution separately and easily. For example,

$$l_{111} = \tfrac{1}{16}\{+25.74 - 48.98 + \cdots + (2 \times 44.18)\} = 0.67, \qquad (9.4.10)$$

and the corresponding contribution to the residual sum of squares is

$$16l_{111}^2 = 7.18. \qquad (9.4.11)$$

For all four, we obtain

$$l_{111} = 0.67, \qquad l_{222} = -0.40, \qquad l_{333} = -0.32, \qquad l_{123} = 0.39, \quad (9.4.12)$$

and a detailed analysis of the residual sum of squares is shown in Table 9.9. (Note that 8, not 16, is the multiplier for the SS calculation for 123.)

None of the mean squares is excessively large compared with the others, nor do they contradict the earlier supposition that $\sigma \doteq 2$, that is, $\sigma^2 \doteq 4$. Thus we have no reason to suspect lack of fit.

Conclusions

What does this investigation allow us to conclude? We can say that:

1. In the immediate neighborhood of its maximum value, elasticity can be represented approximately by a second-degree polynomial in the three predictor variables—concentration 1, concentration 2, and temperature—given by (9.2.4).

2. Examination of individual residual degrees of freedom gives no indication that the fit is inadequate.

3. The overall F value associated with the complete second-degree equation is more than 10 times its 5% significance level, implying that the fitted surface is worthy of interpretation.

4. The maximum point is estimated to be at $C_1^* = 19.4\%$, $C_2^* = 2.5\%$, and $T^* = 146.5°C$, and the estimated elasticity there is $\hat{y}^* = 58.3$.

5. Contour plots indicate some factor dependence between the variables.

EXERCISES

9.1. (*Source:* Optimization of the synthesis of an analogue of Jojoba oil using a fully central composite design, by A. Coteron, N. Sanchez, M. Martinez, and J. Aracil, *Canadian Journal of Chemical Engineering*, **71**, June 1993, 485–488.) An ester with characteristics similar to those of sperm whale oil and jojoba oil is used as a substitute for the two natural oils. Experiments on it involved the three predictor variables: reaction temperature T (°C), initial concentration of catalyst, C (g), and working pressure P (mPa). These were coded (approximately; see Table E9.1) via $x_1 = (T - 140)/24$, $x_2 = (C - 0.64)/0.20$, and $x_3 = (P - 48.4)/46.2$. The response variable was $y =$ yield of ester, oleyl oleate. Table E9.1 shows the randomized run order, the uncoded and coded design levels, and the observed responses. The design was actually performed sequentially in two stages (blocks). In the first stage, the 2^3 design plus 5 center points (run orders 1–13) were used.

Table E9.1. Data from a Study on an Analogue of Jojoba Oil

Block	Run No.	Run Order	T (°C)	C (g)	P (mPa)	x_1	x_2	x_3	y
1	1	6	116	0.44	20	−1	−1	−1	17.0
1	2	4	164	0.44	20	1	−1	−1	44.0
1	3	3	116	0.83	20	−1	1	−1	19.0
1	4	11	164	0.83	20	1	1	−1	46.0
1	5	2	116	0.44	77	−1	−1	1	7.0
1	6	10	164	0.44	77	1	−1	1	55.0
1	7	5	116	0.83	77	−1	1	1	15.0
1	8	13	164	0.83	77	1	1	1	41.0
1	9	12	140	0.64	48.4	0	0	0	29.0
1	10	7	140	0.64	48.4	0	0	0	28.5
1	11	9	140	0.64	48.4	0	0	0	30.0
1	12	1	140	0.64	48.4	0	0	0	27.0
1	13	8	140	0.64	48.4	0	0	0	28.0
2	14	16	100	0.64	48.4	−1.682	0	0	8.0
2	15	18	180	0.64	48.4	1.682	0	0	74.5
2	16	14	140	0.29	48.4	0	−1.682	0	30.0
2	17	19	140	0.98	48.4	0	1.682	0	37.5
2	18	15	140	0.64	2.1	0	0	−1.682	35.0
2	19	17	140	0.64	94.6	0	0	1.682	30.5

(**a**) Fit a plane to these 13 first-stage data points and obtain the analysis of variance table. Evaluate the two-factor interactions (or the corresponding regression coefficients) and the overall measure of curvature, $c = \bar{y}_f - \bar{y}_0$, also called the curvature contrast, where $\bar{y}_f$ is the average response at the n_f factorial points and $\bar{y}_0$ is the average response at the n_0 center points. Compare the results with the corresponding standard errors. Note that the standard deviation of c is $\sigma^2 \{ n_f^{-1} + n_c^{-1} \}$. What do you conclude?

(**b**) Find the sums of squares contributions for the two-factor interactions (or equivalently for the corresponding regression coefficients) and also for the curvature contrast c. Enter the SS into the analysis of variance table in an appropriate way. Is the first-order model adequate?

(**c**) Add runs 14–19 (the star, or axial, points) and fit a full second-order model plus a blocking variable. Choose the blocking variable levels so that blocks are orthogonal to the entire model except for the first column of 1's. Place the blocking variable in the model immediately after the column of 1's so that its contribution is removed right after the intercept. Is the addition of the second-order terms worthwhile?

(**d**) Examine the residuals from the second-order fit and comment on what you see.

(**e**) How would your estimates be biased if the feared ("true") model were a cubic?

(**f**) Estimate all the cubic coefficients or, more accurately, those combinations of cubic coefficients that it is possible to estimate. (Specify what these are; you must account for all the cubic coefficients somewhere in what you say.) Re-divide the residual sum of squares you found in (c) accordingly.

(**g**) How many more terms could be added to the model last fitted?

(**h**) Reviewing everything you have seen, what would you suggest be done next with a view to improving the response levels?

CHAPTER 10

Occurrence and Elucidation of Ridge Systems, I

10.1. INTRODUCTION

In the polynomial elasticity example discussed in Chapter 9, the nature of the fitted surface was explored by plotting contours of estimated elasticity $\hat{y}$ in the space of the three predictor variables, or factors, C_1, C_2, and T. The plots revealed some factor dependence. For example, study of Figure 9.2 shows that the concentration C_1 giving the maximum elasticity is different, depending on the levels of the second concentration C_2 and the temperature T. Examples frequently occur which show factor dependence of a much stronger kind. Also there may be more than two or three factors involved. In this chapter we consider the nature of likely factor dependencies and show, in an elementary way, how they may be elucidated.

Response Functions Near Maxima, Minima, and Stationary Values

Consider, to begin with, an example where there is a single predictor variable "residence time" and the response of interest is the "percentage yield." Thus, in our usual notation we have η = mean yield (%), ξ_1 = residence time (min), and $\eta = f(\xi_1, \boldsymbol{\theta})$. As usual, $\eta = f(\xi_1, \boldsymbol{\theta})$ denotes the true, but presumed unknown, relationship between mean yield η and residence time ξ_1. We shall assume that this relationship contains a maximum and that, in the neighborhood of this maximum,* the relationship can be represented by a graph like that in Figure 10.1.

However, suppose now that two variables are studied, say ξ_1 = residence time and ξ_2 = reaction temperature. It is tempting to assume that the response function can be represented by a surface like a more or less symmetric mound, the contour representation of which might be like that shown in Figure 10.2.

---

*We talk in terms of a maximum, but when the response η passes through a maximum, the response $c - \eta$, where c is any constant, passes through a corresponding minimum. Thus discussion of a maximum includes that of a minimum also, throughout what follows.

Response Surfaces, Mixtures, and Ridge Analyses, Second Edition. By G. E. P. Box and N. R. Draper
Copyright © 2007 John Wiley & Sons, Inc.

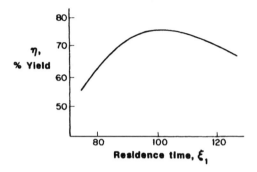

Figure 10.1. A relationship between η and ξ_1 in the region of a maximum.

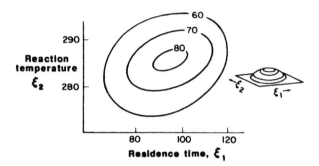

Figure 10.2. Yield contours for a roughly symmetric maximum.

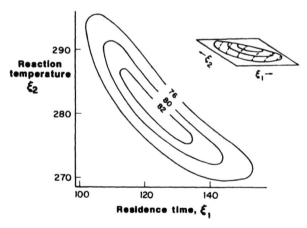

Figure 10.3. Yield contours for an attenuated maximum.

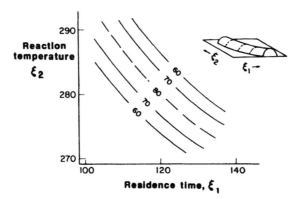

Figure 10.4. Yield contours for a stationary ridge.

As soon as experiments began to be carried out in which actual response surfaces could be roughly plotted, it was found that such a generalization was frequently inadequate. It became clear not only that surfaces were often attenuated in the neighborhood of maxima to form an oblique ridge like that in Figure 10.3, but also that oblique stationary and rising ridge systems like those of Figures 10.4 and 10.5 were of quite common occurrence as well.

Another possibility, which, however, seems to occur much less frequently in practice, is illustrated in Figure 10.6. This surface contains a ridgy *saddle* system, also often called a *col* or *Minimax*.

It will be noted, in Figures 10.2 to 10.5, that any section of the response surfaces taken parallel to either the ξ_1 or the ξ_2 axis will yield a curve like that in Figure 10.1, so that all of these surfaces are entirely built up from such curves.

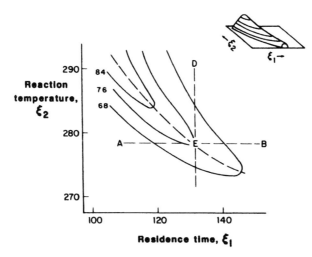

Figure 10.5. Yield contours for a rising ridge.

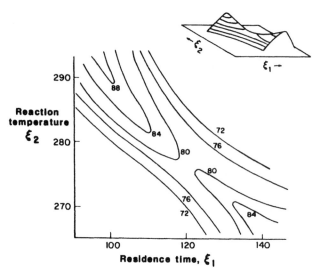

Figure 10.6. Yield contours for a "minimax," "saddle," or "col."

10.2. FACTOR DEPENDENCE

The reason for the occurrence of ridge systems of the kind displayed in Figures 10.3 to 10.5 can be seen when it is remembered that factors like temperature, time, pressure, concentration, and so on, are regarded as "natural" variables only because they happen to be quantities that can be conveniently manipulated and measured. Individual fundamental variables not directly controlled or measured, but in terms of which the behavior of the system could be described more economically (e.g., frequency of a particular type of molecular collision), will often be a function of two or more natural variables.

For this reason, many different combinations of levels of the natural variables may correspond to the best level of a fundamental variable. A similar rationale may be found for many biological and economic systems which display maxima. As an example of the former, consider an investigation of the muscular contraction of a frog's leg resulting from the passage of an electric current. Suppose that the response η = muscle contraction was a function $\eta = f(v)$ only of the voltage v applied and that initially, when v was increased, the contraction increased but, beyond a certain voltage, the muscle was paralyzed so that further increases in v resulted in reduced contraction. The η plotted versus v could give a curve like that of Figure 10.7a, with a maximum muscle contraction of about 74 occurring when a voltage of $v = 100$ was applied. Suppose, however, that this relationship was unknown to the investigator, whose experimental apparatus was set up in such a way that the variables he could actually manipulate were the current ξ_1 and the resistance ξ_2 in the circuit. Suppose finally that the investigator was unaware of the existence of Ohm's law, which would have told him that the voltage applied, $v = \xi_1 \xi_2$, was the product of current and resistance in the circuit. By carrying out experiments in which current ξ_1 and resistance ξ_2 were varied separately, he

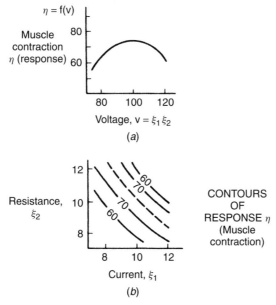

Figure 10.7. Generation of a ridge system. (*a*) Response function in η versus $v = \xi_1 \xi_2$. (*b*) Response function in terms of ξ_1 and ξ_2.

would be exploring a system for which the response surface was a stationary ridge like that shown* in Figure 10.7*b*.

Other functions $v = f(\xi_1, \xi_2)$ can produce ridges with sections that contain a maximum. It will be found, for example, that if b and c are positive constants, the functions $v = a + b\xi_1 + c\xi_2$, $v = a\xi_1^b \xi_2^c$, and $v = a\xi_1^b \exp\{-c/\xi_2\}$ will all produce diagonal ridge systems running from the upper left to the lower right of the diagram, as does the one shown in Figure 10.7*b*, while the functions $v = a + b\xi_1 - c\xi_2$, $v = a\xi_1^b/\xi_2^c$, and $v = a\xi_1^b \exp\{-c\xi_2\}$ all produce ridges running from lower left to upper right. These types of ridge systems are, of course, associated with "interaction" between the predictor variables ξ_1 and ξ_2, the former with a negative two-factor interaction between ξ_1 and ξ_2, the latter with a positive two-factor interaction.

The simplest type of ridge system is produced when v is a linear function of ξ_1 and ξ_2, that is, $v = a + b\xi_1 + c\xi_2$ with any choice of signs. The ridge systems so generated will have parallel straightline contours running in a direction determined by the relative magnitudes of b and c. A section at right angles to these contours will reproduce the original graph of η on v.

We say, in the example quoted above, that there is a "redundancy" of one variable, or that the maximum "possesses 1 df." This is because the apparently two-variables system can in fact be expressed in terms of a single fundamental

---

*Figure 10.7*b* was constructed by drawing contour lines through those points given a constant product $\xi_1 \xi_2$, the appropriate yield being read off from Figure 10.7*a*.

"compound" variable v. The physical and biological sciences abound with examples where, over suitable ranges of the variables, relationships exist similar to those given above.

Frequently, the surface that describes the system, while not of the form shown in Figure 10.4, nevertheless contains a marked component of this type together with an additional component, resulting in an elongated maximum like that shown in Figure 10.3, or a system like that shown in Figure 10.5, where the ridge is steadily rising to higher yields. In the latter "rising ridge" case, the best practical combination of levels to use may be the most extreme point on the ridge that can be attained by the experimenter. Thus, in Figure 10.5, the best conditions could be those using the highest possible temperature (with the appropriate residence time for that temperature).

The examples of Figures 10.3 to 10.6 may be said to show *factor dependence** in the sense that the response function for one factor is not independent of the levels of the other factors. The investigation of factor dependence is of considerable practical importance in response surface studies, for the following reasons.

1. *Alternative Optima.* Where the surface is like that in Figure 10.4, not one, but a whole range of alternative optimum processes, corresponding to points along the crest of the ridge, will be available from which to choose. In practice, some of these processes may be far less costly, or far more convenient to operate, than others. The factors in this system may be said to be compensating in the sense that departure from the maximum response due to change in one variable can be compensated by a suitable change in another variable. The direction of the ridge indicates how much one factor must be changed to compensate for a given change in the other.

2. *Optimizing, or Improving a Second Response.* If we imagine the contours for some auxiliary response η_2 superimposed on those for the major response η_1 we see that (provided the contours of the two systems are not exactly parallel) we could select, for our optimum process, that point near the crest of the ridge for the major response that yielded the most satisfactory value of the auxiliary response.

3. *Directions of Insensitivity.* When the surface is like that in Figure 10.3, the direction of attenuation of the surface indicates those directions in which departures can be made from the optimum process with only small losses in response.

*The idea of factor dependence is analogous to that of stochastic dependence, and diagrams like Figure 10.3 call to mind the familiar contour representation of a bivariate probability surface. Again, there exists the same analogy between the "fundamental variables" we have discussed and the "factors" in factor analysis and principal component analysis.

These analogies are helpful, but it must be emphasized that the two types of dependence are quite distinct and care should be taken to differentiate between them. We are concerned *here* with the deterministic relationship between the mean response η and the levels ξ of a set of predictor variables that can be varied at our choice such as temperature, time, and so on. We do not need any ideas of probability to define *this* relationship. In the analogous stochastic situation, "probability" takes the place of the response, and random variables such as "test scores" take the place of the predictor variables such as temperature and time.

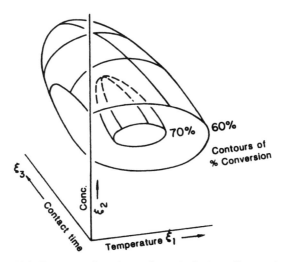

Figure 10.8. Representation of a maximum in three predictor variables.

4. *Yield Improvement by Simultaneous Changes in Several Variables.* The detection of a rising ridge in the surface like that in Figure 10.5 supplies the knowledge that, if the variables are changed together in the direction of the axis of the ridge, then yield improvement is possible even though no improvement may be possible by changing any single variable.

5. *Suggestion of Mechanism or "Natural Law."* Of by no means least importance is the fact that discovery of factor dependence of a particular type may, in conjunction with the experimenter's theoretical knowledge, lead to a better understanding of the basic mechanism of the reaction. Thus, if experimentation with the variables ξ_1 and ξ_2 produced a surface like that of Figure 10.7*b*, we might suspect that some more fundamental variable of the type $v = \xi_1 \xi_2$ existed. (In our frog's leg example, a biologist estimating muscle response might accidently discover Ohm's law!)

More than Two Predictor Variables

So far we have illustrated our discussion with examples in which there were only two predictor variables ξ_1 and ξ_2. The situations that can occur as we add more predictor variables become progressively more complicated. Figure 10.8 shows a contour representation for three predictor variables in the neighborhood of a point maximum (analogous to that in two dimensions shown in Figure 10.2). The contours enclose the maximum point like the skins of an onion enclose its center. Insensitivity of the response to changes in conditions when predictor variables were changed *together* in a certain way would correspond to attenuation of the contours in the direction of the compensatory changes. As an extreme form of such an attenuation, we could imagine a line maximum in the space (i.e., a line such that all sets of conditions on it gave the same maximum yield) surrounded by cylindrical contours of falling yield. We would call this a *line stationary ridge* because it is analogous to the two-dimensional case illustrated in Figure 10.4. Many other possibilities occur, but we shall leave their fuller discussion until later.

It should not be thought that factor dependence is to be expected only in connection with well-defined physical and chemical phenomena, nor that "single variable" redundancy is all that need concern us. Suppose, for example, the problem concerned the making of a cake, the object being to bring some desirable property such as "texture" (which is our response η and which, we suppose, can be measured on some suitable numerical scale) to the "best" level. The natural variables, k in number, whose levels we denote by $\xi_1, \xi_2, \ldots, \xi_k$ might be numerous and might involve, among other things, the amounts of such substances as baking powder (ξ_1), flour (ξ_2), egg white (ξ_3), and citric acid (ξ_4).

Suppose it had happened that the texture η depended in reality on only two fundamental variables, the consistency of the mix ω_1 and its acidity ω_2, and that optimum texture was attained whenever ω_1 and ω_2 were at their optimum levels ω_{10}, ω_{20}, say. For simplicity, we shall make the assumption that ω_1 and ω_2, measured in some suitable way, could be locally approximated by linear functions of the amounts ($\xi_1, \xi_2, \ldots, \xi_k$) of the various ingredients, that is,

$$\text{consistency} \quad \omega_1 = a_1 + b_1\xi_1 + c_1\xi_2 + d_1\xi_3 + \cdots + p_1\xi_k, \quad (10.2.1)$$

$$\text{acidity} \quad \omega_2 = a_2 + b_2\xi_1 + c_2\xi_2 + d_2\xi_3 + \cdots + p_2\xi_k. \quad (10.2.2)$$

Each of the coefficients $b_1, c_1, \ldots, p_1$ in the first equation would measure the change in consistency due to adding a unit amount of the corresponding ingredient. Thus the coefficient might be positive for solid substances like flour, but negative for liquid substances like water. The optimum consistency would thus be obtained on some $(k-1)$-dimensional planar surface in the k-dimensional space of the factors defined by $\omega = \omega_{10}$, that is, such that

$$a_1 + b_1\xi_1 + c_1\xi_2 + d_1\xi_3 + \cdots + p_1\xi_k = \omega_{10}. \quad (10.2.3)$$

Similarly, each of the coefficients $b_2, c_2, \ldots, p_2$ in the second equation would measure the change in acidity due to the addition of a unit of the corresponding ingredient and would be positive for acid substances and negative for alkaline substances. The optimum acidity would thus be obtained on some second $(k-1)$-dimensional planar surface given by $\omega_2 = \omega_{20}$, that is, such that

$$a_2 + b_2\xi_1 + c_2\xi_2 + d_2\xi_3 + \cdots + p_2\xi_k = \omega_{20}. \quad (10.2.4)$$

The intersections of the two planes would give all the ξ levels for which both acidity and consistency were at their best levels, and hence for which optimum texture was attained. Optimum texture would thus be attained on a subspace of $k-2$ dimensions. That is to say, there would be $k-2$ directions at right angles to each other in which we could move while still maintaining optimum texture for the cake. We can express the k variable system in terms of two fundamental "compound" variables ω_1 and ω_2; there is thus a redundancy of $k-2$ variables corresponding to the $k-2$ dimensions in which the maximum is attained. Thus we could say that the maximum "had $k-2$ df."

The surface for this example would again be a generalization of the stationary ridge of Figure 10.4. Rising ridges, attenuated maxima, and combinations of these phenomena all have multidimensional generalizations to which we return later.

10.3. ELUCIDATION OF STATIONARY REGIONS, MAXIMA, AND MINIMA BY CANONICAL ANALYSIS

Let us suppose that a representationally adequate second-degree equation has been fitted* in a near stationary region in which there appears to be some kind of maximum, so that, near the center of the last experimental design, the gradients, as measured by the b_j, $j = 1, 2, \ldots, k$, are not large compared with second-order effects. It is now important to determine the nature of the local surface which we tentatively believe contains a maximum. Some necessary questions are: (a) Is it, in fact, a maximum? (b) If yes, is this maximum most helpfully approximated by a point, a line, or a space (or, approximately speaking, are there 0, 1, 2, or more, df in the maximum)? (c) How can we define the point, line, or space, on which a maximal response is obtained? *Canonical analysis* of the fitted second-degree equation can illuminate these questions.

Canonical Analysis

Canonical analysis is a method of rewriting a fitted second-degree equation in a form in which it can be more readily understood. This is achieved by a rotation of axes which removes all cross-product terms. We call this simplification the *A canonical form*. If desired, this may be accompanied by a change of origin to remove first-order terms as well. We call the result the *B canonical form*.

We first give a brief technical description of the analysis. This is then explained and illustrated.

The *A* Canonical Form

Consider a fitted second-degree model

$$\hat{y} = b_0 + \sum_{j=1}^{k} b_j x_j + \sum\sum_{i \geq j} b_{ij} x_i x_j. \qquad (10.3.1)$$

If we write*

$$\mathbf{x} = \begin{bmatrix} x_1 \\ x_2 \\ \cdots \\ x_k \end{bmatrix}, \qquad \mathbf{b} = \begin{bmatrix} b_1 \\ b_2 \\ \cdots \\ b_k \end{bmatrix}, \qquad \mathbf{B} = \begin{bmatrix} b_{11} & \frac{1}{2}b_{12} & \cdots & \frac{1}{2}b_{1k} \\ \frac{1}{2}b_{12} & b_{22} & \cdots & \frac{1}{2}b_{2k} \\ \cdots & & & \\ \frac{1}{2}b_{1k} & \frac{1}{2}b_{2k} & \cdots & b_{kk} \end{bmatrix} \qquad (10.3.2)$$

---

*Typically, this would be after suitable preliminary experimentation involving (perhaps) steepest ascent to come close to a region of interest, and (possibly) theoretical and empirical knowledge to arrive at suitable transformations of the response and/or the predictor variables.

*Note that, in this discussion, we shall write **b** for a vector of first-order estimated parameters. However, in Chapter 3, **b** was used to denote the full vector of estimates. The reader must be careful not to confuse the two notations.

the fitted equation is

$$\hat{y} = b_0 + \mathbf{x}'\mathbf{b} + \mathbf{x}'\mathbf{B}\mathbf{x}. \tag{10.3.3}$$

Let $\lambda_1, \lambda_2, \ldots, \lambda_i, \ldots, \lambda_k$ be the eigenvalues[†] of the symmetric matrix $\mathbf{B}$, and $\mathbf{m}_1, \mathbf{m}_2, \ldots, \mathbf{m}_i, \ldots, \mathbf{m}_k$ the corresponding eigenvectors[‡] so that, by definition,

$$\mathbf{B}\mathbf{m}_i = \mathbf{m}_i \lambda_i, \qquad i = 1, 2, \ldots, k. \tag{10.3.4}$$

If we standardize each eigenvector so that $\mathbf{m}_i'\mathbf{m}_i = 1$ and if the $k \times k$ matrix $\mathbf{M}$ has $\mathbf{m}_i$ for its ith column, then $\mathbf{M}$ is an orthonormal matrix and the k equations (10.3.4) may be written simultaneously as

$$\mathbf{B}\mathbf{M} = \mathbf{M}\boldsymbol{\Lambda}, \tag{10.3.5}$$

where $\boldsymbol{\Lambda}$ is a diagonal matrix whose ith diagonal element is λ_i. Premultiplying by $\mathbf{M}'(= \mathbf{M}^{-1})$ gives

$$\mathbf{M}'\mathbf{B}\mathbf{M} = \boldsymbol{\Lambda}. \tag{10.3.6}$$

Now by making use of the fact that $\mathbf{M}\mathbf{M}' = \mathbf{I}$, we can write Eq. (10.3.3) as

$$\hat{y} = b_0 + (\mathbf{x}'\mathbf{M})(\mathbf{M}'\mathbf{b}) + (\mathbf{x}'\mathbf{M})\mathbf{M}'\mathbf{B}\mathbf{M}(\mathbf{M}'\mathbf{x}). \tag{10.3.7}$$

If we now write $\mathbf{X} = \mathbf{M}'\mathbf{x}$ and $\boldsymbol{\theta} = \mathbf{M}'\mathbf{b}$, (or, equivalently, $\mathbf{x} = \mathbf{M}\mathbf{X}$ and $\mathbf{b} = \mathbf{M}\boldsymbol{\theta}$) this can be rewritten as

$$\hat{y} = b_0 + \mathbf{X}'\boldsymbol{\theta} + \mathbf{X}'\boldsymbol{\Lambda}\mathbf{X}, \tag{10.3.8}$$

that is, as

$$\hat{y} = b_0 + \theta_1 X_1 + \cdots + \theta_k X_k + \lambda_1 X_1^2 + \cdots + \lambda_k X_k^2. \tag{10.3.9}$$

This constitutes the A canonical form which eliminates cross-product terms by rotating the axes. By differentiating (10.3.9) with respect to $X_1, \ldots, X_k$ we find that the coordinates of the stationary point $(X_{1S}, \ldots, X_{kS})$ in relation to the rotated axes are such that

$$X_{iS} = \frac{-\theta_i}{2\lambda_i}. \tag{10.3.10}$$

The A canonical form has a useful story to tell. The sizes and signs of the λ_i determine the type of second-order surface we have fitted. The θ_i measure the slopes of the surface at the original origin $\mathbf{x} = \mathbf{0}$, in the directions of the (rotated) coordinate axes $X_1, \ldots, X_k$. The values of the X_{iS}'s tell us how far along the various canonical axes is the stationary point S. We illustrate this with a simple constructed example for $k = 2$ variables.

[†]Also called characteristic roots, or latent roots.
[‡]Also called characteristic vectors, or latent vectors.

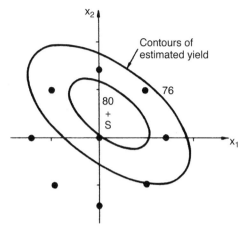

Figure 10.9. Contours $\hat{y} = 76$, and 79 of the fitted equation

$$\hat{y} = 78.8988 + 2.272x_1 + 3.496x_2$$

$$- 2.08x_1^2 - 2.92x_2^2 - 2.88x_1x_2.$$

An Example of Canonical Analysis for $k = 2$ Variables

Suppose the data obtained from a design indicated by dots in Figure 10.9 had led to the fitted second-degree equation

$$\hat{y} = 78.8988 + 2.272x_1 + 3.496x_2 - 2.08x_1^2 - 2.92x_2^2 - 2.88x_1x_2 \quad (10.3.11)$$

and that the usual checks indicated that this equation was representationally adequate. The elliptical contours plotted in Figure 10.9 show that the fitted equation describes a surface with a maximum at a stationary point S which is in the neighborhood of the center of the design. Furthermore, we see that the maximal region is somewhat attenuated in the NW–SE direction. Canonical analysis of the equation makes it possible to deduce this information without graphical representation. For this example,

$$\mathbf{x} = \begin{bmatrix} x_1 \\ x_2 \end{bmatrix}, \quad \mathbf{b} = \begin{bmatrix} 2.272 \\ 3.496 \end{bmatrix}, \quad \mathbf{B} = \begin{bmatrix} -2.08 & -1.44 \\ -1.44 & -2.92 \end{bmatrix}. \quad (10.3.12)$$

The fitted equation in matrix form is therefore

$$\hat{y} = 78.8988 + [x_1, x_2]\begin{bmatrix} 2.272 \\ 3.496 \end{bmatrix} + [x_1, x_2]\begin{bmatrix} -2.08 & -1.44 \\ -1.44 & -2.92 \end{bmatrix}\begin{bmatrix} x_1 \\ x_2 \end{bmatrix}. \quad (10.3.13)$$

The eigenvalues of $\mathbf{B}$ and the corresponding orthonormal eigenvectors are

$$\lambda_1 = -4, \qquad \lambda_2 = -1, \qquad\qquad\qquad (10.3.14)$$

$$\mathbf{m}_1 = \begin{bmatrix} 0.6 \\ 0.8 \end{bmatrix}, \qquad \mathbf{m}_2 = \begin{bmatrix} -0.8 \\ 0.6 \end{bmatrix}. \qquad\qquad (10.3.15)$$

Thus

$$\mathbf{M} = \begin{bmatrix} 0.6 & -0.8 \\ 0.8 & 0.6 \end{bmatrix}, \qquad \Lambda = \begin{bmatrix} -4 & 0 \\ 0 & -1 \end{bmatrix}, \tag{10.3.16}$$

and

$$\theta = \mathbf{M'b} = \begin{bmatrix} 0.6 & 0.8 \\ -0.8 & 0.6 \end{bmatrix} \begin{bmatrix} 2.272 \\ 3.496 \end{bmatrix} = \begin{bmatrix} 4.16 \\ 0.28 \end{bmatrix}, \tag{10.3.17}$$

whereupon

$$X_{1S} = \frac{4.16}{(2 \times 4)} = 0.52,$$

$$\tag{10.3.18}$$

$$X_{2S} = \frac{0.28}{(2 \times 1)} = 0.14.$$

The A canonical form is thus

$$\hat{y} = 78.8988 + 4.16 X_1 + 0.28 X_2 - 4X_1^2 - 1X_2^2, \tag{10.3.19}$$

where

$$X_1 = 0.6x_1 + 0.8x_2,$$

$$X_2 = -0.8x_1 + 0.6x_2. \tag{10.3.20}$$

As illustrated in Figure 10.10, the A canonical reduction refers the equation to the rotated axes (X_1, X_2) which are chosen to lie parallel to the principal axes of the system.

Important quantities to consider in this canonical form are:

1. The signs and relative magnitudes of the eigenvalues λ_i.
2. The size of the coordinates X_{iS} of the stationary point.
3. The nature of the transformation $\mathbf{X} = \mathbf{M'x}$.

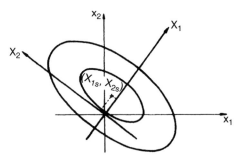

Figure 10.10. A rotation of axes produces canonical form A with fitted equation
$$\hat{y} = 78.8988 + 4.16X_1 + 0.28X_2 - 4X_1^2 - X_2^2.$$

We now discuss these points in detail for our example, in which $k = 2$.

1. For our example, the eigenvalues, which are the coefficients of the quadratic terms X_1^2 and X_2^2 are $\lambda_1 = -4$, $\lambda_2 = -1$. (a) We know, because both signs are negative, that the curvature is negative in both canonical directions and so we must be dealing with a maximum. If λ_1 and λ_2 had both been positive, we would have a minimum, and if they were of different signs, we would have a saddle point. (b) The lengths of the principal axes of the ellipses are proportional to $|\lambda_1|^{-\frac{1}{2}}$ and $|\lambda_2|^{-\frac{1}{2}}$, respectively. Thus, in our example, we know that the fitted surface is somewhat attenuated along the X_2 axis.

2. In the agreed scaling of the response function, distance is preserved by the transformation $\mathbf{X} = \mathbf{M}'\mathbf{x}$, so that we know, from the small magnitudes of X_{1S} and X_{2S}, that the center of the system is close to the center of the design.

3. The transformation $\mathbf{X} = \mathbf{M}'\mathbf{x}$ or, equivalently, $\mathbf{x} = \mathbf{M}\mathbf{X}$ allows us to determine the directions of the canonical axes. For the example, the coordinate X_2 along which the contours are attenuated is defined by $X_2 = -0.8x_1 + 0.6x_2$. The perpendicular axis shown is of course $X_1 = 0.6x_1 + 0.8x_2$.

The B Canonical Form

By equating to zero derivatives with respect to $\mathbf{x}$ in (10.3.3) and $\mathbf{X}$ in (10.3.8), we find the coordinates of the stationary point S to be the solution of the equations

$$-2\mathbf{B}\mathbf{x}_S = \mathbf{b}, \quad \text{or of} \quad -2\mathbf{\Lambda}\mathbf{X}_S = \mathbf{0}, \tag{10.3.21}$$

and the fitted response at this point is

$$\hat{y}_S = b_0 + \tfrac{1}{2}\mathbf{x}_S'\mathbf{b} \quad \text{or, equally,} \quad \hat{y}_S = b_0 + \tfrac{1}{2}\mathbf{X}_S'\mathbf{\theta}. \tag{10.3.22}$$

We now adopt the notation $\tilde{x}_i = x_i - x_{iS}$, $\tilde{X}_i = X_i - X_{iS}$, and the corresponding vector forms $\tilde{\mathbf{x}}$ and $\tilde{\mathbf{X}}$, so that the $\tilde{x}_i$'s and $\tilde{X}_i$'s refer to coordinates measured from the stationary point S. If, in addition to rotating the axes, we also shift the origin to the stationary point S, then, we obtain the form

$$\hat{y} = \hat{y}_S + \tilde{\mathbf{X}}'\mathbf{\Lambda}\tilde{\mathbf{X}}, \tag{10.3.23}$$

that is

$$\hat{y} = \hat{y}_S + \lambda_1\tilde{X}_1^2 + \cdots + \lambda_k\tilde{X}_k^2. \tag{10.3.24}$$

We call this the B canonical form (see Figure 10.11).

For our constructed example with $k = 2$, the equations $-2\mathbf{B}\mathbf{x}_S = \mathbf{b}$ become

$$2\begin{bmatrix} 2.08 & 1.44 \\ 1.44 & 2.92 \end{bmatrix}\begin{bmatrix} x_{1S} \\ x_{2S} \end{bmatrix} = \begin{bmatrix} 2.272 \\ 3.496 \end{bmatrix} \tag{10.3.25}$$

and these have the solution $x_{1S} = 0.2$, $x_{2S} = 0.5$. Also, the estimated response at S, given by the formula $\hat{y}_S = b_0 + \tfrac{1}{2}\mathbf{x}_S'\mathbf{b}$ is

$$\hat{y}_S = 78.8988 + \frac{1}{2}(0.2, 0.5)\begin{bmatrix} 2.272 \\ 3.496 \end{bmatrix} = 80.0. \tag{10.3.26}$$

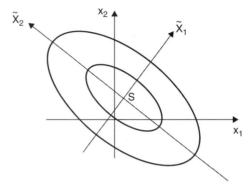

Figure 10.11. A rotation of axes and a shift of origin produces canonical form B with fitted equation

$$\hat{y} = 80 - 4\tilde{X}_1^2 - \tilde{X}_2^2.$$

Thus, finally, moving the origin to S, we express the canonical form $\hat{y} = \hat{y}_S + \tilde{\mathbf{X}}'\boldsymbol{\Lambda}\tilde{\mathbf{X}}$ as

$$\hat{y} = 80 - 4\tilde{X}_1^2 - 1\tilde{X}_2^2 \qquad (10.3.27)$$

with $\tilde{\mathbf{X}} = \mathbf{M}'\tilde{\mathbf{x}} = \mathbf{M}'(\mathbf{x} - \mathbf{x}_S)$, namely,

$$\tilde{X}_1 = 0.6(x_1 - 0.2) + 0.8(x_2 - 0.5),$$
$$\tilde{X}_2 = -0.8(x_1 - 0.2) + 0.6(x_2 - 0.5). \qquad (10.3.28)$$

As we see from Figure 10.11, the B canonical form refers the fitted equation to new axes $(\tilde{X}_1, \tilde{X}_2)$ which are rotated as before, but, in addition, a shift of origin to the stationary point S is made.

Application of Canonical Forms A and B

For two or three predictor variables ($k = 2$ or 3) it is fairly easy to appreciate the nature of various second-degree systems, but for $k > 3$ it becomes more difficult. Canonical analysis makes it possible to understand the system even when k is greater than 3.

In design units, the distance of the center of the system (the stationary point) S from the design center O is given by

$$D = OS = \left\{ \sum_{i=1}^{k} x_{iS}^2 \right\}^{1/2} = \left\{ \sum_{j=1}^{k} X_{jS}^2 \right\}^{1/2}. \qquad (10.3.29)$$

When S is close to O (say, D not much greater than 1), the simpler B canonical form is adequate for understanding the fitted surface. However, particularly with ridge systems, the center S of the fitted system may be remote from the center O of the design. In this case, the analysis is better carried out in terms of the A

canonical form. We first illustrate canonical analysis for the simple system fitted to the elasticity data (see Sections 9.2 and 9.3) using the B canonical form. In Chapter 11, we consider more complicated examples which require use of the A canonical form.

Example 1. *Canonical Analysis for the Polymer Elasticity Data.* We now reexamine a three-variable fitted surface previously discussed in Sections 9.2 and 9.3. It will be recalled that, in that example, the response y, measured elasticity, was considered as a function of three predictor variables, percentage concentrations C_1 and C_2, and reaction temperature T. The variables were coded as

$$x_1 = \frac{(C_1 - 18)}{3}, \qquad x_2 = \frac{(C_2 - 2.7)}{0.4}, \qquad x_3 = \frac{(T - 145)}{10}. \quad (10.3.30)$$

The fitted second-degree equation in a near stationary region was found to be

$$\hat{y} = 57.31 + 1.50x_1 - 2.13x_2 + 1.81x_3$$

$$- 4.69x_1^2 - 6.27x_2^2 - 5.21x_3^2 - 7.13x_1x_2 - 3.27x_1x_3 - 2.73x_2x_3. \quad (10.3.31)$$

The fitted surface was studied in Chapter 9 by plotting contours of $\hat{y}$ (Figure 9.2) in the space of x_1 and x_2 for various fixed values of x_3. The contour plots showed that the stationary point S was at a maximum of the fitted surface. We now look at the corresponding canonical analysis.

Canonical Analysis

We have

$$b_0 = 57.31, \qquad \mathbf{b} = \begin{bmatrix} 1.50 \\ -2.13 \\ 1.81 \end{bmatrix}, \qquad \mathbf{B} = -\begin{bmatrix} 4.690 & 3.565 & 1.635 \\ 3.565 & 6.270 & 1.365 \\ 1.635 & 1.365 & 5.210 \end{bmatrix}. \quad (10.3.32)$$

We thus obtain the coordinates of the stationary point

$$\mathbf{x}_S = -\tfrac{1}{2}\mathbf{B}^{-1}\mathbf{b} = (0.4603, -0.4645, 0.1509)', \quad (10.3.33)$$

at which point the estimated response is

$$\hat{y}_S = b_0 + \tfrac{1}{2}\mathbf{x}_S'\mathbf{b} = 58.29. \quad (10.3.34)$$

The distance from the stationary point S to the center of the design is thus

$$D = \{\mathbf{x}_S'\mathbf{x}_S\}^{1/2} = \{(0.4603)^2 + (-0.4645)^2 + (0.1509)^2\}^{1/2} = 0.6711. \quad (10.3.35)$$

Thus the point S is close to the center of the experimental region, and so we shall proceed directly to the B canonical form. The eigenvalues of $\mathbf{B}$ in order of absolute magnitude, along with their associated orthonormal eigenvectors, are displayed below.

λ_j	-10.03	-4.37	-1.77
Corresponding	0.5899	0.1312	0.7967
$\mathbf{m}_j$	0.7024	0.4034	-0.5865
vector	0.3983	-0.9056	-0.1458

The 3-by-3 lower right portion of this display comprises the matrix $\mathbf{M}$ featured in Eq. (10.3.5) and elsewhere. We obtain $\mathbf{M}^{-1} = \mathbf{M}'$ by transposing $\mathbf{M}$. This, by use of the fact that $\tilde{\mathbf{x}} = \mathbf{x} - \mathbf{x}_S$, leads to the equations

$$\tilde{X}_1 = 0.5899x_1 + 0.7024x_2 + 0.3983x_3 - 0.00537,$$

$$\tilde{X}_2 = 0.1312x_1 + 0.4034x_2 - 0.9056x_3 + 0.2636, \qquad (10.3.36)$$

$$\tilde{X}_3 = 0.7967x_1 - 0.5865x_2 - 0.1458x_3 - 0.6171.$$

The B canonical form is

$$\hat{y} = 58.29 - 10.03\tilde{X}_1^2 - 4.37\tilde{X}_2^2 - 1.77\tilde{X}_3^2. \qquad (10.3.37)$$

This canonical analysis tells us the following important facts, without reference to contour plots:

1. Equation (10.3.35) shows that the center of the fitted system lies close to the center $(0, 0, 0)$ of the experimental region.
2. At the center, the fitted response is $\hat{y} = 58.29$.
3. The negative sign attached to each of the eigenvalues tells us that the fitted response value 58.29 at the center of the fitted surface is a maximum.
4. Because (a) we established in Chapter 9 that the fitted quadratic surface seems to provide a reasonably good explanation of the variation in the data and (b) the appropriate standard errors for the canonical coefficients are roughly of the same size as those of the quadratic coefficients (± 0.5), it seems likely that the *underlying true surface* also contains a maximum. [A more accurate picture of what is known of the location of the true maximum at this stage of experimentation can be gained by the confidence region calculation described in Box and Hunter (1954).]
5. The elements of the eigenvectors, which are also the coefficients in Equations (10.3.36) can provide important information about the relationships of the predictors to the response. This information may be put to purely empirical use, or may possibly aid theoretical conjectures. For example, the coefficients of the eigenvector associated with the smallest (in absolute value) eigenvalue $\lambda_3 = -1.77$ are

$$(0.7967, -0.5865, -0.1458).$$

This suggests that in the $\tilde{X}_3$ direction, which is the direction of greatest elongation of the ellipsoid, the concentrations C_1 and C_2 are, to some extent, compensatory. A reduction in one concentration can be, to some extent, compensated by a corresponding increase in the other.

The last point brings us to the most important practical use of canonical analysis, the detection, exploration, and exploitation of ridge systems. We discuss this further in Chapter 11.

APPENDIX 10A. A SIMPLE EXPLANATION OF CANONICAL ANALYSIS

We now provide a simple geometrical explanation of the mathematics of canonical analysis. We illustrate for $k = 3$ dimensions and give a numerical example for $k = 2$. Suppose we have a fitted second-order equation

$$\hat{y} = b_0 + \mathbf{x'b} + \mathbf{x'Bx}, \tag{10A.1}$$

with a stationary point at $\mathbf{x}_S$. We wish to make a rotation to a coordinate system whose axes lie along the principal axes of the ellipsoidal contours. To do this, we imagine a sphere *of radius unity* centered at S. (Actuallyl any radius choice will serve, but we might as well take it as unity.) The equation of such a sphere is

$$\tilde{\mathbf{x}}'\tilde{\mathbf{x}} = 1, \tag{10A.2}$$

or, equivalently,

$$(\mathbf{x} - \mathbf{x}_S)'(\mathbf{x} - \mathbf{x}_S) = 1. \tag{10A.3}$$

Now consider the response $\hat{y}$ *on the surface of this sphere* (see Figure 10A.1). In general, the response will be stationary at $2k$ points $P_1, P_1'; P_2, P_2'; \ldots; P_k, P'_k$, and the straight lines $P_i SP_i'$, for $i = 1, 2, \ldots, k$, will lie along the k principal axes of the second degree system. We adopt the lines $P_i SP_i'$ as our new coordinate axes. S, of course, is the new center.

The $2k$ stationary points P_i, P_i' may be obtained by using Lagrange's method of undetermined multipliers. Our objective function to be differentiated is

$$F(\mathbf{x}, \lambda) = b_0 + \mathbf{x'b} + \mathbf{x'Bx} - \lambda\{(\mathbf{x} - \mathbf{x}_S)'(\mathbf{x} - \mathbf{x}_S) - 1\}. \tag{10A.4}$$

Equating to zero the derivatives with respect to $\mathbf{x}'$ provides the k simultaneous equations

$$\mathbf{b} + 2\mathbf{Bx} - 2\lambda(\mathbf{x} - \mathbf{x}_S) = \mathbf{0}. \tag{10A.5}$$

Substitution for $\mathbf{b}$ from (10.3.21) provides

$$\mathbf{B}(\mathbf{x} - \mathbf{x}_S) - \lambda(\mathbf{x} - \mathbf{x}_S) = 0, \tag{10A.6}$$

or

$$(\mathbf{B} - \lambda\mathbf{I})\tilde{\mathbf{x}} = 0, \tag{10A.7}$$

where

$$\tilde{\mathbf{x}} = \mathbf{x} - \mathbf{x}_S. \tag{10A.8}$$

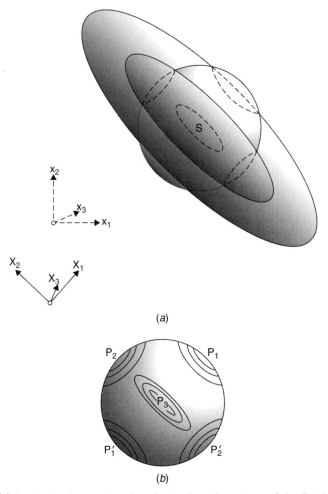

Figure 10A.1. (*a*) $k = 3$. A sphere centered at S cuts through contours of the fitted second degree response system. (*b*) A birds-eye view from the $\tilde{X}_3$ axis of the response contours on the sphere, showing the stationary points.

For this set of homogeneous equations to have a solution, the determinantal equation

$$|\mathbf{B} - \lambda\mathbf{I}| = 0 \qquad (10\text{A}.9)$$

must hold. The solutions λ_i, $i = 1, 2, \ldots, k$, of this determinantal equation are called the eigenvalues of the matrix $\mathbf{B}$. We need consider here only the case where there are k distinct eigenvalues $\lambda_1, \lambda_2, \ldots, \lambda_k$. Theoretical complications, which we consider later, can occur when some of the λ_i are equal to zero or when certain λ_i are equal to one another. However, since we are dealing with experimental data subject to error, neither of these circumstances will ever occur exactly in practice. Now for each λ_i we solve the set of equations

$$(\mathbf{B} - \lambda_i\mathbf{I})\tilde{\mathbf{x}} = 0, \qquad (10\text{A}.10)$$

Table 10A.1. Coordinates of Points in Original and Rotated Systems

Point	Coordinates in the $\tilde{x}$ System				Coordinates in the $\tilde{X}$ System			
	x_1	$\tilde{x}_2$	$\cdots$	$\tilde{x}_k$	$\tilde{X}_1$	$\tilde{X}_2$	$\cdots$	$\tilde{X}_k$
P_1	m_{11}	m_{12}	$\cdots$	m_{1k}	1	0	$\cdots$	0
P_2	m_{21}	m_{22}	$\cdots$	m_{2k}	0	1	$\cdots$	0
$\cdots$	$\cdots$				$\cdots$			
P_k	m_{k1}	m_{k2}	$\cdots$	m_{kk}	0	0	$\cdots$	1
S	0	0	$\cdots$	0	0	0	$\cdots$	0
P_1'	$-m_{11}$	$-m_{12}$	$\cdots$	$-m_{1k}$	-1	0	$\cdots$	0
P_2'	$-m_{21}$	$-m_{22}$	$\cdots$	$-m_{2k}$	0	-1	$\cdots$	0
$\cdots$	$\cdots$							
P'_k	$-m_{k1}$	$-m_{k2}$	$\cdots$	$-m_{kk}$	0	0	$\cdots$	-1

which yields the solution $\tilde{\mathbf{x}} = \mathbf{m}_i$. [These equations are dependent. To solve them, we set one element of $\tilde{\mathbf{x}}$ to any nonzero value (e.g., 1) or some other convenient value, solve any $k - 1$ of the k equations (10A.10), and then normalize this intermediate solution so that the sum of squares of the elements of the solution is 1. The result is $\mathbf{m}_i$ and $\mathbf{m}_i'\mathbf{m}_i = 1$.]

If we take the coordinates of the points $P_1, P_2, \ldots, P_k$ to be, respectively,

$$\tilde{\mathbf{x}} = \mathbf{m}_1, \mathbf{m}_2, \ldots, \mathbf{m}_k \tag{10A.11}$$

then the coordinates of $P_1', P_2', \ldots, P_k'$ will be

$$\tilde{\mathbf{x}} = -\mathbf{m}_1, -\mathbf{m}_2, \ldots, -\mathbf{m}_k. \tag{10A.12}$$

If $SP_1, SP_2, \ldots, SP_k$ are designated as the coordinate axes of the new canonical variables, the correspondence in Table 10A.1 can be set out, where $\mathbf{m}_i = (m_{i1}, m_{i2}, \ldots, m_{ik})'$. The elements of the vector $\mathbf{m}_i = (m_{i1}, m_{i2}, \ldots, m_{ik})'$ are, in fact, the *direction cosines* of the $\tilde{X}_i$ axis in the $\tilde{x}$ space. Furthermore, if we construct a matrix $\mathbf{M}$ whose ith column is $\mathbf{m}_i$, the transformation from old ($\tilde{x}$) to new ($\tilde{X}$) coordinates is given by

$$\tilde{\mathbf{x}} = \mathbf{M}\tilde{\mathbf{X}}. \tag{10A.13}$$

Equivalently, for the respectively parallel coordinate systems x and X, measured from the natural origin, we have the transformation

$$\mathbf{x} = \mathbf{M}\mathbf{X}. \tag{10A.14}$$

Because the transformation is an orthogonal one and $\mathbf{m}_i'\mathbf{m}_j = 0$, $\mathbf{m}_i'\mathbf{m}_i = 1$, $i, j = 1, 2, \ldots, k$, it follows that $\mathbf{M}'\mathbf{M} = \mathbf{M}\mathbf{M}' = \mathbf{I}$, that is, $\mathbf{M}^{-1} = \mathbf{M}'$. Thus the inverse transformations are

$$\tilde{\mathbf{X}} = \mathbf{M}'\tilde{\mathbf{x}} \quad \text{and} \quad \mathbf{X} = \mathbf{M}'\mathbf{x}. \tag{10A.15}$$

From these forms, we see that the jth *row* of **M**, with elements $m_{1j}, m_{2j}, \ldots, m_{kj}$, provides the direction cosines of the $\tilde{x}_j$ axis in the $\tilde{X}$ space (or of the x_j axis in the X space).

Efficient computer programs are available for the actual determination of eigenvalues and eigenvectors. However, to illustrate the above theory, we carry through the calculation for the constructed ($k = 2$) example of Eq. (10.3.11). Using Eq. (10A.9), we obtain the eigenvalues from the solution of the determinantal equation

$$\begin{vmatrix} -2.08 - \lambda & -1.44 \\ -1.44 & -2.92 - \lambda \end{vmatrix} = 0, \qquad (10A.16)$$

namely $(-2.08 - \lambda)(-2.92 - \lambda) - (-1.44)^2 = 0$. This equation has two roots $\lambda_1 = -4$, $\lambda_2 = -1$, and these are the eigenvalues. The corresponding eigenvectors are

$$\mathbf{m}_1 = \begin{bmatrix} 0.6 \\ 0.8 \end{bmatrix}, \qquad \mathbf{m}_2 = \begin{bmatrix} -0.8 \\ 0.6 \end{bmatrix}. \qquad (10A.17)$$

These are obtained by successively substituting in Eqs. (10A.7), namely in $(\mathbf{B} - \lambda\mathbf{I})\tilde{\mathbf{x}} = 0$. For example, putting $\lambda_1 = -4$, Eqs. (10A.7) are

$$(-2.08 + 4)\tilde{x}_1 - 1.44\tilde{x}_2 = 0,$$
$$-1.44\tilde{x}_1 + (-2.92 + 4)\tilde{x}_2 = 0. \qquad (10A.18)$$

These are two dependent equations and so their solution is not unique. If we set $\tilde{x}_1 = 1$, however, we find $\tilde{x}_2 = \frac{4}{3}$ from either equation. This solution $(1, \frac{4}{3})$ is now normalized (uniquely apart from sign) by making the sum of squares of the elements equal to 1 so that

$$\mathbf{m}_1 = \begin{bmatrix} 0.6 \\ 0.8 \end{bmatrix} \qquad (10A.19)$$

as above. Similarly $\lambda_2 = -1$ leads to a solution $\mathbf{m}_2$ as above, whereupon

$$\mathbf{M} = \begin{bmatrix} 0.6 & -0.8 \\ 0.8 & 0.6 \end{bmatrix}. \qquad (10A.20)$$

CHAPTER 11

Occurrence and Elucidation of Ridge Systems, II

11.1. RIDGE SYSTEMS APPROXIMATED BY LIMITING CASES

As we pointed out in Chapter 10, a maximum* in k variables is often most revealingly approximated not by a point but by a line, plane, or hyperplane. Furthermore, the detection, description, and exploitation of such ridge systems is one of the most important uses of response surface techniques.

If we can employ a second-degree polynomial to represent the response in the neighborhood of interest, theoretical systems that approximate ridges occurring in practice can be thought of as limiting cases of the canonical forms considered earlier. We first consider the possibilities for $k = 2$ variables to illustrate the basic ideas.

Stationary Ridge, $k = 2$

Consider the canonical form B for a fitted second-degree equation

$$\hat{y} = \hat{y}_S + \lambda_1 \tilde{X}_1^2 + \lambda_2 \tilde{X}_2^2. \tag{11.1.1}$$

Suppose that λ_1 and λ_2 are both negative, corresponding to a surface like that in Figure 11.1a. Now suppose that λ_1 is kept fixed, while λ_2 is made smaller and smaller in absolute value. Then the contours of the resulting system become more and more drawn out along the $\tilde{X}_2$ axis until, when λ_2 reaches zero, the stationary ridge system of Figure 11.1c is obtained. If we now allow λ_2 to *pass through* zero and to become positive, we shall have an attenuated minimax as in Figure 11.1b. Thus Figure 11.1c can equally well be regarded as a limiting case of Figure 11.1b with the positive λ_2 tending to zero. In practice, a system that is *exactly* of the stationary ridge form is unlikely; experimental error and mild lack of fit can blur the picture somewhat. Thus, for an estimated surface, we would not expect to find

---

*As before, we shall discuss only maxima but parallel considerations apply to minima, also.

Response Surfaces, Mixtures, and Ridge Analyses, Second Edition. By G. E. P. Box and N. R. Draper
Copyright © 2007 John Wiley & Sons, Inc.

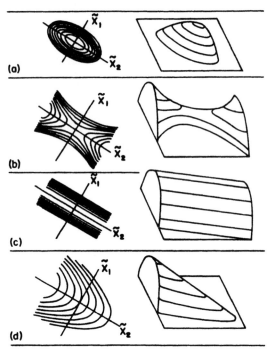

Figure 11.1. Primary and limiting canonical forms of the fitted quadratic equation $\hat{y} = b_0 + b_1 x_1 + b_2 x_2 + b_{11} x_1^2 + b_{22} x_2^2 + b_{12} x_1 x_2$.

a coefficient λ_2 *exactly* zero; rather we might find such a coefficient that was small compared with λ_1 and of about the same size as its standard error. This would imply an attenuation of the surface in the $\tilde{X}_2$ direction. Such an occurrence puts the experimenter on notice that it would be wise to explore the possibility that locally, and approximately, there is a line maximum rather than a point maximum.

Rising Ridge, $k = 2$

A further practical possibility is approximated by a canonical representation of the form

$$\hat{y} = \hat{y}_S + \lambda_1 \tilde{X}_1^2 + \theta_2 \tilde{X}_2 \tag{11.1.2}$$

corresponding to the rising ridge of Figure 11.1d. The system is similar to that of Figure 11.1c, but the crest of the ridge is tipped and has a gradient θ_2 in the $\tilde{X}_2$ direction instead of being level. The contours are parabolas. The system can be imagined as a limiting case either of Figure 11.1a or of Figure 11.1b. It occurs if λ_1 is kept fixed and λ_2 is allowed to become vanishingly small *at the same time as the center of the system is moved to infinity*. In practice, the limiting forms corresponding to Figures 11.1c and 11.1d do not occur exactly, as we have already mentioned; however, these forms may be regarded as reference marks. When we say that an empirical surface is like a limiting case, we mean that it is best thought of as approximated by that case. Approximation by a limiting form of some kind is

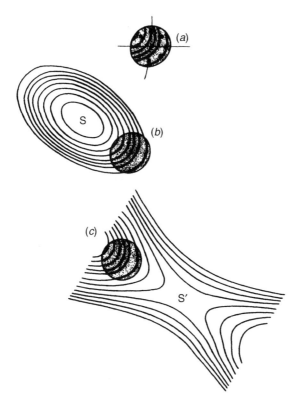

Figure 11.2. When data are taken at emperimental points on a true surface such as (a), the fitted second-degree equation may be part of a system of ellipses centered at S [see (b)] or of a system of hyperbolas centered at S' [see (c)]. In either case, an understanding of what is happening in the (shaded) region of interest is best gained by using canonical form A.

suggested whenever one of the canonical coefficients is close to zero and small in absolute magnitude compared with the other. When this happens, it may also happen that the stationary point for the fitted surface will be remote from the center of the design. The reason for this is that, when the system contains a ridge, steepest ascent will bring the experimental region *close to the ridge*, but the system that represents that ridge may have a remote center.

To see how this can happen, consider Figure 11.2. Suppose that, in the immediate neighborhood of the current experimental region, indicated by shading in Figure 11.2a, the true surface is the rising ridge indicated by the contours. The second-degree equation that is being fitted is, as we have seen, capable of taking on a number of different shapes. Our method of surface fitting will choose that system which most closely represents (in the least-squares sense) the experimental data. It might do this by using part of a system of elliptical contours with a center at S, as shown in Figure 11.2b. Alternatively, it might use part of a system of hyperbolic contours from a minimax system with center at S' as indicated in Figure 11.2c. A small change in the pattern of experimental error could result in a switch from one of these forms to the other. Now obviously we are interested only in approximating the response in the immediate experimental neighborhood, that is, the shaded region in the figure, and either system can do this. By using canonical analysis in the A form, we will see how we can be led to approximately the same canonical form whichever of the apparently quite different fitted representation arise. (Also, see Figure 11.4.)

11.2. CANONICAL ANALYSIS IN THE A FORM TO CHARACTERIZE RIDGE PHENOMENA

In view of the considerations discussed above, when the center S of the fitted system is remote from the region of experimentation, it is most useful to present the reduced equation as canonical form A, namely

$$\hat{y} = b_0 + \theta_1 X_1 + \theta_2 X_2 + \cdots + \theta_k X_k + \lambda_1 X_1^2 + \lambda_2 X_2^2 + \cdots + \lambda_k X_k^2. \quad (11.2.1)$$

For illustration, and further to illuminate the discussion that follows, we show in Figure 11.3 a number of possibilities for *theoretical* second-order surfaces in $k = 3$ dimensions. Table 11.1 shows the values of the θ's and λ's associated with each of

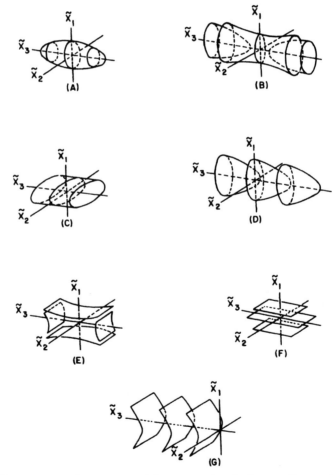

Figure 11.3. Contour systems for some second-degree equations in three predictor variables, including some important limiting forms.

Table 11.1. Characteristics of the Theoretical Second-Degree Equations Whose Contours Are Shown in Figure 11.3, and of the Corresponding Equations Obtained from Data Subject to Error, after Steepest Ascent Has Been Applied

Labels of System in Figure 11.3	Values of $\theta_1\ \theta_2\ \theta_3$ Signs[a] of Diagram as Drawn	$\lambda_1\ \lambda_2\ \lambda_3$	Conditions Implied by the Diagram as Drawn	Type of Contours when a Cross Section is Taken in the Variables Shown[b] (Figures 11.1a–d) $\tilde{X}_1\tilde{X}_2$	$\tilde{X}_1\tilde{X}_3$	$\tilde{X}_2\tilde{X}_3$	System Center in Relation To Fitted Center	Special Features Found in Practice
(A)	0 0 0	− − −	$\|\lambda_2\| > \|\lambda_1\| > \|\lambda_3\|$	(a)	(a)	(a)	Close	—
(B)	0 0 0	− − +	$\|\lambda_2\| > \|\lambda_1\| > \|\lambda_3\|$	(a)	(b)	(b)	Close	When λ_3 is small, true surface may be like (C)
(C)	0 0 0	− − 0	Line of centers on the $\tilde{X}_3$ axis	(a)	(c)	(c)	Indeterminate	λ_3 will have a small value (negative or positive)
(D)	0 0 +	− − 0	Center at infinity on the $\tilde{X}_3$ axis	(a)	(d)	(d)	Remote	λ_3 will have a small value (negative or positive)
(E)	0 0 0	− 0 +	Line of centers on the $\tilde{X}_2$	(c)	(b)	(c)	Indeterminate	When λ_3 is small, true surface may be like (F)
(F)	0 0 0	− 0 0	Plane of centers given by $\tilde{X}_1 = 0$	(c)	(c)	Constant response	Indeterminate	λ_2 and λ_3 will have small values (negative or positive)
(G)	0 0 +	− 0 0	Center at infinity on the $\tilde{X}_3$ axis	(c)	(d)	Parallel straight lines	Remote	λ_2 and λ_3 will have small values (negative or positive)

Characteristics of the Theoretical Canonical Forms: In Practice These Are Only Approximately Attained by the Fitted Surface

Characteristics of Canonically Reduced Fitted Surfaces Approximating the Theoretical Forms

[a] A complete switch of all signs does not alter the contour forms, but reverses the directions in which response increases. "0" indicates that this value is exactly zero in the theoretical canonical representation. For practical cases, see the last column.

[b] The heading $\tilde{X}_1\tilde{X}_2$ indicates the form of the contours in $\tilde{X}_1$ and $\tilde{X}_2$ when a cross-section ($\tilde{X}_3$ = constant) is taken.

these theoretical possibilities and also shows the characteristics to be expected of a corresponding surface *fitted to data subject to error*. (It is assumed that steepest ascent has been applied, so that a near-stationary region has been attained.)

In practice, we may have to deal with complicated ridge systems in many dimensions. Certain quantities calculated from the canonical form A can be of great help in characterizing such systems. It is important to remember that, since we are dealing with data subject to error, we never have to deal with the limiting cases ($\lambda_i = 0$, $\theta_j = \infty$) exactly. What happens in practice is that, within the region of interest, the surface is approximated by a piece of some system in which some of the λ's are small and some of the θ's large.

The A canonical form of Eq. (11.2.1) use the center of the design as its origin; we shall call this point the *design origin O*. Obviously, to identify stationary or rising ridges, it is important to know how close the ridge system is to the design origin. We can find this out as follows. As we have seen in Eq. (10.3.10), the coordinates of the center of the system are given by

$$X_{iS} = -\frac{\theta_i}{2\lambda_i}, \qquad i = 1, 2, \ldots, k. \qquad (11.2.2)$$

These coordinates provide important clues to the nature of the system and should be examined together with the θ's and the λ's.

Distance to the Ridge

Consider now a theoretical system in k variables in which there is a p-dimensional ridge evidenced by p zero eigenvalues. Then let the coordinates of the point R on the ridge which is nearest to the design center O have $(k - p)$ coordinates $\{X_{iS}\}$, with the omitted p coordinates corresponding to the zero eigenvalues. The shortest distance OR to the ridge will thus be

$$OR = \left\{\Sigma X_{iS}^2\right\}^{1/2}, \qquad (11.2.3)$$

where the summation is taken only over those coordinates i corresponding to nonzero eigenvalues. For fitted surfaces, eigenvalues will not be exactly zero and a corresponding summation may be taken omitting subscripts whose eigenvalues are small.

For illustration, consider the situation of Figure 11.4. The rising ridge system depicted by the contours would be approximated either by a system of ellipsoids with contours like those of Figure 11.3A centered on a fairly remote point on the $\tilde{X}_3$ axis such as S, or by a system such as that in Figure 11.3B centered remotely at S'. Thus, the approximating system would have λ_1 and λ_2 both comparatively large

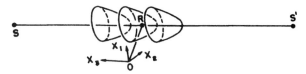

Figure 11.4. A rising ridge may be approximated by ellipsoids centered at S or by hyperboloids centered at S'.

and negative while λ_3 would be close to zero and with a negative [for case (A)] or positive [for case (B)] sign. In either case, the point R on the ridge closest to the design origin O has coordinates $(X_{1S}, X_{2S}, 0)$ and the distance OR is given by $OR = (X_{1S}^2 + X_{2S}^2 + 0^2)^{1/2}$.

Stationarity on the Ridge

For a suspected ridge system with p small eigenvalues $\lambda_{k-p-1}, \ldots, \lambda_k$, a point R having coordinates $(X_{1S}, \ldots, X_{k-p, S}, 0, \ldots, 0)$ will be chosen as being "closest to the ridge." If we now move a distance of one unit in either direction away from R along the ith of the p "small eigenvalue" axes to points R_{i1} and R_{i2}, we can write the three calculated responses in the form

$$
\begin{array}{ll}
\hat{y}_R & \text{at } R, \\
\hat{y}_R + \theta_i + \lambda_i & \text{at } R_{i1}, \\
\hat{y}_R - \theta_i + \lambda_i & \text{at } R_{i2}.
\end{array}
\qquad (11.2.4)
$$

The average of these three responses is $\hat{y}_R + \frac{2}{3}\lambda_i$. The sample variance of these three responses is $\theta_i^2 + \frac{1}{3}\lambda_i^2$. Now, if we have a sample of three observations from a normal distribution, the standard deviation* of the distribution may be approximated by (sample range)$/3^{1/2}$, so that, for the three responses above, the sample range is approximated by

$$
r_i = \left\{ 3\theta_i^2 + \lambda_i^2 \right\}^{1/2}
\qquad (11.2.5)
$$

and this will provide a measure of how much change occurs, in the calculated response, along "near zero eigenvalue" axes. [The reader may wonder why we use (11.2.5) instead of obtaining the actual range of the three responses. This is because we can then avoid working out each of the numbers (11.2.4) to see which is the smallest and largest for each i, in favor of a routine calculation r_i which is of adequate accuracy in practice.]

Summary of Results

It is now possible to summarize the results of the canonical analysis in a table as follows.

Slopes	θ_1	θ_2	$\cdots$	θ_k
Curvatures	λ_1	λ_2	$\cdots$	λ_k
Distances from O of system center	X_{1S}	X_{2S}	$\cdots$	X_{kS}
Changes in $\hat{y}$	r_1	r_2	$\cdots$	r_k

Approximate Standard Errors for θ_i and λ_i

In the rotated axes coordinate system $(X_1, X_2, \ldots, X_k)$, the θ_i represent linear effects and the λ_i are quadratic effects. The second-order designs we use typically have the property (discussed in Section 14.3) of either exact or approximate

---

*For a sample of size two, the formula is (sample range)$/2^{1/2}$ and is *exact*, whether the distribution is normal or not. For a sample of size three, the rule is not very sensitive to the normality assumption.

rotatability. Under these circumstances, $V(\theta_i) = V(b_i)$ and $V(\lambda_i) = V(b_{ii})$, approximately. We have used these results to obtain the "standard errors" for θ_i and λ_i in the examples discussed below. [Also see Bisgaard and Ankenman (1996).]

Example 11.1. Constructed data, $k = 2$. The constructed data analyzed in Section 10.3 gave rise to the fitted equation $\hat{y} = 78.8988 + 2.272x_1 + 3.496x_2 - 2.08x_1^2 - 2.92x_2^2 - 2.88x_1x_2$. We thus find the following.

Description		Subscript		"Standard Error"
		1	2	
Slopes	θ	4.2	0.3	0.3
Curvatures	λ	−4.0	−1.0	0.4
Distances	X_S	0.5	0.1	
Changes, $\hat{y}$	r	8.3	1.1	

The λ's differ significantly from zero, and their sizes show there is more response change in the X_1 (or $\tilde{X}_1$) direction than in X_2 (or $\tilde{X}_2$). We have a point maximum at the center of a set of elliptical contours and we are close to the center of the system. We can take $R = S$ and $OR = OS = [0.5^2 + (-0.1)^2]^{1/2} \doteq 0.5$.

Example 11.2. Polymer elasticity data. (See Section 10.3.) The fitted equation was $\hat{y} = 57.31 + 1.50x_1 - 2.13x_2 + 1.81x_3 - 4.69x_1^2 - 6.27x_2^2 - 5.21x_3^2 - 7.13x_1x_2 - 3.27x_1x_3 - 2.73x_2x_3$.

Description		1	2	3	"Standard Error"
Slopes	θ	0.1	−2.3	2.2	0.5
Curvatures	λ	−10.0	−4.4	−1.8	0.5
Distances	X_S	0.0	−0.3	0.6	
Changes, $\hat{y}$	r	10.0	5.9	4.2	

Although the λ's are considerably different in size, they all apparently differ significantly from zero, so changes in $\hat{y}$ are substantial in all three directions X_1, X_2, X_3. We have a point maximum like Figure 11.3(A), and we are close to the center of the system S. For this situation $R = S$, and $OR = OS = (0^2 + (-0.3^2) + 0.6^2)^{1/2} \doteq 0.7$.

We next consider in detail some other examples which provide interesting ridge systems.

11.3. EXAMPLE 11.3: A CONSECUTIVE CHEMICAL SYSTEM YIELDING A NEAR-STATIONARY PLANAR RIDGE MAXIMUM

Table 11.2 shows data from an investigation of a consecutive chemical system. The products from a batch reaction conducted in an autoclave were analyzed at various conditions of temperature (T), concentration of catalyst (C), and reaction time (t).

Table 11.2. Experimental Data from 19 Runs on a Consecutive Chemical System

Run Number	Levels of Predictor Variables $T(°C)$	$C(\%)$	$t(h)$	First Coding $\dot{x}_1$	$\dot{x}_2$	$\dot{x}_3$	Second Coding x_1	x_2	x_3	Observed Percentage Yield, y
1	162	25	5	-1	-1	-1	-1	-1	-1	45.9
2	162	25	8	-1	-1	1	-1	-1	1	53.3
3	162	30	5	-1	1	-1	-1	1	-1	57.5
4	162	30	8	-1	1	1	-1	1	1	58.8
5	172	25	5	1	-1	-1	1	-1	-1	60.6
6	172	25	8	1	-1	1	1	-1	1	58.0
7	172	30	5	1	1	-1	1	1	-1	58.6
8	172	30	8	1	1	1	1	1	1	52.4
9	167	27.5	6.5	0	0	0	0.01	0.05	0.12	56.9
10	177	27.5	6.5	2	0	0	1.97	0.05	0.12	55.4
11	157	27.5	6.5	-2	0	0	-2.03	0.05	0.12	46.9
12	167	32.5	6.5	0	2	0	0.01	1.87	0.12	57.5
13	167	22.5	6.5	0	-2	0	0.01	-2.16	0.12	55.0
14	167	27.5	9.5	0	0	2	0.01	0.05	1.73	58.9
15	167	27.5	3.5	0	0	-2	0.01	0.05	-2.52	50.3
16	177	20	6.5	2	-3	0	1.97	-3.47	0.12	61.1
17	177	20	6.5	2	-3	0	1.97	-3.47	0.12	62.9
18	160	34	7.5	-1.4	2.6	0.7	-1.41	2.36	0.72	60.0
19	160	34	7.5	-1.4	2.6	0.7	-1.41	2.36	0.72	60.6

The initial planning and analysis were conducted in the coded variables

$$\dot{x}_1 = \frac{(T - 167)}{5}, \qquad \dot{x}_2 = \frac{(C - 27.5)}{2.5}, \qquad \dot{x}_3 = \frac{(t - 6.5)}{1.5}, \quad (11.3.1)$$

whose values will be found in the table under the heading "first coding."

It will be seen that the first 15 runs recorded in the table were made in accordance with a central composite design in the "first coding" units. The four additional runs were then made for confirmation purposes after a preliminary analysis of the initial findings. Subsequently, consideration of the *likely* mechanism underlying the process indicated (Box and Youle, 1955) that it would be more advantageous to work with other predictor variables, namely the reciprocal of the absolute temperature $(T + 273)$, the logarithm of the reaction time t, and the function $f(C)$ of concentration, where

$$f(C) = \ln\left[\frac{C - 4}{\ln(2C - 4) - \ln C}\right]. \quad (11.3.2)$$

If, for our convenience we recode these variables in such a way that the upper and lower levels of the factorial design remain at -1 and $+1$, we obtain the coding

$$x_1 = 88 - \frac{38,715}{(T + 273)},$$

$$x_2 = -38.2056 + 10.5124f(C), \quad (11.3.3)$$

$$x_3 = -7.84868 + 4.25532\ln t.$$

The fitted second-degree equation is then

$$\hat{y} = 59.16 + 2.01x_1 + 1.00x_2 + 0.67x_3 - 2.01x_1^2 - 0.72x_2^2$$
$$- 1.01x_3^2 - 2.79x_1x_2 - 2.18x_1x_3 - 1.16x_2x_3. \qquad (11.3.4)$$

Canonical Analysis

For this example, we find that $b_0 = 59.16$,

$$\mathbf{b} = \begin{bmatrix} 2.01 \\ 1.00 \\ 0.67 \end{bmatrix}, \quad \mathbf{B} = -\begin{bmatrix} 2.01 & 1.39 & 1.09 \\ 1.39 & 0.72 & 0.58 \\ 1.09 & 0.58 & 1.01 \end{bmatrix} \qquad (11.3.5)$$

and solution of the equations $2\mathbf{B}\mathbf{x}_S = -\mathbf{b}$ gives

$$\mathbf{x}_S = (-0.05, 0.90, -0.13)', \qquad (11.3.6)$$

which is close to the center $(0, 0, 0)$ of the experimental region. At $\mathbf{x}_S$, the fitted value is

$$\hat{y}_S = b_0 + \tfrac{1}{2}\mathbf{x}_S'\mathbf{b} = 59.52. \qquad (11.3.7)$$

The eigenvalues in order of absolute magnitude, and their associated eigenvectors are as follows:

$$\lambda_1 = -3.51, \quad \mathbf{m}_1 = (0.762, 0.473, 0.443)',$$
$$\lambda_2 = -0.42, \quad \mathbf{m}_2 = (0.291, 0.361, -0.886)', \qquad (11.3.8)$$
$$\lambda_3 = 0.18, \quad \mathbf{m}_3 = (0.579, -0.804, -0.138)'.$$

Thus the canonical form is

$$\hat{y} = 59.52 - 3.51\tilde{X}_1^2 - 0.42\tilde{X}_2^2 + 0.18\tilde{X}_3^2, \qquad (11.3.9)$$

where

$$\tilde{X}_1 = 0.762x_1 + 0.473x_2 + 0.443x_3 - 0.330,$$
$$\tilde{X}_2 = 0.291x_1 + 0.361x_2 - 0.886x_3 - 0.426, \qquad (11.3.10)$$
$$\tilde{X}_3 = 0.579x_1 - 0.804x_2 - 0.138x_3 + 0.731.$$

We see that the eigenvalues λ_2 and λ_3 are both small in absolute magnitude compared with λ_1. Moreover, the standard errors of the λ's will be, very roughly, of the same magnitude as those of the quadratic coefficients in the fitted equation. These latter are about 0.4. Thus the fitted surface can be realistically approximated by the limiting system

$$\hat{y} = \hat{y}_S + \lambda_1\tilde{X}_1^2, \qquad (11.3.11)$$

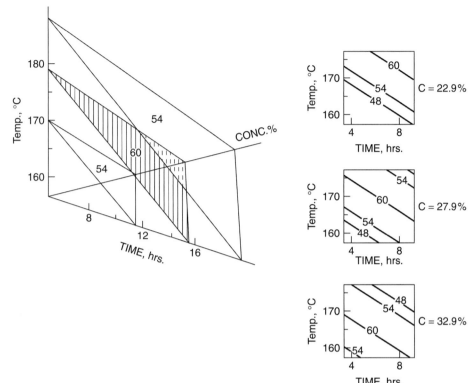

Figure 11.5. Contours of empirical yield surface, with sections at three levels of concentration.

where, approximately, $\hat{y}_S = 60$, $\lambda_1 = -3.51$, and $\tilde{X}_1 = 0.76x_1 + 0.47x_2 + 0.44x_3 - 0.33$. The contours for such a stationary plane ridge system are shown in Figure 11.5.

The associated tabular display is as follows.

Description		1	2	3	"Standard Error"
Slopes	θ	2.3	0.3	0.3	0.3
Curvatures	λ	-3.5	-0.4	0.2	0.4
Distances	X_S	0.3	0.4	-0.7	
Changes, $\hat{y}$	r	5.3	0.7	0.6	

Only λ_1 is substantially and significantly different from zero. The values of θ and λ, taken together with those of r, indicate that very little is happening in any direction except X_1. We have a ($p = 2$) two-dimensional ridge like Figure 11.3 F. The distance OR is quite short, namely, $OR = (0.3^2)^{1/2} = 0.3$. See Eq. (11.2.3) on page 346.

Direct Fitting of the Canonical Form

When originally performed (Box, 1954), this analysis led to the realization that kinetic theory could readily produce a surface which would be approximated by a stationary ridge of this type. We discuss this development later. For the moment, it is sufficient to say that some definite scientific interest is associated with a model of the form of Equation (11.3.11). As a result, this type of canonical form was fitted directly by maximum likelihood. The model, which is now nonlinear in the parameters, can be written

$$y = \gamma_0 - \lambda(\alpha_{11}x_1 + \alpha_{12}x_2 + \alpha_{13}x_3 + \alpha_{10})^2 + \varepsilon \qquad (11.3.12a)$$

with the restriction $\alpha_{11}^2 + \alpha_{12}^2 + \alpha_{13}^2 = 1$. For fitting purposes it is preferable to cast the model in unrestricted form by writing $\gamma_{1j} = \alpha_{1j}\lambda^{1/2}$, $j = 0, 1, 2, 3$, whereupon the model becomes

$$y = \gamma_0 - (\gamma_{11}x_1 + \gamma_{12}x_2 + \gamma_{13}x_3 + \gamma_{10})^2 + \varepsilon \qquad (11.3.12b)$$

and has five independent parameters γ_0, γ_{11}, γ_{12}, γ_{13}, and γ_{10}. On the assumption that the errors ε are normally and independently distributed with constant variance, maximum likelihood estimates of the parameters that appear nonlinearly in Eq. (11.3.12b) are obtained by application of iterative nonlinear least squares, briefly described in Appendix 3D. The resulting fitted equation, recast in the form of (11.3.11), is

$$\hat{y} = 59.93 - 3.76X^2,$$

where

$$X = 0.780x_1 + 0.426x_2 + 0.458x_3 - 0.320. \qquad (11.3.13)$$

The constants in this equation are all readily interpreted; $\hat{y}_S = 59.93$ is the estimated response on the fitted maximal plane, while the constants 0.780, 0.426, and 0.458 are the direction cosines of a line perpendicular to this plane. The value 0.320 measures the distance in design units of the nearest point on the fitted plane to the design origin. The quantity $\lambda_1 = -3.76$ measures the quadratic fall-off in a direction perpendicular to the fitted maximal plane. The residual sum of squares after fitting the canonical form (11.3.13) is 45.2. There are five adjustable coefficients in the new equation, and thus an approximate analysis of variance (approximate because the canonical model is nonlinear in the parameters) may now be obtained as in Table 11.3.

The analysis shows that the canonical form, with only five adjustable constants, explains the data almost as well as does the full quadratic equation with 10 adjustable constants. (Remember that the table entries for df are one fewer in each case, due to elimination of the mean.)

Exploitation of the Canonical Form

When this investigation was completed, great practical interest was associated with the fact that, for this system, there was not a single-point maximum but, locally at least, a plane of near-maxima. This allowed considerable choice of operating conditions (Box, 1954). In particular (see Figure 11.6), it allowed conditions to be chosen so that two other criteria (the level of a certain impurity, and cost) attained their best levels.

Table 11.3. Approximate Analysis of Variance Table for the Fitted Canonical Form Compared with the Full Second-Degree Fit, Section 11.3

Sources		SS	df	MS	F
Full second-degree equation after eliminating mean	Canonical form after eliminating mean	355.7	4	88.92	27.53
		380.2	9	42.24	
	Remainder Residual from nonlinear model	24.5	5	4.90	
			45.2	14	3.23
Residual from full second-degree model		20.7	9	2.30	
Total after eliminating mean		400.9	18		

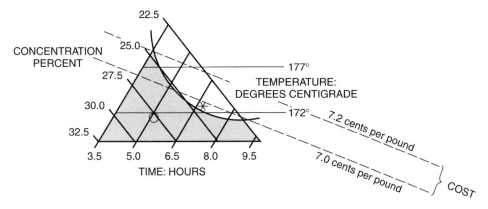

Figure 11.6. Alternative process giving near-maximum yield, with cost contours superimposed. In the shaded region, the by product is unacceptably high. The asterisk indicates the process chosen for future operation.

11.4. EXAMPLE 11.4: A SMALL REACTOR STUDY YIELDING A RISING RIDGE SURFACE

Table 11.4 shows data for an experiment performed sequentially in four blocks. (The blocking sequence has some interesting design aspects that are discussed more fully in Section 15.3.) The series of runs was part of a pilot plant study on a small continuous reactor in which the predictor variables were

r = flow rate in liters per hour,
c = concentration of catalyst,
T = temperature,

Table 11.4. Data for Small Reactor Study (Example 11.4)

Block	Run	x_1	x_2	x_3	Concentration of Product, y	Block Total, T_i
1	1	-1	-1	1	40.0	
	2	1	-1	-1	18.6	
	3	-1	1	-1	53.8	
	4	1	1	1	64.2	
	5	0	0	0	53.5	
	6	0	0	0	52.7	282.8
2	7	-1	-1	-1	39.5	
	8	1	-1	1	59.7	
	9	-1	1	1	42.2	
	10	1	1	-1	33.6	
	11	0	0	0	54.1	
	12	0	0	0	51.0	280.1
3	13	$-\sqrt{2}$	0	0	43.0	
	14	$\sqrt{2}$	0	0	43.9	
	15	0	$-\sqrt{2}$	0	47.0	
	16	0	$\sqrt{2}$	0	62.8	
	17	0	0	$-\sqrt{2}$	25.6	
	18	0	0	$\sqrt{2}$	49.7	272.0
4	19	$-\sqrt{2}$	0	0	39.2	
	20	$\sqrt{2}$	0	0	46.3	
	21	0	$-\sqrt{2}$	0	44.9	
	22	0	$\sqrt{2}$	0	58.1	
	23	0	0	$-\sqrt{2}$	27.0	
	24	0	0	$\sqrt{2}$	50.7	266.2

$$1101.1$$
$$(\Sigma y_i)^2/n = 50{,}517.55$$

and the response variable was $y =$ concentration of product. The runs were randomized within blocks, but are shown in a standard order in Table 11.4.

The stated objective was to find, and explore, conditions giving higher concentrations of product. (This part of the investigation was undertaken after application of the steepest ascent method had resulted in some progress toward higher values of y, but had also indicated the need for a more detailed examination of the response surface.) Flow rate and catalyst concentration were varied on a log scale, and temperature on a linear scale in this design; the actual coding was as follows:

$$x_1 = \frac{(\ln r - \ln 3)}{\ln 2}, \qquad x_2 = \frac{(\ln c - \ln 2)}{\ln 2}, \qquad x_3 = \frac{(T - 80)}{10}. \quad (11.4.1)$$

The choice of relative ranges for the variables was based on the chemist's guess that a halving of the flow rate, a doubling of the concentration of catalyst, and a

Table 11.5. Analysis of Variance Table for Second-Degree Fit with Blocks, Example 11.4

Source	SS		df		MS		F
First-degree equation	1,406.8 ⎫		3 ⎫				
		3004.2		9	333.8		94.03
Extra from second-degree equation $\mid b_0$	1,597.4 ⎭		6 ⎭				
Blocks	28.8		3				
Lack of fit	11.8 ⎫		3 ⎫		3.93 ⎫		
		39.0		11		3.55	
Error	27.2 ⎭		8 ⎭		3.40 ⎭		
Total (after eliminating mean)	3,072.0		23				
Mean (b_0)	50,517.6		1				
Total	50,889.6		24				

10-degree increase in temperature would each individually about double the reaction rate.

The second-degree equation fitted to the 24 runs by least squares was

$$\hat{y} = 51.796 + 0.745x_1 + 4.813x_2 + 8.013x_3 - 3.833x_1^2$$

$$+ 1.217x_2^2 - 6.258x_3^2 + 0.375x_1x_2 + 10.350x_1x_3 - 2.825x_2x_3. \quad (11.4.2)$$

The resulting analysis of variance is given in Table 11.5. For additional detail, see Appendix 11A. For a plot of the surface, see Figure 11.7. See also Figure 12.1a.

No evidence of lack of fit appears. Furthermore, the observed F ratio of 94.03 with 9 and 11 df is over 32 times the percentage point (2.90) needed for significance at the $\alpha = 0.05$ test level. Thus, on the basis of the Box–Wetz criterion discussed in Section 8.2, the equation merits further interpretation.

Canonical Analysis

For this example, $b_0 = 51.796$,

$$\mathbf{b} = \begin{bmatrix} 0.7446 \\ 4.8133 \\ 8.0125 \end{bmatrix}, \quad \mathbf{B} = \begin{bmatrix} -3.8333 & 0.1875 & 5.1750 \\ 0.1875 & 1.2167 & -1.4125 \\ 5.1750 & -1.4125 & -6.2583 \end{bmatrix}, \quad (11.4.3)$$

and the solutions of the equations $2\mathbf{Bx}_S = -\mathbf{b}$ and $\hat{y}_S = b_0 + \frac{1}{2}\mathbf{x}'_S\mathbf{b}$ give the coordinates of the stationary point and the corresponding response as

$$\mathbf{x}_S = (25.78, 15.48, 18.46)', \quad \hat{y}_S = 172.61. \quad (11.4.4)$$

Obviously the fitted equation can have no relevance at this very remote stationary point. Nevertheless its location is of importance as a "construction point" in the analysis that follows. The eigenvalues in order of absolute magnitude, and the

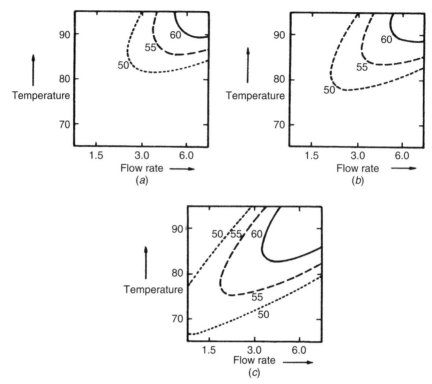

Figure 11.7. Small reactor study. Sections of fitted response surface. (a) Slice of fitted surface at $x_2 = -1$ (i.e., catalyst concentration 1 mole percent). (b) Slice of fitted surface at $x_2 = 0$ (i.e., catalyst concentration 2 mole percent). (c) Slice of fitted surface at $x_2 = 1$ (i.e., catalyst concentration 4 mole percent).

associated eigenvectors are as follows:

$$\lambda_1 = -10.4893, \qquad \mathbf{m}_1 = (-0.6123, 0.1044, 0.7837)',$$

$$\lambda_2 = 1.7109, \qquad \mathbf{m}_2 = (-0.2969, 0.8884, -0.3502)', \qquad (11.4.5)$$

$$\lambda_3 = -0.0965, \qquad \mathbf{m}_3 = (0.7328, 0.4471, 0.5129)'.$$

The B canonical form is thus

$$\hat{y} = 172.61 - 10.4893\tilde{X}_1^2 + 1.7109\tilde{X}_2^2 - 0.0965\tilde{X}_3^2, \qquad (11.4.6)$$

where, from equations $\tilde{\mathbf{X}} = \mathbf{M}'\tilde{\mathbf{x}} = \mathbf{M}'(\mathbf{x} - \mathbf{x}_S)$, we have

$$\tilde{X}_1 = -0.6123x_1 + 0.1044x_2 + 0.7837x_3 - 0.3015,$$

$$\tilde{X}_2 = -0.2969x_1 + 0.8884x_2 - 0.3502x_3 + 0.3649, \qquad (11.4.7)$$

$$\tilde{X}_3 = 0.7328x_1 + 0.4471x_2 + 0.5129x_3 - 35.2796.$$

This canonical form is (as we have seen) centered in a region very remote from the one in which experiments were actually conducted, and of course the response equation has no relevance in that remote region. What we now must do is to determine what the canonical equation tells us about the estimated response *in the region in which it is relevant*—that is, primarily, the region where the experiments themselves were conducted centered at $(x_1, x_2, x_3) = (0, 0, 0)$. We do this by referring the canonical form to a new origin R situated as close as possible to the design origin. Setting x_1, x_2, and $x_3 = 0$ in Eqs. (11.4.7) we find that the design origin has $\tilde{X}$ coordinates

$$\tilde{X}_1 = -0.3015, \qquad \tilde{X}_2 = 0.3649, \qquad \tilde{X}_3 = -35.2796. \qquad (11.4.8)$$

[Note that this implies that $(X_{1S}, X_{2S}, X_{3S}) = (0.3015, -0.3649, 35.2796)$.] Consider the plane $\tilde{X}_1 = 0$ associated with the first eigenvalue $\lambda_1 = -10.49$. The point R on this plane and closest to the design origin is that for which $\tilde{X}_1 = 0$, $\tilde{X}_2 = 0.3649$, $\tilde{X}_3 = -35.2796$. If coordinates referred to this new local region are denoted by dotted X's, then $\tilde{X}_2 = \dot{X}_2 + 0.3649$, $\tilde{X}_3 = \dot{X}_3 - 35.2796$. [Because $\dot{X}_1 = \tilde{X}_1$ we do not use $\dot{X}_1$.] Substitution in the canonical equation gives

$$\hat{y} = 172.61 - 10.4893\tilde{X}_1^2 + 1.7109\left(\dot{X}_2 + 0.3649\right)^2 - 0.0965\left(\dot{X}_3 - 35.2796\right)^2. \qquad (11.4.9)$$

After expansion, along with dropping the squared terms in $\dot{X}_2$ and $\dot{X}_3$ on the grounds that they have relatively small local influence compared with the squared term in $\tilde{X}_1$, we obtain the approximating form which roughly describes the behavior of the fitted function in the neighborhood where it is relevant as

$$\hat{y} = 57.71 - 10.49\tilde{X}_1^2 + 1.25\dot{X}_2 + 6.53\dot{X}_3. \qquad (11.4.10)$$

In this approximating canonical form, the quadratic effect is entirely confined to $\tilde{X}_1$ and the coefficients of $\dot{X}_2$ and $\dot{X}_3$ indicate the orientation of the straight line contours of the rising ridge on the plane $\tilde{X}_1 = 0$.

Now $\dot{X}_2 = \mathbf{m}_2'\mathbf{x}$ and $\dot{X}_3 = \mathbf{m}_3'\mathbf{x}$. Thus the first-order part of the model is

$$1.25\dot{X}_2 + 6.53\dot{X}_3 = 4.41x_1 + 4.03x_2 + 2.91x_3$$

$$= 6.645(0.664x_1 + 0.606x_2 + 0.438x_3)$$

$$= 6.645X, \qquad (11.4.11)$$

where, two lines up, the coefficients of x_1, x_2, and x_3 have been normalized so that their sum of squares is unity. Equation (11.4.10) is therefore equivalent to

$$\hat{y} = 57.71 - 10.49\tilde{X}_1^2 + 6.645X. \qquad (11.4.12)$$

In tabular form, we can display:

Description		1	2	3	"Standard Error"
Slopes	θ	6.3	1.3	6.8	0.47
Curvatures	λ	-10.5	1.7	-0.1	0.55
Distances	X_S	0.3	-0.4	35.3	
Changes, $\hat{y}$	r	15.1	2.8	11.8	

The eigenvalue λ_1 is dominant and there are considerable response changes in the X_1 direction (where the surface is mostly quadratic) and in the X_3 direction (mostly linear); not much is happening in the X_2 direction. We thus have a rising ridge of the type shown in Figure 11.3G.

Direct Fitting of the Canonical Form

In this example, as in the previous one, it is possible to derive a simple kinetic relationship that would produce a rising ridge of the sort found here. We discuss in Appendix 3D how the nonlinear mechanistic model may be fitted directly to the data. For the moment, we consider the result of fitting, by maximum likelihood, a reduced model of the form of (11.4.12). The postulated model is

$$y = \gamma_0 - \lambda_1(\alpha_{11}x_1 + \alpha_{12}x_2 + \alpha_{13}x_3 + \alpha_{10})^2$$
$$+ \lambda_2(\alpha_{21}x_1 + \alpha_{22}x_2 + \alpha_{23}x_3) + \varepsilon, \qquad (11.4.13)$$

where $\alpha_{11}^2 + \alpha_{12}^2 + \alpha_{13}^2 = 1 = \alpha_{21}^2 + \alpha_{22}^2 + \alpha_{23}^2$. Alternatively, we can write this as a seven-parameter model

$$y = \hat{y}_S - (L_{11}x_1 + L_{12}x_2 + L_{13}x_3)^2 + c_1x_1 + c_2x_2 + c_3x_3 + \varepsilon \quad (11.4.14)$$

and drop the restrictions. If we fit the seven-parameter form and then convert it to the form (11.4.13), we obtain

$$\hat{y} = 54.07 - 10.812\{-0.612x_1 + 0.104x_2 + 0.784x_3 - 0.301\}^2$$
$$+ 6.923(0.683x_1 + 0.597x_2 + 0.420x_3). \qquad (11.4.15)$$

Interpreting this equation, we see that $\hat{y}_{S'} = 54.07$ is the estimated response at S', the point nearest to the origin on the rising planar ridge. The quantities $-0.612, 0.104, 0.784$ are the direction cosines of lines perpendicular to this plane. The size of the constant -0.301 measures the shortest distance of the planar ridge from the design origin. The constant -10.812 measures the rate of quadratic fall-off as we move away from the plane of the ridge. The values 0.683, 0.597, and 0.420 are the direction cosines of the line of steepest ascent up the planar ridge. Finally, 6.923 is the linear rate of increase of yield as we move up the ridge. The analysis of variance table is shown as Table 11.6.

For an approximate test of the remainder sum of squares, we compare the ratio (remainder mean square)/(residual mean square) = 3.64 with the upper 5% point of an $F(3, 11)$ distribution, namely 3.59. We see that it is just significant, which

Table 11.6. Approximate Analysis of Variance Table for Second-Degree and Canonical Form Fitted Equations

Source		SS	df	MS	F
Full second-degree equation after allowance for mean	Canonical form, after allowance for mean	2965.4	6	494.23	
		3004.2	9	333.8	
	Remainder	38.8	3	12.93	3.64
Blocks		28.8	3		
Residual		39.0	11	3.55	
Total, corrected for mean		3072.0	23		

implies that some variation is not explained by the fitted canonical form compared with the full second-order model. However, the fitted canonical form model does pick up a proportion $2965.4/3072 = 0.9653$ of the variation about the mean using only six additional parameters, whereas the full second-order model picks up a proportion $3004.2/3072 = 0.9779$, using nine additional parameters, so that, from a practical point of view, the canonical form does provide a worthwhile and parsimonious representation.

Interpretation

In this particular investigation, the maximum temperature attainable with the apparatus used was 100°C, and the highest flow rate possible was 12 L/h. Further experimentation and study showed, as might be expected from this analysis, that a product concentration of about 75 was attainable with the temperature just below 100°C, with a high catalyst concentration of about 5%, and with the maximum permissible flow rate. The most important fact revealed by this response surface study was that even higher concentrations of product were likely to be produced at temperatures higher than 100°C. Further experimentation was therefore conducted in which the reaction was carried out under pressure in an autoclave where temperatures above 100°C were possible.

11.5. EXAMPLE 11.5: A STATIONARY RIDGE IN FIVE VARIABLES

To assist geometric understanding, we have so far illustrated canonical analysis for examples with only two or three predictor variables. The technique becomes even more valuable when there are more than three predictors, for it enables the basic situation to be understood, even though it cannot be easily visualized geometrically. In this next example (Box, 1954), we shall see a two-dimensional ridge system embedded in the five-dimensional space of five predictor variables.

The chemical process under study had two stages, and the five factors considered were the temperatures (T_1, T_2) and the times (t_1, t_2) of reaction at the two stages, along with the concentration (C) of one of the reactants at the first stage. Evidence was available which suggested that the times of reaction were best varied on a logarithmic scale, and it was known that the second reaction time needed to be varied over a wide range. A preliminary application of the steepest ascent procedure had brought the experimenter close to a near-stationary region, and a second-order model was now to be fitted and examined. The coding used in the experiments was

$$x_1 = \frac{(T_1 - 122.5)}{7.5},$$

$$x_2 = \left\{ \frac{2(\log t_1 - \log 5)}{\log 2} \right\} + 1,$$

$$x_3 = \frac{(C - 70)}{10},$$ (11.5.1)

$$x_4 = \frac{(T_2 - 32.5)}{7.5},$$

$$x_5 = \left\{ \frac{2(\log t_2 - \log 1.25)}{\log 5} \right\} + 1.$$

Thirty-two observations of percentage yield were available in all, and these are given in Table 11.7. A second-degree equation fitted to the 32 observed responses produced the estimated coefficients shown (rounded to three places of decimals):

$$b_0 = 68.717, \qquad \mathbf{b} = \begin{bmatrix} 3.258 \\ 1.582 \\ 1.161 \\ 3.474 \\ 1.488 \end{bmatrix},$$

$$\mathbf{B} = \begin{bmatrix} -1.610 & -0.952 & 1.051 & -0.176 & -0.022 \\ & -1.354 & 0.299 & -0.084 & -0.549 \\ & & -2.584 & -1.775 & -0.376 \\ & & & -2.335 & 0.201 \\ \text{symmetric} & & & & -1.421 \end{bmatrix}.$$ (11.5.2)

The standard computations led to the canonical form

$$\hat{y} = 72.51 - 4.46\tilde{X}_1^2 - 2.62\tilde{X}_2^2 - 1.78\tilde{X}_3^2 - 0.40\tilde{X}_4^2 - 0.04\tilde{X}_5^2, \quad (11.5.3)$$

where $\mathbf{x}_S = (2.52, -1.10, 1.27, -0.32, 0.53)'$, and

$$\tilde{\mathbf{X}} = \begin{bmatrix} -0.284 & -0.138 & 0.743 & 0.589 & 0.026 \\ -0.639 & -0.599 & -0.003 & -0.434 & -0.213 \\ -0.254 & 0.255 & 0.255 & -0.383 & 0.820 \\ -0.369 & 0.730 & 0.188 & -0.222 & -0.496 \\ 0.558 & -0.157 & 0.601 & -0.518 & -0.185 \end{bmatrix} \mathbf{x} + \begin{bmatrix} -0.203 \\ -0.925 \\ 0.077 \\ -1.684 \\ 2.408 \end{bmatrix}.$$

(11.5.4)

Table 11.7. Thirty-Two Observations for Example 11.5

Run Number	x_1	x_2	x_3	x_4	x_5	Response, y
1	−1	−1	−1	−1	1	49.8
2	1	−1	−1	−1	−1	51.2
3	−1	1	−1	−1	−1	50.4
4	1	1	−1	−1	1	52.4
5	−1	−1	1	−1	−1	49.2
6	1	−1	1	−1	1	67.1
7	−1	1	1	−1	1	59.6
8	1	1	1	−1	−1	67.9
9	−1	−1	−1	1	−1	59.3
10	1	−1	−1	1	1	70.4
11	−1	1	−1	1	1	69.6
12	1	1	−1	1	−1	64.0
13	−1	−1	1	1	1	53.1
14	1	−1	1	1	−1	63.2
15	−1	1	1	1	−1	58.4
16	1	1	1	1	1	64.3
17	3	−1	−1	1	1	63.0
18	1	−3	−1	1	1	63.8
19	1	−1	−3	1	1	53.5
20	1	−1	−1	3	1	66.8
21	1	−1	−1	1	3	67.4
22	1.23	−0.56	−0.03	0.69	0.70	72.3
23	0.77	−0.82	1.48	1.88	0.77	57.1
24	1.69	−0.30	−1.55	−0.50	0.62	53.4
25	2.53	0.64	−0.10	1.51	1.12	62.3
26	−0.08	−1.75	0.04	−0.13	0.27	61.3
27	0.78	−0.06	0.47	−0.12	2.32	64.8
28	1.68	−1.06	−0.54	1.50	−0.93	63.4
29	2.08	−2.05	−0.32	1.00	1.63	72.5
30	0.38	0.93	0.25	0.38	−0.24	72.0
31	0.15	−0.38	−1.20	1.76	1.24	70.4
32	2.30	−0.74	1.13	−0.38	0.15	71.8

(or $\tilde{\mathbf{X}} = \mathbf{M}'\tilde{\mathbf{x}}$ where $\mathbf{M}'$ is the 5×5 matrix shown). The center of the system lies within the immediate region of experimentation. Also, the canonical variables $\tilde{X}_4$ and $\tilde{X}_5$ play a comparatively minor role in describing the function, their coefficients λ_4 and λ_5 being of the same order of magnitude as their standard errors. At least locally, most of the change in response can be well described by the canonical variables $\tilde{X}_1$, $\tilde{X}_2$, and $\tilde{X}_3$. Thus, the system in five variables has, approximately, a two-dimensional maximum. The sets of conditions that yield, approximately, the same maximum response result of about 72.51 are those in the plane defined by the equations $\tilde{X}_1 = 0$, $\tilde{X}_2 = 0$, and $\tilde{X}_3 = 0$. These three equations are readily transformed to three equations in the five unknowns x_1, x_2, x_3, x_4, and x_5 and thence into three equations in the five natural (uncoded) variables T_1, T_2, t_1, t_2, and C. We have 2 df in the choice of conditions in the sense that, within the local region considered, if we choose values for any two of the predictor variables, we

Table 11.8. Sets of Conditions that Produce, Approximately, the Same Maximum Response Values of About 72.51, for Example 11.5. All these sets of conditions lie on the plane $\bar{X}_1 = \bar{X}_2 = \bar{X}_3 = 0$ in the five-dimensional space of the predictor variables. Rough contours of $t_1 + t_2$, the total reaction time, are sketched over the figures.

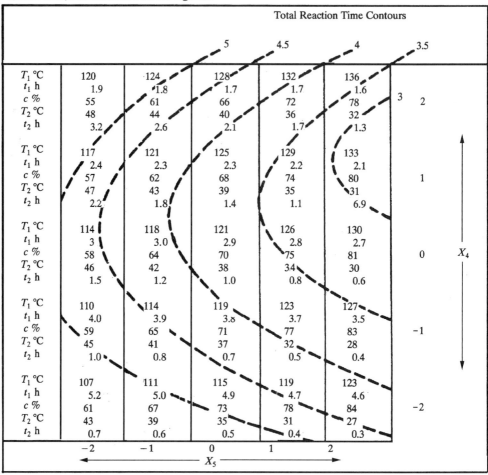

Total Reaction Time Contours

	-2	-1	0	1	2	
T_1 °C	120	124	128	132	136	3 2
t_1 h	1.9	1.8	1.7	1.7	1.6	
c %	55	61	66	72	78	
T_2 °C	48	44	40	36	32	
t_2 h	3.2	2.6	2.1	1.7	1.3	
T_1 °C	117	121	125	129	133	1
t_1 h	2.4	2.3	2.3	2.2	2.1	
c %	57	62	68	74	80	
T_2 °C	47	43	39	35	31	
t_2 h	2.2	1.8	1.4	1.1	6.9	
T_1 °C	114	118	121	126	130	0 X_4
t_1 h	3	3.0	2.9	2.8	2.7	
c %	58	64	70	75	81	
T_2 °C	46	42	38	34	30	
t_2 h	1.5	1.2	1.0	0.8	0.6	
T_1 °C	110	114	119	123	127	-1
t_1 h	4.0	3.9	3.8	3.7	3.5	
c %	59	65	71	77	83	
T_2 °C	45	41	37	32	28	
t_2 h	1.0	0.8	0.7	0.5	0.4	
T_1 °C	107	111	115	119	123	-2
t_1 h	5.2	5.0	4.9	4.7	4.6	
c %	61	67	73	78	84	
T_2 °C	43	39	35	31	27	
t_2 h	0.7	0.6	0.5	0.4	0.3	
	-2	-1	0	1	2	

X_5

Contour labels (top): 5, 4.5, 4, 3.5

obtain three equations in three unknowns that can be solved to give appropriate levels for the remaining three predictor variables. To demonstrate the practical implications to the experimenter of these findings, a table of approximate alternatives, which covered the ranges of interest, was calculated as shown in Table 11.8. Sketched over the table are rough contours of total reaction time, $t_1 + t_2$.

It must be understood that any analysis of this kind is, almost invariably, a tentative one, subject to practical confirmation. Estimation errors, and biases arising from the inability of the second degree equation adequately to represent the system will almost guarantee that, in practice, the findings are not very precise. If the tentative results look at all useful and interesting, therefore, confirmatory runs are essential. In particular, these should be made near conditions likely to be of economic importance. In the present example, the most intriguing possibility was that of achieving the maximum yield of about 73% *at high throughput.* The conditions at the top right-hand corner of Table 11.8 suggest that, with suitable levels for the concentration and the temperatures, a near-maximum yield can be obtained with a total reaction time of only 1.6 + 1.3 = 2.9 h. This would allow almost double the throughput possible with the conditions at the bottom left-hand corner of Table 11.8, where the total reaction time is 5.2 + 0.7 = 5.9 h. Subsequent experimental work, therefore, was directed to investigating the possibilities of the high-throughput process.

11.6. THE ECONOMIC IMPORTANCE OF THE DIMENSIONALITY OF MAXIMA AND MINIMA

The example in the foregoing section illustrates the possible great economic value of the study of the "degrees of freedom" in maxima and minima. Three general facts of importance to be kept in mind are:

1. The natural parsimony of practical systems guarantees that multidimensional maxima and minima will be often best approximated by subspaces (e.g., lines, planes, or hyperplanes) rather than by points. When we consider chemical or biological systems, for example, it is difficult, if not impossible, to think of experimental variables which behave independently of one another and so yield symmetrical maxima. It is much more likely that the system is controlled essentially by a few (and possibly even only one) functions of the predictor variables, in the neighborhood of the maximum.

2. Classical systems of experimentation such as the one-factor-at-a-time method will normally lead the experimenter close to only one set of "best" conditions on a stationary or rising ridge and will, further, mislead him into believing these conditions are:

   i. The *only* maximal conditions (which they are not when the ridge is stationary).

   ii. Necessarily maximal (which they are not when the ridge is rising).

3. When there are degrees of freedom in the maximum, they can almost always be profitably exploited (e.g., to give an equally high yield at less cost and/or greater purity).

These three facts have, in turn, these implications:

   a. Even processes that have been thoroughly explored by the one factor at a time method can often be dramatically improved by the application of response surface methods.

   b. An understanding of the geometrical redundancies in the fitted response function can enable the predictor variables to be adjusted to maintain optimal response levels as prices, specifications, and requirements change.

c. Knowledge of the geometry of maxima can sometimes help the experimenter to postulate possible theoretical mechanisms that would describe the system under study more fundamentally. (We discussed this aspect more fully in Chapter 12 of the first edition.)

11.7. A METHOD FOR OBTAINING A DESIRABLE COMBINATION OF SEVERAL RESPONSES

When several response variables $y_1, y_2, \ldots, y_m$ have been represented by fitted equations $\hat{y}_1, \hat{y}_2, \ldots, \hat{y}_m$ based on the same set of coded input variables $x_1, x_2, \ldots, x_k$, the question often arises of where in the x space the best overall set of response values might be obtained.

When there are only two or three important input variables, it is possible to solve this problem by looking at the response contours for the different fitted responses $\hat{y}_1, \hat{y}_2, \ldots, \hat{y}_m$ simultaneously, for example, by overlays of various contour diagrams. When the problem can be expressed as one of maximizing or minimizing one response subject to constraints in the others, linear programming methods can sometimes be used advantageously. A different approach which is sometimes useful if carefully applied, is to introduce an overall criterion of desirability. One interesting variant on this method is due to G. C. Derringer and R. Suich, and is described in "Simultaneous optimization of several response variables," *Journal of Quality Technology*, **12**, 1980, 214–219. Suppose we choose, for each response, levels $A \leq B \leq C$ such that the product is unacceptable if $y < A$ or $y > C$ and such that the desirability d of the product increases between A and B and decreases between B and C. Define

$$d = \begin{cases} \left\{ \dfrac{(\hat{y} - A)}{(B - A)} \right\}^s, & \text{for } A \leq \hat{y} \leq B, \\[2ex] \left\{ \dfrac{(\hat{y} - C)}{(B - C)} \right\}^t, & B \leq \hat{y} \leq C, \end{cases}$$

and $d = 0$ outside (A, C). By choosing the powers s and t in various ways, we can attribute various levels of desirability to various values of $\hat{y}$ as shown in Figure 11.8. This is a two-sided choice with B the most desirable level of $\hat{y}$. A one-sided choice can be made in either of two ways:

1. Let $A = B$ and define $d = 1$ for $\hat{y} \leq B$.

2. Let $B = C$ and define $d = 1$ for $\hat{y} \geq B$. This second choice is shown in Figure 11.9.

If we write $D = (d_1 d_2 d_3 \ldots d_m)^{1/m}$, where d_i is the desirability function for the ith response, $i = 1, 2, \ldots, m$, we have defined an overall desirability function whose value is specified at each point of the x space. The maximum of D can be sought to give the most desirable point in the x space for all the responses simultaneously. Of course any selected combination of the d_i can be defined as the

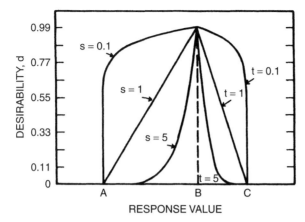

Figure 11.8. Two-sided desirability functions.

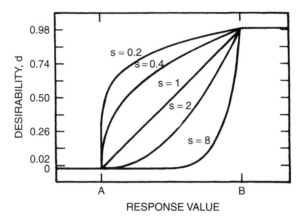

Figure 11.9. One-sided desirability functions.

desirability function; D is simply a specific choice suggested by Derringer and Suich. For examples of this technique, see the quoted reference and also "A statistical methodology for designing elastomer formulations to meet performance specifications," by G. C. Derringer, *Kautschuk + Gummi Kunststoffe*, **36**, 1983, 349–352.

A variety of other work on more than one response can be found in the following references: Allen et al. (1975), Ames et al. (1997), Antony (2001), Box and Draper (1965), Box, et al. (1973), Chang and Shivpuri (1994), Chiao and Hamada (2001), Ch'ng et al. (2005), Copeland and Nelson (1996), Del Castillo (1996), Del Castillo and Montgomery (1993), Del Castillo et al. (1996, 1997, 1999), Ding et al., (2004), Elsayed and Chen (1993), Fan (2000), Khuri and Conlon (1981), Kim and Lin (1998), Lin and Tu (1995), Logothetis and Haigh (1998), Luner (1994), Mays (2001), Miró-Quesado et al. (2004), Myers and Carter (1973), Phadke and Dehnad (1988), Roth and Stewart (1969), Semple (1997), Tang and Xu (2002), Tong et al. (2001–2002), and Vining and Myers (1990).

APPENDIX 11A. CALCULATIONS FOR THE ANALYSIS OF VARIANCE, TABLE 11.5

The "error" sum of squares is comprised of the sums of squares arising from various comparisons, all of which have a zero expectation if the true model is of cubic, or smaller, order. These 8 df arise as follows.

1. There are six comparisons between pairs of runs $(13, 19), (14, 20), \ldots, (18, 24)$. The associated sum of squares is $\frac{1}{2}(43.0 - 39.2)^2 + \frac{1}{2}(43.9 - 46.3)^2 + \cdots + \frac{1}{2}(49.7 - 50.7)^2 = 24.83$. Adjusting this for the sum of squares for the difference between blocks 3 and 4, we obtain an error sum of squares with 5 df of $24.83 - 2.803 = 22.027$.

2. In each of blocks 1 and 2, there is a comparison between the mean of the 2^{3-1} points and the mean of the center points. The difference between these two comparisons $\mathbf{c'y}$, say, has a sum of squares $(\mathbf{c'y})^2/\mathbf{c'c}$, with 1 df. Now $\mathbf{c'} = (\frac{1}{4}, \frac{1}{4}, \frac{1}{4}, \frac{1}{4}, -\frac{1}{2}, -\frac{1}{2}; -\frac{1}{4}, -\frac{1}{4}, -\frac{1}{4}, -\frac{1}{4}, \frac{1}{2}, \frac{1}{2}, \mathbf{0}'; \mathbf{0}')$, $\mathbf{c'y} = -0.15$, $\mathbf{c'c} = 1.5$, so SS $= (-0.15)^2/1.5 = 0.015$.

3. Runs (5.6) and runs $(11, 12)$ provide two error contrasts yielding a sum of squares with 2 df of $\frac{1}{2}(53.5 - 52.7)^2 + \frac{1}{2}(54.1 - 51.0)^2 = 5.125$.

The three portions sum to 27.167, rounded to 27.2 in Table 11.5.

EXERCISES

11.1. (*Source:* The influence of temperature on the scouring of raw wool, by C. C. Grove, *American Dyestuffs Reporter*, **54**, 1965, 13–16.) A two-factor four-level design was used to investigate the response $y =$ "consumption of detergent in lb/100 lb of raw wool" in a wool-scouring experiment. Four bowls were used. In each of the first two (15 ft long, 1 ft wide, capacity 170 gal each), four levels of temperature—50°C, 55°C, 60°C, and 65°C—were investigated. The third bowl (wash, 10 ft by 1 ft, 100 gal) was maintained at 50°C, and the fourth (rinse, 10 ft by 1 ft, 100 gal) at 40°C. The rate of detergent addition was determined by frequent checks on the variable $z =$ "the residual grease content of the wool" which "was kept at approximately 0.6%." The table shows the 16 design points and the results. The temperatures have been coded to $-3, -1, 1, 3$ via $x_i = $ (temperature$_i - $57.5)/2.5, for bowls $i = 1, 2$, the y's are in the body of Table E11.1, and the z's are below the y's in parentheses.

Fit a second-order response surface to these data and check whether use of second-order terms is necessary. Check whether your chosen surface is worthy of further interpretation; if so, plot the contours. If a second-order surface is retained, also perform the canonical reduction. If low levels of detergent consumption are desirable, what combination of first and second bowl temperatures would you recommend?

In your checks of residuals, plot residuals versus z values. Does the presence of the one atypical z value (0.81) alter your conclusions above?

Table E11.1

x_2	x_1			
	-3	-1	1	3
-3	2.038	0.890	0.475	0.528
	(0.81)	(0.56)	(0.56)	(0.60)
-1	1.192	0.689	0.424	0.369
	(0.66)	(0.58)	(0.56)	(0.58)
1	0.592	0.561	0.333	0.413
	(0.59)	(0.60)	(0.60)	(0.61)
3	0.523	0.517	0.318	0.393
	(0.59)	(0.54)	(0.57)	(0.53)

11.2. Perform a canonical reduction on the fitted second-order equations:
(a) $\hat{y} = 60 + 0.35x_1 + 0.22x_2 + 0.12x_1^2 - 0.63x_2^2 - 3.67x_1x_2$.
(b) $\hat{y} = \frac{1}{4}\{(12 + 2\sqrt{3}) - (10 + 2\sqrt{3})x_1 - (14 + 2\sqrt{3})x_2 + 5x_1^2 + 7x_2^2 + 2\sqrt{3}x_1x_2\}$.

11.3. Perform a canonical analysis on the following second-order fitted surfaces:
(a) $\hat{y} = 60 - 44x_1 - 30x_2 + 8x_1^2 + 3x_2^2 + 12x_1x_2$.
(b) $\hat{y} = 23 - 404x_1 + 396x_2 + 101x_1^2 + 101x_2^2 - 198x_1x_2$.
(c) $\hat{y} = 75 + 206x_1 + 0.85x_2 - 4x_1x_2$.

11.4. Perform a canonical reduction on the fitted equation

$$\hat{y} = b_0 + b_1x_1 + b_2x_2 + b_{11}x_1^2 + b_{22}x_2^2 + b_{12}x_1x_2$$

for each of the sets of b's given in Table E11.4. In each case find and plot the canonical axes, state the type of surface found, and draw rough contours. (Accurate plots can also be made if desired.)

11.5. (*Source:* Simultaneous optimization of several response variables, by G. Derringer and R. Such, *Journal of Quality Technology*, **12**, October 1980, 214–219. Adapted with the permission of the copyright holder, the Ameri-

Table E11.4

Problem Number	1	2	3	4
b_0	75.6	26.7	76.4	29.8
b_1	-7.6	-3.1	-1.7	-3.8
b_2	1.5	-10.8	-1.7	-0.2
b_{11}	2.5	0.3	1.3	-0.4
b_{22}	-0.5	3.8	1.7	-0.6
b_{12}	-2.6	2.2	-0.9	1.0

Table E11.5

Compound Number	x_1	x_2	x_3	Y_1	Y_2	Y_3	Y_4
1	−1	−1	1	102	900	470	67.5
2	1	−1	−1	120	860	410	65
3	−1	1	−1	117	800	570	77.5
4	1	1	1	198	2294	240	74.5
5	−1	−1	−1	103	490	640	62.5
6	1	−1	1	132	1289	270	67
7	−1	1	1	132	1270	410	78
8	1	1	−1	139	1090	380	70
9	−1.633	0	0	102	770	590	76
10	1.633	0	0	154	1690	260	70
11	0	−1.633	0	96	700	520	63
12	0	1.633	0	163	1540	380	75
13	0	0	−1.633	116	2184	520	65
14	0	0	1.633	153	1784	290	71
15	0	0	0	133	1300	380	70
16	0	0	0	133	1300	380	68.5
17	0	0	0	140	1145	430	68
18	0	0	0	142	1090	430	68
19	0	0	0	145	1260	390	69
20	0	0	0	142	1344	390	70

$x_1 = \text{(phr silica} - 1.2)/0.5.$
$x_2 = \text{(phr silane} - 50)/10.$
$x_3 = \text{(phr sulfur} - 2.3)/0.5.$

can Society for Quality Control, Inc.) Table E11.5 shows four columns of response data obtained in the development of a tire tread compound on the four responses: PICO Abrasion Index, Y_1; 200% modulus, Y_2; elongation at break, Y_3; hardness, Y_4. Each column was taken at the 20 sets of conditions shown, where x_1, x_2, x_3 are coded levels of the variables X_1 = hydrated silica level, X_2 = silane coupling agent level, and X_3 = sulfur. (The coding is given below the table; phr = parts per hundred.) Fit a full second-order model to each set of data individually. From a practical viewpoint, it is desirable that $Y_1 > 120$, $Y_2 > 1000$, $600 > Y_3 > 400$, and $75 > Y_4 > 60$. What values of (x_1, x_2, x_3) achieve this, based on the fitted surfaces? If this question seems too difficult to handle, consider this simpler question: Which (x_1, x_2, x_3) points in the actual data satisfy all the restrictions? Reduce to canonical form those fitted second-order surfaces that have significant curvature.

11.6. (*Source:* Kinetics of catalytic isomerisation of *n*-pentane, by N. L. Carr, *Industrial and Engineering Chemistry*, **52**, 1960, 391–396. The responses are rounded. Adapted by permission of the copyright holder, the American Chemical Society.) Fit a second-order model to the data in Table E11.6a. Test for lack of fit. If no lack of fit is shown, test whether or not second-order terms are needed, and also whether or not all the *x*'s should be retained. Make your choice of the fitted model to use (if any) on the

Table E11.6a

x_1	x_2	x_3	y_1	y_2
-1	-1	-1	19.2	4.39
1	-1	-1	29.5	5.43
-1	1	-1	21.6	4.65
1	1	-1	33.8	5.82
-1	-1	1	36.1	6.01
1	-1	1	86.6	9.30
-1	1	1	37.4	6.11
1	1	1	44.8	6.70
-1.68	0	0	27.3	5.22
1.68	0	0	45.0	6.71
0	-1.68	0	20.3	4.51
0	1.68	0	30.0	5.47
0	0	-1.68	23.9	4.89
0	0	1.68	42.3	6.50
0	0	0	33.7	5.81
0	0	0	26.8	5.17
0	0	0	28.7	5.35
0	0	0	26.7	5.17

Table E11.6b

x_1	x_2	x_3	y_1	y_2
0	-1.31	-0.38	33.3	5.77
-1.93	1.75	-1.57	16.3	4.03
1.17	-1.29	-1.07	31.8	5.64
-0.49	3.56	-1.53	28.0	5.29
-0.50	0.18	-1.61	17.9	4.23
-1.55	3.68	-1.52	22.5	4.74

basis of these tests. Then, use your selected model to predict responses at the additional points given in Table E11.6b, and check the values obtained against the responses actually recorded.

If canonical reduction is justified for either or both responses, perform the additional analysis.

11.7. (*Source:* The effects of various levels of sodium citrate, glycerol, and equilibrium time on survival of bovine spermatazoa after storage at $-79°C$, by R. G. Cragle, R. M. Myers, R. K. Waugh, J. S. Hunter, and R. L. Anderson, *Journal of Dairy Science,* **38**(5), 1955, 508–514.) A procedure for storing bovine semen involved the addition of $\xi_1\%$ sodium citrate and $\xi_2\%$ glycerol, and maintainance at $5°C$ for an equilibration time of ξ_3 h before freezing. It was desired to know how these three factors affected the

Table E11.7. Percentage Survival Rates of Motile Spermatozoa (y) from a Designed Experiment

x_1	x_2	x_3	y
-1	-1	-1	57
1	-1	-1	40
-1	1	-1	19
1	1	-1	40
-1	-1	1	54
1	-1	1	41
-1	1	1	21
1	1	1	43
0	0	0	63
-2	0	0	28
2	0	0	11
0	-2	0	2
0	2	0	18
0	0	-2	56
0	0	2	46

percentage survival rate of motile spermatozoa. The coding

$$x_1 = \frac{(\xi_1 - 3)}{0.7}, \qquad x_2 = \frac{(\xi_2 - 8)}{3}, \qquad x_3 = \frac{(\xi_3 - 16)}{6},$$

was used and a three-factor cube plus star ($\alpha = 2$) plus one center-point design was used to provide the data shown in Table E11.7. Fit a second-order response surface to these data and provide all the usual regression analyses. Reduce the selected surface to canonical form, interpret the results, and form conclusions about the best levels of (ξ_1, ξ_2, ξ_3).

11.8. (*Source:* Composite designs in agricultural research, by P. Robinson and K. F. Nielsen, *Canadian Journal of Soil Science*, **40**, 1960, 168–176.) Tomato plants in half-gallon glazed pots, grown in North Gower clay loam, received various combinations of three nutrients, nitrogen (N), phosphorus (P_2O_5), and potassium (K_2O). The 33 runs exhibited in Table E11.8 show the levels of the coded variables $x_1 = (N - 150)/100$, $x_2 = (P_2O_5 - 240)/160$, and $x_3 = (K_2O - 210)/140$, the integers being in units of pounds per acre. Replicate response values of yields of plant tops were measured, as shown.

The first 15 runs form a composite design. The remaining 18 runs are the additional runs necessary to provide a complete 3^3 factorial "enclosed within" the composite set of treatments. (Another experiment was performed using Castor silt loam in which the composite design was "enclosed within" the factorial. This is described in a subsequent exercise.) One purpose of this experiment was to compare the relative efficacies of the two designs, composite versus 3^3 factorial. (Note that the comparison is somewhat biased in favor of the 3^3 factorial in that its 27 points are generally "more spread out" than the 15 points of the composite design, as measured

Table E11.8

	Coded Levels			Yields of Tops (gm. dm./pot)	
	x_1	x_2	x_3	Replicate 1	Replicate 2
	−1.5	0	0	5.00	5.14
	1.5	0	0	12.71	14.29
	0	−1.5	0	5.52	4.20
	0	1.5	0	12.59	10.43
	0	0	−1.5	9.96	12.24
	0	0	1.5	10.23	10.38
	0	0	0	9.78	9.40
	−1	−1	−1	6.82	7.83
	−1	−1	1	7.08	6.21
	−1	1	−1	8.85	8.58
	−1	1	1	6.97	6.07
	1	−1	−1	9.49	7.80
	1	−1	1	7.80	8.85
	1	1	−1	16.25	10.83
	1	1	1	15.48	14.56
	−1	−1	0	5.38	6.78
	−1	0	−1	5.90	7.42
	−1	0	0	6.44	6.80
	−1	0	1	5.68	5.22
	−1	1	0	7.03	7.70
	0	−1	−1	9.23	8.50
	0	−1	0	6.52	9.82
	0	−1	1	8.08	7.31
	0	0	−1	8.86	11.56
	0	0	1	12.62	11.63
	0	1	−1	13.68	10.58
	0	1	0	11.90	11.50
	0	1	1	13.00	14.03
	1	−1	0	8.38	9.48
	1	0	−1	13.89	12.58
	1	0	0	12.60	14.79
	1	0	1	13.23	13.51
	1	1	0	14.40	15.58
SS (composite)	12.5	12.5	12.5		
SS (3^3)	18	18	18		

by $\sum_u x_{iu}^2$; these values are shown at the bottom of the table. The factorial points are distant $3^{1/2} = 1.732$ or $2^{1/2} = 1.414$ from the origin, while the axial points are 1.5 units distant, as far as individual points are concerned.) Perform the following calculations.

(a) Fit a second-order response surface to the 30 observations from the replicated composite design.

(b) Fit a second-order response surface to the 54 observations from the replicated 3^3 factorial design.

Table E11.9a. A Replicated Composite Design (30 Runs)

Coded Levels				Yields of Tops (gm. dm./pot)	
x_1	x_2	x_3		Replicate 1	Replicate 2
-1	-1	-1		3.70	3.80
-1	-1	1		2.94	2.11
-1	1	-1		4.94	3.92
-1	1	1		2.31	2.03
1	-1	-1		5.68	5.29
1	-1	1		3.31	3.09
1	1	-1		10.57	11.83
1	1	1		16.82	9.84
0	0	0		5.86	7.24
-1.5	0	0		2.69	2.03
1.5	0	0		13.20	10.09
0	-1.5	0		0.91	0.82
0	1.5	0		6.50	7.09
0	0	-1.5		6.85	7.28
0	0	1.5		6.01	5.53
12.5	12.5	12.5	SS[a]		

[a] For two replicates, SS = 25. This compares with 40.5 for the corresponding 3^3 design which is thus more "widely spread."

(c) Fit a second-order response surface to all 66 observations.

In (a), (b), and (c) perform all the usual response surface analyses and tests, and also answer this question:

(d) Do we need to retain variable x_3 in the model? (*Hint:* Get the extra sum of squares for all b's with a 3 subscript.)

(e) For a *reduced* second-order model in x_1 and x_2 only, repeat (a), (b), and (c).

(f) Plot the three surfaces you obtain in (e) in the square $-1.5 \leq x_i \leq 1.5$, $i = 1, 2$.

(g) State your conclusions about the form of the response surfaces and the comparative efficacies of the two designs, composite and 3^3.

11.9. (*Source:* Same as the foregoing exercise.) Tables E11.9a and E11.9b are for tomatoes grown in Castor silt loam, and give response values for a replicated composite design and for a 3^3 design which "encloses" it, and has wider spread, as measured by second moments of the design. Notice also that the 3^3 observations designated by superscript a's are merely values repeated from replicate 1 of the doubled composite design, and not new observations. Thus, when the designs are combined, these runs must be included *only once.*

(a) Fit a second-order response surface to the 30 observations from the replicated composite design.

Table E11.9b. A 3^3 Factorial Design (27 Runs)

	Coded Levels			Yields of Tops (gm. dm./pot)
x_1	x_2	x_3		Replicate 1
− 1.5	− 1.5	− 1.5		0.80
− 1.5	− 1.5	0		0.87
− 1.5	− 1.5	1.5		0.92
− 1.5	0	− 1.5		2.53
− 1.5	0	0		2.69[a]
− 1.5	0	1.5		1.66
− 1.5	1.5	− 1.5		2.78
− 1.5	1.5	0		2.02
− 1.5	1.5	1.5		1.61
0	− 1.5	− 1.5		0.71
0	− 1.5	0		0.91[a]
0	− 1.5	1.5		0.52
0	0	− 1.5		6.85[a]
0	0	0		5.86[a]
0	0	1.5		6.01[a]
0	1.5	− 1.5		6.00
0	1.5	0		6.50[a]
0	1.5	1.5		6.19
1.5	− 1.5	− 1.5		0.33
1.5	− 1.5	0		0.42
1.5	− 1.5	1.5		0.05
1.5	0	− 1.5		6.91
1.5	0	0		13.20[a]
1.5	0	1.5		10.88
1.5	1.5	− 1.5		7.70
1.5	1.5	0		11.19
1.5	1.5	1.5		12.43
40.5	40.5	40.5	SS	

[a] Response values are those of replicate 1 in the composite design.

(b) Fit a second-order response surface to the 27 observations from the 3^3 factorial design.

(c) Fit a second-order response surface to all 50 observations. (*Reminder:* Observations with superscript *a*'s attached should be omitted for this.)

In (a), (b), and (c) perform all the usual response surface analyses.

(d) Plot the three surfaces you obtain in (a), (b), and (c), for slices at $x_3 = -1.5, 0, 1.5$ and for $-1.5 \le x_i \le 1.5$, $i = 1, 2$.

(e) State your conclusions about the form of the response surfaces and the comparative efficacies of the two designs, composite and 3^3.

(f) Perform a canonical analysis on all three surfaces and compare the three canonical forms.

Table E11.10

| Coefficient | \multicolumn{4}{Time (Days after Planting)} | | | |
	14	28	42	58
b_0	0.3021^b	2.3507^b	3.0303^b	7.4846^b
b_1	-0.3587^b	-0.3802^b	-0.0893	-0.5334^a
b_2	0.0378	0.3087^b	0.2757^b	-0.2484
b_3	0.0099	0.7634^b	0.7188^b	0.9903^b
b_{11}	0.1178^b	-0.1622^a	-0.1424	-0.4302^a
b_{22}	-0.0367	0.0509	-0.0003	-0.0277
b_{33}	-0.0103	-0.0189	0.0948	-0.0027
b_{12}	-0.0294	-0.0162	0.0218	0.3681
b_{13}	0.0404	0.1610	0.2549	-0.0544
b_{23}	0.0057	0.0616	0.2300	0.0681
R^2	0.825	0.815	0.743	0.517

[a] Significant at 0.05 probability level.
[b] Significant at 0.01 probability level.
R is the multiple correlation coefficient.

11.10. (*Source:* The influence of fertilizer placement and rate of nitrogen on fertilizer phosphorus utilization by oats as studied using a central composite design, by M. H. Miller and G. C. Ashton, *Canadian Journal of Soil Science*, **40**, 1960, 157–167.)

Three predictor variables—X_1 = horizontal distance from seed (inches), X_2 = depth below seed (inches), and X_3 = rate of nitrogen applied (pounds per acre)—were examined. The variables were coded as $x_1 = X_1 - 2$, $x_2 = X_2 - 1$, and $x_3 = (X_3 - 10)/5$ and a 38-point design consisting of a doubled (replicated) cube ($\pm 1, \pm 1, \pm 1$), a doubled star ($\pm 2, 0, 0$), ($0, \pm 2, 0$), ($0, 0, \pm 2$), and 10 center points ($0, 0, 0$) was performed. The response data for y_t = fertilizer phosphorus absorption by oats in milligrams per pot (taken at $t = 14, 28, 42, 58$ days after planting) are not given in the paper, but the estimated regression coefficients for the second-order surfaces fitted are given, and they are reproduced in Table E11.10. The authors state that these coefficients are determined from the equations

$$b_0 = 0.0945945(0y) - 0.027027\Sigma(iiy),$$

$$b_i = 0.03125(iy),$$

$$b_{ii} = -0.027027(0y) + 0.015625(iiy) + 0.005489\Sigma(iiy),$$

$$b_{ij} = 0.0625(ijy),$$

where, for example, $(ijy) = \sum_{u=1}^{n} x_{iu}x_{ju}y_u$, and $\Sigma(iiy) = (11y) + (22y) + (33y)$.

(a) Confirm that these equations are correct.

(b) Use the information in the equations and in the table to fit reduced equations of the following forms via least squares:

i. $y_{14} = \beta_0 + \beta_1 x_1 + \beta_{11} x_1^2 + \varepsilon$.

ii. $y_{28} = \beta_0 + \beta_1 x_1 + \beta_2 x_2 + \beta_3 x_3 + \beta_{11} x_1^2 + \beta_{12} x_1 x_2 + \beta_{13} x_1 x_3 + \varepsilon$.

iii. $y_{42} = \beta_0 + \beta_1 x_1 + \beta_2 x_2 + \beta_3 x_3 + \varepsilon$.

iv. $y_{58} = \beta_0 + \beta_1 x_1 + \beta_2 x_2 + \beta_3 x_3 + \beta_{11} x_1^2 + \beta_{12} x_1 x_2 + \beta_{13} x_1 x_3 + \varepsilon$.

(c) Plot the above surfaces for $x_3 = -2, -1, 0, 1, 2$ and comment upon what these show.

(d) Suppose we code time as a fourth predictor $x_4 = (t - 35)/7$ so that the levels of x_4 become $-3, -1, 1, 3.2857$. Do we have adequate information to fit a full second-order model in x_1, x_2, x_3, and x_4? If so, fit such a model and determine which coefficients are "significant" at the 0.05 and 0.01 levels.

(e) Fit a reduced model retaining all "significant" coefficients plus any "nonsignificant" coefficients relating to terms of equal or lesser order in x's whose estimated coefficients were significant. Examples: (i) If b_{11} were significant, retain $\beta_{11}, \beta_{12}, \beta_{13}, \beta_{14}, \beta_1$; (ii) If b_{12} were significant, retain $\beta_{11}, \beta_{12}, \beta_{13}, \beta_{14}, \beta_{22}, \beta_{23}, \beta_{24}, \beta_1, \beta_2$.

(f) Perform a canonical reduction on your reduced model and interpret the results.

11.11. (*Source:* A statistical approach to catalyst development, Part I: The effect of process variables in the vapour phase oxidation of napthalene, by N. L. Franklin, P. H. Pinchbeck, and F. Popper, *Transactions of the Institution of Chemical Engineers*, **34**, 1956, 280–293. See also **36**, 1958, 259–269.) Table E11.11 provides 80 observations on the variables

ξ_1 = air-to-naphthalene ratio (L/pg),

ξ_2 = contact time (s),

ξ_3 = bed temperature (°C),

Y_P = percentage mole conversion of naphthalene to phthalic anhydride,

Y_N = percentage mole conversion of naphthalene to naphthoquinone,

Y_M = percentage mole conversion of naphthalene to maleic anhydride,

Y_C = percentage mole conversion of naphthalene by complete oxidation to CO and CO_2,

which arose in an investigation of the oxidation of napthalene to phthalic anhydride. For a full description, the reader should consult the references given above. Fit, to the response variable Y, by least squares, a full quadratic model in the predictor variables

$$X_1 = \log \xi_1, \qquad X_2 = \log(10\xi_2), \qquad X_3 = 0.01(\xi_3 - 330),$$

Table E11.11

ξ_1	ξ_2	ξ_3	Y_P	Y_N	Y_M	Y_C
7	2.35	404	63.3	2.2	4.7	27.5
5	2.35	406	61.6	2.5	4.2	26.3
7	2.75	403	60.9	1.9	4.7	28.3
9	1.88	403	67.0	1.1	5	24.5
11	1.38	400	73.0	1.9	3.8	14.8
13	0.84	400	75.8	3.4	3	18
15.2	0.31	402	72.4	8.2	3.6	2
13	0.60	405	75.3	3	5	19.9
15	0.84	400	75.2	3.2	3.1	19
15	0.60	399	74.9	2.7	3.8	17.5
10	0.50	407	77.5	2.4	4.8	4
8	0.38	428	80.0	1.2	4	0
7.2	0.30	428	75.8	1.9	5.2	0
10	0.85	402	72.5	1.7	4.3	11.8
16.5	0.40	400	76.4	4	2	6
18	0.80	395	77.5	3.8	2.8	9
20	1.00	397	73.7	3.1	3.3	11.4
10	0.60	405	75.5	2.4	4.8	4
12.3	0.59	405	78.0	3	5	0
10	0.86	404	72.5	1.7	4.3	11.8
12.9	0.84	400	75.8	3.4	3	18
10.4	0.62	385	67.8	12.6	4.2	2
12.3	0.60	385	69.5	12	3.9	1
10	0.88	386	71.2	6.4	4.1	0
12.5	0.89	383	77.0	5.3	4.4	4.3
9.05	0.46	421	80.0	2.5	4.1	6.6
7.9	0.40	413	79.6	3.7	4	0
21	0.50	408	83.2	2	4.4	1.3
40.2	0.20	406	78.0	6.8	2.5	7
10.7	0.50	405	75.0	4.6	4.3	4.3
7	0.59	405	74.5	5.7	5.4	0
29	0.29	404	84.2	3.7	2.4	7.3
25.3	0.79	404	75.8	1.4	5.7	9
25	0.39	404	83.8	1.5	3	7
9.45	1.63	404	65.2	1.3	4.4	25.8
9	1.42	404	70.5	1.9	4.8	20
25	0.60	403	82.4	1.6	2.9	10.7
10.2	0.30	403	62.5	14.5	3.7	1
7.25	0.90	403	74.2	4.7	5.3	5.3
5.75	0.435	403	66.0	11.7	6.9	0
40.2	0.43	401	81.4	3.7	4.1	8.3
12.5	1.07	401	73.0	2.2	1.8	22

Table E11.11. (*Continued*)

ξ_1	ξ_2	ξ_3	Y_P	Y_N	Y_M	Y_C
10.7	1.70	401	71.0	2.2	4.1	22
16.2	0.80	400	74.7	1.8	2.8	10.5
12.1	1.46	400	69.0	2.1	3	18.6
10	0.61	400	74.0	6.8	5.4	0
11.5	0.73	398	79.0	4.4	5.3	1
10.25	0.87	397	77.5	4.1	3.3	0.5[a]
28.5	0.50	396	79.2	3.1	5.9	2.6
12.15	0.87	395	76.4	2.9	3.3	4.7
25.7	0.45	392	75.0	2.5	3.4	7.1
12.4	1.20	392	79.0	2.3	5.2	9.6
12.15	0.59	392	72.0	8.9	3.4	1.4
25.3	0.66	391	83.5	1.6	2.3	8.3
19.4	1.05	391	82.5	1.4	3.3	7.2
24.7	1.6	390	75.0	0.5[a]	3.5	0
16.7	0.87	390	82.5	2.8	5.3	6.3
15.5	1.19	390	77.3	1.6	4.5	9.8
40.2	0.605	389	78.5	2.7	2.9	16.5
31.8	1.21	389	78.5	1.3	2.5	14.7
25.4	0.905	389	84.0	1.8	2.5	10
30.4	0.355	383	75.4	7.6	2.3	7.8
22.7	0.51	383	80.5	4	3.9	4.2
71.5	0.50	382	80.4	4.1	2.6	11.4
39.5	0.50	382	83.0	7.3	3.4	8.8
12.4	1.19	382	72.1	2.9	2.6	7
58.5	0.60	381	87.0	1.9	2.4	10.4
16	1.20	381	78.8	2.9	3.8	8.6
81	0.70	380	84.1	1.3	5.2	9
29.6	0.71	380	82.7	3.6	3.9	7.1
25.3	1.01	380	81.0	2.2	3.7	8
25	1.63	380	77.5	0.5[a]	4	15.7
16	0.88	380	80.0	4	3.7	7.6
39.5	1.21	379	81.5	1.8	2.7	15
19.7	1.08	379	78.5	1.8	3.3	2.2
38	0.81	378	82.5	2.1	3.75	8.8
30.9	1.23	373	77.5	2	2.8	11.6
76.5	0.80	365	86.7	3.1	3.8	0
42	2.35	365	74.2	0.5[a]	3.4	24
103	1.25	334	87.5	4.6	3.5	10.5

[a]Actually recorded as " < 1.0"; the value 0.5 was arbitrarily inserted.

where log denotes logarithm to the base 10. Carry out the usual analysis of variance and checks of residuals. Apply the methods of canonical reduction to elucidate the type of surface represented by the fitted model, and come to appropriate conclusions. (This exercise can be performed for any or all of the four response variables shown.)

11.12. Reduce, to canonical form, the second-order fitted equation $\hat{y} = 67.711 + 1.944x_1 + 0.906x_2 + 1.069x_3 - 1.539x_1^2 - 0.264x_2^2 - 0.676x_3^2 - 3.088x_1x_2 - 2.188x_1x_3 - 1.212x_2x_3$.

Table E11.13

Problem Number	1	2	3	4
b_0	52	43	79	71
b_1	2.01	-1.67	-0.39	1.73
b_2	1.60	-0.69	0.27	0.90
b_3	-0.46	-0.21	0.69	0.41
b_{11}	0.38	2.85	0.99	-0.25
b_{22}	-1.04	1.58	0.11	-1.90
b_{33}	-4.25	1.57	0.02	-1.85
b_{12}	-0.82	-0.18	0.22	-3.28
b_{13}	-3.18	-0.92	-0.10	-1.71
b_{23}	-4.99	-0.97	0.01	0.37

11.13. Perform a canonical analysis on the fitted equation

$$\hat{y} = b_0 + b_1 x_1 + b_2 x_2 + b_3 x_3 + b_{11} x_1^2 + b_{22} x_2^2 + b_{33} x_3^2$$
$$+ b_{12} x_1 x_2 + b_{13} x_1 x_3 + b_{23} x_2 x_3$$

for each of the sets of b's given in Table E11.13. In each case find and plot the canonical axes, state the type of surface found, and draw rough contours.

11.14. (*Source:* Response surface for dry modulus of rupture and drying shrinkage, by H. Hackney and P. R. Jones, *American Ceramics Society Bulletin*, **46**, 1967, 745–749.) Carry out a canonical reduction on the fitted surfaces:

(a) $\hat{y} = 6.88 + 0.0325 x_1 + 0.2588 x_2 - 0.1363 x_3$
$$-0.1466 x_1^2 - 0.0053 x_2^2 + 0.1359 x_3^2$$
$$+0.1875 x_1 x_2 + 0.2050 x_1 x_3 - 0.1450 x_2 x_3.$$

(b) $\hat{y} = 2.714 - 0.0138 x_1 - 0.1300 x_2 - 0.1525 x_3$
$$+0.0333 x_1^2 - 0.0342 x_2^2 + 0.4833 x_3^2$$
$$+0.175 x_1 x_2 + 0.28 x_1 x_3 + 0.5625 x_2 x_3.$$

11.15. (*Source:* Subjective responses in process investigation, by H. Smith and A. Rose, *Industrial and Engineering Chemistry*, **55**(7), 1963, 25–28. Adapted by permission of the copyright holder, the American Chemical Society.) Three predictor variables, the amounts of flour, water, and shortening were varied in a pie crust recipe. Three responses—flakiness, gumminess (toughness), and specific volume (in cm^3/g)—were measured. The fitted second-order surface obtained for flakiness in terms of coded predictors was

$$\hat{y} = 6.89462 + 0.06323 x_1 - 0.12318 x_2 + 0.15162 x_3$$

$$- 0.11544 x_1^2 - 0.03997 x_2^2 - 0.11544 x_3^2$$

$$+ 0.09375 x_1 x_2 - 0.34375 x_1 x_3 - 0.03125 x_{23}.$$

Table E11.6. Extra Trials Completing a Central Composite Design

Trial Number	Factor Levels				Response y_2, DMP
	x_1, NH$_3$	x_2, T	x_3, H$_2$O	x_4, P	
18	−1.4	0	0	0	31.1
19	1.4	0	0	0	28.1
20	0	−1.4	0	0	17.5
21	0	1.4	0	0	49.7
22	0	0	−1.4	0	49.9
23	0	0	1.4	0	34.2
24	0	0	0	−1.4	31.1
25	0	0	0	1.4	43.1

Reduce this equation to canonical form. If flakiness should exceed 7, in what region of (x_1, x_2, x_3) should the crust be made? [*Note:* The original paper is concerned also with limits on the other two responses. Moreover, because the observed F for (regression given b_0) for flakiness is only 1.1 times the $\alpha = 0.05$ percentage point, there is some question as to how well established the fitted surface is, and thus whether canonical reduction is justified. In fact, the original authors did not use canonical reduction, but simply plotted response contours and looked for desirable regions. Properties observed could then be checked with confirmatory runs.]

11.16. (*Source:* Maximum data through a statistical design, by C. D. Chang, O. K. Kononenko, and R. E. Franklin, Jr., *Industrial and Engineering Chemistry*, **52**, November 1960, 939–942. Material reproduced and adapted by permission of the copyright holder, the American Chemical Society. The data were obtained under a grant from the Sugar Research Foundation, Inc., now the World Sugar Research Organization, Ltd.) Consider the data for y_2(DMP) in Exercise 4.16. Remember to count "Trial 17" as six individual trials, each with response 32.8, for the surface fitting and to count, as pure error sum of squares, the quantity $(6 - 1)s_2^2 = 5(9.24) = 46.2$ in the analysis of variance table. Combine the previous data with those of trials 18–25 given in Table E11.16. Fit a full second-order surface to all 30 data points and check for lack of fit. If no lack of fit exists, is the regression significant and, if so, what part(s), first order, second order, both? Is it worthwhile to canonically reduce the fitted surface? Repeat the entire exercise on the response

$$w_i = \sin^{-1}\left(\frac{y_i}{100}\right)^{1/2},$$

and compare and contrast the two analyses. What would you report to the experimenter?

11.17. (*Source:* Evaluation of variables in the pressure-kier bleaching of cotton, by J. J. Gaido and H. D. Terhune, *American Dyestuffs Reporter*, **50**, October

16, 1961, 23–26 and 32). Four predictor variables, coded here as

$x_1 = (\text{temperature } °F - 235)/10,$

$x_2 = (\text{time in minutes } - 60)/20,$

$x_3 = (\text{albone hydrogen peroxide concentration } \% - 3)/0.2,$

$x_4 = (\text{NaOH concentration } \% - 0.3)/0.1,$

were examined in a cotton-bleaching experiment. Three responses, coded

Table E11.17

Run Number	Coded Predictor Levels				Response Values		
	x_1	x_2	x_3	x_4	y_1	y_2	y_3
1	1.3	−2	2	2	8.2	1.2	1.8
2	1.3	−1	3	−1	8.3	1.8	1.5
3	1.3	−1	−3	1	8.0	1.6	1.5
4	1.3	−2	−2	−2	4.0	5.6	1.2
5	0.3	−1.5	2	2	8.4	2.0	1.8
6	0.3	−2.25	3	−1	4.6	4.7	1.2
7	0.3	−2.25	−3	1	5.2	1.5	1.4
8	0.3	−1.5	−2	−2	4.3	3.0	0.9
9	−0.3	3	2	2	9.4	1.2	2.2
10	−0.3	0	3	−1	6.9	3.6	1.5
11	−0.3	0	−3	1	7.7	1.7	1.5
12	−0.3	3	−2	−2	7.0	1.7	1.5
13	−1.3	−0.75	2	2	7.9	2.2	1.6
14	−1.3	1.5	3	−1	7.9	2.5	1.5
15	−1.3	1.5	−3	1	7.7	1.8	1.4
16	−1.3	−0.75	−2	−2	2.2	27.0	0.9
17	1.5	−2	0	0	6.9	1.0	1.4
18	−1.5	6	0	0	8.5	1.1	1.5
19	1.0	1	0	0	8.8	1.0	1.2
20	−1.0	−0.75	0	0	6.3	1.4	1.2
21	0	0	5	1	9.0	0.7	1.6
22	0	0	−5	−1	6.1	1.2	1.3
23	0	0	−1	3	8.4	0.6	1.8
24	0	0	1	−3	5.8	1.5	1.1
25	0	−1.5	0	0	5.7	2.3	1.3
26	0	−1.5	0	0	6.7	2.0	1.5
27	0	0	0	0	8.4	1.8	1.5
28	0	0	0	0	8.3	1.9	1.7
29	0	0	0	0	8.3	0.8	1.6
30	0	0	0	0	7.9	1.3	1.6
31	0	−1.5	0	0	6.4	1.4	1.4
32	1.3	−2	2	5	7.4	2.4	2.0
33	1.3	−1	2	2	7.8	2.4	2.1
34	1.3	−1	2	2	8.7	1.5	2.3
35	1.3	1	2	2	9.6	1.1	2.6
Sums	5.2	−9.25	8	11	250.9	90.5	54.1

here as

$$y_1 = \text{whiteness \% } - 80,$$

$$y_2 = \text{absorbency (s)},$$

$$y_3 = \text{fluidity (rhes)},$$

were recorded. The results from 35 experimental runs appear in Table E11.17.

To each response, fit a second-order surface in the x's. Test for lack of fit, and if no lack of fit is found, test for the need for quadratic terms and/or the need for specific x variables. Note especially the peculiarity of the response data in run 16. The authors remark that this sample is "obviously underbleached" and has an exceptionally high absorbency; thus you should treat this observation, certainly in its y_1 and y_2 values, at least, as a potential outlier.

Whatever is your final choice of fitted surface for each response, examine whether or not it is worth further interpretation and, if it is, plot or canonically reduce it for that interpretation. What are your overall conclusions?

11.18. (*Source:* Effect of raw-material ratios on absorption of whiteware compositions, by W. C. Hackler, W. W. Kriegel, and R. J. Hader, *Journal of the American Ceramic Society*, **39**(1), 1956, 20–25.) A study of four whiteware raw material components was carried out in the following way. First, three ratios of the four ingredients were formed. This avoided the problem arising from the fact that the components necessarily totaled 100% in each run. The three ratios were coded to variables x_1, x_2, and x_3. A fourth predictor

Table E11.18

Coded Levels of the Variables			Value of $100y$ when $x_4 =$		
x_1	x_2	x_3	-1	0	1
-1	-1	-1	1484	881	636
-1	-1	1	1354	726	392
-1	1	-1	1364	746	376
-1	1	1	1286	684	380
1	-1	-1	1171	642	383
1	-1	1	1331	778	550
1	1	-1	1133	597	358
1	1	1	1136	604	362
0	0	0	1304	756	419
-2	0	0	1577	882	455
2	0	0	1070	534	342
0	-2	0	1312	796	488
0	2	0	1278	711	411
0	0	-2	1281	706	418
0	0	2	1238	682	359

variable, firing temperature, was coded and named x_4. The 45 responses obtained are shown in Table E11.18. The response variable y is the "fired absorption" of the whiteware "adjusted for location in the kiln and averaged for the two batches and two firings at each temperature." (For details of the adjustments, see pp. 21–22 of the source reference.)

Fit a second-order model to the data. Use the extra sum of squares principle to investigate if second-order terms are needed, if x_1 is needed, if x_2 is needed, and so on. When you have decided what model to adopt, check whether canonical reduction is worthwhile. If so, perform the canonical reduction and comment on the shape of the surface. If reduced absorption were desirable, what settings of the x's would you recommend for future work?

11.19. (*Source:* Starch vinylation, by J. W. Berry, H. Tucker, and A. J. Deutshman, *Industrial and Engineering Chemistry Process Design and Development,* **2**(4),

Table E11.19

		Coded Values			
x_1	x_2	x_3	x_4	x_5	y
−1	1	−1	1	−1	0.94
−1	1	1	1	1	0.84
1	1	−1	−1	−1	0.75
0	0	0	0	0	0.84
−1	−1	−1	1	1	0.46
1	1	1	1	−1	0.66
−1	−1	1	−1	1	0.24
0	0	0	0	0	0.82
1	1	1	−1	1	0.60
1	−1	1	1	1	0.74
1	−1	−1	−1	1	0.26
1	−1	−1	1	−1	0.10
1	−1	1	−1	−1	0.80
−1	1	1	−1	−1	0.80
0	0	0	0	0	0.84
1	1	−1	1	1	0.78
−1	1	−1	−1	1	0.41
−1	−1	−1	−1	−1	0.25
−1	−1	1	1	−1	0.75
0	0	0	0	0	0.83
−2	0	0	0	0	0.73
0	2	0	0	0	0.72
0	0	−2	0	0	0.73
0	0	0	0	2	0.43
0	−2	0	0	0	0.30
2	0	0	0	0	0.82
0	0	0	−2	0	0.36
0	0	2	0	0	0.89
0	0	0	0	0	0.84
0	0	0	2	0	0.88
0	0	0	0	−2	0.95
0	0	0	0	0	0.84

1963, 318–322. Also see correspondence by I. Klein and D. I. Marshall, **3**(3), 1964, 287–288. Adapted by permission of the copyright holder, the American Chemical Society.) A study of starch vinylation involved five input variables and two responses (only one of which is presented here). Table E11.19 shows a (coded) 2^{5-1} fractional factorial ($x_1 x_2 x_3 x_4 x_5 = -1$) plus 10 axial points distant ± 2 units from the origin, plus 6 center points, and the 32 responses that were observed.

Fit a second-order surface to these data and check for lack of fit and for the needed presence of second-order terms or of any of the x's.

Perform a canonical reduction on the fitted surface you decide to adopt, and interpret the results.

Table E11.20

Run Number	x_1	x_2	x_3	x_4	x_5	y
1	−2	0	0	0	0	67
2	2	0	0	0	0	238
3	0	−2	0	0	0	197
4	0	2	0	0	0	174
5	0	0	−2	0	0	221
6	0	0	2	0	0	152
7	0	0	0	−2	0	195
8	0	0	0	2	0	154
9	0	0	0	0	−2	213
10	0	0	0	0	2	136
11	0	0	0	0	0	221
12	0	0	0	0	0	221
13	−1	−1	−1	−1	1	196
14	1	1	−1	−1	1	198
15	1	1	1	−1	−1	192
16	−1	−1	1	−1	−1	203
17	−1	−1	−1	1	−1	193
18	1	1	−1	1	−1	211
19	−1	−1	1	1	1	76
20	1	1	1	1	1	218
21	0	0	0	0	0	229
22	0	0	0	0	0	225
23	0	0	0	0	0	211
24	1	−1	−1	−1	−1	210
25	−1	1	−1	−1	−1	166
26	1	−1	1	−1	1	251
27	−1	1	1	−1	1	92
28	1	−1	−1	1	1	221
29	−1	1	−1	1	1	81
30	1	−1	1	1	−1	238
31	−1	1	1	1	−1	114
32	0	0	0	0	0	226
33	0	0	0	0	0	234
34	0	0	0	0	0	235

11.20. (*Source:* Statistically designed experiments, by G. C. Derringer, *Rubber Age*, **101**, 1969, November, 66–76.) Five major compounding ingredients in a silica filled rubber compound, coded to $x_1, \ldots, x_5$, were studied via a five-factor central composite design. The design was split into three blocks as indicated in Table E11.20.

Block 1: Ten axial points plus two center points.
Block 2: Half of a 2^{5-1} design ($x_1 x_2 x_3 x_4 x_5 = 1$), such that $x_1 x_2 = 1$, plus three center points.
Block 3: Half of a 2^{5-1} design ($x_1 x_2 x_3 x_4 x_5 = 1$), such that $x_1 x_2 = -1$, plus three center points.

The 2^{5-1} design was blocked on the signs of $x_1 x_2$ because it was known from previous work that the interaction between factors 1 and 2 was insignificant. The observed response shown is $y =$ (tensile strength of the compound, psi)$/10$.

Fit to the data a full second-order model in $x_1, \ldots, x_5$ with two blocking variables z_1 and z_2. Many choices of z_1 and z_2 levels are possible, but values $z_1 = (0, 1, 0)$ and $z_2 = (0, 0, 1)$ for blocks $(1, 2, 3)$, respectively, are suggested. Check for lack of fit and whether or not a reduced model is possible. If appropriate, perform a canonical reduction of the fitted equation you finally select. (*Note:* Blocks are not orthogonal to the model. Various orders of terms in the analysis of variance table are possible. We suggest the following sequence: b_0; blocks given b_0; first-order given b_0 and blocks; second-order given preceding items; residual, split up into lack of fit and pure error. Pure error is calculated within blocks only and has thus $1 + 2 + 2 = 5$ df.)

11.21. (*Source:* A statistical approach to catalyst development. Part II: The integration of process and catalyst variables in the vapour phase oxidation of napthalene, by N. L. Franklin, P. H. Pinchbeck, and F. Popper, *Transactions of the Institution of Chemical Engineers*, **36**, 1958, 259–269.) Two five-predictor-variable fitted response surfaces for yield and purity were obtained as follows.

$$\text{yield} = -110.2 + 88.1 X_1 + 119.8 X_2$$
$$+ 152.1 X_3 - 10.7 X_4 + 27.6 X_5$$
$$- 15.8 X_1^2 - 10.8 X_2^2 - 32.9 X_3^2$$
$$- 6.5 X_4^2 - 7.9 X_5^2 - 14.1 X_1 X_2$$
$$- 28.9 X_1 X_3 + 0.2 X_1 X_4 - 6.3 X_1 X_5$$
$$- 52.2 X_2 X_3 - 1.9 X_2 X_4 - 30.8 X_2 X_5$$
$$+ 5.2 X_3 X_4 + 13.2 X_3 X_5 + 19.4 X_4 X_5.$$

$$\text{purity} = -292.90 + 137.36X_1 + 221.37X_2$$
$$+ 247.75X_3 + 20.44X_4 + 46.14X_5$$
$$- 17.91X_1^2 - 48.35X_2^2 - 38.65X_3^2$$
$$- 9.17X_4^2 - 14.09X_5^2$$
$$- 13.23X_1X_2 - 47.90X_1X_3$$
$$- 6.04X_1X_4 - 10.90X_1X_5$$
$$- 81.24X_2X_3 - 11.63X_2X_4$$
$$- 18.74X_2X_5 - 12.66X_3X_4$$
$$- 15.10X_3X_5 + 27.95X_4X_5.$$

Reduce these surfaces to canonical form and indicate their main features. Does it appear possible to improve yield and purity at the same time (to the extent that the contours can be relied upon as accurate approximations to the underlying true surfaces)? Sketch the contours in the (X_4, X_5) plane for the following three sets of conditions (where log = logarithm to the base 10):

(a) $X_1 = \log 321,$ $X_2 = \log 7,$ $X_3 = 0.5.$
(b) $X_1 = \log 1001,$ $X_2 = \log 7,$ $X_3 = 0.7.$
(c) $X_1 = \log 321,$ $X_2 = \log 7,$ $X_3 = 0.7.$

11.22. (*Source:* An investigation of some of the relationships between copper, iron and molybdenum in the growth and nutrition of lettuce: II. Response surfaces of growth and accumulations of Cu and Fe, by D. P. Moore, M. E. Harward, D. D. Mason, R. J. Hader, W. L. Lott, and W. A. Jackson, *Soil Science Society American Proceedings*, 1957, **21**, 65–74.) In a study of how copper (Cu), molybdenum (Mo), and iron (Fe) affected lettuce growth, a group of experimenters coded CU, Mo, and Fe to x_1, x_2, and x_3, as in Table E11.22a, and ran the design shown in Table E11.22b four times over. The four sets of experimental runs used two different sources of nitrogen (N), which was not one of the variables evaluated but was part of the combinations of chemicals given to the plants, and two different sources of Fe. The two nitrogen sources were denoted by NO_3^- and $NH_4^+ + NO_3^-$ and

Table E11.22a. Codings of Cu, Mo, and Fe to Design Variables x_1, x_2, and x_3

Level No.	x_i	Cu	Mo	Fe
1	−1.68	0.0002	0.0002	0.0025
2	−1	0.0013	0.0013	0.064^a
3	0	0.02	0.02	0.25
4	1	0.31	0.31	3.88
5	1.68	2.00	2.00	25.00

^{a}Through an error in calculation discovered after the experiment was in progress, the second level of Fe actually corresponded to a coded value −0.5, not the intended −1.

Table E11.22b. Observed Yields (grams dry weight) of Lettuce Tops per Culture After Six Weeks' Growth in Solutions Containing the Variables Cu, Mo, and Fe. The four sets of data used four different source combinations of N and Fe. Also shown are degrees of freedom (df) and mean squares from the original reference.

x_1	x_2	x_3	Source 1	Source 2	Source 3	Source 4
-1	-1	-0.5	18.43	15.31	24.01	21.42
1	-1	-0.5	0.82	1.00	7.52	15.92
-1	1	-0.5	18.63	13.44	22.20	22.81
-1	-1	1	7.98	5.11	6.98	14.90
1	1	-0.5	1.03	1.05	9.36	14.95
1	-1	1	1.84	4.60	5.51	7.83
-1	1	1	14.74	3.64	6.21	19.90
1	1	1	2.24	7.70	4.56	4.68
1.68	0	0	0.30	0.36	0.19	0.20
-1.68	0	0	16.16	12.19	15.35	17.65
0	1.68	0	17.16	14.84	21.12	18.16
0	-1.68	0	21.66	20.26	24.40	25.39
0	0	1.68	0.84	26.86	0.65	11.99
0	0	-1.68	8.32	4.16	13.61	7.37
0	0	0	13.48	15.64	—	22.22
0	0	0	20.77	16.08	23.43	19.49
0	0	0	18.56	17.01	—	22.76
0	0	0	18.08	16.53	29.50	24.27
0	0	0	23.39	13.22	27.50	27.88
0	0	0	20.56	15.42	25.18	27.53

Source	df	Mean Squares	Mean Squares	Mean Squares	Mean Squares
Mean	1	3001.0	2518.2	3968.8	6031.6
Linear	3	179.4	62.0	197.2	114.8
Quadratic	6	103.6	78.8	168.6	112.4
Lack of fit	5	22.0	69.6	6.2	10.8
Pure error	5 (*3)	11.2	1.8	7.0*	10.5
Total df	20	20	20	*18	20

Sources:
1. NO_3^- and Fe^{2+}.
2. NO_3^- and Fe^{3+}.
3. $NH_4^+ + NO_3^-$, and Fe^{2+}
4. $NH_4^+ + NO3^-$, and Fe^{3+}.

the two Fe sources by Fe^{2+} and Fe^{3+}. (Note that, because of a minor calculation error, a level of $x_3 = -0.5$ was actually used, instead of the intended $x_3 = -1$. Although this makes the design asymmetric and prevents the design from being an exactly rotatable one as was originally intended, the data will still support the fit of a second-order surface.) Four sets of responses were observed, as shown in Table E11.22b. Also shown, in Table E11.22c, are the authors' fitted regression coefficients and their standard errors.

Table E11.22c. Regression Coefficients and Standard Errors for Yield of Tops (grams dry weight per culture) from Original Reference

Coefficient	Source 1	Source 2	Source 3	Source 4
b_0	18.75, 1.4	15.85, 0.5	26.11, 1.3	24.34, 1.3
b_1	6.23, 0.9	$-3.87, 0.4$	$-4.86, 0.7$	$-4.57, 0.9$
b_2	$-0.22, 0.9$	$-0.80, 0.4$	$-0.51, 0.7$	$-0.82, 0.9$
b_3	$-1.30, 1.1$	2.78, 0.4	$-3.77, 0.9$	$-0.03, 1.0$
b_{11}	$-4.58, 0.9$	$-4.95, 0.4$	$-7.07, 0.8$	$-5.28, 0.9$
b_{22}	$-0.63, 0.9$	$-0.98, 0.4$	$-1.77, 0.8$	$-0.75, 0.9$
b_{33}	$-5.27, 0.9$	$-1.62, 0.4$	$-6.99, 0.8$	$-5.35, 0.9$
b_{12}	$-0.79, 1.2$	0.81, 0.5	0.43, 0.9	$-1.31, 1.2$
b_{13}	2.28, 1.6	4.93, 0.6	4.25, 1.2	$-1.29, 1.5$
b_{23}	1.48, 1.6	0.83, 0.6	$-0.14, 1.2$	0.66, 1.5

Work out the actual coding formulas indicated by Table E11.22a. Use the first set of responses, namely the NO_3^- and Fe^{2+} combination, and confirm the second-order model fitting analysis shown for it. (Minor numerical differences may be found, using current more accurate computer programs.) If the surface appears to fit the data well, perform a canonical analysis on your fitted model. What sort of surface is the one you fitted? Where is the stationary point of the contour system? Is there an attainable maximum response to be found in, or close to, the design region? What levels of CU, Mo, and Fe would you recommend for future lettuce growing, based on what this experiment shows, assuming the only requirement were to maximize yield?

(The other three sets of data provide additional exercises, but solutions for them are not provided.)

11.23. (*Source:* Investigation of calcium and available phosphorus requirements for laying hens by response surface methodology, by W. B. Roush, M. Mylet, J. L. Rosenberger, and J. Derr, *Poultry Science*, 1986, **65**, 964–970.) Twenty-seven observations were taken of the "percentage shell" found in eggs produced by hens on regulated diets of calcium and phosphorus. There were nine diets and three observations at each diet, shown in Table 11.23. Fit, to these data, a second-order surface

$$y = \beta_0 + \beta_1 x_1 + \beta_2 x_2 + \beta_{11} x_1^2 + \beta_{22} x_2^2 + \beta_{12} x_1 x_2 + \varepsilon$$

Table E11.23. Shell Weight of Eggs as a Percentage of Egg Weight. Three observations were taken at each of nine sets of conditions

Phosphorus = 0.65%	8.16, 7.85, 8.13	8.70, 8.99, 8.92	9.04, 8.83, 8.97
Phosphorus = 0.50%	8.84, 8.33, 8.55	8.88, 9.02, 8.95	9.03, 9.14, 8.97
Phosphorus = 0.35%	8.47, 8.34, 8.50	8.92, 8.80, 8.91	8.71, 9.51, 9.05
	Calcium = 2.50%	Calcium = 3.75%	Calcium = 5.00%

Table E11.24a. Crater Surface Data

x_1	x_2	y	x_1	x_2	y
1	3	1	4	6	116
1	4	27	4	7	85
1	5	43	4	8	28
1	6	37	5	1	12
1	7	12	5	2	38
2	2	5	5	3	60
2	3	44	5	4	78
2	4	81	5	5	90
2	5	91	5	6	90
2	6	79	5	7	72
2	7	39	5	8	35
2	8	2	5	9	1
3	1	2	6	2	14
3	2	32	6	3	35
3	3	71	6	4	52
3	4	103	6	5	62
3	5	120	6	6	64
3	6	104	6	7	52
3	7	71	6	8	25
3	8	16	7	3	1
4	1	12	7	4	15
4	2	45	7	5	20
4	3	77	7	6	20
4	4	104	7	7	18
4	5	126	7	8	2

where $x_1 = (\text{Calcium} - 3.75)/1.25$ and $x_2 = (\text{Phosphorus} - 0.50)/0.15$. Carry out the usual statistical analyses, including a test for lack of fit. Estimate the point (x_1, x_2) where the response predicted by the fitted equation attains its maximum value, and find that value. Translate the point of maximum response into the (Calcium, Phosphorus) space.

11.24. (*Source:* Mathematical models of carbide tool crater surfaces, by S. M. Wu and R. N. Meyer, *International Journal of Machine Tool Design Research*, **7**, 1967, 445–463. Adapted with permission from Elsevier, the copyright holder.) In a study of crater surfaces in metal created by a carbide cutting tool, the 50 observations in Table E11.24a were obtained. [The spatial coordinates (x_1, x_2) are in units of 1/100th of an inch; that is, they need to be divided by 100 to be stated in inches. The response values, labeled y, are in units of 1/1000th of an inch, that is, they need to be divided by 1000 to be stated in inches, as they are in the original paper.] Fit response surfaces of second and third order to the data. Are the extra cubic terms worthwhile? Plot the two surfaces and compare them. Does your visual comparison support your previous conclusion about the value of adding the third-order terms?

11.25. (*Source:* Multiple regression and response surface analysis of the effects of calcium chloride and cysteine on heat-induced whey protein gelation, by

Table E11.25a. Predictor Variable Levels and Their Coded Values

Cysteine:	2.6	8.0	21.0	34.0	39.4
$CaCl_2$:	2.5	6.5	16.2	25.9	29.9
Coded level:	-1.414	-1	0	1	1.414

Coding: $x_1 = (Cysteine - 21)/13$, $x_2 = (CaCl_2 - 16.2)/9.7$.

Table E11.25b. Experimental Coded Values and Values of Four Responses:
y_1 = hardness (kg), y_2 = cohesiveness, y_3 = springiness (mm),
y_4 = compressible water (g)

No.	x_1	x_2	y_1	y_2	y_3	y_4
1	-1	-1	2.48	0.55	1.95	0.22
2	1	-1	0.91	0.52	1.37	0.67
3	-1	1	0.71	0.67	1.74	0.57
4	1	1	0.41	0.36	1.20	0.69
5	-1.414	0	2.28	0.59	1.75	0.33
6	1.414	0	0.35	0.31	1.13	0.67
7	0	-1.414	2.14	0.54	1.68	0.42
8	0	1.414	0.78	0.51	1.51	0.57
9	0	0	1.50	0.66	1.80	0.44
10	0	0	1.66	0.66	1.79	0.50
11	0	0	1.48	0.66	1.79	0.50
12	0	0	1.41	0.66	1.77	0.43
13	0	0	1.58	0.66	1.73	0.47

R. H. Schmidt, B. L. Illingworth, J. D. Deng, and J. A. Cornell, *Journal of Agricultural and Food Chemistry*, **27**, 1979, 529–532. Adapted with the permission of the copyright holder, the Americal Chemical Society.) Table 11.25a shows the actual levels (in mM) and the coded levels of two predictor variables systeine (x_1) and calcium chloride, $CaCl_2$, (x_2) used in an experiment to determine their effects on four properties of a food gelatin. The experiment is described in the source reference. A central composite design with five center points was run with the results shown in Table E11.25b. Fit a second-order response surface to each of the responses individually via least squares; check for lack of fit and plot the residuals. Sketch the fitted contours and comment on what you see.

(a) What types of surfaces result?

(b) Is it possible to improve both cohesiveness (y_2) and springiness (y_3) together?

(c) If it is not possible to improve cohesiveness and springiness at the same time, what conditions would provide a reasonable compromise between the two?

CHAPTER 12

Ridge Analysis for Examining Second-Order Fitted Models, Unrestricted Case

12.1. INTRODUCTION

Suppose we investigate the behavior of a response variable y over a specified region of interest by fitting a second-order response surface. Ridge analysis provides a simple way of following the locus of, for example, a maximum response, moving outwards from the origin of the predictor variable space or, indeed, moving from any other point (a selected "focus"). Because ridge analysis does not require one to be able to view the fitted regression surface as a whole, it may be applied even when it is difficult to draw or visualize the surface, for example, in four or more dimensions. The calculation of a ridge trace also helps one to assess and understand the typically complex interplay between the input variables as the response improves. Use of this tool can supplement, or replace, the canonical reduction of a second-order surface.

In so-called "standard" ridge regression, the subject of this chapter, we examine second-order surfaces without linear restrictions. (The imposition of linear restrictions, such as a mixture restriction, for example, is discussed in Chapter 19. More generally, we may wish to explore a subspace defined by other linear restrictions on the predictors.)

Historical Notes

Ridge analysis was first introduced in the context of general response surface methodology by A. E. Hoerl (1959, 1962, 1964). It was further investigated by Draper (1963), who proved several results that A. E. Hoerl had suggested without proof. A wide-ranging discussion is given by R. W. Hoerl (1985). Applications in mixture experiments were discussed by R. W. Hoerl (1987) and Draper and

Response Surfaces, Mixtures, and Ridge Analyses, Second Edition. By G. E. P. Box and N. R. Draper
Copyright © 2007 John Wiley & Sons, Inc.

Pukelsheim (2002, 2003). See also Peterson, Cahya and Del Castillo (2002), Miró-Quesada et al. (2004), and Peterson (1989, 1993, 2003, 2004, 2005).

12.2. STANDARD RIDGE ANALYSIS ABOUT THE ORIGIN $x = 0$

Basic Method, Focus at the Origin

In its original, unrestricted form (A. E. Hoerl, 1959, 1962, 1964), ridge analysis was used on a second-order fitted response to obtain a set of paths, going outwards from the origin $\mathbf{x}' = (x_1, x_2, \ldots, x_q) = (0, 0, \ldots, 0)$ of the factor space. Suppose, in the x-space where we have fitted a second-order response function, we (mentally) construct a sphere of radius R about the origin, which is usually the coded center of experimentation that provided the data used in the fitting. On that sphere there will be a point with the highest response value. As the radius R is changed, all such points will mark out a locus, or path, of highest response when moving from the origin. Similarly, there will also be a path of lowest response. If we extend this idea to points on the spheres where the response is stationary—that is, a minimum or maximum *locally but NOT globally*—more paths will come to light. In fact, there are always $2q$ such paths. Usually, but not always, only the paths that provide the maximum response (path of steepest ascent) and the minimum response (path of steepest descent) on spheres of increasing radius R, centered at the origin, are likely to be interesting to the experimenter. For that reason, these two paths receive the most emphasis in what follows. Secondary paths could well be of interest in certain practical situations; for example, they might provide good, though not optimum, response values at lower cost. Secondary paths typically do not start at the origin, but appear suddenly when certain radii values R (which depend on the specific response surface under study) are attained. As we shall see, we shall always know which of the $2q$ paths we are on. The basic ridge analysis method is as follows. Suppose the fitted second-order surface is

$$\hat{y} = b_0 + b_1 x_1 + b_2 x_2 + \cdots + b_q x_q + b_{11} x_1^2 + b_{22} x_2^2 + \cdots + b_{qq} x_q^2$$

$$+ b_{12} x_1 x_2 + b_{13} x_1 x_3 + \cdots + \beta_{q-1,q} x_{q-1} x_q. \tag{12.2.1}$$

This can be written in matrix form as

$$\hat{y} = b_0 + \mathbf{x}'\mathbf{b} + \mathbf{x}'\mathbf{B}\mathbf{x}, \tag{12.2.2}$$

where $\mathbf{x}' = (x_1, x_2, \ldots, x_q)$, $\mathbf{b}' = (b_1, b_2, \ldots, b_q)$ and where

$$\mathbf{B} = \begin{bmatrix} b_{11} & \frac{1}{2}b_{12} & \cdots & \frac{1}{2}b_{1q} \\ \cdots & b_{22} & \cdots & \frac{1}{2}b_{2q} \\ \cdots & \cdots & \cdots & \cdots \\ \text{sym} & \cdots & \cdots & b_{qq} \end{bmatrix} \tag{12.2.3}$$

is a symmetric matrix containing all second-order coefficients. [Note the halves; it is easy to forget them!] The stationary (or "turning") values of (12.2.1) subject to being on a sphere, radius r, centered at the origin $(0, 0, \ldots, 0)$, with equation

$$\mathbf{x}'\mathbf{x} \equiv x_1^2 + x_2^2 + \cdots + x_q^2 = R^2 \qquad (12.2.4)$$

are obtained by considering the Lagrangian function

$$F = b_0 + \mathbf{x}'\mathbf{b} + \mathbf{x}'\mathbf{Bx} - \lambda(\mathbf{x}'\mathbf{x} - R^2), \qquad (12.2.5)$$

where λ is a Lagrangian multiplier. [For the method of "Lagrange's undetermined multipliers," see Draper and Smith (1998, pp. 231–233).] Differentiating (12.2.5) with respect to $\mathbf{x}$ (which can be achieved by differentiating with respect to $x_1, x_2, \ldots, x_q$ in turn and rewriting those equations in matrix form) gives

$$\frac{\partial F}{\partial \mathbf{x}} = \mathbf{b} + 2\mathbf{Bx} - 2\lambda\mathbf{x}. \qquad (12.2.6)$$

Setting (12.2.6) equal to a zero vector leads to the matrix equation

$$2(\mathbf{B} - \lambda\mathbf{I})\mathbf{x} = -\mathbf{b}. \qquad (12.2.7)$$

We can now select a value for λ. If $(\mathbf{B} - \lambda\mathbf{I})^{-1}$ exists, which will happen as long as λ is not an eigenvalue of $\mathbf{B}$, we can obtain a solution $\mathbf{x}$ for a stationary point of $\hat{y}$ on a sphere of radius R by solving

$$\mathbf{x} = -\tfrac{1}{2}(\mathbf{B} - \lambda\mathbf{I})^{-1}\mathbf{b}. \qquad (12.2.8)$$

With that value of $\mathbf{x}$ in hand, we can then find the radius R, from (12.2.4), associated with the solution $\mathbf{x}$ from (12.2.8). Both R and $\mathbf{x}$ are functions of λ, and their numerical values depend on the particular value of λ selected. Note that when $\lambda = 0$, the solution becomes $\mathbf{x} = -\tfrac{1}{2}\mathbf{B}^{-1}\mathbf{b}$, namely *the stationary point of the second-order contour system*, so the stationary point is always on one of the ridge traces.

How do we know on which of the possible $2q$ stationary paths we have determined a point? The key lies in the values of the eigenvalues of $\mathbf{B}$, $\mu_1, \mu_2, \ldots, \mu_q$, say. (In using this notation, we always assume that these are ordered from most negative to most positive—that is, in ascending order with due regard to sign. Words like lower and higher, or up and down, are always used in that sense here.)

The theory in Draper (1963) tells us that if we select values of λ from $+\infty$ downwards as far as the highest eigenvalue μ_q, we shall always be on the "maximum $\hat{y}$" path. Values of λ from $-\infty$ upwards to the lowest eigenvalue μ_1 put us on the "minimum $\hat{y}$" path. Intermediate paths lie in the rages of λ between the eigenvalues of $\mathbf{B}$. When λ is $+\infty$ or $-\infty$, the value of R is zero. (Look at Figure 12.1.) As we move λ back down from $+\infty$ to the largest eigenvalue μ_q of $\mathbf{B}$, R rises to $+\infty$. As we move λ forward from $-\infty$ to the smallest eigenvalue μ_1 of $\mathbf{B}$, R again rises to $+\infty$. In between each successive pair of eigenvalues, R falls from

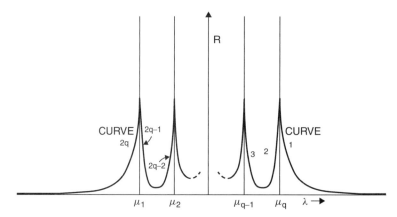

Figure 12.1. General: How the radius R depends on the choice of λ.

$+\infty$ to some value above zero and rises again to $+\infty$, each "loop" corresponding to *two* paths of stationary values of $\hat{y}$. This means that there are $2q$ ridge paths in all, some of which may not appear at low R values of interest. In fact, because we are usually interested only in the smaller values of R that define the region around the origin (or, later, around a selected focal point), the upper portion of Figure 12.1 is usually of little practical interest. However, it is important to appreciate what the behavior of R is, in general, so that one knows what to concentrate on and what to ignore.

Note: There *can* in theory exist second-order surfaces for which R does *not* go to ∞ at an eigenvalue. Such surfaces will have coefficients that obey an unlikely restriction that causes this to happen. In practical cases stemming from real data, this possibility is extraordinarily remote. It is more likely to occur in made-up examples. This incidental point is explored further in Exercise 12.1.

We next provide a small-scale worked example in (x_1, x_2) to illustrate ridge analysis. Further examples in higher dimensions are given in the exercises to this chapter, and their solutions.

12.3. WORKED EXAMPLE OF RIDGE ANALYSIS
ABOUT THE ORIGIN x = 0

This example was first used by A. E. Hoerl (1959). Consider the response surface in two factors with equation

$$\hat{y} = 80 + 0.1x_1 + 0.2x_2 + 0.2x_1^2 + 0.1x_2^2 + x_1x_2. \qquad (12.3.1)$$

So

$$b_0 = 80, \quad \mathbf{B} = \begin{bmatrix} 0.2 & 0.5 \\ 0.5 & 0.1 \end{bmatrix}, \quad \mathbf{b} = \begin{bmatrix} 0.1 \\ 0.2 \end{bmatrix}. \qquad (12.3.2)$$

Equation (12.2.8) becomes

$$(0.2 - \lambda)x_1 + 0.5x_2 = -0.05,$$
$$0.5x_1 + (0.1 - \lambda)x_2 = -0.10, \qquad (12.3.3)$$

with solution

$$x_1 = (9 + 10\lambda)/(2D),$$
$$x_2 = (1 + 20\lambda)/(2D), \qquad (12.3.4)$$

where

$$D = 100\det(\mathbf{B} - \lambda\mathbf{I}) = 100\lambda^2 - 30\lambda - 23. \qquad (12.3.5)$$

Usually we would not obtain specific formulas like (12.3.4). In general, it is easier to perform the computations in matrix form in the computer for individual λ, particularly in higher dimensions. See the solution to Exercise 12.1. The eigenvalues of $\mathbf{B}$ are given by solving the quadratic $D = 0$, which gives

$$\lambda = -0.352 \text{ or } 0.652. \qquad (12.3.6)$$

Suppose we wish to follow the path of the maximum $\hat{y}$ out from the origin $(0, 0)$. We then need to substitute, into the formulas above, values of λ starting from ∞ and working downwards to the larger eigenvalue of 0.652. For each λ, we shall get a point (x_1, x_2), a radius value $R = (x_1^2 + x_2^2)^{1/2}$, and a predicted response $\hat{y}$, obtained by substituting the (x_1, x_2) values into Eq. (12.3.1). Here are a few selected values of these quantities:

λ	x_1	x_2	R	$\hat{y}$
∞	0.00	0.00	0.00	80.00
4.00	0.03	0.06	0.06	80.02
3.00	0.05	0.08	0.09	80.03
1.60	0.14	0.18	0.22	80.08
1.20	0.25	0.29	0.38	80.18
1.00	0.40	0.45	0.60	80.36
0.90	0.58	0.61	0.84	80.64
0.80	1.00	1.00	1.41	81.60
0.78	1.16	1.15	1.64	82.09
0.76	1.39	1.35	1.94	82.86
0.74	1.72	1.65	2.38	84.20
0.72	2.24	2.13	3.09	86.86
0.70	3.20	3.00	4.39	93.47

These $(R, \hat{y})$ values define the curve numbered 1 in Figure 12.2 and $\hat{y}$ in Figure 12.3. Figure 12.4 shows all four ($= 2q$) ridge paths in the (x_1, x_2) space. Figure 12.2, taken from A. E. Hoerl (1959), shows $\hat{y}$ plotted versus the radius R. Curve 4 is developed by using λ values from $-\infty$ to -0.352, and curves 3 and 4 come from

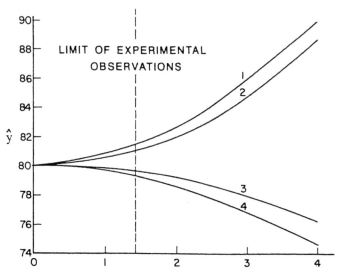

Figure 12.2. Example: Plot of the values of $\hat{y}$ versus radius R on four stationary paths: 1, absolute maximum; 2, smaller local maximum; 3, larger local minimum; 4, absolute minimum.

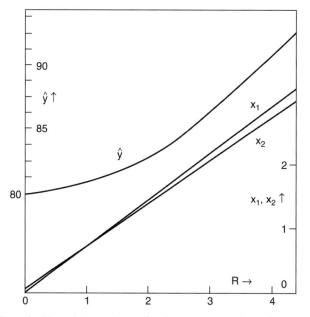

Figure 12.3. Example: Plots of the maximum $\hat{y}$ and its coordinates (x_1, x_2), all versus radius R.

λ values that lie between the two eigenvalues -0.352 and 0.652. For example, a value $\lambda = 0.2$ leads to the point $(-0.22, -0.10)$ for which $R = 0.242$ and $\hat{y} = 79.99$ on curve 3, which is neither a maximum $\hat{y}$, nor a minimum $\hat{y}$ curve. The two loci of intermediate stationary values do not begin until $R = 0.195$, at the point $(-0.196, -0.021)$, corresponding to $\lambda = -0.003$; at this R value, $\partial R/\partial \lambda = 0$; that is, we are at the bottom of the loop of R plotted against λ that lies between the

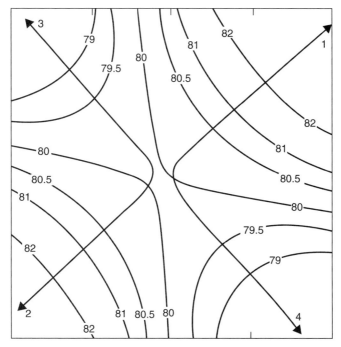

Figure 12.4. Example: The saddle contours and the four stationary paths whose numbers are the same as those in Figure 12.2.

two eigenvalues. (See Figure 12.1.) Because of the scale of the diagram in Figure 12.2, the difference in starting points ($R = 0$ for paths 1 and 4, and $R = 0.195$ for paths 2 and 3) cannot be distinguished, but it can be seen in Figure 12.4. Figure 12.3, a variation of Figure 12.2, shows a plot of the maximum $\hat{y}$ curve (path 1) and the related (x_1, x_2) values, all plotted against the radius R. This is often helpful in assessing the comparative changes that need to be made in the x's to improve the response. In the specific case of Figure 12.3, the two x-lines are very close to being straight lines, showing curvature only near the origin. This is because the center of the system is very close to the origin, and so all of the four paths quickly get on to the major axes and follow those straight lines almost immediately. Often, this would happen outside the immediate region around the origin and so not be so obvious. Both Figures 12.2 and 12.3 have the great advantage that they can be drawn in *two dimensions* for any number of factors q, whereas Figure 12.4 cannot.

As Figure 12.4 shows, the example response surface is a saddle, rising in the first and third quadrants of the (x_1, x_2) plane and falling in the second and fourth quadrants, with ridges oriented approximately 45° to the x-axes and with center very slightly off the origin at ($-\frac{9}{46}, -\frac{1}{46}$). The path of absolute maximum $\hat{y}$ passes from the origin out the first quadrant of the (x_1, x_2) plane, the path of absolute minimum $\hat{y}$ passes from the origin out the fourth quadrant, and the other two paths of stationary points, which are of neither maximum nor minimum $\hat{y}$, pass out the second and third quadrants, starting from the point ($-0.196, -0.021$).

In some examples, the paths of intermediate stationary values are no practical interest; in other examples, they may well be. It is possible that a secondary maximum with near-optimal properties could provide advantages in terms of other responses such as cost, ease of operation, safety, and so on.

We have explored, in our example, only two dimensions (x_1, x_2). This enables us to see the actual response contours and to appreciate the information that ridge analysis provides. When more x-variables are involved, viewing the contours is more difficult, but a ridge analysis can easily be carried out no matter how many dimensions are involved. It can also be carried out from any selected point, as we now discuss.

12.4. RIDGE ANALYSIS AROUND A SELECTED FOCUS f

Ridge analysis does not need to be started from the coded origin $\mathbf{0} = (0, 0, \ldots, 0)'$. Any selected "focal point," or "focus," which we denote here by $\mathbf{f}$, can be used. (For example, $\mathbf{f}$ could be chosen as a central point, perhaps even the exact centroid, of some predefined restricted region of particular interest.) When $\mathbf{f} \neq \mathbf{0}$, the spherical restriction $\mathbf{x}'\mathbf{x} = R^2$ would be replaced by

$$(\mathbf{x} - \mathbf{f})'(\mathbf{x} - \mathbf{f}) = R^2. \tag{12.4.1}$$

We would then consider the Lagrangian function

$$G = b_0 + \mathbf{x}'\mathbf{b} + \mathbf{x}'\mathbf{B}\mathbf{x} - \lambda\{(\mathbf{x} - \mathbf{f})'(\mathbf{x} - \mathbf{f}) - R^2\}, \tag{12.4.2}$$

where λ is a Lagrangian multiplier. Differentiation with respect to $\mathbf{x}$ leads to

$$\frac{\partial G}{\partial \mathbf{x}} = \mathbf{b} + 2\mathbf{B}\mathbf{x} - 2\lambda(\mathbf{x} - \mathbf{f}), \tag{12.4.3}$$

and setting (12.4.3) equal to a zero vector implies that

$$2(\mathbf{B} - \lambda\mathbf{I})\mathbf{x} = -\mathbf{b} - 2\lambda\mathbf{f}. \tag{12.4.4}$$

For many given values of λ (the specific choices will be discussed below), we can write a solution for $\mathbf{x}$ as

$$\mathbf{x} = -\tfrac{1}{2}(\mathbf{B} - \lambda\mathbf{I})^{-1}(\mathbf{b} + 2\lambda\mathbf{f}). \tag{12.4.5}$$

This leads us into the following solution sequence:

1. Choose a value of λ appropriate for the desired path (to be explained below).
2. Obtain $\mathbf{x}$ from (12.4.5). Using that $\mathbf{x}$:
3. Evaluate $R^2 = (\mathbf{x} - \mathbf{f})'(\mathbf{x} - \mathbf{f})$ and thus obtain R.
4. Evaluate $\hat{y} = b_0 + \mathbf{x}'\mathbf{b} + \mathbf{x}'\mathbf{B}\mathbf{x}$
5. Tabulate $\mathbf{x}$, R, and $\hat{y}$.

Then the point **x** will be on the desired path of stationary values and will lie on a sphere of radius R centered at **f**. Again, the eigenvalues of the **B** matrix, namely, the values that result from solving $\det(\mathbf{B} - \lambda\mathbf{I}) = 0$, form the dividing points for the various paths of stationary values. In general, there are q eigenvalues and $2q$ paths; see Figure 12.1 again. Recall that (a) choosing $\lambda > \mu_q$ provides a locus of maximum $\hat{y}$ as R changes, (b) choosing $\lambda < \mu_1$ provides a locus of minimum $\hat{y}$ as R changes, and (c) choosing $\mu_1 < \lambda < \mu_q$ gives intermediate stationary values. Note that we do not actually need the eigenvalues of **B** *to obtain* the paths, but only *to distinguish between* paths. For the loci of the maximum $\hat{y}$ and the minimum $\hat{y}$, which are the most common applications, the eigenvalues are not specifically needed, because choosing λ values decreasing from ∞ gives the path of maximum $\hat{y}$, while using values increasing from $-\infty$ gives the path of minimum $\hat{y}$. However, obtaining the actual eigenvalues is recommended, to avoid the slight possibility of getting onto an unwanted path.

Applying the Above Results to a First-Order Model as a Special Case

The first-order model $\hat{y} = b_0 + b_1 x_1 + b_2 x_2 + \cdots + b_q x_q$ occurs when $\mathbf{B} = \mathbf{0}$. The "eigenvalues of **B**" are thus all zero, and there are only two segments for λ, $-\infty$ to 0 and 0 to $+\infty$. The solution for **x** reduces to

$$\mathbf{x} = \mathbf{f} + (2\lambda)^{-1}\mathbf{b}. \tag{12.4.6}$$

Choosing λ in $[0, \infty]$—that is, positive—gives the straight-line steepest-ascent direction, and choosing λ negative in $[-\infty, 0]$ gives the straight-line steepest-descent direction. Note that $\mathbf{x} - \mathbf{f}$ is proportional to **b**, as it should be.

12.5. WORKED EXAMPLE OF RIDGE ANALYSIS ABOUT A SELECTED FOCUS **f**

We reexamine the surface (12.3.1) previously used, but now choose a focus different from the origin, at $(0.5, 0.5)$. Equation (12.4.5) leads, after reduction, to the solution

$$x_1 = (\lambda^2 + 0.5\lambda + 0.09)/(2\lambda^2 - 0.60\lambda - 0.46),$$

$$x_2 = (\lambda^2 + 0.5\lambda + 0.01)/(2\lambda^2 - 0.60\lambda - 0.46). \tag{12.5.1}$$

When λ exceeds the higher eigenvalue 0.652, these coordinates provide the maximum $\hat{y}$ path in the table, starting from $(0.5, 0.5)$, shown in the table below. As before, we would usually not evaluate specific formulas like (12.5.1). It is easier to perform the computations in matrix form in the computer for individual λ values, particularly in higher dimensions. See the details in Section 12.7 for more on this.

λ	x_1	x_2	R	$\hat{y}$
∞	0.500	0.500	0.000	80.475
4.0	0.621	0.618	0.169	80.685
3.0	0.673	0.668	0.241	80.785
1.6	0.932	0.911	0.597	81.382
1.2	1.253	1.206	1.032	82.337
1.0	1.691	1.606	1.626	84.038
0.9	2.177	2.048	2.283	86.455
0.8	3.324	3.088	3.830	94.377
0.78	3.769	3.492	4.431	98.294

12.6. BEHAVIOR OF RIDGE PATHS IN GENERAL

The $\hat{y}$ values on a ridge path will exhibit one of the following four behaviors. The response will

1. Decrease monotonically, or
2. Increase monotonically, or
3. Pass through a maximum and then decrease monotonically, or
4. Pass through a minimum and then increase monotonically.

In cases 3 and 4, the path changes slope as it passes through the stationary point of the second order response surface at the value $\lambda = 0$. The path that contains the stationary point is determined by the position of $\lambda = 0$ in relation to the eigenvalues of **B**.

12.7. A RIDGE EXPLORATION SOLUTION USING MINITAB

In this section, we illustrate the evaluation of a ridge solution via MINITAB. The fitted second-order equation used for illustration, taken from the solution to Exercise 9.1, is

$$\hat{y} = 30.791 - 2.2912\,B + 17.6x_1 + 0.777x_2 - 1.140x_3$$
$$+ 2.552x_1^2 - 0.099x_2^2 - 0.453x_3^2 - 2.75x_1x_2 + 2.5x_1x_3 - 1.25x_2x_3.$$

The terms $-2.2912\,B$ measures a specific block effect that raises or lowers a particular set of fitted responses to distinguish between blocks and this is dropped from the ridge analysis that follows. In the $\hat{y} = b_0 + \mathbf{x}'\mathbf{b} + \mathbf{x}'\mathbf{Bx}$ notation for a second-order surface, we have $b_0 = 30.791$,

$$\mathbf{b} = \begin{bmatrix} 17.600 \\ 0.777 \\ -1.140 \end{bmatrix} \quad \text{and} \quad \mathbf{B} = \begin{bmatrix} 2.552 & -1.375 & 1.25 \\ -1.375 & -0.099 & -0.625 \\ 1.25 & -0.625 & -0.453 \end{bmatrix}. \quad (12.7.1)$$

The three eigenvalues of the second-order matrix **B** are -0.956, -0.659, and 3.615 so that, for the maximum path, for example, we need values $3.615 < \lambda < \infty$. Note

Table 12.1. Ridge Trace of the Maximum $\hat{y}$ from $(0, 0, 0)$

λ	x_1	x_2	x_3	R	$\hat{y}$
∞	0	0	0	0	30.8
60	0.153	0.003	−0.006	0.153	33.6
30	0.320	−0.002	−0.006	0.320	36.7
20	0.506	−0.015	0.004	0.506	40.4
15	0.714	−0.040	0.022	0.715	44.7
11	1.069	−0.102	0.072	1.076	52.9
10	1.223	−0.134	0.100	1.234	56.7
9	1.431	−0.183	0.141	1.450	62.2
8	1.729	−0.262	0.208	1.761	70.6

that, because 0 does not lie in this interval, the stationary point does not lie on the maximum path. We obtain Table 12.1, which is carried out to roughly $R = 2^{3/4} = 1.682$. (All the data points lie in, or on, a sphere of this radius.)

We see that most of the movement that best improves the response results from changing x_1. The other two variables make smaller contributions. Response diagrams are given in the original source paper.

MINITAB Details

The calculations shown in Table 12.1 were obtained using MINITAB instructions as follows:

```
read 3 1 m1
17.6
0.777
-1.140
read 3 1 m14
0
0
0
read 3 3 m2
2.552 - 1.375 1.25
-1.375 - 0.099 - 0.625
1.25 - 0.625 - 0.453
eigen m2 c24 m24
print c24
print m24
set c19
30.791
end
read 3 3 m3
1 0 0
0 1 0
0 0 1
let k4 = 2
let k3 = -0.5
let k2 = 1000
```

[This keys in the equation and sets $\lambda = 1000$, which represents infinity. The matrix m_{14} contains the coordinates of the starting point of the ridge trace, here set to the origin. The next step is to invoke a MINITAB macro, given below, by typing **%Filename**. The filename can be whatever you choose. [We chose "Fridge" to indicate ridge from a selected focus, here chosen as $(0, 0, 0)$.]

If MINITAB indicates it cannot find the macro and if the macro is in the MACROS folder, the instruction **cd C:\ MTBWIN \ MACROS** should be given. This instruction needs to be invoked only once per session, if at all, and may need to be changed as appropriate for your setup. (See MINITAB Answers Knowledge-base No. 1019 at minitab.com/support/answers for more on this.)

After the macro has performed the calculation, reset k2 = 60 (or whatever you choose to set next) and invoke the macro again, and so on. The macro uses whatever dimensions were used in setting in the data and does not need to be adjusted. The calculations can also be done via a loop with (DO, ENDDO) but, because of the complex relationship between λ and the path points, some interme-diate inspection of results is usually helpful, we have found.

As already mentioned, the instructions above are appropriate when ridge analysis starts at the x-origin $(f_1, f_2, f_3) = (0, 0, 0)$. For a version appropriate when there are four dimensions and the starting point (focus) is (f_1, f_2, f_3, f_4), see below. The pattern shown extends in the obvious way for more x's, with appropriate adjustments made to the input dimensions.

MINITAB Instructions for Four Dimensions and a General Focus Point
(This can be extended to higher dimensions in the obvious way.)

For a four-x example, starting from a focus point (f_1, f_2, f_3, f_4), set in the data like this:

```
read 4 1 m1
b₁
b₂
b₃
b₄
read 4 1 m14
f₁
f₂
f₃
f₄
read 4 4 m2
b₁₁  ½b₁₂  ½b₁₃  ½b₁₄
½b₁₂  b₂₂  ½b₂₃  ½b₂₄
½b₁₃  ½b₂₃  b₃₃  ½b₃₄
½b₁₄  ½b₂₄  ½b₃₄  b₄₄
eigen m2 c24 m24
print c24
print m24
set c19
b₀
end
```

```
read 4 4 m3
1 0 0 0
0 1 0 0
0 0 1 0
0 0 0 1
let k4 = 2
let k3 = − 0.5
let k2 = [selected first λ value]
```

[The lines above key in the equation and set λ. Now invoke the MINITAB macro given below by typing %Filename. If MINITAB indicates it cannot find the macro and if the macro is stored in the macros folder, the instruction cd C:\ MTBWIN \ MACROS is needed. This instruction needs to be invoked only once per session, if at all, and may need to be changed as appropriate for your setup. (See MINITAB Answers Knowledgebase No. 1019 at minitab.com/support/answers.) Next, reset k2 and invoke the macro again, and so on. The macro uses whatever dimensions were used in setting in the data and so does not need to be adjusted.]

GMACRO

```
Filename
mult k2 m14 m15
mult k4 m15 m16
add m16 m1 m17
mult k2 m3 m4
subtract m4 m2 m5
invert m5 m6
mult m6 m17 m7
mult k3 m7 m8
transpose m8 m9
subtract m14 m8 m18
transpose m18 m19
mult m19 m18 m20
copy m20 c20
let c21 = sqrt(c20)
print k2
print c21
print m9
mult m9 m1 m10
mult m9 m2 m11
mult m11 m8 m12
add m10 m12 m13
copy m13 c22
let c23 = c22 + c19
print c23
ENDMACRO
```

[Now reset k2 and invoke the macro again, and so on.]

The following notes explain the purposes of the variables in the input and output.

Input:

m1 is the **b** vector,
m14 is the starting point (focus),
m2 is the **B** matrix,
c19 is b_0,
m3 is the appropriate size unit matrix,
k4 and k3 are fixed constants.

Output:

c24 contains the eigenvalues of **B**,
k2 is λ,
m34 is a column-wise format of a point (x_1, x_2, x_3) on the ridge path for that λ,
c36 is the corresponding R,
c43 is $\hat{y}$ at that point.

The order of the printout can be changed as desired, provided that each quantity printed has been calculated before the point of printout! Unneeded details automatically printed by MINITAB can be ignored or deleted.

EXERCISES

12.1. Consider the second-order (quadratic) surface $\hat{y} = b_0 + b_1 x_1 + b_2 x_2 + b_{11} x_1^2 + b_{22} x_2^2 + b_{12} x_1 x_2$. Show that if

$$b_{22} - b_2 b_{12}/(2b_1) = b_{11} - b_1 b_{12}/(2b_2) = \text{either } \lambda_1 \text{ or } \lambda_2,$$

where λ_1 and λ_2 are the roots of $\det(\mathbf{B} - \lambda\mathbf{I}) = 0$ and where **B** is

$$\mathbf{B} = \begin{bmatrix} b_{11} & \frac{1}{2}b_{12} \\ \frac{1}{2}b_{12} & b_{22} \end{bmatrix},$$

then R in Eq. (12.2.4) does *not* go to $+\infty$ at whichever eigenvalue is satisfied by the condition above. Note that this anomaly is extremely unlikely to occur in a practical situation. In higher dimensions, with more coefficients, the required conditions are even more complicated and thus even less likely to happen.

12.2. (*Source:* A fertilizer production surface with specification of economic optima for corn growth on calcareous Ida silt loam, by E. O. Heady and J. Pesek, *Journal of Farm Economics*, 1954, **36**(3), 466–482.) The yield of corn was measured for a number of combinations of the two variables: 1, pounds of nitrogen per acre; 2, pounds of P_2O_5 per acre. Three models were fitted to the data (which are not given in the paper) and two of the models, called II and III by the authors, were second-order polynomials, as follows:

$$\text{Model II:}\quad \hat{y} = 7.51 + 0.584N + 0.644P - 0.00158N^2$$

$$- 0.00180P^2 + 0.00081NP.$$

$$\text{Model III:}\quad \hat{y} = 5.682 + 6.3512N^{1/2} + 8.5155P^{1/2} - 0.316N$$

$$- 0.417P + 0.34310N^{1/2}P^{1/2}.$$

Note the interesting fact that both of these fitted models are second-order, the first in (N, P) and the second in $N^{1/2}$ and $P^{1/2}$. They simply represent different viewpoints as to how the fertilizers might influence the yield response. The respective R^2 values for the two equations were 0.832 and 0.918. After further examination of the predictions that resulted, the authors eventually threw their weight behind model III, the "square-root" quadratic. Suppose the authors had performed a ridge analysis on both equations, restricting the space of interest to the experimental ranges 0–320 for nitrogen and 0–320 for P_2O_5, how far apart would be their predicted "best $\hat{y}$" locations?

12.3. (*Source:* Robust design and optimization of material handling in an FMS, by J. S. Shang, *International Journal of Product Research*, 1995, **33**(9), 2437–2454.) Investigation of a flexible manufacturing system (FMS) involved two stages: A steepest ascent, followed by the fitting of a five-factor, second-order response surface $\hat{y} = b_0 + \mathbf{x}'\mathbf{b} + \mathbf{x}'\mathbf{B}\mathbf{x}$, in coded x-variables, where $b_0 = 258$, $\mathbf{b}' = (43.7, 38.9, 65.6, 26.8, 68.5, -57.6)$, and

$$\mathbf{B} = \begin{bmatrix} -15.70 & 4.20 & -4.80 & -2.37 & 5.30 & -2.77 \\ 4.20 & 43.00 & 8.00 & 4.77 & -2.23 & 1.33 \\ -4.80 & 8.00 & -21.70 & 0.00 & -0.09 & -1.53 \\ -2.37 & 4.77 & 0.00 & -10.50 & -0.54 & -1.97 \\ 5.30 & -2.23 & -0.09 & -0.54 & -25.80 & -3.57 \\ -2.77 & 1.33 & -1.53 & -1.97 & -3.57 & -37.30 \end{bmatrix}.$$

Find the path of maximum predicted response $\hat{y}$, starting from the origin $(0, 0, 0, 0, 0)$ and determine the point at which a maximum response occurs in, or on, the sphere $\mathbf{x}'\mathbf{x} = 32^{1/2} = 5.657 = 2.378^2$. All the data points lie in or on this sphere. What sort of fitted surface do we have?

12.4. (*Source:* Edible wheat gluten films: Influence of the main process variables on film properties using response surface methodology, by N. Gontard, S. Guilbert, and J. L. Cuq, *Journal of Food Science*, 1992, **57**(1), 190–195, 199.) Table E12.4 shows the results from an experiment to investigate the effects of gluten concentration (G), ethanol (E), and pH on the properties acquired by an edible wheat gluten film. The coded experimental levels shown in the

Table E12.4. Edible Wheat Gluten Film Experiment with Six Responses

No.	x_1	x_2	x_3	y_1	y_2	y_3	y_4	y_5	y_6
1	−1	−1	−1	75.6	62.7	1.05	1.98	17.6	0.71
2	1	−1	−1	12.1	31.5	1.08	2.85	7.6	0.72
3	−1	1	−1	20.0	43.8	0.81	1.17	12.0	0.60
4	1	1	−1	25.0	46.1	0.68	3.13	7.6	0.69
5	−1	−1	1	9.0	49.6	0.90	0.95	4.8	0.67
6	1	−1	1	66.8	39.1	0.74	2.68	5.6	0.64
7	−1	1	1	88.0	75.5	0.84	1.95	10.8	0.64
8	1	1	1	139.2	68.2	1.02	2.12	8.4	0.59
9	−2	0	0	87.6	63.5	0.87	2.57	9.6	0.57
10	2	0	0	93.8	48.3	0.85	3.84	8.8	0.68
11	0	−2	0	55.2	57.2	1.05	1.12	8.8	0.89
12	0	2	0	250.4	100.0	0.97	1.00	10.4	0.79
13	0	0	−2	88.8	49.8	1.12	0.22	12.1	0.71
14	0	0	2	82.4	100.0	0.87	1.76	13.2	0.58
15	0	0	0	59.3	42.8	0.96	1.78	12.4	0.79
16	0	0	0	70.6	48.5	0.86	2.12	12.8	0.82
17	0	0	0	65.0	49.9	0.91	2.03	12.5	0.85
18	0	0	0	67.3	41.8	1.00	1.85	12.2	0.83
19[a]	1	0	−1	34.5	34.1	0.93	2.36	7.3	0.75
[b]				30.1	35.3	0.92	2.56	8.7	0.76

[a] Verification point, after initial response surface analysis.
[b] Values predicted by the fitted model at $(1, 0, -1)$.

table are such that $x_1 = (G - 7.5)/2.5$, $x_2 = (E - 45)/12.5$, $x_3 = pH - 4$. The six property responses are $y_1 =$ opacity, $y_2 =$ solubility, $y_3 =$ water vapor permeability, $y_4 =$ puncture strength, $y_5 =$ puncture deformation, and $y_6 =$ relaxation coefficient. (The various units are described in the source reference.)

(a) Fit second-order models to any or all six responses.

(b) In fact, the authors added a cubic term $x_1 x_2 x_3$ to all the models. If this term were added, how would the extra term change the estimates from the quadratic fit? Which equations, *second-order* or *second-order plus the single cubic term*, would you feel happy interpreting further?

(c) Determine the ridge path of maximum puncture strength, y_4 using the full second order fit. Which coefficients in this model are individually significant? Does a quadratic model involving only x_1 seem reasonable? How different would the ridge path of maximum puncture strength be if only x_1 were in the quadratic model?

12.5. Return to the Hoerl example given in Section 12.3 and find the ridge path details for paths 2, 3 and 4. (Recall that the eigenvalues were −0.352 and 0.652.)

12.6. (*Source:* Multiple regression and response surface analysis of the effects of calcium chloride and cysteine on heat-induced whey protein gelation, by

R. H. Schmidt, B. L. Illingworth, J. C. Deng, and J. A. Cornell, *Journal of Agriculture and Food Chemistry*, **27**, 1979, 529–532.) In Exercise 11.25 you were asked to analyze a set of data involving two predictor variables and four responses. Consider the two responses cohesiveness (y_2) and springiness (y_3). Their fitted equations, of the form $b_0 + \mathbf{x'b} + \mathbf{x'Bx}$, are such that:

For y_2: $b_0 = 0.660$, $\quad \mathbf{b} = \begin{bmatrix} 0.092 \\ 0.010 \end{bmatrix}, \quad \mathbf{B} = \begin{bmatrix} -0.171 & 0.159 \\ 0.159 & -0.098 \end{bmatrix};$

For y_3: $b_0 = 1.776$, $\quad \mathbf{b} = \begin{bmatrix} -0.250 \\ -0.078 \end{bmatrix}, \quad \mathbf{B} = \begin{bmatrix} -0.156 & 0.005 \\ 0.005 & -0.079 \end{bmatrix}.$

Beginning from the coded origin $(x_1, x_2) = (0, 0)$, find the individual ridge paths of maximum response. They should pass through the stationary points determined in the solution to Exercise 11.25 and should pass appropriately through the contours shown on p. 708.

For two further exercises (for which solutions are not provided), find the ridge paths of maximum response for the first and fourth responses in Exercise 11.25. To check your answers make copies of the appropriate response contours on p. 708 and roughly sketch your paths on them.

12.7. (*Source:* Development of techniques for real-time monitoring and control in plasma etching, by K. J. McLaughlin, S. W. Butler, T. F. Edgar, and I. Trachtenberg, *Journal of the Electrochemical Society*, **138**, March 1991, 789–799.) Several possible equations, including transformations of the response variable, were developed from an extensive set of experimental runs describing the etch rate in terms of four predictor variables, power, pressure, H_2, and flow rate. A second-order equation regarded favorably for the silicon dioxide etch rate took the form $b_0 + \mathbf{x'b} + \mathbf{x'Bx}$, such that $b_0 = 350$,

$$\mathbf{b} = \begin{bmatrix} 185.4 \\ -120.1 \\ -21.0 \\ 0 \end{bmatrix}, \quad \text{and} \quad \mathbf{B} = \begin{bmatrix} 11.62 & -16.55 & 14.05 & 0 \\ -16.55 & 14.72 & 0 & 0 \\ 14.05 & 0 & 0 & 0 \\ 0 & 0 & 0 & 0 \end{bmatrix}.$$

Starting from the origin, $(0, 0, 0, 0)$, find the ridge path of maximum response. What feature of the path is particularly noticeable?

12.8. In Exercise 11.22, you fitted a second-order surface and examined it via canonical reduction.

(**a**) Using the same surface, perform a ridge analysis, starting from the origin $(0, 0, 0)$ and following the path of maximum response only, from values $R = 0$ to not more than approximately $R = 1.732$. (Remember that all the design points lie within a sphere of radius $3^{1/2} = 1.732$.) Provide a table which shows sequences of values of λ, x_1, x_2, x_3, R, and $\hat{y}$ and then briefly explain what the maximum path $\hat{y}$ indicates to you. Your answer should be consistent with what you saw in Exercise 11.22, of course.

(b) Now think of this problem slightly differently. Suppose the stationary point has the coordinates $\mathbf{f} = (f_1, f_2, f_3)'$, which you have already worked out in Exercise 11.22. In the Lagrangian problem you solved in (a), suppose you replaced $\mathbf{x}'\mathbf{x} = R^2$ by the equation $(\mathbf{x} - \mathbf{f})'(\mathbf{x} - \mathbf{f}) = R^2$ and reworked the solution. What would you expect the solution to tell you? Explain why. To answer this part, you should not need to do any further computations, but simply bring together the knowledge gained from your previous work.

12.9. (*Source:* Effect of treating *Candida utilis* with acid or alkali, to remove nucleic acids, on the quality of the protein, by I. M. Achor, T. Richardson, and N. R. Draper, *Agricultural and Food Chemistry*, **29**, 1981, 27–33.) In a comprehensive experiment to establish optimum conditions for the removal of nucleic acids from single-cell protein, four factors (yeast concentration, time, temperature, and pH of treatments) were varied in 29 experimental runs in the form of a four-dimensional "cube plus star plus five center points" design. Nineteen response variables were recorded. Table E12.9

Table E12.9. Data from Achor et al. (1981)

No.	x_1	x_2	x_3	x_4	y_1	y_2	y_3	y_4
1	−1	−1	−1	−1	0.7	76.7	82.0	81.7
2	−1	−1	−1	1	10.8	76.7	79.3	75.0
3	1	−1	−1	−1	5.2	75.2	83.2	79.6
4	1	−1	−1	1	7.4	77.2	76.4	76.2
5	−1	1	−1	−1	1.5	76.0	80.7	77.3
6	−1	1	−1	1	11.3	76.0	77.3	78.7
7	1	1	−1	−1	2.4	77.6	81.6	80.4
8	1	1	−1	1	23.0	76.0	77.2	74.4
9	−1	−1	1	−1	5.0	74.0	82.7	80.0
10	−1	−1	1	1	37.7	78.0	79.3	76.3
11	1	−1	1	−1	4.8	74.8	82.8	80.6
12	1	−1	1	1	35.4	74.4	78.8	77.4
13	−1	1	1	−1	6.4	76.0	78.7	79.0
14	−1	1	1	1	37.7	71.3	80.7	76.7
15	1	1	1	−1	4.6	73.6	80.8	80.4
16	1	1	1	1	36.0	73.2	78.0	77.4
17	0	0	0	0	10.8	74.5	81.0	76.0
18	0	0	0	0	11.3	76.0	79.5	75.0
19	0	0	0	0	15.8	74.0	83.5	76.8
20	0	0	0	0	14.5	72.0	81.0	76.0
21	0	0	0	0	11.0	75.5	78.5	78.5
22	0	0	−2	0	1.8	75.5	79.0	81.5
23	0	0	2	0	22.5	70.5	76.5	77.3
24	0	−2	0	0	4.4	76.0	83.0	77.8
25	0	2	0	0	13.5	73.5	83.0	77.8
26	−2	0	0	0	9.0	74.0	83.0	76.5
27	2	0	0	0	10.8	74.7	83.3	82.6
28	0	0	0	−2	1.3	76.0	84.5	78.3
29	0	0	0	2	37.3	71.1	72.5	74.8

Table E12.10. Abrasion Resistance Data

No.	x_1	x_2	x_3	y
1	-1	-1	-1	29.5
2	1	-1	-1	21.7
3	-1	1	-1	24.5
4	1	1	-1	23.7
5	-1	-1	1	51.3
6	1	-1	1	19.4
7	-1	1	1	49.0
8	1	1	1	30.8
9	-2	0	0	42.7
10	2	0	0	16.8
11	0	-2	0	24.6
12	0	2	0	20.9
13	0	0	-2	30.5
14	0	0	2	44.8
15	0	0	0	30.4

shows the design in the usual coded form as well as the results of four of the responses. (Results from the remaining 15 responses are given in the paper.) Fit second-order surfaces to all four responses and, using canonical reduction and/or ridge analysis, determine the position (x_1, x_2, x_3, x_4) in each case of the predicted highest response, within or on a sphere of radius 2. (Note that all the 24 noncentral design points lie on a sphere of radius 2, so that we have no response information outside that sphere.)

12.10. (*Source:* Adapted, with the permission of *Rubber World*, from Statistical methods of compounding—modern and essential tool for the rubber chemist, by J. R. Carney and R. W. Hallman, *Rubber World*, September 1967, 78–83.) Table E12.10 shows 15 observations of NBS abrasion resistance (y), adapted from Figure 5, p. 80, of the source paper. The three predictor variables (x_1, x_2, x_3) shown are coded forms of Sunthene 380, phr Dixie clay, and rubber blend composition. Fit a second-order surface to these data and investigate where, on or within a sphere of radius 2, the highest abrasion resistance is predicted to occur.

12.11. (*Source:* Optimizing the enzymatic maceration of foliole purée from hard pieces of hearts of palm (*Euterpe edulis* Mart.) using response surface analysis, by R. Kitagawa, R. E. Bruns, and T. J. B. De Menezes, *Journal of Food Science*, 1994, **59**(4), 844–848.) A three-dimensional central composite design, with (coded) star distance $\alpha = 3^{1/2} = 1.732$ in two of the dimensions and $\alpha = 1$ in the third, was used to investigate the effects of three predictor variables on the viscosity of foliole purée made from hard pieces of hearts of palm. The variables were coded as follows:

(a) $x_1 = (E - 0.75)/0.25$, where E was a percent enzyme concentration.

(b) $x_2 = (t - 5)/1$, where t was incubation time in hours.

(c) $x_3 = (T - 45)/5$, where T was incubation temperature in degrees centigrade.

Table E12.11a. Experimental Conditions and Viscosity Values for the Central Composite Design with Five Center Points Used to Investigate Hearts of Palm

Run	x_1	x_2	x_3	y_1	y_2
1	-1	-1	-1	160.0	131.8
2	1	-1	-1	91.5	120.0
3	-1	1	-1	66.6	110.6
4	1	1	-1	55.0	75.0
5	-1	-1	1	77.3	130.8
6	1	-1	1	85.6	141.6
7	-1	1	1	101.3	184.0
8	1	1	1	80.0	115.0
9	-1.732	0	0	136.6	103.3
10	1.732	0	0	82.3	55.0
11	0	-1.732	0	121.6	105.0
12	0	1.732	0	93.3	71.7
13	0	0	-1	68.3	128.3
14	0	0	1	73.3	117.7
15	0	0	0	69.9	90.0
16	0	0	0	63.3	95.0
17	0	0	0	75.0	89.2
18	0	0	0	65.0	80.0
19	0	0	0	67.3	90.8

The star distance for the third variable was restricted to one unit because there was no experimental interest in temperatures outside the 40–50°C range. Two responses were observed, both of which were viscosities (cP): y_1 for treatments with cellulase and y_2 for treatments with pectinase. The experiment was devised to investigate whether increased usable yields could be obtained from hearts of palm, using cellulase and pectinase treatments. To achieve this aim, lower values of the viscosities were more desirable. The data are given in Table E12.11a.

Fit second-order response surface models separately to both sets of response data and find approximately where, within the experimental ranges used for the predictor variables, the best (lowest) values of the predicted viscosities are to be found.

12.12. (*Source:* Predicting rubber properties with accuracy, by George C. Derringer, a presentation made to the American Chemical Society, Division of Rubber Chemistry, 1969, in Los Angeles, CA.) An acceleration study of silica filled natural rubber was performed to study the effects of two predictor variables, $X_1 = N$-t-butyl2-benzothiazolesulfenamide, coded as $x_1 = (X_1 - 1.2)/0.56577$, and $X_2 =$ tetraethylthiuram disulfide, coded as $x_2 = (X_2 - 0.6)/0.28288$, on six responses, one of which was tensile strength in pounds per square inch (psi). Because it was not known beforehand whether or not a second order model would be adquate for each response, the design chosen, shown in Table E12.12a, consisted of a third order rotatable design. This was performed in three sequential blocks (each

Table E12.12a. Experimental Data on Rubber Tensile Strength

No.	x_1	x_2	y (psi)	B_1	B_2
1	1	1	3670	-1	1
2	1	-1	3430	-1	1
3	-1	1	3790	-1	1
4	-1	-1	2940	-1	1
5	0	0	3800	-1	1
6	0	0	3860	-1	1
7	0	0	3830	-1	1
8	1.414	0	3670	0	-2
9	-1.414	0	3570	0	-2
10	0	1.414	3770	0	-2
11	0	-1.414	3040	0	-2
12	0	0	3870	0	-2
13	0	0	3690	0	-2
14	0	0	3820	0	-2
15	0.667	0.836	3440	1	1
16	-0.238	1.042	3560	1	1
17	-0.963	0.464	3200	1	1
18	-0.963	-0.464	2860	1	1
19	-0.238	-1.042	3010	1	1
20	0.667	-0.836	3510	1	1
21	1.069	0	3660	1	1

containing seven runs), requiring the addition of two block variables B_1 and B_2 in the model to separate the means of the blocks. Block 1 is a first-order orthogonal (and so first-order rotatable) design; blocks 1 and 2 together form a second-order rotatable design and all three blocks form a third order rotatable design. Note that the seven points of Block 3 all lie on a circle of radius 1.069, a feature not obvious at first sight. In such designs, data are often fitted in sequence to first-, second-, and third-order models, as successive sets of results become available, but here, as in the source reference, all the data will be used together in considering a suitable model. Use the data for tensile strength to fit a model of suitable order and comments on what you see. Where is the highest predicted tensile strength in the experimental region?

12.13. (*Source:* Statistical experimental design: Relationship between injection molding variables and ROVEL physical properties, by C. W. Bawn and V. Ricci, *Proceedings of the 39th Annual Technical Conference, Society of Plastic Engineers*, 1981, 780–782.) Rovel is a high-impact weatherable plastic developed by Uniroyal. There are several grades for extrusion applications. The experimental runs described here were made to investigate the characteristics of a special easy flow grade, intended for injection molding applications. Table E12.13a shows the 27 runs of an experimental design consisting of a cube plus star plus three center points for four variables (x_1, x_2, x_3, x_4).

Table 12.13a. Rovel Design and Four Recorded Responses

Order	No.	x_1	x_2	x_3	x_4	y_1	y_9	y_{13}	y_{15}
B	1	1	1	1	1	5520	138	88	11.8
C	2	1	1	1	-1	5580	149	87	11.4
H	3	1	1	-1	1	5640	143	91	11.1
D	4	1	1	-1	-1	5540	149	91	11.4
I	5	1	-1	1	1	5500	154	85	11.6
E	6	1	-1	1	-1	5550	156	76	11.5
G	7	1	-1	-1	1	5500	157	90	11.4
F	8	1	-1	-1	-1	5510	158	86	11.1
W	9	-1	1	1	1	5950	81	88	7.8
V	10	-1	1	1	-1	5940	80	80	10.1
X	11	-1	1	-1	1	5890	86	92	8.3
U	12	-1	1	-1	-1	5860	89	88	9.8
Y	13	-1	-1	1	1	6050	83	89	10.2
T	14	-1	-1	1	-1	6120	79	64	9.3
Z	15	-1	-1	-1	1	5970	89	91	9.5
S	16	-1	-1	-1	-1	6010	104	92	10.2
A	17	2	0	0	0	5260	156	83	12.3
R	18	-2	0	0	0	6100	34	83	6.5
J	19	0	2	0	0	5770	102	88	11.2
K	20	0	-2	0	0	5870	113	86	11.6
L	21	0	0	2	0	5625	91	89	11.7
M	22	0	0	-2	0	5650	103	95	11.2
O	23	0	0	0	2	5790	117	93	11.7
P	24	0	0	0	-2	5790	90	87	11.3
N	25	0	0	0	0	5750	91	92	11.3
Q	26	0	0	0	0	5700	91	92	10.9
AA	27	0	0	0	0	5720	96	94	11.6

The codings associated with the experimental variables were as follows:

i	Level of x_i	-2	-1	0	1	2
1	Barrel temperature (°F)	300	350	400	450	500
2	Cavity pressure (psi/10^3)	6	7	8	9	10
3	Fill time (seconds)	2	3.5	5	6.5	8
4	Mold temperature (°F)	120	140	160	180	200

The responses shown were:

y_1 = Tensile strength (psi).

y_9 = Gardner drop weight impact at $-20°F$.

y_{13} = Gardner gloss.

y_{15} = Notched izod on $\frac{1}{4}$-inch piece at room temperature.

(The order column indicates the order $A, B, C, \ldots, Z, AA$ in which the experiments were performed.) Fit a second-order surface to the tensile strength data and interpret the equation that results. Find the ridge path of maximum tensile strength, starting at the coded origin $(0, 0, 0, 0)$.

CHAPTER 13

Design Aspects of Variance, Bias, and Lack of Fit

13.1. THE USE OF APPROXIMATING FUNCTIONS

Typically, in a scientific investigation, some response y is measured whose mean value $E(y) = \eta$ is believed to depend on a set of variables $\boldsymbol{\xi} = (\xi_1, \xi_2, \ldots, \xi_k)'$. The exact functional relationship between them,

$$E(y) = \eta = \eta(\boldsymbol{\xi}), \qquad (13.1.1)$$

is usually unknown and possibly unknowable. We often represent the function $\eta(\boldsymbol{\xi})$ as the solution to some set of time- and space-dependent ordinary or partial differential equations. However, we have only to think of the flight of a bird, the fall of a leaf, or the flow of water through a valve, to realize that, even using such equations, we are likely to be able to approximate only the main features of such a relationship. In this book, we employ even cruder approximations using polynomials that exploit local smoothness properties. Approximations of this type are, however, often perfectly adequate locally. Over a short distance, the flight of the bird might be approximated by a straight-line function of time; over somewhat longer distances, a quadratic function might be used. Over narrow ranges, the flow of water through the valve might be similarly approximated by a straight-line function of the valve opening.

Mathematically, this idea may be expressed by saying that, over a limited range of interest $R(\boldsymbol{\xi})$, the main characteristics of a smooth function may be represented by the low-order terms of a Taylor series approximation. It should be understood that the *region of interest* $R(\boldsymbol{\xi})$ will lie within a larger *region of operability* $O(\boldsymbol{\xi})$, defined as the region over which experiments *could* be conducted if desired, using the available apparatus. For example, suppose we are anxious to discover the water temperature ξ which will extract the maximum amount of caffeine η when hot water is added to a particular kind of tea leaves. Suppose, further, that we have a pressure cooker at our disposal. The operability region $O(\boldsymbol{\xi})$ would define the

Response Surfaces, Mixtures, and Ridge Analyses, Second Edition. By G. E. P. Box and N. R. Draper
Copyright © 2007 John Wiley & Sons, Inc.

temperatures at which the cooker could be operated; these might extend from 30°C to 120°C (from room temperature to the temperature attainable when maximum pressure was applied). However, the region of interest $R(\xi)$ typically would be much more limited; initially, at least, $R(\xi)$ would likely be in the immediate vicinity of the boiling point of water—from 95°C to 100°C, say. Depending on what the experiments conducted in this smaller region show, the region of interest might then be changed to cover some other more promising region $R'(\xi)$ within $O(\xi)$, and so on.

It should be carefully noted in this formulation that, in order to best explore some region $R(\xi)$ of current interest, we do not need to perform all the experimental runs within $R(\xi)$. Some runs might be inside and some outside, or all might possibly be outside.

Now, writing $f(\xi)$ for the polynomial approximation, we have

$$E(y) = \eta(\xi) \tag{13.1.2}$$

for all ξ, and

$$\eta(\xi) \doteq f(\xi) \tag{13.1.3}$$

over some limited region of interest $R(\xi)$. The fact that the polynomial is an approximation does not necessarily detract from its usefulness because all models are approximations. Essentially, all models are wrong, but some are useful. However, the approximate nature of the model must always be borne in mind because if $\boldsymbol{\varepsilon} = (\varepsilon_1, \varepsilon_2, \ldots, \varepsilon_n)'$ is a vector of random errors having zero vector mean and if $\mathbf{y} = (y_1, y_2, \ldots, y_n)'$ and $\mathbf{f}(\boldsymbol{\xi}) = \{f(\boldsymbol{\xi}_1), f(\boldsymbol{\xi}_2), \ldots, f(\boldsymbol{\xi}_n)\}$, where $\boldsymbol{\xi}_1, \boldsymbol{\xi}_2, \ldots, \boldsymbol{\xi}_n$ are n observations on $\boldsymbol{\xi}$, the true model is not

$$\mathbf{y} = \mathbf{f}(\boldsymbol{\xi}) + \boldsymbol{\varepsilon} \tag{13.1.4}$$

but

$$\mathbf{y} = \boldsymbol{\eta}(\boldsymbol{\xi}) + \boldsymbol{\varepsilon}, \tag{13.1.5}$$

that is,

$$\mathbf{y} = \mathbf{f}(\boldsymbol{\xi}) + \boldsymbol{\delta}(\boldsymbol{\xi}) + \boldsymbol{\varepsilon}, \tag{13.1.6}$$

where

$$\boldsymbol{\delta}(\boldsymbol{\xi}) = \boldsymbol{\eta}(\boldsymbol{\xi}) - \mathbf{f}(\boldsymbol{\xi}) \tag{13.1.7}$$

is the vector discrepancy, which we would like to be small over $R(\xi)$, between actual and approximate models. In other words, there are *two* types of errors which must be taken into account:

1. Systematic, or bias, errors $\boldsymbol{\delta}(\boldsymbol{\xi}) = \boldsymbol{\eta}(\boldsymbol{\xi}) - \mathbf{f}(\boldsymbol{\xi})$, the difference between the expected value of the response, $E(\mathbf{y}) = \boldsymbol{\eta}(\boldsymbol{\xi})$ and the approximating function $\mathbf{f}(\boldsymbol{\xi})$.

2. Random errors $\boldsymbol{\varepsilon}$.

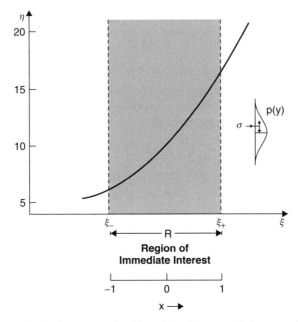

Figure 13.1. A hypothetical unknown relationship existing between $E(y) = \eta$ and ξ. At the right is shown the error distribution $p(y)$ which has standard deviation σ.

Although the above implies that systematic errors $\delta(\xi)$ are always to be expected, there has been, since the time of Gauss (1777–1855), an unfortunate tendency to ignore them and to concentrate only on the random errors ε. This seems to have been because nice mathematical results are possible when this is done. In choosing an experimental design, the ignoring of systematic error is *not* an innocuous approximation and misleading results may well be obtained, as we shall see in due course.

13.2. THE COMPETING EFFECTS OF BIAS AND VARIANCE

Condsider first the elementary case where there is only one important predictor variable ξ. Figure 13.1 represents a typical situation in which the investigator might desire to approximate a relationship

$$E(y) = \eta(\xi) \qquad (13.2.1)$$

over the region (here, interval) of interest R defined by $\xi_- \le \xi \le \xi_+$, by a straight line

$$f(\xi) = \alpha + \beta\xi. \qquad (13.2.2)$$

In practice, of course, the true functional relationship $\eta(\xi)$ indicated by the curve drawn in the figure would be unknown.

Suppose that the errors ε that occur in the observations have a variance σ^2, also unknown in practice, as indicated by the distribution on the right of the figure.

By applying the coding transformation

$$x = \frac{\xi - \frac{1}{2}(\xi_+ + \xi_-)}{(\xi_+ - \xi_-)} \qquad (13.2.3)$$

we can always convert the interval of interest, (ξ_-, ξ_+) in the variable ξ, to the interval $(-1, 1)$ in the standardized variable x. Suppose a particular set of data provides the least-squares fitted equation

$$\hat{y}_x = a + bx. \qquad (13.2.4)$$

The mean squared error associated with estimating η_x by $\hat{y}_x$, standardized for the number, N, of design points and the error variance σ^2 is

$$\frac{NE(\hat{y}_x - \eta_x)^2}{\sigma^2} = \frac{NE\{\hat{y}_x - E(\hat{y}_x) + E(\hat{y}_x) - \eta_x\}^2}{\sigma^2}$$

$$= \frac{NV(\hat{y}_x)}{\sigma^2} + \frac{N\{E(\hat{y}_x) - \eta_x\}^2}{\sigma^2}, \qquad (13.2.5)$$

after reduction. We can represent this symbolically as

$$M_x = V_x + B_x. \qquad (13.2.6)$$

In words, the standardized mean squared error at x is equal to the variance V_x at x plus the squared bias B_x at x.

For the special case of a straight line fitted to observations made at locations $x_1, x_2, \ldots, x_u, \ldots, x_N$, and such that $\bar{x} = 0$, we find that

$$V_x = 1 + \frac{x^2}{\sigma_x^2}, \qquad (13.2.7)$$

where

$$\sigma_x^2 = \sum_{u=1}^{N} \frac{(x_u - \bar{x})^2}{N}, \qquad \bar{x} = \sum_{u=1}^{N} \frac{x_u}{N}. \qquad (13.2.8)$$

Also, it is easily shown that

$$E(\hat{y}_x) = E(a) + E(b)x, \qquad (13.2.9)$$

which, theoretically, could be obtained by fitting a straight line to the *true values* η_u at the design points.

An Illustration Using Symmetric Three Point Designs

Consider the special case of designs consisting of just three points at $(-x_0, 0, x_0)$, so that $\sigma_x^2 = 2x_0^2/3$, or, conversely, $x_0 = 1.225\sigma_x$. Clearly,

$$V_x = 1 + \frac{1.5x^2}{x_0^2}. \qquad (13.2.10)$$

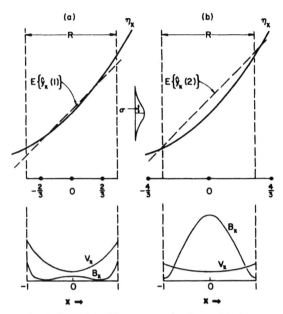

Figure 13.2. Two three-point designs with different spreads, along with the corresponding values of V_x and B_x over the region of interest $-1 \leq x \leq 1$.

In particular, consider the two specific designs shown in Figure 13.2a and 13.2b with $x_0 = \frac{2}{3}$ and $x_0 = \frac{4}{3}$, respectively. Thus design 1 consists of the points $(-\frac{2}{3}, 0, \frac{2}{3})$, and design 2 consists of the points $(-\frac{4}{3}, 0, \frac{4}{3})$. The expected straight lines

$$E(\hat{y}_x) = E(a) + E(b)x \qquad (13.2.11)$$

are drawn in the figures and denoted by $E\{\hat{y}_x(1)\}$ and $E\{\hat{y}_x(2)\}$, where $\hat{y}_x(1)$ and $\hat{y}_x(2)$ are the corresponding fitted lines. For design 1 we have

$$V_x(1) = 1 + \frac{27x^2}{8}, \qquad B_x(1) = \left[E\{\hat{y}_x(1)\} - \eta_x \right]^2, \qquad (13.2.12)$$

and for design 2 we have

$$V_x(2) = 1 + \frac{27x^2}{32}, \qquad B_x(2) = \left[E\{\hat{y}_x(2)\} - \eta_x \right]^2. \qquad (13.2.13)$$

The functions V_x and B_x are plotted below their corresponding designs. From Figure 13.2a we see that the variance error $V_x(1)$ is quite high compared to $B_x(1)$ and to $V_x(2)$. Now, when $\bar{x} = 0$, we have $V_x = 1 + x^2/\sigma_x^2$ and, for the particular type of design shown, $V_x = 1 + 1.5x^2/x_0^2$. Thus we make V_x smaller (except when $x = 0$) by making σ_x^2 larger, that is, by taking x_0 to be larger, and spreading the design points more widely, as is done in design 2. Unfortunately, as we increase x_0, the representational ability of the straight line is strained more severely and thus, as we see from the figure, the squared bias B_x is greatly increased.

13.3. INTEGRATED MEAN SQUARED ERROR

Overall measures of variance and squared bias over any specified region of interest R may be obtained by averaging V_x and B_x over R to provide the quantities

$$V = \frac{\int_R V_x \, dx}{\int_R dx}, \qquad B = \frac{\int_R B_x \, dx}{\int_R dx}. \tag{13.3.1}$$

Denoting the mean squared error integrated over R by M, we can then write

$$M = V + B. \tag{13.3.2}$$

For the specific situations illustrated in Figures 13.2a and b, these equations become

$$
\begin{array}{ll}
2.58 = 2.12 + 0.46 & \text{design 1} \\
5.41 = 1.28 + 4.13 & \text{design 2}
\end{array} \tag{13.3.3}
$$

More generally, if we run a design at levels $(-x_0, 0, x_0)$ and fit a straight line when the true function and error distribution are those shown in Figure 13.1, then the overall integrated mean squared error M and the contributions to M from V and B are shown in Figure 13.3.

This diagram repays careful study. Consider the individual components V and B. The averaged (over R) variance V becomes very large if the spread of the

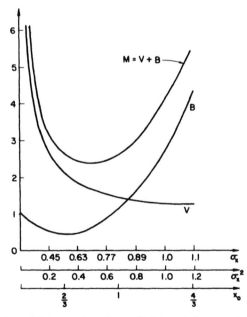

Figure 13.3. The behavior of integrated variance V, integrated squared bias B, and their total $M = V + B$, for three-point designs $(-x_0, 0, x_0)$ as the value of the spread $\sigma_x^2 = 2x_0^2/3$ varies.

design is made very small (that is, as x_0 approaches zero, V approaches infinity). Also, if the design is made very large, V slowly approaches its minimum value of unity (that is, as x_0 approaches infinity, V approaches 1). The average squared bias B on the other hand, has a minimum value when x_0 is about 0.7, and increases for larger or smaller designs. The reason for this behavior is not difficult to see. If x_0 is close to 0.7, bias over the whole interval R can be made small. However, if x_0 is less than about 0.7, so that the design is made smaller, the straight line will be close to the true curve only in the middle of the interval. On the other hand, if x_0 exceeds about 0.7, so that the design is made larger, the straight line will be close to the true curve only nearer the extremes of the interval. The curve M shows the overall effect on averaged mean squared error $M = V + B$. A rather flat minimum for M will be observed near $x_0 = 0.79$, or $\sigma_x = 0.65$. Notice that the design which minimizes averaged mean squared error M in this case is not very different from the design which minimizes averaged squared bias B but is very different from that which minimizes averaged variance V.

The Relative Magnitudes of V and B

In practice, of course, the true relationship $E(y_x) = \eta_x$ would be unknown. To make further progress, we can proceed as follows:

1. Given that we are going to fit a polynomial of degree d_1 (say) to represent the function over some interval R, we can suppose that the true function $E(y_x) = \eta_x$ is a polynomial of degree d_2, greater than d_1. Thus, for example, in Figure 13.1, given that the investigator will use a straight line to represent the function in the interval R, we might assume that the true function was a second-degree polynomial. In this case, then, $d_1 = 1$ and $d_2 = 2$.

2. We need also to say something about the *relative* magnitudes of systematic (bias) and random (variance) errors that we could expect to meet in practical cases. An investigator might typically employ a fitted approximating function such as a straight line, if he believed that the average departure from the truth induced by the approximating function were no worse than that induced by the process of fitting. We shall suppose this to be so and will assume, therefore, that the experimenter will tend to choose the size of his region, R, and the degree of his approximating function in such a way that the integrated random error and the integrated systematic error are about equal. Thus we shall suppose that the situation of typical interest is that where B is roughly equal to V.

13.4. REGION OF INTEREST AND REGION OF OPERABILITY IN k DIMENSIONS

So far we have illustrated the argument for the case where there is only one input variable ($k = 1$). We now consider the extension of these ideas to the general k case.

Readers of this book are already aware of the context in which response surface designs are customarily used. Iterative experimentation results in a sequence of

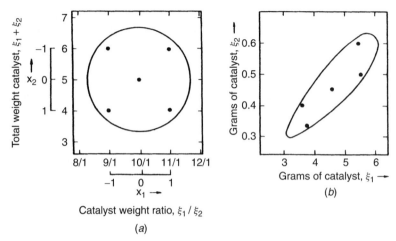

Figure 13.4. The effect of transformation of predictor variables on the region and the design. (a) A 2^2 factorial plus center-point design set in a circular region of interest in the variables $x_1 = \xi_1/\xi_2 - 10$ and $x_2 = \xi_1 + \xi_2 - 5$. (b) The same design and region described in the variables ξ_1 and ξ_2.

designs which move about in the predictor space and which tend to become more complex as the investigation progresses. Also, from one phase to the next, scales, transformations, and the identities of the variables considered will all tend to change.

At a given stage of the investigation then, we suppose there is a region of current interest $R(\xi)$ within which it is hoped that the currently entertained approximating function $f(\xi)$ will provide adequate local representation. When the investigation proceeds to the next stage, both $R(\xi)$ and $f(\xi)$ can change. This region $R(\xi)$ is, as we have said, contained within a much larger region of operability $O(\xi)$, often only vaguely known,* within which experiments *could* be performed using the current experimental apparatus. The approximating function $f(\xi)$ would usually provide a totally inadequate representation over the whole operability region $O(\xi)$.

In routine use of standard designs, both the choice of approximating function $f(\xi)$ and of the selected neighborhood $R(\xi)$ in which $f(\xi)$ is expected to apply are implicit as soon as the experimenter decides on the type of design he intends to employ, the variables he wishes to investigate, the levels of those variables, and the transformations of them he will use. Suppose, for example, that a chemist decides to run a 2^2 factorial with a center point in which he varies the amounts ξ_1 and ξ_2 of two catalysts A and B. Suppose he currently believes that a ratio ξ_1/ξ_2 of $10:1$ by weight is "about right" when a total weight of $\xi_1 + \xi_2 = 5$ g of catalyst is to be used. He further decides to vary the ratio from $9:1$ to $11:1$ and to vary the total catalyst weight from 4 to 6 g. Figure 13.4a shows the design in the coordinate

---

*Occasionally, delineation of certain features of $O(\xi)$ is itself the objective. Thus the experimenter might conceivably want to know: "What is the lowest temperature at which this apparatus will still function?" Sometimes questions of this kind are best put into the context of problems of specification (see Section 1.7). For example, the question "How high a dose of this drug could be given?" might best be studied by jointly considering two separate response surfaces measuring (a) therapeutic value on the one hand and (b) undesirable side effects on the other hand.

system

$$x_1 = \frac{\xi_1}{\xi_2} - 10, \qquad x_2 = \xi_1 + \xi_2 - 5 \qquad (13.4.1)$$

which the experimenter has implicitly chosen to consider. As we have seen in Chapter 4, the design he has chosen is suitable for estimating and checking the applicability of an approximating plane in the immediate neighborhood of the experimental points. More specifically, if a polynomial of degree $d_1 = 1$,

$$\eta = \beta_0 + \beta_1 x_1 + \beta_2 x_2 \qquad (13.4.2)$$

were adequate, one could estimate its β coefficients and also check the adequacy of the approximation by obtaining estimates of β_{12} and $(\beta_{11} + \beta_{22})$ in the polynomial of degree $d_2 = 2$ given by

$$\eta = \beta_0 + \beta_1 x_1 + \beta_2 x_2 + \beta_{11} x_1^2 + \beta_{22} x_2^2 + \beta_{12} x_1 x_2. \qquad (13.4.3)$$

We could express this more formally by saying that the experimenter's choice implies that:

1. For a region of interest $R(\xi)$, approximately represented by the circle in Figure 13.4a, the experimenter hopes that a polynomial of degree $d_1 = 1$ in the variables $\xi_1/\xi_2 - 10$ and $\xi_1 + \xi_2 - 5$ will provide adequate approximation to the true but unknown response function.
2. The experimenter feels that if the first-order polynomial is not adequate, a polynomial of degree $d_2 = 2$ should fit, and he will check the adequacy of the first-degree model by looking at estimates of the representative combinations of second-order effects β_{12} and $(\beta_{11} + \beta_{22})$.

Notice that the choices of design levels and of transformations of the original predictor variables ξ_1 and ξ_2 determine the main characteristics of the region $R(\xi)$. Notice also that the circular region of interest in the (x_1, x_2) space of Figure 13.4a is equivalent to the oval region in the original (ξ_1, ξ_2) space, underlining the facts that the type of design to be used, the forms of the transformations of the variables, and the choice of levels of those transformed variables—or, almost equivalently, the choice of d_1, d_2, and $R(\xi)$—are matters of judgment. At any given stage of an investigation, this judgment will vary from one investigator to another. For example, a second investigator might employ the same design and the same transformations but, instead of varying the total catalyst level 1 g either way about an average of 5 g, might vary it 2 g either way about an average of 7 g. The second investigator would be likely also to choose the levels for catalyst ratio differently. A third investigator might not have set out the design in terms of the ratio ξ_1/ξ_2 of the concentrations and their sum $\xi_1 + \xi_2$, but might have treated ξ_1 and ξ_2 themselves as predictor variables. A fourth might have considered different, or additional, variables at this stage of the investigation.

Such differences would not necessarily have any adverse effect on the *end result* of the investigation. The iterative strategy we have proposed for the exploration of response surfaces is designed to be adaptive and self-correcting. For example, an inappropriate choice of scales or of a transformation can be corrected as the iteration proceeds. However, for a given experimental design, a change of scale can have a major influence on:

1. The variances of estimated effects.
2. The sizes of systematic errors.

The purpose of this discussion is to remind the reader that optimal design properties are critically dependent on this matter of choice of region. Because this choice contains so many arbitrary elements, it is somewhat dubious how far we should push the finer points of optimal theory, which we now discuss.

13.5. AN EXTENSION OF THE IDEA OF A REGION OF INTEREST: THE WEIGHT FUNCTION

Our talk of a *"region of interest R"* thus far has implied that accuracy of prediction is of uniform importance throughout the entire region R and of no interest at all outside R. In some instances, this might not be the case. Sometimes, for example, we might need high accuracy at some point P in the predictor variable space and could tolerate reduced accuracy as we moved away from P. We can express such ideas by introducing a weight function $w(\mathbf{x})$ and minimizing a weighted mean squared error integrated over the whole of the operability region O. With this more general formulation, we have

$$M = V + B \tag{13.5.1}$$

with

$$M = \left(\frac{N}{\sigma^2}\right)\int_O w(\mathbf{x})\, E\{\hat{y}(\mathbf{x}) - \eta(\mathbf{x})\}^2 \, d\mathbf{x}, \tag{13.5.2}$$

$$V = \left(\frac{N}{\sigma^2}\right)\int_O w(\mathbf{x})\, E\{\hat{y}(\mathbf{x}) - E\hat{y}(\mathbf{x})\}^2 \, d\mathbf{x}, \tag{13.5.3}$$

$$B = \left(\frac{N}{\sigma^2}\right)\int_O w(\mathbf{x})\{E\hat{y}(\mathbf{x}) - \eta(\mathbf{x})\}^2 \, d\mathbf{x}. \tag{13.5.4}$$

Two possible weight functions for the $k = 1$ case are shown in Figure 13.5. Note that the vertical scale would be chosen so that $w(\mathbf{x})$ was normalized, that is,

$$\int_O w(\mathbf{x}) = 1.$$

The Importance of "All-Bias" Designs Illustrated for $k = 1$

We now return to the case where there is only one input variable ξ (so that $k = 1$) and where $d_1 = 1$, $d_2 = 2$. We generalize the idea of a region of interest by supposing that we have a symmetric weight function $w(x)$ centered at $x = 0$. The

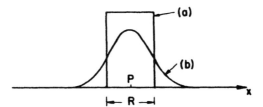

Figure 13.5. Two possible weight functions for $k = 1$. (a) "Uniform over R" type indicating uniform interest over R, no interest outside R. (b) Normal distribution shape, giving greater weight to points nearer P. See Exercise 13.4.

fitted equation is

$$\hat{y}_x = b_0 + b_1 x \tag{13.5.5}$$

while the true model, valid throughout an extensive region O, is

$$\eta_x = \beta_0 + \beta_1 x + \beta_{11} x^2. \tag{13.5.6}$$

Suppose that N runs are to be made at levels $x_1, x_2, \ldots, x_N$, and that $\Sigma x_i = N\bar{x} = 0$. Define

$$m_p = \frac{\Sigma x_u^p}{N}, \qquad \mu_p = \int_O w(x) x^p \, dx, \tag{13.5.7}$$

and note that $m_1 = 0$ and, because of the assumed symmetry, $\mu_p = 0$ for all odd p. Then it is readily shown (see Appendix 13A) that the integrated mean squared error for any design is

$$M = V + B = \left\{ 1 + \frac{\mu_2}{m_2} \right\} + \alpha^2 \left\{ (m_2 - \mu_2)^2 + \left(\mu_4 - \mu_2^2 \right) + \frac{m_3^2 \mu_2}{m_2^2} \right\}, \tag{13.5.8}$$

where

$$\alpha = \frac{N^{1/2} \beta_{11}}{\sigma}. \tag{13.5.9}$$

If M is to be minimized, clearly we should choose $m_3 = 0$ whatever values are taken by α and m_2. One way to do this is to make the design symmetric about its center. The only design characteristic remaining in the reduced expression is m_2, which measures the spread of the x values. Consider, first, the two extreme situations:

The "All-Variance" Case
Here, the bias term B is entirely ignored. (This extreme is the case most often considered by theoretical statisticians.) It is easy to see that the optimal value of m_2 is infinity. The practical interpretation of this is that we are told to spread the design as widely as possible in the operability region O.

Table 13.1. Optimum Spread of Design Points as Measured by $(m_2 / \mu_2)^{1/2}$ for Various $V / (V + B)$ Ratios and Two Types of Weight Functions

		Spread of Design Points as Measured by $(m_2/\mu_2)^{1/2}$	
Design	$V/(V + B)$	For Uniform Weight Function	For Normal Weight Function
All-bias	0	1.00	1.00
	0.2	1.02	1.04
	0.333	1.04	1.09
$V = B$	0.5	1.08	1.16
	0.667	1.14	1.26
	0.8	1.25	1.41
All-variance	1.0	∞	∞

The "All-Bias" Case

Here, the variance term V is entirely ignored and it is assumed that fitting discrepancies arise only from inadequacy of the model. M (with $m_3 = 0$) is minimized when $m_2 = \mu_2$; that is, the design points are restricted to mimic the weight function in location (because $m_1 = \mu_1 = 0$) and spread ($m_2 = \mu_2$) and in third moment ($m_3 = \mu_3 = 0$).

For cases intermediate between these extremes, the minimizing choice of m_2 will depend on the value held by α. Arguing as in Section 13.3, however, we can minimize M (with $m_3 = 0$) with respect to m_2 subject to the constraint that the fraction $V/M = V/(V + B)$ of the mean squared error M accounted for by V has some specific value. As we have seen, special interest attaches to a value of $V/(V + B)$ close to $\frac{1}{2}$ when the contributions from variance and bias are about equal. Table 13.1 shows the optimum spreads of such designs, as measured by the ratio $(m_2/\mu_2)^{1/2}$, for various selected values of the fraction $V/(V + B)$ and for both a uniform weight function and a normal distribution weight function. The relative shapes of these two weight functions are shown in Figure 13.5.

It is very evident that the optimal value of $(m_2/\mu_2)^{1/2}$ for $V = B$, that is, $V/(V + B) = \frac{1}{2}$, is close to that for the all-bias designs $V/(V + B) = 0$ and is dramatically different from that for the all-variance designs, $V/(V + B) = 1$. Furthermore, the function relating $(m_2/\mu_2)^{1/2}$ to $V/(V + B)$ remains extremely flat except when $V/(V + B)$ is close to 1, namely, except when we approach the all-variance case. This suggests that if a simplification is to be made in the design problem, it might be better to ignore the effects of sampling variation rather than those of bias.

13.6. DESIGNS THAT MINIMIZE SQUARED BIAS

The foregoing section indicates that designs that minimize squared bias are of practical importance. We therefore consider next the properties of such designs, using an important general result. Suppose a polynomial model of degree d_1

$$\hat{y}(\mathbf{x}) = \mathbf{x}_1'\mathbf{b}_1 \tag{13.6.1}$$

is fitted to the data, while the true model is a polynomial of degree d_2,

$$\eta(\mathbf{x}) = \mathbf{x}_1'\boldsymbol{\beta}_1 + \mathbf{x}_2'\boldsymbol{\beta}_2. \tag{13.6.2}$$

Thus, for the complete set of N data points we obtain

$$\hat{\mathbf{y}}(\mathbf{x}) = \mathbf{X}_1\mathbf{b}_1,$$
$$\boldsymbol{\eta}(\mathbf{x}) = \mathbf{X}_1\boldsymbol{\beta}_1 + \mathbf{X}_2\boldsymbol{\beta}_2. \tag{13.6.3}$$

Let us now write

$$\mathbf{M}_{11} = N^{-1}\mathbf{X}_1'\mathbf{X}_1, \qquad \mathbf{M}_{12} = N^{-1}\mathbf{X}_1'\mathbf{X}_2,$$

$$\boldsymbol{\mu}_{11} = \int_O w(\mathbf{x})\mathbf{x}_1\mathbf{x}_1'\,d\mathbf{x}, \qquad \boldsymbol{\mu}_{12} = \int_O w(\mathbf{x})\mathbf{x}_1\mathbf{x}_2'\,d\mathbf{x}. \tag{13.6.4}$$

It can now be shown (see Appendix 13B) that, whatever the values of $\boldsymbol{\beta}_1$ and $\boldsymbol{\beta}_2$, a necessary and sufficient condition for the squared bias B to be minimized is that

$$\mathbf{M}_{11}^{-1}\mathbf{M}_{12} = \boldsymbol{\mu}_{11}^{-1}\boldsymbol{\mu}_{12}. \tag{13.6.5}$$

A sufficient (but not necessary) condition for B to be minimized is that

$$\mathbf{M}_{11} = \boldsymbol{\mu}_{11} \quad \text{and} \quad \mathbf{M}_{12} = \boldsymbol{\mu}_{12}. \tag{13.6.6}$$

Now the elements of $\boldsymbol{\mu}_{11}$ and $\boldsymbol{\mu}_{12}$ are of the form

$$\int_O w(\mathbf{x})x_1^{\alpha_1}x_2^{\alpha_2} \cdots x_k^{\alpha_k}\,d\mathbf{x} \tag{13.6.7}$$

and the elements of $\mathbf{M}_{11}$ and $\mathbf{M}_{12}$ are of the form

$$N^{-1}\sum_{u=1}^{N} x_{1u}^{\alpha_1}x_{2u}^{\alpha_2} \cdots x_{ku}^{\alpha_k}. \tag{13.6.8}$$

These typical elements are, respectively, moments of the weight function and moments of the design points of order

$$\alpha = \alpha_1 + \alpha_2 + \cdots + \alpha_k. \tag{13.6.9}$$

Thus, the sufficient condition above states that, up to and including order $d_1 + d_2$, all the moments of the design are equal to all the moments of the weight function.

We have already seen an elementary example in which this result applied when, in the foregoing section, we considered the case $k = 1$, $d_1 = 1$, $d_2 = 2$. There, the all-bias design was obtained by setting $m_2 = \mu_2$ and $m_3 = \mu_3$, where $\mu_3 = 0$, because the weight function was assumed to be symmetric about 0.

We now consider two examples where $k > 1$, and we shall suppose, for illustration, that the weight function is uniform over a sphere of radius 1.

Example 13.1. First-order minimum bias design, $d_1 = 1$, $d_2 = 2$. In this example, we are interested in fitting a plane

$$\hat{y} = b_0 + \sum_{i=1}^{k} b_i x_i \qquad (13.6.10)$$

to describe the response function within a spherical region of interest R of radius 1. The true function over the whole operability region O can be represented by the second-degree polynomial

$$\eta = \beta_0 + \sum_{i=1}^{k} \beta_i x_i + \sum_{i \geq j}^{k} \sum^{k} \beta_{ij} x_i x_j. \qquad (13.6.11)$$

To minimize squared bias, the moments of the design points up to and including order $d_1 + d_2 = 3$ are chosen to match those of a uniform density taken over a k-dimensional sphere. Thus,

$$w(\mathbf{x}) = \begin{cases} c^{-1}, & \text{a constant, within the spherical region } R, \\ 0, & \text{elsewhere}, \end{cases} \qquad (13.6.12)$$

where the value of

$$c = \frac{\{\Gamma(\frac{1}{2})\}^k}{\Gamma(k/2 + 1)}$$

is the volume of a sphere of radius 1.

If we take the origin to be at the center of the spherical region, it is at once obvious from symmetry that all moments of the region of orders 1, 2, and 3 are zero except for the pure second moments. For the ith of these moments we have

$$c \int_R x_i^2 \, d\mathbf{x} = \frac{1}{(k + 2)}. \qquad (13.6.13)$$

Thus, the moments of the design will be equal to the moments of the region if

$$N^{-1} \Sigma x_{iu} = 0,$$

$$N^{-1} \Sigma x_{iu}^2 = \frac{1}{(k + 2)},$$

$$N^{-1} \Sigma x_{iu} x_{ju} = 0, \qquad (13.6.14)$$

$$N^{-1} \Sigma x_{iu}^3 = 0,$$

$$N^{-1} \Sigma x_{iu}^2 x_{ju} = 0,$$

$$N^{-1} \Sigma x_{iu} x_{ju} x_{hu} = 0,$$

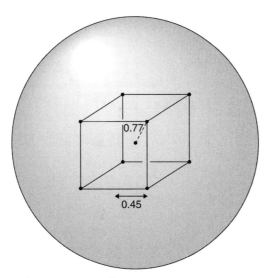

Figure 13.6. A first-order (two-level factorial) design in three factors which minimizes squared bias from second-order terms when the region of interest is a sphere of unit radius.

where all summations are over $u = 1, 2, \ldots, N$. These conditions may be met in a variety of ways and, in particular, are satisfied by any two-level fractional factorial design of resolution IV, appropriately scaled.

We may illustrate the actual placing of the points by using the 2^3 factorial $(\pm a, \pm a, \pm a)$ for the case $k = 3$ as shown in Figure 13.6. When $k = 3$, the condition on the pure second moments becomes

$$0.2 = N^{-1} \sum_{u=1}^{N} x_{iu}^2 = a^2 \tag{13.6.15}$$

so that $a = 0.447$. The radial distance of each point from the center is thus $(3a^2)^{1/2} = 0.7746$. Thus the design points are the vertices of a cube of half-side $a = 0.447$; the vertices lie on a sphere of radius 0.7746 embedded in a spherical region of interest of radius 1. Because the desirable properties of this arrangement are retained when the design and the region are (together) subjected to any linear transformation, we can obtain minimum bias designs for any ellipsoidal region as well.

Example 13.2. Second-order minimum bias designs, $d_1 = 2$, $d_2 = 3$. We again assume a spherical region of interest R of unit radius, and consider the choice of a second-order design given that, over the wider operability region O, a third-degree polynomial model exactly represents the functional relationship. A design that minimizes squared bias may be obtained by choosing the moments of the design, up to and including order $d_1 + d_2 = 5$, to match those of the k-dimensional sphere. This implies that all moments of orders 1, 2, 3, 4, and 5 are zero, except for

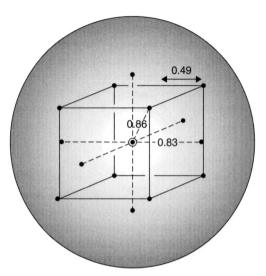

Figure 13.7. A second-order composite rotatable design which minimizes squared bias from third-order terms when a weight function uniform over a spherical region of interest of radius 1 is assumed.

the following, which are evaluated under the assumption of Eq. (13.6.12):

$$\int_O x_i^2 w(\mathbf{x})\, d\mathbf{x} = \frac{1}{k+2};\qquad \int_O x_i^4 w(\mathbf{x})\, d\mathbf{x} = \frac{3}{(k+2)(k+4)};$$

$$\int_O x_i^2 x_j^2 w(\mathbf{x})\, d\mathbf{x} = \frac{1}{(k+2)(k+4)}. \tag{13.6.16}$$

The designs so specified are second-order rotatable.* It follows, in particular, that one class of suitable designs consists of the "cube plus star plus center points" composite design class, in which the "cube" portion is taken as a two-level factorial, or fractional factorial of resolution V. Figure 13.7 shows such a design for $k = 3$. The design consists of a cube of side $2a$, six axial points at a distance αa from the origin, and n_0 center points. Any such design has the required zero moments, and we obtain appropriate values of a, α, and n_0 from the following equations, obtained by equating design moments to region moments.

$$\Sigma x_{iu}^2 = 8a^2 + 2\alpha^2 a^2 = \frac{1}{5}N,$$

$$\Sigma x_{iu}^4 = 8a^4 + 2\alpha^4 a^4 = \frac{3}{35}N,$$

$$\Sigma x_{iu}^2 x_{ju}^2 = 8a^4 = \frac{1}{35}N,$$

$$8 + 6 + n_0 = N. \tag{13.6.17}$$

*A design is rotatable if $V\{\hat{y}(\mathbf{x})\}$ is a function of $r = (x_1^2 + x_2^2 + \cdots + x_k^2)^{1/2}$ only—that is, if the accuracy of prediction of the response depends only on the distance from the center of the design. See Section 14.3.

These equations have the solution $\alpha = 2^{3/4} = 1.682$, $a = 0.4938$, $n_0 = 2.653$. Thus, the design of Figure 13.7 will (very nearly) give minimum average mean squared bias if we set $a = 0.49$ and $\alpha a = 0.83$ and use three center points. The points of the cube will lie a distance $(3a^2)^{1/2} = 0.86$ from the center.

Other Weight Functions, Other Regions of Interest

If more importance were attached to the central portion of the region than to its extremes and a weight function with spherical contours were employed, then the all-bias moment requirements would still call for the use of a second-order rotatable design with scale suitably adjusted. Interest has been shown by some authors in cuboidal regions of interest or, more generally, in regions that can be reduced to cubes by linear transformations. More generally, in that context, we can consider weight functions whose contours are concentric cubes. It is easy to show that the moment requirements for the case $d_1 = 2$, $d_2 = 3$ are again met by suitably chosen (but not rotatable) composite designs. [See Draper and Lawrence (1965c).] Also, see Appendices 6B and 7H.

13.7. ENSURING GOOD DETECTABILITY OF LACK OF FIT

Our aims in choosing a design are necessarily many faceted. In particular, on the one hand when, as inevitably happens, we fit a somewhat inexact model, we do not want our estimate $\hat{y}$ of η to be too far wrong—we want our design to allow *good estimation* of η. On the other hand, when large discrepancies occur which are likely to invalidate the model, we want to know about them so that we can modify the model and, in some cases, augment the data appropriately—we want our design to allow *detection of model inadequacy*. In their general discussion, Box and Draper (1959, 1963) consider a strategy of experimental design in which first a class of designs that minimize mean squared error to satisfy the first requirement is found; then, a further choice is made of a subset of designs that also satisfy the second requirement. It is this second aspect of the choice of design—to give good detection of model inadequacy—that we now consider.

Consider the mechanics of making a test of goodness of fit using the analysis of variance. Suppose we are estimating p parameters, observations are made at $p + f$ distinct points, and repeated observations are made at certain of these points to provide e pure error degrees of freedom, so that the total number of observations is $N = p + f + e$. The resulting analysis of variance table is of the type shown in Table 13.2.

Under the usual assumptions, the expectation of the unbiased pure error mean square is σ^2, the experimental error variance, and the expected value of the lack of fit mean square equals $\sigma^2 + \Lambda^2/f$ where Λ^2 is a noncentrality parameter. The test for goodness of fit is now made by comparing the mean square for lack of fit against the mean square for error, via an $F(f, e)$ test.

In general, the noncentrality parameter takes the form

$$\Lambda^2 = \sum_{u=1}^{N} \{E(\hat{y}_u) - \eta_u\}^2 = E(S_L) - f\sigma^2, \tag{13.7.1}$$

Table 13.2. Skeleton Analysis of Variance Table

Source	df	$E(MS)$
Parameter estimates	p	
Lack of fit	f	$\sigma^2 + \Lambda^2/f$
Pure error	e	σ^2
Total	N	

where S_L is the lack of fit sum of squares. Thus, good detectability of general lack of fit can be obtained by choosing a design that makes Λ^2 large. It turns out that this requirement of good detection of model inadequacy can, like the earlier requirement of good estimation, be achieved by certain conditions on the design moments. Thus, under certain sensible assumptions, it can be shown that a dth-order design would provide high detectability for terms of order $(d + 1)$ if (1) all odd design moments of order $(2d + 1)$ or less are zero and (2) the ratio

$$\frac{\sum_{u=1}^{N} r_u^{2(d+1)}}{\left\{ \sum_{u=1}^{N} r_u^2 \right\}^{d+1}} \tag{13.7.2}$$

is large, where

$$r_u^2 = x_{1u}^2 + x_{2u}^2 + \cdots + x_{ku}^2.$$

In particular, this would require that for a first-order design ($d = 1$) the ratio $\sum r_u^4 / \{\sum r_u^2\}^2$ should be large to provide high detectability of quadratic lack of fit; for a second-order design ($d = 2$), the ratio $\sum r_{iu}^6 / \{\sum r_{iu}^2\}^3$ should be large to provide high detectability of cubic lack of fit. While these general results are of interest, careful attention must be given in practice to the detailed nature of Λ^2, as is now illustrated.

The Nature of the Noncentrality Parameter Λ^2

Suppose that we fit the model

$$\mathbf{y} = \mathbf{X}\boldsymbol{\beta} + \boldsymbol{\varepsilon} \tag{13.7.3}$$

but that the true model is

$$\mathbf{y} = \mathbf{X}\boldsymbol{\beta} + \mathbf{X}_2\boldsymbol{\beta}_2 + \boldsymbol{\varepsilon} = \boldsymbol{\eta} + \boldsymbol{\varepsilon}. \tag{13.7.4}$$

Then (Section 3.10) the estimate $\mathbf{b} = (\mathbf{X}'\mathbf{X})^{-1}\mathbf{X}'\mathbf{y}$ of $\boldsymbol{\beta}$ is biased and

$$E(\mathbf{b}) = \boldsymbol{\beta} + \mathbf{A}\boldsymbol{\beta}_2, \tag{13.7.5}$$

where the bias matrix $\mathbf{A}$ is

$$\mathbf{A} = (\mathbf{X}'\mathbf{X})^{-1}\mathbf{X}'\mathbf{X}_2. \tag{13.7.6}$$

Comparing the formulas for $\mathbf{b}$ and $\mathbf{A}$, we readily see that $\mathbf{A}$ can be regarded as the matrix of regression coefficients of $\mathbf{X}_2$ on $\mathbf{X}$. Now with $\hat{\mathbf{y}} = \mathbf{X}\mathbf{b}$, and $E(\hat{\mathbf{y}}) = \mathbf{X}E(\mathbf{b})$,

$$\Lambda^2 = \{E(\hat{\mathbf{y}}) - \boldsymbol{\eta}\}'\{E(\hat{\mathbf{y}}) - \boldsymbol{\eta}\} = \boldsymbol{\beta}_2'\mathbf{X}_2'(\mathbf{I} - \mathbf{R})'(\mathbf{I} - \mathbf{R})\mathbf{X}_2\boldsymbol{\beta}_2, \tag{13.7.7}$$

where* $\mathbf{R} = \mathbf{X}(\mathbf{X}'\mathbf{X})^{-1}\mathbf{X}'$. Also,

$$(\mathbf{I} - \mathbf{R})\mathbf{X}_2 = \mathbf{X}_2 - \mathbf{X}\mathbf{A}$$

$$= \mathbf{X}_2 - \hat{\mathbf{X}}_2$$

$$= \mathbf{X}_{2\cdot 1}, \tag{13.7.8}$$

say, where the elements of $\hat{\mathbf{X}}_2$ are the "calculated values" obtained by regressing $\mathbf{X}_2$ on $\mathbf{X}$, just as the elements of $\hat{\mathbf{y}}$ are the calculated values obtained via the regression of $\mathbf{y}$ on $\mathbf{X}$. It follows that

$$\Lambda^2 = \boldsymbol{\beta}_2'\mathbf{X}_{2\cdot 1}'\mathbf{X}_{2\cdot 1}\boldsymbol{\beta}_2. \tag{13.7.9}$$

Notice that if the jth column of $\mathbf{X}_2$ was a linear combination of columns of $\mathbf{X}$, then the jth column of $\mathbf{X}_{2\cdot 1} = \mathbf{X}_2 - \hat{\mathbf{X}}_2$ would consist of zeros and no term involving β_{2j} would occur in Λ^2. In such a case, however large the parameter β_{2j} might be, its effect could not be detected by the lack of fit test. Conversely, if the jth column of $\mathbf{X}_2$ were orthogonal to all the columns of $\mathbf{X}$ and to all the other columns of $\mathbf{X}_2$, then the term involving β_{2j} in Λ^2 would be $\beta_{2j}^2\Sigma x_{2j}^2$. In this case, by making the value of Σx_{2j}^2 large, the lack of fit test could be made very sensitive to discrepancies arising from a nonzero values of β_{2j}. We now illustrate these points via an example.

Example 13.3. Suppose $k = 2$, $d_1 = 1$, and $d_2 = 2$ so that initially we fit the model

$$y = \beta_0 + \beta_1 x_1 + \beta_2 x_2 + \varepsilon, \tag{13.7.10}$$

but we bear in mind the possibility that a second-degree approximation

$$y = \beta_0 + \beta_1 x_1 + \beta_2 x_2 + \beta_{11} x_1^2 + \beta_{22} x_2^2 + \beta_{12} x_1 x_2 + \varepsilon \tag{13.7.11}$$

might be needed. Consider the two specific designs whose characteristics are displayed in Table 13.3. See also Figure 13.8. Design A is a 2^2 factorial design with points $(\pm 1, \pm 1)$, and design B is the same design turned through an angle of $45°$, so that the axial distances of the points from the origin are all $2^{1/2}$.

---

*Both $\mathbf{R}$ and $\mathbf{I} - \mathbf{R}$ are symmetric (e.g., $\mathbf{R}' = \mathbf{R}$) and idempotent (e.g., $\mathbf{R}\mathbf{R} = \mathbf{R}^2 = \mathbf{R}$).

Table 13.3. Two First-Order Designs with Associated Matrices and Skeleton Analyses of Variance

Design A		Design B	

$$\begin{array}{ccc} 0 & 1 & 2 \end{array}$$

$$\mathbf{X} = \begin{bmatrix} 1 & -1 & -1 \\ 1 & 1 & -1 \\ 1 & -1 & 1 \\ 1 & 1 & 1 \end{bmatrix}, \quad \begin{array}{ccc} 11 & 22 & 12 \end{array} \quad \mathbf{X}_2 = \begin{bmatrix} 1 & 1 & 1 \\ 1 & 1 & -1 \\ 1 & 1 & -1 \\ 1 & 1 & 1 \end{bmatrix},$$

$$\begin{array}{ccc} 0 & 1 & 2 \end{array}$$

$$\mathbf{X} = \begin{bmatrix} 1 & -2^{1/2} & 0 \\ 1 & 0 & -2^{1/2} \\ 1 & 0 & 2^{1/2} \\ 1 & 2^{1/2} & 0 \end{bmatrix}, \quad \begin{array}{ccc} 11 & 22 & 12 \end{array} \quad \mathbf{X}_2 = \begin{bmatrix} 2 & 0 & 0 \\ 0 & 2 & 0 \\ 0 & 2 & 0 \\ 2 & 0 & 0 \end{bmatrix}$$

$$\mathbf{X}_{2\cdot 1} = \begin{bmatrix} 0 & 0 & 1 \\ 0 & 0 & -1 \\ 0 & 0 & -1 \\ 0 & 0 & 1 \end{bmatrix}$$

$$\mathbf{X}_{2\cdot 1} = \begin{bmatrix} 1 & -1 & 0 \\ -1 & 1 & 0 \\ -1 & 1 & 0 \\ 1 & -1 & 0 \end{bmatrix}$$

$$\mathbf{X}'_{2\cdot 1}\mathbf{X}_{2\cdot 1} = \begin{bmatrix} 0 & 0 & 0 \\ 0 & 0 & 0 \\ 0 & 0 & 4 \end{bmatrix}$$

$$\mathbf{X}'_{2\cdot 1}\mathbf{X}_{2\cdot 1} = \begin{bmatrix} 4 & -4 & 0 \\ -4 & 4 & 0 \\ 0 & 0 & 0 \end{bmatrix}$$

Source	df	E(MS)	Source	df	E(MS)
Parameter estimates	3		Parameter estimates	3	
Lack of fit	1	$\sigma^2 + 4\beta_{12}^2$	Lack of fit	1	$\sigma^2 + 4(\beta_{22} - \beta_{11})^2$

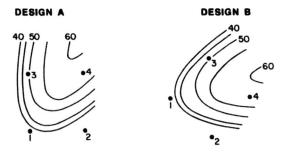

Figure 13.8. Two alternative designs A and B in relation to contours of a possible response function.

Design A is sensitive to lack of fit produced by the *interaction* term in the second-order model, but is completely insensitive to lack of fit produced by the pure *quadratic* terms, while design B has just the opposite sensitivities. The role played by the vectors of $\mathbf{X}_2$, along with their relationships with the vectors of $\mathbf{X}$, is clearly seen. In design A the first two columns of $\mathbf{X}_2$ are identical with the first column of $\mathbf{X}$ and therefore are annihilated when we take residuals, whereas in design B the elements of the final column of $\mathbf{X}_2$ are all zero initially and therefore their residuals are necessarily zero also. These results are to be expected because with design A we would have estimated the interaction term coefficient β_{12} but not the quadratic coefficients β_{11} and β_{22}, while with design B we could have estimated the difference $\beta_{11} - \beta_{22}$ of the quadratic terms, but not the interaction term.

Interim Discussion

With this general theory of tests of lack of fit, and with this specific illustration in mind, we are in a position now to consider some of the more subtle aspects of the problem of checking fit. First of all, we must consider the following dilemma. Suppose we have a design suitable for fitting a polynomial of a certain degree d_1, which we hope provides adequate representation but which we fear may not. Then we could check the fit by running a larger design of order $d_1 + 1$. In the above example, for instance, where $d_1 = 1$, if instead of running either design A or design B we added center points to separate the quadratic effects and combined these two designs, we would obtain a second-order ($d_1 + 1 = 2$) composite design and could then estimate *all* of the coefficients in the second-degree equation. The adequacy of fit test would then consist of checking if the additional second-order terms were negligible. Usually, however, economy in experimentation is important and if, for example, we believed that a first-degree approximation should be adequate, we should not want to expend all the additional runs needed to fit an approximation of second degree. We thus need to seek some principle or principles on which we can choose an intermediate design which supplies some information on the "most feared" discrepancies from the model of order d_1, but is not as comprehensive as a design of order $d_1 + 1$.

If we think in terms of our example, we are thus led to the question "Is it more important to be able to check if β_{12} is small, in which case design A should be used, or is it more important to check if ($\beta_{22} - \beta_{11}$) is small, in which case design B will be appropriate?" However, a little thought reveals this question as somewhat naive. Consider the first-order design A in relation to the response surface whose contours are shown in Figure 13.8a. Obviously, in the situation depicted, serious second-order lack of fit would occur and we would wish to detect it. Now consider design B in relation to the same surface rotated through 45°; this is shown in Figure 13.8b. Since design B is a rotation through 45° of design A, the whole of Figure 13.8b, surface *and* design, is now simply a 45° rotation of Figure 13.8a. Any lack of fit detected by design A for the first surface must therefore be equally detectable by design B for the rotated surface. However, suppose, as might well be the case, that at the time the experiment was being planned, a surface oriented like that in Figure 13.8a was regarded equally as likely to occur as a surface oriented like that in Figure 13.8b. Then there would clearly be nothing to choose between the designs so far as their ability to detect lack of fit was concerned.

To explore the same idea algebraically, consider the second-order terms

$$\beta_{11}x_1^2 + \beta_{22}x_2^2 + \beta_{12}x_1x_2 \qquad (13.7.12)$$

in an equation of second degree for two predictor variables x_1 and x_2. A rotation of axes through 45° corresponds to the transformation to $\dot{x}_1$ and $\dot{x}_2$ via

$$x_1 = \frac{(\dot{x}_1 - \dot{x}_2)}{2^{1/2}}, \qquad x_2 = \frac{(\dot{x}_1 + \dot{x}_2)}{2^{1/2}}. \qquad (13.7.13)$$

In the new coordinate system the second-order terms become

$$\dot{\beta}_{11}\dot{x}_1^2 + \dot{\beta}_{22}\dot{x}_2^2 + \dot{\beta}_{12}\dot{x}_1\dot{x}_2 \qquad (13.7.14)$$

with

$$\dot{\beta}_{11} = \tfrac{1}{2}(\beta_{11} - \beta_{22} + \beta_{12}), \qquad (13.7.15)$$

$$\dot{\beta}_{22} = \tfrac{1}{2}(\beta_{11} + \beta_{22} + \beta_{12}), \qquad (13.7.16)$$

and

$$\dot{\beta}_{12} = \beta_{22} - \beta_{11}, \qquad (13.7.17)$$

the last mentioned being of special relevance to the present discussion. Equivalently, we can write

$$\beta_{12} = \dot{\beta}_{22} - \dot{\beta}_{11}. \qquad (13.7.18)$$

Thus the contents of the expected mean squares, $E(\text{MS})$ in Table 13.3, namely $\sigma^2 + 4\beta_{12}^2$ for design A and $\sigma^2 + 4(\beta_{22} - \beta_{11})^2$ for design B, refer to the same quadratic lack of fit measured in two different orientations that are unlikely to be distinguishable a priori.

Yet another way to understand the comparison between the designs is to consider, for data $\mathbf{y}$ for design A, the contrast

$$\tfrac{1}{2}(y_1 - y_2 - y_3 + y_4). \qquad (13.7.19)$$

If the true model is quadratic, this contrast provides, for design A, an unbiased estimate of β_{12}, that is, of $\dot{\beta}_{22} - \dot{\beta}_{11}$. For data $\dot{\mathbf{y}}$ from design B, the corresponding contrast $\tfrac{1}{2}(\dot{y}_1 - \dot{y}_2 - \dot{y}_3 + \dot{y}_4)$ provides an unbiased estimate of $\beta_{22} - \beta_{11}$, that is, of $\dot{\beta}_{12}$.

This raises the question of whether there are measures of quadratic lack of fit which are invariant under rotation of axes. One measure of this kind is the sum of the quadratic effects

$$\beta_{11} + \beta_{22} + \cdots + \beta_{kk}, \qquad (13.7.20)$$

which becomes $\dot{\beta}_{11} + \dot{\beta}_{22} + \cdots + \dot{\beta}_{kk}$ after rotation. [For the case $k = 2$, this is obvious from Eqs. (13.7.15) and (13.7.16).] This measure is worth practical consideration because we shall often know from the nature of the problem that, while a maximum* showing some degree of ridginess is likely, a minimax is not. This is equivalent to saying that all of the β_{ii}'s are likely to be negative or zero. Consequently, their sum will be a good overall measure of quadratic lack of fit.

Any first-order design with added center points will allow the estimation of $\Sigma \beta_{ii}$ and thus will be sensitive to this kind of quadratic lack of fit. For illustration, suppose that under study are eight input variables whose levels, after suitable transformation and scaling, are denoted by $x_1, x_2, \ldots, x_8$ and that it is expected that, at this stage of experimentation, first-order effects will be dominant. We have seen that first-order arrangements which minimize squared bias are provided by two-level fractional factorial designs of resolution IV or of higher resolution. [See under Eq. (13.6.14).]

*As before, when the same kind of knowledge is available about a minimum, a parallel discussion arises.

Table 13.4. A First-Order Design for $k = 8$ Variables which Checks for Certain Types of Quadratic Bias

x_1	x_2	x_3	x_4	x_5	x_6	x_7	x_8	
1	−1	−1	−1	1	1	1	−1	
1	1	−1	−1	−1	−1	1	1	
1	−1	1	−1	−1	1	−1	1	
1	1	1	−1	1	−1	−1	−1	16-point
1	−1	−1	1	1	−1	−1	1	2_{IV}^{8-4}
1	1	−1	1	−1	1	−1	−1	design
1	−1	1	1	−1	−1	1	−1	with
1	1	1	1	1	1	1	1	generating
−1	1	1	1	−1	−1	−1	1	relation
−1	−1	1	1	1	1	−1	−1	I = 1235
−1	1	−1	1	1	−1	1	−1	= 1246
−1	−1	−1	1	−1	1	1	1	= 1347
−1	1	1	−1	−1	1	1	−1	= 2348
−1	−1	1	−1	1	−1	1	1	
−1	1	−1	−1	1	1	−1	1	
−1	−1	−1	−1	−1	−1	−1	−1	
0	0	0	0	0	0	0	0	n_0
0	0	0	0	0	0	0	0	center
...								points
0	0	0	0	0	0	0	0	

For example, a good design for $d_1 = 1$, $d_2 = 2$, $k = 8$ is that shown in Table 13.4. This consists of a resolution IV fractional factorial design containing $n = 16$ runs, with n_0 added center points for a total of $N = 16 + n_0$ points. On the assumption that a polynomial model of degree $d_2 = 2$ is needed, the design allows estimation of the measure of convexity ($\beta_{11} + \beta_{22} + \cdots + \beta_{88}$) and of seven combinations of two-factor interactions, namely, ($\beta_{12} + \beta_{35} + \beta_{46} + \beta_{78}$), ($\beta_{13} +$

Table 13.5. Breakup of the Residual Sum of Squares and Its Expected Value for the First-Order Design of Table 13.4

	SS	df		Expected Value of Mean Square
Lack of fit, S_L	$n_0 n(\bar{y}_f - \bar{y}_0)^2/N$	1		$\sigma^2 + c^2 n_0 N \left(\sum_{i=1}^{8} \beta_{ii}\right)^2 /n$
	$n\left(\sum_{u=1}^{N} x_{1u} x_{2u} y_u\right)^2 /(c^2 N^2)$	1		$\sigma^2 + c^2 N^2(\beta_{12} + \beta_{35} + \beta_{46} + \beta_{78})^2/n$
	$n\left(\sum_{u=1}^{N} x_{1u} x_{3u} y_u\right)^2 /(c^2 N^2)$	1	8 $\begin{cases} 8\sigma^2 + \Lambda^2 \end{cases}$	$\sigma^2 + c^2 N^2(\beta_{13} + \beta_{25} + \beta_{47} + \beta_{58})^2/n$
	...	...		...
	$n\left(\sum_{u=1}^{N} x_{1u} x_{8u} y_u\right)^2 /(c^2 N^2)$	1		$\sigma^2 + c^2 N^2(\beta_{18} + \beta_{27} + \beta_{35} + \beta_{45})^2/n$
Pure error, S_e	$\sum_{u=1}^{n_0} (y_{0u} - \bar{y}_0)^2$	$n_0 - 1$		σ^2

$\beta_{25} + \beta_{47} + \beta_{68}$), ($\beta_{14} + \beta_{26} + \beta_{37} + \beta_{58}$), ($\beta_{15} + \beta_{23} + \beta_{48} + \beta_{67}$), ($\beta_{16} + \beta_{24} + \beta_{38} + \beta_{57}$), ($\beta_{17} + \beta_{28} + \beta_{34} + \beta_{56}$), and ($\beta_{18} + \beta_{27} + \beta_{36} + \beta_{45}$). Thus the residual sum of squares can be broken up as shown in Table 13.5. In this table,

$$c = \frac{\sum\limits_{u=1}^{n} x_{iu}^2}{N}, \qquad i = 1, 2, \ldots, k$$

$$= \frac{16}{(16 + n_0)}. \tag{13.7.21}$$

Also, $\bar{y}_f$ and $\bar{y}_0$ are the averages of the factorial point responses y_u, $u = 1, 2, \ldots, n$, and of the center points responses y_{0u}, $u = 1, 2, \ldots, n_0$, respectively.

13.8. CHECKS BASED ON THE POSSIBILITY OF OBTAINING A SIMPLE MODEL BY TRANSFORMATION OF THE PREDICTORS

According to the argument outlined above, an efficient design of order d should (a) achieve small mean square error in estimating a polynomial of degree d and (b) allow the adequacy of the dth-degree polynomial to be checked in a parsimonious manner. Parsimony in checking must rely on *a priori* knowledge of what kinds of discrepancies are likely to occur. Knowledge of this kind is utilized when, for example, knowing that a maximum* is more likely to arise than a saddle point, we provide a check of the overall curvature measure $\Sigma \beta_{ii}$ by adding center points to a first-order design.

It is also possible to build in checks for investigating the feasibility of transformations on the predictor variables. Both theoretical knowledge and practical experience frequently indicate that simpler models can be obtained by suitably transforming experimental variables $\xi_1, \xi_2, \ldots, \xi_k$, such as temperature, concentration, pressure, feed rate, and reaction time, and then fitting a polynomial to the transformed variables. We consider first the application of this idea to first-order designs.

Possibility of a First-Order Model in Transformed Predictor Variables

When a curvilinear response relationship exists which is monotonic in the predictor variables over the current region of interest, it may be possible to use a first-order model in which power transformations $\xi_1^{\lambda_1}, \xi_2^{\lambda_2}, \ldots, \xi_k^{\lambda_k}$ are applied to the ξ's.

Assume that, *at worst*, the response may be represented by a second-order degree polynomial in transformed variables, namely by

$$F(\boldsymbol{\xi}, \boldsymbol{\lambda}) = \dot{\beta}_0 + \sum_{i=1}^{k} \dot{\beta}_i \xi_i^{\lambda_i} + \sum_{i=1}^{k} \sum_{j \geq i}^{k} \dot{\beta}_{ij} \xi_i^{\lambda_i} \xi_j^{\lambda_j}. \tag{13.8.1}$$

*Or minimum.

Then a first-degree polynomial model will be appropriate if the λ_i may be chosen so that $\dot{\beta}_{ij} = 0$ for all i and j. In Appendix 13E, we show that this requires that

$$\eta_{ij} = 0, \qquad i \neq j, \tag{13.8.2}$$

$$\eta_{ii} + (S_i/\xi_i)(1 - \lambda_i)\eta_i = 0, \qquad i = 1, 2, \ldots, k, \tag{13.8.3}$$

where

$$\eta_i = \frac{\partial F}{\partial x_i}, \qquad \eta_{ij} = \frac{\partial^2 F}{\partial x_i \, \partial x_j}, \tag{13.8.4}$$

where

$$x_i = \frac{(\xi_i - \xi_{i0})}{S_i}, \qquad i = 1, 2, \ldots, k.$$

(To obtain Eqs. (13.8.2) and (13.8.3), set Eqs. (13E.2a) and (13E.2b) equal to zero.) We define

$$\delta_i = \frac{S_i}{\xi_{i0}}. \tag{13.8.5}$$

Now suppose that a full second-order model

$$\hat{y} = b_0 + \Sigma b_i x_i + \underset{i \leq j}{\Sigma\Sigma} b_{ij} x_i x_j$$

has been fitted to the data from an appropriate design. Then we could approximate the derivatives of Eq. (13.8.4) by the corresponding derivatives of $\hat{y}$, evaluated at $\mathbf{x} = \mathbf{0}$ where necessary, namely by

$$\hat{\eta}_i = b_i, \qquad \hat{\eta}_{ij} = b_{ij}, \qquad i \neq j, \qquad \hat{\eta}_{ii} = 2b_{ii}. \tag{13.8.6}$$

Also, we approximate S_i/ξ_i by δ_i in Eq. (13.8.5). Then (a) the possibility of a first-order representation in transformed variables $\xi_i^{\lambda_i}$ is contraindicated [see Eq. (13.8.2)] if any interaction estimate b_{ij}, $i \neq j$, is significantly different from zero, and (b) supposing such a transformation to be possible, the appropriate transformation parameters are roughly* estimated by

$$\hat{\lambda}_i = 1 + \frac{2b_{ii}}{\delta_i b_i}, \qquad i = 1, 2, \ldots, k. \tag{13.8.7}$$

As always with experimental designs, we can obtain only those features for which we are prepared to pay. However, the above analysis indicates what these features might be:

1. Suppose we are prepared to assume that if the function is not adequately approximated by a first-order model in $\xi_1, \xi_2, \ldots, \xi_k$, it will be by one in

---

*More precise estimates can be found by application of standard nonlinear least squares, fitting the model of Eq. (13.8.1) with $\dot{\beta}_{ij} = 0$ directly to the data.

**Table 13.6. A Design in Four Variables Suitable for Estimating
Power Transformations**

(a) *Design*

x_1	x_2	x_3	x_4
-1	-1	-1	1
0	-1	0	0
1	-1	1	-1
-1	0	1	0
0	0	-1	-1
1	0	0	1
-1	1	0	-1
0	1	1	1
1	1	-1	0

(b) *Association Scheme*

		x_2		
		-1	0	1
	-1	$A\gamma$	$B\beta$	$C\alpha$
x_2	0	$C\beta$	$A\alpha$	$B\gamma$
	1	$B\alpha$	$C\gamma$	$A\beta$

$x_3 = \begin{matrix} -1 & 0 & 1 \\ A & B & C \end{matrix}$ $x_4 = \begin{matrix} -1 & 0 & 1 \\ \alpha & \beta & \gamma \end{matrix}$

$\xi_1^{\lambda_1}, \xi_2^{\lambda_2}, \ldots, \xi_k^{\lambda_k}$. Then designs that provide appropriate checks will allow separate estimation of the β_{ii}. One source of such designs which provide maximum economy in the number of runs to be made are the three-level fractional factorials proposed by Finney. For example, for $k = 4$ and for standardized variables $x_i = (\xi_i - \xi_{i0})/S_i$ chosen at the three levels ± 1 and 0, we might use the fractional factorial design requiring nine runs shown in Table 13.6a.

This design is readily derived by associating the three levels of x_3 and x_4, respectively, with the three Roman (A, B, C) and Greek (α, β, γ) letters in the Graeco–Latin square shown in Table 13.6b. The levels of x_1 and x_2 are shown bordering the square. Regarded as a design to estimate the β_i and β_{ii}, this is a fully saturated arrangement, and must be used with care because ambiguities in interpretation will arise if terms other than those specifically allowed for in the model occur.

2. If we wish to check that a first-order representation is possible in $\xi_1^{\lambda_1}, \xi_2^{\lambda_2}, \ldots, \xi_k^{\lambda_k}$, then we need a design that permits estimation of at least some of the two-factor interactions.

Possibility of a Second-Order Model in Transformed Predictor Variables

We now consider second-order designs that have built in checks sensitive to the possibility that a second-degree approximation in transformed metrics, rather than the original metrics, is needed. For example, the true underlying function may

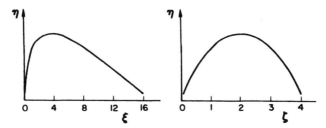

Figure 13.9. An asymmetrical maximum in ξ, on the left, is converted to a symmetrical maximum in ζ, on the right, by the transformation $\zeta = \xi^{1/2}$.

Table 13.7. A Composite Design for $k = 2$ Predictor Variables and Its Associated Estimator Columns; $n_c = 8$, $n_{c0} = 1$, $n_s = 4$, $n_{s0} = 5$, $\alpha = 2$

1	x_1	x_2	x_1^2	x_2^2	$x_1 x_2$	$x_{111}/1.5$	$x_{222}/1.5$	$8CC^a$	Blocks	y
1	-1	-1	1	1	1	1	1	1	-1	37.5
1	-1	-1	1	1	1	1	1	1	-1	38.3
1	1	-1	1	1	-1	-1	1	1	-1	34.7
1	1	-1	1	1	-1	-1	1	1	-1	35.1
1	-1	1	1	1	-1	1	-1	1	-1	27.7
1	-1	1	1	1	-1	1	-1	1	-1	29.2
1	1	1	1	1	1	-1	-1	1	-1	12.2
1	1	1	1	1	1	-1	-1	1	-1	11.4
1	$\cdot$	$\cdot$	$\cdot$	$\cdot$	$\cdot$	$\cdot$		-8	-1	30.1
1	-2	$\cdot$	4	$\cdot$		-2		-1	1	30.2
1	2	$\cdot$	4	$\cdot$		2		-1	1	16.1
1	$\cdot$	-2	$\cdot$	4			-2	-1	1	31.4
1	$\cdot$	2	$\cdot$	4			2	-1	1	16.7
1	$\cdot$	$\cdot$	$\cdot$	$\cdot$				0.8	1	30.5
1	$\cdot$	$\cdot$	$\cdot$	$\cdot$				0.8	1	29.9
1	$\cdot$	$\cdot$	$\cdot$	$\cdot$				0.8	1	29.9
1	$\cdot$	$\cdot$	$\cdot$	$\cdot$				0.8	1	29.8
1	$\cdot$	$\cdot$	$\cdot$	$\cdot$				0.8	1	30.2

^{a}The $8CC$ column provides a "curvature contrast." If the assumptions made about the model being quadratic are true, this contrast has expectation zero. See Box and Draper (1982, Section 3.4) or Eq. (13.8.27).

possess an asymmetrical maximum which, *after suitable transformation of the ξ_i,* may be represented well enough by a quadratic function. (A quadratic function necessarily has a symmetric maximum.) Figure 13.9 illustrates this idea for a single predictor variable ξ, so that $k = 1$. The "suitable transformation" here is $\zeta = \xi^{1/2}$.

It turns out that a parsimonious class of designs of this type consists of the central composite arrangements in which a "cube," consisting of a two-level factorial with coded points $(\pm 1, \pm 1, \dots, \pm 1)$ or a fraction of resolution $R \geq 5$, is augmented by an added "star," with axial points at coded distance α, and by n_0 added center points at $(0, 0, \dots, 0)$. More generally, both the cube and the star might be replicated. A simple example of a design of this type for $k = 2$ is defined by the columns headed x_1 and x_2 in Table 13.7. (We use n_c and n_s for the number of cube and star points and use n_{c0} and n_{s0} for the number of center points associated with them, respectively. Thus $n_0 = n_{c0} + n_{s0}$.) Also shown are

Table 13.8. Analysis of Variance Associated with the Second-Order Model and Its Checks for the Data of Table 13.7

Source		df		SS	MS	F
Mean		1		13,938.934		
Blocks		1		7.347	7.347	
First order extra		2		842.907	421.453	
Second order extra		3		188.142	62.714	
Lack	b_{111}		1	7.701	7.701	16.74
of	b_{222}	3	1	88.923	79.656	29.691	79.656	64.86	173.17
fit	CC		1	1.567	1.567	3.14
Pure error		8		3.657	0.457	
Total		18		15,069.910		

some manufactured data generated by adding random error to response values read from Figure 8.5b, left diagram. The fitted least-squares second-degree equation is

$$\hat{y} = \begin{matrix} 30.59 \\ \pm 0.25 \end{matrix} \begin{matrix} -4.22x_1 \\ \pm 0.17 \end{matrix} \begin{matrix} -5.91x_2 \\ \pm 0.17 \end{matrix} \begin{matrix} -1.66x_1^2 \\ \pm 0.14 \end{matrix} \begin{matrix} -1.44x_2^2 \\ \pm 0.14 \end{matrix} \begin{matrix} -3.41x_1x_2, \\ \pm 0.24 \end{matrix} \quad (13.8.8)$$

where $\pm$ limits beneath each estimated coefficient indicate standard errors, calculated by using the pure error estimate $s_e^2 = 0.457$ to estimate σ^2. An associated analysis of variance table is shown as Table 13.8.

Before accepting the utility of the fitted equations, we would need to be reassured on two questions: (a) Is there evidence from the data of serious lack of fit? If not, (b) is the change in $\hat{y}$, over the experimental region explored by the design, large enough compared with the standard error of $\hat{y}$ to indicate that the response surface is adequately estimated?

The analysis of variance of Table 13.8 sheds light on both these questions. Its use to throw light on the second was discussed in Section 8.2.

Clearly, for this example, it is the marked lack of fit of the second-order model which immediately concerns us. In particular, a need for transformation would be associated with the appearance of third-order terms. Associated with the design of Table 13.7 are four possible third-order columns, namely, those formed by creating entries of the form

$$\left(x_1^3, x_1 x_2^2 \right), \qquad \left(x_2^3, x_2 x_1^2 \right). \qquad (13.8.9)$$

These form two sets of two items, as indicated by the parentheses.

Now suppose these third-order columns are orthogonalized with respect to the lower-order $\mathbf{X}$ vectors. This may be accomplished by regressing them against the first six columns and taking residuals to yield columns x_{111} (from x_1^3), x_{122} (from $x_1 x_2^2$), and so on. Then, in vector notation,

$$\mathbf{x}_{iii} = -3\mathbf{x}_{ijj}, \qquad i \neq j \qquad (13.8.10)$$

and the residual vectors are confounded in two sets of two. Furthermore, the columns $\mathbf{x}_{111}$ and $\mathbf{x}_{222}$ are orthogonal to each other. These vectors, reduced by a

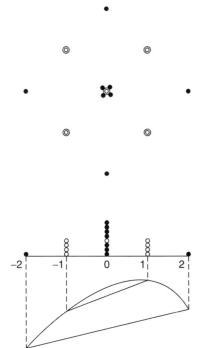

Figure 13.10. Projection of design points on the x_1 axis for the composite design in two factors given in Table 13.7. The contrast c_{31} is an estimate of the difference between the slopes of the two chords.

convenient factor of 1.5 to show their somewhat remarkable basic form, are given in Table 13.7.

Consider now the column $\mathbf{x}_{111}$ in relation to Figure 13.10, which shows the projection of the points of the composite design onto the x_1 axis. Denoting the average of the responses at $x_1 = -\alpha, -1, 1, \alpha$ by $\bar{y}_{-\alpha}$, $\bar{y}_{-1}$, $\bar{y}_1$ and $\bar{y}_\alpha$, respectively, we see that a contrast c_{31} associated with $\mathbf{x}_{111}$ is

$$c_{31} = \frac{1}{36}\mathbf{x}'_{111}\mathbf{y} = \frac{1}{3}\left\{ \frac{\bar{y}_\alpha - \bar{y}_{-\alpha}}{2\alpha} - \frac{\bar{y}_1 - \bar{y}_{-1}}{2} \right\}, \qquad (13.8.11)$$

where, for our example, $\alpha = 2$. The expression in the parentheses is an estimate of the difference in slopes of the two chords joining points equidistant from the design center. For a quadratic response curve, this difference is zero. Thus c_{31} is a natural measure of overall nonquadraticity in the x_1 direction. A corresponding measure in the x_2 direction is, of course, given by $c_{32} = \mathbf{x}'_{222}\mathbf{y}/36$.

The corresponding sums of squares for these contrasts, given in Table 13.8, indicate a highly significant lack of fit. Corresponding plots of the residuals against x_1 and against x_2 show a characteristic pattern. A line joining residuals for observations at $x_i = \alpha$ and $x_i = -\alpha$ slopes up, while the tendency of the remaining residuals is down as x_i is increased. We return to discuss these data later.

General Formulas

In general, a composite design contains:

1. A "cube," consisting of a 2^k factorial, or a 2^{k-p} fractional factorial, made up of points of the type $(\pm 1, \pm 1, \dots, \pm 1)$, of resolution $R \geq 5$ (Box and

Hunter, 1961a, b) replicated r_c (≥ 1) times. There are thus $n_c = r_c 2^{k-p}$ such points (where p may be zero).

2. A "star," that is, $2k$ points $(\pm\alpha, 0, 0, \ldots, 0), (0, \pm\alpha, 0, \ldots, 0), \ldots, (0, 0, 0, \ldots, \pm\alpha)$ on the predictor variable axes, replicated r_s times, so that there are $n_s = 2kr_s$ points in all. We assume $\alpha \neq 1$ [see Draper (1983)].

3. Center points $(0, 0, \ldots, 0)$, n_0 in number, of which n_{c0} are in cube blocks and n_{s0} in star blocks.

It is shown in Appendix 13F that, for any such design, k sets of columns can be isolated with the ith set containing the k columns $x_i x_j^2$, $j = 1, 2, \ldots, k$. This ith set is associated with a single vector $\mathbf{x}_{iii}$ which is orthogonal to the $(k + 1)(k + 2)/2$ columns required for fitting the second-degree equation and is also orthogonal to the $(k - 1)$ similarly constructed vectors $\mathbf{x}_{jjj}$, $j \neq i$.

The elements of these vectors are such that:

For the cube points, $x_{iii} = \phi x_i$, with $\phi = 2r_s\alpha^2(1 - \alpha^2)/(n_c + 2r_s\alpha^2)$.

For the star points, $x_{iii} = \gamma x_i$, with $\gamma = n_c(\alpha^2 - 1)/(n_c + 2r_s\alpha^2)$.

For the center points, $x_{iii} = 0 = x_i$.

Thus, the k estimates of third-order lack of fit, $c_{31}, c_{32}, \ldots, c_{3k}$ are

$$c_{3i} = \frac{\mathbf{x}'_{iii}\mathbf{y}}{\mathbf{x}'_{iii}\mathbf{x}_{iii}} = \frac{1}{\alpha^2 - 1}\left\{\frac{\bar{y}_{\alpha i} - \bar{y}_{-\alpha i}}{2\alpha} - \frac{\bar{y}_{1i} - \bar{y}_{-1i}}{2}\right\} \qquad (13.8.12)$$

with standard deviation

$$\sigma_{c_3} = \frac{1}{\alpha^2 - 1}\left\{\frac{1}{n_c} + \frac{1}{2r_s\alpha^2}\right\}^{1/2}\sigma. \qquad (13.8.13)$$

Also,

$$E(c_{3i}) = \beta_{iii} + (1 - \alpha^2)^{-1}\sum_{j \neq i}^{k}\beta_{ijj}. \qquad (13.8.14)$$

and the contribution to the lack of fit sum of squares is

$$\mathrm{SS}(c_{3i}) = \frac{(\alpha^2 - 1)^2 c_{3i}^2}{\{1/n_c + 1/(2r_s\alpha^2)\}}. \qquad (13.8.15)$$

Note that even if $E(c_{3i}) = 0$, this does not necessarily mean there are no cubic coefficients. A combination of *nonzero* β_{iii} and β_{ijj} could occur for which $\beta_{iii} + (1 - \alpha^2)^{-1}\sum\beta_{ijj} = 0$. It is, of course, impossible to guard against every such possibility unless the full cubic model is fitted.

We now consider more explicitly what relation these lack of fit measures have to the possibility of transformation of the predictor variables. Suppose that, at worst, the response function may be represented by a third-order model in the *trans-*

formed predictor variables. In Appendix 13E we show that the conditions that must then apply if all third-order coefficients of the transformed $\xi_i^{\lambda_i}$ are to be zero are [see Eqs. (13E.5), (13E.3), and (13E.4)]

$$\eta_{ijl} = 0, \qquad \text{all } i \neq j \neq l = 1, 2, \ldots, k; \qquad (13.8.16)$$

$$\eta_{ijj} + \delta_i(1 - \lambda_j)\eta_{ij} = 0, \qquad i \neq j = 1, 2, \ldots, k; \qquad (13.8.17)$$

$$\eta_{iii} + 3\delta_i(1 - \lambda_i)\eta_{ii} + \delta_i^2(1 - \lambda_i)(1 - 2\lambda_i)\eta_i = 0, \qquad i = 1, 2, \ldots, k. \quad (13.8.18)$$

Thus the possibility of second-order representation in the transformed variables is contraindicated if any interaction estimate b_{ijl}, $i \neq j \neq l$, is nonzero.

In practice, the estimation of the transformation (when not contraindicated) is best done using nonlinear least squares directly on the model of Eq. (13.8.1); however, some interesting light on how the curvature measures c_{3i} relate to these transformations is obtained by considering how Eqs. (13.8.17) and (13.8.18) could be used to obtain estimates of the λ_i.

A composite design does not permit all third-order terms to be separately estimated. Suppose, however, that a second-order model augmented with only cubic terms

$$\beta_{111}x_1^3 + \beta_{222}x_2^3 + \cdots + \beta_{kkk}x_k^3, \qquad (13.8.19)$$

was fitted. If the response η could be represented by a full third-order model, the estimates b_i and b_{iii} obtained from the composite design would have expectations

$$E(b_i) = \eta_i - \frac{1}{2}\alpha^2(1 - \alpha^2)^{-1}\sum_{j \neq i}^{k}\eta_{ijj}, \qquad (13.8.20)$$

$$E(b_{iii}) = \frac{1}{6}\eta_{iii} + \frac{1}{2}(1 - \alpha^2)^{-1}\sum_{j \neq i}^{k}\eta_{ijj}. \qquad (13.8.21)$$

If now b_i and b_{iii} are used as estimates of the quantities shown as their expectations then, after appropriate substitutions have been made in Eqs. (13.8.16) to (13.8.18), we obtain the following k equations for the λ_i. (In these equations, $b_{iii} = c_{3i}$.)

$$b_{iii} + \delta_i(1 - \lambda_i)b_{ii} + \frac{1}{6}\delta_i^2(1 - \lambda_i)(1 - 2\lambda_i)b_i$$

$$+ \frac{1}{2}(1 - \alpha^2)^{-1}\left\{1 - \frac{1}{6}\alpha^2\delta_i^2(1 - 2\lambda_i)(1 - \lambda_i)\sum_{j \neq i}^{k}\delta_j(1 - \lambda_j)b_{ij}\right\} = 0,$$

$$i = 1, 2, \ldots, k. \qquad (13.8.22)$$

These equations can be solved iteratively. Guessed values for the λ_i are first substituted in the grouping $(1 - \lambda_i)(1 - 2\lambda_i)$ wherever it occurs and the resulting linear equations solved to provide improved estimates for a second iteration, and so on.

Table 13.9. Analysis of Variance for Second-Order Model in Predictor Variables ln ξ_1 and ξ_2^{-1}

Source	df	SS	MS	F
Mean	1	13,938.934		
Blocks	1	7.347	7.347	
First-order extra	2	552.713	276.357	535.57
Second-order extra	3	565.238	188.413	365.14
Lack of fit $\begin{cases} CC \\ \\ \text{Third order} \end{cases}$ $3\begin{cases} 1 \\ 2.021 \\ 2 \end{cases}$ $\begin{cases} 1.396 \\ 0.674 \\ 0.625 \end{cases}$ $\begin{cases} 1.396 \\ 1.47 \\ 0.313 \end{cases}$ $\begin{cases} 3.05 \\ \\ 0.68 \end{cases}$				
Pure error	8	3.657	0.457	
Total	18	15,069.910		

For the example data, this procedure converges to the values $\hat{\lambda}_1 = -0.23$, $\hat{\lambda}_2 = -0.93$. These may be compared with the values $\hat{\lambda}_1 = 0.09$, $\hat{\lambda}_2 = -0.82$ provided by nonlinear least squares (these are maximum likelihood estimates under the standard normal error assumptions) and with $\lambda_1 = 0$, $\lambda_2 = -1$, the values used to generate the data; see Figure 8.5b.

An analysis of variance for the transformed data is shown in Table 13.9 where, as anticipated, no lack of fit appears.

A Curvature Contrast

We have already considered, indirectly, the curvature contrast

$$c_2 = \bar{y}_c - \bar{y}_{c0}, \tag{13.8.23}$$

which compares the average response at the factorial ("cube") points of a two-level factorial design with the average response at the center of the design. The expected value is

$$E(c_2) = \beta_{11} + \beta_{22} + \cdots + \beta_{kk}, \tag{13.8.24}$$

and this contrast contributes, to the analysis of variance table for a first-order design, a sum of squares

$$\frac{n_0 n_c (\bar{y}_0 - \bar{y}_{c0})^2}{(n_0 + n_c)} \tag{13.8.25}$$

where n_c and n_0 are the number of factorial and center points, respectively. (See Table 13.5 with slightly different notation.) Now when, as in the design of Table 13.7, center points are available both in the factorial block(s) *and* in the star block(s), several (two for our example) such measures are available. Consider, specifically, the two block case for a moment. If the average response at the center of the star is $\bar{y}_{s0}$ and the average over all the star points is $\bar{y}_s$, then the contrast

$$c_2' = \frac{k}{\alpha^2}(\bar{y}_s - \bar{y}_{s0}) \tag{13.8.26}$$

also has expectation $\Sigma \beta_{ii}$. Thus the statistic

$$c_2 - c_2' = \bar{y}_c - \bar{y}_{c0} - \frac{k}{\alpha^2}(\bar{y}_s - \bar{y}_{s0}), \qquad (13.8.27)$$

which is the difference of the two measures of overall curvature, should be zero if the assumptions made about the model being quadratic are true.

From Figure 13.10, we see that the curvature measure c_2 associated with the cube (open circles) is contrasted with c_2' associated with the star (black dots). In general the distance from the center of the design to the cube points is $k^{1/2}$ and that for the star points is α. When, as in our example, $k^{1/2}$ and α are different, a significant value of $c_2 - c_2'$ could indicate (for example) a *symmetric* departure from quadratic fall-off on each side of the maximum, such as we see (for example) in a normal distribution curve.

In general, for two blocks, the standard deviation for $c_2 - c_2'$ is given by

$$\sigma_{c_2-c_2'} = \left\{ \frac{1}{n_c} + \frac{1}{n_{c0}} + \frac{k^2}{\alpha^4}\left[\frac{1}{2kr_s} + \frac{1}{n_{s0}} \right] \right\}^{1/2} \sigma \qquad (13.8.28)$$

and the associated sum of squares for the analysis of variance table entry of Table 13.8 is obtained from

$$\text{SS}(c_2 - c_2') = \frac{(c_2 - c_2')^2}{\{1/n_c + 1/n_{c0} + k/(2r_s\alpha^4) + k^2/(\alpha^4 n_{s0})\}}. \qquad (13.8.29)$$

For our example we find

$$c_2 - c_2' = 1.3925 \pm 0.7520 \qquad (13.8.30)$$

with associated sum of squares 1.567 as shown in Table 13.8. There is clearly no evidence of this sort of lack of fit. When transformed predictors are used, again no lack of fit of this kind is evident, as we see from Table 13.9. No reduction in the degrees of freedom is made for the estimates of λ_1 and λ_2 (Box and Cox 1964, p. 240) because these degrees of freedom are identical to those attributed to third order.

Interaction With Blocks?

When composite designs are run in blocks and if we allow the possibility that effects from the predictor variables could interact with blocks, then the various measures of lack of fit would be confounded with block-effect interactions. Although such contingencies must always be borne in mind, it should be remembered that these particular block–effect interactions are no more likely than any others.

APPENDIX 13A. DERIVATION OF EQ. (13.5.8)

From Eqs. (13.5.3), and (13.5.5), we have

$$V = \left(\frac{N}{\sigma^2}\right)\int_O w(x)\,E\{(b_0 - Eb_0) + (b_1 - Eb_1)x\}^2 \, dx$$

$$= \left(\frac{N}{\sigma^2}\right)\int_O w(x)\Big\{E(b_0 - Eb_0)^2$$

$$\qquad\qquad + 2xE\big[(b_0 - Eb_0)(b_1 - Eb_1)\big] + x^2 E(b_1 - Eb_1)^2\Big\}\,dx$$

$$= \frac{N\{V(b_0) + \mu_2 V(b_1)\}}{\sigma^2}\qquad [\text{because } \mu_1 = 0],$$

$$= 1 + \frac{\mu_2}{m_2}\qquad\qquad\quad [\text{because } \bar{x} = 0].$$

Applying the work of Section 3.10 to Eqs. (13.5.5) and (13.5.6), we find that

$$E(b_0) = \beta_0 + m_2\beta_{11}, \qquad E(b_1) = \beta_1 + \frac{m_3\beta_2}{m_2}.$$

Thus,

$$B = \alpha^2 \int_O w(x)\left(m_2 - x^2 + \frac{m_3 x}{m_2}\right)^2 dx \qquad \left(\text{where } \alpha^2 = \frac{N\beta_{11}^2}{\sigma^2}\right)$$

$$= \alpha^2 \int_O w(x)\left(m_2^2 - 2m_2 x^2 + x^4 + \frac{m_3^2 x^2}{m_2^2} + 2m_3 x - \frac{m_3 x^3}{m_2}\right)dx$$

$$= \alpha^2\left(m_2^2 - 2\mu_2 m_2 + \mu_4 + \frac{m_3^2 \mu_2}{m_2^2}\right)\qquad [\text{because } \mu_p = 0, \text{ for } p \text{ odd}]$$

$$= \alpha^2\left\{(m_2 - \mu_2)^2 + (\mu_4 - \mu_2^2) + \frac{m_3^2 \mu_2}{m_2^2}\right\}.$$

Combination of V and B gives the desired result.

**APPENDIX 13B. NECESSARY AND SUFFICIENT DESIGN
CONDITIONS TO MINIMIZE BIAS**

From (13.5.4), (13.6.3), and (13.5.4), we have

$$B = \left(\frac{N}{\sigma^2}\right)\int_O w(\mathbf{x})\boldsymbol{\beta}_2'(\mathbf{x}_1'\mathbf{A} - \mathbf{x}_2')'(\mathbf{x}_1'\mathbf{A} - \mathbf{x}_2')\boldsymbol{\beta}_2 \, d\mathbf{x}$$

$$= \boldsymbol{\alpha}_2'\boldsymbol{\Delta}\boldsymbol{\alpha}_2,$$

where

$$\boldsymbol{\alpha}_2 = N\boldsymbol{\beta}_2/\sigma^2$$

$$\mathbf{A} = (\mathbf{X}_1'\mathbf{X}_1)^{-1}\mathbf{X}_1'\mathbf{X}_2 = \mathbf{M}_{11}^{-1}\mathbf{M}_{12},$$

$$\boldsymbol{\Delta} = \mathbf{A}'\boldsymbol{\mu}_{11}\mathbf{A} - \boldsymbol{\mu}_{12}'\mathbf{A} - \mathbf{A}'\boldsymbol{\mu}_{12} + \boldsymbol{\mu}_{22}$$

with $\boldsymbol{\mu}_{11}$ and $\boldsymbol{\mu}_{12}$ defined as in Eqs. (13.6.4) and $\boldsymbol{\mu}_{22}$ being similarly defined with appropriate changes in subscripts. Now $\boldsymbol{\Delta}$ can be rewritten as $\boldsymbol{\Delta}_1 + \boldsymbol{\Delta}_2$ where

$$\boldsymbol{\Delta}_1 = \boldsymbol{\mu}_{22} - \boldsymbol{\mu}_{12}'\boldsymbol{\mu}_{11}^{-1}\boldsymbol{\mu}_{12}$$

and

$$\boldsymbol{\Delta}_2 = (\mathbf{M}_{11}^{-1}\mathbf{M}_{12} - \boldsymbol{\mu}_{11}^{-1}\boldsymbol{\mu}_{12})'\boldsymbol{\mu}_{11}(\mathbf{M}_{11}^{-1}\mathbf{M}_{12} - \boldsymbol{\mu}_{11}^{-1}\boldsymbol{\mu}_{12}).$$

It is shown below that both $\boldsymbol{\mu}_{11}$ and $\boldsymbol{\Delta}_1$ are positive semidefinite. It follows that, whatever the value of $\boldsymbol{\beta}_2$, B is minimized if and only if $\mathbf{M}_{11}^{-1}\mathbf{M}_{12} = \boldsymbol{\mu}_{11}^{-1}\boldsymbol{\mu}_{12}$. (If $\mathbf{M}_{11} = \boldsymbol{\mu}_{11}$ and $\mathbf{M}_{12} = \boldsymbol{\mu}_{12}$, than B is also minimized but these conditions are sufficient and not necessary.)

To Show Both $\boldsymbol{\mu}_{11}$ and $\boldsymbol{\Delta}_1$ Are Positive Semidefinite

It is obvious that

$$\int_O w(\mathbf{x})\left[(\mathbf{x}_1', \mathbf{x}_2')\boldsymbol{\beta}\right]^2 dx \geq 0$$

where $\boldsymbol{\beta} = (\boldsymbol{\beta}_1', \boldsymbol{\beta}_2')'$. Writing this as a quadratic form in $\boldsymbol{\beta}$ and performing the integration in the matrix of the quadratic form gives

$$\boldsymbol{\beta}' \begin{bmatrix} \boldsymbol{\mu}_{11} & \boldsymbol{\mu}_{12} \\ \boldsymbol{\mu}_{12}' & \boldsymbol{\mu}_{22} \end{bmatrix} \boldsymbol{\beta} \geq 0,$$

so that both $\boldsymbol{\mu}_{11}$ and the matrix shown are positive semidefinite; but if we define the nonsingular matrix $\mathbf{T}$ as

$$\mathbf{T} = \begin{bmatrix} \mathbf{I}_{11} & -\boldsymbol{\mu}_{11}^{-1}\boldsymbol{\mu}_{12} \\ \mathbf{0} & \mathbf{I}_{22} \end{bmatrix}$$

then

$$\begin{bmatrix} \boldsymbol{\mu}_{11} & \mathbf{0} \\ \mathbf{0} & \boldsymbol{\Delta}_1 \end{bmatrix} = \begin{bmatrix} \boldsymbol{\mu}_{11} & \mathbf{0} \\ \mathbf{0} & \boldsymbol{\mu}_{22} - \boldsymbol{\mu}_{12}'\boldsymbol{\mu}_{11}^{-1}\boldsymbol{\mu}_{12} \end{bmatrix} = \mathbf{T}' \begin{bmatrix} \boldsymbol{\mu}_{11} & \boldsymbol{\mu}_{12} \\ \boldsymbol{\mu}_{12}' & \boldsymbol{\mu}_{22} \end{bmatrix} \mathbf{T}.$$

The right-hand matrix is clearly positive semidefinite. It follows that $\boldsymbol{\Delta}_1$ must be positive semidefinite also.

APPENDIX 13C. MINIMUM BIAS ESTIMATION

In the work we have described, in which designs are sought which minimize integrated variance plus bias, the estimation procedure used was that of least squares. An extra dimension can be created in this work by allowing other, perhaps biased, estimators and treating the least-squares method as a special case of a more general estimation class. Specifically, the use of *minimum bias estimation* was suggested by Karson et al. (1969). In this procedure, the estimator for $\boldsymbol{\beta}$ in the regression model is chosen to minimize the integrated bias B. Provided that $(\mathbf{X}'\mathbf{X})$ is nonsingular,* the estimator that achieves this takes the form

$$\mathbf{b}_1(MB) = \{\mathbf{I}, \boldsymbol{\mu}_{11}^{-1}\boldsymbol{\mu}_{12}\}(\mathbf{X}'\mathbf{X})^{-1}\mathbf{X}'\mathbf{y},$$

where

$$\mathbf{X} = (\mathbf{X}_1, \mathbf{X}_2).$$

This estimator can be written in the alternative form

$$\mathbf{b}_1(MB) = \{\mathbf{I}, \boldsymbol{\mu}_{11}^{-1}\boldsymbol{\mu}_{12}\}\begin{bmatrix} \mathbf{b}_1^* \\ \mathbf{b}_2^* \end{bmatrix}$$

$$= \mathbf{b}_1^* + \boldsymbol{\mu}_{11}^{-1}\boldsymbol{\mu}_{12}\mathbf{b}_2^*,$$

where $\mathbf{b}_1^*$, $\mathbf{b}_2^*$ are the least-squares estimators of $\mathbf{b}_1$ and $\mathbf{b}_2$ *when the latter are estimated together by least squares*. The minimum value of B arising from the estimator $\mathbf{b}_1(MB)$ is

$$\boldsymbol{\alpha}_2'(\boldsymbol{\mu}_{22} - \boldsymbol{\mu}_{12}'\boldsymbol{\mu}_{11}^{-1}\boldsymbol{\mu}_{12})\boldsymbol{\alpha}_2$$

and this is achieved for any experimental design for which

$$\{\mathbf{I}, \boldsymbol{\mu}_{11}^{-1}\boldsymbol{\mu}_{12}\}\begin{pmatrix} \boldsymbol{\beta}_1 \\ \boldsymbol{\beta}_2 \end{pmatrix}$$

is estimable. Note that *if* a design is an "all-bias design" [see Eq. (13.6.5) or Appendix 13B], then

$$\boldsymbol{\mu}_{11}^{-1}\boldsymbol{\mu}_{12} = (\mathbf{X}_1'\mathbf{X}_1)^{-1}\mathbf{X}_1'\mathbf{X}_2 = \mathbf{M}_{11}^{-1}\mathbf{M}_{12} = \mathbf{A},$$

where $\mathbf{A}$ is the usual bias or alias matrix and then

$$\mathbf{b}_1(MB) = \mathbf{b}_1^* + \mathbf{A}\mathbf{b}_2^*$$

$$= (\mathbf{X}_1'\mathbf{X}_1)^{-1}\mathbf{X}_1'\mathbf{y}$$

---

*If it is not, the estimator may still exist. The requirement is then that $(\mathbf{I}, \boldsymbol{\mu}_{11}^{-1}\boldsymbol{\mu}_{12})(\boldsymbol{\beta}_1', \boldsymbol{\beta}_2')'$ be estimable, and the estimator is of the form $\mathbf{T}'\mathbf{Y}$ where $\mathbf{T}'$ must satisfy the condition $\mathbf{T}'(\mathbf{X}_1, \mathbf{X}_2) = (\mathbf{I}, \boldsymbol{\mu}_{11}^{-1}\boldsymbol{\mu}_{12})$.

the usual least-squares estimator for $\boldsymbol{\beta}_1$ estimated alone. The reduction to this arises from the fact that

$$\mathbf{b}^* = \begin{bmatrix} \mathbf{b}_1^* \\ \mathbf{b}_2^* \end{bmatrix} = \begin{bmatrix} \mathbf{X}_1'\mathbf{X}_1 & \mathbf{X}_1'\mathbf{X}_2 \\ \mathbf{X}_2'\mathbf{X}_1 & \mathbf{X}_2'\mathbf{X}_2 \end{bmatrix}^{-1} \begin{bmatrix} \mathbf{X}_1'\mathbf{y} \\ \mathbf{X}_2'\mathbf{y} \end{bmatrix} = \begin{bmatrix} \mathbf{M}_{11} & \mathbf{M}_{12} \\ \mathbf{M}_{12}' & \mathbf{M}_{22} \end{bmatrix} \begin{bmatrix} \mathbf{X}_1'\mathbf{y} \\ \mathbf{X}_2'\mathbf{y} \end{bmatrix}$$

$$= \begin{bmatrix} \mathbf{C}_{11}^{-1} & -\mathbf{M}_{11}^{-1}\mathbf{M}_{12}\mathbf{C}_{22}^{-1} \\ -\mathbf{M}_{22}^{-1}\mathbf{M}_{21}\mathbf{C}_{11}^{-1} & \mathbf{C}_{22}^{-1} \end{bmatrix} \begin{bmatrix} \mathbf{X}_1'\mathbf{y} \\ \mathbf{X}_2'\mathbf{y} \end{bmatrix},$$

where

$$\mathbf{C}_{ii} = \mathbf{M}_{ii} - \mathbf{M}_{ij}\mathbf{M}_{jj}^{-1}\mathbf{M}_{ji}, \quad i, j = 1, 2, \quad i \neq j, \quad \text{and} \quad \mathbf{M}_{21} = \mathbf{M}_{12}'.$$

Premultiplication of $\mathbf{b}^*$ by $(\mathbf{I}, \mathbf{M}_{11}^{-1}\mathbf{M}_{12})$ produces the stated result, after some reduction.

We see immediately some advantages and some drawbacks in minimum bias estimation. Advantages are that it reduces the integrated bias B to its minimum value, and we can then choose a design to minimize the integrated variance V, thus achieving a smaller integrated variance plus bias than would be possible when using least squares; for details, see Karson et al. (1969). A drawback is that for nonsingular* $\mathbf{X}'\mathbf{X}$, we need to be able to estimate *both* $\boldsymbol{\beta}_1$ and $\boldsymbol{\beta}_2$ from the data via least squares to achieve minimum bias. (In the least-squares case, only $\boldsymbol{\beta}_1$ need be estimable.) Moreover, methods of proceeding statistically after the minimum bias estimates have been obtained are unknown. For example, there are no minimum bias estimation equivalents to the analysis of variance table, nor to t or F tests. Our opinion is that minimum bias estimation is an ingenious idea that needs further development before it can be usefully employed.

APPENDIX 13D. A SHRINKAGE ESTIMATION PROCEDURE

Another type of biased estimation procedure was proposed by Kupper and Meydrech (1973, 1974). They suggested using an estimator for $\boldsymbol{\beta}_1$ of form

$$\mathbf{b}_1(KM) = \mathbf{K}\mathbf{b}_1,$$

where $\mathbf{K}$ is a square matrix of the same order as the size of $\mathbf{b}_1$ and $\mathbf{b}_1 = (\mathbf{X}_1'\mathbf{X}_1)^{-1}\mathbf{X}_1'\mathbf{y}$ is the usual least-squares estimator of $\boldsymbol{\beta}_1$. The basic idea is to choose $\mathbf{K}$ in an advantageous manner, to minimize the integrated variance plus bias. In fact, it can be shown that if at least one of the elements in $\boldsymbol{\alpha}_1 = N\boldsymbol{\beta}_1/\sigma^2$ can be bounded, however loosely, a $\mathbf{K}$ can be found which will produce smaller integrated variance plus bias than would the least-squares choice $\mathbf{K} = \mathbf{I}$. Consider, for example, the case of a cuboidal region of interest in k dimensions, and a fitted polynomial of first order, with a true model of second order. Then, if $\alpha_i^2 \leq M_i$ where M_i is some given positive constant, the corresponding diagonal element of

*See previous footnote.

K is chosen as

$$k_i = \frac{m_2 M_i}{(1 + m_2 M_i)},$$

where

$$m_2 = N^{-1} \sum_{u=1}^{N} x_{iu}^2, \qquad i = 1, 2, \ldots, k.$$

Overall, K is chosen as the diagonal matrix whose elements are either k_i as given above for finite M_i, or 1 for $M_i = \infty$. In fact, the integrated variance plus bias will be reduced even if α_i^2 is as large as $2M_i$. For details, see Kupper and Meydrech (1973, 1974).

This procedure produces estimates that are shrunken versions (because $k_i \leq 1$) of the least-squares estimators. It is related in spirit to ridge regression methods and is perfectly appropriate in situations where values of M_i can be realistically settled upon. A disadvantage the method shares with minimum bias estimation is that methods of proceeding after estimation are not known. Again, our opinion is that this is an ingenious idea that needs further development to be of practical value.

APPENDIX 13E. CONDITIONS FOR EFFICACY OF TRANSFORMATION ON THE PREDICTOR VARIABLES

Suppose that an experimental design is run in the k coded variables

$$x_i = \frac{(\xi_i - \xi_{i0})}{S_i}, \qquad i = 1, 2, \ldots, k, \tag{13E.1a}$$

in a situation where the underlying response function can be approximated by a second-degree polynomial $F\{(\xi_i^{\lambda_i})\}$ in the transformed *original* variables $\xi_i^{\lambda_i}$. Thus

$$F(\xi, \lambda) = \dot{\beta}_0 + \sum_{i=1}^{k} \dot{\beta}_i \xi_i^{\lambda_i} + \sum_{i=1}^{k} \sum_{j \geq i}^{k} \dot{\beta}_{ij} \xi_i^{\lambda_i} \xi_j^{\lambda_j}. \tag{13E.1b}$$

Assume all $\lambda_i \neq 0$. [The case when any $\lambda_i = 0$ can be handled as a limiting case using the fact [see, e.g., Box and Cox (1964)] that $(\xi^\lambda - 1)/\lambda$ tends to $\ln \lambda$ as λ tends to zero.]

If Eq. (13E.1b) is to be a suitable representation, then all third derivatives with respect to the $\xi_i^{\lambda_i}$ must vanish identically. Note that

$$\frac{\partial F}{\partial \xi_i^{\lambda_i}} = \frac{\partial F}{\partial \xi_i} \frac{\partial \xi_i}{\partial \xi_i^{\lambda_i}} = F_i \left(\frac{\xi_i^{1-\lambda_i}}{\lambda_i} \right)$$

say, where $F_i = \partial F / \partial \xi_i$. Moreover, because of Eq. (13E.1a),

$$\eta_i = \frac{\partial F}{\partial x_i} = \frac{\partial F}{\partial \xi_i} \frac{\partial \xi_i}{\partial x_i} = F_i S_i$$

so that

$$\frac{\partial F}{\partial \xi_i^{\lambda_i}} = \eta_i S_i^{-1} \left(\frac{\xi_i^{1-\lambda_i}}{\lambda_i} \right).$$

The obvious extensions of these results also follow for the higher derivatives which involve terms of the form

$$\eta_{ij} = \frac{\partial^2 F}{\partial x_i \, \partial x_j} \quad \text{and} \quad \eta_{ijl} = \frac{\partial^3 F}{\partial x_i \, \partial x_j \, \partial x_l}.$$

If we now carry out the appropriate differentiations, we obtain

$$\frac{\partial F}{\partial \xi_i^{\lambda_i}} = \frac{\xi_i^{1-\lambda_i}}{S_i \lambda_i} \eta_i,$$

$$\frac{\partial^2 F}{\partial \xi_i^{\lambda_i} \partial \xi_j^{\lambda_j}} = \frac{\xi_i^{1-\lambda_i} \xi_j^{1-\lambda_j}}{S_i S_j \lambda_i \lambda_j} \eta_{ij},$$

(13E.2a)

$$\frac{\partial^2 F}{\partial^2 (\xi_i^{\lambda_i})} = \frac{\xi_i^{1-2\lambda_i}}{S_i \lambda_i^2} \left[(1 - \lambda_i) \eta_i + \frac{\xi_i}{S_i} \eta_{ii} \right],$$

(13E.2b)

$$\frac{\partial^3 F}{\partial \xi_i^{\lambda_i} \partial^2 (\xi_j^{\lambda_j})} = \frac{\xi_i^{1-\lambda_i} \xi_j^{1-2\lambda_j}}{S_i S_j \lambda_i \lambda_j^2} \left[\frac{\xi_j}{S_j} \eta_{ijj} + (1 - \lambda_j) \eta_{ij} \right] = 0,$$

(13.E.3)

$$\frac{\partial^3 F}{\partial^3 (\xi_i^{\lambda_i})} = \frac{\xi_i^{1-3\lambda_i}}{S_i \lambda_i^3} \left[\left(\frac{\xi_i}{S_i} \right)^2 \eta_{iii} + \frac{3 \xi_i}{S_i} (1 - \lambda_i) \eta_{ii} + (1 - \lambda_i)(1 - 2\lambda_i) \eta_i \right] = 0,$$

(13E.4)

$$\frac{\partial^3 F}{\partial \xi_i^{\lambda_i} \partial \xi_j^{\lambda_j} \partial \xi_l^{\lambda_l}} = \frac{\xi_i^{1-\lambda_i} \xi_j^{1-\lambda_j} \xi_l^{1-\lambda_l}}{S_i S_j S_l \lambda_i \lambda_j \lambda_l} \eta_{ijl} = 0.$$

(13E.5)

Equations (13E.3)–(13E.5) are exact conditions on the λ's. The η_i, η_{ij}, and η_{ijl} also involve the λ's, and cannot be specifically evaluated if the λ's are unknown. If *estimates* of the η_i, η_{ij}, and η_{ijl} are substituted, however, the three equations can be solved to provide estimates of the λ's.

We now assume that the response surface can be approximately represented by a *cubic* polynomial in the coded predictor variables, namely, by

$$\hat{\eta} = b_0 + b_1 x_1 + \cdots + b_k x_k$$

$$+ b_{11} x_1^2 + \cdots + b_{kk} x_k^2$$

$$+ b_{12} x_1 x_2 + \cdots + b_{k-1,k} x_{k-1} x_k$$

$$+ b_{111} x_1^3 + b_{122} x_1 x_2^2 + \cdots$$

$$+ b_{222} x_2^3 + b_{112} x_1^2 x_2 + \cdots$$

$$+ b_{kkk} x_k^3 + b_{11k} x_1^2 x_k + \cdots$$

$$+ b_{123} x_1 x_2 x_3 + \cdots .$$

We can estimate the η_i, η_{ij}, and η_{ijl} by the corresponding derivatives of $\hat{\eta}$ evaluated at the center of the design—that is, at $\mathbf{x} = \mathbf{0}$. In general then,

$$\hat{\eta}_i = b_i, \qquad \hat{\eta}_{ii} = 2b_{ii}, \qquad \hat{\eta}_{iii} = 6b_{iii},$$

$$\hat{\eta}_{ij} = b_{ij}, \qquad \hat{\eta}_{ijj} = 2b_{ijj}, \qquad \hat{\eta}_{ijl} = b_{ijl}, \tag{13E.6}$$

and we would substitute these in Eqs. (13E.3)–(13E.5), at the same time setting $\xi_i = \xi_{i0}$, its value when $x_i = 0$. Note that from Eq. (13E.5) we require that $\eta_{ijl} = 0$, which implies that the assumed transformations $\xi_i^{\lambda_i}$ are not suitable representations unless all three-factor interactions are zero. In practice, then, we would want b_{ijl} to be small and nonsignificant when $i \neq j \neq l$ or else we have a clear indication that the $\xi_i^{\lambda_i}$ will *not* provide a satisfactory second-order representation in (13E.1b). If this aspect is satisfied, however, we now solve the $\frac{1}{2}k(k+1) + k = \frac{1}{2}k(k+3)$ simultaneous equations:

$$\hat{\eta}_{ijj} + \delta_j (1 - \lambda_j) \hat{\eta}_{ij} = 0, \qquad i \neq j = 1, 2, \ldots, k,$$

$$\hat{\eta}_{iii} + 3\delta_i (1 - \lambda_i) \hat{\eta}_{ii} + \delta_i^2 (1 - \lambda_i)(1 - 2\lambda_i) \hat{\eta}_i = 0, \qquad i = 1, 2, \ldots, k \tag{13E.7}$$

where $\delta_i = S_i / \xi_{i0}$, and we have divided through Eqs. (13E.3) and (13E.4) by factors assumed to be nonzero, namely S_i, ξ_{i0}, λ_i, using the $\hat{\eta}$ values from Eq. (13E.6).

Composite Designs

An additional complication arises with composite designs. For such designs, we cannot estimate all the third-order coefficients individually (see Appendix 13F). This means that while Eqs. (13E.7) are still valid, the values in Eq. (13E.6) cannot be used. For a composite design, the column vectors with elements x_i, x_i^3, and

$x_i x_j^2$ are linearly dependent via

$$x_i x_j^2 = \frac{\left(-\alpha^2 x_i + x_i^3\right)}{\left(1 - \alpha^2\right)}.$$

Thus, in the general cubic model, we cannot estimate all $k + k + k(k-1) = k(k+1)$ coefficients in the terms

$$\beta_i x_i + \beta_{iii} x_i^3 + \sum_{j \neq i}^k \beta_{ijj} x_i x_j^2$$

but only the $2k$ coefficients of

$$\left\{ \beta_i - \frac{\alpha^2}{1 - \alpha^2} \sum_{j \neq i}^k \beta_{ijj} \right\} x_i + \left\{ \beta_{iii} + \frac{1}{1 - \alpha^2} \sum_{j \neq i}^k \beta_{ijj} \right\} x_i^3. \qquad (13E.8)$$

It follows that, in the model so reduced,

$$b_i \text{ estimates } \eta_i - \frac{\alpha^2}{1 - \alpha^2} \frac{1}{2} \sum_{j \neq i}^k \eta_{ijj}$$

while

$$b_{iii} \text{ estimates } \frac{1}{6} \eta_{iii} + \frac{1}{1 - \alpha^2} \frac{1}{2} \sum_{j \neq i}^k \eta_{ijj}. \qquad (13E.9)$$

[By examining Eqs. (13.8.14), (13E.6), and the second portion of Eq. (13E.8), we infer that $b_{iii} = c_{3i}$.] Alternatively, if the model is fitted using x_i and the "orthogonalized x_i^3," namely $x_{iii} = x_i^3 - \psi x_i$, the terms of the model are

$$\left\{ \beta_i + \psi \beta_{iii} + \theta \sum_{j \neq i}^k \beta_{ijj} \right\} x_i + \left\{ \beta_{iii} + \frac{1}{1 - \alpha^2} \sum_{j \neq i}^k \beta_{ijj} \right\} \left(x_i^3 - \psi x_i\right),$$

where

$$\theta = \frac{n_c}{\left(n_c + 2r_s \alpha^2\right)}, \qquad \psi = \frac{\left(n_c + 2r_s \alpha^4\right)}{\left(n_c + 2r_s \alpha^2\right)}.$$

In this form, we have that

$$b_i^* \text{ estimates } \eta_i + \frac{1}{6} \psi \eta_{iii} + \frac{1}{2} \theta \sum_{j \neq i}^k \eta_{ijj}$$

and

$$b_{iii}^* \text{ estimates } \frac{1}{6} \eta_{iii} + \frac{1}{1 - \alpha^2} \frac{1}{2} \sum_{j \neq i}^k \eta_{ijj},$$

where b_i^* is the estimated coefficient of x_i and b_{iii}^* is that for $(x_i^3 - \psi x_i)$. (Note that $b_{iii}^* = b_{iii} = c_{3i}$, but that $b_i^* = b_i + \psi b_{iii}$.) We now describe how this affects the estimation of the λ_i. From Eqs. (13E.3) and (13E.4) and setting $\xi_i = \xi_{i0}$ we have

$$\eta_{ijj} + \delta_j(1 - \lambda_j)\eta_{ij} = 0 \tag{13E.10}$$

and

$$\eta_{iii} + 3\delta_i(1 - \lambda_i)\eta_{ii} + \delta_i^2(1 - \lambda_i)(1 - 2\lambda_i)\eta_i = 0. \tag{13E.11}$$

We now combine $\frac{1}{6}$ times Eq. (13E.11) with

$$\left\{ \frac{1}{2(1 - \alpha^2)} - \frac{\alpha^2}{12(1 - \alpha^2)} \delta_i^2 (1 - 2\lambda_i)(1 - \lambda_i) \right\}$$

times Eq. (13E.10) summed over $j \neq i$ to give, for $i = 1, 2, \ldots, k$,

$$\left\{ \frac{1}{6}\eta_{iii} + \frac{1}{2} \frac{1}{1 - \alpha^2} \sum_{j \neq i}^{k} \eta_{ijj} \right\}$$

$$+ \frac{1}{2}\delta_i(1 - \lambda_i)\eta_{ii}$$

$$+ \frac{1}{6}\delta_i^2(1 - \lambda_i)(1 - 2\lambda_i) \left\{ \eta_i - \frac{\alpha^2}{1 - \alpha^2} \frac{1}{2} \sum_{j \neq i}^{k} \eta_{ijj} \right\}$$

$$+ \frac{1}{2} \frac{1}{1 - \alpha^2} \left\{ 1 - \frac{1}{6}\alpha^2\delta_i^2(1 - 2\lambda_i)(1 - \lambda_i) \right\} \sum_{j \neq i}^{k} \delta_j(1 - \lambda_j)\eta_{ij} = 0.$$

Thus, if the third-order model is fitted in terms of x_i and x_i^3 [rather than x_i and $(x_i^3 - \psi x_i)$ as described above], we can substitute appropriate estimates to give the k simultaneous equations for $i = 1, 2, \ldots, k$:

$$b_{iii} + \delta_i(1 - \lambda_i)b_{ii} + \frac{1}{6}\delta_i^2(1 - \lambda_i)(1 - 2\lambda_i)b_i$$

$$+ \frac{1}{2} \frac{1}{1 - \alpha^2} \left\{ 1 - \frac{1}{6}\alpha^2\delta_i^2(1 - 2\lambda_i)(1 - \lambda_i) \right\} \sum_{j \neq i}^{k} \delta_j(1 - \lambda_j)b_{ij} = 0,$$

$$\tag{13E.12}$$

which can be solved for $\lambda_1, \lambda_2, \ldots, \lambda_k$. These awkward equations can be difficult to solve unless some care is applied. We suggest an iterative procedure in which rough estimates of λ_i are used in the grouping $Q_i \equiv (1 - \lambda_i)(1 - 2\lambda_i)$, which occurs in two positions in Eq. (13E.12). The resulting linear equations in $\theta_i = 1 - \lambda_i$ are straightforward to solve, and the results are used in the grouping $(1 - \lambda_i)(1 - 2\lambda_i)$ for a second iteration and so on, until convergence is achieved. To aid convergence, each new iteration can be started from the midpoint of the old and new values, if desired.

Alternative Cubic Case

If the cubic is fitted with estimated terms $b_i^* x_i + b_{iii}^*(x_i^3 - \psi x_i)$ as described above, we obtain b_i and b_{iii} from

$$b_i = b_i^* - \psi b_{iii}^* \qquad \text{and} \qquad b_{iii} = b_{iii}^*.$$

Related Work

The expression in Eq. (13E.9) can be written alternatively in the form

$$\beta_{iii} + (1 - \alpha^2)^{-1} \Sigma \beta_{iij}.$$

Draper and Herzberg (1971, p. 236) show that the sum of squares of these quantities for $i = 1, 2, \ldots, k$ occurs in the expected value of a general measure of lack of fit L_1. Thus, the c_{3i} contrasts essentially provide a split-up of L_1 which permits a more detailed and sensitive analysis. The remaining degrees of freedom pertaining to L_1 can be attributed to other contrasts as already described.

APPENDIX 13F. THE THIRD-ORDER COLUMNS OF THE X MATRIX FOR A COMPOSITE DESIGN

The form of the **X** matrix in the full cubic regression model $\mathbf{y} = \mathbf{X}\boldsymbol{\beta} + \boldsymbol{\varepsilon}$ when the design consists of r_c "cubes" plus r_s "stars" plus n_0 center points is as shown in Table 13F.1. We can denote columns by placing square brackets around the column head; for example $[x_1]$ will denote the x_1 column, and so on. We write $n_c = r_c 2^{k-p}$ for the number of cube points.

All of the cubic columns are orthogonal to all of the other columns with the following exceptions: $[x_i^3]$ is not orthogonal to $[x_i]$, nor to $[x_i x_j^2]$; $[x_i x_j^2]$ is not orthogonal to $[x_i]$, nor to $[x_i^3]$, nor to $[x_i x_l^2]$. The first step is to regress the $[x_i^3]$ and $[x_i x_j^2]$ vectors on the $[x_i]$ and take residuals. Because the columns involved are orthogonal to $[x_0]$, no adjustment for means is needed. We denote the "cube portion" of the $[x_1]$ and $[x_i x_j^2]$ vectors by $\mathbf{c}_i$, as indicated in the table. These two sets of residuals are, where the prime denotes transpose,

$$[x_{iii}] = [x_i^3] - \left\{ \frac{[x_i]'[x_i^3]}{[x_i]'[x_i]} \right\} [x_i]$$

and

$$[x_{ijj}] = [x_i x_j^2] - \left\{ \frac{[x_i]'[x_i x_j^2]}{[x_i]'[x_i]} \right\} [x_i],$$

both of which reduce to multiples $(1 - \alpha^2)m$ and m, respectively, where $m = 2r_s \alpha^2/(n_c + 2r_s \alpha^2)$, of the same vector. For example,

$$[x_{111}]' = [\mathbf{c}_1', d, -d, d, -d, \ldots, d, -d, 0, 0, \ldots, 0](1 - \alpha^2)m,$$

Table 13F.1. The X Matrix for a Cubic Model in k Predictors $x_1, x_2, \ldots, x_k$

x_0	x_1	x_2	$\cdots$	x_k	x_1^2	x_2^2	$\cdots$	x_k^2	$x_1 x_2$	$\cdots$	$x_{k-1}x_k$	x_1^3	$x_1 x_2^2$	$\cdots$	$x_1 x_k^2$	$x_1 x_2 x_3$	$\cdots$
1	$\mathbf{c}_1$	$\mathbf{c}_2$	$\cdots$	$\mathbf{c}_k$													
1																	
1			± 1's			1's				± 1's				± 1's			± 1's	
$\cdot$			2^{k-p} design						interaction patterns				each column same pattern as in x column				
$\cdot$																	
$\cdot$																	
1																	
1	$-\alpha$			0	α^2	0		0				$-\alpha^3$					
1	α			0	α^2	0		0				α^3					
$\cdots$	$\cdots$			$\cdots$	$\cdots$	$\cdots$		$\cdots$				$\cdots$					
1	$-\alpha$		r_s sets	0	α^2	0		0	0			$-\alpha^3$	0		0	0	
1	α		permuted	0	α^2	0		0				α^3					
$\cdots$	$\cdots$		down	$\cdots$	$\cdots$	$\cdots$		$\cdots$				$\cdots$					
			columns	0								0					
	0			0								0					
	0			$\cdots$								$\cdots$					
$\cdot$	$\cdot$			$-\alpha$	$\cdot$			α^2									
				α				α^2									
$\cdot$	$\cdot$			$-\alpha$	$\cdot$			α^2									
1	0			α	0			α^2				0					
1							0			0					0		0
1																	
$\cdots$				0			0			0					0		0
1																	

where

$$d = \frac{n_c}{(2r_s\alpha)}$$

and where there are r_s sets of $(d, -d)$'s in the vector. In general, for $[x_{iii}]'$, c_1' will be replaced by c_i' and the position of the $\pm d$'s will correspond to those of the $\mp\alpha$'s in the corresponding $[x_i]'$ vector. Note that because $c_i'c_j = 0$, $i \neq j$, it is obvious that $[x_{iii}]$ and $[x_{jjj}]$ are orthogonal.

It follows that the k cubic coefficients β_{iii}, β_{ijj} ($j \neq i$, $j = 1, 2, \ldots, k$, otherwise) cannot be estimated individually but only in linear combination, and that an appropriate normalized estimating contrast for this is

$$l_{iii} = \frac{[x_{iii}]'\mathbf{y}}{[x_{iii}]'[x_{iii}]}$$

$$= \frac{\{\mathbf{c}_i\mathbf{y}_1 + d(-r_s\bar{y}_{\alpha i} + r_s\bar{y}_{-\alpha i})\}}{\{n_c(1 - \alpha^2)\}},$$

where $\mathbf{y}_1$ is the portion of $\mathbf{y}$ corresponding to the cube part of the design, and $\bar{y}_{\alpha i}$, $\bar{y}_{-\alpha i}$ are, respectively, the averages of observations taken at the α and $-\alpha$ axial points on the x_i axis. If we similarly denote by $\bar{y}_{1i}$ and $\bar{y}_{-1i}$ the averages of the $n_c/2$ observations in $\mathbf{y}_1$ corresponding to 1 and -1 in $\mathbf{c}_i$, respectively, it follows quickly that $l_{iii} = c_{3i}$ where c_{3i} is given in Eq. (13.8.12). The expected value is

$$E(c_{3i}) = \frac{[x_{iii}]'\mathbf{X}\boldsymbol{\beta}}{[x_{iii}]'[x_{iii}]},$$

where $\mathbf{X}$ is as in Table 13F.1 and the coefficients of $\boldsymbol{\beta}$ correspond to the columns in the obvious manner. Because $[x_{iii}]$ is orthogonal to all columns of $\mathbf{X}$ except the $[x_i^3]$ and $[x_ix_j^2]$ columns, Eq. (13.8.14) emerges almost immediately.

EXERCISES

13.1. Suppose we wish to choose a design to fit a straight line $y = \beta_0 + \beta_1 x + \varepsilon$ over the region of interest $-1 \leq x \leq 1$ and also guard against the possibility that the true model might be a general polynomial of degree $d_2 > 1$. If there were little or no variance error (so that V could effectively be ignored), show that an appropriate "all-bias" design $(x_1, x_2, \ldots, x_N)$ would have moments

$$N^{-1}\sum_{u=1}^{N} x_u^i = \begin{cases} 0, & \text{for } i \text{ odd,} \\ (i + 1)^{-1} & \text{for } i \text{ even,} \end{cases} \quad i \leq d_2 + 1.$$

If $d_2 = 3$ and $N = 6$, find a specific design that satisfies these moment conditions.

Table E13.2

Number of Points at					Value of	
$-b$	$-a$	0	a	b	a	b
1	4	0	4	1	0.478	1.009
1	3	2	3	1	0.567	0.984
1	2	4	2	1	0.790	0.828
2	1	4	1	2	0.777	0.815
2	3	0	3	2	0.358	0.880
2	2	2	2	2	0.452	0.873

13.2. Suppose we wish to choose a design $x_1, x_2, \ldots, x_N$ to fit the model $y = \beta_0 + \beta_1 x + \beta_{11} x^2 + \varepsilon$ over the coded range of interest $-1 \le x \le 1$, but also wish to guard against the possibility that the true model contains an additional term $\beta_{111} x^3$. We furthermore specify that $N = 10$ points should be allocated in some manner to the five levels $x = (-b, -a, 0, a, b)$, where a and b are to be determined. It can be shown (see The choice of a second order rotatable design, by G. E. P. Box and N. R. Draper, *Biometrika*, **50**, 1963, 335–352) that if variance and bias errors are to be roughly the same size, we should choose (approximately) $\Sigma x_u^2 = 3.866$, $\Sigma x_u^4 = 2.494$. Table E13.2 shows six designs that approximately achieve those figures. Evaluate the detectability ratio defined in Section 13.7; this is $\Sigma x_u^6 / \{\Sigma x_u^2\}^3$ but, since Σx_u^2 is approximately constant, evaluate Σx_u^6 for each design instead. Which design would you use? Why?

13.3. Suppose we wish to choose an "all-bias" design to fit a plane $y = \beta_0 + \beta_1 x_1 + \beta_2 x_2 + \cdots \beta_k x_k + \varepsilon$ while guarding against the possibility that the true model might be a second-order polynomial. Show that if $\mathbf{M}$ is any $k \times k$ orthogonal matrix with a first column of all $+1$'s, then the design matrix

$$\mathbf{D} = \begin{bmatrix} \mathbf{M} \\ -\mathbf{M} \end{bmatrix}$$

provides a suitable design.

13.4. (*Source:* Sequential designs for spherical weight functions, by N. R. Draper and W. E. Lawrence, *Technometrics*, **4**, 1967, 517–529.) Consider the density function in k dimensions

$$W(x) = C \exp\left\{ -\tfrac{1}{2} \left[(x_1^2 + x_2^2 + \cdots + x_k^2)/\sigma^2 \right]^{1/(1+\beta)} \right\}, \qquad -1 < \beta \le 1$$

Table E13.5

x_1	x_2	x_3	y	x_1	x_2	x_3	y
-1	-1	-1	353	0	-1.68	0	98
1	-1	-1	361	0	1.68	0	3958
-1	1	-1	2643	0	0	-1.68	1223
1	1	-1	2665	0	0	1.68	1207
-1	-1	1	350				
1	-1	1	354	0	0	0	1216
-1	1	1	2639	0	0	0	1212
1	1	1	2652	0	0	0	1216
				0	0	0	1215
-1.68	0	0	1204	0	0	0	1213
1.68	0	0	1223	0	0	0	1218

where the normalizing constant C is given by

$$C = 2^{-(1/2)k(1+\beta)}\sigma^{-k}(1 + \beta)^{-1}\pi^{-k/2}\frac{\Gamma(k/2)}{\Gamma\{\frac{1}{2}k(1 + \beta)\}}.$$

Show that if $k = 1$, the cases (a) $\beta \to -1$ and (b) $\beta = 0$ give rise to the two weight functions shown in Figure 13.5. In general, $W(x)$ provides a versatile weight function for k dimensions.

13.5. Consider the design and observations y_u given in Table E13.5. Will a second-order model in the x's provide a satisfactory fit to these data? If not, and given that the codings from x's to ξ's are as given below, will a second-order model in $\xi_i^{\lambda_i}$, $i = 1, 2, 3$, provide a satisfactory fit? (In both parts of the question provide all the usual analyses.)

$$\text{Codings: } \xi_1 = 0.85 + 0.09x_1,$$
$$\xi_2 = 50 + 24x_2,$$
$$\xi_3 = 3.5 + 0.9x_3.$$

CHAPTER 14

Variance-Optimal Designs

14.1. INTRODUCTION

We said, in Section 13.7, that our aims in selecting an experimental design must necessarily be multifaceted. The problem of selecting a suitable design is thus a formidable one. Some properties [given by Box and Draper (1976)] of a response surface design—any, all, or some of which might be important, depending on the experimental circumstances—are as follows. The design should:

1. Generate a satisfactory distribution of information throughout the region of interest, R.
2. Ensure that the fitted value at $\mathbf{x}$, $\hat{y}(\mathbf{x})$, be as close as possible to the true value at $\mathbf{x}$, $\eta(\mathbf{x})$.
3. Give good detectability of lack of fit.
4. Allow transformations to be estimated.
5. Allow experiments to be performed in blocks.
6. Allow designs of increasing order to be built up sequentially.
7. Provide an internal estimate of error.
8. Be insensitive to wild observations and to violation of the usual normal theory assumptions.
9. Require a minimum number of experimental runs.
10. Provide simple data patterns that allow ready visual appreciation.
11. Ensure simplicity of calculation.
12. Behave well when errors occur in the settings of the predictor variables, the x's.
13. Not require an impractically large number of levels of the predictor variables.
14. Provide a check on the "constancy of variance" assumption.

Response Surfaces, Mixtures, and Ridge Analyses, Second Edition. By G. E. P. Box and N. R. Draper
Copyright © 2007 John Wiley & Sons, Inc.

These requirements are sometimes conflicting and sometimes confluent, and, in any specific situation, judgment is required to select the design that achieves the best compromise. This judgment is aided by knowledge of the behavior of these design properties. In particular, as we pointed out in Chapter 13, the trade-off between variance and bias is of critical practical importance. Because polynomials, or any other type of empirical function, are approximations whose adequacy depends on the size of the region covered by the design, we must balance off the reduced variance achieved by increasing the size of a design against the extra bias so induced.

Much theoretical work has been done without considering the effect of model inadequacy. The conclusions obtained from such studies can, nevertheless, be of value in providing information for design choice, as long as the possibility of model inadequacy is kept in mind.

14.2. ORTHOGONAL DESIGNS

The work of R. A. Fisher and F. Yates pointed to orthogonality as an important design principle long ago. We first explain and discuss *orthogonality* and then introduce *rotatability* as a logical extension of it.

For simplicity, consider a simple linear regression setup with only two experimental variables ξ_1 and ξ_2, n observations, $u = 1, 2, 3, \ldots, n$, and model

$$y_u = \beta_0 + \beta_1 \tilde{\xi}_{1u} + \beta_2 \tilde{\xi}_{2u} + \varepsilon, \tag{14.2.1}$$

where $\tilde{\xi}_{iu} = \xi_{iu} - \bar{\xi}_i$. If we use least squares to obtain estimates $b_0 = \bar{y}$, b_1, b_2 of β_0, β_1, β_2, respectively, then it is easy to show that the variances and covariances of these estimates are

$$V(b_0) = \frac{\sigma^2}{n}, \tag{14.2.2}$$

$$V(b_1) = \left\{ \frac{1}{(1 - \cos^2 \theta) S_1^2} \right\} \frac{\sigma^2}{n}, \qquad V(b_2) = \left\{ \frac{1}{(1 - \cos^2 \theta) S_2^2} \right\} \frac{\sigma^2}{n}, \tag{14.2.3}$$

$$\text{cov}(b_0, b_1) = 0,$$

$$\text{cov}(b_0, b_2) = 0, \qquad \text{cov}(b_1, b_2) = \left\{ \frac{-\cos \theta}{(1 - \cos^2 \theta) S_1 S_2} \right\} \frac{\sigma^2}{n}, \tag{14.2.4}$$

where

$$S_1^2 = n^{-1} \Sigma \tilde{\xi}_{1u}^2, \qquad S_2^2 = n^{-1} \Sigma \tilde{\xi}_{2u}^2, \qquad \cos \theta = \frac{\Sigma \tilde{\xi}_{1u} \tilde{\xi}_{2u}}{\left\{ (\Sigma \tilde{\xi}_{1u}^2)(\Sigma \tilde{\xi}_{2u}^2) \right\}^{1/2}}, \tag{14.2.5}$$

and all summations are over $u = 1, 2, \ldots, n$. Given that we have decided to make a fixed number of runs, n, and that the experimental error variance $V(\varepsilon_u) = \sigma^2$ is fixed, how should we design an experiment; that is, how should we choose the levels (ξ_{1u}, ξ_{2u}), $u = 1, 2, \ldots, n$? Obviously the choice will depend on the criterion of excellence applied. Suppose we require that the β's be estimated with smallest possible variances. Clearly, $V(b_0)$ will be unaffected by choice of design because σ^2/n is fixed by definition of our problem. Now S_1 and S_2 are measures of spread of the design points in the ξ_1 and ξ_2 directions and θ is the angle between the design vectors $\tilde{\xi}_1 = (\tilde{\xi}_{11}, \tilde{\xi}_{12}, \ldots, \tilde{\xi}_{1n})'$ and $\tilde{\xi}_2 = (\tilde{\xi}_{21}, \tilde{\xi}_{22}, \ldots, \tilde{\xi}_{2n})'$, so that $-1 \leq \cos\theta \leq 1$. Thus we can make $V(b_1)$ and $V(b_2)$ as small as possible by choosing (a) S_1^2 and S_2^2 as large as possible, or (b) $\cos^2\theta$ as small as possible. In practice, we shall need to choose the scale factors S_1 and S_2 together with the location measures $\bar{\xi}_1$ and $\bar{\xi}_2$ so that the design spans the region of interest in the predictor variable space. If this is done and if, in addition, the design is selected so that $\cos\theta = 0$, then for given S_1 and S_2, $V(b_1)$ and $V(b_2)$ will be as small as possible. The condition $\cos\theta = 0$ is the requirement that the n-dimensional design vectors $\tilde{\xi}_1$ and $\tilde{\xi}_2$ are at right angles, that is, *orthogonal* to each other. Since S_1 and S_2 are, by assumption, greater than zero, the condition will be met by making $\sum \tilde{\xi}_{1u}\tilde{\xi}_{2u} = 0$. Notice that $\cos\theta$ is a measure of the correlation between ξ_1 and ξ_2 so that orthogonality implies that the regressors ξ_1 and ξ_2 are uncorrelated. Furthermore, $-\cos\theta$ is the correlation between the two estimates b_1 and b_2; thus, choice of an orthogonal design ensures that these estimates are uncorrelated.

It is possible to show that the models employed in the analysis of randomized blocks, Latin squares, and factorials can all be written in the general linear regression form of Eq. (3.9.1) using indicator or dummy variables. These designs are said to be orthogonal because the dummy X's that carry, for example, the treatment and block contrasts define subspaces that are all orthogonal to one another. This ensures that, for example, the treatment effects from the randomized block design may be estimated independently of the block effects.

Further illustration of the property of orthogonality and, in particular, how it relates to the size of the confidence region for the parameters will be found in Chapter 3; see Figure 3.9 in particular.

Useful arrangements such as factorials may be employed in many different experimental contexts. Therefore it is convenient to write such designs in a standard form in which the predictor variables are located and scaled in a standard way. We often work in terms of the *coded* design variables

$$x_{iu} = \frac{(\xi_{iu} - \bar{\xi}_i)}{S_i}, \tag{14.2.6}$$

where we can choose

$$S_i^2 = n^{-1}\sum(\xi_{iu} - \bar{\xi}_i)^2. \tag{14.2.7}$$

Note that this means that

$$n^{-1}\sum x_{iu}^2 = 1, \tag{14.2.8}$$

where, again, all summations are over $u = 1, 2, \ldots, n$. Although the discussion to this point has involved just two regressors, for simplicity, the arguments would apply generally to k regressors, $k = 1, 2, \ldots$. Consider the model

$$y_u = \beta_0 + \sum_{i=1}^{k} \beta_i x_{iu} + \varepsilon_u. \tag{14.2.9}$$

Then the variances of the estimates b_i of β_i, $i = 1, 2, \ldots, k$ would again attain their smallest values if

$$\Sigma x_{iu} x_{ju} = 0, \qquad \text{for all } i \neq j. \tag{14.2.10}$$

The first-order response surface designs we have discussed have all been orthogonal. In particular, the two-level factorial and fractional factorials discussed in Chapters 4 and 5 are of this type. To illustrate the coding we have adopted, we give below a 2^2 factorial design both in original and coded predictor variables formats.

Temperature, $\xi_1(°C)$	Concentration, $\xi_2(\%)$	x_1	x_2
160	20	-1	-1
180	20	1	-1
160	50	-1	1
180	50	1	1

The appropriate coding here requires $\bar{\xi}_1 = 170$, $\bar{\xi}_2 = 35$, $S_1 = 10$, and $S_2 = 15$, so that

$$x_1 = \frac{(\xi_1 - 170)}{10}, \qquad x_2 = \frac{(\xi_2 - 35)}{15}.$$

First-order orthogonal designs that are not factorials are occasionally of value. In particular, they may be used for the elimination of time trends [see, for example, Box (1952), Box and Hay (1953), and Hill (1960)].

14.3. THE VARIANCE FUNCTION

If it were true that bias could be ignored so that the postulated model function were capable of exactly representing reality, then at $\mathbf{x} = (x_1, x_2, \ldots, x_k)'$

$$E(y_\mathbf{x}) = \eta_\mathbf{x} \tag{14.3.1}$$

and the standardized mean squared error associated with estimating it by $\hat{y}_\mathbf{x}$ would be

$$V_\mathbf{x} = \left(\frac{n}{\sigma^2}\right) V(\hat{y}_\mathbf{x}), \tag{14.3.2}$$

where $V_\mathbf{x}$ will be called the *variance function*. Equivalently, we may consider the *information function* of the design defined by

$$I_\mathbf{x} = V_\mathbf{x}^{-1}. \tag{14.3.3}$$

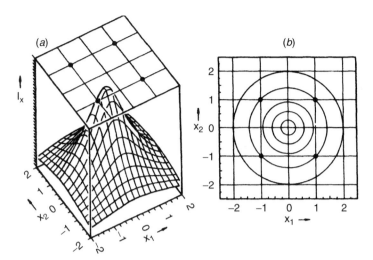

Figure 14.1. (*a*) Information surface for a 2^2 factorial used as a first-order design. (*b*) The corresponding information contours.

For example, for the factorial design illustration in the foregoing section,

$$I_x = \frac{1}{\left(1 + x_1^2 + x_2^2\right)}. \qquad (14.3.4)$$

This is plotted in Figure 14.1*a*. Obviously, if we were prepared to make the unrealistic assumption of no bias, the information function tells us all we can know about the design's ability to estimate the response η. For example, inspection of Figure 14.1*a* shows that a large amount of information on η_x is generated at the design origin and the information falls off rapidly as we move outward from the center. We also notice that for this particular design, setting $\rho = (x_1^2 + x_2^2)^{1/2}$, we have

$$I_x = I_\rho = \frac{1}{\left(1 + \rho^2\right)} \qquad (14.3.5)$$

so that, in units of (x_1, x_2), the function is constant on circles centered at the origin. Thus the information surface is one of rotation. We say that the design is *rotatable*. As far as $V(\hat{y})$ is concerned, any rotation of the original 2^2 factorial design will generate exactly the same information function.

Evidently, rotatable designs are appropriate if we suppose that the experimenter would try to choose the scales and transformations of his predictor variables so that the desired distribution of information is, in his chosen metrics, spherically distributed about the design origin. This seems to be a reasonable goal. When an experimenter chooses a wide scale for one of his variables, it implies that his prior knowledge about the behavior of that variable suggests that a wide range is necessary to make the expected effect of that variable roughly equivalent to that of other variables. Similarly, suppose an experimenter were particularly interested in exploring a diagonal region in which, say, X_1/X_2 varied by only a small percentage, but $X_1 X_2$ varied by a large percentage. This would suggest the use *not* of the original variables X_1 and X_2, but of the transformed variables $\xi_1 = X_1/X_2$ and $\xi_2 = X_1 X_2$. After the design had been appropriately scaled by $x_1 = (\xi_1 - \bar{\xi}_1)/S_1$, $x_2 = (\xi_2 - \bar{\xi}_2)/S_2$, the desired distribution of information could be attained by making the design rotatable in x_1 and x_2. The process, which we must always face, of choosing appropriate scales and transformations for our variables is thus clearly reduced to that of taking account of what is known or suspected about the variables, *a priori*. The design is then superimposed on the chosen system of metrics (in terms of which a spherical distribution of information makes most sense).

So far we have seen that, for a first-order arrangement, orthogonality and rotatability are both desirable. In fact, *for the particular case of first-order designs*, they are equivalent. When we consider designs of higher order, however, this equivalence does not hold. For illustration, consider a 3^2 factorial used as a second-order design for $k = 2$ variables. With the postulated model in the form $\mathbf{y} = \mathbf{X}\boldsymbol{\beta} + \boldsymbol{\varepsilon}$, the $\mathbf{X}$, $\mathbf{X}'\mathbf{X}$, and $(\mathbf{X}'\mathbf{X})^{-1}$ matrices are as follows.

$$
\mathbf{X} =
\begin{array}{c}
\begin{array}{cccccc} 0 & \;\;1 & \;\;2 & 11 & 22 & 12 \end{array} \\
\left[
\begin{array}{cccccc}
1 & -1 & -1 & 1 & 1 & 1 \\
1 & 0 & -1 & 0 & 1 & 0 \\
1 & 1 & -1 & 1 & 1 & -1 \\
1 & -1 & 0 & 1 & 0 & 0 \\
1 & 0 & 0 & 0 & 0 & 0 \\
1 & 1 & 0 & 1 & 0 & 0 \\
1 & -1 & 1 & 1 & 1 & -1 \\
1 & 0 & 1 & 0 & 1 & 0 \\
1 & 1 & 1 & 1 & 1 & 1
\end{array}
\right],
\end{array}
\tag{14.3.6}
$$

$$
\mathbf{X}'\mathbf{X} =
\begin{array}{c}
\begin{array}{ccccccc} & 0 & 1 & 2 & 11 & 22 & 12 \end{array} \\
\begin{array}{c} 0 \\ 1 \\ 2 \\ 11 \\ 22 \\ 12 \end{array}
\left[
\begin{array}{cccccc}
9 & \cdot & \cdot & 6 & 6 & \cdot \\
\cdot & 6 & \cdot & \cdot & \cdot & \cdot \\
\cdot & \cdot & 6 & \cdot & \cdot & \cdot \\
6 & \cdot & \cdot & 6 & 4 & \cdot \\
6 & \cdot & \cdot & 4 & 6 & \cdot \\
\cdot & \cdot & \cdot & \cdot & \cdot & 4
\end{array}
\right],
\end{array}
\tag{14.3.7}
$$

$$
(\mathbf{X}'\mathbf{X})^{-1} = \frac{1}{9}
\begin{array}{c}
\begin{array}{cccccc} 0 & 1 & 2 & 11 & 22 & \;\;\;12 \end{array} \\
\left[
\begin{array}{cccccc}
5 & \cdot & \cdot & -3 & -3 & \cdot \\
\cdot & 1.5 & \cdot & \cdot & \cdot & \cdot \\
\cdot & \cdot & 1.5 & \cdot & \cdot & \cdot \\
-3 & \cdot & \cdot & 4.5 & \cdot & \cdot \\
-3 & \cdot & \cdot & \cdot & 4.5 & \cdot \\
\cdot & \cdot & \cdot & \cdot & \cdot & 2.25
\end{array}
\right].
\end{array}
\tag{14.3.8}
$$

The 3^2 factorial is an orthogonal design in the sense that no two of the estimates for first- and second-order effects are correlated, as is clear from the form of

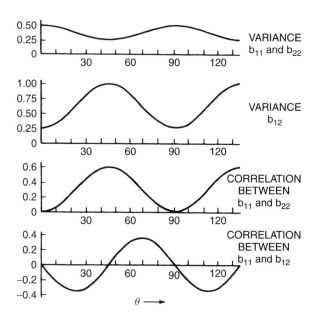

Figure 14.2. Variances of, and correlations between, second-order coefficients estimated from a 3^2 factorial design rotated through an angle θ.

$(\mathbf{X'X})^{-1}$. (The estimate b_0 is, however, inevitably correlated with b_{11} and with b_{22}.) Suppose we are to use the 3^2 factorial to explore a response surface which, it is believed, can be represented by a second-degree expression, but about which little else is known. Now imagine the second-degree surface referred to its principal axes (see Figure 11.1). One thing in particular that will usually not be known will be the orientation of those principal axes, and frequently it will be true that one orientation is as likely as another. Instead of considering rotations of the surface relative to the design, we may equivalently consider rotations of the design relative to the surface.

Figure 14.2 shows how the variances of the second-order terms and the correlations between them change as the 3^2 design is rotated through an angle θ. (For simplicity, we set $\sigma^2 = 1$ here.) Thus, although this 3^2 second-order design is orthogonal (in the sense described) in its initial orientation, it loses this property on rotation, unlike the first-order design. In view of this, it would seem most sensible to choose, if possible, a second-order design for which the variances and covariances of the estimates stay constant as the design is rotated.

Consider the information function for the 3^2 factorial design. Writing $\mathbf{z} = (1, x_1, x_2, x_1^2, x_2^2, x_1 x_2)'$, this is

$$I_x = \left\{ n\mathbf{z}'(\mathbf{X'X})^{-1}\mathbf{z} \right\}^{-1} \tag{14.3.9}$$

$$= \left\{ 5 - 4.5(x_1^2 + x_2^2) + 4.5(x_1^2 + x_2^2)^2 - 6.75x_1^2 x_2^2 \right\}^{-1}, \tag{14.3.10}$$

which is graphed in Figure 14.3a and has its contours plotted in Figure 14.3b. It will be seen that the design generates four pockets of high information which

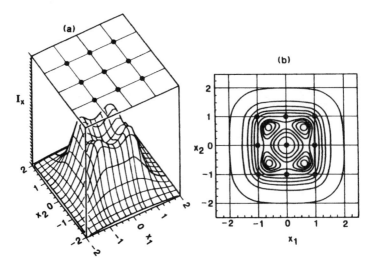

Figure 14.3. (a) Information surface for 3^2 factorial used as a second-order design. (b) The corresponding information contours.

seemingly have little to do with the needs of an experimenter. Rather than attempt to generalize the property of orthogonality to second order, we shall instead generalize the property of rotatability. We shall see that it is possible to choose designs of second and higher orders for which the information contours are spherical. Equivalently, these rotatable designs have the property that the variances and covariances of the effects remain unaffected by rotation. One such rotatable design for two factors, consisting of eight points evenly distributed on a circle with four added center points, is shown in Figure 14.4a together with its information function. The corresponding information contours appear in Figure 14.4b.

A k-dimensional design will be rotatable if the information function I_x depends only on $\rho = (x_1^2 + x_2^2 + \cdots + x_k^2)^{1/2}$, the distance from the origin. Such a design has an $\mathbf{X'X}$ matrix of special form. If we scale the variables so that

$$\sum_{u=1}^{n} x_{iu}^2 = n, \qquad i = 1, 2, \ldots, k, \tag{14.3.11}$$

then, for example, for $k = 2$,

$$n^{-1}\mathbf{X'X} = \begin{array}{c} \\ 0 \\ 1 \\ 2 \\ 11 \\ 22 \\ 12 \end{array} \begin{array}{c} \begin{matrix} 0 & 1 & 2 & 11 & 22 & 12 \end{matrix} \\ \begin{bmatrix} 1 & \cdot & \cdot & 1 & 1 & \cdot \\ \cdot & 1 & \cdot & \cdot & \cdot & \cdot \\ \cdot & \cdot & 1 & \cdot & \cdot & \cdot \\ 1 & \cdot & \cdot & 3\lambda & \lambda & \cdot \\ 1 & \cdot & \cdot & \lambda & 3\lambda & \cdot \\ \cdot & \cdot & \cdot & \cdot & \cdot & \lambda \end{bmatrix} \end{array}. \tag{14.3.12}$$

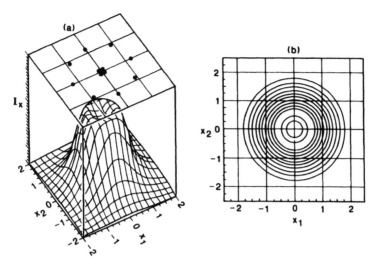

Figure 14.4. (*a*) Information function for a second-order rotatable design consisting of eight points on a circle plus four center points. (*b*) The corresponding information contours.

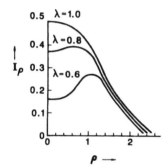

Figure 14.5. Plots of information function versus $\rho = (x_1^2 + x_2^2)^{1/2}$ for various values of λ for a central composite design in $k = 2$ variables.

One adjustable parameter λ is at our disposal. This determines the level of the information function as we move out from the center of the design along any radius vector in the x space. Figure 14.5 shows a plot of I_ρ versus ρ for the indicated values of λ for $k = 2$. The function plotted is

$$I_\rho = \frac{\{4\lambda(2\lambda - 1)\}}{\{8\lambda^2 + 8\lambda(\lambda - 1)\rho^2 + (3\lambda - 1)\rho^4\}}. \tag{14.3.13}$$

[See Box and Hunter (1957, p. 213), for details, especially Eq. (48).] The value attained by λ for a rotatable design is determined by the arrangement of design points and especially by the number n_0 of points placed at the center of the design, as we now explain.

Central Composite Designs

One convenient class of rotatable designs (available for all values of k) is that of the suitably dimensioned central composite designs already discussed in Section

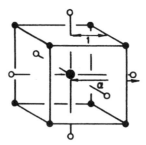

● Factorial (or "cube") points, with
  coordinates $(\pm 1, \pm 1, \pm 1)$

○ Axial (or "star") points, distance α from origin

● Center points. If the design is blocked into cube
  and star portions, some center points would be
  in the cube block, some in the star block.
  See Chapter 15.

Figure 14.6. Central composite design for $k = 3$ variables.

9.2, for $k = 3$, and in Section 13.8. Figure 14.6 shows such a design for $k = 3$. If we scale the eight cube points so that their coordinates are $(\pm 1, \pm 1, \pm 1)$, then, to attain rotatability, the six axial points $(\pm \alpha, 0, 0)$, $(0, \pm \alpha, 0)$, $(0, 0, \pm \alpha)$ must be located at distances $\alpha = 8^{1/4} = 1.682$ from the origin. In general, the cube may be replicated r_c times and the axial points replicated r_s times. In that event, the rotatable values of α would be given by the more general formula $\alpha = (8r_c/r_s)^{1/4}$. For k factors, when a 2^{k-p} fractional factorial, r_c times over, is combined with $2k$ axial points taken r_s times, the appropriate formula is

$$\alpha = \left(\frac{2^{k-p} r_c}{r_s} \right)^{1/4}, \tag{14.3.14}$$

the foregoing special case being that of $k = 3$, $p = 0$.

To ensure that all coefficients in the second-order model are estimable, points are needed at the center of the composite design whenever all the noncentral points fall on a sphere in k dimensions (i.e., on a circle when $k = 2$). When the design is rotatable, their number n_0 then determines λ according to the formula

$$\lambda = \frac{\left(r_c 2^{k-p} + 2kr_s + n_0 \right)}{\left\{ \left(r_c 2^{k-p} \right)^{1/2} + 2r_s^{1/2} \right\}^2}. \tag{14.3.15}$$

(Thus diagrams such as Figure 14.5 could also be constructed and marked using values of n_0 instead of values of λ.) As we have seen, the choice of λ will, in turn, affect the information profile. The question of choosing λ suitably will be discussed in Chapter 15.

Rotatability assures a spherically uniform distribution of information which, in the circumstances we have outlined, seems sensible. However, approximate sphericity is all that will be needed in practice. Whenever the criterion of rotatability is in conflict with some other important consideration (such as, for example, the need to split the design into orthogonal blocks) a moderate departure from exact rotatability may be made to achieve this.

General Conditions for Design Rotatability of First, Second, and Third Order

Suppose there are n design points $(x_{1u}, x_{2u}, \ldots, x_{ku})$, $u = 1, 2, \ldots, n$, in a k factor design. The quantity

$$n^{-1}\Sigma x_{1u}^p x_{2u}^q \cdots x_{ku}^r,$$

where this and subsequent summations are over $u = 1, 2, \ldots, n$, is a design moment of order $p + q + \cdots + r$. If multiplied by n, it is called a "sum of powers" (if only one of $p, q, \ldots, r$ is nonzero) or a "sum of products" (if two or more of $p, q, \ldots, r$ are nonzero) of the same order.

A third-order rotatable design must be such that

$$\Sigma_{iu}^2 = n\lambda_2, \tag{14.3.16}$$

$$\Sigma x_{iu}^4 = 3\Sigma x_{iu}^2 x_{ju}^2 = 3n\lambda_4, \tag{14.3.17}$$

$$\Sigma x_{iu}^6 = 5\Sigma x_{iu}^2 x_{ju}^4 = 15\Sigma x_{iu}^2 x_{ju}^2 x_{lu}^2 = 15n\lambda_6, \tag{14.3.18}$$

where $i \ne j \ne l$, and all other sums of powers and products up to an including order $2d$ (where here $d = 3$) are zero [see Box and Hunter (1957, p. 209)]. These equations serve to define $\lambda_2, \lambda_4, \lambda_6$. For second-order rotatability we need only (14.3.6), (14.3.17), and $2d = 4$; for first-order rotatability, which is simply orthogonality with equal scaling on the axes, we need only (14.3.16) and $2d = 2$. These various assumptions determine appropriate forms for $\mathbf{X'X}$. For a second-order design to be nonsingular (i.e., to provide a nonsingular $\mathbf{X'X}$ matrix), we need

$$\frac{\lambda_4}{\lambda_2^2} > \frac{k}{k+2}. \tag{14.3.19}$$

The (minimum) value $k/(k+2)$ occurs when all the design points lie on a sphere centered at the origin; the addition of center points will then always satisfy (14.3.19). For a third-order design to be nonsingular, both (14.3.19) and

$$\frac{\lambda_6\lambda_2}{\lambda_4^2} > \frac{k+2}{k+4} \tag{14.3.20}$$

must hold. Design points must lie on at least *two* spheres of *nonzero* radii to achieve (14.3.20) (see Draper, 1960b). If this is so, then (14.3.19) is automatically satisfied. (See, for example, Exercise 7.28.)

14.4. THE ALPHABETIC OPTIMALITY APPROACH

A somewhat different approach, which has been very thoroughly explored mathematically, concerns various individual optimal characteristics of the matrix $\mathbf{X'X}$ which occurs in the least-squares fitting of the response function. Such studies can be revealing and are of considerable interest, but they are concerned with optimality of a very narrow kind and their limitations must be clearly understood if the practitioner is not to be misled. To aid this understanding, a brief discussion of some of the main ideas of alphabetic optimality, and a critical appraisal of their application in a response surface context is given here (also see Box 1982).

Aspects of Alphabetic Optimal Design Theory

Consider a response η which is supposed to be an exactly known function $\eta = \mathbf{z}'\boldsymbol{\beta}$ linear in p coefficients $\boldsymbol{\beta}$, where $\mathbf{z} = \{f_1(\boldsymbol{\chi}), \ldots, f_p(\boldsymbol{\chi})\}'$ is a vector of p functions of k experimental variables $\boldsymbol{\chi} = (\chi_1, \chi_2, \ldots, \chi_k)'$. Suppose a design is to be run defining n sets of k experimental conditions given by the $n \times k$ design matrix whose uth row is $\boldsymbol{\chi}'_u = (\chi_{1u}, \chi_{2u}, \ldots, \chi_{ku})$, and yielding n observations $\{y_u\}$, so that

$$\eta_u = \mathbf{z}'_u \boldsymbol{\beta}, \qquad u = 1, 2, \ldots, n, \tag{14.4.1}$$

where $y_u - \eta_u = \varepsilon_u$ is distributed $N(0, \sigma^2)$ and the $n \times p$ matrix $\mathbf{X} = \{\mathbf{z}'_u\}$ consists of n rows with $\mathbf{z}'_u = \{f_1(\boldsymbol{\chi}_u), f_2(\boldsymbol{\chi}_u), \ldots, f_p(\boldsymbol{\chi}_u)\}$ in the uth row. The ε_u are assumed to be independently distributed.

The elements of $\{c_{ij}\} = (\mathbf{X}'\mathbf{X})^{-1}$ are proportional to the variances and covariances of the least-squares estimates $\mathbf{b}$. Within this specification, the problem of experimental design is that of choosing the design $\{\boldsymbol{\chi}'_u\}$ so that the elements c_{ij} are to our liking. Because there are $\frac{1}{2}p(p+1)$ of these, simplification is desirable.

One motivation for simplification is provided by considering the confidence region* for $\boldsymbol{\beta}$:

$$(\boldsymbol{\beta} - \mathbf{b})'\mathbf{X}'\mathbf{X}(\boldsymbol{\beta} - \mathbf{b}) = \text{constant}, \tag{14.4.2}$$

which defines an ellipsoid in p parameters. The eigenvalues $\lambda_1, \lambda_2, \ldots, \lambda_p$ of $(\mathbf{X}'\mathbf{X})^{-1}$ are proportional to the squared lengths of the p principal axes of this ellipsoid. Suppose their maximum, arithmetic mean, and geometric mean are indicated by $\lambda_{\max}$, $\bar{\lambda}$, and $\tilde{\lambda}$. Then it is illuminating to consider the transformation of the $\frac{1}{2}p(p+1)$ elements c_{ij} to an identical number of criteria measuring volume, nonsphericity, and orientation, the meanings of which are easier to comprehend. These are:

1. $D = |\mathbf{X}'\mathbf{X}| = \tilde{\lambda}^{-p}$ (so that $D^{-1/2} = \tilde{\lambda}^{p/2}$ is proportional to the volume of the confidence ellipsoid).
2. $H_1, H_2, \ldots, H_{p-1}$, a set of $p-1$ homogeneous independent functions of order zero in the λ's, which measure the *nonsphericity* or state of ill-conditioning of the ellipsoid. In particular we might choose, for two of these, $H_1 = \bar{\lambda}/\tilde{\lambda}$ and $H_2 = \lambda_{\max}/\tilde{\lambda}$, both of which would take the value unity for a spherical region.
3. $\frac{1}{2}p(p-1)$ independent direction cosines that determine the *orientation* of the orthogonal axes of the ellipsoid.

---

*There are also parallel fiducial and Bayesian rationalizations.

It is traditionally assumed that the $\frac{1}{2}p(p-1)$ elements concerned with orientation of the ellipsoid are of no interest, and attention has been concentrated on particular design criteria that measure in some way or another the sizes of the eigenvalues. All such criteria thus measure some combination of size and sphericity of the confidence ellipsoid. Among these criteria are

$$D = |\mathbf{X}'\mathbf{X}| = \Pi \lambda_i^{-1} = \tilde{\lambda}^{-p},$$

$$A = \Sigma \lambda_i = \text{tr}(\mathbf{X}'\mathbf{X})^{-1} = \sigma^{-2} \Sigma \, \text{var}(b_i) = p \tilde{\lambda} H_1, \qquad (14.4.3)$$

$$E = \max\{\lambda_i\} = \tilde{\lambda} H_2.$$

The D criterion is a function of $\tilde{\lambda}$ only and is thus a measure of region size alone. The A and E criteria both contain the same size measure $\tilde{\lambda}$ multiplied by a measure of nonsphericity.

The desirability of a design, as measured by the D, A, and E criteria, increases as $\tilde{\lambda}$, $\tilde{\lambda} H_1$, and $\tilde{\lambda} H_2$ respectively, are decreased. However, in practical situations, because $\tilde{\lambda}$ is common to all three, each of these criteria will take smaller and hence more desirable values as the ranges for the experimental variables χ_u are taken larger and larger. To cope with this problem, it is usually assumed that the experimental variables χ_u may vary only within some exactly known region in the space of χ, but not outside it. We call this permissible region RO, a notation that (purposely) combines the R and O notations of Chapter 13.

Another characteristic of the problem which makes its study mathematically difficult is the necessary discreteness of the number of runs that can be made at any given location. In a technically brilliant paper, Kiefer and Wolfowitz (1960) dealt with this obstacle by introducing a continuous design measure that determines the *proportion* of runs that should ideally be made at each of a number of points in the χ space. Realizable designs that closely approximated the optimal *measure design(s)* distribution could then be used in practice.

A further important result of Kiefer and Wolfowitz linked the problem of estimating $\boldsymbol{\beta}$ with that of estimating the response η via the property of "G optimality." G optimal designs were defined as those that minimized the maximum value of $V(\hat{y}_x)$ *within RO*. The authors were then able to show, for their measure

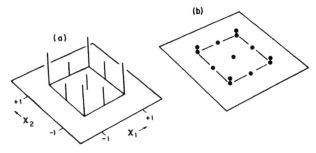

Figure 14.7. (*a*) Fedorov's D/G optimal second-order measure design for the square inner region. The vertical scale indicates the proportion of runs to be made at the various points. (*b*) A realizable design with 13 runs roughly approximating Fedorov's D/G optimal design.

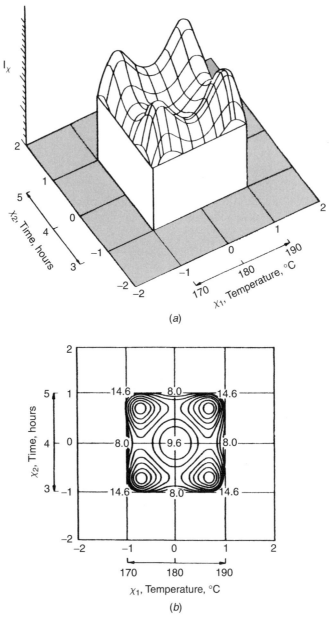

Figure 14.8. (*a*) Information function for a second-order D/G optimal design within the RO region $170 \le \chi_1 \le 190$, $3 \le \chi_2 \le 5$. (*b*) The corresponding contour plots and percentage distribution of the design measure.

designs, the equivalence of G and D optimality. Furthermore, they showed that, for such a design, within the region RO, the maximum value of $nV(\hat{y}_x)/\sigma^2$ was p, and this value was actually attained at each of the design points.

For illustration we consider a second-order measure design in two variables— that is, a design appropriate for the fitting of the second-degree polynomial. Such a design which is both D and G optimal for a square region RO with vertices ($\pm 1, \pm 1$) was given by Fedorov (1972) (see also Herzberg, 1979). The design places 14.6% of the measure at each of the four vertices, 8.0% at each of the midpoints of the edges, and 9.6% at the origin. The distribution measure for this design is shown in Figure 14.7a. A design realizable in practice which would roughly approximate the measure design is shown in Figure 14.7b. This would place two points at each corner of the square region and one point each at the centers of the edges and at the design center.

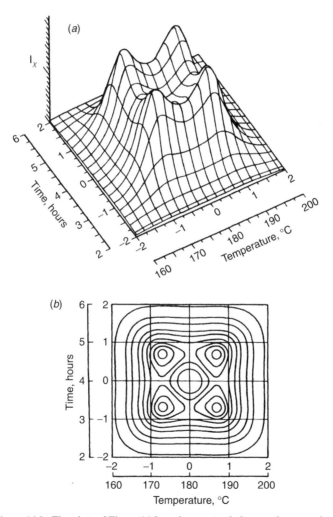

Figure 14.9. The plots of Figure 14.8 are here extended over a larger region.

As we have already seen, a function that represents the overall informational characteristic of the design, and not merely an isolated part of it, is the information function. A plot of this information function for Fedorov's second-order D/G optimal design over the square permissible RO region ($\pm 1, \pm 1$) referred to earlier is shown in Figure 14.8a. The corresponding contour plot and the measure values are shown in Figure 14.8b. To set this design into the kind of scientific context in which it would have to be used, we have supposed that the variables to be studied are χ_1 = temperature (°C) and χ_2 = time (h) and that they have been coded via $x_1 = (\chi_1 - 180)/10$, $x_2 = \chi_2 - 4$. This corresponds to a choice of the RO region in which experimentation is allowed within the limits χ_1 = 170–190°C and χ_2 = 3 to 5 h, but not outside these limits. The extensions of the information function and contour plot over a wider region are shown for comparison in Figure 14.9a and 14.9b. We return to these on pages 478–480.

14.5. ALPHABETIC OPTIMALITY IN THE RESPONSE SURFACE CONTEXT

In the response surface context, a number of questions arise concerning the appropriateness of the specification set out above for alphabetic optimality. These concern:

1. Formulation in terms of the RO region.
2. Distribution of information over a wider region.
3. Sensitivity of criteria to size and shape of the RO region.
4. Ignoring of bias.

As an example, suppose it is desired to study some chemical system, with the object of obtaining a higher value for a response η such as yield which is initially believed to be some function $\eta = g(\chi)$ of k continuous input variables $\chi = (\chi_1, \chi_2, \ldots, \chi_k)'$ such as reaction time, temperature, or concentration. As explained in Chapter 13 and again illustrated in Figure 14.10, it is usually known initially that the system can be operated at some point χ_0 in the space of χ and is expected to be capable of operating over some much more extensive region O called the *operability region*, which,[*] however, is usually unknown or poorly known. Response surface methods of the kind discussed in this book are employed when the nature of the true response function $\eta = g(\chi)$ is also unknown[†] or is inaccessible.

Suppose then that, over some immediate *region of interest R* in the neighborhood of χ_0, it is guessed that a graduating function, such as a dth-degree polynomial in x,

$$\eta_x \doteq z'\beta, \tag{14.5.1}$$

---

[*]A subsidiary objective of the investigation may be to find out more about the operability region O.
[†]Occasionally the true functional form $\eta = g(\chi)$ may be known, or at least conjectured, from knowledge of physical mechanisms. Typically, $g(\chi)$ will then appear as a solution of a set of differential equations that are nonlinear in a number of parameters that may represent physical constants. Problems of nonlinear experimental design then arise which are of considerable interest, although they have received comparatively little attention; see, for example, Box and Lucas (1959), Cochran (1973), and Fisher (1935; see 8th ed., 1966, pp. 218–223). We do not discuss these problems further here, but see Draper and Smith (1998, Chapter 24). Also see Bates and Watts (1987).

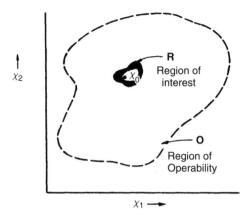

Figure 14.10. The current region of interest R and the region of operability O in the space of two continuous experimental variables χ_1 and χ_2.

might provide a *locally adequate approximation* to the true function $\eta_\chi = g(\chi)$ where, as before, $\mathbf{z}$ is a p-dimensional vector of suitably transformed input variables $\mathbf{z}' = \{f_1(\chi), f_2(\chi), \ldots, f_p(\chi)\}$, and $\boldsymbol{\beta}$ is a vector of coefficients occurring linearly which may be adjusted to approximate the unknown true response function $\eta_\chi = g(\chi)$. Then progress may be achieved by using a sequence of such approximations. For example, when a first-degree polynomial approximation could be employed, it might, via the method of steepest ascent, be used to find a new region of interest R_1 where, say, the yield was higher. Also, a maximum in many variables is often represented by some rather complicated ridge system*; and a second-degree polynomial approximation, when suitably analyzed, might be used to elucidate, describe, and exploit such a system.

Design Information Function

In general, to throw light on the desirable characteristics (1) and (2) of Section 14.1, we again consider the variance function

$$V_\chi = \frac{n \operatorname{var}(\hat{y}_\chi)}{\sigma^2} = n\mathbf{z}'(\mathbf{X}'\mathbf{X})^{-1}\mathbf{z} \qquad (14.5.2)$$

or equivalently the *information function*

$$I_\chi = V_\chi^{-1}. \qquad (14.5.3)$$

If we were to make the unrealistic assumption (made in alphabetic optimality) that the graduating function $\eta = \mathbf{z}'\boldsymbol{\beta}$ is capable of *exactly* representing the true function $g(\chi)$, then the information function would tell us all we could know about the design's ability to estimate η. For example, we have already seen

---

*Empirical evidence suggests this. Also, integration of sets of differential equations which describe the kinetics of chemical systems almost invariably leads to ridge systems; see Box (1954), Box and Youle (1955), Franklin et al. (1956), and Pinchbeck (1957).

information functions and associated information contours for a 2^2 factorial used as a first-order design (Figure 14.1) and for a 3^2 factorial used as a second-order design (Figure 14.3).

Formulation in Terms of the RO Region

As has been pointed out, in response surface studies it is typically true that, at any given stage of an investigation, the current region of interest R is much smaller than the region of operability O, which is, in any case, usually unknown. In particular, it is obvious that this must be so for any investigation in which we allow the possibility that results of one design may allow progress to a *different* unexplored region. Consequently we believe that formulation in terms of an *RO* region which assumes that the region of interest R and the region of operability O are identical is artificial and limiting. In particular, to obtain a good approximation *within R*, it might be sensible to put some experimental points *outside R* and, as long as they are within O, there is no practical reason not to do so. Also, because typically R is only vaguely known, we will want to consider the information function over a wider region, as is done, for example, in Figure 14.9 for Fedorov's second-order D optimal design. We shall soon discuss how the information function for this design compares, over this wider region, with that for the 3^2 factorial, already shown in Figure 14.3.

Distribution of Information over a Wider Region

In the response surface context, the coefficients $\boldsymbol{\beta}$ of a graduating function $\eta_\chi \doteq \mathbf{z}'\boldsymbol{\beta}$, acting as they do merely as adjustments to a kind of mathematical French curve, are not usually of individual interest except insofar as they affect η, in which case only the G optimality criterion among those considered is of direct interest. For response surface studies, however, it is far from clear how desirable is the property of G optimality itself. We illustrate by comparing the information surface for the 3^2 factorial design (Figure 14.3) with that of Fedorov's optimal design (Figure 14.9).

The profiles of Figure 14.11, which were made by taking sections of the surfaces of Figure 14.3 and Figure 14.9, show that neither the D/G optimal design nor the 3^2 design is universally superior. In some subregions one design is slightly better, and in others the other design is slightly better. Both information functions, and particularly that of the D/G optimal design, show a tendency to sag in the middle. This happens for the D/G optimal design because the G optimality characteristic guarantees that (maximized) minima for I_χ, each equal to $1/p$, where p is the number of parameters in the model, occur at every design point, *which must include the center point*. However, this sagging information pattern of the second-order design is not of course a characteristic of the first-order design of Figure 14.1 which is also D/G optimal but contains no center points. Because the information function for a design is a basic characteristic measuring how well we can estimate the response at any given point in the space of the variables, it seems unsatisfactory that the shape of that function should depend so very much on the order of the design under consideration. Indeed, it follows from the Kiefer–Wolfowitz theorem that a second-order design for the $(\pm 1, \pm 1)$ region whose information function did not sag in the middle would necessarily not be D optimal. As we saw earlier, D optimality is only one of many single-valued criteria that might be used

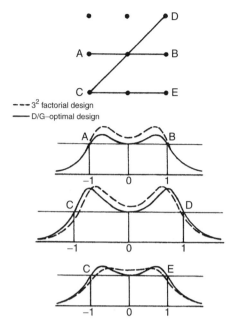

-- - 3^2 factorial design
— D/G-optimal design

Figure 14.11. Profiles of I_x for the second-order D/G optimal design and for the 3^2 factorial design.

in attempts to describe some important characteristic of the $\mathbf{X'X}$ matrix. Others, for example, would be A optimality and E optimality, and these would yield different information profiles. However, because the information function *itself* is the most direct measure of desirability so far as the single issue of variance properties is concerned, our best course would seem to be to choose the design directly by picking a suitable information function, and not indirectly by finding some extremum for A, E, D, or some other arbitrary criterion.

Sensitivity of Criteria to Size and Shape

In the process of scientific investigation, the investigator and the statistician must do a great deal of guesswork. In matching the region of interest R and the degree of complexity of the approximating function, they must try to take into account, for example, that the more flexible second-degree approximating polynomial can be expected to be adequate over a larger region R than the first-degree approximation. Obviously, different experimenters would have different ideas of appropriate locations and ranges for experimental variables. In particular, ranges could easily differ from one experimenter to another by a factor of 2 or more.* In view of this, extreme sensitivity of design criteria to scaling would be disturbing.[†] Such sensitiv-

---

*Over a sequence of designs, initial bad choices of scale and location would tend to be corrected, of course.

[†] In particular, designs can only be fairly compared if they are first scaled to be of the "same size." But how is size to be measured? It was suggested by Box and Wilson (1951) that designs should be judged as being of the same size when their marginal second moments $\Sigma(x_{iu} - \bar{x}_i)^2/n$ were identical. This convention is not entirely satisfactory, but will of course give very different results from those which assume design points to be all included in the same region RO. It is important to be aware that the apparent superiority of one design over another will often disappear if the method of scaling the design is changed. In particular this applies to comparisons such as those made by Nalimov et al. (1970) and Lucas (1976).

ity can be studied as follows. Suppose each dimension of a dth-order experimental design is increased* by a factor c. Then the D criterion is increased by a factor of c^q where

$$q = \frac{2k(k+d)!}{(k+1)!(d-1)!}.$$ (14.5.4)

Equivalently, a confidence region of the same volume as that for a D optimal design can be achieved for a design of given D value by increasing the scale for each variable by a factor of $c = (D_{opt}/D)^{1/q}$, thus increasing the volume occupied by the design in the χ space, by a factor $c^k = (D_{opt}/D)^{k/q}$. For example, the D value for the 3^2 factorial design of Figure 14.3 is 0.98×10^{-2} as compared with a D value of 1.14×10^{-2} for the D optimal design. For $k = 2$, $d = 2$, we find $q = 16$, and $c = (1.14/0.98)^{1/16} = 1.009$. Thus the same value of D (the same volume of a confidence region for the β's) as is obtained for the D optimal design would be obtained from a 3^2 design if *each side of the square region were increased by less than 1%*! Equivalently, the area of the region would be increased by less than 2%. Using the scaling that was used in Figure 14.8 for illustration, we would have to change the temperature by 20.18°C instead of 20°C, and the time by 2 h and 1 min instead of 2 h, for the 3^2 factorial to give the same D value as the D/G optimal design. Obviously, no experimenter can guess to anything approaching this accuracy what are suitable ranges over which to vary these factors. Thus if we limit consideration to alphabetic optimality, the apparent difference between the 3^2 factorial and the D-optimal design is seen, after all, to be entirely trivial, and any practical choice between these designs should be based on other criteria.

Choice of region and choice of information function are, in fact, closely interlinked. For example, *any* set of $N = k + 1$ points in k space which have no coplanarities is obviously a D optimal first-order design for *some*[†] ellipsoidal region. Furthermore the information function for a design of order d is a smooth function whose harmonic average over the n experimental points (which can presumably be regarded as representative of the region of interest) is always $1/p$ wherever we place the points. Thus the problem of design is not so much a question of choosing the design to *increase* total information as it is of *spreading* the total information around in the manner desired.

Scaleless D Optimality

One way to eliminate this massive dependence on scale is to employ the design coding

$$x_{iu} = \frac{(X_{iu} - \bar{X}_i)}{S_i}, \quad \text{where} \quad S_i^2 = n^{-1} \sum_{u=1}^{n} (X_{iu} - \bar{X}_i)^2, \quad i = 1, 2, \ldots, k.$$ (14.5.5)

*A measure of efficiency of a design criterion [see, for example, Atwood (1969) and Box (1968)] is motivated by considering the ratio of the number of runs necessary to achieve the optimal design to the number of runs required for the suboptimal design to obtain the same value of the criterion (supposing fractional numbers of runs to be allowed). In particular for the D criterion, this measure of D efficiency is $(D/D_{opt})^{1/k}$. Equivalently here, to illustrate scale sensitivity, we concentrate attention on the factor c by which each scale would need to be inflated to achieve the same value of the criterion.
[†]Namely, for that region enclosed within the information contour $I_\chi = 1/p$ which must pass through all the $k + 1$ experimental points.

As an example, consider the choice of a first-order design in $k = 2$ variables. In the selected scaling, the design's D criterion takes the form

$$D = n^3(1 - r^2),$$ (14.5.6)

where $nr = \Sigma x_{1u} x_{2u}$. This is maximized when $r = 0$ and the design is orthogonal. In general, for a first-order design in k variables it can be shown that, after scale factors are removed in accordance with Eq. (14.3.11), D optimality requires simply that the design should be orthogonal. This line of argument thus leads to the conclusion that if the criterion of *scaleless D* optimality is applied to the choice of a first-order design, an orthogonal arrangement is best. In practice, the orthogonal scaleless pattern would be related by the experimenter to the experimental variables using the scales and locations that he deemed appropriate. (This is, in fact, traditionally done with factorial and other standard designs.)

Ignoring of Bias

The whole of the above theory rests on the assumption that the graduating polynomial *is* the response function. As we have argued in detail in Chapter 13, such a polynomial must in fact always be regarded as a mathematical French curve which graduates locally the true but unknown response function. Thus, as we have said, there are two sources of error, variance error and bias error, and both should be taken into account in choosing a design. In particular, we have seen that designs which take account of bias tend *not* to place points at the extremes of the region of interest where the credibility of the approximating function is most strained.

Acknowledgment

The diagrams in this chapter were adapted from originals computer-generated by Conrad Fung.

CHAPTER 15

Practical Choice of a Response Surface Design

15.1. INTRODUCTION

We saw in Chapter 14 that the choice of an experimental design requires multifaceted consideration. Like a sailboat, an experimental design has many characteristics whose relative importance differs in different circumstances. In choosing a sailboat, the relative importance of characteristics such as size, speed, sea-worthiness, and comfort will depend greatly on whether we plan to sail on the local pond, undertake a trans-Atlantic voyage, or compete in the America's Cup contest. Some of the characteristics of experimental designs and some of the circumstances that influence their importance are as follows.

Characteristics of Design	Some Relevant Experimental Circumstances
Allows check of fit	Size of the experimental region Smoothness of the response function Complexity of the model
Allows estimation of transformations	Lack of fit that could be corrected by transformation
Permits sequential assembly	Ability to perform runs sequentially Ability to move in the space of the variables
Can be run in blocks	Homogeneity of experimental materials State of control of the process
Provides an independent estimate of error	Number of runs permissible Possibility of large experimental error Existence of reliable prior estimate of error
Robustness of distribution of design points	Possibility of occasional aberrant runs and or observations Nature of the error function
Number of runs required	Cost of making runs
Simplicity of data pattern	Need to visualize data to motivate model

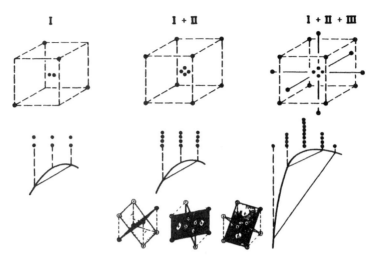

Figure 15.1. An example of sequential assembly, showing checks of linearity and quadraticity. Block I: Regular simplex (see p. 507) plus two center points. Block II: Complementary simplex plus center points. Block III: Six axial points (star).

In choosing a design, we have to take account of the circumstances listed on the right to decide on the relative importance of the design characteristics listed on the left. When we are ignorant about the experimental circumstances, then we need to insure against our fears.

We saw in Section 1.7 that response surface methods are used for various kinds of applications. We mentioned the approximate mapping of a surface, the achievement of a desired specification, and the attainment of best (or at least, good) operating conditions.

Sequential Assembly

In the applications discussed, we have employed first- and second-order designs and have sometimes made suitable transformations of the response and/or the input variables. The designs have often been used sequentially. For example, in investigating a new process, one or more phases of steepest ascent using first-order designs might be necessary before a sufficiently promising region was found within which it would be worth fitting a second-order surface. Because we do not know how such an investigation will proceed until we actually begin to carry it out, a good deal of flexibility in our choice of designs is desirable.

For illustration, Figure 15.1 shows the sequential assembly of a design arranged in three orthogonal blocks, each of six runs, labeled I, II, and III. We first suppose that only the six runs of block I have been performed. Note that these six runs comprise an orthogonal first-order design that also provides a check for overall curvature, obtained by contrasting the average response of the center points with the average response on the cube. Moreover, a single contrast in the center responses is available as a gross check on previous information about experimental error. Suppose now we found that first-order effects were very large compared with their standard errors and that no serious curvature was exhibited. Then it would make sense immediately to explore the indicated direction of steepest ascent. Alternatively, if after analyzing the results from block I there were doubts about the adequacy of a first-degree polynomial model, block II could be performed. It uses the complementary simplex, and while the two parts together (I + II) form a design of only first order, this combination has much greater ability to detect lack of fit due to the addition of contrasts which permit estimation of two-factor interactions. Again, at this point, if first-order terms dominate, there is an opportunity to move to a more promising region. On the other hand, if it is found that the first-order effects are small and the two-factor interactions dominate and/or there is strong evidence of curvature, then block III can be added to produce a composite design (I + II + III) which allows a full second-degree approximating equation to be fitted. The complete design also provides orthogonal checking contrasts for second-order adequacy in each of the three directions. These contrasts can also be regarded as checking the need for transformation in each of the x's. Finally, if it were decided that more information about experimental error was desirable, the replication of the star in a further block IV could furnish this and also provide some increase in the robustness of the design to wild observations.

Robustness

Approaches to the robust design of experiments have been reviewed by Herzberg (1982); see also Herzberg and Andrews (1976). In particular, Box and Draper (1975) suggested that the effects of wild observations could be minimized by making $r = \Sigma r_{uu}^2$ small, where $\mathbf{R} = \{r_{tu}\} = \mathbf{X}(\mathbf{X}'\mathbf{X})^{-1}\mathbf{X}'$. This is equivalent to minimizing $\Sigma r_{uu}^2 - p^2/N = N\sigma^{-4}\,\mathrm{var}\{V(\hat{y})\}$ which takes the value zero when $V(\hat{y}_u) = p/N$ $(u = 1, 2, \ldots, N)$, p being the number of parameters in the model and N the total number of observations. Thus, G optimal designs are optimally robust in this sense.

Number of Runs in an Experimental Design

A good experimental design is one that focuses experimental effort on what is judged important in the particular current experimental context. Suppose that, in addition to estimating the p parameters of the assumed model form, it is concluded that $f \geq 0$ contrasts are needed to check adequacy of fit, $b \geq 0$ further contrasts are needed for blocking, and an estimate of experimental error is needed having $e \geq 0$ df. To obtain independent estimates of all items of interest, we then require a design containing at least $p + f + b + e$ runs. However, the relative importance of checking fit, blocking, and obtaining an independent estimate of error will differ in different circumstances, and the minimum value of N will thus correspondingly differ. But this minimum design will in any case only be adequate if σ^2 is below some critical value. When σ^2 is larger, designs larger than the minimum design will be needed to obtain estimates of sufficient precision. In this circumstance, rather than merely replicating the minimum design, opportunity may be taken to employ a higher-order design allowing the fitting of a more elaborate approximating function which can then cover a wider experimental region. Notice that even when σ^2 is small, designs for which N is larger than p are not necessarily wasteful. This depends on whether the additional degrees of freedom are genuinely used to achieve the experimenter's objectives.

Simple Data Patterns

It has sometimes been argued that we may as well choose points randomly to cover the "design region" or employ some algorithm that distributes them evenly, even though this does not result in a simple data pattern such as is achieved by factorials and composite response surface designs. In favor of this idea, it has been urged that the fitting of a function by least squares to a haphazard set of points is a routine matter, given modern computational devices. This is true, but it overlooks an important attribute of designs that form simple patterns. The statistician's task as a member of a scientific team is a dual one, involving inductive criticism and deductive estimation. The latter involves deducing in the light of the data the consequences of given assumptions (estimating the fitted function), and this can certainly be done with haphazard designs. However, the former involves the questions (a) of what function should be fitted in the first place and (b) of how to examine residuals from the fitted function in an attempt to understand deviations from the initial model, in particular in relation to the predictor variables, in order to be led to appropriate model modification.

Designs such as factorials and composite response surface designs employ patterns of experimental points that allow many such comparisons to be made, both for the original observations and for the residuals from any fitted function. For example, consider a 3^2 factorial design used to elucidate the effects of temperature and concentration on some response such as yield. Intelligent inductive criticism is greatly enhanced by the possibility of being able to plot the original data and residuals against temperature for each individual level of concentration and against concentration for each individual level of temperature.

We now present, in relation to the criteria discussed above, some of the characteristics of various useful response surface designs.

15.2. FIRST-ORDER DESIGNS

First-order designs of great value in response surface methodology are the two-level factorial and fractional factorials discussed in Chapters 4 and 5. In particular, we have discussed in Chapter 6 their use in estimating the direction of steepest ascent. It will be recalled that a convenient curvature check is obtained by adding center points to the factorial points (see Section 6.3). These first-order designs can also play the role of initial building blocks in the construction of second-order designs.

In some situations, Plackett and Burman designs are useful first-order designs (see Section 5.8). They can be used as initial building blocks for some small second-order designs, as described in Section 15.6.

For extreme economy of experimentation, the Koshal-type first-order design can be used (see Section 15.6). Another economical alternative would be to use a regular simplex design, described below in material adapted from Box (1952).

Regular Simplex Designs

We first note that if $\mathbf{O}$ is *any* $N \times N$ *orthogonal* matrix, then $\mathbf{O'O} = \mathbf{OO'} = \mathbf{I}_N$. Suppose now that $\mathbf{Q}$ is an orthogonal $N \times N$ matrix whose first column has all elements equal to $N^{-1/2}$. Consider the $N \times N$ matrix $N^{1/2}\mathbf{Q}$. This has a first column of ones and the sum of squares of any row, or of any column, is N. Pairs of rows or columns are orthogonal. By deleting any $N - k - 1$ columns *except* the first, we can obtain the $N \times (k + 1)$ **X**-matrix of a first-order orthogonal design for k variables in N runs, where $k \le N - 1$. The columns of **X**, other than the first, define the design matrix **D** and the coded variable scales are such that

$$\sum_{u=1}^{N} x_{iu}^2 = N, \qquad i = 1, 2, \ldots, k. \tag{15.2.1}$$

When $k = N - 1$, we obtain the maximum number of factors that can be accommodated and

$$\mathbf{X} = N^{1/2}\mathbf{Q}, \tag{15.2.2}$$

no deletion being necessary. It can be shown (see Box, 1952) that the $k + 1$ design points must be distributed over a k-dimensional sphere, centered at the origin, radius $(N - 1)^{1/2}$, in the x-space, in such a way that the angle subtended at the origin by any pair of points is the same and has cosine $-1/(N - 1)$. Thus the design consists of the vertices of a *regular k-dimensional simplex*.

[*Note:* A *simplex* in k dimensions is the figure formed by joining any $k + 1 = N$ points that do not lie in a $(k - 1)$-dimensional space. For example, any three points not all on the same straight line are the vertices of a simplex (a triangle) in two dimensions. Any four points that do not all lie in the same plane form a simplex (a tetrahedron) in three dimensions. A simplex in k dimensions has $N = k + 1$ points and $k(k + 1)/2$ edges. A regular simplex is one that has all its edges equal. When $k = 2$, it is an equilateral triangle; when $k = 3$, an equilateral tetrahedron, and so on.]

A regular simplex design can be oriented in any manner in the x-space without affecting its orthogonal properties. Thus such a design is first order rotatable (see Section 14.3). For certain values of k, and in certain orientations, the regular simplex becomes a familiar design. For $k = 3$, it can take the form of a 2^{3-1} design defined by either $x_1 x_2 x_3 = 1$ or $x_1 x_2 x_3 = -1$. For $k = 7$, it can take the form of a 2^{7-4} design (see Table 5.4).

One particular type of orientation allows us to write down a regular simplex design for any value of k. We illustrate for $k = 3$. Suppose one face is taken parallel to the plane defined by the x_1 and x_2 axes and one of the sides of that face is taken parallel to the x_1 axis. The design points can then be allocated coordinates as follows, the scaling being chosen so that $\sum_{u=1}^4 x_{iu}^2 = N = 4$, for $i = 1, 2, 3$:

u	x_{1u}	x_{2u}	x_{3u}
1	$-2^{1/2}$	$-(2/3)^{1/2}$	$-1/3^{1/2}$
2	$2^{1/2}$	$-(2/3)^{1/2}$	$-1/3^{1/2}$
3	0	$2(2/3)^{1/2}$	$-1/3^{1/2}$
4	0	0	$3^{1/2}$

Removing a factor from each column, we obtain:

u	x_{1u}	x_{2u}	x_{3u}
1	-1	-1	-1
2	1	-1	-1
3	0	2	-1
4	0	0	3
Scale factor	$2^{1/2}$	$(2/3)^{1/2}$	$1/3^{1/2}$

Extensions to higher values of k are now straightforward. For example, if a fourth factor had to be introduced, we would add an x_{4u} column containing -1, -1, -1, -1, 4, add a row $(5, 0, 0, 0)$ under the $(u, x_{1u}, x_{2u}, x_{3u})$ columns, and either select a scaling factor of $1/5^{1/2}$ for x_{4u} so that $\sum_{u=1}^{5} x_{4u}^2 = 4$ or else rescale all the columns so that $\sum_{u=1}^{5} x_{iu}^2 = N = 5$ $(i = 1, 2, 3, 4)$, or any other selected value. Because the actual experimental levels will have to be coded to design levels in any case, there is no necessity to take $\sum_{u=1}^{N} x_{iu}^2 = N$, $i = 1, 2, \ldots, k$. If desired, the scaling *could* be different for each x-variable but then the design would no longer be rotatable. For the general extension of this method, see Exercise 15.9.

15.3. SECOND-ORDER COMPOSITE DESIGNS

We next consider the features of some of the more useful second-order designs. Consider a composite design consisting of:

1. $n_c = 2^{k-p} r_c$ points with coordinates of the type $(\pm 1, \pm 1, \ldots, \pm 1)$ forming the "cube" part of the design, of resolution at least V.
2. $n_s = 2k r_s$ axial points with coordinates $(\pm \alpha, 0, \ldots, 0)$, $(0, \pm \alpha, \ldots, 0), \ldots, (0, 0, \ldots, \pm \alpha)$ forming the "star" part of the design.
3. n_0 center points at $(0, 0, \ldots, 0)$.

Thus $r_c = n_c / 2^{k-p}$ measures the degree of replication of the chosen cube and $r_s = n_s / (2k)$ is the number of times the star is replicated.

Rotatability

It is shown in Section 14.3 that the design will be rotatable if

$$\alpha = \left(\frac{n_c}{r_s} \right)^{1/4} = \left(\frac{2kn_c}{n_s} \right)^{1/4} \qquad (15.3.1)$$

Orthogonal Blocking

A second-order design is said to be orthogonally blocked if it is divided into blocks in such a manner that block effects do not affect the usual estimates of the parameters of the second-order model. Box and Hunter (1957, pp. 226–234) show that two conditions must be fulfilled to achieve this. Suppose there are n_w runs in the wth block, then:

1. Each block must itself be a first-order orthogonal design, that is,

$$\sum_{u}^{n_w} x_{iu} x_{ju} = 0, \quad i \neq j = 0, 1, 2, \ldots, k \quad \text{all } w. \tag{15.3.2}$$

(*Note:* $x_{0u} = 1$ for all u.)

2. The fraction of the total sum of squares of each variable contributed by every block must be equal to the fraction of the total observations allotted to the block, that is,

$$\frac{\dfrac{\sum_{u}^{n_w} x_{iu}^2}{N}}{\sum_{u=1}^{N} x_{iu}^2} = \frac{n_w}{N}, \quad i = 1, 2, \ldots, k, \quad \text{all } w. \tag{15.3.3}$$

Consider first the division of the design into just two blocks, one consisting of the n_c runs from the cube portion plus n_{c0} center points and the other consisting of the n_s runs from the star portion plus n_{s0} center points. Such blocks obviously satisfy the first condition for orthogonality. The second is satisfied if

$$\frac{n_c}{2r_s \alpha^2} = \frac{n_c + n_{c0}}{n_s + n_{s0}} \tag{15.3.4}$$

that is, if

$$\alpha^2 = \frac{kn_c}{n_s} \left\{ \frac{n_s + n_{s0}}{n_c + n_{c0}} \right\} = k \left\{ \frac{1 + n_{s0}/n_s}{1 + n_{c0}/n_c} \right\} \tag{15.3.5}$$

or

$$\alpha^2 = k \left\{ \frac{1 + p_s}{1 + p_c} \right\}, \tag{15.3.6}$$

where $p_s = n_{s0}/n_s$ and $p_c = n_{c0}/n_c$ are the proportions of center points relative to the noncentral points in the star and cube, respectively. Now p_s and p_c are typically small fractions of about the same magnitude, so that the factor $(1 + p_s)/(1 + p_c)$ is close to unity; hence, for an orthogonally blocked design, α would be close to $k^{1/2}$. In other words, the star points would be roughly the same distance from the design center as the cube points.

Simultaneous Rotatability and Orthogonal Blocking

For a composite design to be both rotatable and orthogonally blocked, Eqs. (15.3.1) and (15.3.6) must be satisfied simultaneously, implying that

$$\frac{r_s}{r_c} = \frac{2^{k-p}}{k^2} \left\{ \frac{1 + p_c}{1 + p_s} \right\}^2. \tag{15.3.7}$$

Table 15.1. Relative Proportions of r_c and r_s Needed to Give Both Rotatability and Orthogonal Blocking when $p_c = p_s$ for a Composite Design which Uses a Full 2^k Factorial

$k =$	2	3	4	5	6	7	8
$\dfrac{r_s}{r_c} = \dfrac{2^k}{k^2} =$	1	0.89	1	1.28	1.78	2.61	4
$\dfrac{r_c}{r_s} = \dfrac{k^2}{2^k} =$	1	1.13	1	0.78	0.56	0.38	0.25
Smallest[a] available cube fractionation	1	1	1	$\frac{1}{2}$	$\frac{1}{2}$	$\frac{1}{2}$	$\frac{1}{4}$

[a]To retain a resolution V cube portion.

Because n_c, n_s, n_{c0}, and n_{s0} must all be integers in practice, it will not always be possible to meet the *exact* requirements for both rotatability and orthogonal blocking. However, neither of these properties is critical. In particular, rotatability introduced to ensure a spherical distribution of information can be relaxed somewhat without appreciable loss, and the number of center points can be adjusted to achieve orthogonal blocking if desired, using Eq. (15.3.5).

Relative Replications of Star and Cube When $p_c = p_s$

If the factor $(1 + p_c)/(1 + p_s) = 1$, then

$$\frac{r_s}{r_c} = \frac{2^{k-p}}{k^2}$$

provides the desirable relative replications of star and cube when $p_c = p_s$. When $p = 0$, substitution of various values of k provides the details shown in Table 15.1.

For example, the entry $r_s/r_c = 4$ for $k = 8$ implies that, if a complete 2^8 factorial were run with proportion of center points for cube and star the same, then four replicates of the star would yield a design which was rotatable and orthogonally blockable. Conversely, if we look at the ratio r_c/r_s, we find the ideal fractional replicate of the cube to be associated with one complete star. So, for $k = 8$, a single star would require a 2^{8-2} design cube portion. When $k = 3, 5, 6,$ or 7, exact achievement of rotatability and orthogonal blocking is not possible for $p_c = p_s$, but the table entries show the smallest fractional cube of resolution V roughly appropriate for each star.

Center Points in Composite Designs

Various criteria have been suggested for deciding the number of center points $n_0 = n_{c0} + n_{s0}$ that should be used in a composite design (for a discussion, see

Table 15.2. Some Useful Second-Order Composite Designs and Their Characteristics

k	2	3	4	5	5	6	7	8
Cube fraction, 2^{-p}	1	1	1	1	$\frac{1}{2}$	$\frac{1}{2}$	$\frac{1}{2}$	$\frac{1}{4}$
r_s	1	1	1	1	1	1	1	1
α for rotatability $(n_c/r_s)^{1/4}$	1.41	1.68	2	2.38	2.00	2.38	2.83	2.83
$k^{1/2}$	1.41	1.73	2	2.24	2.24	2.45	2.65	2.83
n_c	4	8	16	32	16	32	64	64
$n_s = 2k$	4	6	8	10	10	12	14	16
Minimal $\{\ n_0$	2–4	2–4	2–4	2–4	1–4	2–4	2–4	2–4
n_{c0}	1(2)	2	2	2	2	2	2	4
n_{s0}	1(2)	1	1	1	2	1	1	1
Blocks in cube	—	2 × 4	2 × 8	4 × 8	—	2 × 16	4 × 16 8 × 8	4 × 16
Blocking generators	—	**B = 123**	**B = 1234**	**B₁ = 123** **B₂ = 2345**	—	**B = 123**	**B₁ = 1357** **B₂ = 1256** **B₃ = 1234**	**B₁ = 135** **B₂ = 348**
Fractional design generators	—	—	—	—	**I = 12345**	**I = 123456**	**I = 1234567**	**I = 12347** **I = 12568**

Draper, 1982). The criteria include choosing n_0 to:

1. Produce a good profile for the information function.
2. Minimize integrated mean squared error.
3. Give good detectability of third-order lack of fit.
4. Give a robust design insensitive to bad values.

The values of n_0 that satisfy these various criteria do not differ very much, and, in the table of useful designs given above (Table 15.2), convenient compromise values have been chosen. It should be understood that if for some reason (to obtain more information about σ^2, for example, or to achieve equal block sizes, or orthogonal blocking) we wish to introduce more center points than the numbers indicated, nothing will be lost by this except the cost of performing the additional runs.

Smaller Blocks

So far, we have considered the splitting of the composite design into only two blocks, one associated with the cube and one with the star. Smaller blocks can be used if the factorial or fractional factorial part of the design can be further split into blocks that are first-order orthogonal designs. By splitting the n_{c0} center points equally among these smaller blocks, the requirements for orthogonal blocking will remain satisfied. (Naturally, n_{c0} must be a multiple of the number of smaller cube blocks here.) Also, if the star is replicated, then each replication, with appropriate equal allocation of the n_{s0} center points, may be made one of the set of orthogonal blocks.

Useful Second-Order Composite Designs

A short table of useful designs is shown as Table 15.2. We now offer some suggestions for using this table.

If the design is *not* to be run in blocks, the α may as well be chosen to achieve exact rotatability from the row $\alpha = (n_c/r_s)^{1/4}$. If the design is to be blocked, then, if possible, the number of center points should be chosen as indicated, first of all. For example, if $k = 4$, then place one center points in each of the two blocks into which the 2^4 cube is divided and one center point in the star portion. Suppose, however, that $k = 7$ and that the 2^{7-1} cube is to be run in eight blocks of eight runs each. To achieve balance, we could put one center point in each cube block and one in the star block. This would, of course, require nine center points rather than the recommended 2–4 if the design were unblocked. As we have already noted, however, nothing is lost by increasing the number of center points except the cost of the additional runs. In any case, having decided on the allocation of points to the various parts of the blocked design, the best choice of α is now the one that achieves orthogonal blocking, namely,

$$\alpha^2 = k \frac{(1 + n_{s0}/n_s)}{(1 + n_{c0}/n_c)} \, .$$

In general, this will not achieve exact rotatability, although in some instances it happens to, but in all cases it should provide variance contours that are adequately close to the spherical ones that exact rotatability would provide.

Design Examples

Tables 15.3, 15.5, and 15.6 contain examples of useful orthogonally blocked second-order designs for $k = 4$, 5, and 6 factors. For the case $k = 4$ of Table 15.3, orthogonal blocking can be achieved by various combinations of center points, using an appropriate value of α; for examples, see Table 15.4. A similar flexibility is, of course, also achievable for other values of k; however, Tables 15.5 and 15.6 contain a specific example design only.

Note that for $k = 5$ (Table 15.5), blocks I and III (or II and III) can be used as an orthogonally blocked second-order design with the same value of α. This provides a 30-run design for five factors in two blocks consisting of a 2^{5-1} design plus three center points and a star plus one center point. The design is again not rotatable since $\alpha \neq 2^{(k-1)/4} = 2$.

An excellent 24-run, three-factor, second-order, orthogonally blocked design is given in Table 11.4 of Section 11.4.

For other examples of designs that have been used in various applications, see the exercises in Chapters 7 and 11.

Other Effects of Orthogonal Blocking

1. *Pure Error*. Runs that would be repeat runs in an unblocked design are often divided among the blocks. In such a case, these runs are no longer repeat runs

Table 15.3. Orthogonally Blocked Second-Order Designs for Four Factors; with $\frac{1}{2}n_{c0}$ Center Points in Each of Blocks I and II and n_{s0} Center Points in Block III, $\alpha^2 = 4(1 + n_{s0}/8)/(1 + n_{c0}/16)$

Block I (2^{4-1}_{IV} with $x_1x_2x_3x_4 = 1$, plus center points)				Block II (2^{4-1}_{IV} with $x_1x_2x_3x_4 = -1$, plus center points)				Block III (star, axial distance α, plus center points)			
x_1	x_2	x_3	x_4	x_1	x_2	x_3	x_4	x_1	x_2	x_3	x_4
-1	-1	-1	-1	-1	-1	-1	1	$-\alpha$	0	0	0
1	-1	-1	1	1	-1	-1	-1	α	0	0	0
-1	1	-1	1	-1	1	-1	-1	0	$-\alpha$	0	0
1	1	-1	-1	1	1	-1	1	0	α	0	0
-1	-1	1	1	-1	-1	1	-1	0	0	$-\alpha$	0
1	-1	1	-1	1	-1	1	1	0	0	α	0
-1	1	1	-1	-1	1	1	1	0	0	0	$-\alpha$
1	1	1	1	1	1	1	-1	0	0	0	α
0	0	0	0	0	0	0	0	0	0	0	0
$\vdots$	$\vdots$	$\vdots$	$\vdots$	$\vdots$	$\vdots$	$\vdots$	$\vdots$	$\vdots$	$\vdots$	$\vdots$	$\vdots$
0	0	0	0	0	0	0	0	0	0	0	0

Table 15.4. Values of α which Achieve Orthogonal Blocking for Various Selected Center Point Choices for the Four-Factor Design of Table 15.3

n_{c0}	n_{s0}	Center Points in Block			α
		I	II	III	
4	2	2	2	2	2.000[a]
4	3	2	2	3	2.098
4	4	2	2	4	2.191
8	2	4	4	2	1.826
8	3	4	4	3	1.915
8	4	4	4	4	2.000[a]

[a]Note that when $\alpha = 2$, the design is also rotatable.

Table 15.5. An Orthogonally Blocked Second-Order Design for Five Factors

Block I (2^{5-1}, with $x_1x_2x_3x_4x_5 = 1$ plus three center points)					Block II (2^{5-1}, with $x_1x_2x_3x_4x_5 = -1$ plus three center points)					Block III (star, with $\alpha = (88/19)^{1/2} = 2.152110$ [a] plus one center point)				
x_1	x_2	x_3	x_4	x_5	x_1	x_2	x_3	x_4	x_5	x_1	x_2	x_3	x_4	x_5
-1	-1	-1	-1	1	-1	-1	-1	-1	-1	$-\alpha$	0	0	0	0
1	-1	-1	-1	-1	1	-1	-1	-1	1	α	0	0	0	0
-1	1	-1	-1	-1	-1	1	-1	-1	1	0	$-\alpha$	0	0	0
1	1	-1	-1	1	1	1	-1	-1	-1	0	α	0	0	0
-1	-1	1	-1	-1	-1	-1	1	-1	1	0	0	$-\alpha$	0	0
1	-1	1	-1	1	1	-1	1	-1	-1	0	0	α	0	0
-1	1	1	-1	1	-1	1	1	-1	-1	0	0	0	$-\alpha$	0
1	1	1	-1	-1	1	1	1	-1	1	0	0	0	α	0
-1	-1	-1	1	-1	-1	-1	-1	1	1	0	0	0	0	$-\alpha$
1	-1	-1	1	1	1	-1	-1	1	-1	0	0	0	0	α
-1	1	-1	1	1	-1	1	-1	1	-1	0	0	0	0	0
1	1	-1	1	-1	1	1	-1	1	1					
-1	-1	1	1	1	-1	-1	1	1	-1					
1	-1	1	1	-1	1	-1	1	1	1					
-1	1	1	1	-1	-1	1	1	1	1					
1	1	1	1	1	1	1	1	1	-1					
0	0	0	0	0	0	0	0	0	0					
0	0	0	0	0	0	0	0	0	0					
0	0	0	0	0	0	0	0	0	0					

[a]The design is not rotatable because $\alpha \neq 2.378$.

unless they occur in the same block, and the pure error must be calculated on that basis.

2. *Analysis of Variance.* When a design is blocked, the analysis of variance table must contain a sum of squares for blocks. For blocks orthogonal to the model, the appropriate sum of squares for blocks is usually

$$\text{SS(blocks)} = \sum_{w=1}^{m} \frac{B_w^2}{n_w} - \frac{G^2}{N}, \quad \text{with } (m-1) \text{ df},$$

where B_w is the total of the n_w observations in the wth block (there are m blocks

Table 15.6. An Orthogonally Blocked Second-Order Design for Six Factors

Block I $(2^{6-2}$, with $x_5 = -x_1x_2$, $x_6 = -x_3x_4$ plus four center points)						Block II $(2^{6-2}$, with $x_5 = x_1x_2$, $x_6 = x_3x_4$ plus four center points)						Block III (star, with $\alpha = 2(7/5)^{1/2}$ $= 2.366432^a$ plus two center points)					
x_1	x_2	x_3	x_4	x_5	x_6	x_1	x_2	x_3	x_4	x_5	x_6	x_1	x_2	x_3	x_4	x_5	x_6
-1	-1	-1	-1	-1	-1	-1	-1	-1	-1	1	1	$-\alpha$	0	0	0	0	0
1	-1	-1	-1	1	-1	1	-1	-1	-1	-1	1	α	0	0	0	0	0
-1	1	-1	-1	1	-1	-1	1	-1	-1	-1	1	0	$-\alpha$	0	0	0	0
1	1	-1	-1	-1	-1	1	1	-1	-1	1	1	0	α	0	0	0	0
-1	-1	1	-1	-1	1	-1	-1	1	-1	1	-1	0	0	$-\alpha$	0	0	0
1	-1	1	-1	1	1	1	-1	1	-1	-1	-1	0	0	α	0	0	0
-1	1	1	-1	1	1	-1	1	1	-1	-1	-1	0	0	0	$-\alpha$	0	0
1	1	1	-1	-1	1	1	1	1	-1	1	-1	0	0	0	α	0	0
-1	-1	-1	1	-1	1	-1	-1	-1	1	1	-1	0	0	0	0	$-\alpha$	0
1	-1	-1	1	1	1	1	-1	-1	1	-1	-1	0	0	0	0	α	0
-1	1	-1	1	1	1	-1	1	-1	1	-1	-1	0	0	0	0	0	$-\alpha$
1	1	-1	1	-1	1	1	1	-1	1	1	-1	0	0	0	0	0	α
-1	-1	1	1	-1	-1	-1	-1	1	1	1	1	0	0	0	0	0	0
1	-1	1	1	1	-1	1	-1	1	1	-1	1	0	0	0	0	0	0
-1	1	1	1	1	-1	-1	1	1	1	-1	1						
1	1	1	1	-1	-1	1	1	1	1	1	1						
0	0	0	0	0	0	0	0	0	0	0	0						
0	0	0	0	0	0	0	0	0	0	0	0						
0	0	0	0	0	0	0	0	0	0	0	0						
0	0	0	0	0	0	0	0	0	0	0	0						

^{a} The design is not rotatable because $\alpha \neq 2.378$, but is nearly so.

in all) and G is the grand total of all the observations in all of the m blocks. When blocks are not orthogonal to the model, the extra sum of squares principle applies. (It can, of course, be applied in all cases, orthogonally blocked or not, and produces the answer given above in the former case. See Section 3.8.)

15.4. SECOND-ORDER DESIGNS REQUIRING ONLY THREE LEVELS

Second-order composite designs usually require five levels coded $-\alpha$, -1, 0, 1, and α for each of the variables. Circumstances occur, however, where second-order arrangements are required which, for ease of performance, must employ the smallest number of different levels, namely three. One useful class of such designs, which are economical in the number of runs required, is due to Box and Behnken (1960a).

The designs are formed by combining two-level factorial designs with incomplete block designs in a particular manner. This is best illustrated by an example. Table 15.7a shows a balanced incomplete block design for testing $k = 4$ varieties in $b = 6$ blocks of size $s = 2$. If this design were employed in the usual way, varieties 1 and 2 denoted by x_1 and x_2 would be tested in the first block, varieties 3 and 4 in the second, and so on.

Table 15.7. Constructing a Three-Level Second-Order Design for Four Variables

(a) A Balanced Incomplete Block
Design for Four Varieties in Six Blocks   (b) A 2^2 Factorial Design

$k = 4$ varieties

$$
\begin{array}{c}
 \\
 \\
b = 6 \\
\text{blocks}
\end{array}
\begin{array}{c}
1 \\
2 \\
3 \\
4 \\
5 \\
6
\end{array}
\begin{array}{cccc}
x_1 & x_2 & x_3 & x_4 \\
\left[\begin{array}{cccc}
* & * &   &   \\
  &   & * & * \\
\hline
* &   &   & * \\
  & * & * &   \\
\hline
  & * &   & * \\
* &   & * &
\end{array}\right]
\end{array}
\qquad
\begin{array}{cc}
x_i & x_j \\
\left[\begin{array}{cc}
-1 & -1 \\
1 & -1 \\
-1 & 1 \\
1 & 1
\end{array}\right]
\end{array}
$$

Table 15.8. An Incomplete 3^4 Factorial in Three Blocks of Nine Experimental Runs

x_1	x_2	x_3	x_4	
-1	-1	0	0	
1	-1	0	0	
-1	1	0	0	
1	1	0	0	
0	0	-1	-1	Block 1
0	0	1	-1	
0	0	-1	1	
0	0	1	1	
0	0	0	0	
-1	0	0	-1	
1	0	0	-1	
-1	0	0	1	
1	0	0	1	
0	-1	-1	0	Block 2
0	1	-1	0	
0	-1	1	0	
0	1	1	0	
0	0	0	0	
0	-1	0	-1	
0	1	0	-1	
0	-1	0	1	
0	1	0	1	
-1	0	-1	0	Block 3
1	0	-1	0	
-1	0	1	0	
1	0	1	0	
0	0	0	0	

A basis for a three-level design in four variables is obtained by combining this incomplete block design with the 2^2 factorial of Table 15.7(b). The two asterisks in every row of the incomplete block design are replaced by the $s = 2$ columns of the two-level 2^2 design. Wherever an asterisk does not appear, a column of zeros is inserted. The design is completed by the addition of a number of center points $(0, 0, 0, 0)$, about three being desirable with this arrangement. The resulting design is shown in Table 15.8.

This design is a rotatable second-order design* suitable for studying four variables in 27 trials and is capable of being orthogonally blocked in three sets of nine trials, separated by dashed lines in Table 15.8. (Also see Table 15.9 and Exercise 7.23.)

Table 15.9 provides useful designs of this kind for $k = 3, 4, 5, 6, 7$ factors. Note that the design for $k = 6$ is not symmetrical in all factors, because it is based on a *partially* balanced incomplete block design with first associates $(1, 4)$, $(2, 5)$, and $(3, 6)$. The designs for $k = 4$ and 7 are both rotatable. The requirements for orthogonal blocking may be met as follows.

1. Where "replicate sets" can be found in the generating incomplete block design, these provide a basis for orthogonal blocking. These replicate sets are subgroups within which each variety is tested the same number of times.
2. Where the component factorial designs can be divided into blocks which confound only interactions of more than two factors, these can provide a basis for orthogonal blocking.

An illustration of the first method of blocking has already been given in Table 15.8. In Table 15.9, for $k = 4$ and 5, dashed lines indicate the appropriate divisions into replicate sets. Using these divisions, the design for $k = 4$ can be split into three blocks, while the design for $k = 5$ can be split into two blocks. The center points *must* be distributed equally among blocks to retain orthogonality.

The second method may be illustrated with the design for $k = 6$, for which the first method cannot be employed. The basis for the design consists of 48 trials generated from six 2^3 factorial designs. If we were running a single 2^3 factorial design, it could be performed in two sets of four trials, confounding the three-factor interaction with blocks. Trials with levels $(1, 1, 1)$, $(1, -1, -1)$, $(-1, -1, 1)$, $(-1, 1, -1)$ would be included in one set (called the positive set), while trials with levels $(-1, -1, -1)$, $(-1, 1, 1)$, $(1, -1, 1)$, $(1, 1, -1)$ would be included in the other (called the negative set). The complete group of 48 trials can be split into two orthogonal blocks of 24 by allocating one set (either positive or negative) from each of the 2^3 factorial designs to one block, while the remainder can be allocated to the other. This method can also be used for the design for $k = 7$. Three-level designs for $k > 7$ and their blocking arrangements are given in the original reference, Box and Behnken (1960a).

---

*This particular design is, in fact, a rotation of the four variable central composite design with three center points. However, it is not generally true that the present class of designs can be generated from the central composite designs by rotation.

Table 15.9. Some Useful Three-Level Second-Order Response Surface Designs

Number of Factors (k)	Design Matrix	Number of Points

$k = 3$

$$\begin{bmatrix} \pm1 & \pm1 & 0 \\ \pm1 & 0 & \pm1 \\ 0 & \pm1 & \pm1 \\ 0 & 0 & 0 \end{bmatrix}$$

$\left.\right\}$ 12

$\underline{\hspace{3em} 3}$

$N = 15$

$k = 4$

$$\begin{bmatrix} \pm1 & \pm1 & 0 & 0 \\ 0 & 0 & \pm1 & \pm1 \\ 0 & 0 & 0 & 0 \\ \hline \pm1 & 0 & 0 & \pm1 \\ 0 & \pm1 & \pm1 & 0 \\ 0 & 0 & 0 & 0 \\ \hline \pm1 & 0 & \pm1 & 0 \\ 0 & \pm1 & 0 & \pm1 \\ 0 & 0 & 0 & 0 \end{bmatrix}$$

$\left.\right\}$ 8

$\underline{\hspace{2em} 1}$

$\left.\right\}$ 8

$\underline{\hspace{2em} 1}$

$\left.\right\}$ 8

$\underline{\hspace{2em} 1}$

$N = 27$

$k = 5$

$$\begin{bmatrix} \pm1 & \pm1 & 0 & 0 & 0 \\ 0 & 0 & \pm1 & \pm1 & 0 \\ 0 & \pm1 & 0 & 0 & \pm1 \\ \pm1 & 0 & \pm1 & 0 & 0 \\ 0 & 0 & 0 & \pm1 & \pm1 \\ 0 & 0 & 0 & 0 & 0 \\ \hline 0 & \pm1 & \pm1 & 0 & 0 \\ \pm1 & 0 & 0 & \pm1 & 0 \\ 0 & 0 & \pm1 & 0 & \pm1 \\ \pm1 & 0 & 0 & 0 & \pm1 \\ 0 & \pm1 & 0 & \pm1 & 0 \\ 0 & 0 & 0 & 0 & 0 \end{bmatrix}$$

$\left.\right\}$ 20

$\underline{\hspace{2em} 3}$

$\left.\right\}$ 20

$\underline{\hspace{2em} 3}$

$N = 46$

$k = 6$

$$\begin{bmatrix} \pm1 & \pm1 & 0 & \pm1 & 0 & 0 \\ 0 & \pm1 & \pm1 & 0 & \pm1 & 0 \\ 0 & 0 & \pm1 & \pm1 & 0 & \pm1 \\ \pm1 & 0 & 0 & \pm1 & \pm1 & 0 \\ 0 & \pm1 & 0 & 0 & \pm1 & \pm1 \\ \pm1 & 0 & \pm1 & 0 & 0 & \pm1 \\ 0 & 0 & 0 & 0 & 0 & 0 \end{bmatrix}$$

$\left.\right\}$ 48

$\underline{\hspace{2em} 6}$

$N = 54$

$k = 7$

$$\begin{bmatrix} 0 & 0 & 0 & \pm1 & \pm1 & \pm1 & 0 \\ \pm1 & 0 & 0 & 0 & 0 & \pm1 & \pm1 \\ 0 & \pm1 & 0 & 0 & \pm1 & 0 & \pm1 \\ \pm1 & \pm1 & 0 & \pm1 & 0 & 0 & 0 \\ 0 & 0 & \pm1 & \pm1 & 0 & 0 & \pm1 \\ \pm1 & 0 & \pm1 & 0 & \pm1 & 0 & 0 \\ 0 & \pm1 & \pm1 & 0 & 0 & \pm1 & 0 \\ 0 & 0 & 0 & 0 & 0 & 0 & 0 \end{bmatrix}$$

$\left.\right\}$ 56

$\underline{\hspace{2em} 6}$

$n = 62$

15.5. SCREENING DESIGNS WITH THREE LEVELS AND THEIR PROJECTIONS INTO FEWER FACTORS

As we have seen in Chapter 5 in relation to two-level fractional factorial and Plackett–Burman-type designs, screening designs are useful for experiments in which one can make an assumption of *effect sparsity*. In such situations, it is desired to examine *many* factors, but it is anticipated that only a *few* of those factors will have an effect on the response. However, it is not known beforehand which those few may be. A typical analysis for a screening design is to evaluate the estimates of effects associated such a design and then pare down the number of dimensions in some sensible manner, based on one's judgment of the importance of the effects seen. The experimental data can then be reassessed and reanalyzed in terms of those fewer factors. This may suggest the performance of additional runs to resolve ambiguities of interpretation seen in the analysis so far. These extra runs could be made using only the factors that appear to be important, or perhaps in all factors, as seems sensible and/or feasible. Before we choose and run a screening design, then, it is thus useful to know what sorts of designs can arise when projecting it down to various numbers of factors.

Table 15.10. A 27-Run, 13-Variable, Three-Level Screening Design with the Property that a Full Quadratic Model Can Be Fitted to Any Subset of Five Factors

	1	2	3	4	5	6	7	8	9	10	11	12	13
1	−1	−1	−1	−1	−1	0	1	−1	−1	−1	−1	−1	−1
2	−1	0	0	1	1	1	1	1	1	1	0	−1	−1
3	−1	1	1	1	0	1	0	−1	−1	−1	0	1	0
4	−1	−1	0	0	1	0	−1	0	0	−1	0	0	0
5	−1	0	1	1	−1	0	0	0	0	0	1	−1	1
6	−1	1	−1	−1	1	1	−1	1	0	0	−1	1	1
7	−1	−1	−1	0	−1	−1	0	1	1	1	1	1	0
8	−1	0	1	0	0	−1	−1	−1	1	0	−1	0	−1
9	−1	1	0	−1	0	−1	1	0	−1	1	1	0	1
10	0	−1	0	1	0	0	−1	1	−1	0	1	1	−1
11	0	0	−1	−1	0	1	−1	0	1	−1	1	−1	0
12	0	1	0	1	−1	−1	−1	−1	0	1	−1	−1	0
13	0	−1	1	1	1	−1	1	0	1	−1	−1	1	1
14	0	0	−1	0	0	0	1	−1	0	1	0	1	1
15	0	1	−1	0	1	−1	0	0	−1	0	0	−1	−1
16	0	−1	0	−1	−1	1	0	−1	1	0	0	0	1
17	0	0	1	−1	1	0	0	1	−1	1	−1	0	0
18	0	1	1	0	−1	1	1	1	0	−1	1	0	−1
19	1	−1	−1	1	0	1	0	0	0	1	−1	0	−1
20	1	0	0	−1	1	−1	0	−1	0	−1	1	1	−1
21	1	1	1	−1	−1	0	−1	0	1	1	0	1	−1
22	1	−1	1	0	1	1	−1	−1	−1	1	1	−1	1
23	1	0	0	0	−1	1	1	0	−1	0	−1	1	0
24	1	1	−1	1	1	0	1	−1	1	0	1	0	0
25	1	−1	1	−1	0	−1	1	1	0	0	0	−1	0
26	1	0	−1	1	−1	−1	−1	1	−1	−1	0	0	1
27	1	1	0	0	0	0	0	1	1	−1	−1	−1	1

Table 15.11. A 24-Run, Eight-Variable, Three-Level Screening Design with the Property that a Full Quadratic Model Can Be Fitted to Any Subset of Five Factors (Bulutoglu and Cheng, 2003)

	1	2	3	4	5	6	7	8
1	-1	-1	-1	-1	-1	-1	-1	-1
2	-1	-1	0	1	1	0	-1	-1
3	-1	-1	-1	0	1	1	0	-1
4	0	-1	-1	-1	0	1	1	-1
5	1	0	-1	-1	-1	0	1	-1
6	1	1	0	-1	-1	-1	0	-1
7	0	1	1	0	-1	-1	-1	-1
8	-1	0	1	1	0	-1	-1	-1
9	0	0	0	0	0	0	0	0
10	0	0	1	-1	-1	1	0	0
11	0	0	0	1	-1	-1	1	0
12	1	0	0	0	1	-1	-1	0
13	-1	1	0	0	0	1	-1	0
14	-1	-1	1	0	0	0	1	0
15	1	-1	-1	1	0	0	0	0
16	0	1	-1	-1	1	0	0	0
17	1	1	1	1	1	1	1	1
18	1	1	-1	0	0	-1	1	1
19	1	1	1	-1	0	0	-1	1
20	-1	1	1	1	-1	0	0	1
21	0	-1	1	1	1	-1	0	1
22	0	0	-1	1	1	1	-1	1
23	-1	0	0	-1	1	1	1	1
24	1	-1	0	0	-1	1	1	1

Projections of Three-Level Designs

Improvements in computing have led to exhaustive investigations of the projection properties of three-level screening designs. We select for display in Table 15.10 an excellent design constructed by Xu et al. (2004). It is a 13-variable, three-level $(-1, 0, 1)$ screening design in 27 runs. Every five-factor projection will support (i.e., can be used to fit) a full second-order (quadratic) model with 21 terms in the selected five factors. A portion of the original design can be used to screen fewer than 13 variables initially, if desired, merely by choosing as many columns as are needed to accommodate the variables under study. Bulutoglu and Cheng (2003) gave a slightly smaller design with only 24 runs (three fewer), for screening eight or fewer factors. It is shown in Table 15.11. Again, a full second-order model can be fitted to any projected set of five factors.

15.6. DESIGNS REQUIRING ONLY A SMALL NUMBER OF RUNS

We have seen that, if a response function containing p adjustable parameters is to be fitted, we need at least p runs. It sometimes happens that designs containing close to this minimal number of runs are of interest. This could be so for a variety of reasons:

1. Runs might be extremely expensive.
2. The checking of assumptions, the need for an internal error estimate, the need to check fit, and, in particular, the need to consider possible transformations of the predictor variables might not be regarded as important in a particular application.
3. The objective might be to approximate locally, by a polynomial, a function that can be computed *exactly* at any given combination of the input variables; that is, there is no experimental error.

We now discuss some of the possibilities available for such situations.

Small Composite Designs

The composite designs recommended in Section 15.3 for fitting second-order models in k factors all contain cube portions of resolution at least V, plus axial points, plus center points. Of course there must be at least $\frac{1}{2}(k + 1)(k + 2)$ points in the design, this being the number of coefficients to estimate. Hartley (1959) pointed out that the cube portion of the composite design need not be of resolution V. It could, in fact, be of resolution as low as III, provided that two-factor interactions were not aliased with two-factor interactions. (Two-factor interactions could be aliased with main effects, because the star portion provides additional information on the main effects.) This idea permitted much smaller cubes to be used. Westlake (1965) took this idea further by finding even smaller cubes for the $k = 5$, 7, and 9 cases. Table 15.12 shows the numbers of points in the various designs suggested for $2 \leq k \leq 9$.

Westlake (1965) provided three examples of 22-run designs for $k = 5$, one example of a 40-run design for $k = 7$, and one example of a 62-run design for $k = 9$. Subsequent work by Draper (1985b) showed that designs even smaller than Westlake's are possible. Moreover, for $k = 5$ and 9 it is possible to *equal* Westlake's number of runs in a simple manner; and for $k = 7$, simple designs are available with only 42 runs, two more than Westlake's 40. The overall advantage of these suggested designs is that none of the ingenuity shown by Westlake (1965) is needed, thanks to Plackett and Burman (1946), and yet an apparently large

Table 15.12. Points Needed by Some Small Composite Designs

Factors, k	2	3	4	5	6	7	8	9
Coefficients $\frac{1}{2}(k + 1)(k + 2)$	6	10	15	21	28	36	45	55
Points in Box–Hunter (1957) designs	8	14	24	26	44	78	80	146
Hartley's number of points	6	10	16	26	28	46	80	82
Westlake's number of points	—	—	—	22	—	40	—	62

selection of possible designs is immediately available. (As we shall see later, the selection is not as large as first appears!)

The basic method can be simply stated: (a) Use, for the cube portion of the design, k columns of a Plackett and Burman (1946) design. (b) Where repeat runs exist, remove one of each duplicate pair to reduce the number of runs. (c) If $\alpha = k^{1/2}$, center points are mandatory to prevent the design from being singular.

For $k = 5$, for example, we use 5 (of the 11) columns of a 12-run Plackett and Burman (1946) design. There are $\binom{11}{5}$ that is, 462 possible choices, all of which produce nonsingular designs. These require 22 runs, the same number as Westlake's. A detailed examination of the cube portions for the designs shows that there are two basic types. Sixty-six choices of five columns produce a design with a duplicate run that can then be deleted to provide a 21-run design (which beats Westlake's designs by one run). The remaining 396 designs have one pair of runs that are "opposite"—that is, are mirror images, or foldover runs. An example of a duplicate-run design is the one formed by columns $(1, 2, 3, 9, 11)$ of the Plackett and Burman design; runs 3 and 11 are duplicates. Columns $(1, 2, 3, 7, 11)$ produce a design in which runs 2 and 9 are mirror images. [The column numbers are those obtained if the design matrix is written down by (a) following the formula suggested by Plackett and Burman (1946) and (b) cyclically permuting rows by moving the signs to the left. The consequent column numbering is different from that used by Box et al. (1978, p. 398), for which the columns transformation $(1, 2, 3, \ldots, 11) \rightarrow (1, 11, 10, \ldots, 2)$ should be applied. This notational difference arises because Box, Hunter, and Hunter use the Plackett and Burman first row as a column, and they cycle down rather than up.]

There are, as we have said, 462 choices of five columns from the 12-run Plackett and Burman design, but only two basic types. These two essentially different cube portions, are shown in Table 15.13. All others are derivable from these by switching signs in one or more columns and then rearranging rows and columns.

Table 15.13. Two Essentially Different Choices of Five Columns from a 12-Run Plackett and Burman Design

(a) With a Pair of Repeat Runs					(b) With a Mirror-Image Pair of Runs				
−	−	−	−	−	−	−	−	−	−
−	−	−	−	−	+	+	+	+	+
−	−	+	+	+	−	−	+	+	+
−	+	−	+	+	−	+	+	−	+
−	+	+	−	+	+	−	−	+	+
−	+	+	+	−	+	+	−	+	−
+	−	−	+	+	+	+	+	−	−
+	−	+	−	+	+	−	+	−	−
+	−	+	+	−	+	−	−	−	+
+	+	−	−	+	−	+	−	+	−
+	+	−	+	−	−	+	−	−	+
+	+	+	−	−	−	−	+	+	−

Note: All other choices are equivalent to one of these, subject to changes in signs throughout one or more columns, renaming of variables and reordering runs.

For additional details and for the $k = 7$ and 9 cases, see Draper (1985b). For higher dimensions, see the references near the end of Section 5.9, p. 156.

Koshal Designs

A polynomial of degree d in k variables contains $p = \{(k + d)!\}/(d!\,k!)$ coefficients. An example of a design that contains exactly this number of runs for the case $d = 2$ was given by Koshal (1933). He wished to determine the maximum of a likelihood function which could be calculated numerically at any given value, but was difficult to handle analytically. Such designs are occasionally of use in response surface work, particularly in applications like Koshal's; see, for example, Kanemasu (1979). Koshal's design is readily extended to any order d and to any number of predictor variables k.

First-Order Koshal Designs ($d = 1$)
These designs are the well-known "one factor at a time" designs with $N = k + 1$ runs. Thus for $k = 2$, the three constants in a first-order model can be determined using the design

x_1	x_2	
0	0	$d = 1$
1	0	$k = 2$
0	1	$N = p = 3$

where 0, 1 denote the two levels selected for each variable. The generalization to any number of variables k is immediate. For example, the $k = 4$ Koshal first-order design is

x_1	x_2	x_3	x_4	
0	0	0	0	
1	0	0	0	$d = 1$
0	1	0	0	$k = 4$
0	0	1	0	$N = p = 5$
0	0	0	1	

Second-Order Koshal Designs ($d = 2$)
Consider, for $k = 3$ variables, the alternative designs of Table 15.14. Such designs are obviously readily generalized for any k. For designs of type (a), the numbers included in the rows of the design are essentially those of the subscripts of the coefficients of a polynomial of second degree or, equivalently, relate to the partial derivatives entering a Taylor expansion of second order.

It is easy to produce modifications of Koshal designs for various purposes. Design (b) has the possible advantage of providing a less asymmetric pattern than (a). Other variations include the switching of signs in the "interaction" rows.

Table 15.14. Second-Order ($d = 2$) Koshal Designs for $k = 3$ Predictor Variables

(a)			(b)			
x_1	x_2	x_3	x_1	x_2	x_3	
0	0	0	0	0	0	
1	0	0	1	0	0	
0	1	0	0	1	0	
0	0	1	0	0	1	$d = 2$
2	0	0	-1	0	0	$k = 3$
0	2	0	0	-1	0	$N = p = 10$
0	0	2	0	0	-1	
1	1	0	1	1	0	
1	0	1	1	0	1	
0	1	1	0	1	1	

Table 15.15. Third-Order ($d = 3$) Koshal Design for $k = 3$ Predictor Variables

x_1	x_2	x_3	x_1	x_2	x_3	
0	0	0	3	0	0	
1	0	0	0	3	0	
0	1	0	0	0	3	
0	0	1	1	2	0	$d = 3$
2	0	0	2	1	0	$k = 3$
0	2	0	1	0	2	$N = p = 20$
0	0	2	2	0	1	
1	1	0	0	1	2	
1	0	1	0	2	1	
0	1	1	1	1	1	

Higher-Order Koshal Designs

The generalization to higher-order designs is also straightforward. For illustration, consider a third-order design in three variables with $N = p = 20$ runs, shown in Table 15.15. Again, obvious substitutions and sign switchings are possible (for example, to achieve greater symmetry) but we will not pursue these modifications further here.

EXERCISES

15.1. Consider, in $k = 2$ dimensions, the following design: "cube" ($\pm 1, \pm 1$), plus "star" ($\pm \alpha, 0$), ($0, \pm \alpha$), plus four center points. Find the variance function appropriate for the second-order model $y = \beta_0 + \beta_1 x_1 + \beta_2 x_2 +$

$\beta_{11}x_1^2 + \beta_{22}x_2^2 + \beta_{12}x_1x_2 + \varepsilon$, and hence derive the value of α for which the design is second-order rotatable. Show that, in this case, the variance two units from the origin is $9\sigma^2/4$, where σ^2 is the variance of the individual observations.

15.2. (*Source:* Agnes M. Herzberg.) Consider the design in $k = 2$ variables which consists of the nine points $(x_1, x_2) = (0, \pm a), (\pm b, \pm c), (0, 0)$ three times, where $a = 3^{1/2}$, $b = 1.5$, and $c = 3^{1/2}/2$.

    (**a**) Is it possible to estimate a full second-order model in x_1 and x_2 using this design? If so, is the design second-order rotatable?

    (**b**) Evaluate and plot the variance function $NV\{\hat{y}(\mathbf{x})\}/\sigma^2$.

    (**c**) If the true response function were a third degree polynomial, what biases would be induced in the expected values of the second-order estimated coefficients?

15.3. Consider a rotatable design for $k = 3$ consisting of cube, plus star, plus n_0 center points. Find $V(r) = V[\hat{y}(\mathbf{x})]$ as a function of $r = (x_1^2 + \cdots + x_k^2)^{1/2}$, and plot $V(r)$ against r for $n_0 = 2(2)10$. How many center points are needed to achieve $V(0) = V(1)$, approximately? [*Note:* In working this question, keep the cube points fixed at $(\pm 1, \pm 1, \pm 1)$ and *do not* scale so that $\sum_{u=1}^{N} x_{iu}^2 = N$.] How many center points will achieve $V(0) = V(\lambda_2^{1/2})$, approximately, where $\lambda_2 N = \sum_{u=1}^{N} x_{iu}^2$? How many center points will achieve $V(0) = V(k^{1/2})$, approximately?

15.4. Consider the four points of a 2^{3-1} design given by $(1, 1, 1)$, $(1, -1, -1)$, $(-1, 1, -1)$, $(-1, -1, 1)$. If we "add the points in pairs" and attach a multiplier $\frac{1}{2}\alpha$, we get six points: $(\alpha, 0, 0)$ from $\frac{1}{2}\alpha\{(1, 1, 1) + (1, -1, -1)\}$, $(0, \alpha, 0), (0, 0, \alpha), (0, 0, -\alpha), (0, -\alpha, 0), (-\alpha, 0, 0)$, namely the six points of a "star." If we "add the points in threes" with a multiplier of unity, we get the other 2^{3-1} points $(-1, -1, -1), (-1, 1, 1), (1, -1, 1), (1, 1, -1)$. Addition of all four points with any multiplier gives the center point $(0, 0, 0)$. thus we obtain the point locations of a "cube plus star plus center points" composite design for $k = 3$ factors. This is the essence of the "simplex-sum" method of design construction due to G. E. P. Box and D. W. Behnken and described in Simplex-sum designs: A class of second order rotatable designs derivable from those of first order, *Annals of Mathematical Statistics*, **31**, 1960, 838–864.

    Apply this same method to the eight points of a 2^{7-4} design and show that it leads to the 56-point rotatable design in seven factors given by G. E. P. Box and D. W. Behnken in Some new three level designs for the study of quantitative variables, *Technometrics*, **2**, 1960, 455–475 and shown in Table 15.9. [*Note:* In the example above, all the point sets generated were used. Here you will need only the pairs, the 6's, and the sum of all eight, which provides the center points. Note also the mirror image folded property of the r's and the $(N - r)$'s.]

15.5. (*Source:* Mark Thornquist.) Consider the 12-point design below. Confirm that the points of this design form an icosahedron, a regular figure in three dimensions. Confirm also that all the point s lie on a sphere of radius unity, so that center points would be needed to enable inversion of the $\mathbf{X'X}$ matrix, if such a design were used.

$$(1,0,0)$$
$$(a,b,0)$$
$$(a,b\cos\theta,b\sin\theta)$$
$$(a,b\cos2\theta,b\sin2\theta)$$
$$(a,b\cos3\theta,b\sin3\theta)$$
$$(a,b\cos4\theta,b\sin4\theta)$$
$$(-a,-b,0)$$
$$(-a,-b\cos\theta,-b\sin\theta)$$
$$(-a,-b\cos2\theta,-b\sin2\theta)$$
$$(-a,-b\cos3\theta,-b\sin3\theta)$$
$$(-a,-b\cos4\theta,-b\sin4\theta)$$
$$(-1,0,0)$$

$$\theta=\frac{2\pi}{5}$$

$$a=\frac{\cos\theta}{(1+\cos\theta)}$$

$$b=(1-a^2)^{1/2}$$

[Note that the design consists of a single points at $(x_1,x_2,x_3)=(-1,0,0)$, a circle of five equally spaced points at $x_1=-a$, a similar circle but rotated an angle of $\frac{1}{2}\theta$ at $x_1=a$, and a single point at $(1,0,0)$. These points are scaled and rotated forms of the usual representation $(\pm p,\pm q,0)$, $(0,\pm p,\pm q)$, $(\pm q,0,\pm p)$ where $p^2-pq-q^2=0$.]

15.6. Consider, for $k=3$, the cube plus star plus center points composite response surface design. Suppose that the design is split into three blocks.

2^{3-1} ($\mathbf{I}=\mathbf{123}$) plus $\frac{1}{2}c_0$ center points,
2^{3-1} ($\mathbf{I}=-\mathbf{123}$) plus $\frac{1}{2}c_0$ center points,
Star plus s_0 center points.

Confirm that, if $c_0=4$, $s_0=2$, and $\alpha^2=\frac{8}{3}$, the design is orthogonally blocked. Find the formula for values of α^2 that will give an orthogonally blocked design for other values of c_0 and s_0, and tabulate the numerical values of α^2 for $0\le c_0\le 8$, $0\le s_0\le 5$. Is it possible for the design to be rotatable and orthogonally blocked?

15.7. Consider, for $k=4$, the cube plus star plus center points composite response surface design. Suppose that the design is split into three blocks

2^{4-1} ($\mathbf{I}=\mathbf{1234}$) plus $\frac{1}{2}c_0$ center points,
2^{4-1} ($\mathbf{I}=-\mathbf{1234}$) plus $\frac{1}{2}c_0$ center points,
Star plus s_0 center points.

Confirm that if $c_0 = 4$, $x_0 = 2$, and $\alpha = 2$, the design is orthogonally blocked *and* rotatable. Find the formula for values of α^2 that will give an orthogonally blocked design for other values of c_0 and s_0, and tabulate the numerical values of α^2 for $0 \le c_0 \le 8$, $0 \le s_0 \le 5$.

15.8. Consider a composite design for $2 \le k \le 8$ consisting of a cube (2^{k-p} points with $p = 0$ for $k = 2, 3$, and 4; $p = 1$ for $k = 5, 6, 7$, and $p = 2$ for $k = 8$), a replicated star (each of the two with axial distance α), and center points. Divide the cube up into b blocks where $b = 1$ for $k = 2$ and $k = 5$ ($p = 1$), $b = 2$ for $k = 3, 4, 6$ ($p = 1$), $b = 8$ for $k = 7$ ($p = 1$), and $b = 4$ for $k = 8$ ($p = 2$). Each star is itself another block. Consider the cases where there are $A = n_{c0}/b = 0$ or 1 or 2 center points in *each* of the b cube blocks (i.e., no more than 2). Of these 18 situations, which allow a choice of α for which the design can be *both* rotatable *and* orthogonally blocked? Tabulate the α and $B = \frac{1}{2}n_{s0}$ (= number of center points in each star block) values for each case in which it can be achieved.

15.9. Using the method described in Section 15.2, write down regular simplex designs for $k = 4$, 5, and 6, scaling the variables so that $\sum_{u=1}^{k+1} x_{iu}^2 = N = k + 1$, $i = 1, 2, \ldots, k$. Show that the length of an edge is $\sqrt{10}$, $\sqrt{12}$, and $\sqrt{14}$ units respectively in the three cases. Show more generally that, for the general regular simplex design for k variables in $k + 1 = N$ runs obtained

Table E15.10. Data from an Automobile-Door-Closing Study

Run	x_1	x_2	x_3	x_4	y
1	−0.500	−2.000	−2.000	−1.000	−0.500
2	2.182	0.000	0.000	0.000	2.182
3	1.500	2.000	2.000	−1.000	1.500
4	0.500	0.000	0.000	0.000	0.500
5	−1.182	0.000	0.000	0.000	−1.182
6	0.500	0.000	3.364	0.000	0.500
7	0.500	0.000	0.000	0.000	0.500
8	1.500	−2.000	−2.000	1.000	1.500
9	−0.500	−2.000	2.000	−1.000	−0.500
10	0.500	0.000	0.000	−1.682	0.500
11	1.500	2.000	−2.000	−1.000	1.500
12	−0.500	2.000	−2.000	1.000	−0.500
13	0.500	−3.364	0.000	0.000	0.500
14	0.500	0.000	0.000	1.682	0.500
15	1.500	−2.000	2.000	1.000	1.500
16	0.500	0.000	−3.364	0.000	0.500
17	−0.500	2.000	2.000	1.000	−0.500
18	0.500	3.364	0.000	0.000	0.500

by this method, the length of an edge is $(2Nc)^{1/2}$ where

$$\sum_{u=1}^{N} x_{iu}^2 = Nc, \qquad i = 1, 2, \ldots, k.$$

15.10. (*Source:* Response surface designed experiment for door closing effort, by R. E. Chapman and K. Masinda, *Quality Engineering*, 2003, **15**(4), 581–585.) Table E15.10 shows results from a "randomized Draper–Lin small-composite design" conducted for General Motors. Fit a full second-order model to the four factors shown and find the path of minimum predicted response (the effort required to close a door).

CHAPTER 16

Response Surfaces for Mixture Ingredients

16.1. MIXTURE SPACES

In fitting response surfaces so far, we have made no assumptions about links between the x's. We now consider the problem of fitting a response surface when there are q x's and

$$x_1 + x_2 + \cdots + x_q = 1. \tag{16.1.1}$$

This type of problem arises when experiments are performed on a mixture of ingredients. Some examples are (a) blends of bread flours consisting of wheat and various additives; (b) fertilizers consisting of blends of chemicals; (c) wines blended from several sources (and/or types, perhaps) of grapes; (d) paints and dyestuffs; (e) alloys; (f) polymer and fiber blends; (g) concrete; (h) drugs; and (i) gasoline containing additives used to improve performance. We explore what changes this causes to response surface analysis.

Two Ingredients

The simplest situation is when $q = 2$ and $x_1 + x_2 = 1$, for example, for a drink with two ingredients x_1 and x_2. Both x_1 and x_2 must lie in the range $0 \le x_i \le 1$, of course. The extremes of the experimental space occur at $x_1 = 1$, $x_2 = 0$ and $x_1 = 0$, $x_2 = 1$, namely all of one ingredient or the other. In general, x_1 and x_2 will be fractions adding to 1. Geometrically, all possible mixtures fall on the line $x_1 + x_2 = 1$ in the (x_1, x_2) space; the latter is a plane, as depicted in Figure 16.1. Because both x_1 and x_2 must lie between 0 and 1, only that part of the line that includes the points $(0, 1)$ and $(1, 0)$ and falls between them forms the actual mixture space.

Response Surfaces, Mixtures, and Ridge Analyses, Second Edition. By G. E. P. Box and N. R. Draper
Copyright © 2007 John Wiley & Sons, Inc.

509

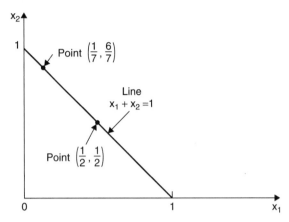

Figure 16.1. A one-dimensional experimental mixture space $x_1 + x_2 = 1$ embedded in the two-dimensional (x_1, x_2) space. Only the region for which $0 \le x_i \le 1$ is valid, as shown.

General Points

It is conventional to call the extreme points "mixtures" even though they contain only one ingredient. However, one should always question whether the response y is a continuous function at these points, or whether "something strange" happens, due to the fact that one ingredient has (or, in more general cases, several ingredients have) been omitted. In most applications, this is not a problem. The response variable y could be of many kinds. For a drink, for example, "taste," "color," or "strength" might be considered. Typically, the response will be continuous variable, capable of being represented via a fitted response equation.

By using fractional ingredients, we circumvent the problem of the size of the mixture batches. A drink, for example, could be 6 oz, 8 oz, or an amount sufficient for a party. Furthermore, if a more general restriction $c_1 t_1 + c_2 t_2 + \cdots + c_q t_q = 1$ in mixture ingredients $(t_1, t_2, \ldots, t_q)$ had initially been specified, we could simply set $c_i t_i = x_i$ and so convert to the form of Eq. (16.1.1).

Three Ingredients

Figure 16.2 shows the triangular mixture space that occurs when $q = 3$. The three "mixtures" $(x_1, x_2, x_3) = (1, 0, 0), (0, 1, 0), (0, 0, 1)$ define the extreme vertices, and they, the lines joining them in pairs, and the inside of the triangle they define form the reduced mixture space. (Outside these boundaries, at least one x_i is negative, which is not possible in a real mixture.) Note that the three sides of the triangle are all mixture spaces in two ingredients, with the third ingredient zero; compare with Figure 16.1. Each interior point has three coordinates, but needs only two coordinate values to specify it, because the third is $1 - $ (sum of the other two). Figure 16.3a shows coordinate lines $x_1 =$ constant "opposite" the $(1, 0, 0)$ vertex. The base line opposite $(1, 0, 0)$ is $x_1 = 0$, and, as x_1 increases, a series of parallel lines is defined, depicted only within the mixture space because we are not interested in the space outside the triangle. The coordinate system $x_2 =$ constant consists of lines parallel to the side $x_2 = 0$ opposite $(0, 1, 0)$. Figure 16.3b shows

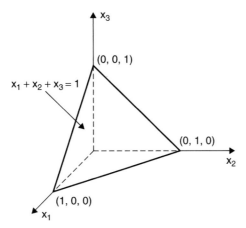

Figure 16.2. A two-dimensional experimental mixture space $x_1 + x_2 + x_3 = 1$ embedded in the three-dimensional (x_1, x_2, x_3) space. Only the region for which $0 \le x_i \le 1$ is valid, as shown.

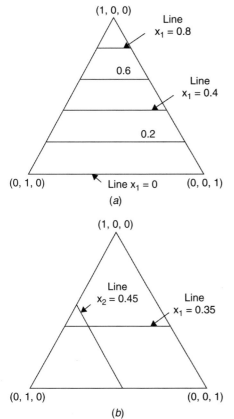

Figure 16.3. (*a*) Parallel coordinate lines $x_1 = 0, 0.2, 0.4, 0.6, 0.8$ in the mixture space $x_1 + x_2 + x_3 = 1$. (*b*) The points $(0.35, 0.45, 0.20)$ is defined by the intersection of the two coordinate lines $x_1 = 0.35$ and $x_2 = 0.45$. Note that $x_3 = 1 - x_1 - x_2 = 0.20$.

the unique point defined by the intersection of the lines $x_1 = 0.35$ and $x_2 = 0.45$. Naturally, $x_3 = 1 - x_1 - x_2 = 0.20$, so the point is $(x_1, x_2, x_3) = (0.35, 0.45, 0.20)$.

Four Ingredients

We cannot draw a four-dimensional space in (x_1, x_2, x_3, x_4), but we can extend mentally our mixture concept by realizing that the mixture space shown in Figure 16.2a can be regarded as the $x_4 = 0$ part of a three-dimensional (four ingredient) space. It follows that the $x_1 = 0$, $x_2 = 0$, and $x_3 = 0$ boundaries must all be of similar character and that the "corner" points $(1, 0, 0, 0)$, $(0, 1, 0, 0)$, $(0, 0, 1, 0)$ and $(0, 0, 0, 1)$ must be vertices of the four ingredient mixture space. Thus the mixture space must be a regular (equal-sided) tetrahedron; see Figure 16.4a. All points inside or on the boundaries are such that $x_1 + x_2 + x_3 + x_4 = 1$. The actual construction of such a space is a useful visualization for the beginner. Draw Figure 16.4b on a piece of paper; cut down to the triangle indicated by the 1's. Fold on the lines 23, 34, and 42 so that the 1's meet in a point defined as $(1, 0, 0, 0)$. The points 2, 3, and 4 are then $(0, 1, 0, 0)$, $(0, 0, 1, 0)$, and $(0, 0, 0, 1)$, respectively. Figure 16.4c shows another way of thinking about the tetrahedron. It can be built up as a series of triangular three-ingredient mixture spaces parallel to any selected base of

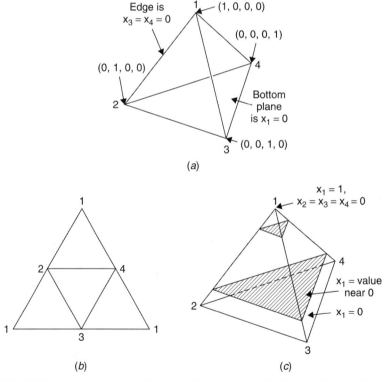

Figure 16.4. (a) The mixture space for four ingredients is an equal-sided tetrahedron (pyramid). (b) Construction of an equal-sided tetrahedron model. (c) Two slices $x_1 = c$ of the four-ingredient mixture space; for one, c is nearer 0, whereas for the other, c is nearer 1.

the tetrahedron. Figure 16.4c shows slices $x_1 = c$, where c ranges from 0 (at the base) to 1 at the 1 vertex. The two shaded slices are at about $x_1 = 0.15$ and $x_1 = 0.85$. The larger the c value, the smaller the slice.

Five Ingredients

The tetrahedron $x_1 + x_2 + x_3 + x_4 = 1$ of Figure 16.4a can be thought of as the $x_5 = 0$ section of the $x_1 + x_2 + x_3 + x_4 + x_5 = 1$ mixture space. When $x_5 = c$, a smaller tetrahedral section is defined with $x_1 + x_2 + x_3 + x_4 = 1 - c$. As c tends to 1, the tetrahedron tends to the single point $(0, 0, 0, 0, 1)$.

All of the mixture spaces we have discussed take the form of a regular simplex, that is, a figure formed by q points equally spaced from one another in a $(q - 1)$-dimensional space. When $q \geq 6$, we have to carry the mental images onward as best we can!

16.2. FIRST-ORDER MIXTURE MODELS

For convenience of writing, we frame our discussion in terms of $q = 3$ ingredients, with n observations of data on them as well as on a response variable y; the generalization to more ingredients will be obvious. We recall that the usual first-order model fitted *in a nonmixture case* is

$$y = \beta_0 + \beta_1 x_1 + \beta_2 x_2 + \beta_3 x_3 + \varepsilon. \tag{16.2.1}$$

This cannot be fitted to mixture data. The **X** matrix for such an (attempted) fit would consist of four vector columns $\mathbf{X} = [\mathbf{1}, \mathbf{x}_1, \mathbf{x}_2, \mathbf{x}_3]$. Because the ingredients add to 1, each element of **1** is equal to the sum of the corresponding elements in $\mathbf{x}_1$, $\mathbf{x}_2$, and $\mathbf{x}_3$. This linear relationship in the columns of **X** causes the **X'X** matrix to be singular. Geometrically, we would need data in a three-dimensional space to fit Eq. (16.2.1), and our mixture space is only two-dimensional (see Figure 16.2). What can we do then? One solution would be to transform the q (here 3) x-coordinates into $q - 1$ (here 2) z-coordinates and fit a model function of the form $\beta_0 + \theta_1 z_1 + \theta_2 z_2 + \cdots + \theta_{q-1} z_{q-1}$ (here $q = 3$). While technically possible, this is not recommended; however, for details of such an approach, see Draper and Smith (1998, Appendix 19A). A preferable solution is to transform the model into a *canonical form*, suggested by Scheffe (1958, 1963), which retains the symmetry in the x's.

Canonical Form of First Order Models

In Eq. (16.2.1) we can rewrite β_0 as $\beta_0(x_1 + x_2 + x_3)$, using the fact that the bracketed portion is equal to 1, to cast the model function in (16.2.1) into the form

$$(\beta_0 + \beta_1)x_1 + (\beta_0 + \beta_2)x_2 + (\beta_0 + \beta_3)x_3 \tag{16.2.2}$$

or

$$\alpha_1 x_1 + \alpha_2 x_2 + \alpha_3 x_3, \tag{16.2.3}$$

where $\alpha_i = \beta_0 + \beta_i$. This canonical form is appealing and popular in practice because of its symmetry in the x's. In fact, however, *any* of the models obtainable by deleting one of the model function terms in (16.2.1) would provide exactly the same fitted values and exactly the same response contours, even though the fitted models would look different.

For general q, the canonical form model function is, by obvious extension,

$$\eta = \beta_1 x_1 + \beta_2 x_2 + \cdots + \beta_q x_q, \tag{16.2.4}$$

reverting to the more usual β-notation, and where $y = \eta + \varepsilon$ is to be fitted by least squares.

Testing for Overall Regression

It would not make sense to test the null hypothesis H_0: all $\beta_i = 0$ in the present context, because such a hypothesis implies a zero response, which is very improbable. Instead, the null hypothesis H_0: $\beta_1 = \beta_2 = \cdots = \beta_q \ (= \beta_0$ say) would be appropriate. The reduced model if H_0 were true would then be $y = \beta_0 + \varepsilon$ because the x's sum to 1. The appropriate F-statistic for the test is

$$F = \left\{ SS(b_1, b_2, \ldots, b_q) - n\bar{y}^2 \right\} / \left\{ (q-1)s^2 \right\} \tag{16.2.5}$$

where s^2 is the residual mean square from the original fit.

Example

Consider the seven observations in Table 16.1. A least-squares fitting of model (16.2.4) with $q = 3$ provides the equation

$$\hat{y} = 10.65 x_1 + 14.17 x_2 + 2.37 x_3. \tag{16.2.6}$$

The total SS (uncorrected because there is no constant term being fitted) is 667.24 (7 df). Subtracting the regression SS (again uncorrected) of 665.97 (3 df) leaves a residual SS of 1.27 (4 df), so that $s^2 = 0.32$. The correction factor $n\bar{y}^2 = 574.22$ is the regression SS when the null hypothesis H_0: $\beta_1 = \beta_2 = \beta_3 \ (= \beta_0$ say) is applied. Thus, to test this H_0, we evaluate $F = \{(665.97 - 574.22)/(3 - 1)\}/0.32 = 143$; this value is highly significant, and so we reject H_0, implying that nonequal component terms β_i are needed for these data. The fitted equation gives rise to the parallel line contour diagram of Figure 16.5.

Table 16.1. Example Data for First-Order Fit

x_1	x_2	x_3	y
1	0	0	10.9
0	1	0	14.4
0	0	1	2.0
0.5	0.5	0	11.6
0.5	0	0.5	6.9
0	0.5	0.5	8.7
0.333	0.333	0.333	8.9

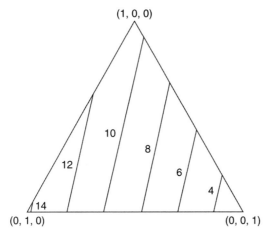

Figure 16.5. Contours of Eq. (16.2.6).

A Second Example

The data of Table 16.2 are given by Snee (1973c, p. 519), and arise from a gasoline blending study whose ingredients $(x_1, x_2, x_3, x_4) =$ (light FCC, alkylate, butane, reformate) yield the values of $y =$ research octane $- 100 = r - 100$ shown. (We subtract 100 from all responses so that the regression analysis is numerically more sensitive to the variation in the response.) A first-order least-squares fit leads to the equation

$$\hat{y} = 3.162x_1 + 4.537x_2 + 3.984x_3 - 2.088x_4. \tag{16.2.7}$$

Of the total SS of 20.790 (9 df), an amount 20.692 (4 df) is picked up by this model and the residual mean square is thus $(20.790 - 20.692)/(9 - 4) = 0.196$. Three of the response contours represented by Eq. (16.2.7) are shown in Figure 16.6. The contours are a series of parallel planes (but they are *not* parallel to the $x_4 = 0$ base of the tetrahedron). For the predicted values of the original response, $\hat{r}$, add 100 to the $\hat{y}$ values shown in Figure 16.6.

The similarity of the first three parameter estimates takes the eye, and Snee suggested considering the null hypothesis H_0: $\beta_1 = \beta_2 = \beta_3$ $(= \beta$, say). The

Table 16.2. Data for a Second Example of First-Order Mixtures Fit

x_1	x_2	x_3	x_4	y	$x_1 + x_2 + x_3$
0.25	0.40	0.100	0.250	2.6	0.750
0.25	0.40	0.030	0.320	2.0	0.680
0.25	0.20	0.100	0.450	1.2	0.550
0.15	0.40	0.030	0.420	1.7	0.580
0.25	0.20	0.030	0.520	0.7	0.480
0.15	0.20	0.100	0.550	0.7	0.450
0.15	0.40	0.100	0.350	1.8	0.650
0.15	0.20	0.030	0.620	0.2	0.380
0.20	0.30	0.065	0.435	1.2	0.565

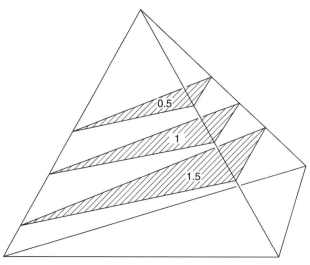

Figure 16.6. Contours of Eq. (16.2.7). The planes are parallel to one another (but not to the $x_4 = 0$ base of the tetrahedron).

reduced model fit is then

$$\hat{y} = 4.096(x_1 + x_2 + x_3) - 2.229x_4, \qquad (16.2.8)$$

and the regression SS is now 20.661 with a residual mean square of $(20.790 - 20.661)/(9 - 2) = 0.0184$. The test statistic for H_0 is $F = (20.692 - 20.661)/(4 - 2) = 0.0155$ and is clearly nonsignificant, indicating that the reduced model is sensible. The contour planes of Eq. (16.2.8) (not drawn) *are* now all parallel to the base; in other respects the situation is similar to that of Figure 16.6. Snee (1973c, pp. 519–520) makes several important points in the following quote (reproduced by permission of the American Statistical Association): "... both equations should be presented to the investigator. It is his decision which equation should be used. Within the region of these data and the octane rating error, the first three components have the same blending behavior. This does not imply, nor is there evidence to support, that these components have the same blending behavior in general. We should keep in mind that the estimated coefficients, called blending values, are the predicted octanes of the pure components [$x_i = 1$, other x's 0] which are far from the nine blends in the study. Hence, we would not expect these blending values to be good approximations to the octanes of the pure components. [More experimental runs would be needed to investigate this.]"

16.3. SECOND-ORDER MIXTURE MODELS

We again choose $q = 3$ ingredients for our initial discussion. The usual second-order response surface model is

$$y = \beta_0 + \beta_1 x_1 + \beta_2 x_2 + \beta_3 x_3 + \beta_{11} x_1^2 + \beta_{22} x_2^2 + \beta_{33} x_3^2$$
$$+ \beta_{12} x_1 x_2 + \beta_{13} x_1 x_3 + \beta_{23} x_2 x_3 + \varepsilon, \qquad (16.3.1)$$

but this (overparameterized for mixtures) model leads to a singular $X'X$ matrix, as in the first-order situation. We again replace β_0 with $\beta_0(x_1 + x_2 + x_3)$, but more replacement than this is needed. Because $x_1 = 1 - x_2 - x_3$, we can multiply by x_1 throughout to give $x_1^2 = x_1 - x_1x_2 - x_1x_3$ and can use this to replace the x_1^2 term in Eq. (16.3.1). Similar substitutions are used for the other two pure quadratic terms. Gathering like terms and renaming coefficients enables us to derive (what we shall call) the S-model (for Scheffé model) canonical form of the model function

$$y = \beta_1 x_1 + \beta_2 x_2 + \beta_3 x_3 + \beta_{12} x_1 x_2 + \beta_{13} x_1 x_3 + \beta_{23} x_2 x_3 + \varepsilon, \quad (16.3.2)$$

where, once again, we revert immediately to the β-notation rather than using α's as an intermediate notation. As in the first-order case, the canonical form is symmetric in all the x's. In general, there are $q(q + 1)/2$ terms in such a model, consisting of q first-order terms and $q(q - 1)/2$ cross-product terms. When $q = 3$, we have $6 = 3 + 3$ terms.

There is also, however, another symmetrical model form that could be used, obtained by alternative substitutions. To obtain this directly from the preceding expression, we replace x_1 by $x_1(x_1 + x_2 + x_3) = x_1^2 + x_1x_2 + x_1x_3$ and use similar expressions for x_2 and x_3. The resulting form, again reverting to the β-notation, is then

$$y = \beta_{11} x_1^2 + \beta_{22} x_2^2 + \beta_{33} x_3^2 + \beta_{12} x_1 x_2 + \beta_{13} x_1 x_3 + \beta_{23} x_2 x_3 + \varepsilon. \quad (16.3.3)$$

Notice that this model consists only of second-order (quadratic) terms! We call this the K-model (Kronecker model) form. The reason for this name is that the Kronecker product $x' \otimes x' = (x_1, x_2, x_3)' \otimes (x_1, x_2, x_3)' = (x_1^2, x_1x_2, x_1x_3; x_2x_1, x_2^2, x_2x_3; x_3x_1, x_3x_2, x_3^2)$ generates all six of the terms in (16.3.3), some of them more than once. (See Appendix 3C for a fuller definition of Kronecker product.)

Example of a Second Order Fit

Consider the data in Table 16.3, simplified and adapted from a larger example given by Busch and Phelen (1999). The three ingredients, x_1, x_2, x_3, are such that

Table 16.3. Data from a $q = 3$ Factor Mixture Experiment

x_1	x_2	x_3	y	e_Q	e_C
1	0	0	40.95	-1.57	-2.33
0	1	0	30.28	-0.06	-0.82
0	0	1	8.25	3.49	2.75
0.5	0.5	0	43.47	-8.03	-3.25
0.5	0	0.5	28.47	-4.42	0.44
0	0.5	0.5	23.98	-2.87	2.00
0.33	0.33	0.34	45.82	5.26	-1.27
0.66	0.17	0.17	55.55	9.53	7.61
0.17	0.66	0.17	44.93	4.86	2.94
0.17	0.17	0.66	19.41	-6.19	-8.07

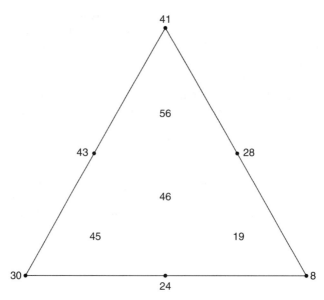

Figure 16.7. A 10-point mixture design for $q = 3$ ingredients; the y-values of Table 16.3 are shown here rounded to integers.

$x_1 + x_2 + x_3 = 1$. The data consist of 10 points (10-mixture ingredient combinations) placed in a symmetrical pattern over the mixture space (see Figure 16.7).

Fitting the model (16.3.2) to these data by least squares provides the fitted equation

$$\hat{y} = 42.523x_1 + 30.337x_2 + 4.755x_3 + 60.30x_1x_2 + 37.02x_1x_3 + 37.20x_2x_3.$$

$$(16.3.4)$$

The associated analysis of variance breakup of the total SS (13,517.6, with 10 df) is as follows.

Source	SS	df	MS	F
b_1, b_2, b_3	12,908.3	3	4,302.8	59.9
Extra for b_{12}, b_{13}, b_{23}	322.1	3	107.4	1.5
Residual	287.2	4	71.8	

This indicates a strong linear component accompanied by a modest second-order curvature, as shown in Figure 16.8a. We shall soon compare these contours with those of an alternative fit, the special cubic, below. The residuals from the quadratic fit (the e_Q values) are listed in Table 16.3.

Appendix 16A contains further examples of second-order (quadratic) response contours in $q = 3$ ingredients. Those diagrams have been taken from Cornell and Ott (1975, pp. 413–415) but are, themselves, adapted versions of more detailed diagrams in Gorman and Hinman (1962, pp. 475–476).

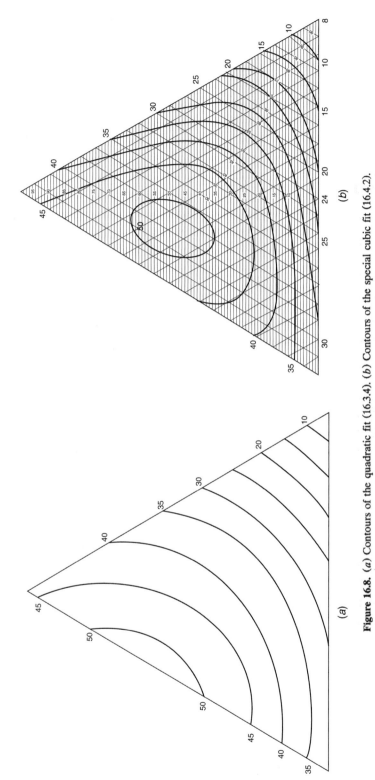

Figure 16.8. (*a*) Contours of the quadratic fit (16.3.4). (*b*) Contours of the special cubic fit (16.4.2).

Points to Note

Regression calculations on mixture data can often be quite sensitive to rounding differences. For example, if x-values such as $(0.33, 0.33, 0.34)$ and $(0.66, 0.17, 0.17)$ are instead entered as $(0.333, 0.333, 0.333)$ and $(0.667, 0.167, 0.167)$, somewhat different coefficients and calculations arise. In general, such differences are usually numerically unimportant, but they can be confusing when results are compared with calculations from other sources. Our x-values shown are those used by the original authors of the material, Busch and Phelan (1999); our y-values and residuals are rounded to two decimal places.

The residuals add to zero even though there is no constant term in the model as fitted. This is because the model *can* be rewritten with a constant term present, if desired. It will be necessary to remove another term (for example, $\beta_1 x_1$); otherwise, singularity will occur in the **X** matrix. Such alternative models give rise to exactly the same fitted values and residuals. Because the residuals sum to zero when the constant term is present, they also must sum to zero when the constant term is removed by a valid re-substitution.

16.4. THIRD-ORDER CANONICAL MIXTURE MODELS

Parallel to the development for first and second order canonical mixture models, we could derive the third-order general canonical cubic which contains q first order terms x_i, $q(q-1)/2$ second-order terms $x_i x_j$, $q(q-1)/2$ cubic terms of the form $x_i x_j (x_i - x_j)$ containing only two x's, and $q(q-1)(q-2)/6$ cubic terms of form $x_i x_j x_k$, containing three x's. The total number of terms is thus $q(q+1)(q+2)/6$. For $q = 4$, for example, the canonical general cubic has 20 terms, as follows:

$$
\begin{aligned}
y = {} & \beta_1 x_1 + \beta_2 x_2 + \beta_3 x_3 + \beta_4 x_4 \\
& + \beta_{12} x_1 x_2 + \beta_{13} x_1 x_3 + \beta_{14} x_1 x_4 + \beta_{23} x_2 x_3 + \beta_{24} x_2 x_4 + \beta_{34} x_3 x_4 \\
& + \alpha_{12} x_1 x_2 (x_1 - x_2) + \alpha_{13} x_1 x_3 (x_1 - x_3) + \alpha_{14} x_1 x_4 (x_1 - x_4) \\
& + \alpha_{23} x_2 x_3 (x_2 - x_3) + \alpha_{24} x_2 x_4 (x_2 - x_4) + \alpha_{34} x_3 x_4 (x_3 - x_4) \\
& + \beta_{123} x_1 x_2 x_3 + \beta_{124} x_1 x_2 x_4 + \beta_{134} x_1 x_3 x_4 + \beta_{234} x_2 x_3 x_4 + \varepsilon, \quad (16.4.1)
\end{aligned}
$$

Third-order models are usually fitted only when a quadratic model appears to be inadequate (and provided there are enough data points to carry out the fit). The full cubic requires $q(q^2 - 1)/6$ more terms than the full quadratic ($q = 3$, four more; $q = 4$, ten more; $q = 5$, twenty more; and so on).

Special Cubic

A reduced version of the full cubic model (16.4.1), usually called the *special cubic*, is obtained by dropping all of the $q(q+1)/2$ α_{ij} terms from the model that is the general form of Eq. (16.4.1), leaving only $q(q^2 + 5)/6$ in general; for $q = 4$, 14 terms remain, for example. The special cubic is typically fitted when it provides an adequate model, in circumstances where the additional α_{ij} terms contribute little. It requires $q(q-1)(q-2)/6$ more terms than the quadratic. For $q = 3$, this is only one more, namely $\beta_{123}x_1x_2x_3$; for $q = 4$, four more; for $q = 5$, ten more; and so on. Two example diagrams of special cubic fits for $q = 3$ ingredients are shown in Appendix 16B. The two cubic equations illustrated have identical first- and second-order terms, but one has a positive and one a negative coefficient attached to the cubic term $x_1x_2x_3$. The diagrams are adapted from those in Gorman and Hinman (1962, p. 477). If we fit (via least squares) the special cubic to the data of Table 16.3, we obtain

$$\hat{y} = 43.276x_1 + 31.096x_2 + 5.500x_3 + 38.12x_1x_2 + 14.56x_1x_3$$

$$+ 14.74x_2x_3 + 357.7x_1x_2x_3. \tag{16.4.2}$$

The addition of the cubic term removes a 1 df sequential sum of squares of size 125.7 from the previous residual SS of 287.2 (4 df). This leaves a new residual SS of $287.2 - 125.7 = 161.5$ (3 df), so that the new $s^2 = 53.83$. The new F-values are then 79.9, 2.00, and 2.34 for first-order, second-order, and third-order terms, respectively. Comparing Eqs. (16.3.4) and (16.4.2), we notice that the first-order coefficients are more or less unchanged, but that introduction of the special cubic term has reduced the coefficients of the three cross-product terms considerably. The residuals from this special cubic fit appear in Table 16.3 under the heading e_C, and the fitted contours are drawn in Figure 16.8b. Although the quadratic and cubic surfaces differ in their details (the quadratic indicates a maximum response on the $x_2 = 0$ boundary, while the cubic, which is slightly more flexible, having one more term available to it, indicates a maximum inside the boundary) the general indications are very similar over the region for both fits. We see from the residuals in Table 16.3 that each model fits better at some locations and worse at others. The major feature of both diagrams is the increase in response as x_2 decreases. It is important that the experimenters see both diagrams in order to form their own assessment and to consider whether additional experimental runs are needed and, if so, where.

Kronecker Forms of the Full Cubic and of Full Polynomials of Higher Order

The alternative Kronecker form of the full cubic contains all distinct x-terms which can be obtained from the triple Kronecker product $\mathbf{x}' \otimes \mathbf{x}' \otimes \mathbf{x}' = (x_1, x_2, \ldots, x_q)'$ $\otimes (x_1, x_2, \ldots, x_q)' \otimes (x_1, x_2, \ldots, x_q)'$. Again, some terms occur more than once (see Draper and Pukelsheim, 1998b). The Kronecker pattern continues for even higher order. For example, all possible fourth-order (quartic) terms are generated from the product $\mathbf{x}' \otimes \mathbf{x}' \otimes \mathbf{x}' \otimes \mathbf{x}'$. This "Kronecker system generation" thus provides a foolproof, albeit tedious, method for obtaining a set of all possible model terms needed for any given order in mixture problems. This set of terms can also be recast in different forms by substitutions using the mixture restriction.

16.5. DESIGN OF EXPERIMENTS FOR CANONICAL POLYNOMIAL MODELS

Now that we have discussed how to fit polynomial models suitable for mixture ingredient regression problems, given the data, we turn to a discussion of some popular types of experimental designs that have been suggested for obtaining the data needed to fit the models.

We have already mentioned that Scheffé (1958, 1963) is responsible for introducing the canonical polynomials. He also suggested some designs that can be used with them. Scheffé's initial ideas were concerned with making it easy to obtain the least-squares estimates by solving some simple equations. Although this aspect is no longer necessary due to computing advances, the designs he suggested are still excellent ones, and they also provide the basic building blocks for larger designs. For the canonical polynomial models, Scheffé suggested *simplex lattice* designs. A $\{q, m\}$ simplex lattice design consists of all valid design points formable by using, for each of the q factors $x_1, x_2, \ldots, x_q$, the $(m + 1)$ equally spaced levels

$$0, \frac{1}{m}, \frac{2}{m}, \ldots, \frac{m-1}{m}, 1. \tag{16.5.1}$$

Naturally, all the points to be formed must satisfy the mixture restriction $x_1 + x_2 + \cdots + x_q = 1$. When $m = 1$, the levels are only 0 and 1, and so the $\{q, 1\}$ designs consist of all vertices of the mixture space. When $m = 2$, the possible levels are 0, 0.5, and 1. Thus we have:

$\{2, 2\}$ simplex lattice (3 points): $(1, 0), (0, 1), (0.5, 0.5)$,

$\{3, 2\}$ simplex lattice (6 points): $(1, 0, 0), (0, 1, 0), (0, 0, 1), (0.5, 0.5, 0)$,

$(0.5, 0, 0.5), (0, 0.5, 0.5)$.

(These six points form part of the design of Table 16.3.) Sometimes the centroid $(\frac{1}{3}, \frac{1}{3}, \frac{1}{3})$ is added to this lattice, also. (This seventh point is also in Table E16.3, with rounded decimals.)

$\{4, 2\}$ simplex lattice (10 points): $(1, 0, 0, 0), (0, 1, 0, 0), (0, 0, 1, 0), (0, 0, 0, 1)$,

$(0.5, 0.5, 0, 0), (0.5, 0, 0.5, 0), (0.5, 0, 0, 0.5)$,

$(0, 0.5, 0.5, 0), (0, 0.5, 0, 0.5), (0, 0, 0.5, 0.5)$.

Sometimes the centroid $(0.25, 0.25, 0.25, 0.25)$ is added to this design.

When $m = 3$, the possible levels are $0, \frac{1}{3}, \frac{2}{3}, 1$. Thus we can form these designs:

$\{2, 3\}$ lattice (4 points): $(1, 0), (0, 1), (\frac{1}{3}, \frac{2}{3}), (\frac{2}{3}, \frac{1}{3})$.

$\{3, 3\}$ lattice (10 points): $(1, 0, 0), (0, 1, 0), (0, 0, 1)$,

$(\frac{1}{3}, \frac{2}{3}, 0), (\frac{2}{3}, \frac{1}{3}, 0), (\frac{1}{3}, 0, \frac{2}{3}), (\frac{2}{3}, 0\frac{1}{3}), (0, \frac{1}{3}, \frac{2}{3})$,

$(0, \frac{2}{3}, \frac{1}{3}), (\frac{1}{3}, \frac{1}{3}, \frac{1}{3})$.

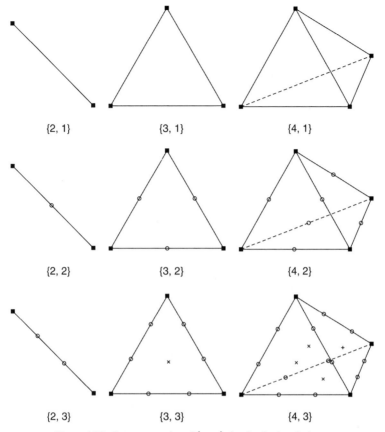

Figure 16.9. Some examples of $\{q, m\}$ simplex lattice designs.

The centroid $(\frac{1}{3}, \frac{1}{3}, \frac{1}{3})$ is already included in this design.

$\{4, 3\}$ lattice (20 points): $(1, 0, 0, 0), (0, 1, 0, 0), (0, 0, 1, 0), (0, 0, 0, 1),$

$$\left(\tfrac{1}{3}, \tfrac{2}{3}, 0, 0\right), \left(\tfrac{2}{3}, \tfrac{1}{3}, 0, 0\right), \left(\tfrac{1}{3}, 0, \tfrac{2}{3}, 0\right), \left(\tfrac{2}{3}, 0, \tfrac{1}{3}, 0\right),$$

$$\left(\tfrac{1}{3}, 0, 0, \tfrac{2}{3}\right), \left(\tfrac{2}{3}, 0, 0, \tfrac{1}{3}\right), \left(0, \tfrac{1}{3}, \tfrac{2}{3}, 0\right), \left(0, \tfrac{2}{3}, \tfrac{1}{3}, 0\right),$$

$$\left(0, \tfrac{1}{3}, 0, \tfrac{2}{3}\right), \left(0, \tfrac{2}{3}, 0, \tfrac{1}{3}\right), \left(0, 0, \tfrac{1}{3}, \tfrac{2}{3}\right), \left(0, 0, \tfrac{2}{3}, \tfrac{1}{3}\right),$$

$$\left(\tfrac{1}{3}, \tfrac{1}{3}, \tfrac{1}{3}, 0\right), \left(\tfrac{1}{3}, \tfrac{1}{3}, 0, \tfrac{1}{3}\right), \left(\tfrac{1}{3}, 0, \tfrac{1}{3}, \tfrac{1}{3}\right), \left(0, \tfrac{1}{3}, \tfrac{1}{3}, \tfrac{1}{3}\right).$$

Sometimes the centroid $(0.25, 0.25, 0.25, 0.25)$ is added to this design.

Some the above designs are pictured in Figure 16.9. A centroid is shown only in the $\{3, 3\}$ case, where it occurs in the basic design. In general, the $\{q, m\}$ lattice contains

$$\frac{(q + m - 1)!}{m! \, (q - 1)!} \tag{16.5.2}$$

design points, while the full canonical polynomials contain q points for a first-order model, $q(q + 1)/2$ for a second order model, $q(q + 1)(q + 2)/6$ for a full third-order model, and so on. It will be found that setting $m = 1, 2$, and 3 in (16.5.2) gives precisely these numbers. In other words, the $\{q, m\}$ simplex lattice provides just enough design points for an exact fit to the mth-order polynomial. Of course, in practice, additional points would be desirable to provide some extra degrees of freedom to check the fit of the model.

Design Points on the Axes

We now see that the design of Table 16.3 and Figure 16.7 is a $\{3, 2\}$ simplex lattice supplemented by a centroid and three symmetrically placed internal points. These three specific internal points are on the *axes* of the mixture space. An axis is defined as the line joining a vertex to the centroid of the opposite face. Thus, for example, the point $(1, 0, \ldots, 0)$ is joined to $(0, \frac{1}{q-1}, \frac{1}{q-1}, \ldots, \frac{1}{q-1})$ in the general q case, and so on for other vertices. A general point on this axis has the form $(\{1 - (q-1)r\}, r, r, \ldots, r)$ where r ranges from 0, at the vertex, to $\frac{1}{q-1}$ at the centroid of the opposite face. Other axes have similar respresentations, with appropriate permutation of x-coordinates. Note that the centroid $(\frac{1}{q}, \frac{1}{q}, \ldots, \frac{1}{q})$ lies on all of the axes, at the location where $r = \frac{1}{q}$.

If an experimenter wished to add axial points to, for example, the $\{3, 2\}$ lattice design where should they be added? The centroid $(\frac{1}{3}, \frac{1}{3}, \frac{1}{3})$ is an obvious choice, followed by a triangle of points, one on each axis, between the vertex and the centroid. A natural choice would be exactly halfway between. This would provide the three points $(\frac{2}{3}, \frac{1}{6}, \frac{1}{6})$, $(\frac{1}{6}, \frac{2}{3}, \frac{1}{6})$, $(\frac{1}{6}, \frac{1}{6}, \frac{2}{3})$. The attractiveness of this design is illustrated in Figure 16.10, where joining up the three design points of the $\{3, 2\}$ simplex lattice that lie in the middle of each of the three sides of the region gives four triangles of identical shape with the "middle one" inverted. The centroid and the three suggested axial points are the centroids of the four triangles. For

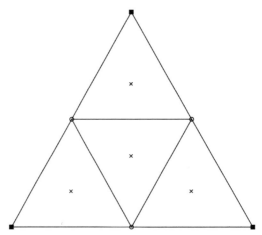

Figure 16.10. The $\{3, 2\}$ simplex lattice design with four extra points, showing symmetrical coverage of the experimental region.

example, the point $(\frac{2}{3}, \frac{1}{6}, \frac{1}{6})$ is the average of the points $(1, 0, 0)$, $(\frac{1}{2}, \frac{1}{2}, 0)$, and $(\frac{1}{2}, 0, \frac{1}{2})$. The mixture region is thus covered very well by these 10 points. Of course, other choices of r might be appropriate in particular cases. A similar situation arises when $q = 4$. Consider the four points $(1, 0, 0, 0)$, $(\frac{1}{2}, \frac{1}{2}, 0, 0)$, $(\frac{1}{2}, 0, \frac{1}{2}, 0)$, and $(\frac{1}{2}, 0, 0, \frac{1}{2})$; their average is $(\frac{5}{8}, \frac{1}{8}, \frac{1}{8}, \frac{1}{8})$, which lies midway between the vertex $(1, 0, 0, 0)$ and the centroid $(\frac{1}{4}, \frac{1}{4}, \frac{1}{4}, \frac{1}{4})$. The reader may wonder whether a similar result applies for general q. See Exercise 16.9.

16.6. SIMPLEX CENTROID DESIGNS AND THE RELATED POLYNOMIALS

The simplex centroid designs of Scheffé (1963) were suggested initially to provide data that would exactly support a particular type of polynomial for q mixture ingredients, namely:

$$
\begin{aligned}
\eta = {} & \beta_1 x_1 + \beta_2 x_2 + \cdots + \beta_q x_q \\
& + \beta_{12} x_1 x_2 + \beta_{13} x_1 x_3 + \cdots + \beta_{q-1,q} x_{q-1} x_q \\
& + \beta_{123} x_1 x_2 x_3 + \beta_{124} x_1 x_2 x_4 + \cdots + \beta_{q-2,q-1,q} x_{q-2} x_{q-1} x_q \\
& + \cdots + \beta_{12\ldots q} x_1 x_2 \ldots x_q.
\end{aligned}
\tag{16.6.1}
$$

In addition to the first-order terms, this polynomial contains all possible product terms in the x's (pairs, threes, fours, etc.) up to and including order q. In all, there are $2^q - 1$ such terms. (This number is obvious once we realize that the product $(1 + x_1)(1 + x_2)\ldots(1 + x_q)$ produces 2^q different terms consisting of all the x-combination terms in Eq. (16.6.1) plus the unneeded single term 1.)

The q-ingredient design suggested by Scheffé (1963) for exactly fitting this polynomial using the same number of design points as coefficients consists of this collection of points:

q points $(1, 0, \ldots, 0), (0, 1, 0, \ldots, 0), \ldots, (0, 0, \ldots, 0, 1)$; that is, all permutations
  of points containing a single 1;

$\binom{q}{2}$ points $(\frac{1}{2}, \frac{1}{2}, 0, \ldots, 0), (\frac{1}{2}, 0, \frac{1}{2}, 0, \ldots, 0), \ldots, (0, 0, \ldots, 0, \frac{1}{2}, \frac{1}{2})$; again, all permutations of this type are used;

$\binom{q}{3}$ points of form $(\frac{1}{3}, \frac{1}{3}, \frac{1}{3}, 0, \ldots, 0)$ plus all other such permutations of three
  thirds; . . . and so on, down to

$\binom{q}{q} = 1$ point $(\frac{1}{q}, \frac{1}{q}, \ldots, \frac{1}{q})$, the overall centroid.

These $2^q - 1$ design points are the vertices and the centroids of all two-dimensional, three-dimensional, . . . , q-dimensional simplexes found in the original mixture space.

For $q = 3$, the seven points are the same as those of the $\{3, 2\}$ simplex lattice plus the centroid $(\frac{1}{3}, \frac{1}{3}, \frac{1}{3})$, and the model (16.6.1) *is* the special cubic. For $q = 4$, the 15 points are the vertices, the midpoints of the edges, the four centroids of the triangular faces, and the overall centroid. Equivalently, we can say that they are the 10 points of the {4, 2} simplex lattice design (see Figure 16.9) plus the four face centroids and the one overall centroid.

In practice, the model of Eq. (16.6.1) is used less frequently than the corresponding canonical polynomials of the type of Eq. (16.3.2) or (16.4.1). Nevertheless, it sometimes provides a useful alternative. Moreover, the designs for model (16.6.1), which has $2^q - 1$ points, can be used with the special cubic, which has $q(q^2 + 5)/6$ coefficients. For example,

> when $q = 4$, there are 15 design points for 14 coefficients; the special cubic lacks only the single $x_1 x_2 x_3 x_4$ term.
>
> when $q = 5$, there are 31 design points for 25 coefficients;
>
> when $q = 6$, there are 63 design points for 41 coefficients;
>
> when $q = 7$, there are 127 design points for 63 coefficients; and so on.

APPENDIX 16A. EXAMPLES OF SECOND-ORDER RESPONSE CONTOURS FOR $q = 3$ MIXTURE INGREDIENTS

The response functions plotted are:

(a) $\hat{y} = 90x_1 + 95x_2 + 100x_3 + 27x_1x_2 + 27x_1x_3 + 27x_2x_3.$
(b) $\hat{y} = 90x_1 + 95x_2 + 100x_3 + 27x_1x_2 + 0x_1x_3 - 27x_2x_3.$

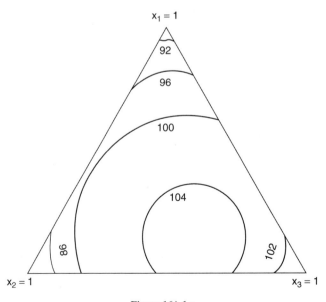

Figure 16A.1*a*

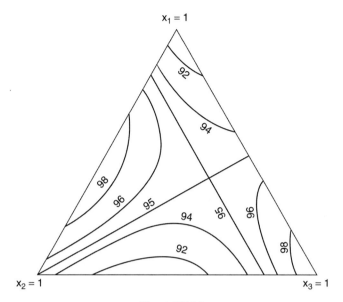

Figure 16A.1*b*

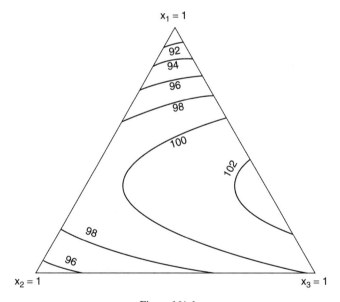

Figure 16A.1*c*

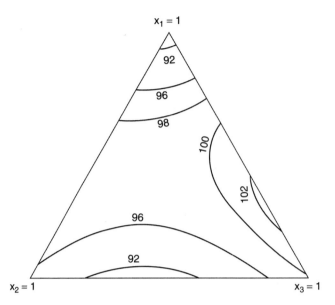

Figure 16A.1*d*

(c) $\hat{y} = 90x_1 + 95x_2 + 100x_3 + 27x_1x_2 + 27x_1x_3 + 0x_2x_3$.

(d) $\hat{y} = 90x_1 + 95x_2 + 100x_3 + 27x_1x_2 + 27x_1x_3 - 27x_2x_3$.

Figures 16A.1*a*–16A.1*d* are from Cornell and Ott (1975, pp. 413–415) but are themselves adapted versions of more detailed diagrams in Gorman and Hinman (1962, pp. 475–476). Their use is by permission of the copyright holder, the American Statistical Association. All rights reserved.

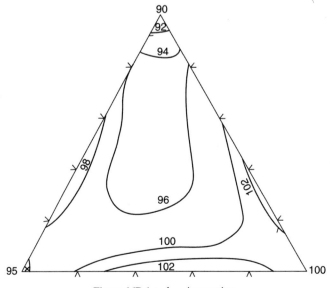

Figure 16B.1*a***.** b_{123} is negative.

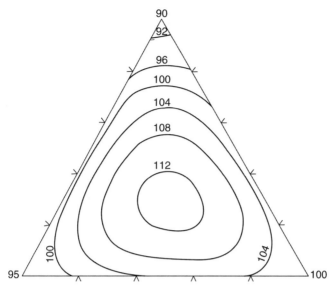

Figure 16B.1b. b_{123} is positive.

APPENDIX 16B. EXAMPLES OF SPECIAL CUBIC RESPONSE CONTOURS FOR $q = 3$ MIXTURE INGREDIENTS

The response functions plotted are:

(a) $\hat{y} = 90x_1 + 95x_2 + 100x_3 + 27x_1x_2 + 27x_1x_3 + 27x_2x_3 - 243x_1x_2x_3$.

(b) $\hat{y} = 90x_1 + 95x_2 + 100x_3 + 27x_1x_2 + 27x_1x_3 + 27x_2x_3 + 243x_1x_2x_3$.

Figures 16B.1a and 16B.1b are adapted versions of more detailed diagrams in Gorman and Hinman (1962, p. 477) used by permission of the copyright holder, the American Statistical Association. All rights reserved.

APPENDIX 16C. EXAMPLES OF A GENERAL CUBIC RESPONSE SURFACE AND A QUARTIC RESPONSE SURFACE FOR $q = 3$ MIXTURE INGREDIENTS

The response functions plotted are:

(a) $\hat{y} = 54.91x_1 + 3.89x_2 + 9.87x_3 - 44.56x_1x_2 - 28.70x_1x_3 + 29.49x_2x_3$
$\quad - 14.33x_1x_2(x_1 - x_2) + 5.13x_1x_3(x_1 - x_3) + 44.64x_2x_3(x_2 - x_3)$
$\quad - 55.54x_1x_2x_3$.

(b) $\hat{y} = 95.03x_1 + 94.97x_2 + 94.80x_3 - 2.00x_1x_2 + 20.47x_1x_3 + 7.93x_2x_3$
$\quad - 0.89x_1x_2(x_1 - x_2) - 4.00x_1x_3(x_1 - x_3) - 10.58x_2x_3(x_2 - x_3)$
$\quad - 6.22x_1x_2(x_1 - x_2)^2 - 31.38x_1x_3(x_1 - x_3)^2 - 18.22x_2x_3(x_2 - x_3)^2$
$\quad - 127.56x_1^2x_2x_3 - 27.64x_1x_2^2x_3 - 17.96x_1x_2x_3^2$

Figures 16C.1a and 16C.1b are taken from Gorman and Hinman (1962, p. 478) by permission of the copyright holder, the American Statistical Association. All rights reserved.

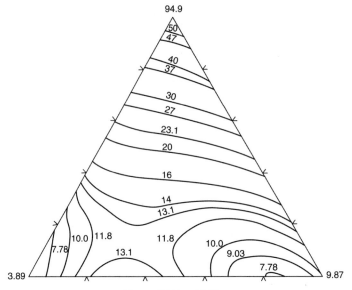

Figure 16C.1a. Cubic.

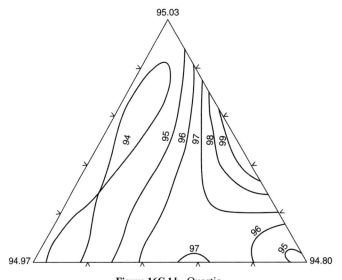

Figure 16C.1b. Quartic.

EXERCISES

16.1 Plot, on triangular graph paper for the space $x_1 + x_1 + x_1 = 1$, the following mixtures of three ingredients: $(x_1, x_2, x_3) = (\frac{1}{3}, \frac{1}{3}, \frac{1}{3})$, $(1, 0, 0)$, $(0.4, 0.4, 0.2)$, $(0.4, 0.2, 0.4)$, $(0.2, 0.4, 0.4)$, $(0.3, 0.3, 0.3)$, $(0.8, 0, 0.2)$, $(0, 0.6, 0.4)$, $(0.5, 0, 0.5)$. Which of these points cannot be plotted and why?

16.2. **(a)** Scheffé's simplex lattice design uses $m + 1$ equally spaced levels of each ingredient. Write down those levels.

(b) Suppose there are three $(k = 3)$ mixture ingredients. Does this mean that there may be $(m + 1)^3$ possible experimental points? Say why or why not.

(c) Using the notation $\{k, m\}$ for the k-ingredient, $m + 1$ level, simplex lattice, plot the points for the $\{3, 2\}$ and $\{3, 3\}$ cases and specify the points for the $\{4, 2\}$ and $\{4, 3\}$ cases.

16.3. Consider the $\{3, 2\}$ simplex lattice design plus centroid, plus three axial points $(1 - 2r, r, r), (r, 1 - 2r, r), (r, r, 1 - 2r)$. Suppose it was decided to choose r so that the sum of squares of distances of the point $(1 - 2r, r, r)$ from the four points $(1, 0, 0)$, $(\frac{1}{2}, \frac{1}{2}, 0)$, $(\frac{1}{2}, 0, \frac{1}{2})$, and $(\frac{1}{3}, \frac{1}{3}, \frac{1}{3})$ was minimized? What value of r achieves this?

16.4. The foregoing exercise can be extended to q dimensions. What value of r minimizes the sum of squares of the distances of the point $(\{1 - \{q - 1\}r\}, r, r, \ldots, r)$ from the $q + 1$ points $(1, 0, \ldots, 0)$, $(\frac{1}{2}, \frac{1}{2}, 0, \ldots, 0)$, $(\frac{1}{2}, 0, \frac{1}{2}, 0, \ldots, 0), \ldots, (\frac{1}{2}, 0, \ldots, 0, \frac{1}{2})$, $(\frac{1}{q}, \frac{1}{q}, \ldots, \frac{1}{q})$?

16.5. Use the data in Table E16.5 to fit a second-order mixture polynomial; it will fit exactly with no df left over for error, of course. Predict the response at $(\frac{1}{3}, \frac{1}{3}, \frac{1}{3})$. Test whether the true response there could be 12, given that an estimate $s^2 = 1$, based on 20 df, is available for the variance of individual observations. [You will need to evaluate the standard error of your prediction at $(\frac{1}{3}, \frac{1}{3}, \frac{1}{3})$ and perform a t-test.]

16.6. Fit a third-order polynomial to the data in Table E16.6. It will fit exactly with no df over for error estimation.

Table E16.5

x_1	x_2	x_3	y
1	0	0	6
0	1	0	12
0	0	1	6
0.5	0.5	0	8
0.5	0	0.5	7
0	0.5	0.5	9

Table E16.6

x_1	x_2	x_3	y
1	0	0	6
0	1	0	12
0	0	1	6
0.333	0.667	0	9
0.667	0.333	0	7
0.333	0	0.667	8
0.667	0	0.333	8
0	0.333	0.667	8
0	0.667	0.333	10
0.333	0.333	0.333	8

16.7. (*Source:* Shrinkage in ternary mixes of container media, by S. Bures, F. A. Pokorny, and G. O. Ware, a paper published in the *Proceedings of the* 1991 *Kansas State University Conference on Applied Statistics in Agriculture*, 43–52, published in 1992.) Plant growth in containers is often studied using an artificial medium rather than natural soil, and the characteristics of the ingredients of such media are often of interest. Here, three mixture ingredients—sand (x_1), pine bark (x_2) and calcined clay (x_3), with different particle size ranges—were mixed together in various individual volume proportions with an initially unit total volume of 1. When the mixture is shaken down, however, some volume shrinkage may occur because smaller particles can occupy space between larger particles. The results in Table E16.7 are 12 of the 66 design points and observations presented in the original paper; the shrinkage numbers (y) shown are proportions adapted from the source paper by rounding to two decimal places. Plot the design points on triangular paper and comment on the design selected here. Fit both a quadratic model and a special cubic model. Is adding the cubic term worthwhile? Choose the model that appears appropriate and plot the contours. What are your conclusions?

Table E16.7. Shrinkage y for Mixtures of Growth Medium Ingredients

x_1	x_2	x_3	y
1	0	0	0.00
0	1	0	0.00
0	0	1	0.00
0.5	0.5	0	0.08
0.5	0	0.5	0.11
0	0.5	0.5	0.02
0.2	0.4	0.4	0.10
0.4	0.2	0.4	0.11
0.4	0.4	0.2	0.11
0.6	0.2	0.2	0.08
0.2	0.6	0.2	0.07
0.2	0.2	0.6	0.11

16.8. Four ingredients are to be investigated in a mixture experiment. Experimenter A suggests the use of a replicated simplex-centroid design. Experimenter B suggests doing, in succession, a simplex centroid for every subset of three ingredients (with the fourth ingredient set to zero) plus a final pair of runs at $(0.25, 0.25, 0.25, 0.25)$. Compare the two designs. Which one is better?

16.9. This exercise generalizes the results at the end of Section 16.5. Find the coordinates of the centroid, C_1, say, of the q-ingredient "corner simplex" defined by the q points $(1, 0, \ldots, 0)$, $(\frac{1}{2}, \frac{1}{2}, 0, 0, \ldots, 0)$, $(\frac{1}{2}, 0, \frac{1}{2}, 0, 0, \ldots, 0)$, $(\frac{1}{2}, 0, 0, \frac{1}{2}, 0, 0, \ldots, 0)$, $\ldots$, $(\frac{1}{2}, 0, 0, \ldots, 0, \frac{1}{2})$. Does C_1 lie exactly in the middle of the axial segment between the two points $(1, 0, \ldots, 0)$ and the overall centroid $(\frac{1}{q}, \frac{1}{q}, \frac{1}{q}, \ldots, \frac{1}{q})$? Your answer will apply to all the other corner simplexes also, of course, with appropriate permutation of coordinate values.

CHAPTER 17

Mixture Experiments in Restricted Regions

17.1. INTRODUCTION

In discussing mixture experiments with q ingredients in the foregoing chapter, we assumed that experiments were feasible anywhere in the mixture space. This is often not the case in practice. Suppose we ask for coffee with cream in a restaurant. We would be astonished if the waitperson poured half a cup of coffee and then expected us to add cream. This is because it is generally accepted that the cream should be a relatively small amount of the total, perhaps 10% or less; in other words, $0 \leq x_1 \leq 0.10$, say, with the 0.10 being an upper bound. In other cases a lower bound may be invoked. For example, a law may specify that a "fruit drink" must contain at least (say) 10% of the juice prominently displayed on the label. Sometimes there are both lower *and* upper bounds, typically in experiments performed by investigators whose experience tells them that the space outside their bounds is of no practical interest.

Whenever bounds are specified on one or more ingredients, the mixture space becomes a restricted region whose shape depends on the specific restrictions. In choosing experimental runs to be performed, one must ensure that all experimental runs are within the restricted space and that sensible coverage of the region is obtained.

Restrictive bounds cannot be chosen arbitrarily; they need to be specified in such a way that the region actually exists. If, for $q = 3$, we specify $x_1 \geq 0.5$, $x_2 \geq 0.30$, and $x_3 \geq 0.25$, no points in the triangular mixture space satisfy these conditions. (*Exercise:* Draw the three lower bound lines on a diagram such as that of Figure 16.3b and see this for yourself.) In what follows, we deal successively with the cases of (1) only lower bounds, (2) only upper bounds, and (3) both lower and upper bounds.

Response Surfaces, Mixtures, and Ridge Analyses, Second Edition. By G. E. P. Box and N. R. Draper

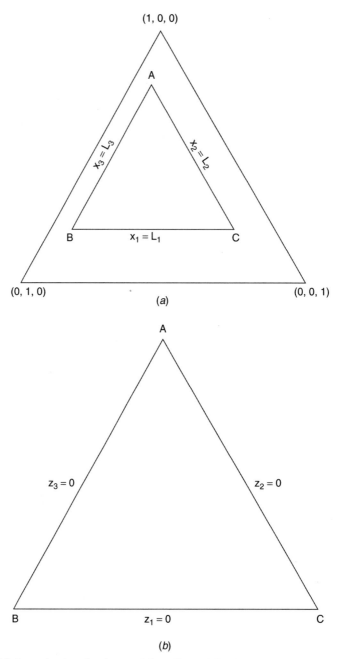

Figure 17.1. (*a*) A restricted region for $q = 3$ ingredients defined by three lower bounds L_1, L_2, and L_3. (*b*) In pseudocomponents, the restricted region can be treated mathematically as though it were the full mixture space.

17.2. LOWER BOUNDS AND PSEUDOCOMPONENTS

Suppose we have $q = 3$ ingredients and there are three lower bounds L_1, L_2, and L_3 such that $x_1 \geq L_1$, $x_2 \geq L_2$, $x_3 \geq L_3$. Let $L = L_1 + L_2 + L_3$. Then we must have $L \leq 1$, or the restricted region does not exist because the restrictions would be mutually incompatible. If $L = 1$, the restricted region consists of one point only, namely (L_1, L_2, L_3), so we specify that L must be less than 1, in general. The restricted region is shown in Figure 17.1a. Note that it is exactly the same shape as the original full region and lies totally within it. The vertex A, which lies at the intersection of the lines $x_2 = L_2$ and $x_3 = L_3$, has coordinates $(1 - L_2 - L_3, L_2, L_3)$. Similarly, we have B at $(L_1, 1 - L_1 - L_3, L_3)$ and C at $(L_1, L_2, 1 - L_1 - L_2)$. In choosing design points, we can apply the same principles discussed in Section 16.5 and can use standard designs to place points at the corners, midpoints of the sides, and so on, of the restricted region, if we wish. To facilitate the application of these design choice principles, we usually convert the coordinates $(x_1, x_2, \ldots, x_q)$ to so-called pseudocoordinates, or pseudocomponents $(z_1, z_2, \ldots, z_q)$, as follows. For the $q = 3$ case,

$$z_1 = \frac{x_1 - L_1}{1 - L}, \qquad z_2 = \frac{x_2 - L_2}{1 - L}, \qquad z_3 = \frac{x_3 - L_3}{1 - L}. \qquad (17.2.1)$$

Equivalently, we can unravel these to the forms

$$x_1 = L_1 + (1 - L)z_1, \qquad x_2 = L_2 + (1 - L)z_2, \qquad x_3 = L_3 + (1 - L)z_3, \qquad (17.2.2)$$

in order to move from the z-metric to the x-metric. We see immediately that when $x_i = L_i$ we have $z_i = 0$, and that $z_1 + z_2 + z_3 = (x_1 + x_2 + x_3 - L_1 - L_2 - L_3)/(1 - L) = 1$, so our restricted region is just an ordinary mixture region, when described in the pseudocomponents z_i; (see Figure 17.1b). The point A now becomes $(z_1, z_2, z_3) = (1, 0, 0)$, B is $(0, 1, 0)$ and C is $(0, 0, 1)$. The extension of pseudocomponents to mixtures with other values of q is done in exactly similar fashion.

Kurotori's Example, $q = 3$

The use of pseudocomponents was introduced by Kurotori (1966), and we now describe the classic example in his paper. In involved the study of a propellant mixture in which the lower bounds of the three ingredients $(x_1, x_2, x_3) = $ (binder, oxidizer, fuel) were $L_1 = 0.20$, $L_2 = 0.40$, and $L_3 = 0.20$, so that $L = 0.80$. The response variable was "modulus of elasticity," the values of which we divided by 1000 for convenience in this representation; the objective was to maximize this response, subject to the restrictions.

Table 17.1 shows the data points (x_1, x_2, x_3) used in the example and the consequent response values denoted by y. The corresponding pseudocomponent coordinates are (z_1, z_2, z_3), where $z_1 = (x_1 - 0.2)/0.2$, $z_2 = (x_2 - 0.4)/0.2$, and

Table 17.1. Data from Kurotori (1966) Together with Pseudocomponent Levels

z_1	z_2	z_3	y	x_1	x_2	x_3
1	0	0	2.35	0.4	0.4	0.2
0	1	0	2.45	0.2	0.6	0.2
0	0	1	2.65	0.2	0.4	0.4
0.5	0.5	0	2.40	0.3	0.5	0.2
0.5	0	0.5	2.75	0.3	0.4	0.3
0	0.5	0.5	2.95	0.2	0.5	0.3
0.33333	0.33333	0.33333	3.00	0.26667	0.46667	0.26667
0.66667	0.16667	0.16667	2.69	0.33333	0.43333	0.23333
0.16667	0.66667	0.16667	2.77	0.23333	0.53333	0.23333
0.16667	0.16667	0.66667	2.98	0.23333	0.43333	0.33333

$z_3 = (x_3 - 0.2)/0.2$, obtained by applying (17.2.1), also appear in the table. The design used is a familiar one, a $\{3, 2\}$ simplex lattice, plus centroid, plus three axial points each standing midway between a vertex and the centroid. (The same design appeared in Section 16.5, and a general form of it was discussed in Exercise 16.9.) In the present case, the three axial points were regarded as extra checkpoints attached to the $\{3, 2\}$ simplex lattice plus centroid design.

Fit of a Quadratic Model

We shall first fit a canonical second-order (quadratic) polynomial to these data. The question arises as to whether it should be fitted in terms of the x's or the z's. It makes little difference and depends on how we want to report our results. The two equations we shall show are, in fact, identical in terms of the contours they represent, and they can be derived from one another by substitution from Eqs. (17.2.1) or (17.2.2), as appropriate. In terms of the z's, we get

$$\hat{y} = 2.339z_1 + 2.433z_2 + 2.640z_3 + 0.373z_1z_2 + 1.387z_1z_3 + 1.976z_2z_3,$$

$$(17.2.3)$$

producing the contours within the z-space shown in Figure 17.2(a). This fit concentrates attention entirely within the restricted region and suggests that an optimum response might be obtained at a point on the $x_1 = 0.20$ boundary. Fitting in terms of the x's, we get instead

$$\hat{y} = -2.732x_1 - 3.341x_2 - 17.261x_3 + 9.32x_1x_2 + 34.69x_1x_3 + 49.42x_2x_3.$$

$$(17.2.4)$$

The contours that give predicted response values of interest (namely, larger values) via this equation are drawn in the full mixture space in Figure 17.2*b*. The curves are identical to those in Figure 17.2*a*, of course, but they are extended outside the restricted region. They indicate that slightly higher response values *might* be obtainable if the lower bound for x_1 could be relaxed. The contours (17.2.3) could also be extended in an identical manner, but note that this would involve substitut-

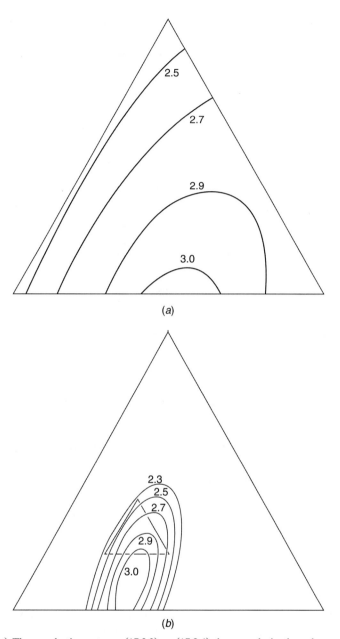

Figure 17.2. (*a*) The quadratic contours (17.2.3) or (17.2.4) drawn only in the relevant space of the pseudocomponents z_j. (*b*) The quadratic contours (17.2.3) or (17.2.4) drawn in the larger mixture space, with the restricted space indicated as the inner triangle.

ing negative values for certain of the z_i, because we would be venturing outside the pseudocomponent simplex.

Whichever equation is fitted, the same basic analysis of variance table is obtained:

Source	SS	df	MS	F
First-order terms	73.025	3	24.342	2617.
Second-order first-order	0.300	3	0.100	10.75
Residual	0.037	4	$s^2 = 0.0093$	
Total	73.362	10		

If we were to break up the 3 df SS for first-order terms into sequential components, we would get (35.661, 21.210, 16.157) for the z fit and get (67.745, 4.077, 1.203) for the x fit; both sets add to 73.025. The cross-product sequential sums of squares are (0.006, 0.096, 0.198), totaling 0.300 in either case. This results from the fact that, for example,

$$z_1 z_2 = (x_1 - L_1)(x_2 - L_2)/(1 - L)^2$$

$$= \{x_1 x_2 - L_1 x_2 - L_2 x_1 + L_1 L_2 (x_1 + x_2 + x_3)\}/(1 - L)^2$$

$$= \{x_1 x_2 - L_1 (1 - L_2) x_2 - L_2 (1 - L_1) x_1 + L_1 L_2 x_3\}/(1 - L)^2.$$

Thus, apart from the scale factor $(1 - L)^2$, which cancels out in the SS calculations, the cross-products survive intact (i.e., $z_1 z_2 \rightarrow x_1 x_2$) but the relationship between z_1 and x_1, for example, is adjusted by extra terms in converting from the z model to the x model.

The first-order terms explain most of the variation in the response; there is, however, a numerically small but statistically significant second-order effect, which provides the curvature in Figure 17.2.

Fit of a Special Cubic

Enough data are available to enable the special cubic model to be fitted by least squares. The two equations obtained are the following:

$$\hat{y} = 2.351 z_1 + 2.446 z_2 + 2.653 z_3 - 0.0063 z_1 z_2 + 1.008 z_1 z_3$$

$$+ 1.597 z_2 z_3 + 6.141 z_1 z_2 z_3, \tag{17.2.5}$$

$$\hat{y} = 50.200 x_1 + 16.991 x_2 + 3.673 x_3 - 153.681 x_1 x_2 - 281.84 x_1 x_3$$

$$- 113.590 x_2 x_3 - 767.62 x_1 x_2 x_3. \tag{17.2.6}$$

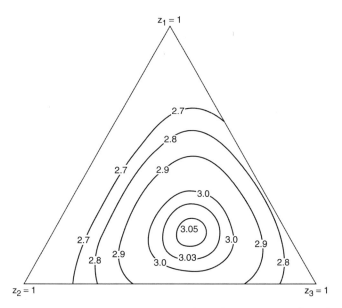

Figure 17.3. The contours of the special cubic equation (17.2.3) or equivalently (17.2.4) are drawn only within the restricted region.

The summary analysis of variance table is the same for both regressions, namely:

Source	df	SS	MS	F	p
Regression	7	73.362	10.480	99572.8	0.000
Error	3	0.000	0.0000		
Total	10	73.362			

The residual standard deviation is actually $s = 0.0103$, derived from a residual sum of squares that rounds to 0.000 at three decimal places. The fit is very close to being an exact one.

These two equations produce the (again identical) contours shown in Figure 17.3, drawn this time only within the restricted region. We see that this equation now predicts a maximum *within* the restricted region. Neither the quadratic nor the cubic can be called "correct." They provide two slightly different interpretations of the data, and the situation must be resolved by further experiments to check whether the predicted best yields can be obtained and, if so, where. In fact, Kurotori (1966, Figure 7, p. 596) concludes that the "desired mixture area" lies in the region where the cubic's $\hat{y} = 3.00$ contour comes nearest to the bottom side of the triangle of restricted space, a position that compromises nicely between the quadratic and the cubic fits.

Kurotori also fitted the special cubic to the data, but in a slightly different manner. He used the data from the {3, 2} simplex lattice plus the observation at the centroid (seven observations in all) to fit the special cubic (seven coefficients). The fit is exact because there are no residual degrees of freedom in such a fit. He then evaluated the predicted values at the three axial point locations and compared them to the responses actually observed at these positions. The differences are very small, as is clear from our fit to all the data, above, where the residual SS is essentially zero.

17.3. UPPER BOUNDS AND PSEUDOCOMPONENTS

Mixture problems whose ingredients have only upper bounds $x_1 \leq U_1, x_2 \leq U_2, \ldots, x_q \leq U_q$, for selected U_i such that $0 \leq U_i \leq 1$, offer more difficulty than those with only lower bounds. The reasons for this are that the restricted experimental region is often not a simplex and, even if it is, it is inverted with respect to the original full simplex region. Figure 17.4a shows, for $q = 3$, a case where the region available for experimentation *is* a simplex (but note the inversion). Figure 17.4b shows an example of the perhaps more typical cases where it is not a simplex; or, more accurately, it is a simplex, parts of which lie outside the original mixture simplex, where experiments cannot be performed.

We follow the path of Crosier (1984), who discusses this situation in detail, and we talk about the "U-simplex" formed by the upper bounds and determine whether it lies fully in the mixture space or not. First, we define $U = U_1 + U_2 + \cdots + U_q$. For $q = 3$, the three upper bounds define the three vertices of the U-simplex (an inverted triangle) as

$$(U_1, U_2, 1 - U_1 - U_2), \quad (U_1, 1 - U_1 - U_3, U_3), \quad \text{and} \quad (1 - U_2 - U_3, U_2, U_3). \tag{17.3.1}$$

The $q = 3$ U-simplex lies entirely within the original mixture space if and only if all its coordinates are positive. This requires that $1 \geq 1 - U_i - U_j \geq 0$ for any choice of i, j, $i \neq j$; subtracting the third upper bound U_k from each side leads to $1 - U \geq -U_k$ or $U - U_k \leq 1$ for $k = 1, 2,$ or 3. A similar sort of argument applies for any q. The general result can be stated concisely as

$$U - U_{\min} \leq 1, \tag{17.3.2}$$

where $U = U_1 + U_2 + \cdots + U_q$, and $U_{\min}$ is the smallest of the q upper bounds. The left-hand side provides the highest value that can be attained and, if (17.3.2) holds, replacement of $U_{\min}$ with any U_i also holds.

For upper bounds $U_1, U_2, \ldots, U_q$, we can use a U-pseudocomponent transformation of the form

$$z_1 = \frac{U_1 - x_1}{U - 1}, \qquad z_2 = \frac{U_2 - x_2}{U - 1}, \qquad z_3 = \frac{U_3 - x_3}{U - 1}. \tag{17.3.3}$$

Note that $z_1 + z_2 + \cdots + z_q = 1$. Points for which any z_i is negative, which would happen if (17.3.2) were not satisfied, would lie outside the original mixture space as in Figure 17.4b. Consider the simpler case of Figure 17.4a, where the restricted region lies entirely within the mixture space. The appropriate U-pseudocomponent transformation is

$$z_1 = (0.50 - x_1)/0.35, \qquad z_2 = (0.45 - x_2)/0.35, \qquad z_3 = (0.40 - x_3)/0.35. \tag{17.3.4}$$

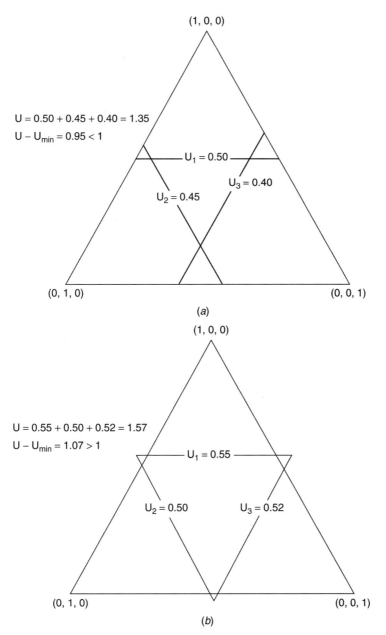

Figure 17.4. For $q = 3$. (*a*) Upper bounds that define a restricted region that is a complete (but inverted) U-simplex *within* the mixture simplex. (*b*) Upper bounds that define a restricted region that is a complete (but inverted) U-simplex, *parts of which lie outside* the mixture simplex.

In the z-coordinates, the experimental space is now the full simplex, and a suitable design can be chosen based on the principles previously discussed. The response data that result can be used to fit a suitable polynomial in the z's. By using the transformation (17.3.4), we can also write the fitted equation in terms of the x's.

Because the U-simplex has an orientation *opposite* to that of the original mixture space simplex, the sign of any corresponding first-order coefficient is reversed. If the response increases as x_i increases, it will decrease as z_i increases, and vice versa. This is clear from the form of Eq. (17.3.4), where positive z_i are related to negative x_i. Except for this reversal, this case is similar to the L-pseudo-component case of the foregoing section. For the situation where the U-pseudo-component simplex is not fully within the mixture space, we recommend using the method of the following section, rather than the pseudocomponent method.

17.2. UPPER AND LOWER BOUNDS

We now consider the case where, for at least some ingredients, there are both lower *and* upper bounds. (Unrestricted x's can be regarded as having a lower bound of 0 and an upper bound of 1 in what follows, or can momentarily be set aside.) Clearly, we need to be able to understand what the restricted region "looks like," even though we cannot properly visualize it in higher dimensions. An important step in doing this is to find the extreme vertices of the region, as explained by McLean and Anderson (1966). Suppose the experimental region is defined by

$$0 \le L_i \le x_i \le U_i \le 1, \qquad \text{for } i = 1, 2, \ldots, q, \qquad (17.4.1)$$

where $L = L_1 + L_2 + \cdots + L_q < 1$, $x_1 + x_2 + \cdots + x_q = 1$, and $U = U_1 + U_2 + \cdots + U_q > 1$. (We also assume that $L_i < U_i$; if any $L_i = U_i$, the problem is exactly the same but in fewer dimensions, because that particular x_i is then held constant.) We need to apply two rules:

1. Write down all possible two-level treatment combinations using the L_i and U_i levels *for all but one factor that is temporarily left blank*. Because each position can be left blank in turn, there are $q2^{q-1}$ such combinations.
2. Reexamine all the $q2^{q-1}$ combinations found in rule 1 and fill in all "admissible blanks." That is, if a blank can be filled in with a factor level that makes the sum of all the factor levels 1 (or the appropriate lesser total if any factors are held constant) and that also satisfies the restrictions (17.4.1), then fill it.

Each "admissible treatment combination" is a vertex of the hyper-polyhedron; however, some vertices may appear more than once, depending on the numbers involved. The duplicates should be dropped. We can use the extreme vertices as design point locations, and this is usually a sensible thing to do. The extreme vertices can also be used to obtain other design points by calculating the centroids of various combinations of the vertices—for example, of edges or, more generally, of multidimensional "faces" that the vertices define.

In general, there will be one-dimensional, two-dimensional, three-dimensional, ..., r-dimensional faces for all $r \leq q - 1$. We identify the various r-dimensional faces (for $2 \leq r \leq q - 1$) by gathering together all the sets of vertices that have $q - r - 1$ ingredient levels identical within the set, while the remaining $r + 1$ ingredient levels vary. The q-dimensional "face" is in fact the hyper-polyhedron itself, defined by all the vertices. Once the faces have been identified, their centroids are obtained by averaging all the x_1 levels, all the x_2 levels, ..., all the x_q levels of all the points defining the face. The example below, taken from McLean and Anderson (1966, pp. 450–452), should make the process clearer. (Used by permission of the American Statistical Association. All rights reserved.)

Example

The chemical constituents of a particular type of flare are magnesium (x_1), sodium nitrate (x_2), strontium nitrate (x_3), and binder (x_4). Ingredients are combined as proportions of weight. Previous engineering experience indicated that the following weight constraints were appropriate.

$$0.40 \leq x_1 \leq 0.60,$$
$$0.10 \leq x_2 \leq 0.50,$$
$$0.10 \leq x_3 \leq 0.50,$$
$$0.03 \leq x_4 \leq 0.08.$$

(17.4.2)

The overall problem was to choose and perform selected experiments and use the resulting data to find the treatment combination (x_1, x_2, x_3, x_4) that gave the maximum response (illumination, measured in candles).

The vertices of the polyhedral space, which consist of all admissible points of the mixture space, are found by applying the two rules. There are $4 \times 2^{4-1} = 32$ candidate locations, but only eight of these can be completed into "admissible treatment combinations," as we see from Table 17.2. The added admissible ingredient values are marked with asterisks (*); blanks left open cannot be completed with values that not only allow the coordinates to sum to 1 but also satisfy the restrictions (17.4.2) as well. (The observation column y will be needed shortly.) The eight extreme vertices and the restricted region are pictured in Figure 17.5, slightly adapted from McLean and Anderson (1966, p. 451). The restricted region looks like an irregular slab paperweight. Each of the six faces is parallel to one of the four faces of the full mixture tetrahedron, in the following pattern:

Face 1234 is parallel to $x_2 = 0$, the simplex base, as drawn, opposite the point $(0, 1, 0, 0)$.

Faces 1256 and 3478 are parallel to $x_1 = 0$, the left face opposite the point $(1, 0, 0, 0)$.

Face 5678 is parallel to $x_3 = 0$, the right face opposite the point $(0, 0, 1, 0)$.

Faces 2468 and 1357 are parallel to $x_4 = 0$, the back face opposite the point $(0, 0, 0, 1)$.

Table 17.2. Application of Two Rules to Find the Extreme Vertices of the Space Defined by (17.4.2); the 32 Candidate Locations and the Actual Eight Vertices, Numbered 1–8

Vertex No.	x_1	x_2	x_3	x_4	y
	0.40	0.10	0.10	—	
	0.40	0.10	0.50	—	
	0.40	0.50	0.10	—	
	0.40	0.50	0.50	—	
	0.60	0.10	0.10	—	
	0.60	0.10	0.50	—	
	0.60	0.50	0.10	—	
	0.60	0.50	0.50	—	
1	0.40	0.10	0.47*	0.03	75
2	0.40	0.10	0.42*	0.08	180
	0.40	0.50	—	0.03	
	0.40	0.50	—	0.08	
3	0.60	0.10	0.27*	0.03	195
4	0.60	0.10	0.22*	0.08	300
	0.60	0.50	—	0.03	
	0.60	0.50	—	0.08	
5	0.40	0.47*	0.10	0.03	145
6	0.40	0.42*	0.10	0.08	230
	0.40	—	0.50	0.03	
	0.40	—	0.50	0.08	
7	0.60	0.27*	0.10	0.03	220
8	0.60	0.22*	0.10	0.08	350
	0.60	—	0.50	0.03	
	0.60	—	0.50	0.08	
	—	0.10	0.10	0.03	
	—	0.10	0.10	0.08	
	—	0.10	0.50	0.03	
	—	0.10	0.50	0.08	
	—	0.50	0.10	0.03	
	—	0.50	0.10	0.08	
	—	0.50	0.50	0.03	
	—	0.50	0.50	0.08	

*The asterisks indicate added admissible ingredient values.

To the extreme vertices, McLean and Anderson added seven additional design point locations. Their particular choices were (a) the six centroids of the six faces of the slab, and (b) the centroid of the eight extreme vertices; these seven points are shown in Table 17.3. The observations y shown in Tables 17.2 and 17.3 result from 15 flares assembled to the specifications indicated by their ingredient values and then tested for their amounts of illumination measured in 1000 candles.

"Faces" Available in the Example, Used and Not Used

Before we carry out a fit of a second-order Scheffé polynomial to these 15 data points, we say something about the r-dimensional "faces" of the polyhedron for

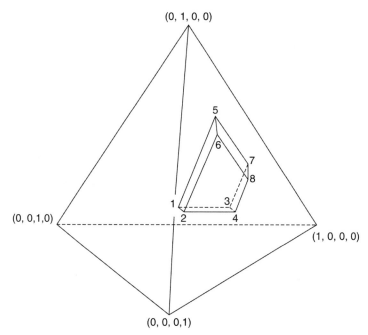

(0, 1, 0, 0)

(0, 0,1,0)

(1, 0, 0, 0)

(0, 0, 0,1)

Figure 17.5. For $q = 4$: The restricted experimental region for the flare experiment, with its extreme vertices numbered 1–8, giving easy identity to the six faces.

Table 17.3. Addition of Face Centroids and Overall Centroid

Point No.	x_1	x_2	x_3	x_4	Points Defining Face	Definition of Face	y
9	0.50	0.1000	0.3450	0.055	1, 2, 3, 4	$x_2 = 0.10$	220
10	0.50	0.3450	0.1000	0.055	5, 6, 7, 8	$x_3 = 0.10$	260
11	0.40	0.2725	0.2725	0.055	1, 2, 5, 6	$x_1 = 0.40$	190
12	0.60	0.1725	0.1725	0.055	3, 4, 7, 8	$x_1 = 0.60$	310
13	0.50	0.2350	0.2350	0.030	1, 3, 5, 7	$x_4 = 0.03$	260
14	0.50	0.2100	0.2100	0.080	2, 4, 6, 8	$x_4 = 0.08$	410
15	0.50	0.2225	0.2225	0.055	1, 2, 3, 4, 5, 6, 7, 8	All vertices	425

this example. Recall that $q = 4$. When $r = 3$, there exists a single three-dimensional face, consisting of all the extreme vertices, points 1–8 in Table 17.2, which have $q - r - 1 = 4 - 3 - 1 = 0$ identical ingredient levels. When $r = 2$, we have $q - r - 1 = 4 - 2 - 1 = 1$, and so the face vertices have one ingredient level identical, as shown in Table 17.3 for the runs 9–14. For example, point 9 is the centroid of the four points 1, 2, 3, 4 that define the face, and all these points have $x_2 = 0.10$. When $r = 1$, we have $q - r - 1 = 4 - 1 - 1 = 2$. The one-dimensional faces are the 12 edges of the polyhedron. As is clear from Figure 17.5, these are obtained as the 12 lines joining the following pairs of points: (1, 2), (1, 3), (1, 5), (2, 4), (2, 6), (3, 4), (3, 7), (4, 8), (5, 6), (5, 7), (6, 8), (7, 8). A check of Table 17.2 shows that each of these pairs of points each has two ingredient levels identical, as the formula $q - r - 1$ specifies. In most circumstances, the midpoints of these edges would also be considered for design points. *In the specific application we are following, however*, they were simply not selected or used.

Fit of a Quadratic Polynomial

When a full 10-term Scheffé quadratic is fitted by least squares to the responses at the 15 selected data points, the following equation is obtained:

$$\hat{y} = -1558x_1 - 2351x_2 - 2426x_3 + 14358x_4$$

$$+ 8300x_1x_2 + 8076x_1x_3 - 6609x_1x_4 + 3214x_2x_3 - 16982x_2x_4 - 17111x_3x_4$$

$$(17.4.3)$$

The analysis of variance table for this fit is as follows:

Source	df	Sum of Squares	Mean Square	F
Mean	1	947527	—	
Extra for first order	3	69465	23155	6.45 (> 5.41)
Extra for second order	6	40461	6743.5	1.88 (< 4.95)
Residual	5	17947	3589.4	
Total	15	1075400		

Most (98.33%) of the variation in the data is picked up by this model; 88.11% by an overall mean, 6.46% more (significant at 5%) by the addition of first-order terms, and the last 3.76% by the addition of second-order terms, which will be retained although they are not significant at 5%.

Two sections of this surface, on the faces $x_4 = 0.03$ and 0.08 and within the restrictions on the other three ingredients, are shown in Figure 17.6. We see that the $\hat{y}$ values increase from front ($x_4 = 0.03$) to back ($x_4 = 0.08$). Smooth contours can be mentally visualized joining up the two sections. Note that the slab is actually quite flat and the distance between the two faces has been greatly exaggerated in the figure, in order to separate the two sections for clarity. The maximum $\hat{y}$ within the restricted region is clearly on the $x_4 = 0.08$ face, at about (0.52, 0.23, 0.17, 0.08). At this location, $\hat{y} = 397.5$. There are, however, some unanswered questions in this example. The residuals from the quadratic fit, in run order and rounded to integers, are (13, 7, 4, -26, 21, 1, -7, -2, -38, -45, -42, 30, -31, 19, and 95). Two features stand out in this listing. Runs 1–8, at the more extreme locations in the space, show the closest fit to the data, and the 15th residual, at the overall centroid of the extreme vertices, is very large compared with the others. It shows as an outlier in a normal plot, or in a "residuals versus fitted values" plot, for example. In the fitting of the model, the corresponding large observation of 425 at this overall centroid is outweighed by the other points. However, such a high value would typically be the focus of further investigation to see if it were reproducible.

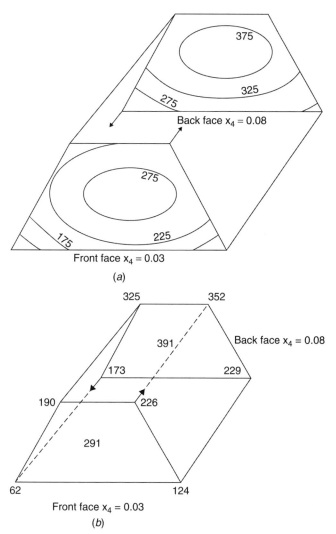

Figure 17.6. (*a*) Two sections of the second-order fitted surface (17.4.3) shown on the faces $x_4 = 0.03$ (front) and $x_4 = 0.08$ (rear). (*b*) Ten of the 15 fitted values $\hat{y}$ are shown at the eight extreme vertices and at the two centroids of the faces; the five other fitted values (258, 232, 280, 291, 330) at design points (9, 11, 12, 13, 15), respectively, have been omitted to avoid crowding the figure.

Advantages of Extreme Vertices Designs

1. The two-rule algorithm enables the experimenter to find the vertices of the restricted region fairly easily, even when there are so many factors that full visualization is impossible. Because the various centroids are then easy to obtain, the entire design can be chosen reasonably quickly.

2. The design is uniquely determined by the constraints.

3. A sequential approach, in which the region of interest is shrunk, expanded or moved, can be employed, because it is easy to choose new designs.

4. Additional or repeat points can easily be added to the design.
5. When the restrictions are such that the restricted region has an embedded simplex shape, this is easy to see, because the hyper-polyhedron has only q distinct vertices.

Disadvantages of Extreme Vertices Designs

1. As k increases, the number of possible extreme vertices increases as does the number of faces and possible centroids. Thus there are many more points to take into consideration in choosing a design. Of course, one does not need to use all the points thus generated. It would be possible to choose just the even-dimensional faces, for example. It might also make sense to use a distance function to check how close together pairs of points are and then omit some points that are "too near" ones already selected. A possible distance function is

$$d_{ij} = \left[\sum_{k=1}^{q} \left\{ (x_{ik} - x_{jk})/(U_k - L_k) \right\}^2 \right]^{1/2}.$$

This function assumes that certain components are more sensitive (in their effect on the response) than others and that the sensitivity for each ingredient is inversely proportional to the width of its constraint interval. This would often be a reasonable assumption.

2. Clustering of design points can occur and steps need to be taken to remove this clustering. A simple example given by Gorman (1966) appears as Exercise 17.3 at the end of this chapter. See also pp. 597–598.

3. Because the design is specifically determined by the various lower and upper bounds, the choice of the constraints plays a large role in selecting the design. Sometimes it makes sense to perform an initial series of experimental runs in order to get information on where the constraints should best be set.

EXERCISES

17.1. Suppose we have a $q = 3$ mixture space restricted by bounds $x_1 \le U_1$, $x_2 \le U_2$, $x_3 \le U_3$, for selected U_i such that $0 < U_i \le 1$ and $U - U_{\min} \le 1$. Show that, if one point of the restricted U-simplex lies on the $x_1 = 0$ boundary of the mixture space, then the other two points are $(U_1, U_2 - U_1, 1 - U_2)$ and $(U_1, U_2, 1 - U_1 - U_2)$. Show further that if all the vertices lie on the $x_i = 0$ boundaries, then $U_1 = U_2 = U_3 = \frac{1}{2}$.

17.2. Extend the second part of Exercise 1 to q dimensions. Show that if all the vertices of the U-simplex lie on the faces of the original mixture simplex, then $U_1 = U_2 = \cdots = U_q = \frac{1}{q-1}$.

Table E17.5. Detergent Formulation Data

No.	Water %	Alcohol %	Urea %	y_1	y_2
1	6	0	0	362.5	36.5
2	0	6	0	78	11.5
3	0	0	6	1630	23.7
4	3	3	0	165	12.9
5	3	0	3	537.5	5.9
6	0	3	3	202.5	5.2
7	1.5	4.5	0	96.5	11.4
8	1.5	0	4.5	560	15.8
9	0	1.5	4.5	310	7.8
10	4.5	1.5	0	230	14.1
11	4.5	0	1.5	455	3.5
12	0	4.5	1.5	108.5	8.5
13	3	1.5	1.5	257.5	7.3
14	1.5	3	1.5	167	8.9
15	1.5	1.5	3	275	3.2

17.3. (*Source:* Discussion of Extreme vertices design of mixture experiments, by R. A. McLean and V. L. Anderson, by J. W. Gorman, *Technometrics*, **8**, 1966, 455–456.) Find the extreme vertices of the restricted region for a four-ingredient mixture problem with restrictions $0 \leq x_1 \leq 0.24$, $0.25 \leq x_2 \leq 0.75$, $0.25 \leq x_3 \leq 0.75$, $0.25 \leq x_4 \leq 0.75$. What is the shape of this region? You may find it helpful to draw a rough diagram of it. Find the centroids of all the nine edges, of all five faces, and of the set of all extreme vertices. Would it make sense to use all 15 of these derived points plus the six vertices, in a 21-point design? What is the obvious drawback? How would you reduce to a more workable 15 point design, for example?

17.4. Suppose we have a three component mixture space $x_1 + x_2 + x_3 = 1$, to which the following restrictions apply: $x_1 \geq 0.2$, $x_2 \geq 0.2$, $x_3 \geq 0.3$. Find

(a) The coordinates of the vertices of the restricted design region.

(b) The equations for the pseudo components in terms of the original components and vice versa.

(c) The original coordinates of a $\{3, 2\}$ design set into the restricted region.

17.5. (*Source:* Use of simplex experimental designs in detergent formulation, by J. P. Narcy and J. Renaud, *Journal of the American Oil Chemists' Society*, 1972, **49**(10), 598–608.) The data in Table E17.5 come from an experiment on detergent formulations. There are three ingredients—water, alcohol, and urea—and the experimental region is a confined one. Use pseudocomponents to translate the predictor variable levels into ranges $0 \leq x_1 \leq 1$, and then fit second-order mixture models to the responses y_1 (viscosity) and y_2 (clear point). Plot the pseudocomponent data on a triangular mixture grid and mark any observations where the fitted quadratic models produce a residual that exceeds its standard error by two standard deviations.

CHAPTER 18

Other Mixture Methods and Topics

18.1. USING RATIOS OF INGREDIENTS AS PREDICTOR VARIABLES

In some mixture problems, it may be desirable to use ratios of the mixture variables as input variables rather than the variables themselves. For example, in glass-making, three basic ingredients are $(x_1, x_2, x_3) =$ (silica, soda, lime). However it may be preferable to work in terms of the silica to soda ratio, $r_1 = x_1/x_2$, instead of the actual proportion of each, and to think of the combined amount of silicon and soda as being controlled by its ratio to lime, which indicates another ratio variable, $r_2 = (x_1 + x_2)/x_3$. Alternatively, either of the ratios x_1/x_3 or x_2/x_3 could be introduced. While the actual numerical details are different for each choice, the basic idea is the same.

If the x's were to lie in the full 0 to 1 range, the r's could vary between 0 and ∞, something that might not be desired. Kenworthy (1963) makes the point that sometimes it is necessary to restrict experimental levels to quite small ranges of the x's. Moreover, if the product is to retain its desired characteristics, omission of any of the ingredients $(x_i = 0)$ may be impossible. In such circumstances, the use of ratios becomes feasible, and may even be preferable, and the permitted ranges of the ratios become reasonable ones. It is then possible to carry out factorial and other types of designs on the ratios to explore the restricted space. The following example is due to Kenworthy (1963).

Example

An experiment is needed to test the weathering characteristics of a synthetic enamel composed entirely of three resins—two alkalyd (x_1 and x_2) and one urea (x_3). The "standard mixture" has the formulation $(0.30, 0.30, 0.40)$ and mixtures which vary 0.10, up or down, need to be investigated. Thus there are three lower bounds $(L_1, L_2, L_3) = (0.20, 0.20, 0.30)$ and three upper bounds $(U_1, U_2, U_3) = (0.40, 0.40, 0.50)$. The region of interest is a hexagon, shown in Figure 18.1. The center of this hexagon is at $(x_1, x_2, x_3) = (0.30, 0.30, 0.40)$ obtained from averaging

Response Surfaces, Mixtures, and Ridge Analyses, Second Edition. By G. E. P. Box and N. R. Draper

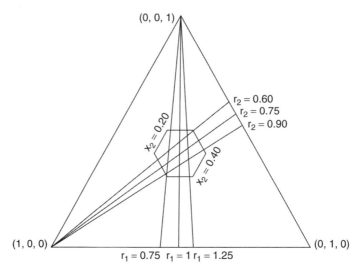

Figure 18.1. The nine design points are at the intersections of the rays.

the respective lower and upper limits, $(L_i + U_i)/2$. Suppose the ratios of interest are

$$r_1 = x_2/x_1 \quad \text{and} \quad r_2 = x_2/x_3. \tag{18.1.1}$$

Any given point (x_1, x_2, x_3) may now be represented in terms of (r_1, r_2); for example the center of the restricted region, $(0.30, 0.30, 0.40)$, translates to $(1, 0.75)$, and so on. The formulas for this particular example are obtained from the two equations above and using the fact that $x_1 + x_2 + x_3 = 1$. This leads, after manipulation, to

$$x_1 = r_2/r, \quad x_2 = r_1 r_2/r, \quad \text{and} \quad x_3 = r_1/r, \quad \text{where } r = r_1 + r_2 + r_1 r_2. \tag{18.1.2}$$

Geometrically, $r_1 = \text{constant}$ is a ray going across the mixture space from the point C $(0, 0, 1)$, while $r_2 = \text{constant}$ is a ray going across the mixture space from the point A $(1, 0, 0)$. In Figure 18.1, the triangular three-ingredient mixture is oriented as in the original reference, a rotation from the usual representation with A at the top. Three rays with equally spaced r_1 values $(0.75, 1, 1.25)$ are shown stemming from the point C while equally spaced r_2 values $(0.60, 0.75, 0.90)$ stem from the point A. The nine intersections of these two sets of rays define the nine points of a 3^2 factorial design in the r-space. In the hexagonal space, they look somewhat irregularly spaced and do not fully cover the region of interest. Table 18.1 shows the coordinates after translation from r's to x's. It is, of course, necessary to make this conversion to set up the experiment.

The outer levels for the ratios must be wide enough apart to satisfactorily cover the region of interest, and the points of intersection of the rays must lie inside the region. The choices of the levels must be made carefully. Figure 18.2 shows an example where the coverage is poor. In higher dimensions, this can make the ratio

Table 18.1. The Nine Intersections of Three Rays of r_1 with Three Rays of r_2 Defines a 3^2 Factorial Design in the Mixture Space (x_1, x_2, x_3)

$r_1 = x_2/x_1$	$r_2 = x_2/x_3 = 0.60$	$r_2 = x_2/x_3 = 0.75$	$r_2 = x_2/x_3 = 0.90$
$r_1 = 0.75$	0.333, 0.250, 0.417	0.364, 0.273, 0.364	0.387, 0.290, 0.323
$r_1 = 1.00$	0.273, 0.273, 0.455	0.300, 0.300, 0.400	0.321, 0.321, 0.357
$r_1 = 1.25$	0.231, 0.288, 0.481	0.255, 0.319, 0.426	0.275, 0.344, 0.382

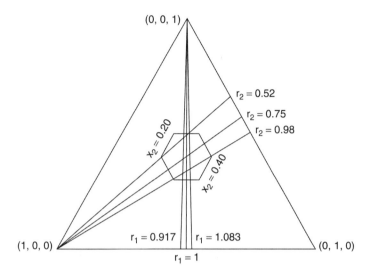

Figure 18.2. A bad choice of ray levels can reduce the coverage of the region.

method for choosing designs more difficult to use because it can be hard to visualize where the intersections lie with respect to the restricted space.

The usual statistical analysis in these ratio problems is to fit a standard (nonmixture)-type polynomial of appropriate order in the r-ratios and interpret that directly. It would also be possible to fit a mixture model directly in the x's, if one could be found that would give a satisfactory fit without too much complication of terms. Fitting directly to the x's effectively deemphasizes the ratios in the analysis, in spite of their use in selecting the design. Much depends on the particular problem under study and the way the experimenter feels about the variables used. Mathematically deriving a fitted polynomial model in the x's *from* the fitted polynomial model in the r's is usually not feasible, because of the x's in the denominators of the r's.

18.2. THE KISSELL AND MARSHALL EXAMPLE

(*Source:* Multi-factor responses of cake quality to basic ingredient ratios by L. T. Kissel and B. D. Marshall, *Cereal Chemistry*, **39**, 1962, 16–30. Adapted with the permission of AACC International.) Kissell and Marshall carried out an ambitious experiment on cakes that used four ratios of five ingredients. We convert their notation to the one used above. The five basic ingredients studied were:

x_1 = baking powder (value at the center of the design = 0.0129)
x_2 = shortening (value at the center of the design = 0.0762)
x_3 = flour (value at the center of the design = 0.2735)
x_4 = sugar (value at the center of the design = 0.3556)
x_5 = water (value at the center of the design = 0.2818)

Note that $x_1 + x_2 + x_3 + x_4 + x_5 = 1$, both generally and, in particular, at the center of the design. The four ratios used were

$$r_1 = x_1/(x_2 + x_3 + x_4 + x_5),$$
$$r_2 = x_2/(x_3 + x_4 + x_5),$$
$$r_3 = x_3/(x_4 + x_5),$$
$$r_4 = x_4/x_5.$$

(18.2.1)

If we multiply these equations through by their denominators, take all the terms to one side, and add the mixture restriction at the top, we get five equations that can be written in matrix form as

$$\begin{bmatrix} 1 & 1 & 1 & 1 & 1 \\ 1 & -r_1 & -r_1 & -r_1 & -r_1 \\ 0 & 1 & -r_2 & -r_2 & -r_2 \\ 0 & 0 & 1 & -r_3 & -r_3 \\ 0 & 0 & 0 & 1 & -r_4 \end{bmatrix} \begin{bmatrix} x_1 \\ x_2 \\ x_3 \\ x_4 \\ x_5 \end{bmatrix} = \begin{bmatrix} 1 \\ 0 \\ 0 \\ 0 \\ 0 \end{bmatrix} \qquad (18.2.2)$$

or

$$\mathbf{Rx} = \mathbf{d}, \qquad (18.2.3)$$

say, so that we can obtain the x's in terms of the r's via

$$\mathbf{x} = \mathbf{R}^{-1}\mathbf{d}. \qquad (18.2.4)$$

Because of the form of $\mathbf{d}$, we need to evaluate only the first column of the inverse $\mathbf{R}^{-1}$. All the other terms of $\mathbf{R}^{-1}$ hit against the 0's in $\mathbf{d}$. The determinant of $\mathbf{R}$ is

$$D = (1 + r_1)(1 + r_2)(1 + r_3)(1 + r_4) \qquad (18.2.5)$$

and Eq. (18.2.4) becomes

$$\mathbf{x} = D^{-1} \begin{bmatrix} r_1(1 + r_2)(1 + r_3)(1 + r_4) & * & * & * & * \\ r_2(1 + r_3)(1 + r_4) & * & * & * & * \\ r_3(1 + r_4) & * & * & * & * \\ r_4 & * & * & * & * \\ 1 & * & * & * & * \end{bmatrix} \begin{bmatrix} 1 \\ 0 \\ 0 \\ 0 \\ 0 \end{bmatrix}, \qquad (18.2.6)$$

where the asterisks denote terms that take no further part in the calculation and so need not be evaluated. The solutions for the x's are then

$$x_1 = r_1/(1 + r_1),$$

$$x_2 = r_2/\{(1 + r_1)(1 + r_2)\},$$

$$x_3 = r_3/\{(1 + r_1)(1 + r_2)(1 + r_3)\}, \qquad (18.2.7)$$

$$x_4 = r_4/\{(1 + r_1)(1 + r_2)(1 + r_3)(1 + r_4)\},$$

$$x_5 = 1/\{(1 + r_1)(1 + r_2)(1 + r_3)(1 + r_4)\}.$$

We see that the r-ratios determine the x's uniquely and vice versa. In choosing ratios, one is usually told to follow the useful rule that each ratio should contain at least one of the components used in another ratio. The acid test that this has been properly done is that the two-way crossover illustrated here by Eqs. (18.2.1) and (18.2.7), which provides a dual identification of each point in the mixture space, should be achievable by the choice made. Note that for q x-components, we need to define $q - 1$ ratios for a complete specification.

Exercise. Go back to the simpler example in Section 18.1 and obtain the solution there in similar fashion. □

Design and Analysis

The design chosen for this experiment was a four-factor central composite design, consisting of (a) a 2^4 full factorial design, the 16 runs of which would usually be coded ($\pm 1, \pm 1, \pm 1, \pm 1$), plus (b) a star, that is, eight axial points, chosen to lie at double the axial distance of the factorial, which would usually be coded ($\pm 2, 0, 0, 0$), $(0, \pm 2, 0, 0)$, $(0, 0, \pm 2, 0)$, $(0, 0, 0, \pm 2)$, plus (c) a center point, usually coded $(0, 0, 0, 0)$. For their convenience, and perhaps to avoid confusing the levels in case minus signs were accidentally omitted, Kissell and Marshall (1962) used levels 1, 2, 3, 4, and 5 instead, simply adding 3 to the usual coding. This is unusual, but perfectly permissible. To attain these five numbers, their r-values were coded in the following way to z's here (but called X's in the paper):

$$z_1 = (r_1 - 0.004)/0.003, \quad \text{or} \quad r_1 = 0.003z_1 + 0.004,$$

$$z_2 = (r_2 - 0.057)/0.009, \quad \text{or} \quad r_2 = 0.009z_2 + 0.057,$$

$$z_3 = (r_3 - 0.339)/0.030, \quad \text{or} \quad r_3 = 0.030z_3 + 0.339, \qquad (18.2.8)$$

$$z_4 = (r_4 - 0.707)/0.185, \quad \text{or} \quad r_4 = 0.185z_4 + 0.707.$$

This design, batter compositions, and the values of two response variables $y_1 =$ cake volume (in cm^3), and top contour score y_2 on a scale of 10 = highly peaked to 1 = extremely fallen are shown in Table 18.2, which is a slightly modified version of Kissell and Marshall's (1962, p. 20) table. A full second-order fit to the

Table 18.2. Experimental Design, Batter Compositions, with Resultant Cake Volumes and Top-Contour Scores

r_1	r_2	r_3	r_4	Baking Powder	Short-ening	Flour	Sugar	Water	Cake Volume (cm^3)	Top Contour Score
3	3	3	3	0.0128	0.0765	0.2734	0.3556	0.2817	571	6.9
2	2	2	2	0.0099	0.0691	0.2626	0.3414	0.3170	489	7.8
4	2	2	2	0.0157	0.0687	0.2611	0.3394	0.3151	556	7.3
2	4	2	2	0.0099	0.0843	0.2583	0.3358	0.3117	483	7.8
4	4	2	2	0.0157	0.0837	0.2568	0.3338	0.3100	562	8.0
2	2	4	2	0.0099	0.0691	0.2898	0.3273	0.3039	502	7.3
4	2	4	2	0.0157	0.0687	0.2881	0.3254	0.3021	586	7.3
2	4	4	2	0.0099	0.0843	0.2850	0.3219	0.2989	505	7.3
4	4	4	2	0.0157	0.0837	0.2833	0.3201	0.2972	583	8.8
2	2	2	4	0.0099	0.0691	0.2627	0.3892	0.2691	529	4.0
4	2	2	4	0.0157	0.0687	0.2611	0.3870	0.2675	463	1.5
2	4	2	4	0.0099	0.0843	0.2294	0.4000	0.2764	472	3.8
4	4	2	4	0.0157	0.0837	0.2280	0.3978	0.2748	416	1.0
2	2	4	4	0.0099	0.0691	0.2897	0.3733	0.2580	565	5.8
4	2	4	4	0.0157	0.0687	0.2881	0.3711	0.2564	514	2.0
2	4	4	4	0.0099	0.0843	0.2850	0.3671	0.2537	557	5.8
4	4	4	4	0.0157	0.0837	0.2833	0.3650	0.2523	512	2.0
1	3	3	3	0.0070	0.0769	0.2750	0.3577	0.2834	493	7.5
5	3	3	3	0.0186	0.0760	0.2718	0.3535	0.2801	528	3.0
3	1	3	3	0.0128	0.0612	0.2780	0.3615	0.2865	565	6.5
3	5	3	3	0.0128	0.0914	0.2689	0.3497	0.2772	570	7.5
3	3	1	3	0.0128	0.0765	0.2454	0.3712	0.2941	532	6.5
3	3	5	3	0.0128	0.0765	0.2991	0.3412	0.2704	595	7.0
3	3	3	1	0.0128	0.0765	0.2734	0.3005	0.3368	490	7.5
3	3	3	5	0.0128	0.0765	0.2734	0.3952	0.2421	454	1.5

y_1 data in terms of the r-ratios gives the following equation:

$$\hat{y}_1 = -69.6 + 194.04r_1 + 10.54r_2 - 18.37r_3 + 231.46r_4 - 16.792r_1^2 - 2.542r_2^2$$
$$- 3.542r_3^2 - 26.417r_4^2 + 1.375r_1r_2 + 2.625r_1r_3 - 32.875r_1r_4 + 5.875r_2r_3$$
$$- 7.125r_2r_4 + 11.375r_3r_4. \tag{18.2.9}$$

(The coefficients shown are very similar to, but differ slightly from, those given by Kissell and Marshall, and the fitted equations have very similar descriptive characteristics. The numerical differences may be ascribable to improvements in the accuracy of regression calculations since the earlier calculations were made.) The model shown explains $R^2 = 0.9063$ or 96.3% of the variation in the data in excess of the mean $\bar{y}$. The normal plot of residuals in Figure 18.3a is well-behaved with a slight indication of a distribution slightly more compact than the normal distribution.

Figures 18.3b and 18.3c show two diagrams constructed for the response surface over the (r_1, r_2) plane, when $r_3 = r_4 = 3$, by Kissell and Marshall (1962,

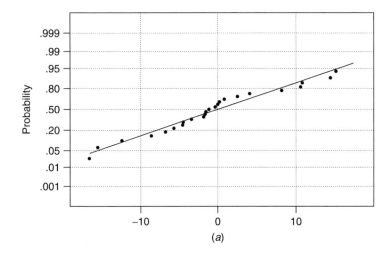

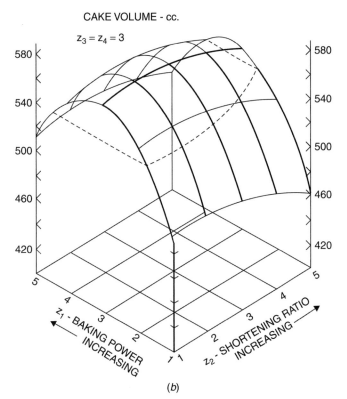

Figure 18.3. (*a*) Normal plot of the residuals from the *r*-fit. (*b*) The *z*-fit response surface over the levels of z_1 and z_2 with $z_3 = z_4 = 3$.

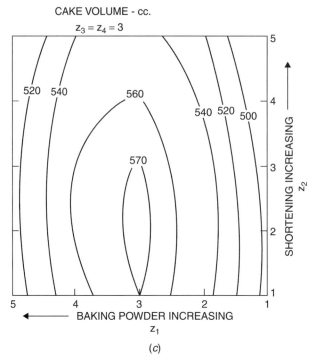

Figure 18.3. (*Continued*) (*c*) The z-fit response surface over the levels of z_1 and z_2 with $z_3 = z_4 = 3$, this time shown as horizontal contour slices of the foregoing figure.

p. 25 and reproduced with the permission of AACC International). Many similar figures can be constructed, and several others are in the same reference, pp. 25–27. By detailed examination of the surface in this way, decisions can be made about the set of conditions needed for a desirable cake and confirming runs can be carried out. See the original paper for further discussion of these issues. Other analyses are also possible; for example, the contours could be converted to canonical form for examination and/or a ridge analysis could be carried out.

Fitting to the Mixture Ingredients

What happens if we fit a model directly to the five mixture ingredients? If we choose the usual 15-term second-order Scheffé mixture model to fit, we obtain the following equation:

$$\begin{aligned}
\hat{y}_1 = {} & -1{,}605{,}003 x_1 + 4487 x_2 + 559 x_3 - 7418 x_4 - 13{,}347 x_5 \\
& + 1{,}731{,}252 x_1 x_2 + 1{,}674{,}333 x_1 x_3 + 1{,}427{,}295 x_1 x_4 \\
& + 1{,}904{,}909 x_1 x_5 - 6202 x_2 x_3 + 912 x_2 x_4 + 7783 x_2 x_5 \\
& + 15718 x_3 x_4 + 4486 x_3 x_5 + 41{,}439 x_4 x_5,
\end{aligned} \tag{18.2.10}$$

which explains 97.7% of the variation in the data unexplained by a model that fits only an overall mean β_0. This is slightly above the 96.3% achieved above by the

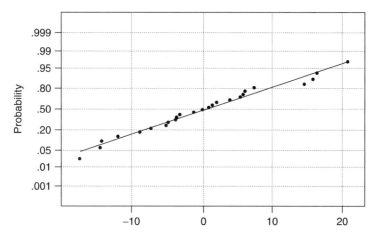

Figure 18.4. Normal plot of the residuals from the x-fit.

ratio fit of Eq. (18.2.9). The normal plot of the residuals in Figure 18.4 gives no cause for anxiety and the model appears to offer a plausible alternative to the ratio fit model.

18.3. MIXTURE DESIGNS AND MODELS WHEN THE TOTAL AMOUNT IS IMPORTANT

The Problem

In experimentation with mixtures, we are often unconcerned with *how much* of a mixture is used in an experimental run. We assume that whether the experimental runs use a thimble-full or a bucket-full of the mixture makes no difference. We are then interested only in the effects caused by combinations of proportions of the ingredients. However, this is not always the case. Consider, for example, a mixture of detergent ingredients, and suppose we are interested in their effects on the cleaning ability of the detergent. If the total mixture amount were varied, would the effectiveness of the detergent vary, even if the proportions of the ingredients were the same? Another example (Cornell, 2003, p. 403) is that of fertilizer, where the total amount typically affects the yield data in addition to the combination of ingredients. For such situations, experimental runs are needed to vary not only the specific mixture ingredients but also the total mixture amount, A. We continue to write, as before, that

$$x_1 + x_2 + \cdots + x_q = 1, \qquad (18.3.1)$$

where the x's represent the *proportions* of ingredients used, and we now also allow A to take various selected values. Thus, for a particular mixture amount A, the ingredient mixture *component amounts* are actually $(a_1, a_2, \ldots, a_q) = (x_1 A, x_2 A, \ldots, x_q A)$, and they add to A. We now need to consider what experimental runs to use and what forms of model can be fitted.

First Thoughts on Choosing a Design and a Suitable Model

A sensible basic way to design an experiment for a mixture amount investigation is to decide on how many values of A are needed, and then to perform the same experiment in terms of the x-values for each level of A. Naturally, the design chosen must allow a suitable model to be fitted. Suppose, for example, that we have three ingredients (x_1, x_2, x_3) and three levels (A_1, A_2, A_3) of the total amount. Three levels of A permit consideration of the possible quadratic effect of A. Suppose further that a Scheffé quadratic model in the x's is thought appropriate at each level of A. Then we can fit a "cross-product" type of model, consisting of terms that can be thought of in either of the following two ways (for our three-ingredient, three-level example):

1. A quadratic polynomial in A with coefficients that each consist of a quadratic polynomial in the x's, namely

$$X(\alpha) + X(\beta)A + X(\gamma)A^2, \tag{18.3.2}$$

where

$$X(\alpha) = \alpha_1 x_1 + \alpha_2 x_2 + \alpha_3 x_3 + \alpha_{12} x_1 x_2 + \alpha_{13} x_1 x_3 + \alpha_{23} x_2 x_3, \tag{18.3.3}$$

and similarly for β and γ. For q ingredients, this "double quadratic" model will have $3q(q + 1)/2$ terms; in our example, $q = 3$ and the formula gives 18 terms.

2. A quadratic polynomial in the x's with coefficients that each consist of a quadratic polynomial in the A's, namely

$$Q(\theta)x_1 + Q(\phi)x_2 + Q(\psi)x_3 + Q(\lambda)x_1 x_2 + Q(\mu)x_1 x_3 + Q(\omega)x_2 x_3, \tag{18.3.4}$$

where

$$Q(\theta) = \theta_0 + \theta_1 A + \theta_2 A^2, \tag{18.3.5}$$

and with five similar expressions involving ϕ, ψ, λ, μ, and ω. Again, there are $3q(q + 1)/2$ terms in general.

Clearly the two models (18.3.2) and (18.3.4) are equivalent once the products are evaluated and the notations are rationalized; each model contains 18 parameters to be estimated. Note that if we believed that A had a linear effect rather than a quadratic, we could omit the A^2 terms. In fact we can test, via the usual regression methods using an extra sums of squares type test, whether or not any set of specific terms are needed in the fitted model. If A has no effect at all, only Eq. (18.3.3) would be needed.

There are many possible other choices of models of this multiplicative type, for example:

(a) A first-order polynomial in the x's with coefficients that each consist of a first-order polynomial in the A's, with $2q$ terms.

(b) A first-order polynomial in the x's with coefficients that each consist of a quadratic polynomial in the A's, with $3q$ terms.

(c) A quadratic polynomial in the x's with coefficients that each consist of a first-order polynomial in the A's, with $q(q + 1)$ terms.

Designs would be selected to match (or overmatch) the amount of detail expressed in the model chosen.

Mixture-Amount Models

The types of models described above are usually called *mixture-amount models*. A disadvantage of all of the forms shown is that when *No* mixture ingredients are present, so that $x_1 = x_2 = \cdots = x_q = 0$, as might be the case when a placebo treatment is part of an experiment, the fitted model will predict a zero response. This may simply not be appropriate in some situations. It is possible to modify the model to accommodate this, as described by Piepel (1988b). It requires a restriction on the parameters of the model. For model (18.3.2), for example, we must take $\alpha_1 = \alpha_2 = \alpha_3$ and $\alpha_{12} = \alpha_{13} = 0$, implying that $X(\alpha)$ is a constant, because its reduced value becomes $\alpha_1(x_1 + x_2 + x_3) = \alpha_1$.

However, there is also another type of model, called the *component-amount model* (in which the "placebo level feature" is automatically in place) that might be more useful in such problems. We describe this next, meanwhile noting that a discussion of mixture-amount and component-amount models is also given by Cornell (2002, pp. 403–418).

Component-Amount Model

We can fit a (nonmixture response surface type of) model directly to the a_i values, $(a_1, a_2, \ldots, a_q) = (x_1 A, x_2 A, \ldots, x_q A)$. For example, we could fit a first-order model

$$\eta = \alpha_0 + \sum_{i=1}^{q} \alpha_i a_i \qquad (18.3.6)$$

containing $(q + 1)$ terms, or a second-order model

$$\eta = \alpha_0 + \sum_{i=1}^{q} \alpha_i a_i + \sum_{i=1}^{q} \alpha_i a_i^2 + \sum \sum_{i<j} \alpha_{ij} a_i a_j \qquad (18.3.7)$$

containing $q + 1 + q + q(q - 1)/2 = (q + 1)(q + 2)/2$ terms.

Note that the numbers of terms in the mixture-amount and component-amount models are not comparable, even for the same order. For example, the general quadratic/quadratic extension of (18.3.2) has $3q(q + 1)/2$ terms, which always (because $q > 1$) exceeds the number of terms $(q + 1)(q + 2)/2$ in the general model, which is the extension of the quadratic (18.3.7). So direct model comparisons are not very useful. Typically, one fits the model(s) that seem(s) sensible in the experimental context, and then he or she makes a choice based on common-

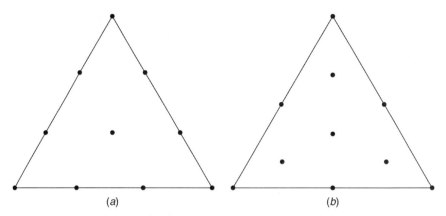

Figure 18.5. Designs for the Claringbold example: (*a*) The 10-point design for observations 1–30 ($B = -1$). (*b*) The 10-point design for observations 31–60 ($B = 1$).

sense examination of the fitted equations. In fitting mixtures models, we are often interested in the overall shape of the response surface in order to find good combinations of ingredients which give rise to a desired response. The interpretation of individual coefficients is often not useful because of the correlations between coefficients and the fact that the model itself can usually be expressed in several ways, which all give rise to the same contours, but whose equations express those contours in different ways. (See Section 16.3, for example.)

Example

We illustrate the applications of mixture-amount and component-amount models using data from Claringbold (1955, p. 181). The design is divided into two blocks, each containing 30 observations. In each block, ten distinct combinations of proportions (x_1, x_2, x_3), where $x_1 + x_2 + x_3 = 1$, of three different hormones were decided upon. In the first block. a 10-point {3, 3} simplex centroid design was selected. (See Figure 18.5*a*.) Three levels of the total amount A were used at each design location, namely 0.75, 1.50, and 3.00, so that the actual component amounts used in the resulting $3 \times 10 = 30$ individual runs are $(a_1, a_2, a_3) = (x_1 A, x_2 A, x_3 A)$. In the second block, a different 10-point design was used (see Figure 18.5*b*, but it was also replicated using the same levels of A as in the first block. In total then, there are 60 runs in two distinct blocks that each contain 30 observations. Each individual experimental combination was injected into 12 mice, and the number of mice, m, that responded to the hormone combination was recorded. We thus have 60 values of m between 0 and 12. The m values are now divided by 12, to provide a response variable that is the *proportion p* of mice affected, which can be assumed to be distributed binomially.

To convert these proportions to data that can be assumed to be normally distributed, at least approximately, the familiar arcsine transformation (see, for example, Draper and Smith, 1998, pp. 292–293) was used by Claringbold. For $1 \le m \le 11$,

$$y = \text{arcsine}(m/12)^{1/2}.$$

Table 18.3. Data from Claringbold (1965)

x_1	x_2	x_3	A	B	m	y
1.000	0.000	0.000	0.75	-1	2	24.09
0.667	0.333	0.000	0.75	-1	0	8.30
0.333	0.667	0.000	0.75	-1	4	35.26
0.000	1.000	0.000	0.75	-1	7	49.80
0.000	0.667	0.333	0.75	-1	2	24.09
0.000	0.333	0.667	0.75	-1	4	35.26
0.000	0.000	1.000	0.75	-1	3	30.00
0.333	0.000	0.667	0.75	-1	3	30.00
0.667	0.000	0.333	0.75	-1	0	8.30
0.333	0.333	0.333	0.75	-1	2	24.09
1.000	0.000	0.000	1.50	-1	5	40.20
0.667	0.333	0.000	1.50	-1	4	35.26
0.333	0.667	0.000	1.50	-1	4	35.26
0.000	1.000	0.000	1.50	-1	7	49.80
0.000	0.667	0.333	1.50	-1	4	35.26
0.000	0.333	0.667	1.50	-1	4	35.26
0.000	0.000	1.000	1.50	-1	6	45.00
0.333	0.000	0.667	1.50	-1	5	40.20
0.667	0.000	0.333	1.50	-1	3	30.00
0.333	0.333	0.333	1.50	-1	3	30.00
1.000	0.000	0.000	3.00	-1	10	65.91
0.667	0.333	0.000	3.00	-1	9	60.00
0.333	0.667	0.000	3.00	-1	9	60.00
0.000	1.000	0.000	3.00	-1	12	81.70
0.000	0.667	0.333	3.00	-1	8	54.94
0.000	0.333	0.667	3.00	-1	7	49.60
0.000	0.000	1.000	3.00	-1	5	40.40
0.333	0.000	0.667	3.00	-1	5	40.40
0.667	0.000	0.333	3.00	-1	9	60.00
0.333	0.333	0.333	3.00	-1	7	49.60
1.000	0.000	0.000	0.75	1	5	40.40
0.500	0.500	0.000	0.75	1	2	24.35
0.000	1.000	0.000	0.75	1	9	60.00
0.000	0.500	0.500	0.75	1	4	35.06
0.000	0.000	1.000	0.75	1	6	45.00
0.500	0.000	0.500	0.75	1	2	24.35
0.667	0.167	0.167	0.75	1	4	35.06
0.167	0.667	0.167	0.75	1	6	45.00
0.167	0.167	0.667	0.75	1	4	35.06
0.333	0.333	0.333	0.75	1	2	24.35
1.000	0.000	0.000	1.50	1	6	45.00
0.500	0.500	0.000	1.50	1	4	35.06
0.000	1.000	0.000	1.50	1	8	54.94
0.000	0.500	0.500	1.50	1	5	40.40
0.000	0.000	1.000	1.50	1	5	40.40
0.500	0.000	0.500	1.50	1	5	40.40
0.667	0.167	0.167	1.50	1	4	35.06

Table 18.3. (*Continued*)

x_1	x_2	x_3	A	B	m	y
0.167	0.667	0.167	1.50	1	6	45.00
0.167	0.167	0.667	1.50	1	4	35.06
0.333	0.333	0.333	1.50	1	5	40.40
1.000	0.000	0.000	3.00	1	9	60.00
0.500	0.500	0.000	3.00	1	10	65.65
0.000	1.000	0.000	3.00	1	10	65.65
0.000	0.500	0.500	3.00	1	8	54.94
0.000	0.000	1.000	3.00	1	12	81.70
0.500	0.000	0.500	3.00	1	7	49.60
0.667	0.167	0.167	3.00	1	7	49.60
0.167	0.667	0.167	3.00	1	7	49.60
0.167	0.167	0.667	3.00	1	6	45.00
0.333	0.333	0.333	3.00	1	5	40.40

The results were recorded in degrees. (They could also have been recorded in radians, for which the divisor $180/\pi = 57.29578$ would be needed. This would just change the scale of the results, not their interpretation.)

The extreme values $m = 0$ and 12 were "smoothed" by Claringbold using special adjustments due to Bartlett and described by Claringbold et al. (1953). When $m = 0$, use $y = \arcsin(1/(4n))^{1/2}$, and when $m = 12$, use $y = 90 - \arcsin(1/(4n))^{1/2}$, when $n = 12$ is the number of mice used for each experimental run.

The data shown in Table 18.3 thus consist of 60 observations of m converted to y and rounded to two places of decimals, in two blocks of size 30. The blocking variable B takes the value -1 for observations 1–30, and 1 for observations 31–60.

Mixture-Amount Fits

A large number of mixture-amount model fits are possible. Table 18.4 shows a selected group of four of them, all incorporating a blocking term bB, where the blocking variable $B = -1$ or 1, as indicated in Table 18.3. The blocking variable is not orthogonal to the models. However, it is orthogonal within rounding to some model terms, and the blocking variable coefficient estimate has very small correlations with all the estimates of the model parameters. Model 1 is a quadratic fit that does not contain A. Model 2 is a quadratic fit with 1 and A portions. Model 3 has additional A^2 terms and represents the model of Eq. (18.3.2) with Eq. (18.3.3) or, equivalently, Eq. (18.3.4) with Eq. (18.3.5). Model 4 is attained by assessing the patterns of statistically significant coefficients (denoted by asterisks) in models 1–3 and forming the conclusion that Model 4 is a sensible compromise choice. This can be confirmed formally by the usual "extra sum of squares" tests. Note that all four models carry with them the implication that, when all the x's are set to zero, $\hat{y} = -2.65$ when $B = -1$ and $\hat{y} = 2.65$ when $B = 1$. In other words, it implies an overall average response of zero with block effects balancing each other above and below zero. No intercept is being fitted, so there are no R^2 values available.

Table 18.4. Selected Mixture-Amount Models for Claringbold (1955) Data

Coefficient of	Model 1	Model 2	Model 3	Model 4
B	2.65	2.65*	2.65*	2.65*
x_1	45.44*	17.25*	16.10	17.45*
x_2	60.27*	44.03*	65.98*	43.63*
x_3	47.56*	30.77*	39.02*	34.98*
x_1x_2	−50.92*	−66.70*	−97.31	−50.92*
x_1x_3	−42.26	−20.54	−88.04	−42.26*
x_2x_3	−54.02*	−37.12	−78.23	−54.02*
Ax_1		16.11*	17.69	16.00*
Ax_2		9.28*	−20.83	9.51*
Ax_3		9.59*	−1.71	7.19*
Ax_1x_2		9.01	51.00	
Ax_1x_3		−12.41	80.20	
Ax_2x_3		−9.66	46.70	
A^2x_1			−0.41	
A^2x_2			7.81	
A^2x_3			2.93	
$A^2x_1x_2$			−10.88	
$A^2x_1x_3$			−24.00	
$A^2x_2x_3$			−14.62	
s^2	178	58.5	62.3	57.0
Residual df	53	47	41	50

An asterisk indicates statistical significance at the 5% level or lower.

For model 4, we have an analysis of variance table as in Table 18.5. We see that most of the variation in the data is explained by first-order terms, with some second-order curvature. The mixture amount A has significant interaction with first-order terms, and this needs to be taken into account in interpreting the effects on the mice. Block differences have a minor but significant effect on the results. Figure 18.6 shows selected contour lines derived from the fitted model 4 for each of the three levels of $A = 0.75$, 1.5, and 3, but ignoring block coefficients—that is, by setting $B = 0$.

Table 18.5. Analysis of Variance Table for the Model 4 Fit

Source of Variation	df	Sum of Squares	Mean Square	F Value
Blocks	1	302	302	5.30
First order	3	109,401	36,467	639.77
Extra for Second Order	3	2,045	682	11.96
Extra for A (first order)	3	6,582	2,195	38.51
Residual	50	2,851	57	
Total	60	121,181		

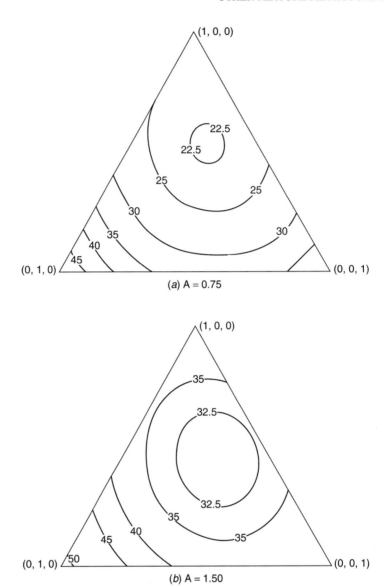

Figure 18.6. Claringbold example. Selected contours defined by the fitted model 4 in Table 18.4, when B is set to zero: (a) At level $A = 0.75$. (b) At level $A = 1.50$.

Component-Amount Fits

Table 18.6 shows three selected component-amount fits to the component quantities $a_i = Ax_i$. Both notations are given in Table 18.6 to allow comparisons with the models in Table 18.4. Model 5 consists only of terms for a first order fit in the a_i. Model 6 is the full quadratic; only one coefficient, indicated by the asterisk, is significant at the 5% level. Adding cubic terms is similarly unpromising. Model 7 is

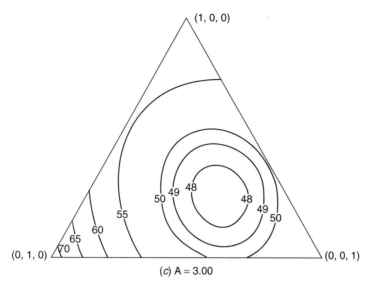

Figure 18.6. (*Continued*) (*c*) At level $A = 3.00$.

Table 18.6. Selected Component-Amount Models for Claringbold (1955) Data

Coefficient of	Coefficient of	Model 5	Model 6	Model 7
Constant	Constant	23.4	24.97	29.12
Blocks	Blocks	2.24	2.49*	2.63*
a_1	Ax_1	9.79*	−2.10	−24.44*
a_2	Ax_2	14.30*	19.01	15.34*
a_3	Ax_3	8.61*	9.48	9.65*
a_1^2	$A^2x_1^2$		5.43	28.13*
a_2^2	$A^2x_2^2$		−0.79	
a_3^2	$A^2x_3^2$		0.86	
a_1a_2	$A^2x_1x_2$		−1.35	
a_1a_3	$A^2x_1x_3$		−1.16	
a_2a_3	$A^2x_2x_3$		−8.57	−7.53*
a_1^3	$A^3x_1^3$			−5.44*
s^2		101.4	77.4	65.2
Residual df		55	49	52
R^2		0.564	0.703	0.699

An asterisk indicates statistical significance at the 5% level or lower.

a compromise model obtained by performing a stepwise regression on all terms up to and including third order, while forcing in the first-order terms (otherwise the a_1 term drops out). The contours produced by Model 7 are, however, quite similar overall to those of Figure 18.6 for model 4, and the stepwise cubic Model 7 ($R^2 = 0.699$, $s^2 = 65.2$) fits somewhat less well than the non-intercept Model 4 ($s^2 = 57.0$). So Model 4 seems to be a sensible choice here.

18.4. DESIGNS FOR COMPONENT-AMOUNT MODELS OBTAINED BY PROJECTION OF STANDARD SIMPLEX LATTICE AND SIMPLEX-CENTROID MIXTURE DESIGNS

As we have already mentioned, one approach to choosing a component-amount design is to perform the same basic design for each level of the total mixture amount A. This may require more runs than the experimenter wishes to perform. For situations where more economy is desirable, Piepel and Cornell (1985) generated designs that were either D_n-optimal or nearly D_n-optimal using the DETMAX program of Mitchell (1974a).

The D_n criterion is generally defined as $|\mathbf{X'X}|/n^p$, where n is the number of runs and p is the number of parameters in the model. Larger values of the D_n criterion give lower values for the variance of the estimated response, and so they are preferred. Larger D_n values also tend to produce designs whose points are more widely spread in the experimental design space. In situations where the values of n and p are not varied, one can simply work with $|\mathbf{X'X}|$ and ignore the n^p.

The candidate points for these designs were chosen from the vertices $(1, 0, \ldots, 0), (0, 1, 0, \ldots, 0), \ldots, (0, 0, \ldots, 0, 1)$, the midpoints of edges $(\frac{1}{2}, \frac{1}{2}, 0, \ldots, 0)$ and permutations, and the face centroids $(\frac{1}{3}, \frac{1}{3}, \frac{1}{3}, 0, \ldots, 0)$ and permutations. The number of points n in the design was varied to provide a number of different designs. This work is very clearly summarized by Cornell (2002, pp. 410–418), and we do not repeat it here.

Table 18.7. Simplex Lattice Mixture Design for $q + r = 4$ Ingredients

Run Number	x_1	x_2	x_3	x_4	Run Number	x_1	x_2	x_3	x_4
1	1	0	0	0	11	1/3	0	0	2/3
2	0	1	0	0	12	2/3	1/3	0	0
3	0	0	1	0	13	2/3	0	1/3	0
4	0	0	0	1	14	2/3	0	0	1/3
5	1/3	1/3	1/3	0	15	0	1/3	2/3	0
6	1/3	1/3	0	1/3	16	0	1/3	0	2/3
7	1/3	0	1/3	1/3	17	0	2/3	1/3	0
8	0	1/3	1/3	1/3	18	0	2/3	0	1/3
9	1/3	2/3	0	0	19	0	0	1/3	2/3
10	1/3	0	2/3	0	20	0	0	2/3	1/3

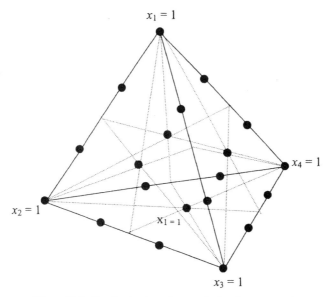

Figure 18.7. Simplex-lattice mixtures design for $q + r = 4$.

Component-Amount Designs Obtained by Projecting Simplex-Lattice Designs and Simplex-Centroid Designs

We next explore work of Prescott and Draper (2004) in which familiar mixture designs in higher dimensions are projected into mixture-amount designs in lower dimensions [see also Hilgers and Bauer (1995) and Hilgers (1999)]. (In the case of some projected designs, discussed subsequently, it is also possible to provide orthogonal blocking.) To obtain projected designs, we simply delete columns of one or more ingredient levels from the original design. Mixture amount designs in q ingredients may be obtained from standard mixture designs (such as simplex-centroid or simplex-lattice designs) in $q + r$ ingredients, by projection into q dimensions. We simply delete any r columns of the original design. For example, suppose we want a mixture amount design for $q = 3$ ingredients. Consider the standard $\{4, 3\}$ simplex-lattice design shown in Table 18.7 and Figure 18.7 which has 20 runs and would normally be used to examine the effects of four ingredients. This design has one run at each of the four vertices, one run at each of the centroids of the four faces, and 12 runs in pairs on the six edges of the tetrahedron forming the mixtures simplex region. The ingredients in each row sum to 1 at present, before projection.

A component-amount design in $q = 3$ ingredients can be obtained from this simplex-lattice design by removing any one column. Removal of column x_4, for example will convert the design into three-ingredient A-slices parallel to the $x_4 = 0$ base of the tetrahedral space. See Figure 18.8. This leads to the three ingredient design of Table 18.8 in which the column headings are now given in terms of $a_i = x_i A$. This projected design has four different levels of the total amount A, namely $0, \frac{1}{3}, \frac{2}{3}$, and 1, with numbers of runs 1, 3, 6, and 10, respectively.

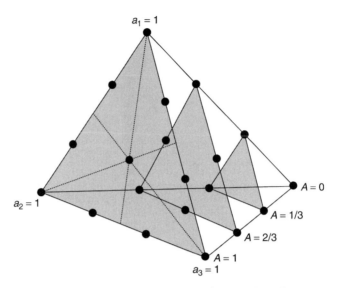

Figure 18.8. Component-amount design for $q = 3$ ingredients.

Table 18.8. Component-Amount Mixture Design for $q = 3$ Ingredients

Run Number	a_1	a_2	a_3	A	Run Number	a_1	a_2	a_3	A
1	1	0	0	1	11	1/3	0	0	1/3
2	0	1	0	1	12	2/3	1/3	0	1
3	0	0	1	1	13	2/3	0	1/3	1
4	0	0	0	0	14	2/3	0	0	2/3
5	1/3	1/3	1/3	1	15	0	1/3	2/3	1
6	1/3	1/3	0	2/3	16	0	1/3	0	1/3
7	1/3	0	1/3	2/3	17	0	2/3	1/3	1
8	0	1/3	1/3	2/3	18	0	2/3	0	2/3
9	1/3	2/3	0	1	19	0	0	1/3	1/3
10	1/3	0	2/3	1	20	0	0	2/3	2/3

An alternative way of writing out the design in terms of the x-levels is shown in Table 18.9; note the rearrangement of runs that separates the design into sections, grouping runs with the same A-levels, as shown in Figure 18.8.

It is clear from Figure 18.8 and Table 18.9 that the 20 runs in the $\{4, 3\}$ simplex-lattice design have been projected into four sets of points in the reduced simplex-lattice space in the following way:

$$\{4, 3\} \rightarrow \{3, 3\} + \{3, 2\} + \{3, 1\} + \{3, 0\},$$

containing points

$$20 \rightarrow 10 + 6 + 3 + 1.$$

Each set of points occurs once. Further projection is possible by removing additional columns. For example, a component-amounts mixtures design in $q = 2$

Table 18.9. The Rearranged Design of Table 18.8 in Terms of the Proportions x_i with Total Amount A Groupings

Run Number	x_1	x_2	x_3	A	Run Number	x_1	x_2	x_3	A
1	1	0	0	1	6	$\frac{1}{3}$	$\frac{1}{3}$	0	$\frac{2}{3}$
2	0	1	0	1	7	$\frac{1}{3}$	0	$\frac{1}{3}$	$\frac{2}{3}$
3	0	0	1	1	8	0	$\frac{1}{3}$	$\frac{1}{3}$	$\frac{2}{3}$
9	$\frac{1}{3}$	$\frac{2}{3}$	0	1	14	$\frac{2}{3}$	0	0	$\frac{2}{3}$
10	$\frac{1}{3}$	0	$\frac{2}{3}$	1	18	0	$\frac{2}{3}$	0	$\frac{2}{3}$
15	0	$\frac{1}{3}$	$\frac{2}{3}$	1	20	0	0	$\frac{2}{3}$	$\frac{2}{3}$
12	$\frac{2}{3}$	$\frac{1}{3}$	0	1	11	$\frac{1}{3}$	0	0	$\frac{1}{3}$
13	$\frac{2}{3}$	0	$\frac{1}{3}$	1	16	0	$\frac{1}{3}$	0	$\frac{1}{3}$
17	0	$\frac{2}{3}$	$\frac{1}{3}$	1	19	0	0	$\frac{1}{3}$	$\frac{1}{3}$
5	$\frac{1}{3}$	$\frac{1}{3}$	$\frac{1}{3}$	1	4	0	0	0	0

Table 18.10. Component-Amount Mixture Design for $q = 2$ Ingredients

Run Number	a_1	a_2	A	Run Number	a_1	a_2	A
1	1	0	1	11	$\frac{1}{3}$	0	$\frac{1}{3}$
2	0	1	1	12	$\frac{2}{3}$	$\frac{1}{3}$	1
3	0	0	0	13	$\frac{2}{3}$	0	$\frac{2}{3}$
4	0	0	0	14	$\frac{2}{3}$	0	$\frac{2}{3}$
5	$\frac{1}{3}$	$\frac{1}{3}$	$\frac{2}{3}$	15	0	$\frac{1}{3}$	$\frac{1}{3}$
6	$\frac{1}{3}$	$\frac{1}{3}$	$\frac{2}{3}$	16	0	$\frac{1}{3}$	$\frac{1}{3}$
7	$\frac{1}{3}$	0	$\frac{1}{3}$	17	0	$\frac{2}{3}$	$\frac{2}{3}$
8	0	$\frac{1}{3}$	$\frac{1}{3}$	18	0	$\frac{2}{3}$	$\frac{2}{3}$
9	$\frac{1}{3}$	$\frac{2}{3}$	1	19	0	0	0
10	$\frac{1}{3}$	0	$\frac{1}{3}$	20	0	0	0

ingredients may be obtained by deleting both the x_3 and x_4 columns from Table 18.7 to give the component-amount design of Table 18.10.

This two-ingredient design, illustrated in Figure 18.9 in the original four ingredient space and in Figure 18.10 in its reduced two dimensions, has the projected 20 runs divided into four runs at each of $A = 0$ and $A = 1$, and 6 runs at each of $A = 1/3$ and $A = 2/3$. The multiplicities are shown point by point in Figure 18.9 and are written next to the point locations in the deflated Figure 18.10. This projection of the $\{4, 3\}$ simplex-lattice design in $q + r = 4$ ingredients to a component-amount design in $q = 2$ ingredients thus produces the relationship

$$\{4, 3\} \rightarrow \{2, 3\} + 2 \times \{2, 2\} + 3 \times \{3, 1\} + 4 \times \{2, 0\}.$$

The corresponding allocation of points has the pattern

$$20 \rightarrow 4 + 2(3) + 3(2) + 4(1).$$

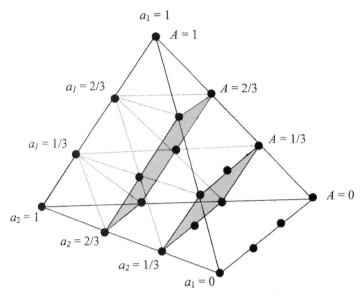

Figure 18.9. Mixture component-amount design for $q = 2$ ingredients in the original four-ingredient space.

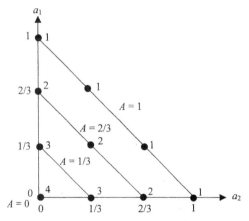

Figure 18.10. Component-amount design in the a_1, a_2 simplex (numbers represent replicates at given point).

The notation indicates that the projected design is made up of the stated number of replicates, 1, 2, 3, and 4, of the smaller simplex-lattice designs $\{2, m\}$, for $m = 3$, 2, 1, and 0, with corresponding mixture amounts given by $A = 1, 2/3, 1/3$, and 0. Of course, there is typically no need to perform all replicates in a projection, especially for just $q = 2$ ingredients, as here.

Higher-Order Base Designs

Similar projected designs may be obtained from higher-order simplex-lattice designs. For example, projecting the $\{5, 3\}$ simplex-lattice design by removing one column gives the relationship

$$\{5,3\} \rightarrow \{4,3\} + \{4,2\} + \{4,1\} + \{4,0\},$$

containing points

$$35 \rightarrow 20 + 10 + 4 + 1.$$

Removal of two columns gives

$$\{5,3\} \rightarrow \{3,3\} + 2 \times \{3,2\} + 3 \times \{3,1\} + 4 \times \{3,0\},$$

with points divided like this:

$$35 \rightarrow 10 + 2 \times 6 + 3 \times 3 + 4 \times 1.$$

Removal of three columns gives

$$\{5,3\} \rightarrow \{2,3\} + 3 \times \{2,2\} + 6 \times \{2,1\} + 10 \times \{2,0\}$$

with the points division

$$35 \rightarrow 4 + 3 \times 3 + 6 \times 2 + 10 \times 1.$$

This final projection indicates that the component-amount design produced by removal of $r = 3$ columns of the $\{5,3\}$ simplex-lattice design with 35 runs is a design containing 35 design points in $q = 2$ ingredients made up of 4 runs at $A = 1$ placed according to the $\{2,3\}$ simplex-lattice, 3 replicates of the 3 runs in $\{2,2\}$ with $A = 2/3$, 6 replicates of the 2 runs in $\{2,1\}$ with $A = 1/3$, and 10 replicates of the single run at $\{2,0\}$ with $A = 0$. Again, not all replicates need to be used.

Table 18.11 shows the component-amount designs with $q = 4$, 3, and 2 resulting from projection of the simplex lattice designs $\{5,3\}$ and $\{5,4\}$ formed by removal of $r = 1$, 2, and 3 columns, respectively.

Generalization to Projections of a $\{q + r, m\}$ Simplex-Lattice Design

This method generalized for $\{q + r, m\}$ simplex lattice designs as follows. The number of design points in these simplex lattice designs is $(q + r + m - 1)!/\{m!(q + r - 1)!\}$. Removal of r columns of the $\{q + r, m\}$ simplex-lattice mixtures design generates a component-amount design with q ingredients consisting of $^{r+k-1}C_k$ replicates of a $\{q, m - k\}$ simplex lattice design at the mixtures amounts $A = (m - k)/m$ for $k = 1, \ldots, m$. The relationship between the initial simplex-lattice mixtures design and the numbers of replicates of the simplex-lattice elements of the mixtures component-amount design is

$$\{q + r, m\} \rightarrow \{q, m\} +{}^{r}C_1\{q, m - 1\} +{}^{r+1}C_2\{q, m - 2\} + \cdots$$
$$+{}^{r+k-1}C_k\{q, m - k\} + \cdots +{}^{r+m-1}C_m\{q, 0\}.$$

Table 18.11. Mixture Component-Amount Designs Obtained by Removing 1, 2, or 3 Columns from the Simplex Lattice Designs {5, 3} and {5, 4}

Initial Design {5, 3}, n = 35 Design Points

Mixture Amount	r = 1 Simplex Component	Runs	Replicates	r = 2 Simplex Component	Runs	Replicates	r = 3 Simplex Component	Runs	Replicates
$A = 1$	{4,3}	20	1	{3,3}	10	1	{2,3}	4	1
$A = 2/3$	{4,2}	10	1	{3,2}	6	2	{2,2}	3	3
$A = 1/3$	{4,1}	4	1	{3,1}	3	3	{2,1}	2	6
$A = 0$	{4,0}	1	1	{3,0}	1	4	{2,0}	1	10

Initial Design {5, 4}, n = 70 Design Points

Mixture Amount	r = 1 Simplex Component	Runs	Replicates	r = 2 Simplex Component	Runs	Replicates	r = 3 Simplex Component	Runs	Replicates
$A = 1$	{4,4}	35	1	{3,4}	15	1	{2,4}	5	1
$A = 3/4$	{4,3}	20	1	{3,3}	10	2	{2,3}	4	3
$A = 1/2$	{4,2}	10	1	{3,2}	6	3	{2,2}	3	6
$A = 1/4$	{4,1}	4	1	{3,1}	3	4	{2,1}	2	10
$A = 0$	{4,0}	1	1	{3,0}	1	5	{2,0}	1	15

Table 18.12. Structure of the Component-Amount Design with $(q + 1)$ Different Levels of Amount A

Amount Level A	Type of Point Structure in Reduced $q - 1$ Space	Number of Design Points
1	The $(q - 1)$-dimensional simplex-centroid design	$2^{q-1} - 1$
$(q - 1)/q$	The $(q - 1)$-dimensional centroid	$\binom{q-1}{0} = 1$
$(q - 2)/(q - 1)$	All $(q - 2)$-dimensional centroids	$\binom{q-1}{1}$
$(q - 3)/(q - 2)$	All $(q - 3)$-dimensional centroids	$\binom{q-1}{2}$
...		...
$(q - r)/(q - r + 1)$	All $(q - r)$-dimensional centroids	$\binom{q-1}{r-1}$
...	(as far as $r = q - 1$, given below)	...
2/3	All two-dimensional centroids (midpoints of edges)	$\binom{q-1}{q-3}$
1/2	All one-dimensional centroids (vertices)	$\binom{q-1}{q-2}$
0	Origin point, for base level of response	1

This relationship provides an equation for the number of design points in the initial simplex-lattice design and for the numbers of replicates of the design points in the smaller simplex-lattice designs that make up the projected mixtures component-amount design. Once again, in any practical application of a mixtures component-amount design it would be up to the experimenter to decide how many replicates of the component parts of the design to use.

Forming Component-Amount Designs from Simplex-Centroid Designs

A q-ingredient simplex-centroid design consists of all vertices of the standard mixtures space plus all centroids obtained by taking these vertices in pairs, threes, ..., q's. It has ${}^qC_1 + {}^qC_2 + \cdots + {}^qC_q = 2^q - 1$ design points in all, where ${}^qC_r$ is the number of ways r items can be selected from q items when order of selection is not relevant. Suppose we project this down to $(q - 1)$ dimensions, dropping any column. We obtain a mixture-amount design with $(q + 1)$ different levels of amount A, with the characteristics shown in Table 18.12.

For example, for $q = 4$ with projection to three dimensions, we obtain the following, using the same order as above: a full seven-point simplex-centroid in three mixture ingredients, namely, $(1, 0, 0)$, $(0, 1, 0)$, $(0, 0, 1)$ $(1/2, 1/2, 0)$, $(1/2, 0, 1/2)$, $(0, 1/2, 1/2)$, $(1/3, 1/3, 1/3)$ with $A = 1$; the centroid point $(1/4, 1/4, 1/4)$ with $A = 3/4$; the three two-ingredient centroids at $(1/3, 1/3, 0)$, $(1/3, 0, 1/3)$, $(0, 1/3, 1/3)$, which have $A = 2/3$; the three vertices $(1/2, 0, 0)$, $(0, 1/2, 0)$, $(0, 0, 1/2)$, with $A = 1/2$; and the origin $(0, 0, 0)$, with $A = 0$. This design has 15 points at five levels of A.

Similarly, for $q = 5$ with projection to four dimensions, we obtain the following: a full simplex-centroid in four mixture ingredients $(1, 0, 0, 0)$, etc., with $A = 1$; the centroid point $(1/5, 1/5, 1/5, 1/5)$ with $A = 4/5$; the four three-ingredient centroids at $(1/4, 1/4, 1/4, 0)$, $(1/4, 1/4, 0, 1/4)$, $(1/4, 0, 1/4, 1/4)$, $(0, 1/4, 1/4, 1/4)$, with $A = 3/4$; the six two-ingredient centroids with points $(1/3, 1/3, 0, 0)$, $(1/3, 0, 1/3, 0)$, $(1/3, 0, 0, 1/3)$, $(0, 1/3, 1/3, 0)$, $(0, 1/3, 0, 1/3)$, $(0, 0, 1/3, 1/3)$, which have $A = 2/3$; the four vertices $(1/2, 0, 0, 0)$, $(0, 1/2, 0, 0)$, $(0, 0, 1/2, 0)$, $(0, 0, 0, 1/2)$ with $A = 1/2$; and the null point $(0, 0, 0, 0)$, with $A = 0$. This design has 31 points at six levels of A.

In projections to lower dimensions, we obtain some repeated subsets of points. For example: for $q = 5$ with projection to three dimensions, we obtain: a full simplex-centroid in three mixture ingredients, namely $(1, 0, 0)$, $(0, 1, 0)$, $(0, 0, 1)$, $(1/2, 1/2, 0)$, $(1/2, 0, 1/2)$, $(0, 1/2, 1/2)$, $(1/3, 1/3, 1/3)$ with 7 points and $A = 1$; replicated (2) centroid points $(1/4, 1/4, 1/4)$, each with $A = 3/4$, 2 points; two two-ingredient centroids at $(1/3, 1/3, 0)$, $(1/3, 0, 1/3)$, $(0, 1/3, 1/3)$, which each have $A = 2/3$ and 6 points; one centroid point $(1/5, 1/5, 1/5)$, with $A = 3/5$; two sets of the three vertices $(1/2, 0, 0)$, $(0, 1/2, 0)$, $(0, 0, 1/2)$, with 6 points and $A = 1/2$; a set of three points $(1/4, 1/4, 0)$, $(1/4, 0, 1/4)$, $(0, 1/4, 1/4)$, with $A = 1/2$; vertices $(1/3, 0, 0)$, $(0, 1/3, 0)$, $(0, 0, 1/3)$, with $A = 1/3$; and the origin $(0, 0, 0)$, with $A = 0$, three times over. We have 31 points at eight levels of A in this design. Compared with the first example projected from $q = 4$, we have an additional 16 design points, but also have 7 levels of A instead of 5, with $A = 3/5$ (one point) and $1/3$ (3 points) added. At other levels in this comparison, there are now some repeated points. There are clearly numerous available possibilities for these projection types. Again, the $D_n = |\mathbf{X}'\mathbf{X}|/n^p$ criterion can be used to choose among competing possibilities for a specific situation. Higher values of D_n are better.

18.5. DESIGNS IN ORTHOGONAL BLOCKS FOR COMPONENT-AMOUNT MODELS

In situations where it is not practicable to carry out all of the experimental runs under similar conditions, it is usually desirable to split the experiment into blocks of runs that are more homogeneous within the blocks. If estimation of the model terms can be made independent of estimation of the block effects by using orthogonal blocking, it is easier to interpret the model. We now consider the construction of mixture component-amount experiments in two or more orthogonal blocks by adapting existing non-component-amount mixture designs via column elimination.

A Design for Two Ingredients in Two Orthogonal Blocks

To illustrate the procedure, we first consider the design of a mixture component-amount experiment in *two* ingredients using two orthogonal blocks. We begin with an orthogonally blocked mixture experiment in *three* ingredients described by Czitrom (1988), based on Latin squares, satisfying the conditions for orthogonality proposed by Nigam (1970) and further refined by John (1984). The original design is shown in Table 18.13, where a, b, and c with $a + b + c = 1$ are the proportional amounts of the three ingredients x_1, x_2, and x_3, and each block contains the common run with coordinates (u, v, w), which is required to prevent the design being singular. Often the centroid point $(1/3, 1/3, 1/3)$ would be used as the

Table 18.13. Block Structure for a Three-Ingredient Mixture Design in Two Orthogonal Blocks

	Block 1			Block 2	
x_1	x_2	x_3	x_1	x_2	x_3
a	b	c	a	c	b
b	c	a	b	a	c
c	a	b	c	b	a
u	v	w	u	v	w

common run, but this is not mandatory. This design, based on two particular 3×3 Latin squares called *mates*, with blocks indexed by -1 and $+1$, allows fitting of the second-order Scheffé mixture model

$$\eta = \beta_1 x_1 + \beta_2 x_2 + \beta_3 x_3 + \beta_{12} x_1 x_2 + \beta_{13} x_1 x_3 + \beta_{23} x_2 x_3 + \delta, \quad (18.5.1)$$

with the block difference, 2δ, estimated orthogonally to all the mixture terms. The determinant of the information matrix for this design and this model is given by

$$|\mathbf{X}'\mathbf{X}| = 96(a - b)^4 (b - c)^4 (c - a)^4 \{(ab + bc + ca) - (uv + uw + vw)\}^2, \quad (18.5.2)$$

which takes its (D-optimum) maximum value of 2.66872×10^{-4} at $a = 0.158$, $b = 0.832$, and $c = 0$ and all other permutations of these values, when $u = v = w = 1/3$—that is, when the common run is taken at the centroid of the simplex.

To obtain a component-amount design for two factors from this three-factor mixture design, we eliminate any one column. For example, if we eliminate the third column, x_3, from Table 18.13, we obtain the two blocks shown in Table 18.14. The quadratic mixtures component-amount model for two ingredients and with a block difference 2δ is given by

$$\eta = \alpha_0 + \alpha_1 a_1 + \alpha_2 a_2 + \alpha_{11} a_1^2 + \alpha_{22} a_2^2 + \alpha_{12} a_1 a_2 + \delta, \quad (18.5.3)$$

Table 18.14. Block Structure for a Two-Ingredient Component-Amount Design in Two Orthogonal Blocks

Block 1			Block 2		
Components		Amounts	Components		Amounts
a_1	a_2	A	a_1	a_2	A
a	b	$a + b$	a	c	$a + c$
b	c	$b + c$	b	a	$b + a$
c	a	$c + a$	c	b	$c + b$
u	v	$u + v$	u	v	$u + v$

which has **X**-matrix given by

$$\mathbf{X} = \begin{pmatrix} 1 & a & b & a^2 & b^2 & ab & -1 \\ 1 & b & c & b^2 & c^2 & bc & -1 \\ 1 & c & a & c^2 & a^2 & ac & -1 \\ 1 & u & v & u^2 & v^2 & uv & -1 \\ 1 & a & c & a^2 & c^2 & ac & 1 \\ 1 & b & a & b^2 & a^2 & ab & 1 \\ 1 & c & b & c^2 & b^2 & bc & 1 \\ 1 & u & v & u^2 & v^2 & uv & 1 \end{pmatrix}.$$

It is clear that this design is orthogonally blocked for the second-order model (18.5.3) because some of the columns of this matrix **X** are a subset of those in the **X** for the original model and design and the others are obviously orthogonal to the blocking column. We next seek the values of the parameters that provide a D-optimal arrangement. The determinant of the information matrix is

$$|\mathbf{X}'\mathbf{X}| = 96(a - b)^4(b - c)^4(a - c)^4\big(ab + bc + ac + u^2 + uv + v^2 - (u + v)\big)^2. \tag{18.5.4}$$

This is maximized when $u = v = 1/3$ so that in this case

$$|\mathbf{X}'\mathbf{X}| = 32(a - b)^4(b - c)^4(a - c)^4(3(ab + bc + ac) - 1)^2/3. \tag{18.5.5}$$

Surprisingly, this is exactly the same as the determinant of the information matrix for the three-ingredient mixture experiment using Scheffé's quadratic mixture model—that is, Eq. (18.5.2) with $u = v = w = 1/3$. It follows that it also takes its maximum value of $|\mathbf{X}'\mathbf{X}| = 2.66872 \times 10^{-4}$ at $a = 0$, $b = 0.168$, $c = 0.832$ and all other permutations of these values. The optimal design with these values of the ingredient amounts is shown in Table 18.15 and illustrated for the ingredient totals, $A = 0.168, 0.667, 0.832$, and 1.0 in Figure 18.11 The two overlapping points are at the same location $(1/3, 1/3)$ but are shown overlapping for illustrative purposes only.

Table 18.15. Optimal Component-Amount Design for Two Ingredients in Two Orthogonal Blocks

Block 1			Block 2		
Components		Amounts	Components		Amounts
a_1	a_2	A	a_1	a_2	A
0	0.168	0.168	0	0.832	0.832
0.168	0.832	1.0	0.168	0	0.168
0.832	0	0.832	0.832	0.168	1.0
0.333	0.333	0.667	0.333	0.333	0.667

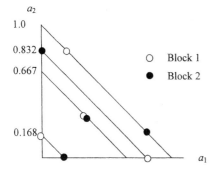

Figure 18.11. Two ingredient orthogonally blocked mixtures component-amount design.

Designs for Three Ingredients in Orthogonal Blocks

A similar approach may be used to produce three-ingredient component-amount designs from four-ingredient mixture designs. Building on the results of John (1984) and Czitrom (1988, 1989), Draper, Prescott et al. (1993) produced an exhaustive investigation of four-ingredient Latin-square-based designs in two orthogonal blocks. They showed that orthogonal block designs could be produced using pairs of (4×4) Latin squares (mates) obtained from the six permutations of each of three of the four available standard Latin squares of side four. Mates were defined as squares giving rise to exactly the same pattern of cross products since this satisfied the condition for block orthogonality. The six permutations of each standard square gave rise to the same cross-product terms but in three specific orders, identifying three pairs of Latin squares as mates. The values of the cross products were different from each standard square but the three sets of patterns were similar. Different orthogonally blocked designs for fitting second-order Scheffé models could be obtained by selecting pairs of mates with one square in block 1 and the other in block 2. Two or more pairs of mates are required to provide sufficient runs to estimate the 11 parameters (including a blocking parameter) in a second-order mixture model, and a common run, usually the centroid at $(1/4, 1/4, 1/4, 1/4)$, is required in each block to avoid singularity. Different combinations of pairs of mates give rise to designs with different properties. We next examine the possible types of designs that can arise.

Designs of Type A

An example of such an orthogonally blocked design, which we call a design of type A, is shown in Table 18.16. The squares in block 1 are SI(2) and SI(3), and their mates in block 2 are SVI(2) and SV(3) in the notation of Draper, Prescott et al. (1993). The cross-product patterns for these pairs of mates are LLKKLL and PQPPQP, respectively, corresponding to the six cross-product terms $\Sigma x_1 x_2$, $\Sigma x_1 x_3$, $\Sigma x_1 x_4$, $\Sigma x_2 x_3$, $\Sigma x_2 x_4$, $\Sigma x_3 x_4$, summed over the columns of the Latin squares used in the design, with $L = ab + ac + bd + cd$, $K = 2(bc + ad)$, $P = ab + ad + bc + cd$, and $Q = 2(ac + bd)$. (Characteristic of designs of Type A is that two runs in block 1 also occur in block 2, permitting the option of eliminating these four runs while retaining orthogonal blocking. Here, the two runs are *dcba* and *cdab*.)

It is shown in Draper, Prescott et al. (1993) that, by combining pairs of mates with different cross-product patterns, several classes of orthogonally blocked

Table 18.16. Design A: Two Pairs of Mates with Different Cross-Product Patterns from Different Standard Squares

Block 1				Block 2			
x_1	x_2	x_3	x_4	x_1	x_2	x_3	x_4
a	b	c	d	a	c	b	d
b	d	a	c	b	a	d	c
c	a	d	b	c	d	a	b
d	c	b	a	d	b	c	a
a	b	c	d	a	d	c	b
b	c	d	a	b	a	d	c
c	d	a	b	c	b	a	d
d	a	b	c	d	c	b	a
u	v	w	z	u	v	w	z

mixture designs may be derived. All the designs within a class have the same algebraic form for the $\det(\mathbf{X}'\mathbf{X})$, and so all give rise to the same maximum determinant, provided that the full design is used, even in the classes where some of the runs are the same in both blocks.

An orthogonally blocked mixtures component-amount design in three ingredients may be obtained by eliminating a column of the four-component design in Table 18.16. In Table 18.17, the x_4 column has been dropped. The corresponding $\mathbf{X}$-matrix for the component-amount model is given below, where the final column of ± 1's corresponds to the blocking variable z.

$$
\mathbf{X} =
\begin{bmatrix}
1 & a & b & c & a^2 & b^2 & c^2 & ab & ac & bc & -1 \\
1 & b & d & a & b^2 & d^2 & a^2 & bd & ab & da & -1 \\
1 & c & a & d & c^2 & a^2 & d^2 & ac & cd & da & -1 \\
1 & d & c & b & d^2 & c^2 & b^2 & cd & bd & bc & -1 \\
1 & a & b & c & a^2 & b^2 & c^2 & ab & ac & bc & -1 \\
1 & b & c & d & b^2 & c^2 & d^2 & bc & bd & cd & -1 \\
1 & c & d & a & c^2 & d^2 & a^2 & cd & ac & da & -1 \\
1 & d & a & b & d^2 & a^2 & b^2 & da & bd & ab & -1 \\
1 & v & v & v & v^2 & v^2 & v^2 & v^2 & v^2 & v^2 & -1 \\
1 & a & c & b & a^2 & c^2 & b^2 & ac & ab & bc & 1 \\
1 & b & a & d & b^2 & a^2 & d^2 & ab & bd & da & 1 \\
1 & c & d & a & c^2 & d^2 & a^2 & cd & ac & da & 1 \\
1 & d & b & c & d^2 & b^2 & c^2 & bd & cd & bc & 1 \\
1 & a & d & c & a^2 & d^2 & c^2 & da & ac & cd & 1 \\
1 & b & a & d & b^2 & a^2 & d^2 & ab & bd & da & 1 \\
1 & c & b & a & c^2 & b^2 & a^2 & bc & ac & ab & 1 \\
1 & d & c & b & d^2 & c^2 & b^2 & cd & bd & bc & 1 \\
1 & v & v & v & v^2 & v^2 & v^2 & v^2 & v^2 & v^2 & 1
\end{bmatrix}.
$$

Table 18.17. A Component-Amount Design of Type A in Three Ingredients in Two Orthogonal Blocks

Block 1				Block 2			
Components			Amount	Components			Amounts
a_1	a_2	a_3	A	a_1	a_2	a_3	A
a	b	c	$a + b + c = 1 - d$	a	c	b	$a + c + b = 1 - d$
b	d	a	$b + d + a = 1 - c$	b	a	d	$b + a + d = 1 - c$
c	a	d	$c + a + d = 1 - b$	c	d	a	$c + d + a = 1 - b$
d	c	b	$d + c + b = 1 - a$	d	b	c	$d + b + c = 1 - a$
a	b	c	$a + b + c = 1 - d$	a	d	c	$a + d + c = 1 - b$
b	c	d	$b + c + d = 1 - a$	b	a	d	$b + a + d = 1 - c$
c	d	a	$c + d + a = 1 - b$	c	b	a	$c + b + a = 1 - d$
d	a	b	$d + a + b = 1 - c$	d	c	b	$d + c + b = 1 - a$
u	v	w	$u + v + w = 1 - z$	u	v	w	$u + v + w = 1 - z$

The corresponding information matrix $\mathbf{X'X}$ has last row and last column consisting of all zeros except for the final diagonal entry, which is 18, showing that the design has the property of block-orthogonality. The design is such that there are at most five levels of the component-amount A. For D-optimality, we require the maxima of the determinant of $\mathbf{X'X}$, or equivalently the maxima of the determinant of $\mathbf{X_0'X_0}$, where $\mathbf{X_0}$ is equal to $\mathbf{X}$ with the final column deleted; see also Chan and Sandhu (1999, Section 2). Algebraic evaluation of the determinant of $\mathbf{X_0'X_0}$ is difficult, because of the complexity of some of the elements of the matrix. However, a simple grid search reveals that the maximum determinant corresponds to points such that $a = 0$, $b = 0$, $c = 0.240$, and $d = 0.760$, with the common run at the original centroid, $u = v = w = \frac{1}{4}$.

These are exactly the same values that produced the optimal designs in this class of four-ingredient mixture designs using Scheffé's quadratic model, but, unlike that case, not all possible combinations of these values give rise to maxima of the determinant of $\mathbf{X'X}$. Numerical computation showed that only four combinations corresponded to maximum values, as shown in Table 18.18. There are in fact only two distinct designs resulting from these four optimal combinations, since solutions 1 and 2 are the same design with the two blocks interchanged, as are the designs produced by solutions 3 and 4. Convenient "nearly optimal" practical values very close to these optimal combinations are $(0, 0, \frac{1}{4}, \frac{3}{4})$ and $(\frac{1}{4}, \frac{3}{4}, 0, 0)$, for which the determinant is $0.12234614 \times 10^{-5}$ compared with the optimal value $0.12397522 \times 10^{-5}$ shown in Table 18.18. The nearly optimal component-amount designs corresponding to these two combinations are shown as Designs A1 and A2 in Tables 18.19 and 18.20. These designs have just three amount levels, $A = \frac{1}{4}, \frac{3}{4}$, and 1, which contain 4, 6, and 8 runs respectively, equally divided between the two orthogonal blocks. The block structures are illustrated in Figures 18.12 and 18.13, where the open circles represent runs in block 1 and the filled circles represent runs in block 2.

As introduced here, in Designs A1 and A2, there is the (sometimes undesirable) practical limitation that the three ingredients occur mostly in two-factor combinations. In practice, the designs can be drawn into the interior of the space by

Table 18.18. Combinations of a, b, c, and d Leading to Type A Optimal Designs

a	b	c	d	$\vert \mathbf{X'X} \vert$	Solutions
0	0	.24	.76	$0.12397552 \times 10^{-5}$	1
0	0	.76	.24	$0.12397552 \times 10^{-5}$	2
0	.24	0	.76	0	
0	.24	.76	0	0	
0	.76	0	.24	0	
0	.76	.24	0	0	
.24	0	0	.76	0	
.24	0	.76	0	0	
.24	.76	0	0	$0.12397552 \times 10^{-5}$	3
.76	0	0	.24	0	
.76	0	.24	0	0	
.76	.24	0	0	$0.12397552 \times 10^{-5}$	4

Table 18.19. A Type A Nearly Optimal, Component-Amount Design for Three Ingredients in Two Orthogonal Blocks Produced from Solutions 1 and 2, Design A1

Block 1				Block 2			
Components			Amounts	Components			Amounts
a_1	a_2	a_3	A	a_1	a_2	a_3	A
0	0	1/4	1/4	0	1/4	0	1/4
0	3/4	0	3/4	0	0	3/4	3/4
1/4	0	3/4	1	1/4	3/4	0	1
3/4	1/4	0	1	3/4	0	1/4	1
0	0	1/4	1/4	0	3/4	1/4	1
0	1/4	3/4	1	0	0	3/4	3/4
1/4	3/4	0	1	1/4	0	0	1/4
3/4	0	0	3/4	3/4	1/4	0	1
1/4	1/4	1/4	3/4	1/4	1/4	1/4	3/4

Table 18.20. A Type A Nearly Optimal, Component-Amount Design for Three Ingredients in Two Orthogonal Blocks Produced from Solutions 3 and 4, Design A2

Block 1				Block 2			
Components			Amounts	Components			Amounts
a_1	a_2	a_3	A	a_1	a_2	a_3	A
1/4	3/4	0	1	1/4	0	3/4	1
3/4	0	1/4	1	3/4	1/4	0	1
0	1/4	0	1/4	0	0	1/4	1/4
0	0	3/4	3/4	0	3/4	0	3/4
1/4	3/4	0	1	1/4	0	0	1/4
3/4	0	0	3/4	3/4	1/4	0	1
0	0	1/4	1/4	0	3/4	1/4	1
0	1/4	3/4	1	0	0	3/4	3/4
1/4	1/4	1/4	3/4	1/4	1/4	1/4	3/4

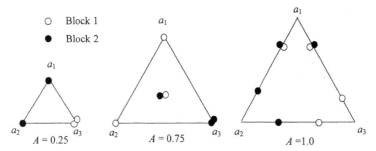

Figure 18.12. Orthogonal block structure for the component-amount design A1.

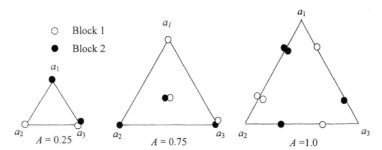

Figure 18.13. Orthogonal block structure for the component-amount design A2.

imposing (the same or different) lower-bound limitations on the ingredients. Positioning points on the boundaries is not only a representational convenience here, but it also produces the highest value of the determinant criterion $D = \det(\mathbf{X}'\mathbf{X})$. Relaxing this latter requirement via restrictions frees the pattern of points to move inside the region. For a useful survey of results about optimal designs in the area of mixtures experiments, see Chan (2000).

Designs of Type B

A different type of design, an example of which appears in Table 18.21, is obtained by choosing pairs of mates from different standard squares but with similar patterns of cross-products. In such *Type B* designs, a characteristic is that no row is repeated, either in the same block or in different blocks. Thus no run reduction is possible if orthogonal blocking is to be maintained. The squares in block 1 are SI(3) and SIII(4), and their mates in block 2 are SV(3) and SVI(4) in the notation of Draper, Prescott et al. (1993). The cross-product patterns for these pairs of mates are PQPPQP and BABBAB respectively, corresponding to the six cross-product terms $\Sigma x_1 x_2$, $\Sigma x_1 x_3$, $\Sigma x_1 x_4$, $\Sigma x_2 x_3$, $\Sigma x_2 x_4$, and $\Sigma x_3 x_4$, summed over the columns of the Latin squares, where $A = 2(ab + cd)$, $B = ac + ad + bc + bd$, P, and Q are as defined earlier.

An orthogonally blocked mixtures component-amount design in three ingredients may be obtained by eliminating any column of the four-component design in Table 18.21. We again choose to eliminate the x_4 column, giving the design in

Table 18.21. Design B: Two Pairs of Mates with Similar Cross-Product Patterns from Different Standard Squares

	Block 1				Block 2		
x_1	x_2	x_3	x_4	x_1	x_2	x_3	x_4
a	b	c	d	a	d	c	b
b	c	d	a	b	a	d	c
c	d	a	b	c	b	a	d
d	a	b	c	d	c	b	a
a	d	b	c	a	c	b	d
b	c	a	d	b	d	a	c
c	a	d	b	c	b	d	a
d	b	c	a	d	a	c	b
u	v	w	z	u	v	w	z

Table 18.22. A Component-Amount Design of Type B in Three Ingredients in Two Orthogonal Blocks

	Block 1				Block 2		
Components			Amounts	Components			Amounts
a_1	a_2	a_3	A	a_1	a_2	a_3	A
a	b	c	$1-d$	a	d	c	$1-b$
b	c	d	$1-a$	b	a	d	$1-c$
c	d	a	$1-b$	c	b	a	$1-d$
d	a	b	$1-c$	d	c	b	$1-a$
a	d	b	$1-c$	a	c	b	$1-d$
b	c	a	$1-d$	b	d	a	$1-c$
c	a	d	$1-b$	c	b	d	$1-a$
d	b	c	$1-a$	d	a	c	$1-b$
u	v	w	$1-z$	u	v	w	$1-z$

Table 18.23. Combinations of a, b, c, and d Leading to Optimal Designs of Type B

a	b	c	d	$|\mathbf{X'X}|$	Solutions
0	0	.24	.76	$0.12397552 \times 10^{-5}$	1
0	0	.76	.24	$0.12397552 \times 10^{-5}$	2
0	.24	0	.76	$0.12397552 \times 10^{-5}$	3
0	.24	.76	0	0	
0	.76	0	.24	$0.12397552 \times 10^{-5}$	4
0	.76	.24	0	0	
.24	0	0	.76	0	
.24	0	.76	0	$0.12397552 \times 10^{-5}$	5
.24	.76	0	0	$0.12397552 \times 10^{-5}$	6
.76	0	0	.24	0	
.76	0	.24	0	$0.12397552 \times 10^{-5}$	7
.76	.24	0	0	$0.12397552 \times 10^{-5}$	8

Table 18.24. A Type B Nearly Optimal Component-Amount Design for Three Ingredients in Two Orthogonal Blocks

	Block 1				Block 2		
	Components		Amounts		Components		Amounts
a_1	a_2	a_3	A	a_1	a_2	a_3	A
0	0	1/4	1/4	0	3/4	1/4	1
0	1/4	3/4	1	0	0	3/4	3/4
1/4	3/4	0	1	1/4	0	0	1/4
3/4	0	0	3/4	3/4	1/4	0	1
0	3/4	0	3/4	0	1/4	0	1/4
0	1/4	0	1/4	0	3/4	0	3/4
1/4	0	3/4	1	1/4	0	3/4	1
3/4	0	1/4	1	3/4	0	1/4	1
1/4	1/4	1/4	3/4	1/4	1/4	1/4	3/4

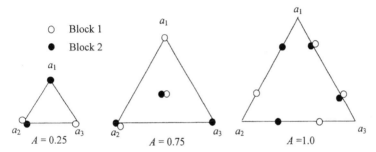

Figure 18.14. Orthogonal block structure for the component-amount design of Type B.

Table 18.22. The determinant of the corresponding information matrix, $\mathbf{X'X}$, of this Type B design is maximized at 8 of the 12 possible combinations of the values $a = 0$, $b = 0$, $c = 0.24$, and $d = 0.76$, with $u = v = w = \frac{1}{4}$ as shown in Table 18.23. The four other possibilities lead to singular designs.

Although there are eight combinations that lead to optimum designs, all give rise to exactly the same block structure for designs of Type B. If we use the more convenient levels $\frac{1}{4}$ and $\frac{3}{4}$ instead of 0.24 and 0.76, we obtain the design shown in Table 18.24 and illustrated in Figure 18.14.

Designs of Type C

A third type of design is obtained by choosing the pairs of mates from the same standard squares but necessarily with different patterns of cross-products. Such a design is shown in Table 18.25. The squares in block 1 are SI(4) and SII(4), and their mates in block 2 are SIV(4) and SV(4) in the notation of Draper, Prescott et al. (1993). The cross-product patterns for these pairs of mates are *ABBBBA* and *BBAABB* respectively, where *A* and *B* are as defined earlier. An orthogonally blocked mixtures component-amount design in three ingredients may be obtained

Table 18.25. Design C: Two Pairs of Mates with Different Cross-Product Patterns from the Same Standard Squares

Block 1				Block 2			
x_1	x_2	x_3	x_4	x_1	x_2	x_3	x_4
a	b	c	d	a	b	d	c
b	a	d	c	b	a	c	d
c	d	b	a	c	d	a	b
d	c	a	b	d	c	b	a
a	c	d	b	a	d	c	b
b	d	c	a	b	c	d	a
c	b	a	d	c	a	b	d
d	a	b	c	d	b	a	c
u	v	w	z	u	v	w	z

Table 18.26. A Component-Amount Design of Type C in Three Ingredients in Two Orthogonal Blocks

Block 1				Block 2			
Components			Amounts	Components			Amounts
a_1	a_2	a_3	A	a_1	a_2	a_3	A
a	b	c	$1-d$	a	b	d	$1-c$
b	a	d	$1-c$	b	a	c	$1-d$
c	d	b	$1-a$	c	d	a	$1-b$
d	c	a	$1-b$	d	c	b	$1-a$
a	c	d	$1-b$	a	d	c	$1-b$
b	d	c	$1-a$	b	c	d	$1-a$
c	b	a	$1-d$	c	a	b	$1-d$
d	a	b	$1-c$	d	b	a	$1-c$
u	v	w	$1-z$	u	v	w	$1-z$

Table 18.27. A Type C Nearly Optimal Component-Amount Design for Three Ingredients in Two Orthogonal Blocks

Block 1				Block 2			
Components			Amounts	Components			Amounts
a_1	a_2	a_3	A	a_1	a_2	a_3	A
0	1/4	0	1/4	0	1/4	3/4	1
1/4	0	3/4	1	1/4	0	0	1/4
0	3/4	1/4	1	0	3/4	0	3/4
3/4	0	0	3/4	3/4	0	1/4	1
0	0	3/4	3/4	0	3/4	0	3/4
1/4	3/4	0	1	1/4	0	3/4	1
0	1/4	0	1/4	0	0	1/4	1/4
3/4	0	1/4	1	3/4	1/4	0	1
1/4	1/4	1/4	3/4	1/4	1/4	1/4	3/4

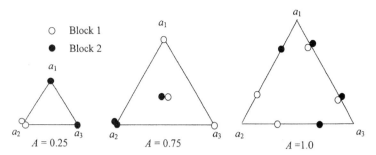

Figure 18.15. Orthogonal block structure for the component-amount design of Type C.

by eliminating any column (for example, the fourth) of the four-component design in Table 18.25 to give the design in Table 18.26. As with type B designs, the determinant of the information matrix, $\mathbf{X'X}$, of this type C design is maximized at 8 of the 12 possible combinations of the values $a = 0$, $b = 0$, $c = 0.24$, and $d = 0.76$. The other four possibilities again give rise to singular designs.

As with the type B designs, although there are 8 combinations leading to optimum designs, all give rise to exactly the same block structure for designs of type C. A nearly optimal version of this design with levels set at $\frac{1}{4}$ and $\frac{3}{4}$ is shown in Table 18.27 and is illustrated in Figure 18.15.

Other Combinations of Pairs of Mates

Other designs may or may not be possible with different combinations of pairs of mates. For example, the pairs of mates SI(4), SIV(4), and SIII(4), SVI(4) may be used to form two blocks similar to the design of type C shown previously, but with cross product patterns *ABBBBA* and *BABBAB*. Projection of this design by elimination of the x_4 column produces a component-amount design with each of the eight optimal solutions giving the identical optimal design. Although this design was produced from pairs of mates from the same standard square and necessarily different cross-product patterns, it is identical to Design A2, as shown in Figure 18.11, which was produced using pairs of mates from different squares and with different cross-product patterns.

A third way of combining pairs of mates from the same standard square uses pairs with cross-product patterns *BBAABB* (mates SII(4) and SV(4)) and *BABBAB* (mates SII(4) and SVI(4)). This combination leads to the same optimal (or near-optimal) design as the initial Type C design shown in Figure 18.15, but with a different orientation (i.e. with a_1 and a_2 interchanged).

18.6. AN EXAMPLE OF A PROJECTION OF A NEARLY *D*-OPTIMAL MIXTURE DESIGN INTO FEWER DIMENSIONS

The data from a bread-making experiment described by Draper, Prescott et al. (1993, p. 270), reproduced with permission from the American Statistical Association in Table 18.28, provide an excellent illustration of an orthogonally blocked

Table 18.28. The Orthogonally Blocked Design and Response Values for the Bread-Making Experiment

	Blocks 1 and 3			Responses from Block			Blocks 2 and 4			Responses from Block	
x_1	x_2	x_3	x_4	1	3	x_1	x_2	x_3	x_4	2	4
0	b	0	d	403	381	0	d	0	b	423	404
b	0	d	0	425	422	b	0	d	0	417	425
0	d	0	b	442	412	0	b	0	d	398	391
d	0	b	0	433	413	d	0	b	0	407	426
0	d	b	0	445	398	0	0	b	d	388	362
b	0	0	d	435	412	b	d	0	0	435	427
0	0	d	b	385	371	0	b	d	0	379	390
d	b	0	0	425	428	d	0	0	b	406	411
u	v	w	z	433	393	u	v	w	z	439	409

nearly D-optimal mixture design that retains orthogonal blocking characteristics when projected into fewer and different dimensions. In the notation of Draper, Prescott et al. (1993), the design consists of the adjoin of squares SI(3) and SV(3) with pattern *PQPPQP*, and of squares SIII(4) and SVI(4), with pattern *BABBAB*. The nearly D-optimal values of (a, b, c, d)—that is, $(0, \frac{1}{4}, 0, \frac{3}{4})$—are employed in the runs of Table 18.28 together with the common centroid $u = v = w = z = \frac{1}{4}$. [Note that the equivalent Table 2 in Draper, Prescott et al. (1993, p. 270) contains two misprinted response values that are corrected here. In block 2, the third and fifth values have been corrected to 398 and 388, replacing the erroneous 388 and 338.]

When a full second-order Scheffé model is fitted to the data, using the orthogonal blocking variables shown in Table 18.29, the fitted equation is

$$\hat{y} = 386.2x_1 + 435.1x_2 + 384.7x_3 + 389.3x_4 + 96.4x_1x_2 + 189.6x_1x_3 + 150.7x_1x_4$$
$$- 36.9x_2x_3 - 29.1x_2x_4 - 56.0x_3x_4 - 3.3z_1 - 6.8z_2 + 4.1z_3. \qquad (18.6.1)$$

An extra sum of squares test indicates it is reasonable to omit the x_2x_3, x_2x_4, and x_3x_4 terms, reducing the refitted model to

$$\hat{y} = 383.4x_1 + 430.3x_2 + 375.2x_3 + 381.6x_4 + 107.8x_1x_2 + 217.9x_1x_3$$
$$+ 169.7x_1x_4 - 3.3z_1 - 6.8z_2 + 4.1z_3. \qquad (18.6.2)$$

Table 18.29. Three Orthogonal Columns of Dummy Variables Used to Separate the Four Blocks

Block	z_1	z_2	z_3
1	-1	-1	1
2	1	-1	-1
3	-1	1	-1
4	1	1	1

Table 18.30. A Comparison of the Four Fitted Equations

Equation Fitted	Extra Sum of Squares Taken Up[a]	Residual Mean Square	Residual df
(18.6.1)	12,490	130.0	23
(18.6.2)	12,298	122.3	26
(18.6.3)	12,164	122.8	27
(18.6.4)	12,164	122.8	27

[a]Out of the possible 15,477 remaining after the fitting of a constant response

Snee (1995) pointed out that the model could sensibly be further simplified by fitting a second-order Scheffé model to the variables x_1, x_2 and $x_3 + x_4$. It will then be found that the cross product term $x_2(x_3 + x_4) = x_2 x_3 + x_2 x_4$ contributes very little and can be dropped to yield the fitted equation

$$\hat{y} = 385.3x_1 + 433.3x_2 + 379.5(x_3 + x_4) + 95.8x_1x_2 + 191.5x_1(x_3 + x_4)$$
$$- 10.2x_2(x_3 + x_4) - 3.3z_1 - 6.8z_2 + 4.1z_3. \tag{18.6.3}$$

An alternative explanation of the data may be obtained if we recall that, because of the mixture constraint, $(x_3 + x_4) = 1 - (x_1 + x_2)$, so that model (18.6.3) is really an equation in the amounts of the first two ingredients x_1 and x_2 and their total amount $A = (x_1 + x_2)$. In hindsight we could have removed the x_3 and x_4 columns from the design and fitted the model:

$$\hat{y} = 379.5 + 197.3x_1 + 43.5x_2 - 191.5x_1^2 + 10.2x_2^2 - 85.5x_1x_2$$
$$- 3.3z_1 - 6.8z_2 + 4.1z_3. \tag{18.6.4}$$

That this provides a relatively simple explanation of the data can be appreciated by comparing the four regression fits (18.6.1) through (18.6.4) in Table 18.30. Eq. (18.6.1) has more parameters than Eqs. (18.6.2)–(18.6.4), yet explains slightly less of the overall variation. Equation (18.6.2) fits only very slightly better than (18.6.3) and (18.6.4), which are simply reparameterizations of each other. This indicates that Snee's suggestion that (18.6.3) or (18.6.4) be adopted is an eminently sensible one. Note that the orthogonal blocking properties of the design are preserved for all four equations, as is clear from the fact that the estimates of the regression coefficients associated with blocks are unchanged throughout. In hindsight, we see that a mixture component-amount analysis *on the response used* could have been conducted in only the first two ingredients. This does not necessarily imply that the other ingredients are not needed in the mixture, however.

Our philosophy in interpreting the results of mixture experiments, or even more general response surface experiments without mixture restrictions, is that it is especially important to understand the overall shape of the fitted surface, either by examining contour plots (for $q = 2$ or 3 ingredients), or by sectional plots in three dimensions ($q > 3$), or by canonical reduction and/or ridge analysis ($q > 4$ usually). The interpretation of individual regression coefficient estimates, while seductive, can be misleading unless performed by an experimenter who understands the physical characteristics of the process being studied, or unless it relates to a very simple model—for example, one of only first order in the ingredients. This is because many mixtures models can be expressed in a multitude of different ways by substitutions that use the mixture restriction, $\sum_{i=1}^{q} x_i = 1$ (see Prescott et al., 2002). Different appearing, but actually equivalent, models may lead to different speculations about the mechanism. However, the various alternative fitted models produce the exact same contours, so no ambiguity can arise from that perspective. (This general philosophy is not shared by all who work in this area.)

Higher Dimensions

Clearly, the fact that the three-factor designs projected from the four factor designs are D-optimal using the same design levels $(0.24, 0.76, 0, 0)$ that made the original four factor designs D-optimal makes the original four-factor designs very versatile ones. Exploratory computations have indicated that this property does not apply in general to higher dimensional original mixture designs, such as those discussed in Prescott et al. (1993) and Lewis et al. (1994). However, it seems likely that a conveniently chosen nearly D-optimal mixtures design will project into nearly D-optimal component-amount designs. Again, because the original designs are orthogonally blocked and the **X**-matrix columns of the projected designs form a subset of the columns for the original designs, the projected designs retain the orthogonal blocking property, under attainable restrictions for the specific choice of blocking variables.

Table E18.1. Component Percentages and Dissolution Values of 14 Tablet Blends

Blend	x_1	x_2	x_3	x_4	y	$\hat{y}$
1	1.883	1.633	0.150	96.334	9.083	9.228
2	2.487	0.	0.025	97.488	4.333	4.624
3	2.500	2.575	0.025	94.900	11.300	11.178
4	2.500	2.450	0.150	94.900	9.633	9.519
5	1.000	3.000	0.025	95.975	9.483	9.206
6	2.500	1.256	0.087	96.157	10.400	10.247
7	2.013	3.000	0.087	94.900	11.167	11.336
8	2.500	2.450	0.150	94.900	9.350	9.519
9	1.000	3.000	0.150	95.850	9.767	9.351
10	1.000	1.413	0.087	97.500	8.783	8.768
11	1.000	3.000	0.025	95.975	8.800	9.206
12	1.000	2.175	0.150	96.675	8.917	9.254
13	1.738	0.738	0.025	97.499	7.283	6.985
14	2.500	0.	0.150	97.350	3.267	3.147
Min.	1.000	0.	0.025	94.900		
Max	2.500	3.000	0.150	97.500		
Average	1.830	1.906	0.092	96.170		
Range	1.500	3.000	0.125	1.300		

EXERCISE

18.1. (*Source:* Two new mixture models: living with collinearity but removing its influence, by J. A. Cornell and J. W. Gorman, *Journal of Quality Technology*, **35**, 2003, 78–88.) The percentage data in Table E18.1 come from an experiment to study how the response y, dissolution of tablets (pills), was affected by the ingredients (x_1, x_2, x_3, x_4) = (binder, disintegrant, surfactant, other ingredients combined). Note that the levels of ingredients x_1, x_2, and x_3 are very tiny compared with those of x_4 and that $x_1 + x_2 + x_3 + x_4 = 100$.

(a) Fit a Scheffé-type second-order mixture model (no intercept term) to these data, and note the very large value of one of the coefficients compared with the response values observed. (The original paper discusses this seeming anomaly, caused by the fact that the experimental design occupies only a small part of the mixture space, while the fitted model provides an interpretation over the whole space; however, we do not follow this route here.)

(b) Next, ignoring the x_4 values, consider the (x_1, x_2, x_3) values to be component-amount values (a_1, a_2, a_3), with total amounts $A = a_1 + a_2 + a_3 = 100 - x_4$, and fit to these a second-order component-amount model of the form (18.3.7), with intercept. Confirm that the fitted values are the same as those obtained in fit (a) but that the coefficients seem "more reasonable" (a somewhat nebulous concept in view of the fact that the same predictions arise).

Ridge Analysis for Examining Second-Order Fitted Models When There Are Linear Restrictions on the Experimental Region

19.1. INTRODUCTION

Suppose we wish to investigate the behavior of a response variable y over a specified region of interest by fitting a second-order response surface. Ridge analysis provides a way of following the locus of, for example, a maximum response, moving outwards from the origin of the predictor variable space or, indeed, from any other selected point. Because this approach does not require one to view the fitted regression surface as a whole, this important technique may be applied even when visualization of the surface is difficult in several dimensions. The calculation of a ridge trace often enables one to assess and understand the typically complex interplay between the input variables as the response improves. Use of this tool can supplement, or can replace, the use of canonical reduction of a second-order surface, depending on the experimenter's perspective.

In so-called "standard" ridge regression, discussed in Chapter 11, we examined a second-order surface without restrictions on the predictor variable space except, perhaps, bounds on individual x's or on the radius of the search. Such simple restrictions are easily taken into account by simply examining the ridge path(s) of interest. In this chapter, we show how ridge analysis can be used in a more general way, enabling the investigation of second-order surfaces based on many predictors when there are linear inequalities restricting the predictor space. A special case of such a problem is the mixture situation where the experimental space is restricted by $x_1 + x_2 + \cdots + x_q = 1$. Other linear restrictions may also apply, either in the mixtures case or in the nonmixtures case. *Any* form of second-order fitted model, whether of Scheffé type, Kronecker type, or something in between, can be accommodated in a very general framework.

19.2. A MOTIVATING EXPERIMENT

We shall use the pharmaceutical mixture example of Anik and Sukumar (1981) to motivate the general theory that will follow. This is an excellent example of a mixture problem which entails *additional* linear inequalities on the mixture ingredients and which is thus clarified by this sort of ridge analysis.

The Anik and Sukumar (1981) paper involved a study of five ingredients, satisfying the usual mixture restriction

$$x_1 + x_2 + x_3 + x_4 + x_5 = 1. \tag{19.2.1}$$

One ingredient, x_5, was held constant at 0.10 (10% of the mixture) so that the remaining four ingredients x_1, x_2, x_3, and x_4, were constrained by the requirement that

$$x_1 + x_2 + x_3 + x_4 = 0.9. \tag{19.2.2}$$

Equation (19.2.2) could be renormalized via $x_i = 0.9u_i$, so that $u_1 + u_2 + u_3 + u_4 = 1$. We do not do this, because it introduces a step that is not needed and would have to be undone in later calculations. (However, such a renormalization can be used to aid the construction of diagrams on triangular graph paper.) The essence of our method is that the ridge paths are obtained directly (and more easily so) without any such additional steps.

Anik and Sukumar (1981) wanted to examine various combinations of the four ingredients in order to fit a quadratic model to a response variable y, solubility, and then seek the maximum response. Each of the four ingredients was restricted to a range within [0, 1] as shown in Table 19.1. hence the authors decided to use an

Table 19.1. Experimental Design Used by Anik and Sukumar (1981) Together with the Lower and Upper Limits that Define the Mixture Space of Interest, and the Response Data Obtained from the Experiment

	Experiment Number	Polyethelyne Glycol 400	Glycerine	Polysorbate 60	Water	Response
Variable		x_1	x_2	x_3	x_4	y
Lower limit		0.10	0.10	0	0.30	—
Upper limit		0.40	0.40	0.08	0.70	—
Vertex	1	0.10	0.10	0	0.70	3.0
Vertex	2	0.10	0.10	0.08	0.62	7.3
"Vertex"	3	0.15	0.40	0	0.35	4.9
"Vertex"	4	0.11	0.40	0.08	0.31	8.4
"Vertex"	5	0.40	0.15	0	0.35	8.6
"Vertex"	6	0.40	0.11	0.08	0.31	12.7
Centroid of 1, 2	7	0.10	0.10	0.04	0.66	5.1
Centroid of 5, 6	8	0.40	0.13	0.04	0.33	10.8
Centroid of 3, 4	9	0.13	0.40	0.04	0.33	6.6
Centroid of 1, 3, 5	10	0.216	0.216	0	0.468	4.4
Centroid of 2, 4, 6	11	0.203	0.203	0.08	0.414	7.9
Centroid of 4, 6	12	0.255	0.255	0.08	0.31	9.4
Centroid of 3, 5	13	0.275	0.275	0	0.35	5.8
Overall centroid	14	0.21	0.21	0.04	0.44	6.3

Note that $x_1 + x_2 + x_3 + x_4 = 0.9$ for each point.

experimental design based on the "extreme vertices" of the restricted region (again, see Table 19.1). This excellent method was first suggested by McLean and Anderson (1966) and is discussed in Chapter 17. Anik and Sukumar (1981) aimed to show how useful a design obtained using extreme vertices can be. To implement this, one generates the extreme points (or "corners") of the region and then selects the design points from (a) vertices, (b) edge (one-dimensional) centroids, (c) face (two-dimensional) centroids, and so on. The last of these groups is the single point represented by the overall centroid, calculated by averaging all the vertices. The method has various subtleties due to the fact that the number of extreme vertices (and consequently of the various centroids) depends on the specific ranges of the x's, which determine the consequent region shape. Anik and Sukumar (1981) were led to use the specific 14-point experimental design shown in Table 19.1. Their selection of design points is further explained below.

Comments on the Design

The experimental design adopted by Anik and Sukumar requires additional explanation to avoid potential confusion. Figure 19.1a shows the triangular sub-

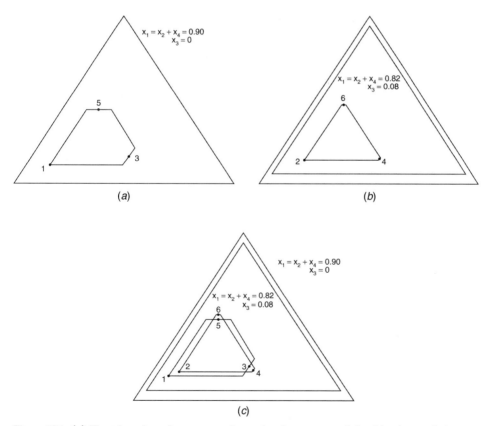

Figure 19.1. (a) The triangular subspace $x_3 = 0$ contains the pentagon defined by the restrictions on the mixture ingredients. (b) The triangular subspace $x_3 = 0.08$ (inner triangle) contains the (different) pentagon defined by the restrictions. (c) The two slices are superimposed as in a view downwards from the $x_3 = 0.90$ vertex of the mixture space. By joining corresponding pentagon vertices, one defines the entire restricted region.

space $x_3 = 0$; within it, the other restrictions create the five-sided figure. The inner triangle and the pentagon in Figure 19.1b play the same respective roles for $x_3 = 0.08$ subspace. The outer triangle of Figure 19.1b is the same triangle as in Figure 19.1a, and this makes the point that the $x_3 = 0.08$ slice of the four-dimensional simplex is smaller than the $x_3 = 0$ slice. Figure 19.1c shows the two slices superimposed as they would be seen in a birds-eye view from the $x_3 = 0.90$ vertex. We further note that for each pentagon, two pairs of vertices are quite close together. Because of this, Anik and Sukumar (1981, p. 898) averaged these close pairs of points, and called the resulting averages "vertices" of their region. Thus, in Table 19.1, "vertex 3" $(0.15, 0.40, 0, 0.35)$ is the average of true vertices $(0.10, 0.40, 0, 0.40)$ and $(0.20, 0.40, 0, 0.30)$; "vertex 4" $(0.11, 0.40, 0.08, 0.31)$ is the average of true vertices $(0.10, 0.40, 0.08, 0.32)$ and $(0.12, 0.40, 0.08, 0.30)$; "vertex 5" $(0.40, 0.15, 0, 0.35)$ is the average of true vertices $(0.40, 0.10, 0, 0.40)$ and $(0.40, 0.20, 0, 0.30)$; and "vertex 6" $(0.40, 0.11, 0.08, 0.31)$ is the average of true vertices $(0.40, 0.10, 0.08, 0.32)$ and $(0.40, 0.12, 0.08, 0.30)$. This sort of reduction is very sensible.

Fitting a Second-Order Model

We shall later revisit this example to illustrate how ridge analysis can be applied to mixture experiments with regions restricted by linear inequalities in the ingredients. For now, we follow the approach of the original authors in fitting a second-order (or quadratic) model via least squares using the data in Table 19.1. A discussion of the various equivalent second-order model forms that can be fitted in a mixture problem is given by Prescott et al. (2002). For purposes of interpretation, it does not matter *which* of the several alternative possible models is fitted, because the resulting response contours will be identical in every case. Thus the ridge paths are also exactly the same if another choice of model form is made. Anik and Sukumar (1981) chose a model with a constant term in it, one of several possibilities. Our choice here will be the Scheffé model

$$y = \beta_1 x_1 + \beta_2 x_2 + \beta_3 x_3 + \beta_4 x_4$$
$$+ \beta_{12} x_1 x_2 + \beta_{13} x_1 x_3 + \beta_{14} x_1 x_4 + \beta_{23} x_2 x_3 + \beta_{24} x_2 x_4 + \beta_{34} x_3 x_4 + \varepsilon$$
$$(19.2.3)$$

[see Chapter 16 or Scheffé (1958, 1963)]. The equation resulting from fitting Eq. (19.2.3) by least squares is

$$\hat{y} = 49.716 x_1 + 8.414 x_2 + 29.95 x_3 + 4.3365 x_4$$
$$- 58.671 x_1 x_2 - 27.83 x_1 x_3 - 74.902 x_1 x_4 + 10.20 x_2 x_3 + 33.81 x_3 x_4.$$
$$(19.2.4)$$

Note that the nonlinear blending term $x_2 x_4$ is missing in Eq. (19.2.4). This happens because, when the Scheffé model is used with the design of Table 19.1, the resulting **X**-matrix is singular. As the reader can confirm, a regression fit of the $x_2 x_4$ column onto the remaining **X**-columns produces an exact fit on the columns

$x_1, x_2, x_1x_3, x_1x_4, x_2x_3$. After rearrangement of terms and factorization, the exact fit equation can be written as

$$(x_1 - x_2)(x_3 + 2x_4 - 0.7) = 0. \tag{19.2.5}$$

For every data point in Table 19.1, either the first or the second factor of (19.2.5) is zero. Because x_2x_4 enters the surface fit in the last position of the terms mentioned above, it is the term eliminated. The contours of the fitted response surface and the associated ridge paths are not affected by which term is eliminated, but substitution of specific numbers into formulas given later in Section 19.4 will, of course, change appropriately. Overall, however, the fitted model is one degree of freedom less flexible in representing the response surface than it could have been with a nonsingular design.

We shall later explore the ridges of this surface in two ways, both covered by the theory to be described in Section 19.4. First, we shall seek the ridges that emanate from a selected "focal point" of the mixture space defined by the boundaries in Table 19.1. Later we shall follow the ridges on the appropriate subspace boundaries.

19.3. STANDARD RIDGE ANALYSIS WITHOUT LINEAR RESTRICTIONS, BRIEF RECAPITULATION OF THE ALGEBRA

Suppose the fitted second-order surface is written as

$$\hat{y} = b_0 + \mathbf{x}'\mathbf{b} + \mathbf{x}'\mathbf{B}\mathbf{x}, \tag{19.3.1}$$

where $\mathbf{x}' = (x_1, x_2, \ldots, x_q)$, $\mathbf{b}' = (b_1, b_2, \ldots, b_q)$ and where

$$\mathbf{B} = \begin{bmatrix} b_{11} & \frac{1}{2}b_{12} & \cdots & \frac{1}{2}b_{1q} \\ \cdots & b_{22} & \cdots & \frac{1}{2}b_{2q} \\ \cdots & \cdots & \cdots & \cdots \\ \text{sym} & \cdots & \cdots & b_{qq} \end{bmatrix} \tag{19.3.2}$$

is a symmetric matrix. Then (19.3.1) is the matrix format for the second-order fitted equation

$$\hat{y} = b_0 + b_1x_1 + b_2x_2 + \cdots + b_qx_q + b_{11}x_1^2 + b_{22}x_2^2 + \cdots + b_{qq}x_q^2$$
$$+ b_{12}x_1x_2 + b_{13}x_1x_3 + \cdots + \beta_{q-1,q}x_{q-1}x_q. \tag{19.3.3}$$

The stationary values of Eq. (19.3.1) subject to being on a sphere, radius r, centered at the origin, and thus with equation

$$\mathbf{x}'\mathbf{x} \equiv x_1^2 + x_2^2 + \cdots + x_q^2 = R^2 \tag{19.3.4}$$

are obtained by considering the Lagrangian function

$$F = b_0 + \mathbf{x}'\mathbf{b} + \mathbf{x}'\mathbf{B}\mathbf{x} - \lambda(\mathbf{x}'\mathbf{x} - R^2). \tag{19.3.5}$$

Differentiating Eq. (19.3.5) with respect to **x** (which can be achieved by differentiating with respect to $x_1, x_2, \ldots, x_q$ in turn and rewriting these equations in matrix form) gives

$$\frac{\partial F}{\partial x} = \mathbf{b} + 2\mathbf{B}\mathbf{x} - 2\lambda\mathbf{x}. \tag{19.3.6}$$

Setting Eq. (19.3.6) equal to a zero vector leads to

$$2(\mathbf{B} - \lambda\mathbf{I})\mathbf{x} = -\mathbf{b}. \tag{19.3.7}$$

We can now select a value for λ. If $(\mathbf{B} - \lambda\mathbf{I})^{-1}$ exists, which will happen as long as λ is not an eigenvalue of **B**, we obtain a solution **x** for a stationary point of $\hat{y}$,

$$\mathbf{x} = -\tfrac{1}{2}(\mathbf{B} - \lambda\mathbf{I})^{-1}\mathbf{b}, \tag{19.3.8}$$

and can then find the radius R, from Eq. (19.3.4), associated with the solution **x** from Eq. (19.3.8). Both R and **x** are functions of λ.

The theory in Draper (1963) tells us that if we select values of λ from $+\infty$ downwards, we shall be on the "maximum $\hat{y}$" path. Values of λ from $-\infty$ upwards give us the "minimum $\hat{y}$" path. Intermediate paths lie in the ranges of λ between the eigenvalues of **B**. For details, review Section 12.2.

19.4. RIDGE ANALYSIS WITH MULTIPLE LINEAR RESTRICTIONS, INCLUDING APPLICATIONS TO MIXTURE EXPERIMENTS

Ridge Analysis Around a Selected Focus

We recall that ridge analysis can be started from any selected "focal point," or "focus," which we again denote by the vector **f**. In mixture experiments, for example, **f** could be chosen as a central point, perhaps even the exact centroid, of some predefined restricted region in which the experimental runs were confined. Wherever it is, it should satisfy all linear restrictions that apply so that the starting point is within the permitted region. (For a special case exception, see Exercise 19.2.) When $\mathbf{f} \neq \mathbf{0}$, Eq. (19.3.4), $\mathbf{x}'\mathbf{x} = R^2$, would be replaced by

$$(\mathbf{x} - \mathbf{f})'(\mathbf{x} - \mathbf{f}) = R^2. \tag{19.4.1}$$

Adding Linear Restrictions

Suppose we wish to perform ridge analysis subject to a set of linear restrictions of the form

$$\mathbf{A}\mathbf{x} = \mathbf{c}, \tag{19.4.2}$$

where **A** is a given $m \times q$ matrix of linearly independent rows, *normalized so that the sum of squares of each row is* 1, and where **c** is a given $m \times 1$ vector. For example, suppose we were investigating a mixture problem with ingredients $x_1, x_2, \ldots, x_q$ restricted by

$$\mathbf{1}'\mathbf{x} = \mathbf{x}'\mathbf{1} = x_1 + x_2 + \cdots + x_q = 1, \tag{19.4.3}$$

where $\mathbf{1}' = (1, 1, \ldots, 1)$, a $1 \times q$ vector. We could choose

$$\mathbf{A} = \frac{1}{q^{1/2}}(1, 1, \ldots, 1) \quad \text{and} \quad \mathbf{c} = \frac{1}{q^{1/2}}. \qquad (19.4.4)$$

If this mixture space were further restricted to the plane

$$(\alpha_1, \alpha_2, \ldots, \alpha_q)\mathbf{x} = \alpha, \qquad (19.4.5)$$

where not all of the pre-specified α's were equal and where $\alpha_1^2 + \alpha_2^2 + \cdots + \alpha_q^2 = 1$, then $m = 2$,

$$\mathbf{A} = \begin{bmatrix} \dfrac{1}{q^{1/2}} & \dfrac{1}{q^{1/2}} & \cdots & \dfrac{1}{q^{1/2}} \\ \alpha_1 & \alpha_2 & \cdots & \alpha_q \end{bmatrix} \quad \text{and} \quad \mathbf{c} = \begin{bmatrix} \dfrac{1}{q^{1/2}} \\ \alpha \end{bmatrix}, \qquad (19.4.6)$$

and so on. Of course, any set of noncontradictory, linearly independent linear restrictions can be adopted. We are not confined only to mixture problems adding to 1. The dimension m of $\mathbf{A}$ must be such that $m < q$ in general. When $m = q$ we are reduced to a single point in the x-space and no paths are possible, or, rather, all paths coalesce into a single point. Note that because the focus $\mathbf{f}$ itself must lie in the restricted space, it must also satisfy Eq. (19.4.2), so that $\mathbf{Af} = \mathbf{c}$.

Generalized Ridge Analysis

Under conditions (19.4.1) and (19.4.2), we now consider the Lagrangian function

$$G = b_0 + \mathbf{x}'\mathbf{b} + \mathbf{x}'\mathbf{Bx} - \lambda\{(\mathbf{x} - \mathbf{f})'(\mathbf{x} - \mathbf{f}) - R^2\} - \theta'(\mathbf{Ax} - \mathbf{c}). \quad (19.4.7)$$

where λ and the elements $(\theta_1, \theta_2, \ldots, \theta_m)$ forming θ' are Lagrangian multipliers. Differentiation with respect to $\mathbf{x}$ leads to

$$\frac{\partial G}{\partial x} = \mathbf{b} + 2\mathbf{Bx} - 2\lambda(\mathbf{x} - \mathbf{f}) - \mathbf{A}'\theta, \qquad (19.4.8)$$

and setting (19.4.8) equal to a zero vector implies that

$$2(\mathbf{B} - \lambda\mathbf{I})\mathbf{x} = \mathbf{A}'\theta - \mathbf{b} - 2\lambda\mathbf{f}. \qquad (19.4.9)$$

For many given values of λ (the specific choices will be discussed below) we can write a solution for $\mathbf{x}$ as

$$\mathbf{x} = \tfrac{1}{2}(\mathbf{B} - \lambda\mathbf{I})^{-1}(\mathbf{A}'\theta - \mathbf{b} - 2\lambda\mathbf{f}). \qquad (19.4.10)$$

This $\mathbf{x}$ must satisfy Eq. (19.4.2), which implies that

$$\mathbf{c} = \tfrac{1}{2}\mathbf{A}(\mathbf{B} - \lambda\mathbf{I})^{-1}\mathbf{A}'\theta - \tfrac{1}{2}\mathbf{A}(\mathbf{B} - \lambda\mathbf{I})^{-1}(\mathbf{b} + 2\lambda\mathbf{f}). \qquad (19.4.11)$$

whereupon

$$\theta = \left\{ A(B - \lambda I)^{-1}A' \right\}^{-1} \left\{ 2c + A(B - \lambda I)^{-1}(b + 2\lambda f) \right\}. \qquad (19.4.12)$$

This leads us into the following solution sequence:

1. Choose values of λ appropriate for the desired path (to be explained below).
2. Solve Eq. (19.4.12) for θ.
3. Obtain x from Eq. (19.4.10).
4. Evaluate the radius R as in Eq. (19.4.1).

Then the point x will be on the desired path of stationary values and will lie on a sphere of radius R. The question to resolve now is whether a chosen value of λ places us on the maximum path, the minimum path, or some intermediate path. To determine this, the eigenvalues of an appropriate matrix are needed; this matrix is *not* B.

Determining the Specific Ridge Paths Under Linear Restrictions

We recall that in the *unrestricted* ridge analysis described in Chapter 12, the matrix of second derivatives

$$\left\{ \frac{\partial F}{\partial x_i\, \partial x_j} \right\} = 2(B - \lambda I) \qquad (19.4.13)$$

was key in determining which path is selected. The eigenvalues of B, namely, the values that result from solving

$$\det(B - \lambda I) = 0, \qquad (19.4.14)$$

formed the dividing points for the various paths of stationary values. In general, there were q eigenvalues and $2q$ paths; recall Figure 12.2. The eigenvalues of B are *not* appropriate for the restricted problem, however. What are needed instead are the eigenvalues of a lower dimension matrix which makes proper allowance for the linear restrictions.

We further recall that if there are m restrictions, as in Eq. (19.4.2), A is a given $m \times q$ matrix with m linearly independent rows of length q, normalized to make the sum of squares of each row equal to 1.

Let T be a $(q - m) \times q$ matrix each of whose $(q - m)$ rows is orthogonal to every row of A and such that $TT' = I_{q-m}$. Another way of saying this is that the columns of A' form a basis for the restriction space, and those of T' form an *orthonormal* basis for the space orthogonal to A'. It follows that

$$\begin{aligned}
TA' &= 0 &&(\text{of size } (q - m) \times m), \\
AT' &= 0 &&(\text{of size } m \times (q - m)), \qquad (19.4.15) \\
TT' &= I_{q-m} &&(\text{of size } (q - m) \times (q - m)).
\end{aligned}$$

The combined matrix,

$$\mathbf{Q} = \begin{bmatrix} \mathbf{A} \\ \mathbf{T} \end{bmatrix}, \tag{19.4.16}$$

is then a $q \times q$ matrix which provides a transformation of the coordinate system $(x_1, x_2, \ldots, x_q)$ into a coordinate system $(z_1, z_2, \ldots, z_q)$ via $\mathbf{z} = \mathbf{Qx}$, whereupon $\mathbf{x} = \mathbf{Q}^{-1}\mathbf{z}$.

If we partition $\mathbf{z} = (z_1, z_2, \ldots, z_m, z_{m+1}, \ldots, z_q)$ into $\mathbf{z}' = (\mathbf{u}, \mathbf{v})$, where $\mathbf{u}' = (z_1, z_2, \ldots, z_m)$ and $\mathbf{v}' = (z_{m+1}, \ldots, z_q)$, then

$$\mathbf{z} = \begin{bmatrix} \mathbf{u} \\ \mathbf{v} \end{bmatrix} = \mathbf{Qx} = \begin{bmatrix} \mathbf{A} \\ \mathbf{T} \end{bmatrix}\mathbf{x} = \begin{bmatrix} \mathbf{Ax} \\ \mathbf{Tx} \end{bmatrix} = \begin{bmatrix} \mathbf{c} \\ \mathbf{Tx} \end{bmatrix} \tag{19.4.17}$$

under the restrictions (19.4.2). Consider the inverse of $\mathbf{Q}$. This is of the form

$$\mathbf{Q}^{-1} = \left[\mathbf{A}'(\mathbf{AA}')^{-1}, \mathbf{T}' \right]. \tag{19.4.18}$$

$\mathbf{AA}'$ is nonsingular because of our assumption below Eq. (19.4.2) that the restrictions are linearly independent. We verify Eq. (19.4.18) by writing

$$\mathbf{QQ}^{-1} = \begin{bmatrix} \mathbf{A} \\ \mathbf{T} \end{bmatrix}\left[\mathbf{A}'(\mathbf{AA}')^{-1}, \mathbf{T}' \right] = \mathbf{I}_q \tag{19.4.19}$$

as a result of conditions (19.4.15). It follows that $\mathbf{Q}^{-1}\mathbf{Q} = \mathbf{I}$ also, because the inverse is unique.

The first quadratic portion of the Lagrangian function (19.4.7) is thus, using $\mathbf{x} = \mathbf{Q}^{-1}\mathbf{z}$, with $\mathbf{z}$ from (19.4.17) and $\mathbf{Q}^{-1}$ from (19.4.18),

$$\mathbf{x}'\mathbf{Bx} = \mathbf{z}'(\mathbf{Q}^{-1})'\mathbf{B}\mathbf{Q}^{-1}\mathbf{z} = [\mathbf{c}', \mathbf{v}']\begin{bmatrix} (\mathbf{AA}')^{-1}\mathbf{A} \\ \mathbf{T} \end{bmatrix}\mathbf{B}\left[\mathbf{A}'(\mathbf{AA}')^{-1}, \mathbf{T}'\right]\begin{bmatrix} \mathbf{c} \\ \mathbf{v} \end{bmatrix} \tag{19.4.20}$$

$$= \left[\mathbf{c}'(\mathbf{AA})^{-1}\mathbf{A} + \mathbf{v}'\mathbf{T} \right]\mathbf{B}\left[\mathbf{A}'(\mathbf{AA}')^{-1}\mathbf{c} + \mathbf{T}'\mathbf{v} \right] \tag{19.4.21}$$

$$= \mathbf{v}'\mathbf{TBT}'\mathbf{v} + 2\mathbf{v}'\mathbf{TBA}'(\mathbf{AA}')^{-1}\mathbf{c} + \mathbf{c}'(\mathbf{AA}')^{-1}\mathbf{ABA}'(\mathbf{AA}')^{-1}\mathbf{c}, \tag{19.4.22}$$

after reduction. This gives us the first quadratic portion in G in terms of $\mathbf{v}$. For the second quadratic portion of G, we can set $\mathbf{B} = \mathbf{I}$ as a special case in Eq. (19.4.22) to give

$$\lambda\mathbf{x}'\mathbf{x} = \lambda\mathbf{v}'\mathbf{v} + \mathbf{0} + \lambda\mathbf{c}'(\mathbf{AA}')^{-1}\mathbf{c}, \tag{19.4.23}$$

after applying the results in Eq. (19.4.15). For G, see Eq. (19.4.7) on p. 601.

Differentiating the transformed version of G twice with respect to $\mathbf{v}$, noting that constants and terms linear in $\mathbf{v}$ then drop out, we obtain

$$\left\{ \frac{\partial G}{\partial v_i \, \partial v_j} \right\} = 2(\mathbf{TBT}' - \lambda\mathbf{I}), \tag{19.4.24}$$

in place of Eq. (19.4.13). The size of the square matrix $(\mathbf{TBT}' - \lambda\mathbf{I})$ is $(q - m)$, not q, because $\mathbf{T}$ is $(q - m) \times q$. When λ is such that Eq. (19.4.24) is positive definite, we have a minimum, while if Eq. (19.4.24) is negative definite, we have a maximum. If Eq. (19.4.24) is indefinite, intermediate stationary values are indicated. In fact, the theory at this point is a complete parallel of that in the unrestricted case. If the eigenvalues of $\mathbf{TBT}'$ are $\mu_1 \leq \mu_2 \leq \cdots \leq \mu_{q-m}$ arranged in order with due regard to sign, then, subject to the restrictions $\mathbf{Ax} = \mathbf{c}$,

(a) Choosing $\lambda > \mu_{q-m}$ provides a locus of maximum $\hat{y}$ as R changes.
(b) Choosing $\lambda \leq \mu_1$ provides a locus of minimum $\hat{y}$ as R changes.
(c) Choosing $\mu_1 \leq \lambda \leq \mu_{q-m}$ gives intermediate stationary values.

As in the unrestricted case, when $\lambda = \mu_i$ exactly for $i = 1, 2, \ldots, q - m$, R is infinite, apart from pathological cases which typically do not occur in practical work.

Note that we do not need these eigenvalues of $\mathbf{TBT}'$ to obtain the paths, but only to distinguish between paths. For the loci of the maximum $\hat{y}$ and the minimum $\hat{y}$, which are the most common applications, the eigenvalues are not necessary at all, because choosing λ values decreasing from ∞ gives the path of maximum $\hat{y}$, while using values increasing from $-\infty$ gives the path of minimum $\hat{y}$. However, knowledge of the actual eigenvalues helps in selecting appropriate λ values for intermediate paths and distinguishing between them.

We next apply these results to the mixture problem described by Anik and Sukumar (1981) to illustrate the technique.

19.5. GENERALIZED RIDGE ANALYSIS OF THE MOTIVATING EXPERIMENT

The foregoing section describes, in a very general context, the calculation details necessary to find the ridge paths as they stream from a selected focus. The important sequence of operations for this lies below Eq. (19.4.12), and the selection of the ridge paths requires the eigenvalues of the matrix $(\mathbf{TBT}' - \lambda\mathbf{I})$. We now apply this theory to the Anik and Sukumar (1981) data set. For this, $q = 4$, and from Eq. (19.2.4) and from Eq. (19.2.2) after normalization, we obtain

$$b_0 = 0, \tag{19.5.1}$$

$$\mathbf{b} = \begin{bmatrix} 49.716 \\ 8.414 \\ 29.95 \\ 4.3365 \end{bmatrix}, \tag{19.5.2}$$

$$\mathbf{B} = \begin{bmatrix} 0 & -29.3355 & -13.915 & -37.451 \\ -29.3355 & 0 & 5.1 & 0 \\ -13.915 & 5.1 & 0 & 16.905 \\ -37.451 & 0 & 16.905 & 0 \end{bmatrix}, \tag{19.5.3}$$

$$\mathbf{A} = \left(\tfrac{1}{2}, \tfrac{1}{2}, \tfrac{1}{2}, \tfrac{1}{2}\right) \quad \text{and} \quad \mathbf{c} = 0.9/2 = 0.45. \tag{19.5.4}$$

Table 19.2. Ridge Paths for the Anik and Sukumar (1981) Data, Applying Only the Mixture Restriction $x_1 + x_2 + x_3 + x_4 = 0.90$

λ	x_1	x_2	x_3	x_4	R	$\hat{y}$
			Path A (Maximum $\hat{y}$)			
∞	0.210	0.210	0.040	0.440	0.000	6.27
2000	0.209	0.207	0.048	0.436	0.010	6.64
1000	0.208	0.204	0.056	0.432	0.020	7.02
750	0.207	0.202	0.062	0.429	0.026	7.27
500	0.206	0.199	0.072	0.423	0.038	7.75
400	0.205	0.196	0.080	0.419	0.048	8.10
300	0.204	0.191	0.092	0.413	0.062	8.66
250	0.203	0.187	0.102	0.408	0.074	9.10
100	0.201	0.152	0.181	0.366	0.170	12.48
62	0.230	0.107	0.243	0.320	0.259	15.40
50	0.441	0.020	0.244	0.195	0.437	21.94
48	0.920	-0.131	0.168	-0.057	0.940	55.58
			Path F (Minimum $\hat{y}$)			
-90	0.248	0.273	-0.194	0.573	0.279	-6.26
-100	0.243	0.266	-0.165	0.556	0.244	-4.55
-200	0.224	0.238	-0.052	0.490	0.109	1.69
-436	0.216	0.223	0.000	0.461	0.048	4.32
-500	0.215	0.221	0.005	0.459	0.041	4.58
-700	0.213	0.218	0.016	0.453	0.029	5.08
-900	0.213	0.216	0.021	0.450	0.023	5.35
$-\infty$	0.210	0.210	0.040	0.440	0.000	6.27

Paths B, C, D, E do not occur within the experimental region.

The First Set of Ridge Paths

We choose, as the focus $\mathbf{f}$ of the ridge system, the centroid of the points 1–6 in Table 19.1, namely $\mathbf{f} = (0.21, 0.21, 0.04, 0.44)'$. The distances from $\mathbf{f}$ to the six points $1, 2, \ldots, 6$ of Table 19.1 are, respectively, 0.0966, 0.763, 0.703, 0.773, 0.703, and 0.804; these values will give some comparative perspective to the R values in Table 19.2. The eigenvalues of $\mathbf{B}$ are not relevant here because of the restriction involving Eq. (19.5.4). Instead, we need the eigenvalues of the matrix $\mathbf{TBT'}$ in Eq. (19.4.24). An appropriate $\mathbf{T}$ [see Eq. (19.4.16)] takes the form

$$\mathbf{T} = \begin{bmatrix} -0.6708204 & -0.2236068 & 0.2236068 & 0.6708204 \\ 0.5 & -0.5 & -0.5 & 0.5 \\ -0.2236068 & 0.6708204 & -0.6708204 & 0.2236068 \end{bmatrix} \quad (19.5.5)$$

The rows of $\mathbf{T}$ consist of the first-, second-, and third-order orthogonal polynomials for four levels, namely $(-3, -1, 1, 3)$, $(1, -1, -1, 1)$, $(-1, 3, -3, 1)$, all normalized so that the sum of the squared elements in each row equals 1. [See, for example, Draper and Smith (1998, p. 466) or the shorter table in Appendix 19A.] The three rows of $\mathbf{T}$ are orthogonal to one another, each row sum of squares equals 1, and all three rows are orthogonal to $\frac{1}{2}\mathbf{1}' = (\frac{1}{2}, \frac{1}{2}, \frac{1}{2}, \frac{1}{2})$, which is the normalized vector of

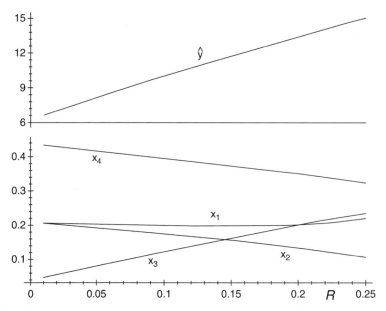

Figure 19.2. The maximum predicted response $\hat{y}$ and its corresponding positional coordinates (x_1, x_2, x_3, x_4) are plotted against R, the distance the point lies from the focus $f = (0.210, 0.210, 0.04, 0.440)'$ in the space $x_1 + x_2 + x_3 + x_4 = 0.90$. The numerical details are given in Table 19.2.

coefficients of the x's in the mixture restriction $x_1 + x_2 + x_3 + x_4 = 0.90$. The three eigenvalues of **TBT'** are $(-20.04, 2.52, 46.87)$ and the radius R becomes infinite when λ takes these eigenvalues. The ridge path of maximum $\hat{y}$ (path A) will be given by choosing λ values from $+\infty$ (where the solution will be $\mathbf{x} = \mathbf{f}$ and where $R = 0$) to 46.87 (where the solution will be $x = +\infty$). The ridge of minimum $\hat{y}$ (path F) will be given by choosing λ values from $-\infty$ (where the solution will be $\mathbf{x} = \mathbf{f}$) to -20.04 (where the solution will be $\mathbf{x} = -\infty$). Other λ values between the eigenvalues will deliver four more paths B, C, D, and E of stationary values of $\hat{y}$.

Table 19.2 shows a selected representative set of values of λ (which we choose initially), of (x_1, x_2, x_3, x_4) on the paths designated, and of the resultant R (from $(\mathbf{x} - \mathbf{f})'(\mathbf{x} - \mathbf{f}) = R^2$) and $\hat{y}$ values, using the formula $\hat{y} = b_0 + \mathbf{x}'\mathbf{b} + \mathbf{x}'\mathbf{Bx}$. Path A, the maximum $\hat{y}$ path begins at the selected focus $\mathbf{f}$, where $R = 0$ and $\hat{y} = 6.27$, and moves quickly (see the x_3 values) to the $x_3 = 0.08$ boundary and beyond, while the values of x_1, x_2, and x_4 change only slowly. This clearly shows the importance of variable x_3 and, unless the range of x_3 can be extended past the $x_3 = 0.08$ value, indicates that further exploration of the fitted surface needs to be carried out on the $x_3 = 0.08$ face of the restricted region.

Figure 19.2, derived from the A path details in Table 19.2, shows how the coordinates x_1, x_2, x_3, and x_4 and the predicted maximum response value $\hat{y}$ change versus R. Such a diagram could be drawn also for any of the ridge paths we provide, and it is considered by many scientists to be the best way to view the ridge results. It enables practitioners to assess and understand the typically complex interplay between the mixture ingredients as the response improves. It also permits the addition of a "cost" curve for the ingredients, or of any other curves measuring selected qualities of the changing mixture. (We provide only this one example, since it duplicates the information in the corresponding table, but such diagrams can be useful at every stage as the ridge trace is followed into further restricted subspaces.) We recall that simple closed form expressions for the dependency of x_i and $\hat{y}$ upon R are not available. However, numerical computer calculations are easily done, and these provide the details for constructing the smooth lines of Figure 19.2. Alternatively, a satisfactory working diagram can be obtained by plotting the selected values given in Table 19.2.

Intermediate paths B and C have no points of practical interest. The x_1 values are negative from the eigenvalue $\lambda = 46.87$ until about $\lambda = 41.5$, where the x_3 value reaches a minimum of about $x_3 = 0.358$, well above the x_3 upper limit for the experimental region. The minimum R value of about 0.379 is attained at about $\lambda = 40$.

Intermediate paths D and E are also of no practical interest, having negative x_1 and x_4 values throughout. Their minimum radius lies beyond the range of R shown in Figure 19.2.

The minimum $\hat{y}$ path F begins, like path A, at the selected focus **f**, where $\lambda = -\infty$, $R = 0$, and $\hat{y} = 6.27$. As might be anticipated from the behavior of path A, path E goes quickly to the (opposite) $x_3 = 0$ boundary, after which it is of no practical interest because x_3 becomes negative. We show selected λ values, to a point where the predicted $\hat{y}$ has turned negative, in Table 19.2.

The Second Set of Ridge Paths

Because we are interested in maximizing $\hat{y}$, we now need to explore the surface on the $x_3 = 0.08$ plane. (Had we been interested in minimizing $\hat{y}$, we would have gone to the $x_3 = 0$ plane instead.)

The theory of Section 19.4 can again be applied, but now with the addition of the linear equality $x_3 = 0.08$. This means that Eq. (19.5.4) is replaced by

$$\mathbf{A} = \begin{bmatrix} \frac{1}{2} & \frac{1}{2} & \frac{1}{2} & \frac{1}{2} \\ 0 & 0 & 1 & 0 \end{bmatrix}, \qquad \mathbf{c} = \begin{bmatrix} 0.45 \\ 0.08 \end{bmatrix}. \tag{19.5.6}$$

Repeating the calculations with these new restrictions requires us to use a new **T**,

$$\mathbf{T} = \begin{bmatrix} 0.267261 & 0.534523 & 0 & -0.801784 \\ 0.771517 & -0.617213 & 0 & -0.154303 \end{bmatrix}, \tag{19.5.7}$$

which leads to the eigenvalues of **TBT'** being $(-0.49, 45.01)$. (The two rows of **T**, orthogonal to each other and also to the rows of **A**, are $(1, 2, 0, -3)$ and $(5, -4, 0, -1)$ renormalized to have sum of squares 1.) There are now four ridge paths which we designate A (maximum $\hat{y}$), B, C and D (minimum $\hat{y}$). A new focus needs to be chosen.

The current restricted region is shown in Figure 19.1*b*. The design points 2, 4, and 6 from Table 19.1 lie on this pentagon and we choose **f** = $(0.203, 0.203, 0.08, 0.413)'$, their centroid. This point lies at distances $R = 0.253, 0.241, 0.241$ from points 2, 4, 6 respectively, and these numbers can be

Table 19.3. Ridge Paths for the Anik and Sukumar (1981) Data, Under the Restrictions $x_3 = 0.08$, $x_1 + x_2 + x_3 = 0.82$

λ	x_1	x_2	x_3	x_4	R	$\hat{y}$
\multicolumn{7}{c}{Path A (Maximum $\hat{y}$)}						
∞	0.203	0.203	0.08	0.414	0.000	8.12
1000	0.207	0.203	0.08	0.410	0.005	8.16
500	0.211	0.202	0.08	0.407	0.010	8.21
200	0.255	0.200	0.08	0.395	0.029	8.41
150	0.236	0.197	0.08	0.387	0.042	8.57
125	0.246	0.194	0.08	0.380	0.055	8.74
100	0.265	0.189	0.08	0.366	0.079	9.10
90	0.279	0.184	0.08	0.357	0.097	9.39
80	0.301	0.177	0.08	0.342	0.124	9.90
75	0.317	0.171	0.08	0.332	0.144	10.32
70	0.341	0.162	0.08	0.317	0.173	10.97
66	0.367	0.152	0.08	0.301	0.205	11.80
65.95	0.368	0.152	0.08	0.300	0.206	11.82
60	0.433	0.127	0.08	0.360	0.287	14.31
55	0.549	0.081	0.08	0.190	0.429	20.13
52	0.698	0.021	0.08	0.101	0.613	30.32
\multicolumn{7}{c}{Path D (Minimum $\hat{y}$)}						
-6	0.158	0.028	0.08	0.634	0.286	6.86
-7	0.156	0.058	0.08	0.606	0.245	7.00
-9	0.154	0.098	0.08	0.568	0.194	7.18
-9.15	0.154	0.100	0.08	0.566	0.191	7.19
-10	0.154	0.111	0.08	0.555	0.176	7.23
-20	0.156	0.168	0.08	0.496	0.101	7.51
-30	0.161	0.184	0.08	0.475	0.077	7.62
-40	0.166	0.192	0.08	0.462	0.063	7.69
-50	0.169	0.196	0.08	0.455	0.054	7.73
-70	0.175	0.200	0.08	0.445	0.043	7.80
-100	0.181	0.202	0.08	0.437	0.033	7.86
-500	0.197	0.204	0.08	0.419	0.008	8.05
-1000	0.200	0.204	0.08	0.416	0.004	8.08
$-\infty$	0.203	0.203	0.08	0.414	0.000	8.12

Paths B and C do not occur within the region $x_3 = 0.08$, $x_1 + x_2 + x_3 = 0.82$.

compared with the values of R that we see on the ridge paths shown in Table 19.3. We recall that $x_3 = 0.08$ throughout, and we show the paths in Figure 19.3. The maximum $\hat{y}$ path crosses the $x_4 = 0.30$ boundary when λ is about 65.95. As in all steepest ascent studies when a boundary is met, one must now move along this boundary. We postpone this for the moment to discuss the other three ridge traces. Neither path B nor path C lies within the restricted region and their details are not given. The minimum $\hat{y}$ path D moves downward until x_1 is about 0.16 and then turns, crossing the $x_2 = 0.10$ boundary at roughly this $x_1 = 0.16$ level (see Figure 19.3).

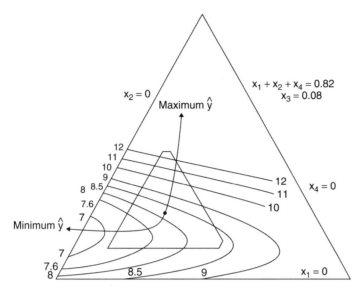

Figure 19.3. The fitted contours defined by Eq. (19.2.4) when $x_3 = 0.08$ are shown in the subspace $x_1 + x_2 + x_3 = 0.82$. The ridge paths of maximum $\hat{y}$ and minimum $\hat{y}$ on spheres of radius R emanating from the focus $f = (0.203, 0.203, 0.08, 0.413)'$ are shown; numerical details are in Table 19.3. The pentagon is the one drawn in Figure 19.1b.

The Third Set of Ridge Paths

To move along the boundary $x_4 = 0.30$, we designate a new focus **f** and need a new matrix **T**. The endpoints of the restricted region along the boundary are the corner points (0.40, 0.12, 0.08, 0.30), near design point 6 in Table 19.1, and (0.12, 0.40, 0.08, 0.30), near design point 4 (see Figure 19.1b or Figure 19.1c). We choose their centroid, namely **f** = (0.26, 0.26, 0.08, 0.30)'. **T** is now a normalized row vector orthogonal to the rows of **A** in **Ax = c**. This equation is now

$$
\begin{bmatrix} 0.5 & 0.5 & 0.5 & 0.5 \\ 0 & 0 & 1 & 0 \\ 0 & 0 & 0 & 1 \end{bmatrix}
\begin{bmatrix} x_1 \\ x_2 \\ x_3 \\ x_4 \end{bmatrix}
=
\begin{bmatrix} 0.45 \\ 0.08 \\ 0.30 \end{bmatrix}. \tag{19.5.8}
$$

Necessarily, **T** = (0.707107, −0.707107, 0, 0), a normalized form of (1, −1, 0, 0), or the similar vector with signs reversed. The sole eigenvalue of the 1×1 matrix **TBT'** is 29.3355, which is $-\frac{1}{2}b_{12}$ where b_{12} is the regression coefficient associated with $x_1 x_2$. Only the path A of maximum $\hat{y}$ ($\lambda > 29.3355$) and the path B of minimum $\hat{y}$ ($\lambda < 29.3355$) exist. On these paths, $x_3 = 0.08$, $x_4 = 0.30$ and so $x_1 + x_2 = 0.52$. Thus we can show the paths most simply by quoting only the x_1 value, as in Table 19.4. Path A is shown only to the point $x_1 = 0.40$ when the first corner point is reached and we attain the maximum predicted response, $\hat{y} = 12.81$, subject to the restrictions. Path B is shown only to $x_1 = 0.12$ when the second

Table 19.4. Ridge Paths for the Anik and Sukumar (1981) Data, Under the Restrictions $x_3 = 0.08$, $x_4 = 0.30$, $x_1 + x_2 = 0.52$

λ	x_1	R	$\hat{y}$
	Path A (Maximum $\hat{y}$)		
∞	0.260	0.000	9.45
1000	0.264	0.006	9.51
500	0.268	0.012	9.58
250	0.278	0.025	9.75
100	0.316	0.079	10.51
90	0.325	0.092	10.72
80	0.338	0.110	11.03
75	0.346	0.122	11.25
70	0.357	0.137	11.53
65	0.371	0.157	11.91
60	0.389	0.182	12.45
57.5	0.400	0.198	12.81
	Path B (Minimum $\hat{y}$)		
1.15	0.120	0.198	8.39
0	0.125	0.190	8.38
-10	0.160	0.142	8.45
-20	0.180	0.113	8.56
-40	0.203	0.081	8.74
-100	0.229	0.043	9.02
-200	0.243	0.024	9.19
-750	0.255	0.007	9.37
$-\infty$	0.260	0.000	9.45

corner point is reached. (It is *not*, of course, the minimum region response, which we would find by exploring the $x_2 = 0.10$ boundary, choosing $\mathbf{f} = (0.25, 0.10, 0.08, 0.47)'$ and $\mathbf{T} = (0.707107, 0, 0, -0.707107)$ or the vector with signs reversed. Moreover, the *global* minimum is on the $x_3 = 0$ face.) Overall, we see that, by a triple application of the ridge analysis technique, we have come to the predicted maximum response in the restricted region, improving from $\hat{y} = 8.09$ in Table 19.2 to $\hat{y} = 11.82$ in Table 19.3 to $\hat{y} = 12.81$ in Table 19.4.

19.6. SOME ADDITIONAL CONSIDERATIONS

We now briefly discuss some points that arise in connection with generalized ridge analysis:

1. The exact choice of the focus $\mathbf{f}$ is not a crucial feature of the restricted steepest ascent/descent procedure we have described. After the first stage in our example, one might have argued that because the path of maximum $\hat{y}$ entered the $x_3 = 0.08$ face of the restricted region at $(0.205, 0.196, 0.08, 0.419)$,

we should start again there. However, steepest ascent is a very flexible procedure, and a rigid method for choosing $\mathbf{f}$ would be inappropriate. Choosing some central point of the region is always safe, barring pathological examples.

2. The formulas given also can be applied to steepest ascent subject to linear restrictions when the model is a first-order mixture model $\hat{y} = b_1 x_1 + b_2 x_2 + \cdots + b_q x_q$. In this case, $b_0 = 0$, $\mathbf{B} = \mathbf{0}$ in Eq. (19.4.7) through Eq. (19.4.12). The "eigenvalues of $\mathbf{B}$" are all zero and, by the choice of $\mathbf{f}$, $\mathbf{Af} = \mathbf{c}$. The solution reduces to

$$\mathbf{x} = \mathbf{f} + (2\lambda)^{-1}\left(\mathbf{I} - \mathbf{A}'(\mathbf{AA}')^{-1}\mathbf{A}\right)\mathbf{b}. \qquad (19.6.1)$$

Choosing λ in $[0, \infty]$ gives the straight-line steepest-ascent direction, and choosing λ in $[-\infty, 0]$ gives the steepest-descent direction. Note that when there are no linear conditions on $\mathbf{x}$, $\mathbf{A} = \mathbf{0}$ and $\mathbf{x}$ is proportional to $\mathbf{b}$ as it should be.

3. A move to a selected focus $\mathbf{f}$ could be accompanied by changing to pseudo-components, if desired. This would involve a preliminary transformation of the form $\mathbf{z} = u\mathbf{x} - \mathbf{v}$ which might improve conditioning for the design used.

4. In our example, the paths of intermediate stationary values were of no practical interest; in other examples, they may well be. It is possible that a secondary maximum, with near-optimal properties, could be in a more distant location in the design space and could provide other advantages in terms of cost, ease of operation, safety, and so on, and might also improve other responses of interest. On the other hand, such locations may fall outside permissible operating conditions, particularly if there are mixture restrictions. Certainly, these other paths must be examined in all cases.

5. One might question whether the stage-by-stage following of the optimum $\hat{y}$ path to, and along, boundaries of the restricted region necessarily leads to the overall optimum. As a specific check of the example of Table 19.4, which gives the maximum $\hat{y} = 12.81$ value at the true vertex $(0.40, 0.12, 0.08, 0.30)$, we can calculate the predicted response values at all ten true vertices of the restricted region. Among these ten $\hat{y}$ values, the second largest is 12.63 and occurs at the vertex $(0.40, 0.10, 0.08, 0.32)$, the vertex closest to the maximum. More generally it would be possible to use the methods of this chapter on any selected subregion, including the faces of the bounding polyhedron. In cases where boundaries cut off the path of the maximum ridge quickly, and where secondary paths begin within the restricted region, it would be possible for the true restricted maximum to lie on another path. In our example, there are no secondary paths within the restricted region, so this cannot occur.

6. The contours of Figure 19.3 are drawn here *only* to show the paths, and thereby display what the method achieves. One does not actually need the contours, as examination of the coordinates in Tables 19.2, 19.3, and 19.4 makes clear. This would be especially important in a high-dimensional mixture space, when contours could be drawn only in sections.

7. The calculations for the Anik and Sukumar (1981) example were carried out in two ways, using MINITAB and MAPLE. A MINITAB routine with numerical details specific to the Anik and Sukumar example, but easily adapted to other examples, is given in Appendix 19B. The macro file attached there is a general one and needs no adaptation.

APPENDIX 19A. ORTHOGONAL AND NORMALIZED ROWS (ORTHOGONAL POLYNOMIALS) FOR $q = 3, 4, 5,$ AND 6 DIMENSIONS

	$q = 3$	
-0.707107	0	0.707107
0.408248	-0.816497	0.408248

	$q = 4$		
-0.670820	-0.223607	0.223607	0.670820
0.5	-0.5	-0.5	0.5
-0.223607	0.670820	-0.670820	0.223607

		$q = 5$		
-0.632456	-0.316228	0	0.316228	0.632456
0.534522	-0.267261	-0.534522	-0.267261	0.534522
-0.316228	0.632456	0	-0.632456	0.316228
0.119523	-0.478091	0.717137	-0.478091	0.119523

			$q = 6$		
-0.597614	0.358569	-0.119523	0.119523	0.358569	0.597614
0.545545	-0.109109	-0.436436	-0.436436	-0.109109	0.545545
-0.372678	0.521749	0.298142	-0.298142	-0.521749	0.372678
0.188982	-0.566947	0.377964	0.377964	-0.566947	0.188982
-0.062994	0.314970	-0.629941	0.629941	-0.314970	0.062994

Comments

If the only restriction is the mixture restriction with $\mathbf{A} = (1, 1, \ldots, 1)$, the tabulated values above provide the easiest choices for the matrix $\mathbf{T}$. Just select the appropriate matrix for the value of q that applies. For a more general $\mathbf{T}$, first try to find a succession of rows with integer coefficients that are orthogonal to the rows of $\mathbf{A}$ and then normalize those rows. If this fails, take any new row vector of size q, regress it against the rows of $\mathbf{A}$ with no intercept, find the residuals, normalize them, and use these as a row vector of $\mathbf{T}$. If the residuals come out to be all zeros, the selected row vector is a linear combination of the rows of $\mathbf{A}$ and another selection needs to be made. Successive performance of this procedure (adding previously found rows to the original $\mathbf{A}$ to form a notional "new $\mathbf{A}$ after the previous computation" each time) always works; however, neater solutions can often be found, as in the main example of this chapter.

In general, many possible candidates for $\mathbf{T}$ matrices exist because there are many possible ways to specify, or span, the space orthogonal to the rows of

A—that is, to the columns of **A**'. Provided that **T** satisfies Eqs. (19.4.15), any choice will do. (When only one dimension remains for **T**, there will be only one choice, however, apart from a change of sign.)

[*Note:* Failure to normalize the rows of **T** will produce the wrong eigenvalues for **TBT**' and therefore give wrong information about the values of λ that need to be examined, even though some of the other calculations will turn out to be correct.]

APPENDIX 19B. ADAPTABLE MINITAB INSTRUCTIONS FOR THE ANIK AND SUKUMAR EXAMPLE

First, enter the data for the specific example; then call the macro, here named FLridge. For another example, change the numerical details and their dimensional read-in specifications to correspond as follows:

m1 is the **b** vector,
m14 is the **f** vector,
m2 is the **B** matrix,
m21 is the **A** matrix,
m22 is the **c** vector,
m43 is the **T** matrix,
c19 is b_0,
m3 is the appropriate size unit matrix,
k4 and k3 are fixed constants.

The instruction cd C:\ MTBWIN \ MACROS is needed if MINITAB indicates it cannot find the macro and if the macro is in the MACROS folder. This instruction needs to be invoked only once per session, if at all, and may need to be changed as appropriate for your setup.

The constant k2 is the chosen λ value, to be successively updated, with the macro recalled for each new value of k2.

The output of interest is as follows:

c48 contains the eigenvalues of **TBT**'
k2 is λ,
m34 is a column-wise format of a point (x_1, x_2, x_3) on the ridge path for that λ,
c36 is the corresponding R,
c43 is $\hat{y}$ at that point.

The order of quantities chosen for printout can be changed. Other unneeded details printed by MINITAB can be ignored or deleted.

```
read 4 1 m1
49.716
8.414
29.95
4.3365
```

```
read 4 1 m14
.21
.21
.04
.44
read 4 4 m2
0 −29.3355 −13.915 −37.451
−29.3355 0 5.1 0
−13.915 5.1 0 16.905
−37.451 0 16.905 0
read 1 4 m21
.5 .5 .5 .5
read 1 1 m22
.45
read 3 4 m43
−0.670820 −0.223607 0.223607 0.670820
0.5 −0.5 −0.5 0.5
−0.223607 0.670820 −0.670820 0.223607
mult m43 m2 m45
transpose m43 m46
mult m45 m46 m47
eigen m47 c48 m48
print c48
print m48
set c19
0
end
read 4 4 m3
1 0 0 0
0 1 0 0
0 0 1 0
0 0 0 1
let k4 = 2
let k3 = −0.5
cd C:\MTBWIN\MACROS

let k2 = 50
%FLridge.mac
```

Here is the macro file to be called:

```
GMACRO
FLridge
mult k2 m14 m15
mult k4 m15 m16
add m16 m1 m17
mult k2 m3 m4
subtract m4 m2 m5
invert m5 m6
mult m6 m17 m7
mult k3 m7 m8
mult m21 m7 m23
mult m22 k4 m24
```

```
add m24 m23 m25
mult m21 m6 m26
transpose m21 m27
mult m26 m27 m28
invert m28 m29
mult m29 m25 m30
mult m27 m30 m31
subtract m17 m31 m32
mult m6 m32 m33
let k5 = 0.5
mult k5 m33 m34
print m34
subtract m14 m34 m35
transpose m35 m36
mult m36 m35 m37
copy m37 c37
let c36 = sqrt(c37)
print c36
transpose m34 m38
mult m38 m1 m39
mult m38 m2 m40
mult m40 m34 m41
add m39 m41 m42
copy m42 c42
let c43 = c42 + c19
print c43
ENDMACRO
```

EXERCISES

19.1. In a letter to the editor of *Technometrics*, quoted with permission from ASA, John Peterson (2003) claimed that fitting the Becker (1968) model of form

$$y = \beta_1 x_1 + \beta_2 x_2 + \beta_3 x_3 + \beta_4 x_4 + \alpha_{12} x_1 x_2/(x_1 + x_2 + h)$$
$$+ \alpha_{13} x_1 x_3/(x_1 + x_3 + h) + \alpha_{14} x_1 x_4/(x_1 + x_4 + h)$$
$$+ \alpha_{23} x_2 x_3/(x_2 + x_3 + h) + \alpha_{34} x_3 x_4/(x_3 + x_4 + h) \quad \text{(E19.1a)}$$

to the Anik and Sukumar data of Table 19.1, where $h = 0.0001$, "gives a 25% reduction in mean squared error" compared with fitting the model

$$y = \beta_1 x_1 + \beta_2 x_2 + \beta_3 x_3 + \beta_4 x_4 + \beta_{12} x_1 x_2 + \beta_{13} x_1 x_3$$
$$+ \beta_{14} x_1 x_4 + \beta_{23} x_2 x_3 + \beta_{34} x_{34}. \quad \text{(E19.1b)}$$

Is this true? If so, what are the practical consequences?

(*Note:* The purpose of the minuscule $h = 0.0001$ is simply to avoid singularity computing problems in data where two x's in a term are simultaneously zero. In fact this does not happen in the Anik and Sukumar data, so that the h could be dropped in this particular example.]

19.2. Suppose we wanted to carry out the first stage of a ridge analysis in an otherwise unconstrained mixture region defined by $x_1 + x_2 + \cdots + x_q = 1$, starting from the centroid $\mathbf{f} = (\frac{1}{q}, \frac{1}{q}, \dots, \frac{1}{q})'$. Show that *for this special case*, we could just as well start from the origin $(0, 0, \dots, 0)$, even though it is not in the mixture space!

19.3. (*Source:* Experiments with mixtures of components having lower bounds, by I. S. Kurotori, *Industrial Quality Control*, **22**, 1966, 592–596. See also Draper and Smith, *Applied Regression Analysis*, Wiley, 1998, pp. 418–419.) An investigation on the modulus of elasticity of a propellant mixture of three ingredients, binder (x_1), oxidizer (x_2) and fuel (x_3), restricted to the region $x_1 \geq 0.2$, $x_2 \geq 0.4$, $x_3 \geq 0.2$ gave rise to the following fitted second-order mixture polynomial: $\hat{y} = -2.756x_1 - 3.352x_2 - 17.288x_3 + 9.38x_1x_2 + 34.76x_1x_3 + 49.49x_2x_3$. [Readers will note that the coefficients here differ slightly from those given in Eq. (17.2.4); this results from rounding the x's here to three decimal places, as was done in the second reference mentioned above.] Find the ridge path of maximum $\hat{y}$ beginning from the centroid $(0.267, 0.467, 0.267)$ of the restricted region and so find the point of maximum response, subject to the restrictions.

19.4. (*Source:* Adapted from Shrinkage in ternary mixtures of container ingredients, by S. Bures, F. A. Pokorny, and G. O. Ware, *Proceedings of the 1991 Kansas State University Conference on Applied Statistics in Agriculture*, 43–52, published in 1992.) Three ingredients $(q = 3)$ restricted to the mixture space $x_1 + x_2 + x_3 = 1$ $(m = 1)$, were investigated, leading to the fitted mixture model $\hat{y} = -0.00658x_1 - 0.00243x_2 + 0.00367x_3 + 0.34265x_1x_2 + 0.47074x_1x_3 + 0.14115x_2x_3$. This was derived from selected data for the proportional shrinkage of a mixture of three ingredients forming an artificial medium for growing plants. See also Exercise 16.7 and Section 20.5. Find the ridge path of maximum $\hat{y}$, beginning at the centroid $(\frac{1}{3}, \frac{1}{3}, \frac{1}{3})$, and so find the point of maximum response, subject to the restrictions of the mixture space.

19.5. [*Source:* Mixture designs for four components in orthogonal blocks, by N. R. Draper, P. Prescott, S. M. Lewis, A. M. Dean, P. W. M. John, and M. G. Tuck, *Technometrics*, **35**(3), 1993, 268–276. Note also a comment by R. D. Snee, *Technometrics*, **37**(1), 1995, 131–132, and two data corrections in the response to Snee; and *Applied Regression Analysis*, by N. R. Draper and H. Smith, John Wiley & Sons, 1998, pp. 419–422.] A second-order response surface was fitted to a set of mixture data on bread flours and certain sensible reductions in terms were made to obtain the following equation (where blocking variables have been omitted for the purpose of this exercise): $\hat{y} = b_0 + \mathbf{x}'\mathbf{b} + \mathbf{x}'\mathbf{B}\mathbf{x}$, where $b_0 = 0$, $\mathbf{b}' = (397.6, 444.5, 389.4, 395.8)$, and

$$\mathbf{B} = \begin{bmatrix} 0 & 53.9 & 108.95 & 84.85 \\ 53.9 & 0 & 0 & 0 \\ 108.95 & 0 & 0 & 0 \\ 84.85 & 0 & 0 & 0 \end{bmatrix}.$$

Table E19.7. Data from a Mixture Experiment

x_1	x_2	x_3	y
1	0	0	18.9
0	1	0	15.2
0	0	1	35.0
0.5	0.5	0	16.1
0.5	0	0.5	18.9
0	0.5	0.5	31.2
0.333	0.333	0.333	19.3
0.666	0.167	0.167	18.2
0.167	0.666	0.167	17.7
0.167	0.167	0.666	30.1
0.333	0.333	0.333	19.0

Beginning from the point $(0, 0.333, 0.333, 0.333)$, which lies on the $x_1 = 0$ boundary, find the path of maximum response within the mixture space $x_1 + x_2 + x_3 + x_4 = 1$. Remember that when the path hits a boundary, it must be followed onto that boundary to keep the mixture ingredient values between 0 and 1 at all times.

19.6. In Exercise 6.4, you were essentially asked to see if the point $(x_1, x_2, x_3) = (-0.89, -2.71, 2)$ was on a dog-leg path of steepest ascent, starting from the origin $(0, 0, 0)$, and subject to the restriction $x_1 + x_2 + 0.5x_3 \geq -2.6$. (A transition has been made here from uncoded X's to coded x's.) The fitted response surface was the first-order equation $\hat{y} = 9.925 - 4.10x_1 - 9.25x_2 + 4.90x_3$. Calling on the second-order methods of this chapter to work this problem seems somewhat ponderous, but use them anyway to see if they work.

19.7. (*Source:* Influence of non-ionic surfactants on the physical and chemical properties of a biodegradable pseuodolatex, by S. E. Frisbee and J. W. McGinity, *European Journal of Pharmaceutics and Biopharmaceutics*, **40**, 1994, 355–363.) Table E19.7 shows data from a mixture experiment. The response y is the glass transition temperature of films cast from poly(DL-lactide) (PLA); and the three predictors x_1, x_2, x_3, whose sum is 1, are amounts of nonionic surfactants, Polaxamer 188 NF (Pluronic F68), Ploy-oxyethylene 40 monostearate (Myrj 52-S), and Polyoxyethylene sorbitan fatty acid ester. NF (Tween 60). Conditions that provide lower responses are desirable.

(a) Fit, to the response variable y, a second-order (quadratic) mixtures model in the three x-variables and find the path of minimum response, starting from the centroid $(\frac{1}{3}, \frac{1}{3}, \frac{1}{3})$ and ending at a boundary of the mixture region.

(b) Suppose, instead, we decide to drop any terms that are non-significant at the 0.05 level in (a) and refit the model to the remaining terms, as did the original authors. Is the path of minimum response, starting from the centroid $(\frac{1}{3}, \frac{1}{3}, \frac{1}{3})$, much changed?

CHAPTER 20

Canonical Reduction of Second-Order Fitted Models Subject to Linear Restrictions

20.1. INTRODUCTION

We saw in Chapter 11 that canonical reduction of second-order response surfaces can be a useful technique for finding the form and shape of surfaces and, often, for discovering redundancies that enable the surface to be expressible in a simpler form with fewer canonical predictor variables than there are original predictor variables. Canonical reduction of models subject to linear restrictions is more difficult, but also possible. An important special application is when the predictor variables are mixture ingredients that must sum to a constant. Other linear restrictions may also be encountered in such problems, as well, as illustrated in Chapter 19. A possible difficulty in interpretation that arises in restricted cases is that the stationary point may fall outside the permissible restricted space. For example, in a mixtures situation the stationary point on the plane $x_1 + x_2 + \cdots + x_q = 1$ may have one or more ingredients negative. In such a situation, some or all of the canonical axes may not pass through the allowable experimental region. In general, one has to investigate the situation to discover if useful conclusions emerge or if a more detailed ridge analysis is needed. In this chapter, techniques for performing a canonical reduction subject to linear restrictions are given, and two mixture examples are canonically reduced, to illustrate what canonical reduction can and cannot provide. In what follows, there is some repetition of work in foregoing chapters, in order to make the technical details self-contained in one location.

Response Surfaces, Mixtures, and Ridge Analyses, Second Edition. By G. E. P. Box and N. R. Draper
Copyright © 2007 John Wiley & Sons, Inc.

20.2. CANONICAL REDUCTION WITH NO RESTRICTIONS (RECAPITULATION)

Canonical reduction is a method of rewriting a fitted second-order equation in a form in which it can be more readily understood. There are two main stages. First, a specific rotation of axes that removes all cross-product terms produces the so-called A canonical form. Then an appropriate change of origin removes all first-order terms as well and gives rise to the so-called B canonical form. This can be easily interpreted to understand the main features of the fitted surface. The general fitted second-order response surface can be written

$$\hat{y} = b_0 + \mathbf{x'b} + \mathbf{x'Bx}, \tag{20.2.1}$$

where $\mathbf{x'} = (x_1, x_2, \ldots, x_q)$ and $\mathbf{b'} = (b_1, b_2, \ldots, b_q)$ and where

$$\mathbf{B} = \begin{bmatrix} b_{11} & \frac{1}{2}b_{12} & \cdots & \frac{1}{2}b_{1q} \\ \cdots & b_{22} & \cdots & \frac{1}{2}b_{2q} \\ \cdots & \cdots & \cdots & \cdots \\ \text{sym} & \cdots & \cdots & b_{qq} \end{bmatrix} \tag{20.2.2}$$

is a symmetric matrix. Then Eq. (20.2.1) is the matrix format for the second-order fitted equation

$$\hat{y} = b_0 + b_1 x_1 + b_2 x_2 + \cdots + b_q x_q + b_{11} x_1^2 + b_{22} x_2^2 + \cdots + b_{qq} x_q^2$$
$$+ b_{12} x_1 x_2 + b_{13} x_1 x_3 + \cdots + b_{q-1,q} x_{q-1} x_q. \tag{20.2.3}$$

Differentiation of the fitted equation (20.2.1) with respect to $\mathbf{x}$ and setting the result equal to a zero vector leads to $\mathbf{b} + 2\mathbf{Bx} = \mathbf{0}$, and the solution,

$$\mathbf{x}_s = -\tfrac{1}{2}\mathbf{B}^{-1}\mathbf{b}, \tag{20.2.4}$$

defines the stationary point of Eq. (20.2.1). This is also the center of the quadratic system. Suppose $\lambda_1, \lambda_2, \ldots, \lambda_q$ are the eigenvalues of $\mathbf{B}$ and $\mathbf{m}_1, \mathbf{m}_2, \ldots, \mathbf{m}_q$ are corresponding normalized and mutually orthogonal eigenvectors. Then the matrix $\mathbf{M} = \{\mathbf{m}_1, \mathbf{m}_2, \ldots, \mathbf{m}_q\}$ has $\mathbf{M'} = \mathbf{M}^{-1}$. Because the eigenvalues and vectors are defined by $\mathbf{Bm}_i = \mathbf{m}_i \lambda_i$, $i = 1, 2, \ldots, q$, we can write $\mathbf{BM} = \mathbf{M\Lambda}$, where $\mathbf{\Lambda} = \text{diagonal}(\lambda_1, \lambda_2, \ldots, \lambda_q)$. Premultiplying by $\mathbf{M'} = \mathbf{M}^{-1}$ leads to $\mathbf{M'BM} = \mathbf{\Lambda}$. Using this result, and by insertion of $\mathbf{MM'} = \mathbf{I}$ into Eq. (20.2.1), we obtain

$$\hat{y} = b_0 + (\mathbf{x'M})(\mathbf{M'b}) + (\mathbf{x'M})\mathbf{M'BM}(\mathbf{M'x}). \tag{20.2.5}$$

We can now let $\mathbf{X} = \mathbf{M'x}$ and $\mathbf{\theta} = \mathbf{M'b}$ (or equivalently, $\mathbf{x} = \mathbf{MX}$ and $\mathbf{b} = \mathbf{M\theta}$) and express Eq. (20.2.5) as

$$\hat{y} = b_0 + \mathbf{X'\theta} + \mathbf{X'\Lambda X}. \tag{20.2.6}$$

This constitutes the "A canonical form" in which the axes have been rotated to remove cross-product terms, but the new variables $\mathbf{X}$ are measured from the original origin.

The "B canonical form" is obtained by moving the origin to the stationary point (20.2.4), by substituting new variables $\mathbf{W} = \mathbf{X} - \mathbf{M}'\mathbf{x}_s = \mathbf{X} - \mathbf{M}'(-\frac{1}{2}\mathbf{B}^{-1}\mathbf{b})$, which implies that $\mathbf{X} = \mathbf{W} - \frac{1}{2}\mathbf{M}'\mathbf{B}^{-1}\mathbf{b}$. After some algebra, this leads to

$$\hat{y} = b_0 - \tfrac{1}{4}\mathbf{b}'\mathbf{B}^{-1}\mathbf{b} + \mathbf{W}'\mathbf{\Lambda}\mathbf{W},$$

or

$$\hat{y} = \hat{y}_s + \mathbf{W}'\mathbf{\Lambda}\mathbf{W}, \tag{20.2.7}$$

where

$$\hat{y}_s = b_0 + \tfrac{1}{2}\mathbf{b}'\mathbf{x}_s = b_0 - \tfrac{1}{4}\mathbf{b}'\mathbf{B}^{-1}\mathbf{b} \tag{20.2.8}$$

is the predicted response at the stationary point. Equation (20.2.7) is the B canonical form, most frequently employed in practice. For examples, see Chapters 10 and 11.

20.3. ADDING LINEAR RESTRICTIONS

Suppose we wish to perform canonical reduction subject to a set of linear restrictions of the form

$$\mathbf{A}\mathbf{x} = \mathbf{c}, \tag{20.3.1}$$

where $\mathbf{A}$ is a given $m \times q$ matrix of linearly independent rows, *normalized so that the sum of squares of each row is* 1, and where $\mathbf{c}$ is a given $m \times 1$ vector. For example, suppose we were investigating a mixture problem with ingredients $x_1, x_2, \ldots, x_q$ restricted by

$$\mathbf{1}'\mathbf{x} = \mathbf{x}'\mathbf{1} = x_1 + x_2 + \cdots + x_q = 1, \tag{20.3.2}$$

where $\mathbf{1}' = (1, 1, \ldots, 1)$, a $1 \times q$ vector. We could choose

$$\mathbf{A} = \frac{1}{q^{1/2}}(1, 1, \ldots, 1) \quad \text{and} \quad \mathbf{c} = \frac{1}{q^{1/2}}. \tag{20.3.3}$$

If this mixture space were further restricted to the plane

$$(\alpha_1, \alpha_2, \ldots, \alpha_q)\mathbf{x} = \alpha, \tag{20.3.4}$$

where all the α_i's were prespecified and $\alpha_1^2 + \alpha_2^2 + \cdots + \alpha_q^2 = 1$; then $m = 2$,

$$\mathbf{A} = \begin{bmatrix} \dfrac{1}{q^{1/2}} & \dfrac{1}{q^{1/2}} & \cdots & \dfrac{1}{q^{1/2}} \\ \alpha_1 & \alpha_2 & \cdots & \alpha_q \end{bmatrix}, \quad \mathbf{c} = \begin{bmatrix} \dfrac{1}{q^{1/2}} \\ \alpha \end{bmatrix}, \tag{20.3.5}$$

and so on. Any set of noncontradictory, linearly independent linear restrictions can be adopted. We are not confined only to mixture problems with ingredients adding to 1. (In other words, restrictions could occur in nonmixtures problems also.) In general, the dimension m of $\mathbf{A}$ must be such that $m < q$. When $m = q$, the restrictions define a single point in the x-space.

20.4. CANONICAL REDUCTION UNDER LINEAR RESTRICTIONS

Let $\mathbf{T}$ be any $(q - m) \times q$ matrix, each of whose $(q - m)$ rows is orthogonal to every row of $\mathbf{A}$, and such that $\mathbf{TT}' = \mathbf{I}_{q-m}$. Another way of saying this is that the columns of $\mathbf{A}'$ form a basis for the restriction space, and those of $\mathbf{T}'$ form an *orthonormal* basis for the space orthogonal to $\mathbf{A}'$. It follows that

$$\mathbf{TA}' = \mathbf{0} \qquad (\text{of size } (q - m) \times m),$$

$$\mathbf{AT}' = \mathbf{0} \qquad (\text{of size } m \times (q - m)), \qquad (20.4.1)$$

$$\mathbf{TT}' = \mathbf{I}_{q-m} \qquad (\text{of size } (q - m) \times (q - m)).$$

The combined matrix,

$$\mathbf{Q} = \begin{bmatrix} \mathbf{A} \\ \mathbf{T} \end{bmatrix}, \qquad (20.4.2)$$

is then a $q \times q$ matrix which provides a transformation of the coordinate system $(x_1, x_2, \ldots, x_q)$ into a coordinate system $(z_1, z_2, \ldots, z_q)$ via $\mathbf{z} = \mathbf{Qx}$, whereupon $\mathbf{x} = \mathbf{Q}^{-1}\mathbf{z}$.

If we partition $\mathbf{z}' = (z_1, z_2, \ldots, z_m, z_{m+1}, \ldots, z_q)$ into $\mathbf{z}' = (\mathbf{u}', \mathbf{v}')$, where $\mathbf{u}' = (z_1, z_2, \ldots, z_m)$ and $\mathbf{v}' = (z_{m+1}, \ldots, z_q)$, then

$$\mathbf{z} = \begin{bmatrix} \mathbf{u} \\ \mathbf{v} \end{bmatrix} = \mathbf{Qx} = \begin{bmatrix} \mathbf{A} \\ \mathbf{T} \end{bmatrix} \mathbf{x} = \begin{bmatrix} \mathbf{Ax} \\ \mathbf{Tx} \end{bmatrix} = \begin{bmatrix} \mathbf{c} \\ \mathbf{Tx} \end{bmatrix} \qquad (20.4.3)$$

under the restrictions (20.3.1). Thus the transformation fixes the first m coordinates at the desired restricted values but leaves free the remaining $q - m$ coordinates which specify points in the space restricted by $\mathbf{Ax} = \mathbf{c}$. Consider the inverse of $\mathbf{Q}$. This is of the form

$$\mathbf{Q}^{-1} = \left[\mathbf{A}'(\mathbf{AA}')^{-1}, \mathbf{T}' \right]. \qquad (20.4.4)$$

$\mathbf{AA}'$ is nonsingular because of our assumption below Eq. (20.3.1) that the restrictions are linearly independent. It is easy to verify that $\mathbf{QQ}^{-1} = \mathbf{I}_q$ because of conditions (20.4.1). It follows that $\mathbf{Q}^{-1}\mathbf{Q} = \mathbf{I}$ also, because the inverse is unique. The quadratic portion of the fitted model function (20.2.1) is thus, using $\mathbf{x} = \mathbf{Q}^{-1}\mathbf{z}$, with $\mathbf{z}$ from (20.4.3) and $\mathbf{Q}^{-1}$ from (20.4.4),

$$\mathbf{x}'\mathbf{Bx} = \mathbf{z}'(\mathbf{Q}^{-1})'\mathbf{B}\mathbf{Q}^{-1}\mathbf{z} = [\mathbf{c}',\mathbf{v}']\begin{bmatrix} (\mathbf{A}\mathbf{A}')^{-1}\mathbf{A} \\ \mathbf{T} \end{bmatrix}\mathbf{B}[\mathbf{A}'(\mathbf{A}\mathbf{A}')^{-1},\mathbf{T}']\begin{bmatrix} \mathbf{c} \\ \mathbf{v} \end{bmatrix} \quad (20.4.5)$$

$$= [\mathbf{c}'(\mathbf{A}\mathbf{A}')^{-1}\mathbf{A} + \mathbf{v}'\mathbf{T}]\mathbf{B}[\mathbf{A}'(\mathbf{A}\mathbf{A}')^{-1}\mathbf{c} + \mathbf{T}'\mathbf{v}] \quad (20.4.6)$$

$$= \mathbf{v}'\mathbf{TBT}'\mathbf{v} + 2\mathbf{v}'\mathbf{TBA}'(\mathbf{A}\mathbf{A}')^{-1}\mathbf{c} + \mathbf{c}'(\mathbf{A}\mathbf{A}')^{-1}\mathbf{ABA}'(\mathbf{A}\mathbf{A}')^{-1}\mathbf{c}, \quad (20.4.7)$$

after reduction. This gives us the quadratic portion of $\hat{y}$ in terms of $\mathbf{v}$ and $\mathbf{c}$. For the linear portion of $\hat{y}$, we obtain

$$\mathbf{x}'\mathbf{b} = \mathbf{z}'(\mathbf{Q}^{-1})'\mathbf{b} = [\mathbf{c}',\mathbf{v}']\begin{bmatrix} (\mathbf{A}\mathbf{A}')^{-1}\mathbf{A} \\ \mathbf{T} \end{bmatrix}\mathbf{b} = \mathbf{c}'(\mathbf{A}\mathbf{A})^{-1}\mathbf{Ab} + \mathbf{v}'\mathbf{Tb}. \quad (20.4.8)$$

The transformed form of (20.2.1) is now b_0 + Eq. (20.4.8) + Eq. (20.4.7), namely,

$$\hat{y} = d_0 + \mathbf{v}'\{\mathbf{Tb} + 2\mathbf{TBA}'(\mathbf{A}\mathbf{A}')^{-1}\mathbf{c}\} + \mathbf{v}'\mathbf{TBT}'\mathbf{v}, \quad (20.4.9)$$

where

$$d_0 = b_0 + \mathbf{c}'(\mathbf{A}\mathbf{A}')^{-1}\mathbf{Ab} + \mathbf{c}'(\mathbf{A}\mathbf{A}')^{-1}\mathbf{ABA}'(\mathbf{A}\mathbf{A}')^{-1}\mathbf{c}. \quad (20.4.10)$$

Differentiating this transformed version (20.4.9) of $\hat{y}$ once with respect to $\mathbf{v}$ and setting the result to zero, we obtain the stationary point in the restricted space from

$$2\mathbf{TBT}'\mathbf{v} + 2\mathbf{TBA}'(\mathbf{A}\mathbf{A}')^{-1}\mathbf{c} + \mathbf{Tb} = \mathbf{0}, \quad (20.4.11)$$

with solution

$$\mathbf{v}_s = -(\mathbf{TBT}')^{-1}\{\tfrac{1}{2}\mathbf{Tb} + \mathbf{TBA}'(\mathbf{A}\mathbf{A}')^{-1}\mathbf{c}\}. \quad (20.4.12)$$

Substituting Eq. (20.4.12) into Eq. (20.4.9) and canceling several terms leads to the predicted response at the restricted stationary point as

$$\hat{y}_s = d_0 - \mathbf{v}_s'(\mathbf{TBT}')\mathbf{v}_s, \quad (20.4.13)$$

where d_0 is given in (20.4.10). [We note that, in the no-restrictions case with $\mathbf{T} = \mathbf{I}$, $\mathbf{A} = \mathbf{0}$, $\mathbf{c} = \mathbf{0}$, so that $\mathbf{v} = \mathbf{x}$, Eq. (20.4.12) reduces to Eq. (20.2.4), as it should. Use the same specialization, Eq. (20.4.13) reduces to Eq. (20.2.8).] Because

$$\mathbf{x} = \mathbf{Q}^{-1}\mathbf{z} = \mathbf{A}'(\mathbf{A}\mathbf{A}')^{-1}\mathbf{c} + \mathbf{T}'\mathbf{v}, \quad (20.4.14)$$

we can obtain the stationary point in the space subject to the restrictions as

$$\mathbf{x}_s = \mathbf{A}'(\mathbf{A}\mathbf{A}')^{-1}\mathbf{c} + \mathbf{T}'\mathbf{v}_s \quad (20.4.15)$$

in terms of the original coordinates. This point is not usually, of course, the stationary point in the full x-space, which is often unattainable because of the restrictions and thus would no longer be relevant anyway. Note that $\mathbf{A}\mathbf{x}_s = \mathbf{A}\mathbf{A}'(\mathbf{A}\mathbf{A}')^{-1}\mathbf{c} + \mathbf{A}\mathbf{T}'\mathbf{v}_s = \mathbf{c}$, by applying (20.4.1), as it should, because $\mathbf{x}_s$ is in the restricted space. It is also clear that $\mathbf{x}_s$ is the same whatever choice is made for $\mathbf{T}$. For if another choice, $\mathbf{T}_2 = \mathbf{P}\mathbf{T}$ is made, where $\mathbf{P}$ is a square $(m - q) \times (m - q)$ nonsingular matrix such that $\mathbf{T}_2$ satisfies Eq. (20.4.1), it will be found that the $\mathbf{P}$'s cancel out when $\mathbf{T}_2$ is inserted throughout in place of $\mathbf{T}$. The result of Eq. (20.4.15) does, of course, depend on $\mathbf{A}$, $\mathbf{c}$, and the coefficients of the fitted model.

By replacing the curly bracket in Eq. (20.4.9) using Eq. (20.4.12), we can rewrite (20.4.9) in the form

$$\hat{y} = d_0 - 2\mathbf{v}'\mathbf{T}\mathbf{B}\mathbf{T}'\mathbf{v}_s + \mathbf{v}'\mathbf{T}\mathbf{B}\mathbf{T}'\mathbf{v}. \tag{20.4.16}$$

Subtracting Eq. (20.4.13) from Eq. (20.4.16) and then factorizing the result gives

$$\hat{y} - \hat{y}_s = \mathbf{v}_s'(\mathbf{T}\mathbf{B}\mathbf{T}')\mathbf{v}_s - 2\mathbf{v}'\mathbf{T}\mathbf{B}\mathbf{T}'\mathbf{v}_s + \mathbf{v}'\mathbf{T}\mathbf{B}\mathbf{T}'\mathbf{v}$$

$$= (\mathbf{v} - \mathbf{v}_s)'\mathbf{T}\mathbf{B}\mathbf{T}'(\mathbf{v} - \mathbf{v}_s). \tag{20.4.17}$$

Suppose we now move the origin to the stationary point of Eq. (20.4.12) by choosing new coordinate values $\mathbf{Z} = \mathbf{v} - \mathbf{v}_s$ and substituting $\mathbf{v} = \mathbf{Z} + \mathbf{v}_s$ into Eq. (20.4.17). This leads immediately to the form

$$\hat{y} = \hat{y}_s + \mathbf{Z}'\mathbf{T}\mathbf{B}\mathbf{T}'\mathbf{Z}, \tag{20.4.18}$$

where $\hat{y}_s$ is defined in Eq. (20.4.13). There are no first-order terms in $\mathbf{Z}$ in this equation, which means that we can convert to canonical form by repeating the same steps as used previously between Eqs. (20.2.4) and (20.2.5), but with the role of $\mathbf{B}$ now being given to the matrix $\mathbf{T}\mathbf{B}\mathbf{T}'$. Suppose that $\lambda_1, \lambda_2, \ldots, \lambda_{q-m}$ are the $q - m$ eigenvalues of $\mathbf{T}\mathbf{B}\mathbf{T}'$ and let $\mathbf{M} = \{\mathbf{m}_1, \mathbf{m}_2, \ldots, \mathbf{m}_{q-m}\}$ denote a corresponding matrix of orthonormal eigenvectors. Write $\mathbf{\Lambda} = \mathbf{diagonal}\{\lambda_1, \lambda_2, \ldots, \lambda_{q-m}\}$. Note that we employ the same sort of notation given previously between Eqs. (20.2.4) and (20.2.5); however the dimensions are reduced by the number of linear restrictions, and so are now $q - m$ rather than q. The rotated orthogonal axes are now of the form

$$\mathbf{W} = \mathbf{M}'\mathbf{Z} = \mathbf{M}'(\mathbf{v} - \mathbf{v}_s) = \mathbf{M}'\mathbf{T}(\mathbf{x} - \mathbf{x}_s), \tag{20.4.19}$$

applying (1) $\mathbf{Z} = \mathbf{v} - \mathbf{v}_s$ [see below Eq. (20.4.17)] and (2) the bottom portion of Eq. (20.4.3). The surface can now be expressed as

$$\hat{y} = \hat{y}_s + \mathbf{W}'\mathbf{\Lambda}\mathbf{W} = \hat{y}_s + \sum_{i=1}^{q-m} \lambda_i W_i^2, \tag{20.4.20}$$

using steps similar to those described between Eq. (20.2.4) and Eq. (20.2.5). The stationary point lies where all $W_i = 0$. Individual axes are determined by the conditions that $q - m - 1$ (that is, all but one) of the W_i are zero.

The numerical coefficients attached to the elements of $\mathbf{x}$ in Eq. (20.4.19) are the rows of $\mathbf{M}'\mathbf{T}$. However $\mathbf{M}'\mathbf{T}\mathbf{A}' = \mathbf{0}$; see Eq. (20.4.1). Thus the evaluation of the product $\mathbf{M}'\mathbf{T}\mathbf{A}'$, which should be $\mathbf{0}$ within rounding error, provides a useful numerical check that the calculations (20.4.19) have been correctly performed. Specifically, in examples where **A** contains a row $\left(\frac{1}{q^{1/2}}, \frac{1}{q^{1/2}}, \ldots, \frac{1}{q^{1/2}}\right)$, the coefficients of $x_1, x_2, \ldots, x_q$, in each W_i will add to zero. In applications with linear restrictions, canonical reduction can give rise to various possible outcomes. Our first concern is typically whether or not the stationary point lies within the appropriate mixture space, which would be the full mixture region if only the mixture restriction applied. If it does, we can then investigate the shape taken by the fitted surface within the restricted space. It will then also be informative to evaluate the coordinates where the new W-axes intersect and exit the boundaries of the restricted region. When the stationary point is outside the region of interest, the behavior of the surface around the stationary point is usually not relevant, particularly when the stationary point is far away. However, it may still be useful to see if any of the canonical axes passes through the restricted region and, if any do, to determine predicted values of the fitted response surface along such axes. In cases where canonical reduction is not fruitful, performing a full ridge analysis is then the best course of action (see Chapter 19). We next provide two illustrative examples of canonical reduction in mixture problems.

20.5. THREE INGREDIENTS MIXTURE EXAMPLE

(This example is adapted from Shrinkage in ternary mixtures of container ingredients, by S. Bures, F. A. Pokorny, and G. O. Ware, *Proceedings of the 1991 Kansas State University Conference on Applied Statistics in Agriculture*, 43–52, published in 1992. Also see Exercises 16.7 and 19.4.)

We illustrate some of the details above via a three ingredient ($q = 3$) example in the usual mixture space $x_1 + x_2 + x_3 = 1$ ($m = 1$), leading to a canonical reduction in $q - m = 2$ dimensions. Consider the fitted mixture model in three ingredients $\hat{y} = -0.00658x_1 - 0.00243x_2 + 0.00367x_3 + 0.34265x_1x_2 + 0.47074x_1x_3 + 0.14115x_2x_3$, derived from selected data for the proportional shrinkage of a mixture of three ingredients forming an artificial medium for growing plants. We have $q = 3$, $b_0 = 0$, $\mathbf{b}' = (-0.00658, -0.00243, 0.00367)$,

$$\mathbf{B} = \begin{bmatrix} 0 & 0.171325 & 0.23537 \\ 0.171325 & 0 & 0.070575 \\ 0.23537 & 0.070575 & 0 \end{bmatrix},$$

$$\mathbf{A} = \left(\frac{1}{3^{1/2}}, \frac{1}{3^{1/2}}, \frac{1}{3^{1/2}}\right) = (0.577350, 0.577350, 0.577350), \quad \mathbf{c} = \frac{1}{3^{1/2}} = 0.577350,$$

$$\mathbf{T} = \begin{bmatrix} -\dfrac{1}{2^{1/2}} & 0 & \dfrac{1}{2^{1/2}} \\ \dfrac{1}{6^{1/2}} & -\dfrac{2}{6^{1/2}} & \dfrac{1}{6^{1/2}} \end{bmatrix} = \begin{bmatrix} -0.707107 & 0 & 0.707107 \\ 0.408248 & -0.816497 & 0.408248 \end{bmatrix}.$$

The rows of **T** are simply the normalized orthogonal polynomials of first and second order. (See Appendix 19A.) Then from Eq. (20.4.12), $\mathbf{v}'_s = (0.0104, 0.3882)$ so that, from Eq. (20.4.15), $\mathbf{x}'_s = (0.484, 0.016, 0.499)$, which is a point just inside the mixture space near the $x_2 = 0$ boundary (see Figure 20.1). The two eigenvalues

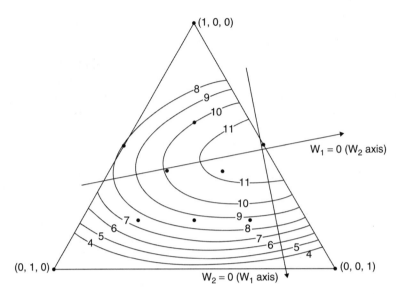

Figure 20.1. Contours of the Bures et al. example mixtures response surface together with the canonical axes.

of **TBT'** are $(-0.2550, -0.0632)$, with a corresponding matrix of eigenvectors of

$$\mathbf{M} = \begin{bmatrix} 0.9474 & 0.3200 \\ -0.3200 & 0.9474 \end{bmatrix}.$$

The canonical form of the fitted surface is now

$$\hat{y} = \hat{y}_s - 0.2550W_1^2 - 0.0632W_2^2,$$

where $\hat{y}_s = 11.62$. This form indicates (see Table 11.1) a set of elliptical contours centered at the point $\mathbf{x}_s$. The negative signs indicate a *maximum* response of 11.62 at the stationary point $W_1 = W_2 = 0$. The lengths of the major axes of the ellipses are in the ratio of $(0.255)^{-1/2} : (0.0632)^{-1/2}$, which is approximately $1:2$ (see Figure 20.1). The $q - m = 3 - 1 = 2$ canonical axes in the space $x_1 + x_2 + x_3 = 1$ are given via (20.4.19) as

$$W_1 = -0.801x_1 + 0.261x_2 + 0.539x_3 + 0.114,$$
$$W_2 = 0.160x_1 - 0.774x_2 + 0.613x_3 - 0.371.$$

We note that the x-coefficients in each W add to zero, within rounding error, as explained in the new paragraph below Eq. (20.4.20). By setting $W_1 = 0$, then setting each $x_i = 0$ in turn, $i = 1, 2, 3$, while maintaining the mixture restriction, we obtain points of intersection with the mixture boundaries with coordinates

$$(0, 2.35, -1.35), (0.49, 0, 0.51), \text{ and } (0.35, 0.65, 0),$$

indicating that the W_2 axis, defined by $W_1 = 0$, meets the $x_1 = 0$ boundary well outside the mixture region (indicated by the negative value -1.35) but intersects the $x_2 = 0$ and $x_3 = 0$ boundaries at valid mixture points.

Similarly, when we let $W_2 = 0$, and then set each $x_i = 0$ in turn, $i = 1, 2, 3$, again while maintaining the mixture restriction, we obtain the intersections

$$(0, 0.17, 0.83), (0.53, 0, 0.47), (1.23, -0.23, 0),$$

so that the W_1 axis cuts across the $x_1 = 0$ and $x_2 = 0$ boundaries within the mixture space restrictions; however, the W_1 axis meets the $x_3 = 0$ boundary outside the mixture space. By substituting for the centroid $(x_1, x_2, x_3) = (\frac{1}{3}, \frac{1}{3}, \frac{1}{3})$ in the W-equations above, we find that it lies at $(W_1, W_2) = (0.114, -0.371)$. This enables us to determine which direction of each axis is the positive one in the canonical equation. (These directions, the choices of which are made by the convention adopted in the program used to obtain eigenvalues and eigenvectors, are needed to interpret the locations of points on the response surface in the restricted space.) All these remarks are illustrated in Figure 20.1.

20.6. FIVE INGREDIENTS MIXTURE EXAMPLE

(*Source:* Multi-factor responses of cake quality to basic ingredient ratios, by L. T. Kissel and B. D. Marshall, *Cereal Chemistry*, **39**, 1962, 16–30. See also Section 18.2.)

This example gave rise to a fitted second-order response surface of the form (20.2.1) with $b_0 = 0$,

$$\mathbf{b} = \begin{bmatrix} -1605003 \\ 4487 \\ 559 \\ -7418 \\ -13347 \end{bmatrix},$$

$$\mathbf{B} = \begin{bmatrix} 0 & 1731252 & 1674333 & 1427295 & 1904909 \\ 1731252 & 0 & -6202 & 912 & 7783 \\ 1674333 & -6202 & 0 & 15718 & 4486 \\ 1427295 & 912 & 15718 & 0 & 41439 \\ 1904909 & 7783 & 4486 & 41439 & 0 \end{bmatrix},$$

$$\mathbf{A} = \left[\frac{1}{\sqrt{5}}, \frac{1}{\sqrt{5}}, \frac{1}{\sqrt{5}}, \frac{1}{\sqrt{5}}, \frac{1}{\sqrt{5}} \right], \quad \mathbf{c} = \frac{1}{\sqrt{5}} = 0.447214,$$

$$
\mathbf{T} = \begin{bmatrix}
-\dfrac{2}{10^{1/2}} & -\dfrac{1}{10^{1/2}} & 0 & \dfrac{1}{10^{1/2}} & \dfrac{2}{10^{1/2}} \\
\dfrac{2}{14^{1/2}} & -\dfrac{1}{14^{1/2}} & -\dfrac{2}{14^{1/2}} & \dfrac{1}{14^{1/2}} & \dfrac{2}{14^{1/2}} \\
-\dfrac{1}{10^{1/2}} & \dfrac{2}{10^{1/2}} & 0 & -\dfrac{2}{10^{1/2}} & \dfrac{1}{10^{1/2}} \\
\dfrac{1}{70^{1/2}} & -\dfrac{4}{70^{1/2}} & \dfrac{6}{70^{1/2}} & -\dfrac{4}{70^{1/2}} & \dfrac{1}{70^{1/2}}
\end{bmatrix}.
$$

$$
= \begin{bmatrix}
-0.632456 & -0.316228 & 0 & 0.316228 & 0.632456 \\
0.534522 & -0.267261 & -0.534522 & -0.267261 & 0.534522 \\
-0.316228 & 0.632456 & 0 & -0.632456 & 0.316228 \\
0.119523 & -0.478091 & 0.717137 & -0.478091 & 0.119523
\end{bmatrix}.
$$

Again the rows of **T** are normalized orthogonal polynomials, here of orders 1 to 4. See Appendix 19A. The stationary point is at $\mathbf{x}_s = (0.335, -1.872, 9.084, -3.783, -2.763)$ well outside the mixture region. None of the major axes passes through the region of interest, which is a subregion of the mixture space defined by $x_1 + x_2 + x_3 + x_4 + x_5 = 1$. Canonical reduction tells us relatively little here, and it would be necessary to use ridge analysis to negotiate the restricted region instead. See Chapter 19.

Answers to Exercises

CHAPTER 3

3.1. **(a)** $\hat{Y} = -1434.544 + 57.38\xi$. Note that the b_0 value is attained at $\xi = 0$, well away from the data, and is not practically meaningful there.

(b) $\xi_0 = 25.10$, $S = 0.05$; $x = -2, -1, 0, 1, 2$.

(c) $\hat{Y} = 5.694 + 2.869x$. The fitted values and residuals are identical to those in (a).

(d) The models are identical via substitution of $x = (\xi - 25.1)/0.05$.

(e) The residuals, $1.594, -0.465, -1.764, -1.453, 2.088$, which correctly sum to zero, indicate curvature around the fitted line, obvious from the plot. Either transformation of Y or use of a quadratic in x might be sensible.

(f) $\hat{y} = 0.6281 + 0.2360x$.

x	y		$\hat{y}$		$e = y - \hat{y}$	
-2	0.1903		0.1561		0.0342	
-1	0.3729		0.3921		-0.0192	
0	0.5944		0.6281		-0.0337	
1	0.8519		0.8641		-0.0122	
2	1.1310		1.1001		0.0309	
0	3.1405		3.1405		0.0	Sum
10	2.533474	$=$	2.529508	$+$	0.003778	SS

(The SS equality has a rounding error of 0.000188.) Taking logs does not remove the curvature pattern.

(g) Apply formula (3.4.4).

Response Surfaces, Mixtures, and Ridge Analyses, Second Edition. By G. E. P. Box and N. R. Draper
Copyright © 2007 John Wiley & Sons, Inc.

629

(h) In addition to the columns in (f), we need the following, where $\eta_0 = 0.6 + 0.25x$.

η_0	$y - \eta_0$	$\hat{y} - \eta_0$	
0.10	0.0903	0.0561	
0.35	0.0229	0.0421	
0.60	−0.0056	0.0281	
0.85	0.0019	0.0141	
1.10	0.0310	0.0001	
3.00	0.1405	0.1405	Sum
	96.74	59.08	SS $\times 10^4$

Source	SS $\times 10^4$	df	MS	F
Model	$\Sigma(\hat{y} - \eta_0)^2 = 59.08$	2	29.54	2.346
Residual	$\Sigma(y - \hat{y})^2 = 37.78$	3	12.59	
Total	$\Sigma(y - \eta_0)^2 = 96.74$	5		

We do not reject H_0: $\beta_0 = 0.6$, $\beta_1 = 0.25$ at the $\alpha = 0.10$ level. (Note round-off error effect in SS column.)

3.2.

$$\mathbf{b} = \begin{bmatrix} 5 & 0 & 10 \\ 0 & 10 & 0 \\ 10 & 0 & 34 \end{bmatrix}^{-1} \begin{bmatrix} 28.47 \\ 28.69 \\ 69.75 \end{bmatrix} = \begin{bmatrix} 34/70 & 0 & -10/70 \\ 0 & 0.1 & 0 \\ -10/70 & 0 & 5/70 \end{bmatrix} \begin{bmatrix} 28.47 \\ 28.69 \\ 69.75 \end{bmatrix}$$

$\hat{Y} = 3.864 + 2.869x + 0.915x^2$.

Y	$\hat{Y}$	$e = Y - \hat{Y}$
1.55	1.786	−0.236
2.36	1.910	0.450
3.93	3.864	0.066
7.11	7.648	−0.538
13.52	13.262	0.258

The quadratic requires an extra parameter but the residuals are less systematic. (Of course in practice one cannot make much out of just five observations! The data do, however, illustrate the calculations and considerations involved in this sort of situation.)

3.3.

$$\hat{\theta} = \left(\Sigma z^2\right)^{-1}(\Sigma zy) = 5^{-1}(45) = 9, \qquad \eta_0 = 8.$$

$$\Sigma(\hat{y} - \eta_0)^2 = 5, \qquad \Sigma(y - \hat{y})^2 = 10, \qquad F = \frac{(5/1)}{(10/4)} = 2.0.$$

$\text{Prob}(F(1, 4) > 2.0)$ exceeds 0.10, so do not reject H_0.

3.4. (a) $\hat{y} = 43.40704z_1$. It is being assumed that the model passes through the $(z_1, y) = (0, 0)$ origin—that is, that with no additives the vehicle starts immediately.

(b) $\hat{y} = 31.79864z_1 + 14.34827z_2$.

(c) $z_{2 \cdot 1} = z_2 - (\Sigma z_1 z_2 / \Sigma z_1^2) z_1 = z_2 - 0.809045 z_1$.
The $z_{2 \cdot 1}$ entries are $z'_{2 \cdot 1} = (3.57286, 0.76382, 2.95477, -1.85427, 0.33668, -2.47236)$ and the vector is orthogonal to z_1 of course. (You may wish to check this.)
$\hat{y} = 43.40704z_1 + 14.34828z_{2 \cdot 1}$.
$\hat{\theta} = \hat{\theta}_1 + 0.809045\hat{\theta}_2$.

(d)

Source	SS/100	df	MS/100
z_1 only	$\hat{\theta}^2 \Sigma z_1^2 = 374950$	1	374950
Extra for z_2	$\hat{\theta}_2^2 \Sigma z_{2 \cdot 1}^2 = 6535$	1	6535
Residual	102	4	26
Total	3815.92	6	

(Note rounding error in table.)
Clearly both variables play a significant role.

(e) Alias "matrix" is $(z'_1 z_1)^{-1} z'_1 z_2 = 161/199 = 0.809045$, so that $E(\theta_1) = \theta_1 + 0.809045\theta_2$. This should be identical to the relationship in (c) and it is.

(f) Region is defined by the interior of the ellipse

$$[(\theta_1 - 31.80), (\theta_2 - 14.35)] \begin{bmatrix} 199 & 161 \\ 161 & 162 \end{bmatrix} \begin{bmatrix} \theta_1 - 31.80 \\ \theta_2 - 14.35 \end{bmatrix}$$

$$= 2(0.26)6.94 = 3.61.$$

3.5. (a) $x = (X - 30)/12$.

(b) $\hat{Y} = 9.471 - 6.3500x$.

Source	SS	df	MS	F
b_0	1255.91	1		
$b_1|b_0$	1209.67	1		
Lack of fit	374.15	3	124.72	186.1
Pure error	6.07	9	0.67	
Total	2845.80	14		

Clear lack of fit is shown. The residuals indicate curvature. Possible alternatives are $Y = \beta_0 + \beta_1 x + \beta_{11} x^2 + \varepsilon$ and $y \equiv \ln Y = \beta_0 + \beta_1 x + \varepsilon$.

(c) $\hat{Y} = 3.326 - 6.3500x + 2.8678x^2$.
$\hat{y} = 1.5219 - 0.86634x$.

The residuals from the first equation produce (in order of increasing x) three pluses, three minuses, five pluses, and three minuses, a disturbing pattern that indicates possible *systematic* departure from the fitted model. The second fitted equation looks preferable.

(d) This discovery raises the possibility of serial correlation between the response values in a column, requiring estimation of the correlation(s) and use of generalized least squares, or an alternative analysis. For some pertinent remarks see, for example, Problems in the analysis of growth and wear curves, by G. E. P. Box, *Biometrics*, **6**, 1950, 362–389.

3.6.   The bias results follow from a direct application of the formula $E(\mathbf{b}) = \boldsymbol{\beta} + \mathbf{A}\boldsymbol{\beta}_2$, with

$$\mathbf{X} = \begin{bmatrix} 1 & x_1 \\ 1 & x_2 \\ \cdots & \cdots \\ 1 & x_n \end{bmatrix}, \quad \mathbf{X}_2 = \begin{bmatrix} x_1^2 \\ x_2^2 \\ \cdots \\ x_n^2 \end{bmatrix}, \quad \mathbf{X}'\mathbf{X} = \begin{bmatrix} n & n\bar{x} \\ n\bar{x} & nc \end{bmatrix}, \quad \mathbf{X}'\mathbf{X}_2 = \begin{bmatrix} nc \\ nd \end{bmatrix},$$

$$\mathbf{A} = (\mathbf{X}'\mathbf{X})^{-1}\mathbf{X}'\mathbf{X}_2.$$

(a) No, because if so, then $c^2 = \bar{x}d$ and $d = \bar{x}c$, implying that $c = \bar{x}^2$, obtained by substituting the second required condition into the first. It follows that the condition $c - \bar{x}^2 = 0$ would have to be satisfied The left-hand side is the sum of squares of the x's about their mean, which is zero only if all x's are identical, giving a useless design with only one level, inadequate for fitting a straight line.

(b) Yes. Any design symmetric about zero has $\bar{x} = d = 0$. For example, a three-level design with n_1, n_2, n_1 observations at $x = -1, 0, 1$, respectively, would have this property; there are many others. For such designs, $E(b_0) = \beta_0 + c\beta_{11}$ and $E(b_1) = \beta_1$.

CHAPTER 4

4.1.   See Appendix 4A.

4.2.   Real effects (in descending order) appear to be 4 (estimate 24), 2 (21), and 1 (18). Replotting the remaining contrasts, we obtain an estimate of 5.25 for the standard deviation of the effects; that is, $\hat{\sigma} = \frac{1}{2}(5.25)(16)^{1/2}$ is the estimated standard deviation of individual observations.

4.3.   The only real effect appears to be D with a value of 66.7. Replotting the remaining contrasts gives a standard error (effect) = 8; that is, $\hat{\sigma} = \frac{1}{2}(8)(16)^{1/2} = 16$ for the estimated standard deviation of individual observations.

4.4.   **(a)** (i) $ME_i = (\Sigma y$'s at upper level of factor $i - \Sigma y$'s at lower level$)/2^{k-1}$.
(ii) $b_i = $ (same)$/2^k$.
Thus $ME_i = 2b_i$.

(b) (i) (Two-factor interaction)$_{ij} = (TFI)_{ij} = \{(\Sigma y$'s for which factor i level times factor j level is $+1) - (\Sigma y$'s for $\ldots$ is $-1)\}/2^{k-1}$.
(ii) $b_{ij} = $ (same)$/2^k$.
Thus $(TFI)_{ij} = 2b_{ij}$.

4.5.   We first add a column of 1's and define the four columns of $\mathbf{X}$ as the columns of $(1, x_1, x_2, x_3)$. The single column of $\mathbf{X}_2$ is defined as the values of x_4 shown. Applying the formula $E(\mathbf{b}) = \boldsymbol{\beta} + (\mathbf{X}'\mathbf{X})^{-1}\mathbf{X}'\mathbf{X}_2\boldsymbol{\beta}_2$, where $\boldsymbol{\beta} = (\beta_0, \beta_1, \beta_2, \beta_3)'$, $\boldsymbol{\beta}_2 = \beta_4$, and $\mathbf{b}$ estimates $\boldsymbol{\beta}$ in the model, we find $(\mathbf{X}'\mathbf{X})^{-1}\mathbf{X}'\mathbf{X}_2 = (0, 2, 3, -2)'$ so that b_0 is unbiased, while

$$E(b_1) = \beta_1 + 2\beta_4,$$

$$E(b_2) = \beta_2 + 3\beta_4,$$

$$E(b_3) = \beta_3 - 2\beta_4.$$

4.6.   The effects are

$$I \leftarrow \quad 9.125, \quad 1 \leftarrow -2.75, \quad 2 \leftarrow -3.25, \quad 3 \leftarrow -0.75,$$
$$12 \leftarrow -0.25, \quad 13 \leftarrow \quad 0.25, \quad 23 \leftarrow \quad 1.75, \quad 123 \leftarrow \quad 0.75.$$

4.7.   Center point observations added to a 2^k design have no effect whatsoever on the factorial effects; they must be included in the mean calculation, however. Also, a comparison can be made between the factorial and center point averages to give an idea of whether curvature is present. (See Section 6.3.)

4.8.   **(a)** $x_1 = (\%C - 0.5)/0.4$,
$x_2 = (\%Mn - 0.8)/0.6$,
$x_3 = (\%Ni - 0.15)/0.05$.

(c) $\bar{y} = 552.5 \quad 1 \leftarrow -575, \quad 12 \leftarrow 10, \quad 123 \leftarrow -10,$
$2 \leftarrow -\,90, \quad 13 \leftarrow 15,$
$3 \leftarrow -\,65, \quad 23 \leftarrow 10.$

(d) $V(\text{effect}) = 4\sigma^2/n = 4(65)^2/8 = (45.962)^2$, so that standard error (effect) $= 46$, approximately. Only the 1-effect is larger than $\pm 2(46) = \pm 92$.

(e) The 1-effect is negative, so increasing the %C level decreases the start temperature.

4.9.   With the coding $x_1 = (X_1 - 85)/25$, $x_2 = (X_2 - 1760)/750$, $x_3 = (X_3 - 4.65)/2.65$, the design is recognizable as a replicated 2^3 experiment with runs $(\pm 1, \pm 1, \pm 1)$. The mean is $\bar{y} = 29.59375$ and the factorial effects, rounded to two decimal places are

$$1 \leftarrow 15.01, \qquad 12 \leftarrow 0.06, \qquad 123 \leftarrow -1.09$$
$$2 \leftarrow 11.73, \qquad 13 \leftarrow -0.69,$$
$$3 \leftarrow 9.54, \qquad 23 \leftarrow 1.34,$$

each with variance $4\sigma^2/16 = \sigma^2/4$. We estimate σ^2 from the repeat runs; each pair [e.g., 12.2, 12.5 from the $(-, -, -)$ runs, etc.] provides a 1-df estimate of σ^2 via the special formula for a pair of repeat runs, for example, $\frac{1}{2}(12.2 - 12.5)^2 = 0.045$. The eight such estimates can be averaged to give $s^2 = 1.8081$ based on 8 df. [Remember that, in general, the weighted average $(v_1 s_1^2 + v_2 s_2^2 + \cdots + v_r s_r^2)/(v_1 + v_2 + \cdots + v_r)$ is needed, but here all the individual degrees of freedom are $v_i = 1$ so that the general formula implies simple averaging in such a case.] The standard error (effect) = $\{1.8081/4\}^{1/2} = 0.672$ and only the three main effects are significant. All are positive, so that increases in X_1, X_2, and X_3 lead to increased penetration rates. In drilling, however, there are other responses to consider such as the wear on the drill, and the consequent overall cost of a succession of drilling jobs. Thus, increasing the levels of the predictors will improve the response, but it is not necessarily the best overall practical policy. Additional information is needed to determine this. See, for example, Tool life testing by response surface methodology, by S. M. Wu, *J. Engineering Industry, Trans. ASME*, **B-86**, 1964, 105–116. See also a paper by S. M. Wu and R. N. Meyer in the November 1965 issue.

4.10.

$$\bar{y} = 58.25,$$

$1 \leftarrow 0.25,$	$12 \leftarrow 0.5,$	$123 \leftarrow 1.75,$
$2 \leftarrow -17.25,$	$13 \leftarrow -1.,$	$124 \leftarrow -0.75,$
$3 \leftarrow 9.75,$	$14 \leftarrow -1.,$	$134 \leftarrow 0.25,$
$4 \leftarrow 21.75,$	$23 \leftarrow -7.5,$	$234 \leftarrow 0.25,$
	$24 \leftarrow 1.,$	$1234 \leftarrow 0.00,$
	$34 \leftarrow -0.5.$	

We have no estimate for σ, but it is clear from the sizes of the main effects that factors 2, 3, 4 are important, and that the 23 interaction is probably operating, also. It is tempting to drop variable 1 entirely. If we did this, we would have a replicated 2^3 design in variables 2, 3, and 4, and could use the ignored effects to estimate σ^2.

For a regression treatment that provides an estimate of σ^2, see Exercise 4.13.

4.11. The results are $\bar{y} = 4.071$,

$$1 \leftarrow 2.99, \qquad 12 \leftarrow -0.14, \qquad 123 \leftarrow -0.11,$$
$$2 \leftarrow -0.06, \qquad 13 \leftarrow -1.08,$$
$$3 \leftarrow -1.93, \qquad 23 \leftarrow 0.04.$$

$s_e^2 = 16.06 = 1.004$. Standard error (effect) = $(4s_e^2/24)^{1/2} = 0.41$, so the 1, 3, and 13 effects all exceed 2 standard errors in size. We rearrange the data in a two-way table of averages of six observations in factors 1 and 3:

$$\begin{array}{ccc} & 2.15 & 4.07 \\ \uparrow & & \\ (3) & 3 \qquad 7.07 & (1) \rightarrow \end{array}$$

Increased tear resistance is clearly attained by the " + " paper, but tearing is greatly affected by the direction of tear.

4.12.
$$\hat{y} = 29.594 - 7.506x_1 + 5.869x_2 + 4.769x_3.$$

Note that $b_0 = \bar{y}$ and the b_i are twice the factorial effects.

Source	SS	df	MS	F
b_0	14,012.64	1	—	
b_1	901.50	1	901.50	382.80
b_2	551.08	1	551.08	234.00
b_3	363.85	1	363.85	154.50
Lack of fit	13.79	4	3.448	1.906
Pure error	14.47	8	1.809	
Total	15,857.33	16		

There is no lack of fit so $s^2 = (13.79 + 14.47)/(4 + 8) = 2.355$ estimates σ^2. The regression coefficients b_1, b_2, b_3 are all highly significant.

4.13. $\hat{y} = 58.25 - 8.625x_2 + 4.875x_3 + 10.875x_4$
$\qquad - 3.75x_2x_3 + 0.5x_2x_4 - 0.25x_3x_4.$

Source	SS	df	MS	F
b_0	54,289.00	1	(same as	
b_2	1,190.25	1	SS's)	442.47
b_3	380.25	1		141.36
b_4	1,892.25	1		703.44
b_{23}	225.00	1		83.64
b_{24}	4.00	1		1.49
b_{34}	1.00	1		0.37
Residual	24.25	9	$s^2 = 2.69$	
Total	58,006.00	16		

The b_2, b_3, b_4, and b_{23} coefficients are significant.

4.14. Two of the factors, x_1 = type of paper and x_3 = direction of tear, are qualitative and would usually not be suitably extrapolated or interpolated. The only quantitative factor, x_2 = humidity, has little effect on the response.

4.15. (a) The fitted equation is $\hat{y} = 54.5 - 0.625x_1 + 1.375x_2 + 3.375x_3$ with analysis of variance table as below.

Source	SS	df	MS	F
First order	109.375	3	36.485	9.85
Lack of fit	26.875	5	5.375	5.86
Pure error	2.750	3	0.917	
Total, corrected	139.000	11		

Lack of fit is not significant; $s^2 = (26.875 + 2.75)/(5 + 3) = 3.703$. First-order terms are significant. A splitup of the first-order sum of squares into its three component pieces of $3.125 + 15.125 + 91.125$ for b_1, b_2, and b_3, respectively, shows that only b_3 is significant. A model $\hat{y} = 54.5 + 3.375x_3$ explains $R^2 = 0.6556$ of the variation about the mean. (Because pure error cannot ever be explained, the maximum possible value of R^2 is 0.9802.)

(b) A set of 12 meaningful orthogonal contrasts consists of the following columns: I(twelve 1's); x_1, x_2, x_3 (as shown); x_1x_2, x_1x_3, x_2x_3, $x_1x_2x_3$; a column generated from $x_i^2 - \bar{x_i^2}$, formed from any x_i^2 column, then made orthogonal to the **I** column, and consisting of eight $\frac{1}{3}$'s followed by four $-\frac{2}{3}$'s; three pure error contrast columns which have zeros

Table A4.16a. Factorial Effects and *CC* Contrasts for the Original Data

Main Effect or Interaction	Response Number			
	1	2	3	4
1	2.39*	-3.24	11.10*	10.13*
2	-0.66	-2.44	-40.00*	-42.98*
3	0.19	-19.01*	31.45*	12.50*
4	0.61	10.24*	0.45	11.43*
12	-0.59	4.39*	4.30	8.23*
13	1.26	-0.29	5.60*	6.45*
14	0.24	3.16	-7.00*	-3.48
23	1.31	16.01*	-9.90*	7.55*
24	1.39*	5.61*	-1.40	5.48*
34	0.39	0.59	-2.10	-1.00
123	0.99	-6.81*	7.40*	1.70
124	0.61	-10.61*	2.10	-8.03*
134	0.51	-11.34*	-1.85	-12.55*
234	0.86	5.11*	2.45	8.30*
1234	1.09	-4.46*	-4.95	-8.45*
Standard error (effects)	0.52	1.52	1.96	1.43
CC	-2.99*	-6.16*	-6.13*	-15.24*
Standard error (*CC*)	0.50	1.46	1.88	1.37

The asterisks are explained on p. 637. See 4.16.

Table A4.16b. Factorial Effects and *CC* Contrasts for the Transformed Data, After Multiplication by 1000

Main Effect or Interaction	Response Number			
	1	2	3	4
1	94	−25	132	127
2	−28	−15	−460*	−508*
3	−15	−235*	357*	132
4	22	122	5	134
12	−25	59	53	69
13	61	5	63	78
14	−1	32	−77	−31
23	46	178	−83	66
24	56	70	−11	46
34	8	17	−16	−34
123	25	−60	77	13
124	−4	−133	26	97
134	13	−121	31	147
234	32	64	37	116
1234	20	−71	57	−78
Standard error (effects)	51	88	100	85
CC	−101	−88	−81	−147
Standard error (*CC*)	49	84	95	81

The asterisks are explained below. See 4.16.

in the first eight positions and any orthogonal pattern such as, for example,

$$
\begin{array}{ccc}
-1 & -1 & 1 \\
1 & -1 & -1 \\
-1 & 1 & -1 \\
1 & 1 & 1
\end{array}
$$

for their last four entries.

4.16. Table A4.16a shows the factorial effects and the *CC* contrasts, with their respective standard errors. The asterisks indicate contrasts which exceed, in modulus, 2.57 times their standard error. The value 2.57 is the upper $2\frac{1}{2}\%$ point of the t distribution with 5 df. A confused pattern is shown and curvature, as measured by the estimate of ($\beta_{11} + \beta_{22} + \beta_{33} + \beta_{44}$), appears for all responses. The standard errors are $s_i/2$ for the factorial effects and $(11s_i^2/48)^{1/2}$ for the *CC* contrasts which are the differences between two averages of 16 and 6 observations and so have variance $\sigma^2/16 + \sigma^2/6 = 11\sigma^2/48$.

Transformation to $\arcsin(y/100)^{1/2}$ gives the effects in Table A4.16b. We cannot work out the standard errors here using appropriate pure error mean squares because we cannot transform the individual center point observations, which we do not know. As a very crude approximation, we use $q_i = \arcsin(s_i/100)^{1/2}$ and then evaluate $q_i/2$ and $q_i(11/48)^{1/2}$. We see that a much simpler picture emerges after transformation. See page 637.

4.17. A partial answer is provided with the question.

4.18. They concluded that the new sintering method and a thick separator should be used (from the 2^3 experiment) and that the order of rolling should be with negative electrode first (from the 2^2 experiment). They did not consider the direction of rolling to be crucial.

4.19. The regression versions of the main effects (half of the factorial effects) are 218.75, -14.25, and -10; the corresponding 2fi's are -15.25, -20, and 15.5; and the 3fi is 16. Although we have no measure of error to compare these numbers to, it is clear that the only the main effect of factor x_1 is comparatively large and the rest can all be regarded as residual variation. From the point of view of battery life, the higher cost batteries appear worthwhile, while the connector type and battery temperature appear to make little difference.

4.20. A full factorial analysis of these data, performed as a regression calculation, gives rise to the following half-effects:

Main effects: 9.56, 3.01, 2.11, 0.24.
2fi: 0.49, -1.11, -0.36, 0.06, -0.06, 0.19.
3fi: 0.49, 0.44, 0.09, -0.39.
4fi: 0.76.

The analysis of variance table is as follows.

Source	SS	df	MS	F	p
Regression	15	1725.66	115.04	13.30	0.011
Lack of fit	1	11.10	11.10	1.42	0.319
Pure error	3	23.49	7.83		
Total	19	1760.25			

The four repeat center points yield a pure error sum of squares of 23.49 with 3 df, so $s_e^2 = 7.83 = (2.80)^2$. The F-test for lack of fit provides an $F(1, 3)$ value of $11.10/7.83 = 1.42$, not significant. The resulting pooled $s^2 = (11.10 + 23.49)/(1 + 3) = 8.6475$. The estimated variance of an individual effect is $s^2/16 = 0.5405$, providing a standard error for each effect of 0.735. Only the first, second, and third main effects show significance. Note that the one df sum of squares 11.10 is the appropriate sum of squares for the curvature test described in Sections 6.3 and 13.8. The factorial point average is $\bar{y}_f = 55.3875$, the center point average is $\bar{y}_c = 53.525$, and the difference is $\bar{y}_f - \bar{y}_c = 1.8625$. We can think of this as $\mathbf{c'y}$, where

$$\mathbf{c'} = \left(\tfrac{1}{16}, \tfrac{1}{16}, \ldots, \tfrac{1}{16}, -\tfrac{1}{4}, -\tfrac{1}{4}, -\tfrac{1}{4}, -\tfrac{1}{4} \right).$$

The sum of squares attributable to any individual contrast $\mathbf{c'y}$ is given by the formula $(\mathbf{c'y})^2 / \mathbf{c'c}$. Here, this gives $(1.8625)^2 / (\tfrac{5}{16}) = 11.1005$, which is rounded to two decimal places in the analysis of variance table.

CHAPTER 5

5.1. (a) True. The largest value of R is given by the half-fractions $\mathbf{I} = \pm \mathbf{12 \ldots k}$, so that R cannot exceed k. For example, the 2_{III}^{3-1} with generator $\mathbf{I} = \mathbf{123}$ has $k = R = 3$.

(b) False. For example, the 2_{III}^{3-1} generated by $\mathbf{I} = \mathbf{123}$ has $k = 3$, $R = 3$, $p = 1$ so that $3 = k < R + p = 4$.

5.2. (a) Assume that all interactions (i) between 1, 2, and 3 are zero; (ii) between 3, 4, and 5 are zero. The defining relation is $\mathbf{I} = \mathbf{1234} = \mathbf{135} = \mathbf{245}$. Ignoring interactions of three or more factors, we can estimate

Mean	(12) + (34)
1 + (35)	(13) + 24 + 5
2 + (45)	(23) + 14
3 + 15	(123) + 4 + 25

Two main effects are clear, three are confounded. (The interactions in parentheses are assumed to be zero, but are listed for information.)

(b) A better design is given by $\mathbf{I} = \mathbf{123} = \mathbf{345} \ (= \mathbf{1245})$ in which we can estimate

Mean	4 + (35)
1 + (23)	5 + (34)
2 + (13)	14 + 25
3 + (12) + (45)	24 + 15

Here, because of the assumptions, all five main effects are clear and the remaining four two-factor interactions are grouped in two pairs.

5.3. Design (a) has $\mathbf{I} = \mathbf{A}$, while design (b) is of resolution III. So (b) is *much* better.

5.4. (a) First fraction $\mathbf{I} = \mathbf{8} = \mathbf{1234} = \mathbf{125} = \mathbf{146} = \mathbf{247}$.
Folded fraction $\mathbf{I} = -\mathbf{8} = \mathbf{1234} = -\mathbf{125} = -\mathbf{146} = -\mathbf{247}$.

(b) $\mathbf{I} = \mathbf{1234} = \mathbf{1258} = \mathbf{1468} = \mathbf{2478}$, a 2_{IV}^{8-4}.

(c) Suppose design (b) has matrix

$$\mathbf{B} = \begin{bmatrix} \mathbf{A} \\ -\mathbf{A} \end{bmatrix}.$$

The new design will have matrix

$$\begin{bmatrix} \mathbf{B} \\ -\mathbf{B} \end{bmatrix} = \begin{bmatrix} \mathbf{A} \\ -\mathbf{A} \\ -\mathbf{A} \\ \mathbf{A} \end{bmatrix},$$

which is a 2_{IV}^{8-4} design, replicated.

5.5. The 2_{IV}^{8-4} design, $\mathbf{I} = \mathbf{1234} = \mathbf{1246} = \mathbf{1347} = \mathbf{2348}$, in two blocks of eight runs each.

Set $\mathbf{B} = \mathbf{12}$ so that $\mathbf{I} = \mathbf{12B}$ generates the blocking. Then we can obtain estimates of

$$I, (1 + 2B), (2 + 1B), (3 + 5B), (4 + 6B),$$
$$(5 + 3B), (6 + 4B), (7 + 8B), (8 + 7B),$$
$$(B + 12 + 35 + 46 + 78), (13 + 25 + 47 + 68),$$
$$(14 + 26 + 37 + 58),$$
$$(23 + 15 + 48 + 67), (24 + 16 + 38 + 57),$$
$$(34 + 17 + 28 + 56), (45 + 36 + 27 + 18).$$

The main effects are clear if the block variable does not interact.

5.6. The defining relation for the design given in the hint is

$$\mathbf{I} = \mathbf{1235} = \mathbf{1246} = \mathbf{12B_1} = \mathbf{13B_2} = \mathbf{3456}$$
$$= \mathbf{35B_1} = \mathbf{25B_2} = \mathbf{46B_1} = \mathbf{2346B_2}$$
$$= \mathbf{23B_1B_2} = \mathbf{123456B_1} = \mathbf{1456B_2}$$
$$= \mathbf{15B_1B_2} = \mathbf{1346B_1B_2} = \mathbf{2456B_1B_2}.$$

The 16 estimates available are those of the mean I, the six main effects, $(B_1 + 12 + 35 + 46)$, $(B_2 + 13 + 25)$, $(B_1B_2 + 15 + 23)$, $(14 + 26)$, $(16 + 24)$, $(34 + 56)$, (higher-order interactions), (higher-order interactions), and $(36 + 45)$, if block variables do not interact.

It might be slightly better to block the design by $\mathbf{B_1} = \mathbf{134}$, and $\mathbf{B_2} = \mathbf{234}$. This will provide estimates of the mean I, the six main effects, $(B_1B_2 + 12 + 35 + 46)$, $(13 + 25)$, $(15 + 23)$, $(14 + 26)$, $(16 + 24)$, $(34 + 56)$, B_1, B_2, $(36 + 45)$, if block variables do not interact. Two of the block effects have been disassociated from the sets of two-factor interactions. However, the first design will provide two estimates of higher-order interactions which can be used to estimate σ^2, which the second design does not provide.

5.7. Generators (including blocking) are

$$I = 1235 = 1246 = 12B_1 = 13B_2 = 14B_3.$$

Defining relation:

$$I = 1235 = 1246 = 12B_1 = 13B_2 = 14B_3 = 3456 = 35B_1 = 25B_2$$

$$= 2345B_3 = 46B_1 = 2346B_2 = 26B_3 = 23B_1B_2 = 24B_1B_3 = 34B_2B_3$$

$$= 123456B_1 = 1456B_2 = 1356B_3 = 15B_1B_2 = 1345B_1B_3 = 1245B_2B_3$$

$$= 1346B_1B_2 = 16B_1B_3 = 1236B_2B_3 = 1234B_1B_2B_3 = 2456B_1B_2$$

$$= 2356B_1B_3 = 56B_2B_3 = 45B_1B_2B_3 = 36B_1B_2B_3 = 1256B_1B_2B_3.$$

If block variables do not interact, and interactions between three or more factors are ignored, the following estimates are available: I, 1, 2, 3, 4, 5, 6, $(B_1 + 12 + 35 + 46)$, $(B_2 + 13 + 25)$, $(B_3 + 14 + 26)$, $(B_1B_2 + 15 + 23)$, $(B_1B_3 + 16 + 24)$, $(B_2B_3 + 34 + 56)$, (third-order interactions), (third-order interactions), $(B_1B_2B_3 + 36 + 45)$.

When there are eight blocks it is not possible to improve on this estimation setup. For example, if we set $B_1 = 134$, $B_2 = 234$, then $B_1B_2 = 12$. The allocation of B_3 to any of the remaining columns aliases a main effect with a block effect. Thus, the three blocking variables must be allocated to columns that usually estimate two-factor interaction combinations. (Note that, as in a previous example, if we want *four* blocks, we can set $B_1 = 134$ and $B_2 = 234$ without trouble.)

5.8. (a) The defining relation for the initial eight runs is $I = 1234 = 235 = 145$. When a design is folded over, the signs of each variable are reversed, so that the defining relation for the second eight-run fraction is $I = 1234 = -235 = -145$. Combining the two designs (symbolically, "averaging" the two defining relations) gives $I = 1234$ and thus the design is a 2^{5-1} of resolution IV.

(b) A 2_V^{5-1} design $I = 12345$ is usually better because all main effects and two-factor interactions can be separately estimated.

5.9. First write down an eight-run resolution III design in six factors, for example, $I = 124 = 135 = 236$. Add a second fraction in which only the sign of variable 1 is switched, namely, $I = -124 = -135 = 236$. The defining relations of the two designs are, respectively,

$$I = 124 = 135 = 236 = 2345 = 1346 = 1256 = 456,$$

$$I = -124 = -135 = 236 = 2345 = -1346 = -1256 = 456.$$

The joint design thus has $\mathbf{I} = \mathbf{236} = \mathbf{2345} = \mathbf{456}$, and any two of the three indicated words will generate it. The main effect of 1 is associated with four- or five-factor interactions and all two-factor interactions involving factor 1 are associated with three- or four-factor interactions, achieving the desired aim.

5.10. The two defining relationships are

$$\mathbf{I} = \mathbf{124} = \mathbf{135} = \mathbf{236} = \mathbf{2345} = \mathbf{1346} = \mathbf{1256} = \mathbf{456},$$

$$\mathbf{I} = \mathbf{124} = -\mathbf{135} = -\mathbf{236} = -\mathbf{2345} = -\mathbf{1346} = \mathbf{1256} = \mathbf{456},$$

with common parts

$$\mathbf{I} = \mathbf{124} = \mathbf{1256} = \mathbf{456}.$$

If we ignore interactions involving three or more factors, the alias structure is thus

I	**13**
1 = 24	**15 = 26**
2 = 14	**16 = 25**
3	**23**
4 = 12 = 56	**34**
5 = 46	**35**
6 = 45	**36**

We see that variable 3 and all its two-factor interactions are estimable individually. This happens because changing the signs of **1** and **2** produces exactly the same effect here as changing the sign of **3**.

5.11. We first set down a 2^4 factorial design in variables **1, 2, 3,** and **4**. Clearly the columns **12, 13, 14, 23, 24, 123** *cannot* be use to accommodate new variables or these interactions would not be individually estimable. Five choices remain, namely **34, 124, 134, 234,** and **1234**. If, for example, we choose **5 = 34** and **6 = 1234** the defining relation is $\mathbf{I} = \mathbf{345} = \mathbf{12346} = \mathbf{1256}$. However, this aliases **5** with **34** and **15** with **26**, and so is no good. No other pairs of choices work, either; such a design does not exist.

   (*Note:* Some readers interpret this question to mean that two factor interactions not specifically mentioned are zero. If this *were* true, setting **5 = 34** and **6 = 1234** *would* work.)

5.12–
5.14. Solutions are implicit in the questions.

5.15. The design is a 2_{IV}^{4-1} generated by $I = -1234$. We can estimate:

$1 - 234 \leftarrow$	$3.75,$	$I \leftarrow$	$-106.875,$
$2 - 134 \leftarrow$	$5.75,$	$12 - 34 \leftarrow$	$1.75,$
$3 - 124 \leftarrow$	$0.26,$	$13 - 24 \leftarrow$	1.25
$4 - 123 \leftarrow$	$-0.25,$	$14 - 23 \leftarrow$	-1.25

Standard error (effect) $= \{4\sigma^2/8\}^{1/2} = 1$. Factors 1 and 2 appear to be the only influential ones. It is, of course, possible that a pair of interactions may be canceling each other out. One possibility for four extra runs is to do a 2^2 factorial in factors 1 and 2, with factors 3 and 4 held fixed. This would provide a recheck on the main effects of 1 and 2 and give an estimate of the interaction 12 (and of 34 when combined with the results above).

5.16. This 2_{III}^{5-2} design has defining relation $I = 123 = 345 = 1245$. Ignoring interactions of three or more factors, we can estimate the following:

$1 + 23$	$\leftarrow \quad 5.375,$	$5 + 34 \leftarrow$	$0.125,$
$2 + 13$	$\leftarrow \quad 0.625,$	$14 + 25 \leftarrow$	$6.125,$
$3 + 12 + 45$	$\leftarrow -1.125,$	$15 + 24 \leftarrow$	$-0.375,$
$4 + 35$	$\leftarrow \quad 7.125,$	$I \leftarrow$	$20.1875.$

Pure error $s^2 = 1.8125$; standard error(effects) $= (s^2/4)^{1/2} = 0.673$. The important effects are $(1 + 23)$, $(4 + 35)$, $(14 + 25)$. A likely possibility is that factors 1 and 4 and the interaction 14 are large. A 2^2 factorial in 1 and 4 would be a useful four-run followup to check this. Other factors would be held constant.

If eight new runs were permissible, the fold-over design $I = -123 = -345 = 1245$, when combined with the original design would give clear estimates of all main effects and two-factor interactions except for $(12 + 45)$, $(14 + 25)$ and $(15 + 24)$, assuming all interactions of three or more factors are zero.

5.17. This cannot be done. $\mathbf{B}_1$ and $\mathbf{B}_2$ must be set equal to interactions with at least three letters, or blocks are confounded with two-factor interactions immediately. However, if we choose, say, $\mathbf{B}_1 = 123$, then also $\mathbf{B}_1 = 123$ $(-12345) = -45$. Similarly, a choice of $\mathbf{B}_1 = 1234$ implies $\mathbf{B}_1 = -5$.

5.18. (a) The $(\mathbf{B}_1, \mathbf{B}_2) = (+, +)$ block is defined by

$$I = 1234567 = 1357 = 1256$$
$$(= 246 = 347 = 2367 = 145).$$

In other words, we have a 2_{III}^{7-3} design which we can generate from initial columns **1**, **2**, **3**, and **4** (the basic 2^4 design) by choosing $5 = 14$, $6 = 24$, and $7 = 34$.

(b) Estimates available consist of the average response $\bar{y} \rightarrow I$, and those for

$1 + 45,$	$12 + 56,$
$2 + 46,$	$13 + 57,$
$3 + 47,$	$16 + 25,$
$4 + 26 + 37 + 15,$	$17 + 35,$
$5 + 14,$	$23 + 67,$
$6 + 24,$	$27 + 36.$
$7 + 34,$	

5.19. **(a)** Ozzie wanted to find a resolution III design which would be such that variables **3** and **5** would never attain their +levels simultaneously. However, every resolution III design gives a 2^2 factorial (possibly replicated) in any two variables. Hence any eight run resolution III design will give a (replicated) 2^2 factorial in variables **3** and **5**, which means that there will be at least one run for which variables **3** and **5** attain their +levels simultaneously.

(b) The principal generating relation is $I = 124 = 135 = 236$. If we do not rename the variables, the only members of this family which can be used are $I = \pm 124 = -134 = \pm 236$. There are four such fractions. If we are allowed to rename the variables, the question becomes: for what members of the family can we find three variables that do not attain their +levels simultaneously? We see, for example, that in the fraction with generating relation $I = -124 = 135 = 236$ our requirement will be satisfied for the three variables **1**, **2**, and **4**. In fact, if we put a minus sign in front of *any* word in the principal generating relation, our requirement will be satisfied for the three variables in that word. Therefore, the *only* member of the family which *cannot* be used is that associated with the principal generating relation.

(c) Defining relation: $I = 124 = -135 = 236 = -2345 = 1346 = -1256 = -456$

$$
\begin{aligned}
1 + 24 - 35 &\leftarrow \quad 40, \\
2 + 14 + 36 &\leftarrow -10, \\
3 - 15 + 26 &\leftarrow \quad 20, \\
4 + 12 - 56 &\leftarrow \quad 90, \\
5 - 13 - 46 &\leftarrow -60, \\
6 + 23 - 45 &\leftarrow \quad 30, \\
34 - 25 + 16 &\leftarrow \quad 0.
\end{aligned}
$$

(d) One possibility is to switch all the signs and separate out all the main effects from combinations of two-factor interactions. Ozzie was more interested in learning about Miss Freeny's interactions, however, than he was in learning about all the main effects. The fact that the estimates that have the largest magnitude involve either the main effect

of Miss Freeny or her interactions gives justification to his decision to switch the sign of variable **4** in the second fraction. (Note that a similar argument could be made for switching the sign of only variable **5** in the second fraction. However, we have to face the fact that Ozzie was far more interested in Miss Freeny than in the gypsy band.)
The defining relation for the second fraction is

$$I = -124 = -135 = 236 = 2345 = -1346 = -1256 = 456.$$

The calculation matrix for the second fraction is

			Design Matrix					
I	**1**	**2**	**3**	**4**	**5**	**6**	**123**	**y**
+	−	−	−	−	−	+	−	135
+	+	−	−	+	+	+	+	165
+	−	+	−	+	−	−	+	285
+	+	+	−	−	+	−	−	175
+	−	−	+	−	+	−	+	205
+	+	−	+	+	−	−	−	195
+	−	+	+	+	+	+	−	295
+	+	+	+	−	−	+	+	145

The estimates obtained from the second fraction are

$$1 - 24 - 35 \leftarrow -60,$$
$$2 - 14 + 36 \leftarrow 50,$$
$$3 - 15 + 26 \leftarrow 20,$$
$$4 - 12 + 56 \leftarrow 70,$$
$$5 - 13 + 46 \leftarrow 20,$$
$$6 + 23 + 45 \leftarrow -30,$$
$$-34 - 25 + 16 \leftarrow 0.$$

(e) Combining the two pieces gives the following estimates:

$$1 - 35 \leftarrow -10, \qquad\qquad 24 \leftarrow 50,$$
$$2 + 36 \leftarrow 20, \qquad\qquad 14 \leftarrow -30,$$
$$3 - 15 + 26 \leftarrow 20, \qquad\qquad \text{error} \leftarrow 0,$$
$$4 \leftarrow 80, \qquad\qquad 12 - 56 \leftarrow 10,$$
$$5 - 13 \leftarrow -20, \qquad\qquad 46 \leftarrow 40,$$
$$6 + 23 \leftarrow 0, \qquad\qquad 45 \leftarrow -30,$$
$$-25 + 16 \leftarrow 0, \qquad\qquad 34 \leftarrow 0.$$

(f) The second fraction was a wise choice. The main effect 4 of Miss Freeny is large and so are four of her five interactions 24, 14, 46, and 45.

(g) Miss Freeny's presence clearly has a positive effect on bar business, an effect that is enhanced when there are chips available for her to pass around (effect 24 ← 50), the light are low (effect 14 ← −30), the band is not playing (effect 45 ← −30), and Dick is the bartender (effect 46 ← 40). Obviously, the customers were much more attentive to Miss Freeny when the lights were low and they were not distracted by the band; they consumed more potato chips, became thirstier, and drank faster. Unfortunately, Tom the bartender, who was generally more alert than Dick, became so enamored of Miss Freeny that he seldom noticed the empty glasses, hence the positive 46 interaction. Note that this particular explanation is just one of several that are consistent with the data. To appreciate that it was probably the right one, you had to be there.

5.20. It is not obvious at first sight that columns 13 and 16 are identical (i.e., these two variables have been set to the same pattern of levels and so their effects cannot be distinguished). However this emerges immediately if one evaluates the effects using a planar regression model with a constant term and 24 variables. Omitting the redundant column 16, the regression leads to the analysis of variance table shown below.

Source	SS	df	MS	F
Effects	134,851.10	23	5863	2.95
Residual	7,953.86	4	1988	
Total, corrected	142,804.96	27		

Overall, the regression is not significant, but we need to explore the relative sizes of the individual effects. On the basis of our *a priori* assumption, we shall transfer, to the residual, the smallest eight sums of squares, associated with variables 3, 5, 6, 11, 19, 21, 22, and 24 and round off decimals in the sums of squares. This gives the table:

Source	SS	df	MS	F
Effects	129,730	15	8649	7.94*
Residual	13,075	12	1090	
Total, corrected	142,805	27		

*Significant at 0.001

If we use the upper 5% point of the $F(1, 12)$ distribution as a point of comparison, an individual 1 df sum of squares would be declared "large" if $\{SS/1\}/1090 \geq 4.75$, namely if $SS \geq 5178$. By this criterion, factors 4, 8, 14, 15, 17, 20, and 22 are designated as possibly the most effective factors; factor 1 falls below the cutoff but is the next largest. This conclusion is consistent with that of the original author, who added that, while factor 17 "was not studied further," the other seven factors were confirmed as effective in subsequent work.

Portions of the data in this exercise have been used as examples in a number of papers. See, for example, Abraham et al. (1999), Beattie et al.

(2002), Chipman et al. (1977), Holcomb et al. (2003), Li and Lin (2002), Lin (1993a, b; 1995a, b), Lu and Wu (2004), Wang (1995), and Westfall et al. (1998).

5.21. We note that the replicated runs are the 2^{3-1} design for which $x_1 x_2 x_3 = 1$, so that pairs like (b_1, b_{23}) will not be independently estimated. If we fit the full factorial model, the $\mathbf{Z}$ matrix has column heads 1, x_1, x_2, x_3, $x_1 x_2$, $x_1 x_3$, $x_2 x_3$, $x_1 x_2 x_3$ and is a 12×8 matrix. The solution $\mathbf{b} = (\mathbf{Z'Z})^{-1} \mathbf{Z'y}$ is then

$$
\begin{bmatrix}
12 & & & & & & & 4 \\
 & 12 & & & & & 4 & \\
 & & 12 & & & 4 & & \\
 & & & 12 & 4 & & & \\
 & & & 4 & 12 & & & \\
 & & 4 & & & 12 & & \\
 & 4 & & & & & 12 & \\
4 & & & & & & & 12
\end{bmatrix}^{-1}
\begin{bmatrix}
615 \\ 137 \\ -5 \\ 11 \\ 13 \\ 45 \\ 47 \\ 209
\end{bmatrix}
=
\begin{bmatrix}
51.125 \\ 11.375 \\ -1.875 \\ 0.625 \\ 0.875 \\ 4.375 \\ 0.125 \\ 0.375
\end{bmatrix}
$$

Source	SS	df	MS	F
b_0	31,518.75	1	(Same	—
b_1	1,564.08	1	as SS's)	125.13
b_2	2.08	1		0.17
b_3	10.08	1		0.81
$b_{12}\|b_3$	8.17	1		0.65
$b_{13}\|b_2$	204.17	1		16.33
$b_{23}\|b_1$	0.17	1		0.01
$b_{123}\|b_0$	1.50	1		0.12
Residual (pure error)	50.00	4	$s_e^2 = 12.50$	
Total	33,359.00	12		

Note that the residual here consists only of the pure error. For all the b's, $\hat{V}(b) = 0.09375(12.50) = 1.172$, so that standard error $(b) = 1.08$.

5.22. Solution is implicit in the questions asked.

5.23. (Wu and Meyer, 1965).

$$\hat{y} = 6.637 + 0.202 x_1 + 0.169 x_2 + 0.025 x_3 + 0.038 x_4 + 0.015 x_5.$$

Source	SS	df	MS	F
b_1	1.3069	1		
b_2	0.9147	1		
b_3	0.0194	1		
b_4	0.0451	1		
b_5	0.0070	1		
Lack of fit	0.0443	10	0.00443	11.25
Pure error	0.0063	16	0.00039	
Total, corrected	2.3437	31		

Lack of fit is indicated. $R^2 = 0.9784$. Thus, most of the variation in the data is explained by the model, but the variation that remains unexplained is still large compared to the pure error variation shown between pairs of runs at the same experimental conditions.

5.24. It is not possible, because no 2_{IV}^{9-5} exists. However suppose we use, for example, the 2_{III}^{9-5} given by $I = 1235 = 1246 = 1347 = 2348 = 129$. The full defining relation contains 26 more "words" including the following (only) of length three: **789**. Thus all main effects are estimated clear except for those in the combinations:

$$1 + 29$$
$$2 + 19$$
$$7 + 89$$
$$8 + 79$$
$$9 + 12 + 78.$$

So we would need six two-factor interactions to be zero before we could obtain clear main effects.

5.25. We first obtain the defining relation of this eight-run design by multiplying together all possible combinations of generators. Rearranging this gives $I = 235 = 136 = 1256 \ [= 124 = 1237 = 1345 = 2346 = 347 = 157 = 267 = 1234567 = 3567 = 1467 = 2457 = 456]$. All the "words" in the square brackets contain either a **4** or a **7** which we drop. Moreover **1256** is the product of **235** and **136**. Thus the reduced design is generated by $I = 235 = 136$ so that $5 = 23$ and $6 = 13$ columns can be obtained from an initial 2^3 design in variables **1**, **2**, and **3**. The design is a 2_{III}^{5-2} in variables **1**, **2**, **3**, **5**, and **6**.

5.26. Inspection shows the design to be a 2_{III}^{5-2} generated by $I = 123 = 145$ $(= 2345)$. If we ignore interactions involving three or more factors, we obtain estimates as follows.

$$1 + 23 + 45 \leftarrow -1.25^*,$$
$$2 + 13 \qquad \leftarrow -0.25,$$
$$3 + 12 \qquad \leftarrow \ \ \ 0.25,$$
$$4 + 15 \qquad \leftarrow -1.75^*,$$
$$5 + 14 \qquad \leftarrow -2.25^*,$$
$$24 + 35 \qquad \leftarrow -0.25,$$
$$25 + 34 \qquad \leftarrow \ \ \ 0.25.$$

The larger estimates (marked with asterisks) involve factors 1, 4, and 5 and/or their interactions. A sensible next step would be to run a second design in which the signs of all variables were switched, namely $I = -123 = -145 \ (= 2345)$. This would enable the main effects to be split off separately.

5.27.   **(a)** The full defining relation is

$$I = -126 = 135 = -237 = -1234 = -2356 = 1367 = 346$$
$$= -1257 = -245 = 147 = 567 = 1456 = -2467 = 3457$$
$$= -1234567.$$

I	1	2	3	4 = −123	5 = 13	6 = −12	7 = −23	y
+	−	−	−	+	+	−	−	46.1
+	+	−	−	−	−	+	−	55.4
+	−	+	−	−	+	+	+	44.1
+	+	+	−	+	−	−	+	58.7
+	−	−	+	−	−	−	+	56.3
+	+	−	+	+	+	+	+	18.9
+	−	+	+	+	−	+	−	46.4
+	+	+	+	−	+	−	−	16.4

Ignoring all interactions involving three or more factors, we obtain

$$l_I = 42.8 \rightarrow \text{average,} \qquad\qquad l_4 = -0.5 \rightarrow 4 + 36 - 25 + 17,$$
$$l_1 = -10.9 \rightarrow 1 - 26 + 35 + 47, \qquad l_5 = -22.8 \rightarrow 5 + 13 - 24 + 67,$$
$$l_2 = -2.8 \rightarrow 2 - 16 - 37 - 45, \qquad l_6 = -3.2 \rightarrow 6 - 12 + 34 + 57,$$
$$l_3 = -16.6 \rightarrow 3 + 15 - 27 + 46, \qquad l_7 = 3.4 \rightarrow 7 - 23 + 14 + 56,$$

Conclude: Relatively large effects are

$$1 - 26 + 35 + 47,$$
$$3 + 15 - 27 + 46,$$
$$5 + 13 - 24 + 67.$$

All decrease packing time. This can be explained in several ways, any of which *might* be correct:

1, 3, 5 are all important main effects, each causing a decrease in packing time.
Or 1 and 3 are the only important main effects, and they interact.
Or 1 and 5 are the only important main effects, and they interact.
Or 3 and 5 are the only important main effects, and they interact.
Or 1 is the only important main effect, but it interacts with 3 and 5.
Or 3 is the only important main effect, but it interacts with 1 and 5.
Or 5 is the only important main effect, but it interacts with 1 and 3.

Other conclusions might be drawn, but these seem most plausible. Since this experiment leads us to at least seven fairly reasonable but differing conclusions, the advantage of performing the second fractional factorial becomes apparent.

(b) For the second fraction

$$I = 126 = -135 = 237 = -1234 = -2356 = 1367 = -346 = -1257$$
$$= 245 = -147 = -567 = 1456 = -2467 = 3457 = 1234567.$$

I	1	2	3	4 = −123	5 = −13	6 = 12	7 = 23	y
+	+	+	+	−	−	+	+	41.8
+	−	+	+	+	+	−	+	40.1
+	+	−	+	+	−	−	−	61.5
+	−	−	+	−	+	+	−	37.0
+	+	+	−	+	+	+	−	22.9
+	−	+	−	−	−	−	−	34.1
+	+	−	−	−	+	−	+	17.7
+	−	−	−	+	−	+	+	42.7

$$l'_I = 37.2 \to \text{average}, \qquad l'_4 = 9.2 \to 4 - 36 + 25 - 17,$$
$$l'_1 = -2.5 \to 1 + 26 - 35 - 47, \qquad l'_5 = -15.6 \to 5 - 13 + 24 - 67,$$
$$l'_2 = -5.0 \to 2 + 16 + 37 + 45, \qquad l'_6 = -2.3 \to 6 + 12 - 34 - 57,$$
$$l'_3 = 15.8 \to 3 - 15 + 27 - 46, \qquad l'_7 = -3.3 \to 7 + 23 - 14 - 56.$$

(c) Combining (a) and (b), we obtain estimates as follows:

$$\tfrac{1}{2}(l_I + l'_I) = 40.0 \to \text{average}, \qquad l_I - l'_I = 5.6 \to \text{block effect},$$
$$\tfrac{1}{2}(l_1 + l'_1) = -6.7 \to 1, \qquad \tfrac{1}{2}(l_1 - l'_1) = -4.2 \to -26 + 35 + 47,$$
$$\tfrac{1}{2}(l_2 + l'_2) = -3.9 \to 2, \qquad \tfrac{1}{2}(l_2 - l'_2) = 1.1 \to -16 - 37 - 45,$$
$$\tfrac{1}{2}(l_3 + l'_3) = -0.4 \to 3, \qquad \tfrac{1}{2}(l_3 - l'_3) = -16.2 \to 15 - 27 + 46,$$
$$\tfrac{1}{2}(l_4 + l'_4) = 4.4 \to 4, \qquad \tfrac{1}{2}(l_4 - l'_4) = -4.9 \to 36 - 25 + 17,$$
$$\tfrac{1}{2}(l_5 + l'_5) = -19.2 \to 5, \qquad \tfrac{1}{2}(l_5 - l'_5) = -3.6 \to 13 - 24 + 67,$$
$$\tfrac{1}{2}(l_6 + l'_6) = -2.8 \to 6, \qquad \tfrac{1}{2}(l_6 - l'_6) = -0.5 \to -12 + 34 + 57,$$
$$\tfrac{1}{2}(l_7 + l'_7) = 0.1 \to 7, \qquad \tfrac{1}{2}(l_7 - l'_7) = 3.4 \to -23 + 14 + 56.$$

Conclusion: The two most important main effects are 1 and 5, and their interaction is probably the cause of the large effect $15 - 27 + 46 \leftarrow -16.2$. The factor owner would therefore do well to keep the foreman on the job at all times in addition to piping in music.

The experiment also indicates that women pack faster than men, that high temperature slows packing, and that packers 25 or over pack faster than those under 25.

5.28. The missing run from the incomplete design can be "estimated" by making the tentative assumption that a (relatively) high-order interaction estimate is zero. We first complete the design by adding a row of signs $+ - + - - +$ which is chosen to give four pluses and four minuses in each column.

We now inspect the design matrix to find three columns that form a full factorial, for example, columns **1**, **2**, and **3**. It is then seen that **4 = 123**, **5 = 23**, and **6 = 13**. Only the **12** column remains for estimation purposes and it will lead to an estimate of the combination of effects (12 + 34 + 56) assuming all interactions involving three or more factors are zero. All the other six columns provide estimates of (main effect plus two-factor interactions) combinations.

The estimate of (12 + 34 + 56) would be zero if the missing observation m were such that

$$\tfrac{1}{4}(29 - 27 - 42 - 0 + 30 + 0 + 39 - m) = 0,$$

that is, if $m = 29$.

We now evaluate effects other than (12 + 34 + 56) which we have "sacrificed." The defining relation of the design is

$$\mathbf{I = 1234 = 235 = 136 = 145 = 246 = 1256 = 3456.}$$

We obtain

$$1 + 36 + 45 \leftarrow \quad 1.0,$$
$$2 + 35 + 46 \leftarrow -21.0,$$
$$3 + 25 + 16 \leftarrow -\ 1.5,$$
$$4 + 15 + 26 \leftarrow -\ 8.5,$$
$$5 + 23 + 14 \leftarrow \quad 0.5,$$
$$6 + 24 + 13 \leftarrow -19.5.$$

In view of the comparatively large sizes of the second, fourth, and sixth of these, a sensible course of action would be to "switch all signs," to perform the 2_{III}^{6-3} fraction $\mathbf{I = 1234 = -235 = -136}$, and to estimate main effects clear by combining both fractions.

5.29. Not provided.

5.30. $\mathbf{I = -134 = 1235 = -126 = -237}$. The design is a 2_{III}^{7-4}.

5.31. (a) The fitted model is $\hat{y} = 1792.92 + 505.083x_1 - 556.083x_2 + 476.417x_3 + 0.417x_4$ and the coefficients have a standard error of 51.68.

(b) The intercept is again 1792.92 and the regression coefficient estimates are (rounding to integer values) 505, −556, 476, 0, 25, 94, 62, 112, 81, −269, 161. A normal plot of these indicates four unusually high ones, the first three and the last but one, −269.

(c) The bias formula shows an alias matrix of form

$$\frac{1}{3}\begin{bmatrix} 3 & 0 & 0 & 0 & 0 & 0 & 0 & 0 & 0 & 0 & 0 \\ 0 & 3 & 0 & 0 & 0 & 0 & 0 & 0 & -1 & -1 & 1 \\ 0 & 0 & 3 & 0 & 0 & 0 & -1 & -1 & 0 & 0 & -1 \\ 0 & 0 & 0 & 3 & 0 & -1 & 0 & 1 & 0 & -1 & 0 \\ 0 & 0 & 0 & 0 & 3 & -1 & 1 & 0 & -1 & 0 & 0 \\ 0 & 0 & 0 & 0 & 0 & -1 & -1 & 1 & -1 & 1 & -1 \\ 0 & 0 & 0 & 0 & 0 & 1 & -1 & 1 & -1 & -1 & -1 \\ 0 & 0 & 0 & 0 & 0 & -1 & 1 & -1 & 1 & -1 & -1 \\ 0 & 0 & 0 & 0 & 0 & -1 & -1 & -1 & -1 & 1 & 1 \\ 0 & 0 & 0 & 0 & 0 & 1 & 1 & -1 & -1 & -1 & -1 \\ 0 & 0 & 0 & 0 & 0 & 1 & -1 & -1 & 1 & 1 & -1 \\ 0 & 0 & 0 & 0 & 0 & -1 & -1 & -1 & 1 & -1 & 1 \end{bmatrix},$$

indicating that, apart from the overall mean, all estimates actually involve interaction effects. This gives justification for a reanalysis via a model that includes main effects and two-factor interactions.

(d) Apart from the mean 1792.92, the 10 coefficients are (rounded to integers) 280, -556, 304, -140, -225, -55, -36, -253, -328, and 93. The approximate normal plot shows that all the points lie roughly on a straight line, indicating no real effects and confirming the initial expectations. The "effects" seen in the first analysis arose because the high fractionation of the Plackett and Burman design was not properly taken into account.

5.32. The design is a 2_{IV}^{5-1} fractional factorial generated by $I = 12345$. A normal plot of the 15 contrast estimates from the factorial analysis (or from the equivalent regression analysis as in Figure 5.32a, omitting the overall average, shows a split of the main mass of the plot, indicating a possible bias effect. We can also tentatively work with the idea that only two factors 4 and 5 and their interaction 45 are responsible for most of the changes in the data. There are four possible combinations of levels of factors $(4, 5)$, each of which is associated with four observations as follows:

$$(-, -)\ 6.92, 6.33, 6.24, 6.29$$
$$(-, +)\ 6.30, 7.83, 7.72, 7.62$$
$$(+, -)\ 7.70, 8.26, 7.13, 7.36$$
$$(+, +)\ 17.97, 18.14, 18.86, 11.26.$$

The 16th observation, 11.26, looks out of place compared with the rest of its group. If we replace it by the average of the other three in the group, that is, by 18.32, and redo the calculations, the normal plot (Figure 5.32b) realigns most of the contrasts and still shows up the main effects of factors 3 and 4 and their interaction.

(The regression coefficients are ($\bar{y} = 9.496$), 0.323, -0.436, -0.359, 2.589, 2.467, 0.637, 0.403, 0.607, 0.444, -0.568, -0.497, -0.162, -0.471,

−0.391, 2.006. The listing is in the usual order with coefficient subscripts $(0), 1, 2, \ldots, 5, 12, 13, \ldots, 45.$)

5.33. With such unbalanced data, the easiest first calculation to make is to use a stepwise regression program to select variables from the set of variables $x_1, x_2, \ldots, x_{12}$ and the 28 two-factor interactions (cross-products) involving variables $x_1, x_2, \ldots, x_8$. Setting the "alpha to enter" and the "alpha to remove" both at 0.15 leads to a path of calculations in which all the significant variables are added successively but none is removed. The final equation, with the terms shown in the order of entry, is

$$\hat{y} = 7.478 + 1.34x_5 + 0.93x_8 - 0.76x_4 + 0.43x_9$$
$$- 0.61x_5x_6 + 0.35x_1 - 0.35x_7.$$

The final R^2 value is $R^2 = 0.567$. Variables x_2 and x_3 are the only ones not selected at all, although x_6 enters only through its interaction with x_5. Of the dummy variables, only x_9 enters the equation. The highest predicted response at the design levels would be found at $x_5 = x_8 = x_9 = x_1 = 1$, $x_4 = x_7 = -1$, and $x_5x_6 = -1$, which then implies that $x_6 = -1$. This particular combination of levels does not occur in the planned design. New exploration could be around this indicated set of conditions. (If a minimum response were sought, the levels shown above would need to be of the opposite signs, of course.)

By re-running the regression with the variables selected above, but adding a lack of fit test, we obtain the same equation, of course, but with an analysis of variance table as follows:

Source	df	SS	MS	F
Regression	7	154.18	22.03	16.10
Lack of fit	8	15.94	1.99	1.53
Pure error	78	101.70	1.30	
Total, corrected	93			

No lack of fit is indicated. The regression F-value, calculated as

$$22.03/\{(15.94 + 101.7)/(8 + 78)\} = 16.10,$$

after re-pooling lack of fit and pure error, is significant at the 0.001 level.

This experiment was followed by a subsequent study in which variables 5, 6, 7, and 8 were retained (but renumbered as 3, 7, 1 and 8) and four new factors were introduced. Once again a 2_{IV}^{8-4} design was planned, with the 16 runs divided up into eight blocks. The details of this subsequent experiment, plus the resulting analyses, are given in Fractional factorial analysis of growth and weaning success in *Peromyscus maniculatus*, by W. P. Porter and R. L. Busch, *Science, New Series*, **202**, No. 4370 (November 24, 1978), 907–910.

5.34. Each triplet of y's provides 2 df for pure error, 16 in all, from which $s^2 = 13.74$. The overall total of 24 df divides up as 16 for pure error, 1 for overall mean and 7 for 7 estimates of effects. We apply a regression analysis to these data. The overall mean is 58.0775. The seven estimates of the following combinations of main effects and 2fi: $(A + BD + CE + FG)$, $(B + AD + CF + EG)$, $(C + AE + BF + DG)$, $(D + AB + CG + EF)$, $(E + AC + BG + DF)$, $(F + AG + BC + DE)$, $(G + AF + BE + CD)$ are respectively *double* the following regression coefficient estimates (all of which have the same standard error of 0.7567): 11.385, -4.3408, 3.8542, 1.2983, -1.1950, -2.2375, 2.7283. The first three and the last two are the significant effects. Of course, we cannot separate the individual main effects from the 2fi in such a small design, but we can use the regression equation to predict responses. If we drop the two nonsignificant coefficients from the regression we obtain a predictive equation of

$$\hat{y} = 58.0775 + 11.385x_1 - 4.3408x_2 + 3.8542x_3 - 2.2375x_6 + 2.7283x_7.$$

Because of the orthogonality of the columns in the **X**-matrix, the selected coefficients do not change when the two nonsignificant terms are dropped. Plots of the residuals for both fitted models show up the 22nd observation 32.330 as a possible too low outlier. The corresponding predicted value is 40.970 leaving a residual of -8.640, which provides a normal deviate of size -2.85, relatively large.

5.35. We first fit a regression model using the 11 observations from the settings 1–8, and a model with a constant, x_1, x_2, x_3, x_4, x_1x_2 $(= x_3x_4)$, x_1x_3 $(= x_2x_4)$, and x_1x_4 $(= x_2x_3)$. This gives estimates that are half the estimates of these factorial combinations: $(1 + 234)$, $(2 + 134)$, $(3 + 124)$, $(4 + 123)$, $(12 + 34)$, $(13 + 24)$, $(14 + 23)$. The respective numerical values are 0.25, 0, 0.25, 0, 0, -0.25, 0. Since pure error is zero, no test of coefficients is possible here. Assuming "all $\geq$ 3fi $= 0$" thus leads us to believe that the "real" effects are 1, 3 and the combination $(13 + 24)$.

We next add the four semi-fold runs to the regression. This enables us to add a 24 column that is now different from the 13 column, due to the four additional runs. The resulting nonzero estimates are 0.25, 0.25, -0.25, and 0 for the effects 1, 3, 13, and 24, respectively. So factors 1 and 3 have positive effects but 13 is negative. Now the combinations used of factors 1 and 3, and their consequent response values, are as follows:

1	3	Results
$-$	$-$	0 0 0 0 0 0
$+$	$-$	1 1 1 1 1
$-$	$+$	1 1
$+$	$+$	1 1

Clearly, failures occur whenever factor 1 is negative (water) and factor 3 is also negative (regular flour). Avoid this combination and the bread will rise!

Table E5.36a. Results from a 2_{III}^{7-3} Fractional Factorial Experiment

Estimate:	−1.31	12.71	−1.04	1.44	9.89	10.61	−0.04
Main effect:	1	2	3	4	5	6	7

Estimate:	−1.09	−0.19	−0.81	−0.61
2 fi sum:	12 + 35 + 46	13 + 25 + 47	14 + 26 + 37	15 + 23 + 67

Estimate:	2.11	8.41	0.24	−0.56
2 fi sum:	24 + 16 + 57	34 + 17 + 56	27 + 36 + 45	0^a

^{a} Interactions of order greater than 2, 234 + Aliases (145, 136, 127, 256, 357, 467, 1234567).

Table E5.36b. Rearrangement of the 2_{III}^{7-3} Fractional Factorial Experiment, Based on Tentatively Assuming that Only Factors 2, 5, and 6 Matter

x_2	x_5	x_6	Mean Observation	Observations	(Difference)^2/2
−1	−1	−1	57.9	58.7, 57.1	1.28
1	−1	−1	71.05	69.0, 73.1	8.405
−1	1	−1	59.0	59.1, 58.9	0.02
1	1	−1	72.9	69.0, 76.8	30.42
−1	−1	1	60.35	60.5, 60.2	0.045
1	−1	1	73.0	71.7, 74.3	3.38
−1	1	1	79.4	79.7, 79.1	0.18
1	1	1	90.55	90.7, 90.4	0.045

5.36. Via the usual factorial analysis (or the similar regression calculation which gives half of the numbers shown for its regression coefficients) we obtain the estimates of Table E5.36a. The overall mean is 70.52. The independent regression (or effect) columns can be constructed from any one of the interactions in the groups shown; for example, the products 12, 35, or 46 all give the same column and so on.

Via inspection, or via a normal plot of the effects, we can tentatively conclude that 2, 5 and 6 are the important variables. If variables 1, 3, 4, and 7 are deleted, we can reanalyze the design as a (pseudo-) replicated 2^3 factorial in 2, 5, and 6 (see Table E5.36b).

The large difference between the two observations at $(1, 1, −1)$ stands out, casts some doubt on our tentative assumption, and also contributes heavily to the last column. These observations need further investigation. The sum of the last column, 43.775, with 8 df, provides an estimate $s^2 = 43.775/8 = 5.49$ of the basic variance of y, and could be compared with any prior value available from past experience, either via a χ^2 test if σ^2 were "known," or via an F-test if an estimate with specific df were available. Unfortunately this information was not available.

CHAPTER 6

Note: For steepest ascent answers, $\mathbf{b} = (b_1, b_2, \ldots, b_k)'$ and does not contain b_0.

6.1. A suitable coding is $x_1 = (X_1 − 90)/10$, $x_2 = (X_2 − 20)/10$.
$$\hat{y} = 11.60 − 8.5x_1 + 6x_2.$$

Points on the path of steepest ascent have coordinates $(x_1, x_2) = (-8.5\theta, 6\theta)$ for various θ values or $(X_1, X_2) = (90 - 85\theta, 20 + 60\theta)$. Thus, all the runs given are on this path; take $\theta = 0.3, 0.5, 0.6, 0.7, 0.55, 0.65$ in turn.

$$\hat{y} = 70 - 1.5x_1, \quad \text{where now } x_1 = (X_1 - 39)/4.25, \ x_2 = (X_2 - 56)/3.$$

Source	SS	df	MS	F
b_1	9	1		
b_2	0	1		
Lack of fit	41	2	20.50	30.75
Pure error	2	3	0.67	
Total, corrected	52	7		

The lack of fit test F value exceeds the upper 5% F point of 9.55. It would be best to fit a second-order surface at this stage, as a diagram (not provided here) well illustrates.

6.2. $\hat{y} = 12 - 8x_1 + 5.5x_2$. Points $(x_1, x_2) = (-8\theta, 5.5\theta)$, that is, points $(X_1, X_2) = (99 - 80\theta, 17 + 110\theta)$ are on the path of steepest ascent. No value of θ gives $(59, 67)$ so the answer is no.

6.3. $\mathbf{a} = \begin{bmatrix} 4 \\ 5 \\ 6 \end{bmatrix}, \quad \mathbf{b} = \begin{bmatrix} 0.25 \\ 1.00 \\ 0.50 \end{bmatrix}, \quad \mathbf{e} = \mathbf{b} - \left(\dfrac{\mathbf{a}'\mathbf{b}}{\mathbf{a}'\mathbf{a}} \right) \mathbf{a} = \begin{bmatrix} -0.22 \\ 0.42 \\ -0.20 \end{bmatrix}.$

Restriction is $4(35 + 5x_1) + 5(15 + 5x_2) + 6(11 + 5x_3) = 371$, or $4x_1 + 5x_2 + 6x_3 = 18 = -a_0$.

$$\lambda_0 = -\frac{a_0}{\mathbf{a}'\mathbf{b}} = 2,$$

$$\mathbf{x}_0 = \lambda_0 \begin{bmatrix} b_1 \\ b_2 \\ b_3 \end{bmatrix} = \begin{bmatrix} 0.5 \\ 2.0 \\ 1.0 \end{bmatrix}, \quad \mathbf{X}_0 = 5\mathbf{x}_0 + \begin{bmatrix} 35 \\ 15 \\ 11 \end{bmatrix} = \begin{bmatrix} 37.5 \\ 25.0 \\ 16.0 \end{bmatrix}.$$

In coded units, the new path is

$$\begin{bmatrix} 0.5 \\ 2.0 \\ 1.0 \end{bmatrix} + \mu \begin{bmatrix} -0.22 \\ 0.42 \\ -0.20 \end{bmatrix} \quad \text{or} \quad \begin{bmatrix} 35 + 5(0.5 - 0.22\mu) \\ 15 + 5(2.0 + 0.42\mu) \\ 11 + 5(1.0 - 0.20\mu) \end{bmatrix}$$

in uncoded units. This gives the point with coordinates $(36.40, 27.10, 15.00)$ if we set $\mu = 1$. So the answer is yes.

6.4. The restriction is $(x_1 + 8) + x_2 + 14 + (\frac{1}{2}x_3 + 7) = 26.4$, or $2.6 + x_1 + x_2 + \frac{1}{2}x_3 = 0$: $\mathbf{a} = (1, 1, \frac{1}{2})'$, $\mathbf{b} = (-4.10, -9.25, 4.90)'$. $c = \mathbf{a'b}/\mathbf{a'a} = -10.9/2.25 = -4.844$. Thus $\mathbf{e} = \mathbf{b} - c\mathbf{a} = (0.744, -4.406, 7.322)$. Also, $\lambda_0 = -2.6/(-10.9) = 0.2385$, so that the point at which the path of steepest ascent is modified is $\lambda_0(-4.10, -9.25, 4.90)$, or $(-0.978, -2.206, 1.169)$. Thus the modified path is on $x_1 = 0.978 + \mu(0.744)$, $x_2 = -2.206 + \mu(-4.406)$, $x_3 = 1.169 + \mu(7.322)$, or on $(X_1, X_2, X_3) = \{7.022 + \mu(0.744), 11.794 - \mu(4.406), 7.584 + \mu(3.611)\}$. A value of $\mu = 0.114$ gives the point $(7.11, 11.29, 8.00)$; so the answer is yes.

6.5. We have $\mathbf{a} = (1, 1, 1, 1)'$, $\mathbf{b} = (1, 2, 3, 2)'$, $\mathbf{e} = \mathbf{b} - 2\mathbf{a} = (-1, 0, 1, 0)'$. Now, $\lambda_0 = -(-4)/8 = \frac{1}{2}$. Thus in coded units the amended path has coordinates $(x_1, x_2, x_3, x_4) = \frac{1}{2}(1, 2, 3, 2) + \mu(-1, 0, 1, 0)$, or $(X_1, X_2, X_3, X_4) = \{5(0.5 - \mu) + 5, 5(1) + 5, (1.5 + \mu) + 5, 5(1) + 5\} = \{7.5 - 5\mu, 10, 12.5 + 5\mu, 10\}$. When $\mu = 2$, this results in the point $(-2.5, 10, 22.5, 10)$. So the answer is yes.

6.6. If $\hat{y}_1 > 21.5$, we must use (x_1, x_2) values "northeast" of the line $1.5 = x_1 + x_2$ which passes through the points $(1.5, 0)$ and $(0, 1.5)$. If $\hat{y}_2 < 54$, we must use (x_1, x_2) values "southwest" of the line $4 = 4x_1 + 2x_2$, which passes through the points $(1, 0)$ and $(0, 2)$. The lines intersect at $(\frac{1}{2}, 1)$. A pointed segment "northwest" of that point is the production region of (x_1, x_2). For example, all points $(0, x_2)$ where $1.5 \le x_2 \le 2$ lie in the production region.

CHAPTER 7

7.1. (a) $y = \beta_0 + \beta_1 x_1 + \beta_2 x_2 + \varepsilon$.

$$\mathbf{Z'Zb} \equiv \begin{bmatrix} 9 & 0 & 0 \\ 0 & 6 & 0 \\ 0 & 0 & 6 \end{bmatrix} \begin{bmatrix} b_0 \\ b_1 \\ b_2 \end{bmatrix} = \begin{bmatrix} 24.93 \\ 2.42 \\ -1.61 \end{bmatrix} \equiv \mathbf{Z'y}.$$

$$\hat{y} = 2.7700 + 0.4033x_1 - 0.2683x_2.$$

Source	SS	df	MS	F
b_0	69.0561	1		
b_1	0.9761 ⎫ 1.4081	2	0.7041	$F = 115$
b_2	0.4320 ⎭			
Residual	0.0363	6	$s^2 = 0.0061$	
	70.5005	9		

The residuals are, in order, 10^{-2} times $-7, 4, 6, 5, 2, -14, -3, 2, 5$ (sum 0), compared with $s = 0.08$, a figure that is roughly of the same order as the standard error of the residuals.

(b) The F statistic for regression, $F = 115$ is highly significant and $R^2 = 1.4081/(1.4081 + 0.0363) = 0.9749$. All of the residuals are of the same order as $s = 0.08$ except for -0.14 which is less than $2s$, and there are no remarkable features exhibited by the residuals plot in the (x_1, x_2) plane.

Overall we conclude that a good fit to the data has been obtained.

7.2. **(a)**

$$Z = \begin{bmatrix} 1 & -1 & -1 & 1 & 1 & 1 \\ 1 & 0 & -1 & 0 & 1 & 0 \\ 1 & 1 & -1 & 1 & 1 & -1 \\ 1 & -1 & 0 & 1 & 0 & 0 \\ 1 & 0 & 0 & 0 & 0 & 0 \\ 1 & 1 & 0 & 1 & 0 & 0 \\ 1 & -1 & 1 & 1 & 1 & -1 \\ 1 & 0 & 1 & 0 & 1 & 0 \\ 1 & 1 & 1 & 1 & 1 & 1 \end{bmatrix},$$

$$Z'Z = \begin{bmatrix} 9 & 0 & 0 & 6 & 6 & 0 \\ & 6 & 0 & 0 & 0 & 0 \\ & & 6 & 0 & 0 & 0 \\ & & & 6 & 4 & 0 \\ & & & & 6 & 0 \\ \text{symmetric} & & & & & 4 \end{bmatrix}.$$

(b) Replace x_1^2 by $x_1^2 - \frac{2}{3}$ as in hint so that the new column is $\frac{1}{3}, -\frac{2}{3}, \frac{1}{3}, \frac{1}{3}, -\frac{2}{3}, \frac{1}{3}, \frac{1}{3}, -\frac{2}{3}, \frac{1}{3}$. Similarly, replacing x_2^2 by $x_2^2 - \frac{2}{3}$, we obtain a new column $\frac{1}{3}, \frac{1}{3}, \frac{1}{3}, -\frac{2}{3}, -\frac{2}{3}, -\frac{2}{3}, \frac{1}{3}, \frac{1}{3}, \frac{1}{3}$. Other columns of Z are as before. The "new $Z'Z$" is now diagonal with diagonal entries 9, 6, 6, 2, 2, 4.

7.3. From the form described in Exercise 7.2(b) we obtain

$$\begin{bmatrix} b_0' \\ b_1 \\ b_2 \\ b_{11} \\ b_{22} \\ b_{12} \end{bmatrix} = \begin{bmatrix} \frac{1}{9} & & & & & \\ & \frac{1}{6} & & \mathbf{0} & & \\ & & \frac{1}{6} & & & \\ & & & \frac{1}{2} & & \\ & \mathbf{0} & & & \frac{1}{2} & \\ & & & & & \frac{1}{4} \end{bmatrix} \begin{bmatrix} 24.93 \\ 2.42 \\ -1.61 \\ -0.08 \\ 0.07 \\ -0.05 \end{bmatrix} = \begin{bmatrix} 2.7700 \\ 0.4033 \\ -0.2683 \\ -0.0400 \\ 0.0350 \\ -0.0125 \end{bmatrix}$$

Now $b_0' = b_0 + \frac{2}{3}b_{11} + \frac{2}{3}b_{22}$, so that $b_0 = 2.7733$. The fitted model is then $\hat{y} = 2.7733 + 0.4033x_1 - 0.2683x_2 - 0.0400x_1^2 + 0.0350x_2^2 - 0.0125x_1x_2$.

Source	SS	df	MS
b_0	69.0561	1	—
b_1	0.9761	1	0.9761
b_2	0.4320	1	0.4320
b_{11}	0.0032 ⎫	1 ⎫	
b_{22}	0.0025 ⎬ 0.0063	1 ⎬ 3	0.0021
b_{12}	0.0006 ⎭	1 ⎭	
Residual	0.0300	3	$s^2 = 0.0100$
Total	70.5005	9	

Clearly the second-order terms are not significant, so there is no point retaining them in the model. The first order test is the same as in Exercise 7.1.

7.4. In view of Exercise 7.3, we might expect to find that a transformation is unnecessary, and this is indeed the case. Numerical details follow.

x_1	x_2	y	$\hat{y}$	$y - \hat{y}$	$q = \hat{y}^2$	$q - \hat{q}$
-1	-1	2.57	2.6350	-0.0650	6.9432	
0	-1	3.08	3.0383	0.0417	9.2313	
1	-1	3.50	3.4416	0.0584	11.8446	
-1	0	2.42	2.3667	0.0533	5.6013	
0	0	2.79	2.7700	0.0200	7.6729	
1	0	3.03	3.1733	-0.1433	10.0698	
-1	1	2.07	2.0984	-0.0284	4.4033	
0	1	2.52	2.5017	0.0183	6.2585	
1	1	2.95	2.9050	0.0450	8.4390	

We obtain $\hat{q} = 7.8293 + 2.2343x_1 - 1.4864x_2$, with regression SS for this model of $\mathbf{a'Z'q} = 594.7792$. [*Notation:* $\mathbf{a'} = (7.8293, 2.2343, -1.4864)$, $\mathbf{Z}$ is the matrix of coefficients of $(1, x_1, x_2)$ as in Exercise 7.1, and $\mathbf{q}$ is the vector of q values.]

Thus, $\Sigma(q_u - \hat{q}_u)^2 = \Sigma q_u^2 - 594.7792 = 0.1366$. Also $\Sigma(y_u - \hat{y}_u)(q_u - \hat{q}_u) = \Sigma(y_u - \hat{y}_u)q_u = 0.0030$. Thus

$$\hat{\alpha} = \frac{(0.0030)}{0.1366} = 0.0220,$$

$$S_\alpha = \frac{(0.0030)^2}{0.1366} = 0.00007.$$

Table 7.6 can now be displayed:

Source	SS	df	MS	F
Transformation	0.00007	1	0.00007	0.01
Adjusted residual	0.03623	5	0.00725	
Residual	0.0363	6		

It is clear that no transformation is called for.

7.5. **(a)** $y = \beta_0 + \beta_1 x_1 + \beta_2 x_2 + \beta_{11} x_1^2 + \beta_{22} x_2^2 + \beta_{12} x_1 x_2$
$\qquad + \beta_{111} x_1^3 + \beta_{222} x_2^3 + \beta_{112} x_1^2 x_2 + \beta_{122} x_1 x_2^2 + \varepsilon.$
(b) In the appropriate $\mathbf{Z}$ matrix, the (*iii*) column $\equiv$ (*i*) column for $i = 1$ and 2 so that the β_{iii} are not separately estimable. However, the two

(*ijj*) columns *cannot* be expressed as a linear combination of the other columns of $\mathbf{Z}$ and so both β_{122} and β_{112} *are* separately estimable.

(c) For the second-order model with third-order alternative, the alias matrix $\mathbf{A} = (\mathbf{X}_1'\mathbf{X}_1)^{-1}\mathbf{X}_1'\mathbf{X}_2$ has the form

$$\frac{1}{36}\begin{bmatrix} 20 & 0 & 0 & -12 & -12 & 0 \\ 0 & 6 & 0 & 0 & 0 & 0 \\ 0 & 0 & 6 & 0 & 0 & 0 \\ -12 & 0 & 0 & 18 & 0 & 0 \\ -12 & 0 & 0 & 0 & 18 & 0 \\ 0 & 0 & 0 & 0 & 0 & 9 \end{bmatrix}\begin{bmatrix} 0 & 0 & 0 & 0 \\ 6 & 0 & 0 & 4 \\ 0 & 6 & 4 & 0 \\ 0 & 0 & 0 & 0 \\ 0 & 0 & 0 & 0 \\ 0 & 0 & 0 & 0 \end{bmatrix} = \left[\begin{array}{c} \mathbf{0} \\ \hline \begin{array}{cccc} 1 & 0 & 0 & \frac{2}{3} \\ 0 & 1 & \frac{2}{3} & 0 \end{array} \\ \hline \mathbf{0} \end{array}\right]$$

Thus b_1 estimates $\beta_1 + \beta_{111} + \frac{2}{3}\beta_{122}$, b_2 estimates $\beta_2 + \beta_{222} + \frac{2}{3}\beta_{112}$. The other estimates are unbiased.

(d) The question "Are your results consistent with what you found in (b)?" is a somewhat misleading one, and is asked in order to make this point: The alias structure does not in itself enable us to say whether or not a coefficient can be estimated. For example, β_1 is aliased with β_{111} (which *cannot* be estimated) and β_{122} (which *can* be estimated). This point is sometimes misunderstood.

7.6. (a) We use the formula $V(\mathbf{b}) = (\mathbf{Z}'\mathbf{Z})^{-1}\sigma^2$ and then estimate σ^2 by $s^2 = 0.0061$ to get the estimated variances est $V(b_0) = 0.0061/9$, est $V(b_i) = 0.0061/6$, $i = 1, 2$. The confidence bands are given by

$$b_i \pm t\left(v, 1 - \tfrac{1}{2}\alpha\right)\{\text{est } V(b_i)\}^{1/2},$$

where v is the df of s^2 ($= 6$, here). See Eq. (3.12.9), page 64, for an alternative notation. For 95% limits $t(6, 0.975) = 2.45$ so that we obtain

$$\beta_0 : 2.77 \pm 0.06, \qquad \beta_1 : 0.40 \pm 0.08, \qquad \beta_2 : -0.27 \pm 0.08.$$

(b) At $(x_1, x_2) = (0.7, 0.5)$, $\hat{y} = 2.9182$. est $V(\hat{y}_0) = \mathbf{x}_0'(\mathbf{Z}'\mathbf{Z})^{-1}\mathbf{x}_0 s^2$

$$= (1, 0.7, 0.5)\begin{bmatrix} \frac{1}{9} & 0 & 0 \\ 0 & \frac{1}{6} & 0 \\ 0 & 0 & \frac{1}{6} \end{bmatrix}\begin{bmatrix} 1 \\ 0.7 \\ 0.5 \end{bmatrix}(0.0061) = 0.001430$$

$$= (0.0378)^2.$$

The 95% limits for the true mean value of y are thus $2.9182 \pm 2.45(0.0378) = 2.92 \pm 0.09$.

(c) The values of (β_1, β_2) in the 90% joint confidence region are bounded by the "ellipse" (which actually is a circle here)

$$[(b_1 - \beta_1), (b_2 - \beta_2)]\begin{bmatrix} 6 & 0 \\ 0 & 6 \end{bmatrix}\begin{bmatrix} b_1 - \beta_1 \\ b_2 - \beta_2 \end{bmatrix} = 2s^2 F(2, 6, 0.90);$$

see formula (3.12.5). Because the x_1 and x_2 columns are orthogonal to the column of 1's, no adjustment for the means of these columns is needed. This becomes

$$(\beta_1 - 0.4033)^2 + (\beta_2 - 0.2683)^2 = 2(0.0061)3.46/6 = 0.007035$$

$$= (0.084)^2,$$

that is, a circle of radius 0.084.

(d) We can compare the correct 90% confidence region in (c) with the incorrect 90.25% region obtained by taking the intersection of the individual confidence intervals for β_1 and β_2. These latter are spread ± 0.078 about the estimates and so define a square of slightly smaller width than the diameter of the circle. On the whole, they would not mislead, mainly due to the lack of correlation between the estimates b_1 and b_2.

(e) In general,

$$V(\hat{y}) = (1, x_1, x_2)\begin{bmatrix} \frac{1}{9} & 0 & 0 \\ 0 & \frac{1}{6} & 0 \\ 0 & 0 & \frac{1}{6} \end{bmatrix}\begin{bmatrix} 1 \\ x_1 \\ x_2 \end{bmatrix}$$

$$= \tfrac{1}{9} + \tfrac{1}{6}\left(x_1^2 + x_2^2\right).$$

This is obviously constant on circles $x_1^2 + x_2^2 = \rho^2$.

7.7.   By using $(x_i^2 - \tfrac{2}{3})$ instead of x_i^2 in the full second-order model, we can do the least-squares fit in orthogonal form. (*Note:* This works here because of the 3^3 design used; it is not always possible to achieve orthogonalization in this way.) The fitted model can be written

$$\hat{y} = 386.63 + 39.78x_1 + 33.89x_2 + 13.83x_3$$

$$- 205.50x_1^2 + 55.17x_2^2 - 32.67x_3^2$$

$$+ 46.42x_1x_2 - 21.17x_1x_3 + 2.50x_2x_3.$$

Residuals in order: $15, 31, -24, -49, -85, 55, 70, -64, 49,$
$51, 46, -84, -43, 35, 49, 53, -114, 7,$
$-63, 30, -3, -41, 87, -9, 6, 33, -41.$

Check total of residuals is -3, which is zero within rounding error. The largest of these is -114, although it does not appear abnormally large on the overall plot of residuals. The calculations for a transformation test show SS(transformation) $= 83$ (1 df), SS(adjusted residual) $= 76{,}066$ (17 df) with a nonsignificant F. No transformation is needed. The 3^k algorithm leads to the following effects.

Final Column of Algorithm	Divisor	Mean Square	Root Mean Square	Rank of rms	Effect Name
7145	27	1,890,779	—	—	Mean
716	18	24,481	169	25	1
-3700	54	253,519	504	26	11
610	18	20,672	144	23	2
557	12	25,854	161	24	12
757	36	15,918	126	20	112
992	54	18,223	135	22	22
-685	36	13,034	114	19	122
335	108	1,039	32	3	1122
249	18	3,445	59	15	3
-254	12	5,376	73	17	13
-804	36	17,956	134	21	113
30	12	75	9	1	23
-125	8	1,953	44	7	123
-291	24	3,528	59	16	1123
-348	36	3,364	58	14	223
217	24	1,962	44	8	1223
399	72	2,211	47	10	11223
-589	54	6,424	80	18	33
266	36	1,965	44	9	133
-292	108	789	28	2	1133
202	36	1,133	34	4	233
-265	24	2,926	54	12	1233
-479	72	3,187	56	13	11233
368	108	1,254	35	5	2233
413	72	2,369	49	11	12233
-541	216	1,355	37	6	112233

The normal plot of root mean squares indicates two things.

1. The larger effects seem to be associated with variables x_1 and x_2.
2. There appears to be bias in at least one observation.

If we take the seventeenth observation (which is associated with the largest residual) as being the possible culprit, we can replace it by a "missing value estimate" obtained in one of several possible ways. One way is to set equal to zero the highest order interaction (112233). This leads to an "estimate" for y_{17} of 497.25 instead of 362. The revised root mean squares are (3) 540 instead of 504, (4) 176, (6) 81, (7) 153, (9) 6, (19) 117 not 80, (21) 24, (22) 11, (24) 7, (25) 9, (27) set equal to zero and so dropped from the normal plot. The remainder are unchanged. A plot of the 25 revised root mean squares shows the same prominent effects as before. Further analysis is left to the reader, who should check the Davies source reference for features suppressed in our presentation of the data.

7.8. Application of the 3^k algorithm provides the figures below.

Final Column of Algorithm	Divisor	Mean Square	Root Mean Square	Rank of rms	Effect Name
8068	27	2,410,838	—	—	Mean
1243	18	85,836	293	26	1
−635	54	7,467	86	22	11
1048	18	61,017	247	25	2
−179	12	2,607	51	19	12
−371	36	3,823	62	21	112
−782	54	11,324	106	23	22
−95	36	251	16	12	122
253	108	593	24	16	1122
583	18	18,883	137	24	3
−31	12	80	9	7	13
13	36	5	2	3	113
56	12	261	16	13	23
−2	8	1	1	1	123
−130	24	704	27	17	1123
82	36	187	14	11	223
20	24	17	4	5	1223
52	72	38	6	6	11223
−437	54	3,536	59	20	33
−23	36	15	4	4	133
13	108	2	1	2	1133
−188	36	982	31	18	233
−44	24	81	9	8	1233
−104	72	150	12	10	11233
−206	108	393	20	15	2233
−146	72	296	17	14	12233
154	216	110	10	9	112233

The normal plot of the root mean squares shows up as large effects (in order)—1, 2, 3, 22, 11, 112, 33, 12—and suggests that a second-order model in all three factors would do well. We can, in fact, obtain the fitted equation

$$\hat{y} = 367.48 + 69.06x_1 + 58.22x_2 + 32.39x_3$$

$$- 35.28x_1^2 - 43.44x_2^2 + 24.28x_3^2$$

$$- 14.92x_1x_2 - 2.58x_1x_3 + 4.67x_2x_3.$$

Most of these coefficients are found by dividing the first column of the table by the corresponding divisor, except for $b_0 = \hat{y} - \frac{2}{3}(b_{11} + b_{22} + b_{33})$. We can read off the SS from the MS column, because each MS has 1 df, to give this analysis of variance table:

Source	SS	df	MS	F
b_0	2,410,838	1	—	
First order	165,736	3	55,245	122.0
Second order $\|b_0$	25,275	6	4,213	9.3
Residual	7,709	17	$s^2 = 453$	
Total	2,609,558	27		

The major influences are clearly first order, although the quadratic contributions are significant also. The interaction terms are not large, comparatively.

7.9. Application of the 3^k algorithm followed by a plot on normal probability paper of root mean squares shows a well-behaved plot with largest effects being, in order, main effects 3, 1, and 2, followed after a gap, by interactions 13, 123, 23. If a second-order model is fitted we obtain

$$\hat{y} = 326.14 + 131.46x_1 + 109.43x_2 + 177.00x_3$$

$$- 28.32x_1^2 - 21.64x_2^2 + 32.74x_3^2$$

$$+ 43.58x_1x_2 + 75.47x_1x_3 + 66.03x_2x_3.$$

The first-order terms pick up $3,271,619/4,284,176 = 0.7637$ of the variation about the mean, while second-order terms with an extra SS (given b_0) of 471,778, pick up another 0.1101 for a total of 0.8738. However, a lack-of-fit test gives an F value of

$$\frac{\{295,990/17\}}{\{244,025/54\}} = 3.86$$

which exceeds the percentage point $F_{0.05}(17, 54) \approx 1.83$. In the circumstances, it would seem sensible to look for a transformation for y, as indications are that use of the model as fitted could lead to problems. (We do not pursue this here, but leave it to the reader.)

7.10. A full second-order model can be fitted to these data to give $\hat{y} = 219.55 + 5.42x_1 - 2.67x_2 + 75.92x_1^2 + 38.17x_2^2 + 7.13x_1x_2$ with ANOVA table as below.

Source	SS	df	MS	F	
b_0	1,572,946.72	1			
b_1	352.08	1	352.08	1.57	Not significant
b_2	85.33	1	85.33	0.38	Not significant
$b_{11}\vert b_0$	23,053.87	1	23,053.87	102.97	Significant
$b_{22}\vert b_0$	5,827.03	1	5,827.03	26.03	Significant
b_{12}	403.13	1	403.13	1.80	Not significant
Residual	2,686.84	12	$s^2 = 223.90$		
Lack of fit	1,085.34	3	361.72	2.03	Not significant
Pure error	1,601.50	9	177.94		
Total	1,605,355.00	18			

$F(3, 9, 0.95) = 3.86$, $F(1, 12, 0.95) = 4.75$, and $F(1, 12, 0.99) = 9.33$. There is no apparent lack of fit, and the response is primarily one of pure second order in x_1 and x_2. The residuals are given below in the same order downward as the corresponding observations (and then left to right in sequence).

$$
x_1 = \begin{cases} -1 \\ 0 \\ 1 \end{cases}
\quad
\begin{array}{ccc}
-1 & x_2 = 0 & 1 \\
0, 6 & -26, 0 & 14, 7 \\
-2, 12 & 22, -13 & 3, -22 \\
-15, -1 & 7, 9 & -8, 6
\end{array}
$$

The residuals add to -1 (≈ 0) and show no abnormalities worth remarking. An analysis via the 3^k algorithm can also be performed. The conclusions are the same; the (122) interaction turns out to be the third largest effect.

7.11. (Partial solution).

(b) $\hat{y} = 36.25 + 3.487x_1 + 4.621x_2$. The analysis is included in (c).

(c) $\hat{y} = 7.5 + 3.487x_1 + 4.621x_2 - 5.813x_1^2 - 5.063x_2^2 + 7.5x_1x_2$.

Source	SS	df	MS	F
b_0	15,768.75	1		
First order	268.15	2	134.08	13.53
Second order$\|b_0$	542.63	3	180.88	18.25
Lack of fit	42.47	3	14.16	2.50
Pure error	17.00	3	5.67	
Total	16,639.00	12		

No lack of fit is apparent, so we use $s^2 = (42.47 + 17.00)/(3 + 3) = 9.912$ in the F tests for first and second order. Both are significant at the $\alpha = 0.01$ level. When just first order is fitted, lack of fit is shown; the "second order$\|b_0$" SS would be in the lack of fit for the smaller model and the lack of fit F value would be the (significant at $\alpha = 0.05$) value

$$F = \frac{\{(542.63 + 42.47)/6\}}{5.67} = 17.20.$$

(d) Adding 37 to observations 7–12 simulates the situation when there is a block effect between the two sets of runs 1–6 (block 1) and 7–12 (block 2). The blocking variable is orthogonal to the model, so exactly the same predicted equation as in (c) emerges except that b_0 is increased from 36.25 to 54.75. Only 2 df now exist for pure error, 1 df for each pair *within* a block, with pure error sum of squares $\frac{1}{2}(45 - 41)^2 + \frac{1}{2}(42 - 46)^2 = 16$. The "lost df" for pure error becomes a blocks df with SS

$$\frac{B_1^2}{6} + \frac{B_2^2}{6} - \frac{G^2}{12} = 3502.08 \text{ (1 df)}$$

where $B_1 = 226$ is the sum of the observations in the first block and $B_2 = 431$ is the sum for the second block. $G = B_1 + B_2 = 657$ is the grand total. $SS(b_0) = 35,970.75$ now, and total SS = 40,331. The lack of fit SS = 31.39, and lack of fit is not significant. The SS for first- and second-order$\|b_0$ are the same. The basic conclusions are unchanged.

7.12. $\hat{y} = 75 + 2.06x_1 + 0.85x_2 - 4x_1x_2.$

Source	SS	df	MS	F
b_0	78,750.0	1		
b_1, b_2	39.8	2	19.9	33.33
$b_{11}, b_{22}, b_{12}\|b_0$	64.0	3	21.3	35.50
Blocks	350.0	1	350.0	583.33
Lack of fit	0.2	3	0.067	0.07
Pure error	4.0	4	1.0	
Total	79,208.0	14		

There is no lack of fit; $s^2 = 4.2/7 = 0.6$, with 7 df. The blocks difference is large, and worth removing. Both first and second order are significant, but second order is all interaction b_{12}. (See Section 13.7.)

7.13. *All 25 observations:*

$$\hat{y}_{(25)} = 80.59 - 2.940x_1 + 2.12x_2 - 2.157x_1^2 - 1.257x_2^2 - 3.350x_1x_2.$$

Source	SS	df	MS	F
b_1, b_2	656.90	2	328.45	30.13
$b_{11}, b_{22}, b_{12}\|b_0$	1558.61	3	519.54	47.66
Residual	207.05	19	10.90	
Total, corrected	2422.56	24		

Both first- and second-order terms are needed.

Table A7.13. Fitted Values from the 25- and 9-Point Fits[a]

y	$\hat{y}_{(25)}$	$\hat{y}_{(9)}$	$\hat{y}_{(9)}$
60	55		28
67	68		55
76	78	76	
82	85		89
91	90		95
64	65		52
71	75	72	
80	81		85
90	86	91	
90	87		90
68	71	68	
77	77		80
87	81	86	
83	81		85
77	80	77	
71	73		75
80	75	80	
76	75		79
70	73	70	
65	68		55
72	70		75
69	69		73
64	66	64	
59	60		48
55	52		26

[a]The two $\hat{y}_{(9)}$ columns are, respectively, data point predictions and predictions at the unused 16 other locations.

Nine selected observations:

$$\hat{y}_{(9)} = 85.78 - 2.917x_1 + 2.250x_2 - 4.021x_1^2 - 3.396x_2^2 - 7.250x_1x_2.$$

Source	SS	df	MS	F
b_1, b_2	162.83	2	81.42	691.9
$b_{11}, b_{22}, b_{12}|b_0$	436.53	3	145.51	123.66
Residual	3.53	3	1.18	
Total, corrected	602.89	8		

Both first- and second-order terms are needed.

Table A7.13 shows that the nine point fit does very well at predicting at the nine data locations but tends to extrapolate badly at the corners due to its greater curvature. Because the 9-point fit is performed with six parameters and is over a smaller region than the 25-point fit, these features are not surprising ones.

7.14. $\hat{y} = 150.84 + 2.47x_1 - 2.06x_2 - 16.07x_1^2 - 16.30x_2^2 - 6.00x_1x_2.$

Source	SS	df	MS	F
b_1, b_2	87.90	2	43.95	2.24
$b_{11}, b_{22}, b_{12}|b_0$	3328.62	3	1109.54	56.57
Lack of fit	100.67	3	33.557	5.92
Pure error	17.00	3	5.667	
Total, corrected	3534.19	11		

Lack of fit is not significant; $s^2 = 117.67/6 = 19.612$. Second-order terms are significant, so the nonsignificant first-order terms must be retained.

The contours of $V = NV\{\hat{y}(\mathbf{x})\}/\sigma^2$ are given by $V = 2.979 + 0.630x_1^2x_2^2 - 1.383r^2 + 1.579r^4$, where $r^2 = x_1^2 + x_2^2$; see the diagram. Note that, if the

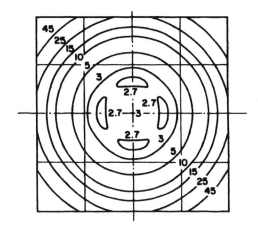

axial points had been at distance $2^{1/2} = 1.414$, rather than 1.5, the contours would have been perfect circles.

[Additional analysis:

$$\frac{(\hat{y}_{\max} - \hat{y}_{\min})}{(ps^2/n)^{1/2}} = \frac{(150.84 - 107.94)}{\{6(19.612)/12\}^{1/2}} = 13.70,$$

(see Section 8.2 for explanation), so that canonical reduction is justified. The reduced form is $\hat{y} = 150.98 - 13.183\tilde{X}_1^2 - 19.187\tilde{X}_2^2$, a simple maximum. The center of the system is close to the origin at $(0.067, -0.051)$. See Chapter 11.]

7.15. A cautious approach is needed here. A first reaction is to fit a second-order surface to the 20 readings as though they were 20 independent data points. (If this is done, lack of fit is exhibited when the lack of fit mean square is compared with a "pure error" mean square based on 11 df and computed from the eight pairs and the four center point readings.) However, the presentation of the data makes it clear that we simply have 10 pairs of duplicate ash readings which probably do not reflect the full experimental error within a pair; moreover, members of each pair are not independent. For this reason, we shall analyze instead the 10 averages (of the pairs), and treat only the averages for runs 1 and 8 as true pure error repeat runs.

The averages are 212, 90, 214, 91, 219.5, 55, 171.5, 165, 139.5, 160.5. The fitted second-order model is $\hat{y} = 149.915 - 57.853x_1 - 0.794x_2 - 5.780x_1^2 + 7.998x_2^2 - 0.250x_1x_2$.

Source	SS	df	MS	F
b_1, b_2	28,454.5	2	14,227.3	174.14
$b_{11}, b_{22}, b_{12}\mid b_0$	970.3	3	323.4	3.96 Not significant
Lack of fit	106.3	3	35.4	0.16 Not significant
Pure error	220.5	1	220.5	
Total, corrected	29,751.6	9		

Now, no lack of fit is indicated; $s^2 = (106.3 + 220.5)/(3 + 1) = 81.7$. Evidently, second-order terms are not needed. If we refit a first-order model, we find that the x_2 term is not needed either, and the model $\hat{y} = 151.8 - 57.853x_1$ appears to adequately describe the data. The residuals indicate nothing to remark upon. Substituting for x_1 in terms of temperature gives the fitted equation $\hat{y} = 541.7 - 0.50307T$. If temperature were difficult to control, so that the settings were unreliable, there would clearly be a problem in predicting dry ash value from this equation. The data thus support the authors' conclusion (p. 44) that "It is recommended that dry ash measurements be eliminated as a control procedure except under special circumstances."

7.16. The two surfaces are very similar, indicating stability from run to run. Two possible directions for increased yields are revealed.

7.17. $\hat{y} = 58.35 - 5.482x_1 - 23.81x_2 + 0.1332x_1^2 + 3.882x_2^2 + 0.3849x_1x_2$.

Source	SS	df	MS	F
First order$\|b_0$	564.66	2	282.33	423.92
Second order$\|b_0$, first	1.13	3	0.377	0.57 Not significant
Lack of fit	2.48	3	0.827	1.28 Not significant
Pure error	17.50	27	0.648	
Total, corrected	585.77	35		

Lack of fit is not significant, and the residuals are unremarkable; $s^2 = (2.48 + 17.50)/(3 + 27) = 0.666$. A full second-order model is not needed. Refitting terms of up to first order provides $\hat{y} = 36.466 - 3.2337x_1 - 6.9835x_2$. Now, $s^2 = 21.11/33 = 0.64$, and standard error$(b_0) = 1.406$, standard error$(b_1) = 0.1101$, standard error$(b_2) = 0.6116$. All coefficients are necessary and this is the model of choice.

7.18. If $G > 1.055$ is substituted in equation (2), we obtain $K < 113.2$. Moreover, we must have $N \geq 0$ and $P \geq 0$.

Figure A7.18 is reproduced from the first source reference with slight modifications. Figure A7.18a shows the surface cut by the $K = 114$ plane, (114 being the next integer up from 113.2 and close enough for our purposes). Figure A7.18b shows the $N = 0$ plane intersecting the surface. Figure A7.18c shows both planes with the $N = 0$ plane regarded as trans-

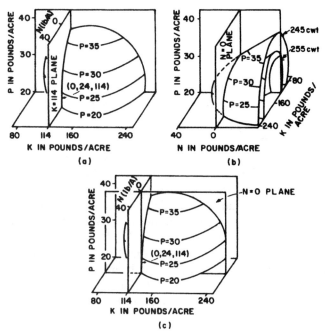

Figure A7.18. Constant yield surfaces showing combinations of N, P, and K fertilization which produce a particular yield. (a) and (c) shows a "front" view of the 245 cwt. potatoes per acre surface, while (b) shows a "side" view of the 245 and 255 cwt. potatoes per acre surfaces. The planes $K = 114$ and $N = 0$ shown cut off the region within which the conditional maximum yield must be sought.

parent. It is clear from the diagram that the response decreases as $N \, (> 0)$ increases and as $K \, (< 114)$ decreases; this is easily substantiated by seeing that the derivatives of $\hat{Y}$ with respect to N and to $-K$ (note the minus direction, which will provide some sign changes) are both < 0 for the ranges of (N, P, K) involved. Thus the maximum response, subject to our limiting conditions, must occur at the intersection of the planes $N = 0$ and $K = 114$. Substituting these values into (1), we obtain

$$\hat{Y} \, (0, P, 114) = 191.4 + 4.55P - 0.0936P^2, \tag{4}$$

which is maximized at $P = -4.55/\{2(-0.0936)\} = 24.3$ or 24 if we round off to the nearest integer. So the best (N, P, K) combination is at the point $(0, 24, 114)$.

We now check if the point $(0, 24, 114)$ lies on or within (3). At the point in question, the left-hand side of (3) has the value 2.669 compared with the right-hand side value of 2.618, so the ratio $(2.669/2.618)^{1/2} = 1.01$ indicates the best point is just outside (1% of the radius outside) the region of experimentation.

Note that the best point on (3) cannot be obtained by substituting $N = 0$, $K = 144$ into (3), because the line so defined does not intersect the sphere (3); the solution is imaginary. We would need to minimize (1) subject to (i) left-hand side of (3) $\leq c^2$, (ii) $N \geq 0$, (iii) $K \leq 114$. This is not an easy problem to solve but could be tackled via the Kuhn–Tucker theorem; see, for example, *Optimization by Vector Space Methods*, by D. G. Luenberger, published by John Wiley & Sons, 1969, pp. 247–253.

7.19. First set of data: $\hat{y} = 14.93 + 0.44046x_1 - 0.49573x_2 - 0.97305x_3 + 0.00994x_3^2 - 0.11748x_1x_3 + 0.07219x_2x_3 + 0.01164x_1x_3^2 - 0.00541x_2x_3^2$. The pure error is very tiny; $s_e^2 = 0.003018$ (14 df). The residual MS = 0.0076 (40 df). Lack of fit MS = $0.262/26 = 0.0101$. Lack of fit $F = 0.0101/0.003018 = 3.347$ which is significant. However the model explains $R^2 = 0.9989$ of the variation about the mean. This is one of those puzzling situations where we might suspect the pure error is "too small." Derringer concludes that while the model "could stand further refinement," it remains a useful predictive tool.

Second set of data: $\hat{y} = 14.19 + 0.5766x_1 - 0.4868x_2 - 0.7944x_3 - 0.0040x_3^2 - 0.2036x_1x_3 + 0.0187x_2x_3 + 0.0206x_1x_3^2 + 0.0030x_2x_3^2$. The pure error is $s_e^2 = 0.00528/5 = 0.00106$, again tiny. The residual MS = $0.0399/11 = 0.0036$. Lack of fit MS = $(0.0399 - 0.00528)/(11 - 5) = 0.00577$ so that the lack of fit $F = 0.00577/0.00106 = 5.44$, which is significant. However, once again, the model explains nearly all the variation in the data about the mean, because $R^2 = 0.9991$. Thus similar conclusions to those for the first set of data may be drawn. It will be found that the model $\hat{y} = 14.26 + 0.5766x_1 - 0.3397x_2 - 0.8304x_3 - 0.2036x_1x_3 + 0.0206x_1x_3^2$, picks up an amount $R^2 = 0.9977$ of the variation about the mean. The terms dropped are the ones that would have been nonsignificant in the full model had not the lack of fit present rendered all the t tests invalid.

(*Note:* Because Derringer has used uncoded ξ_1 and ξ_2 values, his equations are transformed versions of ours, obtained by setting $\xi_1 = 15x_1$

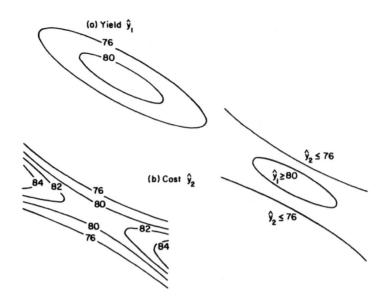

+ 60, $\xi_2 = 15x_2 + 21$. For applications of the models, see the original paper.)

7.20. The region $\hat{y}_1 \geq 80$ is inside an ellipse. The region $\hat{y}_2 \leq 76$ is outside a pair of saddle contours. There is no common region. Other predictor variables would have to be considered to try to find a common region elsewhere. The diagrams make the above solution clear, but it can also be shown algebraically.

7.21. $\hat{y} = 15.8 + 0.525x_1 + 0.925x_2 + 3.808x_3$
$\qquad + 0.019x_1^2 - 0.194x_2^2 + 0.075x_3^2$
$\qquad + 0.7x_1x_2 + 0.025x_1x_3 - 0.025x_2x_3.$

No lack of fit is indicated and second-order terms are not significant as shown in the analysis of variance table. The reduced first-order equation is $\hat{y} = 15.8 + 0.525x_1 + 0.925x_2 + 3.808x_3$. Recomputing $s^2 = 12.359/20 = 0.61795$ we obtain the standard errors of b_1, b_2, b_3 as $(0.61795/16)^{1/2} = 0.20$, so that all terms are needed.

Source	SS	df	MS	F
b_0	5915.760	1		
First order	375.201	3	125.067	302.09
Second order$\mid b_0$	6.558	6	1.093	2.64 Not significant
Lack of fit	3.308	5	0.661	2.39 Not significant
Pure error	2.493	9	0.277	
Total	6303.320	24		

$s^2 = (3.308 + 2.493)/(5 + 9) = 0.414$; this value is used to test for first- and second-order terms in the table.

7.22. The blocking variable is not needed as it removes very little of the variation, so we drop it immediately. The second-order fit is

$$\hat{y} = 6.5961 + 0.2641x_1 + 0.1735x_2 + 0.0781x_3$$
$$- 0.00403x_1^2 - 0.01402x_2^2 - 0.02507x_3^2$$
$$- 0.02128x_1x_2 - 0.01258x_1x_3 - 0.01821x_2x_3.$$

The corresponding analysis of variance table is

Source	SS	df	MS	F
First order	3.3905	3	1.1302	
Second order$|b_0$	0.0320	6	0.0053	
Lack of fit	0.0081	20	0.0004	3.84
Pure error	0.0019	18	0.0001	
Total, corrected	3.4325	47		

There is lack of fit. Nevertheless, $R^2 = 0.9971$, implying that the second-order model explains nearly all the variation in the data about the mean $\bar{y}$. In such circumstances one would often use the model as a predicting tool, in spite of the lack of fit. The technically correct procedure, however, is to see if the lack of fit can be accounted for, and then to recast the model to remove it, unless it is found that the pure error is artificially small for some explainable reason.

The predicted values at the five given locations, obtained from the five corresponding $\exp(\hat{y})$ values, are 838, 707, 661, 846, and 625.

[Additional analysis:

As an exercise, we reduce the equation to the canonical form $\hat{y} = 6.088 - 0.0342\tilde{X}_1^2 - 0.0117\tilde{X}_2^2 + 0.0027\tilde{X}_3^2$ which is of type B (and relatively close to type C) in Table 11.1. The transformation $\tilde{\mathbf{X}} = \mathbf{M}'(\mathbf{x} - \mathbf{x}_S)$ is such that $\mathbf{x}_S = (-16.15, 19.37, -1.43)'$, so that the center is remote, and

$$\mathbf{M} = \begin{bmatrix} 0.3486 & -0.3978 & 0.8487 \\ 0.5321 & -0.6614 & -0.5286 \\ 0.7716 & 0.6359 & -0.0189 \end{bmatrix}$$

See Chapters 10 and 11 for additional explanation.]

7.23. Solution is implicit in the question.

7.24. (e) This is a personal decision. Provided there are no misgivings about the patterns exhibited by the residuals, the first-order model would be a realistic choice.

7.25. A second-order model fitted to the y_1 data reveals a significant lack of fit F value of $(101.26/5)/(0.57/3) = 106.59$. The pure error is extremely small. If it were judged artifically so, and if the results of the lack of fit test were ignored, so that $s^2 = 12.73$ were used to test coefficients, then it would be found that x_3 can be dropped and a quadratic in x_1 and x_2 provides a good fit. The appropriate technical decision, however, is that the model does not fit. For the y_2 data, the lack of fit $F = 20.97$, and $s^2 = 0.4382$, but exactly the same comments otherwise apply.

7.26. The design is not orthogonally blocked, because the ratios $\Sigma x_{iu}^2/$(number of points) are $16/20$ for runs $1\text{--}20$, and $8/8$ for runs $21\text{--}28$, and are unequal. The blocking variable suggested is orthogonal to the column of 1's arising from b_0, however. The fitted second-order model is

$$\hat{y} = 5.8142 + 0.0899z + 0.4688x_1 + 0.2655x_2 + 0.4572x_3$$

$$+ 0.1488x_4 - 0.0565x_1^2 - 0.0190x_2^2 - 0.0335x_3^2$$

$$- 0.0198x_4^2 - 0.0423x_1x_2 - 0.0838x_1x_3 - 0.0284x_1x_4$$

$$- 0.0340x_2x_3 - 0.0094x_2x_4 - 0.0284x_3x_4.$$

The model shows lack of fit, but the tiny amount of variation in the four center points is worrying, and perhaps cause for suspicion. The authors of the source paper argued that "most of the high residuals [high compared

Source	SS × 10^3	df	MS	F
b_0	910,940.2	1		
Blocks	23.5	1	23.5	
First order|blocks	12,514.7	4	3,128.68	
Second order|b_0, blocks	292.4	7	41.77	
Lack of fit	43.6	12	3.63	22.85
Pure error	0.5	3	0.16	
Total	923,814.9	28		

with pure error variation]... are of no practical importance in engineering practice. Therefore, the second-order model is considered to be adequate for the present investigation." A check of the standardized residuals shows that four of the five largest in absolute value lie at star points (trials $21, 22, 27, 28$). It appears possible that the x_1 and x_4 ranges are too wide for a quadratic fit to be fully appropriate. Thus with limitations in those directions, cautious interpretation of the fitted surface may not be too bad. All the x's appear to be needed.

7.27.

Estimated Coefficient	$\hat{y}_0$	$\hat{y}_3$	$\hat{y}_6$	$\hat{y}_9$
b_0	74.18	35.00	25.42	17.7
b_1	46.35	-9.93	-45.69	-22.65
b_2	-13.89	-125.34^a	-115.14^a	-236.54^a
b_3	196.48^a	283.87^a	288.19^a	445.06^a
b_4	-86.13^a	-141.79^a	-123.75^a	-241.75^a
b_5	47.90^a	150.72^a	147.29^a	117.96
b_{11}	-13.56	29.0	43.81	81.7
b_{22}	39.19	114.0	113.3	148.0
b_{33}	103.54	37.8	38.8	93.6
b_{44}	1.69	76.5	73.8	122.3
b_{55}	22.33	52.91	56.96	65.80
b_{12}	36.17	101.07	128.44^a	173.06^a
b_{13}	53.90^a	-1.01	-35.62	-14.44
b_{14}	-54.26^a	47.26	80.00	24.31
b_{15}	21.74	-54.76	-47.81	-163.81^a
b_{23}	3.89	-106.80^a	-92.81	-226.62^a
b_{24}	-28.58	39.30	29.69	153.37
b_{25}	46.68	-63.05	-97.50	-37.87
b_{34}	-106.30^a	-169.86^a	-161.87^a	-279.75^a
b_{35}	27.52	152.99^a	142.19^a	105.25
b_{45}	-29.14	-87.99	-110.94	-158.50

[a]Significant at $\alpha = 0.05$ level.

The table shows the estimated coefficients of second-order fitted models. The pattern of significant coefficients varies from response to response. It would now be possible to look at these four response functions directly or via canonical reduction.

The response data cover several orders of magnitude, however, suggesting that it would be worth checking for a possibly useful transformation, by the Box and Cox (1964) technique, for example. The consequent investigation shows that the ln y transformation is a reasonable choice for all four responses. Moreover, we find, from the analyses of the four second-order fits to ln y, by examining the values of the t statistics stemming from the individual coefficients, that:

(a) Only terms in x_2, x_3, and x_5 appear to be needed for responses y_6 and y_9.

(b) Only terms in x_2, x_3, x_5, and $x_3 x_4$ appear to be needed for responses y_0 and y_3.

When we actually refit the model $\hat{y} = b_0 + b_2 x_2 + b_3 x_3 + b_5 x_5 + b_{34} x_3 x_4$, the term b_{34} becomes nonsignificant only for y_9 so, for uniformity, we reproduce the fitted equations with b_{34} present in all of them in a second table.

Estimated Coefficient	$\ln \hat{y}_0$	$\ln \hat{y}_3$	$\ln \hat{y}_6$	$\ln \hat{y}_9$
b_0	4.4080	4.4579	4.3890	4.8040
b_2	−0.2665	−0.5076	−0.5449	−0.6175
b_3	1.4376	1.5302	1.6209	1.5608
b_5	0.3688	0.5899	0.5581	0.3257
b_{34}	−0.3605	−0.3915	−0.3457	−0.3003
R^2	0.919	0.904	0.877	0.878

Examination of the 20 estimated coefficients in this second table reveals a remarkable similarity in the coefficients line by line, and clearly calls for further conjecture and checking, preferably done with the specialist knowledge of the experimenters. An obvious question is whether or not the same model would provide a satisfactory fit to all 108 data points. Performing this fit we obtain

$$\ln \hat{y} = 4.51473 - 0.48413x_2 + 1.53737x_3 + 0.46064x_5 - 0.34953x_3x_4$$

with $R^2 = 0.870$, a remarkably good fit in the circumstances. The analysis of variance table is as shown. All coefficients are highly significant.

Source	SS	df	MS	F
b_2	16.876	1	16.876	53.87
b_3	170.173	1	170.173	543.16
b_5	20.370	1	20.370	65.02
b_{34}	7.819	1	7.819	24.96
Residual	32.267	103	0.3133	
Total, corrected	247.505	107		

The restriction $800 \leq y_t \leq 1000$ translates into $6.68461 \leq \ln y_t \leq 6.90776$. This implies that

$$2.16988 \leq -0.48413x_2 + 1.53737x_3 + 0.46064x_5 - 0.34953x_3x_4$$

$$\leq 2.39303.$$

Points $(x_1, x_2, x_3, x_4, x_5)$ that lie between the two surfaces so defined are satisfactory ones. (For $y_t \leq 200$, we need $\ln y_t \leq 5.29832$, with similar calculations.) The model underpredicts all the y_t values at $(1, 1, 1, -1, 1)$, which is somewhat surprising. Of the 108 standardized residuals, only five exceed the range $(-2, 2)$ and these are shown in the table below, exhibiting a tendency to overprediction in four of the five cases. Further consideration needs to be given to these details. Nevertheless, all things considered, the five-parameter model does extremely well in predicting the 108 observations.

Coded x's					Value of t			
x_1	x_2	x_3	x_4	x_5	0	3	6	9
-1	-1	1	-1	1	-2.04			
0	0	1	0	0		-2.32	-2.53	
0	1	0	0	0			-2.39	
1	1	1	1	-1				2.43

It is, of course, possible that some suitable mechanistic modeling of the coefficients β_0, β_2, β_3, β_5, and β_{34} in terms of time would be even more effective. This extra pickup of additional variation would, of course, have a price in terms of extra parameters. We remark only that the assumption of quadratic functions—for example, $\beta_{2t} = \beta_{20} + \beta_{21}t + \beta_{21}t^2$, and so on, for the other coefficients—is remarkably ineffective. The reader may wish to consider other alternatives.

Note: In view of the presence of the x_3x_4 interaction in the final model, it might be appropriate to refit, but with an x_4 term added (x_3 is already in). One should also examine two-way tables of ln y averages at the four $(\pm, \pm 1)$ locations in the (x_3, x_4) space. A two-way table can be constructed for each response individually, and for all four together, and the nature of the x_3x_4 interaction can thus be examined. We leave this further exploration to the reader.

7.28.

Estimated Regression Coefficients for Cubic Models

Estimate	$\hat{y}_1$	$\hat{y}_2$	$\hat{y}_3$	$\hat{y}_4$	$\hat{y}_5$	$\hat{y}_6$
b_0	878.34	2363.59	3803.84	666.43	10.81	15.29
b_1	220.93	708.20	582.95	-45.01	1.40	-0.60
b_2	23.02	89.74	38.47	-28.56	-2.37	-2.21
b_{11}	-8.70	1.41	-123.08	-9.17	-0.85	0.97
b_{22}	-73.40	-180.19	-203.64	5.42	0.95	5.28
b_{12}	40.17	79.03	-182.25	-22.30	-0.01	1.19
b_{111}	-42.72	-194.00	-267.07	4.01	0.05	0.29
b_{222}	87.90	184.58	114.60	-4.65	-0.90	-3.93
b_{112}	117.30	257.44	109.97	-14.29	1.16	-1.63
b_{122}	-103.34	-280.67	-237.00	19.09	-0.52	-0.10
a_1	-20.01	-87.17	15.66	7.14	1.34	0.93
a_2	-172.86	-411.44	-297.16	14.29	0.47	-0.07

The specific values of b_0, a_1, a_2 will depend on the particular blocking variable setup selected.

F Ratios for Lack of Fit

Model Order	y_1	y_2	y_3	y_4	y_5	y_6
First	40.7^b	37.9^b	13.7^a	3.5	8.5^a	110.8^b
Second	22.2^b	21.4^b	4.9	1.5	5.6	21.4^b
Third	1.9	4.4	2.2	1.3	3.4	5.3

[a] Lack of fit significant at $\alpha = 0.05$ level.
[b] Lack of fit significant at $\alpha = 0.01$ level.

The test indications are that a cubic equation is needed for responses 1, 2, and 6, a quadratic equation is adequate for responses 3 and 5 and a planar equation is adequate for response 4. (The conclusions in the paper are slightly different, reflecting practical experience of these responses.) Plots of the response surfaces may be found in the source reference.

CHAPTER 8

8.1. First plot θ as abscissa, ϕ as ordinate. Then turn the paper sideways, regard ϕ as abscissa now, and plot log ϕ as ordinate. Returning the paper to its original alignment we can now select θ, read up for ϕ, and then left for log ϕ.

8.2. Since $x^2 + y^2 = r^2$ we have $x^2 + y^2 + 4x + 6y = 12$, or $(x + 2)^2 + (y + 3)^2 = 25$, a circle with major axes parallel to the (x, y) axes, centered at $(-2, -3)$.

8.3. The fitted equation can be written as

$$\hat{y} + 3 \log \xi_3 + 5 \log \xi_4 = 0.703 - 0.517 \log \xi_3 - 0.166 \log \xi_4.$$

The ANOVA table is

Source	SS	df	MS	F
$\hat{\gamma}$	0.2808	1		
$\hat{\delta}_1, \hat{\delta}_2 \mid \hat{\gamma}$	0.0151	2	0.007552	1.41 Not significant
Residual	0.1587	24	0.006611	
Total	0.4546	27		

So we cannot reject $H_0: \delta_1 = \delta_2 = 0$, that is, $H_0: \gamma_1 = -3, \gamma_2 = -5$.

8.4. For both y_1 and y_2, $w = \ln y$ is close to best. This gives rise to the fitted equations

$$\ln \hat{y}_1 = 4.395 + 0.513x_1 - 0.210x_2,$$
$$\ln \hat{y}_2 = 3.739 + 0.572x_1 - 0.237x_2.$$

Source	SS(y_1)	df$_1$	SS(y_2)	df$_2$
b_1, b_2	1.219	2	3.064	2
Lack of fit	0.018	4	0.026	6
Pure error	0.047	4	0.022	4
Total, corrected	1.284	10	3.112	12
	$s_1^2 = 0.0081$	8	$s_2^2 = 0.0049$	10

The equations explain $R_1^2 = 0.949$ and $R_2^2 = 0.985$ of the variation about respective means, there is no lack of fit indicated, and both responses need both x's. The mild irregularities in residuals plots do not seem to call for any remedial action.

8.5. $\hat{\lambda} = -0.8$. The 95% confidence interval is wide, $(-2.4, 0.7)$. Many alternative fits are thus possible. Derringer used $\lambda = 0$ to get $\ln \hat{y} = 4.468 + 0.179 x_1 + 0.068 x_2 + 0.032 x_3$.

Source	SS	df	MS	F
b_1, b_2, b_3	0.3209	3	0.1070	54.87
Lack of fit	0.0150	7	0.0021	3.50
Pure error	0.0006	1	0.0006	
Total, corrected	0.3365	11		

There is no lack of fit and the regression is highly significant with an $R^2 = 0.954$. Predictor x_3 is not significant and, statistically, could be dropped. The ninth residual has a standardized value of -2.36 and may need additional investigation. For the F value for first order, the value $s^2 = (0.0150 + 0.0006)/(7 + 1) = 0.00195$ has been used.

8.6. Applying the transformation of Eq. (8.4.4) for a selection of λ values and performing the corresponding regressions gives these results for the residual sums of squares:

λ:	-1	-0.5	0	0.5	1	1.5	2
RSS:	21,615	9068	5444	5669	7645	10,586	14,306

From a plot of these values, we can estimate the best λ at about 0.25 with an approximate confidence region of about $-0.4 \le \lambda \le 0.75$, which excludes the value 1. It appears that use of a natural log transformation (ln) would make a sensible compromise here. The resulting fit is

$$\ln(\hat{y}) = 5.9212 + 0.2646 x_1 + 0.2266 x_2 + 0.1236 x_3$$

$$- 0.1424 x_1^2 - 0.1757 x_2^2 - 0.0894 x_3^2 - 0.1112 x_1 x_2$$

$$- 0.1112 x_1 x_3 - 0.1112 x_2 x_3.$$

The analysis of variance table indicates that both first- and second-order terms are useful.

Source	df	SS	MS	F
First order	3	2.459	0.820	206.5
Second order	6	0.533	0.089	22.4
Residual	17	0.067	0.004	
Total, corrected	26	3.059		

The eigenvalues of the second-order matrix are all negative, indicating a maximum at the center of the system, which is at $(x_1, x_2, x_3) = (0.693, 0.404, 0.472)$, a point that lies within the experimental region.

8.7. The initial analysis, working out all the factorial effects and normal-plotting them, shows all the effects more or less on a straight line. Bisgaard et al. point out that using an inverse response $1/y$ and repeating the analysis leads to identifying the factors S and B and their interaction SB as most important, permitting a relatively simple explanation. See the paper for additional details.

CHAPTER 9

9.1. Overall comment: The pure error SS in this example is very small and both first- and second-order fits register as inadequate in a lack-of-fit test. This may well be one of those practical examples where the pure error variation does not exhibit the full run-to-run variation and so is "too small." Whether this is so or not, the fitted models explain a large proportion of the variation in the data and the data themselves are interesting.

(a) $\hat{y} = 29.731 + 16x_1 - 0.25x_2 - x_3$.

Source	df	SS	MS	F
Extra, first order	3	2056.5	685.50	
Lack of fit	5	195.81	39.16	31.33
Pure error	4	5.00	1.25	
Total, corrected	12	2257.31		

(b) $b_{12} = -2.75$, $b_{13} = 2.50$, $b_{23} = -1.25$ with common standard error (se) of 1.273, if we estimate σ^2 from the whole 9 df residual mean square $s^2 = (4.723)^2$, as is done by standard regression programs. If s is estimated from pure error only, then se $= 0.301$.

The curvature contrast is $c = 30.5 - 28.2 = 2.3$ with se of $s(13/40)^{1/2}$. If we use $s = (1.25)^{1/2} = 1.118$, then se $= 0.637$. A second-order explanation would seem to be needed, because the curvature contrast is about 3.61 times its standard error.

(c) Suppose we choose the blocking variable $B = 1$ for run orders 1–13 and some other level, B_2 say, for run orders 14–19. Orthogonality is thus automatically achieved for all **X**-columns except for the pure quadratics. To achieve full orthogonality, we need to make $8(1) + 2(1.682)^2 B_2 = 0$, which implies $B_2 = -1.4138652$. The design level 1.682 is a rounded form of $2^{3/4}$. If that level were taken as *exactly* $2^{3/4}$, the second block level would be exactly $-2^{1/2}$. The second-order fit is

$$\hat{y} = 30.791 - 2.2912\,B + 17.6x_1 + 0.777x_2 - 1.140x_3$$

$$+ 2.552x_1^2 - 0.099x_2^2 - 0.453x_3^2 - 2.75x_1x_2 + 2.5x_1x_3 - 1.25x_2x_3.$$

Source	df	SS	MS	F
Blocks, $B\vert b_0$	1	157.09		
First order	3	4238.07	1412.690	
Second order $\vert b_0$	6	221.63	36.938	
Lack of fit	4	128.32	32.080	25.66
Pure error	4	5.00	1.25	
Total, corrected	18	4760.11		

The second-order terms pick up about 62% of the SS left after first order is fitted and seem modestly worthwhile.

(d) The residuals show nothing startling; however, the 18 residuals share only 8 df, so we would not be surprised if the residuals showed some dependence.

(e) The calculation of the alias matrix shows that

$$E(b_i) = \beta_i + 0.58572676(3\beta_{iii} + \beta_{ijj} + \beta_{ill})$$

where $i = 1, 2, 3$ in turn, and where j and l are the other two subscripts that are not i. All other estimates are unbiased.

(f) We next need to evaluate the $\mathbf{Z}_{2\cdot1}$ matrix. This has only four independent columns and shows that four combinations of third-order β's can be evaluated. Three of these involve combinations of the associated third-order betas indicated in (e), but with different coefficients, while the fourth involves only β_{123}. The first three contrasts require the coefficients

$$1 -1\,1 -1\,1 -1\,1 -1 -zz\,000000000$$

$$1\,1 -1 -1\,1\,1 -1 -1\,0\,0 -zz\,0000000$$

$$1\,1\,1\,1 -1 -1 -1 -1\,0000 -zz\,00000$$

multiplying the 19 observations, where $z = 2.3781213$, while the fourth contrast requires the coefficients

$$-1\,1\,1 -1\,1 -1 -1 -1\,100000000000.$$

Any multiples of these rows can also be used (for example, all could be normalized to have sum of squares 1, though this is completely unnecessary). Whatever are chosen, they will give rise to exactly the same sums of squares, via the formula: SS of a contrast of the form $\mathbf{c}'\mathbf{y}$ is given by $(\mathbf{c}'\mathbf{y})^2/\mathbf{c}'\mathbf{c}$. The 1 df sums of squares for these contrasts are, respectively, 47.06, 20.38, 0.38, 60.50, which sum to 128.32 (4 df). Note that this is precisely the value of the lack of fit SS in (c) above, and it uses up all the available model df. We are able to add together the four 1 df sums of squares in this manner only because the corresponding contrasts are mutually orthogonal, of course.

(**g**) None! See the penultimate sentence of (f).

(**h**) A plot or contour diagram of $\hat{y}$ versus the temperature and pressure variables is useful in answering this question. Or, we can perform a ridge analysis to follow the maximum $\hat{y}$ path from the coded origin $(0, 0, 0)$; for this, see Section 12.7.

CHAPTER 11

11.1. $\hat{y} = 0.410469 - 0.112913x_1 - 0.091438x_2 + 0.028766x_1^2 + 0.017328x_2^2 + 0.037593x_1x_2.$

Source	SS	df	MS	F
b_0	6.572814	1		
First order	1.688804	2	0.8444	29.22*
Second order	0.853976	3	0.2846	9.85*
Residual	0.2888915	10	0.0289	
Total	9.404509	16		

*Significant at $\alpha = 0.01$ level.

Both first- and second-order terms appear to be needed, and

$$\frac{(\hat{y}_{\max} - \hat{y}_{\min})}{(ps^2/n)^{1/2}} = 14.3,$$

so further interpretation appears justified. Canonical reduction produces the stationary point $\mathbf{x}_S = (0.819, 1.750)'$, well within the experimental region and, around this point, $\hat{y} = 0.284 + 0.042694\tilde{X}_1^2 + 0.003400\tilde{X}_2^2$, these contours representing a simple minimum form. If low detergent consumption is desirable, the stationary point, which corresponds to temperatures of about (60°C, 62°C), is clearly best as predicted by $\hat{y}$, and elliptical contours of constant $\hat{y}$ values surround this point. A plot would show the canonical axes through $\mathbf{x}_S$ angled at 53.5° and 36.5° to the x_1 axis, elongated roughly

3.5 to 1 in the approximate NW–SE direction.

$$\tilde{\mathbf{X}} = \begin{bmatrix} 0.803 & 0.595 \\ 0.595 & -0.803 \end{bmatrix} (\mathbf{x} - \mathbf{x}_S).$$

A plot of residuals from the quadratic fit versus z is unremarkable except that the z value (0.81) is well separated from the rest. The author of the source reference remarks that "if this lot had been scoured to approximately 0.6% grease as were the others, the detergent consumption at this point would have been even higher than indicated...". In view of the fact that at this point, $(-3, 3)$ in coded coordinates, the fitted surface is relatively high and well away from the minimum, the conclusions about desirable temperatures for low detergent consumption remain intact.

11.2.

	(a)	(b)
Center at	$(0.0266, 0.0971)$	$(1, 1)$
$\hat{y}_S$	60.015	0
λ_1	-2.128	1
λ_2	1.618	2
M	$0.632 \quad 0.775$	$-\frac{1}{2}\sqrt{3} \quad \frac{1}{2}$
	$0.775 \quad -0.632$	$\frac{1}{2} \quad \frac{1}{2}\sqrt{3}$
Type	Saddle	Bowl

11.3. **(a)** Center is at $(2, 1)$. Roots are $\lambda = 12, -1$.

$$\hat{y} = 1 + 12\tilde{X}_1^2 - \tilde{X}_2^2, \qquad \tilde{X}_1 = \{3(x_1 - 2) + 2(x_2 - 1)\}/13^{1/2},$$
$$\tilde{X}_2 = \{-2(x_1 - 2) + 3(x_2 - 1)\}/13^{1/2}.$$

The surface is a saddle, broad in the $\tilde{X}_1$ direction and narrow in the $\tilde{X}_2$ direction. $\tilde{X}_1$ lies in the first quadrant of the $\{(x_1 - 2), (x_2 - 1)\}$ space and $\tilde{X}_2$ lies in the second quadrant.

(b) Center is at $(2, 0)$. Roots are $\lambda = 2, 200$

$$\hat{y} = -381 + 2\tilde{X}_1^2 + 200\tilde{X}_2^2, \qquad \tilde{X}_1 = \{(x_1 - 2) + x_2\}/2^{1/2},$$
$$\tilde{X}_2 = \{-(x_1 - 2) + x_2\}/2^{1/2}.$$

The surface is, analytically, a long, narrow, elliptical bowl but locally looks more like a stationary "river valley" rising steeply in the $\pm\tilde{X}_2$ directions and slightly in the $\pm\tilde{X}_1$ directions. $\tilde{X}_1$ and $\tilde{X}_2$ lie in the first and second quadrants of the $\{(x_1 - 2), x_2\}$ space.

(c) Center is at $(a, b) = (0.2125, 0.515)$. Roots are $\lambda = -2, 2$.

$$\hat{y} = 75.43775 - 2\tilde{X}_1^2 + 2\tilde{X}_2^2, \qquad \tilde{X}_1 = \{(x_1 - a) + (x_2 - b)\}/2^{1/2},$$
$$\tilde{X}_2 = \{-(x_1 - a) + (x_2 - b)\}/2^{1/2}.$$

The surface is a saddle, going down in the first and third quadrants of $\{(x_1 - a), (x_2 - b)\}$ and up in the second and fourth quadrants.

11.4.

	1	2	3	4
x_{1S}	0.978	0.714	0.901	119.0
x_{2S}	-1.043	1.214	0.741	99.0
$\hat{y}_S$	71.1	19.0	75.0	-206.2
λ_1	2.98	4.12	1.99	-1.01
λ_2	-0.98	-0.02	1.01	0.01
$\mathbf{m}_1$	0.937	0.277	-0.545	-0.634
	-0.350	0.961	0.838	0.773
$\mathbf{m}_2$	0.350	0.961	0.838	0.773
	0.937	-0.277	0.545	0.634

Note: $\tilde{\mathbf{X}} = (\mathbf{m}_1, \mathbf{m}_2)'(\mathbf{x} - \mathbf{x}_S)$.
The general surface types are, respectively, (b), (d), (a), and (c) in Figure 11.1.
Specific forms are determined by the sizes and signs of the λ's.

11.5. Note that the value $\alpha = 1.633 = (\frac{8}{3})^{1/2}$ would be an appropriate axial distance for orthogonal blocking if the design were to be divided into (cube plus four center points) plus (star plus two center points). We treat the design as a single block here, however.

	$\hat{Y}_1$	$\hat{Y}_2$	$\hat{Y}_3$	$\hat{Y}_4$
b_0	139.12	1261.11	400.4	68.91
b_1	15.493	268.15	-99.7	-1.41
b_2	17.880	246.50	31.4	4.32
b_3	10.906	139.48	-73.9	1.63
b_{11}	-4.009	-83.55	7.93	1.56
b_{22}	-3.447	-124.79	17.3	0.0577
b_{33}	-1.572	199.17	0.432	-0.317
b_{12}	5.125	69.38	8.75	-1.62
b_{13}	7.125	94.13	6.25	0.125
b_{23}	7.875	104.38	1.25	-0.25

	$\hat{Y}_1$	$\hat{Y}_2$	$\hat{Y}_3$	$\hat{Y}_4$
(3 df) SS(b_i)	9,476.2	2,028,337	218,442.6	310.93
(6 df) SS($b_{ij}|b_0$)	1,472.7	1,081,600	5,509.7	56.24
(5 df) Lack of fit	188.0	1,030,452*	1,442.7	11.85
(5 df) Pure error	126.8	49,941	2,800.0	4.21
(19 df) Total, corrected	11,263.7	4,190,330	228,175.0	383.23

	$\hat{Y}_1$	$\hat{Y}_2$	$\hat{Y}_3$	$\hat{Y}_4$
$\dfrac{\left(\hat{Y}_{\max} - \hat{Y}_{\min}\right)}{\left(ps^2/n\right)^{1/2}}$	25.2	*Lack of fit shown	22.7	18.8
x_{1S}	-1.20	—	14.85	2.79
x_{2S}	-1.37	—	-2.17	4.61
x_{3S}	-2.68	—	-18.72	1.30
$\hat{y}_S$	102.24	—	384.30	77.95
λ_1	-6.68	—	19.20	-0.40
λ_2	-6.31	—	7.22	-0.22
λ_3	3.96	—	-0.75	1.92
Shape	11.1(F)	—	11.1(D)	11.1(B)

$$\tilde{\mathbf{X}} = \mathbf{M}'_i(\mathbf{x} - \mathbf{x}_S),$$

$$\mathbf{M}'_1 = \begin{bmatrix} 0.497 & 0.485 & -0.720 \\ -0.722 & 0.692 & -0.032 \\ 0.483 & 0.536 & 0.693 \end{bmatrix},$$

$$\mathbf{M}'_3 = \begin{bmatrix} 0.383 & 0.919 & 0.094 \\ 0.849 & -0.391 & 0.355 \\ -0.363 & 0.056 & 0.930 \end{bmatrix},$$

$$\mathbf{M}'_4 = \begin{bmatrix} 0.229 & 0.613 & 0.756 \\ -0.333 & -0.681 & 0.653 \\ 0.915 & -0.401 & 0.048 \end{bmatrix}.$$

Ideally, one should express the given inequalities in terms of the fitted surfaces. This would define the boundaries to four desirable regions in the x space and one should now see what, if any, region satisfies all the inequalities. More simply, we can note that the responses to compound numbers 7, 17, and 18 fall within the restrictions, indicating that feasible conditions are possible, and probably between the origin $(0, 0, 0)$ and the cube vertex $(-1, 1, 1)$. Trial values in this region may now be substituted into the response functions to see if the predicted response satisfies the restrictions.

[The original authors used a compound desirability criterion, briefly described in Section 11.7, to pick a desirable point at $(-0.05, 0.145, -0.868)$. They also found that the region around this point was insensitive to small departures from it, a good feature.] (Some details courtesy of J. C. Lu, C. M. Oliveri, and Y. C. Ki.)

11.6. The quadratic model fitted to the y_1 data shows significant lack of fit with $F = 15.23$ (df $= 5, 3$). The sixth observation is associated with a standardized residual of 2.75 and probably needs careful checking. If this observation is dropped, and the surface refitted, the lack of fit becomes nonsignificant. The fitted model is then

$$\hat{y} = 28.933 + 4.4551x_1 + 2.3337x_2 + 6.0363x_3$$

$$+ 2.7233x_1^2 - 1.1741x_2^2 + 1.6427x_3^2$$

$$+ 1.018x_1x_2 - 1.743x_1x_3 + 0.268x_2x_3.$$

Source	SS	df	MS	F
First order	812.50	3	270.83	38.52
Second order$|b_0$	185.38	6	30.90	4.39
Lack of fit	16.90	4	4.23	0.39
Pure error	32.31	3	10.77	
Total, corrected	1047.09	16		

Because there is no lack of fit, we use $s^2 = (16.90 + 32.31)/(4 + 3) = 7.03$ to test first- and second-order terms. Both sets are significant at the $\alpha = 0.05$ level.

At the additional point locations we compare the predicted $\hat{y}_1$'s and the observed y_1's as follows (in the order given in the question).

$\hat{y}_1$:	21.9	16.1	29.4	10.9	20.7	4.5
y_1:	33.3	16.3	31.8	28.0	17.9	22.5

We note that the predictions are particularly bad for the two points well outside the design region, namely the fourth and sixth. The overall message is that the fitted model is likely to be useful only over the region in which the data were taken initially.

The canonical form is $\hat{y} = 16.91 + 3.2415\tilde{X}_1^2 - 1.2613\tilde{X}_2^2 + 1.2117\tilde{X}_3^2$. The center of the system is at $x_S = (-1.682, -0.046, -2.726)'$ and the axes are rotated around this point via $\tilde{X} = M'(x - x_S)$, where

$$M = \begin{bmatrix} -0.8778 & -0.1454 & 0.4565 \\ -0.0869 & 0.9854 & 0.1467 \\ 0.4712 & -0.0891 & 0.8775 \end{bmatrix}.$$

For y_2, no lack of fit is shown for the quadratic model, but only the terms x_1 and x_3 appear effective when the estimated coefficients are compared with their standard errors. For a fit on x_1 and x_3 only, we obtain $\hat{y}_2 = 5.7339 + 0.6298x_1 + 0.7721x_3$. There is no overall lack of fit, and both coefficients are significant.

Source	SS	df	MS	F
b_1, b_3	13.5455	2	6.7727	12.28
Lack of fit	8.0021	12	0.6668	7.30
Pure error	0.2739	3	0.0913	
Total, corrected	21.8214	17		

Regression is tested using $s^2 = (8.0021 + 0.2739)/(12 + 3) = 0.5517$.

At the additional point locations, the predicted $\hat{y}_2$'s and observed y_2 values are:

$\hat{y}_2$:	5.44	3.31	5.64	4.24	4.18	3.58
y_2:	5.77	4.03	5.64	5.29	4.23	4.74

Again, prediction is worst at the two locations well outside the design region, indicating once again the dangers of extrapolation.

11.7.

$$\mathbf{X'X} = \begin{bmatrix} 15 & \mathbf{0} & 16 & 16 & 16 & \mathbf{0} \\ \mathbf{0} & 16\mathbf{I} & \mathbf{0} & \mathbf{0} & \mathbf{0} & \mathbf{0} \\ 16 & \mathbf{0} & 40 & 8 & 8 & \mathbf{0} \\ 16 & \mathbf{0} & 8 & 40 & 8 & \mathbf{0} \\ 16 & \mathbf{0} & 8 & 8 & 40 & \mathbf{0} \\ \mathbf{0} & \mathbf{0} & \mathbf{0} & \mathbf{0} & \mathbf{0} & 8\mathbf{I} \end{bmatrix},$$

$$(\mathbf{X'X})^{-1} = \begin{bmatrix} \frac{7}{9} & 0 & 0 & 0 & -\frac{2}{9} & -\frac{2}{9} & -\frac{2}{9} & 0 & 0 & 0 \\ 0 & \frac{1}{16} & & & & & & & & \\ 0 & & \frac{1}{16} & & & & & & & \\ 0 & & & \frac{1}{16} & & & & & & \\ -\frac{2}{9} & & & & \frac{26}{288} & \frac{17}{288} & \frac{17}{288} & & & \\ -\frac{2}{9} & & & & \frac{17}{288} & \frac{26}{288} & \frac{17}{288} & & & \\ -\frac{2}{9} & & & & \frac{17}{288} & \frac{17}{288} & \frac{26}{288} & & & \\ 0 & & & & & & & \frac{1}{8} & & \\ 0 & & & & & & & & \frac{1}{8} & \\ 0 & & & & & & & & & \frac{1}{8} \end{bmatrix},$$

$$\mathbf{X'y} = \begin{bmatrix} 539 \\ -21 \\ -37 \\ -17 \\ 471 \\ 395 \\ 723 \\ 73 \\ 5 \\ 7 \end{bmatrix}, \qquad \mathbf{b} = \begin{bmatrix} 66.1111 \\ -1.3125 \\ -2.3125 \\ -1.0625 \\ -11.2639 \\ -13.6389 \\ -3.3889 \\ 9.125 \\ 0.625 \\ 0.875 \end{bmatrix}.$$

Note: The **b** vector is not the same as the one in the original paper. We assume the quoted data have been rounded from those that determined the fitted response surface given originally. The analysis of variance table is as follows:

Source	SS	df	MS	F
b_0	19,368.07	1		
b_i	131.19	3	43.73	0.26
$b_{ii}\|b_0$	3,123.00	3	1041.00	6.10
b_{ij}	675.38	3	225.13	1.32
$(b_{ii}, b_{ij}\|b_0$	3,798.38	6	633.06	3.71)
Residual	853.37	5	170.67	
Total	24,151.01	15		

Comments: Only the pure second-order terms alone are significant when compared with the tabular value $F(3, 5, 0.95) = 5.41$. The ratio $6.10/5.41 = 1.13$ is very low, a bad sign (see Section 8.2). All six second-order terms are nonsignificant as a package when compared to $F(6, 5, 0.95) = 4.95$. The inevitable conclusion is that this fitted response surface is a shaky one and that canonical analysis is probably not worthwhile. Overall, this does not seem a very satisfactory fit to this set of data, in spite of an R^2 value of 0.822.

[One wonders whether application of the transformation $W = \sin^{-1}(Y/100)^{1/2}$, useful for transforming proportion or percentage type data, would improve the fit. In fact it does not, as the reader may confirm for him or herself.]

Investigating further, we see that all coefficients with a subscript 3 are small compared to their standard errors:

$$b_3 = -1.0625 \text{ has standard error of } (170.67/16)^{1/2} = 3.266,$$

$$b_{33} = -3.3889 \text{ has standard error of } \{(170.67)(26/288)\}^{1/2} = 3.925,$$

$$b_{13} = 0.625 \text{ has standard error of } (170.67/8)^{1/2} = 4.619,$$

$$b_{23} = 0.875 \text{ has standard error of } (170.67/8)^{1/2} = 4.619.$$

The standard errors are, of course, the square roots of the appropriate diagonal entries in $(\mathbf{X}'\mathbf{X})^{-1}s^2$. The above indicates that x_3 may not be needed in the model. We can check this formally by an extra sum of squares test. First we fit the reduced model

$$\hat{y} = 57.7692 - 1.3125x_1 - 2.3125x_2$$

$$- 9.0481x_1^2 - 11.4231x_2^2 + 9.125x_1x_2$$

with analysis of variance:

Source	SS	df	MS	F	$F(v, 9, 0.95)$
b_0	19,368.07	1	—		
b_1, b_2	113.13	2	56.56	0.51	4.26
$b_{11}, b_{22} \vert b_0$	2,995.79	2	1497.90	13.38	4.26
b_{12}	666.13	1	666.13	5.95	5.12
(All second order$\vert b_0$	3,661.92	3	1220.64	10.90	3.86)
Residual	1,007.89	9	111.99		
	24,151.01	15			

Comparing the two tables, we find the extra sum of squares for b_3, b_{33}, b_{13}, b_{23} given the other b's is

$$1007.89 - 853.37 = 154.52,$$
$$\text{with df} = 9 - 5 = 4.$$

The F statistic for testing H_0: $\beta_3 = \beta_{33} = \beta_{13} = \beta_{23} = 0$ versus H_1: not so, is thus

$$\frac{\{154.52/4\}}{170.67} = 0.23,$$

not significant. We thus drop x_3 from the model. The resulting x_1, x_2 model is significant, but not highly so (see Section 8.2). Nevertheless, if we proceed to a canonical reduction, we obtain (see diagram)

$$\hat{y} = 58.06 - 5.520\tilde{X}_1^2 - 14.948\tilde{X}_2^2,$$

with

$$\tilde{X}_1 = -0.791(x_1 + 0.155) - 0.612(x_2 + 0.163),$$
$$\tilde{X}_2 = -0.612(x_1 + 0.155) + 0.791(x_2 + 0.163).$$

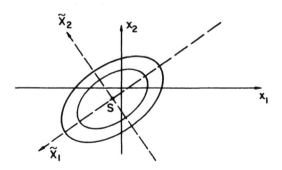

The conclusions then are that a predicted maximum response of 58.06 is obtained at a point estimated as $(x_1, x_2) = (-0.155, -0.163)$ and that the predicted response drops away from that point. In fact, a higher response of 63 was observed at $(0, 0)$. These conclusions must be hedged by a reminder of the unconvincing value of the fitted equation. One wonders if the residual mean square is "too big" but there is no way to check that in these data, because there are no repeats, a lack that is especially felt.

11.8. All three fitted equations can dispense with x_3 as indicated by an extra sum of squares F test. For the whole set of data, the reduced model is then

$$\hat{y} = 10.79 + 2.71x_1 + 1.90x_2 - 0.71x_1^2 - 0.82x_2^2 + 1.26x_1x_2.$$
$$\quad\quad (0.18) \quad\quad (0.18) \quad\quad (0.25) \quad\quad (0.25) \quad\quad (0.25)$$

Standard errors are in parentheses. No lack of fit is indicated.

Source	SS	df	MS	F
b_1, b_2	493.67	2	246.84	155.62
$b_{11}, b_{22}, b_{12}\|b_0$	60.35	3	20.12	12.68
Lack of fit	46.92	27	1.74	1.89
Pure error	48.25	33	1.46	
Total, corrected	649.20	65		

$$s^2 = \frac{(46.92 + 48.25)}{(27 + 33)} = 1.5862.$$

Canonical reduction is justified in view of the fact that

$$\frac{(\hat{y}_{max} - \hat{y}_{min})}{(ps^2/n)^{1/2}} = \frac{(15.13 - 5.13)}{\{6(1.5862)/66\}^{1/2}} = 26.34.$$

The eigenvalues of

$$\mathbf{B} = \begin{bmatrix} -0.7062 & 0.62915 \\ 0.62915 & -0.82440 \end{bmatrix}$$

are -1.3972 and -0.1334. The center of the system is at $\mathbf{x}_S$, where

$$\mathbf{x}'_S = \tfrac{1}{2}(2.7117, 1.9017)\mathbf{B}^{-1} = (9.2074, 8.1804)$$

and $\hat{y}_S = b_0 + \tfrac{1}{2}\mathbf{x}'_S\mathbf{b} = 10.7853 + 20.2617 = 31.0470$. The appropriate transformation is

$$\tilde{\mathbf{X}} = \begin{bmatrix} -0.6732 & 0.7394 \\ 0.7394 & 0.6732 \end{bmatrix} (\mathbf{x} - \mathbf{x}_S)$$

and the surface form is that of an elliptical, cigar-shaped (in a ratio of $3.2:1$) mound pointing out of the positive quadrant at roughly 45 degrees. Because the center is so remote, the shape near the design center is essentially that of a rising ridge. The two other surfaces are technically saddles but with the same rising ridge characteristic near the design center, and it is this feature that dominates the interpretation of the surface. (See Figure 11.2.)

If the two designs are scaled to have the same second moment and the efficiency of the central composite design with respect to the 3^3 factorial is defined as

$$E(b) = \frac{V(b) \text{ for } 3^3 \text{ factorial}}{V(b) \text{ for central composite}} \left(\frac{27}{15}\right),$$

we find $E(b_i) = 1$, $E(b_{ii}) = 0.96$, $E(b_{ij}) = 1.32$. On balance then, the central composite design comes out slightly better. (Details courtesy of M. Lindstrom.)

11.9. All three sets of data show the same basic characteristics and we proceed here only with the full set of 50 observations, to which we fit

$$\hat{y} = 7.120 + 2.026x_1 + 1.936x_2 - 0.023x_3$$

$$- 0.118x_1^2 - 1.299x_2^2 - 0.308x_3^2$$

$$+ 1.187x_1x_2 + 0.390x_1x_3 + 0.208x_2x_3.$$

Source	SS	df	MS	F
First order	479.11	3	159.703	61.53
Second order$\vert b_0$	202.80	6	33.800	13.02
Lack of fit	71.26	25	2.850	1.31
Pure error	32.56	15	2.171	
Total, corrected	785.73	49		

$$s^2 = \frac{(71.26 + 32.56)}{(25 + 15)} = 2.5955.$$

No lack of fit is indicated, and

$$\frac{(\hat{y}_{\max} - \hat{y}_{\min})}{\{ps^2/n\}^{1/2}} = \frac{\{13.165 - (-1.413)\}}{\{10(2.5955)/50\}^{1/2}} = 20.24.$$

Thus, interpretation of the surface is worthwhile. The center of the system is at $\mathbf{x}_S = (-4.316, -1.490, -3.273)'$ and $\hat{y} = 1.342 - 1.546\tilde{X}_1^2 - 0.399\tilde{X}_2^2 + 0.220\tilde{X}_3^2$. The transformation is

$$\tilde{\mathbf{X}} = \begin{bmatrix} 0.382 & -0.924 & 0.0176 \\ -0.362 & -0.132 & 0.923 \\ 0.851 & 0.357 & 0.385 \end{bmatrix} (\mathbf{x} - \mathbf{x}_S).$$

The cross-sectional contours in $(\tilde{X}_1, \tilde{X}_2)$ planes are ellipses, with saddles in other cross sections. When $x_3 = 0$, we obtain the contours shown in the diagram because the center is remote from the experimental region. Similar shapes apply for other x_3 sections. The composite design is slightly more efficient than the factorial for estimating $b_{ii}(E = 1.067)$ but less $(E = 0.768)$ for b_{ij}. (Details courtesy of K. Kim.)

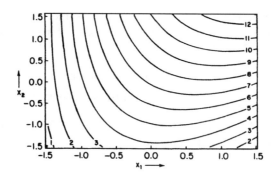

11.10. For this design we have $\sum_u x_{iu}^2 = 32$, $\sum_u x_{iu}^4 = 80$, $\sum_u x_{iu}^2 x_{ju}^2 = 16$.

(a) The portion of the $\mathbf{X}'\mathbf{X}$ matrix stemming from terms in 1, x_1^2, x_2^2, and x_3^2 has inverse

$$
\begin{bmatrix}
38 & 32 & 32 & 32 \\
32 & 80 & 16 & 16 \\
32 & 16 & 80 & 16 \\
32 & 16 & 16 & 80
\end{bmatrix}^{-1}
=
\begin{bmatrix}
P & Q & Q & Q \\
Q & R & S & S \\
Q & S & R & S \\
Q & S & S & R
\end{bmatrix},
$$

where $P = \frac{7}{74}$, $Q = -\frac{1}{27}$, $R = \frac{25}{1184}$, $S = \frac{13}{2368}$. The author's equations now follow directly from setting out $\mathbf{b} = (\mathbf{X}'\mathbf{X})^{-1}\mathbf{X}'\mathbf{y}$.

(b) The next step is to backsolve these equations for $\mathbf{X}'\mathbf{y} = \mathbf{X}'\mathbf{X}\mathbf{b}$ using, successively, each set of b's given in the table. For example, for y_{14}, we obtain

$$
\begin{bmatrix}
(0y) \\
(11y) \\
(22y) \\
(33y)
\end{bmatrix}
=
\begin{bmatrix}
38 & 32 & 32 & 32 \\
32 & 80 & 16 & 16 \\
32 & 16 & 80 & 16 \\
32 & 16 & 16 & 80
\end{bmatrix}
\begin{bmatrix}
0.3021 \\
0.1178 \\
-0.0367 \\
-0.0103
\end{bmatrix}
=
\begin{bmatrix}
13.7454 \\
18.3392 \\
8.4512 \\
10.1408
\end{bmatrix},
$$

$$
(1y) = 32(-0.3587) = -11.4784,
$$

$$
(2y) = 32(0.0378) = 1.2096,
$$

$$
(3y) = 32(0.0099) = 0.3168,
$$

$$
(12y) = 16(-0.0294) = -0.4704,
$$

$$
(13y) = 16(0.0404) = 0.6464,
$$

$$
(23y) = 16(0.0057) = 0.0912.
$$

We actually do not need all these to fit $\hat{y} = b_0 + b_1 x + b_{11} x^2$ for which

$$
\begin{bmatrix} b_0 \\ b_1 \\ b_{11} \end{bmatrix} = \begin{bmatrix} 38 & 0 & 32 \\ 0 & 32 & 0 \\ 32 & 0 & 80 \end{bmatrix}^{-1} \begin{bmatrix} 13.7454 \\ -11.4784 \\ 18.3392 \end{bmatrix}
$$

$$
= \begin{bmatrix} 80 & 0 & -32 \\ 0 & 63 & 0 \\ -32 & 0 & 38 \end{bmatrix} (32 \times 63)^{-1} \begin{bmatrix} 13.7454 \\ -11.4784 \\ 18.3392 \end{bmatrix} = \begin{bmatrix} 0.2544 \\ -0.3587 \\ 0.1275 \end{bmatrix}.
$$

For (ii), we need

$$
\mathbf{b} = \begin{bmatrix} 38 & 0 & 0 & 0 & 32 & 0 & 0 \\ 0 & 32 & 0 & 0 & & & \\ 0 & 0 & 32 & 0 & & \mathbf{0} & \\ 0 & 0 & 0 & 32 & & & \\ 32 & & & & 80 & 0 & 0 \\ 0 & & \mathbf{0} & & 0 & 16 & 0 \\ 0 & & & & 0 & 0 & 16 \end{bmatrix}^{-1} \begin{bmatrix} (0y) \\ (1y) \\ (2y) \\ (3y) \\ (11y) \\ (12y) \\ (13y) \end{bmatrix}
$$

From the original table we can read off $b_1 = -0.3802$, $b_2 = 0.3087$, $b_3 = 0.7634$, $b_{12} = -0.0162$, and $b_{13} = 0.1610$, all these being unchanged. For the remainder we have

$$
\begin{bmatrix} b_0' \\ b_{11}' \end{bmatrix} = \begin{bmatrix} 38 & 32 \\ 32 & 80 \end{bmatrix}^{-1} \begin{bmatrix} (0y) \\ (11y) \end{bmatrix}
$$

$$
= (32 \times 63)^{-1} \begin{bmatrix} 80 & -32 \\ -32 & 38 \end{bmatrix} \begin{bmatrix} (0y) \\ (11y) \end{bmatrix}
$$

$$
= (32 \times 63)^{-1} \begin{bmatrix} 80 & -32 & 0 & 0 \\ -32 & 38 & 0 & 0 \end{bmatrix}
$$

$$
\times \begin{bmatrix} 38 & 32 & 32 & 32 \\ 32 & 80 & 16 & 16 \\ 32 & 16 & 80 & 16 \\ 32 & 16 & 16 & 80 \end{bmatrix} \begin{bmatrix} b_0 \\ b_{11} \\ b_{22} \\ b_{33} \end{bmatrix}
$$

$$
= (32 \times 63)^{-1} \begin{bmatrix} 2016 & 0 & 2048 & 2048 \\ 0 & 2016 & -416 & -416 \end{bmatrix} \begin{bmatrix} b_0 \\ b_{11} \\ b_{22} \\ b_{33} \end{bmatrix}
$$

$$
= \begin{bmatrix} 1 & 0 & 1.015873 & 1.015873 \\ 0 & 1 & -0.206349 & -0.206349 \end{bmatrix} \begin{bmatrix} 2.3507 \\ -0.1622 \\ 0.0509 \\ -0.0189 \end{bmatrix}
$$

$$
= \begin{bmatrix} 2.3832 \\ -0.1688 \end{bmatrix}.
$$

For (iii),

$$
b_0' = (0y)/38
$$

$$
= (38)^{-1}(38, 32, 32, 32) \begin{bmatrix} 3.0303 \\ -0.1424 \\ -0.0003 \\ 0.0948 \end{bmatrix} = 2.9900.
$$

The coefficients $b_1 = -0.0893$, $b_2 = 0.2757$, $b_3 = 0.7188$ are unchanged from before. (iv) The required calculation is exactly parallel to that of part (ii). The fitted equation that emerges is $\hat{y} = 7.4537 - 0.5334x_1 - 0.2484x_2 + 0.9903x_3 - 0.4239x_1^2 + 0.3681x_1x_2 - 0.0544x_1x_3$.

(c) The plots are straightforward and are not shown. (i) is a quadratic independent of x_3 and (iii) is a plane. Sections of (ii) and (iv) for various x_3 are quadratic curves in the (x_1, x_2) plane.

(d) In order to fit a full quadratic model in $x_1, \ldots, x_4$ we would need $(4y)$, $(i4y)$, $i = 1, 2, 3, 4$. We cannot obtain these from the information provided. Thus (e) and (f) cannot be done.

11.11. *FPP-4 responses.*

Estimate	$\hat{Y}_P$	$\hat{Y}_N$	$\hat{Y}_M$	$\hat{Y}_C$
b_0	-314	263.94	42.09	-48.35
b_1	201	-137.05	-35.03	84.61
b_2	288	-155.39	-10.65	-51.01
b_3	418	-290.75	-36.45	16.25
b_{11}	-26.1	19.45	7.60	-16.55
b_{22}	-60.9	24.03	-0.593	28.77
b_{33}	-108	77.46	10.83	1.79
b_{12}	-63.2	36.29	6.13	-11.15
b_{13}	-107	77.06	14.11	-37.70
b_{23}	-163	89.39	7.23	54.92

	$\hat{Y}_P$	$\hat{Y}_N$	$\hat{Y}_M$	$\hat{Y}_C$
(3 df) SS(first order$|b_0$)	1476.0	361.7	15.24	2357.23
(6 df) SS(second order$|b_0$, first)	712.0	138.9	13.54	480.53
(70 df) SS(residual)	484.6	102.9	53.18	1747.53
(79 df) SS(total, corrected)	2672.6	603.4	81.96	4585.30
F, first order	71.07	82.0	6.7	31.5
F, second order	17.14	15.7	3.0	3.2
s^2	6.923	1.47	0.76	24.96
$\dfrac{\left(\hat{Y}_{max} - \hat{Y}_{min}\right)}{\left(ps^2/n\right)^{1/2}}$	21.5	29.5	16.6	19.0
$\hat{y}_0$	81.999	1.389	3.032	9.284
λ_1	-192.3	120.6	17.66	49.29
λ_2	-8.1	3.5	1.98	-30.55
λ_3	5.3	-3.2	-1.83	-4.73

	$\hat{Y}_P$	$\hat{Y}_N$	$\hat{Y}_M$	$\hat{Y}_C$
Approximate surface type	11.1(F)	11.1(F)	11.1(F)	11.1(E)
x_{1S}	1.545	1.081	1.616	1.913
x_{2S}	0.275	1.001	1.053	1.001
x_{3S}	0.963	0.761	0.283	0.269

Transformations matrices $\tilde{\mathbf{X}} = \mathbf{M}'(\mathbf{x} - \mathbf{x}_S)$

$$\mathbf{M}_P = \begin{bmatrix} 0.349 & -0.743 & -0.571 \\ 0.553 & 0.655 & -0.514 \\ 0.756 & -0.137 & 0.640 \end{bmatrix},$$

$$\mathbf{M}_N = \begin{bmatrix} 0.387 & 0.782 & -0.489 \\ 0.446 & -0.623 & -0.643 \\ 0.807 & -0.031 & 0.590 \end{bmatrix},$$

$$\mathbf{M}_M = \begin{bmatrix} 0.605 & -0.774 & 0.185 \\ 0.251 & -0.035 & -0.967 \\ 0.756 & 0.632 & 0.173 \end{bmatrix},$$

$$\mathbf{M}'_C = \begin{bmatrix} 0.226 & -0.802 & -0.553 \\ 0.752 & -0.218 & 0.623 \\ -0.620 & -0.557 & 0.553 \end{bmatrix}.$$

(Details courtesy of W. Lee, R. Luneski, M. S. Chi, and S. K. Ahn.)

11.12. The center is at $(0.060, 0.215, 0.501) = \mathbf{x}'_S$. The reduced form is $\hat{y} = 68.134 - 3.190\tilde{X}_1^2 + 0.780\tilde{X}_2^2 - 0.069\tilde{X}_3^2$. Because λ_3 is relatively small, this is basically a form of the general nature of (E) in Figure 11.3, (E) being viewed as a limiting case of (B). The transformation needed is

$$\tilde{\mathbf{X}} = \begin{bmatrix} 0.7510 & 0.5849 & -0.3063 \\ 0.4883 & -0.8043 & -0.3386 \\ 0.4445 & -0.1047 & 0.8897 \end{bmatrix}'(\mathbf{x} - \mathbf{x}_S).$$

11.13.

Problem number	1	2	3	4
x_{1S}	0.186	0.346	-0.843	-0.196
x_{2S}	-2.431	0.320	0.501	0.430
x_{3S}	1.303	0.267	-19.482	0.244
$\hat{y}_S$	49.94	42.57	72.51	71.07
λ_1	-6.01	3.00	1.01	-3.00
λ_2	1.00	1.99	0.10	-2.00
λ_3	0.10	1.00	0.02	1.00
Approximate surface type in Table 11.1.	E	A	G	B

The transformation to canonical variables $\tilde{\mathbf{X}}$ is $\tilde{\mathbf{X}} = \mathbf{M}_i'(\mathbf{x} - \mathbf{x}_S)$, where $\mathbf{M}_1, \ldots, \mathbf{M}_4$ are as given below.

$$\mathbf{M}_1 = \begin{bmatrix} 0.2426 & 0.8428 & -0.4804 \\ 0.4511 & 0.3404 & 0.8250 \\ 0.8588 & -0.4169 & -0.2977 \end{bmatrix},$$

$$\mathbf{M}_2 = \begin{bmatrix} 0.9459 & 0.2443 & 0.2137 \\ 0.0496 & -0.7593 & 0.6489 \\ -0.3208 & 0.6031 & 0.7303 \end{bmatrix},$$

$$\mathbf{M}_3 = \begin{bmatrix} -0.9914 & 0.1137 & -0.0654 \\ -0.1214 & -0.9842 & 0.1292 \\ 0.0497 & -0.1360 & -0.9895 \end{bmatrix},$$

$$\mathbf{M}_4 = \begin{bmatrix} 0.5552 & 0.0786 & -0.8280 \\ 0.7804 & -0.3935 & 0.4859 \\ 0.2876 & 0.9160 & 0.2798 \end{bmatrix}.$$

11.14. (a) Stationary point is $\mathbf{x}_S = (-1.37756, -2.53382, 0.18873)'$.

$$\hat{y} = 6.52 - 0.235\tilde{X}_1^2 + 0.0392\tilde{X}_2^2 + 0.1798\tilde{X}_3^2,$$

$$\tilde{\mathbf{X}} = \begin{bmatrix} -0.83831 & 0.44259 & 0.31834 \\ 0.49769 & 0.85956 & 0.11598 \\ 0.22188 & -0.25609 & 0.94083 \end{bmatrix}'(\mathbf{x} - \mathbf{x}_S).$$

(b) Stationary point is $\mathbf{x}_S = (2.05894, -0.04153, -0.4145)'$.

$$\hat{y} = 2.73 - 0.00645\tilde{X}_1^2 - 0.16118\tilde{X}_2^2 + 0.65004\tilde{X}_2^2,$$

$$\tilde{\mathbf{X}} = \begin{bmatrix} -0.95524 & -0.03588 & 0.29364 \\ 0.14772 & -0.91784 & 0.36844 \\ 0.25631 & 0.39534 & 0.88205 \end{bmatrix} (\mathbf{x} - \mathbf{x}_S).$$

11.15. $\hat{y} = 6.91082 - 0.28932\tilde{X}_1^2 + 0.07381\tilde{X}_2^2 - 0.05534\tilde{X}_3^2$.
$\mathbf{x}_S = (0.793, -0.429, -0.466)'$. $\tilde{\mathbf{X}} = \mathbf{M}'(\mathbf{x} - \mathbf{x}_S)$ where

$$\mathbf{M} = \begin{bmatrix} -0.7123 & 0.6737 & 0.1966 \\ 0.0903 & 0.3658 & -0.9263 \\ -0.6960 & -0.6421 & -0.3421 \end{bmatrix}.$$

Setting $\hat{y} > 7$ provides a quadratic restriction in x_1, x_2, x_3 and all points satisfying the restriction are suitable. In the canonical form, for example, we can set $\tilde{X}_1 = \tilde{X}_3 = 0$ and choose $\tilde{X}_2 > 1.21^{1/2} = 1.1$. This implies a requirement, for any value of δ, of

$$\mathbf{M}'(\mathbf{x} - \mathbf{x}_S) = \begin{bmatrix} 0 \\ 1.1 + \delta^2 \\ 0 \end{bmatrix}$$

or, because $\mathbf{M}^{-1} = \mathbf{M}'$, of

$$\mathbf{x} = \mathbf{x}_S + (1.1 + \delta^2)\begin{bmatrix} 0.6737 \\ 0.3658 \\ -0.6421 \end{bmatrix}$$

$$= \begin{bmatrix} 1.534 \\ -0.027 \\ -1.172 \end{bmatrix} + \delta^2 \begin{bmatrix} 0.6737 \\ 0.3658 \\ -0.6421 \end{bmatrix},$$

for any δ. Other points giving nonzero values of $\tilde{X}_1$ and $\tilde{X}_3$ are also feasible, of course.

11.16. $\hat{y} = 35.49 - 1.511x_1 + 1.284x_2 - 8.739x_3 + 4.955x_4$
$\qquad - 5.024x_1^2 - 2.983x_2^2 + 1.328x_3^2 - 1.198x_4^2$
$\qquad + 2.194x_1x_2 - 0.144x_1x_3 + 1.581x_1x_4$
$\qquad + 8.006x_2x_3 + 2.806x_2x_4 + 0.294x_3x_4.$

Source	SS	df	MS	F
First order	2088.6	4		
Second order$\vert b_0$	1729.3	10		
Lack of fit	2078.6	10	207.86	22.50
Pure error	46.2	5	9.24	
Total, corrected	5942.7	29		

$R^2 = 0.6475$. Lack of fit is evident. Note that in order to obtain the appropriate corrected total sum of squares we must work out the corrected sum of squares using six center point observations all equal to 32.8, and then add an $(n_0 - 1)s_e^2 = (6 - 1)9.24 = 46.2$. For reasoning, see Draper and Smith (1998, pp. 467–469). The lack of fit casts doubt on the use of the equation for predictive purposes and canonical reduction is probably not justified.

If we now fit a second-order equation to the w_i we are unable to obtain the lack of fit/pure error splitup of the residual sum of squares due to the fact that the six individual center point observations are not available. Moreover, because $R^2 = 0.06632$ is about the same size as in the previous analysis, it looks as if not much improvement has occurred.

The experimenter could be told all of the above, and be asked for the individual replicate runs at the center of the design. Were they available, lack of fit could then be tested on the $\hat{w}$ model.

11.17. $\hat{y}_1 = 8.029 + (0.7543, 0.9139, 0.2177, 0.6499)\mathbf{x} + \mathbf{x}'\mathbf{Bx}$, where $\mathbf{x} = (x_1, x_2, x_3, x_4)'$, and

$$
\mathbf{B} = \begin{bmatrix} -0.203 & -0.136 & -0.0024 & -0.1256 \\ & -0.172 & 0.0066 & -0.0812 \\ & & -0.0187 & -0.0072 \\ \text{symmetric} & & & -0.118 \end{bmatrix}.
$$

Source	SS	df	MS	F
First order $\mid b_0$	66.24	4	16.560	67.04
Second order $\mid$ first, b_0	22.85	10	2.285	9.25
Lack of fit	4.27	15	0.285	2.12
Pure error	0.67	5	0.134	
Total, corrected	94.03	34		

$$
s^2 = \frac{(4.27 + 0.67)}{(15 + 5)} = 0.247(20 \text{ df})
$$

$$
\frac{(\hat{y}_{\max} - \hat{y}_{\min})}{(ps^2/n)^{1/2}} = \frac{(9.61 - 2.43)}{\{14(0.247)/35\}^{1/2}} = 22.84.
$$

So the surface is worthy of interpretation. The stationary point is at $\mathbf{x}_S = (-1.43, 2.67, 0.60, 2.40)'$, and $\hat{y} = 9.55 - 0.40\tilde{X}_1^2 - 0.19\tilde{X}_2^2 - 0.07\tilde{X}_3^2 - 0.02\tilde{X}_4^2$, so that the stationary point represents a maximum. However, this point is somewhat outside the design region. Good whiteness values are obtained as we move from the origin toward the stationary point, nevertheless. The low sixteenth value has a standardized residual of -0.84 and does not appear out of line. (Details courtesy of M. Grassl.)

With observation 16:

$$\hat{y}_2 = 0.585 + (-2.8195, -1.2899, -0.2705, -1.1940)x + x'Bx,$$

$$B = \begin{bmatrix} 2.043 & 0.538 & 0.141 & 0.772 \\ & 0.258 & -0.006 & 0.340 \\ & & 0.008 & 0.085 \\ \text{symmetric} & & & 0.115 \end{bmatrix}.$$

Source	SS	df	MS	F
First order$\|b_0$	141.96	4		
Second order$\|$first, b_0	318.04	10		
Lack of fit	189.34	15	12.623	51.73
Pure error	1.22	5	0.244	
Total, corrected	650.56	34		

Lack of fit is significant, so we cannot proceed with this model. The sixteenth residual is 5.33 times its standard deviation. We now reanalyze the data without this obvious outlier.

Without observation 16.

$$\hat{y}_2 = 1.399 + (-0.2254, -0.3812, 0.0165, -0.2687)x + x'Bx,$$

$$B = \begin{bmatrix} 0.192 & -0.019 & -0.115 & 0.066 \\ & 0.034 & -0.075 & 0.115 \\ & & 0.008 & 0.011 \\ \text{symmetric} & & & 0.088 \end{bmatrix}.$$

Source	SS	df	MS	F
First order$\|b_0$	10.12	4	2.530	2.87
Second order$\|$first, b_0	10.09	10	1.009	1.14
Lack of fit	15.54	14	1.110	4.55
Pure error	1.22	5	0.244	
Total, corrected	36.97	33		

Lack of fit is not significant, and $s^2 = (15.54 + 1.22)/(14 + 5) = 0.882$. Neither first- nor second-order terms are significant and no further interpretation of the fitted surface is called for. The model $\hat{y}_2 = \bar{y}$ appears perfectly adequate. (Some details courtesy of D. Kim.)

$$\hat{y}_3 = 1.589 + (0.1414, 0.1125, 0.0391, 0.1406)x + x'Bx,$$

$$B = \begin{bmatrix} -0.028 & -0.006 & 0.008 & 0.011 \\ & -0.015 & 0.006 & 0.002 \\ & & -0.004 & 0.007 \\ \text{symmetric} & & & -0.013 \end{bmatrix}.$$

Source	SS	df	MS	F
First order$\mid b_0$	3.462	4	0.8655	20.32
Second order$\mid$first, b_0	0.273	10	0.0273	0.64
Lack of fit	0.792	15	0.0528	4.40
Pure error	0.060	5	0.017	
Total, corrected	4.587	34		

Lack of fit is not significant, and $s^2 = (0.792 + 0.060)/(15 + 5) = 0.0426$. Only first-order terms are significant, so that the reduced planar model $\hat{y}_3 = 1.494 + (0.1531, 0.0875, 0.0382, 0.1384)'\mathbf{x}$ can be fitted. This shows no lack of fit, and significant first-order terms, all of which are needed. A sectional contour plot could now be constructed.

11.18. $\hat{y} = 7.277 - 0.667x_1 - 0.353x_2 - 0.092x_3 - 4.330x_4$
$\qquad - 0.0342x_1^2 + 0.0225x_2^2 - 0.1075x_3^2 + 1.3993x_4^2$
$\qquad - 0.0116x_1x_2 + 0.4758x_1x_3 + 0.4294x_1x_4$
$\qquad - 0.0233x_2x_3 - 0.0469x_2x_4 - 0.0175x_3x_4.$

Source	SS	df	MS	F
First order	590.2	4	147.55	607.2
Second order$\mid b_0$	31.9	10	3.19	13.1
Residual	7.3	30	0.243	
Total, corrected	629.4	44		

All x's are needed, and so are second-order terms.

$$\frac{(\hat{y}_{max} - \hat{y}_{min})}{(ps^2/n)^{1/2}} = 39.0 \text{ so canonical reduction is worthwhile.}$$

$$\hat{y} = 8.20 + 1.432\tilde{X}_1^2 - 0.325\tilde{X}_2^2 + 0.151X_3^2 + 0.022\tilde{X}_4^2.$$

The coefficient of $\tilde{X}_4^2$ is close to zero. Ignoring this term leaves a surface of type 11.1(B). In theory, for reduced absorption, we follow the $\tilde{X}_2$ axis in either direction away from $\tilde{\mathbf{X}} = \mathbf{0}$. However, the stationary point is remote at $\mathbf{x}_S = (0.893, 9.899, 0.346, 1.578)'$. Thus, it is more important to plot the contours *in the design region*, for example, by setting $x_4 = -1, 0,$ and 1 in succession. A plot of the response values at each level of x_4 is the next best thing. This was given in the original paper and is reproduced on page 701. The canonical transformation is

$$\tilde{\mathbf{X}} = \begin{bmatrix} 0.1477 & -0.6689 & -0.7254 & 0.0676 \\ -0.0172 & 0.0195 & 0.0714 & 0.9971 \\ 0.0173 & 0.7379 & -0.6738 & 0.0341 \\ 0.9887 & 0.0873 & 0.1214 & 0.0067 \end{bmatrix}'(\mathbf{x} - \mathbf{x}_S).$$

(Some details courtesy of H. Udomon.)

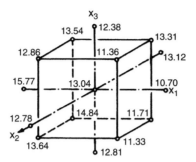

Response y when $x_4 = -1$.

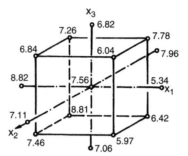

Response y when $x_4 = 0$.

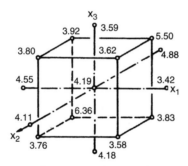

Response y when $x_4 = 1$.

11.19. The authors provide the following fitted regression coefficients.

$$b_0 = 0.845222,$$

$$b_1 = 0.007500, \qquad b_{11} = -0.025222,$$

$$b_2 = 0.125834, \qquad b_{22} = -0.091472,$$

$$b_3 = 0.075001, \qquad b_{33} = -0.016472,$$

$$b_4 = 0.091667, \qquad b_{44} = -0.063972,$$

$$b_5 = -0.073334, \quad b_{55} = -0.046472,$$

$$b_{12} = -0.025000, \quad b_{23} = -0.090000, \quad b_{34} = -0.003750,$$

$$b_{13} = 0.021250, \quad b_{24} = 0.010000, \quad b_{35} = -0.028750,$$

$$b_{14} = -0.088750, \quad b_{25} = -0.020000, \quad b_{45} = 0.091250.$$

$$b_{15} = 0.053750,$$

They also provide this analysis of variance:

Source	df	MS
First order	5	0.1694
Second order $\mid b_0$	15	0.0567
Lack of fit	6	0.0156
Pure error	5	0.0001
Total, corrected	31	

Some practical difficulties now arise. Clearly there is, technically, lack of fit, which would make further analysis of the second-order model invalid. However, the authors argued that the replicated center points exhibit an uncharacteristically low variation, "so the degree of bias was felt to be relatively small" (p. 315). They repooled the lack of fit and pure error sum of squares into a residual sum of squares and estimated σ^2 from that. The resulting standard errors for b_i, b_{ii}, and b_{ij} are then 0.0188, 0.0171, and 0.0231, respectively. Both first- and second-order sets of terms are significant when their mean squares are compared with $s^2 = 0.0085$, and no lack of fit is assumed.

The canonical reduction was carried out by Klein and Marshall. The stationary point is at $\mathbf{x}'_S = (x_{1S}, x_{2S}, x_{3S}, x_{4S}, x_{5S}) = (-2.4495, 0.0429, 2.5937, 0.6375, -2.3912)$ and the canonical form is $\hat{y} = 1.0529 - 0.13168 \tilde{X}_1^2 - 0.11121 \tilde{X}_2^2 + 0.01996 \tilde{X}_3^2 + 0.00000 \tilde{X}_4^2 - 0.02069 \tilde{X}_5^2$. The canonical variables are defined by $\tilde{\mathbf{X}} = \mathbf{M}'(\mathbf{x} - \mathbf{x}_S)$ where

$$\mathbf{M}' = \begin{bmatrix} -0.37932 & 0.34843 & 0.2277 & -0.62911 & 0.53580 \\ 0.27739 & 0.82961 & 0.34969 & 0.31012 & 0.12784 \\ -0.55037 & 0.32512 & -0.63936 & 0.40256 & 0.14337 \\ 0.53409 & -0.11307 & 0.07621 & 0.20054 & 0.80990 \\ 0.24178 & -0.19807 & 0.12837 & 0.52994 & 0.77464 \end{bmatrix}.$$

Note that the predicted response is unaffected by changes in $\tilde{X}_4$, that the coefficients of $\tilde{X}_3^2$ and $\tilde{X}_5^2$ are relatively small, and that the coefficients of $\tilde{X}_1^2$, $\tilde{X}_2^2$, and $\tilde{X}_5^2$ are negative. In the two-dimensional space defined by $\tilde{X}_1 = \tilde{X}_2 = \tilde{X}_5 = 0$, the response is $\hat{y} = 1.5029 + 0.01996 \tilde{X}_3^2$ at any $\tilde{X}_4$. Thus for any chosen $\tilde{X}_4$ value, increases in $\tilde{X}_3^2$ will raise the predicted response, which has been maximized for choice of $\tilde{X}_1$, $\tilde{X}_2$, and $\tilde{X}_5$.

11.20. The fitted quadratic takes the form

$$\hat{y} = 215.7 + 7.4z_1 + 17.4z_2$$

$$+ 40x_1 - 15.08x_2 - 9.58x_3 - 9.92x_4 - 14.50x_5$$

$$- 15.53x_1^2 - 7.28x_2^2 - 7.03x_3^2 - 10.03x_4^2 - 10.03x_5^2$$

$$+ 12.13x_1x_2 + 13.13x_1x_3 + 14.38x_1x_4 + 16.75x_1x_5$$

$$+ 0.75x_2x_3 + 6.75x_2x_4 + 0.38x_2x_5$$

$$- 1.75x_3x_4 - 1.63x_3x_5 - 7.88x_4x_5.$$

Source	SS	df	MS	F
Blocks$\vert b_0$	1,007	2	504	
First order$\vert b_0$, blocks	53,471	5	10,694	
Second order$\vert$above	28,021	15	1,868	
Lack of fit	2,198	6	363	8.07
Pure error	227	5	45	
Total, corrected	84,924	33		

Lack of fit is significant at $\alpha = 0.05$. The source author remarked (his p.75) that "many of the experimental errors, as determined from the replicated center point, were much smaller than previous estimates not reported here. It is suspected that nonhomogeneity of error variance is the probable cause." It is suggested earlier on the same page that "it is reasonable to expect that the difficulty in processing compounds with unusually high pigment or oil levels leads to larger error than more conventional levels of these components." Thus, a reanalysis of the data would need to take into account how the variances of the observations depended on the predictor variables. Further analysis of the model fitted above is not appropriate in the circumstances.

11.21. yield $= -85.25 - 56.88\tilde{X}_1^2 - 21.70\tilde{X}_2^2 + 19.70\tilde{X}_3^2 - 9.79\tilde{X}_4^2 - 5.23\tilde{X}_5^2,$

$$\tilde{\mathbf{X}} = \begin{bmatrix} -0.37 & -0.01 & -0.10 & 0.93 & -0.02 \\ -0.52 & 0.30 & -0.64 & -0.28 & -0.39 \\ -0.77 & -0.32 & 0.43 & -0.25 & 0.25 \\ 0.05 & -0.43 & 0.28 & 0.02 & -0.86 \\ -0.09 & 0.79 & 0.53 & 0.02 & -0.22 \end{bmatrix} \mathbf{X} + \begin{bmatrix} 0.09 \\ 1.38 \\ 0.71 \\ 0.68 \\ -1.25 \end{bmatrix},$$

$$\mathbf{X}_S = (1.15, 1.10, 1.11, 0.49, 0.68)'.$$

In any pair of dimensions involving $\tilde{X}_3$ we have a saddle cross section. In dimensions not involving $\tilde{X}_3$ the contours are elliptical or ellipsoidal.

$$\text{purity} = 68.97 - 93.28\,\tilde{X}_1^2 - 25.67\tilde{X}_2^2 - 21.02\,\tilde{X}_3^2 + 7.35\tilde{X}_4^2 + 4.45\tilde{X}_5^2,$$

$$\tilde{\mathbf{X}} = \begin{bmatrix} -0.28 & -0.04 & 0.78 & 0.53 & -0.19 \\ -0.68 & 0.04 & -0.58 & 0.44 & -0.13 \\ -0.66 & 0.06 & 0.25 & -0.56 & 0.43 \\ -0.08 & 0.66 & 0.06 & -0.32 & -0.67 \\ -0.15 & -0.75 & 0. & -0.33 & -0.56 \end{bmatrix} \mathbf{X} + \begin{bmatrix} -1.03 \\ 1.24 \\ 0.63 \\ 0.78 \\ 1.72 \end{bmatrix},$$

$$\mathbf{X}_S = (1.28, 0.64, 1.31, 1.18, 1.18)'.$$

The contours are elliptical and decreasing outward in $(\tilde{X}_1, \tilde{X}_2, \tilde{X}_3)$ and elliptical and increasing outward in $(\tilde{X}_4, \tilde{X}_5)$. In pairs of dimensions where one is selected from the first bracket and one from the second, the contours are saddles.

A ridge analysis can be used to show that both responses can be simultaneously improved. The sketches requested show:

(a) Two similar saddles for which improvements are obtained by moving from the origin in a southwesterly direction when X_5 is abscissa, X_4 is ordinate.

(b) and **(c)** show very similar characteristics to (a).
(Details courtesy of J. Tort-Martorell).

11.22. The fitted second-order surface has the form $\hat{y} = b_0 + \mathbf{x}'\mathbf{b} + \mathbf{x}'\mathbf{B}\mathbf{x}$, where $b_0 = 18.726$,

$$\mathbf{b} = \begin{bmatrix} -6.234 \\ -0.216 \\ -1.300 \end{bmatrix}, \quad \mathbf{B} = \begin{bmatrix} -4.57410 & -0.39685 & 1.14340 \\ -0.39685 & -0.61300 & 0.73865 \\ 1.14340 & 0.73865 & -5.27560 \end{bmatrix}.$$

The analysis of variance shows a useful regression with nonsignificant lack of fit.

Source	df	SS	MS	F
First order	3	559.43	186.46	11.15*
Extra for second order	6	599.26	99.88	5.97*
Lack of fit	5	110.96	22.19	1.97
Pure error	5	56.23	11.25	
Total, corrected	19	1325.88		

*Using $s^2 = (110.96 + 56.23)/(5 + 5) = 16.719 = (4.09)^2$ with 10 df.

The stationary point is at $(-0.751, -0.041, -0.292)$ relatively close to the origin. The response there, $\hat{y} = b_0 + \frac{1}{2}\mathbf{x}'\mathbf{b} = 21.261$. The eigenvalues of $\mathbf{B}$ are, in increasing order, $-6.24328, -3.73308, -0.48633$. The canonically reduced surface thus takes the form

$$\hat{y} = 21.261 - 6.243X_1^2 - 3.733X_2^2 - 0.486X_3^2,$$

which is an ellipsoidal system with a maximum response of 21.261 at the (stationary point) center of the system. The canonical axes (at the origin, not moved to the stationary point) are given by $\mathbf{X} = \mathbf{M}'\mathbf{x}$, where

$$\mathbf{M} = \begin{bmatrix} -0.582 & -0.146 & 0.800 \\ -0.811 & 0.035 & -0.584 \\ 0.057 & -0.989 & -0.139 \end{bmatrix}.$$

The codings are, where log denotes logarithms to the base 10,

$$x_1 = \{\log \text{Cu} - \log(0.02)\}/2^{1/4}$$

$$x_2 = \{\log \text{Mo} - \log(0.02)\}/2^{1/4}$$

$$x_3 = \{\log \text{Fe} - \log(0.025)\}/2^{1/4}$$

so that the reverse codings are $\text{Cu} = 10^p$, $\text{Mo} = 10^q$, and $\text{Fe} = 10^r$, where

$$p = 2^{1/4}x_{1s} + \log(0.02), \quad q = 2^{1/4}x_{2s} + \log(0.02), \quad r = 2^{1/4}x_{3s} + \log(0.025).$$

Via these formulas, we find that the maximum $\hat{y} = 21.261$ occurs at $(\text{CU}, \text{Mo}, \text{Fe}) = (0.0026, 0.018, 0.112)$.

11.23. $\hat{y} = 8.99593 + 0.33778x_1 - 0.09x_2 - 0.20889x_1^2 - 0.14556x_2^2 + 0.06167x_1x_2$.

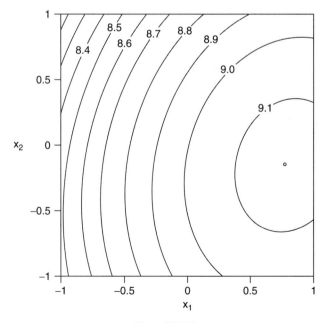

Figure E11.23

Source	df	SS	MS	F
b_1, b_2	2	2.19949	1.099745	28.79*
b_{11}, b_{22}, b_{12}	3	0.43456	0.144853	3.79*
Lack of fit	3	0.17385	0.057950	1.66**
Pure error	18	0.62840	0.034911	
Total, corrected	26	3.43630		

*Significant at level 0.05.
**Not significant.

$$s^2 = (0.17385 + 0.62840)/(3 + 18) = 0.0382 = (0.1955)^2.$$

There is no indication of lack of fit, and both first- and second-order terms are useful. The eigenvalues, which are the coefficients for the second-order terms in the canonical form, are $(-0.2214, -0.1330)$, indicating a predicted maximum at the stationary point $(x_1, x_2) = (0.777619, -0.144422)$, where $\hat{y} = 9.268$. In terms of (Calcium, Phosphorus), the stationary point is at $(1.25x_1 + 3.75, 0.15x_2 + 0.50) = (47, 0.48)$. Figure E11.23 shows a sketch of the contours reproduced from the source paper.

11.24. The fitted quadratic surface is $\hat{y} = -307.73 + 126.38x_1 + 98.847x_2 - 8.203x_1^2 - 6.533x_2^2 + 0.932x_1x_2$, with $R^2 = 0.909$. The fitted cubic is $\hat{y} = -165.07 + 59.859x_1 + 61.732x_2 - 16.401x_1^2 - 6.533x_2^2 - 15.083x_1x_2 + 0.5303x_1^3 - 0.4188x_1^2x_2 + 0.2998x_1x_2^2 + 1.3508x_2^3$, with $R^2 = 0.973$. Selected contours of these response functions appear in the source reference and are reproduced in Figure E11.24. The analysis of variance table for the

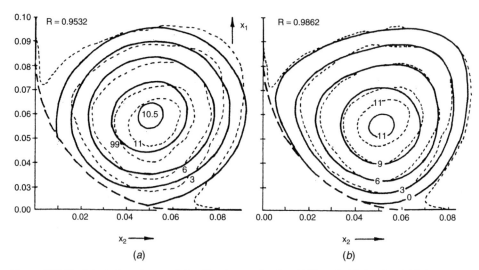

Figure E11.24. Selected contours of the fitted second-order (*left*) and third-order (*right*) response surfaces, reproduced from Wu and Meyer (1967, p. 451) with the permission of Elsevier, the copyright holder.

Table E11.24b. Additional Crater Surface Data

x_1	x_2	y	x_1	x_2	y
2.5	4	90	3.5	6.5	100
2.5	4.5	101	4	4.5	117
2.5	5	104	4	5.5	128
2.5	5.5	101	4	6.5	102
2.5	6	90	4.5	4	95
2.5	6.5	78	4.5	4.5	104
3	4.5	120	4.5	5	116
3	5.5	117	4.5	5.5	118
3	6	90	4.5	6	109
3.5	4	108	4.5	6.5	98
3.5	4.5	121	5	4.5	89
3.5	5	129	5	5.5	93
3.5	5.5	127	5	6.5	88
3.5	6	116			

cubic fit is as follows.

Source	SS	df	MS	F
First order$\mid b_0$	1,459	2	779.5	17.6
Second order extra	57,365	3	19,121.7	430.7
Third order extra	4,138	4	1,034.5	23.3
Residual	1,775	40	44.4	
Total, corrected	64,736	49		

All three groups of terms are significant at the 0.001 level. The accompanying figure is reproduced from the source reference. We see that the value of adding third-order terms is to provide a much better fit to the actual contours, which are the dashed lines shown. In the published paper, an additional 27 grid points were later added to provide detailed mapping of the bottom of the crater. These are given in Table E11.24b. The reader may like to repeat the exercise with these points added, even though no solution is provided!

11.25. (a) All four surfaces have elliptically shaped contours, three with maxima and $\hat{y}_4$ with a minimum; see Figure E11.25, which is a simplified form of the original Figure 3 of the source reference, obtained by omitting the axes which were marked in terms of the uncoded variables. Each circle shown has a radius of 1.414 in the coded variables (x_1, x_2). Thus the design points 1–8 lie *on* each circle, while the design center points 9–13 lie at the center of each circle. The regression coefficients are as shown in Table E11.25c. The residuals plots gave no cause for concern, but lack of fit was seen for the $\hat{y}_1$ surface.

(b) The predicted maxima for $\hat{y}_2$ and $\hat{y}_3$ lie at the stationary points $(-0.574, 0.260)$ and $(-0.819, -0.545)$, respectively, in coded units. Both points lie in the experimental region but at different locations.

(c) However, a workable compromise somewhere between the locations of these two peaks seems possible. Consider, for example the straight line connecting the two stationary points on which there are points with coded coordinates $[\theta_1(-0.574) + \theta_2(-0.819), \theta_1(0.260) + \theta_2(-0.545)]$, where $\theta_1 + \theta_2 = 1$. Points on this line *between* the two stationary points are such that both θ_1 and θ_2 are nonnegative. When $\theta_1 = 1$ we obtain the $\hat{y}_2$ maximum point, and when $\theta_1 = 0$ we obtain the $\hat{y}_3$ maximum point. Table E11.25d shows 11 points (x_1, x_2) equally spaced between the indicated optimal locations for $\hat{y}_2$ and $\hat{y}_3$ individually, and the corresponding predicted responses at these 11 locations. The points were obtained by choosing values $\theta_1 = 1, 0.9, 0.8, \ldots, 0$. Viewing such a table, which can be calculated in any number of dimensions, enables various compromises to be considered. Thus, for example, conditions

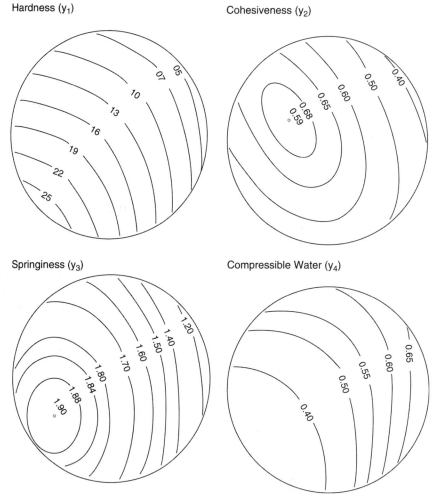

Figure E11.25. Response surface contours for four responses, hardness (y_1), cohesiveness (y_2), springiness (y_3), and compressible water (y_4). The circles indicate the experimental region with a radius 1.414 in the coded x-coordinates. Adapted with the permission of the copyright holder, The American Chemical Society, copyright 1979.

Table E11.25c. Coefficients of the Response Surfaces

b_0	b_1	b_2	b_{11}	b_{22}	b_{12}	R^2	Lack of Fit?
1.526	−0.575	−0.524	−0.171	−0.098	0.318	0.952	Yes, $p = 0.032$
0.660	−0.092	−0.010	−0.096	−0.058	−0.070	0.981	Untested, $s^2 = 0$
1.776	−0.250	−0.078	−0.156	−0.079	0.010	0.977	No
0.468	0.131	0.073	0.026	0.024	−0.083	0.949	No

Table 11.25d. Values of the Responses $\hat{y}_2$ and $\hat{y}_3$ at 11 Locations Equally Spaced on the Line Joining the Two Individual Optima

x_1	x_2	$\hat{y}_2$	$\hat{y}_3$
−0.574	0.260	0.685	1.841
−0.599	0.180	0.685	1.852
−0.623	0.099	0.683	1.862
−0.648	0.019	0.680	1.871
−0.672	−0.062	0.676	1.879
−0.697	−0.143	0.671	1.885
−0.721	−0.223	0.665	1.890
−0.746	−0.304	0.657	1.894
−0.770	−0.384	0.649	1.897
−0.795	−0.465	0.639	1.899
−0.819	−0.545	0.628	1.900

that give a predicted yield for $\hat{y}_2$ of 0.671, down from the predicted maximum of 0.685, provide a predicted response for $\hat{y}_3$ of 1.885, up from the previous corresponding value of 1.841, and so on. If $\hat{y}_2$ were considered more important, a reasonable compromise would probably lie more toward the $\hat{y}_2$ maximum than toward the $\hat{y}_3$ maximum.

In the paper, the authors concluded that "further research . . . is necessary" to build further on what was learned.

[The fitted equations obtained in this exercise are used again in Exercise 12.6.]

CHAPTER 12

12.1. Write out the solution for **x** in (12.2.8) and see that a factor $(\lambda - \lambda_1)$ or $(\lambda - \lambda_2)$, whichever (λ_1 or λ_2) occurs in the situation, appears top and bottom, and so cancels out. Remember that $\det(\mathbf{B} - \lambda\mathbf{I})$, which occurs in the denominator, can be written as $(\lambda - \lambda_1)(\lambda - \lambda_2)$. For a numerical example, consider the case $b_1 = 2^{1/2} - 1$, $b_2 = 1$, $b_{11} = 1$, $b_{22} = 3$, and $b_{12} = 2$. Remember that such an example is *very* unlikely to occur in any practical application of ridge analysis, and you will probably never see one. Also, see Fan (1999).

12.2. In fact it is not necessary to do a full ridge analysis to answer the question reasonably. We first observe that the eigenvalues of **B** are $(-0.00211,$ $-0.00127)$ for model II and $(-0.544, -0.189)$ for model III. Because both sets are negative, both surfaces are mounds with unique maxima. The locations of these maxima are determined by the formula $x_S = -\frac{1}{2}\mathbf{B}^{-1}\mathbf{b}$, which is simply the ridge analysis formula with $\lambda = 0$. Note that the maximum $\hat{y}$ path always uses λ values above the higher eigenvalue, a range that includes zero for both models. So the maximum $\hat{y}$ paths must have rising yields to the stationary point whereupon the response falls again (although the paths always contain the maximum $\hat{y}$ value for the current radius R). For model II, the maximum occurs at the point $(N, P) =$ $(245, 234)$, which is within the experimental region (0-320, 0-320); at that point, $\hat{y} = 154.3$. For model III, the maximum occurs at $(N^{1/2}, P^{1/2}) =$ $(19.9624, 18.3725)$—that is, at $(N, P) = (398, 338)$, a point outside the experimental region, where $\hat{y} = 147.3$. If we follow the maximum $\hat{y}$ path for response III, we find that it passes outside the experimental region for $\lambda = 0.0207$, when $(N, P) = (320, 279)$ and $\hat{y} = 145.95$. While both models tell the same basic story, their conclusions differ slightly, and further experimental runs are needed to resolve the situation. (A diagram of the response surface III appears on p. 473 of the source reference.)

12.3. The six eigenvalues of **B** are (in order from most negative to most positive) -38.85, -29.30, -22.80, -12.81, -8.88, and 44.64, so for the path of maximum response, we explore the range $44.64 < \lambda < \infty$, to a radius of approximately $R \le 2.378$. Note that because five of the eigenvalues are negative, the response falls off as we move away along the corresponding canonical axes. However, the maximum response is *not* at the stationary point because the response increases when we move outwards on the sixth canonical axis. This point is also clear from the fact that setting $\lambda = 0$ (which gives the stationary point) does *not* give a point on the path of maximum response, for which it is necessary that $\lambda > 44.64$, the highest eigenvalue. The maximum response path is shown in Table E12.3.
(We thank Ko-Kung Wang for bringing this paper to our attention.)

Table E12.3. Path of Maximum $\hat{y}$ for the Shang Data

λ	x_1	x_2	x_3	x_4	x_5	x_6	$\hat{y}$	R
∞	0	0	0	0	0	0	258*	0
1000	0.022	0.021	0.032	0.013	0.034	-0.028	266	0.063
100	0.206	0.392	0.290	0.136	0.280	-0.233	340	0.653
70	0.300	0.886	0.424	0.215	0.363	-0.287	411	1.147
60	0.368	1.488	0.531	0.284	0.394	-0.314	517	1.723
58	0.388	1.718	0.566	0.308	0.399	-0.320	565	1.944
57	0.400	1.860	0.587	0.322	0.400	-0.322	597	2.082
56	0.414	2.028	0.611	0.339	0.401	-0.325	636	2.245
55.5	0.421	2.124	0.625	0.348	0.402	-0.326	660	2.388
55	0.429	2.229	0.639	0.358	0.402	-0.327	687	2.440

*Not given in the paper, but imputed from information therein.

Table E12.4a. The Regression Equations for the Six Responses

Coefficient	y_1	y_2	y_3	y_4	y_5	y_6
b_0	53.76	42.62	0.917	2.026	12.027	0.802
b_1	3.93	−4.82	−0.008	0.454	−1.100	0.015
b_2	31.19	8.52	−0.036	−0.021	0.400	−0.026
b_3	9.84	9.29	−0.039	0.103	−0.813	−0.028
b_{11}	3.34	1.76	−0.022	0.335	−0.931	−0.054
b_{22}	18.87	7.43	0.015	−0.201	−0.831	−0.001
b_{33}	2.07	6.51	0.012	−0.219	−0.068	−0.049
b_{12}	7.74	4.59	0.023	−0.059	0.300	0.008
b_{13}	20.94	1.39	0.015	−0.116	1.600	−0.023
b_{23}	24.26	7.41	0.108	0.121	1.800	0.008
b_{123}	−9.39	−3.79	0.063	−0.331	−1.100	−0.013
R^2	0.692	0.856	0.837	0.843	0.728	0.825

Table E12.4b. Summary of Regression Fits

Equation Type	Lack of Fit?	R^2 Value	Significant* Regression?
y_1 quadratic	Yes	0.678	No
y_1 cubic	Yes	0.692	No
y_2 quadratic	Yes	0.838	Yes
y_2 cubic	Yes	0.856	Yes
y_3 quadratic	No	0.708	No
y_3 cubic	No	0.837	Borderline
y_4 quadratic in x_1 only	No	0.631	Yes**
y_4 quadratic	Yes	0.776	No
y_4 cubic	Yes	0.843	Yes
y_5 quadratic	Yes	0.669	No
y_5 cubic	Yes	0.728	No
y_6 quadratic	Yes	0.818	Yes
y_6 cubic	Yes	0.825	No

*At 0.05 level or better, using residual mean square irrespective of lack-of-fit test.
**Non-normal-appearing residual plot.

12.4. Table E12.4a shows all the regression coefficients for the six responses when a full quadratic in (x_1, x_2, x_3) plus a single cubic term $x_1 x_2 x_3$, is fitted to each response. Table E12.4b summarizes the situation with regard to lack-of-fit tests and the significance of the regressions of second-order, labeled "quadratic"—meaning a full quadratic in (x_1, x_2, x_3)—and of the equations labeled "cubic," which, in our context, implies only the addition of the term $x_1 x_2 x_3$. No other cubic terms are in the equations.

We see, as noted by the authors of the source paper, that several equations show lack of fit. As has been mentioned in other exercises, this may be because the repeat runs are not true repeats, or the equations may be inadequate. Nevertheless, even with lack of fit present, the interpretation of a fitted equation is often useful in practice.

Table E12.4c. Path of Maximum $\hat{y}$, Using the Quadratic in x_1

λ	$x_1 = R$	$\hat{y}$
1.35	0.249	1.70
0.90	0.490	1.89
0.74	0.749	2.15
0.66	1.017	2.48
0.62	1.238	2.79
0.59	1.480	3.19
0.57	1.702	3.60
0.55	2.002	4.22

Table E12.4d . Path of Maximum $\hat{y}$, from the Full Quadratic in x_1, x_2, x_3

λ	x_1	x_2	x_3	R	$\hat{y}$
1.35	0.223	−0.010	0.024	0.224	2.15
0.90	0.401	−0.019	0.024	0.402	2.26
0.74	0.561	−0.027	0.018	0.562	2.39
0.66	0.701	−0.035	0.010	0.702	2.51
0.62	0.802	−0.041	0.003	0.803	2.61
0.59	0.898	−0.047	−0.004	0.900	2.71
0.57	0.979	−0.051	−0.011	0.979	2.79
0.55	1.071	−0.057	−0.018	1.073	2.90
0.54	1.125	−0.060	−0.023	1.127	2.97
0.53	1.185	−0.064	−0.028	1.187	3.04
0.50	1.411	−0.078	−0.049	1.414	3.34
0.48	1.616	−0.091	−0.069	1.620	3.65
0.46	1.892	−0.109	−0.096	1.898	4.11
0.45	2.069	−0.120	−0.114	2.076	4.43

An important point to be noted is that the column of the **X**-matrix attributable to the product $x_1x_2x_3$ is orthogonal to all the other columns, and so adding or dropping this cubic term does not change any of the other regression estimates. The sum of squares for the cubic term goes back into the residual sum of squares if the term is removed, of course, and its single degree of freedom also returns in a parallel manner to the residual df.

Specifically on puncture resistance, y_4, we note the following. The full regression in the tables contains only two terms that are individually significant, namely x_1 and x_1^2. (Those details are not shown here.) Suppose we used only those two terms in a separate regression fit and then followed the ridge paths for maximizing response puncture resistance using (i) the quadratic in x_1, and (ii) the full quadratic equation in (x_1, x_2, x_3), how different would the paths be? Tables E12.4c and E12.4d show the two paths. It appears that x_2 and x_3 contribute little.

(We thank Yichi Guo for bringing this paper to our attention.)

12.5. Selected details of paths 2, 3, and 4 are given in the following tables.

2

λ	x_1	x_2	R	$\hat{y}$
0.50	-0.538	-0.423	0.469	80.17
0.52	-0.614	-0.493	0.788	80.24
0.54	-0.717	-0.588	0.927	80.37
0.56	-0.865	-0.723	1.127	80.60
0.58	-1.095	-0.932	1.438	81.05
0.60	-1.500	-1.300	1.985	82.16
0.61	-1.846	-1.614	2.452	83.41
0.62	-2.405	-2.120	3.206	86.04

3

λ	x_1	x_2	R	$\hat{y}$
-0.003	-0.196	-0.021	0.197	79.99
-0.20	-0.269	0.115	0.293	79.98
-0.25	-0.351	0.216	0.413	79.96
-0.30	-0.600	0.500	0.781	79.84
-0.31	-0.721	0.636	0.961	79.74
-0.32	-0.918	0.854	1.254	79.54
-0.33	-1.290	1.267	1.808	78.98
-0.34	-2.258	2.339	3.251	76.53

4

λ	x_1	x_2	R	$\hat{y}$
-100	0	0.001	0.001	80.00
-0.50	0.118	-0.265	0.290	79.94
-0.40	0.500	-0.700	0.860	79.66
-0.39	0.652	-0.870	1.087	79.48
-0.38	0.915	-1.162	1.479	79.10
-0.37	1.480	-1.788	2.321	77.90
-0.368	1.681	-2.010	2.620	77.36
-0.367	1.802	-2.143	2.800	77.00
-0.366	1.941	-2.297	3.007	76.56

12.6. Here is the path of maximum $\hat{y}_2$:

λ	x_1	x_2	R	$\hat{y}$
∞	0.000	0.000	0.000	0.660
10	-0.005	-0.000	0.005	0.660
5	-0.009	-0.001	0.009	0.661
1	-0.042	-0.003	0.042	0.664
0.5	-0.077	-0.004	0.077	0.667

λ	x_1	x_2	R	$\hat{y}$
0.2	-0.156	0.002	0.156	0.672
0.1	-0.238	0.021	0.239	0.677
0.05	-0.330	0.061	0.335	0.680
0.02	-0.436	0.132	0.456	0.684
0*	-0.574	0.261	0.630	0.685
-0.010	-0.700	0.406	0.810	0.684
-0.018	-0.879	0.644	1.089	0.676
-0.020	-0.946	0.740	1.201	0.671
-0.022	-1.029	0.862	1.342	0.663
-0.023	-1.079	0.936	1.428	0.658

*Stationary point and maximum; the response now falls.

Now we have the path of maximum $\hat{y}_3$:

λ	x_1	x_2	R	$\hat{y}$
∞	0.000	0.000	0.000	1.776
50	-0.002	-0.001	0.003	1.777
5	-0.024	-0.008	0.025	1.783
2	-0.058	-0.019	0.061	1.791
1	-0.108	-0.037	0.114	1.804
0.5	-0.191	-0.069	0.203	1.823
0.3	-0.275	-0.107	0.295	1.841
0.1	-0.493	-0.232	0.545	1.876
0*	-0.819	-0.545	0.984	1.900
-0.3	-1.028	-0.901	1.367	1.884
-0.4	-1.127	-1.144	1.606	1.858

*Stationary point and maximum; the response now falls

12.7. The eigenvalues of **B** are -13.178, 0, 6.724, and 32.794, so we search the space $\lambda \geq 32.794$. Here are some points on the path.

λ	x_1	x_2	x_3	x_4	R	$\hat{y}$
∞	0	0	0	0	0	350
1500	0.063	-0.041	-0.006	0	0.075	367
1000	0.095	-0.063	-0.009	0	0.114	376
500	0.194	-0.130	-0.016	0	0.234	403
300	0.334	-0.230	-0.019	0	0.406	444
200	0.524	-0.371	-0.016	0	0.642	503
170	0.632	-0.454	-0.010	0	0.778	539
140	0.798	-0.585	0.005	0	0.989	596
120	0.969	-0.723	0.026	0	1.209	658
100	1.237	-0.944	0.069	0	1.557	763
50	1.719	-1.356	0.171	0	2.196	975

Table E12.8. The Path of Maximum Response

λ	x_1	x_2	x_3	R	$\hat{y}$
∞	0	0	0	0	18.726
500	-0.006	-0.000	-0.001	0.006	18.766
100	-0.030	-0.001	-0.007	0.031	18.917
50	-0.057	-0.002	-0.013	0.059	19.087
30	-0.091	-0.003	-0.021	0.093	19.285
10	-0.218	-0.006	-0.059	0.226	19.958
5	-0.337	-0.009	-0.101	0.352	20.465
4	-0.379	-0.010	-0.118	0.397	20.614
3	-0.432	-0.011	-0.139	0.454	20.783
2	-0.503	-0.013	-0.170	0.531	20.969
1	-0.602	-0.017	-0.215	0.640	21.154
0	-0.751	-0.041	-0.292	0.807	21.261
-0.1	-0.767	-0.051	-0.303	0.829	21.259
-0.4	-0.825	-0.217	-0.360	0.856	21.211
-0.45	-0.819	-0.512	-0.407	1.049	21.106
-0.46	-0.811	-0.706	-0.436	1.160	20.994
-0.47	-0.788	-1.137	-0.498	1.471	20.613
-0.48	-0.687	-2.931	-0.751	9.624	17.052
-0.485	-0.053	-13.923	-2.295	14.111	-74.69
-0.486	2.369	-55.824	-8.179	56.470	-1527
-0.48633	∞	$-\infty$	$-\infty$	∞	$-\infty$

Note that x_4 remains at zero throughout the path; this is because the fitted equation did not contain x_4. The problem could therefore have been treated as a three-x problem instead. However, the four-x ridge analysis takes proper account of this and produces the correct answer. (Other equations in the source paper contain all four predictor variables.)

12.8. (a) The eigenvalues of **B** are, in increasing order, -6.24328, -3.73308, -0.48633. For the maximum $\hat{y}$ path, therefore, we need to choose values of λ from ∞ downwards to -0.48633 and evaluate $(x_1, x_2, x_3)'$ $= \mathbf{x} = -\frac{1}{2}(\mathbf{B} - \lambda\mathbf{I})^{-1}\mathbf{b}$ for each λ followed by $R = (\mathbf{x}'\mathbf{x})^{1/2}$ and $\hat{y} = b_0 + \mathbf{x}'\mathbf{b} + \mathbf{x}'\mathbf{B}\mathbf{x}$. This provides the values in Table E12.8. The path goes from the origin into the stationary point of the response surface as one would anticipate. R starts at zero and rises to infinity at the eigenvalue -0.48633. Note that, after the path passes through the maximum $\hat{y}$ value at $\lambda = 0$, where the solution tells us that we are at the stationary point $\mathbf{x} = -\frac{1}{2}\mathbf{B}^{-1}\mathbf{b}$, the response then declines, again as one would expect. Below $\lambda = -0.45$ and moving closer to the highest eigenvalue -0.48633, the response even goes negative! Of course that is a mathematical fiction, caused by extrapolating the surface too far beyond realistic limits and is of no practical use. We show these results only to indicate what happens to the calculations as the highest eigenvalue is approached.

(b) You would expect the computations to tell you that you are already at the maximum response, which they do, because the maximum path does not move from **f**, apart from small rounding errors that may be generated. This can be shown algebraically. When the derivative of the new Lagrangian function is set equal to zero, we see the equation

$$2(\mathbf{B} - \lambda\mathbf{I})\mathbf{x} = -\mathbf{b} - 2\lambda\mathbf{x}_S.$$

The right-hand side can now be manipulated into the form

$$-\mathbf{B}\mathbf{B}^{-1}\mathbf{b} - 2\lambda\left(-\tfrac{1}{2}\mathbf{B}^{-1}\mathbf{b}\right) = -(\mathbf{B} - \lambda\mathbf{I})\mathbf{B}^{-1}\mathbf{b} = 2(\mathbf{B} - \lambda\mathbf{I})\mathbf{x}_S.$$

It follows that for any value of λ that is not an eigenvalue of **B**, the solution is $\mathbf{x} = \mathbf{x}_S$.

12.9. The coefficients of the four fitted equations are shown below in the format b_0, **x**, **B** corresponding to the predicted value matrix form $\hat{y} = b_0 + \mathbf{x}'\mathbf{b} + \mathbf{x}'\mathbf{B}\mathbf{x}$, followed by the corresponding ridge path tables which include the point of maximum response on a sphere of radius $R = 2$ (or close to 2, when we reach a point where further computing seems pointless). These paths can be extended further if needed, of course, or constructed in more detail, as desired.

First response, y_1

12.680

0.4708
1.4208
6.1125
10.0292

-0.3960	0.3281	-0.6156	0.0281
0.3281	-0.6335	-0.3844	0.5469
-0.6156	-0.3844	0.1665	2.6031
0.0281	0.5469	2.6031	1.9540

λ	x_1	x_2	x_3	x_4	R	$\hat{y}$
∞	0	0	0	0	0	12.68
100	0.002	0.007	0.032	0.052	0.062	13.42
90	0.002	0.008	0.036	0.058	0.069	13.51
70	0.003	0.010	0.046	0.076	0.089	13.77
50	0.004	0.015	0.067	0.108	0.128	14.26
30	0.006	0.025	0.119	0.190	0.226	15.54
20	0.007	0.039	0.193	0.307	0.365	17.50
15	0.006	0.054	0.282	0.443	0.528	19.98
13	0.004	0.064	0.345	0.539	0.643	21.86
10	-0.003	0.089	0.519	0.797	0.955	27.44
9	-0.009	0.103	0.622	0.949	1.140	31.10
8	-0.020	0.121	0.776	1.174	1.413	37.00
7	-0.041	0.150	1.030	1.541	1.860	47.86
6.8	-0.048	0.157	1.101	1.644	1.985	51.19
6.778	-0.049	0.158	1.110	1.656	2.000	51.59

Second response, y_2

74.4

-0.0542
-0.5125
-1.0875
-0.4542

0.2385	0.1531	-0.1219	0.0094
0.1531	0.3385	-0.2156	-0.3844
-0.1219	-0.2156	-0.0990	-0.0594
0.0094	-0.3844	-0.0594	0.0385

λ	x_1	x_2	x_3	x_4	R	$\hat{y}$
∞	0	0	0	0	0	74.40
40	-0.001	-0.006	-0.014	-0.006	0.016	74.42
20	-0.001	-0.013	-0.027	-0.011	0.031	74.44
10	-0.003	-0.025	-0.053	-0.022	0.062	74.48
5	-0.005	-0.047	-0.104	-0.041	0.121	74.55
3	-0.007	-0.074	-0.169	-0.064	0.195	74.64
1	0.005	-0.151	-0.458	-0.147	0.504	74.97
0.8	0.043	-0.109	-0.571	-0.198	0.616	75.09
0.75	0.083	-0.030	-0.627	-0.249	0.681	75.15
0.72	0.155	0.131	-0.697	-0.344	0.803	75.28
0.7	0.338	0.562	-0.840	-0.590	1.218	75.87
0.696	0.439	0.805	-0.915	-0.727	1.486	76.38
0.693	0.567	1.111	-1.009	-0.900	1.839	77.20
0.6922	0.614	1.225	-1.043	-0.964	1.974	77.55
0.69205	0.624	1.249	-1.051	-0.978	2.002	77.63

Third response, y_3

80.7

-0.0542
-0.3958
-0.0375
-2.0625

0.5073	0.0719	-0.0032	-0.3281
0.0719	0.4698	-0.0406	0.2594
-0.0032	-0.0406	-0.8427	0.2844
-0.3281	0.2594	0.2844	-0.6552

λ	x_1	x_2	x_3	x_4	R	$\hat{y}$
∞	0	0	0	0	0	80.70
50	0.000	−0.004	0.000	−0.020	0.021	80.74
10	0.000	−0.023	−0.004	−0.097	0.100	80.91
5	0.007	−0.054	−0.012	−0.186	0.194	81.09
3	0.025	−0.107	−0.026	−0.294	0.315	81.32
1.5	0.130	−0.318	−0.069	−0.546	0.648	81.95
1	0.397	−0.719	−0.124	−0.836	1.178	83.08
0.9	0.568	−0.933	−0.148	−0.966	1.466	83.80
0.85	0.707	−1.091	−0.164	−1.058	1.684	84.40
0.8	0.916	−1.308	−0.186	−1.185	1.997	85.35

Fourth response, y_4

76.46

0.5792
−0.1042
−0.1625
−1.4125

0.7506	0.0031	0.1844	−0.1344
0.0031	0.3131	0.0282	0.2219
0.1844	0.0282	0.7131	0.0781
−0.1344	0.2219	0.0781	0.0006

λ	x_1	x_2	x_3	x_4	R	$\hat{y}$
∞	0	0	0	0	0	76.46
5	0.072	−0.018	−0.019	−0.144	0.163	76.72
3	0.141	−0.040	−0.033	−0.246	0.288	76.93
1.9	0.291	−0.090	−0.052	−0.405	0.509	77.33
1.5	0.469	−0.144	−0.052	−0.537	0.730	77.79
1.3	0.677	−0.197	−0.021	−0.649	0.959	78.32
1.1	1.253	−0.291	0.196	−0.840	1.549	80.06
1.045	1.682	−0.329	0.443	−0.929	1.999	81.76

12.10. The fitted quadratic surface is

$$\hat{y} = 31.378 - 6.9062x_1 - 0.0813x_2 + 4.9812x_3 - 0.285x_1^2 - 2.035x_2^2$$

$$+ 1.690x_3^2 + 2.588x_1x_2 - 5.188x_1x_3 + 1.513x_2x_3,$$

with $R^2 = 0.968$. The analysis of variance table is

Source	SS	df	MS	F
First order	1160.26	3	386.75	34.59*
Extra for second order	510.06	6	85.01	7.60**
Residual	55.91	5	11.182	
Total, corrected	1726.23	14		

*Significant at 0.01 level.
**Significant at 0.05 level.

In the $\hat{y} = b_0 + \mathbf{x}'\mathbf{b} + \mathbf{x}'\mathbf{B}\mathbf{x}$ notation, $b_0 = 31.378$, $\mathbf{b} = (-6.902, -0.0813, 4.9812)'$ and

$$\mathbf{B} = \begin{bmatrix} 0.285 & 1.294 & -2.594 \\ 1.294 & -2.035 & 0.7565 \\ -2.594 & 0.7565 & 1.690 \end{bmatrix}.$$

The eigenvalues of $\mathbf{B}$ are -3.64, -0.66, and 3.14, so the maximum $\hat{y}$ path is associated with values $\lambda > 3.14$. The ridge analysis calculation for this maximum $\hat{y}$ path, beginning from the center of the design at $(0, 0, 0)$, leads

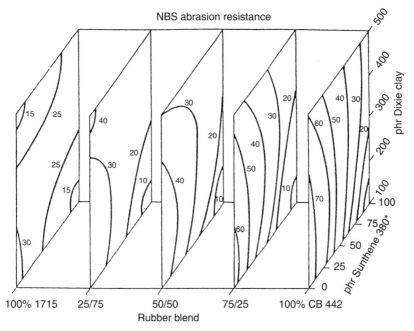

Figure E12.10. Contour slices of a fitted second-order response surface, reproduced from the source paper with the permission of the copyright holder, *Rubber World*.

to the following table:

λ	R	x_1	x_2	x_3	$\hat{y}$
∞	0	0	0	0	31.4
10	0.60	-0.44	-0.03	0.41	37.6
6	1.43	-0.99	-0.07	1.04	49.8
5.5	1.74	-1.18	-0.08	1.27	55.3
5.4	1.82	-1.23	-0.08	1.33	56.7
5.3	1.90	-1.29	-0.09	1.39	58.4
5.19	2.00	-1.35	-0.09	1.47	60.5

So, in or on a sphere of radius 2, an optimum response of about 60.5 is predicted at the coded point $(-1.35, -0.09, 1.47)$. Figure E12.10 shows original contour slices reproduced from Figure 6, page 80 of the source reference. The horizontal axis is the uncoded form of x_3, the vertical axis is the uncoded x_2, and the uncoded x_1 axis slants in from foreground to background. These contours provide a close approximation to our fitted surface. (The source paper did not contain the original data but instead provided fitted values. Our adaptation consisted of adding a set of random normal errors to selected fitted values, whose locations formed the composite design given in the exercise.)

12.11. **(a)** The fitted second-order surface for the first response, written in the form $\hat{y} = b_0 + \mathbf{x}'\mathbf{b} + \mathbf{x}'\mathbf{B}\mathbf{x}$ has the following parameters: $b_0 = 69.399$,

$$\mathbf{b} = \begin{bmatrix} -13.368 \\ -11.466 \\ -2.390 \end{bmatrix} \quad \text{and} \quad \mathbf{B} = \begin{bmatrix} 12.810 & 1.706 & 4.1935 \\ 1.706 & 12.143 & 9.2685 \\ 4.1935 & 9.2685 & -3.471 \end{bmatrix}.$$

The model fits well with $R^2 = 0.910$. The eigenvalues of $\mathbf{B}$ are -8.24, 11.21, and 18.51. To obtain lower responses, we need to explore the λ-range below -8.24. Table E12.11b contains those calculations.

We see that for $\lambda = -10.5$, the optimum path is just about to pass outside the boundary $x_3 = -1$, with an estimated response of $\hat{y} = 48.2$.

Table E12.11b. Ridge Analysis Calculations for the First Response

λ	x_1	x_2	x_3	R	$\hat{y}$
-50	0.104	0.090	-0.002	0.137	67.2
-25	0.175	0.158	-0.047	0.240	65.9
-20	0.205	0.193	-0.088	0.295	65.3
-15	0.254	0.264	-0.201	0.418	63.8
-12	0.316	0.379	-0.427	0.652	60.5
-11	0.356	0.468	-0.615	0.851	57.1
-10.1	0.421	0.626	-0.962	1.222	49.1
-10.5	0.426	0.639	-0.991	1.254	48.2
-10.01	0.431	0.651	-1.016	1.281	47.6
-10	0.432	0.653	-1.022	1.288	47.4

Table E12.11c. Ridge Analysis Calculations for the Second Response

λ	x_1	x_2	x_3	R	$\hat{y}$
-100	0.072	0.042	-0.046	0.095	86.1
-50	0.154	0.097	-0.074	0.197	84.2
-25	0.370	0.256	-0.116	0.465	78.4
-20	0.517	0.369	-0.135	0.650	73.9
-15	0.865	0.647	-0.173	1.094	60.8
-13	1.190	0.910	-0.204	1.512	45.7
-12	1.468	1.138	-0.231	1.872	30.6
-11.5	1.663	1.298	-0.249	2.125	18.7
-11.4	1.709	1.336	-0.253	2.184	15.8
-11.35	1.733	1.355	-0.255	2.214	14.3
-11.3	1.757	1.375	-0.257	2.246	12.7

Uncoding the x-values via $E = 0.75 + 0.25x_1$, $t = 5 + x_2$, and $T = 45 + 5x_3$, provides the approximate optimum response point as $(E, t, T) = (0.86, 5.64, 40.05)$, within the limitations of the experimental region.

(b) The fitted second-order surface for the second response, written in the form $\hat{y} = b_0 + \mathbf{x'b} + \mathbf{x'Bx}$ has the following parameters: $b_0 = 87.933$,

$$
\mathbf{b} = \begin{bmatrix} -13.519 \\ -6.984 \\ 12.340 \end{bmatrix} \quad \text{and} \quad \mathbf{B} = \begin{bmatrix} -2.483 & -6.475 & -0.675 \\ -6.475 & 0.583 & 5.8 \\ -0.675 & 5.8 & 39.067 \end{bmatrix}.
$$

The model fits well with $R^2 = 0.900$. The eigenvalues of $\mathbf{B}$ are -7.81, 5.00, and 39.98. To obtain lower responses, we need to explore the λ-range below -7.81. Table E12.11c contains those calculations.

We see that for $\lambda = -11.35$, the optimum path has just passed outside the boundary $x_1 = 1.732$, with an estimated response of $\hat{y} = 14.3$. Uncoding the x-values via $E = 0.75 + 0.25x_1$, $t = 5 + x_2$, and $T = 45 + 5x_3$ provides the approximate optimum response point as $(E, t, T) = (1.18, 6.36, 43.73)$, within the limitations of the experimental region.

(*Note:* Our calculations for the first response agree quite well with those in the paper. For the second response, however, there are unresolved differences.)

12.12. We placed the two dummy variables into the equation immediately after the constant term because the dummies play the role of allowing a different mean level in each block. If all the data are used to fit a full cubic model, we obtain

$$
\hat{y} = 3710.03 - 148.56B_1 - 54.74B_2 + 583.1x_1 + 38.6x_2 - 123.09x_1^2 - 203.65x_2^2
$$

$$
- 182.27x_1x_2 - 267.16x_1^3 + 109.86x_1^2x_2 - 237.04x_1x_1^2 + 114.53x_2^3,
$$

with $R^2 = 0.966$ and an analysis of variance table as follows.

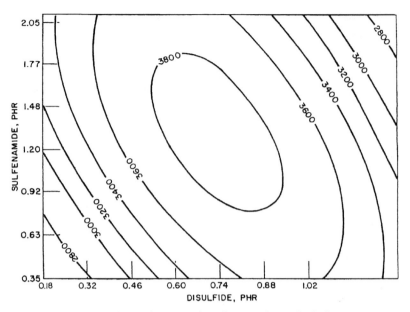

Figure E12.12a. Contours of tensile strength, quadratic fit.

Source	SS	df	MS	F
Blocks	434,981	2	217,490	27.60**
First order	879,636	2	439,818	55.81**
Second order, given first	570,086	3	190,029	24.12**
Third order|first, second	159,245	4	39,811	5.05*
Lack of fit	51,851	5	51,851	2.72[a]
Pure error	19,067	4	19,067	
Total, corrected	2,114,867	20		

[a]Lack of fit is not significant; the remaining F-tests use $s^2 = (51851 + 19067)/(5 + 4) = 7880$ with 9 df as denominator.
*Significant at 0.05 level.
**Significant at 0.01 level.

It thus appears that third-order terms should be included. Figures E12.12(a) and (b) show quadratic and cubic fits. On the one hand, the fits look considerably different; on the other hand, the enclosed areas with desired high tensile strength (see the two *closed* curves marked 3800) have a high degree of overlap. The *nonclosed* 3800 contour of the cubic, however, completely contradicts the curve pattern of diminishing response in the quadratic fit, and so that region would need further investigation if it were of continuing interest. Note that this response is only one of six responses that need to be considered as a group.

Table E12.12b contains data on the five other responses (300% modulus psi; 500% modulus psi; elongation at break %; Mooney scorch@270°F, T_5 min; 95% rheometer cure time@280°F, min.) to be added to the layout of Table E12.12a. Desirable values are, respectively, high, high, low, high, and

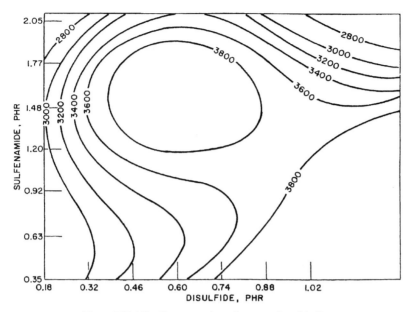

Figure E12.12b. Contours of tensile strength, cubic fit.

Table E12.12b. Additional Responses for Table E12.12a

Number	y_1	y_2	y_3	y_4	y_5	B_1	B_2
1	1140	3040	570	9.4	14.5	−1	1
2	590	1750	720	13.4	28.0	−1	1
3	910	2390	660	7.5	13.0	−1	1
4	530	1500	710	11.9	30.5	−1	1
5	900	2410	660	12.0	16.0	−1	1
6	880	2410	660	10.6	15.0	−1	1
7	870	2330	670	11.3	15.0	−1	1
8	1040	2760	600	12.7	17.5	0	−2
9	650	1830	710	8.2	18.0	0	−2
10	990	2560	630	8.9	13.0	0	−2
11	440	1300	730	20.7	41.5	0	−2
12	840	2240	690	11.2	16.0	0	−2
13	860	2270	660	11.9	16.0	0	−2
14	860	2260	680	11.9	16.5	0	−2
15	870	2400	620	10.1	14.5	1	1
16	730	2000	660	7.9	13.5	1	1
17	570	1590	700	9.4	15.0	1	1
18	490	1350	710	10.5	21.0	1	1
19	460	1290	740	14.5	28.0	1	1
20	620	1880	690	14.7	22.0	1	1
21	870	2440	630	12.3	17.5	1	1

Table E12.13b. Path of Maximum $\hat{y}$

λ	x_1	x_2	x_3	x_4	R	$\hat{y}$
∞	0	0	0	0	0	5723
1000	−0.11	−0.01	0.01	0	0.11	5746
500	−0.21	−0.03	0.01	−0.01	0.22	5770
100	−1.14	−0.53	0.19	−0.12	1.28	6015
90	−1.30	−0.69	0.23	−0.16	1.50	6074
80	−1.53	−0.94	0.30	−0.24	1.84	6169
70	−1.89	−1.42	0.43	−0.39	2.44	6359
60	−2.63	−2.55	0.72	−0.81	3.82	6911

low. These data can be used as exercises or assignments for first-, second-, and third-order fits, either in total, or sequentially, using observations 1–7 for a first-order fit, then 1–14 for a second-order fit, then 1–21 for a third-order fit. Answers are not provided, however!

12.13. The fitted second-order surface for the first response, written in the form $\hat{y} = b_0 + \mathbf{x}'\mathbf{b} + \mathbf{x}'\mathbf{Bx}$ has the following parameters: $b_0 = 5723.33$,

$$
\mathbf{b} = \begin{bmatrix} -213.750 \\ -20.417 \\ 10.000 \\ -3.750 \end{bmatrix} \quad \text{and} \quad \mathbf{B} = \begin{bmatrix} -6.458 & 22.8125 & -11.5625 & 1.5625 \\ 22.8125 & 28.542 & -5.3125 & 7.8125 \\ -11.5625 & -5.3125 & -17.083 & -7.8125 \\ 1.5625 & 7.8125 & -7.8125 & 21.042 \end{bmatrix}.
$$

The model fits well with no lack of fit indicated, and $R^2 = 0.978$. The eigenvalues of $\mathbf{B}$ are -27.00, -10.55, 18.71, and 44.88. The four-dimensional surface is difficult to visualize. In the three axes corresponding to the two negative and either positive eigenvalue, it looks like Figure 11.3(B). To obtain higher responses, we need to explore the λ-range above 44.88. Table E12.13b contains those calculations.

CHAPTER 13

13.1. To obtain the moment conditions, apply the sufficient conditions of Appendix 13B, namely moments of design equal moments of region to order $d_2 + 1$. If $d_2 = 3$, $N = 6$, we need $x_1, x_2, \ldots, x_6$ such that $\Sigma x_u = 0$, $\Sigma x_u^2 = 2$, $\Sigma x_u^3 = 0$, $\Sigma x_u^4 = 1.2$. The first and third of these are satisfied if we choose the x's symmetrically as $-a, -b, -c, c, b, a$, whereupon the second and fourth equations become $a^2 + b^2 + c^2 = 1$ and $a^4 + b^4 + c^4 = 0.6$. A solution is given by

$$
a^2 = \tfrac{1}{2}\{(1 - c^2) \pm (0.2 + 2c^2 - 3c^4)^{1/2}\}, \qquad b^2 = 1 - c^2 - a^2,
$$

for any value of c which leads to positive, nonimaginary solutions for a^2 and b^2. There are many such c values; for example, if $c = 0$, then $a = 0.5257$ and $b = 0.8507$.

13.2. The Σr_u^6 values are proportional to 142, 125, 75, 68, 62, and 60, respectively. Thus, the first design provides the best detectability of cubic lack of fit. However, the second design has experiments at all five levels and the detectability is quite high, so it would also be a good choice. Note that the design with two experiments at each of the five levels is not a particularly good one in the assumed circumstances, despite the fact that it might seem the natural arrangement to select.

13.3. Solution is implicit in question.

13.4. Solution is implicit in question.

13.5. This example is constructed. The surface used was

$$y = 76 + 5X_1 - X_2 + 10X_3 + X_1^2 + 0.5X_2^2 - 7X_3^2 - X_1X_2 - 1.5X_1X_3 + \varepsilon,$$

where $X_1 = \xi_1^{-1}$, $X_2 = \xi_2$, $X_3 = \xi_3^{1/2}$, and $\varepsilon \sim N(0, 4)$. Thus the estimates obtained for the second part of the question should be close to the indicated values.

CHAPTER 15

15.1. The variance function, obtained from the formula $V\{\hat{y}(\mathbf{x})\} = \mathbf{u}'(\mathbf{X}'\mathbf{X})^{-1}\mathbf{u}\sigma^2$, where $\mathbf{u} = (1, x_1, x_2, x_1^2, x_2^2, x_1x_2)$, is of the form σ^2 times

$$P + \left(2Q + \frac{1}{B}\right)r^2 + Rr^4 + \frac{x_1^2 x_2^2 (C - 3D)}{[D(C - D)]},$$

where $P = 4\alpha^4(4 + \alpha^4)/A$, $Q = -4\alpha^4(2 + \alpha^2)/A$, $R = 4(5\alpha^4 - 4\alpha^2 + 8)/A$, $A = 2\alpha^4(16\alpha^4 - 32\alpha^2 + 64)$, $B = 4 + 2\alpha^2$, $C = 4 + 2\alpha^4$, $D = 4$, and $r^2 = x_1^2 + x_2^2$. The design is rotatable if this expression is a function of r^2 only, which happens when $C = 3D$, which implies $\alpha = 2^{1/2}$. Note that, with this value of α, the cube and star points all lie on a circle so that center points *are* essential.

15.2. (a) $\Sigma x_{1u}^2 = 4b^2 = 9$, $\Sigma x_{1u}^4 = 4b^4 = 20.25$

$\Sigma x_{2u}^2 = 2a^2 + 4c^2 = 9$, $\Sigma x_{2u}^4 = 2a^4 + 4c^4 = 20.25$,

$\Sigma x_{1u}^2 x_{2u}^2 = 4b^2c^2 = 6.75$.

All other sums of powers and products up to and including fourth order are zero. It is possible to estimate a full second-order model with this design provided that the $\mathbf{X}'\mathbf{X}$ matrix is nonsingular. In fact

$$(\mathbf{X}'\mathbf{X})^{-1} = \begin{bmatrix} 9 & 0 & 0 & 9 & 9 & 0 \\ 0 & 9 & 0 & 0 & 0 & 0 \\ 0 & 0 & 9 & 0 & 0 & 0 \\ 9 & 0 & 0 & 20.25 & 6.75 & 0 \\ 9 & 0 & 0 & 6.75 & 20.25 & 0 \\ 0 & 0 & 0 & 0 & 0 & 6.75 \end{bmatrix}^{-1}$$

$$= \begin{bmatrix} \frac{1}{3} & 0 & 0 & -\frac{1}{9} & -\frac{1}{9} & 0 \\ 0 & \frac{1}{9} & 0 & 0 & 0 & 0 \\ 0 & 0 & \frac{1}{9} & 0 & 0 & 0 \\ -\frac{1}{9} & 0 & 0 & \frac{5}{54} & \frac{1}{54} & 0 \\ -\frac{1}{9} & 0 & 0 & \frac{1}{54} & \frac{5}{54} & 0 \\ 0 & 0 & 0 & 0 & 0 & \frac{4}{27} \end{bmatrix}.$$

The design is rotatable, because $\Sigma x_{iu}^4 = 3\Sigma x_{1u}^2 x_{2u}^2$.

(b) $V \equiv NV\{\hat{y}(\mathbf{x})\}/\sigma^2 = N\mathbf{x}_0'(\mathbf{X}'\mathbf{X})^{-1}\mathbf{x}_0$ where

$$\mathbf{x}_0' = \left(1, x_1, x_2, x_1^2, x_2^2, x_1 x_2\right).$$

This becomes $V = 3 - r^2 + \frac{5}{6}r^4$, where $r^2 = x_1^2 + x_2^2$. The contours are circles about the origin.

(c) If the true model contained terms $\beta_{111}x_1^3 + \beta_{122}x_1x_2^2 + \beta_{112}x_1^2x_2 + \beta_{222}x_2^3$ in addition to those of second order and below, the alias matrix would be $\mathbf{A} = (\mathbf{X}'\mathbf{X})^{-1}\mathbf{X}'\mathbf{X}_2$ where $\mathbf{X}_2$ was a matrix with columns $x_1^3, x_1x_2^2, x_1^2x_2, x_2^3$ generated from the columns of the second-order $\mathbf{X}$ matrix. It can be shown that

$$\mathbf{A} = \frac{1}{4}\begin{bmatrix} 0 & 0 & 0 & 0 \\ 9 & 3 & 0 & 0 \\ 0 & 0 & 3 & 1 \\ 0 & 0 & 0 & 0 \\ 0 & 0 & 0 & 0 \\ 0 & 0 & 0 & 0 \end{bmatrix}.$$

Thus

$$E(b_1) = \beta_1 + 3(3\beta_{111} + \beta_{122})/4,$$

$$E(b_2) = \beta_2 + (3\beta_{112} + \beta_{222})/4.$$

and the remaining b's are unbiased.

15.3. In general, *for a rotatable design* in k factors, $V(r) = \{P + (2Q + 1/B)r^2 + Rr^4\}\,\sigma^2$ where $P = 2(k + 2)D^2/A$, $Q = -2DB/A$, $R = \{N(k + 1)D - (k - 1)B^2\}/A$, $A = 2D\{N(k + 2)D - kB^2\}$, $B = \Sigma_{u=1}^N x_{iu}^2$, $D =$

$\sum_{u=1}^{N} x_{iu}^2 x_{ju}^2$, and N = total points in design. For the design indicated, $k = 3$, $N = 14 + n_0$, $B = 8 + 4\sqrt{2}$, $D = 8$, and so $P = 1/(n_0 + 0.011775)$, $2Q + 1/B = 0.073223 - 0.682843/(n_0 + 0.011775)$, and

$$R = \frac{(0.117157 + n_0/20)}{(n_0 + 0.011775)},$$

enabling $V(r)$ to be plotted against r for various n_0 values (suggestion). $V(0) = V(1)$ implies that $2Q + 1/B + R = 0$ which implies $n_0 = 4.58$, a result we round to the nearest value, 5. Similar work for $r^2 = B/N$ and for $r^2 = k$ provides $n_0 = 5.55$, and $n_0 = 1.48$, respectively. For additional commentary, see Draper (1982), where the 1.48 values is adjusted to 2, not 1. This is because $|V(0) - V(k^{1/2})|$ is smaller for $n_0 = 2$ than for $n_0 = 1$, even though it is zero at $n_0 = 1.48$.

15.4. Solution is implicit in the question.

15.5. Solution is implicit in the question.

15.6. For orthogonal blocking, we need, for (a) 2^{3-1} blocks and for (b) star block, respectively;

(a) $\dfrac{4}{(8 + 2\alpha^2)} = \dfrac{\left(4 + \frac{1}{2}c_0\right)}{(14 + c_0 + s_0)}$,

(b) $\dfrac{2\alpha^2}{(8 + 2\alpha^2)} = \dfrac{(6 + s_0)}{(14 + c_0 + s_0)}$.

These are satisfied by $c_0 = 4$, $s_0 = 2$, $\alpha^2 = \frac{8}{3}$. From either equation (we really need only one), $\alpha^2 = 4(6 + s_0)/(8 + c_0)$. Some α^2 values are given below.

c_0	\multicolumn{6}{c}{s_0}
	0	1	2	3	4	5
0	3	$\frac{7}{2}$	4	$\frac{9}{2}$	5	$\frac{11}{2}$
2	$\frac{12}{5}$	$\frac{14}{5}$	$\frac{16}{5}$	$\frac{18}{5}$	4	$\frac{22}{5}$
4	2	$\frac{7}{3}$	$\frac{8}{3}$	3	$\frac{10}{3}$	$\frac{11}{3}$
6	$\frac{12}{7}$	2	$\frac{16}{7}$	$\frac{18}{7}$	$\frac{20}{7}$	$\frac{22}{7}$
8	$\frac{3}{2}$	$\frac{7}{4}$	2	$\frac{9}{4}$	$\frac{5}{2}$	$\frac{11}{4}$

For rotatability as well, we need $\alpha = 2^{(k-p)/4}$, that is, $\alpha^2 = 2^{3/2}$. Substituting this in the formula gives $8 + c_0 = 2^{1/2}(6 + s_0)$, which cannot be achieved with integer values of c_0 and s_0. It follows that no design *of this form* (cube plus star plus center points) can be rotatable *and* orthogonally blocked, for $k = 3$. (It can happen for other values of k however, and there are other $k = 3$ designs for which it can be achieved. See Exercise 15.8.)

15.7. For orthogonal blocking we need

$$\frac{8}{(16 + 2\alpha^2)} = \frac{\left(8 + \frac{1}{2}c_0\right)}{(24 + c_0 + s_0)},$$

that is,

$$\alpha^2 = \frac{8(8 + s_0)}{(16 + c_0)}.$$

Clearly $c_0 = 4$, $s_0 = 2$, and $\alpha = 2$ satisfy this and the design is rotatable for this value of α. Other values of α^2 for an orthogonally blocked design are given by substituting in for s_0 and c_0. Note that when $c_0 = 2s_0$, $\alpha = 2$ so that an equal number of center points in *each* block will provide a design that is both rotatable *and* orthogonally blocked.

			s_0			
c_0	0	1	2	3	4	5
0	4	$\frac{9}{2}$	5	$\frac{11}{2}$	6	$\frac{13}{2}$
2	$\frac{32}{9}$	4	$\frac{40}{9}$	$\frac{44}{9}$	$\frac{16}{3}$	$\frac{52}{9}$
4	$\frac{16}{5}$	$\frac{18}{5}$	4	$\frac{22}{5}$	$\frac{24}{5}$	$\frac{26}{5}$
6	$\frac{32}{11}$	$\frac{36}{11}$	$\frac{40}{11}$	4	$\frac{48}{11}$	$\frac{52}{11}$
8	$\frac{8}{3}$	3	$\frac{10}{3}$	$\frac{11}{3}$	4	$\frac{13}{3}$

15.8. For rotatability, $2^{k-p} + 4\alpha^4 = 3(2^{k-p})$, that is, $\alpha^2 = 2^{(k-p-1)/2}$. For orthogonal blocking,

$$\frac{2^{k-p}/b}{2\alpha^2} = \frac{2^{k-p}/b + A}{2k + B},$$

which implies that

$$B = 2^{(k-p-1)/2}\left\{1 + \frac{Ab}{2^{k-p}}\right\} - 2k.$$

We substitute in the various values of $A = 0, 1, 2$, and of k, p, and b. The relevant solutions are those for which B is a positive interger or zero. In fact, of the 18 choices given, only the values $k = 3$, $p = 0$, $A = 2$, $b = 2$, $B = 0$ work. Thus a cube divided into two blocks, each with two center points, plus a replicated star ($\alpha = 2^{1/2}$), each star with no center points, forms a four-block orthogonally blocked rotatable design for $k = 3$ factors. (See Table 11.4.)

15.9. The general simplex design with $\sum_{u=1}^{N} x_{iu}^2 = Nc$ has $N = k + 1$ runs in k columns and the design matrix **D** is an $N \times k$ matrix of form

$$
\begin{array}{c}
 \\
\mathbf{c}_1' \\
\mathbf{c}_2' \\
\mathbf{c}_3' \\
\mathbf{c}_4' \\
\\
\cdots \\
\\
\\
\\
\mathbf{c}_N'
\end{array}
\begin{array}{cccccc}
x_1 & x_2 & x_3 & \cdots & x_i & \cdots & x_k \\
\left[\begin{array}{ccccccc}
-a_1 & -a_2 & -a_3 & \cdots & -a_i & \cdots & -a_k \\
a_1 & -a_2 & -a_3 & & -a_i & & -a_k \\
0 & 2a_2 & -a_3 & & -a_i & & -a_k \\
0 & 0 & 3a_3 & & -a_i & & -a_k \\
& & & & \cdots & & \\
\cdots & & & & ia_i & & \cdots \\
& & & & 0 & & \\
& & & & \cdots & & \\
0 & 0 & 0 & & 0 & & ka_k
\end{array}\right]
\end{array}
$$

where $a_i = \{cN/i(i+1)\}^{1/2}$. Let $\mathbf{c}_u'$ denote the uth row. Then it can be verified that

$$
\mathbf{c}_u' \mathbf{c}_v = \begin{cases} c(N-1) & \text{if } u = v \\ -c & \text{if } u \ne v. \end{cases}
$$

Each $\mathbf{c}_u$ represents a point in k-dimensional space. The distance between the uth and the vth row is r_{uv} where

$$
\begin{aligned}
r_{uv}^2 &= (\mathbf{c}_u - \mathbf{c}_v)'(\mathbf{c}_u - \mathbf{c}_v) \\
&= \mathbf{c}_u' \mathbf{c}_u - \mathbf{c}_u' \mathbf{c}_v - \mathbf{c}_v' \mathbf{c}_u + \mathbf{c}_v' \mathbf{c}_v \\
&= c(N - 1 + 1 + 1 + N - 1) \\
&= 2Nc.
\end{aligned}
$$

The special cases follow immediately. The **X**-matrix is obtained by adjoining a column of one's to the **D** matrix.

15.10. (*Note:* In what follows, we fit a full second-order model and find the ridge path of minimum response, based on that fit, as an exercise. However there are several additional complications, described in the article, which we have ignored here. For those, we suggest reading the original paper. The reader may also question our use of ridge analysis on an equation which, although it has $R^2 = 0.882$, has no coefficients significant at the 0.05 level except for the constant term! In general, we don't recommend doing it.)

The second-order fit has $b_0 = 1.279$, $\mathbf{b}' = (-0.04003, -0.00537, 0.00982, -0.00065)$

$$
\mathbf{B} = \begin{bmatrix}
-0.03726 & 0.00091 & -0.01 & 0.03929 \\
0.00091 & 0.024706 & -0.003125 & 0.003805 \\
-0.01 & -0.003125 & 0.003498 & -0.018125 \\
0.03929 & 0.003805 & -0.018125 & 0.03873
\end{bmatrix}.
$$

The eigenvalues of **B** are -0.0540131, -0.0036962, 0.0242425, and 0.0631408. Their product, the determinant of **B**, is tiny, 3.056×10^{-7}. The path of minimum response is connected to values $-\infty < \lambda < -0.0540131$. Some representative calculations are given below.

λ	x_1	x_2	x_3	x_4	R	$\hat{y}$
$-\infty$	0	0	0	0	0	1.279
-0.1	0.383	0.021	-0.029	-0.111	0.401	1.255
-0.09	0.486	0.024	-0.029	-0.151	0.510	1.246
-0.08	0.666	0.027	-0.026	-0.223	0.703	1.226
-0.07	1.071	0.033	-0.015	-0.388	1.140	1.167
-0.065	1.549	0.039	0.002	-0.585	1.656	1.070
-0.06	2.824	0.053	0.052	-1.113	3.036	0.670
-0.059	3.386	0.059	0.076	-1.346	3.645	0.428
-0.0585	3.761	0.063	0.092	-1.502	4.051	0.245

CHAPTER 16

16.1. The sixth point cannot be plotted because the three ingredients do not add to 1.

16.2. (a) $0, 1/m, 2/m, \ldots, (m-1)/m, 1$.

(b) No, because only combinations of levels adding to 1 can be used. In general, there are $\binom{k+m-1}{m}$ of these; for $k = 3$, this reduces to $\binom{m+2}{m} = (m+2)(m+1)/2 < (m+1)^3$.

(c) See Figure 16.9.

16.3. The sum of squares of the required distances is formed by $6r^2 + \{6r^2 - 3r + \frac{1}{2}\} + \{6r^2 - 3r + \frac{1}{2}\} + \{6r^2 - 4r + \frac{2}{3}\} = 24r^2 - 10r + \frac{5}{3}$. This is minimized by $r = 5/24 = 0.208$. So the three axial points are at $(0.583, 0.208, 0.208)$, $(0.208, 0.583, 0.208)$, $(0.208, 0.208, 0.583)$. By symmetry, each of these points minimizes a similar sum of squares derived from its own local neighbors.

16.4. The general sum of squares of the distances is $q(q-1)r^2 + (q-1)\{q(q-1)r^2 - qr + \frac{1}{2}\} + (q-1)(qr-1)^2/q$, and this is minimized by $r = (q+2)/[2q(q+1)]$. The first axial point is then at $([\frac{1}{2} + (q^2+q)^{-1}], (q+2)/[2q(q+1)], (q+2)/[2q(q+1)], \ldots, (q+2)/[2q(q+1)])$, with permutation of the first coordinate for the remaining $q-1$ axial points. So we get, as first axial point, $(0.667, 0.333)$ for $q = 2$; $(0.584, 0.208, 0.208)$ for $q = 3$; $(0.550, 0.150, 0.150, 0.150)$ for $q = 4$; $(0.532, 0.117, 0.117, 0.117, 0.117)$ for $q = 5$; and so on. As q increases, the first coordinate of this first axial point tends to 0.500. At first sight, it seems peculiar that the other coordinates tend to zero as q increases, but one must remember that there are $q-1$ of them, an increasing number, and the sum of all the q coordinates is always 1, as is easily confirmed.

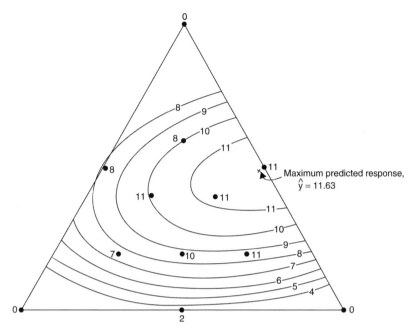

Figure E16.7. Design points, response values, and quadratic fit contours. All numbers shown have been multiplied by 100 to reduce clutter.

16.5. The fitted equation is $\hat{y} = 6x_1 + 12x_2 + 6x_3 - 4x_1x_2 - 4x_1x_3 + 0x_2x_3$. At the centroid $(\frac{1}{3}, \frac{1}{3}, \frac{1}{3})$, $\hat{y} = 8 - 8/9 = 7.11$ with standard error given by 0.7934 (from a regression printout). The t_{20}-value is $(7.11 - 12)/0.7934 = -6.16$, very significant. So it is unlikely that the centroid response could be as high as 12.

16.6. $\hat{y} = 6x_1 + 12x_2 + 6x_3 - 4.5x_1x_2 + 9x_1x_3 + 0x_2x_3$
$\quad - 54x_1x_2(x_1 - x_2) + 0x_1x_3(x_1 - x_3) + 0x_2x_3(x_2 - x_3) - 13.5x_1x_2x_3.$

16.7. The quadratic fitted surface is $\hat{y} = -0.00658x_1 - 0.00243x_2 + 0.00367x_3 + 0.34265x_1x_2 + 0.47074x_1x_3 + 0.14115x_2x_3$. Note that the zero responses at the corners of the region cause this fitted model to depend mostly on the cross-product terms. Of the total sum of squares 0.076500 (12 df), the quadratic model picks up 0.074972 (6 df) leaving a residual sum of squares of 0.001528 (6 df). The special cubic fit is barely better, the cubic term picking up only another 0.000285 (1 df), a nonsignificant amount. So it seems reasonable to choose the quadratic model. A contour plot is shown in Figure E16.7. When x_1 and x_3 are roughly equal, higher shrinkage values occur as x_2 decreases. See also Figure 20.1, where the same contours appear with their canonical axes.

16.8. The simplex centroid design for $q = 4$ consists of four sets of points:

S_1, consisting of $(1, 0, 0, 0)$ plus permutations. (4 points),
S_2, consisting of $(\frac{1}{2}, \frac{1}{2}, 0, 0)$ plus permutations. (6 points),

S_3, consisting of $(\frac{1}{3}, \frac{1}{3}, \frac{1}{3}, 0)$ plus permutations. (4 points),

S_4, consisting of $(\frac{1}{4}, \frac{1}{4}, \frac{1}{4}, 0)$ plus permutations. (1 point).

So, because of the replication, A's design can be thought of as $2S_1 + 2S_2 + 2S_3 + 2S_4$, and contains $2(4) + 2(6) + 2(4) + 2(1) = 30$ points. If B's design is written out, it is quickly seen to be $3S_1 + 2S_2 + S_3 + 2S_4$, and contains $3(4) + 2(6) + 1(4) + 2(1) = 30$ points, also. So the designs are actually almost identical, although conceived differently. One could argue that design B has too many single ingredient design points, with $3S_1$ and S_3, and that design A, with $2S_1$ and $2S_3$, is thus slightly preferable. One could also argue that, because design A is slightly more spread out, its D-criterion, that is, its value of $D = \{\det(\mathbf{X}'\mathbf{X})\}/n^p$, which is a general measure of the goodness of the design, will be slightly higher, and thus will give estimates with lower standard errors, for the usual polynomial models.

16.9. For the corner simplex, add all corresponding coordinates and divide by the number of points, q. For the other two points, add all corresponding coordinates and divide by 2. Both calculations give the same point, which has coordinates $\left(\dfrac{q+1}{2q}, \dfrac{1}{2q}, \dfrac{1}{2q}, \ldots, \dfrac{1}{2q}\right)$. So the answer is yes.

CHAPTER 17

17.1. The restricted region is an inverted triangle with coordinates $(1 - U_2 - U_3, U_2, U_3)$ $(U_1, 1 - U_1 - U_3, U_3)$ $(U_1, U_2, 1 - U_1 - U_2)$. Only the first point can have $x_1 = 0$, implying that $1 - U_2 = U_3$. The result follows immediately.

17.2. All points are of the form $(1 - U_2 - U_3 - \cdots - U_q, U_2, U_3, \ldots, U_q)$, plus similar sorts of permutations. If this first point has $x_1 = 0$, then we get $1 + U_1 = U$, by subtracting and adding back U_1 and rearranging. Similar results apply for the other U_i, for $2 \le U_i \le q$, so that all the U_i must be equal; but the coordinates of each point must also sum to 1, and the first coordinate of the first point above must be 0. This implies that $U_1 = U_2 = \cdots = U_q = \dfrac{1}{q-1}$.

17.3. The six extreme vertices are $(0.24, 0.25, 0.25, 0.26)$, $(0.24, 0.25, 0.26, 0.25)$, $(0.24, 0.26, 0.25, 0.25)$, $(0, 0.50, 0.25, 0.25)$, $(0, 0.25, 0.50, 0.25)$, $(0, 0.25, 0.25, 0.50)$. The first, second, and third of these are very close together, and it would usually not be sensible to include all three points in a design. As they stand, the six points form a simplex with one tip cut off. Note that if the x_1 restriction is re-specified as $0 \le x_1 \le 0.25$, the first three points move and coalesce into the single point $(0.25, 0.25, 0.25, 0.25)$, the centroid of the region, and the restricted region is now a simplex with four vertices. These four points generate various centroids as follows: six averages of pairs, giving $(0, 0.375, 0.375, 0.250)$, $(0, 0.375, 0.250, 0.375)$

$(0, 0.250, 0.375, 0.375)$, $(0.125, 0.375, 0.250, 0.250)$, $(0.125, 0.250, 0.375, 0.250)$, $(0.125, 0.250, 0.250, 0.375)$; four averages of triples, giving $(0, 0.333, 0.333, 0.333)$, $(0.083, 0.333, 0.333, 0.250)$, $(0.083, 0.333, 0.250, 0.333)$, $(0.083, 0.250, 0.333, 0.333)$; and one average of the four vertices at $(0.0625, 0.3125, 0.3125, 0.3125)$. This gives 15 design points in all, in the modified restricted region. If the original restriction on x_1 were unbreakable, we could simply substitute the average $(0.24, 0.253, 0.253, 0.253)$ of the three close points for the point $(0.25, 0.25, 0.25, 0.25)$ in the 15 point design. Various approximations of this sort often make sensible design choices.

17.4. (a) $(0.5, 0.2, 0.3)$, $(0.2, 0.5, 0.3)$, $(0.2, 0.2, 0.6)$.

(b) $z_1 - (x_1 - 0.2)/0.3$, $x_1 = 0.3z_1 + 0.2$; $z_2 = (x_2 - 0.2)/0.3$, $x_2 = 0.3z_2 + 0.2$; $z_3 = (x_3 - 0.3)/0.3$, $x_3 = 0.3z_3 + 0.3$.

(c) $(0.5, 0.2, 0.3)$, $(0.2, 0.5, 0.3)$, $(0.2, 0.2, 0.6)$; $(0.35, 0.35, 0.30)$, $(0.35, 0.20, 0.45)$, $(0.20, 0.35, 0.45)$. If a centroid were added, it would be at $(0.30, 0.30, 0.40)$.

17.5. Translating the data into pseudocomponent levels leads to Table E17.5a. Both second-order fits explain most of the variation in the data. For the y_1 data, the regression coefficients are $(413, 141, 1370, -366, -1640$ and $-2460)$; observation 3 is flagged as "too large" and observation 9 as "too small." A plot on triangular paper indicates that the corner observation 1630 (No. 3) is probably out of line. For the y_2 data, the regression coefficients are $(29.92, 13.47, 24.49, -36.06, -86.40$ and $-51.60)$; observation 1 appears "too large" and observation 11 appears "too small." A plot on triangular paper indicates that the corner observation 36.5 (No. 1) is probably the culprit.

Table E17.5a. Detergent Formulation Data in Pseudocomponent Form

No.	Water %	Alcohol %	Urea %	y_1	y_2
1	1	0	0	362.5	36.5
2	0	1	0	78	11.5
3	0	0	1	1630	23.7
4	0.5	0.5	0	165	12.9
5	0.5	0	0.5	537.5	5.9
6	0	3	3	202.5	5.2
7	0.25	0.75	0	96.5	11.4
8	0.25	0	0.75	560	15.8
9	0	0.25	0.75	310	7.8
10	0.75	0.25	0	230	14.1
11	0.75	0	0.25	455	3.5
12	0	0.75	0.25	108.5	8.5
13	0.5	0.25	0.25	257.5	7.3
14	0.25	0.5	0.25	167	8.9
15	0.25	0.25	0.5	275	3.2

CHAPTER 18

18.1. **(a)** Scheffé's second-order mixture model (no intercept)

Predictor (No Constant)	Coefficient	SE Coefficient	T	P
c1	-88.30	60.13	-1.47	0.216
c2	-113.94	22.48	-5.07	0.007
c3	$-30930.$	8204	-3.77	0.020
c4	-0.05794	0.03379	-1.72	0.161
c12	1.6122	0.5814	2.77	0.050
c13	300.75	82.57	3.64	0.022
c14	0.9444	0.6264	1.51	0.206
c23	310.47	82.50	3.76	0.020
c24	1.2090	0.2384	5.07	0.007
c34	309.97	82.17	3.77	0.020

$s = 0.4599.$

Analysis of Variance

Source	df	SS	MS	F	P
Regression	10	1124.33	112.43	531.56	0.000
Residual error	4	0.85	0.21		
Lack of fit	2	0.57	0.29	2.10	0.323
Pure error	2	0.27	0.14		
Total	14	1125.18			

(b) Component-amount model (a complete quadratic) with intercept term included. The regression equation is

$$C5 = -5.79 + 6.19\ C1 + 7.02\ C2 + 67.7\ C3 - 0.944\ C6 - 1.21\ C7$$
$$- 310\ C8 - 0.541\ C10 - 10.2\ C11 - 0.71\ C13$$

Predictor	Coefficient	SE Coefficient	T	P
Constant	-5.794	3.379	-1.72	0.161
a1	6.189	2.697	2.29	0.083
a2	7.023	1.499	4.69	0.009
a3	67.71	17.21	3.93	0.017
a11	-0.9444	0.6264	-1.51	0.206
a22	-1.2090	0.2384	-5.07	0.007
a33	-309.97	82.17	-3.77	0.020
a12	-0.5412	0.3998	-1.35	0.247
a13	-10.166	3.969	-2.56	0.063
a23	-0.713	2.474	-0.29	0.788

$s = 0.4599$, R-Sq $= 98.8\%$.

Analysis of Variance

Source	df	SS	MS	F	P
Regression	9	68.7374	7.6375	36.11	0.002
Residual error	4	0.8461	0.2115		
Lack of fit	2	0.5728	0.2864	2.10	0.323
Pure error	2	0.2733	0.1366		
Total	13	69.5835			

The fitted values are the same for both equations and are already given in the question; see the last column of Table E18.1.

This table can be fitted to the original data (as shown) or to data that have been "shifted" and "scaled" in some selected manner. Taking an analogy from pseudocomponents, for example, we could choose to transform using $u_i = \{a_i - \min(a_i)\}/\text{range}(a_i)$ for all i, so that the transformed variables range from 0 to 1. The estimated coefficients would then have a simple interpretation.

CHAPTER 19

19.1. Yes, it is true. However, the calculations show that both models fit enormously well and the large sounding "25% reduction" is of little practical importance in this context. Some details of the calculations follow. It is interesting to see how much some of the "corresponding" regression coefficients change, another warning of the possible dangers in interpreting individual coefficients in mixture situations.

Estimate subscripts:	1	2	3	4	12	13	14	23	34
E19.1a:	37.48	9.70	24.67	3.67	−17.35	−13.59	−38.68	−8.22	48.71
E19.1b:	49.72	8.41	29.95	4.34	−58.67	−27.83	−74.90	10.20	33.81

ANOVA:	df	SS E19.1a	MS	SS E19.1b	MS
Regression:	9	821.703		821.676	
Residual:	5	0.077	0.015	0.104	0.021
Total:	14	821.780		821.780	

After seeing these calculations, one wonders what would happen if the term $\alpha_{24} x_2 x_4/(x_2 + x_4 + h)$ were added to Eq. (E19.1a) to form Eq. (E19.1c). Note that although $\beta_{24} x_2 x_4$ *cannot* be added to the original Eq. (E19.1b), because of the linear dependency that arises, the ratio term *can* be added to Eq. (E19.1a) because no linear dependency occurs! Here are the results. There are bewildering changes in several of the corresponding previous coefficients and the additional term takes out roughly 1 df worth of the previous residual sum of squares.

Estimate subscripts:	1	2	3	4	12	13	14	23	34	24
E19.1c:	150.1	−103.0	24.7	3.67	−17.37	−33.58	−384.1	11.77	48.68	345.4

ANOVA:	df	SS E19.1c	MS
Regression:	10	821.714	
Residual:	4	0.066	0.016
Total:	14	821.780	

19.2. *In this special case*, the restriction becomes $R^2 = (\mathbf{x} - \mathbf{f})'(\mathbf{x} - \mathbf{f}) = \mathbf{x}'\mathbf{x} - 2\mathbf{f}'\mathbf{x} + \mathbf{f}'\mathbf{f} = \mathbf{x}'\mathbf{x} - 2/q + \frac{1}{q} = \mathbf{x}'\mathbf{x} - \frac{1}{q}$. Thus $\mathbf{x}'\mathbf{x} = R^2 + \frac{1}{q}$, essentially a redefinition of the radius value to be about the origin, even though it is outside the permissible mixture region. The physical meaning of this is that any sphere centered at the origin $(0, 0, \ldots, 0)$ eventually expands so that its intersection with the mixture space is a subsphere, in $q - 1$ dimensions, centered at the mixture space centroid. This is shown for $q = 2$ and 3 in Figure E19.2.

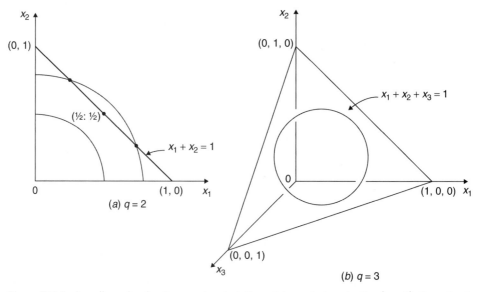

Figure E19.2. A q-dimensional sphere centered at the origin projects onto the $(q - 1)$-dimensional mixture space $x_1 + x_2 + \cdots + x_q = 1$ as a sphere of $q - 1$ dimensions; (*a*) For $q = 2$, a circle centered at the origin intersects the mixture space in two points equidistant from the centroid $(\frac{1}{2}, \frac{1}{2})$; (*b*) For $q = 3$, a sphere centered at the origin intersects the mixture space in a circle with center at the centroid $(\frac{1}{3}, \frac{1}{3}, \frac{1}{3})$.

Table E19.3. Ridge Path of Maximum Response

λ	x_1	x_2	x_3	R	$\hat{y}$
200	0.262	0.466	0.272	0.007	2.91
100	0.258	0.466	0.276	0.013	2.91
40	0.248	0.467	0.285	0.026	2.95
30	0.243	0.469	0.288	0.032	2.97
8	0.207	0.487	0.306	0.074	3.029
7	0.202	0.490	0.307	0.080	3.036
6.6	0.200	0.492	0.308	0.082	3.038

19.3. We have $q = 3$, $b_0 = 0$, $\mathbf{b}' = (-2.756, -3.352, -17.288)$,

$$\mathbf{B} = \begin{bmatrix} 0 & 4.69 & 17.38 \\ 4.69 & 0 & 24.745 \\ 17.38 & 24.745 & 0 \end{bmatrix},$$

$\mathbf{A} = \left(\dfrac{1}{3^{1/2}}, \dfrac{1}{3^{1/2}}, \dfrac{1}{3^{1/2}} \right) = (0.577350, 0.577350, 0.577350)$, $\quad \mathbf{c} = \dfrac{1}{3^{1/2}} = 0.577350$, and $\mathbf{f}' = (0.267, 0.467, 0.267)$. A suitable $\mathbf{T}$ is

$$\mathbf{T} = \begin{bmatrix} -\dfrac{1}{2^{1/2}} & 0 & \dfrac{1}{2^{1/2}} \\ \dfrac{1}{6^{1/2}} & -\dfrac{2}{6^{1/2}} & \dfrac{1}{6^{1/2}} \end{bmatrix} = \begin{bmatrix} -0.707107 & 0 & 0.707107 \\ 0.408248 & -0.816497 & 0.408248 \end{bmatrix}.$$

The rows of $\mathbf{T}$ are simply the normalized orthogonal polynomials of first and second-order; see Appendix 19A. The two eigenvalues of $\mathbf{TBT}'$ are $(-27.319, -3.891)$, so that for a maximum response ridge path, we investigate the range $-3.891 \le \lambda \le \infty$. Some selected calculations are shown in Table E19.3.

The path hits the border $x_1 = 0.2$ at the point $(0.200, 0.492, 0.308)$, where the predicted response is 3.038. We now have to move along this border to find the overall maximum subject to the restrictions. This can be done in a couple of ways.

1. Perform another ridge calculation using

$$\mathbf{A} = \begin{bmatrix} 0.577350 & 0.577350 & 0.577350 \\ 1 & 0 & 0 \end{bmatrix}, \quad \mathbf{c} = \begin{bmatrix} 0.577350 \\ 0.2 \end{bmatrix}.$$

A suitable $\mathbf{T}$ is then

$$\mathbf{T} = [0 \quad -0.707107 \quad 0.707107],$$

or the similar vector with signs reversed; or, more simply,

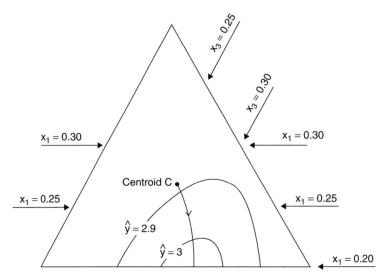

Figure E19.3. The first ridge path of maximum response runs from the restricted region centroid C, at $(0.267, 0.467, 0.267)$, to the lower boundary $x_1 = 0.20$.

2. Write $\hat{y}$ in a reduced form by setting $x_1 = 0.2$ and $x_3 = 0.8 - x_2$, which leads to the function $\hat{y} = -49.49x_2^2 + 48.452x_2 - 8.82$, which can then be maximized with respect to x_2.

Either method leads to the solution $(x_1, x_2, x_3) = (0.200, 0.490, 0.310)$, where $\hat{y} = 3.039$. So there is very little motion from where the first path hit the $x_2 = 0.2$ boundary. See Figure E19.3, derived from Figure 19.7, p. 419 of the Wiley book *Applied Regression Analysis* by N. R. Draper and H. Smith, 1998.

19.4. We have $q = 3$, $b_0 = 0$, $\mathbf{b}' = (-0.00658, -0.00243, 0.00367)$,

$$\mathbf{B} = \begin{bmatrix} 0 & 0.171325 & 0.23537 \\ 0.171325 & 0 & 0.070575 \\ 0.23537 & 0.070575 & 0 \end{bmatrix},$$

$$\mathbf{A} = \left(\frac{1}{3^{1/2}}, \frac{1}{3^{1/2}}, \frac{1}{3^{1/2}} \right) = (0.577350, 0.577350, 0.577350), \quad \mathbf{c} = \frac{1}{3^{1/2}} = 0.577350,$$ and $\mathbf{f}' = (0.333, 0.333, 0.333)$. A suitable $\mathbf{T}$ is

$$\mathbf{T} = \begin{bmatrix} -\dfrac{1}{2^{1/2}} & 0 & \dfrac{1}{-2^{1/2}} \\ \dfrac{1}{6^{1/2}} & -\dfrac{2}{6^{1/2}} & \dfrac{1}{6^{1/2}} \end{bmatrix}$$

$$= \begin{bmatrix} -0.707107 & 0 & 0.707107 \\ 0.408248 & -0.816497 & 0.408248 \end{bmatrix}.$$

Table E19.4. Ridge Path of Maximum Response

λ	x_1	x_2	x_3	R	$\hat{y}$
∞	0.333	0.333	0.333	0.000	0.104
10	0.336	0.331	0.333	0.004	0.105
3	0.342	0.325	0.333	0.012	0.105
1	0.355	0.310	0.334	0.032	0.107
0.5	0.371	0.291	0.338	0.057	0.107
0.1	0.422	0.201	0.377	0.165	0.113
0	0.484	0.016	0.499	0.388	0.116
-0.004	0.490	-0.004	0.514	0.413	0.116

The rows of **T** are simply the normalized orthogonal polynomials of first and second-order; see Appendix 19A. The two eigenvalues of **TBT**$'$ are $(-0.2550, -0.0632)$, so that for a maximum response ridge path, we investigate the range $-0.0632 \leq \lambda \leq \infty$. Some calculations are in Table E19.4. The path crosses the $x_2 = 0$ border at approximately $(0.49, 0.51)$. Moving along this border, using either of the methods described in the answer to the preceding exercise, gives a maximum $\hat{y} = 0.1163$ at about $(0.489, 0, 0.511)$.

19.5. For the first path, the whole mixture space is available. We have **A** $= (0.5, 0.5, 0.5, 0.5)$, **c** $= 0.5$. A suitable choice for **T** is the $q = 4$ suggestion from Appendix 19A, namely,

-0.670820	-0.223607	0.223607	0.670820
0.5	-0.5	-0.5	0.5
-0.223607	0.670820	-0.670820	0.223607

The eigenvalues of **TBT**$'$ are $(-132.473, 0, 8.623)$, so we search $\lambda \geq 8.623$. The path is given in Table E19.5a. As soon as the path hits the $x_4 = 0$ boundary, we need to follow the path *on* that boundary.

Table E19.5a. First Ridge Path of Maximum Response

λ	x_1	x_2	x_3	x_4	R	$\hat{y}$
∞	0	0.333	0.333	0.333	0	410
2000	0.027	0.333	0.320	0.321	0.033	414
1000	0.051	0.332	0.308	0.309	0.061	418
250	0.149	0.336	0.258	0.257	0.183	431
100	0.242	0.356	0.214	0.188	0.307	441
50	0.302	0.406	0.187	0.105	0.412	446
30	0.326	0.490	0.170	0.014	0.509	449
29	0.327	0.498	0.168	0.007	0.517	449
28.1	0.327	0.506	0.167	0.000	0.525	450

Table E19.5b. Second Ridge Path of Maximum Response

λ	x_1	x_2	x_3	x_4	R	$\hat{y}$
∞	0.327	0.506	0.167	0	0	450
500	0.336	0.510	0.153	0	0.017	450.0
200	0.344	0.519	0.137	0	0.037	450.3
50	0.346	0.582	0.072	0	0.123	451.4
30	0.330	0.661	0.009	0	0.222	452.7
28.6	0.327	0.673	0.000	0	0.236	452.9

We now have

$$\mathbf{A} = \begin{bmatrix} 0.5 & 0.5 & 0.5 & 0.5 \\ 0 & 0 & 0 & 1 \end{bmatrix}, \qquad \mathbf{c} = \begin{bmatrix} 0.5 \\ 0 \end{bmatrix}.$$

We find a $\mathbf{T}$ by noting that if we choose the fourth column of it as $(0, 0)'$, the other three columns can be chosen as the normalized orthogonal polynomials for $q = 3$. Thus, we select

$$\mathbf{T} = \begin{bmatrix} -0.707107 & 0 & 0.707107 & 0 \\ 0.408248 & -0.816497 & 0.408248 & 0 \end{bmatrix}.$$

The new $\mathbf{f} = (0.327, 0.506, 0.167, 0.000)'$, the previous stopping point. The calculations give the path in Table E19.5b. We could repeat this sort of procedure to find the best point on the $x_3 = x_4 = 0$ edge. We leave that as an exercise. A simple alternative is to reexpress the response in terms of only x_1 and x_2, substitute $x_2 = 1 - x_1$, and maximize the resulting function of x_1. This provides a solution $x_1 = (b_1 - b_2 + 2b_{12})/(4b_{12}) = 0.282$. So the best point is at $(0.282, 0.718, 0, 0)$, where $\hat{y} = 453.1$. If the result had been outside the range $0 \le x_1 \le 1$, the point of exit, 0 or 1, would have been the appropriate answer for x_1.

19.6. We have $b_0 = 9.925$, $\mathbf{b}' = (-4.10, -9.25, 4.90)$, $\mathbf{B} = \mathbf{0}$, and the initial $\mathbf{f}' = (0, 0, 0)$. Because $\mathbf{B} = \mathbf{0}$, the only eigenvalue is 0, and so the initial path of steepest ascent (maximum response) lies on the straight line given by setting $\mathbf{B} = \mathbf{0}$ in Eq. (12.2.8), that is, $\mathbf{x} = \mathbf{b}/(2\lambda)$, as $\lambda \ge 0$ varies. [Alternatively, see Eq. (19.6.1) and set $\mathbf{f} = \mathbf{0}$ and $\mathbf{A} = \mathbf{0}$.] The line hits the restrictive plane when its coordinates satisfy it, that is, when $b_1 + b_2 + 0.5b_3 = 2\lambda(-2.6)$. This implies that $\lambda = (-4.1 - 9.25 + 0.5 \times 4.90)/(-5.2) = -10.90/(-5.2) = 2.096$, whereupon the point of impact, which will be the focus on the redirected second path, is $\mathbf{x}_f = \mathbf{b}/(4.192) = (-0.978, -2.207, 1.169)$.

The second leg of the dog-leg steepest ascent path is on the restrictive plane $\mathbf{x}'\mathbf{b} = \mathbf{b}'\mathbf{x} = -2.6$. For convenience, we normalize this into the form $\mathbf{A}\mathbf{x} = \mathbf{c}$ by applying the divisor $(1^2 + 1^2 + 0.5^2)^{1/2} = 1.5$ throughout. This leads to $\mathbf{A} = (0.6667, 0.6667, 0.3333)$ and $\mathbf{c} = -1.7333$. From Eq. (19.6.1), using the fact that $\mathbf{A}\mathbf{A}' = 1$, we have that $\mathbf{x} = \mathbf{f} + (2\lambda)^{-1}(\mathbf{I} - \mathbf{A}'\mathbf{A})\mathbf{b}$, which reduces to $x_1 = -0.978 + 0.3722/\lambda$, $x_2 = -2.207 - 2.2028/\lambda$, $x_3 = 1.169$

$+ 3.6611/\lambda$. To achieve $x_3 = 2$, which is the third coordinate of the point given in the exercise, we need to have $\lambda = 4.40567$. Using this value leads to $x_1 = -0.894$, $x_2 = -2.707$, or, after rounding to two decimal places, $(x_1, x_2, x_3) = (-0.89, -2.71, 2)$.

19.7.   **(a)** The six-term full quadratic has $\mathbf{b}' = (19.404, 14.787, 35.916)$, and

$$\mathbf{B} = \begin{bmatrix} 0 & -4.569 & -17.4315 \\ -4.569 & 0 & 9.946 \\ -17.4315 & 9.946 & 0 \end{bmatrix}.$$

We also have $\mathbf{A} = (0.57735, 0.57735, 0.57735)$ and $\mathbf{c} = 0.57735$. We choose $\mathbf{T}$ from Appendix 19A, $q = 3$ to complement the standard mixture restriction of total amount 1. The lowest eigenvalue of $\mathbf{TBT}'$ is -11.798, so we search the λ-interval below this, starting from $-\infty$. As shown in the first table, the lowest response in the mixture region is predicted to be about 14.6 at about $(0.45, 0.55, 0)$.

λ	x_1	x_2	x_3	R	$\hat{y}$
$-\infty$	0.333	0.333	0.333	0	20.69
-200	0.363	0.334	0.303	0.043	19.93
-100	0.387	0.337	0.275	0.079	19.33
-50	0.423	0.353	0.225	0.142	18.39
-40	0.436	0.365	0.200	0.171	17.98
-30	0.450	0.394	0.156	0.221	17.31
-20	0.455	0.507	0.038	0.363	15.37
-19	0.452	0.536	0.012	0.398	14.85
-18.6	0.450	0.551	-0.001	0.416	14.58

(b) Only one regression coefficient in the full model, namely the term $b_{12} = -9.138$, is not significant at the 0.05 level. Dropping the corresponding term and re-fitting leads to $\mathbf{b}' = (18.496, 13.879, 36.060)$, and

$$\mathbf{B} = \begin{bmatrix} 0 & 0 & -17.594 \\ 0 & 0 & 9.784 \\ -17.594 & 9.784 & 0 \end{bmatrix}.$$

$\mathbf{A}$, $\mathbf{c}$, and $\mathbf{T}$ are the same as before. The lowest eigenvalue of the new $\mathbf{TBT}'$ is -13.416, so we search the λ-interval below this value, starting from $-\infty$. As shown in the first table, the lowest response in the mixture region is predicted to be about 14.58 at about $(0.45, 0.55, 0)$.

The second table indicates that the lowest response in the mixture region is predicted to be about 15.85 at about $(0.43, 0.57, 0)$. It is clear that dropping the only nonsignificant term from the first regression and then re-fitting the reduced quadratic does not materially affect the conclusion. In general, we do not recommend dropping terms in this way.

λ	x_1	x_2	x_3	R	$\hat{y}$
$-\infty$	0.333	0.333	0.333	0	20.69
-200	0.362	0.331	0.307	0.039	20.44
-100	0.386	0.331	0.283	0.072	19.94
-50	0.422	0.335	0.243	0.126	19.21
-30	0.452	0.350	0.196	0.183	18.54
-20	0.465	0.409	0.125	0.258	17.76
-17	0.449	0.497	0.054	0.344	16.82
-16.1	0.431	0.563	0.006	0.411	15.98
-16.01	0.428	0.572	0.000	0.421	15.85

For another analysis of these data, based on the use of models suggested by Becker (1968), see Rajagopal and Del Castillo (2005).

Tables

Table A. Tail Area of the Unit Normal Distribution

z	0.00	0.01	0.02	0.03	0.04	0.05	0.06	0.07	0.08	0.09
0.0	0.5000	0.4960	0.4920	0.4880	0.4840	0.4801	0.4761	0.4721	0.4681	0.4641
0.1	0.4602	0.4562	0.4522	0.4483	0.4443	0.4404	0.4364	0.4325	0.4286	0.4247
0.2	0.4207	0.4168	0.4129	0.4090	0.4052	0.4013	0.3974	0.3936	0.3897	0.3859
0.3	0.3821	0.3783	0.3745	0.3707	0.3669	0.3632	0.3594	0.3557	0.3520	0.3483
0.4	0.3446	0.3409	0.3372	0.3336	0.3300	0.3264	0.3228	0.3192	0.3156	0.3121
0.5	0.3085	0.3050	0.3015	0.2981	0.2946	0.2912	0.2877	0.2843	0.2810	0.2776
0.6	0.2743	0.2709	0.2676	0.2643	0.2611	0.2578	0.2546	0.2514	0.2483	0.2451
0.7	0.2420	0.2389	0.2358	0.2327	0.2296	0.2266	0.2236	0.2206	0.2177	0.2148
0.8	0.2119	0.2090	0.2061	0.2033	0.2005	0.1977	0.1949	0.1922	0.1894	0.1867
0.9	0.1841	0.1814	0.1788	0.1762	0.1736	0.1711	0.1685	0.1660	0.1635	0.1611
1.0	0.1587	0.1562	0.1539	0.1515	0.1492	0.1469	0.1446	0.1423	0.1401	0.1379
1.1	0.1357	0.1335	0.1314	0.1292	0.1271	0.1251	0.1230	0.1210	0.1190	0.1170
1.2	0.1151	0.1131	0.1112	0.1093	0.1075	0.1056	0.1038	0.1020	0.1003	0.0985
1.3	0.0968	0.0951	0.0934	0.0918	0.0901	0.0885	0.0869	0.0853	0.0838	0.0823
1.4	0.0808	0.0793	0.0778	0.0764	0.0749	0.0735	0.0721	0.0708	0.0694	0.0681
1.5	0.0668	0.0655	0.0643	0.0630	0.0618	0.0606	0.0594	0.0582	0.0571	0.0559
1.6	0.0548	0.0537	0.0526	0.0516	0.0505	0.0495	0.0485	0.0475	0.0465	0.0455
1.7	0.0446	0.0436	0.0427	0.0418	0.0409	0.0401	0.0392	0.0384	0.0375	0.0367
1.8	0.0359	0.0351	0.0344	0.0336	0.0329	0.0322	0.0314	0.0307	0.0301	0.0294
1.9	0.0287	0.0281	0.0274	0.0268	0.0262	0.0256	0.0250	0.0244	0.0239	0.0233
2.0	0.0228	0.0222	0.0217	0.0212	0.0207	0.0202	0.0197	0.0192	0.0188	0.0183
2.1	0.0179	0.0174	0.0170	0.0166	0.0162	0.0158	0.0154	0.0150	0.0146	0.0143
2.2	0.0139	0.0136	0.0132	0.0129	0.0125	0.0122	0.0119	0.0116	0.0113	0.0110
2.3	0.0107	0.0104	0.0102	0.0099	0.0096	0.0094	0.0091	0.0089	0.0087	0.0084
2.4	0.0082	0.0080	0.0078	0.0075	0.0073	0.0071	0.0069	0.0068	0.0066	0.0064
2.5	0.0062	0.0060	0.0059	0.0057	0.0055	0.0054	0.0052	0.0051	0.0049	0.0048
2.6	0.0047	0.0045	0.0044	0.0043	0.0041	0.0040	0.0039	0.0038	0.0037	0.0036
2.7	0.0035	0.0034	0.0033	0.0032	0.0031	0.0030	0.0029	0.0028	0.0027	0.0026
2.8	0.0026	0.0025	0.0024	0.0023	0.0023	0.0022	0.0021	0.0021	0.0020	0.0019
2.9	0.0019	0.0018	0.0018	0.0017	0.0016	0.0016	0.0015	0.0015	0.0014	0.0014
3.0	0.0013	0.0013	0.0013	0.0012	0.0012	0.0011	0.0011	0.0011	0.0010	0.0010
3.1	0.0010	0.0009	0.0009	0.0009	0.0008	0.0008	0.0008	0.0008	0.0007	0.0007
3.2	0.0007	0.0007	0.0006	0.0006	0.0006	0.0006	0.0006	0.0005	0.0005	0.0005
3.3	0.0005	0.0005	0.0005	0.0004	0.0004	0.0004	0.0004	0.0004	0.0004	0.0003
3.4	0.0003	0.0003	0.0003	0.0003	0.0003	0.0003	0.0003	0.0003	0.0003	0.0002
3.5	0.0002	0.0002	0.0002	0.0002	0.0002	0.0002	0.0002	0.0002	0.0002	0.0002
3.6	0.0002	0.0002	0.0001	0.0001	0.0001	0.0001	0.0001	0.0001	0.0001	0.0001
3.7	0.0001	0.0001	0.0001	0.0001	0.0001	0.0001	0.0001	0.0001	0.0001	0.0001
3.8	0.0001	0.0001	0.0001	0.0001	0.0001	0.0001	0.0001	0.0001	0.0001	0.0001
3.9	0.0000	0.0000	0.0000	0.0000	0.0000	0.0000	0.0000	0.0000	0.0000	0.0000

Table B. Probability Points of the
t Distribution with ν
Degrees of Freedom

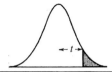

					Tail Area Probability					
ν	0.4	0.25	0.1	0.05	0.025	0.01	0.005	0.0025	0.001	0.0005
1	0.325	1.000	3.078	6.314	12.706	31.821	63.657	127.32	318.31	636.62
2	0.289	0.816	1.886	2.920	4.303	6.965	9.925	14.089	22.326	31.598
3	0.277	0.765	1.638	2.353	3.182	4.541	5.841	7.453	10.213	12.924
4	0.271	0.741	1.533	2.132	2.776	3.747	4.604	5.598	7.173	8.610
5	0.267	0.727	1.476	2.015	2.571	3.365	4.032	4.773	5.893	6.869
6	0.265	0.718	1.440	1.943	2.447	3.143	3.707	4.317	5.208	5.959
7	0.263	0.711	1.415	1.895	2.365	2.998	3.499	4.029	4.785	5.408
8	0.262	0.706	1.397	1.860	2.306	2.896	3.355	3.833	4.501	5.041
9	0.261	0.703	1.383	1.833	2.262	2.821	3.250	3.690	4.297	4.781
10	0.260	0.700	1.372	1.812	2.228	2.764	3.169	3.581	4.144	4.587
11	0.260	0.697	1.363	1.796	2.201	2.718	3.106	3.497	4.025	4.437
12	0.259	0.695	1.356	1.782	2.179	2.681	3.055	3.428	3.930	4.318
13	0.259	0.694	1.350	1.771	2.160	2.650	3.012	3.372	3.852	4.221
14	0.258	0.692	1.345	1.761	2.145	2.624	2.977	3.326	3.787	4.140
15	0.258	0.691	1.341	1.753	2.131	2.602	2.947	3.286	3.733	4.073
16	0.258	0.690	1.337	1.746	2.120	2.583	2.921	3.252	3.686	4.015
17	0.257	0.689	1.333	1.740	2.110	2.567	2.898	3.222	3.646	3.965
18	0.257	0.688	1.330	1.734	2.101	2.552	2.878	3.197	3.610	3.922
19	0.257	0.688	1.328	1.729	2.093	2.539	2.861	3.174	3.579	3.883
20	0.257	0.687	1.325	1.725	2.086	2.528	2.845	3.153	3.552	3.850
21	0.257	0.686	1.323	1.721	2.080	2.518	2.831	3.135	3.527	3.819
22	0.256	0.686	1.321	1.717	2.074	2.508	2.819	3.119	3.505	3.792
23	0.256	0.685	1.319	1.714	2.069	2.500	2.807	3.104	3.485	3.767
24	0.256	0.685	1.318	1.711	2.064	2.492	2.797	3.091	3.467	3.745
25	0.256	0.684	1.316	1.708	2.060	2.485	2.787	3.078	3.450	3.725
26	0.256	0.684	1.315	1.706	2.056	2.479	2.779	3.067	3.435	3.707
27	0.256	0.684	1.314	1.703	2.052	2.473	2.771	3.057	3.421	3.690
28	0.256	0.683	1.313	1.701	2.048	2.467	2.763	3.047	3.408	3.674
29	0.256	0.683	1.311	1.699	2.045	2.462	2.756	3.038	3.396	3.659
30	0.256	0.683	1.310	1.697	2.042	2.457	2.750	3.030	3.385	3.646
40	0.255	0.681	1.303	1.684	2.021	2.423	2.704	2.971	3.307	3.551
60	0.254	0.679	1.296	1.671	2.000	2.390	2.660	2.915	3.232	3.460
120	0.254	0.677	1.289	1.658	1.980	2.358	2.617	2.860	3.160	3.373
∞	0.253	0.674	1.282	1.645	1.960	2.326	2.576	2.807	3.090	3.291

Table C. Probability Points of the χ^2 Distribution with ν Degrees of Freedom

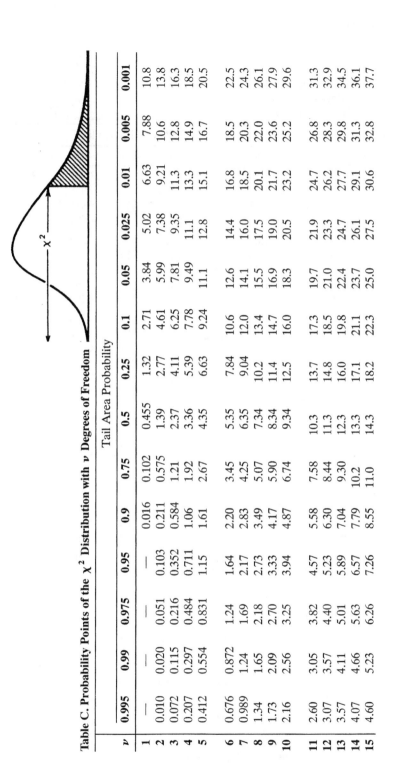

Tail Area Probability

ν	0.995	0.99	0.975	0.95	0.9	0.75	0.5	0.25	0.1	0.05	0.025	0.01	0.005	0.001
1	—	—	—	—	0.016	0.102	0.455	1.32	2.71	3.84	5.02	6.63	7.88	10.8
2	0.010	0.020	0.051	0.103	0.211	0.575	1.39	2.77	4.61	5.99	7.38	9.21	10.6	13.8
3	0.072	0.115	0.216	0.352	0.584	1.21	2.37	4.11	6.25	7.81	9.35	11.3	12.8	16.3
4	0.207	0.297	0.484	0.711	1.06	1.92	3.36	5.39	7.78	9.49	11.1	13.3	14.9	18.5
5	0.412	0.554	0.831	1.15	1.61	2.67	4.35	6.63	9.24	11.1	12.8	15.1	16.7	20.5
6	0.676	0.872	1.24	1.64	2.20	3.45	5.35	7.84	10.6	12.6	14.4	16.8	18.5	22.5
7	0.989	1.24	1.69	2.17	2.83	4.25	6.35	9.04	12.0	14.1	16.0	18.5	20.3	24.3
8	1.34	1.65	2.18	2.73	3.49	5.07	7.34	10.2	13.4	15.5	17.5	20.1	22.0	26.1
9	1.73	2.09	2.70	3.33	4.17	5.90	8.34	11.4	14.7	16.9	19.0	21.7	23.6	27.9
10	2.16	2.56	3.25	3.94	4.87	6.74	9.34	12.5	16.0	18.3	20.5	23.2	25.2	29.6
11	2.60	3.05	3.82	4.57	5.58	7.58	10.3	13.7	17.3	19.7	21.9	24.7	26.8	31.3
12	3.07	3.57	4.40	5.23	6.30	8.44	11.3	14.8	18.5	21.0	23.3	26.2	28.3	32.9
13	3.57	4.11	5.01	5.89	7.04	9.30	12.3	16.0	19.8	22.4	24.7	27.7	29.8	34.5
14	4.07	4.66	5.63	6.57	7.79	10.2	13.3	17.1	21.1	23.7	26.1	29.1	31.3	36.1
15	4.60	5.23	6.26	7.26	8.55	11.0	14.3	18.2	22.3	25.0	27.5	30.6	32.8	37.7

16	5.14	5.81	6.91	7.96	9.31	11.9	15.3	19.4	23.5	26.3	28.8	32.0	34.3	39.3
17	5.70	6.41	7.56	8.67	10.1	12.8	16.3	20.5	24.8	27.6	30.2	33.4	35.7	40.8
18	6.26	7.01	8.23	9.39	10.9	13.7	17.3	21.6	26.0	28.9	31.5	34.8	37.2	42.3
19	6.84	7.63	8.91	10.1	11.7	14.6	18.3	22.7	27.2	30.1	32.9	36.2	38.6	43.8
20	7.43	8.26	9.59	10.9	12.4	15.5	19.3	23.8	28.4	31.4	34.2	37.6	40.0	45.3
21	8.03	8.90	10.3	11.6	13.2	16.3	20.3	24.9	29.6	32.7	35.5	38.9	41.4	46.8
22	8.64	9.54	11.0	12.3	14.0	17.2	21.3	26.0	30.8	33.9	36.8	40.3	42.8	48.3
23	9.26	10.2	11.7	13.1	14.8	18.1	22.3	27.1	32.0	35.2	38.1	41.6	44.2	49.7
24	9.89	10.9	12.4	13.8	15.7	19.0	23.3	28.2	33.2	36.4	39.4	43.0	45.6	51.2
25	10.5	11.5	13.1	14.6	16.5	19.9	24.3	29.3	34.4	37.7	40.6	44.3	46.9	52.6
26	11.2	12.2	13.8	15.4	17.3	20.8	25.3	30.4	35.6	38.9	41.9	45.6	48.3	54.1
27	11.8	12.9	14.6	16.2	18.1	21.7	26.3	31.5	36.7	40.1	43.2	47.0	49.6	55.5
28	12.5	13.6	15.3	16.9	18.9	22.7	27.3	32.6	37.9	41.3	44.5	48.3	51.0	56.9
29	13.1	14.3	16.0	17.7	19.8	23.6	28.3	33.7	39.1	42.6	45.7	49.6	52.3	58.3
30	13.8	15.0	16.8	18.5	20.6	24.5	29.3	34.8	40.3	43.8	47.0	50.9	53.7	59.7

Source: Adapted from Table 8, Percentage points of the χ^2 distribution, in E. S. Pearson and H. O. Hartley (Eds.) (1966), *Biometrika Tables for Statisticians*, Vol. 1, 3rd ed., Cambridge University Press. Used by permission of the Biometrika Trustees.

Table D. Percentage Points of the F Distribution: Upper 25% Points

ν_2 \ ν_1	1	2	3	4	5	6	7	8	9	10	12	15	20	24	30	40	60	120	∞
1	5.83	7.50	8.20	8.58	8.82	8.98	9.10	9.19	9.26	9.32	9.41	9.49	9.58	9.63	9.67	9.71	9.76	9.80	9.85
2	2.57	3.00	3.15	3.23	3.28	3.31	3.34	3.35	3.37	3.38	3.39	3.41	3.43	3.43	3.44	3.45	3.46	3.47	3.48
3	2.02	2.28	2.36	2.39	2.41	2.42	2.43	2.44	2.44	2.44	2.45	2.46	2.46	2.46	2.47	2.47	2.47	2.47	2.47
4	1.81	2.00	2.05	2.06	2.07	2.08	2.08	2.08	2.08	2.08	2.08	2.08	2.08	2.08	2.08	2.08	2.08	2.08	2.08
5	1.69	1.85	1.88	1.89	1.89	1.89	1.89	1.89	1.89	1.89	1.89	1.89	1.88	1.88	1.88	1.88	1.87	1.87	1.87
6	1.62	1.76	1.78	1.79	1.79	1.78	1.78	1.78	1.77	1.77	1.77	1.76	1.76	1.75	1.75	1.75	1.74	1.74	1.74
7	1.57	1.70	1.72	1.72	1.71	1.71	1.70	1.70	1.69	1.69	1.68	1.68	1.67	1.67	1.66	1.66	1.65	1.65	1.65
8	1.54	1.66	1.67	1.66	1.66	1.65	1.64	1.64	1.63	1.63	1.62	1.62	1.61	1.60	1.60	1.59	1.59	1.58	1.58
9	1.51	1.62	1.63	1.63	1.62	1.61	1.60	1.60	1.59	1.59	1.58	1.57	1.56	1.56	1.55	1.54	1.54	1.53	1.53
10	1.49	1.60	1.60	1.59	1.59	1.58	1.57	1.56	1.56	1.55	1.54	1.53	1.52	1.52	1.51	1.51	1.50	1.49	1.48
11	1.47	1.58	1.58	1.57	1.56	1.55	1.54	1.53	1.53	1.52	1.51	1.50	1.49	1.49	1.48	1.47	1.47	1.46	1.45
12	1.46	1.56	1.56	1.55	1.54	1.53	1.52	1.51	1.51	1.50	1.49	1.48	1.47	1.46	1.45	1.45	1.44	1.43	1.42
13	1.45	1.55	1.55	1.53	1.52	1.51	1.50	1.49	1.49	1.48	1.47	1.46	1.45	1.44	1.43	1.42	1.42	1.41	1.40
14	1.44	1.53	1.53	1.52	1.51	1.50	1.49	1.48	1.47	1.46	1.45	1.44	1.43	1.42	1.41	1.41	1.40	1.39	1.38
15	1.43	1.52	1.52	1.51	1.49	1.48	1.47	1.46	1.46	1.45	1.44	1.43	1.41	1.41	1.40	1.39	1.38	1.37	1.36
16	1.42	1.51	1.51	1.50	1.48	1.47	1.46	1.45	1.44	1.44	1.43	1.41	1.40	1.39	1.38	1.37	1.36	1.35	1.34
17	1.42	1.51	1.50	1.49	1.47	1.46	1.45	1.44	1.43	1.43	1.41	1.40	1.39	1.38	1.37	1.36	1.35	1.34	1.33
18	1.41	1.50	1.49	1.48	1.46	1.45	1.44	1.43	1.42	1.42	1.40	1.39	1.38	1.37	1.36	1.35	1.34	1.33	1.32
19	1.41	1.49	1.49	1.47	1.46	1.44	1.43	1.42	1.41	1.41	1.40	1.38	1.37	1.36	1.35	1.34	1.33	1.32	1.30

20	1.40	1.49	1.48	1.47	1.45	1.44	1.43	1.42	1.41	1.40	1.39	1.37	1.36	1.35	1.34	1.33	1.32	1.31	1.29
21	1.40	1.48	1.48	1.46	1.44	1.43	1.42	1.41	1.40	1.39	1.38	1.37	1.35	1.34	1.33	1.32	1.31	1.30	1.28
22	1.40	1.48	1.47	1.45	1.44	1.42	1.41	1.40	1.39	1.39	1.37	1.36	1.34	1.33	1.32	1.31	1.30	1.29	1.28
23	1.39	1.47	1.47	1.45	1.43	1.42	1.41	1.40	1.39	1.38	1.37	1.35	1.34	1.33	1.32	1.31	1.30	1.28	1.27
24	1.39	1.47	1.46	1.44	1.43	1.41	1.40	1.39	1.38	1.38	1.36	1.35	1.33	1.32	1.31	1.30	1.29	1.28	1.26
25	1.39	1.47	1.46	1.44	1.42	1.41	1.40	1.39	1.38	1.37	1.36	1.34	1.33	1.32	1.31	1.29	1.28	1.27	1.25
26	1.38	1.46	1.45	1.44	1.42	1.41	1.39	1.38	1.37	1.37	1.35	1.34	1.32	1.31	1.30	1.29	1.28	1.26	1.25
27	1.38	1.46	1.45	1.43	1.42	1.40	1.39	1.38	1.37	1.36	1.35	1.33	1.32	1.31	1.30	1.28	1.27	1.26	1.24
28	1.38	1.46	1.45	1.43	1.41	1.40	1.39	1.38	1.37	1.36	1.34	1.33	1.31	1.30	1.29	1.28	1.27	1.25	1.24
29	1.38	1.45	1.45	1.43	1.41	1.40	1.38	1.37	1.36	1.35	1.34	1.32	1.31	1.30	1.29	1.27	1.26	1.25	1.23
30	1.38	1.45	1.44	1.42	1.41	1.39	1.38	1.37	1.36	1.35	1.34	1.32	1.30	1.29	1.28	1.27	1.26	1.24	1.23
40	1.36	1.44	1.42	1.40	1.39	1.37	1.36	1.35	1.34	1.33	1.31	1.30	1.28	1.26	1.25	1.24	1.22	1.21	1.19
60	1.35	1.42	1.41	1.38	1.37	1.35	1.33	1.32	1.31	1.30	1.29	1.27	1.25	1.24	1.22	1.21	1.19	1.17	1.15
120	1.34	1.40	1.39	1.37	1.35	1.33	1.31	1.30	1.29	1.28	1.26	1.24	1.22	1.21	1.19	1.18	1.16	1.13	1.10
∞	1.32	1.39	1.37	1.35	1.33	1.31	1.29	1.28	1.27	1.25	1.24	1.22	1.19	1.18	1.16	1.14	1.12	1.08	1.00

Source: M. Merrington and C. M. Thompson. Tables of percentage points of the inverted beta (*F*) distribution. *Biometrika*, **33**, 1943, 73. Used by permission of the Biometrika Trustees.

Table D (*Continued*). **Percentage Points of the F Distribution: Upper 10% Points**

v_2 \ v_1	1	2	3	4	5	6	7	8	9	10	12	15	20	24	30	40	60	120	∞
1	39.86	49.50	53.59	55.83	57.24	58.20	58.91	59.44	59.86	60.19	60.71	61.22	61.74	62.00	62.26	62.53	62.79	63.06	63.33
2	8.53	9.00	9.16	9.24	9.29	9.33	9.35	9.37	9.38	9.39	9.41	9.42	9.44	9.45	9.46	9.47	9.47	9.48	9.49
3	5.54	5.46	5.39	5.34	5.31	5.28	5.27	5.25	5.24	5.23	5.22	5.20	5.18	5.18	5.17	5.16	5.15	5.14	5.13
4	4.54	4.32	4.19	4.11	4.05	4.01	3.98	3.95	3.94	3.92	3.90	3.87	3.84	3.83	3.82	3.80	3.79	3.78	3.76
5	4.06	3.78	3.62	3.52	3.45	3.40	3.37	3.34	3.32	3.30	3.27	3.24	3.21	3.19	3.17	3.16	3.14	3.12	3.10
6	3.78	3.46	3.29	3.18	3.11	3.05	3.01	2.98	2.96	2.94	2.90	2.87	2.84	2.82	2.80	2.78	2.76	2.74	2.72
7	3.59	3.26	3.07	2.96	2.88	2.83	2.78	2.75	2.72	2.70	2.67	2.63	2.59	2.58	2.56	2.54	2.51	2.49	2.47
8	3.46	3.11	2.92	2.81	2.73	2.67	2.62	2.59	2.56	2.54	2.50	2.46	2.42	2.40	2.38	2.36	2.34	2.32	2.29
9	3.36	3.01	2.81	2.69	2.61	2.55	2.51	2.47	2.44	2.42	2.38	2.34	2.30	2.28	2.25	2.23	2.21	2.18	2.16
10	3.29	2.92	2.73	2.61	2.52	2.46	2.41	2.38	2.35	2.32	2.28	2.24	2.20	2.18	2.16	2.13	2.11	2.08	2.06
11	3.23	2.86	2.66	2.54	2.45	2.39	2.34	2.30	2.27	2.25	2.21	2.17	2.12	2.10	2.08	2.05	2.03	2.00	1.97
12	3.18	2.81	2.61	2.48	2.39	2.33	2.28	2.24	2.21	2.19	2.15	2.10	2.06	2.04	2.01	1.99	1.96	1.93	1.90
13	3.14	2.76	2.56	2.43	2.35	2.28	2.23	2.20	2.16	2.14	2.10	2.05	2.01	1.98	1.96	1.93	1.90	1.88	1.85
14	3.10	2.73	2.52	2.39	2.31	2.24	2.19	2.15	2.12	2.10	2.05	2.01	1.96	1.94	1.91	1.89	1.86	1.83	1.80
15	3.07	2.70	2.49	2.36	2.27	2.21	2.16	2.12	2.09	2.06	2.02	1.97	1.92	1.90	1.87	1.85	1.82	1.79	1.76
16	3.05	2.67	2.46	2.33	2.24	2.18	2.13	2.09	2.06	2.03	1.99	1.94	1.89	1.87	1.84	1.81	1.78	1.75	1.72
17	3.03	2.64	2.44	2.31	2.22	2.15	2.10	2.06	2.03	2.00	1.96	1.91	1.86	1.84	1.81	1.78	1.75	1.72	1.69
18	3.01	2.62	2.42	2.29	2.20	2.13	2.08	2.04	2.00	1.98	1.93	1.89	1.84	1.81	1.78	1.75	1.72	1.69	1.66
19	2.99	2.61	2.40	2.27	2.18	2.11	2.06	2.02	1.98	1.96	1.91	1.86	1.81	1.79	1.76	1.73	1.70	1.67	1.63

20	2.97	2.59	2.38	2.25	2.16	2.09	2.04	2.00	1.96	1.94	1.89	1.84	1.79	1.77	1.74	1.71	1.68	1.64	1.61
21	2.96	2.57	2.36	2.23	2.14	2.08	2.02	1.98	1.95	1.92	1.87	1.83	1.78	1.75	1.72	1.69	1.66	1.62	1.59
22	2.95	2.56	2.35	2.22	2.13	2.06	2.01	1.97	1.93	1.90	1.86	1.81	1.76	1.73	1.70	1.67	1.64	1.60	1.57
23	2.94	2.55	2.34	2.21	2.11	2.05	1.99	1.95	1.92	1.89	1.84	1.80	1.74	1.72	1.69	1.66	1.62	1.59	1.55
24	2.93	2.54	2.33	2.19	2.10	2.04	1.98	1.94	1.91	1.88	1.83	1.78	1.73	1.70	1.67	1.64	1.61	1.57	1.53
25	2.92	2.53	2.32	2.18	2.09	2.02	1.97	1.93	1.89	1.87	1.82	1.77	1.72	1.69	1.66	1.63	1.59	1.56	1.52
26	2.91	2.52	2.31	2.17	2.08	2.01	1.96	1.92	1.88	1.86	1.81	1.76	1.71	1.68	1.65	1.61	1.58	1.54	1.50
27	2.90	2.51	2.30	2.17	2.07	2.00	1.95	1.91	1.87	1.85	1.80	1.75	1.70	1.67	1.64	1.60	1.57	1.53	1.49
28	2.89	2.50	2.29	2.16	2.06	2.00	1.94	1.90	1.87	1.84	1.79	1.74	1.69	1.66	1.63	1.59	1.56	1.52	1.48
29	2.89	2.50	2.28	2.15	2.06	1.99	1.93	1.89	1.86	1.83	1.78	1.73	1.68	1.65	1.62	1.58	1.55	1.51	1.47
30	2.88	2.49	2.28	2.14	2.05	1.98	1.93	1.88	1.85	1.82	1.77	1.72	1.67	1.64	1.61	1.57	1.54	1.50	1.46
40	2.84	2.44	2.23	2.09	2.00	1.93	1.87	1.83	1.79	1.76	1.71	1.66	1.61	1.57	1.54	1.51	1.47	1.42	1.38
60	2.79	2.39	2.18	2.04	1.95	1.87	1.82	1.77	1.74	1.71	1.66	1.60	1.54	1.51	1.48	1.44	1.40	1.35	1.29
120	2.75	2.35	2.13	1.99	1.90	1.82	1.77	1.72	1.68	1.65	1.60	1.55	1.48	1.45	1.41	1.37	1.32	1.26	1.19
∞	2.71	2.30	2.08	1.94	1.85	1.77	1.72	1.67	1.63	1.60	1.55	1.49	1.42	1.38	1.34	1.30	1.24	1.17	1.00

Source: M. Merrington and C. M. Thompson. Tables of percentage points of the inverted beta (*F*) distribution. *Biometrika*, **33**, 1943, 73. Used by permission of the Biometrika Trustees.

Table D (*Continued*). **Percentage Points of the *F* Distribution: Upper 5% Points**

ν_2 \ ν_1	1	2	3	4	5	6	7	8	9	10	12	15	20	24	30	40	60	120	∞
1	161.4	199.5	215.7	224.6	230.2	234.0	236.8	238.9	240.5	241.9	243.9	245.9	248.0	249.1	250.1	251.1	252.2	253.3	254.3
2	18.51	19.00	19.16	19.25	19.30	19.33	19.35	19.37	19.38	19.40	19.41	19.43	19.45	19.45	19.46	19.47	19.48	19.49	19.50
3	10.13	9.55	9.28	9.12	9.01	8.94	8.89	8.85	8.81	8.79	8.74	8.70	8.66	8.64	8.62	8.59	8.57	8.55	8.53
4	7.71	6.94	6.59	6.39	6.26	6.16	6.09	6.04	6.00	5.96	5.91	5.86	5.80	5.77	5.75	5.72	5.69	5.66	5.63
5	6.61	5.79	5.41	5.19	5.05	4.95	4.88	4.82	4.77	4.74	4.68	4.62	4.56	4.53	4.50	4.46	4.43	4.40	4.36
6	5.99	5.14	4.76	4.53	4.39	4.28	4.21	4.15	4.10	4.06	4.00	3.94	3.87	3.84	3.81	3.77	3.74	3.70	3.67
7	5.59	4.74	4.35	4.12	3.97	3.87	3.79	3.73	3.68	3.64	3.57	3.51	3.44	3.41	3.38	3.34	3.30	3.27	3.23
8	5.32	4.46	4.07	3.84	3.69	3.58	3.50	3.44	3.39	3.35	3.28	3.22	3.15	3.12	3.08	3.04	3.01	2.97	2.93
9	5.12	4.26	3.86	3.63	3.48	3.37	3.29	3.23	3.18	3.14	3.07	3.01	2.94	2.90	2.86	2.83	2.79	2.75	2.71
10	4.96	4.10	3.71	3.48	3.33	3.22	3.14	3.07	3.02	2.98	2.91	2.85	2.77	2.74	2.70	2.66	2.62	2.58	2.54
11	4.84	3.98	3.59	3.36	3.20	3.09	3.01	2.95	2.90	2.85	2.79	2.72	2.65	2.61	2.57	2.53	2.49	2.45	2.40
12	4.75	3.89	3.49	3.26	3.11	3.00	2.91	2.85	2.80	2.75	2.69	2.62	2.54	2.51	2.47	2.43	2.38	2.34	2.30
13	4.67	3.81	3.41	3.18	3.03	2.92	2.83	2.77	2.71	2.67	2.60	2.53	2.46	2.42	2.38	2.34	2.30	2.25	2.21
14	4.60	3.74	3.34	3.11	2.96	2.85	2.76	2.70	2.65	2.60	2.53	2.46	2.39	2.35	2.31	2.27	2.22	2.18	2.13
15	4.54	3.68	3.29	3.06	2.90	2.79	2.71	2.64	2.59	2.54	2.48	2.40	2.33	2.29	2.25	2.20	2.16	2.11	2.07
16	4.49	3.63	3.24	3.01	2.85	2.74	2.66	2.59	2.54	2.49	2.42	2.35	2.28	2.24	2.19	2.15	2.11	2.06	2.01
17	4.45	3.59	3.20	2.96	2.81	2.70	2.61	2.55	2.49	2.45	2.38	2.31	2.23	2.19	2.15	2.10	2.06	2.01	1.96
18	4.41	3.55	3.16	2.93	2.77	2.66	2.58	2.51	2.46	2.41	2.34	2.27	2.19	2.15	2.11	2.06	2.02	1.97	1.92
19	4.38	3.52	3.13	2.90	2.74	2.63	2.54	2.48	2.42	2.38	2.31	2.23	2.16	2.11	2.07	2.03	1.98	1.93	1.88

20	4.35	3.49	3.10	2.87	2.71	2.60	2.51	2.45	2.39	2.35	2.28	2.20	2.12	2.08	2.04	1.99	1.95	1.90	1.84
21	4.32	3.47	3.07	2.84	2.68	2.57	2.49	2.42	2.37	2.32	2.25	2.18	2.10	2.05	2.01	1.96	1.92	1.87	1.81
22	4.30	3.44	3.05	2.82	2.66	2.55	2.46	2.40	2.34	2.30	2.23	2.15	2.07	2.03	1.98	1.94	1.89	1.84	1.78
23	4.28	3.42	3.03	2.80	2.64	2.53	2.44	2.37	2.32	2.27	2.20	2.13	2.05	2.01	1.96	1.91	1.86	1.81	1.76
24	4.26	3.40	3.01	2.78	2.62	2.51	2.42	2.36	2.30	2.25	2.18	2.11	2.03	1.98	1.94	1.89	1.84	1.79	1.73
25	4.24	3.39	2.99	2.76	2.60	2.49	2.40	2.34	2.28	2.24	2.16	2.09	2.01	1.96	1.92	1.87	1.82	1.77	1.71
26	4.23	3.37	2.98	2.74	2.59	2.47	2.39	2.32	2.27	2.22	2.15	2.07	1.99	1.95	1.90	1.85	1.80	1.75	1.69
27	4.21	3.35	2.96	2.73	2.57	2.46	2.37	2.31	2.25	2.20	2.13	2.06	1.97	1.93	1.88	1.84	1.79	1.73	1.67
28	4.20	3.34	2.95	2.71	2.56	2.45	2.36	2.29	2.24	2.19	2.12	2.04	1.96	1.91	1.87	1.82	1.77	1.71	1.65
29	4.18	3.33	2.93	2.70	2.55	2.43	2.35	2.28	2.22	2.18	2.10	2.03	1.94	1.90	1.85	1.81	1.75	1.70	1.64
30	4.17	3.32	2.92	2.69	2.53	2.42	2.33	2.27	2.21	2.16	2.09	2.01	1.93	1.89	1.84	1.79	1.74	1.68	1.62
40	4.08	3.23	2.84	2.61	2.45	2.34	2.25	2.18	2.12	2.08	2.00	1.92	1.84	1.79	1.74	1.69	1.64	1.58	1.51
60	4.00	3.15	2.76	2.53	2.37	2.25	2.17	2.10	2.04	1.99	1.92	1.84	1.75	1.70	1.65	1.59	1.53	1.47	1.39
120	3.92	3.07	2.68	2.45	2.29	2.17	2.09	2.02	1.96	1.91	1.83	1.75	1.66	1.61	1.55	1.50	1.43	1.35	1.25
∞	3.84	3.00	2.60	2.37	2.21	2.10	2.01	1.94	1.88	1.83	1.75	1.67	1.57	1.52	1.46	1.39	1.32	1.22	1.00

Source: M. Merrington and C. M. Thompson, Tables of percentage points of the inverted beta (F) distribution. *Biometrika*, **33**, 1943, 73. Used by permission of the Biometrika Trustees.

Table D (*Continued*). **Percentage Points of the F Distribution: Upper 1% Points**

ν_2 \ ν_1	1	2	3	4	5	6	7	8	9	10	12	15	20	24	30	40	60	120	∞
1	4052	4999.50	5403	5625	5764	5859	5928	5982	6022	6056	6106	6157	6209	6235	6261	6287	6313	6339	6366
2	98.50	99.00	99.17	99.25	99.30	99.33	99.36	99.37	99.39	99.40	99.42	99.43	99.45	99.46	99.47	99.47	99.48	99.49	99.50
3	34.12	30.82	29.46	28.71	28.24	27.91	27.67	27.49	27.35	27.23	27.05	26.87	26.69	26.60	26.50	26.41	26.32	26.22	26.13
4	21.20	18.00	16.69	15.98	15.52	15.21	14.98	14.80	14.66	14.55	14.37	14.20	14.02	13.93	13.84	13.75	13.65	13.56	13.46
5	16.26	13.27	12.06	11.39	10.97	10.67	10.46	10.29	10.16	10.05	9.89	9.72	9.55	9.47	9.38	9.29	9.20	9.11	9.02
6	13.75	10.92	9.78	9.15	8.75	8.47	8.26	8.10	7.98	7.87	7.72	7.56	7.40	7.31	7.23	7.14	7.06	6.97	6.88
7	12.25	9.55	8.45	7.85	7.46	7.19	6.99	6.84	6.72	6.62	6.47	6.31	6.16	6.07	5.99	5.91	5.82	5.74	5.65
8	11.26	8.65	7.59	7.01	6.63	6.37	6.18	6.03	5.91	5.81	5.67	5.52	5.36	5.28	5.20	5.12	5.03	4.95	4.86
9	10.56	8.02	6.99	6.42	6.06	5.80	5.61	5.47	5.35	5.26	5.11	4.96	4.81	4.73	4.65	4.57	4.48	4.40	4.31
10	10.04	7.56	6.55	5.99	5.64	5.39	5.20	5.06	4.94	4.85	4.71	4.56	4.41	4.33	4.25	4.17	4.08	4.00	3.91
11	9.65	7.21	6.22	5.67	5.32	5.07	4.89	4.74	4.63	4.54	4.40	4.25	4.10	4.02	3.94	3.86	3.78	3.69	3.60
12	9.33	6.93	5.95	5.41	5.06	4.82	4.64	4.50	4.39	4.30	4.16	4.01	3.86	3.78	3.70	3.62	3.54	3.45	3.36
13	9.07	6.70	5.74	5.21	4.86	4.62	4.44	4.30	4.19	4.10	3.96	3.82	3.66	3.59	3.51	3.43	3.34	3.25	3.17
14	8.86	6.51	5.56	5.04	4.69	4.46	4.28	4.14	4.03	3.94	3.80	3.66	3.51	3.43	3.35	3.27	3.18	3.09	3.00
15	8.68	6.36	5.42	4.89	4.56	4.32	4.14	4.00	3.89	3.80	3.67	3.52	3.37	3.29	3.21	3.13	3.05	2.96	2.87
16	8.53	6.23	5.29	4.77	4.44	4.20	4.03	3.89	3.78	3.69	3.55	3.41	3.26	3.18	3.10	3.02	2.93	2.84	2.75
17	8.40	6.11	5.18	4.67	4.34	4.10	3.93	3.79	3.68	3.59	3.46	3.31	3.16	3.08	3.00	2.92	2.83	2.75	2.65
18	8.29	6.01	5.09	4.58	4.25	4.01	3.84	3.71	3.60	3.51	3.37	3.23	3.08	3.00	2.92	2.84	2.75	2.66	2.57
19	8.18	5.93	5.01	4.50	4.17	3.94	3.77	3.63	3.52	3.43	3.30	3.15	3.00	2.92	2.84	2.76	2.67	2.58	2.49
20	8.10	5.85	4.94	4.43	4.10	3.87	3.70	3.56	3.46	3.37	3.23	3.09	2.94	2.86	2.78	2.69	2.61	2.52	2.42
21	8.02	5.78	4.87	4.37	4.04	3.81	3.64	3.51	3.40	3.31	3.17	3.03	2.88	2.80	2.72	2.64	2.55	2.46	2.36
22	7.95	5.72	4.82	4.31	3.99	3.76	3.59	3.45	3.35	3.26	3.12	2.98	2.83	2.75	2.67	2.58	2.50	2.40	2.31
23	7.88	5.66	4.76	4.26	3.94	3.71	3.54	3.41	3.30	3.21	3.07	2.93	2.78	2.70	2.62	2.54	2.45	2.35	2.26
24	7.82	5.61	4.72	4.22	3.90	3.67	3.50	3.36	3.26	3.17	3.03	2.89	2.74	2.66	2.58	2.49	2.40	2.31	2.21
25	7.77	5.57	4.68	4.18	3.85	3.63	3.46	3.32	3.22	3.13	2.99	2.85	2.70	2.62	2.54	2.45	2.36	2.27	2.17
26	7.72	5.53	4.64	4.14	3.82	3.59	3.42	3.29	3.18	3.09	2.96	2.81	2.66	2.58	2.50	2.42	2.33	2.23	2.13
27	7.68	5.49	4.60	4.11	3.78	3.56	3.39	3.26	3.15	3.06	2.93	2.78	2.63	2.55	2.47	2.38	2.29	2.20	2.10
28	7.64	5.45	4.57	4.07	3.75	3.53	3.36	3.23	3.12	3.03	2.90	2.75	2.60	2.52	2.44	2.35	2.26	2.17	2.06
29	7.60	5.42	4.54	4.04	3.73	3.50	3.33	3.20	3.09	3.00	2.87	2.73	2.57	2.49	2.41	2.33	2.23	2.14	2.03
30	7.56	5.39	4.51	4.02	3.70	3.47	3.30	3.17	3.07	2.98	2.84	2.70	2.55	2.47	2.39	2.30	2.21	2.11	2.01
40	7.31	5.18	4.31	3.83	3.51	3.29	3.12	2.99	2.89	2.80	2.66	2.52	2.37	2.29	2.20	2.11	2.02	1.92	1.80
60	7.08	4.98	4.13	3.65	3.34	3.12	2.95	2.82	2.72	2.63	2.50	2.35	2.20	2.12	2.03	1.94	1.84	1.73	1.60
120	6.85	4.79	3.95	3.48	3.17	2.96	2.79	2.66	2.56	2.47	2.34	2.19	2.03	1.95	1.86	1.76	1.66	1.53	1.38
∞	6.63	4.61	3.78	3.32	3.02	2.80	2.64	2.51	2.41	2.32	2.18	2.04	1.88	1.79	1.70	1.59	1.47	1.32	1.00

Source: M. Merrington and C. M. Thompson, Tables of percentage points of the inverted beta (*F*) distribution. *Biometrika*, **33**, 1943, 73. Used by permission of the Biometrika Trustees.

Table D (*Continued*). Percentage Points of the *F* Distribution: Upper 0.1% Points

v_2 \ v_1	1	2	3	4	5	6	7	8	9	10	12	15	20	24	30	40	60	120	∞
1	4053*	5000*	5404*	5625*	5764*	5859*	5929*	5981*	6023*	6056*	6107*	6158*	6209*	6235*	6261*	6287*	6313*	6340*	6366*
2	998.5	999.0	999.2	999.2	999.3	999.3	999.4	999.4	999.4	999.4	999.4	999.4	999.4	999.5	999.5	999.5	999.5	999.5	999.5
3	167.0	148.5	141.1	137.1	134.6	132.8	131.6	130.6	129.9	129.2	128.3	127.4	126.4	125.9	125.4	125.0	124.5	124.0	123.5
4	74.14	61.25	56.18	53.44	51.71	50.53	49.66	49.00	48.47	48.05	47.41	46.76	46.10	45.77	45.43	45.09	44.75	44.40	44.05
5	47.18	37.12	33.20	31.09	29.75	28.84	28.16	27.64	27.24	26.92	26.42	25.91	25.39	25.14	24.87	24.60	24.33	24.06	23.79
6	35.51	27.00	23.70	21.92	20.81	20.03	19.46	19.03	18.69	18.41	17.99	17.56	17.12	16.89	16.67	16.44	16.21	15.99	15.75
7	29.25	21.69	18.77	17.19	16.21	15.52	15.02	14.63	14.33	14.08	13.71	13.32	12.93	12.73	12.53	12.33	12.12	11.91	11.70
8	25.42	18.49	15.83	14.39	13.49	12.86	12.40	12.04	11.77	11.54	11.19	10.84	10.48	10.30	10.11	9.92	9.73	9.53	9.33
9	22.86	16.39	13.90	12.56	11.71	11.13	10.70	10.37	10.11	9.89	9.57	9.24	8.90	8.72	8.55	8.37	8.19	8.00	7.81
10	21.04	14.91	12.55	11.28	10.48	9.92	9.52	9.20	8.96	8.75	8.45	8.13	7.80	7.64	7.47	7.30	7.12	6.94	6.76
11	19.69	13.81	11.56	10.35	9.58	9.05	8.66	8.35	8.12	7.92	7.63	7.32	7.01	6.85	6.68	6.52	6.35	6.17	6.00
12	18.64	12.97	10.80	9.63	8.89	8.38	8.00	7.71	7.48	7.29	7.00	6.71	6.40	6.25	6.09	5.93	5.76	5.59	5.42
13	17.81	12.31	10.21	9.07	8.35	7.86	7.49	7.21	6.98	6.80	6.52	6.23	5.93	5.78	5.63	5.47	5.30	5.14	4.97
14	17.14	11.78	9.73	8.62	7.92	7.43	7.08	6.80	6.58	6.40	6.13	5.85	5.56	5.41	5.25	5.10	4.94	4.77	4.60
15	16.59	11.34	9.34	8.25	7.57	7.09	6.74	6.47	6.26	6.08	5.81	5.54	5.25	5.10	4.95	4.80	4.64	4.47	4.31
16	16.12	10.97	9.00	7.94	7.27	6.81	6.46	6.19	5.98	5.81	5.55	5.27	4.99	4.85	4.70	4.54	4.39	4.23	4.06
17	15.72	10.66	8.73	7.68	7.02	6.56	6.22	5.96	5.75	5.58	5.32	5.05	4.78	4.63	4.48	4.33	4.18	4.02	3.85
18	15.38	10.39	8.49	7.46	6.81	6.35	6.02	5.76	5.56	5.39	5.13	4.87	4.59	4.45	4.30	4.15	4.00	3.84	3.67
19	15.08	10.16	8.28	7.26	6.62	6.18	5.85	5.59	5.39	5.22	4.97	4.70	4.43	4.29	4.14	3.99	3.84	3.68	3.51
20	14.82	9.95	8.10	7.10	6.46	6.02	5.69	5.44	5.24	5.08	4.82	4.56	4.29	4.15	4.00	3.86	3.70	3.54	3.38
21	14.59	9.77	7.94	6.95	6.32	5.88	5.56	5.31	5.11	4.95	4.70	4.44	4.17	4.03	3.88	3.74	3.58	3.42	3.26
22	14.38	9.61	7.80	6.81	6.19	5.76	5.44	5.19	4.99	4.83	4.58	4.33	4.06	3.92	3.78	3.63	3.48	3.32	3.15
23	14.19	9.47	7.67	6.69	6.08	5.65	5.33	5.09	4.89	4.73	4.48	4.23	3.96	3.82	3.68	3.53	3.38	3.22	3.05
24	14.03	9.34	7.55	6.59	5.98	5.55	5.23	4.99	4.80	4.64	4.39	4.14	3.87	3.74	3.59	3.45	3.29	3.14	2.97
25	13.88	9.22	7.45	6.49	5.88	5.46	5.15	4.91	4.71	4.56	4.31	4.06	3.79	3.66	3.52	3.37	3.22	3.06	2.89
26	13.74	9.12	7.36	6.41	5.80	5.38	5.07	4.83	4.64	4.48	4.24	3.99	3.72	3.59	3.44	3.30	3.15	2.99	2.82
27	13.61	9.02	7.27	6.33	5.73	5.31	5.00	4.76	4.57	4.41	4.17	3.92	3.66	3.52	3.38	3.23	3.08	2.92	2.75
28	13.50	8.93	7.19	6.25	5.66	5.24	4.93	4.69	4.50	4.35	4.11	3.86	3.60	3.46	3.32	3.18	3.02	2.86	2.69
29	13.39	8.85	7.12	6.19	5.59	5.18	4.87	4.64	4.45	4.29	4.05	3.80	3.54	3.41	3.27	3.12	2.97	2.81	2.64
30	13.29	8.77	7.05	6.12	5.53	5.12	4.82	4.58	4.39	4.24	4.00	3.75	3.49	3.36	3.22	3.07	2.92	2.76	2.59
40	12.61	8.25	6.60	5.70	5.13	4.73	4.44	4.21	4.02	3.87	3.64	3.40	3.15	3.01	2.87	2.73	2.57	2.41	2.23
60	11.97	7.76	6.17	5.31	4.76	4.37	4.09	3.87	3.69	3.54	3.31	3.08	2.83	2.69	2.55	2.41	2.25	2.08	1.89
120	11.38	7.32	5.79	4.95	4.42	4.04	3.77	3.55	3.38	3.24	3.02	2.78	2.53	2.40	2.26	2.11	1.95	1.76	1.54
∞	10.83	6.91	5.42	4.62	4.10	3.74	3.47	3.27	3.10	2.96	2.74	2.51	2.27	2.13	1.99	1.84	1.66	1.45	1.00

*Multiply these entries by 100.

Source: Reproduced with permission of the Biometrika Trustees from *Biometrika Tables for Statisticians*, eds. E. S. Pearson and H. O. Hartley, Cambridge University Press, 2nd ed., 1958, and also from *Statistical Tables for Biological*, *Agricultural and Medical Research* by R. A. Fisher and F. Yates, Longman Group, London, 5th ed., 1957, with permission of the publishers.

Bibliography

Aboukalam, M. A. F., and Al-Shiha, A. A. (2001). A robust analysis for unreplicated factorial experiments. *Computational Statistics and Data Analysis*, **36**, 31–46.

Abraham, G., Chipman, H., and Vijayan, K. (1999). Some risks in the construction and analysis of supersaturated designs. *Technometrics*, **41**(2), 135–141.

Abraham, T. P., and Rao, V. Y. (1966). An investigation on functional models for fertilizer-response surfaces. *Journal of the Indian Society for Agricultural Statistics*, **18**, 45–61.

Achor, I. M., Richardson, T., and Draper, N. R. (1981). Effect of treating *Candida utilis* with acid or alkali, to remove nucleic acids, on the quality of the protein. *Agriculture and Food Chemistry*, **29**, 27–33.

Addelman, S. (1961). Irregular fractions of the 2^n factorial experiments. *Technometrics*, **3**(4), 479–496.

Addelman, S. (1962a). Orthogonal main-effect plans for asymmetrical factorial experiments. *Technometrics*, **4**(1), 21–46.

Addelman, S. (1962b). Symmetrical and asymmetrical fractional factorial plans. *Technometrics*, **4**(1), 47–57.

Addelman, S. (1962c). Augmenting factorial plans to accommodate additional two-level factors. *Biometrics*, **18**(3), 308–322.

Addelman, S. (1963). Techniques for constructing fractional replicate plans. *Journal of the American Statistical Association*, **58**, 45–71.

Addelman, S. (1964a). Some two-level factorial plans with split-plot confounding. *Technometrics*, **6**(3), 253–258.

Addelman, S. (1964b). Designs for the sequential application of factors. *Technometrics*, **6**(4), 365–370.

Addelman, S. (1965). The construction of a 2^{17-9} resolution V plan in eight blocks of 32. *Technometrics*, **7**(3), 439–443.

Addelman, S. (1969). Sequences of two-level fractional factorial plans. *Technometrics*, **11**(3), 477–509.

Adhikary, B. (1965). On the properties and construction of balanced incomplete block designs with variable replicates. *Calcutta Statistical Association Bulletin*, **14**, 36–64.

Adhikary, B. (1966). Some types of *m*-associate PBIB association schemes. *Calcutta Statistical Association Bulletin*, **15**, 47–74.

Adhikary, B. (1967). Group divisible designs with variable replications. *Calcutta Statistical Association Bulletin*, **16**, 73–92.

Adhikary, B. (1972). A note on restricted Kronecker product method of constructing statistical designs. *Calcutta Statistical Association Bulletin*, **21**, 193–196.

Adhikary, B. (1973). On generalized group divisible designs. *Calcutta Statistical Association Bulletin*, **22**, 75–88.

Adhikary, B., and Panda, R. (1977). Restricted Kronecker product method of constructing second order rotatable designs. *Calcutta Statistical Association Bulletin*, **26**, 61–78.

Adhikary, B., and Panda, R. (1980). Construction of group divisible rotatable designs. *Lecture Notes in Mathematics*, **885**, 171–184.

Adhikary, B., and Panda, R. (1981). Construction of GDSORD from balanced block designs with variable replication. *Calcutta Statistical Association Bulletin*, **30**, 129–137.

Adhikary, B., and Panda, R. (1983). On group divisible response surface designs. *Journal of Statistical Planning and Inference*, **7**, 387–405.

Adhikary, B., and Panda, R. (1984). Group divisible third order rotatable designs (GDTORD). *Sankhya, Series B*, **46**, 135–146.

Adhikary, B., and Panda, R. (1985). Group divisible response surface (GDRS) designs of third order. *Calcutta Statistical Association Bulletin*, **34**, 75–87.

Adhikary, B., and Panda, R. (1986). Construction of nearly D-efficient third order rotatable designs using PBIB designs. *Proceedings of the Symposium on Optimization, Design of Experiments and Graph Theory*, I. I. T. Bombay, pp. 208–217.

Adhikary, B., and Panda, R. (1990). Sequential method of conducting rotatable designs. *Sankhya, Series B*, **52**, 212–218.

Adhikary, B., and Sinha, B. K. (1976). On group divisible rotatable designs. *Calcutta Statistical Association Bulletin*, **25**, 79–93.

Agarwal, S. C., and Das, M. N. (1987). A note on construction and application of balanced *n*-ary designs. *Sankhya, Series B*, **49**(2), 192–196.

Aggrawal, M. L., and Dey, A. (1983). Orthogonal resolution IV designs for some asymmetrical factors. *Technometrics*, **25**(2), 197–199.

Aggrawal, M. L., and Kaul, R. (1999). Combined array approach for optimal designs. *Communications in Statistics, Theory and Methods*, **28**(11), 2655–2670.

Aguirre-Torres, V., and Perez-Trejo, M. E. (2001). Outliers and the use of the rank transformation to detect active effects in unreplicated 2^f experiments. *Communications in Statistics, Simulation and Computation*, **30**(3), 637–663.

Aia, M. A., Goldsmith, R. L., and Mooney, R. W. (1961). Precipitating stoichiometric $CaHPO_4 \cdot 2H_2O$. *Industrial and Engineering Chemistry*, **53**, 55–57.

Aitchison, J. (1982). The statistical analysis of compositional data. *Journal of the Royal Statistical Society, Series B*, **44**, 139–160; discussion: pp. 161–177. **M**

Aitchison, J., and Bacon-Shone, J. (1984). Log contrast models for experiments with mixtures. *Biometrika*, **71**, 323–330. **M**

Aitchison, J., and Bacon-Shone, J. (1999). Convex linear combinations of compositions. *Biometrika*, **86**(2), 351–364. **M**

Aitchison, J., and Greenacre, M. (2002). Biplots of compositional data. *Applied Statistics*, **51**(4), 375–392. **M**

Alamprese, C., Foschino, R., Rossi, M., Pompei, C., and Savani, L. (2002). Survival of Lactobacillus johnsonii La1 and influence of its addition in retail-manufactured ice cream produced with different sugar and fat concentrations. *International Dairy Journal*, **12**, 201–208.

Albert, A. E. (1961). The sequential design of experiments for infinitely many states of nature. *Annals of Mathematical Statistics*, **32**, 774–799.

Alessi, J., and Power, J. F. (1977). Residual effects of nitrogen fertilization on dryland spring wheat in the Northern Plains. I. Wheat yield and water use. *Agronomy Journal*, **69**, 1007–1011.

Alessi, J., and Power, J. F. (1978). Residual effects of nitrogen fertilization on dryland spring wheat in the Northern Plains. II. Fate of fertilizer N. *Agronomy Journal*, **70**, 282–286.

Allen, D. J., Reimers, H. J., Fauerstein, I. A., and Mustard, J. F. (1975). The use and analysis of multiple responses in multicompartment cellular systems. *Biometrics*, **31**, 921–929.

Allen, T. T., and Bernshteyn, M. A. (2003). Supersaturated designs that maximize the probability of identifying active factors. *Technometrics*, **45**(1), 90–97.

Allen, T. T., and Yu, L. (2002). Low cost response surface methods from simulation optimization. *Quality and Reliability in Engineering International*, **18**, 5–7.

Allen, T. T., Bernshteyn, M. A., and Kabiri-Bamoradian, K. (2003). Constructing meta-models for computer experiments. *Journal of Quality Technology*, **35**(3), 264–274.

Allen, T. T., Yu, L., and Schmitz, J. (2003). An experimental design criterion for minimizing meta-model prediction errors applied to die casting process design. *Applied Statistics*, **52**(1), 103–117.

Altan, S., and Raghavarao, D. (1996). Nested Youden square designs. *Biometrika*, **83**, 242–245.

Ames, A. E., Mattucci, N., MacDonald, S., Szonyi, G., and Hawkins, D. M. (1997). Quality loss functions for optimization across multiple response surfaces. *Journal of Quality Technology*, **29**(3), 339–346.

Andere-Rendon, J., Montgomery, D. C., and Rollier, D. A. (1997). Design of mixture experiments using Bayesian D-optimality. *Journal of Quality Technology*, **29**(4), 451–463. **M**

Anderson, C. M., and Wu, C. F. J. (1995). Measuring location effects from factorial experiments with a directional response. *International Statistical Review*, **63**, 345–363.

Anderson, C. M., and Wu, C. F. J. (1996). Dispersion measures and analysis for factorial directional data with replicates. *Applied Statistics*, **45**(1), 47–61.

Anderson, M. J. (2003). Augmented ruggedness testing to prevent failures. *Quality Progress*, **36**(5), May, 38–45.

Anderson, R. L. (1953). Recent advances in finding best operating conditions. *Biometrics*, **48**, 789–798.

Anderson, R. L., and Nelson, L. (1975). A family of models involving intersecting straight lines and concomitant experimental designs useful in evaluating response to fertilizer nutrients. *Biometrics*, **31**(2), 303–318.

Anderson-Cook, C. M. (1998). Designing a first experiment: A project for design of experiment courses. *American Statistician*, **52**(4), 338–342.

Anderson-Cook, C. M. (2005). How to choose the appropriate design. *Quality Progress*, **38**(10) October, pp. 80–82.

Anderson-Cook, C. M., and Ozol-Godfrey, A. (2005). Using fraction of design space plots for informative comparisons between designs. In *Response Surface Methodology and Related Topics* (A. I. Khuri, ed.), 379–407. Hackensack, NJ: World Scientific Publishing.

Andrews, D. F., and Herzberg, A. M. (1979). The robustness and optimality of response surface designs. *Journal of Statistical Planning and Inference*, **3**, 249–258.

Anik, S. T., and Sukumar, L. (1981). Extreme vertexes design in formulation development: Solubility of butoconazole nitrate in a multicomponent system. *Journal of Pharmaceutical Sciences*, **70**(8), 897–900.

Anjaneyulu, G. V. S. R., Dattatreyarao, A. V., and Narasimham, V. L. (1998). Group-divisible second order slope rotatable designs. *Gujarat Statistical Review*, **25**, 3–8.

Anjaneyulu, G. V. S. R., Varma, D. N., and Narasimham, V. L. (1998). A note on second order slope rotatable designs over all directions. *Communications in Statistics, Theory and Methods*, **26**, 1477–1479.

Ankenman, B. E. (1999). Design of experiments with two- and four-level factors. *Journal of Quality Technology*, **31**(4), 363–375.

Ankenman, B. E., Liu, H., Karr, A. F., and Picka, J. D. (2002). A class of experimental designs for estimating a response surface and variance components. *Technometrics*, **44**(1), 45–54.

Annis, D. H. (2005). Rethinking the paper helicopter: Combining statistical and engineering knowledge. *American Statistician*, **59**(4), 320–326.

Anscombe, F. J. (1959). Quick analysis methods for random balance screening experiments. *Technometrics*, **1**(2), 195–209. See also Budne (1959) and Satterthwaite (1959).

Anscombe, F. J. (1961). Examination of residuals. *Proceedings of the Fourth Berkeley Symposium in Mathematical Statistics and Probability*, **1**, 1–36.

Anscombe, F. G., and Tukey, J. W. (1963). The examination and analysis of residuals. *Technometrics*, **5**(2), 141–160.

Antony, J. (2001). Simultaneous optimization of multiple quality characteristics in manufacturing processes using Taguchi's quality loss function. *International Journal of Advanced Manufacturing Technology*, **17**(2), 134–138.

Arap-Koske, J. K. (1987). A fourth order rotatable design in four dimensions. *Communications in Statistics, Theory and Methods*, **16**(9), 2747–2753.

Arap-Koske, J. K. (1989a). The variance function of the difference between two estimated fourth order response surfaces. *Journal of Statistical Planning and Inference*, **23**, 263–266.

Arap-Koske, J. K. (1989b). Rotatable design of order four in four dimensions. *International Journal of Mathematics Education in Science and Technology*, **20**(3), 482–493.

Arap-Koske, J. K. (1989c). On construction of a fourth order rotatable design with three factors. *International Journal of Mathematics Education in Science and Technology*, **20**(6), 809–814.

Arap-Koske, J. K. (1989d). A simpler way of obtaining non-singularity conditions of rotatablity. *Communications in Statistics, Theory and Methods*, **18**(7), 2489–2500.

Arap-Koske, J. K. (1990). Response surface designs with missing observations. *Biometry for Development*, **1**, 51–54.

Arap-Koske, J. K. (1992). A fourth order rotatable design in two dimensions. *East African Journal of Sciences, Series B*, **1**, 47–52.

Arap-Koske, J. K., and Patel, M. S. (1985). Conditions for fourth order rotatability in *k* dimensions. *Communications in Statistics, Theory and Methods*, **14**(6), 1343–1351.

Arap-Koske, J. K., and Patel, M. S. (1986). A fourth order rotatable design in three dimensions. *Communications in Statistics, Theory and Methods*, **15**, 3435–3444.

Arap-Koske, J. K., and Patel, M. S. (1987). Construction of fourth order rotatable designs with estimation of corresponding response surface. *Communications in Statistics, Theory and Methods*, **16**(5), 1361–1376.

Arap-Koske, J. K., and Patel, M. S. (1989). A simpler way of obtaining non-singularity conditions of rotatability. *Communications in Statistics, Theory and Methods*, **18**, 2489–2500.

Aravamuyhan, R., and Yayin, I. (1993). Application of response surface methodology for the maximization of Concora crush resistance of paperboard. *Quality Engineering*, **6**(1), 1–20.

Arbogast, P. G., and Bedrick, E. J. (2004). Model–checking techniques for linear models with parametric variance functions. *Technometrics*, **46**(4), 404–410.

Armbrust, D. V. (1968). Windblown soil abrasive injury to cotton plants. *Agronomy Journal*, **60**, 622–625.

Arthanari, T. S. (2005). A game theory application in robust design. *Quality Engineering*, **17**(2), 291–300.

Ash, A., and Hedayat, A. (1978). An introduction to design optimality with an overview of the literature. *Communications in Statistics, Theory and Methods*, **7**, 1295–1325.

Atkinson, A. C. (1969a). A test for discriminating between models. *Biometrika*, **56**, 337–347.

Atkinson, A. C. (1969b). Constrained maximization and the design of experiments. *Technometrics*, **11**(3), 616–618.

Atkinson, A. C. (1970a). A method of discriminating between models. *Journal of the Royal Statistical Society, Series B*, **32**, 323–345, discussion: pp. 345–353.

Atkinson, A. C. (1970b). The design of experiments to estimate the slope of a response surface. *Biometrika*, **57**, 319–328.

Atkinson, A. C. (1972). Planning experiments to detect inadequate regression models. *Biometrika*, **59**, 275–293.

Atkinson, A. C. (1973a). Multifactor second order designs for cuboidal regions. *Biometrika*, **60**, 15–19.

Atkinson, A. C. (1973b). Testing transformations to normality. *Journal of the Royal Statistical Society, Series B*, **35**, 473–479.

Atkinson, A. C. (1975). Planning experiments for model testing and discrimination. *Mathematik Operationsforchung und Statistik*, **6**, 253–267.

Atkinson, A. C. (1981). A comparison of two criteria for the design of experiments for discriminating between models. *Technometrics*, **23**(3), 301–305.

Atkinson, A. C. (1982). Developments in the design of experiment. *International Statistical Review*, **50**, 161–177.

Atkinson, A. C. (1985). *Plots, Transformations and Regression*, Oxford: Oxford University Press.

Atkinson, A. C. (1996). The usefulness of optimum experimental designs. *Journal of the Royal Statistical Society, Series B*, **58**(1), 59–76; discussion: pp. 95–111.

Atkinson, A. C. (2003). Horwitz's rule, transforming both sides and the design of experiments for mechanistic models. *Applied Statistics*, **52**(3), 261–278.

Atkinson, A. C. (2005). Robust optimum designs for transformation of the responses in a multivariate chemical kinetic model. *Technometrics*, **47**(4), 478–487.

Atkinson, A. C. (2006). Generalized linear models and response transformation. In *Response Surface Methodology and Related Topics* (A. I. Khuri, ed.), pp. 173–202. Hackensack, NJ: World Scientific Publishing.

Atkinson, A. C., and Bailey, R. A. (2001). One hundred years of the design of experiments on and off the pages of Biometrika. *Biometrika*, **88**(1), 53–97.

Atkinson, A. C., and Bogacka, B. (1997). Compound D- and D_s-optimum designs for determining the order of a chemical reaction. *Technometrics*, **39**(4), 347–356.

Atkinson, A. C., and Cook, R. D. (1997). Designing for a response transformation parameter. *Journal of the Royal Statistical Society, Series B*, **59**(1), 111–124.

Atkinson, A. C., and Cox, D. R. (1974). Planning experiments for discriminating between models. *Journal of the Royal Statistical Society, Series B*, **36**(3), 321–334; discussion: pp. 335–348.

Atkinson, A. C., and Donev, A. N. (1989). The construction of exact D-optimum experimental designs with application to blocking response surface designs. *Biometrika*, **76**, 515–526.

Atkinson, A. C., and Donev, A. N. (1996a). *Optimum Experimental Design*. London: Oxford University Press (Clarendon).

Atkinson, A. C., and Donev, A. N. (1996b). Experimental designs optimally balanced for trend. *Technometrics*, **38**(4), 333–341.

Atkinson, A. C., and Federov, V. V. (1975a). The design of experiments for discriminating between two rival models. *Biometrika*, **62**, 57–70.

Atkinson, A. C., and Federov, V. V. (1975b). Optimal design: Experiments for discriminating between several models. *Biometrika*, **62**, 289–303.

Atkinson, A. C., and Haines, L. M. (1996). Designs for nonlinear and generalized linear models. *Handbook of Statistics*. Amsterdam: North-Holland Publ.

Atkinson, A. C., and Hunter, W. G. (1968). The design of experiments for parameter estimation. *Technometrics*, **10**(2), 271–289.

Atkinson, A. C., Chaloner, K., Herzberg, A. M., and Juritz, J. (1993). Optimum experimental designs for properties of a compartmental model. *Biometrics*, **49**, 325–337.

Atkinson, A. C., Demetrio, C. G. B., and Zocchi, S. (1995). Optimum dose levels when males and females differ in response. *Applied Statistics*, **44**, 213–226.

Atkinson, A. C., Pronzaato, L., and Wynn, H. P. (1998). *Advances in Model-Oriented Data Analysis and Experimental Design*. Heidelberg: Physica Verlag.

Atwood, C. L. (1969). Optimal and efficient design of experiments. *Annals of Mathematical Statistics*, **40**, 1570–1602.

Atwood, C. L. (1973). Sequences converging to D-optimal designs of experiments. *Annals of Statistics*, **1**, 342–352.

Atwood, C. L. (1975). Estimating a response surface with an uncertain number of parameters assuming normal errors. *Journal of the American Statistical Association*, **70**, 613–617.

Atwood, C. L. (1976). Convergent design sequences for sufficiently regular optimality criteria. *Annals of Statistics*, **4**, 1124–1138.

Avery, P. J., and Masters, G. A. (1999). Advice on the design and analysis of sensory evaluation experiments using a case-study with cooked pork. *The Statistician*, **48**(3), 349–359.

Azais, J.-M., Monod, H., and Bailey, R. A. (1998). The influence of design on validity and efficiency of neighbour methods. *Biometrics*, **54**(4), 1374–1387.

Azeredo, M. B. de V., da Silva, S. S., and Rekab, K. (2003). Improve molded part quality. *Quality Progress*, **36**(7), July, 72–76.

Baasel, W. D. (1965). Exploring response surfaces to establish optimum conditions. *Chemical Engineering*, **72**, 147–152.

Babu, B. Re. V., and Narasimham, V. L. (1991). Construction of second order slope rotatable designs through balanced incomplete block designs. *Communications in Statistics, Theory and Methods*, **20**, 2467–2478.

Bacon, D. W. (1970). Making the most of a "one-shot" experiment. *Industrial and Engineering Chemistry*, **62**, 27–34.

Bacon, D. W., and Watts, D. G. (1974). Estimating the transition between two intersecting straight lines. *Biometrika*, **58**, 525–534.

Bagchi, B., and Bagchi, S. (2001). Optimality of partial geometric designs. *Annals of Statistics*, **29**(2), 577–594.

Bagchi, T. P., and Templeton, J. G. C. (1993). Constrained optimization: An effective alternative to to Taguchi's two steps to robust design. *Opsearch*, **30**(3).

Bailey, R. A. (1977). Patterns of confounding in factorial designs. *Biometrika*, **64**, 597–603.

Bailey, R. A. (1985). Factorial design and Abelian groups. *Linear Algebra Applications*, **70**, 349–368.

Bailey, R. A., and Druilhet, P. (2004). Optimality of neighbor-balanced designs for total effects. *Annals of Statistics*, **32**(4), 1650–1661.

Bailey, R. A., Gilchrist, F. H. L., and Patterson, H. D. (1977). Identification of effects and confounding patterns in factorial designs. *Biometrika*, **64**, 347–354.

Bain, W. A., and Batty, J. E. (1956). Inactivation of adrenaline and noradrenaline by human and other mammalian liver in vitro. *British Journal of Pharmacology and Chemotherapy*, **11**, 52–57.

Baird, B. L., and Mason, D. D. (1959). Multivariable equations describing fertility-corn yield response surfaces and their agronomic and economic interpretation. *Agronomy Journal*, **51**, 152–156.

Baker, A. E., Doerry, W. T., Kulp, K., and Kemp, K. (1988). A response-surface analysis of the oxidative requirements of no-time doughs. *Cereal Chemistry*, **65**(4), 367–372.

Baker, F. D., and Bargmann, R. E. (1985). Orthogonal central composite designs of the third order in the evaluation of sensitivity and plant growth simulation models. *Journal of the American Statistical Association*, **80**, 574–579.

Balkin, S. D., and Lin, D. K. J. (1998). A graphical comparison of supersaturated designs. *Communications in Statistics, Theory and Practice*, **27**, 1289–1303.

Banerjee, K. S. (1950). A note on the fractional replication of factorial arrangements. *Sankhya*, **10**, 87–94.

Banerjee, K. S., and Federer, W. T. (1963). On estimates for a fraction of a complete factorial experiment as orthogonal linear combinations of the observations. *Annals of Mathematical Statistics*, **34**, 1068–1078.

Banerjee, K. S., and Federer, W. T. (1967). On a special subset giving irregular fractional replicate of a 2^n factorial experiment. *Journal of the Royal Statistical Society, Series B*, **29**, 292–299.

Banks, D. (1993). Is industrial statistics out of control? *Statistical Science*, **8**, 356–377; discussion: pp. 378–409.

Barker, T. B. (1994). *Quality by Experimental Design*, 2nd ed. New York: Dekker.

Barker, T. B. (2001-2002). Deconfounding two-factor interactions with a minimum number of extra runs. *Quality Engineering*, **14**(2), 187–193.

Barnett, J., Czitrom, V., John, P. W. M., and Leon, R. V. (1997). Using fewer wafers to resolve confounding in screening experiments. In *Statistical Case Studies for Industrial Process Improvement*, (V. Citrom and P. D. Spagon, eds.), pp. 235–250. Philadelphia: SIAM.

Barnett, V., and Haworth, J. (1998). A fractional factorial design for bench-mark testing of a Bayesian method for multilocation. *The Statistician*, **47**(4), 617–628.

Barrentine, L. B. (1999). *An Introduction to Design of Experiments: A Simplified Approach*. Milwaukee, WI: ASQ Quality Press.

Barton, R. R. (1998). Design-plots for factorial and fractional-factorial designs. *Journal of Quality Technology*, **30**(1), 40–54.

Barton, R. R. (1999). *Graphical Methods for the Design of Experiments*. New York: Springer-Verlag.

Bates, D. M., and Watts, D. G. (1987). *Nonlinear Regression Analysis and Its Applications*. New York: Wiley.

Bates, R. A., and Wynn, H. P. (2000). Adaptive radial basis function emulators for robust design. In *Evolutionary Design and Manufacture* (E. I. C. Parmee, ed.), pp.343–350. New York: Springer-Verlag.

Bates, R. A., Buck, R. J., Riccomagno, E., and Wynn, H. P. (1996). Experimental design and observation for large systems. *Journal of the Royal Statistical Society, Series B*, **58**, 77–94; discussion: pp. 95–111.

Bates, R. A., Giglio, B., and Wynn, H. P. (2003). A global selection procedure for polynomial interpolators. *Technometrics*, **45**(3), 246–254.

Baumert, L., Golomb, S. W., and Hall, M. (1962). Discovery of an Hadamard matrix of order 92. *American Mathematics Society Bulletin*, **68**, 237–238.

Bawn, C. W., and Ricci, V. (1981). Statistical experimental design: Relationship between injection molding variables and ROVEL physical properties. *Proceedings of the 39th Annual Technical Conference, Society of Plastic Engineers*, pp. 780–782.

Beattie, S. D., Fong, D. K. H., and Lin, D. K. J. (2002). A two-stage Bayesian model selection strategy for supersaturated designs. *Technometrics*, **44**(1), 55–63.

Beauchamp, J. J., and Cornell, R. G. (1969). Spearman simultaneous estimation for a compartmental model. *Technometrics*, **11**(3), 551–560.

Beaver, R. J. (1977). Weighted least squares response surface fitting in factorial paired comparisons. *Communications in Statistics, Theory and Methods*, **6**, 1275–1287.

Becker, N. G. (1968). Models for the response of a mixture. *Journal of the Royal Statistical Society, Series B*, **30**(2), 349–358. **M**

Becker, N. G. (1969). Regression problems when the predictor variables are proportions. *Journal of the Royal Statistical Society, Series B*, **31**(1), 107–112. **M**

Becker, N. G. (1970). Mixture designs for a model linear in the proportions. *Biometrika*, **57**(2), 329–338. **M**

Beckman, R. J. (1996). Plotting p^k factorial or p^{k-n} fractional factorial data. *American Statistician*, **50**(2), 170–174.

Beder, J. H. (1989). The problem of confounding in two-factor experiments. *Communications in Statistics, Theory and Methods*, **18**(8), 2165–2188; see also Beder (1994). **23**(7), 2129–2130.

Beder, J. H. (1998). Conjectures about Hadamard matrices. *Journal of Statistical Planning and Inference*, **72**, 7–14.

Behnken, D. W., and Draper, N. R. (1972). Residuals and their variance patterns. *Technometrics*, **14**(1), 101–111.

Belloto, R. J., Jr., Dean, A. M., Moustafa, A. M., Gouda, M. W., and Sokolski, T. D. (1985). Statistical techniques applied to solubility predictions and pharmaceutical formulations; an approach to problem solving using mixture response surface methodology. *International Journal of Pharmacology (AMST)*, **233**, 195–208. **M**

Bellsley, D. A. (1991). Conditioning Diagnostics, *Collinearity and Weak Data in Regression*. New York: Wiley.

Bemensdorfer, J. L. (1979). Approving a process for production. *Journal of Quality Technology*, **11**(1), 1–12.

Bendell, A., Disney, J., and McCollin, C. (1999). The future role of statistics in quality engineering and management. *The Statistician*, **48**(3), 299–313; discussion: pp. 313–326.

Bennett, C. A., and Franklin, N. L. (1954). *Statistical Analysis in Chemistry and the Chemical Industry*. New York: Wiley.

Benski, H. C. (1989). Use of a normality test to identify significant effects in factorial designs. *Journal of Quality Technology*, **21**(3), 174–178.

Bentley, J., and Schneider, T. J. (2000). Statistics and archaeology in Israel. *Computational Statistics and Data Analysis*, **32**(3-4), 465–483.

Berding, C., Kleider, W., and Gossl, R. (1996). DOPE: D-optimal planning and experimentation. The tool for experimental design in research and development. *Computational Statistics and Data Analysis*, **21**(6), 705–710.

Berg, C. (1960). Optimization in process development. *Chemical Engineering Progress*, **36**(8), 42–47.

Berger, P. D. (1972). On Yates' order in fractional factorial designs. *Technometrics*, **14**(4), 971–972.

Bergerud, W. A. (1996). Displaying factor relationships in experiments. *American Statistician*, **50**(3), 228–233.

Bergman, B., and Hynen, A. (1997). Dispersion effects from unreplicated designs in the 2^{k-p} series. *Technometrics*, **39**(2), 191–198.

Berk, K. N., and Picard, R. R. (1991). Significance tests for saturated orthogonal arrays. *Journal of Quality Technology*, **23**(2), 79–89.

Berry, J. W., Tucker, H., and Deutshman, A. J. (1963). Starch vinylation. *Industrial and Engineering Chemistry Process Design and Developments*, **2**(4), 318–322.

Berube, J., and Nair, V. N. (1998). Exploiting the inherent structure in robust parameter design experiments. *Statistica Sinica*, **8**, 43–66.

Biles, W. E. (1975). A response surface method for experimental optimization of multi-response processes. *Industrial and Engineering Chemistry Process Design and Development*, **14**, 152–158.

Bingham, D. R., and Li, W. (2002). A class of optimal robust parameter designs. *Journal of Quality Technology*, **34**(3), 244–259.

Bingham, D. R., and Sitter, R. R. (1999a). Minimum-aberration two-level fractional factorial split-plot designs. *Technometrics*, **41**(1), 62–70.

Bingham, D. R., and Sitter, R. R. (1999b). Some theoretical results for fractional factorial split-plot designs. *Annals of Statistics*, **27**(4), 1240–1255.

Bingham, D. R., and Sitter, R. R. (2001). Design issues in fractional factorial split-plot experiments. *Journal of Quality Technology*, **33**(1), 2–15.

Bingham, D. R., and Sitter, R. R. (2003). Fractional factorial split-plot designs for robust parameter experiments. *Technometrics*, **45**(1), 80–89.

Bingham, D. R., Schoen, E. D., and Sitter, R. R. (2004). Designing fractional factorial split-plot experiments with few whole-plot factors. *Applied Statistics*, **53**(2), 325–339; corrections: (2005), **54**(5), 955–958.

Bingham, T. C. (1997). An approach to developing multi-level fractional factorial designs. *Journal of Quality Technology*, **29**(4), 370–380.

Birnbaum, A. (1959). On the analysis of factorial experiments without replication. *Technometrics*, **1**(4), 343–357.

Bisgaard, S. (1991). Teaching statistics to engineers. *American Statistician*, **45**(4), 274–283.

Bisgaard, S. (1993). A method for identifying defining contrasts for 2^{k-p} experiments. *Journal of Quality Technology*, **25**(1), 28–35.

Bisgaard, S. (1994a). Blocking generators for small 2^{k-p} designs. *Journal of Quality Technology*, **26**(4), 288–296.

Bisgaard, S. (1994b). A note on the definition of resolution for blocked 2^{k-p} designs. *Technometrics*, **36**(3), 308–312.

Bisgaard, S. (1996). A comparative analysis of the performance of Taguchi's linear graphs for the design of two-level fractional factorials. *Applied Statistics*, **45**(3), 311–322.

Bisgaard, S. (1997). Designing experiments for tolerancing assembled products. *Technometrics*, **39**(2), 142–152.

Bisgaard, S. (1998). Making sure the design fits the problem—an example. *Quality Engineering*, **10**, 771–775.

Bisgaard, S. (2000). The design and analysis of $2^{k-p} \times 2^{q-r}$ split plot experiments. *Journal of Quality Technology*, **32**(1), 39–56.

Bisgaard, S., and Ankenman, B. (1996). Standard errors for the eigenvalues in second order response surface models. *Technometrics*, **38**(3), 238–246.

Bisgaard, S., and Fuller, H. T. (1994). Analysis of factorial experiments with defects and defectives as the response. *Quality Engineering*, **7**(2), 429–443.

Bisgaard, S., and Fuller, H. T. (1995). Sample size estimates for 2^{k-p} designs with binary responses. *Journal of Quality Technology*, **27**(4), 344–354.

Bisgaard, S., and Gertsbakh, I. (2000). 2^{k-p} experiments with binary responses: inverse binomial sampling. *Journal of Quality Technology*, **32**(2), 148–156.

Bisgaard, S., and Kulahci, M. (2001a). Switching one column for Plackett-Burman designs. *Journal of Applied Statistics*, **28**, 943–949.

Bisgaard, S., and Kulahci, M. (2001b). Robust product design: saving trials with split plot confounding. *Quality Engineering*, **13**, 525–530.

Bisgaard, S., and Steinberg, D. M. (1997). The design and analysis of $2^{k-p} \times s$ prototype experiments. *Technometrics*, **39**(1), 52–62.

Bisgaard, S., and Sutherland, M. (2003–2004). Split plot experiments: Taguchi's Ina tile experiment reanalyzed. *Quality Engineering*, **16**(1), 157–164.

Bisgaard, S., and Weiss, P. (2002–2003). Experimental design applied to identify defective parts in assembled products. *Quality Engineering*, **15**(2), 347–350.

Bisgaard, S., Fuller, H. T., and Barrios, E. (1996). Two-level factorials run as split-plot experiments. *Quality Engineering*, **8**, 705–708.

Bisgaard, S., Hunter, W. G., and Pallesen, L. (1984). Economic selection of quality of manufactured product. *Technometrics*, **26**(1), 9–18.

Bisgaard, S., Vivacqua, C. A., and de Pinho, L. S. (2005). Not all models are polynomials! *Quality Engineering*, **17**, 181–186.

Bissell, A. F. (1989). Interpreting mean squares in saturated fractional designs. *Journal of Applied Statistics*, **16**(1), 7–18.

Bissell, A. F. (1992). Mean squares in saturated fractional designs revisited. *Journal of Applied Statistics*, **19**(2), 351–366.

Bjerke, F., Aastveit, A. H., Stroup, W. W., Kirkhus, B., and Naes, T. (2004). Design and analysis of storing experiments: A case study. *Quality Engineering*, **16**(4), 591–611.

Bjorkestøl, K., and Tormod, N. (2005). A discussion of alternative ways of modeling and interpreting mixture data. *Quality Engineering*, **17**(4), 509–533. **M**

Block, R., and Mee, R. (2005). Resolution IV designs with 128 runs. *Journal of Quality Technology*, **37**(4), 282–293. Corrections, **38**(2), 196.

Blomkvist, O., Hynen, A., and Bergman, B. (1997). A method to identify dispersion effects from unreplicated multilevel experiments. *Quality and Reliability Engineering International*, **13**, 525–530.

Böckenholt, U. (1989). Analyzing optima in the exploration of multiple response surfaces. *Biometrics*, **45**, 1001–1008.

Bohrer, R., Chow, W., Faith, R., Joshi, V. M., and Wu, C. F. J. (1981). Multiple three-decision rules for factorial simple effects: Bonferroni wins again. *Journal of the American Statistical Association*, **76**, 119–124.

Bole, J. B., and Freyman, S. (1975). Response of irrigated field and sweet corn to nitrogen and phosphorus fertilizers in southern Alberta. *Canadian Journal of Soil Science*, **55**, 137–143.

Bolker, H. I. (1965). Delignification by nitrogen compounds. *Industrial and Engineering Chemistry Product Research and Development*, **4**, 74–79.

Bonett, D. G., and Woodward, J. A. (1993). Analysis of simple main effects in fractional factorial experimental designs of resolution V. *Communications in Statistics, Theory and Methods*, **22**(6), 1585–1593.

Booth, K. H. V., and Cox, D. R. (1962). Some systematic saturated designs. *Technometrics*, **4**(4), 489–495.

Borkowski, J. J. (1995a). Finding maximum G-criterion values for central composite designs on the hypercube. *Communications in Statistics, Theory and Methods*, **24**(8), 2041–2058.

Borkowski, J. J. (1995b). Spherical prediction-variance properties of central composite designs and Box-Behnken designs. *Technometrics*, **37**(4), 399–410.

Borkowski, J. J. (2003a). A comparison of prediction variance criteria for response surface designs. *Journal of Quality Technology*, **35**(1), 70–77.

Borkowski, J. J. (2003b). Using a genetic algorithm to generate small exact response surface designs. *Journal of Probability and Statistical Science*, **1**(1), 65–88.

Borkowski, J. J. (2006). Graphical methods for assessing the prediction capability of response surface designs. In *Response Surface Methodology and Related Topics* (A. I. Khuri, ed.), pp. 349–378. Hackensack, NJ: World Scientific Publishing.

Borkowski, J. J., and Lucas, J. M. (1997). Designs of mixed resolution for process robustness studies. *Technometrics*, **39**(1), 63–70.

Borkowski, J. J., and Valeroso, E. S. (2001). Comparison of design optimality criteria of reduced models for response surface designs in the hypercube. *Technometrics*, **43**(4), 468–477.

Borror, C. M., and Montgomery, D. C. (2002). Mixed resolution designs as alternatives to Taguchi inner/outer array designs for robust design problems. *Quality and Reliability Engineering International*, **16**, 117–127.

Borror, C. M., Heredia-Langner, A., and Montgomery, D. C. (2000). Generalized linear models in the analysis of industrial experiments. *Journal of Propagations in Probability and Statistics* (*International Edition*), **2**, 127–144.

Borror, C. M., Montgomery, D. C., and Myers, R. H. (2002). Evaluation of statisticsl designs for experiments involving noise variables. *Journal of Quality Technology*, **34**(1), 54–70.

Bose, M. (1996). Some efficient incomplete block sequences. *Biometrika*, **83**(4), 956–961.

Bose, R. C. (1947). Mathematical theory of the symmetrical factorial design. *Sankhya*, **8**, 107–166.

Bose, R. C., and Bush, K. A. (1952). Orthogonal arrays of strength two and three. *Annals of Mathematical Statistics*, **23**, 508–524.

Bose, R. C., and Carter, R. L. (1959). Complex representation in the construction of rotatable designs. *Annals of Mathematical Statistics*, **30**, 771–780.

Bose, R. C., and Draper, N. R. (1959). Second order rotatable designs in three dimensions. *Annals of Mathematical Statistics*, **30**, 1097–1112.

Bose, R. C., and Kishen, K. (1940). On the problem of confounding in the general symmetrical factorial design. *Sankhya*, **5**, 21–26.

Bose, R. C., and Kuebler, R. R. (1960). A geometry of binary sequences associated with group alphabets for information theory. *Annals of Mathematical Statistics*, **31**, 113–139.

Bose, R. C., and Srivastava, J. N. (1964). On a bound useful in the theory of factorial design and error correcting codes. *Annals of Mathematical Statistics*, **35**, 408–414.

Bose, R. C., Clatworthy, W. H., and Shrikande, S. S. (1954). Tables of partially balanced designs with two associate classes. *North Carolina Agricultural Experiment Station, Technical Bulletin*, **107**.

Bose, R. C., Chakravati, I. M., and Knuth, D. E. (1960). On methods of constructing sets of mutually orthogonal Latin squares using a computer. I. *Technometrics*, **2**(4), 507–516.

Bose, R. C., Chakravati, I. M., and Knuth, D. E. (1961). On methods of constructing sets of mutually orthogonal Latin squares using a computer. II. *Technometrics*. **3**(1), 111–117.

Boussouel, N., Mathieu, F., Benoit, V., Linder, M., Revol-Junelles A.-M., and Millière, J.-B. (1999). Response surface methodology, an approach to predict the effects of a lactoperoxidase system, Nisin, alone or in combination, on *Listeria monocytogenes* in skim milk. *Journal of Applied Microbiology*, **86**, 642–652.

Box, G. E. P. (1950). Problems in the analysis of growth and wear curves. *Biometrics*, **6**(4), 362–389.

Box, G. E. P. (1952). Multifactor designs of first order. *Biometrika*, **39**, 49–57 (see also p. 189, note by Tocher).

Box, G. E. P. (1953). Non-normality and tests on variances. *Biometrika*, **40**, 318–335.

Box, G. E. P. (1954). The exploration and exploitation of response surfaces: Some general considerations and examples. *Biometrics*, **10**, 16–60.

Box, G. E. P. (1955). Contribution to the discussion, Symposium on Interval Estimation. *Journal of the Royal Statistical Society, Series B*, **16**, 211–212.

Box, G. E. P. (1957). Evolutionary operation: A method for increasing industrial productivity. *Applied Statistics*, **6**, 3–23.

Box, G. E. P. (1958). Use of statistical methods in the elucidation of basic mechanisms. *Bulletin of the International Statistic Institute*, **36**, 215–225.

Box, G. E. P. (1959a). Discussion of "Random balance experimentation" by F. E. Satterthwaite and "The application of random balance design" by T. A. Budne. *Technometrics*, **1**, 174–180.

Box, G. E. P. (1959b). Answer to query: Replication of non-center points in the rotatable and near-rotatable central composite design. *Biometrics*, **15**, 133–135.

Box, G. E. P. (1960a). Some general considerations in process optimization. *Journal of the Royal Society of Basic Engineering*, **82**, 113–119.

Box, G. E. P. (1960b). Fitting empirical data. *Annals of the New York Academy of Sciences*, **86**, 792–816.

Box, G. E. P. (1963). The effects of errors in the factor levels and experimental design. *Technometrics*, **5**(2), 247–262.

Box, G. E. P. (1966a). A simple system of evolutionary operation subject to empirical feedback. *Technometrics*, **8**(1), 19–26.

Box, G. E. P. (1966b). A note on augmented designs. *Technometrics*, **8**(1), 184–188.

Box, G. E. P. (1966c). Use and abuse of regression. *Technometrics*, **8**(4), 625–629.

Box, G. E. P. (1968). Response surfaces. Article under "Experimental Design" in *The International Encyclopedia of the Social Sciences*, 254–259. New York: MacMillan and Free Press.

Box, G. E. P. (1976). Science and statistics. *Journal of the American Statistical Association*, **71**, 791–799.

Box, G. E. P. (1982). Choice of response surface design and alphabetic optimality. Utilitas Mathematica, **21B**, 11–55.

Box, G. E. P. (1985a). Discussion of off-line quality control, parameter design and the Taguchi method. *Journal of Quality Technology*, 17(4), 198–206.

Box, G. E. P. (1985b). *The Collected Works* (G. C. Tiao, ed.), Vols. 1 and 2. Belmont, CA: Wadsworth.

Box, G. E. P. (1988). Signal-to-noise ratios, performance criteria, and transformations. *Technometrics*, **30**(1), 1–17; discussion: pp. 19–41.

Box, G. E. P. (1990). Finding bad values in factorial designs. *Quality Engineering*, **3**(3), 405–410.

Box, G. E. P. (1991). A simple way to deal with missing observations from designed experiments. *Quality Engineering*, **3**(2), 249–254.

Box, G. E. P. (1992). Teaching engineers experimental design with a paper helicopter. *Quality Engineering*, **4**, 453–459.

Box, G. E. P. (1993). Quality improvement—the new industrial revolution. *International Statistical Review*, **61**, 3–19; discussion: pp. 21–26.

Box, G. E. P. (1994). Statistics and quality improvement. *Journal of the Royal Statistical Society, Series A*, **157**, 209–229.

Box, G. E. P. (1996). Split-plot experiments. *Quality Engineering*, **8**, 515–520.

Box, G. E. P. (1998). Signal-to-noise ratios, performance criteria and transformations. *Technometrics*, **30**(1), 1– 17; discussion: pp. 18–40.

Box, G. E. P. (1999a). Statistics as a catalyst to learning by scientific method. Part I—an example. *Journal of Quality Technology*, 31(1), 1–15.

Box, G. E. P. (1999b). Statistics as a catalyst to learning by scientific method. Part II—a discussion. *Journal of Quality Technology*, 31(1), 16–29.

Box, G. E. P. (1999–2000). The invention of the composite design. *Quality Engineering*, **12**(1), 119–122.

Box, G. E. P. (2000). *Box on Quality and Discovery with Design, Control and Robustness* (G. Tiao, S. Bisgaard, W. Hill, D. Pena, and S. Stigler, eds.). New York: Wiley.

Box, G. E. P. (2001–2002). Comparisons, absolute values, and how I got to go to the Folies Bergères. *Quality Engineering*, **14**(1), 167–169.

Box, G. E. P., and Behnken, D. W. (1960a). Some new three-level designs for the study of quantitative variables. *Technometrics*, **2**(4), 455–475.

Box, G. E. P., and Behnken, D. W. (1960b). Simplex-sum designs: A class of second order rotatable designs derivable from those of first order. *Annals of Mathematical Statistics*, **31**, 838–864.

Box, G. E. P., and Bisgaard, S. (1993). What can you find out from 12 experimental runs. *Quality Engineering*, **5**, 663–668.

Box, G. E. P., and Cox, D. R. (1964). An analysis of transformations. *Journal of the Royal Statistical Society, Series B*, **26**, 211–243; discussion: pp. 244–252.

Box, G. E. P., and Draper, N. R. (1959). A basis for the selection of a response surface design. *Journal of the American Statistical Association*, **54**, 622–652.

Box, G. E. P., and Draper, N. R. (1963). The choice of a second order rotatable design. *Biometrika*, **50**, 335–352.

Box, G. E. P., and Draper, N. R. (1965). The Bayesian estimation of common parameters from several responses. *Biometrika*, **52**, 355–365.

Box, G. E. P., and Draper, N. R. (1968). Isn't my process too variable for EVOP? *Technometrics*, **10**(3), 439–444.

Box, G. E. P., and Draper, N. R. (1969). *Evolutionary Operation*. New York: Wiley.

Box, G. E. P., and Draper, N. R. (1970). EVOP—makes a plant grow better. *Industrial Engineering*, April, pp. 31–33, condensed in *Management Review*, July 1970, pp. 22–25.

Box, G. E. P., and Draper, N. R. (1975). Robust designs. *Biometrika*, **62**, 347–352.

Box, G. E. P., and Draper, N. R. (1980). The variance function of the difference between two estimated responses. *Journal of the Royal Statistical Society, Series B*, **42**, 79–82.

Box, G. E. P., and Draper, N. R. (1982). Measures of lack of fit for response surface designs and predictor variable transformations. *Technometrics*, **24**(1), 1–8, see also Box and Draper (1983).

Box, G. E. P., and Draper, N. R. (1983). Letter to Editor, *Technometrics*, **25**(2), p. 217.

Box, G. E. P., and Draper, N. R. (1987). *Empirical Model-Building and Response Surfaces*. New York: Wiley.

Box, G. E. P., and Fung, C. A. (1983). Some considerations in estimating data transformations. Mathematical Research Center Technical Summary Report No. 2609, University of Wisconsin—Madison.

Box, G. E. P., and Hau, I. (2001). Experimental designs when there are one or more factor constraints. *Journal of Applied Statistics*, **28**(8), 973–989.

Box, G. E. P., and Hay, W. A. (1953). A statistical design for the efficient removal trends occurring in a comparative experiment with an application in biological assay. *Biometrics*, **9**, 304–319.

Box, G. E. P., and Hill, W. J. (1967). Discrimination among mechanistic models. *Technometrics*, **9**(1), 57–71.

Box, G. E. P., and Hill, W. J. (1974). Correcting inhomogeneity of variance with power transformation weighting. *Technometrics*, **16**(3), 385–389.

Box, G. E. P., and Hunter, J. S. (1954). A confidence region for the solution of a set of simultaneous equations with an application to experimental design. *Biometrika*, **41**, 190–199.

Box, G. E. P., and Hunter, J. S. (1957). Multifactor experimental designs for exploring response surfaces. *Annals of Mathematical Statistics*, **28**, 195–241.

Box, G. E. P., and Hunter, J. S. (1959). Condensed calculations for evolutionary operation programs. *Technometrics*, **1**(1), 77–95.

Box, G. E. P., and Hunter, J. S. (1961a). The 2^{k-p} fractional factorial designs. Part 1. *Technometrics*, **3**(3), 311–351. (Also reprinted in the *Technometrics* 2000 historical issue, **42**(1), 28–47.)

Box, G. E. P., and Hunter, J. S. (1961b). The 2^{k-p} fractional factorial designs. Part 2. *Technometrics*, **3**(4), 449–458.

Box, G. E. P., and Hunter, W. G. (1962). A useful method for model building. *Technometrics*, **4**(3), 301–318.

Box, G. E. P., and Hunter, W. G. (1965a). Sequential design of experiments for nonlinear models. *Proceedings of the IBM Scientific Computing Symposium in Statistics*, pp. 113–137.

Box, G. E. P., and Hunter, W. G. (1965b). The experimental study of physical mechanisms. *Technometrics*, **7**(1), 23–42.

Box, G. E. P., and Jones, S. (1992). Split plots for robust product experimentation. *Journal of Applied Statistics*, **19**(1), 3–26.

Box, G. E. P., and Jones, S. (2000–2001). Split plots for robust product and process experimentation. *Quality Engineering* **13**(1), 127–134.

Box, G. E. P., and Liu, P. Y. T. (1999). Statistics as a catalyst to learning by scientific method. Part I—an example. *Journal of Quality Technology*, **31**(1), 1–15.

Box, G. E. P., and Lucas, H. L. (1959). Design of experiments in nonlinear situations. *Biometrika*, **46**, 77–90.

Box, G. E. P., and Luceño, A. (1997). *Statistical Control by Monitoring and Feedback Adjustment*. New York: Wiley.

Box, G. E. P., and Luceño, A. (2000). Six sigma, process drift, capability indices, and feedback adjustment. *Quality Engineering*, **12**(3), 297–302.

Box, G. E. P., and Luceño, A. (2001). Feedforward as a supplement to feedback adjustment to allow for feedstock changes.

Box, G. E. P., and Meyer, R. D. (1986a). An analysis for unreplicated fractional factorials. *Technometrics*, **28**(1), 11–18.

Box, G. E. P., and Meyer, R. D. (1986b). Dispersion effects from fractional designs. *Technometrics*, **28**(1), 19–27.

Box, G. E. P., and Meyer, R. D. (1993). Finding the active factors in fractionated screening experiments. *Journal of Quality Technology*, **25**(2), 94–105.

Box, G. E. P., and Tidwell, P. W. (1962). Transformation of the independent variables. *Technometrics*, **4**(4), 531–550.

Box, G. E. P., and Tyssedal, J. (1996). Projective properties of certain orthogonal arrays. *Biometrika*, **83**(4), 950–955.

Box, G. E. P., and Tyssedal, J. (2001). Sixteen run designs of high projectivity for factor screening. *Communications in Statistics, Simulation and Computation*, **30**(2), 217–228.

Box, G. E. P., and Wetz, J. (1973). Criteria for judging adequacy of estimation by an approximating response function. *Technical Report*—No. 9. University of Wisconsin, Statistics Department.

Box, G. E. P., and Wilson, K. B. (1951). On the experimental attainment of optimum conditions. *Journal of the Royal Statistical Society Series B*, **13**, 1–38; discussion: pp. 38–45.

Box, G. E. P., and Youle, P. V. (1955). The exploration and exploitation of response surfaces: An example of the link between the fitted surface and the basic mechanism of the system. *Biometrics*, **11**, 287–323.

Box, G. E. P., Youdon, W. J., Kempthorne, O., Tukey, J. W., and Hunter, J. S. (1959). Discussion of the papers of Messrs. Satterthwaite and Budne. *Technometrics*, **1**(2), 157–193.

Box, G. E. P., Hunter, W. G., MacGregor, J. F., and Erjavec, J. (1973). Some problems associated with the analysis of multiresponse data. *Technometrics*, **15**(1), 33–51.

Box, G. E. P., Hunter, W. G., and Hunter, J. S. (1978). *Statistics for Experimenters. An Introduction to Design, Data Analysis and Model Building*. New York: Wiley.

Box, G. E. P., Bisgaard, S., and Fung, C. A. (1988). An explanation and critique of Taguchi's contributions to quality engineering. *Quality and Reliability Engineering International*, **4**, 123–131.

Box, G. E. P., Hunter, J. S., and Hunter, W. G. (2005). *Statistics for Experimenters: Design, Innovation and Discovery*, 2nd ed. New York: Wiley.

Box, J. F. (1978). *R. A. Fisher: The Life of a Scientist*. New York: Wiley.

Box, M. J. (1971). Simplified experimental design. *Technometrics*, **13**(1), 19–31.

Box, M. J., and Draper, N. R. (1971). Factorial designs, the det $\mathbf{X'X}$ criterion, and some related matters. *Technometrics*, **13**(4), 731–742; corrections: **14**(2), 511; **15**(2), 430.

Box, M. J., and Draper, N. R. (1972). Estimation and design criteria for multiresponse non-linear models with non-homogeneous variance. *Applied Statistics*, **21**, 13–24.

Box, M. J., and Draper, N. R. (1974). Minimum point second order designs. *Technometrics*, **16**(4), 613–616. See also letters to Editor, **17**(2), p. 282.

Boyd, D. A. (1972). Some recent ideas on fertilizer response curves. *Proceedings of the 9th International Congress of the Potash Institute*, 461–473.

Bradley, R. A. (1958). Determination of optimum operating conditions by experimental methods. Part I. *Industrial Quality Control*, **15**(4), 16–20.

Brand, D., Pandey, A., Rodriguez, L., Roussos, S., Brand, I., and Soccol, C. (2001). Packed bed column fermenter and kinetic modeling for upgrading the nutritional quality of coffee husk in solid-state fermentation. *Biotechnology Progress*, **17**(6), 1065–1070.

Brannigan, M. (1981). An adaptive piecewise polynomial curve fitting procedure for data analysis. *Communications in Statistics, Theory and Methods*, **10**, 1823–1848.

Brenneman, W. A., and Nair, V. N. (2001). Methods for identifying dispersion effects in unreplicated factorial experiments: A critical analysis and proposed strategies. *Technometrics*, **43**(4), 388–405.

Brien, C. J., and Bailey, R. A. (2006). Multiple randomizations. *Journal of the Royal Statistical Society, Series B*, **68**(4), 571–599, discussion 599–609.

Brien, C. J., and Payne, R. W. (1999). Tiers, structure formulae and the analysis of complicated experiments. *The Statistician*, **48**(1) 41–52.

Brinkley, P., Meyer, K., and Lu, J. (1996). Combined generalized linear modeling nonlinear programming approach to robust process design—a case study in circuit board quality improvement. *Applied Statistics*, **45**, 99–110.

Broman, K., Speed, T., and Tigges, M. (1998). Estimation of antigen-responsive T cell frequencies in PBMC from human subjects. *Statistical Science*, **13**(1) 4–8; discussion: pp. 23–29.

Brooks, S. H., and Mickey, M. R. (1961). Optimum estimation of gradient direction in steepest ascent experiments. *Biometrics*, **17**, 48–56.

Brownlee, K. A. (1949). *Industrial Experimentation*. New York: Chemical Publishing Co.

Brownlee, K. A., Kelly, B. K., and Loraine, P. K. (1948). Fractional replication arrangements for factorial experiments with factors at two levels. *Biometrika*, **35**, 268–282.

Bruen, A. A., Haddad, L., and Wehlau, D. L. (1998). Binary codes and caps. *Journal of Combination and Design*, **6**, 275–284.

Brunner, E., Domhof, S., and Langer, F. (2002). *Nonparametric Analysis of Longitudinal Data in Factorial Experiments*. New York: Wiley.

Buchanan, R. L., and Phillips, J. G. (1990). Response surface model for predicting the effects of temperature, pH, sodium chloride content, sodium nitrate concentration and atmosphere on the growth of listeria monocytogenes. *Journal of Food Protection*, **53**, 370–376.

Budne, T. A. (1959). The application of random balance designs. *Technometrics*, **1**(2), 139–155 (discussion, 157–193). See also Satterthwaite (1959) and Anscombe (1959).

Bullington, K. E., Hool, J. N., and Maghsoodloo, S. (1990). A simple method for obtaining resolution IV designs for use with Taguchi's orthogonal arrays. *Journal of Quality Technology*, **22**(4), 260–264.

Bullington, R. G., Lovin, S., Miller, D. M., and Woodall, W. H. (1993). Improvement of an industrial thermostat using designed experiments. *Journal of Quality Technology*, **25**(4), 262–270.

Bulutoglu, D. A., and Cheng, C.-S. (2003). Hidden projection properties of some nonregular fractional factorial designs. *Annals of Statistics*, **31**(3), 1012–1026.

Bulutoglu, D. A., and Cheng, C.-S. (2004). Construction of $E(s^2)$-optimal supersaturated designs. *Annals of Statistics*, **32**(4), 1662–1678.

Burdick, D. S., and Naylor, T. H. (1969). Response surface methods in economics. *Revue of the International Statistical Institute*, **37**, 18–35.

Bures, S., Pokorny, F. A., and Ware, G. O. (1992). Shrinkage in ternary mixtures of container ingredients. *Proceedings of the 1991 Kansas State University Conference on Applied Statistics in Agriculture*, pp. 43–52. **M**

Burgess, L., Kreher, D. L., and Street, D. J. (1999). Small orthogonal main effect plans with four factors. *Communications in Statistics, Theory and Methods*, **28**(10), 2441–2464.

Burridge, J., and Sabastiani, P. (1994). D-optimal designs for generalized linear models with variance proportional to the square of the mean. *Biometrika*, **81**(2), 295–304.

Bursztyn, D., and Steinberg, D. M. (2001). Rotation designs for experiments in high-bias situations. *Journal of Statistical Planning and Inference*, **97**, 399–414.

Bursztyn, D., and Steinberg, D. M. (2002). Rotation designs: Orthogonal first-order designs with higher-order projectivity. *Journal of Applied Stochastic Models in Business and Industry*, **18**, 197–206.

Busch, J. W., and Phelan, P. L. (1999). Mixture models of soybean growth and herbivore performance in response to nitrogen-sulphur-phosphorus nutrient interactions. *Ecological Entomology*, **24**(2), 132–145. **M**

Butler, N. A. (2003). Some theory for constructing minimum aberration fractional factorial designs. *Biometrika*, **90**, 233–238.

Butler, N. A. (2004a). Construction of two-level split-plot fractional factorial designs for multistage processes. *Technometrics*, **46**(4), 445–451.

Butler, N. A. (2004b). Classification of efficient two-level fractional factorial designs of resolution IV or more. *Journal of Statistical Planning and Inference*, **131**, 145–159.

Butler, N. A., and Mead, R. (2001). A general method of constructing $E(s^2)$-optimal supersaturated designs. *Journal of the Royal Statistical Society*, Series B, **63**(3), 621–632.

Butler, T. A., Sikora, L. J., Steinhilber, P. M. and Douglas, L. W. (2001). Compost age and sample storage effects on maturity indicators of biosolids compost. *Journal of Environmental Quality*, **30**, 2141–2148.

Buxton, J. R. (1991). Some comments on the use of response variable transformations in empirical modeling. *Applied Statistics*, **40**, 391–400.

Cain, M., and Owen, R. J. (1990). Regressor levels for Bayesian predictive response. *Journal of the American Statistical Association*, **85**, 228–231.

Cantell, B., and Ramirez, J. G. (1994). Robust design of a polysilicon deposition process using a split-plot analysis. *Quality and Reliability Engineering International*, **10**, 123–132.

Carlson, R., and Carlson, J. E. (2005). *Design and Optimization in Organic Synthesis*, 2nd ed., revised and enlarged. Amsterdam: Elsevier.

Carlyle, W. M., Montgomery, D. C., and Runger, G. C. (2000). Optimization problems and methods in quality control and improvement. *Journal of Quality Technology*, **32**(1), 1–17; discussion: pp. 18–31.

Carney, J. R., and Hallman, R. W. (1967). Statistical methods of compounding—modern and essential tool for the rubber chemist. *Rubber World*, September, pp. 78–83.

Carr, N. L. (1960). Kinetics of catalytic isomerization of *n*-pentane. *Industrial Engineering Chemistry*, **52**, 391–396.

Carrion, J., Sansó, A., and Ortuño, M. (1999) Response surfaces estimates for the Dickey Fuller unit root test with structural breaks. *Economics Letters*, **63**(3), 279–283.

Carter, W. H., Jr., and Wampler, G. L. (1980). Survival analysis of drug combinations using a hazards model with time-dependent covariates. *Biometrics*, **36**, 537–546.

Carter, W. H., Jr., Chinchilla, V. M., Campbell, E. D., and Wampler, G. L. (1984). Confidence interval about the response at the stationary point of a response surface, with an application to preclinical cancer therapy. *Biometrics*, **40**, 1125–1130.

Carter, W. H., Jr., Chinchilla, V. M., Myers, R. H., and Campbell, E. D. (1986). Confidence intervals and an improved ridge analysis of response surfaces. *Technometrics*, **28**(4), 339–346.

Carter, W. H., Jr., Chinchilla, V. M., and Campbell, E. D. (1990). A large-sample confidence region useful in characterizing the stationary point of a quadratic response surface. *Technometrics*, **32**(4), 425–435.

Casler, M. D., and Undersander, D. J. (2000). Forage yield precision, experimental design, and cultivar mean separation for alfalfa cultivar trials. *Agronomy Journal*, **92**, 1064–1071.

Chai, F. S., Mukerjee, R., and Suen, C.-Y. (2002). Further results on orthogonal array plus one plans. *Journal of Statistical Planning and Inference*, **106**, 287–301.

Chakravati, I. M. (1956). Fractional replication in asymmetrical factorial designs and partially balanced arrays. *Sankhya*, **17**, 143–164.

Chakravati, I. M. (1961). On some methods of construction of partially balanced arrays. *Annals of Mathematical Statistics*, **32**, 1181–1185.

Challoner, K., and Larntz, K. (1989). Optimal Bayesian design applied to logistic regression experiments. *Journal of Statistical Planning and Inference*, **21**, 191–208.

Challoner, K., and Verdinelli, I. (1995). Bayesian experimental design: A review. *Statistical Science*, **10**, 273–304.

Chan, L. K., and Mak, T. K. (1995). A regression approach for discovering small variation around a target. *Applied Statistics*, **44**(3), 369–377.

Chan, L.-Y. (1992). D-optimal design for a quadratic log contrast model for experiments with mixtures. *Communications in Statistics, Theory and Methods*, **21**(10), 2909–2930. **M**

Chan, L.-Y. (2000). Optimal designs for experiments with mixtures: A survey. *Communications in Statistics, Theory and Methods*, **29**(9 and 10), 2281–2312. **M**

Chan, L.-Y. and Sandhu, M. K. (1999). Optimal orthogonal block designs for a quadratic mixture model for three components, *Journal of Applied Statistics*, **26**, 19–34. **M**

Chan, L.-Y., Fang, K. T., and Mukerjee, R. (2001). A characterization for orthogonal arrays of strength two via a regression model. *Statistics and Probability Letters*, **54**, 189–192.

Chan, L.-Y., Ma, C.-X., and Goh, T. N. (2003). Orthogonal arrays for experiments with lean designs. *Journal of Quality Technology*, **35**(2), 123–138.

Chang, C. D., and Kononenko, O. K. (1962). Sucrose-modified phenolic resins as plywood adhesives. *Adhesives Age*, **5**(7), July, 36–40.

Chang, C. D., Kononenko, O. K., and Franklin, R. E., Jr. (1960). Maximum data through a statistical design. *Industrial and Engineering Chemistry*, **52**, 939–942.

Chang, S., and Shivpuri, R. (1994). A multiple-objective decision-making approach for assessing simultaneous improvement in die life and casting quality in a die casting process. *Quality Engineering*, **7**(2), 371–383.

Chapman, R. E., and Masinda, K. (2003). Response surface designed experiment for door closing effort. *Quality Engineering*, **15**(4), 581–585.

Chasalow, S. D., and Brand, R. J. (1995). Generation of simplex lattice points. *Applied Statistics*, **44**(4), 534–545.

Chatfield, C. (2002). Confessions of a pragmatic statistician. *The Statistician*, **51**(1), 1–20.

Chatterjee, S. K., and Mandal, N. K. (1981). Response surface designs for estimating the optimal point. *Calcutta Statistical Association Bulletin*, **30**, 145–169.

Chen, H., and Cheng, C.-S. (1999). Theory of optimal blocking of 2^{n-m} designs. *Annals of Statistics*, **27**(6), 1948–1973.

Chen, H., and Cheng, C.-S. (2006). Doubling and projection: A method of constructing two-level designs of resolution IV. *Annals of Statistics*, **34**(1), 546–558.

Chen H., and Hedayat, A. S. (1996). 2^{n-m} designs with weak minimum aberration. *Annals of Statistics*, **24**, 2536–2548.

Chen H., and Hedayat, A. S. (1998). 2^{n-m} designs with resolution III or IV containing clear two-factor interactions. *Journal of Planning and Inference*, **75**, 147–158.

Chen, J. (1992). Some results on 2^{n-k} fractional factorial designs and search for minimum aberration designs. *Annals of Statistics*, **20**, 2124–2141.

Chen, J. (1998). Intelligent search for 2^{13-6} and 2^{14-7} minimum aberration designs. *Statistica Sinica*, **8**, 1265–1270.

Chen, J., and Lin, D. K. L. (1998). On the identifiability of supersaturated designs. *Journal of Statistical Planning and Inference*, **72**, 99–107.

Chen, J., and Wu, C. F. J. (1991). Some results on s^{n-k} fractional factorial designs with minimum aberration or optimal moments. *Annals of Statistics*, **19**, 1028–1041.

Chen, J., Sun, D. X., and Wu, C. F. J. (1993). A catalogue of two-level and three-level fractional factorial designs with small runs. *International Statistical Review*, **61**, 131–145.

Chen, M. H., and Wang, P. C. (2001). Multilevel factorial designs with minimum numbers of level changes. *Communications in Statistics, Theory and Methods*, **30**(5), 875–885.

Chen, Q. H., He, G. Q., and Ali, M. A. M. (2002). Optimization of medium composition for the production of elastase by Bacillus sp. EL31410 with response surface methodology. *Enzyme and Microbial Technology*, **30**, 667–672.

Chen, R.-B., and Huang, M.-N. L. (2000). Exact D-optimal designs for weighted polynomial regression model. *Computational Statistics and Data Analysis*, **33**(2), 137–149.

Cheng, C.-S. (1977). Optimality of some weighing designs and 2^n fractional factorial designs. *Annals of Statistics*, **8**, 436–446.

Cheng, C.-S. (1978a). Optimality of certain asymmetric experimental designs. *Annals of Statistics*, **6**, 1239–1261.

Cheng, C.-S. (1978b). Optimal design for the elimination of multi-way heterogeneity. *Annals of Statistics*, **6**, 1262–1272.

Cheng, C.-S. (1983). Construction of optimal balanced incomplete block designs for correlated observations. *Annals of Statistics*, **11**, 240–246.

Cheng, C.-S. (1987). An application of the Kiefer-Wolfowitz equivalence theorem to a problem in Hadamard transform optics. *Annals of Statistics*, **15**, 1593–1603.

Cheng, C.-S. (1989). Some orthogonal main-effect plans for asymmetrical factorials. *Technometrics*, **31**(4), 475–477.

Cheng, C.-S. (1995a). Complete class results for the moment matrices of designs over permutation-invariant sets. *Annals of Statistics*, **23**(1), 41–54.

Cheng, C.-S. (1995b). Some projection properties of orthogonal arrays. *Annals of Statistics*, **23**(4), 1223–1233.

Cheng, C.-S. (1997). $E(s^2)$-optimal supersaturated designs. *Statistica Sinica*, **7**(4), 929–939.

Cheng, C.-S. (1998a). Some hidden projection properties of orthogonal arrays with strength three. *Biometrika*, **85**(2) 491–495.

Cheng, C.-S. (1998b). Projectivity and resolving power. *Journal of Combination and Information System Science*, **23**, 47–58.

Cheng, C.-S., and Jacroux, M. (1988). The construction of trend-free run orders of two-level factorial designs. *Journal of the American Statistical Association*, **83**, 1152–1158.

Cheng, C.-S., and Li, C.-C. (1993). Constructing orthogonal fractional factorial designs when some factor level-combinations are debarred. *Technometrics*, **35**(3), 277–283.

Cheng, C.-S., and Mukerjee, R. (1998). Regular fractional factorial designs with minimum aberration and maximum estimation capacity. *Annals of Statistics*, **26**(6) 2289–2300.

Cheng, C.-S., and Mukerjee, R. (2001). Blocked regular fractional factorial designs with maximum estimation capacity. *Annals of Statistics*, **29**(2), 530–548; correction: **30**(3), 925–926.

Cheng, C.-S., and Mukerjee, R. (2003). On regular-fractional factorial experiments in row-column designs. *Journal of Statistical Planning and Inference*, **114**, 3–20.

Cheng, C.-S., and Steinberg, D. M.(1991). Trend robust two-level factorial designs. *Biometrika*, **78**, 325–336.

Cheng, C.-S., and Tang, B. (2001). Upper bounds on the number of columns in supersaturated designs. *Biometrika*, **84**(4), 1169–1174.

Cheng, C.-S., and Tang, B. (2005). A general theory of minimum aberration and its application. *Annals of Statistics*, **33**(2), 944–958.

Cheng, C.-S., Martin, R. J., and Tang, B. (1998). Two-level factorial designs with extreme numbers of level changes. *Annals of Statistics*, **26**(4), 1522–1539.

Cheng, C.-S., Steinberg, D. M., and Sun, D. X. (1999). Minimum aberration and model robustness for two-level fractional factorial designs. *Journal of the Royal Statistical Society B*, **61**(1), 85–93.

Cheng, S-W., and Wu, C. F. J. (1980). Balanced repeated measurements designs. *Annals of Statistics*, **8**, 1272–1283.

Cheng, S-W., and Wu, C. F. J. (2001). Factor screening and response surface exploration. *Statistica Sinica*, **11**, 553–604, with discussion.

Cheng, S-W., and Wu, C. F. J. (2002). Choice of optimal blocking schemes in two-level and three-level designs. *Technometrics*, **44**(3), 269–277.

Cheng, S-W., and Ye, K. Q. (2004). Geometric isomorphism and minimum aberration for factorial designs with quantitative factors. *Annals of Statistics*, **32**(5), 2168–2185.

Cheng, S-W., Li, W., and Ye, K. Q. (2004). Blocked nonregular two-level factorial designs. *Technometrics*, **46**(3), 269–279.

Cheng, T.-C. (2005). Robust regression diagnostics with data transformations. *Computational Statistics and Data Analysis*, **49**(3), 875–891.

Chernoff, H. (1953). Locally optimal designs for estimating parameters. *Annals of Mathematical Statistics*, **24**, 586–602.

Chernoff, H. (1959). Sequential design of experiments. *Annals of Mathematical Statistics*, **30**, 755–770.

Chernoff, H. (1999). Gustav Elfving's impact on experiment design. *Statistical Science*, **14**(2), 201–205.

Chew, V., ed. (1958). *Experimental Designs in Industry*. New York: Wiley.

Chiao, C.-H., and Hamada, M. (2001). Analyzing experiments with correlated multiple responses. *Journal of Quality Technology*, **33**(4), 451–465.

Chipman, H. (1998). Handling uncertainty in analysis of robust design experiments. *Journal of Quality Technology*, **30**(1), 11–17.

Chipman, H., and Hamada, M. S. (1996). Discussion. *Technometrics*, **38**, 317–320.

Chipman, H., Hamada, M., and Wu, C. J. F. (1997). A Bayesian variable-selection approach for analyzing designed experiments with complex aliasing. *Technometrics*, **39**(4), 372–381.

Ch'ng, C. K., Quah, S. H., and Low, H. C. (2005). A new approach for multiple response optimization. *Quality Engineering*, **17**(4), 621–626.

Cho, B. R. (1994). *Optimization Issues in Quality Engineering*. Norman: University of Oklahoma.

Cho, B. R., and Leonard, M. S. (1997). Identification and extensions of quasiconvex quality loss functions. *International Journal of Reliability and Quality Safety Engineering*, **4**, 191–204.

Cho, B. R., Kim, Y. J., Kimbler, D. L., and Phillips, M. D. (2000). An integrated joint optimization procedure for robust and tolerance design. *International Journal of Product Research*, **38**, 2309–2325.

Chopra, D. V. (1991). Further investigations on balanced arrays. *Computational Statistics and Data Analysis*, **12**, 231–237.

Chow, W. M. (1962). A note on the calculation of certain constrained maxima. *Technometrics*, **4**(1), 135–137.

Christensen, R., and Huzurbazar, A. V. (1996). A note on augmenting resolution III designs. *American Statistician*, **50**(2), 175–177.

Christians, N. F., Martin, D. P., and Karnock, K. J. (1981). The interrelationship among nutrient elements applied to calcareous sand greens. *Agronomy Journal*, **73**, 929–933.

Chu, C., and Hougan, O. A. (1961). Optimum design of a catalytic nitric oxide reactor. *Chemical Engineering Progress*, **57**, June, 51–58.

Claringbold, P. J. (1955). Use of the simplex design in the study of the joint action of related hormones. *Biometrics*, **11**, 174–185. **M**

Claringbold, P. J., Biggers, J. D., and Emmens, C. W. (1953). The angular transformation in quantal analysis. *Biometrics*, **9**, 467–484. **M**

Clark, J. B., and Dean, A. M. (2001). Equivalence of fractional factorial designs. *Statistica Sinica*, **11**, 537–547.

Clark, V. (1965). Choice of levels in polynomial regression with one or two variables. *Technometrics*, **7**(3), 325–333.

Clarke, G. M., and Kempson, R. E. (1997). *Introduction to the Design and Analysis of Experiments*. London: Arnold. (New York: Wiley.)

Clatworthy, W. H., Connor, W. S., Deming, L. S., and Zelen, M. (1957). Fractional factorial experimental designs for factors at two levels. *U. S. Department of Commerce, National Bureau of Standards, Applied Mathematics Series*, No. 48.

Clayton, C. A., Goldberg, M. M., and Potter, B. B. (1997). Design and analysis of an experiment for assessing cyanide in gold mining wastes. *Chemometrics and Intelligent Laboratory Systems*, **36**, 181–193. **M**

Cobb, G. W. (1998). *Introduction to Design and Analysis of Experiments*. New York: Springer.

Cochran, W. G. (1973). Experiments for nonlinear functions. *Journal of the American Statistical Association*, **68**, 771–781.

Cochran, W. G., and Cox, G. M. (1957) *Experimental Design*, New York: Wiley.

Coffman, C. B., and Genter, W. A. (1977). Responses of greenhouse-grown cannabis sativa L. to nitrogen, phosphorus, and potassium. *Agronomy Journal*, **69**, 832–836.

Colbourne, C. J., and Dinitz, J. H. (1996). *The CRC Handbook of Combinatorial Designs*. Boca Raton, FL: CRC Press.

Coleman, D. E., and Montgomery, D. C. (1993). A systematic approach to planning for a designed industrial experiment. *Technometrics*, **35**(1), 1–12; discussion: pp. 13–27.

Collins, B. J. (1989). Quick confounding. *Technometrics*, **31**(1), 107–110.

Colyer, D., and Kroth, E. H. (1968). Corn yield response and economic optima for nitrogen treatments and plant population over a seven-year period. *Agronomy Journal*, **60**, 524–529.

Connor, W. S., and Young, S. (1961). Fractional factorial designs for experiments with factors at two and three levels. *U. S. Department of Commerce, National Bureau of Standards, Applied Mathematics Series*, No. 58.

Connor, W. S., and Zelen, M. (1961). Fractional factorial experiment designs for factors at three levels. *U. S. Department of Commerce, National Bureau of Standards, Applied Mathematics Series*, No. 54.

Conrad, K. L., and Jones, P. R. (1965). Factorial design of experiments in ceramics, IV. Effect of composition, firing rate, and firing temperature. *American Ceramics Society Bulletin*, **44**, 616–619.

Constantine, G. M. (1987). *Combinatorial Theory and Statistical Design.* New York: Wiley.

Cook, R. D. (1977). Detection of influential observations in linear regression. *Technometrics*, **19**(1), 15–18.

Cook, R. D., and Nachtsheim, C. J. (1980). A comparison of algorithms for constructing exact D-optimal designs. *Technometrics*, **22**(3), 315–324.

Cook, R. D., and Nachtsheim, C. J. (1989). Computer-aided blocking of factorial and response surface designs. *Technometrics*, **31**(3), 339–346; see also Cook and Nachtsheim (1990).

Cook, R. D., and Nachtsheim, C. J. (1990). Letters to the Editor: Response to James M. Lucas, *Technometrics*, **32**(3), 363–365.

Cook, R. D., and Nachtsheim, C. J. (1994). Reweighting to achieve elliptically contoured covariates in regression. *Journal of the American Statistical Association*, **89**, 592–599.

Cook, R. D., and Thibodeau, L. A. (1980). Marginally restricted D-optimal designs. *Journal of the American Statistical Association*, **75**, 366–371.

Cook, R. D., and Weisberg, S. (1982). *Residuals and Influence in Regression.* London: Chapman & Hall.

Cook, R. D., and Weisberg, S. (1999). Graphs in statistical analysis: is the medium the message? *American Statistician*, **53**(1), 29–37.

Cook, R. D., and Wong, W. K. (1994). On the equivalence of constrained and compound optimal designs. *Journal of the American Statistical Association*, **89**, 687–692.

Cooper, H. R., Hughes, I. R., and Mathews, M. E. (1977). Application of response surface methodology to the evaluation of whey protein gel systems. *New Zealand Journal of Dairy Science Technology*, **12**, 248–252.

Cooper, J., Milas, J., and Wallis, W. D. (1978). Hadamard equivalence. *Lecture Notes in Mathematics*, **686**, 126–135.

Cooper, R. L. (1977). Response of soybean cultivars to narrow rows and planting rates under weed-free conditions. *Agronomy Journal*, **69**, 89–91.

Copeland, K. A. F., and Nelson, P. R. (1996). Dual response optimization via direct function minimization. *Journal of Quality Technology*, **28**(3), 331–336.

Cornelius, P. L., Templeton, W. C., and Taylor, T. H. (1979). Curve fitting by regression on smoothed singular vectors. *Biometrics*, **35**, 849–859.

Cornell, J. A. (1975). Some comments on designs for Cox's mixture polynomials. *Technometrics*, **17**(1), 25–35. **M**

Cornell, J. A. (1977). Weighted versus unweighted estimates using Scheffé's mixture model for symmetrical error variance patterns. *Technometrics*, **19**(3), 237–247. **M**

Cornell, J. A. (1988). Analyzing data from mixture experiments containing process variables: A split-plot approach. *Journal of Quality Technology*, **20**, 2–23. **M**

Cornell, J. A. (1990). Embedding mixture experiments inside factorial experiments. *Journal of Quality Technology*, **22**, 265–276. **M**

Cornell, J. A. (1995). Fitting models to data from mixture experiments containing other factors. *Journal of Quality Technology*, **27**(1), 13–33. **M**

Cornell, J. A. (1998). Mixture experiment research. Are we done? *American Statistical Association Proceedings of the Section on Physical and Engineering Sciences*, pp. 94–99. **M**

Cornell, J. A. (2000). Fitting a slack-variable model to mixture data: Some questions raised. *Journal of Quality Technology*, **32**(2), 133–147. **M**

Cornell, J. A. (2002). *Experiments with Mixtures*, 3rd ed. (*Designs, Models, and the Analysis of Mixture Data*.) New York: Wiley. **M**

Cornell, J. A., and Good, I. J. (1970). The mixture problem for categorized components. *Journal of the American Statistical Association*, **65**(1), 339–355. **M**

Cornell, J. A., and Gorman, J. W. (1978). On the detection of an additive blending component in multicomponent mixtures. *Biometrics*, **34**, 251–263. **M**

Cornell, J. A., and Gorman, J. W. (1984). Fractional design plans for process variables in mixture experiments. *Journal of Quality Technology*, **16**(1), 20–38. **M**

Cornell, J. A., and Gorman, J. W. (2003). Two new mixture models: Living with collinearity but removing its influence. *Journal of Quality Technology*, **35**(1), 78–88. **M**

Cornell, J. A., and Khuri, A. I. (1979). Obtaining constant prediction variance on concentric triangles for ternary mixture systems. *Technometrics*, **21**(2), 147–157. **M**

Cornell, J. A., and Montgomery, D. C. (1996a). Interaction models as alternatives to low-order polynomials. *Journal of Quality Technology*, **28**(2), 163–176.

Cornell, J. A., and Montgomery, D. C. (1996b). Fitting models to data: Interaction versus polynomial? Your choice! *Communications in Statistics, Theory and Methods*, **25**(11), 2531–2555.

Cornell, J. A., and Ott, L.(1975). The use of gradients to aid in the interpretation of mixture response surfaces. *Technometrics*, **17**(4), 409–424. **M**

Cornell, J. A., and Ramsey, P.J. (1998). A generalized mixture model for categorized-components problems with an application to a photoresist-coating experiment. *Technometrics*, **40**(1), 48–61. **M**

Coster, D. C. (1993a). Tables of minimum cost, linear trend-free run sequences for two- and three-level fractional factorial designs. *Computational Statistics and Data Analysis*, **16**, 325–336.

Coster, D. C. (1993b). Trend-free run orders of mixed-level fractional factorial designs. *Annals of Statistics*, **21**, 2072–2086.

Coster, D. C., and Cheng, C. S. (1988). Minimum cost trend-free run orders of fractional factorial designs. *Annals of Statistics*, **16**, 1188–1205.

Cote, R., Manson, A. R., and Hader, R. J. (1973). Minimum bias approximation of a general regression model. *Journal of the American Statistical Association*, **68**, 633–638.

Coteron, A., Sanchez, N., Martinez, M., and Aracil, J. (1993). Optimization of the synthesis of an analogue of Jojoba oil using a fully central composite design. *Canadian Journal of Chemical Engineering*, **71**, June, 485–488.

Cotter, S. C. (1979). A screening design for factorial experiments with interactions. *Biometrika*, **66**, 317–320.

Cox, D. R. (1958). *Planning of Experiments*. New York: Wiley.

Cox, D. R. (1971). A note on the polynomial response functions for mixtures. *Biometrika*, **58**(1), 155–159. **M**

Cox, F. R., and Reid, P. H. (1965). Interaction of plant population factors and level of production on the yield and grade of peanuts. *Agronomy Journal*, **57**, 455–457.

Cox, F. R., and Reid, N. (2000). *The Theory of the Design of Experiments*. Boca Raton, FL: Chapman & Hall/CRC.

Coxeter, H. S. M. (1948). *Regular Polytopes*. London: Methuen.

Coxeter, H. S. M. (1973). *Regular Polytopes*, 3rd ed. New York: Dover Books.

Cragle, R. G., Myers, R. M., Waugh, P. K., Hunter, J. S., and Anderson, R. L. (1955). The effects of various levels of sodium citrate, glycerol, and equilibration time on survival of bovine spermatozoa after storage at $-79°C$. *Journal of Dairy Science*, **38**(5), 508–514.

Crosier, R. B. (1984). Mixture experiments: geometry and pseudocomponents. *Technometrics*, **26**(3), 209–216. **M**

Crosier, R. B. (1986). The geometry of constrained mixture experiments. *Technometrics*, **28**(2), 95–102. **M**

Crosier, R. B. (1991). Symmetry in mixture experiments. *Communications in Statistics, Theory and Methods*, **20**, 1911–1935. **M**

Crosier, R. B. (1996). Symmetric orientations for simplex designs. *Journal of Quality Technology*, **28**, 148–152. **M**

Crosier, R. B. (2000). Some new two-level saturated designs. *Journal of Quality Technology*, **32**(2), 103–110.

Crosson, L. S., and Protz, R. (1973). Prediction of soil properties from stereorthophoto measurements of landform properties. *Canadian Journal of Soil Science*, **53**, 259–262.

Crowder, S. V., Jensen, K. L., Stephenson, W. R., and Vardeman, S. B. (1988). An interactive program for the analysis of data from two-level factorial experiments via probability plotting. *Journal of Quality Technology*, **20**(2), 140–148.

Crowther, E. M., and Yates, F. (1941). Fertiliser policy in wartime. *Empire Journal of Experimental Agriculture*. **19**, 77–97.

Curnow, R. N. (1972). The number of variables when searching for an optimum. *Journal of the Royal Statistical Society, Series B*, **34**, 461–476; discussion: pp. 477–481.

Curtin, A. C., De Angelis, M., Cipriani, M., Corbo, M. R., McSweeney, P. L. H., and Gobbetti, M. (2001). Amino acid catabolism in cheese-related bacteria: selection and study of the effects of pH, temperature and NaCl by quadratic response surface methodology. *Journal of Applied Microbiology*, **91**, 312–321.

Curtis, G. J., and Hornsey, K. G. (1972). Competition and yield compensation in relation to breeding sugar beet. *Journal of Agricultural Science*, **79**, 115–119.

Czitrom, V. (1988). Mixture experiments with process variables: D-optimal orthogonal experimental designs. *Communications in Statistics, Theory and Methods*, **17**(1), 105–121. **M**

Czitrom, V. (1989). Experimental designs for four mixture components with process variables, *Communications in Statistics, Theory and Methods*, **18**(12), 4561–4581. **M**

Czitrom, V. (1992). Note on a mixture experiment with process variables. *Communications in Statistics, Simulation and Computation*, **21**(2), 493–498. **M**

Czitrom, V. (1999). One-factor-at-a-time versus designed experiments. *American Statistician*, **53**(2), 126–131.

Czitrom, V., and Spagon, P. D. (1997). *Statistical Case Studies for Industrial Process Improvement*. Philadelphia: The American Statistical Association and The Society for Industrial and Applied Mathematics.

Dalal, S. R., and Mallows, C. L. (1998). Factor-covering designs for testing software. *Technometrics*, **40**(3), 234–243.

Dale, R. F., and Shaw, R. H. (1965). Effect on corn yields of moisture stress and stand at two fertility levels. *Agronomy Journal*, **57**, 475–479.

Daniel, C. (1956). Fractional replication in industrial research. *Proceedings of the 3rd Berkeley Symposium on Mathematical Statistics and Probability*, Vol. 8, pp. 87–98.

Daniel, C. (1959). Use of half-normal plots in interpreting factorial two-level experiments. *Technometrics*, **1**(4), 311–341.

Daniel, C. (1962). Sequences of fractional replicates in the 2^{p-q} series. *Journal of the American Statistical Association*, **57**, 403–429.

Daniel, C. (1973). One-at-a-time plans. *Journal of the American Statistical Association*, **68**, 353–360.

Daniel, C. (1976). *Applications of Statistics to Industrial Experimentation*. New York: Wiley.

Daniel, C., and Riblett, E. W. (1954). A multifactor experiment. *Industrial and Engineering Chemistry*, **46**, 1465–1468.

Daniel, C., and Wilcoxon, F. (1966). Factorial 2^{p-q} plans robust against linear and quadratic trends. *Technometrics*, **8**(2), 259–278.

Daniel, C., and Wood, F. S. (1980). *Fitting Equations to Data*. New York: Wiley.

Darroch, J. N., and Waller, J. (1985). Additivity and interaction in three-component experiments with mixtures. *Biometika*, **72**, 153–163. **M**

Das, A. D. (1976). A note on the construction of asymmetrical rotatable designs with blocks. *Journal of the Indian Society of Agricultural Statistics*, **28**, 91–96.

Das, M. N. (1960). Fractional replicates as asymmetrical factorial designs. *Journal of the Indian Society of Agricultural Statistics*, **12**, 159–174.

Das, M. N. (1961). Construction of rotatable designs from factorial designs. *Journal of the Indian Society of Agricultural Statistics*, **13**, 169–194.

Das, M. N. (1963). On construction of second order rotatable designs through BIB designs with blocks of unequal sizes. *Calcutta Statistical Association Bulletin*, **12**, 31–46.

Das, M. N., and Dey, A. (1967). Group divisible rotatable designs. *Annals of the Institute of Statistical Mathematics* (*Tokyo*), **19**, 331–347.

Das, M. N., and Dey, A. (1970). On blocking second order rotatable designs. *Calcutta Statistical Association Bulletin*, **19**, 75–85.

Das, M. N., and Gill, B. S. (1974). Blocking rotatable designs for agricultural experimentation. *Journal of the Indian Society for Agricultural Statistics*, **26**, 125–138.

Das, M. N., and Giri, N. C. (1979). *Design and Analysis of Experiments*. New Delhi: Wiley Eastern.

Das, M. N., and Mehta, J. S. (1968). Asymmetric rotatable designs and orthogonal transformations. *Technometrics*, **10**(2), 313–322.

Das, M. N., and Narasimham, V. L. (1962). Construction of rotatable designs through BIB designs. *Annals of Mathematical Statistics*, **33**, 1421–1439.

Das, M. N., and Nigam (1966). On a method of construction of rotatable designs with smaller number of points controlling the number of levels. *Calcutta Statistical Association Bulletin*, **15**, 147–157.

David, H. A. (1952). Upper 5 and 1% points of the maximum F-ratio. *Biometrika*, **39**, 422–424.

Davies, O. L., ed. (1978). *Design and Analysis of Industrial Experiments*, 4th ed. New York: Longman.

Dawid, A. P., and Sebastiani, P. (1999). Coherent dispersion criteria for optimal experimental design. *Annals of Statistics*, **27**(1), 65–81.

Dawson, K. S., Eller, T. J., and Carter, W. H. Jr. (2006). Response surface methods and their application in the treatment of cancer with drug combinations: some reflections. In *Response Surface Methodology and Related Topics* (A. I. Khuri, ed.), pp. 151–171. Hackensack, NJ: World Scientific Publishing.

Dean, A. M., and Draper, N. R. (1999). Saturated main-effect designs for factorial experiments. *Statistics and Computing*, **9**, 179–185.

Dean, A. M., and Lewis, S. M. (1992). Multidimensional designs for two-factor experiments. *Journal of the American Statistical Association*, **87**, 1158–1165.

Dean, A. M., and Lewis, S. M. (2002). Comparison of group screening strategies for factorial experiments. *Computational Statistics and Data Analysis*, **39**, 287–297.

Dean, A. M., and Voss, D. (1999). *Design and Analysis of Experiments*. New York: Springer.

Dean, A. M., Lewis, S. M., Draper, N. R., and Prescott, P. (1992). Use of a symbolic algebra computer system to investigate the properties of mixture designs in orthogonal blocks. In *Computational Statistics* (Y. Dodge and J. Whittaker, eds.), Vol. 2, pp. 215–220. Heidelberg: Physica-Verlag. **M**

De Baun, R. M. (1956). Block effects in the determination of optimum conditions. *Biometrics*, **12**, 20–22.

De Baun, R. M. (1959). Response surface designs for three factors at three levels. *Technometrics*, **1**(1), 1–8.

Deckman, D. A., and Van Winkle, M. (1959). Perforated plate column studies by the Box method of experimentation. *Industrial Engineering Chemistry*, **51**, 1015–1018.

De Cock, D., and Stufken, J. (2000). On finding mixed orthogonal arrays of strength 2 with many two-level factors. *Statistics and Probability Letters*, **50**, 383–388.

DeFeo, P., and Myers, R. H. (1992). A new look at experimental design robustness. *Biometrika*, **79**, 375–380.

de Launey, W. (1986). A survey of generalized Hadamard matrices and difference matrices $D(k, \lambda, G)$ with large k. *Utilitas Mathematica*, **30**, 5–29.

de Launey, W. (1987). On difference matrices, transversal designs, resolvable transversal designs and large sets of mutually orthogonal F-squares. *Journal of Statistical Planning and Inference*, **16**, 107–125.

Del Castillo, E. (1996). Multiresponse process optimization via constrained confidence regions. *Journal of Quality Technology*, **28**(1), 61–70.

Del Castillo, E., and Cahya, S. (2001). A tool for computing confidence regions on the stationary point of a response surface. *American Statistician*, **55**(4), 358–365.

Del Castillo, E., and Montgomery, D. C. (1993). A nonlinear programming solution to the dual response problem. *Journal of Quality Technology*, **25**(3), 199–204.

Del Castillo, E., Montgomery, D. C., and McCarville, D. (1996). Modified desirability functions for multiple response optimization. *Journal of Quality Technology*, **28**(3), 337–345.

Del Castillo, E., Fan, S. K., and Semple, J. (1997). The computation of global optima in dual response systems. *Journal of Quality Technology*, **29**(3), 347–353.

Del Castillo, E., Fan, S. K., and Semple, J. (1999). Optimization of dual response systems: A comprehensive procedure for degenerate and nondegenerate problems. *European Journal of Operational Research*, **112**, 174–186.

Del Castillo, E., Hadi, A. S., Conejo, A., and Fernandez-Canteli, A. (2004). A general method for local sensitivity analysis with application to regression models and other optimization problems. *Technometrics*, **46**(4), 430–444.

De Lury, D. B. (1960). *Values and Integrals of the Orthogonal Polynomials up to N = 26*. Toronto, Canada: University of Toronto Press.

de Mooy, C. J., and Pesek, J. (1966). Nodulation responses of soybeans to added phosphorus, potassium and calcium salts. *Agronomy Journal*, **58**, 275–280.

Denes, J., and Keedwell, A. D. (1991). *Latin Squares. New Developments in the Theory and Applications*. Amsterdam: Elsevier.

Deng, L.-Y., and Tang, B. (1999). Generalized resolution and minimum aberration criteria for Plackett-Burman and other nonregular factorial designs. *Statistica Sinica*, **9**, 1071–1082.

Deng, L.-Y., and Tang, B. (2002). Design selection and classification for Hadamard matrices using generalized minimum aberration criteria. *Technometrics*, **44**(2), 173–184.

Deng, L.-Y., Lin, D. K. J., and Wang, J. N. (1996a). Marginally oversaturated designs. *Communications in Statistics, Theory and Methods*, **25**(11), 2557–2573.

Deng, L.-Y., Lin, D. K. J., and Wang, J. N. (1996b). A measure of multifactor nonorthogonality. *Statistics and Probability Letters*, **28**, 203–209.

Deng, L.-Y., Lin, D. K. J., and Wang, J. N. (1999). On resolution rank criterion for supersaturated design. *Statistica Sinica*, **9**, 605–610.

Derringer, G. C. (1969a). Sequential method for estimating response surfaces. *Industrial and Engineering Chemistry*, **61**, 6–13.

Derringer, G. C. (1969b). Statistically designed experiments. *Rubber Age*, **101**, November, 66–76.

Derringer, G. C. (1974a). Variable shear rate viscosity of SBR-filler-plasticizer systems. *Rubber Chemistry and Technology*, **47**, 825–836.

Derringer, G. C. (1974b). An empirical model for viscosity of filled and plasticized elastomer compounds. *Journal of Applied Polymer Science*, **18**, 1083–1101.

Derringer, G. C. (1983). A statistical methodology for designing elastomer formulations to meet performance specifications. *Kautschukt und Gummi Kunststoffe*, **36**(5), 349–352.

Derringer, G. C. (1994). A balancing act: Optimizing a products properties. *Quality Progress*, **27**(6), June, pp. 51–58.

Derringer, G. C., and Suich, R. (1980). Simultaneous optimization of several response variables. *Journal of Quality Technology*, **12**(4), 214–219.

Dette, H. (1990). A generalization of D- and D_1-optimal designs in polynomial regression. *Annals of Statistics*, **18**, 1784–1804.

Dette, H. (1992a). Experimental designs for a class of weighted polynomial regression models. *Computational Statistics and Data Analysis*, **14**, 359–373.

Dette, H. (1992b). Optimal designs for a class of polynomials of odd or even degree. *Annals of Statistics*, **20**, 238–259.

Dette, H. (1993a). On a mixture of the D- and D_1-optimal designs in polynomial regression. *Journal of Statistical Planning and Inference*, **38**, 105–124.

Dette, H. (1993b). Elfving's theorem for D-optimality. *Annals of Statistics*, **21**, 753–766.

Dette, H. (1993c). A note on E-optimal designs for weighted polynomial regression. *Annals of Statistics*, **21**, 767–771.

Dette, H. (1994). Discrimination designs for for polynomial regression on compact intervals. *Annals of Statistics*, **22**, 890–903.

Dette, H. (1995). Optimal designs for identifying the degree of a polynomial regression. *Annals of Statistics*, **23**, 1248–1266.

Dette, H. (1996). A note on Bayesian C- and D-optimal designs in nonlinear regression models. *Annals of Statistics*, **24**, 1225–1234.

Dette, H. (1997a). Designing experiments with respect to 'standardized' optimality criteria. *Journal of the Royal Statistical Society, Series B*, **59**, 97–110.

Dette, H. (1997b). E-optimal designs for regression models with quantitative factors—a reasonable choice? *Canadian Journal of Statistics*, **25**(4), 531–543.

Dette, H. (2003). Efficient design of experiments in the Monod model. *Journal of the Royal Statistical Society, Series B*, **65**(3), 725–742.

Dette, H., and Franke, T. (2000). Constrained D- and D1-optimal designs for polynomial regression. *Annals of Statistics*, **28**(6), 1702–1727.

Dette, H., and Franke, T. (2001). Robust designs for polynomial regression by maximizing a minimum of D- and D_1-efficiencies. *Annals of Statistics*, **29**(4), 1024–1049.

Dette, H., and Haines, L. M. (1994). E-optimal designs for linear and nonlinear models with two parameters. *Biometrika*, **81**, 739–754.

Dette, H., and Haller, G. (1998). Optimal designs for the identification of the order of a Fourier regression. *Annals of Statistics*, **26**(4) 1496–1521.

Dette, H., and Kwiecen, R. (2002). A comparison of sequential and non-sequential designs for discrimination between nested regression models. Technical Report.

Dette, H., and Melas, V. B. (2003). Optimal designs for estimating individual coefficients in Fourier regression models. *Annals of Statistics*, **31**(5), 1669–1692.

Dette, H., and O'Brien, T. E. (1999). Optimality criteria for regression models based on predicted variance. *Biometrika*, **86**(1) 93–106.

Dette, H., and Roeder, I. (1997). Optimal discrimination designs for multifactor experiments. *Annals of Statistics*, **25**, 1161–1175; correction: (1998), **26**, 449.

Dette, H., and Studden, W. J. (1993). Geometry of E-optimality. *Annals of Statistics*, **21**, 416–433.

Dette, H., and Studden, W. J. (1995). Optimal designs for polynomial regression when the degree is not known. *Statistica Sinica*, **5**, 459–473.

Dette, H., and Wong, W. K. (1995a). On G-efficiency calculation for polynomial models. *Annals of Statistics*, **23**, 2081–2101.

Dette, H., and Wong, W. K. (1995b). Recurrence moment formulas for D-optimal designs. *Scandinavian Journal of Statistics, Theory and Applications*, **22**(4), 505–512.

Dette, H., and Wong, W. K. (1996). Robust optimal extrapolation designs. *Biometrika*, **83**, 667–680.

Dette, H., and Wong, W. K. (1998). Bayesian D-optimal designs on a fixed number of design points for heteroscedastic polynomial models. *Biometrika*, **85**(4), 869–882.

Dette, H., and Wong, W. K. (1999a). Optimal designs when the variance is a function of the mean. *Biometrics*, **55**(3), 925–929.

Dette, H., and Wong, W. K. (1999b). E-optimal designs for the Michaelis-Menten model. *Statistical Probability Letters*, **44**, 405–408.

Dette, H., Heiligers, B., and Studden, W. J. (1995). Minimax designs in linear regression models. *Annals of Statistics*, **23**, 30–40.

Dette, H., Haines, L. M., and Imhof, L. (1999). Optimal designs for rational models and weighted polynomial regression. *Annals of Statistics*, **27**(4), 1272–1293.

Dette, H., Melas, V. B., and Pepelyshev, A. (2004a). Optimal designs for estimating individual coefficients in a polynomial regression—A functional approach. *Journal of Statistical Planning and Inference*, **118**, 201–219.

Dette, H., Melas, V. B., and Pepelyshev, A. (2004b). Optimal designs for a class of nonlinear regression models. *Annals of Statistics*, **32**(5), 2142–2167.

Dette, H., Melas, V. B., and Pepelyshev, A. (2005). Optimal designs for three-dimensional shape analysis with spherical harmonic descriptors. *Annals of Statistics*, **33**(6), 2758–2788.

Dey, A. (1970). On response surface designs with equi-spaced doses. *Calcutta Statistical Association Bulletin*, **19**, 135–144.

Dey, A. (1985). *Orthogonal Fractional Factorial Designs*. Calcutta: Wiley Eastern.

Dey, A., and Das, M. N. (1970) On blocking second order rotatable designs. *Calcutta Statistical Association Bulletin*, **19**, 75–85.

Dey, A., and Kulshreshtha, A. C. (1973). Further second order rotatable designs. *Journal of the Indian Society of Agricultural Statistics*, **25**, 91–96.

Dey, A., and Mukerjee, R. (1999). *Fractional Factorial Plans*. New York: Wiley.

Dey, A., and Nigam, A. K. (1968). Group divisible rotatable designs—some further considerations. *Annals of the Institute of Statistical Mathematics* (*Tokyo*), **20**, 477–481.

Dey, A and Suen, C.-Y. (2002). Optimal fractional factorial plans for main effects and specified two-factor interactions: A projective geometric approach. *Annals of Statistics*, **30**(5), 1512–1523.

Diamond, N. T. (1991). The use of a class of foldover designs as search designs. *Australian Journal of Statistics*, **33**, 159–166.

Diamond, N. T. (1995). Some properties of a foldover design. *Australian Journal of Statistics*, **37**, 345–352.

Diamond, W. J. (1989). *Practical Experimental Designs*, 2nd ed. New York: Van Nostrand-Reinhold.

Diaz-Garcia, J. A., and Ramos-Quiroga, R. (2001). An approach to optimization in response surface methodology. *Communications in Statistics, Theory and Methods*, **30**(5), 827–835.

Diaz-Garcia, J. A., Ramos-Quiroga, R., and Cabrera-Vicencio, E. (2005). Stochastic programming methods in the response surface methodology. *Computational Statistics and Data Analysis*, **49**(3), 837–848.

Dickinson, A. W. (1974). Some run orders requiring a minimum number of factor level changes for the 2^4 and 2^5 main effect plans. *Technometrics*, **16**(1), 31–37.

Dillon, J. L. (1977). *The Analysis of Response in Crop and Livestock Production*, 2nd ed. Oxford: Pergamon.

Ding, R., Lin, D. K. J., and Wei, D. (2004). Dual-response surface optimization: A weighted MSE approach. *Quality Engineering*, **16**(3), 377–385.

Dingstad, G. Egelandsdal, B., and Naes, T. (2003). Modelling methods for crosses mixture experiments. A case study from sausage production. *Chemometrics and Intelligent Laboratory Systems*, **66**(2), 175–190. **M**

Dobson, A. J. (2002). *An Introduction to Generalized Linear Models*. Boca Raton, FL: Chapman & Hall/CRC.

Doehlbert, D. H. (1970). Uniform shell designs. *Journal of the Royal Statistical Society*, Series C, **19**, 231–239.

Donev, A. N. (1989). Design of experiments with both mixture and qualitative factors. *Journal of the Royal Statistical Society, Series B*, **51**(2), 297–302. **M**

Donev, A. N., and Atkinson, A. C. (1988). An adjustment algorithm for the construction of exact D-optimum experimental designs. *Technometrics*, **30**(4), 429–433.

Drain, D. (1997). *Handbook of Experimental Methods for Process Improvement*. New York: Chapman & Hall.

Draper, N. R. (1960a). Second order rotatable designs in four or more dimensions. *Annals of Mathematical Statistics*, **31**, 23–33.

Draper, N. R. (1960b). Third order rotatable designs in three dimensions. *Annals of Mathematical Statistics*, **31**, 865–874.

Draper, N. R. (1960c). A third order rotatable design in four dimensions. *Annals of Mathematical Statistics*, **31**, 875–877.

Draper, N. R. (1961a). Third order rotatable designs in three dimensions: Some specific designs. *Annals of Mathematical Statistics*, **32**, 910–913.

Draper, N. R. (1961b). Missing values in response surface designs. *Technometrics*, **3**(3), 389–398.

Draper, N. R. (1962). Third order rotatable designs in three factors: Analysis. *Technometrics*, **4**(2), 219–234.

Draper, N. R. (1963). "Ridge analysis" of response surfaces. *Technometrics*, **5**(4), 469–479.

Draper, N. R. (1982). Center points in response surface designs. *Technometrics*, **24**(2), 127–133.

Draper, N. R. (1983). Cubic lack of fit for three-level second order designs (letter to the editor). *Technometrics*, **25**(2), 217.

Draper, N. R. (1984). Schläflian rotatability. *Journal of the Royal Statististical Society*, Series B, **46**, 406–411.

Draper, N. R. (1985a). Small composite designs. *Technometrics*, **27**(2), 173–180.

Draper, N. R. (1985b). Response transformation: An example. *Journal of Food Science*, **50**, 523–525.

Draper, N. R. (2007). Central composite designs. *Encyclopedia of Statistics in Quality and Reliability*, (F. Ruggeri, R. Kenett, and F. Faltin, eds.) Hoboken, NJ: Wiley.

Draper, N. R. (2007). Ridge analysis. *Encyclopedia os Statistics in Quality and Reliability*, (F. Ruggeri, R. Kenett, and F. Faltin, eds.) Hoboken, NJ: Wiley.

Draper, N. R. (2007). Rotatable designs and rotatability. *Encyclopedia of Statistics in Quality and Reliability*, (F. Ruggeri, R. Kenett, and F. Faltin, eds.) Hoboken, NJ: Wiley.

Draper, N. R. (1999). Comment on Cook and Weisberg, 1999 (letter to the editor with response). *American Statistician*, **53**(3), 295–296.

Draper, N. R., and Beggs, W. J. (1971). Errors in the factor levels and experimental design. *Annals of Mathematical Statistics*, **42**, 46–58.

Draper, N. R., and Guttman, I. (1980). Incorporating overlap effects from neighboring units into response surface models. *Applied Statistics*, **29**, 128–134.

Draper, N. R., and Guttman, I. (1986). Response surface designs in flexible regions. *Journal of the American Statistical Association*, **81**, 1089–1094.

Draper, N. R., and Guttman, I. (1987). A comparison of alternative tests for lack of fit. *Technometrics*, **29**(1), 91–93.

Draper, N. R., and Guttman, I. (1988). An index of rotatability. *Technometrics*, **30**(1), 105–111.

Draper, N. R., and Guttman, I. (1992). Treating bias as variance for experimental design purposes. *Annals of the Institute of Statistical Mathematics*, **44**, 659–671.

Draper, N. R., and Guttman, I. (1996). Confidence intervals versus regions. *The Statistician*, **44**, 399–403.

Draper, N. R., and Guttman, I. (1997). Performing two-level factorial and fractional factorial designs in blocks of size two. *Journal of Quality Technology*, **29**(1), 71–75.

Draper, N. R., and Guttman, I. (2000). Assessing prior information for designed experiments. *Communications in Statistics, Theory and Methods*, **29**(9–10), 255–269.

Draper, N. R., and Herzberg, A. M. (1968). Further second order rotatable designs. *Annals of Mathematical Statistics*, **39**, 1995–2001.

Draper, N. R., and Herzberg, A. M. (1971). On lack of fit. *Technometrics*, **13**(2), 231–241. Correction: **14**(1), 245.

Draper, N. R., and Herzberg, A. M. (1973). Some designs for extrapolation outside a sphere. *Journal of the Royal Statistical Society*, Series B, 35, 268–276.

Draper, N. R., and Herzberg, A. M. (1979a). An investigation of first-order and second-order designs for extrapolation outside a hypersphere. *Canadian Journal of Statistics*, **7**, 97–101.

Draper, N. R., and Herzberg, A. M. (1979b). Designs to guard against outliers in the presence or absence of model bias. *Canadian Journal of Statistics*, **7**, 127–135.

Draper, N. R., and Herzberg, A. M. (1985a). Fourth order rotatability. *Communications in Statistics*, **B14**, 515–528.

Draper, N. R., and Herzberg, A. M. (1985b). A ridge regression sidelight. *American Statistician*, **41**, 282–283.

Draper, N. R., and Hunter, W. G. (1966). Design of experiments for parameter estimation in multiresponse situations. *Biometrika*, **53**, 525–533.

Draper, N. R., and Hunter, W. G. (1967a). The use of prior distributions in the design of experiments for parameter estimation in non-linear situations. *Biometrika*, **53**, 147–153.

Draper, N. R., and Hunter, W. G. (1967b). The use of prior distributions in the design of experiments for parameter estimation in non-linear situations: Multiresponse case. *Biometrika*, **54**, 662–665.

Draper, N. R., and Hunter, W. G. (1969). Transformations: some examples revisited. *Technometrics*, **11**(1), 23–40.

Draper, N. R., and John, J. A. (1988). Response surface designs for quantitative and qualitative variable. *Technometrics*, **30**(4), 423–428.

Draper, N. R., and John, J. A. (1998). Response surface designs where levels of some factors are difficult to change. *Australian and New Zealand Journal of Statistics*, **40**, 487–495.

Draper, N. R., and Joiner, B. L. (1984). Residuals with one degree of freedom. *American Statistician*, **38**, 55–57.

Draper, N. R., and Lawrence, W. E. (1965a). Designs which minimize model inadequacies: cuboidal regions of interest. *Biometrika*, **52**, 111–118.

Draper, N. R., and Lawrence, W. E. (1965b). Mixture designs for three factors. *Journal of the Royal Statistical Society, Series B*, **27**(3), 450–465. **M**

Draper, N. R., and Lawrence, W. E. (1965c). Mixture designs for four factors. *Journal of the Royal Statistical Society, Series B*, **27**(3), 473–478. **M**

Draper, N. R., and Lawrence, W. E. (1966). The use of second order "spherical" and "cuboidal" designs in the wrong regions. *Biometrika*, **53**, 596–599.

Draper, N. R., and Lawrence, W. E. (1967). Sequential designs for spherical weight functions. *Technometrics*, **9**, 517–529.

Draper, N. R., and Lin, D. J. K. (1990a). Capacity considerations for two-level fractional factorial designs. *Journal of Statistical Planning and Inference*, **24**, 25–35.

Draper, N. R., and Lin, D. J. K. (1990b). Small response surface designs. *Technometrics*, **32**(2), 187–194.

Draper, N. R., and Lin, D. J. K. (1990c). Connections between two-level designs of resolution III* and V. *Technometrics*, **32**(3), 283–288.

Draper, N. R., and Lin, D. J. K. (1995). Characterizing projected designs—repeat and mirror image runs. *Communications in Statistics, Theory and Methods*, **24**, 775–795.

Draper, N. R., and Lin, D. J. K. (1996). Response surface designs. In *Handbook of Statistics* (S. Ghosh and C. R. Rao, eds.), Vol. 13, pp. 343–375. Amsterdam: Elsevier.

Draper, N. R., and Mezaki, R. (1973). On the adequacy of a multi-site model suggested for the catalytic hydrogenation of olefins. (Letter to the Editors.) *Journal of Catalysis*, **28**(1), 179–181.

Draper, N. R., and Mitchell, T. J. (1967). The construction of saturated 2_R^{k-p} designs. *Annals of Mathematical Statistics*, **38**, 1110–1126.

Draper, N. R., and Mitchell, T. J. (1968). Construction of the set of 256-run designs of resolution ≥ 5 and the set of even 512-run designs of resolution ≥ 6 with special reference to the unique saturated designs. *Annals of Mathematical Statistics*, **39**, 246–255.

Draper, N. R., and Mitchell, T. J. (1970). Construction of a set of 512-run designs of resolution ≥ 5 and a set of even 1024-run designs of resolution ≥ 6. *Annals of Mathematical Statistics*, **41**, 876–887.

Draper, N. R., and Pukelsheim, F. (1990). Another look at rotatability. *Technometrics*, **32**(2), 195–202.

Draper, N. R., and Pukelsheim, F. (1994). On third-order rotatability. *Metrika*, **41**, 137–161.

Draper, N. R., and Pukelsheim, F. (1996). An overview of design of experiments. *Statistical Papers*, **37**, 1–32.

Draper, N. R., and Pukelsheim, F. (1997). Kshirsager and Cheng's rotatability measure (letter to the editor). *Australian Journal of Statistics*, **39**(2), 236–238.

Draper, N. R., and Pukelsheim, F. (1998a). Polynomial representations for response surface modeling. *IMS Lecture Notes—Monograph Series*, **34**, 199–212.

Draper, N. R., and Pukelsheim, F. (1998b). Mixture models based on homogeneous polynomials. *Journal of Statistical Planning and Inference*, **71**, 303–311. **M**

Draper, N. R., and Pukelsheim, F. (1999). Kiefer ordering of simplex designs for first- and second-degree mixture models. *Journal of Statistical Planning and Inference*, **79**, 325–384. **M**

Draper, N. R., and Pukelsheim, F. (2000). Ridge analysis of mixture response surfaces. *Statistics and Probability Letters*, **48**, 131–140. **M**

Draper, N. R., and Pukelsheim, F. (2002). Generalized ridge analysis under linear restrictions, with particular applications to mixture experiments problems. *Technometrics*, **44**(3), 250–259. See also Draper and Pukelsheim (2003a).

Draper, N. R., and Pukelsheim, F. (2003). Canonical reduction of second order fitted models subject to linear restrictions. *Statistics and Probability Letters*, **63**, 401–410. **M**

Draper, N. R., and Sanders, E. (1988). Designs for minimum bias estimation. *Technometrics*, **30**(3), 319–325.

Draper, N. R., and Smith, H. (1998). *Applied Regression Analysis*, 3rd ed. New York: Wiley. **M**

Draper, N. R., and St John, R. C. (1975). D-optimality for regression designs: A review. *Technometrics*, **17**(1), 15–23.

Draper, N. R., and St. John, R. C. (1977a). A mixtures model with inverse terms. *Technometrics*, **19**(1), 37–46. **M**

Draper, N. R., and St. John, R. C. (1977b). Designs in three and four components for mixtures models with inverse terms. *Technometrics*, **19**(2), 117–130. **M**

Draper, N. R., and Stoneman, D. M. (1964). Estimating missing values in unreplicated two-level factorial and fractional factorial designs. *Biometrics*, **20**, 443–458.

Draper, N. R., and Stoneman, D. M. (1966). Alias relationships for two-level Plackett and Burman designs. *Technical Report—University of Wisconsin, Statistics Department*, No. 96.

Draper, N. R., and Stoneman, D. M. (1968a). Response surface designs for factors at two and three levels and at two and four levels. *Technometrics*, **10**(1), 177–192.

Draper, N. R., and Stoneman, D. M. (1968b). Factor changes and linear trends in eight-run two-level factorial designs. *Technometrics*, **10**(2), 301–311.

Draper, N. R., and Van Nostrand, R. C. (1979). Ridge regression and James-Stein estimation: Review and comments. *Technometrics*, **21**(4), 451–466.

Draper, N. R., and Ying, L. H. (1994). A note on slope rotatability over all directions. *Journal of Statistical Planning and Inference*, **41**, 113–119.

Draper, N. R., Guttman, I., and Lipow, P. (1977). All-bias designs for spline functions joined at the axes. *Journal of the American Statistical Association*, **72**, 424–429.

Draper, N. R., Gaffke, N., and Pukelsheim, F. (1991). First and second order rotatability of experimental designs, moment matrices, and information surfaces. *Metrika*, **38**, 129–161.

Draper, N. R., Gaffke, N., and Pukelsheim, F. (1993). Rotatability of variance surfaces and moment matrices. *Journal of Statistical Planning and Inference*, **36**, 347–356.

Draper, N. R., Prescott, P., Lewis, S. M., Dean, A. M., John, P. W. M., and Tuck, M. G., (1993), Mixture designs for four components in orthogonal blocks. *Technometrics*, **35**(3), 268–276; corrections: (1994), **36**(2), 234. **M**; see also comment by Snee (1995) and two data corrections given in the response to Snee.

Draper, N. R., Davis, T. P., Pozueta, L., and Grove, D. (1994). Isolation of degrees of freedom for Box-Behnken designs. *Technometrics*, **36**(3), 283–291.

Draper, N. R., Heiligers, B., and Pukelsheim, F. (1996). On optimal third-order rotatable designs. *Annals of the Institute of Statistical Mathematics*, **48**, 395–402.

Draper, N. R., Heiligers, B., and Pukelsheim, F. (2000). Kiefer ordering of simplex design for second degree mixture models with four or more ingredients. *Annals of Statistics*, **28**, 578–590. **M**

DuMochel, W., and Jones, B. (1994). A simple Bayesian modification of D-optimal designs to reduce dependence on an assumed model. *Technometrics*, **36**(1), 37–47.

Duthie, A. I. (1993). A class of E-optimal hypercubic designs. *Communications in Statistics, Theory and Methods*, **22**(3), 717–722.

Dyke, G. V. (1988). *Comparative Experiments with Field Crops*. London: Griffin.

Dykstra, O. (1956). Partial duplication of factorial experiments. *Technometrics*, **1**(1), 63–75.

Dykstra, O. (1960). Partial duplication of response surface designs. *Technometrics*, **2**(2), 185–195.

Dykstra, O. (1971). The augmentation of experimental data to maximize det $\mathbf{X'X}$. *Technometrics*, **13**(3), 682–688.

Easterling, R. G. (2004). Teaching experimental design. *American Statistician*, **58**(3), 244–252.

Edmondson, R. N. (1991). Agricultural response surface experiments based on four-level factorial designs. *Biometrics*, **47**, 1435–1448.

Edmondson, R. N. (1993). Systematic row-and-column designs balanced for low order polynomial interactions between rows and columns. *Journal of the Royal Statistical Society, Series B*, **55**, 707–723.

Edmondson, R. N. (1994). Fractional factorial designs for factors with a prime number of quantitative levels. *Journal of the Royal Statistical Society, Series B*, **56**(4), 611–622.

Efromovich, S. (1998). Simultaneous sharp estimation of functions and their derivatives. *Annals of Statistics*, **26**(1), 273–278.

Efron, B. (1998). R. A. Fisher in the 21st century. *Statistical Science*, **13**(2), 95–114; discussion: pp. 114–122.

Eik, K., and Hanway, J. J. (1965). Some factors affecting development and longevity of leaves of corn. *Agronomy Journal* **57**, 7–12.

Ellerton, R. R. W., and Tsai, W.-Y. (1979). Minimum bias estimation and the selection of polynomial terms for response surfaces. *Biometrics*, **35**, 631–635.

Ellis, S. R. M., Jeffreys, G. V., and Wharton, J. T. (1964). Raschig synthesis of hydrazine. Investigation of chloramines formation reaction. *Industrial and Engineering Chemistry Proceedings of Design and Development*, **3**, 18–22.

Elsayed, E., and Chen, A. (1993). Optimal levels of process parameters for products with multiple characteristics. *International Journal of Product Research*, **31**(5), 1117–1132.

Emanuel, J. T., and Palanisamy, M. (2000). Sequential experimentation using two-level fractional factorials. *Quality Engineering*, **12**(3), 335–346.

Engel, B., and Walstra, P. (1991). Increasing precision or reducing expense in regression experiments by using information from a concomitant variable. *Biometrics*, **47**, 13–20; see also Engel and Walstra (1996).

Engel, B., and Walstra, P. (1996). *Biometrics*, **52**, 371–372.

Engel, J. (1992). Modelling variation in industrial experiments. *Applied Statistics*, **41**(3), 579–593.

Engel, J., and Huele, A. F. (1996a). Taguchi parameter design by second-order response surfaces. *Quality and Reliability Engineering International*, **12**, 95–100.

Engel, J., and Huele, A. F. (1996b). A generalized linear modeling approach to robust design. *Technometrics*, **38**(4), 365–373.

Erickson, R. V., Fabian, V., and Marik, J. (1995). An optimum design for estimating the first derivative. *Annals of Statistics*, **23**(4), 1234–1247.

Estes, G. O., Koch, D. W., and Breutsch, T. F. (1973). *Agronomy Journal*, **65**, 972–975.

Etzioni, R., and Kadane, J. B. (1993). Optimal experimental design for another's analysis. *Journal of the American Statistical Association*, **88**, 1404–1411.

Evangelaras, H., and Koukouvinos, C. (2003a). On the use of Hadamard matrices in factorial designs. *Utilitas Mathematica*, **64**, 45–63.

Evangelaras, H., and Koukouvinos, C. (2003b). Screening properties and design selection of certain two-level designs. *Journal of Modern and Applied Statistical Methods*, **2**, 87–107.

Evangelaras, H., and Koukouvinos, C. (2004a). Inequivalent projections of some D-optimal designs. *Utilitas Mathematica*, **65**, 83–96.

Evangelaras, H., and Koukouvinos, C. (2004b). Another look at projection properties of Hadamard matrices. *Communications in Statistics, Theory and Methods*, **33**, 1607–1620.

Evangelaras, H., and Koukouvinos, C. (2004c). Combined arrays with minimum number of runs and maximum estimation efficiency. *Communications in Statistics, Theory and Methods*, **33**, 1621–1628.

Evangelaras, H., and Koukouvinos, C. (2006). A comparison between the Groebner bases approach and hidden projection properties in factorial designs. *Computational Statistics and Data Analysis*, **50**(1), 77–88.

Evangelaras, H., Georgiou, S., and Koukouvinos, C. (2003). Inequivalent projections of Hadamard matrices of orders 16 and 20. *Metrika*, **57**, 29–35.

Evangelaras, H., Kotsireas, I., and Koukouvinos, C. (2003). Application of Gröbner bases to the analysis of certain two or three level factorial designs. *Advances and Applications in Statistics*, **3**, 1–13.

Evangelaras, H., Georgiou, S., and Koukouvinos, C. (2004). Evaluation of inequivalent projections of Hadamard matrices of order 24. *Metrika*, **59**, 51–73.

Evangelaras, H., Kolaiti, E., and Koukouvinos, C. (2004). Regular fractional factorial designs: resolution, defining relations and blocking via coding theory. *Advances and Applications in Statistics*, **4**, 45–58.

Evangelaras, H., Koukouvinos, C., Dean, A. M., and Dingus, C. A. (2005). Projection properties of certain three-level orthogonal arrays. *Metrika*, **62**, 241–257.

Evans, J. W. (1979). Computer augmentation of experimental designs to maximize det $\mathbf{X'X}$. *Technometrics*, **21**(3), 321–330.

Evans, J. W., and Manson, A. R. (1978). Optimal experimental design in two dimensions using minimum bias estimation. *Journal of the American Statistical Association*, **73**, 171–176.

Fan, S.-K. S. (1999). Computational schema on ridge analysis. *Communications in Statistics, Simulation and Computation*, **28**, 767–783.

Fan, S.-K. S. (2000). A generalized global optimization algorithm for dual response systems. *Journal of Quality Technology*, **32**(4), 444–456.

Fan, S-. K., and Chaloner, K. (2006). Design for a trinomial response to dose. In *Response Surface Methodology and Related Topics* (A. I. Khuri, ed.), pp. 225–250. Hackensack, NJ: World Scientific Publishing.

Fang, K. T., and Lin, D. K. J. (2003). Uniform experimental designs and their applications in industry. In *Handbook of Statistics* (R. Khattree and C. R. Rao, eds.), Vol. 22, pp. 131–170. Amsterdam: Elsevier.

Fang, K.-T., and Mukerjee, R. (2000). A connection between uniformity and aberration in regular fractions of two-level factorials. *Biometrika*, **87**(1), 193–198.

Fang, K.-T., and Mukerjee, R. (2004). Optimal selection of augmented pairs designs for response surface modeling. *Technometrics*, **46**(2), 147–152.

Fang, K.-T., Lin, D.K.J., and Ma, C. X. (2000). On the construction of multi-level supersaturated designs. *Journal of Statistical Planning and Inference*, **86**, 239–252.

Fang, K.-T., Lin, D. K. J., Winker, P., and Zhang, Y. (2000). Uniform design: Theory and application. *Technometrics*, **42**(3), 237–248.

Fang, K.-T., Lin, D. K. J., and Ma, C. X. (2001). Wrap-around L_2 discrepancy of random sampling. Latin hypercube and uniform designs. *Journal of Complexity*, **17**, 608–624.

Fang, K.-T., Lin, D. K. J., and Qiu, H. (2003). A note on optimal foldover design. *Statistics and Probability Letters*, **62**, 245–250.

Fang, Z., and Wiens, D.P. (1999). Robust extrapolation designs and weights for biased regression models with heteroscedastic errors. *Canadian Journal of Statistics*, **27**(4), 751–770.

Fearn, T. (1983). A misuse of ridge regression in the calibration of a near infrared reflectance instrument. *Applied Statistics*, **32**(1), 73–79; however, see Hoerl et al. (1985).

Federer, W. T. (1999). *Statistical Design and Analysis for Intercropping Experiments*. Vol. 2. *Three or More Crops*. New York: Springer.

Federer, W. T., and Meredith, M. P. (1992). Covariance analysis for split-plot and split-block designs. *American Statistician*, **46**(2), 155–162.

Federov, V. V. (1972). *Theory of Optimal Experiments*. New York: Academic Press.

Federov, V. V., and Hackl, P. (1997). Model-Oriented Design of Experiments . New York: Springer.

Feeny, A. E., and Landwehr, J. M. (1995). Graphical analysis for a large designed experiment. *Technometrics*, **37**(1), 1–14.

Fellman, J. (1999). Gustav Elfving's contribution to the emergence of the optimal experimental design theory. *Statistical Science*, **14**(2) 197–200.

Ferrante, G. R. (1962). Laboratory evaluation of new creaseproofing agents using statistical techniques. *American Dyestuff Reporter*, **51**, January 22, 41–43.

Ferrer, A. J., and Romero, R. (1993). Small samples estimation of dispersion effects from unreplicated data. *Communications in Statistics, Simulation and Computation*, **222**, 975–995.

Ferrer, A. J., and Romero, R. (1995). A simple method to study dispersion effects from non-necessary replicated data in industrial context. *Quality Engineering*, **7**, 747–755.

Feuell, H. J., and Wagg, R. E. (1949). Statistical methods in detergent investigation. *Research*, **2**, 334–337.

Fieller, E. C. (1955). Some problems in interval estimation. *Journal of the Royal Statistical Society, Series B*, **16**, 175–185.

Filliben, J. J., and Li, K.-C. (1997). A systematic approach to the analysis of complex interaction patterns in two-level factorial designs. *Technometrics*, **39**(3), 286–297.

Finney, D. J. (1945). The fractional replication of factorial arrangements. *Annals of Eugenics*, **12**, 291–301.

Firth, D., and Hinde, J. P. (1997a). On Bayesian D-optimum design criteria and the equivalence theorem in nonlinear models. *Journal of the Royal Statistical Society, Series B*, **59**, 793–797.

Firth, D., and Hinde, J. P. (1997b). Parameter neutral optimum design for nonlinear models. *Journal of the Royal Statistical Society, Series B*, **59**, 799–811.

Fisher, R. A. (1921a). Studies in crop variation. I. An examination of the yield of dressed grain from Broadbalk. *Journal of Agricultural Science*, **11**, 107–135.

Fisher, R. A. (1921b). Studies in crop variation. II. The manorial response of different potato varieties. *Journal of Agricultural Science*, **13**, 311–320.

Fisher, R. A. (1924). The influence of rainfall on the yield of wheat at Rothamsted. *Philosophical Transactions, Series B*, **213**, 89–142.

Fisher, R. A. (1935). See (1966).

Fisher, R. A. (1942). The theory of confounding in factorial experiments, in relation to the theory of groups. *Annals of Eugenics*, **11**, 341–353.

Fisher, R. A.(1945). A system of confounding for factors with more than two alternatives giving completely orthogonal cubes and higher powers. *Annals of Eugenics*, **12**, 283–290.

Fisher, R. A. (1963). *Statistical Methods for Research Workers*, 13th ed. Edinburgh: Oliver & Boyd.

Fisher, R. A. (1966). *The Design of Experiments*, 8th ed. Edinburgh: Oliver & Boyd.

Fisher, R. A., and Yates, F. (1957). *Statistical Tables for Biological, Agricultural and Medical Research*, 5th ed. London: Longman Group.

Fisher, R. A., and Yates, F. (1963). *Statistical Tables for Biological, Agricultural and Medical Research*, 6th ed. Edinburgh: Longman Group.

Floros, J. D., and Chinman, M. S. (1988) Seven factor response surface optimization of a double stage lye (NaOH) peeling process for pimiento peppers. *Journal of Food Science*, **53**, 631–638.

Fontana, R., Pistone, G., and Rogantin, M. P. (2000). Classification of two-level factorial fractions. *Journal of Statistical Planning and Inference*, **87**, 149–172.

Ford, I., Torsney, B., and Wu, C. F. J. (1992). The use of a canonical form in the construction of locally optimal designs for nonlinear problems. *Journal of the Royal Statistical Society, Series B*, **54**, 569–583.

Frankel, S. A. (1961) Statistical design of experiments for process development of MBT. *Rubber Age*, **89**, 453–458.

Franklin, M. F. (1984). Constructing tables of minimum aberration p^{n-m} designs. *Technometrics*, **26**(3), 225–232.

Franklin, M. F. (1985). Selecting defining contrasts and confounded effects in p^{n-m} factorial experiments. *Technometrics*, **27**(2), 165–172.

Franklin, M. F., and Bailey, R. A. (1977). Selecting defining contrasts and confounded effects in two-level experiments. *Applied Statistics*, **26**, 321–326.

Franklin, N. L., Pinchbeck, P. H., and Popper, F. (1956). A statistical approach to catalyst development. Part I. The effect of process variables on the vapour phase oxidation of naphthalene. *Transactions of the Institution of Chemical Engineers*, **34**, 280–293.

Franklin, N. L., Pinchbeck, P. H., and Popper, F. (1958), A statistical approach to catalyst development. Part II. The integration of process and catalyst variables in the vapour phase oxidation of napthalene. *Transactions of the Institution of Chemical Engineers*, **36**, 259–269.

Freckleton, R. P., and Watkinson, A. R. (2000). Designs for greenhouse studies of interactions between plants: an analytical perspective. *Journal of Ecology*, **88**, 386–391.

Freeman, M. F., and Tukey, J. W. (1950). Transformations related to the angular and square root. *Annals of Mathematical Statistics*, **21**, 607–611.

Freeny, A. E., and Landwehr, J. M. (1995). Graphical analysis for a large designed experiment. *Technometrics*, **37**(1), 1–14.

Fries, A., and Hunter, W. G. (1980). Minimum aberration 2^{k-p} designs. *Technometrics*, **22**(4), 601–608.

Frigon, N.L., and Mathews, D. (1997). *Practical Guide to Experimental Design*. New York: Wiley.

Frisbee, S. E., and McGinity, J. W. (1994). Influence of nonionic surfactants on the physical and chemical properties of a biodegradable pseudolatex. *European Journal of Pharmaceutics and Biopharmaceutics*, **40**, 355–363.

Gad, S. C. (1999). *Statistics and Experimental Design for Toxicologists*. Boca Raton: CRC Press.

Gaffke, N., and Heiligers, B. (1996). Approximate designs for polynomial regression: Invariance, admissibility and optimality. In Handbook of Statistics (S. Ghosh and C. R. Rao, eds.), Vol. 13, pp. 1149–1199. Amsterdam: Elsevier.

Gaffke, N., and Kraft, O. (1982). Exact D-optimum designs for quadratic regression. *Journal of the Royal Statistical Society, Series B*, **44**, 394–397.

Gaido, J. J., and Terhune, H. D. (1961). Evaluation of variables in the pressure-kier bleaching of cotton. *American Dyestuff Reporter*, **50**, October 16, 23–26, 32.

Galil, Z., and Kiefer, J. C. (1977a). Comparison of rotatable designs for regression on balls, I (Quadratic). *Journal of Statistical Planning and Inference*, **1**, 27–40.

Galil, Z., and Kiefer, J. C. (1977b). Comparison of designs for quadratic regression on cubes, I (Quadratic). *Journal of Statistical Planning and Inference*, **1**, 121–132.

Galil, Z., and Kiefer, J. C. (1977c). Comparison of Box-Draper and D-optimum designs for experiments with mixtures. *Technometrics*, **19**(4), 441–444. **M**

Galil, Z., and Kiefer, J. C. (1977d). Comparison of simplex designs for quadratic mixture models. *Technometrics*, **19**(4), 445–453. **M**

Galil, Z., and Kiefer, J. C. (1979). Extrapolation designs and ϕ_p-optimum designs for cubic regression on the q-ball. *Journal of Statistical Planning and Inference*, **3**, 27–38.

Galil, Z., and Kiefer, J. C. (1980). Time and space saving computer methods, related to Mitchell's DETMAX for finding D-optimum designs. *Technometrics*, **22**(3), 301–313.

Galil, Z., and Kiefer, J. C. (1982). Construction methods for D-optimum weighing designs when $n = 3 \pmod 4$. *Annals of Statistics*, **10**, 502–510.

Ganju, J., and Lucas, J. M. (1997). Bias in test statistics when restrictions in randomization are caused by factors. *Communications in Statistics, Theory and Methods*, **26**, 47–63.

Ganju, J., and Lucas, J. M. (1999). Detecting randomization restrictions caused by factors. *Journal of Statistical Planning and Inference*, **81**, 129–140.

Ganju, J., and Lucas, J. M. (2000). Analysis of unbalanced data from an experiment with random block effects and unequally spaced factor levels. *American Statistician*, **54**(1), 5–11.

Ganju, J., and Lucas, J. M. (2005). Randomized and random run order experiments. *Journal of Statistical Planning and Inference*, **133**, 199–210.

Ganju, J., and Lucas, J. M. (2006). Random run order, randomization and inadvertent split-plots in response surface experiments. In *Response Surface Methodology and Related Topics* (A. I. Khuri, ed.), pp. 47–64. Hackensack, NJ: World Scientific Publishing.

Gardiner, D. A., and Cowser, K. (1961). Optimization of radionuclide removal from low-level process wastes by the use of response surface methods. *Health Physics*, **5**, 70–78.

Gardiner, D. A., Grandage, A. H. E., and Hader, R. J. (1959). Third order rotatable designs for exploring response surfaces. *Annals of Mathematical Statistics*, **30**, 1082–1096.

Gardiner, W. P., and Gettinby, G. (1998). *Experimental Design Techniques in Statistical Practice*: *A Practical Software-Based Approach*. Chichester, UK: Ellis Horwood.

Gartner, N. H. (1976). On the application of a response surface methodology to traffic signal settings. *Transportation Research*, **10**, 59–60.

Gates, C. E. (1995). What really is experimental error in block designs? *American Statistician*, **49**, 362–363; see also Gates (1996).

Gates, C. E. (1996). Comment and reply. *American Statistician*, **50**, 386–387.

Gelman, A. (2000). Should we take measurements at an intermediate design point? *Biostatistics*, **1**(1), 27–34.

Gelman, A. (2005). Analysis of variance—why it is more important than ever. *Annals of Statistics*, **33**(1), 1–31; discussion: pp. 31–53.

Gennings, C., Chinchilli, V. M., and Carter, W. H., Jr., (1989). Response surface analysis with correlated data: A nonlinear model approach. *Journal of the American Statistical Association*, **84**, 805–809.

Gennings, C., Dawson, K. S., Carter, W. H., Jr., and Myers, R. H. (1990). Interpreting plots of a multidimensional dose-response surface in a parallel coordinate system. *Biometrics*, **46**, 719–735.

George, K. C., and Das, M. N. (1966). A type of central composite response surface designs. *Journal of the Indian Society of Agricultural Statistics*, **18**, 21–29.

Georgiou, S., and Koukouvinos, C. (2003). On equivalence of Hadamard matrices and projection properties. *Ars Combinatoria*, **69**, 79–95.

Georgiou, S., Koukouvinos, C., and Stylianou, S. (2002). On good matrices, skew Hadamard matrices and optimal designs. *Computational Statistics and Data Analysis*, **41**(1), 171–184.

Georgiou, S., Koukouvinos, C., Mitrouli, M., and Seberry, J. (2002a). Necessary and sufficient conditions for two variable orthogonal designs in order 44: Addendum. *Journal of Combinatorial Mathematics and Combinatorial Computing*, **34**, 59–64.

Georgiou, S., Koukouvinos, C., Mitrouli, M., and Seberry, J. (2002b). Necessary and sufficient conditions for three and four variable orthogonal designs in order 36. *Journal of Statistical Planning and Inference*, **106**, 329–352.

Georgiou, S., Koukouvinos, C., and Seberry, J. (2003). Hadamard matrices, orthogonal designs and construction algorithms. In *Designs 2002: Further Computational and Constructive Design Theory* (W. D. Wallis, ed.), pp. 133–205. Norwell, MA: Kluwer Academic Publishers.

Georgiou, S., Koukouvinos, C., and Stylianou, S. (2004). Construction of new skew Hadamard matrices and their use in screening experiments. *Computational Statistics and Data Analysis*, **45**(3), 423–429.

Geramita, A. V., and Seberry, J. (1979). *Orthogonal Designs: Quadratic Forms and Hadamard Matrices*. New York: Dekker.

Geramita, A. V., Geramita, J. M., and Seberry, J. (1976). Orthogonal designs. *Linear and Multilinear Algebra*, **3**, 281–306.

Ghosh, S. (1979). On robustness of designs against incomplete data. *Sankhya*, **3 & 4**, 204–208.

Ghosh, S. (1980). On main effect plus one plans for 2^m factorials. *Annals of Statistics*, **8**(4), 922–930.

Ghosh, S. (1982). Robustness of BIBD against the unavailability of data. *Journal of Statistical Planning and Inference*, **6**, 29–32.

Ghosh, S. (1989). On two methods of identifying influential sets of observations. *Statistics and Probability Letters*, **7**, 241–245.

Ghosh, S. (1991). *Statistical Design and Analysis of Industrial Experiments*. New York: Marcel Dekker.

Ghosh, S., and Al-Sabah, W. S. (1996). Efficient composite designs with small number of runs. *Journal of Statistical Planning and Inference*, **53**, 117–132.

Ghosh, S., and Burns, C. (2002). Comparisons of four new classes of search designs for factor screening experiments with factors at three levels. *Australian and New Zealand Journal of Statistics*, **44**(3), 901–910.

Ghosh, S., and Derderian, E. (1995). Determination of robust design against noise factors and in presence of signal factors. Communications in Statistics, Simulation and Computation, **24**(2), 309–326.

Ghosh, S., and Divecha, J. (1997). Two associate class partially balanced incomplete block designs and partial diallel crosses. *Biometrika*, **84**(1), 245–248.

Ghosh, S., and Duh, Y.-J. (1991). Adjusting residuals in estimation of dispersion effects. *Australian Journal of Statistics*, **33**(1), 65–74.

Ghosh, S., and Duh, Y.-J. (1992). Determination of optimum experimental conditions using dispersion main effects and interactions of factors in replicated factorial experiments. *Journal of Applied Statistics*, **19**, 367–378.

Ghosh, S., and Fairchild, L. D. (2000). Inference on interactions between treatments and subjects within groups in a two-period crossover trial. *Journal of Statistical Planning and Inference*, **88**(2), 301–317.

Ghosh, S., and Lagergren, E. (1991). Dispersion models and estimation of dispersion effects in replicated factorial experiments. *Journal of Statistical Planning and Inference*, **26**(3), 253–262.

Ghosh, S., and Liu, T. (1999). Optimal mixture designs for four components in two orthogonal blocks. *Journal of Statistical Planning and Inference*, **78**, 219–228. **M**

Ghosh, S., and Rao, C.R., eds. (1996). *Handbook of Statistics*, Vol. 13: *Design and Analysis of Experiments*. Amsterdam: Elsevier.

Ghosh, S., and Talebi, H. (1993). Main effect plans with additional search property for 2m factorial experiments. *Journal of Statistical Planning and Inference*, **36**(2–3), 367–384.

Ghosh, S., and Teschmacher, L. (2002). Comparison of search designs using search probabilities. *Journal of Statistical Planning and Inference*, **104**(2), 439–458.

Ghosh, S., and Zhang, X. D. (1987). Two new series of search designs for 3^m factorial experiments. *Utilitas Mathematica*, **32**, 245–254.

Ghoudi, K., and McDonald, D. (2000). Cramer-von Mises regression. *Canadian Journal of Statistics*, **28**(4), 689–714.

Giddings, A.P., Bailey, T. G., and Moore J. T. (2001). Optimality analysis of facility location problems using response surface methodology. *International Journal of Physical Distribution and Logistic Management*, **31**, 38–52.

Giesbrecht, F. G., and Gumpertz, M. L. (2004). *Planning, Construction, and Statistical Analysis of Comparative Experiments*. Hoboken, NJ: Wiley.

Giglio, B., Riccomagno, E., and Wynn, H. P. (2000). Gröbner bases strategies in regression. *Journal of Applied Statistics*, **27**, 923–938.

Gill, B. S., and Das, M. N. (1974). Response surface designs for conduct of agricultural experimentation. *Journal of the Indian Society of Agricultural Statistics*, **26**, 19–32.

Gill, E. P., Murray, W., and Wright, M. H. (1981). *Practical Optimization*. London: Academic Press.

Gilmour, S. G., and Draper, N. R. (2003). Confidence regions around the ridge of optimal response on fitted second-order response surfaces. *Technometrics*, **45**(3), 333–339; see also Gilmour and Draper (2004).

Gilmour, S. G., and Draper, N. R. (2004) Response to Letter to the Editor. *Technometrics*, **46**(3), 358.

Gilmour, S. G., and Mead, R. (1995). Stopping rules for sequences of factorial designs. *Applied Statistics*, **44**(3), 343–355.

Gilmour, S. G., and Mead, R. (1996). Fixing a factor in the sequential design of two-level fractional factorial experiments. *Journal of Applied Statistics*, **23**(1), 21–29.

Gilmour, S. G., and Ringrose, T.J. (1999). Controlling processes in food technology by simplifying the canonical form of fitted response surfaces. *Applied Statistics*, **48**(1), 91–101.

Gilmour, S. G., and Trinca, L. A. (2000). Some practical advice on polynomial regression analysis from blocked response surface designs. *Communications in Statistics, Theory and Methods*, **29**(9–10), 2157–2180.

Gilmour, S. G., and Trinca, L. A. (2003). Row-column response surface designs. *Journal of Quality Technology*, **35**(2), 184–193.

Gilmour, S. G., and Trinca, L. A. (2006). Response surface experiments on processes with high variation. In *Response Surface Methodology and Related Topics*, (A. I. Khuri, ed.), pp. 19–46. Hackensack, NJ: World Scientific Publishing.

Ginebra, J., and Clayton, M. K. (1995). Response surface bandits. *Journal of the Royal Statistical Society, Series B*, **57**(4), 771–784.

Giovannitti-Jensen, A., and Myers, R. H. (1989). Graphical assessment of the prediction capability of response surface designs. *Technometrics*, **31**(2), 159–171.

Gliner, J. A., and Morgan, G. A. (2000). *Research Methods in Applied Settings: An Integrated Approach to Design and Analysis*. Mahwa, NJ: Erlbaum.

Godolphin, J. D. (2004). Simple pilot procedures for the avoidance of disconnected experimental designs. *Applied Statistics*, **53**(1), 133–147.

Goel, B. S., and Nigam, A. K. (1979). Sequential exploration in mixture experiments. *Journal of the Indian Society of Agricultural Statistics*, **21**, 277–285. **M**

Goel, P. K., and Ginebra, J. (2003). When is one experiment 'always better than' another? *The Statistician*, **52**(4), 515–537.

Goh, D., and Street, D. J. (1998). Projective properties of small Hadamard matrices and fractional factorial designs. *Journal of Combinatorial Mathematics and Computing*, **28**, 141–148.

Goh, T. N. (1993). Taguchi methods: some technical, cultural and pedagogical perspectives. *Quality and Reliability Engineering International*, **9**, 185–202.

Goh, T. N. (1996). Economical experimentation via 'lean design.' *Quality and Reliability Engineering International*, **12**, 383–388.

Goldfarb, H. B., and Montgomery, D. C. (2006). Graphical methods for comparing response surface designs for experiments with mixture components. In *Response Surface Methodology and Related Topics*, (A. I. Khuri, ed.), pp. 329–348. Hackensack, NJ: World Scientific Publishing. **M**

Goldfarb, H. B., Borror, C. M., and Montgomery, D. C. (2003). Mixture-process variable experiments with noise variables. *Journal of Quality Technology*, **35**(4), 393–405. **M**

Goldfarb, H. B., Anderson-Cook, C. M., Borror, C. M., and Montgomery, D. C. (2004). Fraction of design space plots for assessing mixture and mixture-process designs. *Journal of Quality Technology*, **36**(2), 169–179. **M**

Goldfarb, H. B., Borror, C. M., Montgomery, D. C., and Anderson-Cook, C. M. (2004a). Three-dimensional variance dispersion graphs for mixture-process experiments. *Journal of Quality Technology*, **36**(1), 109–124. **M**

Goldfarb, H. B., Borror, C. M., Montgomery, D. C., and Anderson-Cook, C. M. (2004b). Evaluating mixture-process designs with control and noise variables. *Journal of Quality Technology*, **36**(3), 245–262. **M**

Goldfarb, H. B., Borror, C. M., Montgomery, D. C., and Anderson-Cook, C. M. (2005). Using genetic algorithms to generate mixture process experimental designs involving control and noise variables. *Journal of Quality Technology*, **37**(1), 60–74. **M**

Gontard, N., Guilbert, S,, and Cuq, J. L. (1992). Edible wheat gluten films: Influence of the main process variables on film properties using response surface methodology. *Journal of Food Science*, **57**(1), 190–195, 199.

Goos, P. (2002). *The Optimal Design of Blocked and Split-Plot Experiments*. Secaucus, NJ: Springer-Verlag.

Goos, P., and Donev, A. N. (2006). The D-optimal design of blocked experiments with mixture components. *Journal of Quality Technology*, **38**(4), 319–332. **M**

Goos, P., and Vandenbroek, M. (2001a). Optimal split-plot designs. *Journal of Quality Technology*, **33**(4), 436–450.

Goos, P., and Vandenbroek, M. (2001b). D-optimal response surface designs in the presence of random block effects. *Computational Statistics and Data Analysis*, **37**(4), 433–453.

Goos, P., and Vandenbroek, M. (2003). D-optimal split-plot designs with given numbers and sizes of whole plots. *Technometrics*, **45**(3), 235–245.

Goos, P., and Vandenbroek, M. (2004). Outperforming completely randomized designs. *Journal of Quality Technology*, **36**(1), 12–26.

Goos, P., Langhans, I., and Vandenbroek, M. (2006). Practical inference from industrial split-plot designs. *Journal of Quality Technology*, **38**(2), 162–179.

Goos, P., Kobilinsky, A., O'Brien, T. E., and Vandebroek (2005). Model-robust and model-sensitive designs. *Computational Statistics and Data Analysis*, **49**(1), 201–216.

Gopalan, R., and Dey, A. (1976). On robust experimental designs. *Sankhya, Series B*, **38**, 297–299.

Gorman, J. W., (1966). Discussion of "Extreme vertices design of mixture experiments" by R. A. McLean and V. L. Anderson. *Technometrics*, **8**(3), 455–456. **M**

Gorman, J. W., and Cornell, J. A. (1982). A note on fitting equations to freezing point data exhibiting eutetics for binary and ternary mixture systems. *Technometrics*, **24**(3), 229–239. **M**

Gorman, J. W., and Cornell, J. A. (1985). A note on model reduction for experiments with both mixture components and process variables. *Technometrics*, **27**(3), 243–247. **M**

Gorman J. W., and Hinman, J. E. (1962). Simplex-lattice designs for multi-component systems. *Technometrics*, **4**(4), 463–487. **M**

Goswami, K. K. (1992). On the construction of orthogonal factorial designs of resolution IV. *Communications in Statistics, Theory and Methods*, **21**(12), 3561–3570.

Gouda, M., Thakur, M., and Karanth, N. (2001). Optimization of the multienzyme system for sucrose biosensor by response surface methodology. *World Journal of Microbiology and Biotechnology*, **17**(6), 595–600.

Greenfield, A. A. (1976). Selection of defining contrasts in two-level experiments. *Applied Statistics*, **25**, 64–67.

Grego, J. M. (1993). Generalized linear models and process variation. *Journal of Quality Technology*, **25** (4), 288–295.

Grego, J. M., Lewis, J. F., and Craney, T. A. (2000). Quantile plots for mean effects in the presence of variance effects for 2^{k-p} designs. *Communications in Statistics, Simulation and Computation*, **29**(4) 1109–1133.

Grize, Y. L. (1991). Plotting scaled effects from unreplicated orthogonal experiments. *Journal of Quality Technology*, **23**(3), 205–212.

Grove, C. C. (1965). The influence of temperature on the scouring of raw wool. *American Dyestuff Reporter*, **54**, 13–16.

Grove, D. M., and Davis, T. P. (1991). Taguchi's idle column method. *Technometrics*, **33**(3), 349–353.

Grove, D. M., and Davis, T. P. (1992). *Engineering Quality and Experimental Design*. New York: Wiley.

Gunst, R. F and McDonald, G. C. (1996). The importance of outcome dynamics, simple geometry, and pragmatic statistical arguments in exposing deficiencies of experimental design strategies. *American Statistician*, **50**, 44–50.

Gunter, B. (1993). Through a funnel slowly with ball bearing and insight to teach experimental design. *American Statistician*, **47**, 265–269.

Gupta, B. C. (1992). On the existence of main effect plus k plans for 2^m factorials and tables for main effect plus 1 and 2 plans for 2^7 factorials. *Communications in Statistics, Theory and Methods*, **21**(4), 1137–1143.

Gupta, T. K., and Dey, A. (1975). On some new second order rotatable designs. *Annals of the Institute of Statistical Mathematics* (Tokyo), **27**, 167–175.

Gupta, V. K., and Nigam, A. K. (1982). On a model useful for approximating fertilizer response relationships. *Journal of the Indian Society of Agricultural Statistics*, **34**(3), 61–74.

Gupta, V. K., and Singh, R. (1989). On E-optimal block designs. *Biometrika*, **76**, 184–188.

Haaland, P. D., and O'Connell, M. R. (1985). Inference for effect-saturated fractional factorials. *Technometrics*, **37**(1), 82–93.

Haaland, P. D., McMillan, N., Nycha, D., and Welch, W. (1994). Analysis of space-filling designs. In *Computing Science and Statistics. Computationally Intensive Statistical Methods. Proceedings of the 26th Symposium on the Interface*, pp. 111-120. Fairfax Station, VA: Interface Foundation of North America.

Hackler, W. C., Kriegel, W. W., and Hader, R. J. (1956). Effect of raw-material ratios on absorption of whiteware compositions. *Journal of the American Ceramic Society*, **39**(1), 20–25.

Hackney, H., and Jones, P. R. (1967). Response surface for dry modulus of rupture and drying shrinkage. *American Ceramic Society Bulletin*, **46**(8), 745–749.

Hadamard, J. (1893). Resolution d'une question relative aux determinants. *Bulletin des Sciences Mathamatiques*, **17**, 240–246.

Hader, R. J., and Park, S. H. (1978). Slope-rotatable central composite designs. *Technometrics*, **20**(4), 413–417.

Hader, R. J., Harward, M. E., Mason, D. D., and Moore, D. P. (1957). An investigation of some of the relationships of copper, iron and molybdenum in the growth and nutrition of lettuce: I. Experimental design and statistical methods for characterizing the response surface. *Soil Science Society of America Proceedings*, **21**, 59–64.

Hahn, G. J. (1976a). Process improvement using evolutionary operation. *Chemical Technology*, **6**, 204–206.

Hahn, G. J. (1976b). Process improvement using simplex evolutionary operation. *Chemical Technology*, **6**, 343–345.

Hahn, G. J. (1984). Experimental design in the complex world. *Technometrics*, **26**(1), 19–31.

Hahn, G. J. (1989). Statistics-aided manufacturing: A look into the future. *American Statistician*, **43**, 74–79.

Hahn, G. J., and Hoerl, R. (1998). Key challenges for statisticians in business and industry. *Technometrics*, **40**(3), 195–200; discussion, pp. 201–213.

Hahn, G. J., and Meeker, W. Q. (1993). Assumptions for statistical inference. *American Statistician*, **47**, 1–11.

Hahn, G. J., Feder, P. I., and Meeker, W. Q. (1976). The evaluation and comparison of experimental designs for fitting regression functions. *Journal of Quality Technology*, **8**(3), 140–157.

Hahn, G. J., Feder, P. I., and Meeker, W. Q. (1978a). Evaluating the effect of incorrect specification of a regression model. *Journal of Quality Technology*, **10**(2), 61–72.

Hahn, G. J., Feder, P. I., and Meeker, W. Q. (1978b). Evaluating the effect of incorrect specification of a regression model. Part 2. Further example and discussion. *Journal of Quality Technology*, **10**(3), 93–98.

Hahn, G. J., Morgan, C. B., and Schmec, J. (1981). The analysis of a fractional factorial experiment with censored data using iterative least squares. *Technometrics*, **23**(1), 33–36.

Haines, L. M. (1987). The application of the annealing algorithm to the construction of the exact optimal designs for linear-regression models. *Technometrics*, **29**(4), 439–437.

Haines, L. M. (2006). Evaluating the performance of non-standard designs: The San Cristobal design. In *Response Surface Methodology and Related Topics*, (A. I. Khuri, ed), pp. 251–281. Hackensack, NJ: World Scientific Publishing.

Hall, M., Jr. (1961). Hadamard matrix of order 16. *Jet Propulsion Laboratory, Research Summary*, **1**, 21–26.

Hall, M., Jr. (1965). Hadamard matrix of order 20. *California Institute of Technology, Technical Report—Jet Propulsion Laboratory*, 32–761.

Hall, P., and Rau, C. (2000). Tracking a smooth fault line in a response surface. *Annals of Statistics*, **28**(3), 713–733.

Hamada, M., and Nelder, J. A. (1997). Generalized linear models for quality-improvement experiments. *Journal of Quality Technology*, **29**(3), 292–304.

Hamada, M., and Wu, C. J. F. (1992). Analysis of designed experiments with complex aliasing. *Journal of Quality Technology*, **24**(3), 130–137.

Hamada, M., and Wu, C. J. F. (1995). The treatment of related experimental factors by sliding levels. *Journal of Quality Technology*, **27**(1), 45–55.

Hamada, M., and Wu, C. J. F. (2000). *Experiments: Planning, Analysis and Parameter Design Optimization*. New York: Wiley.

Hamada, M., Martz, H. F., Reese, C. S., and Wilson, A. G. (2001). Finding near-optimal Bayesian experimental designs via genetic algorithms. *American Statistician*, **55**(3), 175–181.

Hammons, A. R., Kumar, P. V., Calderbank, A. R., Sloane, N. J. A., and Sole, P. (1994). The Z_4-linearity of Kerdock, Preparata, Goethals, and related codes. *IEEE Transactions on Information Theory*, **40**, 301–319.

Hare, L. B. (1979). Designs for mixture experiments involving process variables. *Technometrics*, **21**(2), 159–173. **M**

Hare, L. B. (1988). In the soup: A case study to identify contributors to filling variability. *Journal of Quality Technology*, **20**(1), 36–43.

Hare, L. B., and Brown, P. L. (1977). Plotting response surface contours for component mixtures. *Journal of Quality Technology*, **9**(4), 193–197. **M**

Harrington, E. C., Jr. (1965). The desirability function. *Industrial Quality Control*, **21**, 494–498.

Harris, N. (1988). Taguchi methods. Special issue of *Quality and Reliability Engineering*. Chichester, UK: Wiley.

Hartley, H. O. (1950). The maximum F-ratio as a short-cut test for heterogeneity of variance. *Biometrika*, **37**, 308–312.

Hartley, H. O. (1959). Smallest composite design for quadratic response surfaces. *Biometrics*, **15**, 611–624.

Harville, D. A. (1977). *Matrix Algebra from a Statistician's Perspective*. New York: Springer.

Haslin, C., and Pellegrini, M. (2001). Culture medium composition for optimal thallus regeneration in the Red Alga Asparagopsis armata Harvey (Rhodophyta, Bonnemaisoniaceae). *Botanica Marina*. **44**(1), 23–30.

Hassemer, D. J., Weibe, D. A., and Kramer, T. (1989). The Wisconsin cholesterol study: An assessment of laboratory performance. *Clinical Chemistry Abstracts*, **35**(6), 1068.

He, Z., Studden, W. J., and Sun, D. (1996). Optimal designs for rational models. *Annals of Statistics*, **24**, 2128–2147.

Heady, E. O., and Pesek, J. (1954). A fertilizer production surface with specification of economic optima for corn growth on calcareous Ida silt loam. *Journal of Farm Economics*, **36**(3), 466–482.

Healy, M. J. R. (1995). Frank Yates, 1902–1994. The work of a statistician. *International Statistical Review*, **63**(3), 271–288..

Heapt, L. A., Robertson, J. A., McBeath, D. K., von Maydell, U. M., Love, H. C., and Webster, G. R. (1976a). Development of a barley yield equation for Central Alberta. 1. Effects of soil and fertilizers, N and P. *Canadian Journal of Soil Science*, **56**, 233–247.

Heapt, L. A., Robertson, J. A., McBeath, D. K., von Maydell, U. M., Love, H. C., and Webster, G. R. (1976b). Development of a barley yield equation for Central Alberta. 2. Effects of soil moisture stress. *Canadian Journal of Soil Science*, **56**, 249–256.

Heavlin, W. D. (2003). Designing experiments for causal networks. *Technometrics*, **45**(2), 115–129.

Hebble, T. L., and Mitchell, T. J. (1972). 'Repairing' response surface designs. *Technometrics*, **14**(3), 767–779.

Hedayat, A. S. (1990). New properties of orthogonal arrays and their statistical applications. In *Statistical Design and Analysis of Industrial Experiments*, (S. Ghosh, ed.), pp. 407–422. New York: Dekker.

Hedayat, A. S., and John, P. W. M. (1974). Resistant and susceptible BIB designs. *Annals of Statistics*, **2**, 148–158.

Hedayat, A. S., and Majumdar, D. (1985). Families of A-optimal block designs for comparing test treatments with a control. *Annals of Statistics*, **13**, 757–767.

Hedayat, A. S., and Stufken, J. (1989). On the maximum number of constraints in orthogonal arrays. *Annals of Statistics*, **17**, 448–451.

Hedayat, A. S., and Stufken, J. (1999). Compound orthogonal arrays. *Technometrics*, **41**(1), 57–61.

Hedayat, A. S., and Wallis, W. D. (1978). Hadamard matrices and their applications. *Annals of Statistics*, **6**, 1184–1238.

Hedayat, A. S., and Yang, M. (2005). Optimal and efficient crossover designs for comparing test treatments with a control treatment. *Annals of Statistics*, **33**(2), 915–943.

Hedayat, A. S., Pu, K., and Stufken, J. (1992). On the construction of asymmetrical orthogonal arrays. *Annals of Statistics*, **20**, 2142–2152.

Hedayat, A. S., Stufken, J., and Su, G. (1997). On the construction and existence of orthogonal arrays with three levels. *Annals of Statistics*, **25**, 2044–2053.

Hedayat, A. S., Sloan, N. J. A., and Stufken, J. (1999). *Orthogonal Arrays: Theory and Applications*. New York: Springer-Verlag.

Heiberger, R. M. (1991). *Computation for the Analysis of Designed Experiments*. New York: Wiley.

Heiberger, R. M., Bhaumik, D, K., and Holland, B. (1993). Optimal data augmentation strategies for additive models. *Journal of the American Statistical Association*, **88**, 926–938.

Heiligers, B. (1994). E-optimal designs in weighted polynomial regression. *Annals of Statistics*, **22**, 917–929.

Heise, M. A., and Myers, R. H. (1996). Optimal designs for bivariate logistic regression. *Biometrics*, **52**, 613–624.

Heller, N. B., and Staats, G. E. (1973). Response surface optimization when experimental factors are subject to costs and constraints. *Technometrics*, **15**(1), 113–123.

Henderson, H. V., and Searle, S. R. (1981). The vec-permutation matrix, the vec operator and Kronecker products: A review. *Linear and Multilinear Algebra*, **9**, 271–288.

Henderson, H. V., Pukelsheim, F., and Searle, S. R. (1983). On the history of the Kronecker product. Linear and Multilinear Algebra, **14**, 113–120.

Henderson, M. S., and Robinson, D. L. (1982). Environmental influences on fiber component concentrations of warm-season perennial grasses. *Agronomy Journal*, **74**, 573–579.

Hendry, D. G., Mayo, F. R., and Schetzle, D. (1968). Stability of butadiene polyperoxide. *Industrial Engineering Chemistry Product Research and Development*, **7**, 145–151.

Heo, G., Schmuland, B., and Wiens, D. P. (2001). Restricted minimax robust designs for misspecified regression models. *Canadian Journal of Statistics*, **29**(1), 117–128.

Heredia-Langner, A., Loredo, E. N., Montgomery, D. C., and Griffen, A. H. (2000). Optimization of a bonded leads process using statistically designed experiments. *Robotics and Computer Integrated Manufacturing*, **16**, 377–382.

Heredia-Langner, A., Carlyle, W. M., Montgomery, D. C., Borror, C. M., and Runger, G. C. (2003). Genetic algorithms for the construction of D-optimal designs. *Journal of Quality Technology*, **35**(1), 28–46.

Heredia-Langner, A., Montgomery, D. C., Carlyle, W. M., and Borror, C. M. (2004). Model-robust designs: a genetic algorithm approach. *Journal of Quality Technology*, **36**(3), 263–279.

Hermanson, H. P. (1961). The fertility of some Minnesota peat soils. Ph.D. thesis, University of Minnesota. (University Microfilms, Ann Arbor, MI, No. 61-5846.)

Hermanson, H. P. (1965). Maximization of potato yield under constraint. *Agronomy Journal*, **57**, 210–213.

Hermanson, H. P., Gates, C. E., Chapman, J. W., and Farnham, R. S. (1964). An agronomically useful three-factor response surface design based on dodecahedron symmetry. *Agronomy Journal*, **56**, 14–17.

Herzberg, A. M. (1964). Two third order rotatable designs in four dimensions. *Annals of Mathematical Statistics*, **35**, 445–446.

Herzberg, A. M. (1966). Cylindrically rotatable designs. *Annals of Mathematical Statistics*, **37**, 242–247.

Herzberg, A. M. (1967a). Cylindrically rotatable designs of types 1, 2 and 3. *Annals of Mathematical Statistics*, **38**, 167–176.

Herzberg, A. M. (1967b). A method for the construction of second order rotatable designs in k dimensions. *Annals of Mathematical Statistics*, **38**, 177–180.

Herzberg, A. M. (1967c). The behavior of the variance function of the difference between two estimated responses. *Journal of the Royal Statistical Society, Series B*, **29**, 174–179.

Herzberg, A. M. (1979). Are theoretical designs applicable? *Operations Research Verpharen / Methods of Operations Research*, **30**, 68–76.

Herzberg, A. M. (1982a). The robust design of experiments: A review. *Serdica Bulgaricae Mathematicae Publicationes*, **8**, 223–228.

Herzberg, A. M. (1982b). The design of experiments for correlated error structures: layout and robustness. *Canadian Journal of Statistics*, **10**, 133–138.

Herzberg, A. M., and Andrews, D. F. (1976). Some considerations in the optimal design of experiments in non-optimal situations. *Journal of the Royal Statistical Society, Series B*, **38**, 284–289.

Herzberg, A. M., and Andrews, D. F. (1978). The robustness of chain block designs and coat-of-mail designs. *Communications in Statistics, Theory and Methods*, **7**, 479–485.

Herzberg, A. M., and Cox, D. R. (1969). Recent work on the design of experiments: A bibliography and a review. *Journal of the Royal Statistical Society, Series A*, **132**, 29–67. **R**

Herzberg, A. M., and Cox, D. R. (1972). Some optimal designs for interpolation and extrapolation. *Biometrika*, **59**, 551–561.

Herzberg, A. M., and Traves, W. N. (1994). An optimal experimental design for the Haar regression model. *Canadian Journal of Statistics*, **22**(3), 357–364.

Herzberg, A. M., Garner, C. W. L., and Springer, B. G. F. (1973). Kiss-precise sequential rotatable designs. *Canadian Mathematical Bulletin*, **16**, 207–217.

Higham, J. (1998). Row-complete Latin squares of every composite order exist. *Journal of Combination and Design*, **6**, 63–77.

Hilgers, R-D., (1999), Design efficiency and estimation for component amount models, *Journal of the Indian Society of Agricultural Statistics*, **41**, 783–798. **M**

Hilgers, R-D., and Bauer, P. (1995), Optimal designs for mixture amount experiments, *Journal of Statistical Planning and Inference*, **48**, 241–246. **M**

Hill, H. M. (1960). Experimental designs to adjust for time trends. *Technometrics*, **2**(1), 67–82.

Hill, W. J., and Hunter, W. G. (1966). A review of response surface methodology: A literature survey. *Technometrics*, **8**(4), 571–590. **R**

Hillyer, M. J., and Roth, P. M. (1972). Planning of experiments when the experimental region is constrained. Application of linear transformations to factorial design. *Chemical Engineering* Science, **27**, 187–197.

Hinkelmann, K., and Kempthorne, O. (1994). *Design and Analysis of Experiments*, Vol. 1. *Introduction to Experimental Design*. New York: Wiley.

Hinkelmann, K., and Kempthorne, O. (2005). *Design and Analysis of Experiments*, Vol, 2. Hoboken, NJ: Wiley.

Hodnett, G. E. (1956). The use of response curves in the analysis and planning of series of experiments with fertilizers. *Empire Journal of Experimental Agriculture*, **24**, 205–212.

Hoel, P. G. (1968). On testing for the degree of a polynomial. *Technometrics*, **10**(4), 757–767.

Hoel, P. G., and Jennrich, R. I. (1979). Optimal designs for dose response experiments in cancer research. *Biometrika*, **66**, 307–316.

Hoerl, A. E. (1959). Optimum solution of many variables equations. *Chemical Engineering Progress*, **55**(11), 69–78.

Hoerl, A. E. (1960). Statistical analysis of an industrial production problem. *Industrial Engineering Chemistry*, **52**, 513–514.

Hoerl, A. E. (1962). Application of ridge analysis to regression problems. *Chemical Engineering* Progress, 58(3), 54–59.

Hoerl, A. E. (1964). Ridge analysis. *Chemical Engineering Progress, Symposium Series*, **60**, 67–77.

Hoerl, A. E., and Kennard, R. W. (1970a). Ridge regression: Biased estimation for nonorthogonal problems. *Technometrics*, **12**(1), 55–67.

Hoerl, A. E., and Kennard, R. W. (1970b). Ridge regression: Applications to nonorthogonal problems. *Technometrics*, **12**(1), 69–82; correction: **12**(3), 723.

Hoerl, A. E., Kennard, R. W., and Hoerl, R. W. (1985). Practical use of ridge regression: A challenge met. *Applied Statistics*, **34**(2), 114–120.

Hoerl, R. W. (1985). Ridge analysis 25 years later. *American Statistician*, **39**, 186–192. **R**

Hoerl, R. W. (1987). The application of ridge techniques to mixture data: Ridge analysis. *Technometrics*, **29**(2), 161–172. **M**

Hoerl, R. W., Hooper, J. H., Jacobs, P. J., and Lucas, J. M. (1993). Skills for industrial statisticians to survive and prosper in the emerging quality environment. *American Statistician*, **47**, 280–292.

Hoeting, J. A. (1998). Sandbars in the Colorado river: An environmental consulting project. *Statistical Science*, **13**(1), 9-13; discussion: pp. 23–29.

Hoke, A. T. (1974). Economical second-order designs based on irregular fractions of the 3^n factorial. *Technometrics*, **16**(3), 375–384.

Holcomb, D. R., and Carlyle, W. M. (2002). Some notes on the construction and evaluation of supersaturated designs. *Quality and Reliability Engineering International*, **18**, 299–304.

Holcomb, D. R., Montgomery, D. C., and Carlyle, W. M. (2003). Analysis of supersaturated designs. *Journal of Quality Technology*, **35**(1), 13–27.

Holliday, T., Lawrance, A. J., and Davis, T. P. (1998). Engine-mapping experiments: A two-stage regression approach. *Technometrics*, **40**(2), 120–126.

Holms, A. G. (1998). Design of experiments as expansible sequences of orthogonal blocks with crossed-classification block effects. *Technometrics*, **40**(3), 244–253.

Holt, R. F., and Timmons, D. R. (1968). Influence of precipitation, soil water and plant population interactions on corn grain yields. *Agronomy Journal*, **60**, 379–381.

Hooke, R., and Jeeves, T. A. (1961). Direct search solution of numerical and statistical problems. *Journal of the Association of Computing Machinery*, **8**, 212–229.

Hooper, P. M. (1993). Nearly orthogonal randomized designs. *Journal of the Royal Statistical Society Series B*, **55**(1), 221–236.

Hopkins, H. S., and Jones, P. R. (1965). Factorial design of experiments in ceramics, III. Effects of firing rate, firing temperature, particle size distribution, and thickness. *American Ceramic Society Bulletin*, **44**, 502–505.

Houtman, A. M., and Speed, T. P. (1983). Balance in designed experiments with orthogonal block structure. *Annals of Statistics*, **11**, 1069–1085.

Hu, I. (1998). On sequential designs in nonlinear problems. *Biometrika*, **85**(2), 496–503.

Huang, D., and Allen, T. T. (2005). Design and analysis of variable fidelity experimentation applied to engine valve heat treatment process design. *Journal of the Royal Statistical Society C*, **54**(2), 443–463.

Huang, P., Chen, D., and Voelkel, J.O. (1998). Minimum-aberration two-level split-plot designs. *Technometrics*, **40**(4), 314–326.

Huang, S., Quall. K., and Moss, R. (1998). The optimization of a laboratory processing procedure for southern-style Chinese steamed bread. International *Journal of Food Science and Technology*, **33**, 345–357.

Huang, Y.-C., and Wong, W.-K. (1998). Sequential construction of multiple-objective optimal designs. *Biometrics*, **54**(4), 1388–1397.

Hubele, N. F., Beaumariage, T., Baweja, G., Hong, S.-C., and Chu, R. (1994). Using experimental design to assess the capability of a system. *Journal of Quality Technology*, **26**(1), 1–11.

Huda, S. (1981). A method for constructing second order rotatable designs. *Calcutta Statistical Association Bulletin*, **30**, 139–144.

Huda, S. (1982a). Some third order rotatable designs in three dimensions. *Annals of the Institute of Statistical Mathematics* (*Tokyo*), **34**, 365–371.

Huda, S. (1982b). Some third order rotatable designs. *Biometrical Journal*, **24**, 257–263.

Huda, S. (1982c). Cylindrically rotatable designs of type 3: Further considerations. *Biometrical Journal*, **24**, 469–475.

Huda, S. (1983a). Two third order rotatable designs in four dimensions. *Technical Report, Statistics and Mathematics Division, International Statistical Institute, Calcutta*, No. 33/82.

Huda, S. (1983b). The m-grouped cylindrically rotatable designs of types $(1, 0, m - 1)$, $(0, 1, m - 1)$, $(1, 1, m - 1)$ and $(0, 0, m)$. *Technical Report, Statistics and Mathematics Division, International Statistical Institute, Calcutta*. No. 6/83.

Huda, S. (2006). Design of experiments for estimating differences between responses and slopes of the response. In *Response Surface Methodology and Related Topics* (A. I. Khuri, ed.), pp. 427–446. Hackensack, NJ: World Scientific Publishing.

Huda, S., and Khan, I. H. (1993). On some A-optimal second order designs over cubic regions. Communications in Statistics, Simulation and Computation, **22**(1), 99–115.

Huda, S., and Mukerjee, R. (1984). Minimizing the maximum variance of the difference between two estimated responses. *Biometrika*, **71**, 381–385.

Huda, S., Khan, I. H., and Mukerjee, R. (1995). On optimal designs with restricted circular string property. *Computational Statistics and Data Analysis*. **19**(1), 75–83.

Hung, C. C., Lin, H. C., and Shih, H.C. (2002). Response surface methodology applied to silicon trench etching in Cl2/HBr/O2 using transformer coupled plasma technique. *Solid-State Electronics*, **46**, 791–795.

Hunter G. B., Hodi, F. S., and Eager, T. W. (1982). High-cycle fatigue of weld repaired cast Ti-6AI-4V. *Metallic Transactions*, **13A**, 1589–1594.

Hunter, J. S. (1959). Determination of optimum conditions by experimental methods [in three parts]. *Industrial Quality Control*, **15**(6), 16–24; **15**(7), 7–15; **15**(8), 6–14.

Hunter, J. S. (1960). Some applications of statistics to experimentation. *Chemical Engineering* Progress, Symposium Series, **56**(31), 10–26.

Hunter, J. S. (1964). Sequential factorial estimation. *Technometrics*, **6**(1), 41–55.

Hunter, J. S. (1985). Statistical design applied to product design. *Journal of Quality Technology*, **17**(4), 210–221.

Hunter, W. G., and Kittrell, J. R. (1966). EVOP—A review. *Technometrics*, **8**(3), 389–397.

Hunter, W. G., and Reiner, A. M. (1965). Designs for discriminating between two rival models. *Technometrics*, **7**(3), 307–323.

Hunter, W. G., Hill, W. J., and Wichern, D. W. (1968). A joint design criterion for the dual problem of model discrimination and parameter estimation. *Technometrics*, **10**(1). 145–160.

Huor, S. S., Ahmed, E. M., Rao, P. V., and Cornell, J. A. (1980). Formulation and sensory evaluation of a fruit punch containing watermelon citrullus lanatus juice. *Journal of Food Science*, **45**, 809–813. **M**

Huwang, L., Wu, C. F. J., and Yen, C. H. (2002). The idle column method: design construction, properties and comparisons. *Technometrics*, **44**(4). 347–355.

Hwang, S., Lee, Y., and Yang, K. (2001). Maximization of acetic acid production in partial acidogenesis of swine wastewater. *Biotechnology and Bioengineering*, **75**(5), 521–529.

Hyodo, Y. (1994). Characterization of information matrices for balanced two-level fractional factorial designs of odd resolution derived from two-symbol simple arrays. *Communications in Statistics, Theory and Methods*, **23**(7), 1859–1874.

Iida, T. (1994). A construction method of two-level supersaturated design derived from L12. *Japanese Journal of Applied Statistics*, **23**(3), 147–153.

Imhof, L. A. (2001). Maximin designs for exponential growth models and heteroscedastic polynomial models. *Annals of Statistics*, **29**(2), 561–576.

Imhof, L. A., and Studden, W. J. (2001). E-optimal designs for rational models. *Annals of Statistics*, **29**(3), 763–783.

Irizarry, Maria De Los A., Wilson, J. R., and Trevino, J. (2001). A flexible simulation tool for manufacturing-cell design. II: Response surface analysis and case study. *Institute of Industrial Engineers Transactions*, **33**, 837–846.

Ito, N., Leon, J. S., and Longyear, J. Q. (1981). Classification of 3-(24, 12, 5) designs and 24-dimensional Hadamard matrices. *Journal of Combinatorial Theory, Series A*, **31**, 66–93.

Iversen, P., and Marasinghe, M. (2001). Dynamic graphical tools for teaching experimental design and analysis concepts. *American Statistician*, **55**(4), 345–351.

Jacroux, M. (1989). The A-optimality of block designs for comparing test treatments with a control. *Journal of the American Statistical Association*, **84**, 310–317.

Jacroux, M. (1992). A note on the determination and construction of minimal orthogonal main effect plans. *Technometrics*, **34**(1), 92–96.

Jacroux, M. (1993). On the construction of minimal partially replicated orthogonal main effect plans. *Technometrics*, **35**(1), 32–36.

Jacroux, M., and Ray, R. S. (1990). On the construction of trend-free run orders of treatments. *Biometrika*, **77**, 187–191.

Jacroux, M., Majumdar, D., and Shah, K. R. (1997). On the determination and construction of optimal block designs in the presence of linear trends. *Journal of the American Statistical Association*, **92**, 375–382, 437.

Jamora, J., Ki, S., and Khee, C. (2001). Extrusion puffing of pork meat-defatted soy flour corn starch blends to produce snack-like products. *Journal of Food Science and Nutrition*. **6**(3), 163–169.

Jang, D.-H. (2002). A graphical method for evaluating slope-rotatability in axial directions for second order response surface designs. *Computational Statistics and Data Analysis*. **39**, 343–349.

Jang, D.-H., and Na, H.-J. (1996). A graphical method for evaluating mixture designs with respect to the slope. *Communications in Statistics, Theory and Methods*, **25**(5), 1043–1058. M

Jang, D.-H., and Park, S. H. (1993). A measure and a graphical method for evaluating slope-rotatability in response surface designs. *Communications in Statistics, Theory and Methods*, **22**(7), 1849–1863.

Jeffreys, G. V., and Whorton, J. T. (1965). Raschig synthesis of hydrazine. *Industrial and Engineering Chemistry Process Design and Development*, **4**, 71–76.

Jha, M. P., Kumar, A., and Bapat, B. R. (1981). Some investigations on response to fertilizer and determination of optimum dose using soil test values. *Journal of the Indian Society of Agricultural Statistics*, **33**, 60–70.

Jing, I., and Siebert, K. (2001). Development of a method for assessing haze-active protenin in beer by dye-binding. *Journal of the American Society of Brewing Chemists*, **59**(4), 172–182.

Jiricka, M., Hrma, P., and Vienna, J. D. (2003). The effect of composition on spinel crystals equilibrium in low silica high-level waste glasses. *Journal of Non-Crystal Solids*, **319**(3), 280–288. **M**

John, J. A. (2001a). Updating formula in an analysis of variance model. *Biometrika*, **88**(4), 1175–1178.

John, J. A. (2001b). Recursive formulae for the average efficiency factor in block and row-column designs. *Journal of the Royal Statistical Society, Series B*, **62**(3), 575–583.

John, J. A., and Draper, N. R. (1980). An alternative family of transformations. *Applied Statistics*, **29**, 190–197.

John, J. A., and Quenouille, M. H. (1977). *Experiments: Design and Analysis*. London: Griffin.

John, J. A., and Williams, E. R. (1995). *Cyclic and Computer Generated Designs*, 2nd ed. London: Chapman & Hall.

John, J. A., and Williams, E. R. (1997). The construction of efficient two-replicate row-column designs for use in field trials. *Applied Statistics*, **46**(2), 207–214.

John, P. W. M. (1961). Three quarter replicates of 2^4 and 2^5 designs. *Biometrics*, **17**, 319–321.

John, P. W. M. (1962). Three quarter replicates of 2^n designs. *Biometrics*, **18**, 172–184.

John, P. W. M. (1964). Blocking of $3(2^{n-k})$ designs. *Technometrics*, **6**(4), 371–376.

John, P. W. M. (1966). Augmenting 2^{n-1} designs. *Technometrics*, **8**(3), 469–480.

John, P. W. M. (1969). Some non-orthogonal fractions of 2^n designs. *Journal of the Royal Statistical Society, Series B*, **31**, 270–275.

John, P. W. M. (1971). *Statistical Design and Analysis of Experiment*. New York: Macmillan.

John, P. W. M. (1976). Robustness of balanced incomplete block designs. *Annals of Statistics*, **4**, 960–962.

John, P. W. M. (1979). Missing points in 2^n and 2^{n+k} factorial designs. *Technometrics*, **21**(2), 225–228.

John, P. W. M. (1984). Experiments with mixtures involving process variables. *Technical Report—University of Texas, Austin, Center for Statistical Sciences*, No. 8. **M**

John, P. W. M. (1990). Time trends and factorial experiments. *Technometrics*, **32**(3), 275–282.

Johnson, A. F. (1966). Properties of second order designs: Effects of transformation or truncation on prediction variance. *Applied Statistics*, **15**, 48–50.

Johnson, M. E., and Nachtheim, C. J. (1983). Some guidelines for constructing exact D-optimal designs on convex design spaces. *Technometrics*, **25**(3), 271–277.

Johnson, M. E., Moore, L. M., and Ylvisaker, D. (1990). Minimax and maximin distance designs. *Journal of Statistical Planning and Inference*, 36, 131–148.

Johnson, J. A., Widener, S., Gitlow, H., and Popovich, (2006). A "Six Sigma" black belt case study: G. E. P. Box's paper helicopter experiment Part A. *Quality Engineering*, **18**(4), 413–430.

Johnson, J. A., Widener, S., Gitlow, H., and Popovich, (2006). A "Six Sigma" black belt case study: G. E. P. Box's paper helicopter experiment Part B. *Quality Engineering*, **18**(4), 431–442.

Johnson, N. L., and Leone, F. C. (1977). *Statistics and Experimental Design in Engineering and the Physical Sciences, Volume 1, 2nd ed.*, New York: Wiley.

Johnson, R., Clapp, T., and Baqai, N. (1989). Understanding the effect of confounding in design of experiments: A case study in high speed weaving. *Quality Engineering*, **1**(4), 501–508.

Joiner, B. L., and Campbell, C. (1976). Designing experiments when run order is important. *Technometrics*, **18**(3), 249–259.

Jones, B. (1976). An algorithm for deriving optimal block designs. *Technometrics*, **18**(4), 451–458.

Jones, B. (1979). Algorithms to search for optimum row-and-column designs. *Journal of the Royal Statistical Society, Series B*, **41**, 210–216.

Jones, B., and Eccleston, J. A. (1980). Exchange and interchange procedures to search for optimum designs. *Journal of the Royal Statistical Society, Series B*, **42**, 238–243.

Jones, E. R., and Mitchell, T. J. (1978). Design criteria for detecting model inadequacy. *Biometrika*, **65**, 541–551.

Jordan D. L., Barnes, S., Bogle, C. R., Naderman, G. C., Roberson, G. T., and Johnson, P. D. (2001). Peanut response to tillage and fertilization. *Agronomy Journal*, **93**, 1064–1071.

Jordan L. (1977). Optimal designs for polynomial fitting. *Estadistica*, **31**, 385–394.

Jordan L., and Kempthorne, O. (1977). Min-average bias estimable designs. *Estadistica*, **31**, 66–79.

Joseph, V. R. (2006). A Bayesian approach to the design and analysis of fractionate experiments. *Technometrics*, **48**(2), 219–229.

Joseph, V. R., and Wu, C. F. J. (2002). Robust parameter design of multiple target systems. *Technometrics*, **44**(4), 338–346.

Ju, H. L., and Lucas, J. M. (2002). L^k factorial experiments with hard-to-change and easy-to-change factors. *Journal of Quality Technology*, **34**(4), 411–421.

Juan, J., and Peña, D. (1992). A simple method to identify significant effects in unreplicated two-level factorial designs. *Communications in Statistics, Theory and Methods*, **21**(5), 1383–1403.

Jun, H., and Yong, H. (2001). Changes in color parameters of clarified apple and carrot blend juice using response surface methodology. *Food Science and Biotechnology*, **10**(6), 673–676.

Jung, J. S., and Yum, B. J. (1996). Construction of exact D-optimal designs by tabu search. *Computational Statistics and Data Analysis*, **21**(2), 181–191.

Juusola, J. A., Bacon, D. W., and Downie, J. (1972). Sequential statistical design strategy in an experimental kinetic study. *Canadian Journal of Chemical Engineering*, **50**, 796–801.

Kackar, R. N. (1985). Off-line quality control, parameter designs, and the Taguchi methods. *Journal of Quality Technology*, **17**(4), 176–188; discussion: pp. 189–209.

Kackar, R. N., and Tsui, K.-L. (1990). Interaction graphs: Graphical aids for planning experiments. *Journal of Quality Technology*, **22**(1), 1–14.

Kackar, R. N., Lagergren, E. S., and Filliben, J. J. (1991). Taguchi's fixed element arrays are fractional factorials. *Journal of Quality Technology*, **23**(2), 107–116.

Kalaba, R., and Tesfatsion, L. (1996). A multicriteria approach to model specification and estimation. *Computational Statistics and Data Analysis*, **21**(2), 193–214.

Kalish, L. A. (1990). Efficient design for estimation of median lethal dose and quantal dose-response curves. *Biometrics*, **46**, 737–748.

Kamimura, E., Medieta, O., Rodrigues, M., and Maugeri, F. (2001). Studies on lipase-affinity adsorption using response-surface analysis. *Biotechnology and Applied Biochemistry*, **33**(3), 153–159.

Kanemasu, H. (1979). A statistical approach to efficient use and analysis of simulation models. *Bulletin of the International Statistical Institute*, **48**, 573–604.

Karlin, S., and Studden, W. J. (1966a). Optimal experimental designs. *Annals of Mathematical Statistics*, **37**, 783–815.

Karson, M. J. (1970). Design criterion for minimum bias estimation of response surfaces. *Journal of the American Statistical Association*, **65**, 1565–1572.

Karson, M. J., and Spruill, M. L. (1975). Design criteria and minimum bias estimation. *Communications in Statistics, Theory and Methods*, **4**, 339–355.

Karson, M. J., Manson, A. R., and Hader, R. J. (1969). Minimum bias estimation and experimental design for response surfaces. *Technometrics*, **11**(3), 461–475.

Kasperski, W. J., Schneider, H., and Johnson, K. L. (1995). Analyzing unreplicated factorials using the smallest n contrasts. *Journal of Quality Technology*, **27**(4), 377–386.

Ke, W., and Tang, B. (2003). Selecting 2^{m-p} designs using a minimum aberration criterion when some two-factor interactions are important. *Technometrics*, **45**(4), 352–360.

Kelly, K. (2001). Planting data and foliar fungicide effects on yield components and grain traits of winter wheat, *Agronomy Journal*, **93**, 380–389.

Kempthorne, O. (1947a). A note on differential responses in blocks. *Journal of Agricultural Science*, **37**, 245–248.

Kempthorne, O. (1947b). A simple approach to confounding and fractional replication in factorial experiments. *Biometrika*, **34**, 255–272.

Kempthorne, O. (1952). *The Design and Analysis of Experiments*. New York: Wiley.

Kenward, M. G., and Rogers, J. H. (1997). Small sample inference for fixed effects from restricted maximum likelihood. *Biometrics*, **53**, 983–997.

Kenworthy, O. O. (1963). Factorial experiments with mixtures using ratios. *Industrial Quality Control*, **19**, 24–26. **M**

Kerr, K. F. (2006). Efficient 2^k factorial designs for blocks of size 2 with microarray applications. *Journal of Quality Technology*, **38**(4), 309–318.

Kerr, M. K. (2001). Bayesian-optimal fractional factorials. *Statistica Sinica*, **11**, 605–630.

Keuhl, R. O. (2000). *Design of Experiments: Statistical Principles of Research Design and Analysis*, 2nd ed. Pacific Grove, CA: Brooks/Cole.

Khattree, R. (1996). Robust parameter design: A response surface approach. *Journal of Quality Technology*, **28**(2), 187–198.

Khosla, R., Alley, M. M., and Davis, P. H. (2000). Nitrogen management in no-tillage grain sorghum production. *Agronomy Journal*, **92**, 321–328.

Khuri, A. I. (1988). A measure of rotatability for response surface designs. *Technometrics*, **30**(1), 95–104.

Khuri, A. I. (1990). Analysis of multiresponse experiments: A review. In *Statistical Design and Analysis of Industrial Experiments* (S. Ghosh, ed.), pp. 231–246. New York: Dekker.

Khuri, A. I. (1991). Blocking with rotatable designs. *Calcutta Statistical Association Bulletin*, **41**, 81–98.

Khuri, A. I. (1992a). Response surface models with random block effects. *Technometrics*, **34**(1), 26–37.

Khuri, A. I. (1992b). Diagnostic results concerning a measure of rotatability. *Journal of the Royal Statistical Society, Series B*, **54**, 253–267.

Khuri, A. I. (1994). Effect of blocking on the estimation of a response surface. *Journal of Applied Statistics*, **21**, 305–316.

Khuri, A. I. (1996). Response surface models with mixed effects. *Journal of Quality Technology*, **28**(2), 177–186.

Khuri, A. I. (2006). Mixed response surface models with heterogeneous within-block error variances. *Technometrics*, **48**(2), 206–218.

Khuri, A. I., ed. (2006). *Response Surface Methodology and Related Topics*. Hackensack, NJ: World Scientific Publishing. (17 individual papers)

Khuri, A. I., and Conlon, M. (1981). Simultaneous-optimization of multiple responses represented by polynomial regression functions. *Technometrics*, **23**(4) 363–375.

Khuri, A. I., and Cornell, J. A. (1977). Secondary design considerations for minimum bias estimation. *Communications in Statistics, Theory and Methods*, **6**(7), 631–647.

Khuri, A. I., and Cornell, J. A., eds. (1996). *Response Surfaces: Designs and Analyses*, 2nd ed. New York: Dekker.

Khuri, A. I., and Cornell, J. A. (1998). Lack of fit revisited. *Journal of Combinatorics, Information and System Sciences*, **23**, 193–208.

Khuri, A. I., and Mukhopadhyay, S. (2006). GLM designs: the dependence on unknown parameters dilemma. In *Response Surface Methodology and Related Topics*, (A. I. Khuri. ed.), pp. 203–223. Hackensack, NJ: World Scientific Publishing.

Khuri, A. I., and Myers, R. H. (1979). Modified ridge analysis. *Technometrics*, **21**(4) 467–473.

Khuri, A. I., and Valerosa, E. S. (1999). Optimization methods in multiresponse surface methodology. In *Statistical Process Monitoring and Optimization* (S. H. Park and G. G. Vining, eds.). New York: Dekker.

Khuri, A. I., Kim H. J., and Um, Y. H. (1996). Quantile plots of the prediction variance for response surface designs. *Computational Statistics and Data Analysis*, **22**(4), 395–407.

Khuri, A. I., Harrison, J. M., and Cornell, J. A. (1999). Using quantile plots of the prediction variance for comparing designs for a constrained mixture region: an application involving a fertilizer experiment. *Journal of the Royal Statistical Society, Series C (Applied Statistics)*, **48**(4), 521–532. **M**

Kiefer, J. C. (1958). On the nonrandomized optimality and the randomized nonoptimality of symmetrical designs. *Annals of Mathematical Statistics*, **29**, 675–679.

Kiefer, J. C. (1959). Optimum experimental designs. *Journal of the Royal Statistical Society, Series B*, **21**, 272–304.

Kiefer, J. C. (1960). Optimum experimental designs V, with applications to systematic and rotatable designs. *Proceedings of the Fourth Berkeley Symposium*, **1**, 381–405.

Kiefer, J. C. (1961). Optimum designs in regression problems. II. *Annals of Mathematical Statistics*, **32**(2), 298–325.

Kiefer, J. C. (1974). General equivalence theory for optimum designs (approximate theory). *Annals of Statistics*, **2**, 849–879.

Kiefer, J. C. (1975). Optimal design: variation in structure and performance under change of criterion. *Biometrika*, **62**, 277–288.

Kiefer, J. C. (1978). Asymptotic approach to families of design problems. *Communications in Statistics, Theory and Methods*, **7**, 1347–1362.

Kiefer, J. C. (1984). The publications and writings of Jack Kiefer. *Annals of Statistics*, **12**, 424–430. ["The complete bibliography . . . prepared through the efforts of Roger Farrell and Ingram Olkin." The quotation is from page 403 of the cited journal.]

Kiefer, J. C. (1985). *Jack Carl Kiefer Collected Papers III, Design of Experiments*, (L. D. Brown, I. Olkin, J. Sacks, and H. P. Wynn, eds.) New York: Springer-Verlag. (Note: This is probably the best source for Kiefer's writings on experimental design.)

Kiefer, J. C., and Studden, W. J. (1976). Optimal designs for larger degree polynomial regression. *Annals of Statistics*, **4**, 1113–1123.

Kiefer, J. C., and Wolfowitz, J. (1959). Optimum designs in regression problems. *Annals of Mathematical Statistics*, **30**(2), 271–294.

Kiefer, J. C., and Wolfowitz, J. (1960). The equivalence of two extremum problems. *Canadian Journal of Mathematics*, **12**, 363–366.

Kim, H. J., Um, Y. H., and Khuri, A. (1996). Quantile plots of the average slope variance for response surface designs. *Communications in Statistics, Simulation and Computation*, **25**(4), 995–1014.

Kim, K.-J., and Lin, D. K. J. (1998). Dual response surface optimization: A fuzzy modeling approach. *Journal of Quality Technology*, **30**(1), 1–10.

Kim, K.-J., and Lin, D. K. J. (2000). Simultaneous optimization of mechanical properties of steel by maximizing exponential desirability functions. *Applied Statistics*, **49**(3), 311–325.

Kim, W. B., and Draper, N. R. (1994). Choosing a design for straight line fits to two correlated responses. *Statistica Sinica*, **4**, 275–280.

Kim, Y. J., and Cho, B. R. (2002). Development of priority-based robust design. *Quality Engineering*, **14**(3), 355–363.

Kimura, H. (1989). New Hadamard matrices of order 24. *Graphs and Combinatorics*, **5**, 236–242.

Kiran, K., Manohar, B., and Divakar, S. (2001). A central composite rotatable design analysis of lipase catalyzed synthesis of lauroyl lactic acid at bench-scale level. *Enzyme and Microbial Technology*, **29**(2–3), 122–128.

Kishen, K. (1948). On fractional replication of the general symmetrical factorial design. *Journal of the Indian Society of Agricultural Statistics*, **1**, 91–106.

Kissell, D. E., and Burnett, E. (1979). Response of Coastal bermuda grass to tillage and fertilizers on an eroded Grayland soil. *Agronomy Journal*, **71**, 941–944.

Kissell, L. T. (1959). A lean-formula cake method for varietal evaluation and research. *Cereal Chemistry*, **36**, 168–175. **M**

Kissell, L. T. (1967). Optimization of white layer cake formulations by a multiple-factor experimental design. *Cereal Chemistry*, **44**, 253–268. **M**

Kissell, L. T., and Marshall, B. D. (1962). Multi-factor responses of cake quality to basic ingredient ratios. *Cereal Chemistry*, **39**, 16–30. **M**

Kitawaga, R., Brubs, R. E., and De Menezes, T. J. B. (1994). Optimizing the enzymatic maceration of foliole purée from hard pieces of hearts of palm (*Euterpe edulis* Mart.) using response surface analysis. *Journal of Food Science*, **59**(4), 844–848.

Kitagawa, T., and Mitome, M. (1958). *Tables for the Design of Factorial Experiments*. New York: Dover.

Kitsos, C. P., Titterington, D. M., and Torsney, B. (1988). An optimal design problem in rhythmometry. *Biometrics*, **44**, 657–671.

Klç, M., Bayraktar, E., Ate, S., and Mehmetoglu, Ü. (2002). Investigation of extractive citric acid fermentation using response-surface methodology. *Process Biochemistry*, **37**, 759–767.

Klein, I., and Marshall, D. I. (1964). Note on starch vinylation by Berry, Tucker and Deutchman. *Industrial and Engineering Chemistry Process Design and Development*, **3**, 287–288.

Knapp, W. R., and Knapp, J. S. (1978). Response of winter wheat to date of planting and fall fertilization. *Agronomy Journal*, **70**, 1048–1052.

Ko, Y.-H., Kim, K.-J., and Jun, C.-H. (2005). A new loss-function-based method for multiresponse optimization. *Journal of Quality Technology*, **37**(1), 50–59.

Koehler, T. L. (1960). How statistics apply to chemical processes. *Chemical Engineering*, **67**, 142–152.

Köksoy, O., and Doganaksoy, N. (2003). Joint optimization of mean and standard deviation using response surface methods. *Journal of Quality Technology*, **35**(3), 239–252.

Koons, G. F. (1989). Effect of sinter composition on emissions: A multi-component, highly constrained mixture experiment. *Journal of Quality Technology*, **21**(4), 261–267. M

Kopas, D. A., and McAllister, P. R. (1992). Process improvement exercises for the chemical industry. *American Statistician*, **46**, 34–41.

Koshal, R. S. (1933). Application of the method of maximum likelihood to the improvement of curves fitted by the method of moments. *Journal of the Royal Statistical Society, Series A*, **96**, 303–313.

Koukouvinos, C. (1998). Weighing experiments with $n = 2$ (mod 4) observations. *Computational Statistics and Data Analysis*, **27**(1), 27–31.

Koukouvinos, C., and Seberry, J. (2000). New orthogonal designs and sequences with two and three variables in order 28. *Ars Combinatoria*, **54**, 97–108.

Koukouvinos, C., Mitrouli, M. Seberry, J., and Karabelas, P. (1996). On sufficient conditions for some orthogonal designs and sequences with zero autocorrelation function. *Australian Journal of Combinatorics*, **13**, 197–216.

Koukouvinos, C., Platis, N., and Seberry, J. (1996). Necessary and sufficient conditions for some two variable orthogonal designs in order 36. *Congressus Numerantium*, **114**, 129–140.

Kowalski, S. M. (2002). 24 run split plot experiments for robust parameter design. *Journal of Quality Technology*, **34**(4), 399–410.

Kowalski, S. M., and Potcner, K. J. (2003). How to recognize a split-plot experiment. *Quality Progress*, **36**(11), November, 60–66.

Kowalski, S. M., Cornell, J. A., and Vining, G. G. (2000). A new model and class of designs for mixtures experiments with process variables. *Communications in Statistics: Theory and Methods*, **29**, 2255–2280.

Kowalski, S. M., Cornell, J. A., and Vining, G. G. (2002). Split-plot designs and estimation methods for mixture experiments with process variables. *Technometrics*, **44**(1), 72–79. M

Kowalski, S. M., Landman, D., and Simpson, J. (2003). Design of experiments enhances race car performance. *Scientific Computing and Instrumentation Online*, `www.scimag.com/ scripts/`.

Kowalski, S. M., Vining, C. G., Montgomery, D. C., and Borror, C. M. (2006). Modifying a central composite design to model the process mean and variance when there are hard to change factors. *Applied Statistics*, **55**(5), 615–630.

Kowalski, S. M., Borror, C. M., and Montgomery, D. C. (2005). A modified path of steepest ascent for split-plot experiments. *Journal of Quality Technology*, **37**(1), 75–83.

Krishna, C., and Nokes, S. (2001). Predicting vegetative inoculum performance to maximize phytase production in solid-state fermentation using response surface methodology. *Journal of Industrial Microbiology and Biotechnology.* **26**(3), 161–170.

Krose, J. F., and Mastrangelo, C. M. (2001). Comparing methods for multi-response design problem. *Quality and Reliability International*, **17**, 323–331.

Krouse, D. P. (1994). Patterned matrices and the misspecification robustness of 2^{k-p} designs. *Communications in Statistics, Theory and Methods*, **23**(11), 3285–3301.

Kuhfeld, W. F. (2005). Difference schemes via computerized searches. *Journal of Statistical Planning and Inference*, **127**(1-2), 341–346.

Kuhfeld, W. F., and Tobias, R. D. (2005). Large factorial designs for product engineering and marketing research applications. *Technometrics*, **47**(2), 132–141.

Kuhfeld, W. F., Tobias, R. D., and Garratt, M. (1994). Efficient experimental design with marketing applications. *Journal of Marketing Research*, **31**, 545–557.

Kuhn, A. M. (2003). Optimizing response surface experiments with noise factors using confidence regions. *Quality Engineering*, **15**(3), 419–426.

Kuhn, A. M., Carter, W. H., and Myers, R. H. (2000). Incorporating noise factors into experiments with censored data. *Technometrics*, **42**(4), 376–383.

Kulahci, M., and Bisgaard, S. (2005). The use of Plackett-Burman designs to construct split-plot designs. *Technometrics*, **47**(4), 495–501.

Kulahci, M., and Box, G. E. P. (2003). Catalysis of discovery and development in engineering and industry. *Quality Engineering*, **15**(3), 513–517.

Kulahci, M., Ramırez, J. G., and Tobias, R. (2006). Split-plot fractional designs: Is minimum aberration enough? *Journal of Quality Technology*, **38**(1), 56–64.

Kulshreshtha, A. C. (1969). Fitting of response surface in the presence of a concomitant variate. *Calcutta Statistical Association Bulletin*, **18**, 123–131.

Kunert, J. (1997). On the use of the factor-sparsity assumption to get an estimate of the variance in saturated designs. *Technometrics*, **39**(1), 81–90.

Kunert, J., and Martin, R. J. (1987). On the optimality of finite Williams II(a) designs. *Annals of Statistics*, **15**, 1604–1628.

Kunert, J., and Martin, R. J. (2000). On the determination of optimal designs for an interference model. *Annals of Statistics*, **28**(6), 1728–1742.

Kupper, L. L. (1972). A note on the admissibility of a response surface design. *Journal of the Royal Statistical Society, Series B*, **34**, 28–32.

Kupper, L. L. (1973). Minimax designs for Fourier series and spherical harmonics regressions: A characteristic of rotatable arrangements. *Journal of the Royal Statistical Society, Series B*, **35**, 493–500.

Kupper, L. L., and Meydrech, E. F. (1973). A new approach to mean squared error estimation of response surfaces. *Biometrika*, **60**, 573–579.

Kupper, L. L., and Meydrech, E. F. (1974). Experimental design considerations based on a new approach to mean squared error estimation of response surfaces. *Journal of the American Statistical Association*, **69**, 461–463.

Kurotori, I. S. (1966). Experiments with mixtures of components having lower bounds. *Industrial Quality Control*, **22**, May, 592–596. **M**

Kushner, H. B. (1997). Optimality and efficiency of two-treatment repeated measurements designs. *Biometrika*, **84**(2), 455–468.

Lam, C., and Tonchev, V. D. (1996). Classification of affine-resolvable 2-(27, 9, 4) designs. *Journal of Statistical Planning and Inference*, **56**, 187–1202.

Lamb, R. H., Boos, D. D., and Brownie, C. (1996). Testing for effects on variance in experiments with factorial treatment structure and nested errors. *Technometrics*, **38**(2), 170–177.

Lambrakis, D. P. (1968a). Experiments with mixtures: A generalization of the simplex lattice design. *Journal of the Royal Statistical Society, Series B*, **30**(1), 123–136. **M**

Lambrakis, D. P. (1968b). Experiments with *p*-component mixtures. *Journal of the Royal Statistical Society, Series B*, **30**(1), 137–144. **M**

Lambrakis, D. P. (1969a). Experiments with mixtures: An alternative to the simplex-lattice design. *Journal of the Royal Statistical Society, Series B*, **31**(2), 234–245. **M**

Lambrakis, D. P. (1969b). Experiments with mixtures: Estimated regression function of the multiple-lattice design. *Journal of the Royal Statistical Society, Series B*, **31**(2), 276–284. **M**

Langsrud, O. (2001). Identifying significant effects in fractional factorial multiresponse experiments. *Technometrics*, **43**(4), 415–424.

Langsrud, O., and Naes, T. (1998). A unified framework for significance testing in fractional factorials. *Computational Statistics and Data Analysis*, **28**(4), 413–431.

Lasdon, L., Fox, R., and Ratner, M. (1974). Nonlinear optimization using generalized reduced gradient method. *Revue Française d'Automatique Informatique Recherche Operationnelle*, **8**, 73–103.

Läuter, E. (1974). Experimental design in a class of models. *Mathematik Operations-forschung Statistik*, **5**, 379–398.

Law, R., and Watkinson, A. R. (1987). Response surface analysis of two-species competition: An experiment on phleum arenarium and vulpia fasciculate. *Journal of Ecology*, **75**, 871–886.

Lawless, J. F., and Wang, P. (1976). A simulation study of ridge and other regression estimators. *Communications in Statistics, Theory and Methods*, **5**, 307–323.

Lawrance, A. J. (1996). A design of experiments workshop as an introduction to statistics. *American Statistician*, **50**(2), 156–158.

Lawson, J., Grimshaw, S., and Burt, J. (1998). A quantitative method for identifying active contrasts in unreplicated factorial designs based on the half-normal plot. *Computational Statistics and Data Analysis*, **26**(4), 425–436.

Lawson, J., Liddle, S. W., and Meade, D. (2006). Finding bad values in factorial designs— revisited. *Quality Engineering*, **18**(4), 491–501.

Lawson, J. S. (1988). A case study of effective use of statistical experimental design in a smoke stack industry. *Journal of Quality Technology*, **20**(1), 51–62.

Lawson, J. S. (2003). One-step screening and process optimization experiments. *American Statistician*, **57**(1), 15–20.

Laycock, P. J., and Rowley, P. J. (1995). A method for generating and labeling all regular fractions or blocks for q^{n-m} designs. *Journal of the Royal Statistical Society, Series B*, **57**(1), 191–204.

Leakey, C. L. A. (1972). The effect of plant population and fertility level on yield and its components in two determinate cultivars of *Phaseolus vulgaris*. *Journal of Agricultural Science, Cambridge*, **79**, 259–267.

Ledolter, J. (1995). Projects in introductory statistics courses. *American Statistician*, **49**(4), 364–367.

Ledolter, J., and Swersey, A. (1997). Dorian Shanian's variables search procedure: A critical assessment. *Journal of Quality Technology*, **39**(3), 237–247.

Ledolter, J., and Swersey, A. (2006). Using a fractional factorial design to increase direct mail response at Mother Jones magazine. *Quality Engineering*, **18**(4), 469–475.

Lee, G.-D., and Kwon, J.-H. (1998). The use of response surface methodology to optimize the Maillard reaction to produce melanoidins with high antioxidative and antimutagenic activities. *International Journal of Food Science and Technology*, **33**, 375–383.

Lee, Y., and Nelder, J. A. (1998). Joint modeling of mean and dispersion. *Technometrics*, **40**(2), 168–171.

Lee, Y., and Nelder, J. A. (2003). Robust designs via generalized linear models. *Journal of Quality Technology*, **35**(1), 2–12.

Leitnaker, M. G., and Mee, R. W. (2001). Analytic use of two-level factorials in incomplete blocks to examine the stability of factor effects. *Quality Engineering*, **14**(1), 49–58.

Lenth, R. V. (1989). Quick and easy analysis of unreplicated factorials. *Technometrics*, **31**(4), 469–473.

Leon, R. V., and Wu, C. J. F. (1992). A theory of performance measures in parameter design. *Statistica Sinica*, **2**, 335–358.

Leon, R. V., Shoemaker, A. C., and Kackar, R. N. (1987). Performance measures independent of adjustment: An explanation and extension of Taguchi's signal-to-noise ratio. *Technometrics*, **29**(3), 253–265; discussion: pp. 266–285.

Letsinger, J. D., Myers, R. H., and Lentner, M. (1996). Response surface methods for bi-randomization structures. *Journal of Quality Technology*, **28**(4), 381–397.

Lewis, G.A., Mathieu, D., and Phan-Tan-Luu, R. (1999). *Pharmaceutical Experimental Design*. New York: Dekker.

Lewis, S. L., Montgomery, D. C., and Myers, R. H. (1999–2000). The analysis of designed experiments with non-normal responses. *Quality Engineering*, **12**(2), 225–243.

Lewis, S. L., Montgomery, D. C., and Myers, R. H. (2001a). Examples of designed experiments with nonnormal responses. *Journal of Quality Technology*, **33**(3), 265–278.

Lewis, S. L., Montgomery, D. C., and Myers, R. H. (2001b). Confidence interval coverage for designed experiments analyzed with generalized linear models. *Journal of Quality Technology*, **33**(3), 279–292.

Lewis, S. M., and Dean, A. M. (2001). Detection of interactions in experiments with large numbers of factors. *Journal of the Royal Statistical Society, Series B*, **63**(4), 633–672.

Lewis, S. M., Dean, A. M., Draper, N. R., and Prescott, P. (1994). Mixture designs for q components in orthogonal blocks. *Journal of the Royal Statistical Society, Series B*, **56**, 457–467. **M**

Leyshon, A. J., and Sheard, R. W. (1974). Influence of short-term flooding on the growth and plant nutrient composition of barley. *Canadian Journal of Statistics*, **54**, 463–473.

Li, C., Bai, J., Cai, Z., and Ouyang, F. (2002). Optimization of a cultural medium for bacteriocin production by Lactococcus lactis using response surface methodology. *Journal of Biotechnology*, **93**, 27–34.

Li, H., and Mee, R. W. (2002). Better foldover fractions for Resolution III 2^{k-p} designs. *Technometrics*, **44**(3), 278–283.

Li, K. C. (1992). On principal Hessian directions for data visualization and dimension reduction: Another application of Stein's lemma. *Journal of the American Statistical Association*, **87**, 1025–1039.

Li, R., and Lin, D. K. J. (2002). Data analysis in supersaturated designs. *Statistics and Probability Letters*, **59**, 135–144.

Li, W., and Lin, D. K. J. (2003a). Analysis methods for supersaturated designs: Some comparisons. *Journal of Data Science*, **1**, 103–121.

Li, W., and Lin, D. K. J. (2003b). Optimal foldover plans for two-level fractional factorial designs. *Technometrics*, **45**(2), 142–149.

Li, W., and Mee, R. W. (2002). Better foldover plans for resolution III 2^{k-p} designs. *Technometrics*, **44**(4), 278–283.

Li, W., and Nachtsheim, C. J. (2000). Model-robust factorial designs. *Technometrics*, **42**(4), 345–352.

Li, W., Lin, D. K. J., and Ye, K. Q. (2003). Optimal foldover plans for two-level nonregular orthogonal designs. *Technometrics*, **45**(4), 347–351.

Li, W. W., and Wu, C. J. F. (1997). Columnwise-pairwise algorithms with applications to the construction of supersaturated designs. *Technometrics*, **39**(2), 171–179.

Li, Y. (2000). Constructions of generalized minimum aberration designs through Hadamard matrices and orthogonal arrays. Ph.D. thesis, University of Memphis, Memphis, TN.

Liao, C. T. (1999). Orthogonal three-level parallel flats designs for user-specified resolution. *Communications in Statistics, Theory and Methods*, **28**(8), 1945–1960.

Liao, C. T. (2000). Indentification of dispersion effects from unreplicated 2^{n-k} fractional factorial designs. *Computational Statistics and Data Analysis*, **33**(2), 291–298.

Liao, C. T., and Iyer, H. K. (2000). Optimal 2^{n-p} fractional factorial designs for dispersion effects under a location-dispersion model. *Communications in Statistics, Theory and Methods*, **29**(4), 823–835.

Liao, C. T., Iyer, H. K., and Vecchia, D. F. (1996). Construction of orthogonal two-level designs of user-specified resolution where $N \neq 2^k$. *Technometrics*, **38**(4), 342–353.

Lim, Y. B. (1990). D-optimal design for cubic polynomial regression on the q-simplex. *Journal of Statisticsl Planning and Inference*, **25**, 141–152. **M**

Lim, Y. B., and Studden, W. J. (1988). Efficient D_s-optimal designs for multivariate polynomial regression on the q-cube. *Annals of Statistics*, **16**, 1225–1240.

Lin, D. K. J. (1993a). A new class of supersaturated designs. *Technometrics*, **35**(1), 28–31; see also the exchange of letters in Lin (1995b).

Lin, D. K. J. (1993b). Another look at first-order saturated designs: The p-efficient designs. *Technometrics*, **35**(3), 284–292.

Lin, D. K. J. (1995a). Generating systematic supersaturated designs. *Technometrics*, **37**(2), 213–225.

Lin, D. K. J. (1995b). Exchange of letters. *Technometrics*, **37**(3), 358–359.

Lin, D. K. J. (1998). Spotlight interaction effects in main-effect designs: A supersaturated design approach. *Quality Engineering*, **11**, 133–139.

Lin, D. K. J. (1999). Supersaturated designs (Update). *Encyclopedia of Statistical Sciences (Update)*, **3**, 727–731.

Lin, D. K. J. (2000). Recent developments in supersaturated designs. In *Statistical Process Monitoring and Optimization* (S. H. Park and G. G. Vining, eds.), pp. 305–320. New York: Dekker.

Lin, D. K. J. (2003). Industrial experimentation for screening. In *Handbook of Statistics* (R. Khattree and C. R. Rao, eds.) Vol. 22, pp. 33–73. Amsterdam: Elsevier.

Lin, D. K. J., and Chang, J. Y. (2001). A note on cyclic orthogonal designs. *Statistica Sinica*, **11**, 549–552.

Lin, D. K. J., and Draper, N. R. (1992). Projection properties of Plackett and Burman designs. *Technometrics*, **34**(4), 423–428.

Lin, D. K. J., and Draper, N. R. (1993). Generating alias relationships for two-level Plackett-Burman designs. *Computational Statistics and Data Analysis*, **15**, 147–157.

Lin, D. K. J., and Draper, N. R. (1995). Screening properties of certain two-level designs. *Metrika*, **42**, 99–118.

Lin, D. K. J., and Lam, A. W. (1997). Connections between two-level factorials and Venn diagrams. *American Statistician*, **51**(1), 49–51.

Lin, D. J. K., and Peterson, J. J. (2006). Statistical inference for response surface optima. In *Response Surface Methodology and Related Topics* (A. I. Khuri, ed.), pp. 65–88. Hackensack, NJ: World Scientific Publishing.

Lin, D. K. J., and Tu, W. (1995). Dual response surface optimization. *Journal of Quality Technology*, **27**(1), 34–39.

Lin, H. F., Myers, R. H., and Ye, K. Y. (2000). Bayesian two-stage optimal design for mixture models. *Journal of Statistical Computation and Simulation*, **66**, 209–231. **M**

Lin, K. M., and Kackar, R. N. (1986). Optimizing the wave soldering process. *Electronic Packaging and Production*, February, pp. 108–115.

Lin, T., and Chananda, B. (2003–2004). Quality improvement of an injection-molded product using design of experiments: A case study. *Quality Engineering*, **16**(1), 99–104.

Lin, T., and Sanders, S. (2006). A sweet way to learn DoE. *Quality Progress*, **39**(2), February, p. 88.

Lind, E. E., Goldin, J., and Hickman, J. B. (1960). Fitting yield and cost response surfaces. *Chemical Engineering Progress*, **56**, 62.

Lindsay, B. G. (1989a). On the determinants of moment matrices. *Annals of Statistics*, **17**, 711–721.

Lindsay, B. G. (1989b). Moment matrices: Applications in mixtures. *Annals of Statistics*, **17**, 722–740.

Lindsey, J. K. (1972). Fitting response surfaces with power transformations. *Applied Statistics*, **21**, 234–247.

Lindsey, J. K. (1999). Response surfaces for overdispersion in the study of the conditions for fish eggs hatching. *Biometrics*, **55**(1), 149–155.

Liu, C., and Sun, D. X. (2000). Analysis of interval-censored data from fractionated experiments using covariance adjustment. *Technometrics*, **42**(4), 353–365.

Liu, S., and Neudecker, H. (1995). A V-optimal design for Scheffé's polynomial model. *Statistics and Probability Letters*, **23**, 253–258. **M**

Liu, S. X. (1997). Restricted optimal designs for linear and quadratic polynomial regressions. *Communications in Statistics, Theory and Methods*, **26**(1), 33–46.

Liu, Y., and Dean, A. M. (2004). k-circulant supersaturated designs. *Technometrics*, **46**(1), 32–43.

Loeppky, J. L., and Sitter, R. R. (2002). Analyzing unreplicated blocked or split-plot fractional factorial designs. *Journal of Quality Technology*, **34**(3), 229–243.

Logothetis, N., and Haigh, A. (1998). Characterizing and optimizing multi-response processes by the Taguchi method. *Quality and Reliability Engineering International*, **4**(2), 159–169.

Loh, W.-Y. (1992). Identification of active contrasts in unreplicated factorial experiments. *Computational Statistics and Data Analysis*, **14**, 135–148.

Loh, W.-Y. (1996). A combinatorial central limit theorem for randomized orthogonal array sampling designs. *Annals of Statistics*, **24**(3), 1209–1224.

Lorenzen, T. J., and Anderson, V. L. (1993). *Design of Experiments: A No-Name Aproach*. New York: Dekker.

Loughin, T. M. (1998a). On type I error rates in the analysis of unreplicated 2^3-factorials. *The Statistician*, **47**(1), 207–210.

Loughin, T. M. (1998b). Calibration of the Lenth test for unreplicated factorial designs. *Journal of Quality Technology*, **30**(2), 171–175.

Loughin, T. M., and Noble, W. (1997). A permutation test for effects in an unreplicated factorial design. *Technometrics*, **39**(2), 180–190.

Lu, X., and Meng, Y. (2000). A new method in the construction of two-level supersaturated designs. *Journal of Statistical Planning and Inference*, **86**, 229–238.

Lu, X., and Wu, X. (2004). A strategy of searching active factors in supersaturated screening experiments. *Journal of Quality Technology*, **36**(4), 392–399.

Lu, X., Li, W., and Xie, M. (2006). A class of nearly orthogonal arrays. *Journal of Quality Technology*, **38**(2), 148–161.

Lucas, J. M. (1974). Optimum composite designs. *Technometrics*, **16**(4), 561–567.

Lucas, J. M. (1976). Which [second-order] response surface design is best?: A performance comparison of several types. *Technometrics*, **18**(4), 411–417.

Lucas, J. M. (1977). Design efficiencies for varying numbers of center points. *Biometrika*, **64**, 145–147.

Lucas, J. M. (1994). How to achieve a robust process using response surface methodology. *Journal of Quality Technology*, **26**(4), 248–260.

Luenberger, D. G. (1969). *Optimization by Vector Space Methods*. New York: Wiley.

Luftig, J. T., and Jordan, V. S. (1996). *Design of Experiments in Quality Engineering*. New York: McGraw-Hill.

Lulu, M., and Rao, R. L. (1990). Parameter design for a wave solder process. *Quality Engineering*, **2**, 301–318.

Lund, R. E. (1982). Plans for blocking, and fractions of, nested cube designs. *Communications in Statistics, Theory and Methods*, **11**, 2287–2296.

Lund, R. E., and Linnell, M. G. (1982). Description and evaluation of a nested cube experimental design. *Communications in Statistics, Theory and Methods*, **11**, 2297–2313.

Luner, J. (1994). Achieving continuous improvement with the dual response approach: A demonstration of the Roman catapult. *Quality Engineering*, **6**, 691–705.

Luque-Garcia, J. L., Velasco, J., Dobarganes, M.C., and Luque de Castro, M.D. (2002). Fast quality monitoring of oil from prefried and fried foods by focused microwave-assisted Soxhlet extraction. *Food Chemistry*, **76**, 241–248.

Lv, C., Wang, W., Tang, Y., Wang, L., and Yang, S. (2002). Effect of cholesterol bioavailability-improving factors on cholesterol oxidase production by a mutant Brevibacterium sp. DGCDC-82. *Process Biochemistry*, **37**, 901–907.

Lynch, R. O. (1993). Minimum detectable effects for 2^{k-p} experimental plans. *Journal of Quality Technology*, **25**(1), 12–17.

Ma, C. X., and Fang, K. T. (2001). A note on generalized aberration in factorial designs. *Metrika*, **53**, 85–93.

Ma, C. X., Fang, K. T., and Erkki, L. (2000). A new approach in constructing orthogonal and nearly orthogonal arrays. *Metrika*, **50**, 255–268.

Ma, C. X., Fang, K. T., and Lin, D. K. J. (2001). On the isomorphism of fractional factorial designs. *Journal of Complexity*, **17**, 86–97.

Ma, C. X., Fang, K. T., and Lin, D. K. J. (2003). A note on uniformity and orthogonality. *Journal of Statistical Planning and Inference*, **113**, 323–334.

MacKay, D. C., Entz, T., and Carefoot, J. M. (1989). Factors affecting leaf nutrient concentrations of potatoes on irrigated brown and dark brown chernozemic soils. *Canadian Journal of Plant Science*, **69**, 591–600.

MacWilliams, F. J., and Sloane, N. J. A. (1977). *The Theory of Error-Correcting Codes*. Amsterdam: North-Holland Publ..

Maghsoodloo, S., and Hool, J. N. (1976). On response surface methodology and its computer aided teaching. *American Statistician*, **30**, 140–144.

Mallows, C. (1998). The zeroth problem. *American Statistician*, **52**(1), 1–9.

Mansson, R., and Prescott, P. (2002). Missing observations in Youden square designs. *Computational Statistics and Data Analysis*, **40**(2), 329–338.

Marazzi, A., and Yohai, V. J. (2006). Robust Box-Cox transformations based on minimum residual autocorrelation. *Computational Statistics and Data Analysis*, **50**(10), 2752–2768.

Marcus, M., and Minc, H. (1964). *A Survey of Matrix Theory and Matrix Inequalities*. Boston: Prindle, Weber & Schmidt.

Mardia, K. V. (1989). Shape analysis of triangles through directional techniques. *Journal of the Royal Statistical Society, Series B*, **51**, 449–485.

Margolin, B. H. (1968). Orthogonal main effect $2^n 3^m$ designs and two-factor interaction aliasing. *Technometrics*, **10**(3), 559–573.

Margolin, B. H. (1969a). Results on factorial designs of resolution IV for the 2^n and $2^n 3^m$ series. *Technometrics*, **11**(3), 431–444.

Margolin, B. H. (1969b). Orthogonal main effect plans permitting estimation of all two factor interactions for the $2^n 3^m$ factorial series of designs. *Technometrics*, **11**(4), 747–762.

Margolin, B. H. (1976). Design and analysis of factorial experiments via interactive computing in APL. *Technometrics*, **18**(2), 135–150.

Marquardt, D. W. (1963). An algorithm for least squares estimation of nonlinear parameters. *Journal of the Society for Industrial and Applied Mathematics*, **11**, 431–441.

Marquardt, D. W., and Snee, R. D. (1974). Test statistics for mixture models. *Technometrics*, **16**(4), 533–537. **M**

Marquardt, D. W., and Snee, R. D. (1975). Ridge regression in practice. *The American Statistician*, **29**, 3–19.

Martin, R.J., and Eccleston, J.A. (1998). Variance-balanced change-over designs for dependent observations. *Biometrika*, **85**(4), 883–892.

Mason, R. L., Gunst, R. F., and Hess, J. L. (2003). *Statistical Design and Analysis of Experiments*. New York: Wiley.

Mathew, T., and Sinha, B. K. (1992). Exact and optimum tests in unbalanced split-plot designs under mixed and random models. *Journal of the American Statistical Association*, **87**, 192–200.

Matthews, S. (1973). The effect of time of harvest on the viability and preemergence mortality in soil of pea (*Pisum sativum L.*) seeds. *Annals of Applied Biology*, **73**, 211–219.

Mays, D. P. (1999a). Near-saturated two-stage designs with dispersion effects. *Journal of Statistical Computation and Simulation*, **63**, 13–15.

Mays, D. P. (1999b). Optimal central composite designs in the presence of dispersion effects. *Journal of Quality Technology*, **31**(4), 398–407.

Mays, D. P. (2001). The impact of correlated responses and dispersion effects on optimal three level factorial designs. *Communications in Statistics, Simulation and Computation*, **30**(1), 185–194.

Mays, D. P., and Easter, S. M. (1997). Optimal response surface design in the presence of dispersion effects. *Journal of Quality Technology*, **29**(1), 59–70.

Mays, D. P., and Myers, R. H. (1996). Bayesian approach for the design and analysis of a two level factorial experiment in the presence of dispersion effects. *Communications in Statistics, Theory and Methods*, **25**(7), 1409–1428.

Mays, D. P., and Myers, R. H. (1997). Design and analysis for a two-level factorial experiment in the presence of variance heterogeneity. *Computational Statistics and Data Analysis*, **26**(2), 219–233.

Mays, D. P., and Schwartz, K. R. (1998). Two-stage central composite designs with dispersion effects. *Journal of Statistical Computation and Simulation*, **61**, 191–218.

McCullagh, P. (2000). Invariance and factorial models. *Journal of the Royal Statistical Society, Series B*, **62**(2), 209–238; discussion: pp. 238–256.

McCullagh, P., and Nelder, J. A. (1988). *Generalized Linear Models*, 2nd ed. Boca Raton, FL: Chapman & Hall/CRC.

McDaniel, W. R., and Ankenman, B. E. (2000). A response surface test bed. *Quality and Reliability in Engineering International*al, **16**, 363–372.

McDonald, G. C., and Studden, W. J. (1990). Design aspects of regression-based ratio estimation. *Technometrics*, **32**(4), 417–424.

McGrath, R. N. (2003). Separating location and dispersion effects in unreplicated fractional factorial designs. *Journal of Quality Technology*, **35**(3), 306–316.

McGrath, R. N., and Lin, D. K. J. (2001a). Confounding of location and dispersion effects in unreplicated fractional factorials. *Journal of Quality Technology*, **33**(2), 129–139.

McGrath, R. N., and Lin, D. K. J. (2001b). Testing multiple dispersion effects in unreplicated fractional factorial designs. *Technometrics*, **43**(4), 406–414.

McKay, M. D., Beckman, R. J., and Conover, W. J. (2000). A comparison of three methods for selecting values of imput variables in the analysis of output from a computer code. *Technometrics*, **42**(1), 55–61.

McKean, J. W., Sheather, S. J., and Hettmansperger, T. P. (1994). Robust and high-break-down fits of polynomial models. *Technometrics*, **36**(4), 409–415.

McKee, B., and Kshirsagar, A. M. (1982). Effect of missing plots in some response surface designs. *Communications in Statistics, Theory and Methods*, **11**, 1525–1549.

McLaughlin, K. J., Butler, S. W., Edgar, T. F., and Trachtenberg, I. (1991). Development of techniques for real-time monitoring and control in plasma etching. *Journal of the Electrochemical Society*, **138**, 789–799.

McLean R. A., and Anderson V. A. (1966). Extreme vertices design of mixture experiments. *Technometrics*, **8**(3), 447–454; discussion by J. W. Gorman, pp. 455–456. **M**

McLean, R. A., and Anderson, V. L. (1984). *Applied Factorial and Fractional Factorial Designs*, New York: Dekker.

McLeod, R. G., and Brewster, J. F. (2004). The design of blocked fractional factorial split-plot experiments. *Technometrics*, **46**(2), 135–146.

McLeod, R. G., and Brewster, J. F. (2005). Blocked fractional factorial split-plot experiments for robust parameter design. *Journal of Quality Technology*, **38**(3), 267–279.

Mead, R. (1970). Plant density and crop yield. *Applied Statistics*, **19**, 64–81.

Mead, R. (1979). Competition experiments. *Biometrics*, **35**, 41–54.

Mead, R. (1988). *The Design of Experiments*: *Statistical Principles for Practical Applications*. Cambridge, UK: Cambridge University Press.

Mead, R. (1990). The non-orthogonal design of experiments. *Journal of the Royal Statistical Society, Series A*, **153**, 151–178; discussion: pp. 178–201.

Mead, R., and Freeman, K. H. (1973). An experiment game. *Applied Statistics*, **22**, 1–6.

Mead, R., and Pike, D. J. (1975). A review of response surface methodology from a biometrics viewpoint. *Biometrics*, **31**, 803–851. **R**

Mead, R., and Riley, J. (1981). A review of statistical ideas relevant to intercropping research. *Journal of the Royal Statistical Association, Series A*, **144**, 462–487; discussion: pp. 487–509. **R**

Mee, R. W.(2001). Noncentral composite designs. *Technometrics*, **43**(1), 34–43.

Mee, R. W. (2002).Three-level simplex designs and their use in sequential experimentation. *Journal of Quality Technology*, **34**(2), 152–164.

Mee, R. W. (2004). Efficient two-level designs for estimating all main effects and two-factor interactions. *Journal of Quality Technology*, **36**(4), 400–412; corrections: (2005), **37**(1), 90.

Mee, R. W., and Bates, R. L. (1998). Split-lot designs: experiments for multistage batch processes. *Technometrics*, **40**(2), 127–140.

Mee, R. W., and Peralta, M. (2000). Semifolding 2^{k-p} designs. *Technometrics*, **42**(2), 122–134.

Mehrabi, Y., and Matthews, J.N.S. (1998). Implementable Bayesian designs for limiting dilution assays. *Biometrics*, **54**(4), 1398–1406.

Mehta, J. S., and Das, M. N. (1968). Asymmetric rotatable designs and orthogonal transformations. *Technometrics*, **10**(2), 313–322.

Mendes, A., Carvalho, J., Duarte, M., Duran, N., and Bruns, R. (2001). Factorial design and response surface optimization of crude violacein for Chromobacterium violaceum production. *Biotechnology Letters*, **23**(23), 1963–1969.

Mendoza, J. L., Toothaker, L. E., and Crain, B. R. (1976). Necessary and sufficient conditions for F ratios in the $L \times J \times K$ factorial design with two repeated factors. *Journal of the American Statistical Association*, **71**, 992–993.

Merchant, A., McCann, M., and Edwards, D. (1998). Improved multiple comparisons with a control in response surface analysis. *Technometrics*, **40**(4), 297–303.

Merrington, M., and Thompson, C. M. (1943). Tables of percentage points of the inverted beta (F) distribution. *Biometrika*, **33**, 73–88.

Meydrech, E. F., and Kupper, L. L. (1976). On quadratic approximation of a cubic response. *Communications in Statistics, Theory and Methods*, **5**, 1205–1213.

Meyer, D. L. (1963). Response surface methodology in education and psychology. *Journal of Experimental Education*, **31**, 329–336.

Meyer, R. D., and Box, G. E. P. (1992). Finding the active factors in fractionated screening experiments. *Journal of Quality Technology*, **25**, 94–105.

Meyer, R. D., Steinberg, D. M., and Box, G. E. P. (1996). Follow-up designs to resolve confounding in multifactor experiments. *Technometrics*, **38**(4), 303–313; discussion: pp. 314–332.

Meyer, R. K., and Nachtsheim, C. J. (1995). The coordinate exchange algorithm for constructing exact optimal experimental designs. *Technometrics*, **37**(1), 60–69.

Michaels, S. E. (1964). The usefulness of experimental designs. *Applied Statistics*, **13**(3), 221–228; discussion: pp. 228–235.

Michaels, S. E., and Pengilly, P. J. (1963). Maximum yield for specified cost. *Applied Statistics*, **12**, 189–193.

Mikaeili, F. (1989). D-optimum design for cubic without 3-way effect on the simplex. *Journal of Statistical Planning and Inference*, **21**, 107–115.

Mikaeili, F. (1993). D-optimum design for full cubic on the q-simplex. *Journal of Statistical Planning and Inference*, **35**, 121–130.

Millard, S. P., and Krause, A., editors. (2001). *Applied Statistics in the Pharmaceutical Industry with Case Studies in S-plus*. New York: Springer.

Miller, A. (1997). Strip-plot configurations of fractional factorials. *Technometrics*, **39**(2), 153–161.

Miller, A. (2005). The analysis of unreplicated factorial experiments using all possible comparisons. *Technometrics*, **47**(1), 51–63.

Miller, A., and Sitter, R. R. (2001). Using the folded-over 12-run Plackett-Burman design to consider interactions. *Technometrics*, **43**(1), 44–55.

Miller, A., and Sitter, R. R. (2004). Choosing columns from the 12-run Plackett-Burman design. *Statistics and Probability Letters*, **67**, 193–201.

Miller, A., and Sitter, R. R. (2005). Using folded-over nonorthogonal designs. *Technometrics*, **47**(4), 502–513.

Miller, A., and Wu, C. J. F. (1996). Parameter design for signal-response systems: A different look at Taguchi's dynamic parameter system. *Statistical Science*, **11**(2), 122–136.

Miller, I. (1974). Statistical design for experiments in combination therapy. *Cancer Chemotherapy Reports*, **4**, 151–156.

Miller, M. H., and Ashton, G. C. (1960). The influence of fertilizer placement and rate of nitrogen on fertilizer phosphorus utilization by oats as studied using a central composite design. *Canadian Journal of Soil Science*, **40**, 157–167.

Miró-Quesada, G., and Del Castillo, E. (2004). Two approaches for improving the dual response method in robust parameter design. *Journal of Quality Technology*, **36**(2), 154–168.

Miró-Quesada, G., and Del Castillo, E. (2006). A search method for the exploration of new regions in robust parameter design. In *Response Surface Methodology and Related Topics*, (A. I. Khuri, ed.), pp. 89–121. Hackensack, NJ: World Scientific Publishing.

Miró-Quesada, G., Del Castillo, E., and Peterson, J. P. (2004). A Bayesian approach for multiple response surface optimization in the presence of noise variables. *Journal of Applied Statistics*, **31**, 251–270.

Mirreh, H. F., and Ketcheson, J. W. (1973). Influence of soil water, matric potential, and resistance to penetration on corn root elongation. *Canadian Journal of Soil Science*, **53**, 383–388.

Mislevy, P., Kalmbacher, R. S., and Martin, F. G. (1981). Cutting management of the tropical legume American jointvetch. *Agronomy Journal*, **73**, 771–775.

Mitchell, T. J. (1974a). An algorithm for the construction of D-optimal experimental designs. *Technometrics*, **16**, 203–210.

Mitchell, T. J. (1974b). Computer construction of "D-optimal" first order designs. *Technometrics*, **20**(4), 211–220

Mitchell, T. J. (2000). An algorithm for the construction of "D-optimal" experimental designs. *Technometrics*, **42**(1), 48–54.

Mitchell, T. J., and Bayne, C. K. (1978). D-optimal fractions of three-level fractional designs. *Technometrics*, **20**(4), 369–380; discussion: pp. 381–383.

Mitchell, T. J., Morris, M. D., and Ylvisaker, D. (1995). Two-level fractional factorials and Bayesian prediction. *Statistical Sinica*, **5**, 559–573.

Mitra, R. K. (1981). On G-efficiency of some second order rotatable designs. *Biometrical Journal*, **23**, 749–757.

Moerbeek, M. (2005). Robustness properties of A-, D-, and E-optimal designs for polynomial growth models with autocorrelated errors. *Computational Statistics and Data Analysis*, **48**(4), 765–778.

Moerbeek, M., Van Breukelen, G. J. P., and Berger, M. P. F. (2001). Optimal experimental designs for multilevel logistics models. *The Statistician*, **50**(1), 17–30.

Montepiedra, G., and Yeh, A. B. (1997). A two-stage strategy for the construction of D-optimal experimental designs. *Communications in Statistics, Simulation and Computation*, **27**, 377–402.

Montepiedra, G., and Yeh, A. B. (2004). Two-stage designs for identification and estimation of polynomials. *Computational Statistics and Data Analysis*, **46**(3), 531–546.

Montepiedra, G., Myers, D., and Yeh, A. B. (1998). Application of genetic algorithms to the construction of exact D-optimal designs. *Journal of Applied Statistics*, **25**, 817–826.

Montgomery, D. C. (1990-1991). Using fractional factorial designs for robust process development. *Quality Engineering*, **3**, 193–205.

Montgomery, D. C. (1992). The use of statistical process control and design of experiments in product and process improvement. *IIE Transactions*, **24**, 4–17.

Montgomery, D. C. (1999). Experimental design for product and process design and development (with comments). *The Statistician*, **48**(2), 159–173; discussion: 173–177, see also Montgomery (2000).

Montgomery, D. C. (2000). Reply to comment. *The Statistician*, **49**, 107–110.

Montgomery, D. C. (2005). *Design and Analysis of Experiments*, 6th ed. Hoboken, NJ: Wiley.

Montgomery, D. C., and Borror, C. M. (2001). Discussion of "Factor screening and response surface exploration" by Cheng and Wu. *Statistica Sinica*, **11**, 591–595.

Montgomery, D. C., and Runger, G. C. (1996). Foldovers of 2^{k-p} resolution IV experimental designs. *Journal of Quality Technology*, **28**(4), 446–450.

Montgomery, D. C., and Voth, S. R. (1994). Multicollinearity and leverage in mixture experiments. *Journal of Quality Technology*, **26**(2), 96–108. **M**

Montgomery, D. C., Talavage, J. J., and Mullen, C. J. (1972). A response surface approach to improving traffic signal settings in a street network. *Transportation Research*, **6**, 69–80.

Montgomery, D. C., Borror, C. M., and Stanley, J. D. (1997). Some cautions in the use of Plackett-Burman designs. *Quality Engineering*, **10**(2), 371–381.

Montgomery, D. C., Loredo, Jearkpaporn, E. N., and Testik, M. C. (2002). Experimental designs for constrained regions. *Quality Engineering*, **14**(4), 587–601.

Mooney, R. W., Comstock, A. J., Goldsmith, R. L., and Meisenhalter, G. J. (1960). Predicting chemical composition in precipitation of calcium hydrogen orthophosphate. *Industrial and Engineering Chemistry*, **52**(5), 427–428.

Moore, D. P., Harward, M. E., Mason, D. D. Hader, R. J., Lott, W. L., and Jackson, W. A. (1957). An investigation of some of the relationships of copper, iron and molybdenum in the growth and nutrition of lettuce. II. *Soil Science Society of America Proceedings*, **21**, 65–74.

Moorehead, D. H., and Himmelblau, D. M. (1962). Optimization of operating conditions in a packed liquid-liquid extracting column. *Industrial and Engineering Fundamentals*, pp. 68–72.

Moorehead, P. R., and Wu, C. F. J. (1998). Cost-driven parameter design. *Technometrics*, **40**(1), 111–119.

Morgan, J. P. (1998). Orthogonal collections of Latin squares. *Technometrics*, **40**(4), 327–333.

Morgan, J. P., and Uddin, N. (1991). Two-dimensional design for correlated errors. *Annals of Statistics*, **19**, 2160–2182.

Morgan, J. P., and Uddin, N. (1996). Optimal blocked main effects plans with nested rows and columns and related designs. *Annals of Statistics*, **24**, 1185–1208.

Morris, M. D. (1991). Factorial sampling plans for preliminary computational experiments. *Technometrics*, **33**(2), 161–174.

Morris, M. D. (2000a). Three *Technometrics* experimental design classics. *Technometrics*, **42**(1), 26–27.

Morris, M. D. (2000b). A class of three-level experimental designs for response surface modeling. *Technometrics*, **42**(2), 111–121.

Morris, M. D., and Mitchell, T. J. (1983). Two-level multifactor designs for detecting the presence of interactions. *Technometrics*, **25**(4), 345–355.

Morris, M. D., and Solomon, A. D. (1995). Design and analysis for an inverse problem arising from an advection-dispersion process. *Technometrics*, **37**(3), 293–302.

Mosby, J. F., and Albright, L. A. (1966). Alkylation of isobutene with 1-butane using sulfuric acid as catalyst at high rates of agitation. *Industrial and Engineering Chemistry Product Research and Development*, **5**, 183–190.

Moskowitz, J. R., and Chandler, J. W. (1977). Eclipse-developing products from concepts via consumer ratings. *Food Product Development*, **11**, 50–60.

Mukerjee, R. (1999). On the optimality of orthogonal array plus one run plans. *Annals of Statistics*, **27**(1), 82–93.

Mukerjee R., and Huda, S. (1985). Minimax second- and third-order designs to estimate the slope of a response surface. *Biometrika*, **72**, 173–178.

Mukerjee, R., and Wu, C. F. J. (1995). On the existence of saturated and nearly saturated asymmetrical orthogonal arrays. *Annals of Statistics*, **23**(6), 2102–2115.

Mukerjee, R., and Wu, C. F. J. (1999). Blocking in regular fractional factorials: a projective geometric approach. *Annals of Statistics*, **27**(4), 1256–1271.

Mukerjee, R., and Wu, C. F. J. (2001). Minimum aberration designs for mixed factorials in terms of complimentary sets. *Statistica Sinica*, **11**, 225–239.

Mukerjee R., Chan, L. Y., and Fang, K. T. (2000). Regular fractions of mixed factorials with maximum estimation capacity. *Statistica Sinica*, **10**, 1117–1132.

Mukhopadhyay, A. C. (1969). Operability region and optimum rotatable designs. *Sankhya, Series B*, **31**, 75–84.

Mukhopadhyay, S. (1981). Group divisible rotatable designs which minimize the mean square bias. *Journal of the Indian Society of Agricultural Statistics*, **33**, 40–46.

Müller, C. H. (1997). *Robust Planning and Analysis of Experiments*. New York: Springer.

Müller, W. G., and Pázman, A. (2003). Measures for designs in experiments with correlated errors. *Biometrika*, **90**, 423–434.

Murty, J. S., and Das, M. N. (1968). Design and analysis of experiments with mixtures. *Annals of Mathematical Statistics*, **39**(5), 1517–1539. **M**

Murty, V. N., and Studden, W. J. (1972). Optimal designs for estimating the slope of a polynomial regression. *Journal of the American Statistical Association*, **67**, 869–873.

Myers, R. H. (1991). Response surface methodology in quality improvement. *Communications in Statistics, Theory and Methods*, **20**(2), 457–476.

Myers, R. H. (1999). Response surface methodology—current status and future directions. *Journal of Quality Technology*, **31**(1), 30–44; discussion: pp. 45–74.

Myers, R. H., and Carter, W. H. (1973). Response surface techniques for dual response systems. *Technometrics*, **15**(2), 301–317.

Myers, R. H., and Lahoda, S. J. (1975). A generalization of the response surface mean square error criterion with a specific application to the slope. *Technometrics*, **17**(4), 481–486.

Myers, R. H., and Montgomery, D. C. (1997). A tutorial on generalized linear models. *Journal of Quality Technology*, **29**(3), 274–291.

Myers, R. H., and Montgomery, D. C. (2002). *Response Surface Methodology, Process and Product Optimization Using Designed Experiments*, 2nd ed. New York: Wiley.

Myers, R. H., Khuri, A. I., and Carter, W. H. (1989). Response surface methodology: 1966–1988. *Technometrics*, **31**(2), 137–157. **R**

Myers, R. H., Khuri, A. I., and Vining, G. (1992). Response surface alternatives to the Taguchi robust parameter design approach. *American Statistician*, **46**(2), 131–139; see also Myers et al. (1993).

Myers, R. H., Vining, G. G., Giovannitti-Jensen, A., and Myers, S. L. (1992). Variance dispersion properties of second-order response surface designs. *Journal of Quality Technology*, **24**(1), 1–11.

Myers, R. H., Khuri, A. J., and Vining, G. (1993). Comment by Lucas. *American Statistician*, **47**(1), 86.

Myers, W. R., Myers, R. H., and Carter, W. H. (1994). Some alphabetic optimal designs for the logistic regression model. *Journal of Statistical Planning and Inference*, **42**(1), 57–77.

Myers, R. H., Kim, Y., and Griffiths, K. (1997). Response surface methods and the use of noise variables. *Journal of Quality Technology*, **29**(4), 429–440.

Myers, R. H., Montgomery, D. C., and Vining, G. G. (2002). *Generalized Linear Models with Applications in Engineering and the Sciences*. New York: Wiley.

Myers, R. H., Montgomery, D. C., Vining, G. G., Borror, C. M., and Kowalski, S. M. (2004). Response surface methodology: a retrospective and literature survey. *Journal of Quality Technology*, **36**(1), 53–77. **R**

Naes, T., Bjerke, F., and Faergestad, E. M. (1998a). A comparison of design and analysis techniques for mixtures. *Food Quality and Preferences*, **10**, 209–217. **M**

Naes, T., Faergestad, E. M., and Cornell, J. (1998b). A comparison of methods for analyzing data from a three component mixture experiment in the presence of variation created by two process variables. *Chemometrics and Intelligent Laboratory Systems*, **41**, 221–235. **M**

Naiman, D. Q. (1990). Volumes of tubular neighborhoods of spherical polyhedra and statistical inference. *Annals of Statistics*, **18**, 685–716.

Nair, V. N., ed. (1992). Taguchi's parameter design: A panel discussion. *Technometrics*, **34**(2), 127–161.

Nair, V. N., and Pregibon, D. (1988). Analyzing dispersion effects from replicated factorial experiments. *Technometrics*, **30**(3), 247–257.

Nalimov, V. V., Golikova, T. I., and Mikeshina, N. G. (1970). On practical use of the concept of D-optimality. *Technometrics*, **12**(4), 799–812.

Namini, H., and Ghosh, S. (1990). On robustness of $2^m \times 3^n$ fractional factorial resolution III designs against the unavailability of data. *Computational Statistics and Data Analysis*, **10**, 11–16.

Narasimham, V. L., Rao, P. R., and Rao, K. N. (1983). Construction of second order rotatable designs through a pair of balanced incomplete block designs. *Journal of the Indian Society of Agricultural Statistics*, **35**, 36–40.

Narcy, J. P., and Renaud, J. (1972). Use of simplex experimental designs in detergent formulation. *Journal of the American Oil Chemists' Society*, **49**(10), 598–608.

National Bureau of Standards Applied Mathematics Series. See authors: Clatworthy, W. H., Connor, W. S., Deming, L. S., and Zelen, M. (1957). Fractional factorial experimental designs for factors at two levels. (No. 48); Connor, W. S., and Young, S. (1961). Fractional factorial designs for experiments with factors at two and three levels. (No. 58); Connor, W. S., and Zelen, M. (1961). Fractional factorial experiment designs for factors at three levels. (No. 54).

Navalón, A., Prieto, A., Araujo, L., and Vílchez, J. L. (2002). Determination of oxadiazon residues by headspace solid-phase microextraction and gas chromatography-mass spectrometry. *Journal of Chromatography A*, **946**, 239–245.

Nelder, J. A. (1962). New kinds of systematic designs for spacing experiments. *Biometrics*, **18**, 283–307.

Nelder, J. A. (1966a). Evolutionary experimentation in agriculture. *Journal of Scientific Field Agriculture*, **17**, 7–9.

Nelder, J. A. (1966b). Inverse polynomials, a useful group of multi-factor response functions. *Biometrics*, **22**, 128–141.

Nelder, J. A. (1998). The selection of terms in response-surface models—How strong is the weak-heredity principle? *American Statistician*, **52**(4), 315–318.

Nelder, J. A. (2000). Functional marginality and response surface fitting. *Journal of Applied Statistics*, **27**(1), 109–112.

Nelder, J. A., and Lane, P. W. (1995). The computer analysis of factorial experiments: In memorian—Frank Yates. *American Statistician*, **49**(4), 382–385; see also Nelder and Lane (1996).

Nelder, J. A., and Lane, P. W. (1996). Comment by Randall and reply. *American Statistician*, **50**(4), 382.

Nelder, J. A., and Lee, Y. (1991). Generalized linear models for the analysis of Taguchi-type experiments. *Applied Stochastic Models and Data Analysis*, **7**, 107–120.

Nelder, J. A., and Mead, R. (1965). A simplex method for function minimization. *Computing Journal*, **7**, 308–313.

Nelder, J. A., and Wedderburn, R. W. M. (1972). Generalized linear models. *Journal of the Royal Statistical Association, Series A*, **135**, 370–384.

Nelson, B. J., Montgomery, D. C., Elias, R. J., and Maass, E. (2000). A comparison of several design augmentation strategies. *Quality and Reliability Engineering International*, **16**, 435–449.

Nelson, L. S. (1982). Analysis of two-level factorial experiments. *Journal of Quality Technology*, **14**(2), 95–98.

Nelson, L. S. (1985). What do low F ratios tell you? *Journal of Quality Technology*, **17**(4), 237–238.

Nelson, L. S. (1989). A stabilized normal probability plotting technique. *Journal of Quality Technology*, **21**(3), 213–215.

Nelson, L. S. (1995). Using nested designs. II. Confidence limits for standard deviations. *Journal of Quality Technology*, **27**(3), 265–267.

Nesterov, Y., and Nemirovskii, A. (1994). *Interior-Point Polynomial Algorithms in Convex Programming*. Philadelphia: Society for Industrial and Applied Mathematics.

Neudecker, H. (1969). Some theorems on matrix differentiation with special reference to Kronecker matrix products. *Journal of the American Statistical Association*, **64**, 953–963.

Neumaier, A., and Seidel, J. J. (1990). Measures of strength 2e, and optimal designs of degree e. MS.

Nguyen, N.-K. (1993). An algorithm for constructing optimal resolvable incomplete block designs. *Communications in Statistics, Simulation and Computation*, **22**(3), 911–923.

Nguyen, N.-K. (1994). Construction of optimal block designs by computer. *Technometrics*, **36**(3), 300–307.

Nguyen, N.-K. (1996a). An algorithmic approach to generating supersaturated designs. *Technometrics*, **38**(1), 69–73.

Nguyen, N.-K. (1996b). A note on the construction of near-orthogonal arrays with mixed levels and economic run sizes. *Technometrics*, **38**(3), 279–283.

Nguyen, N.-K., and Dey, A. (1989). Computer aided construction of D-optimal 2^m fractional factorial designs of resolution V. *Australian Journal of Statistics*, **31**, 111–117.

Nguyen, N.-K., and Miller, A. J. (1992). A review of some exchange algorithms for constructing discrete D-optimal designs. *Computational Statistics and Data Analysis*, **14**, 489–498.

Nguyen, N.-K., and Miller, A. J. (1997). 2^m fractional factorial designs of resolution V with high A-efficiency, $7 \leq m \leq 10$. *Journal of Statistical Planning and Inference*, **59**, 379–384.

Nigam, A. K. (1967). On third order rotatable designs with smaller number of levels. *Journal of the Indian Society of Agricultural Statistics*, **19**(2), 36–41.

Nigam, A. K. (1970). Block designs for mixture experiments. *Annals of Mathematical Statistics*, **41**(6), 1861–1869. **M**

Nigam, A. K. (1973). Multifactor mixture experiments. *Journal of the Royal Statistical Society, Series B*, **35**, 51–56. **M**

Nigam, A. K. (1974). Some designs and models for mixture experiments for the sequential exploration of response surfaces. *Journal of the Indian Society for Agricultural Statistics*, **26**, 120–124. **M**

Nigam, A. K. (1976). Corrections to blocking conditions for mixture experiments. *Annals of Mathematical Statistics*, **47**, 1294–1295. **M**

Nigam, A. K. (1977). A note on four and six level second order rotatable designs. *Journal of the Indian Society of Agricultural Statistics*, **29**, 89–91.

Nigam, A. K., and Dey, A. (1970). Four and six level second order rotatable designs. *Calcutta Statistical Association Bulletin*, **19**, 155–157.

Nigam, A. K., and Gupta, V. K. (1985). Construction of orthogonal main-effect plans using Hadamard matrices. *Technometrics*, **27**(1), 37–40.

Nigam, A. K., Gupta, S. C., and Gupta, S. (1983). A new algorithm for extreme vertices designs for linear mixtures models. *Technometrics*, **25**(4), 367–371. **M**

Njui, F., and Patel, M. S. (1988). Fifth order rotatability. *Communications in Statistics, Theory and Methods*, **17**, 833–848.

Nobile, A., and Green, P. J. (2000). Bayesian analysis of factorial experiments by mixture modelling. *Biometrika*, **87**(1), 15–35.

Noh, H. G., Song, M. S., and Park, S. H. (2004). An unbiased method for constructing multilabel classification trees. *Computational Statistics and Data Analysis*, **47**(1), 149–164.

Nordström, K. (1999). The life and work of Gustav Elfving. *Statistical Science*, **14**(2), 174–196.

Norton, C. J., and Moss, T. E. (1963). Oxidative dealkylation of alkylaromatic hydrocarbons. *Industrial and Engineering Chemistry Process Design and Development*, **2**, 140–147; also see Norton and Moss (1964).

Norton, C. J., and Moss, T. E. (1964). Selective vapor-phase catalytic oxidation of alkylaromatic hydrocarbons to their parent homologs. *Industrial and Engineering Chemistry Process Design and Development*, **3**, 23–32.

Notz, W. (1981). Saturated designs for multivariate cubic regression. *Journal of Statistical Planning and Inference*, **5**, 37–43.

Notz, W. (1982). Minimal point second order designs. *Journal of Statistical Planning and Inference*, **6**, 47–58.

Novak, J., Lynn, R.O., and Harrington, E. C. (1962). Process scale-up by sequential experimentation and mathematical optimization. *Chemical Engineering Progress*, **58**(2), 55–59.

O'Brien, T. E., and Funk, G. M. (2003). A gentle introduction to optimal design for regression functions. *American Statistician*, **57**(4), 265–267.

O'Connell, M. Haaland, P., Hardy, S., and Nychka, D. (1995). Nonparametric regression, kriging and process optimization. *Statistical Modelling*, (Seeber et al., eds.) pp. 207–214. New York: Springer-Verlag.

O'Donnell, E. M., and Vining, G. G. (1997). Mean squared error of prediction approach to the analysis of a combined array. *Journal of Applied Statistics*, **24**, 733–746.

Odulaja, A., and Kayode, G. O. (1987). Response of cowpea (*vigna unguiculata*) to spacing in the savanna and rainforest zones of Nigeria: Response surface analysis. *Exploratory Agriculture*, **23**, 63–68.

Ohm, H. W. (1976). Response of 21 oat cultivars to nitrogen fertilization. *Agronomy Journal*, **68**, 773–775.

Oldford, R. W. (1995). A physical device for demonstrating confounding, blocking and the role of randomization in uncovering a causal relationship. *American Statistician*, **49**(2), 210–216; see also Oldford (1996).

Oldford, R. W. (1996). Comment by Pearl and reply. *American Statistician*, **50**(3), 387–388.

Olguin, J., and Fearn, T. (1997). A new look at half-normal plots for assessing the significance of contrasts for unreplicated factorials. *Applied Statistics*, **46**(4), 449–462; see also Olguin and Fearn (1998).

Olguin, J., and Fearn, T. (1998). Letters to the Editors. *Applied Statistics*, **47**(3), 447–448.

Ologunde, O. O., and Sorensen, R. C. (1982). Influence of concentrations of K and Mg in nutrient solutions on sorghum. *Agronomy Journal*, **74**, 41–46.

Olsson, D. M., and Nelson, L. S. (1975a). The Nelder-Mead simplex procedure for function minimization. *Technometrics*, **17**(1), 45–51; see also Olssoon and Nelson (1975b).

Olsson, D. M., and Nelson, L. S. (1975b). *Technometrics*, **17**(3), 393–394.

O'Neill, J. C., Borror, C. M., Eastman, P. Y., Fradkin, D. G., James, M. P., Marks, A. P., and Montgomery, D. C. (2000). Optimal assignment of samples to treatments for robust design. *Quality and Reliability International*, **16**, 417–422.

Ophir, S., El-Gad, V., and Schneider, M. (1988). A case study of the use of an experimental design in preventing shorts in nickel-cadmium cells. *Journal of Quality Technology*, **20**(1), 44–50.

Ortiz, F., Simpson, J. R., Pignatiello, J., and Heredia-Langner, A. (2004). A genetic algorithm approach to multiple-response optimization. *Journal of Quality Technology*, **36**(4), 432–450.

Ott, L., and Cornell, J. A. (1974). A comparison of methods which utilize the integrated mean square error criterion for constructing response surface designs. *Communications in Statistics, Theory and Methods*, **3**(11), 1053–1068.

Ott, L., and Mendenhall, W. (1972). Designs for estimating the slope of a second order linear model. *Technometrics*, **14**(2), 341–354.

Ott, L., and Mendenhall, W. (1973). Designs for comparing the slopes of two second order response curves. *Communications in Statistics, Theory and Methods*, **1**(3), 243–260.

Ott, L., and Meyers, R. H. (1968). Optimal experimental designs for estimating the independent variable in regression. *Technometrics*, **10**(4), 811–823.

Ottenbacher, K. J. (1996). The power of replications and replications of power. *American Statistician*, **50**(3), 271–275.

Ottman, M. J., Doerge, T. A., and Martin, E. C. (2000). Durum grain quality as affected by nitrogen fertilization near athesis and irrigation during grain fill. *Agronomy Journal*, **92**, 1035–1041.

Pajak, T. F., and Addelman, S. (1975). Minimum full sequences of 2^{n-m} resolution III plans. *Journal of the Royal Statistical Society, Series B*, **37**(1), 88–95.

Paley, R. E. A. C. (1933). On orthogonal matrices. *Journal of Mathematical Physics*, **12**, 311–320.

Pan, G. (1999). The impact of unidentified location effects on dispersion-effects identification from unreplicated factorial designs. *Technometrics*, **41**(4), 313–326.

Pan, G., and Santner, T. J. (1998). Selection and screening procedures to determine optimal product designs. *Journal of Statistical Planning and Inference*, **67**, 311–330.

Panda, R. N., and Das, R. N. (1994). First order rotatable designs with correlated errors (FORDWCE). *Calcutta Statistical Association Bulletin*, **44**, 83–101.

Panneton, B., Philion, H., Thériault, R., and Khelifi, M. (2000). Spray chamber evaluation of air-assisted spraying on broccoli. *Crop Science*, **40**, 444–448.

Park, S. H. (1977). Selection of polynomial terms for response surface experiments. *Biometrics*, **33**, 225–229.

Park, S. H. (1978a). Experimental designs for fitting segmented polynomial regression models. *Technometrics*, **20**(2), 151–154.

Park, S. H. (1978b). Selecting contrasts among parameters in Scheffé's mixture models: Screening components and model reduction. *Technometrics*, **20**(3), 273–279. **M**

Park, S. H. (1987). A class of multifactor designs for estimating the slope of response surfaces. *Technometrics*, **29**(4), 449–453.

Park, S. H. (1996). *Robust Design and Analysis for Quality Engineering*. London: Chapman & Hall.

Park, S. H. (2006). Concepts of slope-rotatability for second order response surface designs. In *Response Surface Methodology and Related Topics* (A. I. Khuri, ed.), pp. 409–426. Hackensack, NJ: World Scientific Publishing.

Park, S. H., and Jang, D. H. (1999a). A graphical method for evaluating the effect of blocking in response surface designs. *Communications in Statistics, Simulation and Computation*, **28** (2), 369–380.

Park, S. H., and Jang, D. H. (1999b). Measures for evaluating the effect of blocking in response surface designs. *Communications in Statistics, Theory and Methods*, **28**(7), 1599–1616.

Park, S. H., and Kim, H. J. (1982). Axis-slope-rotatable designs for experiments with mixtures. *Journal of the Korean Statistical Society*, **11**, 36–44. **M**

Park, S. H., and Kim, H. J. (1992). A measure of slope-rotatabilty for second order response surface experimental designs. *Journal of Applied Statistics*, **19**, 391–404.

Park, Y.-J., Richardson, D. E., Montgomery, D. C., Ozol-Godfrey, A., Borror, C. M., Anderson-Cook, C. M. (2005). Prediction variance properties of second order designs for cuboidal regions. *Journal of Quality Technology*, **37**(4), 253–266.

Parks, W. L., and Walker, W. M. (1969). Effect of soil potassium, potassium fertilizer and method of fertilizer placement upon corn yields. *Soil Science Society of America Proceedings*, **33**(3), 427–429.

Patel, M. S., and Arap-Kaske, J. K. (1985). Conditions for fourth order rotatability in k dimensions. *Communications in Statistics, Theory and Methods*, **14**, 1343–1351.

Paterson, L. J. (1988). Some recent work on making incomplete-block designs available as a tool for science. *International Statistical Review*, **56**, 129–138.

Paterwardhan, V. S., and Eckert, R. E. (1981). Maximization of the conversion to m-toluene-sulfonic in liquid-phase sulfonation. *Industrial and Engineering Chemistry Process Design and Development*, **20**, 82–85.

Patnaik, P. B. (1949). The non-central χ^2- and F-distributions and their applications. *Biometrika*, **36**, 202–232.

Payne, R. W. (1997). Inversion of matrices with contents subject to modulo arithmetic. *Applied Statistics*, **46**(2), 295–298.

Payne, R. W. (1998). Design keys, pseudo-factors and general balance. *Computional Statistics and Data Analysis*, **29**(2), 217–229.

Pearce, S. C. (1975). Row and column designs. *Applied Statistics*, **24**, 60–74.

Pearson, E. S., and Hartley, H. O. (1958). *Biometrika Tables for Statisticians*, 2nd ed., Vol. 1. New York: Cambridge University Press.

Pearson, E. S., and Hartley, H. O. (1966). *Biometrika Tables for Statisticians*, 3rd ed., Vol. 1. New York: Cambridge University Press.

Pedreira, C. G. S., Sollenberger, L. E., and Mislevy, P. (2000). Botanical composition, light interception, and carbohydrate reserve status of grazed 'florakirk' bermudagrass. *Agronomy Journal*, **92**, 194–199.

Peixoto, J. L. (1990). A property of well-formulated polynomial regression models. *American Statistician*, **44**(1), 26–30.

Peixoto, J. L., and Diaz-Saiz, J. (1998). On the construction of hierarchical polynomial regression models. *Estadistica, Journal of the Inter-American Statistical Institute*, **48**, 175–210.

Pesotan, H., and Raktoe, B. L. (1981). Further results on invariance and randomization in fractional replication. *Annals of Statistics*, **9**, 418–423.

Pesotchinsky, L. (1975). D-optimum and quasi-D-optimum second order designs on a cube. *Biometrika*, **62**, 335–340.

Peterson, J. J. (1989). First and second order derivatives having applications to estimation of response surface optima. *Statistics and Probability Letters*, **8**, 29–34.

Peterson, J. J. (1993). A general approach to ridge analysis with confidence intervals. *Technometrics*, **35**(2), 204–214. **M**

Peterson, J. J. (2003). Letter to the Editor. *Technometrics*, **45**(2), 185.

Peterson, J. J. (2004). A posterior predictive approach to multiple response surface optimization. *Journal of Quality Technology*, **36**(2), 139–153.

Peterson, J. J. (2005). Ridge analysis with noise variables. *Technometrics*, **47**(3), 274–283. **M**

Peterson, J. J., Cahya, S., and Del Castillo, E. (2002). A general approach to confidence regions for optimal factor levels of response surfaces. *Biometrics*, **58**(2), 422–431; see also Peterson et al. (2004).

Peterson, J. J., Cahya, S., and Del Castillo, E. (2004). Letter to the Editor. *Technometrics*, **46**(3), 355–357.

Petres, J., and Czukor, B. (1989a). Investigation of the effects of extrusion cooking on antinutritional factors in soyabeans employing response surface analysis. Part 1. Effect of extrusion cooking on trypsin-inhibitor activity. *Nahrung*, **33**, 275–281.

Petres, J., and Czukor, B. (1989b). Investigation of the effects of extrusion cooking on antinutritional factors in soyabeans employing response surface analysis. Part 2. Effect of extrusion cooking on urease and hemagglutinin activity. *Nahrung*, **33**, 729–736.

Pettit, A. N. (1996). Infinite estimates with fractional factorial experiments. *The Statistician*, **45**(2), 197–206.

Phadke, M. S. (1989). *Quality Engineering using Robust Design*. Englewood Cliffs, NJ: Prentice-Hall.

Phadke, M. S., and Dehnad, K. (1988). Optimization of product and process design for quality and cost. *Quality and Reliability Engineering International*, **4**(2), 159–169.

Phifer, L. H., and Maginnis, J. B. (1960). Dry ashing of pulp and factors which influence it. *Tappi*, **43**(1), 38–44.

Piegorsch, W. W., Weinberg, C. R., and Margolin, B. H. (1988). Exploring simple independent action in multifactor tables of proportions. *Biometrics*, **44**, 595–603.

Piepel, G. F. (1982). Measuring component effects in constrained mixture experiments. *Technometrics*, **24**(1), 29–39. **M**

Piepel, G. F. (1983a). Defining consistent constraint regions in mixture experiments. *Technometrics*, **25**(1), 97–101. **M**

Piepel, G. F. (1983b). Calculating centroids in constrained mixture experiments. *Technometrics*, **25**(3), 279–284. **M**

Piepel, G. F. (1988a). Programs for generating extreme vertices and centroids of linearly constrained experimental regions. *Journal of Quality Technology*, **20**(2), 125–139. **M**

Piepel, G. F. (1988b). A note on models for mixture-amount experiments when the total amount takes a zero value. *Technometrics*, **30**(4), 449–450. **M**

Piepel, G. F. (1990). Screening designs for constrained mixture experiments derived from classical screening designs. *Journal of Quality Technology*, **22**(1), 23–33. **M**

Piepel, G. F. (1991). Screening designs for constrained mixture experiments derived from classical screening designs an addendum. *Journal of Quality Technology*, **23**(2), 96–101. **M**

Piepel, G. F. (1997). Survey of software with mixture experiment capabilities. *Journal of Quality Technology*, **29**(1), 76–85. **M**

Piepel, G. F. (2006). 50 years of mixture experiment research: 1955–2004. In *Response Surface Methodology and Related Topics* (A. I. Khuri, ed.), pp. 283–327. Hackensack, NJ: World Scientific Publishing. (Contains 360 references!) **M**

Piepel, G. F., and Cornell, J. A. (1985). Models for mixture experiments when the response depends on the total amount. *Technometrics*, **27**(3), 219–227. **M**

Piepel, G. F., and Cornell, J. A. (1994). Mixture experiment approaches: Examples, discussion, and recommendations. *Journal of Quality Technology*, **26**(3), 177–196. **M**

Piepel, G. F., Cooley, S. K., Peeler, D. K, Vienna, J. D., and Edwards, T. B. (2002). Augmenting a waste glass mixture experiment study with additional glass components and experimental runs. *Quality Engineering*, **15**(1), 91–111. **M**

Piepel, G. F., Hicks, R. D., Szychowski, J. M., and Loeppky, J. L. (2002). Methods for assessing curvature and interaction in mixture experiments. *Technometrics*, **44**(2), 161–172. **M**

Piepel, G. F., Szychowski, J. M., and Loeppky, J. L. (2002). Augmenting Scheffé linear mixture models with squared and/or cross-product terms. *Journal of Quality Technology*, **34**(3), 297–314. **M**

Pigeon, J. G., and McAllister, P. R. (1989). A note on partially replicated orthogonal main-effect plans. *Technometrics*, **31**(2), 249–251.

Pignatiello, J. J. (1993). Strategies for robust multiresponse quality engineering. *IIE Transactions*, **25**, 5–15.

Pignatiello, J. J., and Ramberg, J. S. (1991). Top ten triumphs and tragedies of Genichi Taguchi. *Quality Engineering*, **4**, 211–225.

Pinchbeck, P. H. (1957). The kinetic implications of an empirically fitted yield surface for the vapour phase oxidation of naphthalene to pthalic anhydride. *Chemical Engineering Science*, **6**, 105–111.

Pinthus, M. J. (1972). The effect of chlormequat seed-parent treatment on the resistance of wheat seedlings to turbutryne and simazine. *Weed Research*, **12**, 241–247.

Pistone, G., and Wynn, H. P. (1996). Generalized confounding with Gröbner bases. *Biometrika*, **83**(3), 653–666.

Plackett, R. L. (1946). Some generalizations in the multifactorial design. *Biometrika*, **33**, 328–332.

Plackett, R. L., and Burman, J. P. (1946). The design of optimum multifactorial experiments. *Biometrika*, **33**, 305–325, 328–332.

Plackett, R. L., and Hewlett, P. S. (1952). Quantal responses to mixtures of poisons. *Journal of the Royal Statistical Society*, Series B, **14**(2), 141–161. **M**

Pledger, M. (1996). Observable uncontrollable factors in parameter design. *Journal of Quality Technology*, **28**(2), 153–162.

Ponce de Leon, A. C., and Atkinson, A. C. (1991). Optimum experimental design for discriminating between two rival models in the presence of prior information. *Biometrika*, **78**, 601–608.

Porter, W. P., and Busch, R. L. (1978). Fractional factorial analysis of growth and weaning success in Peromyscus maniculatus. *Science, New Series*, **202**, November 24, 1978, 907–910.

Prairie, R. R., and Zimmer, W. J. (1964). 2^p factorial experiments with the factors applied sequentially. *Journal of the American Statistical Association*, **59**, 1205–1216.

Prairie, R. R., and Zimmer, W. J. (1968). Fractional replications of 2^p factorial experiments with the factors applied sequentially. *Journal of the American Statistical Association*, **63**, 644–652.

Prasad, M. (1976). Response of sugarcane to filter press mud and N, P, and K fertilizers. I. Effect on sugarcane yield and sucrose content. II. Effects on plant composition and soil chemical properties. *Agronomy Journal*, **68**, 539–546.

Preece, D. A. (1990). R. A. Fisher and experimental design: A review. *Biometrics*, **46**, 925–935.

Prescott, P. (1998). Nearly optimal orthogonally blocked designs for a quadratic mixture model with q components. *Communications in Statistics, Theory and Methods*, **27**(10), 2559–2580.

Prescott, P. (2004). Modelling in mixture experiments including interactions with process variables. *Quality Technology and Quantitative Management*, **1**(1), 87–103. **M**

Prescott, P., and Draper, N. R. (1998). Mixture designs for constrained components in orthogonal blocks. *Journal of Applied Statistics*, **25**(5), 613–638. **M**

Prescott, P., and Draper, N. R. (2004). Mixture component-amount designs via projections, including orthogonally blocked designs. *Journal of Quality Technology*, **36**(4), 413–431.

Prescott, P., and Mansson, R. (2004). Robustness of diallel cross designs to the loss of one or more observations. *Computational Statistics and Data Analysis*, **47**(1), 91–109.

Prescott, P., Draper, N. R., Dean, A. M., and Lewis, S. M. (1993). Mixture designs for five components in orthogonal blocks. *Journal of Applied Statistics*, **20**, 105–117. **M**

Prescott, P., Draper, N. R., Dean, A. M., and Lewis, S. M. (1997). Further properties of mixture designs for five components in orthogonal blocks. *Journal of Applied Statistics*, **24**, 147–156. **M**

Prescott, P., Dean, A. M., Draper, N. R., and Lewis, S. M. (2002). Mixture experiments: Ill-conditioning and quadratic model specification. *Technometrics*, **44**(3), 260–268. **M**

Pronzato, L. (2000). Adaptive optimization and D-optimum experimental design. *Annals of Statistics*, **28**(6), 1743–1761.

Pukelsheim, F. (1993). *Optimal Design of Experiments*. New York: Wiley.

Pukelsheim. F., and Rieder, S. (1992). Efficient rounding of approximate designs. *Biometrika*, **79**, 763–770.

Pukelsheim. F., and Rosenberger, J. L. (1993). Experimental designs for model discrimination. *Journal of the American Statistical Association*, **88**, 642–649.

Pukelsheim. F., and Studden, W. J. (1993). E-optimal designs for polynomial regression. *Annals of Statistics*, **21**, 402–415.

Pukelsheim. F., and Titterington, D. M. (1983). General differential and Lagrangian theory for optimal experimental design. *Annals of Statistics*, **11**, 1614–1625.

Pukelsheim. F., and Torsney, B. (1991). Optimal designs for experimental designs on linearly independent support points. *Annals of Statistics*, **19**, 1060–1068

Puri, S., Beg, Q. K., and Gupta, R. (2002). Optimization of alkaline protease production from Bacillus sp. by response surface methodology. *Current Microbiology*, **44**, 286–290.

Puri, Y. P., Miller, M. F., Sah, R. N., Baghott, K. G., Freres-Castel, E., and Meyer, R. D. (1988). Response surface analysis of the effects of seeding rates, N-rates and irrigation frequencies on durum wheat I. Grain yield and yield components. *Fertilizer Research*, **17**, 197–218.

Quesada, G. M., and Del Castillo, E. (2004). A dual-response approach to the multivariate robust parameter design problem. *Technometrics*, **46**(2), 176–187.

Radson, D., and Herrin, G. D. (1995). Augmenting a factorial experiment when one factor is an uncontrollable random variable: A case study. *Technometrics*, **37**(1), 70–81.

Raghavarao, D. (1959). Some optimum weighing designs. *Annals of Mathematical Statistics*, **30**, 295–303.

Raghavarao, D. (1963). Construction of second order rotatable designs through incomplete block designs. *Journal of the Indian Statistical Association*, **1**, 221–225.

Raghavarao, D. (1971). *Constructions and Combinatorial Problems in Design of Experiments*. New York: Wiley.

Rajagopal, R., and Del Castillo, E. (2005). Model-robust process optimization using Bayesian model averaging. *Technometrics*, **47**(2), 152–163.

Rajagopal, R., Del Castillo, E., and Peterson, J. J. (2005). Model- and distribution-robust process optimization with noise factors. *Journal of Quality Technology*, **37**(3), 210–222; corrections: (2006), **38**(1), 83.

Rao, C. R. (1946). On hypercubes of strength d leading to confounded designs in factorial experiments. *Bulletin of the Calcutta Mathematical Society*, **38**, 67–78.

Rao, C. R. (1947). Factorial experiments derivable from combinatorial arrangements of arrays. *Journal of the Royal Statistical Society, Series B*, **9**, 128–139.

Rao, C. R. (1950). The theory of fractional replication in factorial experiments. *Sankhya*, **10**, 84–86.

Rao, C. R. (1973a). *Linear Statistical Inference and Its Applications*, 2nd ed. New York: Wiley.

Rao, C. R. (1973b). Some combinatorial problems of arrays and applications to design of experiments. In *A Survey of Combinatorial Theory* (J. N. Srivastava et al., eds.), Chapter 29. Amsterdam: North Holland/American Elsevier.

Ravindra, M. R., and Chattopadhyay, P. K. (2000). Optimisation of osmotic preconcentration and fluidized bed drying to produce dehydrated quick-cooking potato cubes. *Journal of Food Engineering*, **41**, 5–11.

Rechtschaffner, R. L. (1967). Saturated fractions of 2^n and 3^n factorial designs. *Technometrics*, **9**(4), 569–575.

Reid, D. (1972). The effects of the long-term application of a wide range of nitrogen rates on the yields from perennial ryegrass swards with and without white clover. *Journal of Agricultural Science*, **79**, 291–301.

Remmenga, M. D., and Johnson, D. E. (1995). A comparison of inference procedures in unbalanced split-plot designs. *Journal of Statistical Computation and Simulation*, **51**, 353–367.

Remmers, E. G., and Dunn, G. G. (1961). Process improvement of a fermentation product. *Industrial Engineering Chemistry*, **53**, 743–745.

Reyes, M., Parra, I., Milan, C., and Zazueta, N. (2002). A response surface methodology approach to optimize pretreatments to prevent enzymatic browning in potato (*Solanum tuberosim L*) cubes. *Journal of the Science of Food and Agriculture*, **82**(1), 69–79.

Ribeiro, J. L. D., Fogliatto, F. S., and ten Caten, C. S. (2001). Minimizing manufacturing and quality costs in multiresponse optimization. *Quality Engineering*, **13**(4), 559–569.

Riccomagno, E., Schwabe, R., and Wynn, H. P. (1997). Latticebased D-optimum design for Fourier regression. *Annals of Statistics*, **25**(6), 2313–2327.

Richert, S. H., Morr, C. V., and Cooney, C. M. (1974). Effect of heat and other factors upon foaming properties of whey protein concentrates. *Journal of Food Science*, **39**, 42–48.

Riedell, W. E., Beck, D. L., and Schumacher, T. E. (2000). Corn response to fertilizer placement treatments in an irrigated no-till system. *Agronomy Journal*, **92**, 316–320.

Ringwald, T. J., and Forth, S. A. (2005). Simplifying multivariate second-order surfaces by fitting constrained models using automatic differentiation. *Technometrics*, **47**(3), 249–259.

Robinson, G. K. (1993). Improving Taguchi's packaging of fractional factorial designs. *Journal of Quality Technology*, **25**(1), 1–11.

Robinson, L. W. (2003). Concise experimental designs. *Quality Engineering*, **15**(3), 403–406.

Robinson, P., and Nielsen, K. F. (1960). Composite designs in agricultural research. *Canadian Journal of Soil Science*, **40**, 168–176.

Robinson, T. J., and Wulff, S. S. (2006). Response surface approaches to robust parameter design. In *Response Surface Methodology and Related Topics* (A. I. Khuri, ed.), pp. 123–157. Hackensack, NJ: World Scientific Publishing.

Robinson, T. J., Myers, R. H., and Montgomery, D. C. (2004). Analysis considerations in industrial split-plot experiments with non-normal responses. *Journal of Quality Technology*, **36**(2), 180–192.

Robinson, T. J., Wulff, S. S., Montgomery, D. C., and Khuri, A. I. (2006). Robust parameter design using generalized linear mixed models. *Journal of Quality Technology*, **38**(1), 65–75.

Rohan, V. **M.**, and Jones, G. (2000). Efficient run orders for a two-factor response surface experiment on a correlated process. *Communications in Statistics, Theory and Methods*, **29**(3), 593–609.

Romano, D., Varetto, M., and Vicario, G. (2004). Multiresponse robust design: A general framework based on combined array. *Journal of Quality Technology*, **36**(1), 27–37.

Roquemore, K. G. (1976a). Hybrid designs for quadratic response surfaces. *Technometrics*, **18**(4), 419–423; see also Roquemore (1976b).

Roquemore, K. G. (1976b). Errata. *Technometrics*, **19**(1), 106.

Rosenbaum, P. R. (1994). Dispersion effects from fractional factorials in Taguchi's method of quality design. *Journal of the Royal Statistical Society, Series B*, **56**, 641–652.

Rosenbaum, P. R. (1996). Some useful compound dispersion experiments in quality design. *Technometrics*, **38**(4), 354–364.

Rosenbaum, P. R. (1999). Blocking in compound dispersion experiments. *Technometrics*, **41**(2), 125–134.

Rosenberger, J., and Kalish, L. (1978). Optimal designs for estimation of parameters of the logistic function (abstract). *Biometrics*, **34**, 746.

Ross, E., Damron, B. L., and Harms, R. H. (1972). The requirement for inorganic sulfate in the diet of chicks for optimum growth and feed efficiency. *Poultry Science*, **51**, 1606–1612.

Roth, P. M., and Stewart, R. A. (1969). Experimental studies with multiple responses. *Applied Statistics*, **18**, 221–228.

Roush, W. B., Mylet, M., Rosenberger, J. L., and Derr, J. (1986). Investigation of calcium and available phosphors requirements for laying hens by response surface methodology. *Poultry Science*, **65**, 964–970.

Roy, R. K. (2001). *Design of Experiments Using the Taguchi Approach: Sixteen Steps to Product and Process Improvement*. New York: Wiley.

Royle, J. A. (2002). Exchange algorithms for constructing large spatial designs. *Journal of Statistical Planning and Inference*, **100**, 121–134.

Ruggoo, A., and Vandebroek, M. (2004). Bayesian sequential D-D optimal model-robust designs. *Computational Statistics and Data Analysis*, **47**(4), 655–673.

Ryan, T. P. (2000). *Statistical Methods for Quality Improvement*. New York: Wiley.

Ryan, T. P., and Schwertman, N. C. (1997). Survey of software with mixture experiment capabilities. *Journal of Quality Technology*, **29**(1), 76–85. **M, R**

Sa, P., and Edwards, D. (1993a). Tables of the Casella and Strawderman (1980) critical points for use in multiple comparisons with a control in response surface methodology. *Communications in Statistics, Simulation and Computation*, **22**(3), 765–777.

Sa, P., and Edwards, D. (1993b). Multiple comparisons with a control in response surface methodology. *Technometrics*, **35**(4), 436–445.

Sacks, J., and Ylvisaker, D. (1978). Linear estimation for approximately linear models. *Annals of Statistics*, **5**, 1122–1137.

Sacks, J., Schiller, S. B., and Welch, W. J. (1989). Designs for computer experiments. *Technometrics*, **31**(1), 41–47.

Sacks, J., Welch, W. J., Mitchell, T. J., and Wynn, H. P. (1989). Design and analysis of computer experiments. *Statistical Science*, **4**(4), 409–423; discussion: pp. 423–435.

Safizadeh, M.H. (2002). Minimizing the bias and variance of the gradient estimate in RSM simulation studies. *European Journal of Operational Research*, **136**, 121–135.

Saha, G. M., and Das, A. R. (1973). Four level second order rotatable designs from partially balanced arrays. *Journal of the Indian Society for Agricultural Statistics*, **25**, 97–102.

Sahni, N. S., Isaksson, T., and Naes, T. (2001). The use of experimental design methodology and multivariate analysis to determine critical control points in a process. *Chemometrics and Intelligent Laboratory Systems*, **56**, 105–121.

Sahni, N. S., Aastveit, A., and Naes, T. (2005). In-line process and product control using spectroscopy and multivariate calibration. *Journal of Quality Technology*, **37**(1), 1–20. **M**

Saklar, S., Katnas, S., and Ungan, S. (2001). Determination of optimum hazelnut roasting conditions. *International Journal of Food Science and Technology*, **36**, 271–281.

Saltelli, A. Andres, T. H., and Homma, T. (1995). Sensitivity analysis of model output. Performance of the iterated fractional factorial design method. *Computational Statistics and Data Analysis*, **20**(4), 387–407.

Salter, K. C., and Fawcett, R. F. (1993). The ART test of interaction: A robust and powerful rank test of interaction in factorial models. *Communications in Statistics, Simulation and Computation*, **22**(1), 137–153.

Sánchez-Lafuente, C., Furlanetto, S., Fernández-Arévalo, M., Alvarez-Fuentes, J., Rabasco, A. M., Faucci, M. T., Pinzauti, S., and Mura, P. (2002). Didanosine extended-release matrix tablets: Optimization of formulation variables using statistical experimental design. *International Journal of Pharmaceutics*, **237**, 107–118.

Sanders, D., Leitnaker, M., and McLean, R. (2001). Randomized complete block designs in industrial studies. *Quality Engineering*, **14**(1), 1–8.

Sarma, G. S., and Ravindram, M. (1975). Studies in dealkylation—statistical analysis of process variables. *Journal of the Indian Institute of Science*, **58**, 67–83.

Sasse, C. E., and Baker, D. H. (1972). The phenylalanine and tyrosine requirements and their interrelationship for the young chick. *Poultry Science*, **51**, 1531–1535.

Sathe, Y. S., and Shenoy, R. G. (1989). A-optimal weighing designs when $N \equiv$ (mod 4). *Annals of Statistics*, **17**, 1906–1915.

Satterthwaite, F. E. (1959). Random balance experimentation. *Technometrics*, **1**(2), 111–137; discussion: pp. 157–193; see also Budne (1959) and Anscombe (1959).

Satyabrata, P. (1978). Some methods of construction of second order rotatable designs. *Calcutta Statistical Association Bulletin*, **27**, 127–140.

Saxena, S. K., and Nigam, A. K. (1973). Symmetric-simplex block designs for mixtures. *Journal of the Royal Statistical Society*, Series B, **35**(3), 466–472. **M**

Saxena, S. K., and Nigam, A. K. (1977). Restricted exploration of mixtures by symmetric-simplex design. *Technometrics*, **19**(1), 47–52. **M**

Schabenberger, O., Grégoire, T. G., and Kong, F. (2000). Collections of simple effects and their relationship to main effects and interactions in factorials. *American Statistician*, **54**(3), 210–214.

Scheffé H. (1958). Experiments with mixtures. *Journal of the Royal Statistical Society*, Series B, **20**, 344–360; correction: (1959), **21**, 238. **M**

Scheffé, H. (1959). *The Analysis of Variance*. New York: Wiley.

Scheffé, H. (1963). The simplex-centroid design for experiments with mixtures. *Journal of the Royal Statistical Society*, Series B, **25**, 235–251; discussion: pp. 251–263. **M**

Schepers, A. W., Thibault, J., and Lacroix, C. (2002). Lactobacillus helveticus growth and lactic acid production during pH-controlled batch cultures in whey permeate/yeast extract medium. Part I. Multiple factor kinetic analysis. *Enzyme and Microbial Technology*, **30**, 176–186.

Schmidt, R. H., Illingworth, B. L., Deng, J. C., and Cornell, J. A. (1979). Multiple regression and response surface analysis of the effects of calcium chloride and cysteine on heat-induced whey protein gelation. *Journal of Agricultural and Food Chemistry*, **27**, 529–532.

Schneider, H., Kasperski, W. J., and Weissfield, L. (1993). Finding significant effects for unreplicated fractional factorials using the smallest n contrasts. *Journal of Quality Technology*, **25**(1), 18–27.

Schoen, E. D. (1999). Designing fractional factorial two-level experiments with nested error structures. *Journal of Applied Statistics*, **26**, 495–508.

Schoen, E. D., and Kaul, E. A. A. (2000). Three robust scale estimators to judge unreplicated experiments. *Journal of Quality Technology*, **32**(3), 276–283.

Schoen, E. D., and Wolff, K. (1997). Design and analysis of a fractional $4^1 3^1 2^5$ split-plot experiment. *Journal of Applied Statistics*, **24**, 409–419.

Schoney, R. A., Bay, T. F., and Moncrief, J. F. (1981). Use of computer graphics in the development and evaluation of response surfaces. *Agronomy Journal*, **73**, 437–442.

Schroeder, E. C., Nair, K. P. C., and Cardeilhac, P. T. (1972). Response of broiler chicks to a single dose of Aflatoxin. *Poultry Science*, **51**, 1552–1556.

Scibilia, B., Kobi, A. Chassagnon, R., and Barreau, A. (2002). Minimal design augmentation schemes to resolve complex aliasing in industrial experiments. *Quality Engineering*, **14**(4), 523–529.

Scrucca, L. (2001). A review and computer code for assessing the structural dimension of a regression model: Uncorrelated 2D views. *Computational Statistics and Data Analysis*, **36**, 163–177.

Searle, S. R. (1982). *Matrix Algebra Useful for Statistics*. New York: Wiley.

Sebastiani, P., and Wynn, H. P. (2000). Maximum entropy sampling and optimal Bayesian experimental design. *Journal of the Royal Statistical Society*, Series B, **62**(1), 145–157.

Seheult, A. (1978). Minimum bias or least squares estimation? *Communications in Statistics, Theory and Methods*, **7**, 277–283.

Seidel, J. J. (2001). Definitions for spherical designs. *Journal of Statistical Planning and Inference*, **95**(1-2), 307–313.

Semple, J. (1997). Optimality conditions and solutions procedures for nondegenerate dual response systems. *IIE Transactions*, **29**, 743–752.

Senanayake, S. P. J. N., and Shahidi, F. (2002). Lipase-catalyzed incorporation of docosahexaenoic acid (DHA) into borage oil: Optimization using response surface methodology. *Food Chemistry*, **77**, 115–123.

Sethuraman, V. S., Raghavarao, D., and Sinha, B. K. (2006). Optimal s^n factorial designs when observations within-blocks are correlated. *Computational Statistics and Data Analysis*, **50**(10), 2855–2862.

Setodji, C. M., and Cook, R. D. (2004). K-means inverse regression. *Technometrics*, **46**(4), 421–429.

Sexton, C. J., Anthony, D. K., Lewis, S. M., Please, C. P., and Keane, A. J. (2006). Design of experiments algorithms for assembled products. *Journal of Quality Technology*, **38**(4), 298–308.

Sexton, C. J., Lewis, S. M., and Please, C. P. (2001). Experiments for derived factors with application to hydraulic gear pumps. *Applied Statistics*, **50**(2), 155–170.

Seymour, P. D., and Zaslavsky, N. N. (1984). Averaging sets: A generalization of mean values and spherical designs. *Advances in Mathematics*, **52**, 213–240.

Shah, K., Bose, M., and Raghavarao, D. (2005). Universal optimality of Patterson's crossover designs. *Annals of Statistics*, **33**(6), 2854–2872.

Shah, K. R., and Sinha, B. K. (1989). *Theory of Optimal Designs*. New York: Springer-Verlag.

Shah, K. R., and Sinha, B. K. (1989). On the choice of optimality criteria in comparing statistical designs. *Canadian Journal of Statistics*, **17**, 345–348.

Shang, J. S. (1995). Robust design and optimization of material handling in an FMS. International Journal of Product Research, **33**(9), 2437–2454.

Sharma, G. P., and Prasad, S. (2002). Dielectric properties of garlic (*Allium sativum L.*) at 2450 MHz as function of temperature and moisture content. *Journal of Food Engineering*, **52**, 343–348.

Sharma, M. L. (1976). Interaction of water potential and temperature effects on germination of three semi-arid plant species. *Agronomy Journal*, **68**, 390–394.

Sheldon, F. R. (1960). Statistical techniques applied to production situations. *Industrial and Engineering Chemistry*, **52**, 507–509.

Shelton, J. T., Khuri, A. I., and Cornell, J. A. (1983). Selecting check points for testing lack of fit in response surface models. *Technometrics*, **25**(4), 357–365.

Shen, Z., Mishra, V., Imison, B., Palmer, M., and Fairclough, R. (2002). Use of adsorbent and supercritical carbon dioxide to concentrate flavor compounds from orange oil. *Journal of Agricultural and Food Chemistry*, **50**, 154–160.

Shirafuji, M. (1959). A two stage sequential design in response surface analysis. *Bulletin of Mathematical Statistics*, **8**, 115–126.

Shoemaker, A. C., and Tsui, K.-L. (1993). Response model analysis for robust design experiments. Communications in Statistics, Simulation and Computation, **22**(4), 1037–1064.

Shoemaker, A. C., Tsui, K.-L., and Wu, C. J. F (1991). Economical experimentation methods for robust design. *Technometrics*, **33**(4), 415–427; correction: **34**(4), p. 502.

Short, T. G., Ho, T. Y., Minto, C. F., Schnider, T. W., and Shafer, S. L. (2002). Efficient trial design for eliciting a pharmacokinetic-pharmacodynamic model-based response surface describing the interaction between two intravenous anesthetic drugs. *Anesthesiology*, **96**, 400–408.

Shukla, G. K., and Subrahmanyam, G. S. V. (1999). A note on an exact test and confidence interval for competition and overlap effects. *Biometrics*, **55**(1), 273–276.

Si, B. C., and Kachanoski, R. G. (2000). Estimating soil hydraulic properties during constant flux infiltration. *Soil Science Society of America Journal*, **64**, 439–449.

Sihota, S. S., and Banerjee, K. S. (1981). On the algebraic structures in the construction of confounding plans in mixed factorial designs on the lines of White and Hultquist. *Journal of the American Statistical Association*, **76**, 996–1101.

Silvey, S. D. (1980). *Optimal Design*. London: Chapman & Hall.

Silvey, S. D., and Titterington, D. M. (1973). A geometrical approach to optimal design theory. *Biometrikka*, **60**, 21–32.

Silvey, S. D., Titterington, D. M., and Torsney, B. (1978). An algorithm for optimal designs on a finite design space. *Communications in Statistics, Theory and Methods*, **7**, 1379–1389.

Singh, M. (1979). Group divisible second order rotatable designs. *Biometrical Journal,*, **21**, 579–589.

Singh, M., and Hinklemann, K. (1995). Partial diellel crosses in incomplete blocks. *Biometrics*, **51**(4), 1302–1314.

Singh, M., Dey, A., and Mitra, R. K. (1979). An algorithm for the choice of optimal response surface designs. *Journal of the Indian Society for Agricultural Statistics*, **31**, 50–54.

Sinha, K. (1992). Construction of balanced treatment incomplete block designs. *Communications in Statistics, Theory and Methods*, **21**(5), 1377–1382.

Sitter, R. R. (1992). Robust designs for binary data. *Biometrics*, **48**(4), 1145–1155.

Sitter, R. R., and Torsney, B. (1995). Optimal designs for binary response experiments with two design variables. *Statistica Sinica*, **5**, 405–419.

Sitter, R. R., and Wu, C. F. J. (1993). Optimal designs for binary response experiments—Fieller, D and A criteria. *Scandinavian Journal of Statistics*, **20**, 329–341.

Sitter, R. R., Chen, J., and Feder, M. (1997). Fractional resolution and minimum aberration in blocked 2^{n-k} designs. *Technometrics*, **39**(4), 382–390.

Sloane, N. J. A. (1993). Covering arrays and intersecting codes. *Journal of Combinatorial Design*, **1**, 51–63.

Sloane, N. J. A. (2004). A library of orthogonal arrays. URL: http://www.research.att.com/njas/oadir/

Smith, D., and Smith, R. R. (1977). Response of red clover to increasing rates of topdressed potassium fertilizer. *Agronomy Journal*, **69**, 45–48.

Smith, H., and Rose, A. (1963). Subjective response in process investigations. *Industrial and Engineering Chemistry*, **55**(7), 25–28.

Smith, W. F., and Beverly, T. A. (1997). Generating linear and quadratic Cox mixture model. *Journal of Quality Technology*, **29**(2), 211–219. **M**

Smith, W. F., and Cornell, J. A. (1993). Biplot displays for looking at multiple response data in mixture experiments. *Technometrics*, **35**(4), 337–350. **M**

Smith, W. L. (2005). *Experimental Design for Formulation*. ASA-SIAM Series on Statistics and Applied Probability. Philadelphia: SIAM Books. **M**

Snee, R. D. (1971). Design and analysis of mixture experiments. *Journal of Quality Technology*, **3**(4), 159–169. **M**

Snee, R. D. (1973a). Some aspects of nonorthogonal data analysis. Part I. Developing prediction equations. *Journal of Quality Technology*, **5**(2), 67–69.

Snee, R. D. (1973b). Some aspects of nonorthogonal data analysis. Part II. Comparison of means. *Journal of Quality Technology*, **5**(3), 109–122.

Snee, R. D. (1973c). Techniques for the analysis of mixture data. *Technometrics*, **15**(3), 517–528. **M**

Snee, R. D. (1975). Experimental designs for quadratic models in constrained mixture spaces. *Technometrics*, **17**(2), 149–159.

Snee, R. D. (1977). Validation of regression models: methods and examples. *Technometrics*, **19**(4), 415–428.

Snee, R. D. (1979). Experimental designs for mixture systems with multicomponent restraints. *Communications in Statistics, Theory and Methods*, **8**(4), 303–326. **M**

Snee, R. D. (1981). Developing blending models for gasoline and other mixtures. *Technometrics*, **23**(2), 119–130. **M**

Snee, R. D. (1985). Computer-aided design of experiments—some practical experiences. *Journal of Quality Technology*, **17**(4), 222–236.

Snee, R. D. (1995). Comment on Draper, Prescott, Lewis, Dean, John and Tuck (1993): Understanding formulation systems. *Technometrics*, **37**, 131–132. **M**

Snee, R. D., and Hoerl, R. W. (2004). Statistical leadership. *Quality Progress*, **37**(10), October.

Snee, R. D., and Hoerl, R. W. (2005). *Six Sigma Beyond the Factory Floor*. Upper Saddle River, NJ: Pearson Prentice Hall.

Snee, R. D., and Marquardt, D. W. (1974). Extreme vertices designs for linear mixture models. *Technometrics*, **16**(3), 399–408. **M**

Snee, R. D., and Marquardt, D. W. (1976). Screening concepts and designs for experiments with mixtures. *Technometrics*, **18**(1), 19–29. **M**

Snee, R. D., and Marquardt, D. W. (1984). Collinearity diagnostics depend on the domain of prediction, the model and the data. *Annals of Statistics*, **38**, 83–87.

Snee, R. D., and Rayner, A. A. (1982). Assessing the accuracy of mixture model regression calculations. *Journal of Quality Technology*, **14**(2), 67–79. **M**

Snee, R. D., Hare, L. B., and Trout, J. R. (1985). *Experimenting with a Large Number of Variables*. Milwaukee, WI: ASQ Quality Press.

Springall, A. (1973). Response surface fitting using a generalization of the Bradley-Terry paired comparison model. *Applied Statistics*, **22**, 59–68.

Spruill, M. C., and Tu, R. (2001). Locally asymptotically optimal designs for testing in logistic regression. *Annals of Statistics*, **29**(4), 1050–1057.

Spurrier, J. D. (1993). Comparison of simultaneous confidence bands for quadratic regression over a finite interval. *Technometrics*, **35**(3), 315–320. **M**

Sreenath, P. R. (1989). Construction of some balanced incomplete block designs with nested rows and columns. *Biometrika*, **76**, 399–402.

Srivastava, J. N., and Chopra, D. V. (1971). Balanced optimal 2^m fractional factorial designs of resolution V, $m = 6$. *Technometrics*, **13**(2), 257–269.

Srivastava, J. N., and Ghosh, S. (1980). Enumeration and representation of non-isomorphic bipartite graphs. *Annals of Discrete Mathematics*, **6**, 315–332.

Srivastava, J. N., and Ghosh, S. (1996). On nonorthogonality and nonoptimality of Addelman's main-effect plans satisfying the condition of proportional frequencies. *Statistics and Probability Letters*, **26**, 51–60.

Stablein, D. M., Carter, W. H,, Jr., and Wampler, G. L. (1983). Confidence regions for constrained optima in response surface experiments. *Biometrics*, **39**, 759–763.

Stauber, M. S., and Burt, O. R. (1973). Implicit estimate of residual nitrogen under fertilised range conditions in the Northern Great Plains. *Agronomy Journal*, **65**, 897–901.

Steinberg, D. M. (1985). Model robust response surface designs: Scaling two-level factorials. *Biometrika*, **72**, 513–526.

Steinberg, D. M. (1988). Factorial experiments with time trends. *Technometrics*, **30**(3), 259–269.

Steinberg, D. M., and Bursztyn, D. (1994). Dispersion effects in robust-design experiments with noise factors. *Journal of Quality Technology*, **26**(1), 12–20.

Steinberg, D. M., and Bursztyn, D. (1998). Noise factors, dispersion effects and robust design. *Statistica Sinica*, **8**, 67–85.

Steinberg, D. M., and Hunter, W. G. (1984). Experimental design: Review and comment. *Technometrics*, **26**(2), 71–97; discussion: pp. 98–130. **R**

Steiner, S. H., and Hamada, M. (1997). Making mixtures robust to noise and mixing measurement errors. *Journal of Quality Technology*, **29**(4), 441–450. **M**

Stephenson, W. R. (1991). A computer program for the quick and easy analysis of unreplicated factorials. *Journal of Quality Technology*, **23**(1), 63–67.

Stephenson, W. R., Hulting, F. L., and Moore, K. (1989). Posterior probabilities for identifying active effects in unreplicated experiments. *Journal of Quality Technology*, **21**(3), 202–212.

Stidley, C.A., and Schrader, R. M. (1999). The analysis of mixed and random effects models for equireplicate variance-balanced block designs. *Communications in Statistics, Theory and Methods*, **28**(10), 2509–2526.

Stigler, S. M. (1971). Optimal experimental design for polynomial regression. *Journal of the American Statistical Association*, **66**, 311–318.

St. John, R. C., and Draper, N. R. (1975). D-optimality for regression designs: A review. *Technometrics*, **17**(1), 15–23. **R**

St. John, R. C., and Draper, N. R. (1977). Designs in three and four components for mixtures models with inverse terms. *Technometrics*, **19**(2), 117–130. **M**

Stone, R. A. (1998). The blind paper cutter: teaching about variation, bias, stability, and process control. *American Statistician*, **52**(3), 244–247.

Stone, R. L., Wu, S. M., and Tiemann, T. D. (1965). A statistical experimental design and analysis of the extraction of silica from quartz by digestion in sodium hydroxide solutions. *Transactions of the Society of Mining Engineers*, **232**, 115–124.

Studden, W. J. (1968). Optimal designs on Tchebychev points. *Annals of Mathematical Statistics*, **39**, 1435–1447.

Studden, W. J. (1971a). Optimal designs for multivariate polynomial extrapolation. *Annals of Mathematical Statistics*, **42**, 828–832.

Studden, W. J. (1971b). Elfving's theorem and optimal designs for quadratic loss. *Annals of Mathematical Statistics*, **42**, 1613–1621.

Studden, W. J. (1982). Some robust type D-optimal designs in polynomial regression. *Journal of the American Statistical Association*, **77**, 916–921.

Studden, W. J. (1989). Note on some ϕ_p-optimal designs for polynomial regression. *Annals of Statistics*, **17**, 618–623.

Studden, W. J., and Tsay, J. Y. (1976). Remez's procedure for finding optimal designs. *Annals of Statistics*, **4**, 1271–1279.

Stufken, J. (1987). A-optimal block designs for comparing test treatments with a control. *Annals of Statistics*, **15**, 1629–1638.

Suen, C. Y. (1989a). A class of orthogonal main effects plans. *Journal of Statistical Planning and Inference*, **21**, 391–394.

Suen, C. Y. (1989b). Some resolvable orthogonal arrays with two symbols. *Communications in Statistics, Theory and Methods*, **18**, 3875–3881.

Suen, C. Y. (2003). Construction of mixed orthogonal arrays by juxtaposition. *Statistics and Probability Letters*, **15**, 161–163.

Suen, C. Y., Cheng, H., and Wu, C. J. F. (1977). Some identities on q^{n-m} designs with application to minimum aberration designs. *Annals of Statistics*, **25**(3), 1176–1188.

Sulieman, H., McLellan, P. J., and Bacon, D. W. (2001). A profile-based approach to parametric sensitivity analysis of nonlinear regression models. *Technometrics*, **43**(4), 425–433.

Sulieman, H., McLellan, P. J., and Bacon, D. W. (2004). A profile-based approach to parametric sensitivity in multiresponse regression models. *Computational Statistics and Data Analysis*, **45**(4), 721–740.

Sullivan, T. W., and Al-Timini, A. A. (1972). Safety and toxicity of dietary and organic arsenicals relative to performances of young turkeys. 3. Nitarsone. *Poultry Science*, **51**, 1582–1586.

Sun, D. X., and Wu, C. J. F. (1994). Interaction graphs for three-level fractional factorial designs. *Journal of Quality Technology*, **26**(4), 297–307.

Sun, D. X., Wu, C. J. F., and Chen, Y. (1997). Optimal blocking schemes for 2^n and 2^{n-p} designs. *Technometrics*, **39**(3), 298–307.

Swanson, A. M., Geissler, J. J., Magnino, P. J., and Swanson, J. A. (1967). Use of statistics and computers in the selection of optimum food processing. *Food Technology*, **21**(11), 99–102.

Swanson, C. L., Maffsiger, T., Russel, C., Hofreiter, B., and Rist, C. (1964). Xanthation of starch by a continuous process. *Industrial and Engineering Chemistry Process Design and Development*, **3**, 22–27.

Szegö, G. (1975). *Orthogonal Polynomials*, 4th ed. Providence, RI: American Mathematical Society.

Sztendur, E. W., and Diamond, N. T. (2002). Extensions to confidence region calculations for the path of steepest ascent. *Journal of Quality Technology*, **34**(3), 289–296.

Tack, L., and Vandebroek, M. (2002). An adjustment algorithm for optimal run orders in design of experiments. *Computational Statistics and Data Analysis*, **40**(3), 559–577.

Tack, L., and Vandebroek, M. (2003). Semiparametric exact optimal run orders. *Journal of Quality Technology*, **35**(2), 168–183.

Taguchi, G. (1987). *System of Experimental Design: Engineering Methods to Optimize Quality and Minimize Cost*. White Plains, NY: Quality Resources.

Taguchi, G. (1991). *Introduction to Quality Engineering*, White Plains, NY: UNIPUB/Kraus International.

Taguchi, G., and Wu, Y. (1985). *Introduction to Off-Line Quality Control*, Dearborn, MI: Central Japan Quality Control Association, American Supplier Institute.

Tang, B. (1993). Orthogonal array-based Latin hypercubes. *Journal of the American Statistical Association*, **88**, 1392–1397.

Tang, B. (1994). A theorem for selecting OA-based Latin hypercubes using a distance criterion. *Communications in Statistics, Theory and Methods*, **23**(7), 2047–2058.

Tang, B. (2001). Theory of J-characteristics for fractional factorial designs and projection justification of minimum G_2-aberration. *Biometrika*, **88**(2), 401–407.

Tang, B., and Deng, L.-Y. (1999). Minimium G_2-aberration for nonregular fractional factorial designs. *Annals of Statistics*, **27**(6), 1914–1926.

Tang, B., and Wu, C. J. F. (1996). Characterization of minimum aberration 2^{n-k} designs in terms of their complementary designs. *Annals of Statistics*, **24**(6), 2549–2559.

Tang, B., and Wu, C. J. F. (1997). A method for constructing supersaturated designs and its $E(s^2)$ optimality. *Canadian Journal of Statistics*, **25**(2), 191–201.

Tang, B., Ma, F., Ingram, D., and Wang, H. (2002). Bounds on the maximum number of clear two-factor interactions for 2^{m-p} designs of resolution III and IV. *Canadian Journal of Statistics*, **30**, 127–136.

Tang, D. I. (1993). Minimax regression designs under uniform departure models. *Annals of Statistics*, **21**, 434–446.

Tang, L. C and Xu, K. (2002). A unified approach for dual response surface optimization. *Journal of Quality Technology*, **34**(4), 437–447.

Tasaka, M. (1968). Optimum designs for polynomial approximation. *Bulletin of Mathematical Statistics*, **13**, 25–39.

Taubman, S. B., and Addelman, S. (1978). Some two-level multiple response factorial plans. *Journal of the American Statistical Association*, **73**, 607–612.

Taylor, S. (1998). Setting up computer assisted personal interviewing in the Australian longitudinal study of ageing. *Statistical Science*, **13**(1), 14–18; discussion: pp. 23–29.

Terhune, H. D. (1963). A statistical analysis of the effect of four variable operating conditions on bleaching results. *American Dyestuff Reporter*, **52**(7), April 1, 33–38.

Terman, G. L., Khasawneh, F. E., Allen, S. E., and Engelstadt, O. P. (1976). Yield-nutrient absorption relationships as affected by environmental growth factors. *Agronomy Journal*, **68**, 107–110.

Thaker, P. J. (1962). Some infinite series of second order rotatable designs. *Journal of the Indian Society for Agricultural Statistics*, **14**, 110–120.

Thaker, P. J., and Das, M. N. (1961). Sequential third order rotatable designs for up to eleven factors. *Journal of the Indian Society for Agricultural Statistics*, **13**, 218–231.

Thomas, E. V. (1994). Evaluating the ignition sensitivity of thermal-battery heat pellets. *Technometrics*, **36**(3), 273–282.

Thomas, E. V., and Ge, N. (2000). Development of robust multivariate calibration models. *Technometrics*, **42**(2), 168–177.

Thomas, J. R., and McLean, D. M. (1967). Growth and mineral composition of squash (*cucurbita pepo L.*) as affected by N, P, K and tobacco ring spot virus. *Agronomy Journal*, **59**, 67–69.

Thompson, T. L., Doerge, T. A., and Godin, R. E. (2002). Subservice drip irrigation and fertigation of broccoli. I: Yield, quality and nitrogen uptake. II: Agronomic, economic and environmental outcomes. *Soil Science Society of America Journal*, **66**, 178–192.

Thompson, W. O. (1973). Secondary criteria in the selection of minimum bias designs in two variables. *Technometrics*, **15**(2), 319–328.

Thompson, W. O., and Myers, R. H. (1968). Response surface designs for mixture problems. *Technometrics*, **10**(4), 739–756. **M**

Tiahrt, K. J., and Weeks, P. L. (1970). A method of constrained randomization for 2^n factorials. *Technometrics*, **12**(3), 471–486.

Tian, Y., and Herzberg, A. M. (2006). A-minimax and D-minimax robust optimal designs for approximately linear Haar-wavelet models. *Computational Statistics and Data Analysis*, **50**(10), 2942–2951.

Tiao, G., Bisgaard, S., Hill, W., Pena, D., and Stigler, S., eds. (2000). (George E. P.) *Box on Quality and Discovery with Design, Control and Robustness*. New York: Wiley.

Tidwell, P. W. (1960). Chemical process improvement by response surface methods. *Industrial Engineering Chemistry*, **52**, 510–512.

Tinsson, W. (2001). Prediction of the variations of the mean response by using experimental design with quantitative factors and random block effects. *Communications in Statistics, Theory and Methods*, **30**(2), 209–228.

Titterington, D. M. (1975). Optimal design: some geometrical aspects of D-optimality. *Biometrika*, **62**, 313–320.

Tjur, T. (1991). Block designs and electrical networks. *Annals of Statistics*, **19**, 1010–1027.

Tobias, R. (1996). Saturated second-order two-level designs: An empirical approach. *SAS Institute Technical Report*.
[http://support.sas.com/rnd/app/papers/minres5.pdf]

Toman, B. (1994). Bayes optimal designs for two- and three-level factorial experiments. *Journal of the American Statistical Association*, **89**, 937–946.

Toman, B. (1996). Bayesian experimental design for multiple hypothesis testing. *Journal of the American Statistical Association*, **91**, 185–190.

Tong, L.-I., Wang, C.-H., Houng, J.-Y., and Chen, J.-Y. (2001-2002). Optimizing dynamic multiresponse problems using the dual response method. *Quality Engineering*, **14**(1), 115–125.

Torres, V. A. (1993). A simple analysis of unreplicated factorials with possible abnormalities. *Journal of Quality Technology*, **25**(3), 183–187.

Toutenberg, H. (2002). *Statistical Analysis of Designed Experiments*. New York: Springer-Verlag.

Tracy, D. S., and Jinadasa, K. G. (1989). Partitioned Kronecker products of matrices and applications. *Canadian Journal of Statistics*, **17**, 107–120.

Tranter, R. L., ed. (2000). *Design and Analysis in Chemical Research*. Sheffield, UK: Sheffield Academic Press.

Trinca, L. A., and Gilmour, S. G. (1998). Variance dispersion graphs for comparing blocked response surface designs. *Journal of Quality Technology*, **30**(4), 314–327.

Trinca, L. A., and Gilmour, S. G. (1999). Difference variance dispersion graphs for comparing response surface designs with applications in food technology. *Applied Statistics*, **48**(4), 441–455.

Trinca, L. A., and Gilmour, S. G. (2000). An algorithm for arranging response surface designs in small blocks. *Computational Statistics and Data Analysis*, **33**(1), 25–43; correction: (2002), **40**(3), 475.

Trinca, L. A., and Gilmour, S. G. (2001). Multi-stratum response surface design. *Technometrics*. **43**(1), 25–33.

Tsai, P.-W., Gilmour, S. G., and Mead, R. (2000). Projective three-level main effects designs robust to model uncertainty. *Biometrika*, **87**(2) 467–475.

Tsui, K.-L. (1994). Avoiding unnecessary bias in robust design analysis. *Computational Statistics and Data Analysis*, **18**, 535–546.

Tsui, K.-L. (1996). A critical look at Taguchi's modeling approach for robust design. *Journal of Applied Statistics*, **23**, 81–99.

Tuck, M. G., Lewis, S. M., and Cottrell, J. I. L. (1993). Response surface methodology and Taguchi: A quality improvement study from the milling industry. *Applied Statistics*, **42**, 671–681.

Tukey, J. W. (1949). One degree of freedom for non-additivity. *Biometrics*, **5**, 232–242.

Tukey, J. W. (1957). On the comparative anatomy of transformations. *Annals of Mathematical Statistics*, **28**, 602–632.

Ture, T. E. (1994). Optimal row-column designs for multiple camparisons with a control: A complete catalog. *Technometrics*, **36**(3), 292–299.

Turiel, T. P. (1988a). A Fortran program to generate fractional factorial experiments. *Journal of Quality Technology*, **20**(1), 63–72.

Turiel, T. P. (1988b). A computer program to determine defining contrasts and factor combinations for two-level fractional factorial designs of resolution III, IV, and V. *Journal of Quality Technology*, **20**(4), 267–272.

Tweedie, R. (1998). Consulting: real problems, real interactions, real outcomes. *Statistical Science*, **13**(1), 1–3; discussion: pp. 23–29.

Tweedie, R., and Hall, N. (1998). Queueing at the tax office. *Statistical Science*, **13**(1), 18–23; discussion: pp. 23–29.

Tyagi, B. N. (1964a). A note on the construction of a class of second order rotatable designs. *Journal of the Indian Statistical Association*, **2**, 52–54.

Tyagi, B. N. (1964b). On construction of second and third order rotatable designs through pair-wise balanced and doubly balanced designs. *Calcutta Statistical Association Bulletin*, **13**, 150–162.

Uddin, N. (1900). Some series constructions for minimal size equi-neighbourhooded balanced incomplete block designs with nested rows and columns. *Biometrika*, **77**, 829–833.

Uddin, N., and Morgan, J. P. (1990). Some constructions for balanced incomplete block designs with nested rows and columns. *Biometrika*, **77**, 193–202.

Uddin, N., and Morgan, J. P. (1997a). Efficient block designs for settings with spatially correlated errors. *Biometrika*, **84**, 443–454.

Uddin, N., and Morgan, J. P. (1997b). Universally optimal designs with blocksize $p \times 2$ and correlated observations. *Annals of Statistics*, **25**(3), 1189–1207.

Umland, A. W., and Smith, W. N. (1959). The use of Lagrange multipliers with response surfaces. *Technometrics*, **1**(3), 289–292.

Underwood, W. M. (1962). Experimental methods for designing extrusion screws. *Chemical Engineering Progress*, **58**(1), 59–65.

Vaithiyalingam, S., and Khan, M .A. (2002). Optimization and characterization of controlled release multi-particulate beads formulated with a customized cellulose acetate butyrate dispersion. *International Journal of Pharmaceutics*, **234**, 179–193.

Vandenbrande, W. (2005). Design of experiments for dummies. *Quality Progress*, **38**(4), April, 59–65.

Van der Vaart, H. R. (1960). On certain types of bias in current methods of response surface estimation. *Bulletin of the International Statistical Institute*, **37**, 191–203.

Van der Vaart, H. R. (1961). On certain characteristics of the distribution of the latent roots of a symmetric random matrix under general conditions. *Annals of Mathematical Statistics*, **32**, 864–873.

Vanhonacker, W. R. (1996). Meta-analysis and response surface extrapolation: A least squares approach. *American Statistician*, **50**(4), 294–299.

Vanleeuwen, D. M., Birkes, D. S., and Seely, J. F. (1999). Balance and orthogonality in designs for mixed classified models. *Annals of Statistics*, **27**(6), 1927–1947.

Vatsala, C. N., Saxena, C. D., and Haridas Rao, P. H. (2001). Optimization of ingredients and process conditions for the preparation of puri using response surface methodology. *International Journal of Food Science and Technology*, **36**, 407–414.

Vaughan, T. S. (1993). Experimental design for response surface gradient estimation. *Communications in Statistics, Theory and Methods*, **22**(6), 1535–1555.

Venter, J. H., and Steel, S. J. (1996). A hypothesis-testing approach toward identifying active contrasts. *Technometrics*, **38**(2), 161–169.

Venter, J. H., and Steel, S. J. (1998). Identifying active contrasts by stepwise testing. *Technometrics*, **40**(4), 304–313.

Verdinelli, I. (2000). A note on Bayesian design for the normal linear model with unknown error variance. *Biometrika*, **87**(1), 222–227.

Vienna, J. D., Hrma, P., Crum, J. V., and Mika, M. (2001). Liquidus temperature-composition model for multi-component glasses in the Fe, Cr, Ni and Mn spinel primary Phase field. *Journal of Non-Crystal Solids*, **292**(1-3), 1–24. **M**

Vining, G. G. (1993). A computer program for generating variance dispersion graphs. *Journal of Quality Technology*, **25**(1), 45–58.

Vining, G. G. (1998). A compromise approach to multiresponse optimization. *Journal of Quality Technology*, **30**(4), 309–313.

Vining, G.G., and Bohn, L. L. (1998). Response surfaces for the mean and variance using a nonparametric approach. *Journal of Quality Technology*, **30**(3), 282–291.

Vining, G. G., and Myers, R. H. (1990). Combining Taguchi and response surface philosophies: A dual response approach. *Journal of Quality Technology*, **22**(1), 38–45.

Vining, G. G., and Myers, R. H. (1991). A graphical approach for evaluating response surface designs in terms of the mean squared error of prediction. *Technometrics*, **33**(3), 315–326.

Vining, G. G., and Schaub, D. (1996). Experimental designs for estimating both mean and variance functions. *Journal of Quality Technology*, **28**(2), 135–147. Correction, **28**(4), p. 496.

Vining, G. G., Cornell, J. A., and Myers, R. H. (1993). A graphical approach for evaluating mixture designs. *Applied Statistics*, **42**, 127–138. **M**

Vining, G. G., Kowalski, S. M., and Montgomery, D. C. (2005) . Response surface designs within a split-plot structure. *Journal of Quality Technology*, **37**(2), 115–129.

Vooelkel, J. G. (2005). The efficiencies of fractional factorial designs. *Technometrics*, **47**(4), 488–494.

Vohra, A., and Satyanarayana, T. (2002). Statistical optimization of the medium components by response surface methodology to enhance phytase production by Pichia anomala. *Process Biochemistry*, **37**, 999–1004.

Voss, D. T., and DeMartino, S. (1994). Spatial interpolation for photographic reproduction of color. *Technometrics*, **36**(3), 249–259.

Voss, R., and Pesek, J (1965). Geometrical determination of uncontrolled-controlled factor relationships affecting crop yield. *Agronomy Journal*, **57**, 460–463.

Voss, R., and Pesek, J (1967). Yield of corn grain as affected by fertilizer rates and environmental factors. *Agronomy Journal*, **59**, 567–572.

Vuchkov, I. N., and Boyadjieva, L. N. (1992). Quality improvement through design of experiments with both product parameters and external noise factors. In *Model Oriented Data Analysis* (V. Fedorov, W. G. Muller, and I. N. Vuchkov, eds.), pp. 195–212. Heidelberg: Physica-Verlag.

Vuchkov, I. N., Damgaliev, O. L., and Yontchev, Ch. A. (1981). Sequentially generated second order quasi D-optimal designs for experiments with mixture and process variables. *Technometrics*, **23**(3), 233–238. **M**

Walker, C. E., and Parkhurst, A. M. (1984). Response surface analysis of bake-lab data with a personal computer. *Research*, **29**, 662–666.

Walker, W. M., and Carmer, S. G. (1967). Determination of input levels for a selected probability of response on a curvilinear regression function. *Agronomy Journal*, **59**, 161–162.

Walker, W. M., and Long, O. M. (1966). Effect of selected soil fertility parameters on soybean yields. *Agronomy Journal*, **58**, 403–405.

Walker, W. M., and Pesek, J. (1967). Yield of Kentucky bluegrass (*poa pratensis*) as a function of its percentage of nitrogen, phosp

Walker, W. M., Pesek, J., and Heady, E. O. (1963). Effect of nitrogen, phosphorus and pot

Walkowiak, R., and Kala, R. (2000). Two-phase nonlinear regression with smooth transition. *Communications in Statistics, Simulation and Computation*, **29**(2), 385–397.

Wallace, D. L. (1958). Intersection region confidence procedures with an application to the location of the maximum in quadratic regression. *Annals of Mathematical Statistics*, **29**, 455–475.

Wang, J. C. (1996). Mixed difference matrices and the construction of orthogonal arrays. *Statistics and Probability Letters*, **28**, 121–126.

Wang, J. C., and Wu, C. F. J. (1991). An approach to the construction of asymmetrical orthogonal arrays. *Journal of the American Statistical Association*, **86**, 450–456.

Wang, J. C., and Wu, C. F. J. (1992). Nearly orthogonal arrays with mixed levels and small runs. *Technometrics*, **34**(4), 409–422.

Wang, J. C., and Wu, C. F. J. (1995). A hidden projection property of Plackett-Burman and related designs. *Statistica Sinica*, **5**, 235–250.

Wang, P. C. (1988). A simple method for analyzing binary data from orthogonal arrays. *Journal of Quality Technology*, **20**(4), 230–232.

Wang, P. C. (1995). Comments on "A new class of supersaturated design," by D. K. J. Lin. *Technometrics*, **37**(3), 358–359.

Wang, P. C., and Jan, H. W. (1995). Designing two-level experiments using orthogonal arrays when the run order is important. *The Statistician*, **44**(3), 379–388.

Wang, P. C., and Lin, D. F. (2001). Dispersion effects in signal-response data from fractional factorial experiments. *Computational Statistics and Data Analysis*, **38**, 95–111.

Wang, P. C., Chen, M. H., and Wei, S. C. (1995). To analyze binary data from mixed-level orthogonal arrays. *Computational Statistics and Data Analysis*, **20**(6), 689–697.

Wang, S. G., and Nyquist, H. (1991). Effects on the eigenstructure of a data matrix when deleting an observation. *Computational Statistics and Data Analysis*, **11**, 179–188.

Wang, W., and Voss, D. T. (2001). Control of error rates in adaptive analysis of orthogonal saturated designs. *Annals of Statistics*, **29**(4), 1058–1065.

Wang, Y., Lin, D. K. J., and Fang, K.-T. (1995). Designing outer array points. *Journal of Quality Technology*, **27**(3), 226–241.

Warncke, D. D., and Barber, S. A. (1973). Ammonium and nitrate uptake by corn, as influenced by nitrogen concentration and NH_4^+/NO_3^- ratio. *Agronomy Journal*, **65**, 950–953.

Wasiloff, E., and Hargitt, C. (1999). Using design of experiment to determine AA battery life. *Quality Progress*, **32**(3), March, 67–71.

Watson, G. S. (1969). Linear regression on proportions. *Biometrics*, **25**, 585–588.

Watts, D. G. (1995). Understanding canonical analysis. *Journal of Quality Technology*, **27**(1), 40–44.

Watts, D. G. (1997). Explaining power using two-level factorial designs. *Journal of Quality Technology*, **29**(3), 305–306.

Watts, D. G., and Bacon, D. W. (1974). Using a hyperbola as a transition model to fit two-regime straight line data. *Technometrics*, **16**(3), 369–373.

Webb, D., Lucas, J. M., and Borkowski, J. J. (2004). Factorial experiments when factors are not necessarily reset. *Journal of Quality Technology*, **36**(1), 1–11.

Webb, S. R. (1968a). Saturated sequential factorial designs. *Technometrics*, **10**(2), 291–299.

Webb, S. R. (1968b). Nonorthogonal designs of even resolution. *Technometrics*, **10**(3), 535–550.

Webb, S. R. (1971). Small incomplete factorial experiment designs for two and three level factors. *Technometrics*, **13**(2), 243–256.

Weerakkody, G. J. (1992). A note on the recovery of inter-block information in balanced incomplete block designs. *Communications in Statistics, Theory and Methods*, **21**(4), 1125–1136.

Wegman, E. J. (1990). Hyperdimensional data analysis using parallel coordinates. *Journal of the American Statistical Association*, **85**, 664–675.

Wei, F., and Notz, W. I. (1992). D-optimal weighing designs and optimal chemical balance weighing designs for estimating the total weight. *Communications in Statistics, Theory and Methods*, **21**(3), 667–678.

Weissert, F. C., and Cundiff, R. R. (1963). Compounding Diene/NR blends for truck tires. *Rubber Age*, **92**, 881–887.

Welch, L. F., Adams, W. E., and Carmon, J. L. (1963). Yield response surfaces, isoquants and economic fertilizer optima for Coastal bermudagrass. *Agronomy Journal*, **55**, 63–67.

Welch, L. F., Johnson, P. E., McKibben, G. E., Boone, L. V., and Pendleton, J. W. (1966). Relative efficiency of broadcast versus banded potassium for corn. *Agronomy Journal*, **58**, 618–621.

Welch, W. J. (1982). Branch and bound search for experimental designs based on D-optimality and other criteria. *Technometrics*, **24**(1), 41–48.

Welch, W. J. (1983). A mean squared error criterion for the design of experiments. *Biometrika*, **70**, 205–213.

Welch, W. J. (1984). Computer-aided design of experiments for response estimation. *Technometrics*, **26**(3), 217–224.

Welch, W. J. (1985). ACED: Algorithms for the construction of experimental designs. *American Statistician*, **39**, 146.

Welch, W. J., Yu, T. K., Kang, S. M., and Sacks, J. (1990). Computer experiments for quality control by parameter design. *Journal of Quality Technology*, **22**(1), 15–22.

Welch, W. J., Buck, R. J., Sacks, J., Wynn, H. P., Mitchell, T. J., and Morris, M. D. (1992). Screening, prediction and computer experiments. *Technometrics*, **34**(1), 15–25.

Wen, L.-F., Rodis, P., and Wasserman, B. P. (1990). Starch fragmentation and protein insolubilization during twin-screw extrusion of corn meal. *Cereal Chemistry*, **67**, 268–275.

Westfall, P. H., Young, S. S., and Lin, D. K. J. (1998). Forward selection error control in the analysis of supersaturated designs. *Statistica Sinica*, **8**(1), 101–117.

Westlake, W. J. (1962). A numerical analysis problem in constrained quadratic regression analysis. *Technometrics*, **4**(3), 426–430.

Westlake, W. J. (1965). Composite designs based on irregular fractions of factorials. *Biometrics*, **21**, 324–336.

Wetherill, G. B. (1963). Sequential estimation of quantal response curves. *Journal of the Royal Statistical Society, Series B*, **25**, 1–38; discussion: pp. 39–48.

Wheeler, D. J. (1988). *Understanding Industrial Experimentation*. Knoxville, TN: Statistical Process Controls.

Whitaker, D., Triggs, C. M., and John, J. A. (1990). Construction of block designs using mathematical programming. *Journal of the Royal Statistical Society, Series B*, **52**, 497–503.

Whittle, P. (1965). Some general results in sequential design. *Journal of the Royal Statistical Society, Series B*, **27**, 371–387; discussion: pp. 387–394.

Whitwell, J. C., and Morbey, G. K. (1961). Reduced designs of resolution five. *Technometrics*, **3**(4), 459–477.

Wichern, D. W., and Churchill, G. A. (1978). A comparison of ridge estimators. *Technometrics*, **20**(3), 301–310.

Wiens, D. P. (1991). Designs for approximately linear regression: Two optimality properties of uniform designs. *Statistics and Probability Letters*, **12**, 217–221.

Wiens, D. P. (1993). Designs for approximately linear regression: Maximizing the minimum coverage probability of confidence ellipsoids. *Canadian Journal of Statistics*, **21**, 59–70.

Wiens, D. P. (1996). Robust sequential designs for approximately linear models. *Canadian Journal of Statistics*, **24**(1), 67–79.

Wiens, D. P. (1998). Minimax robust designs and weights for approximately specified regression models with heteroscedastic errors. *Journal of the American Statistical Association*, **93**, 1440–1450.

Wiens, D. P., and Zhou, J. (1997). Robust designs based on the infinitesimal approach. *Journal of the American Statistical Association*, **92**, 1503–1511.

Wild, C. J., and Pfannkuch, M. (1999). Statistical thinking in empirical enquiry. *International Statistical Review*, **67**(3), 223–248; discussion: pp. 248–265.

Willey, R. W., and Heath, S. B. (1969). The quantitative relationships between plant population and crop yield. *Advances in Agronomy*, **21**, 281–321.

Williams, D. A. (1970). Discrimination between regression models to determine the pattern of enzyme synthesis in synchronous cell cultures. *Biometrics*, **26**, 23–32.

Williams, E. R., and John, J. A. (1996a). A note on optimality in lattice square designs. *Biometrika*, **83**, 709–713.

Williams, E. R., and John, J. A. (1996b). Row-column factorial designs for use in agricultural field trials. *Applied Statistics*, **45**(1), 39–46.

Williams, E. R., and John, J. A. (2000). Updating the average efficiency factor in α-designs. *Biometrika*, **87**(3),695–699.

Williams, K. R. (1968). Designed experiments. *Rubber Age*, **100**, August, 65–71.

Williamson, J. (1944). Hadamard's determinant theorem and the sum of four squares. *Duke Mathematical Journal*, **11**, 65–81.

Wilson, J. H., Clowes, M. St. J., and Allison, J. C. S. (1973). Growth and yield of maize at different altitudes in Rhodesia. *Annals of Applied Biology*, **73**, 77–84.

Wishart, J. (1938). Growth rate determination in nutrition studies with the bacon pig, and their analysis. *Biometrika*, **30**, 16–28.

Wishart, J. (1939). Statistical treatment of animal experiments. *Journal of the Royal Statistical Society*, Series B, **6**, 1–22.

Wolfinger, R. D., and Tobias, R. D. (1998). Joint estimation of location, dispersion, and random effects in robust design. *Technometrics*, **40**(1), 62–71.

Wong, W. K. (1994). Comparing robust properties of A, D, E and G-optimal designs. *Computational Statistics and Data Analysis*, **18**, 441–448.

Wong, W. K., and Cook, R. D. (1993). Heteroscedastic G-optimal designs. *Journal of the Royal Statistical Society*, Series B, **55**, 871–880.

Wood, C. L., and Cady, F. B. (1981). Intersite transfer of estimated response surfaces. *Biometrics*, **37**, 1–10.

Wood, J. T. (1974). An extension of the analysis of transformations of Box and Cox. *Applied Statistics*, **23**, 278–283.

Woodall, W. H., Koudelik, R., Tsui, K.-L., Kim, S. B., Stoumbos, Z. G., and Carvounis, C. P. (2003). A review and analysis of the Mahalanobis-Taguchi system. *Technometrics*, **45**(1), 1–15; discussion: pp. 16–30.

Wooding, W. M. (1973). The split-plot design. *Journal of Quality Technology*, **5**(1), 16–33.

Woods, D. C., Lewis, S. M., Eccleston, J. A., and Russell, K. G. (2006). Designs for generalized linear models with several variables and model uncertainty. *Technometrics*, **48**(2), 284–292.

Wu, C. J. F. (1985). Efficient sequential designs with binary data. *Journal of the American Statistical Association*, **80**, 974–984.

Wu, C. J. F. (1989). Construction of $2^m 4^n$ designs via a grouping scheme. *Annals of Statistics*, **17**, 1880–1885.

Wu, C. J. F. (1993). Construction of supersaturated designs through partially aliased interactions. *Biometrika*, **80**(3), 661–669.

Wu, C. F. J., and Chen, Y. (1992). A graph-aided method for planning two-level experiments when certain interactions are important. *Technometrics*, **34**(2), 162–175.

Wu, C. F. J., and Ding, Y. (1998). Construction of response surface design for qualitative and quantitative factors. *Journal of Statistical Planning and Inference*, **71**, 331–348.

Wu, C. F. J., and Hamada, M. (2000). *Experiments*: *Planning, Analysis, and Parameter Design Optimization*. New York: Wiley.

Wu, C. F. J., and Zhang, R. (1993). Minimum aberration designs with two-level and four-level factors. *Biometrika*, **80**, 203–209.

Wu, C. F. J., and Zhu, Y. (2003). Optimal selection of single arrays for parameter design experiments. *Statistica Sinica*, **13**, 1179–1199.

Wu, H. (1997). Optimal exact designs on a circle or circular arc. *Annals of Statistics*, **25**(5), 2027–2043.

Wu, H., and Wu, C. F. J. (2002). Clear two-factor interactions and minimum aberration. *Annals of Statistics*, **30**(5), 1496–1511.

Wu, S. M. (1964a). Tool-life testing by response surface methodology. Part I. *Journal of Engineering for Industry, ASME Transactions, Series B*, **86**, May, 105–110.

Wu, S. M. (1964b). Tool-life testing by response surface methodology, Part II. *Journal of Engineering for Industry, ASME Transactions, Series B*. **86**, 111–116.

Wu, S. M., and Meyer, R. N. (1964). Cutting tool temperature-predicting equation by response surface methodology. *Journal of Engineering for Industry, ASME Transactions, Series B*, **86**, May, 150–156.

Wu, S. M., and Meyer, R. N. (1965). A first order five-variable cutting tool temperature equation and chip equivalent. *Journal of Engineering for Industry, ASME Transactions, Series B*, **87**, November, 395–400.

Wu, S.M., and Meyer, R. N. (1967). Mathematical models of carbide tool crater surfaces. *International Journal of Machine Tool Design Research*, **7**, 445–463.

Wynn, H. P. (1970). The sequential generation of D-optimum experimental designs. *Annals of Mathematical Statistics*, **41**(5), 1655–1664.

Wynn, H. P. (1972). Results in the theory and construction of D-optimum experimental designs. *Journal of the Royal Statistical Society, Series B*, 34, 133–147; discussion: pp. 170–185.

Wynn, H. P. (1984). Jack Kiefer's contributions to experimental design. *Annals of Statistics*, **12**, 416–423.

Xu, H. (2002). An algorithm for constructing orthogonal and nearly orthogonal arrays with mixed levels and small runs. *Technometrics*, **44**(4). 356–368.

Xu, H. (2003). Minimum moment aberration for non-regular designs and supersaturated designs. *Statistica Sinica*, 13, 691–708.

Xu, H., and Deng, L.-Y. (2005). Moment aberration projection for nonregular fractional factorial designs. *Technometrics*, **47**(2), 121–131.

Xu, H., and Wu, C. F. J. (2001). Generalized minimum aberration for asymmetrical fractional factorial designs. *Annals of Statistics*, **29**(4), 1066–1077. [This is a corrected version of **29**(2), 549–560.]

Xu, H., and Wu, C. J. F. (2005). Construction of optimal multi-level supersaturated designs. *Annals of Statistics*, **33**(6), 2811–2836.

Xu, H., Cheng, S.-W., and Wu, C. F. J. (2004). Optimal projective three-level designs for factor screening and interaction detection. *Technometrics*, **46**(3), 280–292.

Yamada, S., and Lin, D. K. J. (1997). Supersaturated designs including an orthogonal base. *Canadian Journal of Statistics*, **25**(2), 203–213.

Yamada, S., and Lin, D. K. J. (1999). Three-level supersaturated designs. *Statistics and Probability Letters*, **45**, 31–39.

Yamada, S., and Lin, D. K. J. (2002). Construction of mixed-level supersaturated designs. *Metrika*, **56**, 205–214.

Yamada, S., and Matsui, T. (2002). Optimality of mixed–level supersaturated designs. *Journal of Statistical Planning and Inference*, **104**, 459–468.

Yamada, S., Ikebe, Y., Hashiguchi, H., and Niki, N. (1999). Construction of three-level supersaturated designs. *Journal of Statistical Planning and Inference*, **81**, 183–193.

Yamada, S., Matsui, M., Matsui, T., Lin, D. K. J., and Takahashi, T. (2006). A general construction method for mixed-level supersaturated design. *Computational Statistics and Data Analysis*, **50**(1), 254–265.

Yang, C. H. (1966). Some designs of maximal $(+1, -1)$ determinant of order $n \equiv 2(\bmod 4)$. *Mathematical Computations*, **20**, 147–148.

Yang, C. H. (1968). On the designs of maximal $(+1, -1)$ matrices of order $n \equiv 2(\bmod 4)$. *Mathematical Computations*, **22**, 174–180.

Yang, Y. J and Draper, N.R. (2003). Two-level factorial and fractional factorial designs in blocks of size two. *Journal of Quality Technology*, **35**(3), 294–305.

Yang, Y. J., and Draper, N. R. (2005). Two-level factorial and fractional factorial designs in blocks of size two. Part 2. In *Response Surface Methodology and Related Topics* (A. I. Khuri, ed.), pp. 1–18. Hackensack, NJ: World Scientific Publishing.

Yates, F. (1935). Complex experiments. *Journal of the Royal Statistical Society*, Series B, **2**, 181–223; discussion: pp. 223–247.

Yates, F. (1937). *The Design and Analysis of Factorial Experiments*. London: Imperial Bureau of Soil Science.

Yates, F. (1967). A fresh look at the basic principles of the design and analysis of experiments. In *Proceedings of the Fifth Berkeley Symposium in Mathematical Statistics and Probability*, *IV*, pp. 777–790. Berkeley and Los Angeles: University of California Press.

Yates, F. (1970). *Experimental Design: Selected Papers of Frank Yates*. Darien, CT: Hafner.

Ye, K. Q. (1998). Orthogonal column Latin hypercubes and their application in computer experiments. *Journal of the American Statistical Association*, **93**(444), 1430–1439.

Ye, K. Q. (2003). Indicator function and its application in two-level factorial designs. *Annals of Statistics*, **31**(3), 984–994.

Ye, K. Q., and Hamada, M. (2000). Critical values of the Lenth method for unreplicated factorial designs. *Journal of Quality Technology*, **32**(1), 57–66.

Ye, K. Q., and Li, W. (2003). Some properties of blocked and unblocked foldovers of 2^{k-p} designs. *Statistica Sinica*, **13**, 403–408.

Ye, K. Q., Hamada, M., and Wu, C. F. J. (2001). A step-down Lenth method for analyzing unreplicated factorial designs, *Journal of Quality Technology*. **33**(2), 140–152.

Yin, X., and Seymour, L. (2005). Standard errors for the multiple roots in quadratic response surface models. *Technometrics*, **47**(3), 260–263.

Ying, L. H., Pukelsheim, F., and Draper, N. R. (1995a). Slope rotatability over all directions designs. *Journal of Applied Statistics*, **22**, 331–341.

Ying, L. H., Pukelsheim, F., and Draper, N. R. (1995b). Slope rotatability over all directions designs for $k \geq 4$. *Journal of Applied Statistics*, **22**, 343–354.

Yong, J., Myung, H., Gee, D., Ji, H., and Ok, M. (2001). Establishment on the preparation condition of pumpkin honey kochujang by response surface methodology. *Journal of the Korean Society of Food Science and Nutrition*, pp. 1102–1107.

Young, J. (1996). Blocking, replication, and randomization—the key to effective experimentation: A case study. *Quality Engineering*, **9**, 269–277.

Zahren, A., Anderson-Cook, C. M., and Myers, R. H. (2003). Fraction of design space to assess prediction capability of response surface designs. *Journal of Quality Technology*, **35**(4), 377–386.

Zahren, A., Anderson-Cook, C. M., Myers, R. H., and Smith, E. P. (2003). Modifying 2^2 factorial designs to accommodate a restricted design space. *Journal of Quality Technology*, **35**(4), 387–392.

Zhang, A., Fang, K.-T., Li, R., and Sudjianto, A. (2005). Majorization framework for balanced lattice designs. *Annals of Statistics*, **33**(6), 2837–2853.

Zhang, Y. S., Lu, Y., and Pang, S. (1999). Orthogonal arrays obtained by orthogonal decompositions of projection matrices. *Statistica Sinica*, **9**, 595–604.

Zhang, Y. S., Pang, S., and Wang, Y. (2001). Orthogonal arrays obtained by generalized Hadamard product. *Discrete Mathematics*, **238**, 151–170.

Zhang, Y. S., Duan, L., Lu, Y., and Zheng, Z. (2002). Construction of generalized Hadamard matrices. *Journal of Statistical Planning and Inference*, **104**, 239–258.

Zhang, Y. S., Weiguo, L., Meixia, M., and Zheng, Z. (2005). Orthogonal arrays obtained by generalized Kronecker product. *Journal of Statistical Planning and Inference*. to appear.

Zhou, J. (2001a). Integer-valued, minimax robust designs for approximately linear models with correlated errors. *Communications in Statistics, Theory and Methods*, **30**(1), 21–39.

Zhou, J. (2001b). A robust criterion for experimental designs for serially correlated observations. *Technometrics*, **43**(4), 462–467.

Zhou, X., Joseph, L., Wolfson, D. B., and Bélisle, P. (2003). A Bayesian A-optimal and model robust design criterion. *Biometrics*, **59**, 1082–1088.

Zhu, Y. (2003). Structure function for aliasing patterns in 2^{l-n} designs with multiple groups of factors. *Annals of Statistics*, **31**(3), 995–1011.

Zocchi, S. S., and Atkinson, A. C. (1999). Optimum experimental designs for multinomial logistic models. *Biometrics*, **55**(2), 437–444.

Zondagh, I. B., Holmes, Z. A., Rowe, K. E., and Schrumpf, D. E. (1986). The use of a central composite rotatable design with response surface analysis to predict turkey breast and thigh meat quality characteristics. *Poultry Science*, **65**, 520–5

Author Index

851

Subject Index

WILEY SERIES IN PROBABILITY AND STATISTICS

ESTABLISHED BY WALTER A. SHEWHART AND SAMUEL S. WILKS

Editors: *David J. Balding, Noel A. C. Cressie, Nicholas I. Fisher,*
Iain M. Johnstone, J. B. Kadane, Geert Molenberghs, David W. Scott,
Adrian F. M. Smith, Sanford Weisberg
Editors Emeriti: *Vic Barnett, J. Stuart Hunter, David G. Kendall,*
Jozef L. Teugels

The *Wiley Series in Probability and Statistics* is well established and authoritative. It covers many topics of current research interest in both pure and applied statistics and probability theory. Written by leading statisticians and institutions, the titles span both state-of-the-art developments in the field and classical methods.

Reflecting the wide range of current research in statistics, the series encompasses applied, methodological and theoretical statistics, ranging from applications and new techniques made possible by advances in computerized practice to rigorous treatment of theoretical approaches.

This series provides essential and invaluable reading for all statisticians, whether in academia, industry, government, or research.

† ABRAHAM and LEDOLTER · Statistical Methods for Forecasting
  AGRESTI · Analysis of Ordinal Categorical Data
  AGRESTI · An Introduction to Categorical Data Analysis, *Second Edition*
  AGRESTI · Categorical Data Analysis, *Second Edition*
  ALTMAN, GILL, and McDONALD · Numerical Issues in Statistical Computing for the
    Social Scientist
  AMARATUNGA and CABRERA · Exploration and Analysis of DNA Microarray and
    Protein Array Data
  ANDĚL · Mathematics of Chance
  ANDERSON · An Introduction to Multivariate Statistical Analysis, *Third Edition*
* ANDERSON · The Statistical Analysis of Time Series
  ANDERSON, AUQUIER, HAUCK, OAKES, VANDAELE, and WEISBERG ·
    Statistical Methods for Comparative Studies
  ANDERSON and LOYNES · The Teaching of Practical Statistics
  ARMITAGE and DAVID (editors) · Advances in Biometry
  ARNOLD, BALAKRISHNAN, and NAGARAJA · Records
* ARTHANARI and DODGE · Mathematical Programming in Statistics
* BAILEY · The Elements of Stochastic Processes with Applications to the Natural
    Sciences
  BALAKRISHNAN and KOUTRAS · Runs and Scans with Applications
  BALAKRISHNAN and NG · Precedence-Type Tests and Applications
  BARNETT · Comparative Statistical Inference, *Third Edition*
  BARNETT · Environmental Statistics
  BARNETT and LEWIS · Outliers in Statistical Data, *Third Edition*
  BARTOSZYNSKI and NIEWIADOMSKA-BUGAJ · Probability and Statistical Inference
  BASILEVSKY · Statistical Factor Analysis and Related Methods: Theory and
    Applications
  BASU and RIGDON · Statistical Methods for the Reliability of Repairable Systems
  BATES and WATTS · Nonlinear Regression Analysis and Its Applications

*Now available in a lower priced paperback edition in the Wiley Classics Library.
†Now available in a lower priced paperback edition in the Wiley–Interscience Paperback Series.

BECHHOFER, SANTNER, and GOLDSMAN · Design and Analysis of Experiments for Statistical Selection, Screening, and Multiple Comparisons

BELSLEY · Conditioning Diagnostics: Collinearity and Weak Data in Regression

† BELSLEY, KUH, and WELSCH · Regression Diagnostics: Identifying Influential Data and Sources of Collinearity

BENDAT and PIERSOL · Random Data: Analysis and Measurement Procedures, *Third Edition*

BERRY, CHALONER, and GEWEKE · Bayesian Analysis in Statistics and Econometrics: Essays in Honor of Arnold Zellner

BERNARDO and SMITH · Bayesian Theory

BHAT and MILLER · Elements of Applied Stochastic Processes, *Third Edition*

BHATTACHARYA and WAYMIRE · Stochastic Processes with Applications

BILLINGSLEY · Convergence of Probability Measures, *Second Edition*

BILLINGSLEY · Probability and Measure, *Third Edition*

BIRKES and DODGE · Alternative Methods of Regression

BLISCHKE AND MURTHY (editors) · Case Studies in Reliability and Maintenance

BLISCHKE AND MURTHY · Reliability: Modeling, Prediction, and Optimization

BLOOMFIELD · Fourier Analysis of Time Series: An Introduction, *Second Edition*

BOLLEN · Structural Equations with Latent Variables

BOLLEN and CURRAN · Latent Curve Models: A Structural Equation Perspective

BOROVKOV · Ergodicity and Stability of Stochastic Processes

BOULEAU · Numerical Methods for Stochastic Processes

BOX · Bayesian Inference in Statistical Analysis

BOX · R. A. Fisher, the Life of a Scientist

BOX and DRAPER · Response Surfaces, Mixtures, and Ridge Analyses, *Second Edition*

* BOX and DRAPER · Evolutionary Operation: A Statistical Method for Process Improvement

BOX and FRIENDS · Improving Almost Anything, *Revised Edition*

BOX, HUNTER, and HUNTER · Statistics for Experimenters: Design, Innovation, and Discovery, *Second Editon*

BOX and LUCEÑO · Statistical Control by Monitoring and Feedback Adjustment

BRANDIMARTE · Numerical Methods in Finance: A MATLAB-Based Introduction

BROWN and HOLLANDER · Statistics: A Biomedical Introduction

BRUNNER, DOMHOF, and LANGER · Nonparametric Analysis of Longitudinal Data in Factorial Experiments

BUCKLEW · Large Deviation Techniques in Decision, Simulation, and Estimation

CAIROLI and DALANG · Sequential Stochastic Optimization

CASTILLO, HADI, BALAKRISHNAN, and SARABIA · Extreme Value and Related Models with Applications in Engineering and Science

CHAN · Time Series: Applications to Finance

CHARALAMBIDES · Combinatorial Methods in Discrete Distributions

CHATTERJEE and HADI · Regression Analysis by Example, *Fourth Edition*

CHATTERJEE and HADI · Sensitivity Analysis in Linear Regression

CHERNICK · Bootstrap Methods: A Practitioner's Guide

CHERNICK and FRIIS · Introductory Biostatistics for the Health Sciences

CHILÈS and DELFINER · Geostatistics: Modeling Spatial Uncertainty

CHOW and LIU · Design and Analysis of Clinical Trials: Concepts and Methodologies, *Second Edition*

CLARKE and DISNEY · Probability and Random Processes: A First Course with Applications, *Second Edition*

* COCHRAN and COX · Experimental Designs, *Second Edition*

CONGDON · Applied Bayesian Modelling

CONGDON · Bayesian Models for Categorical Data

CONGDON · Bayesian Statistical Modelling

*Now available in a lower priced paperback edition in the Wiley Classics Library.

†Now available in a lower priced paperback edition in the Wiley–Interscience Paperback Series.

CONOVER · Practical Nonparametric Statistics, *Third Edition*
COOK · Regression Graphics
COOK and WEISBERG · Applied Regression Including Computing and Graphics
COOK and WEISBERG · An Introduction to Regression Graphics
CORNELL · Experiments with Mixtures, Designs, Models, and the Analysis of Mixture
    Data, *Third Edition*
COVER and THOMAS · Elements of Information Theory
COX · A Handbook of Introductory Statistical Methods
* COX · Planning of Experiments
CRESSIE · Statistics for Spatial Data, *Revised Edition*
CSÖRGŐ and HORVÁTH · Limit Theorems in Change Point Analysis
DANIEL · Applications of Statistics to Industrial Experimentation
DANIEL · Biostatistics: A Foundation for Analysis in the Health Sciences, *Eighth Edition*
* DANIEL · Fitting Equations to Data: Computer Analysis of Multifactor Data,
    Second Edition
DASU and JOHNSON · Exploratory Data Mining and Data Cleaning
DAVID and NAGARAJA · Order Statistics, *Third Edition*
* DEGROOT, FIENBERG, and KADANE · Statistics and the Law
DEL CASTILLO · Statistical Process Adjustment for Quality Control
DeMARIS · Regression with Social Data: Modeling Continuous and Limited Response
    Variables
DEMIDENKO · Mixed Models: Theory and Applications
DENISON, HOLMES, MALLICK and SMITH · Bayesian Methods for Nonlinear
    Classification and Regression
DETTE and STUDDEN · The Theory of Canonical Moments with Applications in
    Statistics, Probability, and Analysis
DEY and MUKERJEE · Fractional Factorial Plans
DILLON and GOLDSTEIN · Multivariate Analysis: Methods and Applications
DODGE · Alternative Methods of Regression
* DODGE and ROMIG · Sampling Inspection Tables, *Second Edition*
* DOOB · Stochastic Processes
DOWDY, WEARDEN, and CHILKO · Statistics for Research, *Third Edition*
DRAPER and SMITH · Applied Regression Analysis, *Third Edition*
DRYDEN and MARDIA · Statistical Shape Analysis
DUDEWICZ and MISHRA · Modern Mathematical Statistics
DUNN and CLARK · Basic Statistics: A Primer for the Biomedical Sciences,
    Third Edition
DUPUIS and ELLIS · A Weak Convergence Approach to the Theory of Large Deviations
EDLER and KITSOS · Recent Advances in Quantitative Methods in Cancer and Human
    Health Risk Assessment
* ELANDT-JOHNSON and JOHNSON · Survival Models and Data Analysis
ENDERS · Applied Econometric Time Series
† ETHIER and KURTZ · Markov Processes: Characterization and Convergence
EVANS, HASTINGS, and PEACOCK · Statistical Distributions, *Third Edition*
FELLER · An Introduction to Probability Theory and Its Applications, Volume I,
    Third Edition, Revised; Volume II, *Second Edition*
FISHER and VAN BELLE · Biostatistics: A Methodology for the Health Sciences
FITZMAURICE, LAIRD, and WARE · Applied Longitudinal Analysis
* FLEISS · The Design and Analysis of Clinical Experiments
FLEISS · Statistical Methods for Rates and Proportions, *Third Edition*
† FLEMING and HARRINGTON · Counting Processes and Survival Analysis
FULLER · Introduction to Statistical Time Series, *Second Edition*
† FULLER · Measurement Error Models

*Now available in a lower priced paperback edition in the Wiley Classics Library.
†Now available in a lower priced paperback edition in the Wiley–Interscience Paperback Series.

*Now available in a lower priced paperback edition in the Wiley Classics Library.

†Now available in a lower priced paperback edition in the Wiley–Interscience Paperback Series.

*Now available in a lower priced paperback edition in the Wiley Classics Library.

†Now available in a lower priced paperback edition in the Wiley–Interscience Paperback Series.

---

*Now available in a lower priced paperback edition in the Wiley Classics Library.

†Now available in a lower priced paperback edition in the Wiley–Interscience Paperback Series.

*Now available in a lower priced paperback edition in the Wiley Classics Library.

†Now available in a lower priced paperback edition in the Wiley–Interscience Paperback Series.

---

*Now available in a lower priced paperback edition in the Wiley Classics Library.

†Now available in a lower priced paperback edition in the Wiley–Interscience Paperback Series.